Springer Proceedings in Mathematics & Statistics

Volume 321

Springer Proceedings in Mathematics & Statistics

This book series features volumes composed of selected contributions from workshops and conferences in all areas of current research in mathematics and statistics, including operation research and optimization. In addition to an overall evaluation of the interest, scientific quality, and timeliness of each proposal at the hands of the publisher, individual contributions are all refereed to the high quality standards of leading journals in the field. Thus, this series provides the research community with well-edited, authoritative reports on developments in the most exciting areas of mathematical and statistical research today.

More information about this series at http://www.springer.com/series/10533

Alberto Facchini · Marco Fontana ·
Alfred Geroldinger · Bruce Olberding
Editors

Advances in Rings, Modules and Factorizations

Graz, Austria, February 19–23, 2018

Springer

Editors
Alberto Facchini
Dipartimento di Matematica
"Tullio Levi-Civita"
Università degli Studi di Padova
Padua, Italy

Alfred Geroldinger
Institut für Mathematik und
Wissenschaftliches Rechnen
Universität Graz
Graz, Austria

Marco Fontana
Dipartimento di Matematica e Fisica
Università degli Studi Roma Tre
Rome, Italy

Bruce Olberding
Department of Mathematical Sciences
New Mexico State University
Las Cruces, NM, USA

ISSN 2194-1009 ISSN 2194-1017 (electronic)
Springer Proceedings in Mathematics & Statistics
ISBN 978-3-030-43415-1 ISBN 978-3-030-43416-8 (eBook)
https://doi.org/10.1007/978-3-030-43416-8

Mathematics Subject Classification (2010): 13A15, 13A50, 13B21, 13F15, 13G05, 13H99, 14H99, 16D25, 20M25

This Springer imprint is published by the registered company Springer Nature Switzerland AG
The registered company address is: Gewerbestrasse 11, 6330 Cham, Switzerland

Preface

This volume is occasioned by the conference *Rings and Factorizations* that took place at the University of Graz (Austria), February 19–23, 2018. The meeting featured five days of lectures on a range of topics that reflected new research trends in areas such as ideal theory (especially Prüfer, Krull, and Mori rings); topological methods in ring theory; rings of integer-valued polynomials; module theory and direct-sum decompositions; and factorization and divisibility theory in rings and semigroups. The conference continues a long tradition of international conferences in commutative ring theory, module theory and factorization theory that have been held in Austria, France, Italy, South Korea, Morocco and the United States.

The volume consists of invited, refereed research and expository articles from leading researchers in the fields represented at the conference. Most of the contributors to the volume were speakers at the conference. The diverse list of topics in the volume, like that of the lectures at the conference, represents areas of research that are expanding and transcending traditional boundaries between fields of study.

The conference *Rings and Factorizations* was organized by Alfred Geroldinger, Jun Seok Oh, Salvatore Tringali, and Qinghai Zhong. It was supported by the University of Graz, Institute for Mathematics and Scientific Computing, NAWI Graz, Federal State of Styria, and by the Austrian Science Fund FWF (Project Numbers P28864-N35 and W1230 Doctoral Program Discrete Mathematics). We thank all our sponsors. Without their assistance and support, the conference would not have been possible.

Finally, we thank the authors for their contributions, the referees for their work, and the editorial staff at Springer for their guidance and patience in directing this volume to its publication.

Padua, Italy — Alberto Facchini
Rome, Italy — Marco Fontana
Graz, Austria — Alfred Geroldinger
Las Cruces, NM, USA — Bruce Olberding
September 2019

Contents

Commutative Rings Whose Principal Ideals Have Unique Generators

P. N. Ánh, Keith A. Kearnes, and Ágnes Szendrei

Abstract We investigate the class of commutative unital rings in which principal ideals have unique generators. We prove that this class forms a finitely axiomatizable, relatively ideal distributive quasivariety, and also that it equals the quasivariety generated by the class of integral domains with trivial unit group.

Keywords Divisibility · Relatively distributive quasivariety

2010 Mathematics Subject Classification 13A05 · 13A15 · 08C15

1 Introduction

What can be said about the class of commutative rings in which, if a differs from b, the set of elements divisible by a differs from the set of elements divisible by b? Equivalently, what can be said about the class of rings where $a \neq b$ implies $(a) \neq (b)$? In this paper we show that this class is a relatively ideal distributive quasivariety, and we give a set of axioms for the quasivariety. Along the way we learn that this quasivariety is exactly the quasivariety of commutative rings generated by the class of integral domains with trivial unit group.

P. N. Ánh (✉)
Rényi Institute of Mathematics, Hungarian Academy of Sciences,
Pf. 127, 1364 Budapest, Hungary
e-mail: anh.pham.ngoc@renyi.mta.hu

K. A. Kearnes · Á. Szendrei
Department of Mathematics, University of Colorado, Boulder, CO 80309-0395, USA
e-mail: Kearnes@Colorado.EDU

Á. Szendrei
e-mail: Szendrei@Colorado.EDU

A. Facchini et al. (eds.), *Advances in Rings, Modules and Factorizations*,
Springer Proceedings in Mathematics & Statistics 321,
https://doi.org/10.1007/978-3-030-43416-8_1

2 The Quasivariety of Rings Whose Principal Ideals have Unique Generators

Our goal in this section is to describe the class of commutative rings whose principal ideals have unique generators. The main result is that this class is a relatively ideal distributive quasivariety, so let us explain now what that means. (For more details about relatively congruence distributive/modular quasivarieties, we refer to [1–3].)

A *quasi-identity* in the language of commutative rings is a universally quantified implication of the form

$$(s_1 = t_1) \wedge \cdots \wedge (s_n = t_n) \to (s_0 = t_0)$$

where s_i and t_i are ring terms (= "words", or "polynomials"). We allow $n = 0$, in which case the quasi-identity reduces to an identity: $s_0 = t_0$ (universally quantified). To emphasize this last point: identities are special quasi-identities.

A *variety* is a class axiomatized by identities. A *quasivariety* is a class axiomatized by quasi-identities. For an example of the former, the class of commutative rings is a variety. For an example of the latter, the class of rings axiomatized by the identities defining commutative rings together with the quasi-identity $(x^2 = 0) \to (x = 0)$ is the quasivariety of reduced commutative rings (rings with no nonzero nilpotent elements).

If $\mathcal{Q}$ is a quasivariety of commutative rings, $R \in \mathcal{Q}$, and $I \lhd R$ is an ideal of R, then I is a *$\mathcal{Q}$-ideal* (or a *relative* ideal) if $R/I \in \mathcal{Q}$. For example, if $\mathcal{Q}$ is the quasivariety of commutative reduced rings and $R \in \mathcal{Q}$, then I is a relative ideal of R exactly when I is a semiprime ideal of R.

The collection of $\mathcal{Q}$-ideals of some $R \in \mathcal{Q}$, when ordered by inclusion, forms an algebraic lattice. It is not a sublattice of the ordinary ideal lattice, but it is a subset of the ordinary ideal lattice that is closed under arbitrary meet.

A quasivariety $\mathcal{Q}$ of commutative rings is *relatively ideal distributive* if the $\mathcal{Q}$-ideal lattice of any member of $\mathcal{Q}$ satisfies the distributive law:

$$I \wedge (J \vee K) = (I \wedge J) \vee (I \wedge K).$$

Here, the meet operation is just intersection ($I \wedge J = I \cap J$) while the join operation depends on $\mathcal{Q}$; all that can be said is that $I \vee J$ is the least $\mathcal{Q}$-ideal that contains $I \cup J$ (or, equivalently, contains $I + J$).

It is interesting to find that some particular quasivariety of rings is relatively ideal distributive. Any distributive algebraic lattice is isomorphic to the lattice of open sets of a topology defined on the set of meet irreducible lattice elements. Therefore, if $\mathcal{Q}$ is relatively ideal distributive, then to each member of $\mathcal{Q}$ there is a naturally associated topological space, its $\mathcal{Q}$-spectrum. It is possible to treat a member $R \in \mathcal{Q}$ as a ring of functions defined over its $\mathcal{Q}$-spectrum. It turns out that the quasivariety of commutative reduced rings, mentioned earlier as an example, is relatively ideal distributive, and for this $\mathcal{Q}$ the $\mathcal{Q}$-spectrum of any $R \in \mathcal{Q}$ is just the ordinary prime spectrum of R.

The main result of this section is that the class of commutative rings whose principal ideals have unique generators is a relatively ideal distributive quasivariety, and for which we provide an axiomatization.

Theorem 1. *Let $\mathcal{Q}$ be the class of all commutative rings (with 1) having the property that each principal ideal has a unique generator. Let $\mathcal{D}$ be the class of domains in $\mathcal{Q}$.*

(1) *$\mathcal{Q}$ is a quasivariety. It is exactly the class of rings axiomatized by the quasi-identity $(xyz = z) \to (yz = z)$ along with the identities defining the variety of all commutative rings. All rings in $\mathcal{Q}$ have trivial unit group and are reduced. Such rings are $\mathbb{F}_2$-algebras.*
(2) *$\mathcal{D}$ is exactly the class of domains with trivial unit group.*
(3) *$\mathcal{Q}$ consist of the subrings of products of members of $\mathcal{D}$ (we write $\mathcal{Q} = \mathsf{SP}(\mathcal{D})$).*
(4) *$\mathcal{Q}$ is a relatively ideal distributive quasivariety.*
(5) *The class of locally finite algebras in $\mathcal{Q}$ is the class of Boolean rings. This class is the largest subvariety of $\mathcal{Q}$.*

Proof. We argue the first two claims of Item (1) together. Namely, we show that $R \in \mathcal{Q}$ if and only if R belongs to the quasivariety of commutative rings satisfying $(xyz = z) \to (yz = z)$.

For the "if" part, let R be a commutative ring satisfying $(xyz = z) \to (yz = z)$. Choose $r \in R$ and assume that $(r) = (s)$ for some s. Then $s = qr$ and $r = ps$ for some $p, q \in R$. Since $pqr = r$, the quasi-identity yields $qr = r$, or $s = r$. Thus, $(r) = (s)$ implies $r = s$, showing that R satisfies the unique generator property for principal ideals. Conversely, for "only if", suppose that R does not satisfy $(xyz = z) \to (yz = z)$. R must have elements p, q, r such that $pqr = r$ and $qr \neq r$, Then $(r) = (qr)$ and $qr \neq r$, so R does not have the unique generator property.

For the second to last statement of Item (1), suppose that $R \in \mathcal{Q}$ and that u is a unit in R. Then $(u) = R = (1)$, so by the unique generator property $u = 1$. Also, to see that R is reduced, assume that $n \in R$ satisfies $n^2 = 0$. Then $1 + n$ is a unit (with inverse $1 - n$), so $1 + n = 1$, so $n = 0$.

For the final statement of Item (1), the fact that any $R \in \mathcal{Q}$ is an $\mathbb{F}_2$-algebra follows from the fact that -1 is a unit, so $1 = -1$. Then the prime subring of R is isomorphic to $\mathbb{F}_2$, which is enough to establish that R is an $\mathbb{F}_2$-algebra.

For Item (2), if $D \in \mathcal{D}$, then D is a domain by definition, and it has trivial unit group by Item (1). Conversely, suppose that D is a domain with trivial unit group. If $(a) = (b)$ in D, then a and b must differ by a unit, hence $a = b$, showing that D has the unique generator property, so D is a domain in $\mathcal{Q}$, yielding $D \in \mathcal{D}$.

In order to establish Item (3) we first prove a claim.

Claim 2. *If $R \in \mathcal{Q}$ and $S \subseteq R$ is a subset, then the annihilator $A = ann(S)$ is a $\mathcal{Q}$-ideal (meaning that $R/A \in \mathcal{Q}$).*

Proof of claim. For this we must verify that R/A satisfies the quasi-identity $(xyz = z) \to (yz = z)$. Equivalently, we must show that if $x, y, z \in R$ and $xyz \equiv z \pmod{A}$, then $yz \equiv z \pmod{A}$. We begin: If $xyz \equiv z \pmod{A}$, then $(xyz - z) \in A$, so $(xyz - z)s = 0$ for any $s \in S$. This means that $xy(zs) = (zs)$ for any $s \in S$. Applying the quasi-identity from Item (1) with zs in place of z we derive that $yzs = zs$, or $(yz - z)s = 0$ for any $r \in I$. Hence $yz \equiv z \pmod{A}$, as desired.

Next we argue that if $R \in \mathcal{Q}$ is not a domain, then R has disjoint nonzero $\mathcal{Q}$-ideals I and J. If R is not a domain, then there exist nonzero r and s such that $rs = 0$. Take $I = \mathrm{ann}(r)$ and $J = \mathrm{ann}(I)$. I is nonzero since it contains s, and J is nonzero since it contains r. Both I and J are $\mathcal{Q}$-ideals by Claim 2. If $t \in I \cap J$, then $t^2 \in IJ = \{0\}$, so t is nilpotent. According to Item (1), any $R \in \mathcal{Q}$ is reduced, so $t = 0$. Thus I and J are indeed disjoint nonzero $\mathcal{Q}$-ideals.

The argument for Item (3) is completed by noting that any quasivariety $\mathcal{Q}$ is expressible as $\mathsf{SP}(\mathcal{K})$ where $\mathcal{K}$ is the subclass of relatively subdirectly irreducible members of $\mathcal{Q}$. This is a version of Birkhoff's subdirect representation theorem, stated for quasivarieties, and it holds for quasivarieties because relative ideal/congruence lattices are algebraic. The previous paragraph shows that the only members of $\mathcal{Q}$ that could possibly be relatively subdirectly irreducible are the domains. (That is, R not a domain $\Rightarrow$ R has disjoint nonzero $\mathcal{Q}$-ideals $\Rightarrow$ R is not relatively subdirectly irreducible.)

To prove Item (4), we refer to general criteria from [3] for proving that a quasivariety is relatively congruence distributive. Specifically, we will use Theorems 4.1 and 4.3 of that paper, along with some of the remarks between those theorems.

Here is a summary of what we are citing. From Theorem 4.1 of [3], a quasivariety is relatively congruence modular if and only if it satisfies the "extension principle" and the "relative shifting lemma". From remarks following the proof of Theorem 4.1, the "extension principle" can be replaced by the "weak extension principle". From Theorem 2.1 of that paper, the "relative shifting lemma" can be replaced by the "existence of quasi-Day terms". Finally, from Theorem 4.3 of that paper, a quasivariety is relatively congruence distributive if and only if it is relatively congruence modular and no member has a nonzero abelian congruence.

What this reduces to in our setting is this: to prove that our quasivariety $\mathcal{Q}$ is relatively ideal distributive (Item (4)) it suffices to show that $\mathcal{Q}$

(i) has "quasi-Day terms",
(ii) satisfies the "weak extension principle", and
(iii) has no member with a nontrivial abelian congruence (i.e., with a nonzero ideal A satisfying $A^2 = 0$).

Condition (i) holds since $\mathcal{Q}$ has ordinary Day terms, in fact a Maltsev term. (More explicitly, the singleton set $\Sigma_s := \{(p(w,x,y,z), q(w,x,y,z))\}$ where $p(w,x,y,z) := w - x + y$ and $q(w,x,y,z) := z$ meets the defining conditions from Theorem 2.1(2) of [3] for "quasi-Day terms".)

Condition (iii) holds since if $A \lhd R \in \mathcal{Q}$ and $A^2 = 0$, then the elements of A are nilpotent. As argued in the proof of Item (1), the only nilpotent element in R is 0, hence $A = 0$.

Condition (ii) means that if $R \in \mathcal{Q}$ has disjoint ideals I and J, then I and J can be extended to $\mathcal{Q}$-ideals $\overline{I} \supseteq I$ and $\overline{J} \supseteq J$ that are also disjoint. To prove that Condition (ii) holds we modify an argument from above: If R has ideals I and J such that $I \cap J = 0$, then $IJ = 0$. The $\mathcal{Q}$-ideal $\overline{J} = \text{ann}(I)$ contains J, the $\mathcal{Q}$-ideal $\overline{I} = \text{ann}(\overline{J})$ contains I, both are $\mathcal{Q}$-ideals, and $\overline{I} \cap \overline{J} = 0$ (since the elements in this intersection square to zero and R is reduced). This shows that disjoint ideals I and J may be extended to disjoint $\mathcal{Q}$-ideals.

For Item (5), to show that a locally finite ring in $\mathcal{Q}$ is a Boolean ring it suffices to show that any finite ring $F \in \mathcal{Q}$ is Boolean. (The reason this reduction is permitted is that the property of being a Boolean ring is expressible by the identity $x^2 = x$, and a locally finite structure satisfies a universal sentence if and only if its finite substructures satisfy the sentence.)

So choose a finite $F \in \mathcal{Q}$. As F has trivial unit group, and $1 + \text{rad}(F) \subseteq U(F)$, we get that F must be semiprimitive. Since F is finite it must be a product of fields. Since F has only trivial units, each factor field must have size 2, so F is Boolean.

Conversely, if B is any Boolean ring, then multiplication is a semilattice operation, so $xyz \leq yz \leq z$ in the semilattice order for any $x, y, z \in B$. If, in B, we have first = last ($xyz = z$), then we must have middle = last ($yz = z$). Hence $B \in \mathcal{Q}$.

To complete the proof of Item (5) we must show that if $\mathcal{V}$ is a variety and $\mathcal{V} \subseteq \mathcal{Q}$, then $\mathcal{V}$ consists of Boolean rings. For this it suffices to show that if $R \in \mathcal{Q}$ is not Boolean (i.e., R has an element r satisfying $r \neq r^2$), then $R \notin \mathcal{V}$. This holds because $\langle r^2 \rangle \subsetneq \langle r \rangle$ by the unique generator property, so $r/\langle r^2 \rangle$ is a nonzero nilpotent element of $R/\langle r^2 \rangle$, establishing that some homomorphic image of R is not in $\mathcal{Q}$. □

By substituting $z = 1$ in the quasi-identity $(xyz = z) \to (yz = z)$ we obtain the consequence $(xy = 1) \to (y = 1)$, which expresses that the unit group is trivial. Since a consequence can be no stronger than the original statement, this is enough to deduce that the quasivariety of commutative rings with trivial unit group contains the quasivariety of commutative rings whose principal ideals have unique generators. This containment is proper, and the following example describes a commutative ring satisfying $(xy = 1) \to (y = 1)$ but not $(xyz = z) \to (yz = z)$.

Example 3. Let R be the commutative $\mathbb{F}_2$-algebra presented by

$$\langle X, Y, Z \mid XYZ = Z \rangle.$$

That is, R is the quotient of the polynomial ring $\mathbb{F}_2[X, Y, Z]$ by the ideal $(XYZ - Z)$.

We may view the relation $XYZ - Z = 0$ as a reduction rule $XYZ \to Z$ to produce a normal form for elements of R. This single rule is applied as follows: choose a monomial of the form $XYZW$ (W is a product of variables) of an element in a coset of $(XYZ - Z) \subseteq \mathbb{F}_2[X, Y, Z]$ and replace $XYZW$ by ZW. That is, if each of X, Y, Z appear in a monomial, we delete one instance of X and one instance of Y from that monomial.

The Diamond Lemma applies to show that there is a normal form for elements of R, and the elements in normal form are exactly the polynomials over $\mathbb{F}_2$ in the generators X, Y, Z where no monomial is divisible by each of X, Y, and Z.

Note that each application of the reduction rule reduces the X-degree and the Y-degree of some monomial, but does not alter the Z-degree of any monomial. This is enough to prove that the unit group of R is trivial. For if R had a unit u with inverse v, then the Z-degree of the product $uv = 1$ is zero, but it is also the sum of the Z-degrees of u and v. Hence the normal form of a unit must be Z-free. But then u and v would then be inverse units in the subring $\mathbb{F}_2[X, Y]$, where all elements are in normal form. Now one can argue in this subring, using X-degree and Y-degree, to conclude that none of X, Y, Z appear in the normal form of a unit. We are left with $u = v = 1$ as the only possibility.

Notice also that $YZ - Z$ is in normal form, so $YZ - Z \neq 0$ in R. This shows that R fails to satisfy $(xyz = z) \rightarrow (yz = z)$, but does satisfy $(xy = 1) \rightarrow (y = 1)$. In particular, the fact that $XYZ = Z$ while $YZ \neq Z$ means that $(YZ) = (Z)$, while $YZ \neq Z$, so the principal ideal (Z) does not have a unique generator.

3 Some Related Quasivarieties

We saw in the previous section that the class of commutative rings whose principal ideals have unique generators is the quasivariety generated by the class of domains with trivial unit group. We also saw that this quasivariety is relatively ideal distributive, and that it is axiomatized by the quasi-identity $(xyz = z) \rightarrow (yz = z)$.

In this section we will show that the quasivariety $\mathcal{Q}_{|n}$ generated by those domains D whose unit group $U(D)$ is cyclic of order dividing n is also relatively ideal distributive, and we shall provide an axiomatization for $\mathcal{Q}_{|n}$.

Write $\mathcal{D}_{|n}$ for the class of domains whose unit group is cyclic of order dividing n.

Theorem 4. *By definition, we have that $\mathcal{Q}_{|n}$ is the quasivariety generated by $\mathcal{D}_{|n}$.*

(1) $\mathcal{Q}_{|n}$ *is axiomatized* by

  (a) *the identities defining commutative rings,*
  (b) *the quasi-identity* $(x^2 = 0) \rightarrow (x = 0)$, *which expresses that the only nilpotent element is* 0, *and*
  (c) *the quasi-identity* $(xyz = z) \rightarrow (y^n z = z)$.

(2) $\mathcal{Q}_{|n}$ *is a relatively ideal distributive quasivariety.*

Proof. To prove Item (1), let $\mathcal{K}$ be the quasivariety axiomatized by the sentences in (a), (b), and (c). It is easy to see that $\mathcal{D}_{|n}$ satisfies the quasi-identities in (a), (b), and (c), so $\mathcal{D}_{|n} \subseteq \mathcal{K}$, and therefore $\mathcal{Q}_{|n} \subseteq \mathcal{K}$.

Conversely, we must show that $\mathcal{K} \subseteq \mathcal{Q}_{|n}$ For this, we need the analogue of Claim 2 for $\mathcal{K}$:

Claim 5. *If $R \in \mathcal{K}$ and $S \subseteq R$ is a subset, then the annihilator $A = ann(S)$ is a $\mathcal{K}$-ideal.*

Proof of claim. Our goal is to prove that $R/A \in \mathcal{K}$, so we must prove that R/A is a commutative ring satisfying $(x^2 = 0) \to (x = 0)$ and $(xyz = z) \to (y^n z = z)$. It is clear that R/A is a commutative ring (identities are preserved under quotients), so we only need to verify that R/A satisfies $(x^2 = 0) \to (x = 0)$ and $(xyz = z) \to (y^n z = z)$. For the second of these, the proof is exactly like the proof of Claim 2, while for the first there is an extra idea. We prove the first only.

To prove that R/A satisfies $(x^2 = 0) \to (x = 0)$, we must show that R satisfies $x^2 \equiv 0 \pmod{A}$ implies $x \equiv 0 \pmod{A}$. If $x^2 \equiv 0 \pmod{A}$, or $x^2 \in A$, then $x^2 s = 0$ for all $s \in S$. This implies $(xs)^2 = (x^2 s)s = 0$ for all $s \in S$. (This is the "extra idea".) But R satisfies $(x^2 = 0) \to (x = 0)$, so from $(xs)^2 = 0$ we deduce $xs = 0$ for all $s \in S$. This proves that $x \in A$ or $x \equiv 0 \pmod{A}$.

We will use Claim 5 the same way we used Claim 2 in the proof of Theorem 1. If $R \in \mathcal{K}$ is not a domain, then there exist nonzero r and s such that $rs = 0$. Take $I = \text{ann}(r)$ and $J = \text{ann}(I)$. I is nonzero since it contains s, and J is nonzero since it contains r. By Claim 5, I and J are $\mathcal{K}$-ideals. Any element in $I \cap J$ must square to zero, so since $\mathcal{K}$ satisfies axiom (b) we get $I \cap J = \{0\}$. Thus, if R is not a domain, then it has a pair of nonzero, disjoint, $\mathcal{K}$-ideals. This is enough to guarantee that R is not subdirectly irreducible relative to $\mathcal{K}$.

In the contrapositive form, we have shown that any relatively subdirectly irreducible member of $\mathcal{K}$ is a domain. Hence $\mathcal{K}$ is generated by its subclass of domains.

But if $D \in \mathcal{K}$ is a domain, then by substituting $z = 1$ in the quasi-identity (1)(c) we obtain that D satisfies $(xy = 1) \to (y^n = 1)$. This implies that the unit group $U(D)$ of D is a cyclic group of order dividing n. The reason for this is that $U(D)$ is an abelian group satisfying $x^n = 1$, hence $U(D)$ is a locally finite abelian group. If $U(D)$ is not cyclic, then it contains a finite noncyclic subgroup $G \subseteq U(D)$. But now G is a finite noncyclic subgroup of the field of fractions of D, and we all know that the multiplicative group of a field contains no finite noncyclic subgroup. This shows that the domains in $\mathcal{K}$ lie in $\mathcal{D}_{|n}$, so $\mathcal{K}$ is contained in the quasivariety generated by $\mathcal{D}_{|n}$, which is $\mathcal{Q}_{|n}$.

Item (2) of this theorem is proved exactly like Item (4) of Theorem 1. □

Observations 6. A quick test to rule out that some nonzero ring R belongs to some quasivariety $\mathcal{Q}_{|n}$ is to show that the prime subring of R does not belong to $\mathcal{Q}_{|n}$. Since the prime subring of R is isomorphic either to $\mathbb{Z}$ or to $\mathbb{Z}_k$ for some $k > 1$, and since the units of $\mathbb{Z}$ and $\mathbb{Z}_k$ are easy to determine, it is not hard to derive some consequences.

Namely, $\mathbb{Z}_k$ satisfies the quasi-identity in Theorem 4(1)(b) if and only if k is square-free, and hence $k = p_1 \dots p_m$ and $\mathbb{Z}_k \cong \mathbb{Z}_{p_1} \times \cdots \times \mathbb{Z}_{p_m}$ for distinct primes $p_1, \dots, p_m$. Now, for such a k, $\mathbb{Z}_k$ satisfies the quasi-identity in Theorem 4(1)(c) if and only if $p_i - 1$ divides n for each i. Therefore, these conditions on k are necessary for R to belong to $\mathcal{Q}_{|n}$ whenever the prime subring of R is isomorphic to $\mathbb{Z}_k$. Similarly, $\mathbb{Z}$ satisfies the quasi-identity in Theorem 4(1)(c) if and only if n is even. Hence, if the prime subring of R is isomorphic to $\mathbb{Z}$, then for R to belong to $\mathcal{Q}_{|n}$ the number n must be even.

These considerations imply, in particular, that if n is odd, then $\mathbb{Z}_k \notin \mathcal{Q}_{|n}$ unless $k = 2$, and $\mathbb{Z} \notin \mathcal{Q}_{|n}$. Hence, each ring in $\mathcal{Q}_{|n}$ may be thought of as an $\mathbb{F}_2$-algebra.

Notice also that, when n is odd, the axiom $(x^2 = 0) \to (x = 0)$ from Theorem 4(1)(b) is a consequence of the axiom $(xyz = z) \to (y^n z = z)$ from Theorem 4(1)(c). For if R satisfies $(xyz = z) \to (y^n z = z)$ for some odd n, and some $r \in R$ satisfies $r^2 = 0$, then as observed in the previous paragraph the characteristic of R must be 2, so $(1 + r)^2 = 1$. This implies that $1 + r$ is a unit of order dividing 2 (and also n), so necessarily $1 + r = 1$, which implies that $r = 0$.

We can use Observation 6 to show that not all the quasivarieties $\mathcal{Q}_{|n}$ are distinct, in particular

Theorem 7. *If p is an odd prime, then $\mathcal{Q}_{|p} = \mathcal{Q}_{|1}$ unless p is a Mersenne prime.*

Proof. To prove that $\mathcal{Q}_{|p} = \mathcal{Q}_{|1}$ when p is an odd non-Mersenne prime, it will suffice to show that these quasivarieties contain the same domains. We always have $\mathcal{Q}_{|m} \subseteq \mathcal{Q}_{|n}$ when $m \mid n$, from the definition of these quasivarieties, so we must show that any domain $D \in \mathcal{Q}_{|p}$ is contained in $\mathcal{Q}_{|1}$ (i.e., has a trivial unit group).

Choose $D \in \mathcal{Q}_{|p}$. From Observation 6, we know (since p is odd) that D is an $\mathbb{F}_2$-algebra. Suppose that $\theta \in D$ is a nontrivial unit. Since θ has finite multiplicative order, and the prime subring of D is finite, the subring $S \subseteq D$ generated by θ is finite. S is a subring of a domain itself, hence it is a field, and $U(S) = S^{\times}$. S belongs to $\mathcal{Q}_{|p}$, so $S^{\times}$ has order dividing p, and it must therefore be that $|S^{\times}| = p$. This shows that S is a finite field of characteristic 2 and of cardinality $|S| = p + 1$. We derive that $p + 1 = 2^s$ for some s, or $p = 2^s - 1$. This completes the proof that $\mathcal{Q}_{|p} = \mathcal{Q}_{|1}$ unless p is a Mersenne prime.

The primality of p did not play a big role in the proof. The same argument shows that if n is any odd number, then $\mathcal{Q}_{|n} = \mathcal{Q}_{|1}$ unless n is divisible by some number $x > 1$ of the form $x = 2^s - 1$. So, for example, $\mathcal{Q}_{|55} = \mathcal{Q}_{|25} = \mathcal{Q}_{|1}$. But if n is divisible by some number $x > 1$ of the form $x = 2^s - 1$, then $\mathcal{Q}_{|n}$ will contain some finite fields that are not in $\mathcal{Q}_{|1}$.

Acknowledgements This material is based upon work supported by the National Research, Development and Innovation Office NKFIH K119934, the Vietnam Institute for Advanced Study in Mathematics (VIASM), the Vietnamese Institute of Mathematics, the National Science Foundation grant no. DMS 1500254, the Hungarian National Foundation for Scientific Research (OTKA) grant no. K115518, and the National Research, Development and Innovation Fund of Hungary (NKFI) grant no. K128042.

References

1. Kearnes, K.A.: Relatively congruence distributive subquasivarieties of a congruence modular variety. Bull. Austral. Math. Soc. **41**(1), 87–96 (1990). https://doi.org/10.1017/S0004972700017871
2. Kearnes K.A.: Relatively congruence modular quasivarieties of modules. In: Czelakowski, J. (eds.) Don Pigozzi on Abstract Algebraic Logic, Universal Algebra, and Computer Science. Outstanding Contributions to Logic, vol. 16, pp. 221–232. Springer, Cham (2018). https://doi.org/10.1007/978-3-319-74772-9_8
3. Kearnes, K., McKenzie, R.: Commutator theory for relatively modular quasivarieties. Trans. Amer. Math. Soc. **331**(2), 465–502 (1992). https://doi.org/10.2307/2154123

On Monoids of Ideals of Orders in Quadratic Number Fields

Johannes Brantner, Alfred Geroldinger, and Andreas Reinhart

Abstract We determine the set of catenary degrees, the set of distances, and the unions of sets of lengths of the monoid of nonzero ideals and of the monoid of invertible ideals of orders in quadratic number fields.

Keywords Orders · Quadratic number fields · Sets of lengths · Sets of distances · Catenary degree

2010 Mathematics Subject Classification 11R11 · 11R27 · 13A15 · 13F15 · 20M12 · 20M13

1 Introduction

Factorization theory for Mori domains and their semigroups of ideals splits into two cases. The first and best understood case is that of Krull domains (i.e., of completely integrally closed Mori domains). The arithmetic of a Krull domain depends only on the class group and on the distribution of prime divisors in the classes, and it can be studied—at least to a large extent—with methods from additive combinatorics. The link to additive combinatorics is most powerful when the Krull domain has a finite class group and when each class contains at least one prime divisor (this holds true,

J. Brantner · A. Geroldinger (✉)
Institut Für Mathematik und Wissenschaftliches Rechnen, Karl-Franzens-Universität Graz, NAWI Graz, Heinrichstraße 36, 8010 Graz, Austria
e-mail: alfred.geroldinger@uni-graz.at
URL: https://imsc.uni-graz.at/geroldinger

J. Brantner
e-mail: j.brantner@gmx.at

A. Reinhart
Department of Mathematical Sciences, New Mexico State University, Las Cruces, NM 88003, USA
e-mail: andreas.reinhart@uni-graz.at

A. Facchini et al. (eds.), *Advances in Rings, Modules and Factorizations*, Springer Proceedings in Mathematics & Statistics 321,
https://doi.org/10.1007/978-3-030-43416-8_2

among others, for rings of integers in number fields). Then sets of lengths, sets of distances, and of catenary degrees of the domain can be studied in terms of zero-sum problems over the class group. Moreover, we obtain a variety of explicit results for arithmetical invariants in terms of classical combinatorial invariants (such as the Davenport constant of the class group) or even in terms of the group invariants of the class group. We refer to [15] for a description of the link to additive combinatorics and to the recent survey [32] discussing explicit results for arithmetical invariants.

Let us consider Mori domains that are not completely integrally closed but have a nonzero conductor toward their complete integral closure. The best investigated classes of such domains are weakly Krull Mori domains with finite v-class group and C-domains. For them there is a variety of abstract arithmetical finiteness results but in general there are no precise results. For example, it is well-known that sets of distances and of catenary degrees are finite but there are no reasonable bounds for their size. The simplest not completely integrally closed Mori domains are orders in number fields. They are one-dimensional noetherian with nonzero conductor, finite Picard group, and all factor rings modulo nonzero ideals are finite. Thus they are weakly Krull domains and C-domains. Although there is recent progress for seminormal orders, for general orders in number fields there is no characterization of half-factoriality (for progress in the local case see [26]) and there is no information on the structure of their sets of distances or catenary degrees (neither for orders nor for their monoids of ideals).

In the present paper, we focus on monoids of ideals of orders in quadratic number fields and establish precise results for their set of distances $\Delta(\cdot)$ and their set of catenary degrees $\mathrm{Ca}(\cdot)$. Orders in quadratic number fields are intimately related to quadratic irrationals, continued fractions, and binary quadratic forms and all these areas provide a wealth of number theoretic tools for the investigation of orders. We refer to [25] for a modern presentation of these connections and to [9, 29] for recent progress on the arithmetic and ideal theoretic structure of quadratic orders.

Let $\mathcal{O}$ be an order in a quadratic number field, $\mathcal{I}^*(\mathcal{O})$ be the monoid of invertible ideals, and $\mathcal{I}(\mathcal{O})$ be the monoid of nonzero ideals (note that $\mathcal{I}(\mathcal{O})$ is not cancellative if $\mathcal{O}$ is not maximal). Since $\mathcal{I}^*(\mathcal{O})$ is a divisor-closed submonoid of $\mathcal{I}(\mathcal{O})$, the set of catenary degrees and the set of distances of $\mathcal{I}^*(\mathcal{O})$ are contained in the respective sets of $\mathcal{I}(\mathcal{O})$. We formulate a main result of this paper and then we compare it with related results in the literature.

Theorem 1.1. *Let $\mathcal{O}$ be an order in a quadratic number field K with discriminant d_K and conductor $\mathfrak{f} = f\mathcal{O}_K$ for some $f \in \mathbb{N}_{\geq 2}$.*

1. *The following statements are equivalent:*
   (a) *$\mathcal{I}(\mathcal{O})$ is half-factorial.*
   (b) $\mathsf{c}\big(\mathcal{I}(\mathcal{O})\big) = 2$.
   (c) $\mathsf{c}\big(\mathcal{I}^*(\mathcal{O})\big) = 2$.
   (d) *$\mathcal{I}^*(\mathcal{O})$ is half-factorial.*
   (e) *f is squarefree and all prime divisors of f are inert.*

2. *Suppose that* $\mathcal{I}^*(\mathcal{O})$ *is not half-factorial.*
   (a) *If* f *is squarefree, then* $\mathsf{Ca}(\mathcal{I}(\mathcal{O})) = [1, 3]$, $\mathsf{Ca}(\mathcal{I}^*(\mathcal{O})) = [2, 3]$, $\Delta(\mathcal{I}(\mathcal{O})) = \Delta(\mathcal{I}^*(\mathcal{O})) = \{1\}$.
   (b) *Suppose that* f *is not squarefree.*
      (i) *If* $\mathsf{v}_2(f) \notin \{2, 3\}$ *or* $d_K \not\equiv 1 \mod 8$, *then* $\mathsf{Ca}(\mathcal{I}(\mathcal{O})) = [1, 4]$, $\mathsf{Ca}(\mathcal{I}^*(\mathcal{O})) = [2, 4]$, *and* $\Delta(\mathcal{I}(\mathcal{O})) = \Delta(\mathcal{I}^*(\mathcal{O})) = [1, 2]$.
      (ii) *If* $\mathsf{v}_2(f) \in \{2, 3\}$ *and* $d_K \equiv 1 \mod 8$, *then* $\mathsf{Ca}(\mathcal{I}(\mathcal{O})) = [1, 5]$, $\mathsf{Ca}(\mathcal{I}^*(\mathcal{O})) = [2, 5]$, *and* $\Delta(\mathcal{I}(\mathcal{O})) = \Delta(\mathcal{I}^*(\mathcal{O})) = [1, 3]$.

We say that a cancellative monoid H is *weakly Krull* if $\bigcap_{P \in \mathfrak{X}(H)} H_P = H$ and $\{P \in \mathfrak{X}(H) \mid a \in P\}$ is finite for each $a \in H$ (where $\mathfrak{X}(H)$ denotes the set of height-one prime ideals of H). Moreover, a cancellative monoid H is called *weakly factorial* if every nonunit of H is a finite product of primary elements of H. Let all notation be as in Theorem 1.1, and recall that $\mathcal{I}^*(\mathcal{O})$ is a weakly factorial C-monoid, and that for every atomic monoid H with $\Delta(H) \neq \emptyset$ we have $\min \Delta(H) = \gcd \Delta(H)$.

There is a characterization (due to Halter-Koch) when the order $\mathcal{O}$ is half-factorial [16, Theorem 3.7.15]. This characterization and Theorem 1.1 or [30, Corollary 4.6] show that the half-factoriality of $\mathcal{O}$ implies the half-factoriality of $\mathcal{I}^*(\mathcal{O})$. Consider the case of seminormal orders whence suppose that $\mathcal{O}$ is seminormal. Then f is squarefree (this follows from an explicit characterization of seminormal orders given by Dobbs and Fontana in [10, Corollary 4.5]). Moreover, $\mathcal{I}^*(\mathcal{O})$ is seminormal and if $\mathcal{I}^*(\mathcal{O})$ is not half-factorial, then its catenary degree equals three by [18, Theorems 5.5 and 5.8]. Clearly, this coincides with 2.(a) of the above theorem. Among others, Theorem 1.1 shows that the sets of distances and of catenary degrees are intervals and that the minimum of the set of distances equals 1. We discuss some analogous results and some results which are in sharp contrast to this. If H is a Krull monoid with finite class group, then H is a weakly Krull C-monoid and if there are prime divisors in all classes, then the sets $\mathsf{Ca}(H)$ and $\Delta(H)$ are intervals [23, Theorem 4.1]. On the other hand, for every finite set $S \subset \mathbb{N}$ with $\min S = \gcd S$ (resp. every finite set $S \subset \mathbb{N}_{\geq 2}$) there is a finitely generated Krull monoid H such that $\Delta(H) = S$ (resp. $\mathsf{Ca}(H) = S$) [21] resp. [11, Proposition 3.2]. Just as the monoids of ideals under discussion, every numerical monoid is a weakly factorial C-monoid. However, in contrast to them, the set of distances need not be an interval [8], its minimum need not be 1 [5, Proposition 2.9], and a recent result of O'Neill and Pelayo [28] shows that for every finite set $S \subset \mathbb{N}_{\geq 2}$ there is a numerical monoid H such that $\mathsf{Ca}(H) = S$.

We proceed as follows. In Section 2 we summarize the required background on the arithmetic of monoids. In Section 3 we do the same for orders in quadratic number fields and we provide an explicit description of (invertible) irreducible ideals in orders of quadratic number fields (Theorem 3.6). In Section 4 we give the proof of Theorem 1.1. Based on this result we establish a characterization of those orders $\mathcal{O}$ with $\min \Delta(\mathcal{O}) > 1$ (Theorem 4.14) which allows us to give the first explicit examples of orders $\mathcal{O}$ with $\min \Delta(\mathcal{O}) > 1$. Our third main result (given in Theorem 5.2) states that unions of sets of lengths of $\mathcal{I}(\mathcal{O})$ and of $\mathcal{I}^*(\mathcal{O})$ are intervals.

2 Preliminaries on the Arithmetic of Monoids

Let $\mathbb{N}$ be the set of positive integers, $\mathbb{P} \subset \mathbb{N}$ the set of prime numbers, and for every $m \in \mathbb{N}$, we denote by

$$\varphi(m) = \left|(\mathbb{Z}/m\mathbb{Z})^\times\right| \quad Euler's\ \varphi\text{-}function.$$

For $a, b \in \mathbb{Q} \cup \{-\infty, \infty\}$, $[a, b] = \{x \in \mathbb{Z} \mid a \le x \le b\}$ denotes the discrete interval between a and b. Let $L, L' \subset \mathbb{Z}$. We denote by $L + L' = \{a + b \mid a \in L,\ b \in L'\}$ their *sumset*. A positive integer $d \in \mathbb{N}$ is called a *distance* of L if there exists a $k \in L$ such that $L \cap [k, k + d] = \{k, k + d\}$, and we denote by $\Delta(L)$ the *set of distances* of L. If $\emptyset \ne L \subset \mathbb{N}$, we denote by $\rho(L) = \sup L / \min L \in \mathbb{Q}_{\ge 1} \cup \{\infty\}$ the *elasticity* of L. We set $\rho(\{0\}) = 1$ and $\max \emptyset = \min \emptyset = \sup \emptyset = 0$. All rings and semigroups are commutative and have an identity element.

2.1 Monoids

Let H be a multiplicatively written commutative semigroup. We denote by $H^\times$ the group of invertible elements of H. We say that H is reduced if $H^\times = \{1\}$ and we denote by $H_{\text{red}} = \{aH^\times \mid a \in H\}$ the associated reduced semigroup of H. An element $u \in H$ is said to be cancellative if $au = bu$ implies that $a = b$ for all $a, b \in H$. The semigroup H is said to be

- *cancellative* if every element of H is cancellative.
- *unit-cancellative* if $a, u \in H$ and $a = au$ implies that $u \in H^\times$.

By definition, every cancellative semigroup is unit-cancellative. All semigroups of ideals, that are studied in this paper, are unit-cancellative but not necessarily cancellative.

Throughout this paper, a monoid means a commutative unit-cancellative semigroup with identity element.

Let H be a monoid. A submonoid $S \subset H$ is said to be *divisor-closed* if $a \in S$ and $b \in H$ with $b \mid a$ implies that $b \in S$. An element $u \in H$ is said to be

- *prime* if $u \notin H^\times$ and, for all $a, b \in H$, $u \mid ab$ and $u \nmid a$ implies $u \mid b$.
- *primary* if $u \notin H^\times$ and, for all $a, b \in H$, $u \mid ab$ and $u \nmid a$ implies $u \mid b^n$ for some $n \in \mathbb{N}$.
- *irreducible* (or an *atom*) if $u \notin H^\times$ and, for all $a, b \in H$, $u = ab$ implies that $a \in H^\times$ or $b \in H^\times$.

The monoid H is said to be *atomic* if every $a \in H \setminus H^\times$ is a product of finitely many atoms. If H satisfies the ascending chain condition (ACC) on principal ideals, then H is atomic [12, Lemma 3.1].

2.2 Sets of Lengths

For a set P, we denote by $\mathcal{F}(P)$ the free abelian monoid with basis P. Every $a \in \mathcal{F}(P)$ is written in the form

$$a = \prod_{p \in P} p^{\mathsf{v}_p(a)} \text{ with } \mathsf{v}_p(a) \in \mathbb{N}_0 \quad \text{and} \quad \mathsf{v}_p(a) = 0 \text{ for almost all } p \in P.$$

We call $|a| = \sum_{p \in P} \mathsf{v}_p(a)$ the length of a and $\operatorname{supp}(a) = \{p \in P \mid \mathsf{v}_p(a) > 0\} \subset P$ the support of a. Let H be an atomic monoid. The free abelian monoid $\mathsf{Z}(H) = \mathcal{F}(\mathcal{A}(H_{\text{red}}))$ denotes the *factorization monoid* of H and

$$\pi \colon \mathsf{Z}(H) \to H_{\text{red}} \quad \text{satisfying} \quad \pi(u) = u \text{ for all } u \in \mathcal{A}(H_{\text{red}})$$

denotes the *factorization homomorphism* of H. For every $a \in H$,

$$\begin{aligned} \mathsf{Z}_H(a) = \mathsf{Z}(a) = \pi^{-1}(aH^\times) \quad & \text{is the } \textit{set of factorizations} \text{ of } a \text{ and} \\ \mathsf{L}_H(a) = \mathsf{L}(a) = \{|z| \mid z \in \mathsf{Z}(a)\} \quad & \text{is the } \textit{set of lengths} \text{ of } a. \end{aligned}$$

For a divisor-closed submonoid $S \subset H$ and an element $a \in S$, we have $\mathsf{Z}(S) \subset \mathsf{Z}(H)$ whence $\mathsf{Z}_S(a) = \mathsf{Z}_H(a)$, and $\mathsf{L}_S(a) = \mathsf{L}_H(a)$. We denote by

- $\mathcal{L}(H) = \{\mathsf{L}(a) \mid a \in H\}$ the *system of sets of lengths* of H and by
- $\Delta(H) = \bigcup_{L \in \mathcal{L}(H)} \Delta(L) \subset \mathbb{N}$ the *set of distances* of H.

The monoid H is said to be *half-factorial* if $\Delta(H) = \emptyset$ and if H is not half-factorial, then $\min \Delta(H) = \gcd \Delta(H)$.

2.3 Distances and Chains of Factorizations

Let two factorizations $z, z' \in \mathsf{Z}(H)$ be given, say

$$z = u_1 \cdot \ldots \cdot u_\ell v_1 \cdot \ldots \cdot v_m \quad \text{and} \quad z' = u_1 \cdot \ldots \cdot u_\ell w_1 \cdot \ldots \cdot w_n \,,$$

where $\ell, m, n \in \mathbb{N}_0$ and all $u_i, v_j, w_k \in \mathcal{A}(H_{\text{red}})$ such that $v_j \ne w_k$ for all $j \in [1, m]$ and all $k \in [1, n]$. Then $\mathsf{d}(z, z') = \max\{m, n\}$ is the *distance* between z and z'. If $\pi(z) = \pi(z')$ and $z \ne z'$, then

$$1 + \big||z| - |z'|\big| \le \mathsf{d}(z, z') \text{ resp. } 2 + \big||z| - |z'|\big| \le \mathsf{d}(z, z') \text{ if } H \text{ is cancellative} \tag{2.1}$$

(see [12, Proposition 3.2] and [16, Lemma 1.6.2]). Let $a \in H$ and $N \in \mathbb{N}_0$. A finite sequence $z_0, \ldots, z_k \in \mathsf{Z}(a)$ is called an N-chain of factorizations (concatenating z_0 and z_k) if $\mathsf{d}(z_{i-1}, z_i) \le N$ for all $i \in [1, k]$. For $z, z' \in \mathsf{Z}(H)$ with $\pi(z) = \pi(z')$, we set $\mathsf{c}(z, z') = \min\{N \in \mathbb{N}_0 \mid z$ and z' can be concatenated by an N-chain of factorizations from $\mathsf{Z}(\pi(z))\}$. Then, for every $a \in H$,

$$\mathsf{c}(a) = \sup\{\mathsf{c}(z, z') \mid z, z' \in \mathsf{Z}(a)\} \in \mathbb{N}_0 \cup \{\infty\} \quad \text{is the } \textit{catenary degree} \text{ of } a.$$

Clearly, a has unique factorization (i.e., $|\mathsf{Z}(a)| = 1$) if and only if $\mathsf{c}(a) = 0$. We denote by

$$\mathrm{Ca}(H) = \{\mathsf{c}(a) \mid a \in H, \mathsf{c}(a) > 0\} \subset \mathbb{N} \quad \text{the } \textit{set of catenary degrees} \text{ of } H,$$

and then

$$\mathsf{c}(H) = \sup \mathrm{Ca}(H) \in \mathbb{N}_0 \cup \{\infty\} \quad \text{is the catenary degree of } H.$$

We use the convention that $\sup \emptyset = 0$ whence H is factorial if and only if $\mathsf{c}(H) = 0$. Note that $\mathsf{c}(a) = 0$ for all atoms $a \in H$. The restriction to positive catenary degrees in the definition of $\mathrm{Ca}(H)$ simplifies the statement of some results whence it is usual to restrict to elements with positive catenary degrees. If H is cancellative, then Equation (2.1) implies that $\min \mathrm{Ca}(H) \geq 2$ and

$$2 + \sup \Delta(H) \leq \mathsf{c}(H) \quad \text{if } H \text{ is not factorial.}$$

If $H = \coprod_{i \in I} H_i$, then a straightforward argument shows that

$$\mathrm{Ca}(H) = \bigcup_{i \in I} \mathrm{Ca}(H_i) \quad \text{whence} \quad \mathsf{c}(H) = \sup\{\mathsf{c}(H_i) \mid i \in I\}. \tag{2.2}$$

2.4 *Semigroups of Ideals*

Let R be a domain. We denote by $\mathsf{q}(R)$ its quotient field, by $\mathfrak{X}(R)$ the set of minimal nonzero prime ideals of R, and by $\overline{R}$ its integral closure. Then $R \setminus \{0\}$ is a cancellative monoid,

- $\mathcal{I}(R)$ is the semigroup of nonzero ideals of R (with usual ideal multiplication),
- $\mathcal{I}^*(R)$ is the subsemigroup of invertible ideals of R, and
- $\mathrm{Pic}(R)$ is the Picard group of R.

For every $I \in \mathcal{I}(R)$, we denote by $\sqrt{I}$ its radical and by $\mathcal{N}(I) = (R : I) = |R/I| \in \mathbb{N} \cup \{\infty\}$ its norm.

Let S be a Dedekind domain and $R \subset S$ a subring. Then R is called an *order* in S if one of the following two equivalent conditions hold:

- $\mathsf{q}(R) = \mathsf{q}(S)$ and S is a finitely generated R-module.
- R is one-dimensional noetherian and $\overline{R} = S$ is a finitely generated R-module.

Let R be an order in a Dedekind domain $S = \overline{R}$. We analyze the structure of $\mathcal{I}^*(R)$ and of $\mathcal{I}(R)$.

Since R is noetherian, Krull's intersection theorem holds for R whence $\mathcal{I}(R)$ is unit-cancellative [20, Lemma 4.1]. Thus $\mathcal{I}(R)$ is a reduced atomic monoid with identity R and $\mathcal{I}^*(R)$ is a reduced cancellative atomic divisor-closed submonoid. For the sake of clarity, we will say that an ideal of R is an ideal atom if it is an atom of the monoid $\mathcal{I}(R)$. If $I, J \in \mathcal{I}^*(R)$, then $I \mid J$ if and only if $J \subset I$. The prime elements of $\mathcal{I}^*(R)$ are precisely the invertible prime ideals of R. Every ideal is a product of primary ideals belonging to distinct prime ideals (in particular, $\mathcal{I}^*(R)$ is a weakly factorial monoid). Thus every ideal atom (i.e., every $I \in \mathcal{A}(\mathcal{I}(R))$ is primary, and if $\sqrt{I} = \mathfrak{p} \in \mathfrak{X}(R)$, then I is $\mathfrak{p}$-primary. Since $\overline{R}$ is a finitely generated R-module, the conductor $\mathfrak{f} = (R : \overline{R})$ is nonzero, and we set

$$\mathcal{P} = \{\mathfrak{p} \in \mathfrak{X}(R) \mid \mathfrak{p} \not\supset \mathfrak{f}\} \quad \text{and} \quad \mathcal{P}^* = \mathfrak{X}(R) \setminus \mathcal{P}.$$

Let $\mathfrak{p} \in \mathfrak{X}(R)$. We denote by

$$\mathcal{I}_{\mathfrak{p}}^*(R) = \{I \in \mathcal{I}^*(R) \mid \sqrt{I} \supset \mathfrak{p}\} \quad \text{and} \quad \mathcal{I}_{\mathfrak{p}}(R) = \{I \in \mathcal{I}(R) \mid \sqrt{I} \supset \mathfrak{p}\}$$

the set of invertible $\mathfrak{p}$-primary ideals of R and the set of $\mathfrak{p}$-primary ideals of R. Clearly, these are monoids and, moreover,

$$\mathcal{I}_{\mathfrak{p}}(R) \subset \mathcal{I}(R), \quad \mathcal{I}_{\mathfrak{p}}^*(R) \subset \mathcal{I}_{\mathfrak{p}}(R), \quad \text{and} \quad \mathcal{I}_{\mathfrak{p}}^*(R) \subset \mathcal{I}^*(R)$$

are divisor-closed submonoids. Thus $\mathcal{I}_{\mathfrak{p}}^*(R)$ is a reduced cancellative atomic monoid, $\mathcal{I}_{\mathfrak{p}}(R)$ is a reduced atomic monoid, and if $\mathfrak{p} \in \mathcal{P}$, then $\mathcal{I}_{\mathfrak{p}}^*(R) = \mathcal{I}_{\mathfrak{p}}(R)$ is free abelian. Since R is noetherian and one-dimensional,

$$\alpha : \mathcal{I}(R) \to \coprod_{\mathfrak{p} \in \mathfrak{X}(R)} \mathcal{I}_{\mathfrak{p}}(R), \quad \text{defined by} \quad \alpha(I) = (I_{\mathfrak{p}} \cap R)_{\mathfrak{p} \in \mathfrak{X}(R)} \tag{2.3}$$

is a monoid isomorphism which induces a monoid isomorphism

$$\alpha_{|\mathcal{I}^*(R)} : \mathcal{I}^*(R) \to \coprod_{\mathfrak{p} \in \mathfrak{X}(R)} \mathcal{I}_{\mathfrak{p}}^*(R). \tag{2.4}$$

3 Orders in Quadratic Number Fields

The goal of this section is to prove Theorem 3.6 which provides an explicit description of (invertible) ideal atoms of an order in a quadratic number field. These results are essentially due to Butts and Pall (see [6] where they are given in a different style), and they were summarized without proof by Geroldinger and Lettl in [19].

Unfortunately, that presentation is misleading in one case (namely, in case $p = 2$ and $d_K \equiv 5 \mod 8$). Thus we restate the results and provide a full proof.

First we put together some facts on orders in quadratic number fields and fix our notation which remains valid throughout the rest of this paper. For proofs, details, and any undefined notions, we refer to [25]. Let $d \in \mathbb{Z} \setminus \{0, 1\}$ be squarefree, $K = \mathbb{Q}(\sqrt{d})$ be a quadratic number field,

$$\omega = \begin{cases} \sqrt{d}, & \text{if } d \equiv 2, 3 \mod 4; \\ \frac{1+\sqrt{d}}{2}, & \text{if } d \equiv 1 \mod 4. \end{cases} \quad \text{and} \quad d_K = \begin{cases} 4d, & \text{if } d \equiv 2, 3 \mod 4; \\ d, & \text{if } d \equiv 1 \mod 4. \end{cases}$$

Then $\mathcal{O}_K = \mathbb{Z}[\omega]$ is the ring of integers and d_K is the discriminant of K. For every $f \in \mathbb{N}$, we define

$$\varepsilon \in \{0, 1\} \text{ with } \varepsilon \equiv f d_K \mod 2\,, \quad \eta = \frac{\varepsilon - f^2 d_K}{4}\,, \quad \text{and} \quad \tau = \frac{\varepsilon + f\sqrt{d_K}}{2}.$$

Then

$$\mathcal{O}_f = \mathbb{Z} \oplus f\omega\mathbb{Z} = \mathbb{Z} \oplus \tau\mathbb{Z}$$

is an order in $\mathcal{O}_K$ with conductor $\mathfrak{f} = f\mathcal{O}_K$, and every order in $\mathcal{O}_K$ has this form. With the notation of Section 2.4 we have

$$\mathcal{P}^* = \{\mathfrak{p} \in \mathfrak{X}(\mathcal{O}_f) \mid \mathfrak{p} \supset \mathfrak{f}\} = \{p\mathbb{Z} + f\omega\mathbb{Z} \mid p \in \mathbb{P}, p \mid f\}.$$

If $\alpha = a + b\sqrt{d} \in K$, then $\overline{\alpha} = a - b\sqrt{d}$ is its conjugate, $\mathcal{N}_{K/\mathbb{Q}}(\alpha) = \alpha\overline{\alpha} = a^2 - b^2 d$ is its norm, and $\mathrm{tr}(\alpha) = \alpha + \overline{\alpha} = 2a$ is its trace. For an $I \in \mathcal{I}(\mathcal{O}_f)$, $\overline{I} = \{\overline{\alpha} \mid \alpha \in I\}$ denotes the conjugate ideal. A simple calculation shows that

$$\mathcal{N}_{K/\mathbb{Q}}(r + \tau) = r^2 + \varepsilon r + \eta \quad \text{for each } r \in \mathbb{Z}.$$

If $\mathcal{O}$ is an order and $I \in \mathcal{I}^*(\mathcal{O})$, then $(\mathcal{O}_K : I\mathcal{O}_K) = (\mathcal{O} : I)$ and if $a \in \mathcal{O} \setminus \{0\}$, then

$$(\mathcal{O} : a\mathcal{O}) = (\mathcal{O}_K : a\mathcal{O}_K) = |\mathcal{N}_{K/\mathbb{Q}}(a)|$$

(see [17, Pages 99 and 100] and note that the factor rings $\mathcal{O}_K / I\mathcal{O}_K$ and $\mathcal{O}/I$ need not be isomorphic). For $p \in \mathbb{P}$ and for $a \in \mathbb{Z}$ we denote by $\left(\frac{a}{p}\right) \in \{-1, 0, 1\}$ the *Kronecker symbol* of a modulo p. A prime number $p \in \mathbb{Z}$ is called

- *inert* if $p\mathcal{O}_K \in \mathrm{spec}(\mathcal{O}_K)$.
- *split* if $p\mathcal{O}_K$ is a product of two distinct prime ideals of $\mathcal{O}_K$.
- *ramified* if $p\mathcal{O}_K$ is the square of a prime ideal of $\mathcal{O}_K$.

An odd prime

$$p \text{ is } \begin{cases} \text{inert} & \text{if } \left(\frac{d_K}{p}\right) = -1; \\ \text{split} & \text{if } \left(\frac{d_K}{p}\right) = 1; \\ \text{ramified} & \text{if } \left(\frac{d_K}{p}\right) = 0. \end{cases} \quad \text{and 2 is } \begin{cases} \text{inert} & \text{if } d_K \equiv 5 \mod 8; \\ \text{split} & \text{if } d_K \equiv 1 \mod 8; \\ \text{ramified} & \text{if } d_K \equiv 0 \mod 2. \end{cases}$$

Proposition 3.1. *Let p be a prime divisor of f, $\mathcal{O} = \mathcal{O}_f$, and $\mathfrak{p} = p\mathbb{Z} + f\omega\mathbb{Z}$.*

1. *The primary ideals with radical $\mathfrak{p}$ are exactly the ideals of the form*

$$\mathfrak{q} = p^{\ell}(p^m\mathbb{Z} + (r+\tau)\mathbb{Z})$$

with $\ell, m \in \mathbb{N}_0$, $\ell + m \geq 1$, $0 \leq r < p^m$ and $\mathcal{N}_{K/\mathbb{Q}}(r+\tau) \equiv 0 \mod p^m$. Moreover, $\mathcal{N}(\mathfrak{q}) = p^{2\ell+m}$.

2. *A primary ideal $\mathfrak{q} = p^{\ell}(p^m\mathbb{Z} + (r+\tau)\mathbb{Z})$ is invertible if and only if*

$$\mathcal{N}_{K/\mathbb{Q}}(r+\tau) \not\equiv 0 \mod p^{m+1}.$$

Proof. 1. Let $\mathfrak{q}$ be a $\mathfrak{p}$-primary ideal in $\mathcal{O}$. By [25, Theorem 5.4.2] there exist nonnegative integers ℓ, m, r such that $\mathfrak{q} = \ell(m\mathbb{Z} + (r+\tau)\mathbb{Z})$, $r < m$ and $\mathcal{N}_{K/\mathbb{Q}}(r+\tau) \equiv 0 \mod m$. Since $\mathfrak{q}$ is nonzero and proper, we have $\ell m > 1$. We prove, that ℓm is a power of p. First observe that $\mathfrak{q} \subset \sqrt{\mathfrak{q}} = \mathfrak{p}$ implies that $p \mid \ell m$. Assume to the contrary that there exists another rational prime $p' \neq p$ dividing ℓm, say $\ell m = p's$. But then $p's \in \mathfrak{q}$, $s \notin \mathfrak{q}$ and $p' \notin \mathfrak{p} = \sqrt{\mathfrak{q}}$. A contradiction to $\mathfrak{q}$ being primary. Conversely, assume that $\mathfrak{q} = p^{\ell}(p^m\mathbb{Z} + (r+\tau)\mathbb{Z})$ for integers $\ell, m \in \mathbb{N}_0$, $\ell + m \geq 1$, $0 \leq r < p^m$ and $\mathcal{N}_{K/\mathbb{Q}}(r+\tau) \equiv 0 \mod p^m$. By [25, Theorem 5.4.2], $\mathfrak{q}$ is an ideal of $\mathcal{O}$. Since $p \in \sqrt{\mathfrak{q}}$ and $\mathfrak{p}$ is the only prime ideal in $\mathcal{O}$ containing p we obtain that $\sqrt{\mathfrak{q}} = \bigcap_{\mathfrak{a} \in \operatorname{spec}(\mathcal{O}), \mathfrak{a} \supset \mathfrak{q}} \mathfrak{a} = \mathfrak{p}$. The nonzero prime ideal $\mathfrak{p}$ is maximal, since $\mathcal{O}$ is one-dimensional. Therefore, $\mathfrak{q}$ is $\mathfrak{p}$-primary. It follows from [25, Theorem 5.4.2] that $\mathcal{N}(\mathfrak{q}) = p^{2\ell+m}$.

2. By [25, Theorem 5.4.2], $\mathfrak{q} = p^{\ell}(p^m\mathbb{Z} + (r+\tau)\mathbb{Z})$ is invertible if and only if $\gcd\left(p^m, 2r+\varepsilon, \frac{\mathcal{N}_{K/\mathbb{Q}}(r+\tau)}{p^m}\right) = 1$. Since $p \mid f$ and $\mathcal{N}_{K/\mathbb{Q}}(r+\tau) = \frac{1}{4}((2r+\varepsilon)^2 - f^2 d_K)$, this is the case if and only if $p \nmid \frac{\mathcal{N}_{K/\mathbb{Q}}(r+\tau)}{p^m}$, that is $\mathcal{N}_{K/\mathbb{Q}}(r+\tau) \not\equiv 0 \mod p^{m+1}$. □

If $x \in \mathbb{Z}$ and $y \in \mathbb{N}$, then let $\operatorname{rem}(x, y)$ be the unique $z \in [0, y-1]$ such that $y \mid x - z$. Let p be a prime divisor of f. Note that $\mathrm{v}_p(0) = \infty$, and if $\emptyset \neq A \subset \mathbb{N}_0$, then $\min(A \cup \{\infty\}) = \min A$. We set

$$P_{f,p} = p\mathbb{Z} + f\omega\mathbb{Z}, \quad \mathcal{I}_p^*(\mathcal{O}_f) = \mathcal{I}_{P_{f,p}}^*(\mathcal{O}_f), \mathcal{I}_p(\mathcal{O}_f) = \mathcal{I}_{P_{f,p}}(\mathcal{O}_f), \quad \text{and}$$
$$\mathcal{M}_{f,p} = \{(x, y, z) \in \mathbb{N}_0^3 \mid z < p^y, \mathrm{v}_p(z^2 + \varepsilon z + \eta) \geq y\}.$$

Let $* : \mathcal{M}_{f,p} \times \mathcal{M}_{f,p} \to \mathcal{M}_{f,p}$ be defined by $(u, v, w) * (x, y, z) = (a, b, c)$, where

$$a = u + x + g,\ b = v + y + e - 2g,$$

$$c = \operatorname{rem}\left(h - t\frac{h^2 + \varepsilon h + \eta}{p^g}, p^b\right),\ g = \min\{v, y, \mathrm{v}_p(w + z + \varepsilon)\},$$

$$e = \min\{g, \mathrm{v}_p(w - z), \mathrm{v}_p(w^2 + \varepsilon w + \eta) - v, \mathrm{v}_p(z^2 + \varepsilon z + \eta) - y\},$$

$$t \in \mathbb{Z} \text{ is such that } t\frac{w + z + \varepsilon}{p^g} \equiv 1 \mod p^{\min\{v,y\}-g}, \text{ and } h = \begin{cases} z & \text{if } y \geq v \\ w & \text{if } v > y \end{cases}.$$

Let $\xi_{f,p} : \mathcal{M}_{f,p} \to \mathcal{I}_p(\mathcal{O}_f)$ be defined by $\xi_{f,p}(x, y, z) = p^x(p^y\mathbb{Z} + (z + \tau)\mathbb{Z})$.

Proposition 3.2. *Let p be a prime divisor of f and $I, J \in \mathcal{I}_p(\mathcal{O}_f)$.*

1. *$(\mathcal{M}_{f,p}, *)$ is a reduced monoid and $\xi_{f,p}$ is a monoid isomorphism.*
2. *If $w, z \in \mathbb{Z}$ are such that $\mathrm{v}_p(w^2 + \varepsilon w + \eta) > 0$ and $\mathrm{v}_p(z^2 + \varepsilon z + \eta) > 0$, then $\mathrm{v}_p(w + z + \varepsilon) > 0$ and $\mathrm{v}_p(w - z) > 0$.*
3. *$\mathcal{N}(I)\mathcal{N}(J) \mid \mathcal{N}(IJ)$ and $\mathcal{N}(IJ) = \mathcal{N}(I)\mathcal{N}(J)$ if and only if I is invertible or J is invertible. If I and J are proper, then $IJ \subset p\mathcal{O}_f$.*
4. *If $I \in \mathcal{A}(\mathcal{I}_p(\mathcal{O}_f))$, then there is some $I' \in \mathcal{A}(\mathcal{I}_p^*(\mathcal{O}_f))$ such that $\mathcal{N}(IJ) \mid \mathcal{N}(I'J)$. If $I \in \mathcal{A}(\mathcal{I}_p(\mathcal{O}_f))$ is not invertible, then $\mathcal{N}(I) \mid \mathcal{N}(I')$ and $\mathcal{N}(I) < \mathcal{N}(I')$ for some $I' \in \mathcal{A}(\mathcal{I}_p^*(\mathcal{O}_f))$.*
5. *If $I \in \mathcal{A}(\mathcal{I}_p^*(\mathcal{O}_f))$, then $\overline{I} \in \mathcal{A}(\mathcal{I}_p^*(\mathcal{O}_f))$ and $I\overline{I} = \mathcal{N}(I)\mathcal{O}_f$.*

Proof. 1. Let $(u, v, w), (x, y, z) \in \mathcal{M}_{f,p}$. Set $g = \min\{v, y, \mathrm{v}_p(w + z + \varepsilon)\}$ and $e = \min\{g, \mathrm{v}_p(w - z), \mathrm{v}_p(w^2 + \varepsilon w + \eta) - v, \mathrm{v}_p(z^2 + \varepsilon z + \eta) - y\}$. Note that $\gcd(p^{\min\{v,y\}}, w + z + \varepsilon) = p^g$, and hence there are some $s, t \in \mathbb{Z}$ such that $sp^{\min\{v,y\}} + t(w + z + \varepsilon) = p^g$. This implies that $t\frac{w+z+\varepsilon}{p^g} \equiv 1 \mod p^{\min\{v,y\}-g}$. Set $a = u + x + g$, $b = v + y + e - 2g$ and let $h = z$ if $y \geq v$ and $h = w$ if $v > y$. Finally, set $c = \operatorname{rem}(h - t\frac{h^2+\varepsilon h+\eta}{p^g}, p^b)$. First we show that c does not depend on the choice of t. Let $t' \in \mathbb{Z}$ be such that $t'\frac{w+z+\varepsilon}{p^g} \equiv 1 \mod p^{\min\{v,y\}-g}$. Then $p^{\min\{v,y\}-g} \mid t - t'$. Note that $\min\{v, y\} + \mathrm{v}_p(h^2 + \varepsilon h + \eta) \geq v + y + e$, and hence $p^b \mid (t - t')\frac{h^2+\varepsilon h+\eta}{p^g}$. Consequently, $c = \operatorname{rem}(h - t'\frac{h^2+\varepsilon h+\eta}{p^g}, p^b)$.

Next we show that $(a, b, c) \in \mathcal{M}_{f,p}$. It is clear that $(a, b, c) \in \mathbb{N}_0^3$ and $c < p^b$. It remains to show that $\mathrm{v}_p(c^2 + \varepsilon c + \eta) \geq b$. Without restriction we can assume that $v \leq y$. Then $h = z$. Set $k = z - t\frac{z^2+\varepsilon z+\eta}{p^g}$. There is some $r \in \mathbb{Z}$ such that $c = k + rp^b$. Since $c^2 + \varepsilon c + \eta = k^2 + \varepsilon k + \eta + mp^b$ for some $m \in \mathbb{Z}$, it is sufficient to show that $\mathrm{v}_p(k^2 + \varepsilon k + \eta) \geq b$.

Observe that $k^2 + \varepsilon k + \eta = \frac{z^2+\varepsilon z+\eta}{p^{2g}}(p^{2g} - tp^g(2z + \varepsilon) + t^2(z^2 + \varepsilon z + \eta)) = \frac{z^2+\varepsilon z+\eta}{p^{2g}}(sp^{v+g} + tp^g(w - z) + t^2(z^2 + \varepsilon z + \eta))$. Note that $g + \mathrm{v}_p(w - z) = \min\{v + \mathrm{v}_p(w - z), \mathrm{v}_p(w + z + \varepsilon) + \mathrm{v}_p(w - z)\} = \min\{v + \mathrm{v}_p(w - z), \mathrm{v}_p(w^2 + \varepsilon w + \eta - (z^2 + \varepsilon z + \eta))\} \geq \min\{v + \mathrm{v}_p(w - z), \mathrm{v}_p(z^2 + \varepsilon z + \eta), \mathrm{v}_p(w^2 + \varepsilon w + \eta)\} \geq v$. Moreover, we have $\mathrm{v}_p(z^2 + \varepsilon z + \eta) \geq y + e$. Therefore, $\mathrm{v}_p(k^2 + \varepsilon k + \eta) \geq \mathrm{v}_p(z^2 + \varepsilon z + \eta) - 2g + \min\{v + g, g + \mathrm{v}_p(w - z), \mathrm{v}_p(z^2 + \varepsilon z + \eta)\} \geq y + e - 2g + v = b$.

Now we prove that $p^u(p^v\mathbb{Z}+(w+\tau)\mathbb{Z})p^x(p^y\mathbb{Z}+(z+\tau)\mathbb{Z}) = p^a(p^b\mathbb{Z}+(c+\tau)\mathbb{Z})$. (Note that this can be shown by using [25, Theorem 5.4.6].) Set $I = p^u(p^v\mathbb{Z}+(w+\tau)\mathbb{Z})p^x(p^y\mathbb{Z}+(z+\tau)\mathbb{Z})$. Without restriction let $v \le y$. Note that $(w+\tau)(z+\tau) = wz-\eta+(w+z+\varepsilon)\tau$. Set $\alpha = p^v(z+\tau)$ and $\beta = wz-\eta+(w+z+\varepsilon)\tau$. We infer that $I = p^{u+x}(p^{v+y}\mathbb{Z}+p^y(w+\tau)\mathbb{Z}+\alpha\mathbb{Z}+\beta\mathbb{Z})$.

Moreover, $p^y(w+\tau)\mathbb{Z}+\alpha\mathbb{Z} = p^y(w-z)\mathbb{Z}+\alpha\mathbb{Z}$. Observe that $s\alpha+t\beta = p^g z - t(z^2+\varepsilon z+\eta)+p^g\tau$. Set $k = z - t\frac{z^2+\varepsilon z+\eta}{p^g}$. Then $s\alpha+t\beta = p^g(k+\tau)$. We have $\alpha - p^v(k+\tau) = tp^{v-g}(z^2+\varepsilon z+\eta)$ and $(w+z+\varepsilon)(k+\tau)-\beta = sp^{v-g}(z^2+\varepsilon z+\eta)$. Set $r = p^{v-g}(z^2+\varepsilon z+\eta)$. Consequently, $\alpha\mathbb{Z}+\beta\mathbb{Z} = sr\mathbb{Z}+tr\mathbb{Z}+p^g(k+\tau)\mathbb{Z} = r\mathbb{Z}+p^g(k+\tau)\mathbb{Z}$, since $\gcd(s,t)=1$. Putting these facts together gives us $I = p^{u+x}(p^{v+y}\mathbb{Z}+p^y(w-z)\mathbb{Z}+r\mathbb{Z}+p^g(k+\tau)\mathbb{Z})$.

We have $\gcd(p^{v+y}, p^y(w-z), r) = p^\ell$ with $\ell = \min\{v+y, y+\mathsf{v}_p(w-z), v-g+\mathsf{v}_p(z^2+\varepsilon z+\eta)\}$ and $p^{v+y}\mathbb{Z}+p^y(w-z)\mathbb{Z}+r\mathbb{Z} = p^\ell\mathbb{Z}$. Note that $\ell = v+y-g+\min\{g, \mathsf{v}_p(w-z)-v+g, \mathsf{v}_p(z^2+\varepsilon z+\eta)-y\}$ and $\mathsf{v}_p(w-z)-v+g = \min\{\mathsf{v}_p(w-z), \mathsf{v}_p(w-z)+\mathsf{v}_p(w+z+\varepsilon)-v\} = \min\{\mathsf{v}_p(w-z), \mathsf{v}_p(w^2+\varepsilon w+\eta-(z^2+\varepsilon z+\eta))-v\}$, and hence $\ell = v+y-g+\min\{g, \mathsf{v}_p(w-z), \mathsf{v}_p(w^2+\varepsilon w+\eta-(z^2+\varepsilon z+\eta))-v, \mathsf{v}_p(z^2+\varepsilon z+\eta)-y\}$.

CASE 1: $\mathsf{v}_p(w^2+\varepsilon w+\eta) \ge \mathsf{v}_p(z^2+\varepsilon z+\eta)$. Then $\mathsf{v}_p(w^2+\varepsilon w+\eta)-v \ge \mathsf{v}_p(z^2+\varepsilon z+\eta)-y$ and $\mathsf{v}_p(w^2+\varepsilon w+\eta-(z^2+\varepsilon z+\eta))-v \ge \mathsf{v}_p(z^2+\varepsilon z+\eta)-y$.

CASE 2: $\mathsf{v}_p(z^2+\varepsilon z+\eta) > \mathsf{v}_p(w^2+\varepsilon w+\eta)$. Then $\mathsf{v}_p(w^2+\varepsilon w+\eta-(z^2+\varepsilon z+\eta))-v = \mathsf{v}_p(w^2+\varepsilon w+\eta)-v$.

In any case we have $\min\{\mathsf{v}_p(w^2+\varepsilon w+\eta-(z^2+\varepsilon z+\eta))-v, \mathsf{v}_p(z^2+\varepsilon z+\eta)-y\} = \min\{\mathsf{v}_p(w^2+\varepsilon w+\eta)-v, \mathsf{v}_p(z^2+\varepsilon z+\eta)-y\}$. Obviously, $\ell = v+y+e-g$ and $I = p^{u+x+g}(p^{v+y+e-2g}\mathbb{Z}+(z-t\frac{z^2+\varepsilon z+\eta}{p^g}+\tau)\mathbb{Z})$. Consequently, $I = p^a(p^b\mathbb{Z}+(c+\tau)\mathbb{Z})$.

So far we know that $*$ is an inner binary operation on $\mathcal{M}_{f,p}$. It follows from Proposition 3.1.1 that $\xi_{f,p}$ is surjective. It follows from [25, Theorem 5.4.2] that $\xi_{f,p}$ is injective. It is clear that $(\mathcal{I}_p(\mathcal{O}_f), \cdot)$ is a reduced monoid. We have shown that $\xi_{f,p}$ maps products of elements of $\mathcal{M}_{f,p}$ to products of elements of $\mathcal{I}_p(\mathcal{O}_f)$. It is clear that $(0,0,0)$ is an identity element of $\mathcal{M}_{f,p}$ and $\xi_{f,p}(0,0,0) = \mathcal{O}_f$. Therefore, $(\mathcal{M}_{f,p}, *)$ is a reduced monoid and $\xi_{f,p}$ is a monoid isomorphism.

2. Let $w, z \in \mathbb{Z}$ be such that $\mathsf{v}_p(w^2+\varepsilon w+\eta) > 0$ and $\mathsf{v}_p(z^2+\varepsilon z+\eta) > 0$. Then $p \mid z^2+\varepsilon z+\eta = \frac{1}{4}((2z+\varepsilon)^2 - f^2 d_K)$, and hence $p \mid 2z+\varepsilon$. Moreover $p \mid w^2+\varepsilon w+\eta-(z^2+\varepsilon z+\eta) = (w+z+\varepsilon)(w-z)$, and thus $p \mid w+z+\varepsilon$ or $p \mid w-z$. Since $p \mid 2z+\varepsilon$, we infer that $p \mid w+z+\varepsilon$ if and only if $p \mid w-z$. Consequently, $\min\{\mathsf{v}_p(w+z+\varepsilon), \mathsf{v}_p(w-z)\} > 0$.

3. By 1., there are $(u,v,w), (x,y,z), (a,b,c) \in \mathcal{M}_{f,p}$ such that $I = p^u(p^v\mathbb{Z}+(w+\tau)\mathbb{Z})$, $J = p^x(p^y\mathbb{Z}+(z+\tau)\mathbb{Z})$, and $IJ = p^a(p^b\mathbb{Z}+(c+\tau)\mathbb{Z})$ with $a = u+x+g$, $b = v+y+e-2g$, $g = \min\{v, y, \mathsf{v}_p(w+z+\varepsilon)\}$ and $e = \min\{g, \mathsf{v}_p(w-z), \mathsf{v}_p(w^2+\varepsilon w+\eta)-v, \mathsf{v}_p(z^2+\varepsilon z+\eta)-y\}$. It follows by Proposition 3.1.1 that $\mathcal{N}(I) = p^{2u+v}$, $\mathcal{N}(J) = p^{2x+y}$, and $\mathcal{N}(IJ) = p^{2a+b} = p^{2(u+x)+v+y+e}$. It is obvious that $\mathcal{N}(I)\mathcal{N}(J) \mid \mathcal{N}(IJ)$. Moreover, $\mathcal{N}(IJ) = \mathcal{N}(I)$

$\mathcal{N}(J)$ if and only if $e = 0$. We infer by 2. that $e = 0$ if and only if $v = 0$ or $y = 0$ or $\mathrm{v}_p(w^2 + \varepsilon w + \eta) = v$ or $\mathrm{v}_p(z^2 + \varepsilon z + \eta) = y$, which is the case if and only if I is invertible or J is invertible by Proposition 3.1.2. If I and J are proper, then $u + v > 0$ and $x + y > 0$, and hence $a > 0$ by 2. This implies that $IJ \subset p(p^b\mathbb{Z} + (c+\tau)\mathbb{Z}) \subset p\mathcal{O}_f$.

4. Let $I \in \mathcal{A}(\mathcal{I}_p(\mathcal{O}_f))$. Without restriction let I be not invertible. We have $I = p^b\mathbb{Z} + (r+\tau)\mathbb{Z}$ for some $(0, b, r) \in \mathcal{M}_{f,p}$ and $b < \mathrm{v}_p(r^2 + \varepsilon r + \eta)$. Set $c = \mathrm{v}_p(r^2 + \varepsilon r + \eta)$ and $I' = p^c\mathbb{Z} + (r+\tau)\mathbb{Z}$. Then $I' \in \mathcal{A}(\mathcal{I}_p^*(\mathcal{O}_f))$, $\mathcal{N}(I) \mid \mathcal{N}(I')$, and $\mathcal{N}(I) < \mathcal{N}(I')$ by Proposition 3.1. There is some $(x, y, z) \in \mathcal{M}_{f,p}$ such that $J = p^x(p^y\mathbb{Z} + (z+\tau)\mathbb{Z})$. Then $\mathcal{N}(I'J) = p^{c+2x+y}$ and $\mathcal{N}(IJ) = p^{b+2x+y+e}$ with $e = \min\{b, y, \mathrm{v}_p(r+z+\varepsilon), \mathrm{v}_p(r-z), c-b, \mathrm{v}_p(z^2+\varepsilon z+\eta) - y\} \le c - b$. Therefore, $\mathcal{N}(IJ) \mid \mathcal{N}(I'J)$.

5. Let $I \in \mathcal{A}(\mathcal{I}_p^*(\mathcal{O}_f))$. If $I = p\mathcal{O}_f$, then $\overline{I} = p\mathcal{O}_f$ and $\mathcal{N}(I) = p^2$ by Proposition 3.1.1. Therefore, $I\overline{I} = \mathcal{N}(I)\mathcal{O}_f$. Now let $I \neq p\mathcal{O}_f$. There is some $(0, m, r) \in \mathcal{M}_{f,p}$ such that $I = p^m\mathbb{Z} + (r+\tau)\mathbb{Z}$. Set $s = p^m - r - \varepsilon$. It follows that $\overline{I} = p^m\mathbb{Z} + (r+\overline{\tau})\mathbb{Z} = p^m\mathbb{Z} + (r+\varepsilon-\tau)\mathbb{Z} = p^m\mathbb{Z} + (s+\tau)\mathbb{Z}$. Observe that $s^2 + \varepsilon s + \eta = r^2 + \varepsilon r + \eta + p^m(p^m - (2r+\varepsilon))$. Since $p \mid r^2 + \varepsilon r + \eta = \frac{1}{4}((2r+\varepsilon)^2 - f^2 d_K)$, we have $\mathrm{v}_p(2r+\varepsilon) > 0$, and hence $\mathrm{v}_p(p^m(p^m - (2r+\varepsilon))) > m$. Since $\mathrm{v}_p(r^2 + \varepsilon r + \eta) = m$, we infer that $\mathrm{v}_p(s^2 + \varepsilon s + \eta) = m$, and thus $(0, m, s) \in \mathcal{M}_{f,p}$. Therefore, $\overline{I} \in \mathcal{A}(\mathcal{I}_p^*(\mathcal{O}_f))$. Note that $\min\{m, \mathrm{v}_p(r+s+\varepsilon)\} = m$, and thus $I\overline{I} = p^m\mathcal{O}_f = \mathcal{N}(I)\mathcal{O}_f$ by 1. and Proposition 3.1.1. □

Proposition 3.3. *Let p be a prime divisor of f and $f' = p^{\mathrm{v}_p(f)}$. Set $\mathcal{O} = \mathcal{O}_f$, $\mathcal{O}' = \mathcal{O}_{f'}$, $P = P_{f,p}$ and $P' = P_{f',p}$. For $g \in \mathbb{N}$ let $\varphi_{g,p} : \mathcal{I}_p(\mathcal{O}_g) \to \mathcal{I}((\mathcal{O}_g)_{P_{g,p}})$ be defined by $\varphi_{g,p}(I) = I_{P_{g,p}}$ and $\zeta_{g,p} : \mathcal{I}((\mathcal{O}_g)_{P_{g,p}}) \to \mathcal{I}_p(\mathcal{O}_g)$ be defined by $\zeta_{g,p}(J) = J \cap \mathcal{O}_g$.*

1. *$\mathcal{O}_P = \mathcal{O}'_{P'}$.*
2. *$\varphi_{f,p}$ and $\zeta_{f,p}$ are mutually inverse monoid isomorphisms.*
3. *There is a monoid isomorphism $\delta : \mathcal{I}_p(\mathcal{O}) \to \mathcal{I}_p(\mathcal{O}')$ such that $\delta(p\mathcal{O}) = p\mathcal{O}'$ and $\delta_{|\mathcal{I}_p^*(\mathcal{O})} : \mathcal{I}_p^*(\mathcal{O}) \to \mathcal{I}_p^*(\mathcal{O}')$ is a monoid isomorphism.*

Proof. 1. It is clear that $\mathcal{O} \subset \mathcal{O}'$ and $P' \cap \mathcal{O} = P$. Therefore, $\mathcal{O}_P \subset \mathcal{O}'_{P'}$. Observe that $\mathcal{O} \setminus P = (\mathbb{Z} \setminus p\mathbb{Z}) + f\omega\mathbb{Z}$ and $\mathcal{O}' \setminus P' = (\mathbb{Z} \setminus p\mathbb{Z}) + f'\omega\mathbb{Z}$. It remains to show that $\{f'\omega\} \cup \{x^{-1} \mid x \in (\mathbb{Z} \setminus p\mathbb{Z}) + f'\omega\mathbb{Z}\} \subset \mathcal{O}_P$. Since $\frac{f}{f'} f'\omega = f\omega \in \mathcal{O}$ and $\frac{f}{f'} \in \mathbb{Z} \setminus p\mathbb{Z} \subset \mathcal{O} \setminus P$, we have $f'\omega \in \mathcal{O}_P$. Therefore, $\mathcal{O}' \subset \mathcal{O}_P$. Now let $a \in \mathbb{Z} \setminus p\mathbb{Z}$ and $b \in \mathbb{Z}$. Observe that $a + bf'\overline{\omega} \in \mathcal{O}' \subset \mathcal{O}_P$. Since $\omega + \overline{\omega}, \omega\overline{\omega} \in \mathbb{Z}$, we have $(a + bf'\omega)(a + bf'\overline{\omega}) = a^2 + abf'(\omega + \overline{\omega}) + b^2(f')^2\omega\overline{\omega} \in \mathbb{Z} \setminus p\mathbb{Z} \subset \mathcal{O} \setminus P$. Therefore, $\frac{1}{a+bf'\omega} = \frac{a+bf'\overline{\omega}}{(a+bf'\omega)(a+bf'\overline{\omega})} \in \mathcal{O}_P$.

2. It is clear that $\varphi_{f,p}$ is a well-defined monoid homomorphism. Note that $\zeta_{f,p}$ is a well-defined map (since every nonzero proper ideal J of $\mathcal{O}_P$ is P_P-primary, and hence $J \cap \mathcal{O}$ is P-primary). Moreover, $\zeta_{f,p}(\mathcal{O}_P) = \mathcal{O}$. Now let $J_1, J_2 \in \mathcal{I}(\mathcal{O}_P)$. Observe that $J_1J_2 \cap \mathcal{O}$ and $(J_1 \cap \mathcal{O})(J_2 \cap \mathcal{O})$ coincide locally (note that both are either P-primary or not proper). Therefore, $J_1J_2 \cap \mathcal{O} = (J_1 \cap \mathcal{O})(J_2 \cap \mathcal{O})$, and hence $\zeta_{f,p}$ is a monoid homomorphism. If $J \in \mathcal{I}(\mathcal{O}_P)$, then $(J \cap \mathcal{O})_P = J$. Therefore,

$\varphi_{f,p} \circ \zeta_{f,p} = \mathrm{id}_{\mathcal{I}(\mathcal{O}_P)}$. If I is a P-primary ideal of $\mathcal{O}$, then $I_P \cap \mathcal{O} = I$. This implies that $\zeta_{f,p} \circ \varphi_{f,p} = \mathrm{id}_{\mathcal{I}_p(\mathcal{O})}$.

3. Set $\delta = \zeta_{f',p} \circ \varphi_{f,p}$. Then $\delta : \mathcal{I}_p(\mathcal{O}) \to \mathcal{I}_p(\mathcal{O}')$ is a monoid isomorphism by 1. and 2. Furthermore, we have by 1. that $\delta(p\mathcal{O}) = \zeta_{f',p}(\varphi_{f,p}(p\mathcal{O})) = \zeta_{f',p}(p\mathcal{O}_P) = \zeta_{f',p}(p\mathcal{O}'_{P'}) = p\mathcal{O}'_{P'} \cap \mathcal{O}' = p\mathcal{O}'$.

Since $\mathcal{O}$ is noetherian, we have $\mathcal{I}_p^*(\mathcal{O})$ is the set of cancellative elements of $\mathcal{I}_p(\mathcal{O})$. It follows by analogy that $\mathcal{I}_p^*(\mathcal{O}')$ is the set of cancellative elements of $\mathcal{I}_p(\mathcal{O}')$. Therefore, $\delta(\mathcal{I}_p^*(\mathcal{O})) = \mathcal{I}_p^*(\mathcal{O}')$, and hence $\delta_{|\mathcal{I}_p^*(\mathcal{O})}$ is a monoid isomorphism. □

Lemma 3.4. *Let p be a prime number, let $k \in \mathbb{N}_0$, let $c, n \in \mathbb{N}$ be such that* $\gcd(c, p) = 1$ *and for each $\ell \in \mathbb{N}$ let* $g_\ell = |\{y \in [0, p^\ell - 1] \mid y^2 \equiv c \mod p^\ell\}|$.

1. *If $p \neq 2$, then $p^k c$ is a square modulo p^n if and only if $k \geq n$ or ($k < n$, k is even and $(\frac{c}{p}) = 1$).*
2. *$2^k c$ is a square modulo 2^n if and only if one of the following conditions holds.*
   (a) *$k \geq n$.*
   (b) *k is even and $n = k + 1$.*
   (c) *k is even, $n = k + 2$ and $c \equiv 1 \mod 4$.*
   (d) *k is even, $n \geq k + 3$ and $c \equiv 1 \mod 8$.*
3. *If $\ell \in \mathbb{N}$, then* $g_\ell = \begin{cases} 4 & \text{if } p = 2, \ell \geq 3, c \equiv 1 \mod 8 \\ 2 & \text{if } (p \neq 2, (\frac{c}{p}) = 1) \text{ or } (p = 2, \ell = 2, c \equiv 1 \mod 4) \\ 1 & \text{if } p = 2, \ell = 1 \\ 0 & \text{else} \end{cases}$.

Proof. Note that $p^k c$ is a square modulo p^n if and only if $k \geq n$ or ($k < n$, k is even and c is a square modulo p^{n-k}).

1. Let $p \neq 2$. It remains to show that if $\ell \in \mathbb{N}$, then c is a square modulo p^ℓ if and only if $(\frac{c}{p}) = 1$. If $\ell \in \mathbb{N}$ and c is a square modulo p^ℓ, then c is a square modulo p, and hence $(\frac{c}{p}) = 1$. Now let $(\frac{c}{p}) = 1$. It suffices to show by induction that c is a square modulo p^ℓ for all $\ell \in \mathbb{N}$. The statement is clearly true for $\ell = 1$. Now let $\ell \in \mathbb{N}$ and let $x \in \mathbb{Z}$ be such that $x^2 \equiv c \mod p^\ell$. Without restriction let $\mathrm{v}_p(x^2 - c) = \ell$. Note that $p \nmid x$, and hence $2bx \equiv -1 \mod p$ for some $b \in \mathbb{Z}$. Set $y = x + b(x^2 - c)$. Then $y^2 \equiv c \mod p^{\ell+1}$.

2. It remains to show that if $\ell \in \mathbb{N}$, then c is a square modulo 2^ℓ if and only if $\ell = 1$ or ($\ell = 2$ and $c \equiv 1 \mod 4$) or ($\ell \geq 3$ and $c \equiv 1 \mod 8$). Let $\ell \in \mathbb{N}$ and let c be a square modulo 2^ℓ. If $\ell = 2$, then c is a square modulo 4 and $c \equiv 1 \mod 4$. Moreover, if $\ell \geq 3$, then c is a square modulo 8 and $c \equiv 1 \mod 8$.

Clearly, if $\ell = 1$ or ($\ell = 2$ and $c \equiv 1 \mod 4$), then c is a square modulo 2^ℓ. Now let $c \equiv 1 \mod 8$. It is sufficient to show by induction that c is a square modulo 2^ℓ for each $\ell \in \mathbb{N}_{\geq 3}$. The statement is obviously true for $\ell = 3$. Now let $\ell \in \mathbb{N}_{\geq 3}$ and let $x \in \mathbb{Z}$ be such that $x^2 \equiv c \mod 2^\ell$. Without restriction let $\mathrm{v}_2(x^2 - c) = \ell$. Set $y = x + 2^{\ell-1}$. Then $y^2 \equiv c \mod 2^{\ell+1}$.

3. Let $\ell \in \mathbb{N}$. By 1. and 2., it is sufficient to consider the case $g_\ell > 0$. Let $g_\ell > 0$. Observe that $g_\ell = |\{y \in [0, p^\ell - 1] \mid y^2 \equiv 1 \mod p^\ell\}| = |\{y \in (\mathbb{Z}/p^\ell\mathbb{Z})^\times \mid \mathrm{ord}(y) \leq 2\}|$. If $p = 2$ and $\ell = 1$, then $(\mathbb{Z}/p^\ell\mathbb{Z})^\times$ is trivial, and hence $g_\ell = 1$. If $(p = 2, \ell = 2$ and $c \equiv 1 \mod 4)$ or $(p \neq 2$ and $(\frac{c}{p}) = 1)$, then $(\mathbb{Z}/p^\ell\mathbb{Z})^\times$ is a cyclic group of even order, and thus $g_\ell = 2$. Finally, if $p = 2$, $\ell \geq 3$ and $c \equiv 1 \mod 8$, then $(\mathbb{Z}/2^\ell\mathbb{Z})^\times \cong \mathbb{Z}/2\mathbb{Z} \times C_{2^{\ell-2}}$ is the product of two cyclic groups of even order. Consequently, $g_\ell = 4$. □

Lemma 3.5. *Let p be a prime number, $a, m \in \mathbb{N}$, $c = \frac{a}{p^{\mathrm{v}_p(a)}}$, $M = \{x \in [0, p^m - 1] \mid \mathrm{v}_p(x^2 - a) = m\}$, $N = |M|$ and for each $\ell \in \mathbb{N}$ let $g_\ell = |\{y \in [0, p^\ell - 1] \mid y^2 \equiv c \mod p^\ell\}|$.*

1. *If $m < \mathrm{v}_p(a)$, then* $N = \begin{cases} \varphi(p^{m/2}) & \text{if } m \text{ is even} \\ 0 & \text{if } m \text{ is odd} \end{cases}.$
2. *Let $m = \mathrm{v}_p(a)$.*
   (a) *If a is a square modulo p^{m+1}, then* $N = \begin{cases} p^{m/2-1}(p-2) & \text{if } p \neq 2 \\ 2^{m/2-1} & \text{if } p = 2 \end{cases}.$
   (b) *If a is not a square modulo p^{m+1}, then $N = p^{\lfloor m/2 \rfloor}$.*
3. *If $m > \mathrm{v}_p(a)$ and a is not a square modulo p^m, then $N = 0$.*
4. *If $k \in \mathbb{N}$ is such that $m = k + \mathrm{v}_p(a)$ and a is a square modulo p^m, then $N = p^{\mathrm{v}_p(a)/2-1}(pg_k - g_{k+1})$.*

Proof. 1. Let $m < \mathrm{v}_p(a)$. Observe that $M = \{x \in [0, p^m - 1] \mid 2\mathrm{v}_p(x) = m\}$. Clearly, if m is odd, then $N = 0$. Now let m be even. We have $M = \{p^{m/2}y \mid y \in [0, p^{m/2} - 1], \gcd(y, p) = 1\}$, and thus $N = |\{y \in [0, p^{m/2} - 1] \mid \gcd(y, p) = 1\}| = \varphi(p^{m/2})$.

2. Note that $M = \{x \in [0, p^m - 1] \mid 2\mathrm{v}_p(x) \geq m, x^2 \not\equiv a \mod p^{m+1}\}$ and $|\{x \in [0, p^m - 1] \mid 2\mathrm{v}_p(x) \geq m\}| = p^{\lfloor m/2 \rfloor}$. Set $M' = \{x \in [0, p^m - 1] \mid x^2 \equiv a \mod p^{m+1}\}$. Then $M' = \{x \in [0, p^m - 1] \mid 2\mathrm{v}_p(x) \geq m, x^2 \equiv a \mod p^{m+1}\}$ and $N = p^{\lfloor m/2 \rfloor} - |M'|$. If a is not a square modulo p^{m+1}, then $M' = \emptyset$, and hence $N = p^{\lfloor m/2 \rfloor}$. Now let a be a square modulo p^{m+1}. Then $M' \neq \emptyset$, and thus m is even. Observe that $M' = \{x \in [0, p^m - 1] \mid 2\mathrm{v}_p(x) = m, x^2 \equiv a \mod p^{m+1}\} = \{p^{m/2}y \mid y \in [0, p^{m/2} - 1], y^2 \equiv c \mod p\}$. Therefore, $|M'| = |\{y \in [0, p^{m/2} - 1] \mid y^2 \equiv c \mod p\}| = p^{m/2-1}|\{y \in [0, p - 1] \mid y^2 \equiv c \mod p\}|$.

If $p \neq 2$, then $N = p^{\lfloor m/2 \rfloor} - |M'| = p^{m/2} - 2p^{m/2-1} = p^{m/2-1}(p - 2)$ by Lemma 3.4.3. Moreover, if $p = 2$, then $N = 2^{\lfloor m/2 \rfloor} - |M'| = 2^{m/2} - 2^{m/2-1} = 2^{m/2-1}$ by Lemma 3.4.3.

3. This is obvious.

4. Let $k \in \mathbb{N}$ be such that $m = k + \mathrm{v}_p(a)$ and let a be a square modulo p^m. It follows by Lemma 3.4 that $\mathrm{v}_p(a)$ is even. Set $r = \mathrm{v}_p(a)/2$ and for $\theta \in \{0, 1\}$ set $M_\theta = \{x \in [0, p^m - 1] \mid 2\mathrm{v}_p(x) = \mathrm{v}_p(a), x^2 \equiv a \mod p^{m+\theta}\}$. Then $M = \{x \in [0, p^m - 1] \mid \mathrm{v}_p(x) = r, \mathrm{v}_p(x^2 - a) = m\} = M_0 \setminus M_1$. Since $\{x \in [0, p^m - 1] \mid \mathrm{v}_p(x) = r\} = \{p^r y \mid y \in [0, p^{k+r} - 1], \gcd(y, p) = 1\}$, we infer that $M_\theta = \{p^r y \mid y \in$

$[0, p^{k+r} - 1], y^2 \equiv c \mod p^{k+\theta}\}$. Therefore, $|M_\theta| = |\{y \in [0, p^{k+r} - 1] \mid y^2 \equiv c \mod p^{k+\theta}\}| = p^{r-\theta}|\{y \in [0, p^{k+\theta} - 1] \mid y^2 \equiv c \mod p^{k+\theta}\}| = p^{r-\theta} g_{k+\theta}$. This implies that $N = |M_0| - |M_1| = p^r g_k - p^{r-1} g_{k+1} = p^{r-1}(p g_k - g_{k+1})$. □

Theorem 3.6. *Let $\mathcal{O}$ be an order in a quadratic number field K with conductor $\mathfrak{f} = f\mathcal{O}_K$ for some $f \in \mathbb{N}_{\geq 2}$, p be a prime divisor of f, and $\mathfrak{p} = \mathcal{P}_{f,p}$.*

1. *The primary ideals with radical $\mathfrak{p}$ are exactly the ideals of the form*

$$\mathfrak{q} = p^\ell(p^m\mathbb{Z} + (r + \tau)\mathbb{Z})$$

with $\ell, m \in \mathbb{N}_0$, $\ell + m \geq 1$, $0 \leq r < p^m$, and $\mathcal{N}_{K/\mathbb{Q}}(r + \tau) \equiv 0 \mod p^m$. Moreover, $\mathcal{N}(\mathfrak{q}) = p^{2\ell+m}$.

2. *A primary ideal $\mathfrak{q} = p^\ell(p^m\mathbb{Z} + (r + \tau)\mathbb{Z})$ is invertible if and only if*

$$\mathcal{N}_{K/\mathbb{Q}}(r + \tau) \not\equiv 0 \mod p^{m+1}.$$

3. *A primary ideal $\mathfrak{q}$ with radical $\mathfrak{p}$ is an ideal atom if and only if $\mathfrak{q} = p\mathcal{O}$ or $\mathfrak{q} = p^m\mathbb{Z} + (r + \tau)\mathbb{Z}$ with $m \in \mathbb{N}$ and $p^m \mid \mathcal{N}_{K/\mathbb{Q}}(r + \tau)$.*
4. *Table 1 gives the number of invertible ideal atoms of the form $p^m\mathbb{Z} + (r + \tau)\mathbb{Z}$ with norm p^m; this number is 0 if m is not listed in the table.*

Table 1 Number of nontrivial invertible $\mathfrak{p}$-primary ideal atoms

m	$2h$ $1 \leq h < v_p(f)$	$2v_p(f)$	$2v_p(f) + 1$	$> 2v_p(f) + 1$
p is inert	$\varphi\left(p^{m/2}\right)$	$p^{v_p(f)}$		0
p is ramified			$p^{v_p(f)}$	
p splits		$p^{v_p(f)-1}(p-2)$	$2\varphi\ p^{v_p(f)}$	

5. *The number of ideal atoms with radical $\mathfrak{p}$ is finite if and only if the number of invertible ideal atoms with radical $\mathfrak{p}$ is finite if and only if p does not split.*

Proof. 1. and 2. are an immediate consequence of Proposition 3.1.

3. In 1. we have seen, that all $\mathfrak{p}$-primary ideals of $\mathcal{O}$ are of the form $\mathfrak{q} = p^\ell(p^m\mathbb{Z} + (r + \tau)\mathbb{Z})$. If both ℓ and m are greater than 0, then $\mathfrak{q}$ is not an ideal atom. Indeed, $\mathfrak{q} = (p\mathcal{O})^\ell(p^m\mathbb{Z} + (r + \tau)\mathbb{Z})$ is a nontrivial factorization. It remains to be proven that $p\mathcal{O}$ and $p^m\mathbb{Z} + (r + \tau)\mathbb{Z}$ are ideal atoms.

Assume that there exist proper ideals $\mathfrak{a}_1, \mathfrak{a}_2$ of $\mathcal{O}$ such that $p\mathcal{O} = \mathfrak{a}_1\mathfrak{a}_2$. Since $p\mathcal{O}$ is $\mathfrak{p}$-primary, we have $\mathfrak{a}_1$ and $\mathfrak{a}_2$ are $\mathfrak{p}$-primary. Using this information, we deduce that $p\mathcal{O} \subset \mathfrak{p}^2$, implying

$$p \in p\mathcal{O} \subset \mathfrak{p}^2 = (p^2, pf\omega, f^2\omega^2) = p(p, f\omega, \frac{f}{p}\omega f\omega) = p(p, f\omega) = p\mathfrak{p}.$$

Therefore, $1 \in \mathfrak{p}$, a contradiction.

Assume that there exist proper ideals $\mathfrak{a}_1, \mathfrak{a}_2$ of $\mathcal{O}$ such that $p^m\mathbb{Z} + (r+\tau)\mathbb{Z} = \mathfrak{a}_1\mathfrak{a}_2$. Note that $\mathfrak{a}_1$ and $\mathfrak{a}_2$ are $\mathfrak{p}$-primary. By Proposition 3.2.3, it follows that $p^m\mathbb{Z} + (r+\tau)\mathbb{Z} \subset p\mathcal{O}$, a contradiction to $r+\tau \notin p\mathcal{O}$.

4. By 1. and 3., the nontrivial $\mathfrak{p}$-primary ideal atoms of norm p^m are all $\mathfrak{q} = p^m\mathbb{Z} + (r+\tau)\mathbb{Z}$ with $m \in \mathbb{N}$, $0 \le r < p^m$ and $\mathcal{N}_{K/\mathbb{Q}}(r+\tau) \equiv 0 \mod p^m$. By 2., an ideal of this form is invertible if and only if $\mathcal{N}_{K/\mathbb{Q}}(r+\tau) \not\equiv 0 \mod p^{m+1}$.

Thus if we want to count the number of invertible $\mathfrak{p}$-primary ideal atoms of the form $\mathfrak{q} = p^m\mathbb{Z} + (r+\tau)\mathbb{Z}$, we have to count the number of solutions $r \in [0, p^m - 1]$ of the equation

$$\mathsf{v}_p(\mathcal{N}_{K/\mathbb{Q}}(r+\tau)) = m. \tag{3.1}$$

Set $N = |\{r \in [0, p^m-1] \mid \mathsf{v}_p(\mathcal{N}_{K/\mathbb{Q}}(r+\tau)) = m\}|$ and $a = \begin{cases} (\frac{f}{2})^2 d_K & \text{if } p = 2 \\ f^2 d_K & \text{if } p \neq 2 \end{cases}$.

Next we show that $N = |\{r \in [0, p^m-1] \mid \mathsf{v}_p(r^2 - a) = m\}|$. Note that $\mathcal{N}_{K/\mathbb{Q}}(r+\tau) = \frac{(2r+\varepsilon)^2 - f^2 d_K}{4}$ for each $r \in [0, p^m-1]$. If $p = 2$, then $\varepsilon = 0$, and hence $\mathcal{N}_{K/\mathbb{Q}}(r+\tau) = r^2 - a$. Now let $p \neq 2$. Then $\mathsf{v}_p(\mathcal{N}_{K/\mathbb{Q}}(r+\tau)) = \mathsf{v}_p((2r+\varepsilon)^2 - a)$ for each $r \in [0, p^m-1]$. Let $f : \{r \in [0, p^m-1] \mid \mathsf{v}_p(r^2-a) = m\} \to \{r \in [0, p^m-1] \mid \mathsf{v}_p((2r+\varepsilon)^2 - a) = m\}$ and $g : \{r \in [0, p^m-1] \mid \mathsf{v}_p((2r+\varepsilon)^2 - a) = m\} \to \{r \in [0, p^m-1] \mid \mathsf{v}_p(r^2-a) = m\}$ be defined by $f(r) = \begin{cases} \frac{r-\varepsilon}{2} & \text{if } r-\varepsilon \text{ is even} \\ \frac{r+p^m-\varepsilon}{2} & \text{if } r-\varepsilon \text{ is odd} \end{cases}$ and $g(r) = \operatorname{rem}(2r+\varepsilon, p^m)$ for each $r \in [0, p^m-1]$. Observe that f and g are well-defined injective maps. Therefore, $N = |\{r \in [0, p^m-1] \mid \mathsf{v}_p(r^2-a) = m\}|$ in any case. Set $c = \frac{a}{p^{\mathsf{v}_p(a)}}$ and for $\ell \in \mathbb{N}$ set $g_\ell = |\{y \in [0, p^\ell - 1] \mid y^2 \equiv c \mod p^\ell\}|$. If $m < \mathsf{v}_p(a)$, then the statement follows immediately by Lemma 3.5.1. Therefore, let $m \ge \mathsf{v}_p(a)$. In what follows we use Lemmas 3.4 and 3.5 without further citation.

CASE 1: $p = 2$ and 2 is inert. We have $\mathsf{v}_2(a) = 2\mathsf{v}_2(f) - 2$, $c \equiv d_K \equiv 5 \mod 8$, $g_1 = 1$, $g_2 = 2$ and $g_3 = 0$. If $m = \mathsf{v}_2(a)$, then a is a square modulo 2^{m+1}, and hence $N = 2^{m/2-1} = \varphi(2^{m/2})$. If $m = \mathsf{v}_2(a) + 1$, then a is a square modulo 2^m, and thus $N = 2^{\mathsf{v}_2(a)/2-1}(2g_1 - g_2) = 0$. If $m = \mathsf{v}_2(a) + 2$, then a is a square modulo 2^m, whence $N = 2^{\mathsf{v}_2(a)/2-1}(2g_2 - g_3) = 2^{\mathsf{v}_2(a)/2+1} = 2^{\mathsf{v}_2(f)}$. Finally, let $m \ge \mathsf{v}_2(a) + 3$. Then a is not a square modulo 2^m, and hence $N = 0$.

CASE 2: $p = 2$ and 2 is ramified. Note that $\mathsf{v}_2(a) \in \{2\mathsf{v}_2(f), 2\mathsf{v}_2(f)+1\}$. First let $\mathsf{v}_2(a) = 2\mathsf{v}_2(f)$. Then $a = f^2 d$ with $c \equiv d \equiv 3 \mod 4$, $g_1 = 1$ and $g_\ell = 0$ for each $\ell \in \mathbb{N}_{\ge 2}$. If $m = \mathsf{v}_2(a)$, then a is a square modulo 2^{m+1}, and thus $N = 2^{m/2-1} = 2^{\mathsf{v}_2(f)-1} = \varphi(2^{\mathsf{v}_2(f)})$. If $m = \mathsf{v}_2(a) + 1$, then a is a square modulo 2^m, and hence $N = 2^{\mathsf{v}_2(a)/2-1}(2g_1 - g_2) = 2^{\mathsf{v}_2(f)}$. Finally, let $m \ge \mathsf{v}_2(a) + 2$. Then a is not a square modulo 2^m, and thus $N = 0$.

Now let $\mathsf{v}_2(a) = 2\mathsf{v}_2(f) + 1$. If $m = \mathsf{v}_2(a)$, then a is not a square modulo 2^{m+1}, and hence $N = 2^{\lfloor m/2 \rfloor} = 2^{\mathsf{v}_2(f)}$. If $m > \mathsf{v}_2(a)$, then a is not a square modulo 2^m, and thus $N = 0$.

CASE 3: $p = 2$ and 2 splits. Observe that $\mathrm{v}_2(a) = 2\mathrm{v}_2(f) - 2$, $c \equiv d_K \equiv 1 \mod 8$, $g_1 = 1, g_2 = 2$ and $g_\ell = 4$ for each $\ell \in \mathbb{N}_{\geq 3}$. If $m = \mathrm{v}_2(a)$, then a is a square modulo 2^{m+1}, and hence $N = 2^{m/2-1} = \varphi(2^{m/2})$. Now let $m > \mathrm{v}_2(a)$ and set $k = m - \mathrm{v}_2(a)$. Note that a is a square modulo 2^m, and hence $N = 2^{\mathrm{v}_2(a)/2-1}(2g_k - g_{k+1})$. If $m < \mathrm{v}_2(a) + 3$, then $N = 0$. Finally, let $m \geq \mathrm{v}_2(a) + 3$. Then $N = 2^{\mathrm{v}_2(a)/2+1} = 2^{\mathrm{v}_2(f)} = 2\varphi(2^{\mathrm{v}_2(f)})$.

CASE 4: $p \neq 2$ and p is inert. We have $\mathrm{v}_p(a) = 2\mathrm{v}_p(f)$, $(\frac{c}{p}) = (\frac{d_K}{p}) = -1$ and $g_\ell = 0$ for each $\ell \in \mathbb{N}$. If $m = \mathrm{v}_p(a)$, then a is not a square modulo p^{m+1}, and hence $N = p^{\lfloor m/2 \rfloor} = p^{\mathrm{v}_p(f)}$. If $m > \mathrm{v}_p(a)$, then a is not a square modulo p^m, and thus $N = 0$.

CASE 5: $p \neq 2$ and p is ramified. It follows that $\mathrm{v}_p(a) = 2\mathrm{v}_p(f) + 1$. If $m = \mathrm{v}_p(a)$, then a is not a square modulo p^{m+1}, and thus $N = p^{\lfloor m/2 \rfloor} = p^{\mathrm{v}_p(f)}$. If $m > \mathrm{v}_p(a)$, then a is not a square modulo p^m, and thus $N = 0$.

CASE 6: $p \neq 2$ and p splits. Note that $\mathrm{v}_p(a) = 2\mathrm{v}_p(f)$, $(\frac{c}{p}) = (\frac{d_K}{p}) = 1$ and $g_\ell = 2$ for each $\ell \in \mathbb{N}$. If $m = \mathrm{v}_p(a)$, then a is a square modulo p^{m+1}, and hence $N = p^{m/2-1}(p-2) = p^{\mathrm{v}_p(f)-1}(p-2)$. If $m > \mathrm{v}_p(a)$, then a is a square modulo p^m, and thus $N = p^{\mathrm{v}_p(a)/2-1}(pg_k - g_{k+1}) = 2p^{\mathrm{v}_p(f)-1}(p-1) = 2\varphi(p^{\mathrm{v}_p(f)})$.

5. It is an immediate consequence of 4. that the number of invertible ideal atoms with radical $\mathfrak{p}$ is finite if and only if p does not split. It remains to show that $\mathcal{A}(\mathcal{I}_p(\mathcal{O}))$ is finite if and only if $\mathcal{A}(\mathcal{I}_p^*(\mathcal{O}))$ is finite. It follows from [1, Theorem 4.3] that $\mathcal{I}(\mathcal{O}_\mathfrak{p})$ is a finitely generated monoid if and only if $\mathcal{I}^*(\mathcal{O}_\mathfrak{p})$ is a finitely generated monoid. Therefore, Proposition 3.3.2 implies that $\mathcal{I}_p(\mathcal{O})$ is a finitely generated monoid if and only if $\mathcal{I}_p^*(\mathcal{O})$ is a finitely generated monoid. Observe that $\mathcal{I}_p(\mathcal{O})$ and $\mathcal{I}_p^*(\mathcal{O})$ are atomic monoids. Therefore, $\mathcal{A}(\mathcal{I}_p(\mathcal{O}))$ is finite if and only if $\mathcal{I}_p(\mathcal{O})$ is a finitely generated monoid if and only if $\mathcal{I}_p^*(\mathcal{O})$ is a finitely generated monoid if and only if $\mathcal{A}(\mathcal{I}_p^*(\mathcal{O}))$ is finite. □

4 Sets of Distances and Sets of Catenary Degrees

The goal in this section is to prove Theorem 1.1. The proof is based on the precise description of ideals given in Theorem 3.6. We proceed in a series of lemmas and propositions and use all notation on orders as introduced at the beginning of Section 3. In particular, $\mathcal{O} = \mathcal{O}_f$ is an order in a quadratic number with conductor $f\mathcal{O}_K$ for some $f \in \mathbb{N}_{\geq 2}$.

Proposition 4.1. *Let H be a reduced atomic monoid and suppose there is a cancellative atom $u \in \mathcal{A}(H)$ such that for each $a \in H \setminus H^\times$ there are $n \in \mathbb{N}_0$ and $v \in \mathcal{A}(H)$ such that $a = u^n v$.*

1. *For all $n, m \in \mathbb{N}_0$ and $v, w \in \mathcal{A}(H)$ such that $u^n v = u^m w$, it follows that $n = m$ and $v = w$.*
2. *For all $n \in \mathbb{N}_0$ and $v \in \mathcal{A}(H)$, it follows that* $\max \mathsf{L}(u^n v) = n + 1$.

3. $\mathsf{c}(H) = \sup\{\mathsf{c}(w \cdot y, u^n \cdot v) \mid n \in \mathbb{N}$ *and* $v, w, y \in \mathcal{A}(H)$ *such that* $wy = u^n v\}$.
4. *If H is half-factorial, then* $\mathsf{c}(H) \le 2$.
5. $\sup \Delta(H) = \sup\{\ell - 2 \mid \ell \in \mathbb{N}_{\ge 3}$ *such that* $\mathsf{L}(vw) \cap [2, \ell] = \{2, \ell\}$ *for some* $v, w \in \mathcal{A}(H)\}$.

Proof. 1. Let $n, m \in \mathbb{N}_0$ and $v, w \in \mathcal{A}(H)$ be such that $u^n v = u^m w$. Without restriction let $n \le m$. Since u is cancellative, we infer that $v = u^{m-n} w$. Since $v \in \mathcal{A}(H)$, we have $n = m$, and thus $v = w$.

2. It is clear that $n + 1 \in \mathsf{L}(u^n v)$ for all $n \in \mathbb{N}_0$ and $v \in \mathcal{A}(H)$. Therefore, it is sufficient to show by induction that for all $n \in \mathbb{N}_0$ and $v \in \mathcal{A}(H)$, $\max \mathsf{L}(u^n v) \le n + 1$. Let $n \in \mathbb{N}_0$ and $v \in \mathcal{A}(H)$. If $n = 0$, then the assertion is obviously true. Now let $n > 0$ and $z \in \mathsf{Z}(u^n v)$. Then there are some $z', z'' \in \mathsf{Z}(H) \setminus \{1\}$ such that $z = z' \cdot z''$. There are some $m', m'' \in \mathbb{N}_0$ and $w', w'' \in \mathcal{A}(H)$ such that $\pi(z') = u^{m'} w'$ and $\pi(z'') = u^{m''} w''$. There are some $\ell \in \mathbb{N}$ and $y \in \mathcal{A}(H)$ such that $w'w'' = u^\ell y$. We infer that $u^n v = u^{m'+m''+\ell} y$, and thus $n = m' + m'' + \ell$ by 1. Since $m', m'' < n$, it follows by the induction hypothesis that $|z'| \le m' + 1$ and $|z''| \le m'' + 1$. Consequently, $|z| \le m' + m'' + 2 \le m' + m'' + \ell + 1 = n + 1$.

3. Set $k = \sup\{\mathsf{c}(w \cdot y, u^n \cdot v) \mid n \in \mathbb{N}_0$ and $v, w, y \in \mathcal{A}(H)$ such that $wy = u^n v\}$. Since $\mathsf{c}(H) = \sup\{\mathsf{c}(z, z') \mid a \in H, z, z' \in \mathsf{Z}(a)\}$, it is obvious that $k \le \mathsf{c}(H)$. It remains to show by induction that for all $n \in \mathbb{N}_0$ and $v \in \mathcal{A}(H)$, it follows that $\mathsf{c}(u^n v) \le k$. Let $n \in \mathbb{N}_0$ and $v \in \mathcal{A}(H)$. Since $\mathsf{c}(v) = 0$, we can assume without restriction that $n > 0$. Since $\mathsf{c}(u^n v) = \sup\{\mathsf{c}(z, u^n \cdot v) \mid z \in \mathsf{Z}(u^n v)\}$, it remains to show that $\mathsf{c}(z, u^n \cdot v) \le k$ for all $z \in \mathsf{Z}(u^n v)$. Let $z \in \mathsf{Z}(u^n v)$.

CASE 1: For all $w, y \in \mathcal{A}(H) \setminus \{u\}$, we have $w \cdot y \nmid z$. There are some $m \in \mathbb{N}$ and $w \in \mathcal{A}(H)$ such that $z = u^m \cdot w$. We infer by 1. that $z = u^n \cdot v$, and thus $\mathsf{c}(z, u^n \cdot v) = 0 \le k$.

CASE 2: There are some $w, y \in \mathcal{A}(H) \setminus \{u\}$ such that $w \cdot y \mid z$. Set $z' = \frac{z}{w \cdot y}$. There exist $m \in \mathbb{N}$ and $a \in \mathcal{A}(H)$ such that $wy = u^m a$. We infer that $m \le n$ and $u^n v = \pi(z) = \pi(w \cdot y)\pi(z') = u^m a \pi(z')$, and thus $a\pi(z') = u^{n-m} v$. Observe that $\mathsf{c}(z, u^m \cdot a \cdot z') \le \mathsf{c}(w \cdot y, u^m \cdot a) \le k$. Since $n - m < n$, it follows by the induction hypothesis that $\mathsf{c}(u^m \cdot a \cdot z', u^n \cdot v) \le \mathsf{c}(a \cdot z', u^{n-m} \cdot v) \le k$, and hence $\mathsf{c}(z, u^n \cdot v) \le k$.

4. Let H be half-factorial, $n \in \mathbb{N}$ and $v, w, y \in \mathcal{A}(H)$ be such that $wy = u^n v$. We infer that $n = 1$, and thus $\mathsf{c}(w \cdot y, u^n \cdot v) \le \mathsf{d}(w \cdot y, u \cdot v) \le 2$. Therefore, $\mathsf{c}(H) \le 2$ by 3.

5. Set $N = \sup\{\ell - 2 \mid \ell \in \mathbb{N}_{\ge 3}$ such that $\mathsf{L}(vw) \cap [2, \ell] = \{2, \ell\}$ for some $v, w \in \mathcal{A}(H)\}$. It is obvious that $N \le \sup \Delta(H)$. It remains to show that $k \le N$ for each $k \in \Delta(H)$. Let $k \in \Delta(H)$. Then there are some $a \in H$ and $r, s \in \mathsf{L}(a)$ such that $r < s$, $\mathsf{L}(a) \cap [r, s] = \{r, s\}$, and $k = s - r$. Let $z \in \mathsf{Z}(a)$ with $|z| = r$ be such that $\mathsf{v}_u(z) = \max\{\mathsf{v}_u(z') \mid z' \in \mathsf{Z}(a)$ with $|z'| = r\}$. Since $r < \max \mathsf{L}(a)$, it follows by 2., that there are some $v, w \in \mathcal{A}(H) \setminus \{u\}$ such that $v \cdot w \mid z$. There are some $n \in \mathbb{N}$ and $y \in \mathcal{A}(H)$ such that $vw = u^n y$. Since $\mathsf{v}_u(z)$ is maximal among all factorizations of a of length r, we have $n \ge 2$. Consequently, there is some $\ell \in \mathsf{L}(vw)$ such that

$2 < \ell \le n+1$ and $\mathsf{L}(vw) \cap [2, \ell] = \{2, \ell\}$. Note that $r + \ell - 2 \in \mathsf{L}(a)$, and thus $s \le r + \ell - 2$. This implies that $k \le \ell - 2 \le N$. □

Theorem 3.6 implies that, for all prime divisors p of f, $\mathcal{I}_p^*(\mathcal{O}_f)$ and $\mathcal{I}_p(\mathcal{O}_f)$ are reduced atomic monoids satisfying the assumption in Proposition 4.1.

Lemma 4.2. *Let p be a prime divisor of f.*

1. $\mathsf{Z}(pP_{f,p}) = \{A \cdot P_{f,p} \mid A = P_{f,p}$ *or* $A \in \mathcal{A}(\mathcal{I}_p^*(\mathcal{O}_f))$ *such that* $\mathcal{N}(A) = p^2\}$ *and* $1 \in \mathrm{Ca}(\mathcal{I}_p(\mathcal{O}_f))$.
2. *If $I, J \in \mathcal{A}(\mathcal{I}_p^*(\mathcal{O}_f))$ are such that $\mathcal{N}(I) = p^2$ and $\mathcal{N}(J) > p^2$, then $IJ = pL$ for some $L \in \mathcal{A}(\mathcal{I}_p^*(\mathcal{O}_f))$.*
3. $2 \in \mathrm{Ca}(\mathcal{I}_p^*(\mathcal{O}_f))$.

Proof. 1. Note that $\{I \in \mathcal{I}_p(\mathcal{O}_f) \mid \mathcal{N}(I) = p\} = \{P_{f,p}\}$. First we show that $\mathsf{Z}(pP_{f,p}) = \{A \cdot P_{f,p} \mid A = P_{f,p}$ or $A \in \mathcal{A}(\mathcal{I}_p^*(\mathcal{O}_f))$ such that $\mathcal{N}(A) = p^2\}$.

Let $z \in \mathsf{Z}(pP_{f,p})$. It follows from Proposition 4.1.2 that $|z| \le 2$, and hence $|z| = 2$. Consequently, $z = A \cdot B$ for some $A, B \in \mathcal{A}(\mathcal{I}_p(\mathcal{O}_f))$. By Proposition 3.2.1 there are some $(u, v, w), (x, y, t) \in \mathcal{M}_{f,p}$ such that $A = p^u(p^v\mathbb{Z} + (w + \tau)\mathbb{Z})$ and $B = p^x(p^y\mathbb{Z} + (t + \tau)\mathbb{Z})$. Set $g = \min\{v, y, \mathrm{v}_p(w + t + \varepsilon)\}$ and $e = \min\{g, \mathrm{v}_p(w - t), \mathrm{v}_p(w^2 + \varepsilon w + \eta) - v, \mathrm{v}_p(t^2 + \varepsilon t + \eta) - y\}$. We infer by Proposition 3.2.1 that $u + x + g = 1$ and $v + y + e - 2g = 1$. Note that $g \in \{0, 1\}$. If $g = 0$, then $u + x = v + y = 1$, and thus ($A = p\mathcal{O}_f$ and $B = P_{f,p}$) or ($A = P_{f,p}$ and $B = p\mathcal{O}_f$). Now let $g = 1$. Then $u = x = 0$, $v, y \ge 1$, $v + y + e = 3$, and $e \in \{0, 1\}$. If $e = 1$, then $v = y = 1$, and thus $A = B = P_{f,p}$. Now let $e = 0$. Then ($v = 1$ and $y = 2$) or ($v = 2$ and $y = 1$). Without restriction let $v = 2$ and $y = 1$. Then $B = P_{f,p}$, $\mathcal{N}(A) = p^v = p^2$, and $\mathcal{N}(A)\mathcal{N}(B) = p^3 = \mathcal{N}(pP_{f,p}) = \mathcal{N}(AB)$. Since B is not invertible, it follows by Proposition 3.2.3 that A is invertible.

To prove the converse inclusion note that $P_{f,p} = p\mathbb{Z} + (r + \tau)\mathbb{Z}$ for some $(0, 1, r) \in \mathcal{M}_{f,p}$. By Proposition 3.2.1 we have $P_{f,p}^2 = p^a(p^b\mathbb{Z} + (c + \tau)\mathbb{Z}$ with $(a, b, c) \in \mathcal{M}_{f,p}$, $a = \min\{1, \mathrm{v}_p(2r + \varepsilon)\}$ and $b = 2 + e - 2a$ with $e = \min\{a, \mathrm{v}_p(r^2 + \varepsilon r + \eta) - 1\}$. By Proposition 3.2.3 we have $a > 0$, and thus $a = b = e = 1$. Consequently, $P_{f,p}^2 = pP_{f,p}$. Now let $A \in \mathcal{A}(\mathcal{I}_p^*(\mathcal{O}_f))$ be such that $\mathcal{N}(A) = p^2$. It follows by Proposition 3.2.3 that $\mathcal{N}(AP_{f,p}) = \mathcal{N}(A)\mathcal{N}(P_{f,p}) = p^3$ and $AP_{f,p} = pI$ for some $I \in \mathcal{I}_p(\mathcal{O}_f)$. We infer that $\mathcal{N}(I) = p$, and hence $I = P_{f,p}$.

Observe that $\mathsf{d}(z', z'') \le 1$ for all $z', z'' \in \mathsf{Z}(pP_{f,p})$ and $(p\mathcal{O}_f) \cdot P_{f,p}$ and $P_{f,p}^2$ are distinct factorizations of $pP_{f,p}$. Therefore, $1 = \mathsf{c}(pP_{f,p}) \in \mathrm{Ca}(\mathcal{I}_p(\mathcal{O}_f))$.

2. Let $I, J \in \mathcal{A}(\mathcal{I}_p^*(\mathcal{O}_f))$ be such that $\mathcal{N}(I) = p^2$ and $\mathcal{N}(J) > p^2$. Without restriction we can assume that $I \ne p\mathcal{O}_f$. There are some $(0, 2, r), (0, k, s) \in \mathcal{M}_{f,p}$ such that $I = p^2\mathbb{Z} + (r + \tau)\mathbb{Z}$ and $J = p^k\mathbb{Z} + (s + \tau)\mathbb{Z}$. Since I and J are invertible, we have $\mathrm{v}_p(r^2 + \varepsilon r + \eta) = 2$ and $\mathrm{v}_p(s^2 + \varepsilon s + \eta) = k > 2$. Therefore, $\mathrm{v}_p(r + s + \varepsilon) + \mathrm{v}_p(r - s) = \mathrm{v}_p(r^2 + \varepsilon r + \eta - (s^2 + \varepsilon s + \eta)) = 2$, and thus $\mathrm{v}_p(r + s + \varepsilon) = 1$, by Proposition 3.2.2. Therefore, $\min\{2, k, \mathrm{v}_p(r + s + \varepsilon)\} = 1$, and hence $IJ = pL$ for some $L \in \mathcal{A}(\mathcal{I}_p^*(\mathcal{O}_f))$ by Proposition 3.2.1.

3. We distinguish two cases.

CASE 1: $p \neq 2$ or $\mathsf{v}_p(f) \geq 2$ or $d \not\equiv 1 \mod 8$. It follows from Theorem 3.6 that there is some $I \in \mathcal{A}(\mathcal{I}_p^*(\mathcal{O}_f))$ such that $\mathcal{N}(I) = p^2$ and $I \neq p\mathcal{O}_f$. We have $I\overline{I} = (p\mathcal{O}_f)^2$, and hence $\mathsf{L}(I\overline{I}) = \{2\}$. Since $I \cdot \overline{I}$ and $(p\mathcal{O}_f) \cdot (p\mathcal{O}_f)$ are distinct factorizations of $I\overline{I}$, we have $2 = \mathsf{c}(I\overline{I}) \in \mathrm{Ca}(\mathcal{I}_p^*(\mathcal{O}_f))$.

CASE 2: $p = 2$, $\mathsf{v}_p(f) = 1$ and $d \equiv 1 \mod 8$. By Proposition 3.3.3 we can assume without restriction that $f = 2$. By Theorem 3.6 there is some $I \in \mathcal{A}(\mathcal{I}_2^*(\mathcal{O}_f))$ such that $\mathcal{N}(I) = 8$. There is some $(0, 3, r) \in \mathcal{M}_{f,2}$ such that $I = 8\mathbb{Z} + (r + \tau)\mathbb{Z}$. We have $\mathsf{v}_2(r^2 - d) = 3$, and hence $\mathsf{v}_2(r) = 0$. Therefore, $\min\{3, \mathsf{v}_2(2r)\} = 1$, and thus $I^2 = 2J$ for some $J \in \mathcal{A}(\mathcal{I}_2^*(\mathcal{O}_f))$. Consequently, $\mathsf{L}(I^2) = \{2\}$. Since $I \cdot I$ and $(2\mathcal{O}_f) \cdot J$ are distinct factorizations of I^2, it follows that $2 = \mathsf{c}(I^2) \in \mathrm{Ca}(\mathcal{I}_p^*(\mathcal{O}_f))$. □

Proposition 4.3. *Let p be an odd prime divisor of f such that $\mathsf{v}_p(f) \geq 2$.*

1. *There is a $C \in \mathcal{A}(\mathcal{I}_p^*(\mathcal{O}_f))$ such that $\mathsf{L}(C^2) = \{2, 3\}$ whence $1 \in \Delta(\mathcal{I}_p^*(\mathcal{O}_f))$ and $3 \in \mathrm{Ca}(\mathcal{I}_p^*(\mathcal{O}_f))$. Moreover, if ($p \neq 3$ or $d \not\equiv 2 \mod 3$ or $\mathsf{v}_p(f) > 2$), then there are $I, J, L \in \mathcal{A}(\mathcal{I}_p^*(\mathcal{O}_f))$ such that $I^2 = p^2 J$ and $J^2 = p^2 L$.*
2. *If $|\mathrm{Pic}(\mathcal{O}_f)| \leq 2$ and ($p \neq 3$ or $d \not\equiv 2 \mod 3$ or $\mathsf{v}_p(f) > 2$), then there is a nonzero primary $a \in \mathcal{O}_f$ such that $2, 3 \in \mathsf{L}(a)$ whence $1 \in \Delta(\mathcal{O}_f)$.*

Proof. 1. By Proposition 3.3.3 there is a monoid isomorphism $\delta : \mathcal{I}_p^*(\mathcal{O}_f) \to \mathcal{I}_p^*(\mathcal{O}_{\frac{f}{2^{\mathsf{v}_2(f)}}})$ such that $\delta(p\mathcal{O}_f) = p\mathcal{O}_{\frac{f}{2^{\mathsf{v}_2(f)}}}$. Therefore, we can assume without restriction that f is odd.

CLAIM: $\mathsf{L}(I^2) = \{2, 3\}$ for some $I \in \mathcal{A}(\mathcal{I}_p^*(\mathcal{O}_f))$, $1 \in \Delta(\mathcal{I}_p^*(\mathcal{O}_f))$, $3 \in \mathrm{Ca}(\mathcal{I}_p^*(\mathcal{O}_f))$ and if $\mathsf{v}_p(p^4 + f^2 d) = 4$, then $I^2 = p^2 J$ and $J^2 = p^2 L$ for some $I, J, L \in \mathcal{A}(\mathcal{I}_p^*(\mathcal{O}_f))$.

For $r \in \mathbb{N}_0$ set $k = \mathsf{v}_p(\mathcal{N}_{K/\mathbb{Q}}(r + \tau))$ and $I = p^k\mathbb{Z} + (r + \tau)\mathbb{Z}$. Let $k > 0$ and $r < p^k$. Then $I \in \mathcal{A}(\mathcal{I}_p^*(\mathcal{O}_f))$. Moreover, $I^2 = p^a(p^b\mathbb{Z} + (c + \tau)\mathbb{Z})$ with $a = \min\{k, \mathsf{v}_p(2r + \varepsilon)\}$, $b = 2(k - a)$ and $c = \mathrm{rem}(r - t\frac{\mathcal{N}_{K/\mathbb{Q}}(r+\tau)}{p^a}, p^b)$ for each $t \in \mathbb{Z}$ with $t\frac{2r+\varepsilon}{p^a} \equiv 1 \mod p^{k-a}$. Set $J = p^b\mathbb{Z} + (c + \tau)\mathbb{Z}$. Then $I^2 = p^a J$ and if $b > 0$, then $J \in \mathcal{A}(\mathcal{I}_p^*(\mathcal{O}_f))$. In particular, if $a = 2$ and $b > 0$, then $I, J \in \mathcal{A}(\mathcal{I}_p^*(\mathcal{O}_f))$ and $\mathsf{L}(I^2) = \{2, 3\}$, and hence $1 \in \Delta(I^2) \subset \Delta(\mathcal{I}_p^*(\mathcal{O}_f))$ and $3 = \mathsf{c}(I^2) \in \mathrm{Ca}(\mathcal{I}_p^*(\mathcal{O}_f))$. Observe that $J^2 = p^{a'}(p^{b'}\mathbb{Z} + (c' + \tau)\mathbb{Z})$ with $a' = \min\{b, \mathsf{v}_p(2c + \varepsilon)\}$, $b' = 2(b - a')$ and $c' \in \mathbb{N}_0$ such that $c' < p^{b'}$. Set $L = p^{b'}\mathbb{Z} + (c' + \tau)\mathbb{Z}$. Then $J^2 = p^{a'} L$ and if $b' > 0$, then $L \in \mathcal{A}(\mathcal{I}_p^*(\mathcal{O}_f))$.

CASE 1: $d \not\equiv 1 \mod 4$. Set $r = p^2$. We have $\mathcal{N}_{K/\mathbb{Q}}(r + \tau) = p^4 - f^2 d$, $k \geq 4$, $a = 2$, $b = 2(k - 2) > 0$, $r < p^k$, and $t = \frac{p^{k-2}+1}{2}$ satisfies the congruence. Therefore, $c = \mathrm{rem}(p^2 - \frac{(p^{k-2}+1)(p^4 - f^2 d)}{2p^2}, p^{2(k-2)}) = \frac{p^4 + f^2 d + p^{k-2} f^2 d - p^{k+2} + 2\ell p^{2(k-1)}}{2p^2}$ for some $\ell \in \mathbb{Z}$. For the rest of this case let $\mathsf{v}_p(p^4 + f^2 d) = 4$. It follows that $\mathsf{v}_p(c) = 2$, and hence $a' = \min\{2(k - 2), \mathsf{v}_p(2c)\} = 2$ and $b' = 4(k - 3) > 0$.

CASE 2: $d \equiv 1 \mod 4$. Set $r = \frac{p^2 - 1}{2}$. Observe that $\mathcal{N}_{K/\mathbb{Q}}(r + \tau) = \frac{p^4 - f^2 d}{4}$, $k \geq 4$, $a = 2$, $b = 2(k - 2) > 0$, $r < p^k$, and $t = 1$ satisfies the congruence. Consequently, $2c + \varepsilon = 2\mathrm{rem}(\frac{p^2 - 1}{2} - \frac{p^4 - f^2 d}{4p^2}, p^{2(k-2)}) + 1 = \frac{p^4 + f^2 d + 4\ell p^{2(k-1)}}{2p^2}$ for some

$\ell \in \mathbb{Z}$. For the rest of this case let $\mathsf{v}_p(p^4 + f^2 d) = 4$. We infer that $a' = \min\{2(k-2), \mathsf{v}_p(2c+\varepsilon)\} = 2$. Moreover, $b' = 4(k-3) > 0$. This proves the claim.

Note that if $g \in \mathbb{N}$ with $\mathsf{v}_p(g) = \mathsf{v}_p(f)$, then there is a monoid isomorphism $\alpha : \mathcal{I}_p^*(\mathcal{O}_f) \to \mathcal{I}_p^*(\mathcal{O}_g)$ such that $\alpha(p\mathcal{O}_f) = p\mathcal{O}_g$ by Proposition 3.3.3. By the claim it remains to show that if ($p \neq 3$ or $d \not\equiv 2 \mod 3$ or $\mathsf{v}_p(f) > 2$), then there is some odd $g \in \mathbb{N}$ such that $\mathsf{v}_p(g) = \mathsf{v}_p(f)$ and $\mathsf{v}_p(p^4 + g^2 d) = 4$.

Let ($p \neq 3$ or $d \not\equiv 2 \mod 3$ or $\mathsf{v}_p(f) > 2$). Furthermore, let $\mathsf{v}_p(p^4 + f^2 d) > 4$. This implies that $\mathsf{v}_p(f) = 2$ and $p \nmid d$. Without restriction we can assume that $\mathsf{v}_p(p^4 + (p^2)^2 d) > 4$. We have $\mathsf{v}_p(1+d) > 0$, and hence $p \neq 3$. Set $g = (p-2)p^2$. Then $\mathsf{v}_p(g) = \mathsf{v}_p(f)$. Assume that $\mathsf{v}_p(p^4 + g^2 d) > 4$. Then $p^5 \mid p^4 + (p-2)^2 p^4 d - p^4(1+d)$, and thus $p \mid (p-2)^2 - 1 = p^2 - 4p + 3$. It follows that $p = 3$, a contradiction.

2. Let $|\mathrm{Pic}(\mathcal{O}_f)| \le 2$ and let $p \neq 3$ or $d \not\equiv 2 \mod 3$ or $\mathsf{v}_p(f) > 2$. By 1. there are some $I, J, L \in \mathcal{A}(\mathcal{I}_p^*(\mathcal{O}_f))$ such that $I^2 = p^2 J$ and $J^2 = p^2 L$. We infer that I^2 is principal, and hence J and L are principal. Consequently, there are some $u, v \in \mathcal{A}(\mathcal{O}_f)$ such that $J = u\mathcal{O}_f$, $L = v\mathcal{O}_f$ and $u^2 = p^2 v$. Note that u^2 is primary. Since $p \in \mathcal{A}(\mathcal{O}_f)$, we have $2, 3 \in \mathsf{L}(u^2)$. Therefore, $1 \in \Delta(\mathcal{O}_f)$. □

Proposition 4.4. *Let p be a prime divisor of f such that $\mathsf{v}_p(f) \ge 2$. Then there are $I, J \in \mathcal{A}(\mathcal{I}_p^*(\mathcal{O}_f))$ such that $\mathsf{L}(IJ) = \{2, 4\}$ whence $2 \in \Delta(\mathcal{I}_p^*(\mathcal{O}_f))$ and $4 \in \mathrm{Ca}(\mathcal{I}_p^*(\mathcal{O}_f))$.*

Proof. CASE 1: $p \neq 2$ or $\mathsf{v}_p(f) > 2$ or $d \not\equiv 1 \mod 8$. By Theorem 3.6 there is some $I \in \mathcal{A}(\mathcal{I}_p^*(\mathcal{O}_f))$ such that $\mathcal{N}(I) = p^4$. Set $J = \overline{I}$. We infer that $IJ = (p\mathcal{O}_f)^4$, and hence $\{2, 4\} \subset \mathsf{L}(IJ) \subset \{2, 3, 4\}$. Assume that $3 \in \mathsf{L}(IJ)$. Then there are some $A, B, C \in \mathcal{A}(\mathcal{I}_p^*(\mathcal{O}_f))$ such that $IJ = ABC$ and $\mathcal{N}(A) \le \mathcal{N}(B) \le \mathcal{N}(C)$. Again by Theorem 3.6 we have $\mathcal{N}(L) \in \{p^2\} \cup \{p^n \mid n \in \mathbb{N}_{\ge 4}\}$ for all $L \in \mathcal{A}(\mathcal{I}_p^*(\mathcal{O}_f))$. This implies that $\mathcal{N}(A) = \mathcal{N}(B) = p^2$ and $\mathcal{N}(C) = p^4$. It follows by Lemma 4.2.2 that $ABC = p^2 L$ for some $L \in \mathcal{A}(\mathcal{I}_p^*(\mathcal{O}_f))$. Consequently, $L = p^2\mathcal{O}_f$, a contradiction. We infer that $\mathsf{L}(IJ) = \{2, 4\}$ whence $2 \in \Delta(\mathcal{I}_2^*(\mathcal{O}_f))$ and $4 \in \mathrm{Ca}(\mathcal{I}_2^*(\mathcal{O}_f))$.

CASE 2: $p = 2$, $\mathsf{v}_p(f) = 2$ and $d \equiv 1 \mod 8$. Since $\mathcal{I}_2^*(\mathcal{O}_4) \cong \mathcal{I}_2^*(\mathcal{O}_f)$ by Proposition 3.3.3, we can assume without restriction that $f = 4$. We set

$$w = \begin{cases} 6 & \text{if } d \equiv 1 \mod 16 \\ 2 & \text{if } d \equiv 9 \mod 16 \end{cases} \quad \text{and} \quad z = \begin{cases} 18 & \text{if } d \equiv 1 \mod 32 \\ 22 & \text{if } d \equiv 9 \mod 32 \\ 2 & \text{if } d \equiv 17 \mod 32 \\ 6 & \text{if } d \equiv 25 \mod 32 \end{cases}.$$

In any case, we have $\mathsf{v}_2(\mathcal{N}_{K/\mathbb{Q}}(w+\tau)) = 5$ and $\mathsf{v}_2(\mathcal{N}_{K/\mathbb{Q}}(z+\tau)) = 6$. Set $I = 32\mathbb{Z} + (w+\tau)\mathbb{Z}$ and $J = 64\mathbb{Z} + (z+\tau)\mathbb{Z}$. Then $I, J \in \mathcal{A}(\mathcal{I}_2^*(\mathcal{O}_4))$ and Proposition 3.2.1 implies that $IJ = 2^a(2^b\mathbb{Z} + (c+\tau)\mathbb{Z})$ with $a = \min\{5, 6, \mathsf{v}_2(w+z)\}$, $b = 5 + 6 - 2a$ and $c \in \mathbb{N}_0$ such that $c < 2^b$. Observe that $\mathsf{v}_2(w+z) = 3$, and thus $a = 3$ and $b = 5$. Set $L = 32\mathbb{Z} + (c+\tau)\mathbb{Z}$. Then $L \in \mathcal{A}(\mathcal{I}_2^*(\mathcal{O}_4))$ and $IJ = (2\mathcal{O}_4)^3 L$. We infer that $\{2, 4\} \subset \mathsf{L}(IJ) \subset \{2, 3, 4\}$, by Proposition 4.1.2.

Assume that $3 \in \mathsf{L}(IJ)$. Then there are some $A, B, C \in \mathcal{A}(\mathcal{I}_2^*(\mathcal{O}_4))$ such that $IJ = ABC$ and $\mathcal{N}(A) \leq \mathcal{N}(B) \leq \mathcal{N}(C)$. It follows by Theorem 3.6 that $\mathcal{N}(U) \in \{4\} \cup \{2^n \mid n \geq 5\}$ for all $U \in \mathcal{A}(\mathcal{I}_2^*(\mathcal{O}_4))$. Since $\mathcal{N}(A)\mathcal{N}(B)\mathcal{N}(C) = \mathcal{N}(I)\mathcal{N}(J) = 2048$, we infer that $\mathcal{N}(A) = \mathcal{N}(B) = 4$ and $\mathcal{N}(C) = 128$. It follows by Lemma 4.2.2 that $ABC = 4D$ for some $D \in \mathcal{A}(\mathcal{I}_2^*(\mathcal{O}_4))$. This implies that $D = 2L$, a contradiction. Consequently, $\mathsf{L}(IJ) = \{2, 4\}$, and thus $2 \in \Delta(\mathcal{I}_2^*(\mathcal{O}_4))$ and $4 = \mathsf{c}(IJ) \in \mathrm{Ca}(\mathcal{I}_2^*(\mathcal{O}_4))$. □

Proposition 4.5. *Suppose that one of the following conditions hold:*

(a) $\mathrm{v}_2(f) \geq 5$ *or* ($\mathrm{v}_2(f) = 4$ *and* $d \not\equiv 1 \mod 4$).
(b) $\mathrm{v}_2(f) = 3$ *and* $d \equiv 2 \mod 4$.
(c) $\mathrm{v}_2(f) = 2$ *and* $d \equiv 1 \mod 4$.

Then there are $I, J \in \mathcal{A}(\mathcal{I}_2^*(\mathcal{O}_f))$ *with* $\mathsf{L}(IJ) = \{2, 3\}$ *whence* $1 \in \Delta(\mathcal{I}_2^*(\mathcal{O}_f))$ *and* $3 \in \mathrm{Ca}(\mathcal{I}_2^*(\mathcal{O}_f))$. *If* $|\mathrm{Pic}(\mathcal{O}_f)| \leq 2$, *then there is a nonzero primary* $a \in \mathcal{O}_f$ *with* $2, 3 \in \mathsf{L}(a)$ *whence* $1 \in \Delta(\mathcal{O}_f)$.

Proof. CASE 1: $\mathrm{v}_2(f) \geq 5$ or ($\mathrm{v}_2(f) = 4$ and $d \not\equiv 1 \mod 4$). We show that there are some $A, B, I, J, L \in \mathcal{A}(\mathcal{I}_2^*(\mathcal{O}_f))$ such that $A^2 = 32I$, $B^2 = 16J$ and $IJ = 4L$. Set $k = \mathrm{v}_2(\mathcal{N}_{K/\mathbb{Q}}(16 + \tau))$ and $A = 2^k\mathbb{Z} + (16 + \tau)\mathbb{Z}$. Then $k \geq 8$, $A \in \mathcal{A}(\mathcal{I}_2^*(\mathcal{O}_f))$ and $A^2 = 32(2^{2k-10}\mathbb{Z} + (c + \tau)\mathbb{Z})$ with $(5, 2k - 10, c) \in \mathcal{M}_{f,2}$ and $\mathrm{v}_2(c) \geq 3$. Set $I = 2^{2k-10}\mathbb{Z} + (c + \tau)\mathbb{Z}$. Then $I \in \mathcal{A}(\mathcal{I}_2^*(\mathcal{O}_f))$. Set $B = 64\mathbb{Z} + (8 + \tau)\mathbb{Z}$. Then $B \in \mathcal{A}(\mathcal{I}_2^*(\mathcal{O}_f))$ and $B^2 = 16(16\mathbb{Z} + (4 + \tau)\mathbb{Z})$. Set $J = 16\mathbb{Z} + (4 + \tau)\mathbb{Z}$. Then $B^2 = 16J$, $J \in \mathcal{A}(\mathcal{I}_2^*(\mathcal{O}_f))$ and $IJ = 4L$ with $L \in \mathcal{A}(\mathcal{I}_2^*(\mathcal{O}_f))$.

CASE 2: $\mathrm{v}_2(f) = 3$ and $d \equiv 2 \mod 4$. We show that $AB = 2I$, $AC = 2I'$, $BC = 8I''$, $B^2 = 16J$, $IJ = 4L$, $I'J = 4L'$, $I''J = 4L''$ for some $A, B, C, I, I', I'', J, L, L', L'' \in \mathcal{A}(\mathcal{I}_2^*(\mathcal{O}_f))$. By Proposition 3.3.3, we can assume without restriction that $f = 8$. Set $A = 4\mathbb{Z} + (2 + \tau)\mathbb{Z}$, $B = 64\mathbb{Z} + (8 + \tau)\mathbb{Z}$ and $C = 128\mathbb{Z} + \tau\mathbb{Z}$. Then $A, B, C \in \mathcal{A}(\mathcal{I}_2^*(\mathcal{O}_f))$, $AB = 2(64\mathbb{Z} + (40 + \tau)\mathbb{Z})$, $AC = 2(128\mathbb{Z} + (64 + \tau)\mathbb{Z})$, $B^2 = 16(16\mathbb{Z} + (12 + \tau)\mathbb{Z})$ and $BC = 8(128\mathbb{Z} + (c + \tau)\mathbb{Z})$ with $(3, 7, c) \in \mathcal{M}_{f,2}$ and $\mathrm{v}_2(c) = 4$. Furthermore, $(64\mathbb{Z} + (40 + \tau)\mathbb{Z})(16\mathbb{Z} + (12 + \tau)\mathbb{Z}) = 4(64\mathbb{Z} + (56 + \tau)\mathbb{Z})$, $(128\mathbb{Z} + (64 + \tau)\mathbb{Z})(16\mathbb{Z} + (12 + \tau)\mathbb{Z}) = 4(128\mathbb{Z} + (r + \tau)\mathbb{Z})$ with $(2, 7, r) \in \mathcal{M}_{f,2}$ and $(128\mathbb{Z} + (c + \tau)\mathbb{Z})(16\mathbb{Z} + (12 + \tau)\mathbb{Z}) = 4(128\mathbb{Z} + (s + \tau)\mathbb{Z})$ with $(2, 7, s) \in \mathcal{M}_{f,2}$. Set $J = 16\mathbb{Z} + (12 + \tau)\mathbb{Z}$. In particular, if $I \in \{64\mathbb{Z} + (40 + \tau)\mathbb{Z}, 128\mathbb{Z} + (64 + \tau)\mathbb{Z}, 128\mathbb{Z} + (c + \tau)\mathbb{Z}\}$, then $I, J \in \mathcal{A}(\mathcal{I}_2^*(\mathcal{O}_f))$ and $IJ = 4L$ for some $L \in \mathcal{A}(\mathcal{I}_2^*(\mathcal{O}_f))$.

CASE 3: $\mathrm{v}_2(f) = 2$ and $d \equiv 1 \mod 4$. We show that $A^2 = 4I$ and $I^2 = 4L$ for some $A, I, L \in \mathcal{A}(\mathcal{I}_2^*(\mathcal{O}_f))$. By Proposition 3.3.3, we can assume without restriction that $f = 4$. First let $d \equiv 1 \mod 8$. If $d \equiv 1 \mod 16$, then set $A = 32\mathbb{Z} + (6 + \tau)\mathbb{Z}$ and if $d \equiv 9 \mod 16$, then set $A = 32\mathbb{Z} + (2 + \tau)\mathbb{Z}$. In any case, we have $A \in \mathcal{A}(\mathcal{I}_2^*(\mathcal{O}_f))$ and $A^2 = 4(64\mathbb{Z} + (c + \tau)\mathbb{Z})$ with $(2, 6, c) \in \mathcal{M}_{f,2}$ and $\mathrm{v}_2(c) = 1$. Set $I = 64\mathbb{Z} + (c + \tau)\mathbb{Z}$. Then $I \in \mathcal{A}(\mathcal{I}_2^*(\mathcal{O}_f))$, $A^2 = 4I$ and $I^2 = 4(256\mathbb{Z} + (r + \tau)\mathbb{Z})$ with $(2, 8, r) \in \mathcal{M}_{f,2}$.

Now let $d \equiv 5 \mod 8$. Set $A = 16\mathbb{Z} + (2 + \tau)\mathbb{Z}$. Then $A \in \mathcal{A}(\mathcal{I}_2^*(\mathcal{O}_f))$ and $A^2 = 4(16\mathbb{Z} + (c + \tau)\mathbb{Z})$ with $(2, 4, c) \in \mathcal{M}_{f,2}$ and $\mathrm{v}_2(c) = 1$. Set $I = 16\mathbb{Z} + (c + \tau)\mathbb{Z}$. Then $A^2 = 4I$ and $I^2 = 4(16\mathbb{Z} + (z + \tau)\mathbb{Z})$ with $(2, 4, z) \in \mathcal{M}_{f,2}$.

Using the case analysis above we can find $I, J, L \in \mathcal{A}(\mathcal{I}_2^*(\mathcal{O}_f))$ such that $IJ = 4L$. In particular, $\mathsf{L}(IJ) = \{2, 3\}$, $1 \in \Delta(\mathcal{I}_p^*(\mathcal{O}_f))$ and $3 = \mathsf{c}(IJ) \in \mathrm{Ca}(\mathcal{I}_p^*(\mathcal{O}_f))$. Now let $|\mathrm{Pic}(\mathcal{O}_f)| \le 2$. Observe that if $A, B, C \in \mathcal{A}(\mathcal{I}_2^*(\mathcal{O}_f))$, then A^2 is principal and $\{AB, AC, BC\}$ contains a principal ideal of $\mathcal{O}_f$. In any case we can choose I, J, L to be principal. There are some $u, v, w \in \mathcal{A}(\mathcal{O}_f)$ such that $I = u\mathcal{O}_f$, $J = v\mathcal{O}_f$, $L = w\mathcal{O}_f$ and $uv = 4w$. Note that uv is primary. Since $2 \in \mathcal{A}(\mathcal{O}_f)$, we have $2, 3 \in \mathsf{L}(uv)$, and thus $1 \in \Delta(\mathcal{O}_f)$. □

Proposition 4.6. *Let p be a prime divisor of f. Then the following statements are equivalent:*

(a) *$\mathcal{I}_p^*(\mathcal{O}_f)$ is half-factorial.*
(b) *$\mathcal{I}_p(\mathcal{O}_f)$ is half-factorial.*
(c) *$\mathsf{c}(\mathcal{I}_p^*(\mathcal{O}_f)) = 2$.*
(d) *$\mathsf{c}(\mathcal{I}_p(\mathcal{O}_f)) = 2$.*
(e) *$\mathsf{v}_p(f) = 1$ and p is inert.*

Proof. (a) ⇒ (e) If $\mathsf{v}_p(f) > 1$ or p is not inert, then there is some $I \in \mathcal{A}(\mathcal{I}_p^*(\mathcal{O}_f))$ such that $\mathcal{N}(I) > p^2$ by Theorem 3.6.4. Set $k = \mathsf{v}_p(\mathcal{N}(I))$. Then $k \ge 3$ and $I\overline{I} = (p\mathcal{O}_f)^k$ by Proposition 3.2.5. Since $\overline{I} \in \mathcal{A}(\mathcal{I}_p^*(\mathcal{O}_f))$, we have $2, k \in \mathsf{L}(I\overline{I})$.

(e) ⇒ (b) Observe that $\mathcal{N}(A) \in \{p, p^2\}$ for each $A \in \mathcal{A}(\mathcal{I}_p(\mathcal{O}_f))$, and thus $\mathcal{A}(\mathcal{I}_p(\mathcal{O}_f)) = \{P_{f,p}\} \cup \{A \in \mathcal{A}(\mathcal{I}_p^*(\mathcal{O}_f)) \mid \mathcal{N}(A) = p^2\}$. Let $I \in \mathcal{I}_p(\mathcal{O}_f) \setminus \{\mathcal{O}_f\}$. There are some $k \in \mathbb{N}_0$ and $J \in \mathcal{A}(\mathcal{I}_p(\mathcal{O}_f))$ such that $I = p^k J$. Let $z \in \mathsf{Z}(I)$. Then $z = (\prod_{i=1}^n I_i) \cdot P_{f,p}^\ell$ with $\ell, n \in \mathbb{N}_0$ and $I_i \in \mathcal{A}(\mathcal{I}_p^*(\mathcal{O}_f))$ for each $i \in [1, n]$. Note that $|z| = n + \ell$. It is sufficient to show that $n + \ell = k + 1$.

CASE 1: I is invertible. Then J is invertible and $\ell = 0$. It follows that $p^{2n} = \mathcal{N}(\prod_{i=1}^n I_i) = \mathcal{N}(I) = \mathcal{N}(p^k J) = p^{2k+2}$ by Proposition 3.2.3, and thus $n + \ell = n = k + 1$.

CASE 2: I is not invertible. Then $J = P_{f,p}$ and $\ell > 0$. It follows from Lemma 4.2 that $P_{f,p}^\ell = p^{\ell-1} P_{f,p}$. Consequently,

$$p^{2(n+\ell)-1} = \mathcal{N}(\prod_{i=1}^n I_i)\mathcal{N}(p^{\ell-1}P_{f,p}) = \mathcal{N}(I) = \mathcal{N}(p^k P_{f,p}) = p^{2k+1}$$

by Proposition 3.2.3, and hence $n + \ell = k + 1$.

(b) ⇒ (d) Since $\mathcal{I}_p^*(\mathcal{O}_f)$ is a cancellative divisor-closed submonoid of $\mathcal{I}_p(\mathcal{O}_f)$ and not factorial, we infer by Proposition 4.1.4 that

$$2 \le \mathsf{c}(\mathcal{I}_p^*(\mathcal{O}_f)) \le \mathsf{c}(\mathcal{I}_p(\mathcal{O}_f)) \le 2.$$

(d) ⇒ (c) Note that $\mathcal{I}_p^*(\mathcal{O}_f)$ is a divisor-closed submonoid of $\mathcal{I}_p(\mathcal{O}_f)$, and thus $\mathsf{c}(\mathcal{I}_p^*(\mathcal{O}_f)) \le \mathsf{c}(\mathcal{I}_p(\mathcal{O}_f)) = 2$. Since $\mathcal{I}_p^*(\mathcal{O}_f)$ is not factorial, we infer that $\mathsf{c}(\mathcal{I}_p^*(\mathcal{O}_f)) = 2$.

(c) $\Rightarrow$ (a) Since $\mathcal{I}_p^*(\mathcal{O}_f)$ is cancellative and not factorial, it follows that $2 + \sup \Delta(\mathcal{I}_p^*(\mathcal{O}_f)) \le \mathsf{c}(\mathcal{I}_p^*(\mathcal{O}_f)) = 2$, and thus $\sup \Delta(\mathcal{I}_p^*(\mathcal{O}_f)) = 0$. Consequently, $\Delta(\mathcal{I}_p^*(\mathcal{O}_f)) = \emptyset$, and hence $\mathcal{I}_p^*(\mathcal{O}_f)$ is half-factorial. □

Lemma 4.7. *Let p be a prime divisor of f, $|\mathrm{Pic}(\mathcal{O}_f)| \le 2$, $I, J, L \in \mathcal{A}(\mathcal{I}_p^*(\mathcal{O}_f))$.*

1. *If J is principal and $IJ = p^2L$, then $1 \in \Delta(\mathcal{O}_f)$.*
2. *If I and J are not principal and $IJ = pL$, then $1 \in \Delta(\mathcal{O}_f)$.*

Proof. Note that if $|\mathrm{Pic}(\mathcal{O}_f)| > 1$, then it follows from [16, Corollary 2.11.16] that there is some invertible prime ideal P of $\mathcal{O}_f$ that is not principal. Observe that $p \in \mathcal{A}(\mathcal{O}_f)$. Also note that if I is not principal, then PI is principal, and hence PI is generated by an atom of $\mathcal{O}_f$, since PI has no nontrivial factorizations in $\mathcal{I}^*(\mathcal{O}_f)$.

1. Let J be principal and $IJ = p^2L$. There is some $v \in \mathcal{A}(\mathcal{O}_f)$ such that $J = v\mathcal{O}_f$.

CASE 1: I is principal. Then L is principal, and hence there are some $u, w \in \mathcal{A}(\mathcal{O}_f)$ such that $I = u\mathcal{O}_f$, $L = w\mathcal{O}_f$ and $uv = p^2w$. We infer that $2, 3 \in \mathsf{L}(uv)$, and thus $1 \in \Delta(\mathcal{O}_f)$.

CASE 2: I is not principal. Then L is not principal and $|\mathrm{Pic}(\mathcal{O}_f)| > 1$, and thus there are some $u, w \in \mathcal{A}(\mathcal{O}_f)$ such that $PI = u\mathcal{O}_f$, $PL = w\mathcal{O}_f$ and $uv = p^2w$. It follows that $2, 3 \in \mathsf{L}(uv)$, and thus $1 \in \Delta(\mathcal{O}_f)$.

2. Let I and J not be principal and $IJ = pL$. Then L is principal and $|\mathrm{Pic}(\mathcal{O}_f)| > 1$, and hence there are some $u, v, w, y \in \mathcal{A}(\mathcal{O}_f)$ such that $PI = u\mathcal{O}_f$, $PJ = v\mathcal{O}_f$, $P^2 = w\mathcal{O}_f$, $L = y\mathcal{O}_f$ and $uv = pwy$. Therefore, $2, 3 \in \mathsf{L}(uv)$, and hence $1 \in \Delta(\mathcal{O}_f)$. □

Proposition 4.8. *Let p be a prime divisor of f.*

1. *If $\mathsf{v}_p(f) \ge 2$ or p is not inert, then there are $I, J \in \mathcal{A}(\mathcal{I}_p^*(\mathcal{O}_f))$ such that $\mathsf{L}(IJ) = \{2, 3\}$ whence $1 \in \Delta(\mathcal{I}_p^*(\mathcal{O}_f))$ and $3 \in \mathrm{Ca}(\mathcal{I}_p^*(\mathcal{O}_f))$.*
2. *Suppose that $\mathcal{O}_f$ is not half-factorial and that one of the following conditions holds:*
   (i) *$|\mathrm{Pic}(\mathcal{O}_f)| \ge 3$ or $\mathsf{v}_p(f) \ge 2$ or p does split.*
   (ii) *p is inert and there is some $C \in \mathcal{A}(\mathcal{I}_p^*(\mathcal{O}_f))$ that is not principal.*
   (iii) *p is ramified and there is some principal $C \in \mathcal{A}(\mathcal{I}_p^*(\mathcal{O}_f))$ such that $\mathcal{N}(C) = p^3$.*
   (iv) *f is a squarefree product of inert primes.*

   Then $1 \in \Delta(\mathcal{O}_f)$.

Proof. We prove 1. and 2. simultaneously. Set $G = \mathrm{Pic}(\mathcal{O}_f)$. Let $\mathcal{B}(G)$ be the monoid of zero-sum sequences of G. It follows by [16, Theorem 6.7.1.2] that if $|G| \ge 3$, then $1 \in \Delta(\mathcal{B}(G))$. We infer by [16, Proposition 3.4.7 and Theorems 3.4.10.3 and 3.7.1.1] that there exists an atomic monoid $\mathcal{B}(\mathcal{O}_f)$ such that $\Delta(\mathcal{B}(\mathcal{O}_f)) = \Delta(\mathcal{O}_f)$ and $\mathcal{B}(G)$ is a divisor-closed submonoid of $\mathcal{B}(\mathcal{O}_f)$. In particular, if $|G| \ge 3$, then

$1 \in \Delta(\mathcal{O}_f)$. Thus, for the second assertion we only need to consider the case $|G| \le 2$. By Propositions 4.3 and 4.5 we can restrict to the following cases.

CASE 1: $p = 2$ and $((\mathsf{v}_2(f) \in \{3, 4\}$ and $d \equiv 1 \mod 4)$ or $(\mathsf{v}_2(f) \in \{2, 3\}$ and $d \equiv 3 \mod 4))$. If $(\mathsf{v}_2(f) = 4$ and $d \equiv 1 \mod 4)$ or $(\mathsf{v}_2(f) = 3$ and $d \equiv 3 \mod 4)$, then set $I = 16\mathbb{Z} + (4 + \tau)\mathbb{Z}$. If $\mathsf{v}_2(f) = 3$ and $d \equiv 1 \mod 4$, then set $I = 16\mathbb{Z} + \tau\mathbb{Z}$. Finally, if $\mathsf{v}_2(f) = 2$ and $d \equiv 3 \mod 4$, then there is some $I \in \mathcal{A}(\mathcal{I}_2^*(\mathcal{O}_f))$ such that $\mathcal{N}(I) = 32$ by Theorem 3.6. In any case, it follows that $I \in \mathcal{A}(\mathcal{I}_2^*(\mathcal{O}_f))$.

It is a consequence of Proposition 3.2.1 and Theorem 3.6 that there are some $A, J \in \mathcal{A}(\mathcal{I}_2^*(\mathcal{O}_f))$ and $\ell \in \mathbb{N}$ such that $A^2 = \ell J$ with values according to the following table. Let $k \in \{1, 3, 5, 7\}$ be such that $d \equiv k \mod 8$. Note that $I = 2^a\mathbb{Z} + (r + \tau)\mathbb{Z}$ and $J = 2^b\mathbb{Z} + (s + \tau)\mathbb{Z}$ with $(0, a, r), (0, b, s) \in \mathcal{M}_{f,2}$.

$\mathsf{v}_2(f)$	k	$\mathcal{N}(A)$	ℓ	$\mathcal{N}(J)$	$\mathsf{v}_2(r)$	$\mathsf{v}_2(s)$
4	1	512	16	1024	2	3
4	5	256	16	256	2	3
3	1	128	8	256	∞	2
3	5	64	8	64	∞	2
3	3 or 7	128	16	64	2	≥ 4
2	3 or 7	32	8	16	2	≥ 3

Since $\mathsf{v}_2(r + s) = 2$ in any case, we infer that $IJ = 4L$ for some $L \in \mathcal{A}(\mathcal{I}_2^*(\mathcal{O}_f))$. Now let $|G| \le 2$. We have J is principal, and hence $1 \in \Delta(\mathcal{O}_f)$ by Lemma 4.7.1.

CASE 2: $p = 2$, $\mathsf{v}_2(f) = 2$ and $d \equiv 2 \mod 4$. Set $A = 32\mathbb{Z} + \tau\mathbb{Z}$ and $B = 32\mathbb{Z} + (8 + \tau)\mathbb{Z}$. Then $A, B \in \mathcal{A}(\mathcal{I}_2^*(\mathcal{O}_f))$ and $AB = 8I$ for some $I \in \mathcal{A}(\mathcal{I}_2^*(\mathcal{O}_f))$ with $I = 16\mathbb{Z} + (r + \tau)\mathbb{Z}$, $(0, 4, r) \in \mathcal{M}_{f,2}$, and $\mathsf{v}_2(r) = 2$. Therefore, we have $AI = 4J$ and $BI = 4L$ for some $J, L \in \mathcal{A}(\mathcal{I}_2^*(\mathcal{O}_f))$. Now let $|G| \le 2$. Since $\{A, B, I\}$ contains a principal ideal of $\mathcal{O}_f$, we infer by Lemma 4.7.1 that $1 \in \Delta(\mathcal{O}_f)$.

CASE 3: $p = 3$, $\mathsf{v}_3(f) = 2$ and $d \equiv 2 \mod 3$. First let $d \not\equiv 1 \mod 4$. Set $I = 81\mathbb{Z} + \tau\mathbb{Z}$ and $J = 81\mathbb{Z} + (9 + \tau)\mathbb{Z}$. Then $I, J \in \mathcal{A}(\mathcal{I}_3^*(\mathcal{O}_f))$ and $IJ = 9L$ for some $L \in \mathcal{A}(\mathcal{I}_3^*(\mathcal{O}_f))$ with $L = 81\mathbb{Z} + (r + \tau)\mathbb{Z}$, $(0, 4, r) \in \mathcal{M}_{f,3}$, and $\mathsf{v}_3(r) = 2$. It follows that $IL = 9A$ for some $A \in \mathcal{A}(\mathcal{I}_3^*(\mathcal{O}_f))$.

Now let $d \equiv 1 \mod 4$. By Proposition 3.3.3 we can assume without restriction that f is odd. Set $I = 81\mathbb{Z} + (4 + \tau)\mathbb{Z}$ and $J = 81\mathbb{Z} + (13 + \tau)\mathbb{Z}$. Then $I, J \in \mathcal{A}(\mathcal{I}_3^*(\mathcal{O}_f))$ and $IJ = 9L$ for some $L \in \mathcal{A}(\mathcal{I}_3^*(\mathcal{O}_f))$. There is some $(0, 4, r) \in \mathcal{M}_{f,3}$ such that $L = 81\mathbb{Z} + (r + \tau)\mathbb{Z}$. Since $\mathsf{v}_3(2r + 1) \ge 2$, we have $IL = 9A$ for some $A \in \mathcal{A}(\mathcal{I}_3^*(\mathcal{O}_f))$ or $JL = 9A$ for some $A \in \mathcal{A}(\mathcal{I}_3^*(\mathcal{O}_f))$.

In any case if $|G| \le 2$, then $\{I, J, L\}$ contains a principal ideal of $\mathcal{O}_f$, and hence $1 \in \Delta(\mathcal{O}_f)$ by Lemma 4.7.1.

CASE 4: $\mathsf{v}_p(f) = 1$ and p splits. By Theorem 3.6 there is some $I \in \mathcal{A}(\mathcal{I}_p^*(\mathcal{O}_f))$ such that $\mathcal{N}(I) = p^3$. There is some $(0, 3, r) \in \mathcal{M}_{f,p}$ such that $I = p^3\mathbb{Z} + (r + \tau)\mathbb{Z}$. Observe that $\mathsf{v}_p(2r + \varepsilon) = 1$. We infer that $I^2 = pJ$ for some $J \in \mathcal{A}(\mathcal{I}_p^*(\mathcal{O}_f))$

and $I\overline{I} = p^2L$ with $\overline{I} \in \mathcal{A}(\mathcal{I}_p^*(\mathcal{O}_f))$ and $L = p\mathcal{O}_f \in \mathcal{A}(\mathcal{I}_p^*(\mathcal{O}_f))$. Now let $|G| \leq 2$. We infer by Lemma 4.7 that $1 \in \Delta(\mathcal{O}_f)$.

CASE 5: $\mathsf{v}_p(f) = 1$ and p is ramified. By Theorem 3.6 there is some $C \in \mathcal{A}(\mathcal{I}_p^*(\mathcal{O}_f))$ such that $\mathcal{N}(C) = p^3$. Note that $C\overline{C} = p^3\mathcal{O}_f$ and $\overline{C} \in \mathcal{A}(\mathcal{I}_p^*(\mathcal{O}_f))$. Now let C be principal. It follows by Lemma 4.7.1 that $1 \in \Delta(\mathcal{O}_f)$.

Cases 1-5 show that there are some $I, J, L \in \mathcal{A}(\mathcal{I}_p^*(\mathcal{O}_f))$ such that $IJ = p^2L$. In particular, $\mathsf{L}(IJ) = \{2, 3\}$, $1 \in \Delta(\mathcal{I}_p^*(\mathcal{O}_f))$ and $3 = \mathsf{c}(IJ) \in \mathrm{Ca}(\mathcal{I}_p^*(\mathcal{O}_f))$. This proves 1. For the rest of this proof let $\mathcal{O}_f$ be not half-factorial and $|G| \leq 2$.

CASE 6: $\mathsf{v}_p(f) = 1$, p is inert and there is some $C \in \mathcal{A}(\mathcal{I}_p^*(\mathcal{O}_f))$ that is not principal. We have $C^2 = pL$ for some $L \in \mathcal{A}(\mathcal{I}_p^*(\mathcal{O}_f))$, and thus $1 \in \Delta(\mathcal{O}_f)$ by Lemma 4.7.2.

CASE 7: f is a squarefree product of inert primes. Then $\mathcal{I}_p^*(\mathcal{O}_f)$ is half-factorial by Proposition 4.6. If G is trivial, then $\mathcal{O}_f$ is half-factorial, a contradiction. Note that $\mathcal{O}_f$ is seminormal by [10, Corollary 4.5]. It follows from [18, Theorem 6.2.2.(a)] that $1 \in \Delta(\mathcal{O}_f)$. □

Lemma 4.9. *Let p be a prime divisor of f, $k \in \mathbb{N}_{\geq 2}$, and $N = \sup\{\mathsf{v}_p(\mathcal{N}(A)) \mid A \in \mathcal{A}(\mathcal{I}_p^*(\mathcal{O}_f))\}$. If $\ell \in \mathbb{N}$ and $A \in \mathcal{I}_p(\mathcal{O}_f))$ is both a product of k atoms and a product of ℓ atoms, then $\ell \leq \frac{kN}{2}$.*

Proof. Let $\ell \in \mathbb{N}$ and suppose that a product of k atoms can be written as a product of ℓ atoms and set $P = P_{f,p}$. There are some $a, b \in \mathbb{N}_0$, $I_i \in \mathcal{A}(\mathcal{I}_p(\mathcal{O}_f)) \setminus \{P\}$ for each $[1, b]$ and $J_j \in \mathcal{A}(\mathcal{I}_p(\mathcal{O}_f))$ for each $j \in [1, k]$ such that $\ell = a + b$ and $\prod_{j=1}^k J_j = P^a \prod_{i=1}^b I_i$. Note that $p^2 \mid \mathcal{N}(I_i)$ for each $i \in [1, b]$.

CASE 1: $a = 0$. Then $b = \ell$. It follows by induction from Proposition 3.2.4 that there are $J_j' \in \mathcal{A}(\mathcal{I}_p^*(\mathcal{O}_f))$ for each $j \in [1, k]$ such that $\mathcal{N}(\prod_{j=1}^k J_j) \mid \mathcal{N}(\prod_{j=1}^k J_j')$. Set $M = \mathrm{lcm}\{\mathcal{N}(J_j') \mid j \in [1, k]\}$. Then $p^{2\ell} \mid \prod_{i=1}^\ell \mathcal{N}(I_i) \mid \mathcal{N}(\prod_{i=1}^\ell I_i) = \mathcal{N}(\prod_{j=1}^k J_j) \mid \mathcal{N}(\prod_{j=1}^k J_j') = \prod_{j=1}^k \mathcal{N}(J_j') \mid M^k$. This implies that $2\ell \leq k\mathsf{v}_p(M) \leq kN$, and thus $\ell \leq \frac{kN}{2}$.

CASE 2: $a > 0$. By Lemma 4.2 we have $P^a = p^{a-1}P$, and thus $\mathcal{N}(P^a) = p^{2a-1}$. Note that $\prod_{j=1}^k J_j$ is not invertible, and hence one member of the product, say J_1, is not invertible. Observe that $\mathsf{v}_p(\mathcal{N}(J_1)) \leq N - 1$ by Proposition 3.2.4. We infer by induction from Proposition 3.2.4 that there are $J_j' \in \mathcal{A}(\mathcal{I}_p^*(\mathcal{O}_f))$ for each $j \in [2, k]$ such that $\mathcal{N}(\prod_{j=1}^k J_j) \mid \mathcal{N}(J_1 \prod_{j=2}^k J_j')$. Set $M = \mathrm{lcm}\{\mathcal{N}(J_j') \mid j \in [2, k]\}$. Then $p^{2\ell-1} \mid \mathcal{N}(P^a) \prod_{i=1}^b \mathcal{N}(I_i) \mid \mathcal{N}(P^a \prod_{i=1}^b I_i) = \mathcal{N}(\prod_{j=1}^k J_j) \mid \mathcal{N}(J_1 \prod_{j=2}^k J_j') = \mathcal{N}(J_1) \prod_{j=2}^k \mathcal{N}(J_j') \mid \mathcal{N}(J_1)M^{k-1}$. This implies that $2\ell - 1 \leq \mathsf{v}_p(\mathcal{N}(J_1)) + (k-1)\mathsf{v}_p(M) \leq kN - 1$, and hence $\ell \leq \frac{kN}{2}$. □

Lemma 4.10. *Let p be a prime divisor of f. For every $I \in \mathcal{A}(\mathcal{I}_p^*(\mathcal{O}_f))$, we set $\mathsf{v}_I = \mathsf{v}_p(\mathcal{N}(I))$, and let $\mathcal{B} = \{\mathsf{v}_A \mid A \in \mathcal{A}(\mathcal{I}_p^*(\mathcal{O}_f))\}$.*

1. *For all $I \in \mathcal{A}(\mathcal{I}_p^*(\mathcal{O}_f))$, we have $\mathsf{c}(I \cdot \overline{I}, (p\mathcal{O}_f)^{\mathsf{v}_I}) \leq 2 + \sup \Delta(\mathcal{B})$.*

2. *Let* $p = 2$, $d \equiv 1 \mod 8$, *and* $\mathsf{v}_p(f) \geq 4$. *Then* $\mathsf{c}(I \cdot \overline{I}, (p\mathcal{O}_f)^{\mathsf{v}_I}) \leq 4$ *for all* $I \in \mathcal{A}(\mathcal{I}_p^*(\mathcal{O}_f))$.

Proof. 1. It is sufficient to show by induction that for all $n \in \mathbb{N}_{\geq 2}$ and $I \in \mathcal{A}(\mathcal{I}_p^*(\mathcal{O}_f))$ with $\mathsf{v}_I = n$, it follows that $\mathsf{c}(I \cdot \overline{I}, (p\mathcal{O}_f)^n) \leq 2 + \sup \Delta(\mathcal{B})$. Let $n \in \mathbb{N}_{\geq 2}$ and $I \in \mathcal{A}(\mathcal{I}_p^*(\mathcal{O}_f))$ be such that $\mathsf{v}_I = n$. If $n = 2$, then $\mathsf{c}(I \cdot \overline{I}, (p\mathcal{O}_f)^2) \leq \mathsf{d}(I \cdot \overline{I}, (p\mathcal{O}_f)^2) \leq 2 \leq 2 + \sup \Delta(\mathcal{B})$. Now let $n > 2$. Note that $2 = \mathsf{v}_{p\mathcal{O}_f} \in \mathcal{B}$, and hence there is some $k \in \mathcal{B}$ such that $2 \leq k < n$ and $\mathcal{B} \cap [k, n] = \{k, n\}$. Observe that $n - k \in \Delta(\mathcal{B})$. Furthermore, there is some $J \in \mathcal{A}(\mathcal{I}_p^*(\mathcal{O}_f))$ such that $k = \mathsf{v}_J$. Note that $J\overline{J} = (p\mathcal{O}_f)^k$, and thus $I\overline{I} = (p\mathcal{O}_f)^{n-k} J\overline{J}$. By the induction hypothesis, we infer that $c((p\mathcal{O}_f)^{n-k} \cdot J \cdot \overline{J}, (p\mathcal{O}_f)^n) \leq c(J \cdot \overline{J}, (p\mathcal{O}_f)^k) \leq 2 + \sup \Delta(\mathcal{B})$. Since $\mathsf{d}(I \cdot \overline{I}, (p\mathcal{O}_f)^{n-k} \cdot J \cdot \overline{J}) \leq 2 + (n - k) \leq 2 + \sup \Delta(\mathcal{B})$, it follows that $\mathsf{c}(I \cdot \overline{I}, (p\mathcal{O}_f)^n) \leq 2 + \sup \Delta(\mathcal{B})$.

2. By Proposition 3.3.3 we can assume without restriction that $f = 2^{\mathsf{v}_2(f)}$. We show by induction that for all $n \in \mathbb{N}_{\geq 2}$ and $I \in \mathcal{A}(\mathcal{I}_2^*(\mathcal{O}_f))$ with $\mathsf{v}_I = n$, we have $\mathsf{c}(I \cdot \overline{I}, (2\mathcal{O}_f)^n) \leq 4$. Let $n \in \mathbb{N}_{\geq 2}$ and $I \in \mathcal{A}(\mathcal{I}_2^*(\mathcal{O}_f))$ be such that $\mathsf{v}_I = n$. If $n = 2$, then $\mathsf{c}(I \cdot \overline{I}, (2\mathcal{O}_f)^2) \leq \mathsf{d}(I \cdot \overline{I}, (2\mathcal{O}_f)^2) \leq 2 \leq 2 + \sup \Delta(\mathcal{B})$. Next let $n > 2$. Observe that $2 = \mathsf{v}_{2\mathcal{O}_f} \in \mathcal{B}$, and hence there is some $k \in \mathcal{B}$ such that $2 \leq k < n$ and $\mathcal{B} \cap [k, n] = \{k, n\}$. There is some $J \in \mathcal{A}(\mathcal{I}_2^*(\mathcal{O}_f))$ such that $k = \mathsf{v}_J$. Note that $J\overline{J} = (2\mathcal{O}_f)^k$, and hence $I\overline{I} = (2\mathcal{O}_f)^{n-k} J\overline{J}$. By the induction hypothesis, we have $c((2\mathcal{O}_f)^{n-k} \cdot J \cdot \overline{J}, (2\mathcal{O}_f)^n) \leq c(J \cdot \overline{J}, (2\mathcal{O}_f)^k) \leq 4$.

CASE 1: $n \neq 2\mathsf{v}_2(f) + 1$. It follows from Theorem 3.6 that $n - k \leq 2$. Since $\mathsf{d}(I \cdot \overline{I}, (2\mathcal{O}_f)^{n-k} \cdot J \cdot \overline{J}) \leq 4$, we infer that $\mathsf{c}(I \cdot \overline{I}, (2\mathcal{O}_f)^n) \leq 4$.

CASE 2: $n = 2\mathsf{v}_2(f) + 1$. By Theorem 3.6 we have $n - k = 3$. Set $A = 16\mathbb{Z} + (4 + \tau)\mathbb{Z}$, $B = 2^{n-3}\mathbb{Z} + (2^{n-5} + \tau)\mathbb{Z}$, and $C = 2^{n-3}\mathbb{Z} + (2^{n-4} + \tau)\mathbb{Z}$. Then $A, B, C \in \mathcal{A}(\mathcal{I}_2^*(\mathcal{O}_f))$ and $ABC = 2^{n-5}A(16\mathbb{Z} + (12 + \tau)\mathbb{Z}) = (2\mathcal{O}_f)^{n-1}$. Observe that $\mathsf{d}(I \cdot \overline{I}, (2\mathcal{O}_f) \cdot A \cdot B \cdot C) \leq 4$ and $\mathsf{d}((2\mathcal{O}_f) \cdot A \cdot B \cdot C, (2\mathcal{O}_f)^{n-k} \cdot J \cdot \overline{J})) \leq 4$. Therefore, $\mathsf{c}(I \cdot \overline{I}, (2\mathcal{O}_f)^n) \leq 4$. □

Proposition 4.11. *Let* p *be a prime divisor of* f *and set* $\mathcal{B} = \{\mathsf{v}_p(\mathcal{N}(A)) \mid A \in \mathcal{A}(\mathcal{I}_p^*(\mathcal{O}_f))\}$.

1. $\sup \Delta(\mathcal{I}_p(\mathcal{O}_f)) \leq \sup \Delta(\mathcal{B})$ *and* $\mathsf{c}(\mathcal{I}_p(\mathcal{O}_f)) \leq 2 + \sup \Delta(\mathcal{B})$.
2. *Let* $p = 2$, $d \equiv 1 \mod 8$, *and* $\mathsf{v}_p(f) \geq 4$. *Then* $\sup \Delta(\mathcal{I}_2(\mathcal{O}_f)) \leq 2$ *and* $\mathsf{c}(\mathcal{I}_2(\mathcal{O}_f)) \leq 4$.

Proof. 1. First we consider the case that $\mathsf{v}_p(f) = 1$ and p is inert. It follows from Theorem 3.6 that $\sup \Delta(\mathcal{B}) = 0$. Proposition 4.6 implies that $\sup \Delta(\mathcal{I}_p(\mathcal{O}_f)) = 0$ and $\mathsf{c}(\mathcal{I}_p(\mathcal{O}_f)) = 2$. Now let $\mathsf{v}_p(f) \geq 2$ or p not inert. Observe that $\sup \Delta(\mathcal{B}) \geq 1$ by Theorem 3.6. Let $I, J \in \mathcal{A}(\mathcal{I}_p(\mathcal{O}_f))$. There are some $n \in \mathbb{N}$ and $L \in \mathcal{A}(\mathcal{I}_p(\mathcal{O}_f))$ such that $IJ = p^n L$.

By Proposition 4.1, it remains to show that $\mathsf{c}(I \cdot J, (p\mathcal{O}_f)^n \cdot L) \leq 2 + \sup \Delta(\mathcal{B})$ and if $\ell \in \mathbb{N}_{\geq 3}$ is such that $\mathsf{L}(IJ) \cap [2, \ell] = \{2, \ell\}$, then $\ell - 2 \leq \sup \Delta(\mathcal{B})$. Set $N = \sup \mathcal{B}$. Since a product of two atoms of $\mathcal{I}_p(\mathcal{O}_f)$ can be written as a product of $n + 1$ atoms, Lemma 4.9 implies that $n + 1 \leq N$. If $n = 1$, then $\mathsf{d}(I \cdot J, (p\mathcal{O}_f) \cdot L) \leq 2 \leq$

$2 + \sup \Delta(\mathcal{B})$ and there is no $\ell \in \mathbb{N}_{\geq 3}$ with $\mathsf{L}(IJ) \cap [2, \ell] = \{2, \ell\}$. Now let $n \geq 2$ and $\ell \in \mathbb{N}_{\geq 3}$ be such that $\mathsf{L}(IJ) \cap [2, \ell] = \{2, \ell\}$.

CASE 1: $n \in \mathcal{B}$. Then $A\overline{A} = (p\mathcal{O}_f)^n$ for some $A \in \mathcal{A}(\mathcal{I}_p^*(\mathcal{O}_f))$. Therefore, $\mathsf{c}(A \cdot \overline{A} \cdot L, (p\mathcal{O}_f)^n \cdot L) \leq \mathsf{c}(A \cdot \overline{A}, (p\mathcal{O}_f)^n) \leq 2 + \sup \Delta(\mathcal{B})$ by Lemma 4.10.1. Moreover, $\mathsf{d}(I \cdot J, A \cdot \overline{A} \cdot L) \leq 3 \leq 2 + \sup \Delta(\mathcal{B})$, and thus $\mathsf{c}(I \cdot J, (p\mathcal{O}_f)^n \cdot L) \leq 2 + \sup \Delta(\mathcal{B})$ and $\ell - 2 = 1 \leq \sup \Delta(\mathcal{B})$.

CASE 2: $n \notin \mathcal{B}$. Note that $n \geq 3$. It follows by Theorem 3.6 that $\mathsf{v}_p(f) \geq 2$ and $\sup \Delta(\mathcal{B}) \geq 2$.

CASE 2.1: $p \neq 2$ or $d \not\equiv 1 \mod 8$ or $n \neq 2\mathsf{v}_p(f)$. Since $n \leq N$, it follows from Theorem 3.6 that $n - 1 = \mathcal{N}(A)$ for some $A \in \mathcal{A}(\mathcal{I}_p^*(\mathcal{O}_f))$, and hence $A\overline{A} = (p\mathcal{O}_f)^{n-1}$. We infer that $\mathsf{c}((p\mathcal{O}_f) \cdot A \cdot \overline{A} \cdot L, (p\mathcal{O}_f)^n \cdot L) \leq \mathsf{c}(A \cdot \overline{A}, (p\mathcal{O}_f)^{n-1}) \leq 2 + \sup \Delta(\mathcal{B})$ by Lemma 4.10.1. Moreover, we have $\mathsf{d}(I \cdot J, A \cdot \overline{A} \cdot (p\mathcal{O}_f) \cdot L) \leq 4 \leq 2 + \sup \Delta(\mathcal{B})$, and thus $\mathsf{c}(I \cdot J, (p\mathcal{O}_f)^n \cdot L) \leq 2 + \sup \Delta(\mathcal{B})$ and $\ell - 2 \leq 2 \leq \sup \Delta(\mathcal{B})$.

CASE 2.2: $p = 2$, $d \equiv 1 \mod 8$ and $n = 2\mathsf{v}_p(f)$. We infer by Theorem 3.6 that $\sup \Delta(\mathcal{B}) = 3$. By Theorem 3.6 there is some $A \in \mathcal{A}(\mathcal{I}_2^*(\mathcal{O}_f))$ such that $n - 2 = \mathcal{N}(A)$, and thus $A\overline{A} = (2\mathcal{O}_f)^{n-2}$. This implies that $\mathsf{c}((2\mathcal{O}_f)^2 \cdot A \cdot \overline{A} \cdot L, (2\mathcal{O}_f)^n \cdot L) \leq \mathsf{c}(A \cdot \overline{A}, (2\mathcal{O}_f)^{n-2}) \leq 2 + \sup \Delta(\mathcal{B})$ by Lemma 4.10.1. Observe that $\mathsf{d}(I \cdot J, A \cdot \overline{A} \cdot (2\mathcal{O}_f)^2 \cdot L) \leq 5 = 2 + \sup \Delta(\mathcal{B})$, and hence $\mathsf{c}(I \cdot J, (2\mathcal{O}_f)^n \cdot L) \leq 2 + \sup \Delta(\mathcal{B})$ and $\ell - 2 \leq 3 = \sup \Delta(\mathcal{B})$.

2. By Proposition 3.3.3 we can assume without restriction that $f = 2^{\mathsf{v}_2(f)}$. Let $I, J \in \mathcal{A}(\mathcal{I}_2(\mathcal{O}_f))$. There are some $n \in \mathbb{N}$ and $L \in \mathcal{A}(\mathcal{I}_2(\mathcal{O}_f))$ such that $IJ = 2^n L$. It follows from Lemma 4.9 that $n + 1 \leq \sup \mathcal{B}$. By Proposition 4.1, it is sufficient to show that $\mathsf{c}(I \cdot J, (2\mathcal{O}_f)^n \cdot L) \leq 4$ and if $\ell \in \mathbb{N}_{\geq 3}$ is such that $\mathsf{L}(IJ) \cap [2, \ell] = \{2, \ell\}$, then $\ell - 2 \leq 2$. The assertion is trivially true for $n = 1$. Let $n \geq 2$ and let $\ell \in \mathbb{N}_{\geq 3}$ be such that $\mathsf{L}(IJ) \cap [2, \ell] = \{2, \ell\}$.

CASE 1: $n \in \mathcal{B}$. There is some $A \in \mathcal{A}(\mathcal{I}_2^*(\mathcal{O}_f))$ such that $A\overline{A} = (2\mathcal{O}_f)^n$. It follows by Lemma 4.10.2 that $\mathsf{c}(A \cdot \overline{A} \cdot L, (2\mathcal{O}_f)^n \cdot L) \leq \mathsf{c}(A \cdot \overline{A}, (2\mathcal{O}_f)^n) \leq 4$. Furthermore, $\mathsf{d}(I \cdot J, A \cdot \overline{A} \cdot L) \leq 3$, and thus $\mathsf{c}(I \cdot J, (2\mathcal{O}_f)^n \cdot L) \leq 4$ and $\ell - 2 \leq 1$.

CASE 2: $n \notin \mathcal{B}$ and $n \neq 2\mathsf{v}_2(f)$. It follows by Theorem 3.6 that there is some $A \in \mathcal{A}(\mathcal{I}_2^*(\mathcal{O}_f))$ such that $A\overline{A} = (2\mathcal{O}_f)^{n-1}$. We infer by Lemma 4.10.2 that $\mathsf{c}((2\mathcal{O}_f) \cdot A \cdot \overline{A} \cdot L, (2\mathcal{O}_f)^n \cdot L) \leq \mathsf{c}(\cdot A \cdot \overline{A}, (2\mathcal{O}_f)^{n-1}) \leq 4$. Furthermore, $\mathsf{d}(I \cdot J, (2\mathcal{O}_f) \cdot A \cdot \overline{A} \cdot L) \leq 4$, and thus $\mathsf{c}(I \cdot J, (2\mathcal{O}_f)^n \cdot L) \leq 4$ and $\ell - 2 \leq 2$.

CASE 3: $n = 2\mathsf{v}_2(f)$. By Theorem 3.6 there is some $D \in \mathcal{A}(\mathcal{I}_2^*(\mathcal{O}_f))$ such that $D\overline{D} = (2\mathcal{O}_f)^{n-2}$. Set $A = 16\mathbb{Z} + (4 + \tau)\mathbb{Z}$, $B = 2^{n-2}\mathbb{Z} + (2^{n-4} + \tau)\mathbb{Z}$ and $C = 2^{n-2}\mathbb{Z} + (2^{n-3} + \tau)\mathbb{Z}$. Then $A, B, C \in \mathcal{A}(\mathcal{I}_2^*(\mathcal{O}_f))$ and $ABC = 2^{n-4}A(16\mathbb{Z} + (12 + \tau)\mathbb{Z}) = (2\mathcal{O}_f)^n$. This implies that $\mathsf{c}((2\mathcal{O}_f)^2 \cdot D \cdot \overline{D} \cdot L, (2\mathcal{O}_f)^n \cdot L) \leq \mathsf{c}(D \cdot \overline{D}, (2\mathcal{O}_f)^{n-2}) \leq 4$ by Lemma 4.10.2. Moreover, $\mathsf{d}(A \cdot B \cdot C \cdot L, (2\mathcal{O}f)^2 \cdot D \cdot \overline{D} \cdot L) \leq 4$ and $\mathsf{d}(I \cdot J, A \cdot B \cdot C \cdot L) \leq 4$. Consequently, $\mathsf{c}(I \cdot J, (2\mathcal{O}_f)^n \cdot L) \leq 4$ and $\ell - 2 \leq 2$. □

Proposition 4.12. *Let* $\mathsf{v}_2(f) \in \{2, 3\}$ *and* $d \equiv 1 \mod 8$. *Then* $3 \in \Delta(\mathcal{I}_2^*(\mathcal{O}_f))$ *and* $5 \in \mathrm{Ca}(\mathcal{I}_2^*(\mathcal{O}_f))$.

Proof. We distinguish two cases.

CASE 1: $v_2(f) = 2$. By Theorem 3.6 there is some $I \in \mathcal{A}(\mathcal{I}_2^*(\mathcal{O}_f))$ such that $\mathcal{N}(I) = 32$. Set $J = \overline{I}$. Then $IJ = 32\mathcal{O}_f$, and hence $\{2, 5\} \subset \mathsf{L}(IJ) \subset [2, 5]$. Again by Theorem 3.6 we have $\mathcal{N}(L) \in \{4\} \cup \{2^n \mid n \in \mathbb{N}_{\geq 5}\}$ for all $L \in \mathcal{A}(\mathcal{I}_2^*(\mathcal{O}_f))$. Note that if $A, B, C, D \in \mathcal{A}(\mathcal{I}_2^*(\mathcal{O}_f))$, then $\mathcal{N}(ABCD) \in \{256\} \cup \mathbb{N}_{\geq 2048}$. Since $\mathcal{N}(IJ) = 1024$, we have $4 \notin \mathsf{L}(IJ)$. Assume that $3 \in \mathsf{L}(IJ)$. Then there are some $A, B, C \in \mathcal{A}(\mathcal{I}_2^*(\mathcal{O}_f))$ such that $IJ = ABC$ and $\mathcal{N}(A) \leq \mathcal{N}(B) \leq \mathcal{N}(C)$. Therefore, $\mathcal{N}(A) = \mathcal{N}(B) = 4$ and $\mathcal{N}(C) = 64$. We infer by Lemma 4.2.2 that $ABC = 4L$ for some $L \in \mathcal{A}(\mathcal{I}_2^*(\mathcal{O}_f))$, and hence $L = 8\mathcal{O}_f$, a contradiction. We have $\mathsf{L}(IJ) = \{2, 5\}$, and thus $3 \in \Delta(\mathcal{I}_2^*(\mathcal{O}_f))$ and $5 = \mathsf{c}(IJ) \in \mathrm{Ca}(\mathcal{I}_2^*(\mathcal{O}_f))$.

CASE 2: $v_2(f) = 3$. By Proposition 3.3.3 we can assume without restriction that $f = 8$. By Theorem 3.6 there are some $I, J \in \mathcal{A}(\mathcal{I}_2^*(\mathcal{O}_f))$ such that $\mathcal{N}(I) = 128$ and $\mathcal{N}(J) = 16$. We have $I\overline{I} = 128\mathcal{O}_f$ and $J\overline{J} = 16\mathcal{O}_f$, and hence $I\overline{I} = 8J\overline{J}$. This implies that $\{2, 5\} \subset \mathsf{L}(I\overline{I})$. It follows from Theorem 3.6 that $\mathcal{N}(L) \in \{4, 16\} \cup \{2^n \mid n \in \mathbb{N}_{\geq 7}\}$ for all $L \in \mathcal{A}(\mathcal{I}_2^*(\mathcal{O}_f))$.

First assume that $3 \in \mathsf{L}(I\overline{I})$. Then there exist $A, B, C \in \mathcal{A}(\mathcal{I}_2^*(\mathcal{O}_f))$ such that $I\overline{I} = ABC$, and $\mathcal{N}(A) \leq \mathcal{N}(B) \leq \mathcal{N}(C)$. Therefore, $(\mathcal{N}(A), \mathcal{N}(B), \mathcal{N}(C)) \in \{(4, 16, 256), (4, 4, 1024)\}$. If $(\mathcal{N}(A), \mathcal{N}(B), \mathcal{N}(C)) = (4, 16, 256)$, then it follows by Lemma 4.2.2 that $AB = 2D$ for some $D \in \mathcal{A}(\mathcal{I}_2^*(\mathcal{O}_f))$ with $\mathcal{N}(D) = 16$. We infer that $DC = 64\mathcal{O}_f$, and hence $C = 4\overline{D}$, a contradiction. Now let $(\mathcal{N}(A), \mathcal{N}(B), \mathcal{N}(C)) = (4, 4, 1024)$. Then $ABC = 4D$ for some $D \in \mathcal{A}(\mathcal{I}_2^*(\mathcal{O}_f))$ by Lemma 4.2.2, and thus $D = 32\mathcal{O}_f$, a contradiction. Consequently, $3 \notin \mathsf{L}(I\overline{I})$.

Next assume that $4 \in \mathsf{L}(I\overline{I})$. Then there exist $A, B, C, D \in \mathcal{A}(\mathcal{I}_2^*(\mathcal{O}_f))$ such that $I\overline{I} = ABCD$, and $\mathcal{N}(A) \leq \mathcal{N}(B) \leq \mathcal{N}(C) \leq \mathcal{N}(D)$.

Then $(\mathcal{N}(A), \mathcal{N}(B), \mathcal{N}(C), \mathcal{N}(D)) \in \{(4, 4, 4, 256), (4, 16, 16, 16)\}$.

If $(\mathcal{N}(A), \mathcal{N}(B), \mathcal{N}(C), \mathcal{N}(D)) = (4, 4, 4, 256)$, then $ABCD = 8E$ for $E \in \mathcal{A}(\mathcal{I}_2^*(\mathcal{O}_f))$ by Lemma 4.2.2, and hence $E = 16\mathcal{O}_f$, a contradiction. Now let $(\mathcal{N}(A), \mathcal{N}(B), \mathcal{N}(C), \mathcal{N}(D)) = (4, 16, 16, 16)$. By Lemma 4.2.2 there is some $E \in \mathcal{A}(\mathcal{I}_2^*(\mathcal{O}_f))$ with $\mathcal{N}(E) = 16$ such that $AB = 2E$. Therefore, $ECD = 64\mathcal{O}_f$, and hence $CD = 4\overline{E}$. There are some $(0, 4, r), (0, 4, s) \in \mathcal{M}_{f,2}$ such that $C = 16\mathbb{Z} + (r + \tau)\mathbb{Z}$ and $D = 16\mathbb{Z} + (s + \tau)\mathbb{Z}$. We have $v_2(r^2 - 16d) = v_2(s^2 - 16d) = 4$. Since $d \equiv 1 \mod 8$, this implies that $v_2(r), v_2(s) \geq 3$. Therefore, $\min\{4, v_2(r + s + \varepsilon)\} \in \{3, 4\}$, and hence $CD = 8F$ for some $F \in \mathcal{A}(\mathcal{I}_2^*(\mathcal{O}_f))$. We infer that $\overline{E} = 2F$, a contradiction. Consequently, $4 \notin \mathsf{L}(I\overline{I})$.

Therefore, 2 and 5 are adjacent lengths of $I\overline{I}$, and hence $3 \in \Delta(\mathcal{I}_2^*(\mathcal{O}_f))$. Note that $\mathsf{c}(\mathcal{I}_2^*(\mathcal{O}_f)) \leq 5$ by Proposition 4.11.1 and Theorem 3.6. Moreover, since $\mathcal{I}_2^*(\mathcal{O}_f)$ is a cancellative monoid, we have $5 \leq 2 + \sup \Delta(\mathsf{L}(I\overline{I})) \leq \mathsf{c}(I\overline{I}) \leq 5$, and thus $5 = \mathsf{c}(I\overline{I}) \in \mathrm{Ca}(\mathcal{I}_2^*(\mathcal{O}_f))$. □

Lemma 4.13. *Let $H \in \{\mathcal{I}(\mathcal{O}_f), \mathcal{I}^*(\mathcal{O}_f)\}$. For every prime divisor p of f, we set $H_p = \mathcal{I}_p(\mathcal{O}_f)$ if $H = \mathcal{I}(\mathcal{O}_f)$ and $H_p = \mathcal{I}_p^*(\mathcal{O}_f)$ if $H = \mathcal{I}^*(\mathcal{O}_f)$.*

1. *H is half-factorial if and only if H_p is half-factorial for every $p \in \mathbb{P}$ with $p \mid f$.*
2. *If H is not half-factorial, then* $\sup \Delta(H) = \sup\{\sup \Delta(H_p) \mid p \in \mathbb{P}$ *with* $p \mid f\}$.
3. $\mathsf{c}(H) = \sup\{\mathsf{c}(H_p) \mid p \in \mathbb{P}$ *with* $p \mid f\}$.

Proof. By Eqs. 2.3 and 2.4, we have

$$\mathcal{I}^*(\mathcal{O}_f) \cong \coprod_{P \in \mathfrak{X}(\mathcal{O}_f)} \mathcal{I}^*_P(\mathcal{O}_f) \quad \text{and} \quad \mathcal{I}(\mathcal{O}_f) \cong \coprod_{P \in \mathfrak{X}(\mathcal{O}_f)} \mathcal{I}_P(\mathcal{O}_f).$$

Thus the assertions are easy consequences (see [16, Propositions 1.4.5.3 and 1.6.8.1]). □

Proof (ProofofTheorem 1.1). 1. This is an immediate consequence of Proposition 4.6 and Lemma 4.13.

2. First, suppose that f is squarefree. By 1., we have f is not a product of inert primes. It follows from Lemma 4.13, Proposition 4.11.1, and Theorem 3.6 that $\mathsf{c}(\mathcal{I}^*(\mathcal{O})) \le \mathsf{c}(\mathcal{I}(\mathcal{O})) \le 3$ and $\sup \Delta(\mathcal{I}^*(\mathcal{O})) \le \sup \Delta(\mathcal{I}(\mathcal{O})) \le 1$. By Lemma 4.2 and Proposition 4.8.1, it follows that $1 \in \Delta(\mathcal{I}^*(\mathcal{O}))$, $1 \in \mathrm{Ca}(\mathcal{I}(\mathcal{O}))$ and $[2,3] \subset \mathrm{Ca}(\mathcal{I}^*(\mathcal{O}))$, and thus $\mathrm{Ca}(\mathcal{I}(\mathcal{O})) = [1,3]$, $\mathrm{Ca}(\mathcal{I}^*(\mathcal{O})) = [2,3]$, and $\Delta(\mathcal{I}(\mathcal{O})) = \Delta(\mathcal{I}^*(\mathcal{O})) = \{1\}$.

Now we suppose that f is not squarefree and we distinguish two cases.

CASE 1: $\mathsf{v}_2(f) \notin \{2,3\}$ or $d_K \not\equiv 1 \mod 8$. By Lemma 4.13, Proposition 4.11, and Theorem 3.6 it follows that $\mathsf{c}(\mathcal{I}^*(\mathcal{O})) \le \mathsf{c}(\mathcal{I}(\mathcal{O})) \le 4$ and $\sup \Delta(\mathcal{I}^*(\mathcal{O})) \le \sup \Delta(\mathcal{I}(\mathcal{O})) \le 2$. We infer by Lemma 4.2 and Propositions 4.4 and 4.8 that $[1,2] \subset \Delta(\mathcal{I}^*(\mathcal{O}))$, $1 \in \mathrm{Ca}(\mathcal{I}(\mathcal{O}))$, and $[2,4] \subset \mathrm{Ca}(\mathcal{I}^*(\mathcal{O}))$, and hence $\mathrm{Ca}(\mathcal{I}(\mathcal{O})) = [1,4]$, $\mathrm{Ca}(\mathcal{I}^*(\mathcal{O})) = [2,4]$, and $\Delta(\mathcal{I}(\mathcal{O})) = \Delta(\mathcal{I}^*(\mathcal{O})) = [1,2]$.

CASE 2: $\mathsf{v}_2(f) \in \{2,3\}$ and $d_K \equiv 1 \mod 8$. We infer by Lemma 4.13, Proposition 4.11.1, and Theorem 3.6 that $\mathsf{c}(\mathcal{I}^*(\mathcal{O})) \le \mathsf{c}(\mathcal{I}(\mathcal{O})) \le 5$ and sup $\Delta(\mathcal{I}^*(\mathcal{O})) \le \sup \Delta(\mathcal{I}(\mathcal{O})) \le 3$. Lemma 4.2 and Propositions 4.4, 4.8 and 4.12 imply that $[1,3] \subset \Delta(\mathcal{I}^*(\mathcal{O}))$, $1 \in \mathrm{Ca}(\mathcal{I}(\mathcal{O}))$ and $[2,5] \subset \mathrm{Ca}(\mathcal{I}^*(\mathcal{O}))$. Consequently, $\mathrm{Ca}(\mathcal{I}(\mathcal{O})) = [1,5]$, $\mathrm{Ca}(\mathcal{I}^*(\mathcal{O})) = [2,5]$, and $\Delta(\mathcal{I}(\mathcal{O})) = \Delta(\mathcal{I}^*(\mathcal{O})) = [1,3]$. □

Based on the results of this section we derive a result on the set of distances of orders. Let $\mathcal{O}$ be a non-half-factorial order in a number field. Then the set of distances $\Delta(\mathcal{O})$ is finite. If $\mathcal{O}$ is a principal order, then it is easy to show that $\min \Delta(\mathcal{O}) = 1$ (indeed much stronger results are known, namely, that sets of lengths of almost all elements—in a sense of density—are intervals, see [16, Theorem 9.4.11]). The same is true if $|\mathrm{Pic}(\mathcal{O})| \ge 3$ or if $\mathcal{O}$ is seminormal [24, Theorem 1.1]. However, it was unknown so far whether there exists an order $\mathcal{O}$ with $\min \Delta(\mathcal{O}) > 1$. In the next result of this section we characterize all non-half-factorial orders in quadratic number fields with $\min \Delta(\mathcal{O}) > 1$ which allows us to give the first explicit examples of orders $\mathcal{O}$ with $\min \Delta(\mathcal{O}) > 1$. A characterization of half-factorial orders in quadratic number fields is given in [16, Theorem 3.7.15].

Let $\mathcal{O}$ be an order in a quadratic number field K with conductor $f \in \mathbb{N}_{\ge 2}$. Then the class numbers $|\mathrm{Pic}(\mathcal{O}_K)|$ and $|\mathrm{Pic}(\mathcal{O})|$ are linked by the formula [25, Corollary 5.9.8]

$$|\mathrm{Pic}(\mathcal{O})| = |\mathrm{Pic}(\mathcal{O}_K)| \frac{f}{(\mathcal{O}_K^\times : \mathcal{O}^\times)} \prod_{p \in \mathbb{P}, p \mid f} \left(1 - \left(\frac{d_K}{p}\right) p^{-1}\right), \tag{4.1}$$

and $|\mathrm{Pic}(\mathcal{O})|$ is a multiple of $|\mathrm{Pic}(\mathcal{O}_K)|$.

Since the number of imaginary quadratic number fields with class number at most two is finite (an explicit list of these fields can be found, for example, in [31]), (4.1) shows that the number of orders in imaginary quadratic number fields with $|\mathrm{Pic}(\mathcal{O})| = 2$ is finite. The complete list of non-maximal orders in imaginary quadratic number fields with $|\mathrm{Pic}(\mathcal{O})| = 2$ is given in [27, page 16]. We refer to [25] for more information on class groups and class numbers and end with explicit examples of non-half-factorial orders $\mathcal{O}$ satisfying $\min \Delta(\mathcal{O}) > 1$.

Theorem 4.14. *Let $\mathcal{O}$ be a non-half-factorial order in a quadratic number field K with conductor $f\mathcal{O}_K$ for some $f \in \mathbb{N}_{\geq 2}$. Then the following statements are equivalent:*

(a) $\min \Delta(\mathcal{O}) > 1$.
(b) $|\mathrm{Pic}(\mathcal{O})| = 2$, *$f$ is a nonempty squarefree product of ramified primes times a (possibly empty) squarefree product of inert primes, and for every prime divisor p of f and every $I \in \mathcal{A}(\mathcal{I}_p^*(\mathcal{O}))$, I is principal if and only if $\mathcal{N}(I) = p^2$.*

If these equivalent conditions are satisfied, then K is a real quadratic number field and $\min \Delta(\mathcal{O}) = 2$.

Proof. CLAIM: If $|\mathrm{Pic}(\mathcal{O})| = 2$, p is a ramified prime with $\mathrm{v}_p(f) = 1$, and every $I \in \mathcal{A}(\mathcal{I}_p^*(\mathcal{O}))$ with $\mathcal{N}(I) = p^3$ is not principal, then every $L \in \mathcal{A}(\mathcal{I}_p^*(\mathcal{O}))$ with $\mathcal{N}(L) = p^2$ is principal.

Let $|\mathrm{Pic}(\mathcal{O})| = 2$, let p be a ramified prime with $\mathrm{v}_p(f) = 1$, and suppose that every $I \in \mathcal{A}(\mathcal{I}_p^*(\mathcal{O}))$ with $\mathcal{N}(I) = p^3$ is not principal. By Theorem 3.6 we have $\{\mathcal{N}(J) \mid J \in \mathcal{A}(\mathcal{I}_p^*(\mathcal{O}))\} = \{p^2, p^3\}$. There is some $I \in \mathcal{A}(\mathcal{I}_p^*(\mathcal{O}))$ such that $\mathcal{N}(I) = p^3$. If $J \in \mathcal{A}(\mathcal{I}_p^*(\mathcal{O}))$ with $\mathcal{N}(J) = p^3$, then $IJ = p^2 L$ for some $L \in \mathcal{A}(\mathcal{I}_p^*(\mathcal{O}))$ with $\mathcal{N}(L) = p^2$ (since there are no atoms with norm bigger than p^3). It follows by Theorem 3.6 that $|\{J \in \mathcal{A}(\mathcal{I}_p^*(\mathcal{O})) \mid \mathcal{N}(J) = p^3\}| = |\{L \in \mathcal{A}(\mathcal{I}_p^*(\mathcal{O})) \mid \mathcal{N}(L) = p^2\}| = p$ (note that $\mathcal{N}(p\mathcal{O}) = p^2$). Let $g : \{J \in \mathcal{A}(\mathcal{I}_p^*(\mathcal{O})) \mid \mathcal{N}(J) = p^3\} \to \{L \in \mathcal{A}(\mathcal{I}_p^*(\mathcal{O})) \mid \mathcal{N}(L) = p^2\}$ be defined by $g(J) = L$ where $L \in \mathcal{A}(\mathcal{I}_p^*(\mathcal{O}))$ is such that $\mathcal{N}(L) = p^2$ and $IJ = p^2 L$. Then g is a well-defined bijection. Now let $L \in \mathcal{A}(\mathcal{I}_p^*(\mathcal{O}))$ with $\mathcal{N}(L) = p^2$. There is some $J \in \mathcal{A}(\mathcal{I}_p^*(\mathcal{O}))$ such that $\mathcal{N}(J) = p^3$ and $IJ = p^2 L$. Since $|\mathrm{Pic}(\mathcal{O})| = 2$ and I and J are not principal, we have IJ is principal, and hence L is principal. This proves the claim.

(a) $\Rightarrow$ (b) Observe that if p is an inert prime such that $\mathrm{v}_p(f) = 1$, then $\{\mathcal{N}(J) \mid J \in \mathcal{A}(\mathcal{I}_p^*(\mathcal{O}))\} = \{p^2\}$ by Theorem 3.6. Also note that if p is a ramified prime such that $\mathrm{v}_p(f) = 1$, then $\{\mathcal{N}(J) \mid J \in \mathcal{A}(\mathcal{I}_p^*(\mathcal{O}))\} = \{p^2, p^3\}$ by Theorem 3.6. The assertion now follows by the claim and Proposition 4.8.2.

(b) $\Rightarrow$ (a) Assume to the contrary that $\min \Delta(\mathcal{O}) = 1$. Let $\mathcal{H}$ be the monoid of nonzero principal ideals of $\mathcal{O}$. There is some minimal $k \in \mathbb{N}$ such that $\prod_{i=1}^{k} U_i = \prod_{j=1}^{k+1} U_j'$ with $U_i \in \mathcal{A}(\mathcal{H})$ for each $i \in [1, k]$ and $U_j' \in \mathcal{A}(\mathcal{H})$ for each $j \in [1, k+1]$. Set $\mathcal{Q}_1 = \{P \in \mathfrak{X}(\mathcal{O}) \mid P \text{ is principal}\}$, $\mathcal{Q}_2 = \{P \in \mathfrak{X}(\mathcal{O}) \mid P \text{ is invertible and not principal}\}$, $\mathcal{L} = \{p \in \mathbb{P} \mid p \mid f, p \text{ is ramified}\}$, and $\mathcal{K} = \{\{p, q\} \mid p, q \in \mathcal{L}, p \neq q\}$.

For every prime divisor p of f set $\mathcal{A}_p = \{V \in \mathcal{A}(\mathcal{I}_p^*(\mathcal{O})) \mid \mathcal{N}(V) = p^2\}$, $a_p = |\{i \in [1,k] \mid U_i \in \mathcal{A}_p\}|$ and $a'_p = |\{j \in [1,k+1] \mid U'_j \in \mathcal{A}_p\}|$. For $p \in \mathcal{L}$ set $\mathcal{D}_p = \{V \in \mathcal{A}(\mathcal{I}_p^*(\mathcal{O})) \mid \mathcal{N}(V) = p^3\}$, $\mathcal{B}_p = \{PV \mid P \in \mathcal{Q}_2$ and $V \in \mathcal{D}_p\}$, $b_p = |\{i \in [1,k] \mid U_i \in \mathcal{B}_p\}|$ and $b'_p = |\{j \in [1,k+1] \mid U'_j \in \mathcal{B}_p\}|$. Set $\mathcal{C} = \{PQ \mid P, Q \in \mathcal{Q}_2\}$, $c = |\{i \in [1,k] \mid U_i \in \mathcal{C}\}|$ and $c' = |\{j \in [1,k+1] \mid U'_j \in \mathcal{C}\}|$. If $z \in \mathcal{K}$ is such that $z = \{p,q\}$ with $p,q \in \mathcal{L}$ and $p \neq q$, then set $\mathcal{E}_z = \{VW \mid V \in \mathcal{D}_p, W \in \mathcal{D}_q\}$, $e_z = |\{i \in [1,k] \mid U_i \in \mathcal{E}_z\}|$ and $e'_z = |\{j \in [1,k+1] \mid U'_j \in \mathcal{E}_z\}|$.

Since $|\mathrm{Pic}(\mathcal{O})| = 2$, we have $\mathcal{A}(\mathcal{H}) \subset (\mathcal{A}(\mathcal{I}^*(\mathcal{O})) \cap \mathcal{H}) \cup \{VW \mid V, W \in \mathcal{A}(\mathcal{I}^*(\mathcal{O}))$, V and W are not principal$\}$. As shown in the proof of the claim, $VW \notin \mathcal{A}(\mathcal{H})$ for all $p \in \mathcal{L}$ and $V, W \in \mathcal{D}_p$. We infer that $\mathcal{A}(\mathcal{H}) = \mathcal{Q}_1 \cup \bigcup_{p \in \mathbb{P}, p \mid f} \mathcal{A}_p \cup \bigcup_{p \in \mathcal{L}} \mathcal{B}_p \cup \mathcal{C} \cup \bigcup_{z \in \mathcal{K}} \mathcal{E}_z$.

Since k is minimal, we have $U_i, U'_j \notin \mathcal{Q}_1$ for all $i \in [1,k]$ and $j \in [1,k+1]$. Again since k is minimal and $\mathcal{I}_p^*(\mathcal{O})$ is half-factorial for all inert prime divisors p of f by Proposition 4.6, we have $a_p = a'_p = 0$ for all inert prime divisors p of f. Therefore,

$$k = \sum_{p \in \mathcal{L}} (a_p + b_p) + c + \sum_{z \in \mathcal{K}} e_z \text{ and } k+1 = \sum_{p \in \mathcal{L}} (a'_p + b'_p) + c' + \sum_{z \in \mathcal{K}} e'_z.$$

If $i \in [1,k]$, then $\sum_{P \in \mathcal{Q}_2} \mathsf{v}_P(U_i) = \begin{cases} 2 & \text{if } U_i \in \mathcal{C} \\ 1 & \text{if } U_i \in \bigcup_{p \in \mathcal{L}} \mathcal{B}_p \\ 0 & \text{else} \end{cases}$. This implies that $\sum_{P \in \mathcal{Q}_2} \mathsf{v}_P(\prod_{i=1}^k U_i) = \sum_{i=1}^k \sum_{P \in \mathcal{Q}_2} \mathsf{v}_P(U_i) = \sum_{p \in \mathcal{L}} b_p + 2c$. It follows by analogy that $\sum_{P \in \mathcal{Q}_2} \mathsf{v}_P(\prod_{j=1}^{k+1} U'_j) = \sum_{p \in \mathcal{L}} b'_p + 2c'$. Therefore, $\sum_{p \in \mathcal{L}} b_p + 2c = \sum_{p \in \mathcal{L}} b'_p + 2c'$. Let $r \in \mathcal{L}$.

If $i \in [1,k]$, then $\mathsf{v}_r(\mathcal{N}((U_i)_{P_{f,r}} \cap \mathcal{O})) = \begin{cases} 3 & \text{if } U_i \in \mathcal{B}_r \cup \bigcup_{q \in \mathcal{L} \setminus \{r\}} \mathcal{E}_{\{r,q\}} \\ 2 & \text{if } U_i \in \mathcal{A}_r \\ 0 & \text{else} \end{cases}$.

Consequently,

$$\mathsf{v}_r(\mathcal{N}((\prod_{i=1}^k U_i)_{P_{f,r}} \cap \mathcal{O})) = \sum_{i=1}^k \mathsf{v}_r(\mathcal{N}((U_i)_{P_{f,r}} \cap \mathcal{O})) = 2a_r + 3b_r + 3 \sum_{q \in \mathcal{L} \setminus \{r\}} e_{\{r,q\}}.$$

By analogy we have $\mathsf{v}_r(\mathcal{N}((\prod_{j=1}^{k+1} U'_j)_{P_{f,r}} \cap \mathcal{O})) = 2a'_r + 3b'_r + 3 \sum_{q \in \mathcal{L} \setminus \{r\}} e'_{\{r,q\}}$.

This implies that $2a_r + 3b_r + 3 \sum_{q \in \mathcal{L} \setminus \{r\}} e_{\{r,q\}} = 2a'_r + 3b'_r + 3 \sum_{q \in \mathcal{L} \setminus \{r\}} e'_{\{r,q\}}$. We infer that

$$\sum_{p\in\mathcal{L}}(a'_p-a_p+b'_p-b_p)+c'-c+\sum_{z\in\mathcal{K}}(e'_z-e_z)=1,\ \sum_{p\in\mathcal{L}}(b'_p-b_p)=2(c-c')$$

and $2\sum_{p\in\mathcal{L}}(a'_p-a_p)+3\sum_{p\in\mathcal{L}}(b'_p-b_p)+3\sum_{p\in\mathcal{L}}\sum_{q\in\mathcal{L}\setminus\{p\}}(e'_{\{p,q\}}-e_{\{p,q\}})=0$.

Note that $\sum_{p\in\mathcal{L}}\sum_{q\in\mathcal{L}\setminus\{p\}}(e'_{\{p,q\}}-e_{\{p,q\}})=2\sum_{z\in\mathcal{K}}(e'_z-e_z)$, and hence $\sum_{p\in\mathcal{L}}(a'_p-a_p)=3(c'-c)-3\sum_{z\in\mathcal{K}}(e'_z-e_z)$. Consequently,

$$\begin{aligned}1&=\sum_{p\in\mathcal{L}}(a'_p-a_p+b'_p-b_p)+c'-c+\sum_{z\in\mathcal{K}}(e'_z-e_z)\\&=3(c'-c)-3\sum_{z\in\mathcal{K}}(e'_z-e_z)+2(c-c')+c'-c+\sum_{z\in\mathcal{K}}(e'_z-e_z)\\&=2(c'-c-\sum_{z\in\mathcal{K}}(e'_z-e_z)),\end{aligned}$$

a contradiction.

Now let the equivalent conditions be satisfied. Assume to the contrary that K is an imaginary quadratic number field. Since $\mathcal{O}$ is a non-maximal order with $|\mathrm{Pic}(\mathcal{O})|=2$, it follows from [27, page 16] that $(f,d_K)\in\{(2,-8),(2,-15)\}\cup\{(3,-4),(3,-8),(3,-11),(4,-3),(4,-4),(4,-7),(5,-3),(5,-4),(7,-3)\}$.

Since f is squarefree and divisible by a ramified prime, we infer that $f=2$ and $d_K=-8$. Therefore, $\mathcal{O}=\mathbb{Z}+2\sqrt{-2}\mathbb{Z}$. Set $I=8\mathbb{Z}+2\sqrt{-2}\mathbb{Z}$. Observe that $I\in\mathcal{A}(\mathcal{I}_2^*(\mathcal{O}))$ and $\mathcal{N}(I)=8$. Moreover, $I=2\sqrt{-2}\mathcal{O}$ is principal, a contradiction. Consequently, K is a real quadratic number field.

It remains to show that $\min\Delta(\mathcal{O})=2$. There is some ramified prime p which divides f and there is some $J\in\mathcal{A}(\mathcal{I}_p^*(\mathcal{O}))$ with $\mathcal{N}(J)=p^3$. As shown in the proof of the claim, $J^2=p^2L$ for some $L\in\mathcal{A}(\mathcal{I}_p^*(\mathcal{O}))$. By [16, Corollary 2.11.16], there is some invertible prime ideal P of $\mathcal{O}$ that is not principal. Observe that J is not principal. We have PJ, P^2 and L are principal, and hence there are some $u,v,w\in\mathcal{A}(\mathcal{O})$ such that $PJ=u\mathcal{O}$, $P^2=v\mathcal{O}$, $L=w\mathcal{O}$, and $u^2=p^2vw$. Therefore, $\{2,4\}\subset\mathsf{L}(u^2)$, and since $\min\Delta(\mathcal{O})>1$, we infer that $\min\Delta(\mathcal{O})=2$. □

Proposition 4.15. *Let $\mathcal{O}$ be an order in the quadratic number field K with conductor $f\mathcal{O}_K$ for some $f\in\mathbb{N}_{\geq 2}$ such that $\min\Delta(\mathcal{O})>1$, let g be the product of all inert prime divisors of f and let $\mathcal{O}'$ be the order in K with conductor $g\mathcal{O}_K$. Then $\mathcal{O}'$ is half-factorial and, in particular, $g\in\{1\}\cup\mathbb{P}\cup\{2p\mid p\in\mathbb{P}\setminus\{2\}\}$.*

Proof. Set $\mathcal{Q}_1=\{P\in\mathfrak{X}(\mathcal{O}')\mid P$ is principal$\}$ and $\mathcal{Q}_2=\{P\in\mathfrak{X}(\mathcal{O}')\mid P$ is invertible and not principal$\}$. Observe that $\mathcal{N}(I)=|\mathcal{O}/I|=|\mathcal{O}'/I\mathcal{O}'|=\mathcal{N}(I\mathcal{O}')$ for all $I\in\mathcal{I}^*(\mathcal{O})$. Note that for all inert prime divisors p of f and all $I\in\mathcal{A}(\mathcal{I}_p^*(\mathcal{O}))$ and $J\in\mathcal{A}(\mathcal{I}_p^*(\mathcal{O}'))$, we have $\mathcal{N}(I)=\mathcal{N}(J)=p^2$. Moreover, for all ramified prime divisors p of f, we have $\{\mathcal{N}(I)\mid I\in\mathcal{A}(\mathcal{I}_p^*(\mathcal{O}))\}=\{p^2,p^3\}$. In this proof we will use Theorem 4.14 without further citation.

CLAIM 1: For all prime divisors p of g and all $I \in \mathcal{A}(\mathcal{I}_p^*(\mathcal{O}'))$, it follows that I is principal. Let p be a prime divisor of g and let $I \in \mathcal{A}(\mathcal{I}_p^*(\mathcal{O}'))$. Set $P = P_{f,p}$ and $P' = P_{g,p}$. It follows by Proposition 3.3 that $\mathcal{O}_P = \mathcal{O}'_{P'}$ and that $\delta : \mathcal{I}_p^*(\mathcal{O}) \to \mathcal{I}_p^*(\mathcal{O}')$ defined by $\delta(J) = J_P \cap \mathcal{O}'$ for all $J \in \mathcal{I}_p^*(\mathcal{O})$ is a monoid isomorphism. In particular, we have $\mathcal{A}(\mathcal{I}_p^*(\mathcal{O}')) = \{J_P \cap \mathcal{O}' \mid J \in \mathcal{A}(\mathcal{I}_p^*(\mathcal{O}))\}$. Therefore, there is some $J \in \mathcal{A}(\mathcal{I}_p^*(\mathcal{O}))$ such that $J_P \cap \mathcal{O}' = I$. Note that $\mathcal{N}(I) = p^2 = \mathcal{N}(J) = \mathcal{N}(J\mathcal{O}')$. Since $J\mathcal{O}' \subset J\mathcal{O}'_{P'} \cap \mathcal{O}' = J\mathcal{O}_P \cap \mathcal{O}' = I$, we infer that $I = J\mathcal{O}'$. Since J is a principal ideal of $\mathcal{O}$, it follows that I is principal. This proves Claim 1.

CLAIM 2: If $P \in \mathcal{Q}_2$, p is a ramified prime divisor of f such that $P \cap \mathbb{Z} = p\mathbb{Z}$ and $I \in \mathcal{A}(\mathcal{I}_p^*(\mathcal{O}))$ with $\mathcal{N}(I) = p^3$, then P^2 is principal and $I\mathcal{O}' = P^3$. Let $P \in \mathcal{Q}_2$, p a ramified prime divisor of f such that $P \cap \mathbb{Z} = p\mathbb{Z}$ and $I \in \mathcal{A}(\mathcal{I}_p^*(\mathcal{O}))$ with $\mathcal{N}(I) = p^3$. Since p is ramified, there is some $A \in \mathfrak{X}(\mathcal{O}_K)$ such that $p\mathcal{O}_K = A^2$. Observe that $\mathcal{N}(A^2) = p^2$, and thus $\mathcal{N}(A) = p$. We have $A \cap \mathcal{O}' = P$, $P\mathcal{O}_K = A$ and $\mathcal{N}(P) = \mathcal{N}(A) = p$. Note that since P is invertible, it follows that every P-primary ideal of $\mathcal{O}'$ is a power of P. Therefore, $p\mathcal{O}' = P^k$ for some $k \in \mathbb{N}$, and hence $p^k = \mathcal{N}(P^k) = \mathcal{N}(p\mathcal{O}') = p^2$. Consequently, $k = 2$ and P^2 is principal. Clearly, $I\mathcal{O}'$ is a P-primary ideal of $\mathcal{O}'$, and thus $I\mathcal{O}' = P^m$ for some $m \in \mathbb{N}$. We infer that $p^m = \mathcal{N}(P^m) = \mathcal{N}(I\mathcal{O}') = \mathcal{N}(I) = p^3$, and thus $m = 3$ and $I\mathcal{O}' = P^3$. This proves Claim 2.

CLAIM 3: PQ is principal for all $P, Q \in \mathcal{Q}_2$. Let $P, Q \in \mathcal{Q}_2$.

CASE 1: $P \cap \mathcal{O}$ and $Q \cap \mathcal{O}$ are invertible. Note that $P = (P \cap \mathcal{O})\mathcal{O}'$, $Q = (Q \cap \mathcal{O})\mathcal{O}'$ and $P \cap \mathcal{O}$ and $Q \cap \mathcal{O}$ are not principal. Since $|\mathrm{Pic}(\mathcal{O})| = 2$, we have $(P \cap \mathcal{O})(Q \cap \mathcal{O})$ is a principal ideal of $\mathcal{O}$, and thus $PQ = (P \cap \mathcal{O})(Q \cap \mathcal{O})\mathcal{O}'$ is principal.

CASE 2: ($P \cap \mathcal{O}$ is invertible and $Q \cap \mathcal{O}$ is not invertible) or ($P \cap \mathcal{O}$ is not invertible and $Q \cap \mathcal{O}$ is invertible). Without restriction let $P \cap \mathcal{O}$ be invertible and let $Q \cap \mathcal{O}$ be not invertible. Observe that $P = (P \cap \mathcal{O})\mathcal{O}'$. Moreover, there is some ramified prime q that divides f such that $Q \cap \mathbb{Z} = q\mathbb{Z}$ and there is some $J \in \mathcal{A}(\mathcal{I}_q^*(\mathcal{O}))$ with $\mathcal{N}(J) = q^3$. Observe that $P \cap \mathcal{O}$ and J are not principal. Since $|\mathrm{Pic}(\mathcal{O})| = 2$, it follows that $(P \cap \mathcal{O})J$ is a principal ideal of $\mathcal{O}$. Note that $PQ^3 = (P \cap \mathcal{O})J\mathcal{O}'$ by Claim 2, and thus PQ^3 is principal. Since Q^2 is principal by Claim 2, we infer that PQ is principal.

CASE 3: $P \cap \mathcal{O}$ and $Q \cap \mathcal{O}$ are not invertible. There are ramified primes p and q that divide f such that $P \cap \mathbb{Z} = p\mathbb{Z}$ and $Q \cap \mathbb{Z} = q\mathbb{Z}$. There are some $I \in \mathcal{A}(\mathcal{I}_p^*(\mathcal{O}))$ and $J \in \mathcal{A}(\mathcal{I}_q^*(\mathcal{O}))$ with $\mathcal{N}(I) = p^3$ and $\mathcal{N}(J) = q^3$. Since $|\mathrm{Pic}(\mathcal{O})| = 2$ and I and J are not principal, we have IJ is a principal ideal of $\mathcal{O}$. It follows that $P^3Q^3 = IJ\mathcal{O}'$ by Claim 2, and hence P^3Q^3 is principal. Since P^2 and Q^2 are principal by Claim 2, we have PQ is principal. This proves Claim 3.

Finally, we show that $\mathcal{O}'$ is half-factorial. Set $\mathcal{C} = \{PQ \mid P, Q \in \mathcal{Q}_2\}$ and let $\mathcal{H}$ denote the monoid of nonzero principal ideals of $\mathcal{O}'$. It is an immediate consequence of Claim 1 and Claim 3 that $\mathcal{A}(\mathcal{H}) = \mathcal{Q}_1 \cup \mathcal{C} \cup \bigcup_{p \in \mathbb{P}, p \mid g} \mathcal{A}(\mathcal{I}_p^*(\mathcal{O}'))$.

Let $k, \ell \in \mathbb{N}$ and $I_i, I'_j \in \mathcal{A}(\mathcal{H})$ for each $i \in [1, k]$ and $j \in [1, \ell]$ be such that $\prod_{i=1}^k I_i = \prod_{j=1}^\ell I'_j$. It remains to show that $k = \ell$. Set $b = |\{i \in [1, k] \mid I_i \in \mathcal{Q}_1\}|$,

$b' = |\{j \in [1,\ell] \mid I'_j \in \mathcal{Q}_1\}|$, $c = |\{i \in [1,k] \mid I_i \in \mathcal{C}\}|$, $c' = |\{j \in [1,\ell] \mid I'_j \in \mathcal{C}\}|$ and for each prime divisor p of g set $a_p = |\{i \in [1,k] \mid I_i \in \mathcal{A}(\mathcal{I}_p^*(\mathcal{O}'))\}|$ and $a'_p = |\{j \in [1,\ell] \mid I'_j \in \mathcal{A}(\mathcal{I}_p^*(\mathcal{O}'))\}|$. If p is a prime divisor of g, then $\mathcal{I}_p^*(\mathcal{O}')$ is half-factorial by Proposition 4.6, and hence $a_p = a'_p$ by Claim 1. We have $b = \sum_{i=1}^k \sum_{P\in\mathcal{Q}_1} \mathsf{v}_P(I_i) = \sum_{P\in\mathcal{Q}_1} \mathsf{v}_P(\prod_{i=1}^k I_i) = \sum_{P\in\mathcal{Q}_1} \mathsf{v}_P(\prod_{j=1}^\ell I'_j) = \sum_{j=1}^\ell \sum_{P\in\mathcal{Q}_1} \mathsf{v}_P(I'_j) = b'$.

Moreover, $2c = \sum_{P\in\mathcal{Q}_2} \mathsf{v}_P(\prod_{i=1}^k I_i) = \sum_{P\in\mathcal{Q}_2} \mathsf{v}_P(\prod_{j=1}^\ell I'_j) = 2c'$. Therefore, $k = b + c + \sum_{p\in\mathbb{P}, p|g} a_p = b' + c' + \sum_{p\in\mathbb{P}, p|g} a'_p = \ell$.

The remaining assertion follows from [16, Theorem 3.7.15]. □

Remark 4.16. Let $\mathcal{O}$ be an order in the quadratic number field K with conductor $f\mathcal{O}_K$ for some $f \in \mathbb{N}$ such that $|\mathrm{Pic}(\mathcal{O})| = 2$ and let p be an odd ramified prime such that $\mathsf{v}_p(f) = 1$ and $I \in \mathcal{A}(\mathcal{I}_p^*(\mathcal{O}))$ such that $\mathcal{N}(I) = p^3$ and I not principal. Then every $J \in \mathcal{A}(\mathcal{I}_p^*(\mathcal{O}))$ with $\mathcal{N}(J) = p^3$ is not principal.

Proof. Set $\mathcal{L} = \{J \in \mathcal{A}(\mathcal{I}_p^*(\mathcal{O})) \mid \mathcal{N}(J) = p^3\}$ and $\mathcal{K} = \{L \in \mathcal{A}(\mathcal{I}_p^*(\mathcal{O})) \mid \mathcal{N}(L) = p^2\}$. It follows by the claim in the proof of Theorem 4.14 that for all $J \in \mathcal{L}$ and $L \in \mathcal{K}$, there is a unique $A \in \mathcal{L}$ such that $AJ = p^2L$. By Theorem 3.6 we have $|\mathcal{L}| = |\mathcal{K}| = p$, and hence $|\{(A,J) \in \mathcal{L}^2 \mid AJ = p^2L\}| = p$ for all $L \in \mathcal{K}$. Since p is odd, we infer that for each $L \in \mathcal{K}$ there is some $A \in \mathcal{L}$ such that $A^2 = p^2L$. Consequently, every $L \in \mathcal{K}$ is principal. Now let $J \in \mathcal{L}$. There is some $B \in \mathcal{K}$ such that $IJ = p^2B$, and thus IJ is principal. Therefore, J is not principal. □

Next we show that the assumption that p is odd in Remark 4.16 is crucial.

Example 4.17. Let $\mathcal{O} = \mathbb{Z} + 2\sqrt{-2}\mathbb{Z}$ be the order in the quadratic number field $K = \mathbb{Q}(\sqrt{-2})$ with conductor $2\mathcal{O}_K$. Let $I = 8\mathbb{Z} + 2\sqrt{-2}\mathbb{Z}$ and $J = 8\mathbb{Z} + (4 + 2\sqrt{-2})\mathbb{Z}$. Then 2 is ramified, $|\mathrm{Pic}(\mathcal{O})| = 2$, $I, J \in \mathcal{A}(\mathcal{I}_2^*(\mathcal{O}))$, $\mathcal{N}(I) = \mathcal{N}(J) = 8$, I is principal and J is not principal.

Proof. It is clear that $J \in \mathcal{A}(\mathcal{I}_2^*(\mathcal{O}))$ and $\mathcal{N}(J) = 8$. By the proof of Theorem 4.14, it remains to show that J is not principal. Assume that J is principal. Then there are some $a, b \in \mathbb{Z}$ such that $J = (8a + 4b + 2\sqrt{-2}b)\mathcal{O}$, and hence $8 = \mathcal{N}(J) = |\mathcal{N}_{K/\mathbb{Q}}(8a + 4b + 2\sqrt{-2}b)| = |(8a + 4b)^2 + 8b^2|$. Therefore, $2(2a + b)^2 + b^2 = 1$. It is clear that $|b| \le 1$. If $b = 0$, then $8a^2 = 1$, a contradiction. Therefore, $|b| = 1$ and $2a + b = 0$, a contradiction. □

Lemma 4.18. *Let $d \in \mathbb{N}_{\ge 2}$ be squarefree, let $K = \mathbb{Q}(\sqrt{d})$, let $\mathcal{O}$ be the order in K with conductor $f\mathcal{O}_K$ for some $f \in \mathbb{N}_{\ge 2}$, and let p be a ramified prime with $\mathsf{v}_p(f) = 1$. If ($p \equiv 1 \mod 4$ and $(\frac{d/p}{p}) = -1$) or ($(\frac{p}{q}) = -1$ for some prime q with $q \equiv 1 \mod 4$ and $q \mid df$), then each $I \in \mathcal{A}(\mathcal{I}_p^*(\mathcal{O}))$ with $\mathcal{N}(I) = p^3$ is not principal.*

Proof. Note that if p is odd, then $\{I \in \mathcal{A}(\mathcal{I}_p^*(\mathcal{O})) \mid \mathcal{N}(I) = p^3\} = \{p^3\mathbb{Z} + (p^2k + \frac{\varepsilon p^2 + f\sqrt{d_K}}{2})\mathbb{Z} \mid k \in [0, p-1]\}$. Moreover, if $p = 2$ and d is odd, then $\{I \in \mathcal{A}(\mathcal{I}_p^*$

$(\mathcal{O})) \mid \mathcal{N}(I) = p^3\} = \{8\mathbb{Z} + (2k + f\sqrt{d})\mathbb{Z} \mid k \in \{1, 3\}\}$. Furthermore, if $p = 2$ and d is even, then $\{I \in \mathcal{A}(\mathcal{I}_p^*(\mathcal{O})) \mid \mathcal{N}(I) = p^3\} = \{8\mathbb{Z} + (2k + f\sqrt{d})\mathbb{Z} \mid k \in \{0, 2\}\}$.

CASE 1: $p \equiv 1 \mod 4$ and $(\frac{d/p}{p}) = -1$. Let $I \in \mathcal{A}(\mathcal{I}_p^*(\mathcal{O}))$ be such that $\mathcal{N}(I) = p^3$. Since p is odd, we have $I = p^3\mathbb{Z} + (p^2k + \frac{\varepsilon p^2 + f\sqrt{d_K}}{2})\mathbb{Z}$ for some $k \in [0, p-1]$. Assume that I is principal. Then there are some $a, b \in \mathbb{Z}$ such that $I = (p^3a + p^2bk + \frac{\varepsilon p^2 + f\sqrt{d_K}}{2}b)\mathcal{O}$. We infer that $p^3 = \mathcal{N}(I) = |\mathcal{N}_{K/\mathbb{Q}}(p^3a + p^2bk + \frac{\varepsilon p^2 + f\sqrt{d_K}}{2}b)| = \frac{1}{4}|p^4(2pa + 2bk + \varepsilon b)^2 - f^2b^2d_K|$, and hence $\frac{f^2}{p^2}b^2\frac{d_K}{p} \equiv 4\beta \mod p$ for some $\beta \in \{-1, 1\}$. Since $p \equiv 1 \mod 4$, we have $(\frac{-1}{p}) = 1$, and thus $(\frac{d/p}{p}) = (\frac{d_K/p}{p}) = (\frac{f^2b^2d_K/p^3}{p}) = (\frac{4\beta}{p}) = 1$, a contradiction.

CASE 2: There is some prime q such that $q \equiv 1 \mod 4$, $q \mid df$ and $(\frac{p}{q}) = -1$. Let $I \in \mathcal{A}(\mathcal{I}_p^*(\mathcal{O}))$ be such that $\mathcal{N}(I) = p^3$. First let p be odd. Then $I = p^3\mathbb{Z} + (p^2k + \frac{\varepsilon p^2 + f\sqrt{d_K}}{2})\mathbb{Z}$ for some $k \in [0, p-1]$. Assume that I is principal. Then there are some $a, b \in \mathbb{Z}$ such that $I = (p^3a + p^2bk + \frac{\varepsilon p^2 + f\sqrt{d_K}}{2}b)\mathcal{O}$. This implies that $p^3 = \mathcal{N}(I) = |\mathcal{N}_{K/\mathbb{Q}}(p^3a + p^2bk + \frac{\varepsilon p^2 + f\sqrt{d_K}}{2}b)| = \frac{1}{4}|p^4(2pa + 2bk + \varepsilon b)^2 - f^2b^2d_K|$, and thus $\ell^2 \equiv 4\beta p^3 \mod q$ for some $\ell \in \mathbb{Z}$ and $\beta \in \{-1, 1\}$. Since $q \equiv 1 \mod 4$, we have $(\frac{-1}{q}) = 1$, and hence $(\frac{p}{q})^3 = (\frac{4\beta p^3}{q}) = 1$. Therefore, $(\frac{p}{q}) = 1$, a contradiction.

Now let $p = 2$. Then $I = 8\mathbb{Z} + (2k + f\sqrt{d})\mathbb{Z}$ for some $k \in [0, 3]$. Assume that I is principal. Then there are some $a, b \in \mathbb{Z}$ such that $I = (8a + 2bk + bf\sqrt{d})\mathcal{O}$. Consequently, $8 = \mathcal{N}(I) = |(8a + 2bk)^2 - b^2f^2d|$, and thus $\ell^2 \equiv 8\beta \mod q$ for some $\ell \in \mathbb{Z}$ and $\beta \in \{-1, 1\}$. This implies that $(\frac{2}{q})^3 = (\frac{8\beta}{q}) = 1$. Therefore, $(\frac{2}{q}) = 1$, a contradiction. □

Proposition 4.19. *Let $d \in \mathbb{N}_{\geq 2}$ be squarefree, let $K = \mathbb{Q}(\sqrt{d})$, and let $\mathcal{O}$ be the order in K with conductor $f\mathcal{O}_K$ such that f is a nonempty squarefree product of ramified primes times a squarefree product of inert primes and $|\mathrm{Pic}(\mathcal{O})| = |\mathrm{Pic}(\mathcal{O}_K)| = 2$. If for every ramified prime divisor p of f, we have ($p \equiv 1 \mod 4$ and $(\frac{d/p}{p}) = -1$) or ($(\frac{p}{q}) = -1$ for some prime q with $q \equiv 1 \mod 4$ and $q \mid df$), then* $\min \Delta(\mathcal{O}) = 2$.

Proof. It follows by Lemma 4.18 that for every ramified prime divisor p of f and every $I \in \mathcal{A}(\mathcal{I}_p^*(\mathcal{O}))$ with $\mathcal{N}(I) = p^3$, we have I is not principal. It follows by the claim in the proof of Theorem 4.14 that $I \in \mathcal{A}(\mathcal{I}_p^*(\mathcal{O}))$ is principal if and only if $\mathcal{N}(I) = p^2$. Now let p be an inert prime divisor of f and let $J \in \mathcal{A}(\mathcal{I}_p^*(\mathcal{O}))$. Since $|\mathrm{Pic}(\mathcal{O})| = |\mathrm{Pic}(\mathcal{O}_K)|$, it follows that the group epimorphism $\theta : \mathrm{Pic}(\mathcal{O}) \to \mathrm{Pic}(\mathcal{O}_K)$ defined by $\theta([L]) = [L\mathcal{O}_K]$ for all $L \in \mathcal{I}^*(\mathcal{O})$ is a group isomorphism. Set $P = p\mathcal{O}_K$. Then $J\mathcal{O}_K$ is a P-primary ideal of $\mathcal{O}_K$, and hence $J\mathcal{O}_K$ is a principal ideal of $\mathcal{O}_K$. Since θ is an isomorphism, we infer that J is a principal ideal of $\mathcal{O}$. Now it follows by Theorem 4.14 that $\min \Delta(\mathcal{O}) = 2$. □

Next we provide two counterexamples that show that the additional assumption on the ramified prime divisors of f in Proposition 4.19 is important.

Example 4.20. There is some real quadratic number field K and some order $\mathcal{O}$ in K with conductor $p\mathcal{O}_K$ for some ramified prime p such that $p \equiv 1 \mod 4$, $|\mathrm{Pic}(\mathcal{O})| = |\mathrm{Pic}(\mathcal{O}_K)| = 2$, and $\min \Delta(\mathcal{O}) = 1$.

Proof. Let $\mathcal{O} = \mathbb{Z} + 5\sqrt{30}\mathbb{Z}$ be the order in the real quadratic number field $K = \mathbb{Q}(\sqrt{30})$ with conductor $5\mathcal{O}_K$. Observe that 5 is ramified, $5 \equiv 1 \mod 4$, $|\mathrm{Pic}(\mathcal{O}_K)| = 2$ and $\alpha = 11 + 2\sqrt{30}$ is a fundamental unit of $\mathcal{O}_K$. Since $\alpha \notin \mathcal{O}$ and $(\mathcal{O}_K^\times : \mathcal{O}^\times) \mid 5$, we infer that $(\mathcal{O}_K^\times : \mathcal{O}^\times) = 5$, and hence $|\mathrm{Pic}(\mathcal{O})| = |\mathrm{Pic}(\mathcal{O}_K)| \frac{5}{(\mathcal{O}_K^\times : \mathcal{O}^\times)} = 2$. Let $I = 125\mathbb{Z} + 5\sqrt{30}\mathbb{Z}$. Then $I \in \mathcal{A}(\mathcal{I}_5^*(\mathcal{O}))$ with $\mathcal{N}(I) = 125$. Since $I = (12625 + 2305\sqrt{30})\mathcal{O}$ is principal, we infer by Theorem 4.14 that $\min \Delta(\mathcal{O}) = 1$. □

Example 4.21. There is some real quadratic number field $K = \mathbb{Q}(\sqrt{d})$ with $d \in \mathbb{N}_{\geq 2}$ squarefree and some order $\mathcal{O}$ in K with conductor $p\mathcal{O}_K$ for some odd ramified prime p such that $(\frac{d/p}{p}) = -1$, $|\mathrm{Pic}(\mathcal{O})| = |\mathrm{Pic}(\mathcal{O}_K)| = 2$, and $\min \Delta(\mathcal{O}) = 1$.

Proof. Let $\mathcal{O} = \mathbb{Z} + 7\sqrt{42}\mathbb{Z}$ be the order in the real quadratic number field $K = \mathbb{Q}(\sqrt{42})$ with conductor $7\mathcal{O}_K$. Note that 7 is an odd ramified prime, $(\frac{42/7}{7}) = -1$, $|\mathrm{Pic}(\mathcal{O}_K)| = 2$ and $\alpha = 13 + 2\sqrt{42}$ is a fundamental unit of $\mathcal{O}_K$. We have $\alpha \notin \mathcal{O}$ and $(\mathcal{O}_K^\times : \mathcal{O}^\times) \mid 7$. Therefore, $(\mathcal{O}_K^\times : \mathcal{O}^\times) = 7$, and thus $|\mathrm{Pic}(\mathcal{O})| = |\mathrm{Pic}(\mathcal{O}_K)| \frac{7}{(\mathcal{O}_K^\times : \mathcal{O}^\times)} = 2$. Set $I = 343\mathbb{Z} + 7\sqrt{42}\mathbb{Z}$. Then $I \in \mathcal{A}(\mathcal{I}_7^*(\mathcal{O}))$, $\mathcal{N}(I) = 343$, and $I = (825601 + 127393\sqrt{42})\mathcal{O}$ is principal. Consequently, $\min \Delta(\mathcal{O}) = 1$ by Theorem 4.14. □

Finally, we provide the examples of orders $\mathcal{O}$ in quadratic number fields with $\min \Delta(\mathcal{O}) = 2$.

Example 4.22. Let K be a quadratic number field and $\mathcal{O}$ the order in K with conductor $f\mathcal{O}_K$ such that $(f, d_K) \in \{(2, 60), (3, 60), (5, 60), (6, 60), (10, 60)\} \cup \{(15, 60), (30, 60), (10, 85), (35, 40), (195, 65), (30, 365)\}$.

1. If $(f, d_K) \in \{(2, 60), (3, 60), (5, 60)\}$, then f is a ramified prime.
2. If $(f, d_K) \in \{(6, 60), (10, 60), (15, 60)\}$, then f is the product of two distinct ramified primes.
3. If $(f, d_K) = (30, 60)$, then f is the product of three distinct ramified primes.
4. If $(f, d_K) \in \{(10, 85), (35, 40)\}$, then f is the product of an inert prime and a ramified prime.
5. If $(f, d_K) = (195, 65)$, then f is the product of an inert prime and two distinct ramified primes.
6. If $(f, d_K) = (30, 365)$, then f is the product of two distinct inert primes and a ramified prime.
7. $\min \Delta(\mathcal{O}) = 2$.

Proof. It is straightforward to prove the first six assertions. We prove the last assertion in the case that $d_K = 60$ and $f \in \mathbb{N}_{\geq 2}$ is a divisor of 30. The remaining cases can be proved in analogy by using Proposition 4.19. It is clear that 2, 3, and 5 are ramified primes. Note that $|\mathrm{Pic}(\mathcal{O}_K)| = 2$ (e.g., [25, page 22]) and $\alpha = 4 + \sqrt{15}$ is a fundamental unit of $\mathcal{O}_K$.

We have $\alpha^2 = 31 + 8\sqrt{15}$, $\alpha^3 = 244 + 63\sqrt{15}$, and $\alpha^5 = 15124 + 3905\sqrt{15}$. Moreover, $\alpha^6 = 119071 + 30744\sqrt{15}$, $\alpha^{10} = 457470751 + 118118440\sqrt{15}$, and $\alpha^{15} = 13837575261124 + 3572846569215\sqrt{15}$. Set $k = (\mathcal{O}_K^\times : \mathcal{O}^\times)$. Then k is a divisor of f by (4.1). Observe that $\alpha \notin \mathbb{Z} + 2\sqrt{15}\mathbb{Z}$, $\alpha \notin \mathbb{Z} + 3\sqrt{15}\mathbb{Z}$, $\alpha \notin \mathbb{Z} + 5\sqrt{15}\mathbb{Z}$, $\alpha^2, \alpha^3 \notin \mathbb{Z} + 6\sqrt{15}\mathbb{Z}$, $\alpha^2, \alpha^5 \notin \mathbb{Z} + 10\sqrt{15}\mathbb{Z}$, $\alpha^3, \alpha^5 \notin \mathbb{Z} + 15\sqrt{15}\mathbb{Z}$, and $\alpha^6, \alpha^{10}, \alpha^{15} \notin \mathbb{Z} + 30\sqrt{15}\mathbb{Z}$. This implies that $k = f$, and hence $|\mathrm{Pic}(\mathcal{O})| = \frac{f}{k}|\mathrm{Pic}(\mathcal{O}_K)| = |\mathrm{Pic}(\mathcal{O}_K)| = 2$ by (4.1). We have $5 \equiv 1 \mod 4$ and $(\frac{15/5}{5}) = (\frac{3}{5}) = (\frac{2}{5}) = -1$. We infer by Proposition 4.19 that $\min \Delta(\mathcal{O}) = 2$. □

5 Unions of Sets of Lengths

The goal of this section is to show that all unions of sets of lengths of the monoid of (invertible) ideals in orders of quadratic number fields are intervals (Theorem 5.2). To gather the background on unions of sets of lengths, let H be an atomic monoid with $H \neq H^\times$ and $k \in \mathbb{N}_0$. Then

$$\mathcal{U}_k(H) = \bigcup_{k \in L \in \mathcal{L}(H)} L \qquad \text{denotes the } \textit{union of sets of lengths} \text{ containing } k \text{ and}$$

$$\rho_k(H) = \sup \mathcal{U}_k(H) \qquad \text{is the } \textit{kth elasticity} \text{ of } H.$$

Then, for the *elasticity* $\rho(H)$ of H, we have [12, Proposition 2.7],

$$\rho(H) = \sup\{\rho(L) \mid L \in \mathcal{L}(H)\} = \lim_{k \to \infty} \frac{\rho_k(H)}{k}.$$

Clearly, $\mathcal{U}_0(H) = \{0\}$, $\mathcal{U}_1(H) = \{1\}$ and $\mathcal{U}_k(H)$ is the set of all $\ell \in \mathbb{N}_0$ with the following property:

There are atoms $u_1, \ldots, u_k, v_1, \ldots, v_\ell$ in H such that $u_1 \cdot \ldots \cdot u_k = v_1 \cdot \ldots \cdot v_\ell$.

Let $d \in \mathbb{N}$ and $M \in \mathbb{N}_0$. A subset $L \subset \mathbb{Z}$ is called an AAP (with difference d and bound M) if

$$L = y + \left(L' \cup L^* \cup L''\right) \subset y + d\mathbb{Z},$$

where $y \in \mathbb{Z}$, L^* is a nonempty arithmetical progression with difference d and $\min L^* = 0$, $L' \subset [-M, -1]$, and $L'' \subset \sup L^* + [1, M]$ (with the convention that $L'' = \emptyset$ if L^* is infinite). We say that H satisfies the *Structure Theorem for Unions* if there are $d \in \mathbb{N}$ and $M \in \mathbb{N}_0$ such that $\mathcal{U}_k(H)$ is an AAP with difference d and bound M for all sufficiently large $k \in \mathbb{N}$. If $\Delta(H)$ is finite and the structure theorem for unions holds for some parameter $d \in \mathbb{N}$, then $d = \min \Delta(H)$ [12, Lemma 2.12].

The structure theorem for unions holds for a wealth of monoids and domains (see [2, 13, 34] for recent contributions and see [12, Theorem 4.2] for an example where it does not hold). Since it holds for C-monoids [14], it holds for the monoid of invertible ideals of orders in number fields. In some special cases (including Krull monoids having prime divisors in all classes) all unions of sets of lengths are intervals, in other words the structure theorem for unions holds with $d = 1$ and $M = 0$ [15, Theorem 3.1.3], [18, Theorem 5.8], [33]. In Theorem 5.2 we show that the same is true for the monoids of (invertible) ideals of orders in quadratic number fields.

Proposition 5.1. *Let p be a prime divisor of f and let $N = \sup\{\mathsf{v}_p(\mathcal{N}(A)) \mid A \in \mathcal{A}(\mathcal{I}_p^*(\mathcal{O}_f))\}$.*

1. *If p splits, then $\mathcal{U}_\ell(\mathcal{I}_p(\mathcal{O}_f)) = \mathcal{U}_\ell(\mathcal{I}_p^*(\mathcal{O}_f)) = \mathbb{N}_{\ge 2}$ for all $\ell \in \mathbb{N}_{\ge 2}$.*
2. *If p does not split, then $\mathcal{U}_\ell(\mathcal{I}_p(\mathcal{O}_f)) \cap \mathbb{N}_{\ge \ell} = \mathcal{U}_\ell(\mathcal{I}_p^*(\mathcal{O}_f)) \cap \mathbb{N}_{\ge \ell} = [\ell, \lfloor \frac{\ell N}{2} \rfloor]$ for all $\ell \in \mathbb{N}_{\ge 2}$.*

Proof. We prove 1. and 2. simultaneously. By Proposition 3.3.3 we can assume without restriction that $f = p^{\mathsf{v}_p(f)}$. First we show that both assertions are true for $\ell = 2$. It follows from Theorem 3.6 that $[2, N] = [2, 2\mathsf{v}_p(f)] \cup \{\mathsf{v}_p(\mathcal{N}(A)) \mid A \in \mathcal{A}(\mathcal{I}_p^*(\mathcal{O}_f))\}$. It is obvious that $\mathcal{U}_2(\mathcal{I}_p^*(\mathcal{O}_f)) \subset \mathcal{U}_2(\mathcal{I}_p(\mathcal{O}_f))$. It follows from Lemma 4.9 that $\mathcal{U}_2(\mathcal{I}_p(\mathcal{O}_f)) \subset [2, N]$.

Let $k \in [2, N]$. It remains to show that $k \in \mathcal{U}_2(\mathcal{I}_p^*(\mathcal{O}_f))$. If $k > 2\mathsf{v}_p(f)$, then there is some $I \in \mathcal{A}(\mathcal{I}_p^*(\mathcal{O}_f))$ such that $\mathcal{N}(I) = p^k$. It follows by Proposition 3.2.5 that $I\overline{I} = (p\mathcal{O}_f)^k$, and hence $k \in \mathcal{U}_2(\mathcal{I}_p^*(\mathcal{O}_f))$. Now let $k \le 2\mathsf{v}_p(f)$. By Proposition 4.8.1 we can assume without restriction that $\mathsf{v}_p(f) \ge 2$ and $k \ge 4$.

CASE 1: $d \not\equiv 1 \mod 4$ or ($d \equiv 1 \mod 4$, $p = 2$ and $k \le 2(\mathsf{v}_2(f) - 1)$). We set $a = \mathsf{v}_p(\mathcal{N}_{K/\mathbb{Q}}(p^{k-2} + \tau))$ and $b = \mathsf{v}_p(\mathcal{N}_{K/\mathbb{Q}}(p^{k-2}(p-1) + \tau))$. Observe that if $d \not\equiv 1 \mod 4$, then $a, b \ge \min\{2k - 4, 2\mathsf{v}_p(f)\} \ge k$. Moreover, if $d \equiv 1 \mod 4$, $p = 2$ and $k \le 2(\mathsf{v}_2(f) - 1)$, then $a, b \ge \min\{2k - 4, 2(\mathsf{v}_2(f) - 1)\} \ge k$. Set $I = p^a\mathbb{Z} + (p^{k-2} + \tau)\mathbb{Z}$ and $J = p^b\mathbb{Z} + (p^{k-2}(p-1) + \tau)\mathbb{Z}$. Then $I, J \in \mathcal{A}(\mathcal{I}_p^*(\mathcal{O}_f))$, $\min\{a, b, \mathsf{v}_p(p^{k-2} + p^{k-2}(p-1) + \varepsilon)\} = k - 1$, and $a + b - 2(k-1) > 0$. Therefore, there is some $L \in \mathcal{A}(\mathcal{I}_p^*(\mathcal{O}_f))$ such that $IJ = p^{k-1}L$, and hence $k \in \mathsf{L}(IJ) \subset \mathcal{U}_2(\mathcal{I}_p^*(\mathcal{O}_f))$.

CASE 2: $d \equiv 1 \mod 4$ and $p \ne 2$. We set $a = \mathsf{v}_p(\mathcal{N}_{K/\mathbb{Q}}(\frac{p^{k-2}-1}{2} + \tau))$ and $b = \mathsf{v}_p(\mathcal{N}_{K/\mathbb{Q}}(\frac{p^{k-2}(p^2+p-1)-1}{2} + \tau))$. Note that $a, b \ge \min\{2k - 4, 2\mathsf{v}_p(f)\} \ge k$. Set $I = p^a\mathbb{Z} + (\frac{p^{k-2}-1}{2} + \tau)\mathbb{Z}$ and $J = p^b\mathbb{Z} + (\frac{p^{k-2}(p^2+p-1)-1}{2} + \tau)\mathbb{Z}$. Then $I, J \in \mathcal{A}(\mathcal{I}_p^*(\mathcal{O}_f))$, $\min\{a, b, \mathsf{v}_p(\frac{p^{k-2}-1}{2} + \frac{p^{k-2}(p^2+p-1)-1}{2} + \varepsilon)\} = k - 1$, and $a + b - 2(k-1) > 0$. Consequently, there is some $L \in \mathcal{A}(\mathcal{I}_p^*(\mathcal{O}_f))$ such that $IJ = p^{k-1}L$, and thus $k \in \mathsf{L}(IJ) \subset \mathcal{U}_2(\mathcal{I}_p^*(\mathcal{O}_f))$.

CASE 3: $d \equiv 1 \mod 8$, $p = 2$ and $k \in \{2\mathsf{v}_2(f) - 1, 2\mathsf{v}_2(f)\}$. Set $h = \mathsf{v}_2(f)$. If $h = 2$, then $k = 4$, and hence $k \in \mathcal{U}_2(\mathcal{I}_2^*(\mathcal{O}_f))$ by Proposition 4.4. Now let $h \ge 3$. Note that 2 splits. By Theorem 3.6 there are some $I, J, L \in \mathcal{A}(\mathcal{I}_2^*(\mathcal{O}_f))$ such that $\mathcal{N}(I) = 2^{2h+1}$, $\mathcal{N}(J) = 2^{2h+2}$ and $\mathcal{N}(L) = 16$. By Proposition 3.2.5 we have $L\overline{L} = 16\mathcal{O}_f$, $I\overline{I} = 2^{2h+1}\mathcal{O}_f = 2^{2h-3}L\overline{L}$ and $J\overline{J} = 2^{2h+2}\mathcal{O}_f = 2^{2h-2}L\overline{L}$. We infer that $k \in \{2h-1, 2h\} \subset \mathcal{U}_2(\mathcal{I}_2^*(\mathcal{O}_f))$.

CASE 4: $d \equiv 5 \mod 8$, $p = 2$ and $k \in \{2v_2(f) - 1, 2v_2(f)\}$. Set $h = v_2(f)$. If $h = 2$, then $k = 4$, and thus $k \in \mathcal{U}_2(\mathcal{I}_2^*(\mathcal{O}_f))$ by Proposition 4.4. Now let $h \geq 3$. Set $A = 2^{2h}\mathbb{Z} + (2^{h-1} + \tau)\mathbb{Z}$, $B = 2^{2h}\mathbb{Z} + (2^{2h-2} - 2^{h-1} + \tau)\mathbb{Z}$, and $C = 2^{2h}\mathbb{Z} + (2^{2h-1} - 2^{h-1} + \tau)\mathbb{Z}$. Then $A, B, C \in \mathcal{A}(\mathcal{I}_2^*(\mathcal{O}_f))$, $AB = 2^{2h-2}I$ and $AC = 2^{2h-1}J$ for some $I, J \in \mathcal{A}(\mathcal{I}_2^*(\mathcal{O}_f))$. Therefore, $k \in \{2h - 1, 2h\} \subset \mathcal{U}_2(\mathcal{I}_2^*(\mathcal{O}_f))$.

So far we have proved that both assertions are true for $\ell = 2$. If p splits, then we have $N = \infty$ by Theorem 3.6, and hence $\mathcal{U}_2(\mathcal{I}_p(\mathcal{O}_f)) = \mathcal{U}_2(\mathcal{I}_p^*(\mathcal{O}_f)) = \mathbb{N}_{\geq 2}$. The first assertion now follows easily by induction on ℓ. Now let p not split. Then $N < \infty$. Next we show that 2. is true for $\ell = 3$.

Since $[3, N + 1] = \{1\} + \mathcal{U}_2(\mathcal{I}_p^*(\mathcal{O}_f)) \subset \mathcal{U}_3(\mathcal{I}_p^*(\mathcal{O}_f)) \cap \mathbb{N}_{\geq 3} \subset \mathcal{U}_3(\mathcal{I}_p(\mathcal{O}_f)) \cap \mathbb{N}_{\geq 3} \subset [3, \lfloor \frac{3N}{2} \rfloor]$ by Lemma 4.9 and $N \in \{2v_p(f), 2v_p(f) + 1\}$, it remains to show that $N + m \in \mathcal{U}_3(\mathcal{I}_p^*(\mathcal{O}_f))$ for all $m \in [2, v_p(f)]$. Let $m \in [2, v_p(f)]$. It is sufficient to show that there are some $I, J, L \in \mathcal{A}(\mathcal{I}_p^*(\mathcal{O}_f))$ such that $IJ = p^m L$ and $\mathcal{N}(L) = p^N$, since then $IJ\overline{L} = p^{N+m}\mathcal{O}_f$ by Proposition 3.2.5, and thus $N + m \in \mathcal{U}_3(\mathcal{I}_p^*(\mathcal{O}_f))$.

CASE 1: p is inert. Observe that $N = 2v_p(f)$ by Theorem 3.6. Let $m \in [2, v_p(f)]$. First let $p \neq 2$. If $d \not\equiv 1 \mod 4$, then set $I = p^{2m}\mathbb{Z} + (p^m + \tau)\mathbb{Z}$ and $J = p^{2v_p(f)}\mathbb{Z} + (p^{2v_p(f)-m} + \tau)\mathbb{Z}$. If $d \equiv 1 \mod 4$, then set $I = p^{2m}\mathbb{Z} + (\frac{p^m - 1}{2} + \tau)\mathbb{Z}$ and $J = p^{2v_p(f)}\mathbb{Z} + (\frac{p^{2v_p(f)-m} - 1}{2} + \tau)\mathbb{Z}$. In any case we have $I, J \in \mathcal{A}(\mathcal{I}_p^*(\mathcal{O}_f))$ and $IJ = p^m L$ for some $L \in \mathcal{A}(\mathcal{I}_p^*(\mathcal{O}_f))$ with $\mathcal{N}(L) = p^N$.

Next let $p = 2$. Since 2 is inert, it follows that $d \equiv 5 \mod 8$. If $m < v_2(f) - 1$, then set $I = 2^{2m}\mathbb{Z} + (2^m + \tau)\mathbb{Z}$. If $m = v_2(f) - 1$, then set $I = 2^{2m}\mathbb{Z} + \tau\mathbb{Z}$. Finally, if $m = v_2(f)$, then set $I = 2^{2m}\mathbb{Z} + (2^{m-1} + \tau)\mathbb{Z}$. Set $J = 2^{2v_2(f)}\mathbb{Z} + (2^{v_2(f)-1} + \tau)\mathbb{Z}$. Observe that $I, J \in \mathcal{A}(\mathcal{I}_2^*(\mathcal{O}_f))$ and $IJ = 2^m L$ for some $L \in \mathcal{A}(\mathcal{I}_2^*(\mathcal{O}_f))$ with $\mathcal{N}(L) = 2^N$.

CASE 2: p is ramified. It follows that $N = 2v_p(f) + 1$ by Theorem 3.6. Let $m \in [2, v_p(f)]$. First let $p \neq 2$. Since p is ramified, we have $p \mid d$. If $d \not\equiv 1 \mod 4$, then set $I = p^{2m}\mathbb{Z} + (p^m + \tau)\mathbb{Z}$ and $J = p^{2v_p(f)+1}\mathbb{Z} + (p^{v_p(f)+1} + \tau)\mathbb{Z}$. If $d \equiv 1 \mod 4$, then set $I = p^{2m}\mathbb{Z} + (\frac{p^m - 1}{2} + \tau)\mathbb{Z}$ and $J = p^{2v_p(f)+1}\mathbb{Z} + (\frac{p^{v_p(f)+1} - 1}{2} + \tau)\mathbb{Z}$. We infer that $I, J \in \mathcal{A}(\mathcal{I}_p^*(\mathcal{O}_f))$ and $IJ = p^m L$ for some $L \in \mathcal{A}(\mathcal{I}_p^*(\mathcal{O}_f))$ with $\mathcal{N}(L) = p^N$ in any case.

Now let $p = 2$. Since 2 is ramified, we have $d \not\equiv 1 \mod 4$. If d is even or $m < v_2(f)$, then set $I = 2^{2m}\mathbb{Z} + (2^m + \tau)\mathbb{Z}$. If d is odd and $m = v_2(f)$, then set $I = 2^{2m}\mathbb{Z} + \tau\mathbb{Z}$. If d is even, then set $J = 2^{2v_2(f)+1}\mathbb{Z} + \tau\mathbb{Z}$. If d is odd, then set $J = 2^{2v_2(f)+1}\mathbb{Z} + (2^{v_2(f)} + \tau)\mathbb{Z}$. In any case we have $I, J \in \mathcal{A}(\mathcal{I}_2^*(\mathcal{O}_f))$ and $IJ = 2^m L$ for some $L \in \mathcal{A}(\mathcal{I}_2^*(\mathcal{O}_f))$ with $\mathcal{N}(L) = 2^N$.

Finally, we prove the second assertion by induction on ℓ. Let $\ell \in \mathbb{N}_{\geq 2}$ and let $H \in \{\mathcal{I}_p(\mathcal{O}_f), \mathcal{I}_p^*(\mathcal{O}_f)\}$. Without restriction we can assume that $\ell \geq 4$. We infer by the induction hypothesis that $(\mathcal{U}_{\ell-2}(H) \cap \mathbb{N}_{\geq \ell-2}) + \mathcal{U}_2(H) = [\ell - 2, \lfloor \frac{(\ell-2)N}{2} \rfloor] + [2, N] = [\ell, \lfloor \frac{\ell N}{2} \rfloor]$. Observe that $(\mathcal{U}_{\ell-2}(H) \cap \mathbb{N}_{\geq \ell-2}) + \mathcal{U}_2(H) \subset \mathcal{U}_\ell(H) \cap \mathbb{N}_{\geq \ell}$. It follows by Lemma 4.9 that $\mathcal{U}_\ell(H) \cap \mathbb{N}_{\geq \ell} \subset [\ell, \lfloor \frac{\ell N}{2} \rfloor]$, and thus $\mathcal{U}_\ell(H) \cap \mathbb{N}_{\geq \ell} = [\ell, \lfloor \frac{\ell N}{2} \rfloor]$. □

Theorem 5.2. *Let $\mathcal{O}$ be an order in a quadratic number field K with conductor $f\mathcal{O}_K$ for some $f \in \mathbb{N}_{\geq 2}$.*

1. *If f is divisible by a split prime, then $\mathcal{U}_k(\mathcal{I}(\mathcal{O})) = \mathcal{U}_k(\mathcal{I}^*(\mathcal{O})) = \mathbb{N}_{\geq 2}$ for all $k \in \mathbb{N}_{\geq 2}$.*
2. *Suppose that f is not divisible by a split prime and set $M = \max\{\mathrm{v}_p(f) \mid p \in \mathbb{P}\}$. Then $\mathcal{U}_k(\mathcal{I}(\mathcal{O})) = \mathcal{U}_k(\mathcal{I}^*(\mathcal{O}))$ is a finite interval for all $k \in \mathbb{N}_{\geq 2}$, and for their maxima we have*

   (a) *If $\mathrm{v}_q(f) = M$ for a ramified prime q, then $\rho_k(\mathcal{I}(\mathcal{O})) = \rho_k(\mathcal{I}^*(\mathcal{O})) = kM + \lfloor \frac{k}{2} \rfloor$ for all $k \in \mathbb{N}_{\geq 2}$ and $\rho(\mathcal{I}(\mathcal{O})) = \rho(\mathcal{I}^*(\mathcal{O})) = M + \frac{1}{2}$.*
   (b) *If $\mathrm{v}_q(f) < M$ for all ramified primes q, then $\rho_k(\mathcal{I}(\mathcal{O})) = \rho_k(\mathcal{I}^*(\mathcal{O})) = kM$ for all $k \in \mathbb{N}_{\geq 2}$ and $\rho(\mathcal{I}(\mathcal{O})) = \rho(\mathcal{I}^*(\mathcal{O})) = M$.*

Proof. 1. Let f be divisible by a split prime p and let $k \in \mathbb{N}_{\geq 2}$. Since $\mathcal{I}_p^*(\mathcal{O})$ is a divisor-closed submonoid of $\mathcal{I}^*(\mathcal{O})$ and $\mathcal{I}_p(\mathcal{O})$ is a divisor-closed submonoid of $\mathcal{I}(\mathcal{O})$, it follows from Proposition 5.1.1 that $\mathcal{U}_k(\mathcal{I}(\mathcal{O})) = \mathcal{U}_k(\mathcal{I}^*(\mathcal{O})) = \mathbb{N}_{\geq 2}$.

2. Let $k \in \mathbb{N}_{\geq 2}$ and $\ell \in \mathcal{U}_k(\mathcal{I}(\mathcal{O}))$. There are $I_i \in \mathcal{A}(\mathcal{I}(\mathcal{O}))$ for each $i \in [1, k]$ and $J_j \in \mathcal{A}(\mathcal{I}(\mathcal{O}))$ for each $j \in [1, \ell]$ such that $\prod_{i=1}^k I_i = \prod_{j=1}^\ell J_j$. Note that $\sqrt{I_i}, \sqrt{J_j} \in \mathfrak{X}(\mathcal{O})$ for all $i \in [1, k]$ and $j \in [1, \ell]$. For $P \in \mathfrak{X}(\mathcal{O})$ set $k_P = |\{i \in [1, k] \mid \sqrt{I_i} = P\}|$ and $\ell_P = |\{j \in [1, \ell] \mid \sqrt{J_j} = P\}|$. If p is a prime divisor of f, then set $k_p = k_{P_{f,p}}$ and $\ell_p = \ell_{P_{f,p}}$. Observe that $k = \sum_{P \in \mathfrak{X}(\mathcal{O})} k_P$ and $\ell = \sum_{P \in \mathfrak{X}(\mathcal{O})} \ell_P$. Recall that the P-primary components of $\prod_{i=1}^k I_i$ are uniquely determined, and thus $\ell_P \in \mathcal{U}_{k_P}(\mathcal{I}_P(\mathcal{O}))$ for all $P \in \mathfrak{X}(\mathcal{O})$. If $P \in \mathfrak{X}(\mathcal{O})$ does not contain the conductor, then $\mathcal{I}_P(\mathcal{O})$ is factorial, and hence $\ell_P = k_P$. Also note that if $P \in \mathfrak{X}(\mathcal{O})$ and $k_P \leq 1$, then $\ell_P = k_P$. If p is an inert prime that divides f, then it follows from Proposition 5.1.2 and Theorem 3.6 that $\rho_r(\mathcal{I}_p(\mathcal{O})) = \rho_r(\mathcal{I}_p^*(\mathcal{O})) = r\mathrm{v}_p(f)$ for all $r \in \mathbb{N}_{\geq 2}$. We infer again by Proposition 5.1.2 and Theorem 3.6 that $\rho_r(\mathcal{I}_p(\mathcal{O})) = \rho_r(\mathcal{I}_p^*(\mathcal{O})) = r\mathrm{v}_p(f) + \lfloor \frac{r}{2} \rfloor$ for all ramified primes p that divide f and all $r \in \mathbb{N}_{\geq 2}$.

CASE 1: $\mathrm{v}_q(f) = M$ for some ramified prime q. If $P \in \mathfrak{X}(\mathcal{O})$, then $\ell_P \leq k_P M + \lfloor \frac{k_P}{2} \rfloor$.

Consequently, $\ell = \sum_{P \in \mathfrak{X}(\mathcal{O})} \ell_P \leq (\sum_{P \in \mathfrak{X}(\mathcal{O})} k_P)M + \sum_{P \in \mathfrak{X}(\mathcal{O})} \lfloor \frac{k_P}{2} \rfloor \leq kM + \lfloor \frac{k}{2} \rfloor$. In particular, $\rho_k(\mathcal{I}(\mathcal{O})) \leq kM + \lfloor \frac{k}{2} \rfloor = \max\{\rho_k(\mathcal{I}_p^*(\mathcal{O})) \mid p \in \mathbb{P}, p \mid f\} \leq \rho_k(\mathcal{I}^*(\mathcal{O})) \leq \rho_k(\mathcal{I}(\mathcal{O}))$. This implies that $\rho_k(\mathcal{I}(\mathcal{O})) = \rho_k(\mathcal{I}^*(\mathcal{O})) = \max\{\rho_k(\mathcal{I}_p^*(\mathcal{O})) \mid p \in \mathbb{P}, p \mid f\} = kM + \lfloor \frac{k}{2} \rfloor$.

CASE 2: $\mathrm{v}_q(f) < M$ for all ramified primes q. Note that $\ell_p \leq k_p \mathrm{v}_p(f) + \lfloor \frac{k_p}{2} \rfloor \leq k_p M$ for all ramified primes p that divide f. Therefore, $\ell_P \leq k_P M$ for all $P \in \mathfrak{X}(\mathcal{O})$. This implies that $\ell = \sum_{P \in \mathfrak{X}(\mathcal{O})} \ell_P \leq (\sum_{P \in \mathfrak{X}(\mathcal{O})} k_P)M = kM$. We infer that $\rho_k(\mathcal{I}(\mathcal{O})) \leq kM = \max\{\rho_k(\mathcal{I}_p^*(\mathcal{O})) \mid p \in \mathbb{P}, p \mid f\} \leq \rho_k(\mathcal{I}^*(\mathcal{O})) \leq \rho_k(\mathcal{I}(\mathcal{O}))$, and thus $\rho_k(\mathcal{I}(\mathcal{O})) = \rho_k(\mathcal{I}^*(\mathcal{O})) = \max\{\rho_k(\mathcal{I}_p^*(\mathcal{O})) \mid p \in \mathbb{P}, p \mid f\} = kM$.

By Proposition 5.1.2, we obtain that $\mathcal{U}_k(\mathcal{I}(\mathcal{O})) \cap \mathbb{N}_{\geq k} = \mathcal{U}_k(\mathcal{I}^*(\mathcal{O})) \cap \mathbb{N}_{\geq k}$ is a finite interval. Since the last assertion holds for every $k \in \mathbb{N}_{\geq 2}$, we infer that

$\mathcal{U}_k(\mathcal{I}(\mathcal{O})) = \mathcal{U}_k(\mathcal{I}^*(\mathcal{O}))$ is a finite interval for all $k \in \mathbb{N}_{\geq 2}$. If $\mathsf{v}_q(f) = M$ for some ramified prime q, then

$$\rho(\mathcal{I}(\mathcal{O})) = \rho(\mathcal{I}^*(\mathcal{O})) = \lim_{k\to\infty} \frac{\rho_k(\mathcal{I}(\mathcal{O}))}{k} = \lim_{k\to\infty} M + \frac{1}{k}\left\lfloor \frac{k}{2} \right\rfloor = M + \frac{1}{2}.$$

Finally, let $\mathsf{v}_q(f) < M$ for all ramified primes q. Then

$$\rho(\mathcal{I}(\mathcal{O})) = \rho(\mathcal{I}^*(\mathcal{O})) = \lim_{k\to\infty} \frac{\rho_k(\mathcal{I}(\mathcal{O}))}{k} = \lim_{k\to\infty} \frac{kM}{k} = M. \qquad \square$$

In a final remark we gather what is known on further arithmetical invariants of monoids of ideals of orders in quadratic number fields.

Remark 5.3. Let $\mathcal{O}$ be an order in a quadratic number field K with conductor $f\mathcal{O}_K$ for some $f \in \mathbb{N}_{\geq 2}$.

1. The monotone catenary degree of $\mathcal{I}^*(\mathcal{O})$ is finite by [20, Corollary 5.14]. Precise values for the monotone catenary degree are available so far only in the seminormal case [18, Theorem 5.8].

2. The tame degree of $\mathcal{I}^*(\mathcal{O})$ is finite if and only if the elasticity is finite if and only if f is not divisible by a split prime. This follows from Equations. 2.3 and 2.4, Theorem 5.2, and from [16, Theorem 3.1.5]. Precise values for the tame degree are not known so far.

3. For an atomic monoid H, the set $\{\rho(L) \mid L \in \mathcal{L}(H)\} \subset \mathbb{Q}_{\geq 1}$ of all elasticities was first studied by Chapman et al. and then it found further attention by several authors (e.g., [4, 7], [22, Theorem 5.5], [23, 35]). We say that H is *fully elastic* if for every rational number q with $1 < q < \rho(H)$ there is an $L \in \mathcal{L}(H)$ with $\rho(L) = q$. Since $\mathcal{I}^*(\mathcal{O})$ is cancellative and has a prime element, it is fully elastic by [3, Lemma 2.1]. Since $\mathcal{I}^*(\mathcal{O}) \subset \mathcal{I}(\mathcal{O})$ is divisor-closed and $\rho(\mathcal{I}(\mathcal{O})) = \rho(\mathcal{I}^*(\mathcal{O}))$ by Theorem 5.2, it follows that $\mathcal{I}(\mathcal{O})$ is fully elastic.

4. For an atomic monoid H, let

$$\daleth^*(H) = \{\min(L \setminus \{2\}) \mid 2 \in L \in \mathcal{L}(H) \text{ with } |L| > 1\} \subset \mathbb{N}_{\geq 3}.$$

By definition, we have $\daleth^*(H) \subset 2 + \Delta(H)$ and in [11, 23] the invariant $\daleth^*(H)$ was used as a tool to study $\Delta(H)$. Proposition 4.1.4 shows that, both for $H = \mathcal{I}(\mathcal{O})$ and for $H = \mathcal{I}^*(\mathcal{O})$, we have $\max \daleth^*(H) = 2 + \max \Delta(H)$.

Acknowledgements This work was supported by the Austrian Science Fund FWF, Project Numbers J4023-N35 and P28864-N35. We would like to thank the referee for carefully reading the manuscript and for many suggestions and comments that improved the quality of this paper and simplified the proof of Theorem 3.6.

References

1. Anderson, D.D., Mott, J.L.: Cohen-Kaplansky domains: integral domains with a finite number of irreducible elements. J. Algebra **148**, 17–41 (1992)
2. Baeth, N.R., Smertnig, D.: Arithmetical invariants of local quaternion orders. Acta Arith. **186**, 143–177 (2018)
3. Baginski, P., Chapman, S.T., Crutchfield, C., Kennedy, K.G., Wright, M.: Elastic properties and prime elements. Result. Math. **49**, 187–200 (2006)
4. Barron, T., O'Neill, C., Pelayo, R.: On the set of elasticities in numerical monoids. Semigroup Forum **94**, 37–50 (2017)
5. Bowles, C., Chapman, S.T., Kaplan, N., Reiser, D.: On delta sets of numerical monoids. J. Algebra Appl. **5**, 695–718 (2006)
6. Butts, H.S., Pall, G.: Ideals not prime to the conductor in quadratic orders. Acta Arith. **21**, 261–270 (1972)
7. Chapman, S.T., Holden, M., Moore, T.: Full elasticity in atomic monoids and integral domains. Rocky Mt. J. Math. **36**, 1437–1455 (2006)
8. Colton, S., Kaplan, N.: The realization problem for delta sets of numerical monoids. J. Commut. Algebra **9**, 313–339 (2017)
9. Coykendall, J., Malcolmson, P., Okoh, F.: Inert primes and factorization in extensions of quadratic orders. Houston J. Math. **43**, 61–77 (2017)
10. Dobbs, D.E., Fontana, M.: Seminormal rings generated by algebraic integers. Mathematika **34**, 141–154 (1987)
11. Fan, Y., Geroldinger, A.: Minimal relations and catenary degrees in Krull monoids. J. Commut. Algebra **11**, 29–47 (2019)
12. Fan, Y., Geroldinger, A., Kainrath, F., Tringali, S.: Arithmetic of commutative semigroups with a focus on semigroups of ideals and modules. J. Algebra Appl. **11**, 1750234 (42 pages) (2017)
13. Fan, Y., Tringali, S.: Power monoids: a bridge between factorization theory and arithmetic combinatorics. J. Algebra **512**, 252–294 (2018)
14. Gao, W., Geroldinger, A.: On products of k atoms. Monatsh. Math. **156**, 141–157 (2009)
15. Geroldinger, A.: Additive group theory and non-unique factorizations. Combinatorial Number Theory and Additive Group Theory, Advanced Courses in Mathematics CRM Barcelona, Birkhäuser, pp. 1–86 (2009)
16. Geroldinger, A., Halter-Koch, F.: Non-Unique Factorizations. Algebraic, Combinatorial and Analytic Theory. Pure and Applied Mathematics, vol. 278. Chapman & Hall/CRC (2006)
17. Geroldinger, A., Halter-Koch, F., Kaczorowski, J.: Non-unique factorizations in orders of global fields. J. Reine Angew. Math. **459**, 89–118 (1995)
18. Geroldinger, A., Kainrath, F., Reinhart, A.: Arithmetic of seminormal weakly Krull monoids and domains. J. Algebra **444**, 201–245 (2015)
19. Geroldinger, A., Lettl, G.: Factorization problems in semigroups. Semigroup Forum **40**, 23–38 (1990)
20. Geroldinger, A., Reinhart, A.: The monotone catenary degree of monoids of ideals. Internat. J. Algebra Comput. **29**, 419–458 (2019)
21. Geroldinger, A., Schmid, W.A.: A realization theorem for sets of distances. J. Algebra **481**, 188–198 (2017)
22. Geroldinger, A., Schmid, W.A., Zhong, Q.: Systems of sets of lengths: transfer Krull monoids versus weakly Krull monoids. In: Rings, Polynomials, and Modules, pp. 191–235. Springer, Cham (2017)
23. Geroldinger, A., Zhong, Q.: Sets of arithmetical invariants in transfer Krull monoids. J. Pure Appl. Algebra **223**, 3889–3918 (2019)
24. Geroldinger, A., Zhong, Q.: The set of distances in seminormal weakly Krull monoids. J. Pure Appl. Algebra **220**, 3713–3732 (2016)
25. Halter-Koch, F.: Quadratic Irrationals. Pure and Applied Mathematics, vol. 306. Chapman & Hall/CRC (2013)

26. Kainrath, F.: On local half-factorial orders. In: Arithmetical Properties of Commutative Rings and Monoids. Lecture Notes in Pure and Applied Mathematics, vol. 241, pp. 316–324. Chapman & Hall/CRC (2005)
27. Klaise, J.: Orders in quadratic imaginary fields of small class number (2012). https://warwick.ac.uk/fac/cross_fac/complexity/people/students/dtc/students2013/klaise/janis_klaise_ug_report.pdf
28. O'Neill, C., Pelayo, R.: Realisable sets of catenary degrees of numerical monoids. Bull. Australian Math. Soc. **97**, 240–245 (2018)
29. Peruginelli, G., Zanardo, P.: The lattice of primary ideals of orders in quadratic number fields. Int. J. Number Theory **12**(7), 2025–2040 (2016)
30. Philipp, A.: A precise result on the arithmetic of non-principal orders in algebraic number fields, J. Algebra Appl. **11**, 1250087, 42pp
31. Ribenboim, P.: Classical theory of Algebraic Numbers. Universitext, Springer, New York (2001)
32. Schmid, W.A.: Some recent results and open problems on sets of lengths of Krull monoids with finite class group. In: Multiplicative Ideal Theory and Factorization Theory, pp. 323–352. Springer (2016)
33. Smertnig, D.: Sets of lengths in maximal orders in central simple algebras. J. Algebra **390**, 1–43 (2013)
34. Tringali, S.: Structural properties of subadditive families with applications to factorization theory. Israel J. Math. **234**, 1–35 (2019)
35. Zhong, Q.: On elasticities of v-noetherian monoids. J. Algebra **34**, 145–167 (2019)

UMT-domains: A Survey

Gyu Whan Chang

Abstract Let D be an integral domain, X be an indeterminate over D, and $D[X]$ be the polynomial ring over D. A nonzero prime ideal Q of $D[X]$ is called an *upper to zero* in $D[X]$ if $Q \cap D = (0)$. We say that D is a UMT-domain if each upper to zero in $D[X]$ is a maximal t-ideal of $D[X]$. The notion of UMT-domains was introduced by Houston and Zafrullah in 1989. In this paper, we survey the results on UMT-domains with focus on uppers to zero, Nagata rings, graded integral domains, and constructions of new UMT-domains.

Keywords t-operation · Upper to zero · UMT-domain · Nagata ring · Graded integral domain

1 Introduction

A Prüfer domain is an integral domain whose nonzero finitely generated ideals are invertible. It is well known that Dedekind domains are Prüfer domains, and Prüfer domains are Dedekind domains if and only if it is Noetherian. Hence, the notion of Prüfer domains is a natural generalization of Dedekind domains to non-Noetherian integral domains. It is well known that an integral domain D is a Prüfer domain if and only if D_M is a valuation domain for all maximal ideals M of D. However, note that $D[X]$, the polynomial ring over D, is a Prüfer domain if and only if D is a field. When we study the ideal-theoretic properties of integral domains, the so-called t-operation is very useful. For example, D is a Krull domain if and only if every nonzero ideal of D is t-invertible; a Krull domain is a UFD if and only if it has a trivial divisor class group; and a Krull domain is a Dedekind domain if and only if its Krull dimension is one. A PvMD is an integral domain whose nonzero finitely generated ideals are t-invertible; equivalently, D_P is a valuation domain for all maximal t-ideals P of D

G. W. Chang (✉)
Department of Mathematics Education, Incheon National University,
Incheon 22012, Korea
e-mail: whan@inu.ac.kr

A. Facchini et al. (eds.), *Advances in Rings, Modules and Factorizations*,
Springer Proceedings in Mathematics & Statistics 321,
https://doi.org/10.1007/978-3-030-43416-8_3

[42, Theorem 5]. Hence, a Krull domain and a PvMD are the t-operation analogs of Dedekind domains and Prüfer domains, respectively. Moreover, D is a Krull domain (resp., PvMD) if and only if $D[X]$ is a Krull domain (resp., PvMD). Thus, the class of PvMDs includes Dedekind domains, Prüfer domains, Krull domains, and polynomial rings over a PvMD.

A quasi-Prüfer domain is an integral domain whose integral closure is a Prüfer domain [31, Corollary 6.5.14]; a UMT-domain D is an integral domain such that D_P has the Prüfer integral closure for all maximal t-ideals P of D; and D is a PvMD if and only if D is an integrally closed UMT-domain. Hence, UMT-domains can be considered as the t-operation analog of quasi-Prüfer domains or non-integrally closed PvMDs. The notion of UMT-domains was introduced by Houston and Zafrullah [46] and studied carefully by Fontana, Gabelli, and Houston [30]. In this paper, we survey several properties of UMT-domains. Precisely, this paper consists of five sections containing the introduction. In Section 2, we review the definitions and basic results on star operations (Section 2.1), $*$-invertibility (Section 2.2), t-class groups (Section 2.3), and PvMDs, Nagata rings, and Kronecker function rings (Section 2.4). In Section 3, we give basic properties of uppers to zero, several characterizations of UMT-domains (Section 3.1) and Kaplansky type theorems for uppers to zero (Section 3.2) which lead to the study of a general theory of almost factoriality. Section 4 is devoted to the UMT-domain property of graded integral domains. In Section 4.1, we first introduce some definitions for graded integral domains. We then study the UMT-domain property of graded integral domains in Section 4.2 (general case) and Section 4.3 (when R_H is a UFD). Finally, in Section 5, we introduce the technique of constructing new UMT-domains from old one via semigroup rings and pullback.

2 Definitions Related to the t-operation

Throughout D denotes an integral domain with quotient field K, X is an indeterminate over D, $D[X]$ is the polynomial ring over D, and an overring of D means a subring of K containing D.

2.1 *Star Operations*

Let $\mathbf{F}(D)$ (resp., $\mathbf{f}(D)$) be the set of nonzero (resp., nonzero finitely generated) fractional ideals of D. A *star operation* on D is a function $*$ from $\mathbf{F}(D)$ into $\mathbf{F}(D)$ satisfying the following three properties for all $0 \neq x \in K$ and $I, J \in \mathbf{F}(D)$:

1. $(xD)_* = xD$, $(xI)_* = xI_*$,
2. $I \subseteq I_*$, and $I \subseteq J$ implies $I_* \subseteq J_*$, and
3. $(I_*)_* = I_*$.

Given a star operation $*$ on D, we can construct two new star operations $*_f$ and $*_w$ on D as follows:

- $I_{*_f} = \bigcup\{J_* \mid J \in \mathbf{f}(D) \text{ and } J \subseteq I\}$ and
- $I_{*_w} = \{x \in K \mid xJ \subseteq I \text{ for some } J \in \mathbf{f}(D) \text{ with } J_* = D\}$.

Then $(*_f)_f = *_f$ and $(*_w)_f = (*_f)_w = *_w$. We say that $*$ is of *finite character* if $*_f = *$; so $*_f$ and $*_w$ are of finite character. For $I \in \mathbf{F}(D)$, let $I_d = I$. Then d is the identity function of $\mathbf{F}(D)$, and hence d is a star operation on D with $d = d_f = d_w$. For another well-known examples of star operations, note that if we let $I^{-1} = \{x \in K \mid xI \subseteq D\}$ for $I \in \mathbf{F}(D)$, then $I^{-1} \in \mathbf{F}(D)$. Hence, the v-operation given by $I_v = (I^{-1})^{-1}$ for all $I \in \mathbf{F}(D)$ is a star operation. The t-and w-operations are defined by $t = v_f$ and $w = v_w$, respectively.

Lemma 2.1. *Let D be an integral domain, $I, J \in \mathbf{F}(D)$, $\{I_\alpha\}$ be a nonempty subset of $\mathbf{F}(D)$, and $*$ be a star operation on D.*

1. $I \subseteq I_{*_w} \subseteq I_{*_f} \subseteq I_*$.
2. $I_v = \bigcap\{xD \mid x \in K \text{ and } I \subseteq xD\}$.
3. $I_* \subseteq I_v$, *and hence* $I \subseteq I_{*_f} \subseteq I_t$ *and* $I \subseteq I_{*_w} \subseteq I_w$.
4. $(IJ)_* = (IJ_*)_* = (I_*J_*)_*$.
5. *If* $\bigcap_\alpha I_\alpha \neq (0)$, *then* $\bigcap_\alpha (I_\alpha)_* = (\bigcap_\alpha (I_\alpha)_*)_*$.

Proof. (1) Clear. (2) [39, Theorem 34.1]. (3) If $x \in K$ with $I \subseteq xD$, then $I_* \subseteq (xD)_* = xD$. Thus, $I_* \subseteq I_v$ by (2). (4) and (5) [39, Proposition 32.2].

Let $*$ be a star operation on D. An $I \in \mathbf{F}(D)$ is called a *$*$-ideal* if $I_* = I$. A $*$-ideal of D is said to be of *finite type* if $I = J_*$ for some $J \in \mathbf{f}(D)$; so a d-ideal of finite type is just a nonzero finitely generated ideal. A $*$-ideal is a *maximal $*$-ideal* if it is maximal among proper integral $*$-ideals. Let $*$-Max(D) be the set of maximal $*$-ideals of D. Hence, d-Max(D) := Max(D) is the set of maximal ideals of D. It may happen that $*$-Max$(D) = \emptyset$ even though D is not a field as in the case of a rank-one nondiscrete valuation domain D where v-Max$(D) = \emptyset$. However, we have the following nice properties of star operations of finite character.

Lemma 2.2. *Let D be an integral domain and $*$ be a star operation of finite character on D (e.g., $* = d, t$ or w).*

1. *$*$-Max$(D) \neq \emptyset$ if D is not a field.*
2. *Each maximal $*$-ideal of D is a prime ideal.*
3. *Each proper $*$-ideal of D is contained in a maximal $*$-ideal.*
4. *Each prime ideal of D minimal over a $*$-ideal is a $*$-ideal; so each height-one prime ideal is a $*$-ideal.*
5. *$*$-Max$(D) = *_w$-Max(D).*
6. $D = \bigcap_{P \in *\text{-}Max(D)} D_P$.
7. $I_{*_w} = \bigcap_{P \in *\text{-}Max(D)} ID_P$ *for all* $I \in \mathbf{F}(D)$.

Proof. (1), (2), and (3) This can be proved by an easy argument of Zorn's lemma.

(4) Let I be a $*$-ideal of D and P be a minimal prime ideal of I. It suffices to show that $P_* \subseteq P$. Let $0 \neq x \in P_*$. Then there is a $J \in \mathbf{f}(D)$ such that $J \subseteq P$ and $x \in J_*$. Note that P is minimal over I and J is finitely generated. Hence, there are an integer $n \geq 1$ and $s \in D \setminus P$ so that $sJ^n \subseteq I$, and thus $s(J_*)^n \subseteq (sJ^n)_* \subseteq I_* = I \subseteq P$. Thus, $x \in J_* \subseteq P$.

(5) [5, Theorem 2.16]. (6) This is an easy exercise. (7) [5, Corollary 2.10].

Let $A \subseteq B$ be an extension of integral domains. As in [29], we say that B is *t-linked* over A if $I^{-1} = A$ for $I \in \mathbf{f}(A)$ implies $(IB)^{-1} = B$. It is easy to see that B is t-linked over A if and only if $B = \bigcap_{P \in t\text{-Max}(A)} B_P$ [18, Lemma 3.2], if and only if either $Q \cap A = (0)$ or $Q \cap A \neq (0)$ and $(Q \cap A)_t \subsetneq A$ for all $Q \in t\text{-Max}(B)$ [7, Propositions 2.1].

Lemma 2.3. ([58] or [28, Lemma 2.3]) *Let R be a t-linked overring of an integral domain D. For each $A \in \mathbf{F}(D)$, let*

$$A_{w_D} = \{x \in K \mid xJ \subseteq A \text{ for some } J \in \mathbf{f}(D) \text{ with } J_v = D\}.$$

Then w_D is a star operation of finite character on D.

We mean by $*$-dim$(D) = 1$ that $*$-Max$(D) \neq \emptyset$ and each prime $*$-ideal of D is a maximal $*$-ideal of D. Clearly, if dim$(D) = 1$ (i.e., D is one-dimensional), then t-dim$(D) = 1 = w$-dim(D). For more on basic properties of star operations, the reader can refer to [39, Sections 32 and 34].

2.2 *$*$-Invertibility*

Let $*$ be a star operation on D. We say that an $I \in \mathbf{F}(D)$ is *$*$-invertible* if $(II^{-1})_* = D$. Clearly, if $*_f = *$, then I is $*$-invertible if and only if $II^{-1} \nsubseteq P$ for all $P \in *$-Max(D). Hence, by Lemma 2.2(5), I is $*_f$-invertible if and only if I is $*_w$-invertible. Note that $I = I_d \subseteq I_w \subseteq I_t \subseteq I_v$ for all $I \in \mathbf{F}(D)$; so I is invertible $\Rightarrow I$ is w-invertible $\Leftrightarrow I$ is t-invertible $\Rightarrow I$ is v-invertible.

Lemma 2.4. *[48, Proposition 2.6] Let $*$ be a star operation of finite character on an integral domain D. Then $I \in \mathbf{F}(D)$ is $*$-invertible if and only if $I_* = J_*$ for some $J \in \mathbf{f}(D)$ and ID_P is principal for all $P \in *$-Max(D).*

Let $*_1$ and $*_2$ be star operations. Then $I_{*_1} \subseteq I_{*_2}$ if and only if $(I_{*_2})_{*_1} = (I_{*_1})_{*_2} = I_{*_2}$ for all $I \in \mathbf{F}(D)$. Hence, every v-ideal is a $*$-ideal for any star operation $*$ on D. The next result shows that a $*$-ideal is a v-ideal when it is $*$-invertible.

Proposition 2.5. (cf. [60, Theorem 1.1]) *Let $*_1$ and $*_2$ be star operations on an integral domain D such that $I_{*_1} \subseteq I_{*_2}$ for all $I \in \mathbf{F}(D)$. Then a $*_1$-invertible $*_1$-ideal of D is a $*_2$-invertible v-ideal.*

Proof. Let I be a $*_1$-invertible $*_1$-ideal of D. Clearly, I is $*_2$-invertible. Also, if $x \in I_v$, then $xI^{-1} \subseteq D$, and hence $x \in xD = x(II^{-1})_{*_1} \subseteq I_{*_1}$. Hence, $I_v \subseteq I_{*_1}$, and thus $I_v = I_{*_1} = I$.

We say that D is a $*$-*Dedekind domain* if each nonzero ideal of D is $*$-invertible. We also say that D is a *Prüfer $*$-multiplication domain* (P$*$MD) if each nonzero finitely generated ideal of D is $*_f$-invertible. An integral domain D is a *v-domain* if each nonzero finitely generated ideal of D is v-invertible. Clearly,

• $*$-Dedekind domain $\Rightarrow$ P$*$MD, and

• Dedekind domain $=$ d-Dedekind domain $\Rightarrow$ Prüfer domain $=$ PdMD $\Rightarrow$ P$*$MD $=$ P$*_f$MD $=$ P$*_w$MD $\Rightarrow$ PvMD $\Rightarrow$ v-domain.

Proposition 2.6. *Let D be an integral domain.*

1. *D is a t-Dedekind domain if and only if D is a Krull domain.*
2. *D is a v-Dedekind domain if and only if D is completely integrally closed.*

Proof. (1) [49, Theorem 3.6]. (2) [39, Theorem 34.3].

A Bézout domain (resp., GCD domain) D is an integral domain in which each nonzero finitely generated ideal (resp., I_v for each $I \in \mathbf{f}(D)$) is principal. Hence, by Proposition 2.5,

• Bézout domain $=$ GCD domain $+$ Prüfer domain

and GCD domains are PvMDs.

2.3 t-Class Group and Picard Group

Let $T(D)$ (resp., $Inv(D)$, $Prin(D)$) be the group of t-invertible fractional t-ideals (resp., invertible fractional ideals, nonzero principal fractional ideals) of D under the t-multiplication $I * J = (IJ)_t$. It is obvious that $Prin(D) \subseteq Inv(D) \subseteq T(D)$. The *$t$-class group* of D is the abelian group $Cl(D) = T(D)/Prin(D)$ and the *Picard group* of D is a subgroup $Pic(D) = Inv(D)/Prin(D)$ of $Cl(D)$. It is clear that if D is a Krull domain, then $Cl(D)$ is the divisor class group of D. Also, if D is a Prüfer domain or one-dimensional integral domain, then $Cl(D) = Pic(D)$.

Proposition 2.7. *Let D be an integral domain.*

1. *If each maximal ideal of D is a t-ideal, then $Cl(D) = Pic(D)$.*
2. *D is a GCD domain if and only if D is a PvMD with $Cl(D) = \{0\}$.*
3. *D is a UFD if and only if D is a Krull domain with $Cl(D) = \{0\}$.*

Proof. (1) Clear. (2) [13, Proposition 2]. (3) [35, Proposition 6.1].

It is well known that D is a GCD domain (resp., UFD) if and only if $D[X]$, the polynomial ring over D, is a GCD domain (resp., UFD); so in this case, $Cl(D[X]) = Cl(D)$. The next result shows that $Cl(D)$ (resp., $Pic(D)$) is a subgroup of $Cl(D[X])$ (resp., $Pic(D[X])$).

Lemma 2.8. *Let I be a nonzero fractional ideal of an integral domain D.*

1. $(ID[X])^{-1} = I^{-1}D[X]$.
2. $(ID[X])_v = I_v D[X]$.
3. $(ID[X])_t = I_t D[X]$.
4. $ID[X] \cap K = I$.

Hence, I is a (prime) t-ideal of D (resp., invertible, t-invertible) if and only if $ID[X]$ is a (prime) t-ideal of $D[X]$ (resp., invertible, t-invertible). In particular, $Pic(D) \subseteq Pic(D[X])$ and $Cl(D) \subseteq Cl(D[X])$.

Proof. (1), (2), and (3) [48, Proposition 2.2]. (4) Clear.

For a polynomial $f \in K[X]$, we denote by $c_D(f)$ (simply, $c(f)$) the fractional ideal of D generated by the coefficients of f. For convenience, we mean by $c(f)_* = D$ that $f \neq 0$ and $c(f)_* = D$ for any star operation $*$ on D. Dedekind–Mertens lemma states that if $f, g \in K[X]$ are nonzero, then

$$c(f)^{m+1}c(g) = c(f)^m c(fg)$$

for $m = \deg(g)$ [39, Theorem 28.1]. In particular, if $c(f)$ is $*$-invertible, then $(c(f)c(g))_* = c(fg)_*$.

Proposition 2.9. *Let D be an integral domain. Then the following statements are equivalent.*

1. *D is integrally closed.*
2. $Cl(D) = Cl(D[X])$.
3. $fK[X] \cap D[X] = fc(f)^{-1}[X]$ *for all* $0 \neq f \in D[X]$.
4. $c(fg)_v = (c(f)c(g))_v$ *for all* $0 \neq f, g \in D[X]$.

Proof. (1) $\Leftrightarrow$ (2) [36, Theorem 3.6]. (1) $\Rightarrow$ (4) [39, Proposition 34.8]. (4) $\Rightarrow$ (3) This is true by the proof of [39, Corollary 34.9]. (3) $\Rightarrow$ (1) Let $a, b \in D$, $b \neq 0$ such that $x = \frac{a}{b}$ is integral over D. So if $f = bX - a$, then $Q_f := fK[X] \cap D[X] = f(a,b)^{-1}[X]$ by assumption. Thus, $x \in D$ [36, Lemma 3.5].

2.4 PvMDs, Nagata Rings, and Kronecker Function Rings

Let X be an indeterminate over an integral domain D and $D[X]$ be the polynomial ring over D. For a star operation $*$ on D, let $N_* = \{f \in D[X] \mid c(f)_* = D\}$; then N_* is a saturated multiplicative set of $D[X]$ by Dedekind–Mertens lemma and $N_d \subseteq$

$N_* \subseteq N_v$. We call $D(X) := D[X]_{N_d}$ (resp., $D[X]_{N_*}$) the Nagata ring (resp., $*$-Nagata ring) of D.

Proposition 2.10. *1. $Max(D[X]_{N_*}) = \{P[X]_{N_*} \mid P \in *_f\text{-}Max(D)\}$.*
2. $Pic(D[X]_{N_*}) = \{0\}$.
3. Each maximal ideal of $D[X]_{N_v}$ is a t-ideal, and hence

$$Cl(D[X]_{N_v}) = Pic(D[X]_{N_v}) = \{0\}.$$

Proof. These results appear in Proposition 2.1, Theorem 2.14, and Proposition 2.2 of [48], respectively.

A star operation $*$ on D is said to be *endlich arithmetisch brauchbar* (*e.a.b.*) if $(AB)^* \subseteq (AC)^*$ for all $A, B, C \in \mathbf{f}(D)$ implies $B^* \subseteq C^*$. It is well known that if D admits an *e.a.b.* star operation, then D is integrally closed [39, Corollary 32.8]. Conversely, if D is integrally closed, then the star operation b on D, defined by $I_b = \bigcap\{IV \mid V$ is a valuation overring of $D\}$, is an *e.a.b.* star operation of finite character and b-Max$(D) =$ Max(D) [15, Lemma 3.1].

Theorem 2.11. ([39, Theorem 32.7]) *Let D be an integral domain, $*$ be an e.a.b. star operation on D, X be an indeterminate over D, and*

$$Kr(D, *) = \left\{\frac{f}{g} \mid f, g \in D[X], g \neq 0, \text{ and } c(f) \subseteq c(g)_*\right\}.$$

1. *$Kr(D, *)$ is a Bézout domain,*
2. *$Kr(D, *) \cap K = D$, and*
3. *$fKr(D, *) = c(f)Kr(D, *)$ and $fKr(D, *) \cap K = c(f)_*$ for all $0 \neq f \in D[X]$.*

In this case, $Kr(D, *)$ is called the Kronecker function ring of D with respect to $*$.

Let $*$ be an *e.a.b.* star operation on D. It is clear that $D[X]_{N_*} \subseteq$ Kr$(D, *)$, and next in Corollary 2.13, we study when $D[X]_{N_*} =$ Kr$(D, *)$.

Theorem 2.12. *Let $*$ be a star operation on an integral domain D. Then the following statements are equivalent.*

1. *D is a P$*$MD.*
2. *$*_w$ is an e.a.b. star operation.*
3. *$D[X]_{N_*}$ is a Prüfer domain.*
4. *Every ideal of $D[X]_{N_*}$ is extended from D.*
5. *$c(fg)_{*_w} = (c(f)c(g))_{*_w}$ for all $0 \neq f, g \in D[X]$.*
6. *D_P is a valuation domain for all $P \in *_f$-Max(D).*
7. *D is a PvMD and $*_f = t$ on D.*

Proof. (1) ⇔ (2) ⇔ (3) [32, Theorem 3.1]. (1) ⇔ (4) ⇔ (5) [15, Theorem 3.7]. (1) ⇔ (6) [45, Theorem 1.1]. (1) ⇔ (7) [32, Proposition 3.15].

Arnold showed that if D is integrally closed, then D is a Prüfer domain if and only if $\mathrm{Kr}(D, b) = D(X)$ [12, Theorem 4]. This was generalized to PvMDs by Gilmer as follows: If D is a v-domain (equivalently, the v-operation on D is *e.a.b.*), then D is a PvMD if and only if $\mathrm{Kr}(D, v) = D[X]_{N_v}$ [38, Theorem 2.5].

Corollary 2.13. ([32, Theorem 3.1]) *Let $*$ be an e.a.b star operation on an integral domain D. Then D is a P$*$MD if and only if $D[X]_{N_*} = Kr(D, *)$.*

In [33], Fontana and Loper used an arbitrary star operation $*$ on an integral domain D to define a Kronecker function ring $Kr(D, *)$ as follows.

Lemma 2.14. ([33, Theorem 5.1, Proposition 4.5(2), and Corollary 3.5]) *Let $*$ be a star operation on an integral domain D, and let $Kr(D, *) = \{\frac{f}{g} \mid f, g \in D[X], g \neq 0$, and $(c(f)c(h))_* \subseteq (c(g)c(h))_*$ for some $0 \neq h \in D[X]\}$. Then $Kr(D, *)$ is a Bézout domain with quotient field $K(X)$.*

For more on star operations, P$*$MDs, Nagata rings, and Kronecker function rings, the reader can refer to [15, 32, 34], [39, Sections 32 and 33], [48, 60].

3 UMT-domains

Let D denote an integral domain with quotient field K, $\bar{D}$ be the integral closure of D in K, X be an indeterminate over D, $D[X]$ be the polynomial ring over D, and $N_v = \{f \in D[X] \mid c(f)_v = D\}$. In this section, we study several characterizations of UMT-domains (Section 3.1) and Kaplansky type theorems for uppers to zero in polynomial rings (Section 3.2).

3.1 UMT-domains

Let Q be a nonzero prime ideal of $D[X]$. We say that Q is an *upper to zero* in $D[X]$ if $Q \cap D = (0)$. The next result gives some basic properties of uppers to zero in $D[X]$. For more on uppers to zero in polynomial rings, the reader can refer to Houston's survey article [44].

Proposition 3.1. *Let Q be an upper to zero in $D[X]$.*

1. *$Q = fK[X] \cap D[X]$ for some nonzero irreducible polynomial $f \in K[X]$.*
2. *$htQ = 1$, and hence Q is a prime t-ideal of $D[X]$.*
3. *The following statements are equivalent.*

(a) *Q is a maximal t-ideal of* $D[X]$.
(b) *Q is t-invertible.*
(c) *Q contains a nonzero polynomial* $f \in D[X]$ *with* $c(f)_v = D$.

4. *If D is integrally closed, then Q is a v-ideal.*

Proof. (1) By definition, $Q \cap D = (0)$, and hence $Q_{D\setminus\{0\}}$ is a nonzero prime ideal of $K[X]$. Note that $K[X]$ is a PID; so $Q_{D\setminus\{0\}} = fK[X]$ for some nonzero irreducible polynomial $f \in K[X]$. Thus, $Q = Q_{D\setminus\{0\}} \cap K[X] = fK[X] \cap D[X]$.

(2) $\mathrm{ht}Q = \mathrm{ht}(Q_{D\setminus\{0\}}) = \mathrm{ht}(fK[X]) = 1$.

(3) [46, Theorem 1.4].

(4) By (1), $Q = fK[X] \cap D[X]$ for some $0 \neq f \in D[X]$, and since D is integrally closed, $Q = fc(f)^{-1}[X]$ by Proposition 2.9. Thus, by Lemma 2.8, $Q_v = (fc(f)^{-1}[X])_v = f(c(f)^{-1})_v[X] = fc(f)^{-1}[X] = Q$.

The next corollary shows when an upper to zero in $D[X]$ is a maximal ideal, which can be proved by Proposition 3.1(2) with some other results.

Corollary 3.2. ([19, Corollary 5]) *Let* $f = a_0 + a_1X + \cdots + a_nX^n \in D[X]$ *be irreducible in* $K[X]$ *and* $Q = fK[X] \cap D[X]$. *Then Q is a maximal ideal of* $D[X]$ *if and only if* $(\frac{a_1}{a_0}, \ldots, \frac{a_n}{a_0}) \subseteq P$ *for all nonzero prime ideals P of* $\bar{D}$. *In this case,* $fK[X] \cap \bar{D}[X] = \frac{1}{a_0} f\bar{D}[X]$.

By Proposition 3.1, every upper to zero in $D[X]$ is a prime t-ideal. Following [46], we say that D is a *UMT-domain* if each upper to zero in $D[X]$ is a maximal t-ideal of $D[X]$. (UMT stands for Upper to zero is a Maximal T-ideal.)

Theorem 3.3. ([46, Proposition 3.2]) *An integrally closed domain D is a UMT-domain if and only if D is a PvMD.*

Hence, UMT-domains can be considered as non-integrally closed PvMDs. A *quasi-Prüfer domain* is a UMT-domain in which every maximal ideal is a t-ideal; equivalently, its integral closure is a Prüfer domain [31, Chapter VI].

Theorem 3.4. *The following statements are equivalent for an integral domain D.*

1. *D is a UMT-domain.*
2. $\bar{D}_P$ *is a Prüfer domain for all* $P \in t\text{-}Max(D)$.
3. $\bar{D}_P$ *is a Prüfer domain for all prime t-ideals P of D.*
4. $\bar{D}_P$ *is a Bézout domain for all* $P \in t\text{-}Max(D)$.
5. $\bar{D}_P$ *is a Bézout domain for all prime t-ideals P of D.*
6. *Each t-linked overring of D is a UMT-domain.*

Proof. (1) ⇔ (2) ⇔ (3) [30, Theorem 1.5]. (1) ⇔ (5) [16, Lemma 2.2]. (5) ⇒ (4) ⇒ (2) Clear. (1) ⇒ (6) [30, Proposition 1.2]. (6) ⇒ (1) Clear.

Corollary 3.5. ([30, Proposition 1.2]) *Let D be a UMT-domain. Then* D_S *is a UMT-domain for every multiplicative set S of D.*

Proof. Let I be a nonzero finitely generated ideal of D. Then $(ID_S)^{-1} = I^{-1}D_S$ [39, Theorem 4.4]. Hence, D_S is t-linked over D. Thus, by Theorem 3.4, D_S is a UMT-domain.

We next give some characterizations of UMT-domains via polynomial rings, Nagata rings, and Kronecker function rings.

Theorem 3.6. *The following statements are equivalent for an integral domain D.*

1. *D is a UMT-domain.*
2. *$D[X]$ is a UMT-domain.*
3. *Each prime ideal of $D[X]_{N_v}$ is extended from D.*
4. *$D[X]_{N_v}$ is a quasi-Prüfer domain.*
5. *$\bar{D}[X]_{N_v}$ is a Prüfer domain.*
6. *$Kr(D, w) = \bar{D}[X]_{N_v}$, where $Kr(D, w)$ is as in Lemma 2.14.*
7. *Every upper to zero in $D[X]$ is t-invertible.*
8. *Each upper to zero in $D[X]$ contains an $0 \neq f \in D[X]$ with $c(f)_v = D$.*

Proof. (1) ⇔ (2) [30, Theorem 2.4]. (1) ⇔ (3) [46, Theorem 3.1]. (1) ⇔ (4) ⇔ (5) [23, Corollary 2.4 and Theorem 2.16]. (1) ⇔ (6) [17, Corollary 7]. (1) ⇔ (7) ⇔ (8) Proposition 3.1(3).

Let $D^{[w]} = \bigcap_{P \in t\text{-Max}(D)} \bar{D}_P$. Then $D^{[w]}$ is an integrally closed t-linked overring of D, $D^{[w]} = \bar{D}[X]_{N_v} \cap K$ and $D^{[w]}[X]_{N_v} = \bar{D}[X]_{N_v}$ [28, Theorem 1.3]. In particular, the w_D-operation on $D^{[w]}$ can be defined as in Lemma 2.3. It is easy to see that $A_{w_D} = A\bar{D}[X]_{N_v} \cap K$ for all $A \in \mathbf{F}(D^{[w]})$ [28, Lemma 2.3].

Theorem 3.7. *Let D be an integral domain and w_D be the star operation of finite character on $D^{[w]}$ as in Lemma 2.3. Then the following statements are equivalent.*

1. *D is a UMT-domain.*
2. *$D^{[w]}$ is a Pw_DMD.*
3. *$D^{[w]}$ is a PvMD and t-$Max(D^{[w]}) = \{Q \in Spec(D^{[w]}) \mid Q \cap D \in t\text{-}Max(D)\}$.*

Proof. (1) ⇔ (2) [58, Theorem 4.2]. (1) ⇔ (3) [28, Theorem 2.6].

For another characterization of UMT-domains, recall from [4] that a multiplicative subset S of D is a *t-splitting set* if for each $0 \neq d \in D, dD = (AB)_t$ for some integral ideals A and B of D, where $A_t \cap sD = sA_t$ (equivalently, $(A, s)_t = D$) for all $s \in S$ and $B_t \cap S \neq \emptyset$. It is known that S is a t-splitting set of D if and only if $dD_S \cap D$ is t-invertible for all $0 \neq d \in D$ [4, Corollary 2.3].

Theorem 3.8. ([22, Corollary 2.9]) *An integral domain D is a UMT-domain if and only if $D \setminus \{0\}$ is a t-splitting set in $D[X]$.*

A strong Mori domain (SM domain) is an integral domain which satisfies the ascending chain condition on integral w-ideals. Clearly, Noetherian domains are SM domains, while SM domains need not be Noetherian (e.g., $\mathbb{Q}[\{X_\alpha\}]$ for an infinite set $\{X_\alpha\}$ of indeterminates over the field $\mathbb{Q}$ of rational numbers).

Theorem 3.9. ([28, Corollary 3.2]) *Let D be an SM domain. Then D is a UMT-domain if and only if t-dim$(D) = 1$.*

If D is integrally closed, then each upper to zero in $D[X]$ is a v-ideal by Proposition 3.1(4). In [47], Houston and Zafrullah called D a *UMV-domain* if each upper to zero in $D[X]$ is a maximal v-ideal. It is clear that if D is a UMT-domain, then each upper to zero in $D[X]$ is a maximal v-ideal. Hence, UMT-domains are UMV-domains, while UMV-domains need not be UMT-domains (e.g., see [37, Example 3.1] for v-domains that are not PvMDs and [47, Example 3.5] for a non-integrally closed UMV-domain that is not a UMT-domain).

Theorem 3.10. ([47, Theorem 3.3]) *The following statements are equivalent for an integral domain D.*

1. *D is a v-domain.*
2. *D is an integrally closed UMV-domain.*
3. *D is integrally closed and each upper to zero in $D[X]$ is v-invertible.*

Recall that D is an *S-domain* if $\text{ht}(PD[X]) = 1$ for every prime ideal P of D with $\text{ht}P = 1$ [50, p. 26]. It is easy to see that a UMT-domain is an S-domain; and if t-dim$(D) = 1$ (e.g., dim$(D) = 1$), then D is an S-domain if and only if D is a UMT-domain (cf. [55, Theorem 8]). However, S-domains need not be UMT-domains. For example, if $D = \mathbb{R} + (X, Y)\mathbb{C}[[X, Y]]$, where $\mathbb{C}[[X, Y]]$ is the power series ring over the field $\mathbb{C}$ of complex numbers and $\mathbb{R}$ is the field of real numbers, then D is a 2-dimensional local Noetherian domain [14, Theorem 4 and Corollary 9] whose maximal ideal is a t-ideal. Hence, D is an S-domain [50, Theorem 148] but not a UMT-domain by Theorem 3.9. Moreover, it is clear that $D^{[w]} = \bar{D} = \mathbb{C}[[X, Y]]$, and hence $D^{[w]}$ is a PvMD. Thus, in Theorem 3.7, $D^{[w]}$ being a PvMD does not imply that D is a UMT-domain.

3.2 Kaplansky Type Theorems

It is well known that D is a UFD if and only if every nonzero prime ideal of D contains a nonzero prime element [50, Theorem 5]. Also, by Theorem 3.6 (resp., [23, Theorem 1.1]), we have

Proposition 3.11. *An integral domain D is a UMT-domain (resp., quasi-Prüfer domain) if and only if every upper to zero in $D[X]$ contains a nonzero polynomial $f \in D[X]$ with $c(f)_v = D$ (resp., $c(f) = D$).*

In this subsection, we study this kind of theorem, called Kaplansky type theorem, for uppers to zero in $D[X]$. Note that, nonzero principal ideal $\Rightarrow$ invertible ideal $\Rightarrow$ t-invertible ideal; and prime ideal $\Rightarrow$ primary ideal.

Lemma 3.12. *Let $0 \neq f \in K[X]$ and $Q = fK[X] \cap D[X]$.*

1. *Q is principal if and only if $c(f)_v$ is principal. In this case, $Q = \frac{f}{a}D[X]$ for some $0 \neq a \in D$ with $c(f)_v = aD$.*
2. *[57, Theorem A] f is a prime element of $D[X]$ if and only if f is irreducible over K and $c(f)_v = D$.*

Proof. (1) If $Q = gD[X]$ for some $g \in D[X]$, then $fK[X] = gK[X]$, and hence $f = ag$ for some $0 \neq a \in D$ and $c(g)_v = D$ [10, Lemma 2.1]. Thus, $c(f)_v = aD$. Conversely, if $c(f)_v = bD$ for some $b \in D$, then $c(\frac{f}{b})_v = D$, and thus $Q = \frac{f}{b}D[X]$ by Dedekind–Mertens lemma.

(2) If f is a prime element, then $fK[X]$ is a prime ideal and $fD[X] = fK[X] \cap D[X]$. Hence, f is irreducible over K and $c(f)_v = D$ by (1). The reverse implication follows directly from (1).

We next give a Kaplansky type characterization of GCD domains via uppers to zero in polynomial rings.

Theorem 3.13. ([57, Theorem I] and [24, Theorem 2.2]) *The following statements are equivalent for an integral domain D.*

1. *D is a GCD domain.*
2. *Each upper to zero in $D[X]$ contains a nonzero prime element.*
3. *Each upper to zero in $D[X]$ is principal.*
4. *Every $(aX + b)K[X] \cap D[X]$ for $0 \neq a, b \in D$ is principal.*

Proof. (1) $\Rightarrow$ (3) Let Q be an upper to zero in $D[X]$. Then $Q = fK[X] \cap D[X]$ for some $0 \neq f \in D[X]$, and since $c(f)_v$ is principal, Q is principal by Lemma 3.12.

(2) $\Leftrightarrow$ (3) $\Rightarrow$ (4) Clear.

(4) $\Rightarrow$ (1) Let $f = aX + b$. Then f is irreducible in $K[X]$, and hence $fK[X] \cap D[X]$ is an upper to zero in $D[X]$. Hence, $fK[X] \cap D[X]$ is principal by assumption, and thus $(a, b)_v = c(f)_v$ is principal by Lemma 3.12. Thus, D is a GCD domain.

Corollary 3.14. ([24, Corollary 2.3]) *An integral domain D is a Bézout domain if and only if each upper to zero in $D[X]$ contains a nonzero prime element f with $c(f) = D$.*

A primary element of an integral domain D is a nonzero nonunit $a \in D$ such that $a|bc$ for $b, c \in D$ implies either $a|b$ or $a|c^n$ for some integer $n \geq 1$; equivalently, aD is a primary ideal. Clearly, a prime element is primary.

Theorem 3.15. ([10, Theorem 2.4]) *Each upper to zero in $D[X]$ contains a primary element if and only if D is a UMT-domain with $Cl(D[X])$ torsion.*

Corollary 3.16. ([24, Corollary 2.5]) *Each upper to zero in $D[X]$ contains a nonzero primary element f with $c(f) = D$ if and only if D is a quasi-Prüfer domain with $Cl(D[X])$ torsion.*

Proof. ($\Rightarrow$) This follows directly from Proposition 3.11 and Theorem 3.15. ($\Leftarrow$) Note that a quasi-Prüfer domain is a UMT-domain. Hence, each upper to zero in $D[X]$ contains a primary element f by Theorem 3.15. Then $c(f)_v = D$, and since each maximal ideal of D is a t-ideal, $c(f) = D$.

We say that D is an almost GCD domain (AGCD-domain) if, for each $0 \neq a, b \in D$, there is an integer $n = n(a, b) \geq 1$ such that $(a^n, b^n)_v$ is principal; equivalently, $a^n D \cap b^n D$ is principal. It is known that an integrally closed domain D is an AGCD-domain if and only if D is a PvMD with $Cl(D)$ torsion [59, Theorem 3.9].

Corollary 3.17. *Let D be an integrally closed domain.*

1. *[10, Corollary 2.5] Each upper to zero in $D[X]$ contains a primary element if and only if D is an AGCD-domain.*
2. *[24, Corollary 2.6] Each upper to zero in $D[X]$ contains a primary element f with $c(f) = D$ if and only if D is a Prüfer domain with $Cl(D)$ torsion.*

An integral domain D is called a generalized GCD domain (GGCD domain) if the intersection of two invertible ideals of D is invertible [1]; equivalently, every v-ideal of finite type is invertible [1, Theorem 1]. Clearly, GGCD-domains are PvMDs because PvMDs are integral domains whose v-ideals of finite type are t-invertible and invertible ideals are t-invertible.

Theorem 3.18. ([6, Theorem 15]) *An integral domain D is a GGCD-domain if and only if each upper to zero in $D[X]$ is invertible.*

Proof. This can be proved by the fact that if D is integrally closed, then $fK[X] \cap D[X] = fc(f)^{-1}[X]$ for all $0 \neq f \in D[X]$ by Proposition 2.9.

An integral domain D is an almost GGCD-domain (AGGCD-domain) if for $0 \neq a, b \in D$, there is an integer $n = n(a, b) \geq 1$ such that $a^n D \cap b^n D$ (equivalently, $(a^n, b^n)_v$) is invertible. It is known that D is an AGCD-domain if and only if D is an AGGCD-domain with $Cl(D)$ torsion [51, Theorem 5.1].

Theorem 3.19. ([24, Theorem 2.8]) *Let D be an integrally closed domain.*

1. *Every upper to zero in $D[X]$ contains a t-invertible primary ideal if and only if D is a PvMD.*
2. *Every upper to zero in $D[X]$ contains an invertible primary ideal if and only if D is an AGGCD-domain.*

For another example of UMT-domains, recall that D is called an almost PvMD (APvMD) if, for each $0 \neq a, b \in D$, there exists an integer $n = n(a, b) \geq 1$ such that (a^n, b^n) is t-invertible. Note that $(a^n, b^n)_v$ is t-invertible if and only if (a^n, b^n) is t-invertible; so AGGCD domains are APvMDs.

Theorem 3.20. *Let D be an integral domain with integral closure $\bar{D}$.*

1. *D is an APvMD if and only if D is a UMT-domain and $D \subseteq \bar{D}$ is a root extension, i.e., $a \in \bar{D}$ implies $a^n \in D$ for some integer $n = n(a) \geq 1$.*
2. *D is an AGCD-domain if and only if D is an APvMD with $Cl(D)$ torison.*

Proof. These appear in Theorem 3.8 and Theorem 3.1 of [52], respectively.

In [56], Storch studied almost factorial domains which are Krull domains with torsion divisor class groups [35, Proposition 6.8]. Then, in [59], Zafrullah began to study a general theory of almost factoriality where he defined the notion of AGCD-domains. As we noted in this subsection, we have AGCD-domain ⇒ AGGCD-domain ⇒ APvMD ⇒ UMT-domain.

4 Graded Integral Domains Which Are UMT-domains

In this section, we study the UMT-domain property of graded integral domains. We first review the definitions related with graded integral domains.

4.1 Definitions of Graded Integral Domains

Let Γ be a nonzero torsionless grading monoid, that is, Γ is a nonzero torsionless commutative cancellative monoid (written additively), and $\langle \Gamma \rangle = \{a - b \mid a, b \in \Gamma\}$ be the quotient group of Γ; so $\langle \Gamma \rangle$ is a torsionfree abelian group. It is well known that a cancellative monoid Γ is torsionless if and only if Γ can be given a total order compatible with the monoid operation [54, page 123]. By a (Γ-)graded integral domain $R = \bigoplus_{\alpha \in \Gamma} R_\alpha$, we mean an integral domain graded by Γ. Hence, each nonzero $x \in R_\alpha$ has degree α, i.e., $\deg(x) = \alpha$, $\deg(0) = 0$, and $R_\alpha R_\beta \subseteq R_{\alpha+\beta}$ for all $\alpha, \beta \in \Gamma$. Thus, each nonzero $f \in R$ can be written uniquely as $f = x_{\alpha_1} + \cdots + x_{\alpha_n}$ with $\deg(x_{\alpha_i}) = \alpha_i$ and $\alpha_1 < \cdots < \alpha_n$. A nonzero $x \in R_\alpha$ for every $\alpha \in \Gamma$ is said to be *homogeneous*. Let $H = \bigcup_{\alpha \in \Gamma}(R_\alpha \setminus \{0\})$; so H is the saturated multiplicative set of nonzero homogeneous elements of R. Then R_H, called the *homogeneous quotient field* of R, is a $\langle \Gamma \rangle$-graded integral domain whose nonzero homogeneous elements are units. Hence, R_H is a completely integrally closed GCD domain [3, Proposition 2.1]. For a fractional ideal A of R with $A \subseteq R_H$, let A^h be the fractional ideal of R generated by homogeneous elements in A; so $A^h \subseteq A$. The fractional ideal A is said to be *homogeneous* if $A^h = A$. Let $C(f)$ denote the fractional ideal of R generated by the homogeneous components of $f \in R_H$ and $C(A) = \sum_{f \in A} C(f)$ for a fractional ideal A of R with $A \subseteq R_H$. Clearly, $C(f)$ and $C(A)$ are homogeneous fractional ideals of R. It is known that if $f, g \in R$, then $C(f)^{m+1}C(g) = C(f)^m C(fg)$ for some integer $m \geq 1$ [54].

Examples of torsionless grading monoids include the additive monoids $\mathbb{Z}^+$ and $\mathbb{Q}^+$ of nonnegative integers and nonnegative rational numbers, respectively. Also, $\langle \mathbb{Z}^+ \rangle = \mathbb{Z}$ and $\langle \mathbb{Q}^+ \rangle = \mathbb{Q}$ are the additive groups of integers and rational numbers, respectively. It is clear that the semigroup ring $D[\Gamma]$ of Γ over an integral domain D is a Γ-graded integral domain with $\deg(aX^{\alpha}) = \alpha$ for $0 \neq a \in D$ and $\alpha \in \Gamma$. Also, the polynomial ring $D[X]$ and the Laurent polynomial ring $D[X, X^{-1}]$ are graded integral domains graded by $\mathbb{Z}^+$ and $\mathbb{Z}$, respectively.

Throughout this section, Γ is always a nonzero torsionless grading monoid, $R = \bigoplus_{\alpha \in \Gamma} R_{\alpha}$ is a Γ-graded integral domain such that $R_{\alpha} \neq (0)$ for all $\alpha \in \Gamma$, $\bar{R}$ is the integral closure of R, H is the saturated multiplicative set of nonzero homogeneous elements of R, and $N(H) = \{f \in R \mid f \neq 0 \text{ and } C(f)_v = R\}$; so $N(H)$ is a saturated multiplicative set of R.

4.2 General Case

Our first result shows why we are mainly interested in homogeneous maximal t-ideals when we study the divisibility properties (e.g., PvMDs and Krull domains) of graded integral domains.

Lemma 4.1. *Let Q be a maximal t-ideal of $R = \bigoplus_{\alpha \in \Gamma} R_{\alpha}$.*

1. *If $Q \cap H = \emptyset$, then R_Q is a valuation domain. In particular, if R_H is a UFD, then R_Q is a rank-one DVR.*
2. *If $Q \cap H \neq \emptyset$, then Q is homogeneous.*

Hence, $\{Q \in t\text{-}Max(D) \mid Q \cap H \neq \emptyset\} = \{Q \in t\text{-}Max(D) \mid Q \text{ is homogeneous}\}$.

Proof. (1) [27, Theorem 2]. (2) [9, Lemma 1.2].

The next result is a graded integral domain analog of Theorem 3.4.

Theorem 4.2. ([27, Corollary 7] and [43, Theorem 3.2]) *The following statements are equivalent for a graded integral domain $R = \bigoplus_{\alpha \in \Gamma} R_{\alpha}$.*

1. *R is a UMT-domain.*
2. *$\bar{R}_Q$ is a Prüfer domain for all homogeneous maximal t-ideals Q of R.*
3. *R_Q is a quasi-Prüfer domain for all homogeneous maximal t-ideals Q of R.*
4. *$R_{N(H)}$ is a UMT-domain.*
5. *$R_{H \setminus Q}$ is a UMT-domain and $Q_{H \setminus Q}$ is a t-ideal for all homogeneous maximal t-ideals Q of R.*

Proof. This result follows directly from Theorem 3.4, Lemma 4.1 and the fact that $Q_{N(H)}$ is a maximal t-ideal of $R_{N(H)}$ for all homogeneous maximal t-ideals Q of R [11, Proposition 1.3].

Corollary 4.3. ([25, Lemma 2.7] and [2, Theorem 6.4]) *Let* $R = \bigoplus_{\alpha \in \Gamma} R_\alpha$ *be a graded integral domain. Then the following statements are equivalent.*

1. *R is a PvMD.*
2. R_Q *is a valuation domain for all homogeneous maximal t-ideals Q of R.*
3. *Every nonzero finitely generated homogeneous ideal of R is t-invertible.*

Proof. (1) ⇒ (3) Clear. (3) ⇒ (2) This follows because IR_Q is principal for all nonzero finitely generated homogeneous ideals I of R. (2) ⇒ (1) This is an immediate consequence of Theorem 2.12 and Lemma 4.1 (or Theorems 3.3 and 4.2).

As in [11], we say that R satisfies property (#) if $C(I)_t = R$ implies $I \cap N(H) \neq \emptyset$ for all nonzero ideals I of R; equivalently, $\text{Max}(R_{N(H)}) = \{Q_{N(H)} \mid Q \in t\text{-Max}(D)$ with $Q \cap H \neq \emptyset\}$ [11, Proposition 1.4]. Examples of graded integral domains with property (#) include (i) graded integral domains with a unit of nonzero degree, (ii) $D[\Gamma]$, the monoid domain of Γ over an integral domain D, and (iii) $D[\{X_\alpha\}]$, the polynomial ring over D, [11, Example 1.6].

Corollary 4.4. ([27, Corollary 8]) *Let* $R = \bigoplus_{\alpha \in \Gamma} R_\alpha$ *be a graded integral domain with property* (#). *Then the following statements are equivalent.*

1. *R is a UMT-domain.*
2. $R_{N(H)}$ *is a quasi-Prüfer domain.*
3. $\bar{R}_{N(H)}$ *is a Prüfer domain.*

Let $R = D[X, X^{-1}]$ be the Laurent polynomial ring over an integral domain D. Then R is a $\mathbb{Z}$-graded integral domain and $R_{N(H)} = D[X]_{N_v}$. Hence, Corollary 4.4 is a natural generalization of Theorem 3.6 (1), (4), and (5).

4.3 When R_H Is a UFD

In this subsection, we study the UMT-domain property of graded integral domains $R = \bigoplus_{\alpha \in \Gamma} R_\alpha$ such that R_H is a UFD; so throughout this subsection, we always assume that R_H is a UFD. We first give some examples of graded integral domains R with R_H a UFD.

Example 4.5. Let $R = \bigoplus_{\alpha \in \Gamma} R_\alpha$ be a graded integral domain. Then R_H is a UFD if one of the following conditions are satisfied:

1. [8, Proposition 3.5] $\langle \Gamma \rangle$ satisfies the ascending chain condition on its cyclic subgroups.
2. $R = D[\{X_\alpha\}]$ is the polynomial ring over an integral domain D.
3. [53, Section A.I.4.] $\langle \Gamma \rangle = \mathbb{Z}$ is the additive group of integers.

Let $R = D[X]$ be the polynomial ring over D. Then Q is an upper to zero in $D[X]$ if and only if either $Q = XD[X]$ or $Q = fK[X, X^{-1}] \cap D[X]$ for some prime element $0 \neq f \in K[X]$. Also, $R_H = K[X, X^{-1}]$, and hence if Q is an upper to zero in $D[X]$ with $Q \neq XD[X]$, then $Q = fR_H \cap R$. Motivated by this fact, Chang introduced the notion of graded UMT-domains. As we will see, the notions of UMT-domains and graded UMT-domains are distinct in general.

Definition 4.6. [21, Definition 1.4] Let $R = \bigoplus_{\alpha \in \Gamma} R_\alpha$ be a graded integral domain, and assume that R_H is a UFD.

1. A nonzero prime ideal Q of R is an *upper to zero in* R if $fR_H \cap R$ for some $f \in R_H$. (In this case, f is a nonzero prime element of R_H.)
2. R is a *graded UMT-domain* if every upper to zero in R is a maximal t-ideal.

Clearly, $XD[X]$ is a maximal t-ideal of $D[X]$. Hence, D is a UMT-domain if and only if $D[X]$ is a graded UMT-domain. However, if $D = \mathbb{R} + (y, z)\mathbb{C}[\![y, z]\!]$ (see the remark after Theorem 3.10), then $R = D + XK[X]$ is a graded UMT-domain with property (#) but not a UMT-domain ([20, Corollary 9] and [21, Proposition 4.2]).

Theorem 4.7. ([21, Proposition 1.7]) *Let $R = \bigoplus_{\alpha \in \Gamma} R_\alpha$ be a UMT-domain. Then R is a graded UMT-domain.*

The next result is a graded integral domain analog of Proposition 3.1(3) which is very useful for the study of UMT-domain properties.

Proposition 4.8. *[21, Proposition 1.8] Let Q be a prime t-ideal of $R = \bigoplus_{\alpha \in \Gamma} R_\alpha$ such that $Q \cap H = \emptyset$. Then the following statements are equivalent.*

1. *$C(Q)_t = R$.*
2. *Q is t-invertible.*
3. *Q is a maximal t-ideal.*

In this case, ht$Q = 1$, and hence Q is an upper to zero in R.

Let $f \in D[X]$ be a nonzero polynomial with $c(f)_v = D$. If A is an ideal of $D[X]$ with $f \in A$, then A is t-invertible [46, Proposition 4.1] and $fD[X] = (Q_1^{e_1} \cdots Q_n^{e_n})_t$ for some uppers to zero Q_i in $D[X]$ and integers $e_i \geq 1$ [41, p. 144]. The next result is an extension of these results to graded integral domains.

Corollary 4.9. (cf. [21, Proposition 1.12]) *Let A be a nonzero ideal of $R = \bigoplus_{\alpha \in \Gamma} R_\alpha$ such that $C(A)_t = R$. Then $A_t = (Q_1^{e_1} \cdots Q_n^{e_n})_t$ for some t-invertible uppers to zero Q_i in R and integers $e_i \geq 1$. In particular, A is t-invertible.*

Proof. Let Q be a maximal t-ideal of R with $A \subseteq Q$. Then $R = C(A)_t \subseteq C(Q)_t \subseteq R$ or $C(Q)_t = R$. Hence, $Q \cap H = \emptyset$ by Lemma 4.1, and thus Q is an upper to zero in R that is t-invertible by Proposition 4.8. This also implies that each prime t-ideal of R containing A is t-invertible. Thus, $A_t = (Q_1^{e_1} \cdots Q_n^{e_n})_t$ for some uppers to zero Q_i in R and integers $e_i \geq 1$ [41, Theorem 1.3] and A is t-invertible.

Corollary 4.10. ([21, Corollary 1.13]) *Let* $R = \bigoplus_{\alpha \in \Gamma} R_\alpha$ *and* $0 \neq f \in R$ *with* $C(f)_v = R$. *Then* $fR = (Q_1^{e_1} \cdots Q_n^{e_n})_t$ *for some uppers to zero* Q_i *in* R *and integers* $e_i \geq 1$.

The next result gives a characterization of graded UMT-domains $R = \bigoplus_{\alpha \in \Gamma} R_\alpha$ with property (#). Note that if $R = D[X, X^{-1}]$, then R satisfies property (#) and $R_{N(H)} = D[X]_{N_v}$; hence, the next result is an extension of Theorem 3.6 to graded integral domains with property (#).

Theorem 4.11. *Let* $R = \bigoplus_{\alpha \in \Gamma} R_\alpha$ *be a graded integral domain with property* (#). *Then the following statements are equivalent.*

1. *R is a graded UMT-domain.*
2. *Every upper to zero in R contains an $f \in R$ with $C(f)_v = R$.*
3. *Every prime ideal of $R_{N(H)}$ is extended from a homogeneous ideal of R.*
4. *Every upper to zero in R is t-invertible.*

Proof. (1) ⇔ (2) ⇔ (3) [21, Theorem 2.2]. (2) ⇔ (4) Proposition 4.8.

We now give several characterizations of graded UMT-domains with a unit of nonzero degree, and in this case, graded UMT-domains are UMT-domains.

Theorem 4.12. *Let* $R = \bigoplus_{\alpha \in \Gamma} R_\alpha$ *be a graded integral domain with a unit of nonzero degree. Then the following statements are equivalent.*

1. *R is a graded UMT-domain.*
2. *Every upper to zero in R contains an $f \in Q$ with $C(f)_v = R$.*
3. *Every prime ideal of $R_{N(H)}$ is extended from a homogeneous ideal of R.*
4. *R is a UMT-domain.*
5. *$\bar{R}_{N(H)}$ is a Prüfer domain.*
6. *$R_{N(H)}$ is a UMT-domain.*
7. *$R_{N(H)}$ is a quasi-Prüfer domain.*
8. *Every upper to zero in R is t-invertible.*
9. *Let Q be a nonzero prime ideal of R with $C(Q)_t \subsetneq R$. Then Q is a homogeneous prime t-ideal.*
10. *Let Q be a nonzero prime ideal of R such that $Q \subseteq M$ for some homogeneous maximal t-ideal M of R. Then Q is a homogeneous prime t-ideal.*

Proof. (1)–(7) [21, Theorem 3.5]. (1) ⇔ (8) Theorem 4.11 because graded integral domains with a unit of nonzero degree satisfy property (#). (1) ⇔ (9) ⇔ (10) [21, Corollary 3.10].

Corollary 4.13. ([26, Theorem 2.5]) *Let* $R = \bigoplus_{\alpha \in \Gamma} R_\alpha$ *be a graded integral domain with a unit of nonzero degree. Then R is an integrally closed graded UMT-domain if and only if R is a PvMD.*

Let $R = D[X]$ be the polynomial ring over D. Then, as we note after Definition 4.6, if R is a graded UMT-domain, then R is a UMT-domain even though R does not contain a unit of nonzero degree. The next result explains why this phenominon happens.

Corollary 4.14. ([21, Corollary 3.13]) *Let $R = \bigoplus_{\alpha\in\Gamma} R_\alpha$ be a graded integral domain with a nonzero homogeneous prime element p such that $ht(pR) = 1$ and $deg(p) \neq 0$. Then R is a graded UMT-domain if and only if R is a UMT-domain.*

Proof. Let $S = \{p^n \mid n \geq 1\}$. Then it is easy to see that R is a (graded) UMT-domain if and only if R_S is a (graded) UMT-domain. Thus, the result is an immediate consequence of Theorem 4.12.

5 Constructing New UMT-domains from Old Ones

In this section, we construct new UMT-domains from old UMT-domains via semigroup rings (Section 5.1) and pullbacks (Section 5.2).

5.1 Semigroup Rings

Let D be an integral domain, Γ be a nonzero torsionless grading monoid, and $D[\Gamma]$ be the semigroup ring of Γ over R. Then $D[\Gamma]$ is an integral domain [40, Theorem 8.1] graded by Γ with $\deg(aX^\alpha) = \alpha$ for $0 \neq a \in D$ and $\alpha \in \Gamma$. We say that Γ is a *valuation monoid* if either $g \in \Gamma$ or $-g \in \Gamma$ for every $g \in \langle\Gamma\rangle$. Clearly, $\mathbb{Z}^+$ and $\mathbb{Q}^+$ are valuation monoids.

Lemma 5.1. *Let Γ be a torsionless grading monoid. Then the following statements are equivalent.*

1. *Γ is a valuation monoid.*
2. *$\langle\Gamma\rangle = \Gamma \cup (-\Gamma)$.*
3. *$a + \Gamma \subseteq b + \Gamma$ or $b + \Gamma \subseteq a + \Gamma$ for every $a, b \in \Gamma$.*
4. *Every finitely generated ideal of Γ is principal.*

Proof. This is an easy exercise.

Let $\bar{D}$ be the integral closure of D and $\bar{\Gamma}$ be the integral closure of Γ. Then $\overline{D[\Gamma]} = \bar{D}[\bar{\Gamma}]$ [40, Theorem 12.10] and $\mathrm{Max}(D[\Gamma]_{N(H)}) = \{P[X]_{N(H)} \mid P \in t\text{-}\mathrm{Max}(D)\} \cup \{D[S]_{N(H)} \mid S \in t\text{-}\mathrm{Max}(\Gamma)\}$ [11, Example 1.6].

Theorem 5.2. *The following statements are equivalent for $D[\Gamma]$.*

1. *$D[\Gamma]$ is a UMT-domain.*
2. *$D[\Gamma]_{N(H)}$ is a quasi-Prüfer domain.*
3. *$\overline{D[\Gamma]}_{N(H)}$ is a Prüfer domain.*
4. *D is a UMT-domain and $\bar{\Gamma}_S$ is a valuation monoid for all maximal t-ideals S of Γ.*
5. *$D[\Gamma]_Q$ is a quasi-Prüfer domain for all maximal t-ideals Q of $D[\Gamma]$ with $Q \cap D \neq (0)$ or $X^\alpha \in Q$ for some $\alpha \in \Gamma$.*

Proof. (1) $\Leftrightarrow$ (2) $\Leftrightarrow$ (3) $\Leftrightarrow$ (4) [27, Theorem 17]. (1) $\Leftrightarrow$ (5) Note that Q is a homogeneous maximal t-ideal of $D[\Gamma]$ if and only if either $Q \cap D \neq (0)$ or $X^{\alpha} \in Q$ for some $\alpha \in \Gamma$. Thus, the result follows from Theorem 4.2.

Corollary 5.3. ([27, Corollary 18]) *Let G be a torsionfree abelian group. Then $D[G]$ is a UMT-domain if and only if D is a UMT-domain.*

Proof. Clearly, G has no maximal t-ideal. Thus, the result follows directly from Theorem 5.2.

Corollary 5.4. ([27, Corollary 18]) *Let Γ be a numerical semigroup. Then $D[\Gamma]$ is a UMT-domain if and only if D is a UMT-domain.*

Proof. Clearly, $\bar{\Gamma} = \mathbb{Z}^{+}$ is a valuation monoid. Thus, the result follows directly from Theorem 5.2.

Corollary 5.5. ([30, Theorems 2.4 and 2.5] and [21, Corollary 3.14]) *Let $\{X_{\alpha}\}$ be a nonempty set of indeterminates over D and $N_v = \{f \in D[\{X_{\alpha}\}] \mid c(f)_v = D\}$. Then the following statements are equivalent.*

1. *D is a UMT-domain.*
2. *$D[\{X_{\alpha}\}]$ is a UMT-domain.*
3. *$D[\{X_{\alpha}, X_{\alpha}^{-1}\}]$ is a UMT-domain.*
4. *$D[\{X_{\alpha}\}]_{N_v}$ is a UMT-domain.*
5. *$D[\{X_{\alpha}\}]_{N_v}$ is a quasi-Prüfer domain.*

Proof. For each α, let $\mathbb{Z}_{\alpha} = \mathbb{Z}$ be the additive group of integers; so if $G = \bigoplus_{\alpha} \mathbb{Z}_{\alpha}$, then G is a torsionfree abelian group and the group ring $D[G]$ of G over D is isomorphic to $D[\{X_{\alpha}, X_{\alpha}^{-1}\}]$. Hence, the result follows from Theorem 5.2.

5.2 Pullbacks

Let T be an integral domain, M be a maximal ideal of T, $k = T/M$, D be a subring of k, $\varphi : T \to k$ be the canonical homomorphism, and $R = \varphi^{-1}(D)$ be the pullback of the following diagram:

$$\begin{array}{ccc} R = \varphi^{-1}(D) & \longrightarrow & D \\ \downarrow & & \downarrow \\ T & \xrightarrow{\ \varphi\ } & k = T/M. \end{array}$$

Then T is a t-linked overring of R [30, Proposition 3.1] and M is a prime t-ideal of R. Hence, by Theorem 3.4, if R is a UMT-domain, then T is a UMT-domain. We call R a pullback of type ($\star$). For example, if T is a valuation domain with maximal ideal M such that $T = k + M$ (e.g., $T = k[\![X]\!]$ is the power series ring over k), then $R = D + M$.

Theorem 5.6. ([30, Theorem 3.7]) *Let R be a pullback of type ($\star$). Then R is a UMT-domain if and only if D and T are UMT-domains, M is a maximal t-ideal of T, and k is algebraic over the quotient field of D.*

Corollary 5.7. *Let D be an integral domain with quotient field K, k be an algebraic extension field of K, and X be an indeterminate over k. Then the following statements are equivalent.*

1. *D is a UMT-domain.*
2. *$D + X^n k[X]$ is a UMT-domain for all integers $n \geq 1$.*
3. *$D + X^n k[\![X]\!]$ is a UMT-domain for all integers $n \geq 1$.*

Proof. (1) $\Leftrightarrow$ (2) Clearly, $k[X]$ is a UMT-domain, and hence $k[S]$ for $S = \{0, n, n+1, \dots\}$ is a UMT-domain (by Corollary 5.4) and $X^n k[X]$ is a maximal t-ideal of $k[S]$. Thus, the results follow from Theorem 5.6.

(1) $\Leftrightarrow$ (3) Let $T = k[\![X^n, X^{n+1}, \cdots]\!]$. Then $\bar{T} = k[\![X]\!]$ is a rank-one DVR. Hence, T is a quasi-Prüfer domain and $X^n k[\![X]\!]$ is a maximal t-ideal of T. Thus, by Theorem 5.6, D is a UMT-domain if and only if $D + X^n k[\![X]\!]$ is a UMT-domain.

Corollary 5.8. *Let D be an integral domain with quotient field K, k be an algebraic extension field of K, $k[\mathbb{Q}^+]$ be the semigroup ring of $\mathbb{Q}^+$ over k, and $M = \{g \in k[\mathbb{Q}^+] \mid g(0) = 0\}$. Then D is a UMT-domain if and only if $D + M$ is a UMT-domain.*

Proof. Note that $k[\mathbb{Q}^+]$ is a PvMD by Corollary 4.3 and M is a maximal t-ideal of $k[\mathbb{Q}^+]$ (cf. Lemma 4.1). Thus, the result follows directly from Theorem 5.6.

Acknowledgements This research was supported by Basic Science Research Program through the National Research Foundation of Korea(NRF) funded by the Ministry of Education (2017R1D1A1B06029867).

References

1. Anderson, D.D., Anderson, D.F.: Generalized GCD domains. Comment. Math. Univ. St. Pauli **28**, 215–221 (1979)
2. Anderson, D.D., Anderson, D.F.: Divisorial ideals and invertible ideals in a graded integral domain. J. Algebra **76**, 549–569 (1982)
3. Anderson, D.D., Anderson, D.F.: Divisibility properties of graded domains. Canad. J. Math. **34**, 196–215 (1982)
4. Anderson, D.D., Anderson, D.F., Zafrullah M.: The ring $D + XD_S[X]$ and t-splitting sets, Commutative Algebra. Arab. J. Sci. Eng. Sect. C (26), 3–16 (2001)
5. Anderson, D.D., Cook, S.J.: Two star-operations and their induced lattices. Comm. Algebra **28**, 2461–2475 (2000)
6. Anderson, D.D., Dumitrescu, T., Zafrullah, M.: Quasi-Schreier domains, II. Comm. Algebra **35**, 2096–2104 (2007)

7. Anderson, D.D., Houston, E., Zafrullah, M.: t-linked extensions, the t-class group, and Nagata's theorem. J. Pure Appl. Algebra **86**, 109–124 (1993)
8. Anderson, D.F.: Graded Krull domains. Comm. Algebra **7**, 79–106 (1979)
9. Anderson, D.F., Chang, G.W.: Homogeneous splitting sets of a graded integral domain. J. Algebra **288**, 527–544 (2005)
10. Anderson, D.F., Chang, G.W.: Almost splitting sets in integral domains, II. J. Pure Appl. Algebra **208**, 351–359 (2007)
11. Anderson, D.F., Chang, G.W.: Graded integral domains and Nagata rings. J. Algebra **387**, 169–184 (2013)
12. Arnold, J.T.: On the ideal theory of the Kronecker function ring and the domain $D(X)$. Canad. J. Math. **21**, 558–563 (1969)
13. Bouvier, A.: Le groupe des classes d'un anneaux integre, 107 eme Congres des Societes Savantes, Brest fasc. IV 85–92 (1982)
14. Brewer, J., Rutter, E.: $D + M$ constructions with general overrings. Michigan Math. J. **23**, 33–42 (1976)
15. Chang, G.W.: Prüfer $*$-multiplication domains, Nagata rings, and Kronecker function rings. J. Algebra **319**, 309–319 (2008)
16. Chang, G.W.: Locally pseudo-valuation domains of the form $D[X]_{N_v}$. J. Korean Math. Soc. **45**, 1405–1416 (2008)
17. Chang, G.W.: Kronecker function rings and Prüfer-like domains. Korean J. Math. **20**, 371–379 (2012)
18. Chang, G.W.: The $A + XB[X]$ construction from Prüfer v-multiplication domains. J. Algebra **439**, 417–437 (2015)
19. Chang, G.W.: Uppers to zero in polynomial rings which are maximal ideals. Bull. Korean Math. Soc. **52**, 525–530 (2015)
20. Chang, G.W.: Graded integral domains and Nagata rings II. Korean J. Math. **25**, 215–227 (2017)
21. Chang, G.W.: Graded integral domains and Prüfer-like domains. J. Korean Math. Soc. **54**, 1733–1757 (2017)
22. Chang, G.W., Dumitrescu, T., Zafrullah, M.: t-Splitting sets in integral domains. J. Pure Appl. Algebra **187**, 71–86 (2004)
23. Chang, G.W., Fontana, M.: Uppers to zero in polynomial rings and Prüfer-like domains. Comm. Algebra **37**, 164–192 (2009)
24. Chang, G.W., Kim, H.: Kaplansky-type theorems II. Kyungpook Math. J. **51**, 339–344 (2011)
25. Chang, G.W., Kang, B.G., Lim, J.W.: Prüfer v-multiplication domains and related domains of the form $D + D_S[\Gamma^*]$. J. Algebra **323**, 3124–3133 (2010)
26. Chang, G.W., Kim, H., Oh, D.Y.: Kaplansky-type theorems in graded integral domains. Bull. Korean Math. Soc. **52**, 1253–1268 (2015)
27. Chang, G.W., Sahandi, P.: Graded integral domains which are UMT-domains. Comm. Algebra **46**, 2742–2752 (2018)
28. Chang, G.W., Zafrullah, M.: The w-integral closure of integral domains. J. Algebra **295**, 195–210 (2006)
29. Dobbs, D., Houston, E., Lucas, T., Zafrullah, M.: t-linked overrings and PVMDs. Comm. Algebra **17**, 2835–2852 (1989)
30. Fontana, M., Gabelli, S., Houston, E.: UMT-domains and domains with Prüfer integral closure. Comm. Algebra **25**, 1017–1039 (1998)
31. Fontana, M., Huckaba, J., Papick, I.: Prüfer Domains. Marcel Dekker, New York (1997)
32. Fontana, M., Jara, P., Santos, E.: Prüfer $*$-multiplication domains and semistar operations. J. Algebra Appl. **2**, 1–30 (2003)
33. Fontana, M., Loper, K.A.: Kronecker function rings: a general approach. In: Ideal Theoretic Methods in Commutative Algebra, Lecture Notes in Pure and Applied Mathematics vol. 220, pp. 189-205. Marcel Dekker (2001)

34. Fontana, M., Loper, K.A.: A historical overview of Kronecker function rings, Nagata rings, and related star and semistar operations. In: Brewer, J.W., Glaz, S., Heinzer, W.J., Olberding, B.M. (eds.) Multiplicative Ideal Theory in Commutative Algebra, pp. 169–187. Springer, A Tribute to the Work of Robert Gilmer (2006)
35. Fossum, R.: The Divisor Class Group of a Krull Domain. Springer, New York/Berlin (1973)
36. Gabelli, S.: On divisorial ideals in polynomial rings over Mori domains. Comm. Algebra **15**, 2349–2370 (1987)
37. Gabelli, S., Roitman, M.: Maximal divisorial ideals and t-maximal ideals. JPJ. Algebra Number Theory Appl. **4**, 323–336 (2004)
38. Gilmer, R.: An embedding theorem for HCF-rings. Proc. Cambridge Philos. Soc. **68**, 583–587 (1970)
39. Gilmer, R.: Multiplicative Ideal Theory. Marcel Dekker, New York (1972)
40. Gilmer, R.: Commutative Semigroup Rings. Univ. of Chicago Press, Chicago (1984)
41. Gilmer, R., Mott, J., Zafrullah, M.: t-invertibility and comparability. In: Commutative Ring Theory. Lecture Notes in Pure and Applied Mathematics, vol. 153, pp. 141–150. Marcel Dekker, New York (1994)
42. Griffin, M.: Some results on v-multiplication rings. Canad. J. Math. **19**, 710–722 (1967)
43. Hamdi, H., Sahandi, P.: Uppers to zero in polynomial rings over graded domains and UMt-domains. Bull. Korean Math. Soc. **55**, 187–204 (2018)
44. Houston, E.: Uppers to zero in polynomial rings. In: Brewer, J.W., Glaz, S., Heinzer, W.J., Olberding, B.M. (eds.) Multiplicative Ideal Theory in Commutative Algebra, pp. 243–261. Springer, A Tribute to the Work of Robert Gilmer (2006)
45. Houston, E., Malik, S., Mott, J.: Characterizations of $*$-multiplication domains. Canad. Math. Bull. **27**, 48–52 (1984)
46. Houston, E., Zafrullah, M.: On t-invertibility II. Comm. Algebra **17**, 1955–1969 (1989)
47. Houston, E., Zafrullah, M.: UMV-domains. In: Arithmetical Properties of Commutative Rings and Monoids, Lecture Notes in Pure and Applied Mathematics, vol. 241, pp. 304-315. Chapman & Hall/CRC, New York (2005)
48. Kang, B.G.: Prüfer v-multiplication domains and the ring $R[X]_{N_v}$. J. Algebra **123**, 151–170 (1989)
49. Kang, B.G.: On the converse of a well-known fact about Krull domains. J. Algebra **124**, 284–299 (1989)
50. Kaplansky, I.: Commutative Rings, revised edn. University of Chicago, Chicago (1974)
51. Lewin, R.L.: Almost generalized GCD-domains, In: Andeson, D.D. (eds.) Factorization in Integral Domains. Lecture Notes in Pure and Applied Mathematics, vol. 189, New York, Basel, pp. 371–382. Marcel Dekker, Hong Kong (1997)
52. Li, Q.: On almost Prüfer v-multiplication domains. Algebra Colloq. **19**, 493–500 (2012)
53. Năstăsescu, C., Van Oystaeyen, F.: Graded Ring Theory, Library of Math, vol. 28. North Holland, Amsterdam (1982)
54. Northcott, D.G.: Lessons on Rings, Modules, and Multiplicities. Cambridge University Press, Cambridge (1968)
55. Seidenberg, A.: A note on the dimension theory of rings. Pacific J. Math. **3**, 505–512 (1953)
56. Storch, U.: Fastfaktorielle ringe, Schriftenreihe Math. Inst. Univ. Münster, Heft 36 (1967)
57. Tang, H.T.: Gauss' lemma. Proc. Amer. Math. Soc. **35**, 372–376 (1972)
58. Wang, F.: On induced operations and UMT-domains. Sichuan Shifan Daxue Xuebao Ziran Kexue Ban **27**, 1–9 (2004)
59. Zafrullah, M.: A general theory of almost factoriality. Manuscripta Math. **51**, 29–62 (1985)
60. Zafrullah, M.: Putting t-invertibility to use, Non-Noetherian commutative ring theory, vol. 520, pp. 429–457, Math. Appl., Kluwer Academic Publishers, Dordrecht (2000)

The Apéry Set of a Good Semigroup

Marco D'Anna, Lorenzo Guerrieri, and Vincenzo Micale

Abstract We study the Apéry Set of good subsemigroups of $\mathbb{N}^2$, a class of semigroups containing the value semigroups of curve singularities with two branches. Even if this set in infinite, we show that, for the Apéry Set of such semigroups, we can define a partition in "levels" that allows to generalize many properties of the Apéry Set of numerical semigroups, i.e., value semigroups of one-branch singularities.

Keywords Value semigroups · Algebroid curves · Gorenstein rings · Symmetric semigroups · Apéry Set

MSC 13A18 · 14H99 · 13H99 · 20M25

1 Introduction

The concept of *good semigroup* was formally defined in [1] in order to study value semigroups of Noetherian analytically unramified one-dimensional semilocal reduced rings, e.g., the local rings arising from curve singularities (and from their blowups), possibly with more than one branch; the properties of these semigroups were already considered in [3, 5, 6, 9, 13–15], but it was in [1] that their structure was systematically studied. Similarly to the one-branch case, when the value semigroup is a numerical semigroup, the properties of the rings can be translated and studied at semigroup level. For example, the celebrated result by Kunz (see [18]) that a one-dimensional analytically irreducible local domain is Gorenstein if and only if

M. D'Anna (✉) · L. Guerrieri · V. Micale
Università di Catania, Viale Andrea Doria 6, 95125 Catania, Italy
e-mail: mdanna@dmi.unict.it

L. Guerrieri
e-mail: guelor@guelan.com

V. Micale
e-mail: vmicale@dmi.unict.it

A. Facchini et al. (eds.), *Advances in Rings, Modules and Factorizations*,
Springer Proceedings in Mathematics & Statistics 321,
https://doi.org/10.1007/978-3-030-43416-8_4

its value semigroup is symmetric can be generalized to analytically unramified rings (see [14] and also [6]); in the same way, the numerical characterization of the canonical module in the analytically irreducible case (see [16]) can be given also in the more general case (see [9]).

However good semigroups present some problems that make difficult their study; first of all they are not finitely generated as monoid (even if they can be completely determined by a finite set of elements (see [7, 10, 15])) and they are not closed under finite intersections. Secondly, the behavior of the good ideals of good semigroups (e.g., the ideals arising as values of ideals of the corresponding ring) is not good at all, in the sense that the class of good ideals is not closed under sums and differences (see, e.g., [1, 17]).

Hence, unlike what happens for numerical semigroups (in analogy to analytically irreducible domains), it is not clear how to define the concept of complete intersection good semigroups and also the concepts of embedding dimension and type for these semigroups.

Moreover, in the same paper [1], it is shown that the class of good semigroups is larger than the class of value semigroups and, at the moment, no characterization of value good semigroups is known (while, for the numerical semigroup case, it is easily seen that any such semigroup is the value semigroup of the ring of the corresponding monomial curve). This means that to prove a property for good semigroups it is not possible to take advantage of the nature of value semigroups and it is necessary to work only at semigroup level.

Despite this bad facts, there is a concept quite natural to define that seems very promising in order to study good semigroups and to translate at numerical level other ring concepts: the *Apéry Set*. In general, given any monoid S and any element $s \in S$, the $\mathbf{Ap}(S, s)$ is defined as the set $\{t \in S : t - s \notin S\}$ (where the $-$ is taken in the group generated by S). For studying numerical semigroups, this concept reveals to be very useful and it is also a bridge between semigroup and ring properties, since many important ring properties are stable under quotients with respect to principal ideals generated by a nonzero divisor (x) and the values of the nonzero elements in $R/(x)$ are exactly the elements of $\mathbf{Ap}(S, v(x))$. This strategy was used, e.g., in [4, 8, 11, 12] taking (x) to be a minimal reduction of the maximal ideal that, in this situation, corresponds to an element of minimal nonzero value.

For good semigroups, the notion of $\mathbf{Ap}(S, s)$ was used in [2], in order to obtain an algorithmic characterization of those good semigroups that are value semigroups of a plane singularity with two branches. In that paper, using deeply the structure of the rings associated to plane singularities, it is proved that $\mathbf{Ap}(S, \boldsymbol{e})$ (where $\boldsymbol{e} = (e_1, e_2)$ is the minimal nonzero element in $S \subset \mathbb{N}^2$) can be divided in $e = e_1 + e_2$ subsets, where the integer e corresponds to the multiplicity of the ring.

In this paper we want to investigate the Apéry Set of a good semigroup. Again the problem is that we have to deal with an infinite set; so, first of all, we want to understand if there is a natural partition of it in e subsets, where e is the sum of the components of the minimal nonzero element of S and, in case S is a value semigroup, it represents also the multiplicity of the corresponding ring; to answer to this question we decided to restrict to the good subsemigroups of $\mathbb{N}^2$, otherwise the technicalities would increase too much. After finding a possible partition $\mathbf{Ap}(S) = \bigcup_{i=1}^{e} A_i$, we prove that, if S is the value semigroup of a ring $(R, \mathfrak{m}, k)$, it is possible to choose e elements $\boldsymbol{\alpha}_i$ in the Apéry Set, one for each A_i, so that, taking any element $f_i \in R$ of valuation $v(f_i) = \boldsymbol{\alpha}_i$, the classes $\bar{f_i}$ are a basis of the e-dimensional k-vector space $R/(x)$ (where x is a minimal reduction of $\mathfrak{m}$). This fact make us confident that the definition of the partition is the one we where looking for.

At this point it is natural to investigate if it is possible to generalize the well-known characterization of symmetric numerical semigroups given via their Apéry Set. It turns out that also good symmetric semigroups have $\mathbf{Ap}(S, \boldsymbol{e})$ whose partition satisfies a duality property similar to the duality that holds for the numerical case.

The structure of the paper is the following: after recalling in Section 2 all the preliminary definitions and results needed for the rest of the paper, in Section 3 we define the partition of the Apéry Set of S and we prove that this partition produces exactly $e_1 + e_2$ levels (Theorem 3); successively we deepen the study of the structure of Apéry Set (Theorem 5) and we prove that, in the case of value semigroups, the partition allows to find a basis for $R/(x)$ as explained above (Theorem 6).

In Section 4 we study the properties of the Apéry Set of symmetric good semigroups with particular attention to duality properties of its elements (Proposition 7 and Theorem 8) and in Section 5 we use these results to prove a duality for the levels, characterizing the symmetric semigroups, in analogy to the duality of Apéry Set in the numerical semigroup case (Theorem 9). Finally we deepen this duality showing that we can find sequences of elements, one for each level, that have the same duality properties of the Apéry Set of the numerical semigroups (Theorem 10).

2 Preliminaries

Let $\mathbb{N}$ be the set of nonnegative integers. As usual, $\leq$ stands for the natural partial ordering in $\mathbb{N}^2$: set $\boldsymbol{\alpha} = (\alpha_1, \alpha_2)$, $\boldsymbol{\beta} = (\beta_1, \beta_2)$, then $\boldsymbol{\alpha} \leq \boldsymbol{\beta}$ if $\alpha_1 \leq \beta_1$ and $\alpha_2 \leq \beta_2$. Trough this paper, if not differently specified, when referring to minimal or maximal elements of a subset of $\mathbb{N}^2$, we refer to minimal or maximal elements with respect to $\leq$. Given $\boldsymbol{\alpha}, \boldsymbol{\beta} \in \mathbb{N}^2$, the infimum of the set $\{\boldsymbol{\alpha}, \boldsymbol{\beta}\}$ (with respect to $\leq$) will be denoted by $\boldsymbol{\alpha} \wedge \boldsymbol{\beta}$. Hence

$$\boldsymbol{\alpha} \wedge \boldsymbol{\beta} = (\min(\alpha_1, \beta_1), \min(\alpha_2, \beta_2)).$$

Let S be a submonoid of $(\mathbb{N}^2, +)$. We say that S is a *good semigroup* if

(G1) for all $\boldsymbol{\alpha}, \boldsymbol{\beta} \in S$, $\boldsymbol{\alpha} \wedge \boldsymbol{\beta} \in S$;
(G2) if $\boldsymbol{\alpha}, \boldsymbol{\beta} \in S$ and $\alpha_i = \beta_i$ for some $i \in \{1, 2\}$, then there exists $\boldsymbol{\delta} \in S$ such that $\delta_i > \alpha_i = \beta_i$, $\delta_j \geq \min\{\alpha_j, \beta_j\}$ for $j \in \{1, 2\} \setminus \{i\}$ and $\delta_j = \min\{\alpha_j, \beta_j\}$ if $\alpha_j \neq \beta_j$;
(G3) there exists $\boldsymbol{c} \in S$ such that $\boldsymbol{c} + \mathbb{N}^2 \subseteq S$.

A good subsemigroup of $\mathbb{N}^2$ is said to be *local* if $\mathbf{0} = (0, 0)$ is its only element with a zero component. In the following we will work only with local good semigroups hence we will omit the word local.

Notice that, from condition (G1), if $\boldsymbol{c}$ and $\boldsymbol{d}$ fulfill (G3), then so does $\boldsymbol{c} \wedge \boldsymbol{d}$. So there exists a minimum $\boldsymbol{c} \in \mathbb{N}^2$ for which condition (G3) holds. Therefore we will say that

$$\boldsymbol{c} := \min\{\boldsymbol{\alpha} \in \mathbb{Z}^2 \mid \boldsymbol{\alpha} + \mathbb{N}^2 \subseteq S\}$$

is the *conductor* of S. We denote $\boldsymbol{\gamma} := \boldsymbol{c} - \mathbf{1}$.

In light of [1, Proposition 2.1], value semigroups of Noetherian, analytically unramified, residually rational, one-dimensional, reduced semilocal rings with two minimal primes are good subsemigroups of $\mathbb{N}^2$ and R is local if and only it its value semigroup is local; in the rest of this paper, we will always assume these hypotheses on the rings R unless differently stated.

We give the following technical definitions that are commonly used in the literature about good semigroups:

(1) $\Delta_i(\boldsymbol{\alpha}) := \{\boldsymbol{\beta} \in \mathbb{Z}^2 \mid \alpha_i = \beta_i \text{ and } \alpha_j < \beta_j \text{ for } j \neq i\}$,
(2) $\Delta_i^S(\boldsymbol{\alpha}) := \Delta_i(\boldsymbol{\alpha}) \cap S$,
(3) $\Delta(\boldsymbol{\alpha}) := \Delta_1(\boldsymbol{\alpha}) \cup \Delta_2(\boldsymbol{\alpha})$,
(4) $\Delta^S(\boldsymbol{\alpha}) := \Delta(\boldsymbol{\alpha}) \cap S$.

An element $\boldsymbol{\alpha} \in S$ is said to be *absolute* if $\Delta^S(\boldsymbol{\alpha}) = \emptyset$. By definition of conductor we immediately get $\Delta^S(\boldsymbol{\gamma}) = \emptyset$. Given $\boldsymbol{\alpha}, \boldsymbol{\beta} \in \mathbb{N}^2$, we say that $\boldsymbol{\beta}$ is *above* $\boldsymbol{\alpha}$ if $\boldsymbol{\beta} \in \Delta_1(\boldsymbol{\alpha})$ and that $\boldsymbol{\beta}$ is *on the right* of $\boldsymbol{\alpha}$ if $\boldsymbol{\beta} \in \Delta_2(\boldsymbol{\alpha})$.

Remark 1. Let $\boldsymbol{c} = (c_1, c_2)$ be the conductor of S. By properties (G1) and (G2), we have that if $\boldsymbol{\alpha} = (\alpha_1, c_2) \in S$, then $\Delta_1(\boldsymbol{\alpha}) = \Delta_1^S(\boldsymbol{\alpha})$ (that is, each point $\boldsymbol{\beta} \in \mathbb{N}^2$, above $\boldsymbol{\alpha}$, is in S). Similarly, if $\boldsymbol{\alpha} = (c_1, \alpha_2) \in S$, then $\Delta_2(\boldsymbol{\alpha}) = \Delta_2^S(\boldsymbol{\alpha})$ (that is, each point $\boldsymbol{\beta} \in \mathbb{N}^2$, on the right of $\boldsymbol{\alpha}$, is in S).

The Apéry Set of S with respect to $\boldsymbol{\beta} \in S$ is defined as the set

$$\mathbf{Ap}(S, \boldsymbol{\beta}) = \{\boldsymbol{\alpha} \in S | \boldsymbol{\alpha} - \boldsymbol{\beta} \notin S\}.$$

Property (G1) implies that for a local good semigroup there exists a smallest nonzero element that we will denote by $\boldsymbol{e} = (e_1, e_2)$. We will usually consider the Apéry Set of S with respect to $\boldsymbol{e}$ and, in this case, we will simply write $\mathbf{Ap}(S)$ (see Fig. 1).

Remark 2. By definitions of conductor of S and of $\mathbf{Ap}(S)$, we have

$$\{\boldsymbol{\alpha} \in \mathbb{N}^2 \mid \boldsymbol{\alpha} \geq \boldsymbol{\gamma} + \boldsymbol{e} + \mathbf{1}\} \cap \mathbf{Ap}(S) = \emptyset.$$

Let A be a subset of $\mathbb{N}^2$ and let $\boldsymbol{\alpha}, \boldsymbol{\beta} \in A$, we say that $\boldsymbol{\beta}$ is a consecutive point of $\boldsymbol{\alpha}$ in A if, for every $\boldsymbol{\mu} \in \mathbb{N}^2$ with $\boldsymbol{\alpha} < \boldsymbol{\mu} < \boldsymbol{\beta}$, we necessary have $\boldsymbol{\mu} \notin A$.

Remark 3. Let $\boldsymbol{\alpha}, \boldsymbol{\beta} \in \mathbf{Ap}(S)$. If $\boldsymbol{\beta}$ is a consecutive point of $\boldsymbol{\alpha}$ in S, then $\boldsymbol{\beta}$ is a consecutive point of $\boldsymbol{\alpha}$ in $\mathbf{Ap}(S)$. The converse it is not true.

Let $\boldsymbol{\alpha}, \boldsymbol{\beta} \in A$. The chain of points in A, $\{\boldsymbol{\alpha} = \boldsymbol{\alpha}_1 < \cdots < \boldsymbol{\alpha}_h < \cdots < \boldsymbol{\alpha}_n = \boldsymbol{\beta}\}$, with $\boldsymbol{\alpha}_{h+1}$ consecutive of $\boldsymbol{\alpha}_h$, is called *saturated* of length n. In this case $\boldsymbol{\alpha}$ and $\boldsymbol{\beta}$ are called the initial point and the final point, respectively, of the chain.

A *relative ideal* of a good semigroup S is a subset $\emptyset \neq E \subseteq \mathbb{Z}^2$ such that $E + S \subseteq E$ and $\boldsymbol{\alpha} + E \subseteq S$ for some $\boldsymbol{\alpha} \in S$. A relative ideal E contained in S is simply called an *ideal*. If E satisfies (G1) and (G2), then we say that E is a *good ideal* of S (condition (G3) follows from the definition of good relative ideal).

Proposition 1 ([9, 17]) *All the saturated chains in a good ideal E of S with fixed initial and final points have the same length.*

If $\boldsymbol{\alpha}, \boldsymbol{\beta} \in E$, we denote by $d_E(\boldsymbol{\alpha}, \boldsymbol{\beta})$ the common length of all the saturated chains in the good ideal E with initial point $\boldsymbol{\alpha}$ and final point $\boldsymbol{\beta}$. Moreover, if $E \supseteq F$ are two good ideals, considering $\boldsymbol{m}_E$ and $\boldsymbol{m}_F$ the minimal elements, respectively, of E and F and taking $\boldsymbol{\alpha} \geq \boldsymbol{c}_F$ where $\boldsymbol{c}_F$ is the conductor of F, it is possible to define the following distance function $d(E \setminus F) = d_E(\boldsymbol{m}_E, \boldsymbol{\alpha}) - d_F(\boldsymbol{m}_F, \boldsymbol{\alpha})$ (cf [9, 17] to see that it is a well defined distance). This distance funtion allows to translate many ring properties at semigroup level, since, if $I \supseteq J$ are two fractional ideals of R, it is proved, in the same papers, that the length $\lambda_R(I/J)$ equals $d_{v(R)}(v(I) \setminus v(J))$.

Remark 4. In the numerical semigroup case, that is $S = v(R)$ with $(R, \mathfrak{m})$ a one-dimensional, Noetherian, analitically irreducible, residually rational, local domain, it is well known that, if we denote by x a minimal reduction of $\mathfrak{m}$ (i.e., an element of minimal value e), then $\overline{y} \neq 0$ in $R/(x)$ if and only if $v(y) \in \mathbf{Ap}(S) = \{\omega_0, \dots, \omega_{e-1}\}$. Moreover, if $\omega_i = v(y_i)$, then $\{\overline{y}_0, \dots \overline{y}_{e-1}\}$ are linear independent in the $R/\mathfrak{m}$-vector space $R/(x)$ and so they form a basis for it.

The first part of this remark can be easily generalized as follows, while we will be able to generalize the second part using all the results of the next section (see Theorem 6).

Proposition 2 *Let $y \in R$; then $\overline{y} \neq 0$ in $R/(x)$ if and only if $v(y) \in \mathbf{Ap}(S)$.*

Proof. By definition $\overline{y} \neq 0$ in $R/(x)$ if and only if $y \notin (x)$ that is equivalent to say that $yx^{-1} \notin R$. Since $v(yx^{-1}) = v(y) - v(x)$, if $v(y) \in \mathbf{Ap}(S)$ we immediately get that $yx^{-1} \notin R$, i.e., $\overline{y} \neq 0$ in $R/(x)$. Conversely, assume that $v(y) \notin \mathbf{Ap}(S)$, i.e., $v(yx^{-1}) = v(y) - v(x) = v(r)$, for some $r \in R$. Since R and both its projections on the two minimal primes are residually rational, it follows that there exists an invertible u in R such that $v(yx^{-1} - ur) > v(r)$; moreover, we can choose u in order to increase the first or the second component, as we prefer. Hence, applying repeatedly this argument we obtain, after a finite number of steps, that $v(yx^{-1} - u'r') \geq c$, that implies $yx^{-1} - u'r' \in (R : \overline{R}) \subset R$; therefore $y \in (x)$, a contradiction. □

Remark 5. When S is the value semigroup of a ring $(R, \mathfrak{m})$, it is not difficult to see that an element x is a minimal reduction of $\mathfrak{m}$ if and only if $v(x) = \boldsymbol{e}$; hence, the integer $e = e_1 + e_2$ coincides with the multiplicity of R: in fact $e(R) = \lambda_R(R/(x)) = \dim_{R/\mathfrak{m}}(R/(x))$ (where λ_R denotes the length of an R module); now, using the computation of lengths explained above, it is not difficult to check that $\lambda_R(R/(x)) = d(S \setminus \boldsymbol{e} + S) = e_1 + e_2$.

It is useful to remark, at this point, that $S \setminus \boldsymbol{e} + S = \mathbf{Ap}(S)$. Hence, our first goal is, starting form a good semigroup $S \subseteq \mathbb{N}^2$, to get a partition $\mathbf{Ap}(S)$ in $e_1 + e_2 = e$ levels that should correspond to the e elements of the Apéry Set of a numerical semigroup. After that we would like to find e elements $\boldsymbol{\alpha}_1, \ldots, \boldsymbol{\alpha}_e$ of the Apéry set of S, one in each class of the partition, with the property that, if $S = v(R)$ and $f_i \in R$ are such that $v(f_i) = \boldsymbol{\alpha}_i$, then $\overline{f}_1, \ldots, \overline{f}_e$ are linear independent in the $R/\mathfrak{m}$-vector space $R/(x)$ and so they form a basis for it.

We notice again that, given a good semigroups, it is not known a procedure to see if it is a value semigroup of a ring or not; so we are forced to use semigroup arguments without the possible help of ring techniques.

As in the numerical semigroup case, a symmetric good semigroup has a duality property. Indeed in [14], a good semigroup S is said to be *symmetric* if

$$\boldsymbol{\alpha} \in S \Leftrightarrow \Delta^S(\gamma - \alpha) = \emptyset.$$

Moreover, in the numerical semigroup case, the symmetry of the semigroup S can be characterized by a symmetry of its the Apéry Set: if we order its elements in increasing order $\mathbf{Ap}(S) = \{w_1, \ldots w_e\}$, then S is symmetric if and only if $w_i + w_{e-i+1} = w_e$ for every $i = 1, \ldots, e$.

In this paper we also look for an analogue property for $\mathbf{Ap}(S)$ when S is a symmetric good subsemigroup of $\mathbb{N}^2$.

3 Apéry Set of Good Semigroups in $\mathbb{N}^2$

In order to get the partition of the Apéry Set of S we are looking for, we need to introduce a new relation on $\mathbb{N}^2$ (as it is done in [1]): we say that $(\alpha_1, \alpha_2) \leq\leq (\beta_1, \beta_2)$ if an only if $(\alpha_1, \alpha_2) = (\beta_1, \beta_2)$ or $(\alpha_1, \alpha_2) \neq (\beta_1, \beta_2)$ and $(\alpha_1, \alpha_2) \ll (\beta_1, \beta_2)$, where the last means $\alpha_1 < \beta_1$ and $\alpha_2 < \beta_2$.

We define the following subsets of $\mathbf{Ap}(S)$:

$$B^{(1)} = \{\boldsymbol{\alpha} \in \mathbf{Ap}(S) : \boldsymbol{\alpha} \text{ is maximal with respect to } \leq\leq\},$$

$$C^{(1)} := \{\boldsymbol{\alpha} \in B^{(1)} : \boldsymbol{\alpha} = \boldsymbol{\beta}_1 \wedge \boldsymbol{\beta}_2 \text{ for some } \boldsymbol{\beta}_1, \boldsymbol{\beta}_2 \in B^{(1)} \setminus \{\boldsymbol{\alpha}\}\} \text{ and } D^{(1)} = B^{(1)} \setminus C^{(1)}.$$

Assume $i > 1$ and that $D^{(1)}, \ldots, D^{(i-1)}$ have been defined; we set

$$B^{(i)} = \{\boldsymbol{\alpha} \in \mathbf{Ap}(S) \setminus (\bigcup_{j<i} D^{(j)}) : \boldsymbol{\alpha} \text{ is maximal with respect to } \leq\leq\},$$

$$C^{(i)} := \{\boldsymbol{\alpha} \in B^{(i)} : \boldsymbol{\alpha} = \boldsymbol{\beta}_1 \wedge \boldsymbol{\beta}_2 \text{ for some } \boldsymbol{\beta}_1, \boldsymbol{\beta}_2 \in B^{(i)} \setminus \{\boldsymbol{\alpha}\}\} \text{ and } D^{(i)} = B^{(i)} \setminus C^{(i)}.$$

Clearly, for some $N \in \mathbb{N}_+$, we have $\mathbf{Ap}(S) = \bigcup_{i=1}^{N} D^{(i)}$ and $D^{(i)} \cap D^{(j)} = \emptyset$, for any $i \neq j$. For simplicity we prefer to number the set of the partition in increasing order, so we set $A_i = D^{(N+1-i)}$. Hence

$$\mathbf{Ap}(S) = \bigcup_{i=1}^{N} A_i$$

We want to prove that $N = e_1 + e_2$. We will call the sets A_i *levels* of the Apéry Set (see Fig. 2).

Remark 6. It is straightforward to see that $A_N = \Delta(\boldsymbol{\gamma} + \boldsymbol{e}) = \Delta^S(\boldsymbol{\gamma} + \boldsymbol{e})$ and that $A_1 = \{\mathbf{0}\}$. Moreover, if $\boldsymbol{\alpha}, \boldsymbol{\beta} \in \mathbf{Ap}(S)$, $\boldsymbol{\alpha} \ll \boldsymbol{\beta}$ and $\boldsymbol{\alpha} \in A_i$, then $\boldsymbol{\beta} \in A_j$ for some $j > i$.

Lemma 1. *The sets A_i satisfy the following properties:*

(1) for any $\boldsymbol{\alpha} \in A_i$ there exists $\boldsymbol{\beta} \in A_{i+1}$ such that $\boldsymbol{\alpha} \ll \boldsymbol{\beta}$ or $\boldsymbol{\alpha} = \boldsymbol{\beta}_1 \wedge \boldsymbol{\beta}_2$ with $\boldsymbol{\beta}_1, \boldsymbol{\beta}_2 \in A_{i+1}$ (both cases can happen at the same time);
(2) for every $\boldsymbol{\alpha} \in A_i$ and $\boldsymbol{\beta} \in A_j$, with $j \geq i$, $\boldsymbol{\beta} \not\ll \boldsymbol{\alpha}$;
(3) if $\boldsymbol{\alpha} \in A_i$, $\boldsymbol{\beta} \in \mathbf{Ap}(S)$ and $\boldsymbol{\beta} \geq \boldsymbol{\alpha}$, then $\boldsymbol{\beta} \in A_i \cup \cdots \cup A_N$;
(4) if $\boldsymbol{\alpha} = (\alpha_1, \alpha_2)$, $\boldsymbol{\beta} = (\alpha_1, \beta_2) \in A_i$, with $\alpha_2 < \beta_2$, then for any $\boldsymbol{\delta} = (\delta_1, \delta_2) \in \mathbf{Ap}(S)$ such that $\delta_1 > \alpha_1$ and $\delta_2 \geq \alpha_2$, we get $\boldsymbol{\delta} \in A_{i+1} \cup \cdots \cup A_N$; an analogous statement holds switching the components;
(5) if $\boldsymbol{\alpha} \ll \boldsymbol{\beta} \in \mathbf{Ap}(S)$ and they are consecutive in S, then there exists $i > 0$ such that $\boldsymbol{\alpha} \in A_i$ and $\boldsymbol{\beta} \in A_{i+1}$; if $\boldsymbol{\alpha} < \boldsymbol{\beta} \in \mathbf{Ap}(S)$, they share a component and they are consecutive in S, then there exists $i > 0$ such that either $\boldsymbol{\alpha} \in A_i$ and $\boldsymbol{\beta} \in A_{i+1}$ or $\boldsymbol{\alpha}, \boldsymbol{\beta} \in A_i$;
(6) let $\boldsymbol{\alpha} \in A_i$ and let be $\boldsymbol{\beta}_1, \ldots, \boldsymbol{\beta}_j$ all the elements of $\mathbf{Ap}(S)$, $\boldsymbol{\alpha} < \boldsymbol{\beta}_r$ and consecutive to $\boldsymbol{\alpha}$ in $\mathbf{Ap}(S)$. Then at least one of them belongs to A_{i+1};
(7) $\boldsymbol{\alpha} = (\alpha_1, \gamma_2 + e_2 + 1) \in A_i \Leftrightarrow (\alpha_1, n) \in A_i$ for some $n \geq \gamma_2 + e_2 + 1 \Leftrightarrow (\alpha_1, n) \in A_i$ for all $n \geq \gamma_2 + e_2 + 1$; an analogous statement holds switching the components;
(8) if $\boldsymbol{\alpha} = (\alpha_1, \gamma_2 + e_2 + 1) \in A_i$ and $\boldsymbol{\beta} = (\beta_1, \gamma_2 + e_2 + 1) \in \mathbf{Ap}(S)$, with $\beta_1 < \alpha_1$ and such that for every a, $\beta_1 < a < \alpha_1$, the point $(a, \gamma_2 + e_2 + 1) \notin \mathbf{Ap}(S)$, then $\boldsymbol{\beta} \in A_{i-1}$; an analogous statement holds switching the components. (We could state this property saying that, definitively, consecutive vertical lines (respectively, horizontal lines) of the Apéry Set belong to consecutive levels.)

Proof. Properties (1), (2), and (3) follow immediately by definition.
(4) If $\delta_2 > \alpha_2$, then $\boldsymbol{\alpha} \ll \boldsymbol{\delta}$ and the assertion follows by definition of levels. If $\delta_2 = \alpha_2$, by property (3), $\boldsymbol{\delta} \in A_i \cup \cdots \cup A_N$; if $\boldsymbol{\delta} \in A_i$ we get a contradiction by the definition of levels, since $\boldsymbol{\alpha} = \boldsymbol{\beta} \wedge \boldsymbol{\delta}$.
(5) Let i be such that $\boldsymbol{\alpha} \in A_i$. If $\boldsymbol{\alpha} = (\alpha_1, \alpha_2) \ll \boldsymbol{\beta} = (\beta_1, \beta_2)$ and they are consecutive in S, then there are no elements (a, b) of S such that $\alpha_1 \leq a < \beta_1$ and $\alpha_2 < b$ or $\alpha_1 < a$ and $\alpha_2 \leq b < \beta_2$, since $(a, b) \wedge \boldsymbol{\beta}$ would be a point of S between $\boldsymbol{\alpha}$ and $\boldsymbol{\beta}$; in particular $\boldsymbol{\alpha}$ cannot be obtained as infimum of points of S. Moreover $\boldsymbol{\beta} \in A_j$,

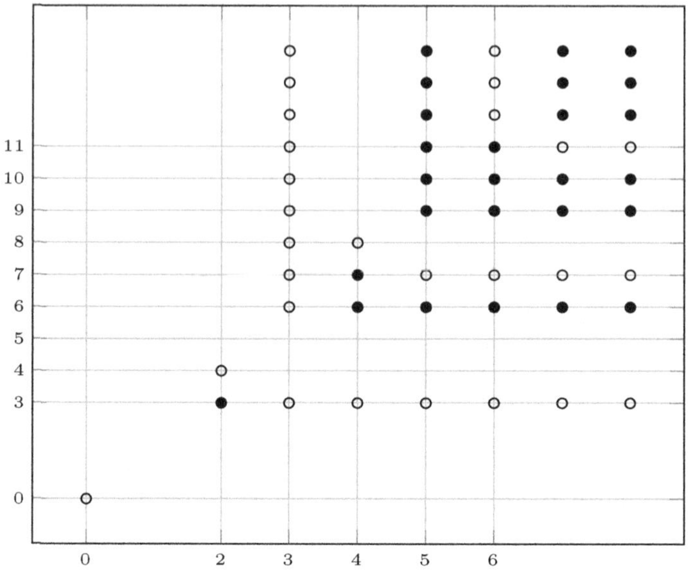

Fig. 1 The value semigroup of $k[[X, Y, Z]]/(X^3 - Z^2) \cap (X^3 - Y^4)$. The elements of the Apéry Set are indicated with $\circ$

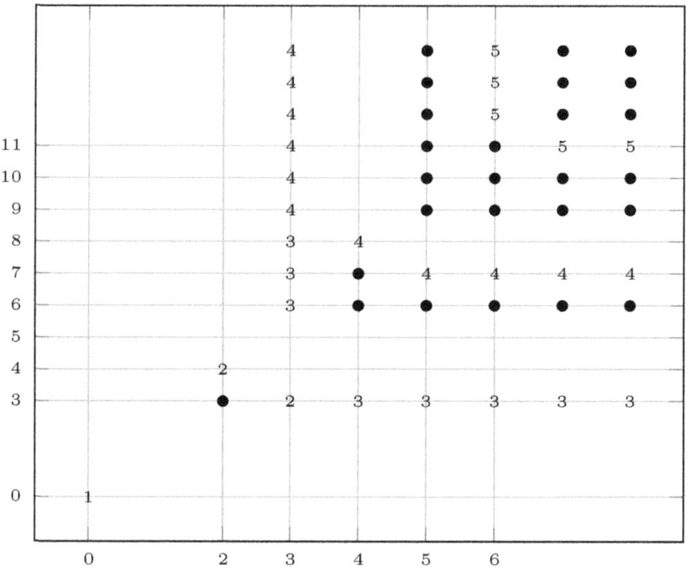

Fig. 2 The Apéry Set of the semigroup in Fig. 1. We mark the elements of the set A_i with the number i.

with $j > i$. By property (3) all the points of $\mathbf{Ap}(S)$ bigger than or equal to $\boldsymbol{\beta}$ belong to A_h, with $h \geq j$; hence, $\boldsymbol{\alpha}$ is maximal in $\mathbf{Ap}(S) \setminus (\bigcup_{h \geq j} A_h)$ with respect to $\leq\leq$ and, by definition of levels, this implies that $j = i + 1$.

Assume now that $\boldsymbol{\alpha}$ and $\boldsymbol{\beta}$ share a component (e.g., $\boldsymbol{\alpha} = (\alpha_1, \alpha_2)$, $\boldsymbol{\beta} = (\alpha_1, \beta_2)$). Since they are consecutive in S, there are no elements (a, b) of S such that $\alpha_1 \leq a$ and $\alpha_2 < b < \beta_2$. Let i be such that $\boldsymbol{\alpha} \in A_i$; by property (3) $\boldsymbol{\beta} \in A_j$ with $j \geq i$ and all the points of $\mathbf{Ap}(S)$ bigger than or equal to $\boldsymbol{\beta}$ belong to A_h, with $h \geq j$; by definition of levels, $\boldsymbol{\beta}$ is maximal in $\mathbf{Ap}(S) \setminus (\bigcup_{h > j} A_h)$ with respect to $\leq\leq$ and therefore also $\boldsymbol{\alpha}$ is maximal in the same set; hence either $j = i$ or $j = i + 1$.

(6) Assume by way of contradiction that for all r, $\boldsymbol{\beta}_r \in A_{h_r}$ with $h_r > i + 1$, set $h = \min\{h_r : r = 1, \ldots, j\}$. Hence $\boldsymbol{\alpha}$ is maximal in $\mathbf{Ap}(S) \setminus (\bigcup_{s \geq h} A_s)$ with respect to $\leq\leq$. In order to have $\boldsymbol{\alpha} \in A_i$, either $h = i + 1$ (that is the thesis) or $h = i + 2$ and $\boldsymbol{\alpha}$ is obtained as infimum of two elements $\boldsymbol{\delta}_1, \boldsymbol{\delta}_2 \in A_{i+1}$. But also in the second case, both $\boldsymbol{\delta}_i$ have to be consecutive to $\boldsymbol{\alpha}$ (otherwise they would be bigger than some elements $\boldsymbol{\beta}_r$, so they could not belong to A_{i+1}); hence, we get a contradiction by the assumption that $h = i + 2$. If, on the other hand, there exists $\boldsymbol{\beta}_r \in A_i$, it necessarily shares a component with $\boldsymbol{\alpha}$. Hence, by property (4) all the other $\boldsymbol{\beta}_s$ belong to $A_{i+1} \cup \cdots \cup A_N$. Now, applying the same argument as above, one of them has to be in A_{i+1}.

(7) Any element $\boldsymbol{\alpha} = (\alpha_1, \alpha_2)$ with $\alpha_2 > \gamma_2$ (respectively, $\alpha_1 > \gamma_1$) belongs to S if and only if $(\alpha_1, \gamma_2 + 1) \in S$ (respectively, $((\gamma_1 + 1, \alpha_2) \in S)$. Hence it is clear that $\boldsymbol{\alpha} = (\alpha_1, \gamma_2 + e_2 + 1) \in \mathbf{Ap}(S)$ if and only if $(\alpha_1, n) \in \mathbf{Ap}(S)$ for some $n \geq \gamma_2 + e_2 + 1$ if and only if $(\alpha_1, n) \in \mathbf{Ap}(S)$ for all (α_1, n) (and the analogous statement holds switching the components).

So we have only to prove that these elements belongs to the same level (we will prove this fact for vertical lines and the corresponding statement for horizontal line is analogous). If not, by property (5), there exist two elements $\boldsymbol{\alpha}_1 = (\alpha_1, \alpha_2) \in A_i$ and $\boldsymbol{\alpha}_2 = (\alpha_1, \alpha_2 + 1) \in A_{i+1}$ consecutive in $\mathbf{Ap}(S)$. Now let $\beta_1 > \alpha_1$ be the smallest integer such that $\boldsymbol{\beta} = (\beta_1, \alpha_2) \in \mathbf{Ap}(S)$; since $\boldsymbol{\alpha}_1 = \boldsymbol{\alpha}_2 \wedge \boldsymbol{\beta}$ and there are no elements (a, α_2) of $\mathbf{Ap}(S)$, with $\alpha_1 < a < \beta_1$, it is clear that either $\boldsymbol{\beta} \in A_i$ or $\boldsymbol{\beta} \in A_{i+1}$; moreover $\boldsymbol{\alpha}_2 \ll (\beta_1, \alpha_2 + 2)$, hence, the last belongs to A_j with $j > i + 1$. Hence, also in the vertical line corresponding to β_1 there are elements on different levels. Iterating the argument we get that the same happens for $\Delta_1(\boldsymbol{\gamma} + \boldsymbol{e}) \subseteq A_N$; a contradiction.

(8) By property (6), alle the elements of S above $\boldsymbol{\alpha}$ are in A_i. Hence $\boldsymbol{\beta} \ll (\alpha_1, \gamma_2 + e_2 + 2)$ and therefore it belongs to A_j with $j < i$. Moreover the hypothesis implies that $\boldsymbol{\beta}$ is maximal in $\mathbf{Ap}(S) \setminus (\bigcup_{h \geq i} A_h)$ with respect to $\leq\leq$ and cannot be obtained as infimum of two other elements maximal in the same set. The thesis follows immediately. □

Next lemma describes global properties of the elements of a good semigroup S and of its Apéry Set.

Lemma 2. *The following assertions hold:*

(1) Let $\boldsymbol{\alpha} \in \mathbb{N}^2$ and assume there is a finite positive number of elements in $\Delta_1^S(\boldsymbol{\alpha}) \cap (\boldsymbol{e} + S)$. Call $\boldsymbol{\delta}$ the maximum of them. Hence $\Delta^S(\boldsymbol{\delta}) \subseteq \mathbf{Ap}(S)$;

(2) *Let* $\boldsymbol{\alpha} \in \mathbf{Ap}(S)$. *If there exists* $\boldsymbol{\beta} \in (\boldsymbol{e} + S) \cap \Delta_1(\boldsymbol{\alpha})$, *then* $\Delta_2^S(\boldsymbol{\alpha}) \subseteq \mathbf{Ap}(S)$*;*
(3) *Let* $\boldsymbol{\alpha} = (a_1, a_2) \in A_i$ *and suppose there exists* $b_2 < a_2$ *such that* $\boldsymbol{\delta} = (a_1, b_2) \in S$ *and* $\Delta_2^S(\boldsymbol{\delta}) \subseteq \mathbf{Ap}(S)$. *Then the minimal element* $\boldsymbol{\beta} = (b_1, b_2)$ *of* $\Delta_2^S(\boldsymbol{\delta})$ *is in* A_j *for some* $j \leq i$. *In particular, if* $\Delta^S(\boldsymbol{\delta}) \subseteq \mathbf{Ap}(S)$ *and* $\boldsymbol{\alpha}$ *is the minimal element of* $\Delta_1^S(\boldsymbol{\delta})$, $\boldsymbol{\beta} \in A_i$.
(4) *Let* $\boldsymbol{\alpha} = (a_1, a_2) \in A_i$ *and suppose there exists* $\boldsymbol{\delta} \in (\boldsymbol{e} + S) \cap \Delta_1(\boldsymbol{\alpha})$. *Then* $\Delta_2^S(\boldsymbol{\alpha}) \subseteq \mathbf{Ap}(S)$ *and the minimal element* $\boldsymbol{\beta}=(b_1, a_2)$ *of* $\Delta_2^S(\boldsymbol{\alpha})$ *is also in* A_i.
(5) *Let* $\boldsymbol{\alpha} \in A_i$ *and assume* $\Delta_1^S(\boldsymbol{\alpha}) \subseteq \mathbf{Ap}(S)$. *Assume also that there exists* $\boldsymbol{\beta} \in \Delta_1^S(\boldsymbol{\alpha}) \cap A_{i+1}$. *Then there exists* $\boldsymbol{\theta} \in (\Delta_1^S(\boldsymbol{\alpha}) \cap A_i) \cup \{\boldsymbol{\alpha}\}$ *such that* $\boldsymbol{\theta} < \boldsymbol{\beta}$ *and* $\Delta^S(\boldsymbol{\theta}) \subseteq \mathbf{Ap}(S)$.
The analogous assertions hold switching the components.

Proof. (1) Since $\boldsymbol{\delta}$ is the maximum of $\Delta_1^S(\boldsymbol{\alpha}) \cap (\boldsymbol{e} + S)$, we have $\Delta_1^S(\boldsymbol{\delta}) \subseteq \mathbf{Ap}(S)$. Now, since $(\boldsymbol{e} + S)$ is a good ideal of S, by property (G2), also $\Delta_2^S(\boldsymbol{\delta}) \subseteq \mathbf{Ap}(S)$.
(2) Assume that there exists $\boldsymbol{\delta} \in \Delta_2(\boldsymbol{\alpha}) \cap (\boldsymbol{e} + S)$. Then again, since $(\boldsymbol{e} + S)$ is a good ideal, by property (G1), $\boldsymbol{\alpha} = \boldsymbol{\beta} \wedge \boldsymbol{\delta} \in (\boldsymbol{e} + S)$ and this is a contradiction.
(3) First we notice that $\Delta_2^S(\boldsymbol{\delta})$ is non-empty since also $\Delta_1^S(\boldsymbol{\delta})$ is non-empty (by axiom (G2)). Now, if $i = N$, the thesis easily follows by Remark 6. For $i < N$, we use the following argument: by definition of $\mathbf{Ap}(S)$ we can always find an element $\boldsymbol{\theta} = (g_1, g_2) \in A_{i+1}$ with $g_1 > a_1$ and $g_2 \geq a_2$. Hence, the fact that $\boldsymbol{\beta}$ is the minimal element of $\Delta_2^S(\boldsymbol{\delta})$ implies that $g_1 \geq b_1$ and this implies $\boldsymbol{\beta} \in A_j$ for $j \leq i + 1$.

If we assume by way of contradiction $\boldsymbol{\beta} \in A_{i+1}$, we would have $g_1 = b_1$ and hence, by axiom (G2), there exists an element $\boldsymbol{\omega} = (h_1, b_2)$ with $h_1 > b_1$. Since $\Delta_2^S(\boldsymbol{\delta}) \subseteq \mathbf{Ap}(S)$, we have $\boldsymbol{\omega} \in \mathbf{Ap}(S)$ and we may assume $\boldsymbol{\omega}$ minimal in $\Delta_2^S(\boldsymbol{\beta})$. Thus, if $\boldsymbol{\omega} \in A_{i+1}$ we have $\boldsymbol{\beta} = \boldsymbol{\theta} \wedge \boldsymbol{\omega} \in A_i$, otherwise we should have $\boldsymbol{\omega} \in A_j$ for some $j > i + 1$. But now we are in the same situation of the hypothesis of the lemma with $\boldsymbol{\theta}, \boldsymbol{\beta}, \boldsymbol{\omega}$ playing the role of $\boldsymbol{\alpha}, \boldsymbol{\delta}, \boldsymbol{\beta}$. In this way, iterating the process, we can create an infinite sequence of elements $\boldsymbol{\omega}^k \in \Delta_2(\boldsymbol{\delta}) \cap A_{i+k}$ and this is a contradiction because the levels of $\mathbf{Ap}(S)$ are in a finite number. The last sentence of the statement follows since, having also $\Delta_1^S(\boldsymbol{\delta}) \subseteq \mathbf{Ap}(S)$, we apply the same result and get the level of $\boldsymbol{\alpha}$ less or equal than the level of $\boldsymbol{\beta}$.
(4) If $i = N$, the thesis is clear by Remark 6. For $i < N$, clearly $\Delta_2^S(\boldsymbol{\alpha})$ is non-empty and it is contained in $\mathbf{Ap}(S)$ by (2). First assume that there exists $\boldsymbol{\theta} = (g_1, g_2) \in A_{i+1}$ such that $\boldsymbol{\theta} \gg \boldsymbol{\alpha}$. Since $\boldsymbol{\beta}$ is the minimal element of $\Delta_2^S(\boldsymbol{\alpha})$, we have $g_1 \geq b_1$. But now, if $g_1 > b_1$, then $\boldsymbol{\theta} \gg \boldsymbol{\beta}$ and hence $\boldsymbol{\beta} \in A_i$. Instead if $g_1 = b_1$ we can find a minimal element $\boldsymbol{\omega} \in \Delta_2(\boldsymbol{\beta}) \cap \mathbf{Ap}(S)$ and $\boldsymbol{\beta} = \boldsymbol{\theta} \wedge \boldsymbol{\omega}$. By (3), $\boldsymbol{\omega} \in A_j$ with $j \leq i + 1$ and thus $\boldsymbol{\beta} \in A_i$.

Now assume that there is no element of A_{i+1} dominating $\boldsymbol{\alpha}$. It follows that $\boldsymbol{\alpha} = \boldsymbol{\theta} \wedge \boldsymbol{\beta}$ with $\boldsymbol{\theta}, \boldsymbol{\beta} \in A_{i+1}$. We can assume by way of contradiction both of these elements to be minimal, otherwise we would have the minimal $\boldsymbol{\beta} \in A_i$. If $\boldsymbol{\theta} \in \Delta_1(\boldsymbol{\alpha})$ is still such that $\boldsymbol{\theta} < \boldsymbol{\delta}$, we have $\Delta_2^S(\boldsymbol{\theta}) \subseteq \mathbf{Ap}(S)$ and we can find in it a minimal element $\boldsymbol{\omega} \in \Delta_2^S(\boldsymbol{\theta})$ which has to be in A_{i+2} by Lemma 1(5) (otherwise we would have an element of A_{i+1} dominating $\boldsymbol{\alpha}$).

Suppose there exists $\omega^1 \in A_{i+2}$ such that $\omega^1 \gg \theta$. Clearly we cannot have neither $\omega^1 \gg \omega$ nor $\omega^1 \in \Delta_1(\omega)$ because this would contradict (3). Hence there exists $\omega^1 \wedge \omega \in \Delta_2^S(\theta)$ and this is a contradiction since ω is minimal in $\Delta_2^S(\theta)$.

Hence, we can assume that there are not elements of A_{i+2} dominating θ. This means that θ is the minimum of two elements $\omega^1, \eta^2 \in A_{i+2}$ and we can iterate the process replacing the elements θ, α, β with ω^1, θ, η^2. If $\Delta_1(\alpha)$ is eventually contained in $e + S$, after a finite number of iteration we will find an element ω^k which is maximal in $\Delta_1(\alpha) \cap \mathbf{Ap}(S)$ and it is dominated by some element of a greater level of $\mathbf{Ap}(S)$ (notice that in this case the elements in $\Delta_1(\alpha)$ are dominated by elements in A_N). This would lead to a contradiction like in the previous paragraph. If $\Delta_1(\alpha) \cap (e + S)$ has a maximum δ, then $\Delta^S(\delta) \subseteq \mathbf{Ap}(S)$ by (1) and our iterative process will end, by replacing the name of the elements and of the levels, in a situation with $\alpha = \theta \wedge \beta$ with $\theta, \beta \in A_{i+1}$ and $\theta > \delta > \alpha$. In this setting, we can find a minimal element $\omega \in \Delta_2(\delta) \cap \mathbf{Ap}(S)$. By (3), $\omega \in A_j$ with $j \leq i + 1$ and this again contradicts the assumption since $\omega \gg \alpha$.

(5) Assume $\Delta_2^S(\alpha) \nsubseteq \mathbf{Ap}(S)$. Hence by (4), we have that the minimal element ω of $\Delta_1^S(\alpha)$ is also in A_i. Hence $\omega < \beta$. If again $\Delta_2^S(\omega) \nsubseteq \mathbf{Ap}(S)$ we find an element $\omega^1 \in \Delta_1^S(\alpha) \cap A_i$ and $\omega^1 < \beta$. Since there is a finite number of elements in $\Delta_1^S(\alpha)$ between α and β, the process must stop to an element $\theta \in \Delta_1^S(\alpha) \cap A_i$ such that $\theta < \beta$ and $\Delta^S(\theta) \subseteq \mathbf{Ap}(S)$. □

Theorem 3. *Let $S \subseteq \mathbb{N}^2$ be a good semigroup and let $e = (e_1, e_2)$ be its minimal nonzero element. Let $\mathbf{Ap}(S) = \bigcup_{i=1}^N A_i$ where the sets A_i are defined as above. Then $N = e_1 + e_2$.*

Proof. We have that $\mathbf{Ap}(S) = S \setminus (e + S)$. Moreover both S and $e + S$ are good ideals so we can compute the distance function $d(S \setminus e + S)$ as $d_S(\mathbf{0}, \gamma + e + \mathbf{1}) - d_{e+S}(e, \gamma + e + \mathbf{1})$; on the other hand we know that $d(S \setminus e + S) = e_1 + e_2$.

Hence, to prove that $N \geq e_1 + e_2$ we show that there exists a saturated chain in S, between $\mathbf{0}$ and $\gamma + e + \mathbf{1}$ that contains exactly one element of every level A_i: if we do not consider the N elements of $\mathbf{Ap}(S)$ in this chain we get a chain (not necessarily saturated) in $e + S$; hence, $d_S(\mathbf{0}, \gamma + e + \mathbf{1}) - N \leq d_{e+S}(e, \gamma + e + \mathbf{1})$, that means

$$e_1 + e_2 = d_S(\mathbf{0}, \gamma + e + \mathbf{1}) - d_{e+S}(e, \gamma + e + \mathbf{1}) \leq N.$$

To construct such a chain we start from $\mathbf{0} \in S \cap \mathbf{Ap}(S)$ and then we choose N elements, one for each level A_i using property (6) of Lemma 1: given $\alpha_i \in A_i$ we choose $\alpha_{i+1} \in A_{i+1}$ consecutive to α_i in $\mathbf{Ap}(S)$; so we get a chain of N elements in $\mathbf{Ap}(S)$, each one consecutive to the previous in $\mathbf{Ap}(S)$. Hence when we saturate this chain in S, we add only elements in $S \setminus \mathbf{Ap}(S) = e + S$, and we obtain the desired chain.

In order to prove that $N \leq e_1 + e_2$ we want to construct a saturated chain in $e + S$ between e and $\gamma + e + \mathbf{1}$ such that, when we saturate it in S, as a chain between $\mathbf{0}$ and $\gamma + e + \mathbf{1}$, we use at least one element for every level A_i (it is clear that we can only add elements of $\mathbf{Ap}(S) = S \setminus (e + S)$: in fact, if we add $n \geq N$ elements in $\mathbf{Ap}(S)$, this would imply $d_S(\mathbf{0}, \boldsymbol{\gamma}+\boldsymbol{e}+\mathbf{1}) = d_{e+S}(\boldsymbol{e}, \boldsymbol{\gamma}+\boldsymbol{e}+\mathbf{1}) + n \geq d_{e+S}(\boldsymbol{e}, \boldsymbol{\gamma}+\boldsymbol{e}+\mathbf{1}) + N$, that is $N \leq d_S(\mathbf{0}, \boldsymbol{\gamma}+\boldsymbol{e}+\mathbf{1}) - d_{e+S}(\boldsymbol{e}, \boldsymbol{\gamma}+\boldsymbol{e}+\mathbf{1}) = e_1 + e_2$.

To construct such a chain we start with $\mathbf{0} \ll \boldsymbol{e}$, that is a saturated chain in S; hence, we can assume that we have constructed a saturated chain in S, say $\boldsymbol{\alpha}_0 = \mathbf{0} < \boldsymbol{\alpha}_1 < \cdots < \boldsymbol{\alpha}_h$, such that, $\boldsymbol{\alpha}_h \leq \boldsymbol{\gamma}+\boldsymbol{e}+\mathbf{1}$, if we delete the elements $\boldsymbol{\alpha}_i \in \mathbf{Ap}(S)$ we get a saturated chain in $\boldsymbol{e}+S$ and every level $A_1, \ldots, A_j$ has at least one representative in it. To apply a recursive argument we need either to stretch the chain adding one or more new elements, the first consecutive to $\boldsymbol{\alpha}_h$ in S and any of the others consecutive to the previous one, or to produce a new chain with the same properties (replacing the last elements of the constructed chain) for which all the levels $A_1, \ldots, A_{j+1}$ have at least one representative in it. The process will end, since the length of a saturated chain is bounded by $d_S(\mathbf{0}, \boldsymbol{\gamma}+\boldsymbol{e}+\mathbf{1})$ (and at the last step we have to touch A_N) and the number of levels is bounded by N. We explain now how to add a new element to the chain in all the different possible cases. Before starting, we observe that if we have $\boldsymbol{\gamma}+\boldsymbol{e}+1 \in \Delta^S(\boldsymbol{\alpha}_h)$ we complete the chain adding all the elements between $\boldsymbol{\alpha}_h$ and $\boldsymbol{\gamma}+\boldsymbol{e}+1$ sharing a coordinate with them (they are all on the same horizontal or on the same vertical line). By Lemma 1(7 and 8) this chain will touch exactly once all the levels of $\mathbf{Ap}(S)$ between $j+1$ and N.

Hence we consider now all the cases in which $\boldsymbol{\alpha}_h \ll \boldsymbol{\gamma}+\boldsymbol{e}+1$. We can start from one of the following two cases: either (case A) $\boldsymbol{\alpha}_h \in \boldsymbol{e}+S$ or (case B) $\boldsymbol{\alpha}_h \in A_j$ (notice that at the beginning of the chain we have $\mathbf{0} \ll \boldsymbol{e}$, so we start from case A).

In both cases if $\boldsymbol{\alpha}_h$ has only one consecutive element $\boldsymbol{\beta}$ in S, necessarily (by axiom (G2)) we have $\boldsymbol{\alpha}_h \ll \boldsymbol{\beta}$ and we should be forced to choose $\boldsymbol{\alpha}_{h+1} = \boldsymbol{\beta}$. If $\boldsymbol{\beta} \in \boldsymbol{e}+S$, the new chain obviously satisfies the requested properties; moreover it has one element more and now we are in case A. On the other hand, assume $\boldsymbol{\beta} \in \mathbf{Ap}(S)$; the condition $\boldsymbol{\alpha}_h \ll \boldsymbol{\beta}$ implies that $\boldsymbol{\alpha}_i \ll \boldsymbol{\beta}$, for all $i = 0, \ldots, h$; in particular, let $\boldsymbol{\alpha}_r \in A_j$ be the last element of the Apéry Set in the chain. Hence, by Remark 6 we know that $\boldsymbol{\beta} \in A_{j+1} \cup \cdots \cup A_N$; if it is in A_{j+1}, we simply add $\boldsymbol{\beta}$ to the chain and proceed in case B. Otherwise, if $\boldsymbol{\beta}$ it is not in A_{j+1}, by Lemma 1(6) there exists another element $\boldsymbol{\delta}$ consecutive in $\mathbf{Ap}(S)$ to $\boldsymbol{\alpha}_r$, such that $\boldsymbol{\delta} \in A_{j+1}$ (notice that, by Lemma 1(5), this situation can happen only in case A). Now $\boldsymbol{\delta}$ has to share a component with one of the elements $\boldsymbol{\alpha}_r, \ldots, \boldsymbol{\alpha}_h$ otherwise, taking infimums, it would create a new point that makes the original chain non-saturated; more precisely, if it is above (respectively, on the right) of some $\boldsymbol{\alpha}_l$ for $r \leq l \leq h$, then $\boldsymbol{\alpha}_{l+1}$ has to be either above or on the right of $\boldsymbol{\alpha}_l$. Now we change the chain substituting $\boldsymbol{\alpha}_m$ with $\boldsymbol{\delta} \wedge \boldsymbol{\alpha}_m$, for every $m \geq l+1$; successively we add to the chain $\boldsymbol{\delta} \wedge \boldsymbol{\beta}$ and all the other elements of S on the vertical (respectively, horizontal) line until we reach $\boldsymbol{\delta}$ (notice that we may have $\boldsymbol{\delta} \wedge \boldsymbol{\alpha}_m = \boldsymbol{\delta}$ for some m and in that case we simply stop to $\boldsymbol{\delta}$ because we have reached it). We show an example of the preceding process in the first picture of Figure 3.

Hence we created a new chain with the requested properties, such that every level $A_1, \ldots, A_{j+1}$ has at least one representative in it and now we are in case B.

It remains to study what happens if $\boldsymbol{\alpha}_h = \boldsymbol{\beta}_1 \wedge \boldsymbol{\beta}_2$ (where both $\boldsymbol{\beta}_i$ are consecutive to $\boldsymbol{\alpha}_h$ in S).

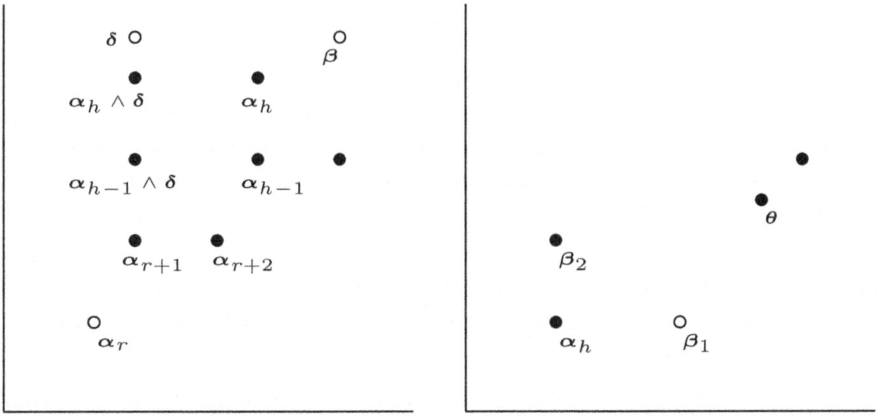

Fig. 3 Explanation of parts of the proof

• Case A. In this case let $\boldsymbol{\alpha}_r \in A_j$ be the last element of the Apéry Set in the chain. The following situations can occur:

(A.1) both $\boldsymbol{\beta}_i \in \mathbf{Ap}(S)$: if at least one $\boldsymbol{\beta}_i$ belongs to A_{j+1}, we set $\boldsymbol{\alpha}_{h+1} := \boldsymbol{\beta}_i$ and we switch to case B. Otherwise, since $\boldsymbol{\alpha}_r \ll \boldsymbol{\beta}_i$, for at least one $i = 1, 2$, we have, for the same i, $\boldsymbol{\beta}_i \in A_{j+2} \cup \cdots \cup A_N$. By Lemma 1(6), there exists another element $\boldsymbol{\delta}$ consecutive in $\mathbf{Ap}(S)$ to $\boldsymbol{\alpha}_r$, such that $\boldsymbol{\delta} \in A_{j+1}$. Using the same argument as above, i.e., replacing the last part of the chain with $\boldsymbol{\delta} \wedge \boldsymbol{\alpha}_m$ (for $r < m \leq h$) and then proceeding on a single line until reaching $\boldsymbol{\delta}$, we obtain the desired result and we switch to case B;
(A.2) both $\boldsymbol{\beta}_i \in \boldsymbol{e} + S$: we can move to any one of them indifferently and proceed in case A;
(A.3) $\boldsymbol{\beta}_1 \in \mathbf{Ap}(S)$ and $\boldsymbol{\beta}_2 \in \boldsymbol{e} + S$: if $\boldsymbol{\beta}_1 \in A_j$ (this can happen only if the last element of the Apéry Set in the chain $\boldsymbol{\alpha}_r$ shares a component with both $\boldsymbol{\alpha}_h$ and $\boldsymbol{\beta}_1$), we set $\boldsymbol{\alpha}_{h+1} = \boldsymbol{\beta}_2$ and we proceed in case A; if $\boldsymbol{\beta}_1 \in A_{j+2}$, we take again $\boldsymbol{\delta}$ consecutive in $\mathbf{Ap}(S)$ to $\boldsymbol{\alpha}_r$ such that $\boldsymbol{\delta} \in A_{j+1}$ and, replacing the last part of the chain with $\boldsymbol{\delta} \wedge \boldsymbol{\alpha}_m$ (the elements $\boldsymbol{\alpha}_m$ are defined as in A.1), we obtain the desired result and we switch to case B.

It remains the hardest situation, i.e., $\boldsymbol{\beta}_1 \in A_{j+1}$; in this case, we have to show that both $\boldsymbol{\beta}_1$ and $\boldsymbol{\beta}_2$ do not have the same element as unique consecutive in S; if this was the case and if $\boldsymbol{\beta}_1$ was the only consecutive of $\boldsymbol{\alpha}_r$ in $\mathbf{Ap}(S)$ belonging to A_{j+1}, it would be impossible to proceed, because either we would skip the level A_{j+1} or we would create a chain such that, if we delete the elements of $\mathbf{Ap}(S)$, we do not get a saturated chain in $\boldsymbol{e} + S$ (for this situation, see the second picture of Figure 3, in which we denote by $\boldsymbol{\theta}$ the unique consecutive element in S of both $\boldsymbol{\beta}_1$ and $\boldsymbol{\beta}_2$ and we consider the two possible chains between $\boldsymbol{\alpha}_h$ and $\boldsymbol{\theta}$).

Let assume that $\boldsymbol{\alpha}_h$ and $\boldsymbol{\beta}_2$ share the first component (the other case is symmetric): So $\boldsymbol{\alpha}_h = (a_1, a_2)$ and $\boldsymbol{\beta}_2 = (a_1, b_2)$, with $b_2 > a_2$. They do not belong to $\mathbf{Ap}(S)$ hence both $\boldsymbol{\alpha}_h - \boldsymbol{e}$ and $\boldsymbol{\beta}_2 - \boldsymbol{e}$ belong to S. Hence, by Property (G2), there must be an element $(c_1, a_2 - e_2) \in S$ and so we have also another point of S, $\boldsymbol{\delta} = (c_1 + e_1, a_2) \in \boldsymbol{e} + S$, on the right of $\boldsymbol{\alpha}_h$. Since $\boldsymbol{\beta}_1$ is also on the right of $\boldsymbol{\alpha}_h$ and it is consecutive to it in S, we have three points in the same horizontal line: $\boldsymbol{\alpha}_h < \boldsymbol{\beta}_1 < \boldsymbol{\delta}$. Now, choosing $\boldsymbol{\delta}$ minimal, we are sure that it is consecutive to $\boldsymbol{\alpha}_h$ in $\boldsymbol{e} + S$. Moreover, we are sure that moving from $\boldsymbol{\beta}_1$ to $\boldsymbol{\delta}$ on the horizontal line, if we meet points of $\mathbf{Ap}(S)$, by Lemma 1(5) we do not skip any level and by Lemma 2(4) and Lemma 1(4) we do not repeat twice the same level. Hence we can stretch the chain up to $\boldsymbol{\delta}$ and proceed in case A.

• Case B. We notice that, in this case, if $\boldsymbol{\beta}_i \in \mathbf{Ap}(S)$ it has to belong either to A_j or to A_{j+1}, since they are consecutive in S to $\boldsymbol{\alpha}_h \in \mathbf{Ap}(S)$ (again by Lemma 1(5)).

The following situations can occur:

(B.1) both $\boldsymbol{\beta}_i \in A_{j+1}$: we can move to any one of them indifferently and proceed in case B;
(B.2) both $\boldsymbol{\beta}_i \in \boldsymbol{e} + S$: this cannot happen by Lemma 2(2);
(B.3) $\boldsymbol{\beta}_1 \in A_j$ and $\boldsymbol{\beta}_2 \in A_{j+1}$: we move to $\boldsymbol{\beta}_2$ and proceed in case B;
(B.4) $\boldsymbol{\beta}_1 \in A_j$ and $\boldsymbol{\beta}_2 \in \boldsymbol{e} + S$: we move to $\boldsymbol{\beta}_2$ and switch to case A;
(B.5) $\boldsymbol{\beta}_1 \in A_{j+1}$ and $\boldsymbol{\beta}_2 \in \boldsymbol{e} + S$: we can assume that $\boldsymbol{\beta}_1$ is on the right of $\boldsymbol{\alpha}_h$ and $\boldsymbol{\beta}_2$ is above it; in this case either there is another element $\boldsymbol{\beta}' \in A_{j+1}$ such that $\boldsymbol{\alpha} = \boldsymbol{\beta}_1 \wedge \boldsymbol{\beta}'$ and $\boldsymbol{\alpha} < \boldsymbol{\beta}_2 < \boldsymbol{\beta}'$ share the first component; we choose $\boldsymbol{\beta}'$ minimal with this property and we move to $\boldsymbol{\beta}_2$ and then to $\boldsymbol{\beta}'$ (considering all possible elements between them, that have to belong to $\boldsymbol{e} + S$) or there exists $\boldsymbol{\beta}'' \in A_{j+1}$ above $\boldsymbol{\beta}_1$ and consecutive to both $\boldsymbol{\beta}_i$; so we move to $\boldsymbol{\beta}_2$ and then to $\boldsymbol{\beta}''$. In both cases we proceed in case B.

The proof is complete. □

From now on we are going to denote the number of levels of the Apéry Set by $e = e(S) = e_1 + e_2$ that, as we noticed in the previous section, coincides with the multiplicity of the ring R, in case $S = v(R)$.

We derive from the proof of the preceding theorem, a sort of converse of property (1) of Lemma 1. We are going to make use of this next result in the last section of this article, while proving a duality property for the levels of the Apéry Set of a symmetric good semigroup.

Proposition 4 *Let $S \subseteq \mathbb{N}^2$ be a good semigroup and let $\mathbf{Ap}(S) = \bigcup_{i=1}^{e} A_i$ be its Apéry Set. Let $\boldsymbol{\alpha} \in A_i$ for $i \geq 2$. Then, there exists $\boldsymbol{\beta} \in A_{i-1}$ such that $\boldsymbol{\beta} \leq \boldsymbol{\alpha}$.*

Proof. Since $A_1 = \{\mathbf{0}\}$, and $\boldsymbol{\alpha} \geq \mathbf{0}$ for every $\boldsymbol{\alpha} \in S$, the thesis is true for $i = 2$ and hence, by induction, we can assume it true for every $j < i$. Assume by way of contradiction that there exists $\boldsymbol{\alpha} \in A_i$ such that $\boldsymbol{\theta} \nleq \boldsymbol{\alpha}$ for every $\boldsymbol{\theta} \in A_{i-1}$. We can further assume that also $\boldsymbol{\delta} \nleq \boldsymbol{\alpha}$ for every $\boldsymbol{\delta} \in A_i$, otherwise we can simply replace $\boldsymbol{\alpha}$ with some element $\boldsymbol{\delta} \leq \boldsymbol{\alpha}$ and minimal in A_i with respect to "$\leq$".

Take $\boldsymbol{\omega} \in \mathbf{Ap}(S)$ such that $\boldsymbol{\omega} \leq \boldsymbol{\alpha}$ and they are consecutive in $\mathbf{Ap}(S)$, hence, $\boldsymbol{\omega} \in A_j$ with $j < i - 1$. We may assume j to be the maximal level of an element of $\mathbf{Ap}(S)$ having $\boldsymbol{\alpha}$ as a consecutive element in $\mathbf{Ap}(S)$. Assume there exists an element $\boldsymbol{\omega}' \in A_j$ such that $\boldsymbol{\omega} \in \Delta^S(\boldsymbol{\omega}')$. Hence, we can find a saturated chain in S between $\boldsymbol{\omega}'$ and $\boldsymbol{\alpha}$ that does not contain any other elements of $\mathbf{Ap}(S)$. Indeed, we can find

$\beta \in S$ such that $\omega' \leq \beta \leq \alpha$, $\beta \in \Delta^S(\omega')$ and it is incomparable with ω (i.e., either $\beta \in \Delta_2^S(\omega')$ and $\omega \in \Delta_1^S(\omega')$ or the converse). If $\beta \in \mathbf{Ap}(S)$, then it has to be in A_j, but this is impossible since $\omega' = \omega \wedge \beta$ would be the minimum of two elements of A_j, and therefore not in A_j.

It follows that we can choose an element $\tilde{\omega} \leq \omega$ minimal with respect to the property of being in A_j (this element could be ω itself) and find a saturated chain in S between $\tilde{\omega}$ and α not containing any other elements of $\mathbf{Ap}(S)$. By inductive hypothesis, $\tilde{\omega} \geq \delta \in A_{j-1}$, and hence we can iterate the preceding process and construct a saturated chain in S between $\mathbf{0}$ and α, containing only one element for every level of $\mathbf{Ap}(S)$ between 1 and j and not containing any element in the levels strictly between j and i. As in the first part of the proof of Theorem 3, we can extend this chain adding a chain in S from α to $\gamma - \boldsymbol{e} + \mathbf{1}$ including only one element for each level of $\mathbf{Ap}(S)$ greater than i. The obtained chain going between $\mathbf{0}$ and $\gamma + \boldsymbol{e} + \mathbf{1}$ contains $h := e - (i - j) + 1$ elements of $\mathbf{Ap}(S)$, thus, removing those elements, we can find a chain in $\boldsymbol{e} + S$ between $\boldsymbol{e}$ and $\gamma + \boldsymbol{e} + \mathbf{1}$ of length

$$d_S(\mathbf{0}, \gamma + \boldsymbol{e} + \mathbf{1}) - h = d_{\boldsymbol{e}+S}(\boldsymbol{e}, \gamma + \boldsymbol{e} + \mathbf{1}) + e - h.$$

Since $j < i - 1$, this length is strictly bigger than $d_{\boldsymbol{e}+S}(\boldsymbol{e}, \gamma + \boldsymbol{e} + \mathbf{1})$ and this is a contradiction. □

As we can observe in all the preceding examples of good semigroups, the first levels of $\mathbf{Ap}(S)$ are finite while the others contain either one or two infinite lines of elements. After formalizing the concept of infinite lines of elements in two definitions, we describe precisely this behavior in the next theorem (Figure 4).

Definition 1. Let $S \subseteq S_1 \times S_2 \subseteq \mathbb{N}^2$ be a good semigroup. Given an element $s_1 \in S_1$, we say that $\Delta_1(s_1, r)$ is an *infinite line* of S if there exists $r \in S_2$ such that $\Delta_1(s_1, r) \subseteq S$. If $\Delta_1(s_1, r) \subseteq \mathbf{Ap}(S)$, we say that $\Delta_1(s_1, r)$ is an infinite line of $\mathbf{Ap}(S)$.

Analogously, given an element $s_2 \in S_2$, $\Delta_2(q, s_2)$ is an *infinite line* of S (resp. $\mathbf{Ap}(S)$) if there exists $q \in S_1$ such that $\Delta_2(q, s_2) \subseteq S$ (resp. $\mathbf{Ap}(S)$). If an infinite line of S is not an infinite line of $\mathbf{Ap}(S)$, then it is an infinite line of $\boldsymbol{e} + S$.

Definition 2. Let $S \subseteq S_1 \times S_2 \subseteq \mathbb{N}^2$ be a good semigroup and let $\mathbf{Ap}(S) = \bigcup_{i=1}^{e} A_i$ be its Apéry Set. For $i = 1, \ldots, e$, we say that

1. A_i contains two infinite lines if there exist two elements $s_1 \in S_1$ and $s_2 \in S_2$, such that, for some $q \in S_1$, $r \in S_2$, $\Delta_1(s_1, r)$, $\Delta_2(q, s_2)$ are infinite lines of $\mathbf{Ap}(S)$ and they are contained in A_i.
2. A_i contains only one infinite line if only one of the previous conditions hold.
3. A_i is finite if it contains a finite number of elements.

Theorem 5. *Let $S \subseteq S_1 \times S_2 \subseteq \mathbb{N}^2$ be a good semigroup, let $\boldsymbol{e} = (e_1, e_2)$ be its minimal nonzero element. Let $\mathbf{Ap}(S) = \bigcup_{i=1}^{e} A_i$ be the Apéry Set of S. Assume $e_1 \geq e_2$. Then,*

(1) The levels $A_e, A_{e-1}, \ldots, A_{e-e_2+1}$ contain two infinite lines.

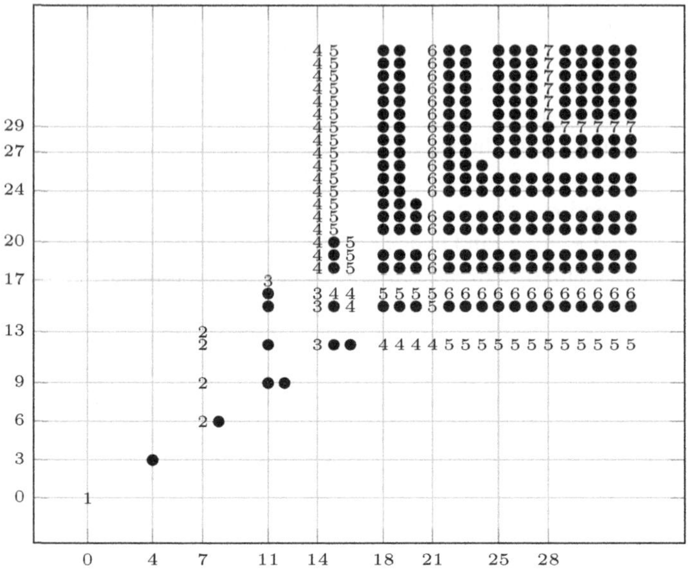

Fig. 4 This is an example of a good semigroup that is not the value semigroup of any ring, see [1, Example 2.16, p. 8]

(2) The levels $A_{e-e_2}, \ldots, A_{e-e_1+1}$ contain only one infinite line of the form $\Delta_1(s_1, r)$ corresponding to some element $s_1 \in S_1$.

(3) The levels $A_{e-e_1}, \ldots, A_2, A_1$ are finite.

If $e_1 \leq e_2$ the correspondent analogous conditions hold.

Proof. First we show that in the projection S_1 of S there are exactly e_1 elements $s_1, \ldots, s_{e_1}$ such that $\Delta_1(s_i, r)$ is an infinite line of $\mathbf{Ap}(S)$ (for some $r \in S_2$). Let $\boldsymbol{c} = (c_1, c_2)$ be the conductor of S. Following the preceding definitions, we have that for every $n \geq c_1$ and sufficiently large $r \in S_2$, $\Delta_1(n, r)$ is an infinite line of S. Moreover, $\Delta_1(n, r) \subseteq \boldsymbol{e} + S$ if and only if also $\Delta_1(n - e_1, r)$ is an infinite line of S, and conversely $\Delta_1(n, r) \subseteq \mathbf{Ap}(S)$ if $\Delta_1(n - e_1, r)$ is not an infinite line of S for any $r \in S_2$. It follows that for every n there exists a unique $m \equiv n \bmod e_1$ such that $\Delta_1(m, r)$ is an infinite line of $\mathbf{Ap}(S)$. With the same argument it can be shown that the analogous situation happens on S_2 and therefore there are e_2 infinite lines of $\mathbf{Ap}(S)$ of the form $\Delta_2(q, t_i)$ corresponding to some elements $t_1, \ldots, t_{e_2} \in S_2$. Now, notice that, by Lemma 1(7), if an infinite line is contained in $\mathbf{Ap}(S)$, then its elements must be contained eventually in a level A_i. Moreover, by Lemma 1(8), A_i cannot contain more than two infinite lines and, if it contains two of them, they must be one of the form $\Delta_1(s_1, r)$ and the other of the form $\Delta_2(q, s_2)$. Applying inductively the definition of the levels A_i, it follows that the levels $A_e, A_{e-1}, \ldots, A_{e-e_j+1}$ contain the e_j infinite lines contained in $\mathbf{Ap}(S)$ and corresponding to the elements of S_j. □

We conclude this section by proving a generalization of Remark 4 holding for good semigroups. We show that if $(R, \mathfrak{m}, k)$ is an analytically unramified one-dimensional local reduced ring, its quotient ring $R/(x)$, where x is a nonzero element having minimal value in the value semigroup $S = v(R)$, can be generate as a k-vector space by a set of e elements having values in all the different levels of the Apéry Set of S.

Theorem 6. *Let $(R, \mathfrak{m}, k)$ be an analytically unramified one-dimensional local reduced ring, having value semigroup $S = v(R)$ and let $x \in R$ such that $v(x) = \boldsymbol{e} = \min(S \setminus \{\boldsymbol{0}\})$. Let $\mathbf{Ap}(S) = \bigcup_{i=1}^{e} A_i$ be the Apéry Set of S. It is possible to construct a chain*

$$\boldsymbol{\alpha}_1 < \boldsymbol{\alpha}_2 < \ldots < \boldsymbol{\alpha}_e \in S$$

such that $\alpha_i \in A_i$ and, for every collection of $f_i \in R$ having $v(f_i) = \alpha_i$,

$$\frac{R}{(x)} = \langle \overline{f_1}, \overline{f_2}, \ldots, \overline{f_e} \rangle_k.$$

Proof. By Remark 5, the dimension over k of $R/(x)$ is e, hence, we only need to show that, for $i = 1, \ldots, e$, we can find elements $\boldsymbol{\alpha}_i$ such that the correspondent f_i are linearly independent over k.

We set $\boldsymbol{\alpha}_1 = \boldsymbol{0}$ and then we define the other elements $\boldsymbol{\alpha}_i$ using the following procedure: in case $\boldsymbol{\alpha}_i \ll \boldsymbol{\beta}$ for some $\boldsymbol{\beta} \in A_{i+1}$, we may simply set $\boldsymbol{\alpha}_{i+1} := \boldsymbol{\beta}$. Otherwise, if $\boldsymbol{\alpha}_i = \boldsymbol{\beta} \wedge \boldsymbol{\delta}$ with $\boldsymbol{\beta} \in \Delta_1^S(\boldsymbol{\alpha}_i) \cap A_{i+1}$ and $\boldsymbol{\delta} \in \Delta_2^S(\boldsymbol{\alpha}_i) \cap A_{i+1}$, by Lemma 2(2), we have $\Delta_h^S(\boldsymbol{\alpha}_i) \subseteq \mathbf{Ap}(S)$ and then we set $\boldsymbol{\alpha}_{i+1} := \boldsymbol{\beta}$ if $h = 2$ and $\boldsymbol{\alpha}_{i+1} := \boldsymbol{\delta}$ if $h = 1$ (if $\Delta^S(\boldsymbol{\alpha}_i) \subseteq \mathbf{Ap}(S)$ we can take indifferently one of them).

Now, taking $f_i \in R$ such that $v(f_i) = \boldsymbol{\alpha}_i \in \mathbf{Ap}(S)$, we clearly get by Proposition 2 that $\overline{f_i}$ is nonzero in $R/(x)$. Then we consider $v(\sum_{i=1}^{e} \lambda_i f_i)$ for $\lambda_i \in k$. Let j be the minimal index such that $\lambda_j \neq 0$. If $\boldsymbol{\alpha}_j \ll \boldsymbol{\alpha}_{j+1}$, we obtain

$$v(\sum_{i=1}^{e} \lambda_i f_i) = v(\lambda_j f_j) = \boldsymbol{\alpha}_j \in \mathbf{Ap}(S)$$

and therefore $\sum_{i=1}^{e} \lambda_i \overline{f_i}$ is nonzero in $R/(x)$. Otherwise, we may assume $\boldsymbol{\alpha}_{j+1} \in \Delta_1^S(\boldsymbol{\alpha}_j)$, and hence, our procedure used to define the $\boldsymbol{\alpha}_i$ implies now that $\Delta_2^S(\boldsymbol{\alpha}_j) \subseteq \mathbf{Ap}(S)$. It follows that

$$v(\sum_{i=1}^{e} \lambda_i f_i) \in \Delta_2^S(\boldsymbol{\alpha}_j) \cup \{\boldsymbol{\alpha}_j\} \subseteq \mathbf{Ap}(S)$$

and thus $\sum_{i=1}^{e} \lambda_i \overline{f_i}$ is nonzero in $R/(x)$. □

4 Symmetric Good Semigroups

In this section we describe more properties of the Apéry Set of a good semigroup in the symmetric case.

Definition 3. A good semigroup S is *symmetric* if, for every $\alpha \in \mathbb{N}^2$, $\alpha \in S$ if and only if $\Delta^S(\gamma - \alpha) = \emptyset$.

Symmetry is an interesting concept because, in the value semigroup case, it is equivalent to the Gorensteiness of the associated ring. Indeed, an analytically unramified one-dimensional local reduced ring is Gorenstein if and only if its value semigroup is symmetric. But more in general, a symmetric good semigroup has other nice properties, that we list in next proposition. Some of them have been already proved in [2, Proposition 3.2]. An interesting fact that we are proving is that it is possible to know the number of absolute elements of a symmetric good semigroup only looking at one of its (numerical) projections (Figure 5).

Remark 7. The projections of a symmetric good semigroup may fail to be symmetric numerical semigroups, as one can see for instance in Figure 6.

Proposition 7 *Let $S \subseteq S_1 \times S_2 \subseteq \mathbb{N}^2$ be a symmetric good semigroup, let $\mathbf{e} = (e_1, e_2)$, $\gamma = (\gamma_1, \gamma_2)$ and $\mathbf{Ap}(S)$ be defined as previously.*

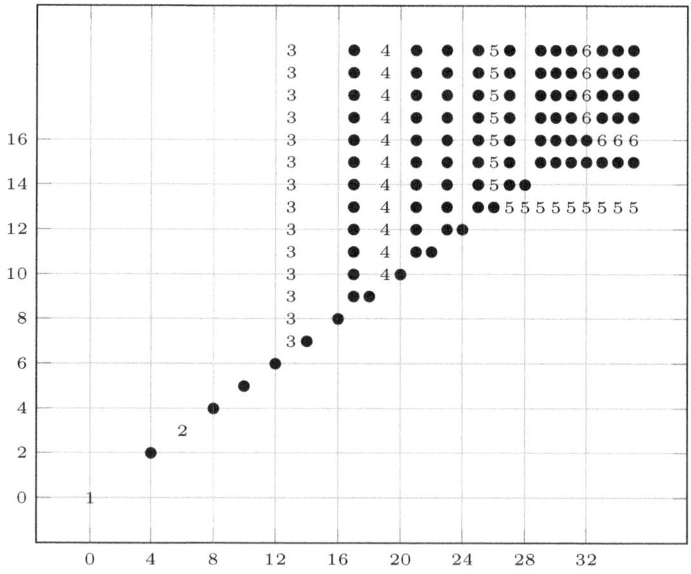

Fig. 5 The symmetric value semigroup of $k[[X, Y]]/(Y^4 - 2X^3Y^2 - 4X^5Y + X^6 - X^7)(Y^2 - X^3)$, see [2, p. 8]. It is possible to observe that the number of absolute elements of this semigroup is $13 = 21 - 8 = 14 - 1$ as predicted by the formula in Proposition 7(2)

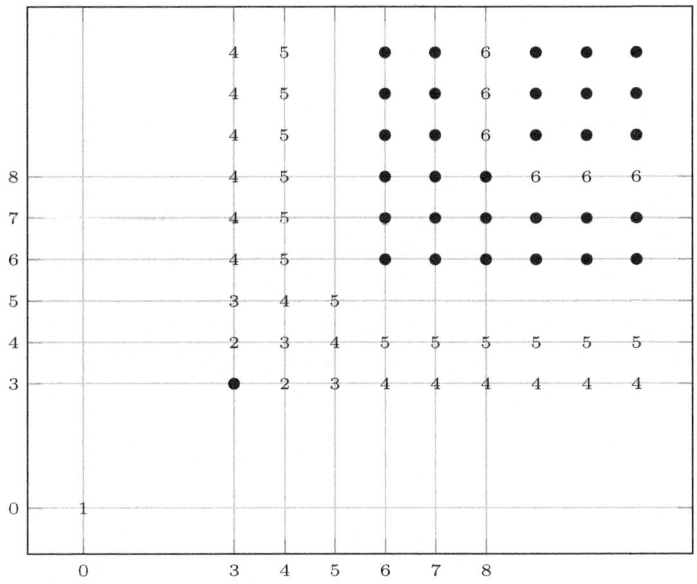

Fig. 6 An example of a symmetric good semigroup whose projections are not symmetric numerical semigroups. It is a good example to check the duality property stated in Theorem 9

(1) If $\boldsymbol{\alpha} \in S$ is a absolute element, then also $\boldsymbol{\gamma} - \boldsymbol{\alpha} \in S$ and it is a absolute element.
(2) The number of the absolute elements of S is $n(S_1) - b(S_1) = n(S_2) - b(S_2)$, where $n(S_i) = |S_i \cap \{0, 1, \ldots, \gamma_i\}|$ and $b(S_i) = |\mathbb{N} \setminus S_i|$.
(3) For $\boldsymbol{\alpha} \in S$, $\boldsymbol{\alpha} \in \mathbf{Ap}(S)$ if and only if $\Delta^S(\boldsymbol{\gamma} + \boldsymbol{e} - \boldsymbol{\alpha}) \neq \emptyset$.
(4) If $\boldsymbol{\alpha} \in \mathbf{Ap}(S)$, then $\Delta^S(\boldsymbol{\gamma} + \boldsymbol{e} - \boldsymbol{\alpha}) \subseteq \mathbf{Ap}(S)$.
(5) Let $\boldsymbol{\alpha} \in \mathbb{N}^2$. If $\Delta^S(\boldsymbol{\alpha}) \subseteq \mathbf{Ap}(S)$ (possibly it is empty), then $\boldsymbol{\gamma} + \boldsymbol{e} - \boldsymbol{\alpha} \in S$.
(6) Let $\boldsymbol{\alpha} \in \mathbf{Ap}(S)$. Then for $i = 1, 2$; $\Delta_i^S(\boldsymbol{\gamma} + \boldsymbol{e} - \boldsymbol{\alpha}) = \emptyset$ if and only if $\Delta_i^S(\boldsymbol{\alpha}) \nsubseteq \mathbf{Ap}(S)$.

Proof. (1) Follows by the definitions of symmetric semigroup and absolute element.
(2) By definition of symmetric good semigroup, we have that $n \notin S_1$ if and only if $\Delta_1^S(n, 0) = \emptyset$, and if and only if $(\gamma_1 - n, \gamma_2 + m) \in S$ for every $m \geq 0$. It follows that the number of elements $s \in S_1$ such that $\Delta_1(s_1, r)$ is an infinite line of S (for some $r \in S_2$) is exactly $b(S_1)$. Call M the number of absolute elements of S. Hence,

$$\gamma_1 = M + 2b(S_1),$$

and, since $\gamma_1 = n(S_1) + b(S_1)$, we obtain $M = n(S_1) - b(S_1)$. In the same way, we can show $M = n(S_2) - b(S_2)$.
(3) An element $\boldsymbol{\alpha} \in S$ is in $\mathbf{Ap}(S)$ if and only if $\boldsymbol{\alpha} - \boldsymbol{e} \notin S$, and this happens by Definition 3 if and only if $\Delta^S(\boldsymbol{\gamma} + \boldsymbol{e} - \boldsymbol{\alpha}) \neq \emptyset$.
(4) Since $\boldsymbol{\alpha} \in \mathbf{Ap}(S)$, $\Delta^S(\boldsymbol{\gamma} - \boldsymbol{\alpha}) = \emptyset$. It follows that $\Delta^S(\boldsymbol{\gamma} + \boldsymbol{e} - \boldsymbol{\alpha}) \subseteq \mathbf{Ap}(S)$.

(5) If $\Delta^S(\boldsymbol{\alpha}) \subseteq \mathbf{Ap}(S)$, then by definition $\Delta^S(\boldsymbol{\alpha} - \boldsymbol{e}) = \emptyset$. and therefore $\boldsymbol{\gamma} + \boldsymbol{e} - \boldsymbol{\alpha} \in S$.
(6) We prove the result for $i = 1$. By (3) and (4), we have that $\Delta^S(\boldsymbol{\gamma} + \boldsymbol{e} - \boldsymbol{\alpha})$ is not empty and contained in $\mathbf{Ap}(S)$. Assuming $\Delta_1^S(\boldsymbol{\gamma} + \boldsymbol{e} - \boldsymbol{\alpha}) = \emptyset$, by axiom (G2) we get $\boldsymbol{\gamma} + \boldsymbol{e} - \boldsymbol{\alpha} \notin S$ and $\Delta_2^S(\boldsymbol{\gamma} + \boldsymbol{e} - \boldsymbol{\alpha}) \neq \emptyset$. By (5), it follows that $\Delta^S(\boldsymbol{\alpha}) \nsubseteq \mathbf{Ap}(S)$ and moreover, since there exists $\boldsymbol{\omega} \in \Delta_2^S(\boldsymbol{\gamma} + \boldsymbol{e} - \boldsymbol{\alpha}) \subseteq \mathbf{Ap}(S)$, we get by (4) $\Delta_2^S(\boldsymbol{\alpha}) \subseteq \Delta_2^S(\boldsymbol{\gamma} + \boldsymbol{e} - \boldsymbol{\omega}) \subseteq \mathbf{Ap}(S)$. Hence $\Delta_1^S(\boldsymbol{\alpha}) \nsubseteq \mathbf{Ap}(S)$.

Conversely, if $\Delta_1^S(\boldsymbol{\alpha}) \nsubseteq \mathbf{Ap}(S)$, there exists $\boldsymbol{\theta} \in \Delta_1^S(\boldsymbol{\alpha}) \cap (\boldsymbol{e} + S)$ and therefore, again by (3) $\Delta_1^S(\boldsymbol{\gamma} + \boldsymbol{e} - \boldsymbol{\alpha}) \subseteq \Delta_1^S(\boldsymbol{\gamma} + \boldsymbol{e} - \boldsymbol{\theta}) = \emptyset$. □

Let $S \subseteq S_1 \times S_2 \subseteq \mathbb{N}^2$ be a good semigroup. We describe now, in the symmetric case, the absolute elements and the infinite lines of $\mathbf{Ap}(S)$ and of $\boldsymbol{e} + S$ in terms of the elements of a single projection, say S_1. For $n \in \mathbb{N}$, we consider the set

$$\Delta_1^S(n, 0) = \{(n, m) \in S \mid m \geq 0\}.$$

This set can be empty, finite or infinite. It is infinite if and only if $\Delta_1^S(n, r)$ is an infinite line of S for some $r \in S_2$; it is finite if and only if $\Delta_1^S(n, 0)$ contains a absolute element of S; it is empty if and only if $n \notin S_1$. The analogous situation holds for the other projection S_2.

Lemma 3. *The set $\Delta_1^S(n, 0) \subseteq \mathbf{Ap}(S)$ if and only if $n \in \mathbf{Ap}(S_1)$.*

Proof. We have $n \in \mathbf{Ap}(S_1)$ if and only if $n - e_1 \notin S_1$ if and only if $\Delta_1^S(n - e_1, 0) = \emptyset$. The result now follows from the definition of $\mathbf{Ap}(S)$. □

Theorem 8. *Let $S \subseteq S_1 \times S_2 \subseteq \mathbb{N}^2$ be a symmetric good semigroup, let $\boldsymbol{e} = (e_1, e_2)$ be its minimal nonzero element. Let $n \in \mathbb{N}$ and define $n' = \gamma_1 + e_1 - n$.*

(1) $\Delta_1^S(n, 0) = \emptyset$ if and only if $\Delta_1^S(n', 0)$ is infinite and eventually contained in $\boldsymbol{e} + S$.
(2) $\Delta_1^S(n, 0)$ is finite with maximal element in $\boldsymbol{e} + S$ if and only if $\Delta_1^S(n', 0)$ is finite with maximal element in $\boldsymbol{e} + S$.
(3) $\Delta_1^S(n, 0) \nsubseteq \mathbf{Ap}(S)$ and it is finite with maximal element in $\mathbf{Ap}(S)$ if and only if $\Delta_1^S(n', 0) \nsubseteq \mathbf{Ap}(S)$ and it is finite with maximal element in $\mathbf{Ap}(S)$.
(4) $\Delta_1^S(n, 0) \subseteq \mathbf{Ap}(S)$ and it is finite if and only if $\Delta_1^S(n', 0)$ is infinite and eventually contained in $\mathbf{Ap}(S)$ but it contains some element of $\boldsymbol{e} + S$.
(5) $\Delta_1^S(n, 0) \subseteq \mathbf{Ap}(S)$ and it is infinite if and only if $\Delta_1^S(n', 0) \subseteq \mathbf{Ap}(S)$ and it is infinite.

All the correspondent statements hold replacing S_1 with S_2.

Proof. (1) Observe that $\Delta_1^S(n, 0) = \emptyset$ if and only if $\Delta^S(n, -m) = \emptyset$ for all $m \geq 0$. This is equivalent by Definition 3 to say that $(\gamma_1 - n, \gamma_2 + m) \in S$. Hence $\Delta_1^S(n, 0) = \emptyset$ if and only if $\Delta_1^S(\gamma_1 - n, 0)$ is infinite, which is equivalent to say that $\Delta_1^S(\gamma_1 + e_1 - n, 0) = \Delta_1^S(n', 0)$ is infinite and contained in $\boldsymbol{e} + S$.
(2) Let $\boldsymbol{\alpha} \in \boldsymbol{e} + S$ be the maximal element of S belonging to $\Delta_1^S(n, 0)$. Hence $\boldsymbol{\alpha}$ is an absolute element and, by Proposition 7(1) $\boldsymbol{\gamma} - \boldsymbol{\alpha}$ is also an absolute element of S. It

follows that $\boldsymbol{\alpha}' = \boldsymbol{\gamma} + \boldsymbol{e} - \boldsymbol{\alpha} \in \boldsymbol{e} + S$. Moreover, since $\boldsymbol{\alpha} \in \boldsymbol{e} + S$, $\Delta^S(\boldsymbol{\alpha}') = \emptyset$ by Proposition 7(3). Thus $\boldsymbol{\alpha}'$ is a absolute element and $\Delta_1^S(n', 0)$ is finite with maximal element in $\boldsymbol{e} + S$. The converse is tautological.
(3) Let $\Delta_1^S(n, 0) \not\subseteq \mathbf{Ap}(S)$ and it is finite with maximal element $\boldsymbol{\alpha} = (n, m) \in \mathbf{Ap}(S)$. As in (2), we have that $\boldsymbol{\alpha}' = \boldsymbol{\gamma} + \boldsymbol{e} - \boldsymbol{\alpha} \in \boldsymbol{e} + S$. But, by exclusion, (1) and (2) imply that $\Delta_1^S(n', 0)$ neither is infinite and eventually contained $\boldsymbol{e} + S$ nor has a maximal in $\boldsymbol{e} + S$. Hence there must exist $\boldsymbol{\beta} = (n', r) \in \mathbf{Ap}(S)$ with $r > m' := \gamma_2 + e_2 - m$. We conclude saying that, since $\Delta_1^S(n, 0) \not\subseteq \mathbf{Ap}(S)$, there must exist an element $\boldsymbol{\delta} = (n, d) \in \boldsymbol{e} + S$ with $d < m$ and hence $\Delta^S(\boldsymbol{\delta}') = \emptyset$. Thus there are only finite elements $(n', q) \in S$ with $q > m'$ and they are all in $\mathbf{Ap}(S)$ by Proposition 7(4), since they are elements of $\Delta^S(\boldsymbol{\alpha}')$. Since there exists at least one of such elements, namely $\boldsymbol{\beta}$, the thesis follows.
(4) Assume that $\Delta_1^S(n, 0) \subseteq \mathbf{Ap}(S)$ and it is finite. Again by (1) and (2) we exclude that $\Delta_1^S(n', 0)$ is infinite and eventually contained $\boldsymbol{e} + S$ or has a maximal in $\boldsymbol{e} + S$. Let $\boldsymbol{\alpha} = (n, m) \in \mathbf{Ap}(S)$ be the maximal element in $\Delta_1^S(n, 0)$. We proceed like in the proof of (3) to say that $\boldsymbol{\alpha}' = \boldsymbol{\gamma} + \boldsymbol{e} - \boldsymbol{\alpha} \in \boldsymbol{e} + S$. If by way of contradiction $\Delta_1^S(n', 0)$ contains a maximal element $\boldsymbol{\theta} \in \mathbf{Ap}(S)$, it would follow by Proposition 7(3 and 5) that $\boldsymbol{\theta}' \in \Delta_1^S(n, 0) \cap (\boldsymbol{e} + S)$ and this is a contradiction.

Conversely, assume $\Delta_1^S(n', 0)$ is infinite and eventually contained in $\mathbf{Ap}(S)$ but it contains some element $\boldsymbol{\theta} \in \boldsymbol{e} + S$. Since $\Delta^S(\boldsymbol{\theta}') = \emptyset$ (by Proposition 7(3)), $\Delta_1^S(n, 0)$ must be finite. We conclude by exclusion, since we characterized in (2) and (3) the other possible cases of a finite $\Delta_1^S(n, 0)$.
(5) It follows since we excluded all the other possible cases in (1),(2),(3), and (4). □

Corollary 1. *Assume the same notations of Theorem 8. Hence, $n \in \mathbf{Ap}(S_1)$ if and only if $\Delta_1^S(n', 0)$ is infinite and eventually contained in $\mathbf{Ap}(S)$.*

5 Duality of the Apéry Set of Symmetric Good Semigroups

The symmetry of a numerical semigroup S can be characterized by the symmetry of its Apéry Set with respect to its largest element: if we order the elements of $\mathbf{Ap}(S)$ in increasing order $\mathbf{Ap}(S) = \{w_1, \ldots, w_e\}$, then S is symmetric if and only if $w_i + w_{e-i+1} = w_e$.

Hence, there is a duality relation associating to each element w_i, the element w_{e-i+1}. In the case of a symmetric good semigroup we do not have this relation by choosing arbitrary elements, one from each level A_i of $\mathbf{Ap}(S)$; but we find a more general duality relation associating the level A_i to the level A_{e-i+1}, and involving both the elements of $\boldsymbol{\alpha} \in A_i$ and the sets $\Delta^S(\boldsymbol{\gamma} + \boldsymbol{e} - \boldsymbol{\alpha})$. After two preparatory lemmas, we define and prove this duality in Theorem 9.

In this Section, we denote as before $\boldsymbol{\alpha}' := \boldsymbol{\gamma} + \boldsymbol{e} - \boldsymbol{\alpha}$.

Lemma 4. *Let $S \subseteq S_1 \times S_2 \subseteq \mathbb{N}^2$ be a symmetric good semigroup. Let* $\mathbf{Ap}(S) = \bigcup_{i=1}^{e} A_i$ *be the Apéry Set of S. If $\boldsymbol{\alpha} \in A_{e-i+1}$, then for every $j < i$,*

$$\Delta^S(\boldsymbol{\alpha}') \cap A_j = \emptyset.$$

Proof. We use induction on i. For $i = 1$, the result is clear. Let $\boldsymbol{\alpha} \in A_{e-i+1}$. We separate the proof in two cases:

Case 1: Assume $\boldsymbol{\alpha} \ll \boldsymbol{\theta}$ for some $\boldsymbol{\theta} \in A_{e-i+2}$. By inductive hypothesis $\Delta^S(\boldsymbol{\theta}') \cap A_j = \emptyset$ for every $j < i - 1$. Since $\boldsymbol{\alpha} \ll \boldsymbol{\theta}$, it follows that $\boldsymbol{\theta}' \ll \boldsymbol{\alpha}'$ and hence for every $\boldsymbol{\delta} \in \Delta^S(\boldsymbol{\alpha}')$ there exists $\boldsymbol{\beta} \in \Delta^S(\boldsymbol{\theta}')$ such that either $\boldsymbol{\beta} \ll \boldsymbol{\delta}$ or there exists $\boldsymbol{\omega} = \boldsymbol{\delta} \wedge \boldsymbol{\beta} \in \Delta^S(\boldsymbol{\theta}') \subseteq \mathbf{Ap}(S)$. In the first case the level of $\boldsymbol{\beta}$ in $\mathbf{Ap}(S)$ is smaller than the level of $\boldsymbol{\delta}$. In the second case, as a consequence of Lemma 2(3), the element $\boldsymbol{\omega}$ is in a level of $\mathbf{Ap}(S)$ smaller than the level of $\boldsymbol{\delta}$. Hence we can find elements in $\Delta^S(\boldsymbol{\theta}')$ in some level smaller than the level of any element of $\Delta^S(\boldsymbol{\alpha}')$. It follows that $\Delta^S(\boldsymbol{\alpha}') \subseteq \bigcup_{j \geq i} A_j$ and hence the thesis.

Case 2: Now assume $\boldsymbol{\alpha} = \boldsymbol{\theta} \wedge \boldsymbol{\delta}$ with $\boldsymbol{\theta} \in \Delta_1^S(\boldsymbol{\alpha}) \cap A_{e-i+2}$ and $\boldsymbol{\delta} \in \Delta_2^S(\boldsymbol{\alpha}) \cap A_{e-i+2}$. Hence $\boldsymbol{\alpha}' \in \Delta_1(\boldsymbol{\theta}') \cap \Delta_2(\boldsymbol{\delta}')$. Assuming $\Delta_1^S(\boldsymbol{\alpha}') \neq \emptyset$ and taking $\boldsymbol{\omega} \in \Delta_1^S(\boldsymbol{\alpha}')$, we can find an element $\boldsymbol{\beta} \in \Delta^S(\boldsymbol{\delta}')$ such that either $\boldsymbol{\beta} \ll \boldsymbol{\omega}$ or there exists $\boldsymbol{\beta}^1 = \boldsymbol{\omega} \wedge \boldsymbol{\beta} \in \Delta^S(\boldsymbol{\delta}') \subseteq \mathbf{Ap}(S)$ (it is possible to have $\boldsymbol{\beta}^1 = \boldsymbol{\alpha}'$). Using the same argument of Case 1, we show that one element among $\boldsymbol{\beta}$ and $\boldsymbol{\beta}^1$ is in a level of $\mathbf{Ap}(S)$ smaller than the level of $\boldsymbol{\omega}$ and therefore we get the same thesis of Case 1. In case $\Delta_2^S(\boldsymbol{\alpha}') \neq \emptyset$ we can use the same argument to find the needed elements in $\Delta^S(\boldsymbol{\theta}')$. □

Lemma 5. *Let $S \subseteq S_1 \times S_2 \subseteq \mathbb{N}^2$ be a symmetric good semigroup. Let* $\mathbf{Ap}(S) = \bigcup_{i=1}^{e} A_i$ *be the Apéry Set of S. If $\boldsymbol{\alpha} \in A_i$, then*

$$\Delta^S(\boldsymbol{\alpha}') \cap A_{e-i+1} \neq \emptyset.$$

Proof. Again we use induction on i. For $i = 1$, the result follows since $A_1 = \{\mathbf{0}\}$ and $\Delta^S(\mathbf{0}') = \Delta^S(\boldsymbol{\gamma} + \mathbf{e}) = A_e$. Let $\boldsymbol{\alpha} \in A_i$. By Lemma 4, we have that $\Delta^S(\boldsymbol{\alpha}') \subseteq \bigcup_{j \geq e-i+1} A_j$. By Proposition 4 we have that $\boldsymbol{\alpha} \geq \boldsymbol{\theta}$ for some $\boldsymbol{\theta} \in A_{i-1}$. We separate the proof in two cases:

Case 1: Assume $\boldsymbol{\alpha} \gg \boldsymbol{\theta}$. By inductive hypothesis, we know that there exists $\boldsymbol{\beta} \in \Delta^S(\boldsymbol{\theta}') \cap A_{e-i+2}$. Using the argument of the proof of Lemma 4 (Case 1) we can show that there exists some element $\boldsymbol{\omega} \in \Delta^S(\boldsymbol{\alpha}')$ which is in a level of $\mathbf{Ap}(S)$ smaller than A_{e-i+2}. Thus we must have $\boldsymbol{\omega} \in A_{e-i+1}$.

Case 2: Now assume $\boldsymbol{\alpha} \in \Delta^S(\boldsymbol{\theta})$. Without loss of generality, we say that $\boldsymbol{\alpha} \in \Delta_1^S(\boldsymbol{\theta})$. Now, if $\Delta_1^S(\boldsymbol{\theta}) \nsubseteq \mathbf{Ap}(S)$, by Proposition 7(6) we have $\Delta_1^S(\boldsymbol{\theta}') = \emptyset$ and therefore $\Delta_2^S(\boldsymbol{\theta}') \neq \emptyset$. Otherwise, if $\Delta_1^S(\boldsymbol{\theta}) \subseteq \mathbf{Ap}(S)$, applying Lemma 2(5) we find an element $\boldsymbol{\omega} \in (\Delta_1^S(\boldsymbol{\theta}) \cap A_{i-1}) \cup \{\boldsymbol{\theta}\}$ such that $\Delta_2^S(\boldsymbol{\omega}) \subseteq \mathbf{Ap}(S)$ and hence again by Proposition 7(6), $\Delta_2^S(\boldsymbol{\omega}') \neq \emptyset$. In both cases we have found an element $\boldsymbol{\omega} \in A_{i-1}$ such that $\boldsymbol{\alpha} \in \Delta_1^S(\boldsymbol{\omega})$ and $\Delta_2^S(\boldsymbol{\omega}') \neq \emptyset$. Proceeding like in Case 2 of Lemma 4 and using the inductive hypothesis, we find an element $\boldsymbol{\beta} \in \Delta^S(\boldsymbol{\alpha}')$ which is in a level of $\mathbf{Ap}(S)$ smaller than A_{e-i+2}. Thus we must have $\boldsymbol{\beta} \in A_{e-i+1}$ as in Case 1. □

Theorem 9. *Let $S \subseteq S_1 \times S_2 \subseteq \mathbb{N}^2$ be a good semigroup. Let $\mathbf{Ap}(S) = \bigcup_{i=1}^{e} A_i$ be the Apéry Set of S. Denote*

$$A_i' = \left(\bigcup_{\omega \in A_i} \Delta^S(\omega') \right) \setminus \left(\bigcup_{\omega \in A_j,\ j<i} \Delta^S(\omega') \right).$$

The following assertions are equivalent:

1. *S is symmetric.*
2. *$A_i' = A_{e-i+1}$ for every $i = 1, \ldots, e$.*

Proof. (1) $\to$ (2): Assume S to be symmetric and notice that in this case by Proposition 7(4), $A_i' \subseteq \mathbf{Ap}(S)$. As a consequence of the definition of the levels, $A_e' = A_1$ and $A_1' = A_e$, thus we can assume by induction $A_j' = A_{e-j+1}$ for $j < i$ and $A_j = A_{e-j+1}'$ for $j > e-i+1$.

We first show $A_i' \subseteq A_{e-i+1}$. Let $\boldsymbol{\delta} \in A_i'$, hence $\boldsymbol{\delta} \in \Delta^S(\boldsymbol{\omega}')$ for some $\boldsymbol{\omega} \in A_i$ and $\boldsymbol{\delta} \notin \bigcup_{\boldsymbol{\theta} \in A_j,\ j<i} \Delta^S(\boldsymbol{\theta}')$. Since $\boldsymbol{\omega} \in A_i$, by Lemma 4, $\boldsymbol{\delta} \notin A_j$ for $j < e-i+1$. By way of contradiction assume $\boldsymbol{\delta} \in A_j$ for some $j > e-i+1$. Thus, by inductive hypothesis $A_j = A_{e-j+1}'$ and $e-j+1 < i$. Hence $\boldsymbol{\delta} \in \Delta^S(\boldsymbol{\theta}')$ for some $\boldsymbol{\theta} \in A_{e-j+1}$, but this is a contradiction since $\boldsymbol{\delta} \in A_i'$ and $e-j+1 < i$.

Now we show the other containment $A_{e-i+1} \subseteq A_i'$. Let $\boldsymbol{\omega} \in A_{e-i+1}$ and take $\boldsymbol{\delta} \in \Delta^S(\boldsymbol{\omega}') \cap A_i$ which does exist by Lemma 5. Hence $\boldsymbol{\omega} \in \Delta^S(\boldsymbol{\delta}')$. We need to prove that $\boldsymbol{\omega} \notin \Delta^S(\boldsymbol{\theta}')$ for every $\boldsymbol{\theta} \in A_j$ with $j < i$. If by way of contradiction, we assume $\boldsymbol{\omega} \in \Delta^S(\boldsymbol{\theta}')$ for some $\boldsymbol{\theta} \in A_j$ with $j < i$, we can take a minimal j such that this happens, and hence by definition of A_j' and by inductive hypothesis, we get $\boldsymbol{\omega} \in A_j' = A_{e-j+1} = A_h$ with $h > e-i+1$. But this is impossible since the levels of $\mathbf{Ap}(S)$ are disjoint. With the same proof it is possible to show that $A_i = A_{e-i+1}'$ and continue with the induction to prove (2).

(2) $\to$ (1): We argue by way of contradiction. Assuming that S is not symmetric, we can find $\boldsymbol{\alpha} \notin S$ such that $\Delta^S(\boldsymbol{\gamma} - \boldsymbol{\alpha}) = \emptyset$. Since there exists a minimal $k \in \mathbb{N}$ such that $\boldsymbol{\alpha} + k\boldsymbol{e} \in S$ and for every k, also $\Delta^S(\boldsymbol{\gamma} - \boldsymbol{\alpha} - k\boldsymbol{e}) = \emptyset$, we may assume, replacing $\boldsymbol{\alpha}$ by $\boldsymbol{\alpha} + k\boldsymbol{e}$, that $\boldsymbol{\alpha} + \boldsymbol{e} \in \mathbf{Ap}(S)$. Assuming $\boldsymbol{\alpha} + \boldsymbol{e} \in A_i$, we show that $\boldsymbol{\alpha} + \boldsymbol{e} \notin A_j'$ for every j, and therefore $A_i \neq A_{e-i+1}'$. Indeed,

$$\emptyset = \Delta^S(\boldsymbol{\gamma} - \boldsymbol{\alpha}) = \Delta^S(\boldsymbol{\gamma} + \boldsymbol{e} - (\boldsymbol{\alpha} + \boldsymbol{e}))$$

and, if $\boldsymbol{\alpha} + \boldsymbol{e} \in \Delta^S(\boldsymbol{\beta}')$ for some $\boldsymbol{\beta} \in \mathbf{Ap}(S)$, we would have $\boldsymbol{\beta} \in \Delta^S(\boldsymbol{\gamma} + \boldsymbol{e} - (\boldsymbol{\alpha} + \boldsymbol{e}))$ and this is a contradiction. □

Corollary 2. *Let $S \subseteq \mathbb{N}^2$ be a symmetric good semigroup and let $\boldsymbol{\alpha} \in A_{e-i+1}$. The minimal elements of $\Delta^S(\boldsymbol{\alpha}')$ with respect to $\leq$ are in A_i.*

Proof. By Lemma 4, for every $j < i$, $\Delta^S(\boldsymbol{\alpha}') \cap A_j = \emptyset$, while by Lemma 5, $\Delta^S(\boldsymbol{\alpha}') \cap A_i \neq \emptyset$. Hence there exists a minimal element $\boldsymbol{\beta}$ of $\Delta^S(\boldsymbol{\alpha}')$ in A_i. If $\boldsymbol{\theta}$ is another minimal element of $\Delta^S(\boldsymbol{\alpha}')$, we clearly have $\boldsymbol{\alpha}' = \boldsymbol{\beta} \wedge \boldsymbol{\theta} \in S$, and hence $\boldsymbol{\theta} \in A_i$ by Lemma 2(3). □

In the next theorem, we provide a specific sequence of elements of a good semigroup S, taken one from each level A_i, behaving like the elements of the Apéry Set of a numerical semigroup with respect to sums. Notice that this sequence may not be the unique having the required property, but we give here a canonical way to construct one.

Theorem 10. *Let $S \subseteq S_1 \times S_2 \subseteq \mathbb{N}^2$ be a symmetric good semigroup and let $\mathbf{Ap}(S) = \bigcup_{i=1}^{e} A_i$ be the Apéry Set of S. Assume $e_1 \geq e_2$.*

1. *If e is even, there exists a sequence of elements $\boldsymbol{\alpha}_1, \boldsymbol{\alpha}_2, \ldots, \boldsymbol{\alpha}_e$ such that*

$$\boldsymbol{\alpha}_i \in A_i$$

and

$$\boldsymbol{\alpha}_i + \boldsymbol{\alpha}_{e-i+1} = \boldsymbol{\alpha}_e.$$

2. *If e is odd, set $e = 2d - 1$. Then, there exists a sequence of elements $\boldsymbol{\alpha}_1, \boldsymbol{\alpha}_2, \ldots, \boldsymbol{\alpha}_e, \boldsymbol{\beta}$ such that*

$$\boldsymbol{\alpha}_i \in A_i, \ \boldsymbol{\beta} \in A_d$$

and

$$\boldsymbol{\alpha}_i + \boldsymbol{\alpha}_{e-i+1} = \boldsymbol{\alpha}_e$$

for $i \neq d$, and moreover

$$\boldsymbol{\alpha}_d + \boldsymbol{\beta} = \boldsymbol{\alpha}_e$$

Proof. Let $\mathbf{Ap}(S_1) = \{\omega_1 = 0, \omega_2, \ldots, \omega_{e_1}\}$ be the Apéry Set of S_1 with elements listed in increasing order. For $i = 1, \ldots, e_1$ set

$$\boldsymbol{\alpha}_i := \min \Delta_1^S(\omega_i, 0).$$

We observe that, defined in this way, $\boldsymbol{\alpha}_i \in A_i$, since, by Corollary 1, the set $\Delta_1^S(\omega_i', 0)$ is eventually contained in $\mathbf{Ap}(S)$ and in particular, by Theorem 5 it is eventually contained in A_{e-i+1}. Hence, we get $\boldsymbol{\alpha}_i \in A_i$ by Corollary 2. Moreover, there exists a minimal $h_i \geq 0$ such that $\boldsymbol{\gamma} + \mathbf{e} - \boldsymbol{\alpha}_i + (0, h_i) \in A_{e-i+1}$.

Call $H := \max\{h_i\}$ and define again for $i = 1, \ldots, e_1$,

$$\boldsymbol{\alpha}_{e-i+1} := \boldsymbol{\gamma} + \mathbf{e} - \boldsymbol{\alpha}_i + (0, H).$$

It follows that $\boldsymbol{\alpha}_{e-i+1} \in A_{e-i+1}$ and that $\boldsymbol{\alpha}_i + \boldsymbol{\alpha}_{e-i+1} = \boldsymbol{\gamma} + \mathbf{e} + (0, H) = \boldsymbol{\alpha}_e$. The second assertion is proved in the same way by defining $\boldsymbol{\beta} := \boldsymbol{\gamma} + \mathbf{e} - \boldsymbol{\alpha}_d + (0, H)$. □

We conclude giving a quite surprising result about symmetric good semigroup with large conductor.

Proposition 11 *Let* $S \subseteq S_1 \times S_2 \subseteq \mathbb{N}^2$ *be a symmetric good semigroup and let* $\mathbf{Ap}(S) = \bigcup_{i=1}^{e} A_i$ *be the Apéry Set of S. Assume* $e_1 \geq e_2$ *and*

$$\gamma_1 > 2f(S_1) + e_1$$

where $f(S_1)$ *denotes the Frobenius number of* S_1*. Then,* $e_1 = e_2$*.*

Proof. Set $\mathbf{Ap}(S_1) = \{\omega_1 = 0, \omega_2, \ldots, \omega_{e_1}\}$ with elements listed in increasing order. In the proof of Theorem 10 is shown that $\boldsymbol{\alpha}_i = \min \Delta_1^S(\omega_i, 0) \in A_i$ and moreover by Theorem 9, any element $\boldsymbol{\theta} = (\gamma_1 + e_1 - \omega_i, t_2)$, with $t_2 \geq \gamma_2 + e_2$, is in A_{e-i+1} that is an infinite level by Corollary 1. Since by assumption,

$$\omega_{e_1} = f(S_1) + e_1 < \gamma_1 + e_1 - (f(S_1) + e_1) = \omega'_{e_1},$$

we get $\boldsymbol{\alpha}_{e_1} \ll \boldsymbol{\theta} \in A_{e-e_1+1}$ and therefore $e_1 < e - e_1 + 1 = e_2 + 1$, since by Theorem 3, $e = e_1 + e_2$. It follows that $e_1 = e_2$. □

References

1. Barucci, V., D'Anna, M., Fröberg, R.: Analytically unramified one-dimensional semilocal rings and their value semigroups. J. Pure Appl. Alg. **147**, 215–254 (2000)
2. Barucci, V., D'Annae, M., Fröberg, R.: The Apéry algorithm for a plane singularity with two branches. Beitrage zur Algebra und Geometrie **46**(1) (2005)
3. Barucci, V., D'Anna, M., Fröberg, R.: The semigroup of values of a one-dimensional local ring with two minimal primes. Comm. Algebra **28**(8), 3607–3633 (2000)
4. Bryant, L.: Goto numbers of a numerical semigroup ring and the gorensteinness of associated graded rings. Comm. Alg. **38**(6), 2092–2128 (2010)
5. Campillo, A., Delgado, F., Gusein-Zade, S.M.: On generators of the semigroup of a plane curve singularity. J. London Math. Soc. **60**(2), 420–430 (1999)
6. Campillo, A., Delgado, F., Kiyek, K.: Gorenstein properties and symmetry for one-dimensional local Cohen-Macaulay rings. Manuscripta Math. **83**, 405–423 (1994)
7. Carvalho, E., Hernandes, M.E.: The semiring of values of an algebroid curve (2017). arXiv:1704.04948v1
8. Cortadellas, T., Jafari, R., Zarzuela, S.: On the Apery sets of monomial curves. Semigroup Forum **86**(2), 289–320 (2013)
9. D'Anna, M.: The canonical module of a one-dimensional reduced local ring. Comm. Algebra **25**, 2939–2965 (1997)
10. D'Anna, M., Garcia Sanchez, P., Micale, V., Tozzo, L.: Good semigroups of N^n. Int. J. Algebra Comput. **28**, 179–206 (2018)
11. D'Anna, M., Jafari, R., Strazzanti, F.: Tangent cones of monomial curves obtained by numerical duplication (2018). arXiv:1803.08302
12. D'Anna, M., Micale, V., Sammartano, A.: When the associated graded ring of a semigroup ring is Complete Intersection. J. Pure Appl. Algebra **217**, 1007–1017 (2013)

13. Delgado, F.: The semigroup of values of a curve singularity with several branches. Manuscripta Math. **59**, 347–374 (1987)
14. Delgado, F.: Gorenstein curves and symmetry of the semigroup of value, Manuscripta Math. **61**, 285–296. Springer, Perspectives (1988)
15. García, A.: Semigroups associated to singular points of plane curves. J. Reine Angew. Math. **336**, 165–184 (1982)
16. Jäger, J.: Längenberechnung und kanonische Ideale in eindimensionalen Ringen. Arch. Math. (Basel) **29**(5), 504–512 (1977)
17. Korell, P., Schulze, M., Tozzo, L.: Duality on value semigroups. arXiv:1510.04072
18. Kunz, E.: The value-semigroup of a one-dimensional Gorenstein ring, Proc. Amer. Math. Soc. **25**, 748–751 (1970)

On Syzygies for Rings of Invariants of Abelian Groups

M. Domokos

Abstract It is well known that results on zero-sum sequences over a finitely generated abelian group can be translated to statements on generators of rings of invariants of the dual group. Here the direction of the transfer of information between zero-sum theory and invariant theory is reversed. First it is shown how a presentation by generators and relations of the ring of invariants of an abelian group acting linearly on a finite-dimensional vector space can be obtained from a presentation of the ring of invariants for the corresponding multiplicity free representation. This combined with a known degree bound for syzygies of rings of invariants yields bounds on the presentation of a block monoid associated to a finite sequence of elements in an abelian group. The results have an equivalent formulation in terms of binomial ideals, but here the language of monoid congruences and the notion of catenary degree is used.

Keywords Affine monoid · Ring of invariants · Catenary degree · Syzygies · Toric variety

1 Introduction

Let G be an abelian group (written multiplicatively) and $G_0 \subseteq G$ a finite subset. Consider the additive monoid $\mathbb{N}_0^{G_0}$ (maps from G_0 into the set of non-negative integers, with pointwise addition). It contains the submonoid

$$\mathcal{B}(G_0) := \{\alpha \in \mathbb{N}_0^{G_0} \mid \prod_{g \in G_0} g^{\alpha(g)} = 1 \in G\} \tag{1}$$

This research was partially supported by National Research, Development and Innovation Office, NKFIH K 119934.

M. Domokos (✉)
Rényi Institute, Budapest, Hungary
e-mail: domokos.matyas@renyi.hu

A. Facchini et al. (eds.), *Advances in Rings, Modules and Factorizations*,
Springer Proceedings in Mathematics & Statistics 321,
https://doi.org/10.1007/978-3-030-43416-8_5

called the *block monoid of* G_0 or the *monoid of product-one sequences over* G_0, see [14, Definition 3.4.1]. It causes no loss of generality in the construction of $\mathcal{B}(G_0)$ if we assume that the group G is generated by G_0. We note that in most of the related literature the group is written additively, and therefore the terminology of *zero-sum sequences* is used. $\mathcal{B}(G_0)$ is a reduced affine monoid (see for example [14, Theorem 3.4.2.1]), write $\mathcal{A}(\mathcal{B}(G_0))$ for the finite set of its atoms.

Our focus is on a similar but more general construction. Fix an m-tuple $\underline{g} = (g_1, \dots, g_m) \in G^m$ of elements of the abelian group G, and set

$$\mathcal{B}(\underline{g}) = \{\alpha \in \mathbb{N}_0^m \mid \prod_{i=1}^{m} g_i^{\alpha_i} = 1 \in G\}. \tag{2}$$

This is a finitely generated submonoid of the additive monoid $\mathbb{N}_0^m$. Write $\mathrm{supp}(\underline{g})$ for the subset $\{g_1, \dots, g_m\}$ of G. In the special case when $g_1, \dots, g_m$ are distinct, the monoid $\mathcal{B}(\underline{g})$ can be identified with $\mathcal{B}(\mathrm{supp}(\underline{g}))$. In general (i.e., when $g_1, \dots, g_m$ are not all distinct) the monoid $\mathcal{B}(\underline{g})$ is different from $\mathcal{B}(\mathrm{supp}(\underline{g}))$. So the construction (2) is indeed a generalization of (1) (however, see Remark 6 in Section 4).

Interest in the monoids $\mathcal{B}(G_0)$ and $\mathcal{B}(\underline{g})$ and their semigroup rings comes from several mathematical topics: factorization theory in monoids, multiplicative ideal theory, zero-sum theory, invariant theory, toric varieties, binomial or toric ideals. For example, it has been long known that results on the atoms in monoids of the form $\mathcal{B}(\underline{g})$ can be reformulated in terms of generators or degree bounds for rings of invariants of abelian groups (see, e.g., [9] for some details and references).

In this paper we shall study presentations of reduced affine monoids, with a particular attention on the monoids $\mathcal{B}(\underline{g})$. By a *monoid* we mean a commutative cancellative semigroup with an identity element. A monoid is *affine* if it is a finitely generated submonoid of a finitely generated free abelian group. We say that a monoid is *reduced* if the identity element is its only unit (invertible element). Recall that by Grillet's theorem, reduced affine monoids are exactly the monoids that are isomorphic to a finitely generated submonoid of the additive monoid $\mathbb{N}_0^k$ for some k (see for example [3, Proposition 2.16]).

Let S be a reduced affine monoid (written multiplicatively) and $\mathcal{A}(S)$ the set of atoms in S. Then $\mathcal{A}(S)$ is finite, and it is a minimal generating set of S, see for example Proposition 1.1.7 in [14]. Denote by $\mathbb{N}_0^{\mathcal{A}(S)}$ the additive monoid of functions from $\mathcal{A}(S)$ into the additive monoid $\mathbb{N}_0$ of non-negative integers; this is a free monoid generated by $|\mathcal{A}(S)|$ elements. Take commuting indeterminates $\{x_a \mid a \in \mathcal{A}(S)\}$, and let M denote the free monoid generated by them (written multiplicatively):

$$M = \{x^\alpha = \prod_{a \in \mathcal{A}(S)} x_a^{\alpha(a)} \mid \alpha \in \mathbb{N}_0^{\mathcal{A}(S)}\}$$

with multiplication $x^\alpha x^\gamma = x^{\alpha+\gamma}$. Consider the unique semigroup homomorphism

$$\pi : M \to S \text{ given by } x_a \mapsto a, \quad a \in \mathcal{A}(S)$$

(called *factorization homomorphism* in [1]). Denote by $\sim_S$ the congruence on M defined by

$$x^\alpha \sim_S x^\gamma \iff \pi(x^\alpha) = \pi(x^\gamma) \in S,$$

and call it the *defining congruence of S* (it is called *the monoid of relations of S* in [1]). The semigroup homomorphism π factors through a monoid isomorphism

$$M/\sim_S \xrightarrow{\cong} S.$$

Formally the congruence $\sim_S$ will be viewed as a subset of $M \times M$. The congruence $\sim_S$ is finitely generated by [25]. By a *presentation* of S we mean a finite subset of $M \times M$ generating the conguence $\sim_S$ (see for example [16, Section I.4] for basic notions related to semigroup congruences).

Now let us summarize the content of the present paper. Our Theorem 1 tells in particular how a presentation of $\mathcal{B}(\underline{g})$ can be derived from a presentation of $\mathcal{B}(\text{supp}(\underline{g}))$. It turns out that in most cases the catenary degree (cf. Definition 1) of $\mathcal{B}(\underline{g})$ coincides with the catenary degree of $\mathcal{B}(\text{supp}(\underline{g}))$ (see Corollary 3). Combining Theorem 1 with a degree bound of Derksen [10] for the defining relations of the ring of invariants of a linearly reductive group, we derive in Theorem 4 degree bounds for a presentation of $\mathcal{B}(\underline{g})$. In order to formulate this result, we introduce the notion of graded catenary degree of a graded monoid (see Definition 2), which is a refinement of (and an upper bound for) the ordinary catenary degree. These results on presentations of monoids have an equivalent formulation in terms of generators of (binomial) ideals of relations of semigroup rings of monoids, this is pointed out in Section 2. A Gröbner basis version of Theorem 1 is given in Theorem 2. Since the semigroup rings of the form $\mathbb{C}[\mathcal{B}(G)]$ are exactly the rings of invariants of abelian groups, the results have relevance for toric varieties; this is expanded a bit in Section 7, and some examples of toric quiver varieties are reviewed. We point out finally that Theorem 2 provides a source of Koszul algebras.

2 Ring Theoretic Characterization of the Catenary Degree

From now on S will be a reduced affine monoid. The *catenary degree* $c(S)$ of the monoid S is a basic arithmetical invariant studied in factorization theory, let us recall its definition. For $\alpha \in \mathbb{N}_0^{\mathcal{A}(S)}$ we set

$$|\alpha| = \sum_{a \in \mathcal{A}(S)} |\alpha(a)|$$

and

$$a^\alpha := \prod_{a \in \mathcal{A}(S)} a^{\alpha(a)} \in S.$$

For $\alpha, \gamma \in \mathbb{N}_0^{\mathcal{A}(S)}$ we write $\gcd(\alpha, \gamma)$ for the greatest common divisor of α, γ in the additive monoid $\mathbb{N}_0^{\mathcal{A}(S)}$ (i.e., $\gcd(\alpha, \gamma)(a) = \min\{\alpha(a), \gamma(a)\}$ for each $a \in \mathcal{A}(S)$).

Definition 1. (see [14, Definition 1.6.1]) *For $\alpha, \gamma \in \mathbb{N}_0^{\mathcal{A}(S)}$ set*

$$d(\alpha, \gamma) := \max\{|\alpha - \gcd(\alpha, \gamma)|, |\gamma - \gcd(\alpha, \gamma)|\}.$$

Given the monoid S as above, we say that α and γ can be connected by a d-chain if there exists a sequence $\alpha^{(0)} = \alpha, \alpha^{(1)}, \ldots, \alpha^{(k)} = \gamma \in \mathbb{N}_0^{\mathcal{A}(S)}$ such that $a^{\alpha^{(j)}} = a^{\alpha^{(j+1)}}$ and $d(\alpha^{(j)}, \alpha^{(j+1)}) \leq d$ for $j = 0, 1, \ldots, k-1$. The catenary degree $c(S)$ *is the minimal non-negative integer d such that if $a^\alpha = a^\gamma$, then α and γ can be connected by a d-chain.*

Remark 1. Note that $c(S) = 0$ if and only if S is a free (or factorial) monoid (i.e., S is isomorphic to the additive monoid $\mathbb{N}_0^m$ for some m), and $c(S)$ is never equal to 1.

Characterizations of the catenary degree and its variants are given in [22, 23]. In particular, [22, Proposition 16] characterizes the catenary degree in terms of the *monoid* of relations. Now extending an observation from [4] we formulate a characterization of the catenary degree in terms of semigroup congruences. Denote by $c'(S)$ the minimal non-negative integer d such that there exists a generating set $\Lambda \subset M \times M$ of the semigroup congruence $\sim_S$, satisfying that for all $(x^\alpha, x^\gamma) \in \Lambda$ we have $|\alpha| \leq d$ and $|\gamma| \leq d$. An explicit description of the semigroup congruence generated by a subset of $M \times M$ can be found for example in [18, page 176].

Proposition 1. *We have $c(S) = c'(S)$.*

Proof. Set

$$\Lambda := \{(x^\alpha, x^\gamma) \mid x^\alpha \sim_S x^\gamma \text{ and } |\alpha| \leq c(S),\ |\gamma| \leq c(S)\}.$$

We claim that Λ generates the semigroup congruence $\sim_S$. Indeed, take a pair $(\alpha, \gamma) \in \mathbb{N}_0^{\mathcal{A}(S)} \times \mathbb{N}_0^{\mathcal{A}(S)}$ such that $x^\alpha \sim_S x^\gamma$. Then α and γ can be connected by a $c(S)$-chain $\alpha^{(0)} = \alpha, \alpha^{(1)}, \ldots, \alpha^{(k)} = \gamma$. For $i = 0, 1, \ldots, k-1$ the pair $(\alpha^{(i)}, \alpha^{(i+1)})$ is of the form $(\beta^{(i)} + \delta^{(i)}, \rho^{(i)} + \delta^{(i)})$, where $\delta^{(i)}, \beta^{(i)}, \rho^{(i)} \in \mathbb{N}_0^{\mathcal{A}(S)}$, $|\beta^{(i)}| \leq c(S)$, $|\rho^{(i)}| \leq c(S)$. Moreover, since S is cancellative, $a^{\alpha^{(i)}} = a^{\alpha^{(i+1)}}$, implies $a^{\beta^{(i)}} = a^{\rho^{(i)}}$, so $x^{\beta^{(i)}} \sim_S x^{\rho^{(i)}}$. Thus $(x^{\beta^{(i)}}, x^{\rho^{(i)}}) \in \Lambda$. It follows that $(x^{\alpha^{(i)}}, x^{\alpha^{(i+1)}}) = (x^{\beta^{(i)}} x^{\delta^{(i)}}, x^{\rho^{(i)}} x^{\delta^{(i)}})$ belongs to the congruence generated by Λ for $i = 0, \ldots, k-1$, implying in turn that $(x^\alpha, x^\gamma) = (x^{\alpha^{(0)}}, x^{\alpha^{(k)}})$ belongs to the congruence generated by Λ. This proves the inequality $c'(S) \leq c(S)$.

The reverse inequality $c(S) \leq c'(S)$ is pointed out in [4, Proposition 3.1].

Fix a commutative ring R (having an identity element) and consider the semigroup rings $R[M]$ and $R[S]$. Note that $R[M] = R[x_a \mid a \in \mathcal{A}(S)]$ is the polynomial ring over R with indeterminates $\{x_a \mid a \in \mathcal{A}(S)\}$. The monoid homomorphism $\pi : M \to S$ extends uniquely to an R-algebra homomorphism

$$\pi_R : R[M] \to R[S], \quad \pi(x^\alpha) = a^\alpha. \tag{3}$$

The statement below is well known, see [18, Proposition 1.5] or [16, Chapter II.7.]:

Proposition 2. *The following conditions are equivalent for a set of pairs $B \subset \mathbb{N}_0^{\mathcal{A}(S)} \times \mathbb{N}_0^{\mathcal{A}(S)}$:*

(1) The semigroup congruence $\sim_S$ is generated by $\{(x^\alpha, x^\gamma) \mid (\alpha, \gamma) \in B\}$.
(2) The ideal $\ker(\pi_R)$ is generated by the binomials $\{x^\alpha - x^\gamma \mid (\alpha, \gamma) \in B\}$.

Remark 2. Condition (1) of Proposition 2 does not depend on the ring R; therefore, a set of binomials generates the ideal $\ker(\pi_R)$ for some ring R if and only if it generates $\ker(\pi_R)$ for any ring R.

Propositions 1 and 2 imply the following ring theoretic characterization of $c(S)$:

Corollary 1. *The catenary degree $c(S)$ is the minimal positive integer d such that the kernel of $\pi_R : R[M] \to R[S]$ is generated (as an ideal) by binomials of degree at most d for some (hence any) commutative ring R (where $R[M]$ is graded in the standard way, namely, the generators x_a have degree 1 and the non-zero scalars in R have degree 0).*

3 Graded Monoids

Let S be a *graded* monoid; that is, S is partitioned into the disjoint union of subsets S_d, $d \in \mathbb{N}_0$, such that $S_d \cdot S_e \subseteq S_{d+e}$. For $s \in S$ write $|s| = d$ if $s \in S_d$. The identity element of S necessarily belongs to S_0. We call a graded monoid *connected graded* if S_0 consists only of the identity element. It seems natural to modify Definition 1 for graded monoids as follows:

Definition 2. *Given a connected graded reduced affine monoid S, for $\alpha, \gamma \in \mathbb{N}_0^{\mathcal{A}(S)}$ set*

$$|\alpha|_{\mathrm{gr}} = \sum_{a \in \mathcal{A}(S)} \alpha(a)|a|$$

and

$$d_{\mathrm{gr}}(\alpha, \gamma) := \max\{|\alpha - \gcd(\alpha, \gamma)|_{\mathrm{gr}}, |\gamma - \gcd(\alpha, \gamma)|_{\mathrm{gr}}\}.$$

We say that α and *γ* can be connected by a chain of weight at most *d if there exists a sequence $\alpha^{(0)} = \alpha, \alpha^{(1)}, \dots, \alpha^{(k)} = \gamma \in \mathbb{N}_0^{\mathcal{A}(S)}$ such that $a^{\alpha^{(j)}} = a^{\alpha^{(j+1)}}$ and $d_{\mathrm{gr}}(\alpha^{(j)}, \alpha^{(j+1)}) \le d$ for $j = 0, 1, \dots, k-1$. The* graded catenary degree *$c_{\mathrm{gr}}(S)$ is the minimal d such that if $a^\alpha = a^\gamma$, then α and γ can be connected by a chain of weight at most d.*

As a straightforward modification of Proposition 1 we get the following.

Proposition 3. *The graded catenary degree $c_{\mathrm{gr}}(S)$ is the minimal non-negative integer d such that there exists a generating set $\Lambda \subset M \times M$ of the semigroup congruence $\sim_S$ satisfying that for all $(x^\alpha, x^\gamma) \in \Lambda$ we have $|\alpha|_{\mathrm{gr}} \leq d$ (note that $x^\alpha \sim_S x^\gamma$ implies $|\alpha|_{\mathrm{gr}} = |\gamma|_{\mathrm{gr}}$).*

The grading of S induces a grading on the semigroup ring $R[S] = \bigoplus_{d=0}^{\infty} R[S]_d$, where $R[S]_d$ is the R-submodule generated by S_d. Lift the grading to M and $R[M]$ by setting the degree of x_a to be equal to the degree $|a|$ of $a \in \mathcal{A}(S)$. Then the map $\pi_R : R[M] \to R[S]$ is a homomorphism of graded algebras, and so $\ker(\pi_R)$ is a homogeneous ideal. Our assumptions on the grading imply that all indeterminates x_a have positive degree. Moreover, $x^\alpha \sim_S x^\gamma$ implies that a^α and a^γ belong to the same homogeneous component of S, and therefore the binomials in $\ker(\pi_R)$ are homogeneous. For an ideal I in $R[M]$ denote by $\mu(I)$ the minimal non-negative integer d such that I is generated by elements of degree at most d; this number is finite for any binomial ideal I.

Corollary 2. *For any connected graded reduced affine monoid we have the equality*

$$c_{\mathrm{gr}}(S) = \mu(\ker(\pi_R)),$$

where $R[M]$ is endowed with the grading that makes π_R a homomorphism of graded algebras.

Proof. Recall that any homogeneous generating system of a homogeneous ideal I contains a minimal (with respect to inclusion) homogeneous generating system, and $\mu(I)$ is the maximal degree of an element in any minimal homogeneous generating system (this follows from the graded Nakayama lemma). The ideal $\ker(\pi_R)$ is generated by homogeneous binomials. Take a minimal set of binomials generating $\ker(\pi_R)$. Then the maximal degree of an element in this set of binomials equals $\mu(\ker(\pi_R))$ on one hand, and it equals $c_{\mathrm{gr}}(S)$ by Propositions 2 and 3, on the other hand.

Remark 3. The notions introduced in this section apply for block monoids. Indeed, $\mathbb{N}_0^m$ is graded by setting the degree of $\alpha = (\alpha_1, \dots, \alpha_m) \in \mathbb{N}_0^m$ to be $\alpha_1 + \cdots + \alpha_m$, and the submonoid $\mathcal{B}(\underline{g})$ inherits this grading.

4 Repetition of Elements

A surjective monoid homomorphism $\theta : T \to B$ between reduced affine monoids T and B is called a *transfer homomorphism* if for any $t \in T, b, c \in B$ with $\theta(t) = bc$, there exist elements $u, v \in T$ such that $t = uv$ and $\theta(u) = b$, $\theta(v) = c$ (see [14, Definition 3.2.1]).

Now let $\underline{g}$ be an m-tuple of elements in an abelian group G and $\mathcal{B}(\underline{g})$, $\mathcal{B}(\mathrm{supp}(\underline{g}))$ the monoids introduced in Section 1. The map

$$\mathbb{N}_0^m \to \mathbb{N}_0^{\mathrm{supp}(\underline{g})}, \quad \alpha \mapsto (g \mapsto \sum_{g_i = g} \alpha_i) \tag{4}$$

restricts to a *transfer homomorphism* $\mathcal{B}(\underline{g}) \to \mathcal{B}(\mathrm{supp}(\underline{g}))$. In factorization theory transfer homomorphisms are used to reduce the computation of arithmetic invariants of monoids to the corresponding invariants of other monoids (frequently of block monoids). In particular, it is known that the catenary degrees of monoids connected by a transfer homomorphism are linked as follows:

Lemma 1. *[14, Theorem 3.2.5.5] Let $\theta : T \to B$ be a transfer homomorphism, where T and B are reduced affine monoids. Then we have the inequalities*

$$c(B) \leq c(T) \leq \max\{c(B), c(T, \theta)\}$$

(see [14, page 171] for the definition of $c(T, \theta)$).

The aim of this section is to refine Lemma 1 for the transfer homomorphism $\mathcal{B}(\underline{g}) \to \mathcal{B}(\mathrm{supp}(\underline{G}))$ given by (4). More precisely, it will be shown how one can get a generating system of the defining congruence of $\mathcal{B}(\underline{g})$ from a given generating system of the defining congruence of $\mathcal{B}(\mathrm{supp}(\underline{g}))$. This will be done in a more general setup.

Assume that S is a (not necessarily connected) graded, reduced, affine monoid, and denote by $\tilde{S}$ the monoid

$$\tilde{S} = \{s[i] \mid s \in S, 0 \leq i \leq |s|\} \text{ with multiplication } s[i] \cdot t[j] = (s \cdot t)[i + j]. \tag{5}$$

So $\tilde{S}$ is a submonoid of the direct product of S and the additive monoid $\mathbb{N}_0$. Obviously the map $\tilde{S} \to S$, $s[i] \mapsto s$ is a transfer homomorphism, and

$$\mathcal{A}(\tilde{S}) = \{a[i] \mid a \in \mathcal{A}(S), \ 0 \leq i \leq |a|\}.$$

This transfer homomorphism $\tilde{S} \to S$ induces a monoid homomorphism

$$\kappa : \mathbb{N}_0^{\mathcal{A}(\tilde{S})} \to \mathbb{N}_0^{\mathcal{A}(S)}, \quad \lambda \mapsto \left(a \mapsto \sum_{i=0}^{|a|} \lambda(a[i]) \right).$$

Set

$$\delta : \mathbb{N}_0^{\mathcal{A}(\tilde{S})} \to \mathbb{N}_0, \quad \lambda \mapsto \sum_{a[i] \in \mathcal{A}(\tilde{S})} i\lambda(a[i]).$$

Note that for $\lambda, \mu \in \mathbb{N}_0^{\mathcal{A}(\tilde{S})}$ we have

$$x^\lambda \sim_{\tilde{S}} x^\mu \iff x^{\kappa(\lambda)} \sim_S x^{\kappa(\mu)} \text{ and } \delta(\lambda) = \delta(\mu). \tag{6}$$

Also we have $\delta(\lambda) \leq |a^{\kappa(\lambda)}|$.

Lemma 2. *Assume that $\alpha, \beta \in \mathbb{N}_0^{\mathcal{A}(\tilde{S})}$ satisfy*

$$\kappa(\alpha) = \kappa(\beta) \quad \text{and} \quad \delta(\alpha) = \delta(\beta).$$

Then $x^\alpha \sim x^\beta$ with respect to the semigroup congruence $\sim$ on the free monoid $\tilde{M} = \{x^\alpha \mid \alpha \in \mathbb{N}_0^{\mathcal{A}(\tilde{S})}\}$ generated by

$$\{(x_{a[k]}x_{b[l]}, x_{a[k+1]}x_{b[l-1]}) \mid a, b \in \mathcal{A}(S),\ 0 \leq k \leq |a| - 1,\ 1 \leq l \leq |b|\}.$$

Proof. Apply induction on $\sum_{a[j] \in \mathcal{A}(\tilde{S})} \alpha(a[j])$. If this number is 1, then clearly $\alpha = \beta$, and so the desired conclusion holds. Assume $\sum_{a[j] \in \mathcal{A}(\tilde{S})} \alpha(a[j]) > 1$.

Case I: x^α and x^β involve a common variable $x_{a[i]}$. Then $x^\alpha = x_{a[i]}x^{\alpha'}$, $x^\beta = x_{a[i]}x^{\beta'}$, and the assumptions on the pair (α, β) in the statement of the lemma hold for the pair (α', β'). By the induction hypothesis we may conclude that $x^{\alpha'} \sim x^{\beta'}$, implying in turn that $x^\alpha \sim x^\beta$.

Case II: x^α and x^β are not divisible by a common variable. Take a variable $x_{a[i]}$ dividing x^α. Then $\kappa(\alpha) = \kappa(\beta)$ implies that x^β is divisible by $x_{a[k]}$ for some $k \neq i$. By symmetry it is sufficient to deal with the case $i > k$. By the assumptions on α, β there must exist an atom $b \in \mathcal{A}(S)$ and integers $j < l$ such that $x_{a[i]}x_{b[j]}$ divides x^α and $x_{a[k]}x_{b[l]}$ divides T^β. We have $x^\alpha = x_{a[i]}x_{b[j]}x^\gamma \sim x_{a[i-1]}x_{b[j+1]}x^\gamma = x^{\alpha'}$. The conditions of the lemma on the pair (α, β) hold also for the pair (α', β). If $i - 1 = k$, then $x^{\alpha'}$ and x^β are divisible by a common variable, and we are back in Case I. Otherwise similarly to the above process we have $x^{\alpha'} \sim x^{\alpha''}$ where $x^{\alpha''}$ is divisible by $x_{a[i-2]}$ and $\kappa(\alpha'') = \kappa(\beta)$ and $\delta(\alpha'') = \delta(\beta)$. After finitely many such steps we get back to Case I.

Remark 4. Lemma 2 says that for the transfer homomorphism $\theta : \tilde{S} \to S, s[i] \mapsto s$ we have $c(\tilde{S}, \theta) \leq 2$ in Lemma 1.

Theorem 1. *Suppose that the congruence $\sim_S$ is generated by $\{(x^\lambda, x^\mu) \mid (\lambda, \mu) \in \Lambda\}$ for some $\Lambda \subset \mathbb{N}_0^{\mathcal{A}(S)} \times \mathbb{N}_0^{\mathcal{A}(S)}$. For each $\lambda \in \mathbb{N}_0^{\mathcal{A}(S)}$ such that $(\lambda, \mu) \in \Lambda$ or $(\mu, \lambda) \in \Lambda$ for some μ, and for each $0 \leq i \leq |a^\lambda|$ choose $\lambda[i] \in \mathbb{N}_0^{\mathcal{A}(\tilde{S})}$ such that $\kappa(\lambda[i]) = \lambda$ and $\delta(\lambda[i]) = i$ (this is clearly possible). Then the congruence $\sim_{\tilde{S}}$ is generated by*

$$\{(x^{\lambda[i]}, x^{\mu[i]}),\ (x_{a[k]}x_{b[l]}, x_{a[k+1]}x_{b[l-1]}) \mid (\lambda, \mu) \in \Lambda,\ 0 \le i \le |a^{\lambda}|,\\ a, b \in \mathcal{A}(S),\ 0 \le k \le |a| - 1,\ 1 \le l \le |b|\}.$$

Proof. The pairs given in the statement do belong to the congruence $\sim_{\tilde{S}}$. Denote by $\sim$ the semigroup congruence on $\tilde{M} = \{x^{\alpha} \mid \alpha \in \mathbb{N}_0^{\mathcal{A}(\tilde{S})}\}$ generated by them. It is sufficient to show that if $x^{\alpha} \sim_{\tilde{S}} x^{\beta}$ for some $\alpha, \beta \in \mathbb{N}_0^{\mathcal{A}(\tilde{S})}$, then $x^{\alpha} \sim x^{\beta}$. By (6) we have $x^{\kappa(\alpha)} \sim_S x^{\kappa(\beta)}$ and $\delta(\alpha) = \delta(\beta)$. Therefore there exists a sequence $(\lambda_j, \mu_j) \in \mathbb{N}_0^{\mathcal{A}(S)} \times \mathbb{N}_0^{\mathcal{A}(S)}$ and $\gamma_j \in \mathbb{N}_0^{\mathcal{A}(S)}$ $(j = 1, \dots, s)$ such that $(\lambda_j, \mu_j) \in \Lambda$ or $(\mu_j, \lambda_j) \in \Lambda$ (implying in particular that $|a^{\lambda_j}| = |a^{\mu_j}|$), $\lambda_1 + \gamma_1 = \kappa(\alpha)$, $\mu_s + \gamma_s = \kappa(\beta)$, and $\mu_j + \gamma_j = \lambda_{j+1} + \gamma_{j+1}$ for $j = 1, \dots, s-1$. Set $d := \delta(\alpha) = \delta(\beta)$. For each $j = 1, \dots, s$ choose a non-negative integer k_j with

$$d - |a^{\gamma_j}| \le k_j \le |a^{\lambda_j}|.$$

This is possible, because

$$d \le |a^{\kappa(\alpha)}| = |a^{\lambda_j + \gamma_j}| = |a^{\lambda_j}| + |a^{\gamma_j}|.$$

Taking into account Lemma 2 we get

$$\begin{aligned} x^{\alpha} \sim x^{\kappa(\alpha)[d]} &= x^{(\lambda_1+\gamma_1)[d]} \sim x^{\lambda_1[k_1]} x^{\gamma_1[d-k_1]} \sim x^{\mu_1[k_1]} x^{\gamma_1[d-k_1]} \sim \\ x^{(\mu_1+\gamma_1)[d]} &= x^{(\lambda_2+\gamma_2)[d]} \sim x^{\lambda_2[k_2]} x^{\gamma_2[d-k_2]} \sim x^{\mu_2[k_2]} x^{\gamma_2[d-k_2]} \sim \\ x^{(\mu_2+\gamma_2)[d]} &= x^{(\lambda_3+\gamma_3)[d]} \sim \dots \sim x^{(\mu_s+\gamma_s)[d]} = x^{\kappa(\beta)[d]} \sim x^{\beta}. \end{aligned}$$

Remark 5. Statements related to Theorem 1 are proved in [21], studying toric ideals associated to nested configurations (see also [26] for some generalization). The construction of $\tilde{S}$ from S (see (5)) can be seen as a special case of the construction of nested configurations. In particular, when S is a submonoid of $\mathbb{N}_0^d$ generated by finitely many elements $\alpha^{(1)}, \dots, \alpha^{(m)}$ such that there exists a $v \in \mathbb{R}^d$ with $\sum_{j=1}^{d} \alpha_j^{(i)} v_j = 1$ for all $i = 1, \dots, m$ (this implies that S can be graded in such a way that each generator has degree 1), the results of [21] apply to the binomial ideal associated to the monoid $\tilde{S}$ and yield a system of generators similar to the one given by Theorem 1.

The monoid $\mathcal{B}(\underline{g})$ can be obtained from the monoid $\mathcal{B}(\mathrm{supp}(\underline{g}))$ by a repeated application of the construction (5), and therefore Theorem 1 can be applied to relate the catenary degree of $\mathcal{B}(\underline{g})$ to the catenary degree of $\mathcal{B}(\mathrm{supp}(\underline{g}))$. Indeed, start with an m-tuple $\underline{g} = (g_1, \dots, g_m) \in G^m$ of not necessarily distinct elements in G, and denote by $\underline{\tilde{g}}$ the $m+1$-tuple $(g_1, \dots, g_m, g_m)$ obtained from $\underline{g}$ by repeating the mth component. Consider the grading on the monoid $\mathcal{B}(\underline{g})$ given by

$$\mathcal{B}(\underline{g})_d = \{\alpha \in \mathcal{B}(\underline{g}) \subseteq \mathbb{N}_0^m \mid \alpha_m = d\}.$$

Proposition 4. *We have $\mathcal{B}(\underline{\tilde{g}}) \cong \tilde{S}$, where $S = \mathcal{B}(\underline{g})$ is endowed with the above grading and $\tilde{S}$ is defined as in* (5).

Proof. A general element of $\tilde{S}$ is of the form $\alpha[i]$ where $\alpha = (\alpha_1, \dots, \alpha_m) \in \mathcal{B}(\underline{g})$ and $0 \le i \le \alpha_m$. The map $\tilde{S} \to \mathcal{B}(\underline{\tilde{g}})$ sending $\alpha[i] \in \tilde{S}$ to $(\alpha_1, \dots, \alpha_{m-1}, \alpha_m - i, i)$ is an isomorphism between the monoids $\tilde{S}$ and $\mathcal{B}(\underline{\tilde{g}})$.

Proposition 1, Theorem 1, and Proposition 4 imply the following:

Corollary 3. *We have $c(\mathcal{B}(\underline{g})) = c(\mathcal{B}(\mathrm{supp}(\underline{g}))$, unless $\mathcal{B}(\mathrm{supp}(\underline{g}))$ is a free monoid and the components $g_1, \dots, g_m$ of $\underline{g}$ are not all distinct. In the latter case we have $c(\mathcal{B}(\mathrm{supp}(\underline{g}))) = 0$ whereas $c(\mathcal{B}(\underline{g})) = 2$.*

Remark 6. Being a finitely generated reduced Krull monoid, $\mathcal{B}(\underline{g}) \cong \mathcal{B}(H_0)$ for some finite subset H_0 in an abelian group H (different from G in general) by [14, Theorem 2.7.14] also when $g_1, \dots, g_m$ are not all distinct. However, the representation of our monoid in the form $\mathcal{B}(\underline{g})$ is fundamental for our discussions.

An easy direct proof of the isomorphism $\mathcal{B}(\underline{g}) \cong \mathcal{B}(H_0)$ can be derived from the following observation. Take $\underline{g} \in G^m$, and suppose that $g_{m-1} = g_m$. Consider the group $H := G \times \mathbb{Z}$, and the sequence

$$\underline{h} := ((g_1, 0), \dots, (g_{m-1}, 0), (g_{m-1}, 1), (0, -1)) \in H^{m+1}.$$

It is easy to see that we have a monoid isomorphism $\mathcal{B}(\underline{g}) \cong \mathcal{B}(\underline{h})$. Now observe that there are less component repetitions in the sequence $\underline{h}$ then the number of component repetitions in $\underline{g}$. Note that the "price" for this manipulation was that we had to extend the group G.

5 Gröbner Bases

In this section we give a Gröbner basis variant of the results of Section 4. As it was explained in Sections 2 and 3, the catenary degree of a monoid can be expressed in terms of generators of a binomial ideal associated to the monoid. Since Gröbner bases are special generating systems of ideals in a polynomial algebra that are at the heart of many algorithms in computational commutative algebra (see for example [7] for Gröbner basis theory), it is worthwhile to translate this notion to the language of semigroup congruences.

Fix an admissible total order $\prec$ on the finitely generated free multiplicative monoid M; that is, $\prec$ is a total order such that for $x, y, z \in M$ with $x \prec y$ we have $xz \prec yz$, and $1 \prec x$ for each $x \in M \setminus \{1\}$. The latter condition ensures that the order $\prec$ is artinian (i.e., there is no infinite strictly descending chain with respect to $\prec$ in M), so any non-empty subset of M contains a unique minimal element. Note also that $\prec$ is a term order of the polynomial ring $\mathbb{F}[M]$ (where $\mathbb{F}$ is a field) in the sense of Gröbner basis theory.

Definition 3. *A finite set* $\Lambda \subset M \times M$ *is a* Gröbner system of the semigroup congruence $\sim$ on M *if the following conditions hold:*

(i) $x \sim y$ *and* $y \prec x$ *for each* $(x, y) \in \Lambda$*;*
(ii) $z \in M$ *is the minimal element in its congruence class with respect to* $\sim$ *if there is no* $(x, y) \in \Lambda$ *such that* x *divides* z *in* M.

The name *Gröbner system* is justified by the relation to Gröbner bases given in Proposition 5 (iii).

Proposition 5. *(i) If* $\Lambda \subset M \times M$ *is a Gröbner system of the semigroup congruence* $\sim$ *then* Λ *generates* $\sim$.
(ii) Every semigroup congruence $\sim$ *on* M *has a Gröbner system.*
(iii) $\Lambda \subset M \times M$ *is a Gröbner system of* $\sim$ *if and only if* $\{x - y \mid (x, y) \in \Lambda\}$ *is a Gröbner basis (satisfying* $y \prec x$ *for each of its elements* $x - y$*) of the ideal* $\ker(\pi_{\mathbb{F}})$*, where* $\mathbb{F}$ *is a field and* $\pi_{\mathbb{F}} : \mathbb{F}[M] \to \mathbb{F}[M/\sim]$ *is induced by the natural surjection* $M \to M/\sim$.

Proof. (i) Denote by $\sim_\Lambda$ the congruence generated by Λ. By assumption it is contained in $\sim$, since for each $(x, y) \in \Lambda$ we have $x \sim y$. To see the reverse inclusion it is sufficient to show that for any $z \in M$ we have $z \sim_\Lambda u$, where u is the minimal element in the $\sim$-congruence class of z. If $z = u$, we are done. Otherwise $u \prec z$, hence, by assumption there exists a pair $(x, y) \in \Lambda$ and $v \in M$ such that $z = xv$. Set $z_1 = yv$. Then $z_1 = yv \prec xv = z$ and $z \sim_\Lambda z_1$. If $z_1 = u$, then we are done. Otherwise repeat the same step for z_1 instead of z (note that $z_1 \sim z \sim u$). We obtain $z_2 \in M$ with $z_2 \prec z_1$ and $z_2 \sim_\Lambda z_1$. If $z_2 = u$ we are done, otherwise repeat the above step with z_2 instead of z_1. Since the order $\prec$ is artinian, in finitely many steps we must end up with $z \sim_\Lambda z_k = u$.

(ii) It is well known that any binomial ideal has a Gröbner basis consisting of binomials, see for example [28, Lemma 8.2.17]. Therefore the statement follows from (iii).

(iii) Suppose $\{x - y \mid (x, y) \in \Lambda\}$ is a Gröbner basis of the ideal $\ker(\pi_{\mathbb{F}})$ (where $y \prec x$ for each $(x, y) \in \Lambda$). It follows that the initial ideal of $\ker(\pi_{\mathbb{F}})$ is generated by $L := \{x \mid \exists y : (x, y) \in \Lambda\}$. Now take any $z \in M$ which is not minimal in its congruence class with respect to $\sim$. Then there is an $u \prec z$ such that $z \sim u$, so $z - u \in \ker(\pi_{\mathbb{F}})$ has initial term z. Therefore there is an $x \in L$ such that x divides z, so condition (ii) of Definition 3 holds for Λ (it is obvious that condition (i) of Definition 3 holds for Λ).

Conversely, assume that Λ is a Gröbner system of $\sim$, and consider the subset $L := \{x - y \mid (x, y) \in \Lambda\}$ in $\mathbb{F}[M]$. Denote by J the ideal generated by the initial terms of the elements in L. Clearly $L \subseteq \ker(\pi_{\mathbb{F}})$, hence, J is contained in the ideal K generated by the initial terms of the ideal $\ker(\pi_{\mathbb{F}})$. By assumption the elements of $M \setminus J$ are all minimal in their congruence class with respect to $\sim$. In particular, they are pairwise incongruent; hence, they are mapped by $\pi_{\mathbb{F}}$ to elements in $\mathbb{F}[M/\sim]$ that are linearly independent over $\mathbb{F}$. On the other hand, $M \setminus J \supseteq M \setminus K$, and the latter is mapped by $\pi_{\mathbb{F}}$ to an $\mathbb{F}$-vector space basis of $\mathbb{F}[M/\sim]$ (see for example [27,

Proposition 1.1]). It follows that $M \setminus J = M \setminus K$, implying in turn that $J = K$. The latter equality means that L is a Gröbner basis of $\ker(\pi_{\mathbb{F}})$.

We keep the notation of Section 4. In particular, S is a reduced, affine, graded monoid, and $\tilde{S}$ is the monoid defined as in (5). Fix an admissible total order $\prec$ on $M = \{x^\alpha \mid \alpha \in \mathbb{N}_0^{\mathcal{A}(S)}\}$. Define an admissible total order (denoted also by $\prec$) on the free monoid $\tilde{M}$ generated by $\{x_a \mid a \in \mathcal{A}(\tilde{S})\}$ as follows. Enumerate the atoms in $\mathcal{A}(S) = \{a_1, \dots, a_n\}$ such that $x_{a_1} \prec x_{a_2} \prec \dots \prec x_{a_n}$. Set $d_i = |a_i|$. For $\lambda, \mu \in \mathbb{N}_0^{\mathcal{A}(\tilde{S})}$ we set $x^\mu \prec x^\lambda \in \tilde{M}$ if

1. $x^{\kappa(\mu)} \prec x^{\kappa(\lambda)}$ in $(M, \prec)$; or
2. $\kappa(\mu) = \kappa(\lambda)$ and the sequence

$$(\mu(a_1[0]), \mu(a_1[1]), \dots, \mu(a_1[d_1]), \mu(a_2[0]), \dots, \mu(a_n[0]), \dots, \mu(a_n[d_n]))$$

is lexicographically greater than

$$(\lambda(a_1[0]), \lambda(a_1[1]), \dots, \lambda(a_1[d_1]), \lambda(a_2[0]), \dots, \lambda(a_n[0]), \dots, \lambda(a_n[d_n])).$$

Theorem 2. *Suppose that* $\{(x^\lambda, x^\mu) \mid (\lambda, \mu) \in \Lambda\}$ *is a Gröbner system of the semigroup congruence* $\sim_S$. *Then* $\Gamma_1 \cup \Gamma_2 \cup \Gamma_3$ *is a Gröbner system of the defining congruence* $\sim_{\tilde{S}}$ *of* $\tilde{S}$, *where*

$$\Gamma_1 = \{(x^\lambda, x^\mu) \mid \lambda, \mu \in \mathbb{N}_0^{\mathcal{A}(\tilde{S})}, \quad (\kappa(\lambda), \kappa(\mu)) \in \Lambda, \quad \delta(\lambda) = \delta(\mu)\}$$

$$\Gamma_2 = \{(x_{a[i]}x_{b[j]}, x_{a[i-1]}x_{b[j+1]}) \mid a, b \in \mathcal{A}(S),\ x_a \prec x_b,\ 0 < i \leq |a|,\ 0 \leq j < |b|\}$$

$$\Gamma_3 = \{(x_{a[i]}x_{a[j]}, x_{a[i-1]}x_{a[j+1]}) \mid a \in \mathcal{A}(S),\ 0 < i \leq j < |a|\}.$$

Proof. Take $(x^\lambda, x^\mu) \in \Gamma_1$. Then $(\kappa(\lambda), \kappa(\mu)) \in \Lambda$, hence, $x^{\kappa(\lambda)} \sim_S x^{\kappa(\mu)}$ and $x^{\kappa(\mu)} \prec x^{\kappa(\lambda)} \in M$. It follows that $x^\mu \prec x^\lambda \in \tilde{M}$. Moreover, $x^{\kappa(\lambda)} \sim_S x^{\kappa(\mu)}$ and $\delta(\lambda) = \delta(\mu)$ imply $x^\lambda \sim_{\tilde{S}} x^\mu$ by (6). Therefore condition (i) of Definition 3 holds for the elements of Γ_1. It obviously holds for the elements of Γ_2 and Γ_3 by definition of the ordering $\prec$ on $\tilde{M}$.

It remains to check that condition (ii) of Definition 3 holds for $\Gamma_1 \cup \Gamma_2 \cup \Gamma_3$. In order to do so, take $\lambda \in \mathbb{N}_0^{\mathcal{A}(\tilde{S})}$ such that $x^\lambda \in \tilde{M}$ is not minimal in its congruence class with respect to $\sim_{\tilde{S}}$.

Assume first that $x^{\kappa(\lambda)} \in M$ is not minimal in its congruence class with respect to $\sim_S$. Then by the assumption of the theorem on Λ, there exist $(\alpha, \beta) \in \Lambda$ and $\gamma \in \mathbb{N}_0^{\mathcal{A}(S)}$ such that $\kappa(\lambda) = \alpha + \gamma$. Clearly there exist $\tilde{\alpha}, \tilde{\gamma} \in \mathbb{N}_0^{\mathcal{A}(\tilde{S})}$ with $\lambda = \tilde{\alpha} + \tilde{\gamma}$, $\kappa(\tilde{\alpha}) = \alpha$, and $\kappa(\tilde{\gamma}) = \gamma$. Also $x^\alpha \sim_S x^\beta$ implies $\sum_{a \in \mathcal{A}(S)} \alpha(a)|a| = \sum_{a \in \mathcal{A}(S)} \beta(a)|a|$. It is easy to infer from this equality the existence of $\tilde{\beta} \in \mathbb{N}_0^{\mathcal{A}(\tilde{S})}$ with $\kappa(\tilde{\beta}) = \beta$ and $\delta(\tilde{\beta}) = \delta(\tilde{\alpha})$. Moreover, $(\alpha, \beta) \in \Lambda$ implies $x^\beta \prec x^\alpha$, hence, $x^{\tilde{\beta}} \prec x^{\tilde{\alpha}}$.

So $(x^{\tilde{\alpha}}, x^{\tilde{\beta}}) \in \Gamma_1$ by definition of Γ_1. Therefore Γ_1 testifies the non-minimality of x^λ as it is required by (ii) of Definition 3.

Suppose next that $x^{\kappa(\lambda)}$ is minimal in its congruence class in M with respect to $\sim_s$, and $x^\lambda \in \tilde{M}$ is not minimal in its congruence class with respect to $\sim_{\tilde{S}}$. It is easy to deduce from condition 2. of the definition of the ordering $\prec$ on $\tilde{M}$ that there must exist $(y, z) \in \Gamma_2 \cup \Gamma_3$ such that y divides x^λ. Consequently, the non-minimality of x^λ is testified by $\Gamma_2 \cup \Gamma_3$ as it is required by (ii) of Definition 3.

Remark 7. The papers [21, 26] mentioned in Remark 7 also give Gröbner bases of the binomial ideals considered there.

We call a Gröbner system Λ *quadratic* if $|\lambda| \leq 2$, $|\mu| \leq 2$ for all $(\lambda, \mu) \in \Lambda$.

Corollary 4. *If the semigroup congruence $\sim_S$ has a quadratic Gröbner system, then the semigroup congruence $\sim_{\tilde{S}}$ also has a quadratic Gröbner system.*

Koszul algebras form an interesting class of rings well studied in commutative algebra, see for example [5]. In general it is a difficult task to decide whether a quadratic algebra (a factor of the multivariate polynomial algebra modulo an ideal generated by homogeneous quadratic elements) is Koszul or not. The significance of Corollary 4 is that it can be used to produce examples of Koszul algebras. Indeed, note that if S has a quadratic Göbner system, then by Proposition 5 (iii) the ideal of relations among the generators of the semigroup algebra $\mathbb{F}[S]$ has a quadratic Gröbner basis, hence is Koszul (see [24] for background on Koszul algebras). Therefore an iterated use of Corollary 4 yields the following:

Corollary 5. *If $\mathcal{B}(\text{supp}(\underline{g}))$ has a quadratic Gröbner system, then $\mathcal{B}(\underline{g})$ also has a quadratic Gröbner system, and hence, the semigroup algebra $\mathbb{F}[\mathcal{B}(\underline{g})]$ is Koszul.*

Example 1. Consider the additive group $\mathbb{Z}/6\mathbb{Z} = \{0, 1, 2, 3, 4, 5\}$, and the monoid $S := \mathcal{B}((1, 2, 3))$. The atoms in this monoid are

$$\mathcal{A}(S) = \{(1, 1, 1),\ (4, 1, 0),\ (3, 0, 1),\ (2, 2, 0),\ (0, 0, 2),\ (0, 3, 0),\ (6, 0, 0)\}.$$

The commuting indeterminates corresponding to the atoms will be denoted by $x_1, x_2, x_3, x_4, x_5, x_6, x_7$, and $\pi : M \to S$ is the monoid homomorphism mapping the generator x_i of the free monoid M to the ith atom in the above list ($i = 1, \ldots, 7$). Denote by $\prec$ the graded reverse lexicographic order on M. That is, $x^\alpha \prec x^\beta$ if $\sum_{i=1}^7 \alpha_i < \sum_{i=1}^7 \beta_i$, or if $\sum_{i=1}^7 \alpha_i = \sum_{i=1}^7 \beta_i$ and for the largest j with $\alpha_j \neq \beta_j$ we have $\alpha_j > \beta_j$ (in particular, $x_7 \prec x_6 \prec \cdots \prec x_1$). We claim that the following is a Gröbner system for S:

$$\Lambda := \{(x_1^2, x_4x_5),\ (x_2^2, x_4x_7),\ (x_3^2, x_5x_7),\ (x_4^2, x_2x_6),\ (x_1x_2, x_3x_4),$$
$$(x_1x_3, x_2x_5),\ (x_2x_3, x_1x_7),\ (x_1x_4, x_3x_6),\ (x_2x_4, x_6x_7)\}.$$

Indeed, for each pair $(x, y) \in \Lambda$ we have $x \sim_S y$ and $y \prec x$. Moreover, it is easy to see that the elements in M not divisible by any of the first components of the pairs in Λ are the following:

$$\{wx_5^i x_6^j x_7^k \mid i, j, k \in \mathbb{N}_0, \ w \in \{1, x_1, x_2, x_3, x_4, x_3x_4\}\}.$$

Now the above elements are mapped by π to distinct elements in S. To see this note that the modulo 6 residue of the first component of $\pi(wx_5^i x_6^j x_7^k)$ uniquely determines w, and then i, j, k are determined by $\pi(wx_5^i x_6^j x_7^k)$. Thus S has a quadratic Gröbner system, and consequently by Corollary 5 the algebra $\mathbb{F}[\mathcal{B}(\underline{g})]$ is Koszul for any $\underline{g}$ with $\mathrm{supp}(\underline{g}) = \{1, 2, 3\} \subset \mathbb{Z}/6\mathbb{Z}$.

6 Relation to Invariant Theory

We need to recall a result from invariant theory (see [11] for an introduction to invariant theory). Let H be a linearly reductive subgroup of the group $GL(V)$ of invertible linear transformations of a finite-dimensional vector space V over an algebraically closed field $\mathbb{F}$. The action of H on V induces an action via graded $\mathbb{F}$-algebra automorphism on the symmetric tensor algebra $S(V)$ of V (graded in the standard way, namely, $V \subset S(V)$ is the degree 1 homogeneous component). Since H is linearly reductive, the algebra $S(V)^H = \{f \in S(V) \mid h \cdot f = f \ \forall h \in H\}$ of polynomial invariants is known to be finitely generated. Let $f_1, \dots, f_n$ be a minimal homogeneous generating system of $S(V)^H$, enumerated so that $\deg(f_1) \geq \deg(f_2) \geq \dots \geq \deg(f_n)$. Consider the $\mathbb{F}$-algebra surjection

$$\varphi : \mathbb{F}[x_1, \dots, x_n] \to S(V)^H \text{ with } x_i \mapsto f_i \quad (i = 1, \dots, n). \tag{7}$$

Endow $\mathbb{F}[x_1, \dots, x_n]$ with the grading given by $\deg(x_i) = \deg(f_i)$, so φ is a homomorphism of graded algebras. Recall that the factor of $S(V)$ modulo the ideal generated by $f_1, \dots, f_n$ is called the *algebra of coinvariants*. It is a finite-dimensional graded vector space when H is finite; in this case write $b(H, V)$ for its top degree (equivalently, all homogeneous elements in $S(V)$ of degree greater than $b(H, V)$ belong to the Hilbert ideal $S(V)f_1 + \dots + S(V)f_n$, and there is a homogeneous element in $S(V)$ of degree $b(H, V)$ not contained in the Hilbert ideal). Denote by s the Krull dimension of $S(V)^H$. Note that $s \leq n$ with equality only if $\ker(\varphi) = \{0\}$.

Theorem 3. (Derksen [10, Theorems 1 and 2])

(i) We have the inequality $\mu(\ker(\varphi)) \leq \sum_{i=1}^{\min\{n,s+1\}} \deg(f_i) - s,$

(ii) When H is finite, we have the inequality $\mu(\ker(\varphi)) \leq 2b(G, V) + 2.$

The *Davenport constant of a finite subset* G_0 of an abelian group G is defined as

$$\mathsf{D}(G_0) = \max\{|\alpha|: \alpha \in \mathcal{A}(\mathcal{B}(G_0))\},$$

where $|\alpha| = \sum_{g\in G_0} \alpha(g)$. When G_0 generates a finite subgroup in G, the *little Davenport constant of* G_0 can be defined as

$$\mathsf{d}(G_0) = \max\{|\alpha|: \alpha \in \mathbb{N}_0^{G_0},\ \forall\gamma \in \mathcal{A}(\mathcal{B}(G_0))\ \exists g \in G_0 \text{ with } \gamma(g) > \alpha(g)\},$$

the maximal length of a sequence over G_0 containing no product-one subsequence (see [14, Proposition 5.1.3.2]).

Now let $\underline{g} = (g_1, \dots, g_m)$ be a sequence of elements from an arbitrary abelian group G, and use the notation developed in Section 4. Consider the following grading of the block monoid $\mathcal{B}(\underline{g})$: for $\alpha \in \mathcal{B}(\underline{g})$ its degree is $|\alpha| = \sum_{i=1}^m \alpha_i$. The graded catenary degree $c_{\mathrm{gr}}(\mathcal{B}(\underline{g}))$ is defined in Definition 2 accordingly. Denote by $r(\mathcal{B}(\underline{g}))$ the rank of the free abelian subgroup in $\mathbb{Z}^m$ generated by $\mathcal{B}(\underline{g})$. Obviously $|\mathcal{A}(\mathcal{B}(\underline{g}))| \geq r(\mathcal{B}(\underline{g}))$ with equality if and only if $\mathcal{B}(\underline{g})$ is a free monoid. Set

$$\mathcal{A}(\mathcal{B}(\mathrm{supp}(\underline{g}))) := \{a_1, \dots, a_n\} \text{ with } |a_1| \geq |a_2| \geq \cdots \geq |a_n|.$$

Theorem 4. *(i) We have the inequalities*

$$c_{\mathrm{gr}}(\mathcal{B}(\underline{g})) \leq \max\{2|a_1|, c_{\mathrm{gr}}(\mathcal{B}(\mathrm{supp}(\underline{g})))\}, \tag{8}$$

and

$$c_{\mathrm{gr}}(\mathcal{B}(\mathrm{supp}(\underline{g}))) \leq \sum_{i=1}^{\min\{n,r+1\}} |a_i| - r, \tag{9}$$

where $r = r(\mathcal{B}(\mathrm{supp}(\underline{g})))$.

(ii) If $g_1, \dots, g_m$ *generate a finite subgroup of* G*, then*

$$c_{\mathrm{gr}}(\mathcal{B}(\underline{g})) \leq 2\mathsf{d}(\mathrm{supp}(\underline{g})) + 2. \tag{10}$$

Proof. We may assume that the components of $\underline{g}$ generate G. So G is a finitely generated abelian group, whence it is isomorphic to $G_1 \times \mathbb{Z}^k$, where G_1 is a finite abelian group, and $\mathbb{Z}^k$ is the free abelian group of rank k. Consider the linear algebraic group $H = G_1 \times T$, where T is the torus $(\mathbb{C}^\times)^k$. For an abelian linear algebraic group A denote by $X(A)$ the group of homomorphisms $A \to \mathbb{C}^\times$ (as algebraic groups). Then $X(G_1) \cong G_1$ and $X(T) \cong \mathbb{Z}^k$, whence $X(H) \cong G_1 \times \mathbb{Z}^k \cong G$. From now on we identify G with $X(H)$. Let V be a $\mathbb{C}$-vector space with basis $x_1, \dots, x_m$, and define an action of H on V via linear transformations by setting $h \cdot x_i = g_i(h)x_i$ for $i = 1, \dots, m$. The algebra $S(V)$ is the polynomial algebra $\mathbb{C}[x_1, \dots, x_m]$. The monomials span 1-dimensional invariant subspaces, and for $\alpha = (\alpha_1, \dots, \alpha_m) \in \mathbb{N}_0^m$ and $h \in H$ we have that

$$h \cdot x^{\alpha} = \left(\prod_{i=1}^{m} g_i(h)^{\alpha_i} \right) x^{\alpha}.$$

It follows that the map $\mathbb{N}_0^m \to S(V)$, $\alpha \mapsto x^{\alpha}$ induces an isomorphism of the semigroup algebras

$$\mathbb{C}[\mathcal{B}(\underline{g})] \xrightarrow{\cong} \mathbb{C}[x_1, \ldots, x_m]^H. \tag{11}$$

The isomorphism (11) is an isomorphism of graded algebras. We may select as homogeneous generators of $S(V)^H$ the monomials $\{x^{\alpha} \mid \alpha \in \mathcal{A}(\mathcal{B}(\underline{g}))\}$. Then the presentation (7) of $S(V)^H$ is identified via (11) with the presentation (3) of the semigroup algebra $\mathbb{C}[\mathcal{B}(\underline{g})]$. So $\mu(\ker(\varphi)) = \mu(\ker(\pi_{\mathbb{C}}))$ (see (3) in Section 2 for the definition of $\pi_{\mathbb{C}} : \mathbb{C}[M] \to \mathbb{C}[\mathcal{B}(\underline{g})]$). By Corollary 2 we know that $\mu(\ker(\pi_{\mathbb{C}})) = c_{\mathrm{gr}}(\mathcal{B}(\underline{g}))$. Thus we have $\mu(\ker(\varphi)) = c_{\mathrm{gr}}(\mathcal{B}(\underline{g}))$. On the other hand applying Theorem 3 (i) for $\mu(\ker(\varphi))$ in the special case when $g_1, \ldots, g_m$ are distinct (i.e., when $\mathcal{B}(\underline{g}) = \mathcal{B}(\mathrm{supp}(\underline{g}))$) we obtain the inequality (9) (by (11) the Krull dimension of $\mathbb{C}[x_1, \ldots, x_m]^H$ equals the Krull dimension of $\mathbb{C}[\mathcal{B}(\underline{g})]$, and the latter coincides with the rank of the free abelian subgroup of $\mathbb{Z}^m$ generated by $\mathcal{B}(\underline{g})$). Combining (9) with Theorem 1 and Proposition 4 we get the inequality (8). The explanation of (11) shows also $b(G, V) = \mathsf{d}(\mathrm{supp}(\underline{g}))$, so Theorem 3 (ii) yields (10).

Remark 8. When $G \cong \mathbb{Z}^k$, the group H in the above proof is an algebraic torus, and the results in [29] give various bounds for $|a_1|$ in Theorem 4 (i). Moreover, [30] characterizes the cases when $\mathbb{C}[\mathcal{B}(\underline{g})] \cong S(V)^H$ (for a torus H) is a polynomial ring, i.e., when $c_{\mathrm{gr}}(\mathcal{B}(\underline{g}) = 0$.

Corollary 6. *For any subset G_0 of a finite abelian group G we have the inequalities*

$$c_{\mathrm{gr}}(\mathcal{B}(G_0)) \leq 2\mathsf{d}(G_0) + 2 \leq 2\mathsf{D}(G) \leq 2|G|.$$

Proof. The first inequality is a special case of Theorem 4 (ii). To see the second inequality note the trivial inequality $\mathsf{d}(G_0) \leq \mathsf{d}(G)$, and the well known equality $\mathsf{d}(G) + 1 = \mathsf{D}(G)$ (cf. [14, Proposition 5.1.3.2]).

It follows immediately from Definitions 1 and 2 that

$$c(S) \leq \frac{1}{\min\{|a| \colon a \in \mathcal{A}(S)\}} c_{\mathrm{gr}}(S). \tag{12}$$

Therefore Theorem 4 implies bounds on the ordinary (not graded) catenary degree. For example, an immediate consequence of Corollary 6 and (12) is the following:

Corollary 7. *Let G_0 be a subset in a finite abelian group G. Then*

$$c(\mathcal{B}(G_0)) \leq \frac{2\mathsf{d}(G_0) + 2}{\min\{|\alpha| \colon \alpha \in \mathcal{A}(\mathcal{B}(G_0))\}}.$$

As an application we recover the following known bound on $c(\mathcal{B}(G))$:

Corollary 8. [14, Theorem 3.4.10.5] *For any finite abelian group G we have the inequality $c(\mathcal{B}(G)) \le \mathsf{D}(G)$.*

Proof. The monoid isomorphism $\mathcal{B}(G) \cong \mathcal{B}(G \setminus \{1_G\}) \times \mathcal{B}(\{1_G\}) \cong \mathcal{B}(G \setminus \{1_G\}) \times \mathbb{N}_0$ implies that $c(\mathcal{B}(G)) = c(\mathcal{B}(G \setminus \{1_G\})$. For a nontrivial group G the minimal degree of an atom in $\mathcal{B}(G \setminus \{1_G\})$ is 2, hence, Corollary 7 gives $c(\mathcal{B}(G \setminus \{1_G\}) \le \frac{2\mathsf{d}(G)+2}{2} = \mathsf{d}(G) + 1 = \mathsf{D}(G)$.

Example 2. It is known that the bound in Corollary 8 is sharp for G with $|G| \ge 3$ if and only if G is cyclic or G is an elementary 2-group by [14, Theorem 6.4.7] (see also [15], where the finite groups with $c(\mathcal{B}(G)) = \mathsf{D}(G) - 1$ are characterized). The inequality $c_{\mathrm{gr}}(\mathcal{B}(G_0)) \le 2\mathsf{d}(G_0) + 2$ in Corollary 6, the inequality in Corollary 7, and the inequalities (8) and (10) in Theorem 4 are also sharp for these groups:

(i) Set $G_0 := \{1, -1\} \subset \mathbb{Z}/n\mathbb{Z}$, $S := \mathcal{B}(G_0)$. Then $\mathcal{A}(S) = \{(n, 0), (0, n), (1, 1)\}$. The defining congruence $\sim_S$ of S is generated by $(x^n_{(1,1)}, x_{(n,0)}x_{(0,n)})$ (a single generator). We have $\mathsf{d}(G_0) = n - 1$ and $c_{\mathrm{gr}}(S) = 2n$, whereas $\min\{|\alpha| \colon \alpha \in \mathcal{A}(S)\} = 2$ and $\max\{|\alpha| \colon \alpha \in \mathcal{A}(S)\} = n$.

(ii) Let $e_1, \dots, e_n$ be a basis of the elementary 2-group of rank $n \ge 2$, and $G_0 := \{e_1 + \dots + e_n, e_1, \dots, e_n\}$, $S := \mathcal{B}(G_0)$. Then $\mathcal{A}(S) = \{a_1 := (1, 1, \dots, 1), a_2 := (2, 0, \dots, 0), a_3 := (0, 2, 0, \dots, 0), \dots, a_{n+2} := (0, \dots, 0, 2)\}$. The defining congruence of S is generated by $(x^2_{a_1}, x_{a_2} \cdots x_{a_{n+2}})$. We have $\mathsf{d}(G_0) = n$ and $c_{\mathrm{gr}}(S) = 2n + 2$, whereas $\min\{|\alpha| \colon \alpha \in \mathcal{A}(S)\} = 2$ and $\max\{|\alpha| \colon \alpha \in \mathcal{A}(S)\} = n + 1$.

7 Relation to Toric Varieties

The quotient construction of toric varieties (cf. [6]) represents a toric variety as the categorical quotient of a Zariski open subset in a vector space endowed with an action of a diagonalizable group (see [8] for background on toric varieties). Rings of invariants are at the basis of quotient constructions in algebraic geometry. In the proof of Theorem 4, we recalled that the ring of invariants $\mathbb{C}[x_1, \dots, x_m]^H$ of a diagonalizable group action is isomorphic to a semigroup ring $\mathbb{C}[\mathcal{B}(\underline{g})]$ of a block monoid. Therefore the results in Sections 4, 5, 6 have relevance for toric varieties.

In more details, the coordinate rings of affine toric varieties with no torus factors are the semigroup rings (over $\mathbb{C}$) of reduced, affine Krull monoids. This class of rings (up to isomorphism) is the same as the class of rings of invariants $\mathbb{C}[x_1, \dots, x_m]^H$, where H is an abelian group, and each variable spans an H-invariant subspace (see for example [3, Corollary 5.19]), which is the same as the class of rings of the form $\mathbb{C}[\mathcal{B}(\underline{g})]$.

Projective toric varieties can be constructed as the projective spectrum of semigroup algebras of reduced affine Krull monoids, see for example [20, Chapter

10], [8, Theorem 14.2.13]. Namely, take $\underline{g} = (g_1, \dots, g_m) \in G^m$ such that $\mathcal{B}(\underline{g}) = \{0\}$, and fix an element $h \in G$. Endow the monoid $\mathcal{B}((\underline{g}, h))$ with the grading given by $\mathcal{B}((\underline{g}, h))_d = \{\alpha \in \mathcal{B}((\underline{g}, h)) \subseteq \mathbb{N}_0^{m+1} \mid \alpha_{m+1} = d\}$, $d = 0, 1, 2, \dots$. Then $\mathbb{C}[\mathcal{B}((\underline{g}, h))]$ becomes a graded algebra, whose projective spectrum is a projective toric variety.

Example 3. We reformulate a result on presentations of homogeneous coordinate rings of projective toric quiver varieties from [12] in the terminology of the present paper. Let Γ be an acyclic quiver (i.e., a finite directed graph having no oriented cycles), with vertex set $\{1, \dots, k\}$ and arrow set $\{e_1, \dots, e_m\}$. For an arrow e_i denote by $s(e_i)$ the starting vertex of e_i, and denote by $t(e_i)$ the terminating vertex of e_i. In the additive group $\mathbb{Z}^k$ consider the elements $g_i = (g_{i1}, \dots, g_{ik})$, $i = 1, \dots, m$ given by

$$g_{ij} = \begin{cases} -1, & \text{if } j = s(e_i) \\ 1 & \text{if } j = t(e_i) \\ 0 & \text{otherwise.} \end{cases}$$

Pick an element $h \in \mathbb{Z}^k$ whose additive inverse is contained in the subgroup of $\mathbb{Z}^k$ generated by $g_1, \dots, g_m$. Then [12, Theorem 9.3] asserts that the catenary degree $c(\mathcal{B}((\underline{g}, h)))$ of $\mathcal{B}((\underline{g}, h))$ is at most 3. Moreover, it is shown in [13] that if we assume in addition that if $r(\mathcal{B}((\underline{g}, h))) \leq 5$ (i.e., the corresponding toric variety has dimension at most 4), then $c(\mathcal{B}((\underline{g}, h))) \leq 2$ with essentially one exception. We mention that presentations of the coordinate ring of affine toric quiver varieties are studied in [19].

Well known open conjectures (of increasing strength) in combinatorial commutative algebra are the following (called sometimes Bøgvad's conjecture; see [27, Conjecture 13.19] or [2]): Given a smooth, projectively normal projective toric variety, its

(i) vanishing ideal is generated by quadratic elements.
(ii) homogeneous coordinate ring is Koszul.
(iii) vanishing ideal has a quadratic Göbner basis.

For a finite subset G_0 of G with $\mathcal{B}(G_0) = \{0\}$ and an element h whose inverse belongs to the subgroup generated by G_0, the monoid $\mathcal{B}(G_0 \cup \{h\})$ is endowed with the grading such that the degree d component consists of the elements α in $\mathcal{B}(G_0 \cup \{h\})$ with $\alpha(h) = d$. Suppose that $\mathcal{B}(G_0 \cup \{h\})$ has a quadratic Gröbner system. According to the above conjecture this is expected to happen when $\mathcal{B}(G_0 \cup \{h\})$ is generated in degree 1 (so $\mathcal{B}(G_0 \cup \{h\})$ is half-factorial in the sense of factorization theory), and the projective spectrum of $\mathbb{C}[\mathcal{B}(G_0 \cup \{h\})]$ is a smooth projective variety. (For instance, in the setup of Example 3 this holds by [17] for almost all choices of h when Γ is a bipartite directed graph with 3 source and 3 sink vertices.) Then for any $\underline{g}$ with $\text{supp}(\underline{g}) = G_0$ we have by Corollary 5 that $\mathcal{B}((\underline{g}, h))$ has a quadratic Gröbner basis, and hence the algebra $\mathbb{C}[\mathcal{B}((\underline{g}, h))]$ is Koszul (although its projective spectrum typically fails to be a smooth projective variety).

Acknowledgements I thank Kálmán Cziszter, Dániel Joó, and Szabolcs Mészáros for several discussions on the topic of this paper. I am also grateful to Alfred Geroldinger for helpful comments on the manuscript.

References

1. Blanco, V., García-Sánchez, P.A., Geroldinger, A.: Semigroup-theoretical characterizations of arithmetical invariants with applications to numerical monoids and Krull monoids. Illinois J. Math. **55**, 1385–1414 (2011)
2. Bruns, W.: The quest for counterexamples in Toric geometry, Proc. CAAG 2010, Ramanujan Math. Soc. Lect. Notes Series No. 17, 1–17 (2013)
3. Bruns, W., Gubeladze, J.: Polytopes, Rings, and K-theory. Springer, New York (2009)
4. Chapman, S.T., García-Sánchez, P.A., Llena, D., Ponomarenko, V., Rosales, J.C.: The catenary and tame degree in finitely generated commutative cancellative monoids. Manuscripta Math. **120**, 253–264 (2006)
5. Conca, A., De Negri, E., Rossi, M.E.: Koszul algebras and regularity. In: Peeva, I. (ed.) Commutative Algebra: Expository Papers Dedicated to David Eisenbud on the Occasion of His 65th Birthday, pp. 285–315. Springer (2013)
6. Cox, D.: The homogeneous coordinate ring of a toric variety. J. Algebr. Geom. **4**, 15–50 (1995)
7. Cox, D., Little, J., O'Shea, D.: Ideals, Varieties, and Algorithms. Springer, New York, Berlin, Heidelberg (1992)
8. Cox, D., Little, J., Schenck, H.: Toric Varieties. Amer. Math. Soc, Providence, RI (2010)
9. Cziszter, K., Domokos, M., Geroldinger, A.: The interplay of invariant theory with multiplicative ideal theory and with arithmetic combinatorics. In: Chapman, S.T., Fontana, M., Geroldinger, A., Olberding, B. (eds.), Multiplicative Ideal Theory and Factorization Theory, pp. 43–95, Springer (2016)
10. Derksen, H.: Degree bounds for syzygies of invariants. Adv. Math. **185**, 207–214 (2004)
11. Derksen, H., Kemper, G.: Computational Invariant Theory, Encyclopaedia of Mathematical Sciences 130. Springer, Berlin, Heidelberg, New York (2002)
12. Domokos, M., Joó, D.: On the equations and classification of toric quiver varieties. Proc. Royal Soc. Edinburgh: Sect. A Math. **146**, 265–295 (2016)
13. Domokos, M., Joó, D.: Toric quiver cells. arXiv:1609.03618
14. Geroldinger, A., Halter-Koch, F.: Non-Unique Factorizations. Algebraic, Combinatorial and Analytic Theory, Pure and Applied Mathematics, Vol. 278, Chapman & Hall/CRC (2006)
15. Geroldinger, A., Zhong, Q.: The catenary degree of Krull monoids II. J. Australian Math. Soc. **98**, 324–354 (2015)
16. Gilmer, R.: Commutative Semigroup Rings. The University of Chicago Press (1984)
17. Haase, C., Paffenholz, A.: Quadratic Gröbner bases for smooth 3×3 transportation polytopes. J. Algebr. Comb. **30**, 477–489 (2009)
18. Herzog, J.: Generators and relations of abelian semigroups and semigroup rings. Manuscripta Math. **3**, 175–193 (1970)
19. Joó, D.: Complete intersection quiver settings with one dimensional vertices. Algebr. Represent. Theory **16**, 1109–1133 (2013)
20. Miller, E., Sturmfels, B.: Combinatorial Commutative Algebra, Graduate Texts in Mathematics, Vol. 227. Springer (2005)
21. Ohsugi, H., Hibi, T.: Toric rings and ideals of nested configurations. J. Commut. Algebra **2**, 187–208 (2010)
22. Philipp, A.: A characterization of arithmetical invariants by the monoid of relations. Semigroup Forum **81**, 424–434 (2010)
23. Philipp, A.: A characterization of arithmetical invariants by the monoid of relations II: the monotone catenary degree and applications to semigroup rings. Semigroup Forum **90**, 220–250 (2015)

24. Polishchuk, A., Positelski, L.: Quadratic Algebras, University Lecture Series 37. AMS, Providence, Rhode Island (2005)
25. Rédei, L.: The theory of Finitely Generated Commutative Semigroups. Pergamon, Oxford-Edinburgh-New York (1965)
26. Shibuta, T.: Gröbner bases of contraction ideals. J. Algebr. Comb. **36**, 1–19 (2012)
27. Sturmfels, B.: Gröbner Bases and Convex Polytopes, University Lecture Series 8. AMS, Providence, Rhode Island (1996)
28. Villareal, R.H.: Monomial Algebras, CRC Press, Chapman and Hall Book, Monographs and Research Notes in Mathematics, Second edition, Boca Raton, Florida (2015)
29. Wehlau, D.: Constructive invariant theory for tori. Ann. Inst. Fourier Grenoble **43**, 1055–1066 (1993)
30. Wehlau, D.: When is a ring of torus invariants a polynomial ring? Manuscripta Math. **82**, 161–170 (1994)

A Bazzoni-Type Theorem for Multiplicative Lattices

Tiberiu Dumitrescu

Abstract We prove a Bazzoni-type theorem for multiplicative lattices thus unifying several ring/monoid theoretic results of this type.

Keywords Prüfer domain · Multiplicative lattice

2010 Mathematics Subject Classification Primary 06F05 · Secondary 13F15

1 Introduction

Let D be an integral domain. Consider the following two assertions:

(i) If I is an ideal of D whose localizations at the maximal ideals are finitely generated, then I is finitely generated.

(ii) Every $x \in D - \{0\}$ belongs to only finitely many maximal ideals of D.

While $(ii) \Rightarrow (i)$ is well-known and easy to prove, Bazzoni [3, p. 630] conjectured that the converse is true for Prüfer domains. Recall that D is a *Prüfer domain* if every finitely generated ideal I of D is locally principal.

Holland et al. [10, Theorem 10] proved Bazzoni's conjecture for Prüfer domains using techniques from lattice-ordered groups theory and McGovern [14, Theorem 11] proved the same result using a direct ring theoretic approach. Halter-Koch [9, Theorem 6.11] proved Bazzoni's conjecture for r-Prüfer monoids (see Section 4). Zafrullah [16, Proposition 5] proved Bazzoni's conjecture for Prüfer v-multiplication domains. Finocchiaro and Tartarone [6, Theorem 4.5] proved Bazzoni's conjecture for almost Prüfer ring extensions (see Section 3). Recently, Chang and Hamdi [4, Theorem 2.4] proved Bazzoni's conjecture for almost Prüfer v-multiplication domains (see Section 4).

T. Dumitrescu (✉)
Facultatea de Matematica si Informatica, University of Bucharest,
14 Academiei Street, 010014 Bucharest, RO, Romania
e-mail: tiberiu@fmi.unibuc.ro

A. Facchini et al. (eds.), *Advances in Rings, Modules and Factorizations*,
Springer Proceedings in Mathematics & Statistics 321,
https://doi.org/10.1007/978-3-030-43416-8_6

The purpose of this paper is to prove a Bazzoni-type theorem for multiplicative lattices (see Section 2), thus unifying the results mentioned above (see Sections 3 and 4). Our standard references are [1, 7, 8].

2 Main Result

We use an abstract ideal theory approach, so we work with multiplicative lattices.

Definition 1. A *multiplicative lattice* is a complete lattice $(L, \leq)$ (with bottom element 0 and top element 1) which is also a multiplicative commutative monoid with identity 1 (the top element) and satisfies $a(\bigvee b_\alpha) = \bigvee ab_\alpha$ for each $a, b_\alpha \in L$.

Let L be a multiplicative lattice. The elements in $L - \{1\}$ are said to be *proper*. Denote by $Max(L)$ the set of maximal elements of L. For $x, y \in L$, set $(y : x) = \bigvee\{a \in L;\ ax \leq y\}$.

We recall some standard terminology.

Definition 2. Let L be a multiplicative lattice and let $x, p \in L$.

(1) p is *prime* if $p \neq 1$ and for all $a, b \in L$, $ab \leq p$ implies $a \leq p$ or $b \leq p$. It follows easily that every maximal element is prime.

(2) x is *compact* if whenever $x \leq \bigvee_{y \in S} y$ with $S \subseteq L$, we have $x \leq \bigvee_{y \in T} y$ for some finite subset T of S.

(3) L is a *C-lattice* if the set L^* of compact elements of L is closed under multiplication, $1 \in L^*$ and every element in L is a join of compact elements.

(4) x is *meet-principal* if $y \wedge zx = ((y : x) \wedge z)x$ for all $y, z \in L$ (in particular $(y : x)x = x \wedge y$).

(5) x is *join-principal* if $y \vee (z : x) = ((yx \vee z) : x)$ for all $y, z \in L$ (in particular $(xy : x) = y \vee (0 : x)$).

(6) x is *cancellative* if for all $y, z \in L$, $xy = xz$ implies $y = z$.

(7) x is *CMP* (ad hoc name) if x is cancellative and meet-principal.

(8) L is a *lattice domain* if $(0 : a) = 0$ for all $a \in L - \{0\}$.

In the sequel, we work with C-lattices and their localization theory. Let L be a C-lattice. For $p \in L$ a prime element and $x \in L$, we set

$$x_p = \bigvee\{a \in L^*;\ ab \leq x \text{ for some } b \in L^* \text{ with } b \not\leq p\}.$$

Then $L_p := \{x_p;\ x \in L\}$ is again a lattice with multiplication $(x_p, y_p) \mapsto (xy)_p = (x_p y_p)_p$, join $\{(b_\alpha)_p\} \mapsto (\bigvee(b_\alpha)_p)_p = (\bigvee b_\alpha)_p$ and meet $\{(b_\alpha)_p\} \mapsto (\bigwedge(b_\alpha)_p)_p$. The next lemma collects several basic properties.

Lemma 3. *Let L be a C-lattice, let $x, y \in L$ and let $p \in L$ be a prime element.*

(1) $x_p = 1$ *if and only if* $x \not\leq p$.

(2) $(x \wedge y)_p = x_p \wedge y_p$.

(3) *If x is compact, then* $(y : x)_p = (y_p : x_p)$.

(4) $x = y$ *if and only if* $x_m = y_m$ *for each* $m \in Max(L)$.

(5) *A cancellative element x is CMP if and only if* $(y : x)x = x \wedge y$ *for all* $y \in L$.

(6) *If x is compact, then x_p is compact in L_p. Conversely, if x_p is compact in L_p, then $x_p = c_p$ for some compact element $c \leq x$.*

(7) *If x and y are CMP elements, then so is xy.*

(8) *If x is compact, y is CMP and $x \leq y$, then $(x : y)$ is compact.*

Proof. For (1–4) see [11, pp. 201–202], for (5) see [15, Lemma 2.10], while (6–7) follow easily from definitions. We prove (8). Note that $(x : y)y = x \wedge y = x$. Suppose that $(x : y) \leq \bigvee_{i \in A} z_i$. Then $x = (x : y)y \leq \bigvee_{i \in A} z_i y$, so $(x : y)y \leq \bigvee_{i \in B} z_i y$ for some finite subset B of A. Cancel y to get $(x : y) \leq \bigvee_{i \in B} z_i$. □

Say that x and $y \in L$ are *comaximal* if $x \vee y = 1$. Clearly, $x \vee yz = 1$ if and only if $x \vee y = 1$ and $x \vee z = 1$. When $t \leq u$ we say that t is *below* u or that u is *above* t.

Lemma 4. *Let L be a C-lattice and $z \in L - \{1\}$ a compact element such that $\{m \in Max(L);\ z \leq m\}$ is infinite. There exists an infinite set $\{a_n;\ n \geq 1\}$ of pairwise comaximal proper compact elements such that $z \leq a_n$ for each n.*

Proof. We may clearly assume that $z = 0$ (just change L by $\{x \in L;\ x \geq z\}$). Say that a proper compact element h is *big* (ad hoc name) if h is below only one maximal element $M(h)$. We separate in two cases.

Case (1): *Every proper compact element is below some big compact element.* We proceed by induction. Suppose that $n \geq 1$ and we already have big compacts $a_1,...,a_n$ such that $M(a_1),...,M(a_n)$ are distinct maximal elements (for $n = 1$ just pick an arbitrary big compact a_1). Let p be a maximal element other than $M(a_1),...,M(a_n)$. There exists a compact element $c \leq p$ such that $c \nleq M(a_i)$ for $1 \leq i \leq n$ (take $c = c_1 \vee \cdots \vee c_n$ where each $c_i \in L^*$ satisfies $c_i \leq p$ and $c_i \nleq M(a_i)$). Then take a big compact element $a_{n+1} \geq c$. This way we construct an infinite set $\{a_n;\ n \geq 1\}$ of big compacts such that all $M(a_n)$'s are distinct. Hence the a_n's are pairwise comaximal.

Case (2): *There exists a proper compact element a_0 which is not below any big compact element.* Clearly every proper compact above a_0 inherits this property. Pick two distinct maximal elements p and q above a_0. As L is a C-lattice, there exist two comaximal compacts $a_1 \leq p$ and $b_1 \leq q$ (note that $p \vee q = 1$, express p and q as joins of compact elements and use the fact that 1 is compact). Repeating this argument for a_1, there exist two comaximal proper compacts $a_2 \geq a_1$ and $b_2 \geq a_1$. Note that b_2 and b_1 are comaximal. Thus we construct inductively an infinite set $\{b_n;\ n \geq 1\}$ of pairwise comaximal proper compact elements. □

In a C-lattice L, we say that an element x is *locally compact* if x_m is compact in L_m for each $m \in Max(L)$. We state our main result which is a Bazzoni-type theorem for C-lattices.

Theorem 5. *Let L be a C-lattice domain satisfying the following two conditions:*

(a) *every nonzero element is above some cancellative compact element, and*

(b) *every compact element $x \neq 1$ has some power x^n below some proper CMP element.*

Then the following conditions are equivalent:

(i) *Every locally compact element of L is compact.*

(ii) *Every nonzero element is below at most finitely many maximal elements.*

Proof. $(ii) \Rightarrow (i)$. Although this part is well-known and easy, we include a proof for the reader's convenience. Let x be a nonzero locally compact element of L and let $a \leq x$ be a nonzero compact element. By (ii), there are only finitely many maximal elements above a, say $m_1,...,m_k$. For each i between 1 and k, pick a compact element $c_i \leq x$ such that $x_{m_i} = (c_i)_{m_i}$. A local check shows that $x = a \vee c_1 \vee \cdots \vee c_k$, so x is compact. Note that this part works for any C-lattice.

$(i) \Rightarrow (ii)$. Deny, so suppose that some nonzero element c is below infinitely many maximal elements. By hypothesis (a), we may assume that c is a cancellative compact element. By Lemma 4 and hypothesis (b), there exist an infinite set $\{b_n;\ n \geq 1\}$ of proper pairwise comaximal CMP elements and integers $k_n \geq 1$ such that $c^{k_n} \leq b_n$ for $n \geq 1$ (k_n minimal with this property). Restricting to a subsequence, we may assume that $k_n \leq k_{n+1}$ for all n. We then have $c^{k_n} \leq b_1 \wedge \cdots \wedge b_n = b_1 \cdots b_n$ for all n.

Claim $(*)$: *The element $a := \bigvee_{n \geq 1}(c^{k_n} : b_1 \cdots b_n)$ is locally compact.*

Pick $m \in \mathrm{Max}(L)$. Since the b_n's are pairwise comaximal, m is above at most one of them. Assume first that $m \geq b_s$. Since each product $b_1 \cdots b_n$ is compact, we get

$$a_m = (\bigvee_{n \geq 1}((c^{k_n})_m : (b_1 \cdots b_n)_m))_m = (c^{k_1} \vee (c^{k_s} : b_s))_m$$

which is compact in L_m, cf. Lemma 3. Similarly, when m is above no b_n, we get $a_m = (c^{k_1})_m$, so a_m is compact in L_m, hence Claim $(*)$ is proved. By (i), a is compact. So $a = \bigvee_{n=1}^{q}(c^{k_n} : b_1 \cdots b_n)$ for some $q \geq 1$. We get

$$(c^{k_{q+1}} : b_1 \cdots b_{q+1}) \leq (c^{k_1} : b_1 \cdots b_q)$$

so multiplying by $b_1 \cdots b_{q+1}$ (which is a CMP element) and taking into account that $c^{k_{q+1}} \leq b_1 \cdots b_{q+1}$, we get

$$c^{k_{q+1}} \leq (c^{k_1} : b_1 \cdots b_q) b_1 \cdots b_{q+1} \leq c^{k_1} b_{q+1}.$$

Since $k_{q+1} \geq k_1$ and c^{k_1} is cancellative, we get $c^{k_{q+1}-k_1} \leq b_{q+1}$, which is a contradiction since k_{q+1} was minimal with $c^{k_{q+1}} \leq b_{q+1}$. □

Recall that a C-lattice domain is a *Prüfer lattice* if every compact element is principal (i.e., meet-principal and join-principal). In a C-lattice domain, every nonzero join-principal element x is cancellative (because $(yx : x) = y \vee (0 : x) = y$ for each y). So in a Prüfer lattice domain every nonzero compact element is CMP.

Corollary 6. *Let L be a C-lattice domain in which every nonzero compact element is CMP (e.g., a Prüfer lattice domain). Then conditions (i) and (ii) of Theorem 5 are equivalent.*

Bazzoni's conjecture for Prüfer domains [10, Theorem 10] (see Introduction) follows from Corollary 6 since the ideal lattice of a Prüfer domain is clearly a Prüfer lattice.

3 Almost Prüfer Extensions

We recall several definitions from [6, 12]. Let $A \subseteq B$ be a commutative ring extension and I an ideal of A. Then I is called *B-regular* if $IB = B$ and I is called *B-invertible* if $IJ = A$ for some A-submodule J of B. Every B-invertible ideal is B-regular, since $A = IJ \subseteq IB$ implies $IB = B$. We say that $A \subseteq B$ is an *almost Prüfer extension* if every finitely generated B-regular ideal of A is B-invertible.

Finocchiaro and Tartarone [6, Theorem 4.5] proved Bazzoni's conjecture for almost Prüfer ring extensions. We state their result and derive it from Corollary 6.

Theorem 7. (Finocchiaro and Tartarone) *If $A \subseteq B$ is an almost Prüfer extension, the following are equivalent:*

(i) Every B-regular locally principal ideal of A is B-invertible.

(ii) Every B-regular ideal of A is contained in only finitely many maximal ideals of A.

Proof. It is well-known and easy to prove that (ii) implies (i), see [6, Corollary 3.5]. We prove the converse. Let L be the set of all B-regular ideals of A together with the zero ideal and order L by inclusion. As shown in [15, Lemma 7.1], L is a C-lattice domain under usual ideal multiplication, where the join is the ideal sum and the meet is the ideal intersection except the case when we get a non-B-regular ideal when we put $\bigwedge = 0$. By [15, Lemma 7.1], the set L^* of compact elements in L is exactly the set of (B-regular) finitely generated ideals of A together with the zero ideal. After this preparation it becomes clear that $[(i) \Rightarrow (ii)]$ follows from Corollary 6 provided we prove the two claims below. Write $x \in L$ as $\widehat{x}$ when considered as an ideal of A.

Claim 1: *Every nonzero compact element of L is a CMP element.*

Let c be a nonzero compact element of L. As $A \subseteq B$ is almost Prüfer, $\widehat{c}$ is a B-invertible ideal, so $\widehat{c}J = A$ for some A-submodule J of B. Then c is clearly cancellative. By Lemma 3, it suffices to show that $(x : c)c = x \wedge c$ for each $x \in L$. Changing x by $x \wedge c$, we may assume that $x \leq c$. We have $\widehat{x} = \widehat{x}J\widehat{c}$, so $x = yc$ where $y \in L$ is such that $\widehat{y} = \widehat{x}J$ (note that $\widehat{x}J \subseteq A$). From $x = yc$ we get $y \leq (x : c)$, so $x = yc \leq (x : c)c \leq x$, thus $(x : c)c = x$.

Claim 2: *Every locally compact element of L is compact.*

Suppose that c is a nonzero locally compact element of L. Let m be a maximal element of L, that is, $\widehat{m}$ is a B-regular maximal ideal of A. So $c_m = \bigvee\{y \in L^*;\ ys \leq c$ for some $s \in L^*, s \nleq m\}$ is compact in the lattice $L_m = \{x_m;\ x \in L\}$. Then $c_m = h_m$ for some $h \in L^*$. Extending these ideals in $A_{\widehat{m}}$, we get $\widehat{c}A_{\widehat{m}} = \widehat{c_m}A_{\widehat{m}} = \widehat{h_m}A_{\widehat{m}} = \widehat{h}A_{\widehat{m}}$. Since $A \subseteq B$ is almost Prüfer, $\widehat{h}$ is B-invertible. By [12, Proposition 2.3], $\widehat{h}A_{\widehat{m}} = \widehat{c}A_{\widehat{m}}$ is a principal ideal of $A_{\widehat{m}}$. Thus $\widehat{c}$ is a locally principal ideal of A. By (i), $\widehat{c}$ is B-invertible, so $\widehat{c}$ is finitely generated, cf. [12, Proposition 2.3]. Thus c is compact in L. □

4 Ideal Systems on Monoids and Integral Domains

Let H be a commutative multiplicative monoid (with zero element 0 and unit element 1) such that every nonzero element of H is cancellative and let $\mathcal{P}(H)$ be the power set of H. A map $r : \mathcal{P}(H) \to \mathcal{P}(H)$, $X \mapsto X_r$, is an *ideal system* on H if the following conditions hold for all $X, Y \in \mathcal{P}(H)$ and $c \in H$:

$$(1)\ cX_r = (cX)_r,\ (2)\ X \subseteq X_r,\ (3)\ X \subseteq Y \text{ implies } X_r \subseteq Y_r,\ (4)\ (X_r)_r = X_r.$$

Then X_r is called the r-*closure* of X and a set of the form X_r is called an r-ideal. An r-ideal I is r-*finite* if $I = Y_r$ for some finite subset Y of I. The ideal system r is called *finitary* if $X_r = \bigcup\{Y_r;\ Y \subseteq X \text{ finite}\}$.

Assume that r is finitary. A proper r-ideal P is *prime* if, for $x, y \in H$, $xy \in P$ implies $x \in P$ or $y \in P$. The *localization of* H *at* P is the fraction monoid $H_P = \{x/t;\ x \in H, t \in H - P\}$ which comes together with the finitary ideal system r_P defined by $\{a_1/s_1, ..., a_n/s_n\}_{r_P} = (\{a_1, ..., a_n\}_r)_P$ for all $a_1, ..., a_n \in H$ and $s_1, ..., s_n \in H - P$. A *maximal* r-*ideal* is a maximal element of the set of proper r-ideals of H. Any proper r-ideal is contained in a maximal one and a maximal r-ideal is prime. A nonzero r-ideal I is r-*invertible* if I is r-finite and r-*locally principal* (i.e., $I_M = \{x/t;\ x \in I, t \in H - M\} = yH_M$ with $y \in I_M$ (depending on M) for all maximal r-ideals M). Next H is an r-*Prüfer monoid* if every nonzero r-finite r-ideal of H is r-invertible. For complete details we refer to [8].

Halter-Koch [9, Theorem 6.11] proved Bazzoni's conjecture for r-Prüfer monoids. We state his result and derive it from Corollary 6.

Theorem 8. (Halter-Koch) *If H is an r-Prüfer monoid for some finitary ideal system r on H, the following are equivalent.*

(i) Every r-locally principal r-ideal of H is r-finite.

(ii) Every nonzero r-ideal of H is contained in only finitely many maximal r-ideals.

Proof. It is well-known and easy to prove that (ii) implies (i). We prove the converse. Let L be the set of all r-ideals of H ordered by inclusion. As shown in [8, Chapter 8], L is a C-lattice domain under r-ideal multiplication $(I, J) \mapsto (IJ)_r$, join $\bigvee\{J_\alpha\} := (\bigcup J_\alpha)_r$ and meet $\bigwedge\{J_\alpha\} := \bigcap J_\alpha$. By [15, Lemma 8.1], the set L^* of compact elements in L is exactly the set of r-finite r-ideals of H. After this preparation it becomes clear that $[(i) \Rightarrow (ii)]$ follows from Corollary 6 provided we prove the two claims below. Write $x \in L$ as $\widehat{x}$ when considered as an r-ideal of H.

Claim 1: *Every nonzero compact element of L is a CMP element.*
Let $c \in L$ be a nonzero compact, in other words $\widehat{c}$ is nonzero r-finite r-ideal. Since H is r-Prüfer, $\widehat{c}$ is r-invertible. So c is a CMP element of L, cf. [15, Lemma 8.2].

Claim 2: *Every locally compact element of L is compact.*
Suppose that c is a locally compact element of L. Let m be a maximal element of L, that is, $\widehat{m}$ is a maximal r-ideal of H. So $c_m = \bigvee\{y \in L^*;\ ys \leq c$ for some $s \in L^*, s \nleq m\}$ is compact in the lattice $L_m = \{x_m;\ x \in L\}$. Then $c_m = h_m$ for some $h \in L^*$. Switching to H, we have $(\widehat{c})_{\widehat{m}} = (\widehat{c_m})_{\widehat{m}} = (\widehat{h_m})_{\widehat{m}} = \widehat{h}_{\widehat{m}}$. Since H is an r-Prüfer monoid, $\widehat{h}_{\widehat{m}}$ is a principal $r_{\widehat{m}}$-ideal of $H_{\widehat{m}}$, cf. [8, Theorem 12.3]. Thus $\widehat{c}$ is an r-locally principal r-ideal. By (i), $\widehat{c}$ is r-finite, thus c is compact in L. □

Next we present an application for integral domains. Let D be an integral domain. The t ideal system on D is defined by $X_t = \bigcup\{Y_v;\ Y \subseteq X$ finite$\}$ for all $X \subseteq D$, where $Y_v = \bigcap\{(aD :_D b);\ a, b \in D, bY \subseteq aD\}$. Clearly t is finitary, so the set $Max_t(D)$ of maximal t-ideals is nonempty, see [8, Chapter 11].

The w ideal system on D is defined by $X_w = \bigcap\{XD_M;\ M \in Max_t(D)\}$ for all $X \subseteq D$, where XD_M is the ideal generated by X in D_M. So a w-ideal is an ideal of the ring D. For $X \subseteq D$, we have $X_w = (XD)_w$ and $X_w D_M = XD_M$ for each $M \in Max_t(D)$. Moreover, w is finitary and the set of maximal w-ideals is exactly $Max_t(D)$. A w-finite ideal has the form $((a_1, ..., a_n)D)_w$ for some a_i's in D. And a nonzero w-ideal is *w-invertible* if it is w-finite and t-locally principal (i.e., ID_M is a principal ideal of D_M for each $M \in Max_t(D)$). For details on the w ideal system we refer to [2, 5].

According to [13], D is an *almost Prüfer v-multiplication domain* (in short *APVMD*) if for every $a_1, ..., a_n \in D - \{0\}$, the ideal $((a_1^k, ..., a_n^k)D)_w$ is w-invertible for some $k \geq 1$. Say that a w-ideal I of D is *t-locally finitely generated*, if ID_M is a finitely generated ideal of D_M for each $M \in Max_t(D)$.

Chang and Hamdi [4, Theorem 2.4] proved Bazzoni's conjecture for APVMDs. We state their result and derive it from Theorem 5.

Theorem 9. (Chang and Hamdi) *For an APVMD D, the following statements are equivalent:*

(i) *Each nonzero t-locally finitely generated w-ideal of D is w-finite.*

(ii) *Every nonzero ideal of D is contained in only finitely many maximal t-ideals.*

Proof. It is well-known and easy to prove that (ii) implies (i), see for instance the proof of (3) $\Rightarrow$ (1) in [4, Theorem 2.4]. We prove that (i) implies (ii). Let L be the set of all w-ideals of D ordered by inclusion. As shown in [8, Chapter 8], L is a C-lattice domain under w-ideal multiplication $(I, J) \mapsto (IJ)_w$, join $\bigvee\{J_\alpha\} := (\bigcup J_\alpha)_w$ and meet $\bigwedge\{J_\alpha\} := \bigcap J_\alpha$. By [15, Lemma 8.1], the set L^* of compact elements in L is exactly the set of w-finite ideals of D. Condition (a) of Theorem 5 holds clearly for L (any nonzero ideal contains a nonzero principal ideal). After this preparation it becomes clear that $[(i) \Rightarrow (ii)]$ follows from Theorem 5 provided we prove the two claims below. Write $x \in L$ as $\widehat{x}$ when considered as a w-ideal of D.

Claim 1: *Every* $d \in L^* - \{1\}$ *has some power* d^n *below some proper CMP element.*
Let $d \in L^* - \{1\}$. Then $\widehat{d} = (a_1, ..., a_k)_w$ for some elements $a_i \in \widehat{d}$. Since D is an APVMD, $(a_1^s, ..., a_k^s)_w$ is w-invertible for some $s \geq 1$. If $(a_1^s, ..., a_k^s)_w = \widehat{f}$ with $f \in L$, then f is a proper CMP element of L, cf. [15, Lemma 8.2]. Moreover $d^{sk} \leq f$, so Claim 1 is proved.

Claim 2: *Every locally compact element of L is compact.*
Suppose that c is a locally compact element of L. Let m be a maximal element of L, that is, $\widehat{m}$ is a maximal w-ideal of D. So $c_m = \bigvee\{y \subset L^*;\ ys \leq c$ for some $s \in L^*, s \nleq m\}$ is compact in the lattice $L_m = \{x_m;\ x \in L\}$. Then $c_m = h_m$ for some $h \in L^*$. Hence $\widehat{h} = (b_1, ..., b_n)_w$ for some elements $b_i \in \widehat{h}$. Switching to D, we have

$$\widehat{c}D_{\widehat{m}} = \widehat{c_m}D_{\widehat{m}} = (\widehat{h_m})D_{\widehat{m}} = \widehat{h}D_{\widehat{m}} = (b_1, ..., b_n)_w D_{\widehat{m}} = (b_1, ..., b_n)D_{\widehat{m}}.$$

Thus $\widehat{c}$ is a t-locally finitely generated ideal of D. By (i), $\widehat{c}$ is w-finite, thus c is compact in L. □

Acknowledgements I thank the referee for the close reading of the manuscript and for the detailed report which heavily improved the quality of this paper.

References

1. Anderson, D.D.: Abstract commutative ideal theory without chain conditions. Algebra Univ. **6**, 131–145 (1976)
2. Anderson, D.D., Cook, S.J.: Two star-operations and their induced lattices. Comm. Algebra **28**, 2461–2475 (2000)
3. Bazzoni, S.: Class semigroups of Prüfer domains. J. Algebra **184**, 613–631 (1996)
4. Chang, G.W. Hamdi, H.: Bazzoni's conjecture and almost Prüfer domains, to appear in Comm. Algebra
5. Fanggui, W., McCasland, R.L.: On w-modules over strong Mori domains. Comm. Algebra **25**, 1285–1306 (1997)
6. Finnochiaro, C., Tartarone, F.: Invertibility of ideals in Prüfer extensions. Comm. Algebra **45**, 4521–4527 (2017)
7. Gilmer, R.: Multiplicative Ideal Theory. Marcel Dekker, New York (1972)
8. Halter-Koch, F.: Ideal System. An Introduction to Multiplicative Ideal Theory, Marcel Dekker, New York (1998)
9. Halter-Koch, F.: Clifford semigroups of ideals in monoids and domains. Forum Math. **21**, 1001–1020 (2009)
10. Holland, W.C., Martinez, J., McGovern, W.W., Tesemma, M.: Bazzoni's conjecture. J. Algebra **320**, 1764–1768 (2008)
11. Jayaram, C., Johnson, E.G.: S-prime elements in multiplicative lattices. Periodica Math. Hungarica **31**, 201–208 (1995)
12. Knebusch, M., Zhang, D.: Manis Valuations and Prüfer Extensions I: A New Chapter in Commutative Algebra. Springer, Berlin, Heidelberg (2002)
13. Lewin, R.L.: Generalizations of GCD-domains and related topics, Ph.D. Thesis, University of Iowa (1991)

14. McGovern, W.W.: Prüfer domains with Clifford class semigroup. J. Comm. Algebra **3**, 551–559 (2011)
15. Olberding, B., Reinhart, A.: Radical factorization in commutative rings, monoids and multiplicative lattices, Math. (2018). arXiv: 1811.00242v1, math.AC
16. Zafrullah, M.: t-invertibility and Bazzoni-like statements. J. Pure Appl. Algebra **214**, 654–657 (2010)

What is the Spectral Category?

María José Arroyo Paniagua, Alberto Facchini, Marino Gran, and George Janelidze

Abstract For a category $\mathcal{C}$ with finite limits and a class $\mathcal{S}$ of monomorphisms in $\mathcal{C}$ that is pullback stable, contains all isomorphisms, is closed under composition, and has the strong left cancellation property, we use pullback stable $\mathcal{S}$-essential monomorphisms in $\mathcal{C}$ to construct a spectral category $\mathrm{Spec}(\mathcal{C}, \mathcal{S})$. We show that it has finite limits and that the canonical functor $\mathcal{C} \to \mathrm{Spec}(\mathcal{C}, \mathcal{S})$ preserves finite limits. When $\mathcal{C}$ is a normal category, assuming for simplicity that $\mathcal{S}$ is the class of all monomorphisms in $\mathcal{C}$, we show that pullback stable $\mathcal{S}$-essential monomorphisms are the same as what we call subobject-essential monomorphisms.

Keywords Spectral category · Normal category · Essential monomorphism

---

M. J. Arroyo Paniagua
Departamento de Matemáticas, Universidad Autónoma Metropolitana, Unidad Iztapalapa, Ciudad de México, México
e-mail: mariajose.mja@gmail.com

A. Facchini (✉)
Dipartimento di Matematica "Tullio Levi-Civita", Università di Padova, Padova, Italy
e-mail: facchini@math.unipd.it

M. Gran
Institut de Recherche en Mathématique et Physique, Université catholique de Louvain, 1348 Louvain-la-Neuve, Belgique
e-mail: marino.gran@uclouvain.be

G. Janelidze
Department of Mathematics and Applied Mathematics, University of Cape Town, Rondebosch 7700, South Africa
e-mail: george.janelidze@uct.ac.za

A. Facchini et al. (eds.), *Advances in Rings, Modules and Factorizations*, Springer Proceedings in Mathematics & Statistics 321,
https://doi.org/10.1007/978-3-030-43416-8_7

1 Introduction

The *spectral category* $\mathrm{Spec}(\mathcal{C})$ of a Grothendieck category $\mathcal{C}$ was introduced by Gabriel and Oberst [11]. According to the Abstract of [11], $\mathrm{Spec}(\mathcal{C})$ is obtained from $\mathcal{C}$ by formally inverting all essential monomorphisms. Although there is no reference to Gabriel and Zisman [12], the definition given in Section 1.2 of [11] is, in fact, a construction based on the fact that the class of essential monomorphisms in $\mathcal{C}$ admits the calculus of right fractions. Indeed, it presents the abelian groups $\mathrm{Hom}_{\mathrm{Spec}(\mathcal{C})}(A, B)$ (for all $A, B \in \mathrm{Ob}(\mathcal{C}) = \mathrm{Ob}(\mathrm{Spec}(\mathcal{C}))$) as directed colimits

$$\mathrm{Hom}_{\mathrm{Spec}(\mathcal{C})}(A, B) = \mathrm{colim}\,\mathrm{Hom}_{\mathcal{C}}(A', B)$$

taken over all subobjects A' of A. It is also easy to see that the spectral category $\mathrm{Spec}(\mathcal{C})$ can equivalently be defined as the quotient category of the category of injective objects in $\mathcal{C}$ modulo the ideal consisting of all morphisms in $\mathcal{C}$ whose kernels are essential monomorphisms. Although this is not mentioned in [11], it is said there that $\mathrm{Spec}(\mathcal{C})$ is a replacement of the *spectrum* of $\mathcal{C}$, which is defined (when $\mathcal{C}$ is the category of modules over a ring) as the collection of isomorphism classes of indecomposable injective objects.

Introducing the spectral category of a Grothendieck category $\mathcal{C}$ can also be motivated by non-functoriality of injective envelopes as follows. For each object C in $\mathcal{C}$, let us fix an injective envelope (=injective hull) $\iota_C : C \to E(C)$ of it. One might expect E to become an endofunctor of $\mathcal{C}$, and ι to become a natural transformation $1_{\mathcal{C}} \to E$. However, there are strong negative results against these expectations:

- According to Proposition 1.12 in [13], E cannot be made a functor even when $\mathcal{C}$ is the category of abelian groups.
- Let R be a ring and $\mathcal{C}$ the category of R-modules. The ring R can be chosen in such a way that not all R-modules are injective, but E can be made an endofunctor of $\mathcal{C}$ (see [13, Exercise 24, p. 48] or [8]), but even in those cases ι will not become a natural transformation $1_{\mathcal{C}} \to E$. This follows from a very general Theorem 3.2 of [2].

On the other hand, the canonical functor $P : \mathcal{C} \to \mathrm{Spec}(\mathcal{C})$, which the spectral category $\mathrm{Spec}(\mathcal{C})$ comes equipped with, nicely plays the roles of both $1_{\mathcal{C}}$ and E, since each object in $\mathrm{Spec}(\mathcal{C})$ is injective, as shown in [11].

In this paper, however, we are not interested in injective objects, and our main aim is to construct $\mathrm{Spec}(\mathcal{C})$ in full generality, when $\mathcal{C}$ is supposed to be an arbitrary category with finite limits. Apart from the Grothendieck category case above, this was already done in the case of an arbitrary abelian category [13, p. 15], and for some nonadditive categories [3].

In fact we begin by taking not just an arbitrary category $\mathcal{C}$ with finite limits, but also any class $\mathcal{S}$ of its monomorphisms that contains all isomorphisms and is pullback stable and closed under composition. We define the spectral category $\mathrm{Spec}(\mathcal{C}, \mathcal{S})$ of the pair $(\mathcal{C}, \mathcal{S})$ to be the category

$$\mathcal{C}[(\mathrm{St}(\mathrm{Mono}_E(\mathcal{C},\mathcal{S})))^{-1}]$$

of fractions of $\mathcal{C}$ for the class $\mathrm{St}(\mathrm{Mono}_E(\mathcal{C},\mathcal{S}))$ of pullback stable $\mathcal{S}$-essential monomorphisms of $\mathcal{C}$. When $\mathcal{S}$ is the class of all monomorphisms in $\mathcal{C}$, we write $\mathrm{Spec}(\mathcal{C},\mathcal{S}) = \mathrm{Spec}(\mathcal{C})$ and call this category the spectral category of $\mathcal{C}$.

We make various observations concerning the spans and fractions involved. The most important one is that the class $\mathrm{St}(\mathrm{Mono}_E(\mathcal{C},\mathcal{S}))$ admits the calculus of right fractions, just as the class of essential monomorphism in an abelian category does.

We point out that the spectral category $\mathrm{Spec}(\mathcal{C})$ has finite limits and that the canonical functor $P : \mathcal{C} \to \mathrm{Spec}(\mathcal{C})$ preserves finite limits. When $\mathcal{C}$ is a normal category [18], assuming for simplicity that $\mathcal{S}$ is the class of all monomorphisms in $\mathcal{C}$, we show that pullback stable $\mathcal{S}$-essential monomorphisms are the same as what we call subobject-essential monomorphisms. These are those monomorphisms $m : M \to A$ in $\mathcal{C}$ such that, for any monomorphism $n : N \to A$, one has that $N = 0$ whenever $M \times_A N = 0$. Finally, when $\mathcal{C}$ is normal, the monoid $\mathrm{End}_{\mathrm{Spec}(\mathcal{C})}(P(A))$ of endomorphisms of an object $P(A)$ in the spectral category is a division monoid whenever A is a *uniform* object (a notion extending the classical one of uniform module in the additive context).

The theory we develop is indeed an extension of what was done in [11] for the case of Grothendieck categories and in [3] for the category of G-groups. Note that there are several papers involving essential monomorphisms in non-abelian contexts (see e.g. [4, 21] and the references therein), although it is not their purpose to introduce spectral categories.

Throughout this paper, $\mathcal{C}$ denotes a category with finite limits.

2 Stabilization of Classes of Morphisms

Let $\mathcal{M}$ be a class of morphisms in $\mathcal{C}$. Following [7], define the stabilization $\mathrm{St}(\mathcal{M})$ of $\mathcal{M}$ as the class of morphisms $m\colon M \to A$ such that, for every pullback diagram of the form

$$\begin{array}{ccc} U & \xrightarrow{u} & X \\ \downarrow & & \downarrow \\ M & \xrightarrow[m]{} & A, \end{array}$$

u is in $\mathcal{M}$. Let us recall that the symbol "St" was used in [15], while in [7] the stabilization of $\mathcal{M}$ was simply denoted by $\mathcal{M}'$. Similar constructions were also used before, of course.

Proposition 1. *The stabilization* St($\mathcal{M}$) *of* $\mathcal{M}$ *has the following properties:*

(a) *The class* St($\mathcal{M}$) *is pullback stable.*
(b) *If* $\mathcal{M}$ *contains all isomorphisms, then so does* St($\mathcal{M}$).
(c) *If* $\mathcal{M}$ *is closed under composition, then so is* St($\mathcal{M}$).
(d) *If* $\mathcal{M}$ *has the right cancellation property of the form*

$$(mm' \in \mathcal{M} \ \&\ m' \in \mathcal{S}) \Rightarrow m \in \mathcal{M}$$

for some pullback stable class $\mathcal{S}$ *of morphisms in* $\mathcal{C}$, *then* St($\mathcal{M}$) *has the same property with respect to the same class* $\mathcal{S}$.
(e) *If* $\mathcal{M}$ *has the weak right cancellation property*

$$(mm' \in \mathcal{M} \ \&\ m' \in \mathcal{M}) \Rightarrow m \in \mathcal{M},$$

then St($\mathcal{M}$) *has the same property.*
(f) St($\mathcal{M}$) *has the left cancellation property of the form*

$$(mm' \in \mathrm{St}(\mathcal{M}) \ \&\ m \in \mathrm{Mono}(\mathcal{C})) \Rightarrow m' \in \mathrm{St}(\mathcal{M}),$$

where Mono($\mathcal{C}$) *denotes the class of all monomorphisms in* $\mathcal{C}$.

Proof. (a) and (b) are obvious.

To prove (c), (d), and (e), use a diagram of the form

$$\begin{array}{ccccc} U' & \longrightarrow & U & \longrightarrow & X \\ \downarrow & & \downarrow & & \downarrow \\ M' & \xrightarrow[m']{} & M & \xrightarrow[m]{} & A, \end{array}$$

where the squares are pullbacks and the unlabeled arrows are the suitable pullback projections.

To prove (f), consider the diagram

$$\begin{array}{ccccc} M' \times_M L & \longrightarrow & L & \xrightarrow{1_L} & L \\ \downarrow & & \downarrow{\scriptstyle l} & & \downarrow{\scriptstyle l} \\ M' & \xrightarrow{m'} & M & \xrightarrow{1_M} & M \\ {\scriptstyle 1_{M'}}\downarrow & & {\scriptstyle 1_M}\downarrow & & \downarrow{\scriptstyle m} \\ M' & \xrightarrow[m']{} & M & \xrightarrow[m]{} & A, \end{array}$$

where $l\colon L \to M$ is an arbitrary morphism and the unlabeled arrows are the pullback projections. Note that all its squares are pullbacks, except for the right-hand bottom square, although it is also a pullback if m is a monomorphism. Therefore, if mm' is in $\mathrm{St}(\mathcal{M})$ and m is a monomorphism, the pullback projection $M' \times_M L \to L$ is in $\mathcal{M}$. This proves the desired implication. □

Remark 1. Properties 1(a)–(c) are mentioned in [7] and 1(d) is 'almost' there, with $\mathcal{E}$ instead of $\mathcal{M}$. Property 1(e) also holds in the main example there, but for the trivial reason that $(mm' \in \mathrm{St}(\mathcal{E})$ & $m \in \mathrm{Mono}(\mathcal{C}))$ implies that m is an isomorphism.

3 Essential and Pullback Stable Essential Monomorphisms

Throughout this paper, we will consider a class $\mathcal{S}$ of monomorphisms in $\mathcal{C}$ that is pullback stable, contains all isomorphisms, is closed under composition, and has the strong left cancellation property

$$mm' \in \mathcal{S} \Rightarrow m' \in \mathcal{S}.$$

According to a well-known definition, a morphism $m\colon M \to A$ from $\mathcal{S}$ is said to be an *$\mathcal{S}$-essential monomorphism*, if a morphism $f\colon A \to B$ from $\mathcal{C}$ is in $\mathcal{S}$ whenever so is fm. When $\mathcal{S}$ is the class of all monomorphisms in $\mathcal{C}$, we will say "essential" instead of "$\mathcal{S}$-essential". The class of all $\mathcal{S}$-essential monomorphisms will be denoted by $\mathrm{Mono}_E(\mathcal{C}, \mathcal{S})$. This class has many "good" properties well-known in the case of an abelian $\mathcal{C}$ with $\mathcal{S}$ being the class of all monomorphisms in $\mathcal{C}$ (see, e.g., any of the following: Section 5 in Chapter II of [10], Section 2 in Chapter III of [19], or Section 15.2 of [20]), and also known in the general case, as briefly mentioned in Remark 9.23 of [1]. The known properties we will need are collected in:

Proposition 2. *The class* $\mathrm{Mono}_E(\mathcal{C}, \mathcal{S})$ *of $\mathcal{S}$-essential monomorphisms*

(a) contains all isomorphisms;
(b) is closed under composition;
(c) has the right cancellation property of the form

$$(mm' \in \mathrm{Mono}_E(\mathcal{C}, \mathcal{S}) \ \&\ m \in \mathcal{S}) \Rightarrow m \in \mathrm{Mono}_E(\mathcal{C}, \mathcal{S});$$

(d) has the weak right cancellation property

$$(mm' \in \mathrm{Mono}_E(\mathcal{C}, \mathcal{S}) \ \&\ m' \in \mathrm{Mono}_E(\mathcal{C}, \mathcal{S})) \Rightarrow m \in \mathrm{Mono}_E(\mathcal{C}, \mathcal{S}),$$

and, in particular, every split monomorphism that belongs to $\mathrm{Mono}_E(\mathcal{C}, \mathcal{S})$ *is an isomorphism.* □

Remark 2. Note the difference between our Proposition 2(c) and Proposition 9.14(3) of [1]: we have omitted the redundant assumption $m' \in \mathcal{S}$.

From Propositions 1 and 2, we immediately obtain:

Theorem 1. *The class* $\mathrm{St}(\mathrm{Mono}_E(\mathcal{C}, \mathcal{S}))$ *of pullback stable* $\mathcal{S}$*-essential monomorphisms in* $\mathcal{C}$

(a) is pullback stable;
(b) contains all isomorphisms;
(c) is closed under composition;
(d) has the right cancellation property of the form

$$(mm' \in \mathrm{St}(\mathrm{Mono}_E(\mathcal{C}, \mathcal{S})) \ \&\ m \in \mathcal{S}) \Rightarrow m \in \mathrm{St}(\mathrm{Mono}_E(\mathcal{C}, \mathcal{S}));$$

(e) has the weak right cancellation property

$$(mm' \in \mathrm{St}(\mathrm{Mono}_E(\mathcal{C}, \mathcal{S})) \ \&\ m' \in \mathrm{St}(\mathrm{Mono}_E(\mathcal{C}, \mathcal{S}))) \Rightarrow m \in \mathrm{St}(\mathrm{Mono}_E(\mathcal{C}, \mathcal{S})),$$

and, in particular, every split monomorphism that belongs to $\mathrm{St}(\mathrm{Mono}_E(\mathcal{C}, \mathcal{S}))$ *is an isomorphism;*
(f) has the left cancellation property of the form

$$(mm' \in \mathrm{St}(\mathrm{Mono}_E(\mathcal{C}, \mathcal{S})) \ \&\ m \in \mathrm{Mono}(\mathcal{C})) \Rightarrow m' \in \mathrm{St}(\mathrm{Mono}_E(\mathcal{C}, \mathcal{S})).$$

□

4 Spans and Fractions

Let $\mathcal{C}$ be a category with pullbacks. The bicategory $\mathrm{Span}(\mathcal{C})$ of spans in $\mathcal{C}$, originally introduced in [5] (motivated by the study of spans of additive categories in [22]) is constructed as follows, omitting obvious coherent isomorphisms:

- The objects (=0-cells) of $\mathrm{Span}(\mathcal{C})$ are the same as the objects of $\mathcal{C}$.
- A morphism (1-cell) $A \to B$ in $\mathrm{Span}(\mathcal{C})$ is a diagram in $\mathcal{C}$ of the form

$$A \xleftarrow{\ x\ } X \xrightarrow{\ f\ } B,$$

  usually written either as the triple (f, X, x) or as the pair (f, x).
- The composite $(g, Y, y)(f, X, x) = (gq, X \times_B Y, xp)$ of $(f, X, x)\colon A \to B$ and $(g, Y, y)\colon B \to C$ is defined via the diagram

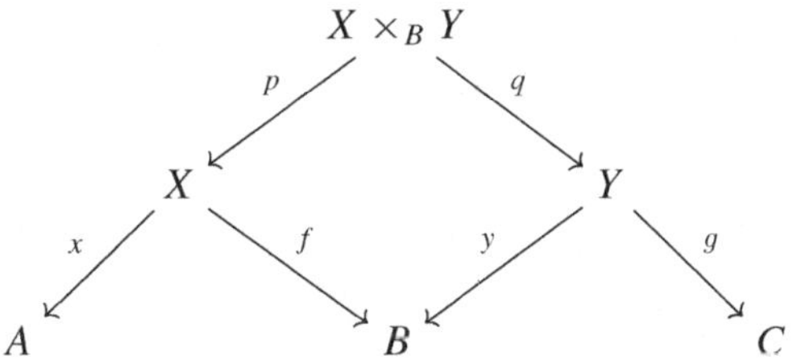

in which $p\colon X \times_B Y \to X$ and $q\colon X \times_B Y \to Y$ are the pullback projections.
- A 2-cell from $(f, X, x)\colon A \to B$ to $(f', X', x')\colon A \to B$ is a morphism $s\colon X \to X'$ with $x's = x$ and $f's = f$, and the 2-cells compose as in $\mathcal{C}$.

More generally, given a pullback stable class $\mathcal{M}$ of morphisms in $\mathcal{C}$ that contains all identity morphisms and is closed under composition—we can then form the bicategory $\mathrm{Span}_{\mathcal{M}}(\mathcal{C})$ as above but requiring its morphisms (f, x) to have x in $\mathcal{M}$. As it was observed in a discussion with Janelidze and Mac Lane [14] (and most probably known before, which is why the content of that discussion was never published), the assignment $(cls(f, x)\colon A \to B) \mapsto (fx^{-1}\colon A \to B)$ (here cls is the abbreviation for "class") determines an isomorphism

$$\Pi(\mathrm{Span}_{\mathcal{M}}(\mathcal{C})) \approx \mathcal{C}[\mathcal{M}^{-1}],$$

in which:

- $\Pi(\mathrm{Span}_{\mathcal{M}}(\mathcal{C}))$ is the Poincaré category of $\mathrm{Span}_{\mathcal{M}}(\mathcal{C})$ (in the sense of [5]), that is, it has the same objects as $\mathrm{Span}_{\mathcal{M}}(\mathcal{C})$, and its hom sets are the sets of connected components of hom categories of $\mathrm{Span}_{\mathcal{M}}(\mathcal{C})$.
- $\mathcal{C}[\mathcal{M}^{-1}]$ is the category of fractions [12] of $\mathcal{C}$ for $\mathcal{M}$.

We are assuming that the reader is familiar with the content of [5]. Repeating here the necessary details from that influential paper would take too much space.

Under the isomorphism above, the functor $\mathcal{C} \to \Pi(\mathrm{Span}_{\mathcal{M}}(\mathcal{C}))$, corresponding to the canonical functor

$$P_{\mathcal{M}}\colon \mathcal{C} \to \mathcal{C}[\mathcal{M}^{-1}],$$

is defined by $(f\colon A \to B) \mapsto (cls(f, 1_A)\colon A \to B)$.

Recall that a class $\mathcal{M}$ is *focal* [6] if it satisfies the following four conditions:

(F_0) For each object $X \in \mathcal{C}$ there exists an $s \in \mathcal{M}$ with codomain X.

(F_1) For all $\xrightarrow{s_1}\ \xrightarrow{s_0}$ with $s_i \in \mathcal{M}$, there exists a morphism f in $\mathcal{C}$ such that the composite $s_0 s_1 f$ is defined and is in $\mathcal{M}$.

(F_2) Each diagram

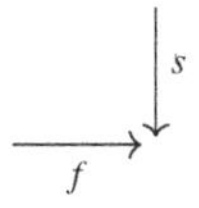

with $s \in \mathcal{M}$ can be completed in a commutative square

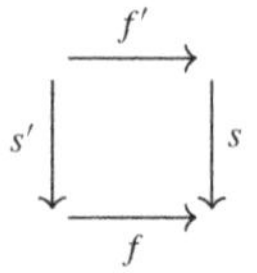

where $s' \in \mathcal{M}$.

(F_3) If a pair (f, g) of parallel morphisms is coequalized by some $s \in \mathcal{M}$, it is also equalized by some $s' \in \mathcal{M}$.

Proposition 3. *If $\mathcal{M}$ is a pullback stable class of morphisms in $\mathcal{C}$ that contains all identity morphisms and is closed under composition, with $\mathcal{M} \subseteq \mathrm{Mono}(\mathcal{C})$, then $\mathcal{M}$ is focal and, moreover, $\mathcal{M}$ admits the calculus of right fractions in the sense of [12].*

Proof. All we need to check is that $\mathcal{M}$ satisfies the condition dual to condition 2.2(d) in Chapter I of [12], i.e., that whenever two parallel morphisms f and g admit a morphism $m \in \mathcal{M}$ with $mf = mg$, they also admit a morphism $n \in \mathcal{M}$ with $fn = gn$. This condition holds trivially because $\mathcal{M} \subseteq \mathrm{Mono}(\mathcal{C})$. □

Remark 3. Note the following levels of generality (in fact there are many more of them, including those suggested by distinguishing sets of morphisms from proper classes of morphisms), where we omitted all required conditions on $\mathcal{M}$ in the first five items:

(a) For an arbitrary class $\mathcal{M}$ of morphisms of $\mathcal{C}$, we can still form the category $\mathcal{C}[\mathcal{M}^{-1}]$ of fractions of $\mathcal{C}$ for $\mathcal{M}$.
(b) As shown in [6], the morphisms of $\mathcal{C}[\mathcal{M}^{-1}]$ can be presented in the form $P_{\mathcal{M}}(f)P_{\mathcal{M}}(x)^{-1}$ with $x \in \mathcal{M}$ if and only if $\mathcal{M}$ satisfies conditions (F_0), (F_1), and (F_2).
(c) In particular, this is the case when $\mathcal{M}$ satisfies the conditions dual to conditions 2.2(a), 2.2(b), and 2.2(c) in Chapter I of [12].
(d) If the equivalent conditions in (b) hold, then the following conditions are equivalent: (d_1) $\mathcal{M}$ is focal; (d_2) $\mathcal{M}$ satisfies condition (F_3), which is the same as condition 2.2(d) in Chapter I of [12]; (d_3) not only can the morphisms of $\mathcal{C}[\mathcal{M}^{-1}]$ be presented as in (b), but also $P_{\mathcal{M}}(f)P_{\mathcal{M}}(x)^{-1} = P_{\mathcal{M}}(f')P_{\mathcal{M}}(x')^{-1}$ if and only there exists a commutative diagram in $\mathcal{C}$ of the form

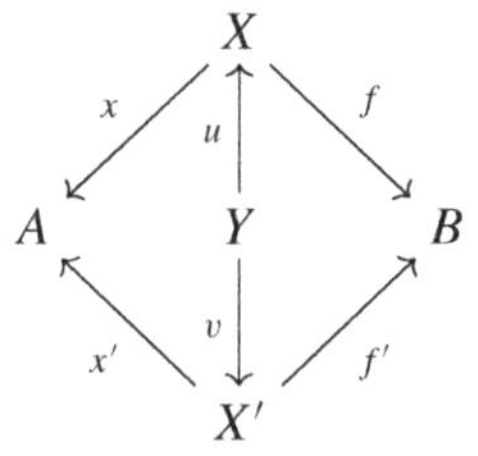

with $xu \in \mathcal{M}$.

(e) In particular, the equivalent conditions (d_1)–(d_3) hold when the class $\mathcal{M}$ admits the calculus of right fractions in the sense of [12].

(f) If $\mathcal{M}$ contains all identity morphisms, is closed under composition, and is pullback stable, then not only are we in the situation (c), but we also have the isomorphism between $\mathcal{C}[\mathcal{M}^{-1}]$ and $\Pi(\mathrm{Span}_{\mathcal{M}}(\mathcal{C}))$ mentioned above.

(g) The situation of Proposition 3. Note, in particular, that in this case the morphisms u and v in the diamond diagram of (d) belong to $\mathcal{M}$. This follows from Proposition 1(f) and the fact that $\mathrm{St}(\mathcal{M}) = \mathcal{M}$ here.

The levels of generality listed above are related as follows:

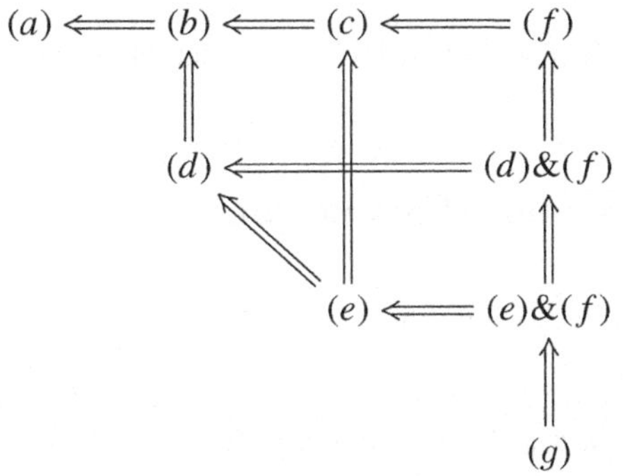

Remark 4. We recall from [12] that already in the situation (d), the equivalent conditions mentioned there imply that the hom sets of $\mathcal{C}[\mathcal{M}^{-1}]$ can be constructed as filtered colimits

$$\hom_{\mathcal{C}[\mathcal{M}^{-1}]}(A, B) = \mathrm{colim}(\hom(M, B)),$$

where the colimit is taken over all $m : M \to A$ in $\mathcal{M}$ (see Page 13 in [12], where the dual construction is described explicitly).

5 The Spectral Category

Let $\mathcal{S}$ be a class of monomorphisms in $\mathcal{C}$ satisfying the conditions required at the beginning of Section 3. Then, as follows from (a)–(c) and (f) of Theorem 1, the class $\mathcal{M} = \mathrm{St}(\mathrm{Mono}_E(\mathcal{C}, \mathcal{S}))$ satisfies the conditions required in Proposition 3. We are ready to give the following:

Definition 1. *The* spectral category $\mathrm{Spec}(\mathcal{C}, \mathcal{S})$ *of* $(\mathcal{C}, \mathcal{S})$ *is the category*

$$\mathcal{C}[(\mathrm{St}(\mathrm{Mono}_E(\mathcal{C}, \mathcal{S})))^{-1}]$$

of fractions of $\mathcal{C}$ *for the class* $\mathrm{St}(\mathrm{Mono}_E(\mathcal{C}, \mathcal{S}))$ *of pullback stable* $\mathcal{S}$*-essential monomorphisms of* $\mathcal{C}$. *When* $\mathcal{S}$ *is the class of all monomorphisms in* $\mathcal{C}$, *we shall simply write* $\mathrm{Mono}_E(\mathcal{C}, \mathcal{S}) = \mathrm{Mono}_E(\mathcal{C})$ *and* $\mathrm{Spec}(\mathcal{C}, \mathcal{S}) = \mathrm{Spec}(\mathcal{C})$, *and say that* $\mathrm{Spec}(\mathcal{C})$ *is the spectral category of* $\mathcal{C}$.

Thanks to the results of [12], specifically Proposition 3.1 and Corollary 3.2 of Chapter I there, our Proposition 3 implies:

Theorem 2. *The spectral category* $\mathrm{Spec}(\mathcal{C}, \mathcal{S})$ *has finite limits. Moreover, the canonical functor*

$$P_{\mathcal{C},\mathcal{S}} = P_{\mathrm{St}(\mathrm{Mono}_E(\mathcal{C},\mathcal{S}))} : \mathcal{C} \to \mathrm{Spec}(\mathcal{C}, \mathcal{S}),$$

defined by $(f : A \to B) \mapsto (cls(f, 1_A) : A \to B)$, *preserves finite limits.* □

6 Subobject-Essential Monomorphisms

Assuming $\mathcal{C}$ to be pointed, we define:

Definition 2. *A monomorphism* $m : M \to A$ *in* $\mathcal{C}$ *is said to be* subobject-essential *if, for a monomorphism* $n : N \to A$, *one has* $M \times_A N = 0 \Rightarrow N = 0$. *The class of all subobject-essential monomorphisms in* $\mathcal{C}$ *will be denoted by* $\mathrm{Mono}_{SE}(\mathcal{C})$.

Recall that a regular epimorphism in a category $\mathcal{C}$ is a morphism that is the coequalizer of two morphisms in $\mathcal{C}$.

A finitely complete category $\mathcal{C}$ is *regular* if any morphism $f : A \to B$ can be factorized as the composite morphism of a regular epimorphism $p : A \to I$ and a monomorphism $m : I \to B$

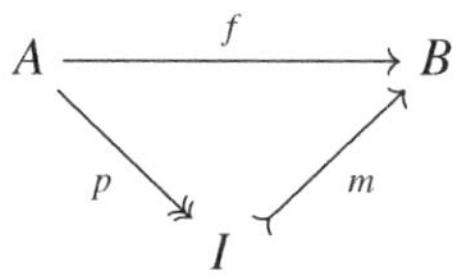

and these factorizations are pullback stable. Following [18], we will call $\mathcal{C}$ *normal* if it is pointed, regular, and any regular epimorphism is a normal epimorphism (i.e., a cokernel of some arrow in $\mathcal{C}$). In such a category, any regular epimorphism is then the cokernel of its kernel and, as a consequence, a morphism in $\mathcal{C}$ is a monomorphism if and only if its kernel is zero.

Remark 5. For a pointed variety $\mathcal{V}$ of universal algebras, being a normal category is the same as being a 0-regular variety in the sense of [9] (see [17] for further explanations and historical remarks about the relationship between the properties of 0-regularity and normality). The algebraic theory of a pointed 0-regular variety $\mathcal{V}$ is characterized by the existence of a unique constant 0 and binary terms $d_1, \ldots, d_n$ such that the identities $d_i(x, x) = 0$ (for $i \in \{1, \ldots, n\}$) and the implication $(d_1(x, y) = 0 \,\&\, \ldots \,\&\, d_n(x, y) = 0) \Rightarrow x = y$ hold. Intuitively, these operations $d_i(x, y)$ can then be thought of as a kind of "generalized subtraction". This implies that the varieties of groups, loops, rings, associative algebras, Lie algebras, crossed modules, and G-groups (for a group G) are all normal. There are also plenty of examples of normal categories that are not varieties, such as the categories of topological groups, cocommutative K-Hopf algebras over a field K, and C^*-algebras, for instance. In general, any semi-abelian category [16] is, in particular, a normal category.

For an object A in $\mathcal{C}$, the smallest and the largest congruence (=effective equivalence relation) on A will be denoted by Δ_A and ∇_A, respectively. Note that equalities like $E = \Delta_A$ should usually be understood as equalities of subobjects (of $A \times A$ in this case).

When $\mathcal{C}$ is normal, it is natural to ask how different subobject-essential monomorphisms are from essential ones (recall that "essential" means $\mathcal{S}$-essential for $\mathcal{S} = \mathrm{Mono}(\mathcal{C})$). Most of this section is devoted to studying various ways to compare them.

Let us begin with the following proposition, well-known in the case of an abelian category $\mathcal{C}$:

Proposition 4. *If $\mathcal{C}$ is normal, then the following conditions on a monomorphism $m : M \to A$ in $\mathcal{C}$ are equivalent:*

(a) m is an essential monomorphism, that is, a morphism $f : A \to B$ is a monomorphism whenever so is fm.
(b) For any congruence E on A, one has $(M \times M) \times_{A \times A} E = \Delta_M \Rightarrow E = \Delta_A$.
(c) For any normal monomorphism $n : N \to A$, one has $M \times_A N = 0 \Rightarrow N = 0$.
(d) For any morphism $f : A \to B$, one has $\mathsf{Ker}(fm) = 0 \Rightarrow \mathsf{Ker}(f) = 0$.

Proof. (a) $\Rightarrow$ (b). Let (E, e_1, e_2) be a congruence on A, and $f : A \to C$ a morphism such that E is the kernel pair of f:

$$\begin{array}{ccc} E & \xrightarrow{e_2} & A \\ {\scriptstyle e_1}\downarrow & & \downarrow{\scriptstyle f} \\ A & \xrightarrow[f]{} & C. \end{array} \tag{1}$$

Consider the commutative diagram

$$\begin{array}{ccccc} M & \longrightarrow & E & \longrightarrow & C \\ {\scriptstyle (1_M, 1_M)}\downarrow & & {\scriptstyle (e_1, e_2)}\downarrow & & \downarrow{\scriptstyle (1_C, 1_C)} \\ M \times M & \xrightarrow[m \times m]{} & A \times A & \xrightarrow[f \times f]{} & C \times C \end{array}$$

where the right-hand square is a pullback by definition of kernel pair, and the left-hand square is a pullback by the assumption $(M \times M) \times_{A \times A} E = \Delta_M$. The fact that the rectangle is a pullback means that fm is a monomorphism. Since m is an essential monomorphism, it follows that f is a monomorphism and $E = \Delta_A$.

(b) ⇒ (c) This follows from the fact that a congruence (E, e_1, e_2) as in (1) is the discrete equivalence relation Δ_A if and only if the normal monomorphism $\ker(f)\colon N \to A$ corresponding to E is $0 \to A$.

(c) ⇒ (d) It suffices to apply the assumption to the pullback

$$\begin{array}{ccc} 0 = \mathsf{Ker}(fm) & \longrightarrow & \mathsf{Ker}(f) \\ \downarrow & & \downarrow \\ M & \xrightarrow[m]{} & A. \end{array}$$

(d) ⇒ (a) This is immediate since, in a normal category, monomorphisms are characterized by the fact that their kernel is 0. □

From Proposition 4, we immediately obtain:

Corollary 1. *Let $\mathcal{C}$ be a normal category. Then:*

(a) Every subobject-essential monomorphism is essential.
(b) If A is an object in $\mathcal{C}$ for which every monomorphism with codomain A is normal, then a monomorphism $m : M \to A$ is subobject-essential if and only if it is essential.
(c) In particular, if $\mathcal{C}$ is abelian, then a monomorphism in $\mathcal{C}$ is subobject-essential if and only if it is essential.

Next, we have:

Proposition 5. *The class* $\mathrm{Mono}_{SE}(\mathcal{C})$ *of subobject-essential monomorphisms in $\mathcal{C}$*

(a) contains all isomorphisms;
(b) is closed under composition;
(c) has the right cancellation property of the form

$$(mm' \in \mathrm{Mono}_{SE}(\mathcal{C}) \;\&\; m \in \mathrm{Mono}(\mathcal{C})) \Rightarrow m \in \mathrm{Mono}_{SE}(\mathcal{C}).$$

(d) If $\mathcal{C}$ is normal, then the class $\mathrm{Mono}_{SE}(\mathcal{C})$ *has the weak right cancellation property*

$$(mm' \in \mathrm{Mono}_{SE}(\mathcal{C}) \;\&\; m' \in \mathrm{Mono}_{SE}(\mathcal{C})) \Rightarrow m \in \mathrm{Mono}_{SE}(\mathcal{C})$$

and, in particular, every split monomorphism that belongs to it is an isomorphism.

(e) It has the left cancellation property of the form

$$(mm' \in \mathrm{Mono}_{SE}(\mathcal{C}) \;\&\; m \in \mathrm{Mono}(\mathcal{C})) \Rightarrow m' \in \mathrm{Mono}_{SE}(\mathcal{C}).$$

(f) If $\mathcal{C}$ is normal, then the class $\mathrm{Mono}_{SE}(\mathcal{C})$ *is pullback stable.*

Proof. (a) is obvious.

(b) and (c): Given monomorphisms $m: M \to A$, $m': M' \to M$ and $n: N \to A$, consider the diagram

$$\begin{array}{ccccc} M' \times_A N & \xrightarrow{m' \times 1} & M \times_A N & \longrightarrow & N \\ \downarrow & & \downarrow & & \downarrow{\scriptstyle n} \\ M' & \xrightarrow[m']{} & M & \xrightarrow[m]{} & A, \end{array}$$

where the unlabeled arrows are the suitable pullback projections. Since both squares in this diagram are pullbacks, we can argue as follows:

- If $m, m' \in \mathcal{M}$, then $M' \times_A N = 0 \Rightarrow M \times_A N = 0 \Rightarrow N = 0$.
- If $mm' \in \mathcal{M}$, then $M \times_A N = 0 \Rightarrow M' \times_A N = 0 \Rightarrow N = 0$, where the first implication holds because $\mathsf{Ker}(m') = 0$.

(d): Suppose mm' and m are in $\text{Mono}_{SE}(\mathcal{C})$. Thanks to (c), we only need to prove that m is a monomorphism. Therefore, since $\mathcal{C}$ is normal, it suffices to prove that m has zero kernel. For, consider the diagram

$$\begin{array}{ccccc} 0 & \longrightarrow & \mathsf{Ker}(m) & \longrightarrow & 0 \\ \downarrow & & \downarrow{\scriptstyle \ker(m)} & & \downarrow \\ M' & \xrightarrow[m']{} & M & \xrightarrow[m]{} & A \end{array}$$

and observe that:

- Its left-hand square is a pullback because so is its right-hand square, and mm' is a monomorphism because it is in $\text{Mono}_{SE}(\mathcal{C})$.
- Since m' is in $\text{Mono}_{SE}(\mathcal{C})$, we have that $\mathsf{Ker}(m) = 0$.

(e): Suppose mm' is in $\text{Mono}_{SE}(\mathcal{C})$ and m is a monomorphism. First notice that, since mm' is a monomorphism, so is m'. After that, consider the diagram

$$\begin{array}{ccccc} M' \times_M L & \longrightarrow & L & \xrightarrow{1_L} & L \\ \downarrow & & \downarrow{\scriptstyle l} & & \downarrow{\scriptstyle ml} \\ M' & \xrightarrow[m']{} & M & \xrightarrow[m]{} & A, \end{array}$$

where the unlabeled arrows are the suitable pullback projections. Since both squares in this diagram are pullbacks, we have

$$M' \times_M L = 0 \Rightarrow M' \times_A L = 0 \Rightarrow L = 0.$$

(f): Given $m\colon M \to A$ from $\mathrm{Mono}_{SE}(\mathcal{C})$, a morphism $x\colon X \to A$, and a monomorphism $u\colon U \to X$, consider the diagram

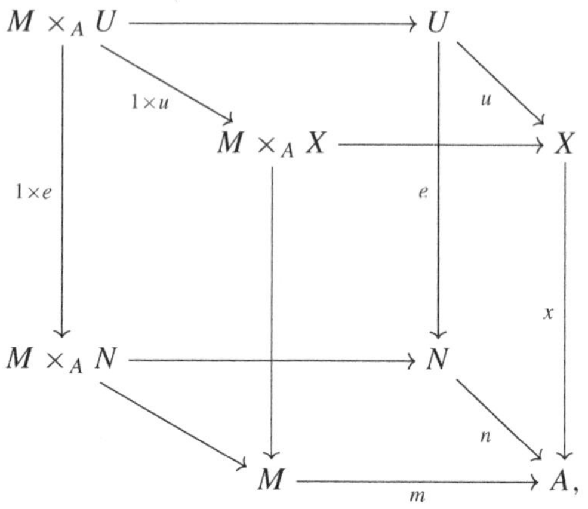

in which $M \times_A U = (M \times_A X) \times_X U$, $xu = ne$ is a (regular epi, mono) factorization of xu, and the unlabeled arrows are the suitable pullback projections. Assuming $M \times_A U = 0$, we have to prove that $U = 0$. Indeed:

- Since e is a regular epimorphism, so is $1 \times e$.
- Since $1 \times e$ is an epimorphism and $M \times_A U = 0$, we have $M \times_A N = 0$.
- Since $M \times_A N = 0$ and m is in $\mathrm{Mono}_{SE}(\mathcal{C})$, we have that $N = 0$.
- Since $N = 0$, we have $xe = ne = 0$, and so u factors through the kernel of x.
- Since u factors through the kernel of x, it also factors through the pullback projection $M \times_A X \to X$.
- Since u factors through the pullback projection $M \times_A X \to X$, and the top part of our diagram is a pullback, the pullback projection $M \times_A U \to U$ is a split epimorphism.

It follows that $U = 0$, as desired. □

Remark 6. For a composable pair (m, m') of monomorphisms, m' can be seen as a pullback of mm' along m (this well-known fact was used in the proof of Proposition 1(f) for the pair (m, l)). This implies that every pullback stable class of monomorphisms has the strong left cancellation property and, in particular, that, in the case of normal $\mathcal{C}$, Proposition 5(e) could be deduced from Proposition 5(f).

Remark 7. In contrast to Theorem 1(f) and Proposition 5(e), the class $\mathrm{Mono}_E(\mathcal{C})$ does not even have, in general, the weak left cancellation property $m,\ mm' \in \mathrm{Mono}_E(\mathcal{C}) \Rightarrow m' \in \mathrm{Mono}_E(\mathcal{C})$. One can easily construct counterexamples in many non-abelian semi-abelian algebraic categories by suitably choosing $m' : M' \to M$ to be a split monomorphism (that is not an isomorphism) and choosing $m : M \to A$ with simple A. For example:

(a) Let $\mathcal{C}$ be the category of groups, A any simple group that has an element a of order pq with relatively prime p and q, M the subgroup of A generated by a, M' the subgroup of A generated by a^p, and $m : M \to A$ and $m' : M' \to M$ the inclusion maps. Then m and mm' are in $\mathcal{M}_E$, but m' is not.
(b) Let $\mathcal{C}$ be the category of rings (commutative or not; we do not require them to have identity element, to make $\mathcal{C}$ semi-abelian), $M' = K$ be a field, $M = K[x]$ the polynomial ring in one variable x over K, $A = K(x)$ the field of fractions of M, and $m : M \to A$ and $m' : M' \to M$ the canonical monomorphisms. Then, again, m and mm' are in $\mathcal{M}_E$, but m' is not.

Remark 8. Although the non-pullback-stability of $\mathrm{Mono}_E(\mathcal{C})$ in the category of groups follows from Remark 7(a), let us give what seems to be the simplest counterexample. Consider the pullback

$$\begin{array}{ccc} 0 & \longrightarrow & S_2 \\ \downarrow & & \downarrow \\ A_3 & \longrightarrow & S_3 \end{array}$$

of monomorphisms, where S_2, S_3, A_3 are the symmetric/alternating groups. Its bottom arrow is an essential monomorphism, while the top one is not. This also shows that $A_3 \to S_3$ is an example of an essential monomorphism that is not subobject-essential.

Theorem 3. *If $\mathcal{C}$ is normal, then* $\mathrm{Mono}_{SE}(\mathcal{C}) = \mathrm{St}(\mathrm{Mono}_E(\mathcal{C}))$*, that is, a morphism in $\mathcal{C}$ is a subobject-essential monomorphism if and only if it is a pullback stable essential monomorphism.*

Proof. The inclusion $\mathrm{Mono}_{SE}(\mathcal{C}) \subseteq \mathrm{St}(\mathrm{Mono}_E(\mathcal{C}))$ follows from Corollary 1(a) and Proposition 5(f). Conversely, let $m : M \to A$ be a pullback stable essential monomorphism in $\mathcal{C}$ and $n : N \to A$ a monomorphism in $\mathcal{C}$ with $M \times_A N = 0$. Then $0 \to N$ is an essential monomorphism because it is a pullback of m. Hence, from the last assertion of Proposition 2(d), $0 \to N$ is an isomorphism, that is, $N = 0$. □

Since every abelian category is normal, we easily get:

Corollary 2. *If $\mathcal{C}$ is abelian then* $\mathrm{Spec}(\mathcal{C})$ *is the same as the spectral category of $\mathcal{C}$ in the usual sense (see [11] and [12]).*

Proof. Having in mind Corollary 1(c), this follows from Theorem 3 and the description of the spectral category of an abelian category given in 2.5(e) of [12, Chapter I]. □

7 Uniform Objects

Let us return to the general situation of Remark 4(a), where $\mathcal{M}$ is an arbitrary class of morphisms in $\mathcal{C}$, but let us assume that $\mathcal{C}$ is pointed and that

$$x \in \mathcal{M} \Rightarrow \mathsf{Ker}(x) = 0.$$

As already observed, in any *normal* category, this property simply says that $\mathcal{M}$ is a class of monomorphisms.

Definition 3. *An object A of $\mathcal{C}$ is said to be $\mathcal{M}$-*uniform *if a morphism $x\colon X \to A$ belongs to $\mathcal{M}$ whenever $X \neq 0$ and $\mathsf{Ker}(x) = 0$.*

The term *uniform* comes from module theory, where a nonzero module M is called a uniform module if every nonzero submodule N of M is an essential submodule. Note that this term was also used in [3] for an analogue notion in the category of G-groups.

Proposition 6. *Let A and B be $\mathcal{M}$-uniform objects in $\mathcal{C}$. Every nonzero morphism $A \to B$ in $\mathcal{C}[\mathcal{M}^{-1}]$ of the form $P_{\mathcal{M}}(f)P_{\mathcal{M}}(x)^{-1}$ with x in $\mathcal{M}$ is an isomorphism.*

Proof. Consider the diagram

$$\begin{array}{ccc} & \mathsf{Ker}(f) & \\ & \Big\downarrow{\scriptstyle \ker(f)} & \\ A \xleftarrow{\;\;x\;\;} & X & \xrightarrow{\;\;f\;\;} B, \end{array}$$

where X is the domain of x (the domain of f).

Suppose $\mathsf{Ker}(f) \neq 0$. Since A is $\mathcal{M}$-uniform and x and $\ker(f)$ have zero kernels, the composite $\mathsf{Ker}(f) \to X \to A$ belongs to $\mathcal{M}$. As x also belongs to $\mathcal{M}$, this implies that $P_{\mathcal{M}}(\ker(f))$ is an isomorphism, and we can write

$$\begin{aligned} P_{\mathcal{M}}(f)P_{\mathcal{M}}(x)^{-1} &= P_{\mathcal{M}}(f)P_{\mathcal{M}}(\ker(f))P_{\mathcal{M}}(\ker(f))^{-1}P_{\mathcal{M}}(x)^{-1} \\ &= P_{\mathcal{M}}(f(\ker(f))P_{\mathcal{M}}(\ker(f))^{-1}P_{\mathcal{M}}(x)^{-1} \\ &= P_{\mathcal{M}}(0)P_{\mathcal{M}}(\ker(f))^{-1}P_{\mathcal{M}}(x)^{-1} \\ &= 0. \end{aligned}$$

That is, we can suppose $\mathsf{Ker}(f) = 0$. If so, then, since B is $\mathcal{M}$-uniform, f belongs to $\mathcal{M}$, which makes $P_{\mathcal{M}}(f)P_{\mathcal{M}}(x)^{-1}$ an isomorphism. □

In order to state the next result, let us recall that a *division monoid* is a nontrivial monoid M with the property that the submonoid $U(M)$ of invertible elements is given by $U(M) = M \setminus \{0\}$. We write $\mathrm{End}_{\mathrm{Spec}(\mathcal{C})}(A)$ for the monoid of endomorphisms of an object A in the spectral category $\mathrm{Spec}(\mathcal{C})$, where $\mathcal{C}$ is a normal category.

Corollary 3. *Let $\mathcal{C}$ be a normal category, and A an $\mathcal{M}$-uniform object in $\mathcal{C}$ for $\mathcal{M}$ being the class of subobject-essential monomorphisms in $\mathcal{C}$. Then the monoid $\mathrm{End}_{\mathrm{Spec}(\mathcal{C})}(A)$ of endomorphisms of A in $\mathrm{Spec}(\mathcal{C})$ is a division monoid.*

Proof. This immediately follows from Proposition 6, by taking into account the fact that the class of subobject-essential monomorphisms coincides with the class of pullback stable essential monomorphisms whenever $\mathcal{C}$ is a normal category (by Theorem 3). □

This last result extends Lemma 5.4 in [3], where the base category $\mathcal{C}$ was the category of G-groups, to the general context of a normal category $\mathcal{C}$.

Acknowledgements The second author was partially supported by Ministero dell' Istruzione, dell' Università e della Ricerca (Progetto di ricerca di rilevante interesse nazionale "Categories, Algebras: Ring-Theoretical and Homological Approaches (CARTHA)") and Dipartimento di Matematica "Tullio Levi-Civita" of Università di Padova (Research program DOR1828909 "Anelli e categorie di moduli"). The third author was supported by a "Visiting scientist scholarship—year 2019" by Università degli Studi di Padova. The fourth author was partially supported by South African NRF.

References

1. Adámek, J., Herrlich, H., Strecker, G.E.: Abstract and concrete categories. The joy of cats, Pure and Applied Math., John Wiley & Sons, Inc., New York (1990)
2. Adámek, J., Herrlich, H., Rosický, J., Tholen, W.: Injective hulls are not natural. Algebra Universalis **48**, 374–388 (2002)
3. Arroyo Paniagua, M.J., Facchini, A.: Category of G-groups and its spectral category, Comm. Algebra **45**(4), 1696–1710 (2017)
4. Banaschewski, B.: Injectivity and essential extensions in equational classes of algebras. "Queen's papers in Pure and Applied Mathematics No. 25" (Proceedings of the Conference on Universal Algebra, Kingston 1969), pp. 131–147. Kingston, Ontario (1970)
5. Bénabou, J.: Introduction to bicategories. In: 1967 Reports of the Midwest Category Seminar. Springer, Berlin, pp. 1–77
6. Bénabou, J.: Some geometric aspects of the calculus of fractions, the European Colloquium of category theory (Tours, 1994). Appl. Categ. Struct. **4**(2), 139–165 (1996)
7. Carboni, A., Janelidze, G., Kelly, G.M., Paré, R.: On localization and stabilization of factorization systems. Appl. Categ. Struct. **5**, 1–58 (1997)
8. Facchini, A.: Injective modules, spectral categories, and applications. In: Jain, S.K., Parvathi, S. (eds.) Noncommutative Rings, Group Rings, Diagram Algebras and their Applications, Contemp. Math. 456, Amer. Math. Soc., Providence, RI, pp. 1–17 (2008)
9. Fichtner, K.: Eine Bermerkung über Mannigfaltigkeiten universeller Algebren mit Idealen. Monatsh. d. Deutsch. Akad. d. Wiss. (Berlin) **12**, 21–25 (1970)
10. Gabriel, P.: Des catégories abéliennes. Bull. Soc. Math. Fr. **90**, 323–448 (1962)
11. Gabriel, P., Oberst, U.: Spektralkategorien und reguläre Ringe im von-Neumannschen Sinn. Math. Z. **92**, 389–395 (1966)
12. Gabriel, P., Zisman, M.: Calculus of Fractions and Homotopy Theory. Springer, Berlin (1967)
13. Goodearl, K.R.: Ring Theory. Nonsingular Rings and Modules. Marcel Dekker Inc., New York and Basel (1976)
14. Janelidze, G., Mac Lane, S.: Private communication
15. Janelidze, G., Tholen, W.: Strongly separable morphisms in general categories. Theory Appl. Categ. **23**(7), 136–149 (2010)
16. Janelidze, G., Márki, L., Tholen, W.: Semi-abelian categories. J. Pure Appl. Algebra **168**, 367–386 (2002)
17. Janelidze, G., Márki, L., Ursini, A.: Ideals and clots in universal algebra and in semi-abelian categories. J. Algebra **307**, 191–208 (2007)

18. Janelidze, Z.: The pointed subobject functor, 3×3 lemmas, and subtractivity of spans. Theory Appl. Categ. **23**(11), 221–242 (2010)
19. Mitchell, B.: Theory of categories. Pure Appl. Math. **17** (1965). Academic Press
20. Schubert, H.: Categories. Springer, New York, Heidelberg (1972)
21. Tholen, W.: Injective objects and cogenerating sets. J. Algebra **73**, 139–155 (1981)
22. Yoneda, N.: On Ext and exact sequences. J. Fac. Sci. Univ. Tokyo Sect. **I**(8), 507–576 (1960)

A Survey on the Local Invertibility of Ideals in Commutative Rings

Carmelo Antonio Finocchiaro and Francesca Tartarone

Abstract Let D be an integral domain. We give an overview on connections between the (t)-finite character property of D (i.e., each nonzero element of D is contained in finitely many (t)-maximal ideals) and problems of local invertibility of ideals.

Keywords Finite character · Invertible ideal · Star operation

2000 Mathematics Subject Classification Primary: 13A15 · 13F05 · Secondary: 13B30 · 13G05

1 Introduction

A well-known characterization of invertible ideals in integral domains states that "a nonzero ideal is invertible if and only if it is finitely generated and locally principal" [8, II §5, Theorem 4].

The condition that the ideal is finitely generated can be dropped down, for instance, if the domain has the finite character on maximal ideals, i.e., each nonzero element is contained in finitely many maximal ideals (see, for instance, the argument provided in the proof of [3, Chapter 7, Exercise 9]. Here the fact that finite character allows to characterize Noetherian domains among locally Noetherian domains is put in light).

C. A. Finocchiaro
Dipartimento di Matematica, Università degli Studi di Padova, Via Trieste 63, 35121 Padova, Italy
e-mail: carmelo@math.unipd.it; cafinocchiaro@unict.it

Dipartimento di Matematica e Informatica, Università degli Studi di Catania, Viale Andrea Doria 6, 95125 Catania, Italy

F. Tartarone (✉)
Dipartimento di Matematica e Fisica, Università degli Studi Roma Tre, Largo San Leonardo Murialdo 1, 00146 Roma, Italy
e-mail: tfrance@mat.uniroma3.it

A. Facchini et al. (eds.), *Advances in Rings, Modules and Factorizations*,
Springer Proceedings in Mathematics & Statistics 321,
https://doi.org/10.1007/978-3-030-43416-8_8

An interesting problem considered, for instance, by S. Glaz and W. Vasconcelos in [22, 23], asks for conditions on a domain D in order to have that flat ideals of D are invertible. This question can be specialized by asking when faithfully flat ideals are invertible.

We recall that for ideals in a domain, projective is equivalent to invertible and faithfully flat is equivalent to locally principal ([2]). Thus the condition "faithfully flat ideals are projective" is exactly "locally principal ideals are invertible".

In [22], the authors conjecture the equivalence between faithfully flat and projective ideals in H-domains (i.e., a domain in which t-maximal ideals are divisorial). This conjecture has been disproved by G. Picozza and F. T. in [32, Example 1.10], but the problem that S. Glaz and W. Vasconcelos posed has also been considered for other classes of domains like the Prüfer ones (i.e., domains whose localization at a prime ideal is a valuation domain).

Bazzoni [6] conjectured that:

In a Prüfer domain D "locally principal ideals are invertible" if and only if D has the finite character on maximal ideals.

Bazzoni's conjecture was proved at the same time by W.C. Holland, J. Martinez, W. Wm. McGovern, M. Tesemma in [27] by using methods (of independent interest) of the theory of lattice-ordered groups and by F. Halter-Koch using the theory of r-Prüfer monoids [25], where r is an ideal system.

After the publication of these papers, a growing interest in this question (in more general contexts) came up.

It was considered a kind of t-version of Bazzoni's conjecture which replaces the finite character with the t-finite character and the local invertibility with the t-local invertibility. In this context, the original conjecture has been generalized to Prüfer v-multiplication domains and to even larger classes of integral domains (cfr. [18, 25, 35]). Section 3 of the paper is completely dedicated to a discussion of the t-local invertibility.

Finally, in Section 4, we present some recent results about the local invertibility and its connections with the finite character on maximal ideals in commutative (not necessarily integral) rings. As seen in Theorem 4.5, Bazzoni's conjecture can be extended to general rings with zero-divisors. In this context, the authors use the concept of Manis valuations and Prüfer extensions in place of Prüfer domains (see [28]).

2 The Prüfer Case

In this section, we will provide a deeper insight into the so-called Bazzoni's conjecture, which states that a Prüfer domain D has finite character if and only if every locally principal of D is invertible (i.e., finitely generated). This conjecture, stated in [4] and [6], was first solved by Holland, Martinez, Mc Govern and Tesemma in [27] and their result was then generalized to several other classes of rings (see, for instance, [18, 25, 35]). We are going to present the main steps of the first proof of Bazzoni's conjecture. It is based on an argument involving some basic tools on lattice-ordered abelian groups. Thus we will recall now some preliminaries for the reader's convenience.

As usual, for any ring R, $\mathrm{Spec}(R)$ denotes the set of all prime ideals of R and, if S is any subset of R, we set

$$V(S) := \{\mathfrak{p} \in \mathrm{Spec}(R) : \mathfrak{p} \supseteq S\}.$$

If $(X, \leq)$ is a partially ordered set and $x_1, \ldots, x_n \in X$, then $\sup(x_1, \ldots, x_n)$ (resp., $\inf(x_1, \ldots, x_n)$) will denote the supremum (resp., the infimum) of $\{x_1, \ldots, x_n\}$ in X, if it exists. Recall that a nonempty and proper subset F of a partially ordered set $(X, \leq)$ is a *filter* (see [12, Definition 14.1]) if it satisfies the following properties:

- given $x, y \in F$, there exists $\inf(x, y)$ and $\sup(x, y) \in F$;
- if $f \in F, x \in X$ and $f \leq x$, then $x \in F$.

Let $\mathcal{F}(X)$ denote the set of all filters on X. If $(X, \leq)$ is a lattice, for any $a \in X$, the set $\{x \in X : x \geq a\}$ is clearly a filter and it is called *a principal filter*.

Now, let $(G, \cdot, \leq)$ be a lattice-ordered abelian group (for short, a ℓ-group). Recall that an ℓ-subgroup H of G is *convex* if, given elements $h, k \in H, g \in G$ such that $h \leq g \leq k$, then $g \in H$. Clearly, the intersection of any nonempty collection of convex ℓ-subgroups of G is still convex, and thus for any subset S of G there exists the smallest convex ℓ-subgroup $\mathrm{Conv}(S)$ of G containing S, and it is called *the convex envelop of* S. If e is the identity element of G, let

$$G^+ := \{g \in G : g \geq e\}$$

be the positive cone of G. If $S \subseteq G^+$, then, by [12, Proposition 7.11(b)],

$$\mathrm{Conv}(S) = \{g \in G : |g| \leq s_1 \cdots s_n, \ \text{ for some } s_1, \ldots, s_n \in S, n \in \mathbb{N}^+\},$$

where $|g| := \sup(g, e) \cdot \sup(g^{-1}, e)$. A convex ℓ-subgroup P of G is *prime* if, whenever $g, h \in G$ and $\inf(g, h) = e$, then either $g \in P$ or $h \in P$. A straightforward application of Zorn's Lemma shows that G admits *minimal prime subgroups* (i.e., prime subgroups which are minimal under inclusion) and every prime subgroup of G contains some minimal prime subgroup [12, Theorem 9.6]. Let $\mathcal{P}(G)$ (resp., $\mathcal{M}(G)$) denote the set of all prime (resp., minimal prime) subgroups of G. For every $g \in G$, let $U(g) := \{P \in \mathcal{M}(G) : g \notin P\}$.

The bridge which links Bazzoni's conjecture and the theory of ℓ-groups is the ideal structure of a Prüfer domain. Let D be an integral domain and let $\mathrm{Inv}(D)$ be the multiplicative group consisting of all invertible fractional ideals of D, endowed with partial order given by the opposite inclusion $\supseteq$. Since D is the identity of $\mathrm{Inv}(D)$, the positive cone $\mathrm{Inv}(D)^+$ of $\mathrm{Inv}(D)$ is just the set of all integral invertible ideals.

Theorem 2.1 ([9, Theorem 2]). *If D is a Prüfer domain, then* $\mathrm{Inv}(D)$ *is an ℓ-group.*

More precisely, given two fractional ideals $I, J \in \mathrm{Inv}(D)$, then they are finitely generated, in particular. Thus $I + J$ is finitely generated too and, since D is Prüfer, $I + J \in \mathrm{Inv}(D)$, proving that $I + J = \inf(I, J)$. Moreover, D is a coherent domain (meaning that the intersection of finitely many finitely generated fractional ideals is finitely generated too), being it Prüfer, by [21, Proposition (25.4)(1)], and thus $I \cap J \in \mathrm{Inv}(D)$, proving that $I \cap J = \sup(I, J)$. Let $\mathcal{I}^\bullet(D)$ denote the set of all nonzero (integral) ideals of D.

Lemma 2.2 (see [27, Lemma 1]). *Let D be a Prüfer domain. The following properties hold.*

(1) The map $\varphi : \mathcal{I}^\bullet(D) \longrightarrow \mathcal{F}(\mathrm{Inv}(D)^+)$ defined by setting

$$\varphi(\mathfrak{i}) := \{\mathfrak{a} \in \mathrm{Inv}(D)^+ : \mathfrak{a} \subseteq \mathfrak{i}\}$$

is a bijection.

(2) For every $\mathfrak{i} \in \mathcal{I}^\bullet(D)$, $\varphi(\mathfrak{i})$ is a principal filter if and only if $\mathfrak{i}$ is invertible (i.e., finitely generated).

Proof. (1). The fact that φ is well defined and injective is trivial. Now, let F be a filter on $\mathrm{Inv}(D)^+$ and set $\mathfrak{i} := \sum_{\mathfrak{a} \in F} \mathfrak{a}$. Thus, by definition, $F \subseteq \varphi(\mathfrak{i})$. Conversely, take an ideal $\mathfrak{b} \in \varphi(\mathfrak{i})$, i.e., $\mathfrak{b}$ is invertible and $\mathfrak{b} \subseteq \mathfrak{i}$. Since $\mathfrak{b}$ is, in particular, finitely generated, there exist ideals $\mathfrak{a}_1, \ldots, \mathfrak{a}_n \in F$ such that $\mathfrak{b} \subseteq \mathfrak{a}_1 + \ldots + \mathfrak{a}_n$, that is, $\mathfrak{b} \geq \inf(\mathfrak{a}_1, \ldots, \mathfrak{a}_n)$. Keeping in in mind that F is a filter it follows $\inf(\mathfrak{a}_1, \ldots, \mathfrak{a}_n) \in F$ and finally $\mathfrak{b} \in F$.

(2) is clear, for the definitions. □

Remark 2.3. Let D be a Prüfer domain and let $\mathrm{Spec}(D)^\bullet$ denote the set of all nonzero prime ideals of D. For any ideal $\mathfrak{p} \in \mathrm{Spec}(D)^\bullet$, set $X_\mathfrak{p} := \{\mathfrak{a} \in \mathrm{Inv}(D)^+ : \mathfrak{a} \not\subseteq \mathfrak{p}\}$ and $\overline{\mathfrak{p}} := \mathrm{Conv}(X_\mathfrak{p})$.

(1) $\overline{\mathfrak{p}}$ is a prime subgroup of $\mathrm{Inv}(D)$. Take invertible ideals $\mathfrak{a}_1, \ldots, \mathfrak{a}_n$ of D such that $\inf(\mathfrak{a}_1, \ldots, \mathfrak{a}_n) := \mathfrak{a}_1 + \ldots + \mathfrak{a}_n = D$. It follows $\mathfrak{a}_i \not\subseteq \mathfrak{p}$, for some i, i.e., $\mathfrak{a}_i \in X_\mathfrak{p} \subseteq \overline{\mathfrak{p}}$. Furthermore, by [12, Proposition 7.11(b)] and keeping in mind that $X_\mathfrak{p}$ is closed under multiplication, we infer

$$\overline{\mathfrak{p}} = \{I \in \mathrm{Inv}(D) : |I| \supseteq \mathfrak{a},\ \text{for some } \mathfrak{a} \in X_\mathfrak{p}\}.$$

(2) Since $|\mathfrak{a}| = \mathfrak{a}$, for any $\mathfrak{a} \in \mathrm{Inv}(D)^+$, it follows that $\overline{\mathfrak{p}} \cap \mathrm{Inv}(D)^+ = X_\mathfrak{p}$.

(3) Consider $G = \mathrm{Inv}(D)$ and the map $\psi : \mathrm{Spec}(D)^\bullet \longrightarrow \mathcal{P}(G)$ defined by setting, $\psi(\mathfrak{p}) := \overline{\mathfrak{p}}$, for every $\mathfrak{p} \in \mathrm{Spec}(D)^\bullet$. Keeping in mind part (2) of the present remark, it easily follows that, for every $\mathfrak{p}, \mathfrak{q} \in \mathrm{Spec}(D)^\bullet$, $\mathfrak{p} \subseteq \mathfrak{q}$ if and only if $\overline{\mathfrak{q}} \subseteq \overline{\mathfrak{p}}$. In particular, ψ is injective and order reversing.

(4) ψ restricts to a bijection of $\mathrm{Max}(D)$ onto $\mathcal{M}(\mathrm{Inv}(D))$. As a matter of fact, let P be a minimal prime subgroup of $\mathrm{Inv}(D)$. If, for every maximal ideal $\mathfrak{m}$ of

D, $X_{\mathfrak{m}} \not\subseteq P$, there is an invertible integral ideal $\mathfrak{a}_{\mathfrak{m}}$ of D such that $\mathfrak{a}_{\mathfrak{m}} \not\subseteq \mathfrak{m}$ and $\mathfrak{a}_{\mathfrak{m}} \notin P$. It follows $\sum_{\mathfrak{m}\in \mathrm{Max}(D)} \mathfrak{a}_{\mathfrak{m}} = D$ and thus (since every ring has the identity element) there are maximal ideals $\mathfrak{m}_1, \ldots, \mathfrak{m}_r$ of D such that $\mathfrak{a}_{\mathfrak{m}_1} + \ldots + \mathfrak{a}_{\mathfrak{m}_r} = D$. Since P is a prime subgroup, there exists $i \in \{1, \ldots, r\}$ such that $\mathfrak{a}_{\mathfrak{m}_i} \in P$, a contradiction. It follows that there is a maximal ideal $\mathfrak{m}$ of D such that $X_{\mathfrak{m}} \subseteq P$ and, since P is a minimal prime subgroup (and, in particular, it is convex), we deduce $\overline{\mathfrak{m}} := \mathrm{Conv}(X_{\mathfrak{m}}) = P$. Conversely, if $\mathfrak{m}$ any maximal ideal of D and P is a minimal prime subgroup of $\mathrm{Inv}(D)$ such that $\overline{\mathfrak{m}} \supseteq P$, take a maximal ideal $\mathfrak{n}$ of D such that $P = \overline{\mathfrak{n}}$ (in view of what we have just proved). By part (3) it follows $\mathfrak{m} = \mathfrak{n}$ and thus $\overline{\mathfrak{m}}$ is minimal.

(5) By the previous parts, for every integral invertible ideal $\mathfrak{a}$ of D, we have

$$\psi(\mathrm{Max}(D) \cap V(\mathfrak{a})) = U(\mathfrak{a}).$$

The following result immediately follows from parts (4, 5) of the previous remark.

Lemma 2.4 ([27, Theorem 2]). *A Prüfer domain D has a finite character if and only if, for every integral invertible ideal $\mathfrak{a}$ of D, the set $U(\mathfrak{a})$ is finite.*

Thus the previous lemma provides the translation of the finite character of a Prüfer domain into a statement in the language of ℓ-groups. The next goal is to provide the translation of the LPI property.

Definition 2.5 ([27, Definition 3]). Let G be an ℓ-group and let F be a filter on G^+. Then F is said to be a *cold filter* provided that, for every $P \in \mathcal{M}(G)$, there exists some element $f \in F$ such that $f + P \leq g + P$, for all $g \in F$ (where $\leq$ is the canonical total order induced by the order of G into the factor group G/P).

Lemma 2.6 ([27, Proposition 5]). *Let D be a Prüfer domain. Then a nonzero ideal $\mathfrak{i}$ of D is locally principal if and only if the filter $\varphi(\mathfrak{i})$ (see Lemma 2.2) is a cold filter.*

Combining Lemmas 2.2 and 2.6, it is clear that for a Prüfer domain D the following conditions are equivalent.

(1) Every nonzero locally principal ideal of D is invertible.
(2) Every cold filter on $\mathrm{Inv}(D)^+$ is principal.

This provides the complete translation of Bazzoni's conjecture into a conjecture regarding ℓ-groups. The key step to show it is to observe that, if G is an ℓ-group such that every cold filter on G^+ is principal, then every element of G^+ is greater than only finitely many mutually disjoint elements ([27, Proposition 7]). Now several relevant results of Conrad [11] are very helpful, together with the Finite Basis Theorem ([12, Theorem 46.12]). These tools lead to the main result.

Theorem 2.7 ([27, Theorem 9]). *Let G be an ℓ-group such that every cold filter on G^+ is principal. Then, for every $g \in G^+$, the set $U(g)$ is finite.*

Thus finally, keeping in mind the previous theorem and Lemmas 2.4, 2.6, the desired conclusion follows.

Corollary 2.8. *A Prüfer domain D has finite character if and only if every nonzero locally principal ideal of D is invertible.*

3 Generalization to Non-Prüfer Domains

We start this section by recalling some basic facts about star operations.

Let D be an integral domain with quotient field K. The set $\mathbf{F}(D)$ denotes the nonzero fractional ideals of D and $\mathbf{f}(D)$ the nonzero finitely generated fractional ideals of D.

A map $\star : \mathbf{F}(D) \to \mathbf{F}(D)$, $I \mapsto I^\star$ is called a *star operation* if the following conditions hold for all $x \in K \setminus \{0\}$ and $I, J \in \mathbf{F}(D)$:

($\star_1$) $(xD)^\star = xD$;
($\star_2$) $I \subseteq J \Rightarrow I^\star \subseteq J^\star$;
($\star_3$) $I \subseteq I^\star$ and $I^{\star\star} := (I^\star)^\star = I^\star$.

Given a star operation $\star$, a nonzero ideal I of D such that $I = I^\star$ is called a *$\star$–ideal.*

Examples of star operations are the d-operation, the v-operation, and the t-operation:

- The *d-operation* is the identity map $I \mapsto I$.
- The *v-operation* is the map:

$$I \mapsto I^v := (D\colon (D\colon I)), \quad \text{where} \quad (D\colon I) := I^{-1} = \{x \in K \mid xI \subseteq D\}.$$

- the *t-operation* is the map:

$$I \mapsto I^t := \bigcup_{J \in \mathbf{f}(D),\, J \subseteq I} J^v.$$

A star operation $\star$ on D is *of finite type* if for all $I \in F(D)$,

$$I^\star = \bigcup \{J^\star : J \subseteq I, J \in \mathbf{f}(D)\}.$$

From the definition, it follows that the t-operation is of finite type.

Definition 3.1. Given a star operation $\star$ on a domain D and $I \in \mathbf{F}(D)$,

I is *$\star$-invertible* if there exists $J \in \mathbf{F}(D)$ such that $(IJ)^\star = D$ (it is easy to see that, in this case, $J = I^{-1}$).

Thus, taking $\star = d$, we find the usual definition of invertible ideal, that is $II^{-1} = D$.

An ideal I is $\star$-finite if there exists a finitely generated ideal J such that $J^\star = I^\star$. If $\star$ is of finite type, J can always be taken inside I.

If $\star$ is a star operation of finite type, the set of $\star$-ideals has (proper) maximal elements called $\star$-maximal ideals (this set is denoted by $\star - \text{Max}(D)$). A $\star$-maximal ideal is a prime ideal and every integral $\star$-ideal is contained in a $\star$-maximal ideal.

A domain D has the *$\star$-finite character* if each $\star$-ideal (equivalently, each nonzero element) of D is contained in finitely many $\star$-maximal ideals.

It is well-known that a nonzero ideal of a domain D is invertible if and only if I is finitely generated and locally principal (see [8, II Section 5, Theorem 4]).

A similar characterization holds for $\star$-invertible ideals.

In fact, an ideal I is $\star$-invertible if and only it is $\star$-finite and $\star$-locally principal (that is, ID_M is principal for each $M \in \star - \text{Max}(D)$) ([26, page 137]).

In the characterization of invertible ideals given above, the hypothesis that I is finitely generated can be dropped down in some classes of domains, called LPI domains, introduced by D. D. Anderson and M. Zafrullah in [1].

A domain D is LPI if every nonzero locally principal ideal is invertible, or equivalently, if every faithfully flat ideal is finitely generated. Thus, LPI domains are exactly the domains in which faithfully flat ideals are projective.

Mori domains, and therefore Noetherian domains are LPI.

The finite character condition on a domain D is sufficient to have that D is LPI. In fact, it is a straightforward exercise to prove that the finite character implies that a locally principal ideal is finitely generated. Nevertheless, this condition is not necessary for the LPI property of D.

For instance, Noetherian domains are LPI but they do not always have the finite character (see $\mathbb{Z}[X]$).

Definition 3.2. A domain D has the t-finite character if each nonzero element $x \in D$ is contained in only finitely many t-maximal ideals.

Noetherian domains have the t-finite character (see [5, Proposition 2.2(b)]) and the t-finite character is a sufficient condition for a general domain D to be LPI ([32, Lemma 1.12]).

Remark 3.3. In Prüfer domains the t-operation is the identity ([21, Theorem 22.1 (3)]), that is, each ideal is a t-ideal. Thus, for this class of domains, the t-finite character coincides with the finite character, which is exactly the property required for Prüfer domains to be LPI in the conjecture by S. Bazzoni.

The above remark brings to consider the t-version of Prüfer domain, the Prüfer v-multiplication domains (PvMD).

We recall that a domain D is a PvMD if each t-finite ideal is t-invertible. Equivalently, if and only if D_P is a valuation domain for each t-prime (or t-maximal) ideal P of D ([29, Theorem 4.3]).

Thus, by replacing the finite character with the t-finite character and the invertibility property for ideals with the t-invertibility property it is possible to generalize S. Bazzoni's conjecture using the t-operation.

A first step in this direction is the definition of t-LPI domains.

Definition 3.4. A domain D is t-LPI if each nonzero t-locally principal t-ideal is t-invertible.

The t-version of Bazzoni's conjecture is then:

"A PvMD is t-LPI if and only if it has the t-finite character."

M. Zafrullah and F. Halter-Koch proved this result almost at the same time using different techniques (see [35, Proposition 5] and [25]). F. Halter-Koch considered the problem in a more general setting involving a general $\star$-operation.

Now, we consider the t-version of Bazzoni's conjecture outside the natural context of PvMD and present contributions that prove the equivalence ($\diamond$)

"t-finite character $\Leftrightarrow$ each nonzero t-locally principal ideal is t-invertible"

for more general classes of domains.

We have seen that the t-finite character is a sufficient condition for a general domain D in order to have that D is t-LPI ([32, Lemma 1.12]).

Conversely, by [18, Example 2.3], if D is not a PvMD, the t-finite character is not a necessary condition to have that D is t-LPI.

Thus, a question that was investigated, for instance, by T. Dumitrescu and M. Zafrullah in [13] and, independently, by C.A. F. - G. Picozza and F.T. in [18], concerns the characterization of classes of domains strictly larger than PvMD verifying condition ($\diamond$) given above.

T. Dumitrescu and M. Zafrullah considered the case of t-Schreier domains, that we define below.

Given a domain D, $\mathrm{Inv}_t(D)$ is the set of the t-invertible t-ideals of D.

A domain D is t-Schreier if $\mathrm{Inv}_t(D)$ is a Riesz group, that is: if every finite intersection of nonzero principal ideals is a direct union of t-invertible t-ideals. For instance, PvMD's are t-Schreier (see [15, Lemma 1.8]). More precisely, an integral domain is a PvMD if and only if it is t-Schreier and v-coherent, by [14, Corollary 6(a)].

Theorem 3.5. [14, Proposition 17] *If D is t-Schreier, then D is t-LPI if and only if D has the t-finite character.*

Thus t-Schreier domains enlarge the class of domains verifying the t-version of Bazzoni's conjecture.

In this direction, we also find the results by C.A. F. - G. Picozza and F.T. about v-coherent domains.

In their paper [18] the authors consider the more general case of a $\star$-operation of finite type, thus including the t-operation.

The general question is then to find a characterization of domains for which the $\star$-finite character is equivalent to $\star$-LPI (that is, every $\star$-locally principal $\star$-ideal is $\star$-invertible).

We have seen as Noetherian domains suggest to use the t-finite character condition in the study of (t)-local invertibility for ideals.

Again, Noetherian domains bring to consider another interesting condition involving (t)-comaximal ideals.

In a Noetherian domain every nonzero nonunit element belongs to only a finite number of *mutually comaximal* proper invertible ideals.

Definition 3.6. Two proper ideals $I, J \subset D$ are t-comaximal if $(I + J)_t = D$. In particular, I, J are t-comaximal if $(I + J)_t = D$, which means that I and J are not contained in a common t-maximal ideal.

Theorem 3.7. [18, Proposition 1.6] *Let D be an integral domain. Then the following conditions are equivalent.*

(i) D has the $t-$finite character.
(ii) Every family of mutually t-comaximal t-finite t-ideals of D with nonzero intersection is finite.

Thus, in order to prove that t-LPI is equivalent to the t-finite character, it would be interesting to see whether the condition (ii) of the above Theorem has connections with the t-LPI property.

Since in Prüfer domains the t-operation is the identity, we can restate Theorem 3.7 as follows:

Corollary 3.8. *A Prüfer domain D has the finite character if and only if each invertible integral ideal of D is contained in at most a finite number of mutually comaximal invertible ideals.*

Corollary 3.8 can be easily extended to PvMD's by replacing comaximality with t-comaximality.

Corollary 3.9. *A PvMD has the t-finite character if and only if each integral t-invertible t-ideal is contained in at most a finite number of mutually t-comaximal t-invertible t-ideals.*

Remark 3.10. Consider the following t-invertibility like conditions for ideals in a domain D:

(1) t-locally t-finite (i.e. I_M is t-finite for each $M \in t-\mathrm{Max}(D)$, with respect to the t-operation of D_M) t-ideals are t-finite;
(2) t-locally principal (i.e., I_M is principal for each $M \in t-\mathrm{Max}(D)$) t-ideals are t-invertible (t-LPI);

We observe that

(a) (1) $\Rightarrow$ (2);
(b) conditions (1)–(2) are equivalent to LPI in the case of Prüfer domains and to t-LPI for PvMD's;
(c) the t-finite character implies conditions (1)–(2).

We recall that a domain D is *v-coherent* if for any nonzero finitely generated ideal I of D, I^{-1} is v-finite (see, for instance, [16, Proposition 3.6] and [30]).

A domain D is *t-locally v-coherent* if D_M is v-coherent, for each $M \in t{-}\mathrm{Max}(D)$.

Important classes of v-coherent domains are Noetherian domains, Mori domains, Prüfer domains, PvMD's, finite conductor domains (i.e., $(x) \cap (y)$ is finitely generated for each $x, y \in A$), coherent domains (i.e., the intersection of two finitely generated ideals is finitely generated).

Using pullback constructions it is possible to give examples of t-locally v-coherent domains which are not v-coherent (cfr. [20]).

Since both Prüfer domains and PvMD's are t-locally v-coherent, a first step in the direction of generalizing Bazzoni's conjecture to any domain is the following theorem.

Theorem 3.11. [18, Theorem 1.11] *Let D be an integral domain which is t-locally v-coherent. Then the following conditions are equivalent.*

(i) D has the t-finite character;
(ii) every family of t-finite, t-comaximal, t-ideals over a nonzero element $a \in D$ is finite;
(iii) every nonzero t-locally t-finite t-ideal is t-finite.

Remark 3.12. (a) A t-locally v-coherent domain is not necessarily a PvMD. In fact any Noetherian domain is t-locally v-coherent (and it is not always a PvMD). Thus, the class of domains considered in Theorem 3.11 is larger than one of the PvMD's.

(b) Condition (iii) of Theorem 3.11 is exactly point (1) of Remark 3.10. Thus it implies the t-LPI property (point (2) of the same Remark) and it is equivalent to t-LPI in PvMD's. In general, we don't know whether these two conditions are equivalent.
Anyway, Theorem 3.11 suggests that a natural statement to generalize to any domain of the t-version of Bazzoni's conjecture should claim the equivalence between the t-finite character and condition (1) of Remark 3.10.

(c) In general, Theorem 3.11 cannot be extended to any finite type star operation. For instance, it fails if we take the identity operation d. In fact, a Noetherian domain does not need to have the finite character on maximal ideals, but each locally finitely generated ideal is finitely generated.

So far, the t-operation seems to be the only star operation of finite type that has an interesting role in the generalization of Bazzoni's conjecture.

In fact there is not an analogue of Theorem 3.11 for a generic $\star$-operation. Here below we give two partial results in this direction.

Independently, the authors in [13, Corollary 3] and in [18, Proposition 1.6] put in connection some families of mutually $\star$-comaximal ideals of D with the $\star$-finite character of D.

The following theorem generalizes Theorem 3.7 to any star operation of finite type.

Theorem 3.13. *Let D be an integral domain and $\star$ a finite type star operation on D. Then the following conditions are equivalent.*

(i) D has the $\star$-finite character.
(ii) Every nonzero element $a \in D$ is contained in at most finitely many proper $\star$-finite, mutually $\star$-comaximal $\star$-ideals of D.

There is not a proven connection between condition (ii) of the theorem above and generalizations of $\star$-LPI condition as it happens for the t-operation by Theorem 3.11.

The following result states that if things "work well" for the t-operation, then there are positive cascade results for finite type star operations.

Proposition 3.14. [18, Proposition 2.2] *Let D be a domain in which each t-locally principal t-ideal is t-finite. Then, for any star operation of finite type, each $\star$-locally principal $\star$-ideal is $\star$-finite. In particular, a locally principal ideal is finitely generated (and so, invertible).*

In Proposition 3.14 the t-operation cannot be replaced by any finite type star operation. For instance, [18, Example 2.3] shows that it does not hold when $\star = d$.

Another interesting class of domains recently studied in this context are the *finitely stable* domains.

Definition 3.15. An ideal I of a domain D is *finitely stable* if I is invertible (or projective) in its endomorphism ring

$$End(I) = (I : I) = \{x \in K \mid xI \subseteq I\}.$$

A domain D is *stable* if each nonzero ideal is stable and D is *finitely stable* if each finitely generated ideal is stable. Obviously, stable domains are finitely stable.

An important result proven by B. Olberding in [31] states that stable domains have the finite character. Moreover, integrally closed stable domains are Prüfer, thus they are LPI.

It is also well-known that integrally closed, finitely stable domains are exactly Prüfer domains, hence they generalize the Prüfer ones.

Moreover, finitely stable domains are a distinct class from v-coherent domains. In fact, $D = K[[X^2, X^3]]$ (where K is any field) is Mori, hence v-coherent, but it is not finitely stable because its maximal ideal is not stable (see [7, Example 1]).

On the other hand, if we take a PvMD that is not Prüfer (e.g., $\mathbb{Z}[X]$), then this is not finitely stable but it is v-coherent.

S. Bazzoni in [7] before, and S. Xing and F. Wang in [34] after, study conditions on finitely stable domains in order to verify the LPI property.

A domain has the *local stability property* if each nonzero ideal that is locally stable is stable.

In [7, Theorem 4.5] a characterization of finitely stable domains with finite character is given.

Theorem 3.16. *Let D be a finitely stable domain. Then D has the finite character if and only if it has the local stability property.*

[7, Lemma 3.2] shows that if D is a finitely stable domain that has the local stability property then it is LPI.

As we know, the question whether a finitely stable LPI domain has the local stability property is still open ([7, Question 4.6]).

In view of Theorem 3.16, if the answer to this question is positive, then this would imply that also for (non integrally closed) finitely stable domain the finite character is equivalent to LPI, as it happens in the Prüfer case.

Other interesting results about the interplay between the finite character and the LPI property for finitely stable domains are given in [34].

First of all, in this paper the authors show that LPI is not a local property. In fact they give an example of a domain D that is LPI, but D_S is not LPI for a suitable multiplicatively closed subset $S \subseteq D$ ([34, Example 2.4]). This fact has no connections with the finite character question, but it is interesting by itself.

Anyway, the main result of [34] gives a characterization of LPI finitely stable involving the finite character.

The authors denote by $\mathcal{T}(D)$ the set of maximal ideals m of D for which there exists a finitely generated ideal I such that m is the only maximal ideal containing I. For each ideal I, $\Omega(I)$ is the set of maximal ideals of D containing I. Thus, the finite character is equivalent to ask that $\Omega(I)$ is finite for each nonzero ideal $I \subseteq D$.

Then, the following results are proven:

Theorem 3.17. [34, Theorem 2.6] *Let D be a finitely stable LPI domain. Then every nonzero element of D is contained in, at most, finitely many ideals of $\mathcal{T}(D)$.*

Thus we have that LPI on finitely stable domains implies the finite character on the subset $\mathcal{T}(D)$ of $\mathrm{Max}(D)$.

From Theorem 3.17 it follows the next corollary:

Corollary 3.18. [34, Corollary 2.7] *Let D be a finitely stable LPI domain. Then, the following conditions are equivalent:*

(i) *D ha the finite character;*
(ii) *$\mathcal{T}(D) \cap \Omega(I) \neq \emptyset$, for each nonzero, finitely generated ideal I of D.*

We observe that Corollary 3.18 does not hold without the LPI hypothesis.

In fact, the following example shows that there exists a Prüfer domain D without the finite character and such that $\mathcal{T}(D) = \emptyset$. In this case D is finitely stable (since it is Prüfer) and point (ii) of Corollary 3.18 trivially holds since $\mathcal{T}(D) = \emptyset$, but the domain has not the finite character.

Example 3.19. Consider the integer valued polynomial ring

$$\mathrm{Int}(\mathbb{Z}) = \{f(X) \in \mathbb{Q}[X] \mid f(\mathbb{Z}) \subseteq \mathbb{Z}\}.$$

It is well-known that Int($\mathbb{Z}$) is a Prüfer domain ([10, Theorem VI.1.7]) and it has not the finite character. In fact, each prime $p \in \mathbb{Z}$ is contained in infinitely many maximal ideals of the type $M_{p,\alpha}$ described in [10, Theorem V.2.7].

We easily see that $\mathcal{T}(\mathrm{Int}(\mathbb{Z})) = \emptyset$. By [10, Theorem V.2.7], the maximal ideals of Int($\mathbb{Z}$) are the $M_{p,\alpha}$, $p \in \mathbb{Z}$, $\alpha \in \widehat{\mathbb{Z}_p}$ (the p-adic completion of $\mathbb{Z}$). Suppose that a maximal ideal $M_{p,\alpha}$ belongs to $\mathcal{T}(\mathrm{Int}(\mathbb{Z}))$ and let I be a finitely generated ideal such that $M_{p,\alpha}$ is the only maximal ideal containing I. Since Int($\mathbb{Z}$) is a Prüfer domain, I is invertible. But

$$I^{-1} \subseteq \bigcap_{\mathcal{M} \in \mathrm{Max}(\mathrm{Int}(\mathbb{Z})), I \nsubseteq \mathcal{M}} \mathrm{Int}(\mathbb{Z})_{\mathcal{M}},$$

where second term is an overring of Int($\mathbb{Z}$) defined by Kaplansky and so-called the *Kaplansky transform* of the ideal I [17, Theorem 3.2.2]. Using the same argument of [33, Proposition 2.2] to see that $\mathfrak{P}^{-1} = \mathrm{Int}(D)$, we can show that $I^{-1} = \mathrm{Int}(\mathbb{Z})$ against the hypothesis that I is a proper ideal.

4 Generalization to Rings with Zero-Divisors

In the present section we will present a generalization of Bazzoni's conjecture to (commutative) rings with zero-divisors. We start with recalling some terminology and preliminaries and we will follow [28]. Let $A \subseteq B$ be a ring extension and let X be an A-submodule of B. We will say that X is *B-regular* if $XB = B$. Furthermore, X is said to be *B-invertible* if $XY = A$, for some A-submodule Y of B. It is worth noting that, in case B is the total ring of fractions $\mathrm{T}(A)$ of A, then X is B-regular if and only if it is regular (i.e., it contains a regular element of A) and X is B-invertible if and only if it is invertible in the sense of Griffin (see [24]). Several known facts about fractional ideals of integral domains extend naturally in this setting: for instance, X is B-invertible if and only if X is B-regular, finitely generated and locally principal [28, Section 2, Proposition 2.3].

Definition 4.1. Let $A \subseteq B$ be a ring extension. We say that *$A \subseteq B$ has finite character* if every B-regular ideal of A is contained only in finitely many maximal ideals of A.

Clearly, if $B = \mathrm{T}(A)$, then $A \subseteq B$ has finite character if and only if every regular ideal of A is contained only in finitely many maximal ideal. The following result extends Theorem 3.13 (in case $\star = d$) to every ring extension.

Proposition 4.2. [19, Corollary 3.3] *For a ring extension $A \subseteq B$ the following conditions are equivalent.*

(1) $A \subseteq B$ has finite character.
(2) For any finitely generated and B-regular ideal $\mathfrak{a}$ of A, every collection of mutually comaximal finitely generated (and B-regular) ideals of A containing $\mathfrak{a}$ is finite.

Finite character ring extensions allows to test relevant properties of ideals locally, as the following result shows.

Proposition 4.3. ([19, Proposition 3.4 and Corollary 3.5]) *Let* $A \subseteq B$ *be a ring extension with finite character and let* $\mathfrak{a}$ *be a* B*-regular ideal of* A*. If* $\mathfrak{a}$ *is locally finitely generated, then* $\mathfrak{a}$ *is finitely generated. In particular,* $\mathfrak{a}$ *is* B*-invertible if and only if it is locally principal.*

According to [28, Chapter 2, Theorem 2.1], we say that a ring extension $A \subseteq B$ is a *Prüfer extension* if $A \subseteq B$ is a flat epimorphism (in the category of rings) and every finitely generated B-regular ideal of A is B-invertible. In case $B = \mathrm{T}(A)$, then the previous definition extends the notion of Prüfer ring given by Griffin (i.e., every regular finitely generated ideal is invertible). In what follows it will suffice to work with the following weaker notion.

Definition 4.4. A ring extension $A \subseteq B$ is said to be an *almost Prüfer* extension if every finitely generated B-regular ideal of A is B-invertible.

Now we are in condition to state Bazzoni's conjecture for rings with zero-divisors.

Theorem 4.5. [19, Theorem 4.5] *For an almost Prüfer extension* $A \subseteq B$ *the following conditions are equivalent.*

(1) $A \subseteq B$ *has finite character.*
(2) Every B*-regular locally principal ideal of* A *is* B*-invertible.*

Keeping in mind the remarks made at the beginning of the present section, the following corollary is now clear.

Corollary 4.6. [19, Corollary 4.6] *Let* A *be a Prüfer ring. Then every regular locally principal ideal of* A *is invertible if and only if every regular element of* A *is contained in only finitely many maximal ideal.*

Finally we list some problems and questions that can motivate further investigation about this topic.

Question 1. Is there a class of integral domains, larger than that of t-locally v-coherent domains, for which the equivalent conditions of Theorem 3.11 hold?

Question 2. Does the statement of Theorem 3.16 admit some generalization for rings with zero-divisors?

Acknowledgements The first and the second authors were partially supported by GNSAGA of Istituto Nazionale di Alta Matematica. The first author was partially supported by Dipartimento di Matematica e Informatica of Università di Catania (Research program "Proprietà algebriche locali e globali di anelli associati a curve e ipersuperfici" PTR 2016–18), and by the research program "Reducing complexity in algebra, logic, combinatorics - REDCOM" "Ricerca Scientifica di Eccellenza 2018" Fondazione Cariverona.

References

1. Anderson, D.D., Zafrullah, M.: Integral domains in which nonzero locally principal ideals are invertible. Commun. Algebra **39**(3), 933–941 (2011)
2. Anderson, D.D., Anderson, D.F.: Some remarks on cancellation ideals. Math. Jpn. **29**(6), 879–886 (1984)
3. Atiyah, M.F., MacDonald, I.G.: Introduction to Commutative Algebra, Addison-Wesley Series in Mathematics (1969)
4. Bazzoni, S.: Clifford regular domains. J. Algebra **238**, 703–722 (2001)
5. Barucci, V., Gabelli, S.: How far is a Mori domain from being a Krull domain? J. Pure Appl. Algebra **2**(45), 101–112 (1987)
6. Bazzoni, S.: Class semigroups of Prüfer domains. J. Algebra **184**, 613–631 (1996)
7. Bazzoni, S.: Finite character of finitely stable domains. J. Pure Appl. Algebra **215**(6), 1127–1132 (2011)
8. Bourbaki, N.: Algèbre Commutative. Chapter 1–7. Hermann Paris, pp. 1961–1965
9. Brewer, J., Klingler, L.: The ordered group of invertible ideals of a Prüfer domain of finite character. Commun. Algebra **33**(11), 4197–4203 (2005)
10. Cahen, P.-J. Chabert, J.-L.: Integer-Valued Polynomials. Surveys and Monographs, vol. 48, Providence (1997) (Am. Math. Soc.)
11. Conrad, P.: Some structure theorems for lattice-ordered groups. Trans. Am. Math. Soc. **99**(2), 212–240 (1961)
12. Darnel, M.: Theory of Lattice-Ordered Groups. Monographs and Textbooks Pure and Applied Mathematics, vol. 187. Marcel Dekker (1995)
13. Dumitrescu, T., Zafrullah, M.: Characterizing domains of finite star-character. J. Pure Appl. Algebra **214**(11), 2087–2091 (2010)
14. Dumitrescu, T., Zafrullah, M.: t-Schreier Domains. Commun. Algebra **39**(3), 808–818 (2011)
15. El Baghdadi, S., Gabelli, S., Zafrullah, M.: Unique representation domains II. J. Pure Appl. Algebra **212**, 376–393 (2008)
16. Fontana, M., Gabelli, S.: On the class group and the local class group of a pullback. J. Algebra **181**(3), 803–835 (1996)
17. Fontana, M., Huckaba, J.A., Papick, I.J.: Prüfer Domains. Pure and Applied Mathematics, vol. 203. Marcel Dekker, New York (1997)
18. Finocchiaro, C.A., Picozza, G., Tartarone, F.: Star-invertibility and t-finite character in integral domains. J. Algebra Appl. **10**(4), 755–769 (2011)
19. Finocchiaro, C.A., Tartarone,F.: Invertibility of ideals in Prüfer extensions. Commun. Algebra **45**(10), 4521–4527 (2017)
20. Gabelli, S., Houston, E.G.: Coherent-like conditions in pullbacks. Mich. Math. J. **44**, 99–123 (1997)
21. Gilmer, R.: Multiplicative Ideal Theory. Marcel Dekker, New York (1972)
22. Glaz, S., Vasconcelos, W.V.: Flat ideals II. Manuscripta Math. **22**(4), 325–341 (1977)
23. Glaz, S., Vasconcelos, W.V.: Flat ideals. III. Commun. Algebra **12**(2), 199–227 (1984)
24. Griffin, M.: Prüfer rings with zero-divisors. J. Reine Angew Math. **239**(240), 55–67 (1970)
25. Halter-Koch, F.: Clifford semigroups of ideals in monoids and domains. Forum Math. **21**, 1001–1020 (2009)
26. Halter-Koch, F.: Ideal Systems. An Introduction to Multiplicative Ideal Theory. Monographs and Textbooks in Pure and Applied Mathematics, vol. 211. Marcel Dekker (1998)
27. Holland, W.C., Martinez, J., McGovern, W.Wm., Tesemma, M.: Bazzoni's conjecture. J. Algebra **320**, 1764–1768 (2008)
28. Knebush, M., Zhang, D.: Manis Valuations and Prüfer Extensions I. Lecture Notes in Mathematics, vol. 1791. Springer (2002)
29. Mott, J., Zafrullah, M.: On Prüfer v-multiplication domains. Manuscripta Math. **35**(1–2), 1–26 (1981)
30. Nour El Abidine, D.: Groupe des classes de certains aimeaux integres et ideaux transformes. These de Doctorat, Universite Claude Bernard, F-Lyon 1 (1992)

31. Olberding, B.: On the structure of stable domains. Commun. Algebra **30**(2), 877–895 (2002)
32. Picozza, G., Tartarone, F.: Flat ideals and stability in integral domains. J. Algebra **324**, 1790–1802 (2010)
33. Tartarone, Francesca: Divisorial prime ideals of Int(D) when D is a Krull-type domain. Commun. Algebra **28**(10), 4647–4663 (2000)
34. Xing, S., Wang, F.: Two questions on domains in which locally principal ideals are invertible. J. Algebra Appl. **16**(6), 1750112 (2017)
35. Zafrullah, M.: t-Invertibility and Bazzoni-like statements. J. Pure Appl. Algebra **210**(5), 654–657 (2010)

Idempotence and Divisoriality in Prüfer-Like Domains

Marco Fontana, Evan Houston, and Mi Hee Park

Abstract Let D be a Prüfer $\star$-multiplication domain, where $\star$ is a semistar operation on D. We show that certain ideal-theoretic properties related to idempotence and divisoriality hold in Prüfer domains, and we use the associated semistar Nagata ring of D to show that the natural counterparts of these properties also hold in D.

Keywords Idempotent ideal · Semistar operation · Prüfer $\star$-multiplication domain · Nagata ring · Divisorial ideal

1 Introduction and Preliminaries

Throughout this work, D will denote an integral domain, and K will denote its quotient field. Recall that Arnold [1] proved that D is a Prüfer domain if and only if its associated Nagata ring $D[X]_N$, where N is the set of polynomials in $D[X]$ whose coefficients generate the unit ideal, is a Prüfer domain. This was generalized

---

The first-named author was partially supported by *GNSAGA* of *Istituto Nazionale di Alta Matematica.*
The second-named author was supported by a grant from the Simons Foundation (#354565).
The third-named author was supported by the National Research Foundation of Korea (NRF) grant funded by the Korea government (MSIP) (No. 2015R1C1A2A01055124).

M. Fontana
Dipartimento di Matematica e Fisica, Università degli Studi Roma Tre,
Largo San Leonardo Murialdo, 1, 00146 Roma, Italy
e-mail: fontana@mat.uniroma3.it

E. Houston (✉)
Department of Mathematics and Statistics, University of North Carolina at Charlotte,
Charlotte, NC 28223, USA
e-mail: eghousto@uncc.edu

M. H. Park
Department of Mathematics, Chung-Ang University, Seoul 06974, Korea
e-mail: mhpark@cau.ac.kr

A. Facchini et al. (eds.), *Advances in Rings, Modules and Factorizations*,
Springer Proceedings in Mathematics & Statistics 321,
https://doi.org/10.1007/978-3-030-43416-8_9

to Prüfer v-multiplication domains (PvMDs) by Zafrullah [16] and Kang [14] and to Prüfer $\star$-multiplication domains (P$\star$MDs) by Fontana, Jara, and Santos [8].

Our goal in this paper is to show that certain ideal-theoretic properties that hold in Prüfer domains transfer in a natural way to P$\star$MDs. For example, we show that an ideal I of a Prüfer domain is idempotent if and only if it is a radical ideal each of whose minimal primes is idempotent (Theorem 2.9), and we use a Nagata ring transfer "machine" to transfer a natural counterpart of this characterization to P$\star$MDs. For another example, in Theorem 3.5 we show that an ideal in a Prüfer domain of finite character is idempotent if and only if it is a product of idempotent prime ideals and, perhaps more interestingly, we characterize ideals that are simultaneously idempotent and divisorial as (unique) products of incomparable divisorial idempotent primes; and we then extend this to P$\star$MDs.

Let us review the terminology and notation. Denote by $\overline{\boldsymbol{F}}(D)$ the set of all nonzero D–submodules of K, and by $\boldsymbol{F}(D)$ the set of all nonzero fractional ideals of D, i.e., $E \in \boldsymbol{F}(D)$ if $E \in \overline{\boldsymbol{F}}(D)$ and there exists a nonzero $d \in D$ with $dE \subseteq D$. Let $\boldsymbol{f}(D)$ be the set of all nonzero finitely generated D–submodules of K. Then, obviously, $\boldsymbol{f}(D) \subseteq \boldsymbol{F}(D) \subseteq \overline{\boldsymbol{F}}(D)$.

Following Okabe-Matsuda [15], a *semistar operation* on D is a map $\star : \overline{\boldsymbol{F}}(D) \to \overline{\boldsymbol{F}}(D),\ E \mapsto E^\star$, such that, for all $x \in K$, $x \neq 0$, and for all $E, F \in \overline{\boldsymbol{F}}(D)$, the following properties hold:

($\star_1$) $(xE)^\star = xE^\star$;
($\star_2$) $E \subseteq F$ implies $E^\star \subseteq F^\star$;
($\star_3$) $E \subseteq E^\star$ and $E^{\star\star} := (E^\star)^\star = E^\star$.

Of course, semistar operations are natural generalizations of star operations–see the discussion following Corollary 2.5 below.

The semistar operation $\star$ is said to have *finite type* if $E^\star = \bigcup\{F^\star \mid F \in \boldsymbol{f}(D), F \subseteq E\}$ for each $E \in \overline{\boldsymbol{F}}(D)$. To any semistar operation $\star$ we can associate a finite-type semistar operation $\star_{_f}$ given by

$$E^{\star_f} := \bigcup\{F^\star \mid F \in \boldsymbol{f}(D), F \subseteq E\}.$$

We say that a nonzero ideal I of D is a *quasi-$\star$-ideal* if $I = I^\star \cap D$, a *quasi-$\star$-prime ideal* if it is a prime quasi-$\star$-ideal, and a *quasi-$\star$-maximal ideal* if it is maximal in the set of all proper quasi-$\star$-ideals. A quasi-$\star$-maximal ideal is a prime ideal. We will denote by $\text{QMax}^\star(D)$ ($\text{QSpec}^\star(D)$) the set of all quasi-$\star$-maximal ideals (quasi-$\star$-prime ideals) of D. While quasi-$\star$-maximal ideals may not exist, quasi-$\star_{_f}$-maximal ideals are plentiful in the sense that each proper quasi-$\star_{_f}$-ideal is contained in a quasi-$\star_{_f}$-maximal ideal. (See [9] for details.) Now we can associate to $\star$ yet another semistar operation: for $E \in \overline{\boldsymbol{F}}(D)$, set

$$E^{\widetilde{\star}} := \bigcap\{ED_Q \mid Q \in \text{QMax}^{\star_f}(D)\}.$$

Then $\widetilde{\star}$ is also a finite-type semistar operation, and we have $E^{\widetilde{\star}} \subseteq E^{\star_f} \subseteq E^{\star}$ for all $E \in \overline{F}(D)$.

Let $\star$ be a semistar operation on D. Set $N(\star) = \{g \in D[X] \mid c(g)^{\star} = D^{\star}\}$, where $c(g)$ is the *content* of the polynomial g, i.e., the ideal of D generated by the coefficients of g. Then $N(\star)$ is a saturated multiplicatively closed subset of $D[X]$, and we call the ring $\mathrm{Na}(D, \star) := D[X]_{N(\star)}$ the *semistar Nagata ring of D with respect to* $\star$. The domain D is called a *Prüfer* $\star$*- multiplication domain* (P$\star$MD) if $(FF^{-1})^{\star_f} = D^{\star_f}$ $(= D^{\star})$ for each $F \in f(D)$ (i.e., each such F is $\star_f$-invertible). (Recall that $F^{-1} = (D : F) = \{u \in K \mid uF \subseteq D\}$.)

In the following two lemmas, we assemble the facts we need about Nagata rings and P$\star$MDs. Most of the proofs can be found in [6, 9] or [5].

Lemma 1.1. *Let* $\star$ *be a semistar operation on D. Then:*

(1) $D^{\star} = D^{\star_f}$.
(2) $\mathrm{QMax}^{\star_f}(D) = \mathrm{QMax}^{\widetilde{\star}}(D)$.
(3) *The map* $\mathrm{QMax}^{\star_f}(D) \to \mathrm{Max}(\mathrm{Na}(D, \star))$, $P \mapsto P\mathrm{Na}(D, \star)$, *is a bijection with inverse map* $M \mapsto M \cap D$.
(4) $P \mapsto P\mathrm{Na}(D, \star)$ *defines an injective map* $\mathrm{QSpec}^{\widetilde{\star}}(D) \to \mathrm{Spec}(\mathrm{Na}(D, \star))$.
(5) $N(\star) = N(\star_f) = N(\widetilde{\star})$ *and (hence)* $\mathrm{Na}(D, \star) = \mathrm{Na}(D, \star_f) = \mathrm{Na}(D, \widetilde{\star})$.
(6) *For each* $E \in \overline{F}(D)$, $E^{\widetilde{\star}} = E\mathrm{Na}(D, \star) \cap K$, *and* $E^{\widetilde{\star}}\mathrm{Na}(D, \star) = E\mathrm{Na}(D, \star)$.
(7) *A nonzero ideal I of D is a quasi-*$\widetilde{\star}$*-ideal if and only if* $I = I\mathrm{Na}(D, \star) \cap D$.

Lemma 1.2. *Let* $\star$ *be a semistar operation on D.*

(1) *The following statements are equivalent.*
    (a) *D is a P*$\star$*MD.*
    (b) $\mathrm{Na}(D, \star)$ *is a Prüfer domain.*
    (c) *The ideals of* $\mathrm{Na}(D, \star)$ *are extended from ideals of D.*
    (d) D_P *is a valuation domain for each* $P \in \mathrm{QMax}^{\star_f}(D)$.
(2) *Assume that D is a P*$\star$*MD. Then:*
    (a) $\widetilde{\star} = \star_f$ *and (hence)* $D^{\star} = D^{\widetilde{\star}}$.
    (b) *The map* $\mathrm{QSpec}^{\star_f}(D) \to \mathrm{Spec}(\mathrm{Na}(D, \star))$, $P \mapsto P\mathrm{Na}(D, \star)$, *is a bijection with inverse map* $Q \mapsto Q \cap D$.
    (c) *Finitely generated ideals of* $\mathrm{Na}(D, \star)$ *are extended from finitely generated ideals of D.*

2 Idempotence

We begin with our basic definition.

Definition 2.1. *Let* $\star$ *be a semistar operation on D. An element* $E \in \overline{F}(D)$ *is said to be* $\star$*-idempotent if* $E^{\star} = (E^2)^{\star}$.

Our primary interest will be in (nonzero) $\star$-idempotent *ideals* of D. Let $\star$ be a semistar operation on D, and let I be a nonzero ideal of D. It is well known that $I^\star \cap D$ is a quasi-$\star$-ideal of D. (This is easy to see: we have

$$(I^\star \cap D)^\star \subseteq I^{\star\star} = I^\star = (I \cap D)^\star \subseteq (I^\star \cap D)^\star,$$

and hence $I^\star = (I^\star \cap D)^\star$; it follows that $I^\star \cap D = (I^\star \cap D)^\star \cap D$.) It, therefore, seems natural to call $I^\star \cap D$ the *quasi-$\star$-closure* of I. If we also call I *$\star$-proper* when $I^\star \subsetneq D^\star$, then it is easy to see that I is $\star$-proper if and only if its quasi-$\star$-closure is a proper quasi-$\star$-ideal. Now suppose that I is $\star$-idempotent. Then

$$(I^\star \cap D)^\star = I^\star = (I^2)^\star = ((I^\star)^2)^\star = (((I^\star \cap D)^\star)^2)^\star = ((I^\star \cap D)^2)^\star,$$

whence $I^\star \cap D$ is a $\star$-idempotent quasi-$\star$-ideal of D. A similar argument gives the converse. Thus a ($\star$-proper) nonzero ideal is $\star$-idempotent if and only if its quasi-$\star$-closure is a (proper) $\star$-idempotent quasi-$\star$-ideal.

Our study of idempotence in Prüfer domains and P$\star$MDs involves the notions of sharpness and branchedness. We recall some notation and terminology.

Given an integral domain D and a prime ideal $P \in \text{Spec}(D)$, set

$$\begin{aligned} \nabla(P) &:= \{M \in \text{Max}(D) \mid M \not\supseteq P\} \text{ and} \\ \Theta(P) &:= \bigcap\{D_M \mid M \in \nabla(P)\}. \end{aligned}$$

We say that P is *sharp* if $\Theta(P) \nsubseteq D_P$ (see [11, Lemma 1] and [3, Section 1 and Proposition 2.2]). The domain D itself is *sharp* (*doublesharp*) if every maximal (prime) ideal of D is sharp. (Note that a Prüfer domain D is doublesharp if and only if each overring of D is sharp [7, Theorem 4.1.7].) Now let $\star$ be a semistar operation on D. Given a prime ideal $P \in \text{QSpec}^{\star_f}(D)$, set

$$\begin{aligned} \nabla^{\star_f}(P) &:= \{M \in \text{QMax}^{\star_f}(D) \mid M \not\supseteq P\} \text{ and} \\ \Theta^{\star_f}(P) &:= \bigcap\{D_M \mid M \in \nabla^{\star_f}(P)\}. \end{aligned}$$

Call P *$\star_f$-sharp* if $\Theta^{\star_f}(P) \nsubseteq D_P$. For example, if $\star = d$ is the identity, then the $\star_f$-sharp property coincides with the sharp property. We then say that D is $\star_f$-(double)sharp if each quasi-$\star_f$-maximal (quasi-$\star_f$-prime) ideal of D is $\star_f$-sharp. (For more on sharpness, see [10, 11, 13], [7, page 62], [3], [4, Chapter 2, Section 3] and [5].)

Recall that a prime ideal P of a ring is said to be *branched* if there is a P-primary ideal distinct from P. Also, recall that the domain D has *finite character* if each nonzero ideal of D is contained in only finitely many maximal ideals of D.

We now prove a lemma that discusses the transfer of ideal-theoretic properties between D (on which a semistar operation $\star$ has been defined) and its associated Nagata ring.

Lemma 2.2. *Let $\star$ be a semistar operation on D.*

(1) *Let $E \in \overline{F}(D)$. Then E is $\widetilde{\star}$-idempotent if and only if $E\mathrm{Na}(D,\star)$ is idempotent. In particular, if D is a P$\star$MD, then E is $\star_f$-idempotent if and only if $E\mathrm{Na}(D,\star)$ is idempotent.*

(2) *Let P be a quasi-$\widetilde{\star}$-prime of D and I a nonzero ideal of D. Then:*

 (a) *I is P-primary in D if and only if I is a quasi-$\widetilde{\star}$-ideal of D and $I\mathrm{Na}(D,\star)$ is $P\mathrm{Na}(D,\star)$-primary in $\mathrm{Na}(D,\star)$.*

 (b) *P is branched in D if and only if $P\mathrm{Na}(D,\star)$ is branched in $\mathrm{Na}(D,\star)$.*

(3) *D has $\star_f$-finite character (i.e., each nonzero element of D belongs to only finitely many (possibly zero) $M \in \mathrm{QMax}^{\star_f}(D)$) if and only if $\mathrm{Na}(D,\star)$ has finite character.*

(4) *Let I be a quasi-$\widetilde{\star}$-ideal of D. Then I is a radical ideal if and only if $I\mathrm{Na}(D,\star)$ is a radical ideal of $\mathrm{Na}(D,\star)$.*

(5) *Assume that D is a P$\star$MD. Then:*

 (a) *If $P \in \mathrm{QSpec}^{\star_f}(D)$, then P is $\star_f$-sharp if and only if $P\mathrm{Na}(D,\star)$ is sharp in $\mathrm{Na}(D,\star)$.*

 (b) *D is $\star_f$-(double)sharp if and only if $\mathrm{Na}(D,\star)$ is (double)sharp.*

Proof. (1) We use Lemma 1.1(6). If $E\mathrm{Na}(D,\star)$ is idempotent, then $E^{\widetilde{\star}} = E\mathrm{Na}(D,\star) \cap K = E^2\mathrm{Na}(D,\star) \cap K = (E^2)^{\widetilde{\star}}$. Conversely, if E is $\widetilde{\star}$-idempotent, then $(E\mathrm{Na}(D,\star))^2 = E^2\mathrm{Na}(D,\star) = (E^2)^{\widetilde{\star}}\mathrm{Na}(D,\star) = E^{\widetilde{\star}}\mathrm{Na}(D,\star) = E\mathrm{Na}(D,\star)$. The "in particular" statement follows because $\star_f = \widetilde{\star}$ in a P$\star$MD (Lemma 1.2(2a)).

(2) (a) Suppose that I is P-primary. Then $ID[X]$ is $PD[X]$-primary. Since P is a quasi-$\widetilde{\star}$-prime of D, $P\mathrm{Na}(D,\star)$ is a prime ideal of $\mathrm{Na}(D,\star)$ (Lemma 1.1(4)), and then, since $\mathrm{Na}(D,\star)$ is a quotient ring of $D[X]$, $I\mathrm{Na}(D,\star)$ is $P\mathrm{Na}(D,\star)$-primary in $\mathrm{Na}(D,\star)$. Also, again using the fact that $ID[X]$ is $PD[X]$-primary (along with Lemma 1.1(6)), we have

$$I^{\widetilde{\star}} \cap D = I\mathrm{Na}(D,\star) \cap D \subseteq ID[X]_{PD[X]} \cap D[X] \cap D = ID[X] \cap D = I,$$

whence I is a quasi-$\widetilde{\star}$-ideal of D. Conversely, assume that I is a quasi-$\widetilde{\star}$-ideal of D and that $I\mathrm{Na}(D,\star)$ is $P\mathrm{Na}(D,\star)$-primary. Then for $a \in P$, there is a positive integer n for which $a^n \in I\mathrm{Na}(D,\star) \cap D = I^{\widetilde{\star}} \cap D = I$. Hence $P = \mathrm{rad}(I)$. It now follows easily that I is P-primary. (b) Suppose that P is branched in D. Then there is a P-primary ideal I of D distinct from P, and $I\mathrm{Na}(D,\star)$ is $P\mathrm{Na}(D,\star)$-primary by (a). Also by (a), I is a quasi-$\widetilde{\star}$-ideal, from which it follows that $I\mathrm{Na}(D,\star) \neq P\mathrm{Na}(D,\star)$. Now suppose that $P\mathrm{Na}(D,\star)$ is branched and that J is a $P\mathrm{Na}(D,\star)$-primary ideal of $\mathrm{Na}(D,\star)$ distinct from $P\mathrm{Na}(D,\star)$. Then it is straightforward to show that $J \cap D$ is distinct from P and is P-primary.

(3) Let ψ be a nonzero element of $\mathrm{Na}(D,\star)$, and let N be a maximal ideal with $\psi \in N$. Then $\psi\mathrm{Na}(D,\star) = f\mathrm{Na}(D,\star)$ for some $f \in D[X]$, and writing $N = M\mathrm{Na}(D,\star)$ for some $M \in \mathrm{QMax}^{\star_f}(D)$ (Lemma 1.1(3)), we must have $f \in MD[X]$

and hence $c(f) \subseteq M$. Therefore, if D has finite $\star_f$-character, then $\mathrm{Na}(D, \star)$ has finite character. The converse is even more straightforward.

(4) Suppose that I is a radical ideal of D, and let $\psi^n \in I\mathrm{Na}(D, \star)$ for some $\psi \in \mathrm{Na}(D, \star)$ and positive integer n. Then there is an element $g \in N(\star)$ with ($g\psi^n$ and hence) $(g\psi)^n \in ID[X]$. Since $ID[X]$ is a radical ideal of $D[X]$, $g\psi \in ID[X]$ and we must have $\psi \in I\mathrm{Na}(D, \star)$. Therefore, $I\mathrm{Na}(D, \star)$ is a radical ideal of $\mathrm{Na}(D, \star)$. The converse follows easily from the fact that $I\mathrm{Na}(D, \star) \cap D = I^{\tilde{\star}} \cap D = I$ (Lemma 1.1(7)).

(5) (a) This is part of [5, Proposition 3.5], but we give here a proof more in the spirit of this paper. Let $P \in \mathrm{QSpec}^{\star_f}(D)$. If P is $\star_f$-sharp, then by [5, Proposition 3.1] P contains a finitely generated ideal I with $I \nsubseteq M$ for all $M \in \nabla^{\star_f}(P)$, and, using the description of $\mathrm{Max}(\mathrm{Na}(D, \star))$ given in Lemma 1.1(3), $I\mathrm{Na}(D, \star)$ is a finitely generated ideal of $\mathrm{Na}(D, \star)$ contained in $P\mathrm{Na}(D, \star)$ but in no element of $\nabla(P\mathrm{Na}(D, \star))$. Therefore, $P\mathrm{Na}(D, \star)$ is sharp in the Prüfer domain $\mathrm{Na}(D, \star)$. For the converse, assume that $P\mathrm{Na}(D, \star)$ is sharp in $\mathrm{Na}(D, \star)$. Then $P\mathrm{Na}(D, \star)$ contains a finitely generated ideal J with $J \subseteq P\mathrm{Na}(D, \star)$ but $J \nsubseteq N$ for $N \in \nabla(P\mathrm{Na}(D, \star))$ [13, Corollary 2]. Then $J = I\mathrm{Na}(D, \star)$ for some finitely generated ideal I of D (necessarily) contained in P by Lemma 1.2(2c), and it is easy to see that $I \nsubseteq M$ for $M \in \nabla^{\star_f}(D)$. Then by [5, Proposition 3.1], P is $\star_f$-sharp. Statement (b) follows easily from (a) (using Lemma 1.2). □

Let D be an almost Dedekind domain with a non-finitely generated maximal ideal M. Then $M^{-1} = D$, but M is not idempotent (since MD_M is not idempotent in the Noetherian valuation domain D_M). Our next result shows that this cannot happen in a sharp Prüfer domain.

Theorem 2.3. *Let D be a Prüfer domain. If D is (d-)sharp and I is a nonzero ideal of D with $I^{-1} = D$, then I is idempotent.*

Proof. Assume that D is sharp. Proceeding contrapositively, suppose that I is a nonzero, non-idempotent ideal of D. Then, for some maximal ideal M of D, ID_M is not idempotent in D_M. Since D is a sharp domain, we may choose a finitely generated ideal J of D with $J \subseteq M$ but $J \nsubseteq N$ for all maximal ideals $N \neq M$. Since ID_M is a non-idempotent ideal in the valuation domain D_M, there is an element $a \in I$ for which $I^2D_M \subsetneq aD_M$. Let $B := J + Da$. Then $I^2D_M \subseteq BD_M$ and, for $N \in \mathrm{Max}(D) \setminus \{M\}$, $I^2D_N \subseteq D_N = BD_N$. Hence $I^2 \subseteq B$. Since B is a proper finitely generated ideal, we then have $(I^2)^{-1} \supseteq B^{-1} \supsetneq D$. Hence $(I^2)^{-1} \neq D$, from which it follows that $I^{-1} \neq D$, as desired. □

Kang [14, Proposition 2.2] proves that, for a nonzero ideal I of D, we always have $I^{-1}\mathrm{Na}(D, v) = (\mathrm{Na}(D, v)) : I)$. This cannot be extended to general semistar Nagata rings; for example, if D is an almost Dedekind domain with non-invertible maximal ideal M and we define a semistar operation $\star$ by $E^\star = ED_M$ for $E \in \overline{F}(D)$, then $(D : M) = D$ and hence $(D : M)\mathrm{Na}(D, \star) = \mathrm{Na}(D, \star) = D[X]_{M[X]} = D_M(X) \subsetneq (D_M : MD_M)D_M(X) = (\mathrm{Na}(D, \star) : M\mathrm{Na}(D, \star))$ (where the proper inclusion holds because MD_M is principal in D_M). At any rate, what we really require is the equality

$(D^\star : E)\mathrm{Na}(D, \star) = (\mathrm{Na}(D, \star) : E)$ for $E \in \overline{\boldsymbol{F}}(D)$. In the next lemma, we show that this holds in a P$\star$MD but not in general. The proof of part (1) of the next lemma is a relatively straightforward translation of the proof of [14, Proposition 2.2] to the semistar setting. In carrying this out, however, we discovered a minor flaw in the proof of [14, Proposition 2.2]. The flaw involves a reference to [12, Proposition 34.8], but this result requires that the domain D be integrally closed. To ensure complete transparency, we give the proof in full detail.

Lemma 2.4. *Let $\star$ be a semistar operation on D. Then:*

(1) $(D^\star : E)\mathrm{Na}(D, \star) \supseteq (\mathrm{Na}(D, \star) : E)$ *for each* $E \in \overline{\boldsymbol{F}}(D)$.
(2) *The following statements are equivalent:*

   (a) $(D^\star : E)\mathrm{Na}(D, \star) = (\mathrm{Na}(D, \star) : E)$ *for each* $E \in \overline{\boldsymbol{F}}(D)$.
   (b) $D^\star = D^{\widetilde{\star}}$.
   (c) $D^\star \subseteq \mathrm{Na}(D, \star)$.

(3) $(D^{\widetilde{\star}} : E)\mathrm{Na}(D, \star) = (\mathrm{Na}(D, \star) : E)$ *for each* $E \in \overline{\boldsymbol{F}}(D)$.
(4) *If D is a P$\star$MD, then the equivalent conditions in (2) hold.*

Proof. (1) Let $E \in \overline{\boldsymbol{F}}(D)$, and let $\psi \in (\mathrm{Na}(D, \star) : E)$. For $a \in E$, $a \neq 0$, we may find $g \in N(\star)$ such that $\psi a g \in D[X]$. This yields $\psi g \in a^{-1}D[X] \subseteq K[X]$, and hence $\psi = f/g$ for some $f \in K[X]$. We claim that $c(f) \subseteq (D^\star : E)$. Granting this, we have $f \in (D^\star : E)D[X]$, from which it follows that $\psi = f/g \in (D^\star : E)\mathrm{Na}(D, \star)$, as desired. To prove the claim, take $b \in E$, and note that $fb \in \mathrm{Na}(D, \star)$. Hence $fbh \in D[X]$ for some $h \in N(\star)$, and so $c(fh)b \subseteq D$. By the content formula [12, Theorem 28.1], there is an integer m for which $c(f)c(h)^{m+1} = c(fh)c(h)^m$. Using the fact that $c(h)^\star = D^\star$, we obtain $c(f)^\star = c(fh)^\star$ and hence that $c(f)b \subseteq c(fh)^\star b \subseteq D^\star$. Therefore, $c(f) \subseteq (D^\star : E)$, as claimed.

(2) Under the assumption in (c), $D^\star \subseteq \mathrm{Na}(D, \star) \cap K = D^{\widetilde{\star}}$ (Lemma 1.1(6)). Hence (c) $\Rightarrow$ (b). Now assume that $D^\star = D^{\widetilde{\star}}$. Then for $E \in \overline{\boldsymbol{F}}(D)$, we have $(D^\star : E)E \subseteq D^\star = D^{\widetilde{\star}} \subseteq \mathrm{Na}(D, \star)$; using (1), the implication (b) $\Rightarrow$ (a) follows. That (a) $\Rightarrow$ (c) follows upon taking $E = D$ in (a).

(3) This follows easily from (2), because $\mathrm{Na}(D, \star) = \mathrm{Na}(D, \widetilde{\star})$ by Lemma 1.1(5).

(4) This follows from (2), since if D is a P$\star$MD, then $D^\star = D^{\widetilde{\star}}$ by Lemma 1.2(2a). □

The conditions in Lemma 2.4(2) need not hold: Let $F \subsetneq k$ be fields, $V = k[[x]]$ the power series ring over V in one variable, and $D = F + M$, where $M = xk[[x]]$. Define a (finite-type) semistar operation $\star$ on D by $A^\star = AV$ for $A \in \overline{\boldsymbol{F}}(D)$. Then $D^\star = V \supsetneq D = D_M = D^{\widetilde{\star}}$.

We can now extend Theorem 2.3 to P$\star$MDs.

Corollary 2.5. *Let $\star$ be a semistar operation on D such that D is a $\star_f$-sharp P$\star$MD, and let I be a nonzero ideal of D with $(D^\star : I) = D^\star$. Then I is $\star_f$-idempotent.*

Proof. By Lemma 2.4(3), we have

$$(\mathrm{Na}(D,\star) : I\mathrm{Na}(D,\star)) = (D^\star : I)\mathrm{Na}(D,\star) = D^\star\mathrm{Na}(D,\star) = \mathrm{Na}(D,\star).$$

Hence $I\mathrm{Na}(D,\star)$ is idempotent in the Prüfer domain $\mathrm{Na}(D,\star)$ by Theorem 2.3. Lemma 2.2(1) then yields that I is $\star_f$-idempotent. □

Many semistar counterparts of ideal-theoretic properties in domains result in equations that are "external" to D, since for a semistar operation $\star$ on D and a nonzero ideal I of D, it is possible that $I^\star \not\subseteq D$. Of course, $\star$-idempotence is one such property. Often, one can obtain a "cleaner" counterpart by specializing from P$\star$MDs to "ordinary" PvMDs. We recall some terminology. Semistar operations are generalizations of *star* operations, first considered by Krull and repopularized by Gilmer [12, Sections 32, 34]. Roughly, a star operation is a semistar operation restricted to the set $\boldsymbol{F}(D)$ of nonzero fractional ideals of D with the added requirement that one has $D^\star = D$. The most important star operation (aside from the d-, or trivial, star operation) is the *v-operation*: For $E \in \boldsymbol{F}(D)$, put $E^{-1} = \{x \in K \mid xE \subseteq D\}$ and $E^v = (E^{-1})^{-1}$. Then v_f (restricted to $\boldsymbol{F}(D)$) is the *t-operation* and $\widetilde{v}$ is the w-operation. Thus a *PvMD* is a domain in which each nonzero finitely generated ideal is t-invertible. Corollary 2.5 then has the following restricted interpretation (which has the advantage of being *internal* to D).

Corollary 2.6. *If D is a t-sharp PvMD and I is a nonzero ideal of D for which $I^{-1} = D$, then I is t-idempotent.*

Our next result is a partial converse to Theorem 2.3.

Proposition 2.7. *Let D be a Prüfer domain such that I is idempotent whenever I is a nonzero ideal of D with $I^{-1} = D$. Then, every branched maximal ideal of D is sharp.*

Proof. Let M be a branched maximal ideal of D. Then $MD_M = \mathrm{rad}(aD_M)$ for some nonzero element $a \in M$ [12, Theorem 17.3]. Let $I := aD_M \cap D$. Then I is M-primary, and since $ID_M = aD_M$, (ID_M and hence) I is not idempotent. By hypothesis, we may choose $u \in I^{-1} \setminus D$. Since $Iu \subseteq D$ and $ID_N = D_N$ for $N \in \mathrm{Max}(D) \setminus \{M\}$, then $u \in \bigcap\{D_N \mid N \in \mathrm{Max}(D), N \neq M\}$. On the other hand, since $u \notin D$, $u \notin D_M$. It follows that M is sharp. □

Now we extend Proposition 2.7 to P$\star$MDs.

Corollary 2.8. *Let $\star$ be a semistar operation on D, and assume that D is a P$\star$MD such that I is $\star_f$-idempotent whenever I is a nonzero ideal of D with $(D^\star : I) = D^\star$. Then, each branched quasi-$\star_f$-maximal ideal of D is $\star_f$-sharp.* (*In particular if D is a PvMD in which I is t-idempotent whenever I is a nonzero ideal of D with $I^{-1} = D$, then each branched maximal t-ideal of D is t-sharp.*)

Proof. Let J be a a nonzero ideal of the Prüfer domain $\mathrm{Na}(D,\star)$ with $(\mathrm{Na}(D,\star):J)=\mathrm{Na}(D,\star)$. By Lemma 1.2(1c), $J=I\mathrm{Na}(D,\star)$ for some ideal I of D. Applying Lemma 2.4(3) and Lemma 1.1(6), we obtain $(D^\star:I)=D^\star$. Hence, by hypothesis, I is $\star_f$-idempotent, and this yields that $J=I\mathrm{Na}(D,\star)$ is idempotent in the Prüfer domain $\mathrm{Na}(D,\star)$ (Lemma 2.2(1)). Now, let M be a branched quasi-$\star_f$-maximal ideal of D. Then, by Lemma 2.2(2), $M\mathrm{Na}(D,\star)$ is a branched maximal ideal of $\mathrm{Na}(D,\star)$. We may now apply Proposition 2.7 to conclude that $M\mathrm{Na}(D,\star)$ is sharp. Therefore, M is $\star_f$-sharp in D by Lemma 2.2(5). □

If P is a prime ideal of a Prüfer domain D, then powers of P are P-primary by [12, Theorem 23.3(b)]; it follows that P is idempotent if and only if PD_P is idempotent. We use this fact in the next result.

It is well known that a proper idempotent ideal of a valuation domain must be prime [12, Theorem 17.1(3)]. In fact, according to [12, Exercise 3, p. 284], a proper idempotent ideal in a Prüfer domain must be a radical ideal. We (re-)prove and extend this fact and add a converse.

Theorem 2.9. *Let D be a Prüfer domain, and let I be an ideal of D. Then I is idempotent if and only if I is a radical ideal each of whose minimal primes is idempotent.*

Proof. The result is trivial for $I=(0)$ and vacuously true for $I=D$. Suppose that I is a proper nonzero idempotent ideal of D, and let P be a prime minimal over I. Then ID_P is idempotent, and we must have $ID_P=PD_P$ [12, Theorem 17.1(3)]. Hence PD_P is idempotent, and therefore, by the comment above, so is P. Now let M be a maximal ideal containing I. Then ID_M is idempotent, hence prime (hence radical). It follows (checking locally) that I is a radical ideal.

Conversely, let I be a radical ideal each of whose minimal primes is idempotent. If M is a maximal ideal containing I and P is a minimal prime of I contained in M, then $ID_M=PD_M$. Since P is idempotent, this yields $ID_M=I^2D_M$. It follows that I is idempotent. □

We next extend Theorem 2.9 to P$\star$MDs.

Corollary 2.10. *Let D be a P$\star$MD, where $\star$ is a semistar operation on D, and let I be a quasi-$\star_f$-ideal of D. Then I is $\star_f$-idempotent if and only if I is a radical ideal each of whose minimal primes is $\star_f$-idempotent. (In particular, if D is a PvMD and I is a t-ideal of D, then I is t-idempotent if and only if I is a radical ideal each of whose minimal primes is t-idempotent.)*

Proof. Suppose that I is $\star_f$-idempotent. Then $I\mathrm{Na}(D,\star)$ is an idempotent ideal in $\mathrm{Na}(D,\star)$ by Lemma 2.2(1). By Theorem 2.9, $I\mathrm{Na}(D,\star)$ is a radical ideal of $\mathrm{Na}(D,\star)$, and hence, by Lemma 2.2(4), I is a radical ideal of D. Now let P be a minimal prime of I in D. Then P is a quasi-$\star_f$-prime of D. By Lemma 1.2(2b) $P\mathrm{Na}(D,\star)$ is minimal over $I\mathrm{Na}(D,\star)$, whence $P\mathrm{Na}(D,\star)$ is idempotent, again by Theorem 2.9. The $\star_f$-idempotence of P now follows from Lemma 2.2(1).

The converse follows by similar applications of Theorem 2.9 and Lemma 2.2. □

Recall that a Prüfer domain is said to be *strongly discrete* (*discrete*) if it has no nonzero (branched) idempotent prime ideals. Since unbranched primes in a Prüfer domain must be idempotent [12, Theorem 23.3(b)], a Prüfer domain is strongly discrete if and only if it is discrete and has no unbranched prime ideals. We have the following straightforward application of Theorem 2.9.

Corollary 2.11. *Let D be a Prüfer domain.*

(1) *If D is discrete, then an ideal I of D is idempotent if and only if I is a radical ideal each of whose minimal primes is unbranched.*
(2) *If D is strongly discrete, then D has no proper nonzero idempotent ideals.*

Let us call a P$\star$MD $\star_f$- *strongly discrete* ($\star_f$- *discrete*) if it has no (branched) $\star_f$-idempotent quasi-$\star_f$-prime ideals. From Lemma 2.2(1, 2), we have the usual connection between a property of a P$\star$MD and the corresponding property of its $\star$-Nagata ring.

Proposition 2.12. *Let $\star$ be a semistar operation on D. Then D is $\star_f$-(strongly) discrete P$\star$MD if and only if* $\mathrm{Na}(D, \star)$ *is a (strongly) discrete Prüfer domain.*

Applying Corollary 2.10 and Lemma 2.2(1, 2), we have the following extension of Corollary 2.11.

Corollary 2.13. *Let D be a domain.*

(1) *Assume that D is a P$\star$MD for some semistar operation $\star$ on D.*
   (a) *If D is $\star_f$-discrete, then a nonzero quasi-$\star_f$-ideal I of D is $\star_f$-idempotent if and only if I is a radical ideal each of whose minimal primes is unbranched.*
   (b) *If D is $\star_f$-strongly discrete, then D has no $\star_f$-proper $\star_f$-idempotent ideals.*
(2) *Assume that D is a PvMD.*
   (a) *If D is t-discrete, then a t-ideal I of D is t-idempotent if and only if I is a radical ideal each of whose minimal primes is unbranched.*
   (b) *If D is t-strongly discrete, then D has no t-proper t-idempotent ideals.*

3 Divisoriality

According to [7, Corollary 4.1.14], if D is a doublesharp Prüfer domain and P is a nonzero, nonmaximal ideal of D, then P is divisorial. The natural question arises: If D is a $\star_f$-doublesharp P$\star$MD and $P \in \mathrm{QSpec}^{\star_f}(D) \setminus \mathrm{QMax}^{\star_f}(D)$, is P necessarily divisorial? Since $\star$ is an arbitrary semistar operation and divisoriality specifically involves the v-operation, one might expect the answer to be negative. Indeed, we give a counterexample in Example 3.4 below. However, in Theorem 3.2 we prove a general result, a corollary of which does yield divisoriality in the "ordinary" PvMD case. First, we need a lemma, the first part of which may be regarded as an extension of [14, Proposition 2.2(2)].

Lemma 3.1. *Let $\star$ be a semistar operation on D. Then*

(1) $(D^{\widetilde{\star}} : (D^{\widetilde{\star}} : E))\mathrm{Na}(D, \star) = (\mathrm{Na}(D, \star) : (\mathrm{Na}(D, \star) : E))$ *for each* $E \in \overline{\boldsymbol{F}}(D)$*, and*

(2) *if I is a nonzero ideal of D, then $I^{\widetilde{\star}}$ is a divisorial ideal of $D^{\widetilde{\star}}$ if and only if $I\mathrm{Na}(D, \star)$ is a divisorial ideal of* $\mathrm{Na}(D, \star)$.

In particular, if D is a P$\star$MD, then $(D^{\star} : (D^{\star} : E))\mathrm{Na}(D, \star) = (\mathrm{Na}(D, \star) : (\mathrm{Na}(D, \star) : E))$ *for each $E \in \overline{\boldsymbol{F}}(D)$; and, for a nonzero ideal I of D, $I^{\star_f}$ is divisorial in $D^{\star}$ if and only if $I\mathrm{Na}(D, \star)$ is divisorial in* $\mathrm{Na}(D, \star)$.

Proof. Set $\mathcal{N} = \mathrm{Na}(D, \star)$. For (1), applying Lemma 2.4, we have

$$(D^{\widetilde{\star}} : (D^{\widetilde{\star}} : E))\mathcal{N} = (\mathcal{N} : (D^{\widetilde{\star}} : E)) = (\mathcal{N} : (\mathcal{N} : E)).$$

(2) Assume that I is a nonzero ideal of D. If $I^{\widetilde{\star}}$ is divisorial in $D^{\widetilde{\star}}$, then (using (1))

$$(\mathcal{N} : (\mathcal{N} : I\mathcal{N})) = (D^{\widetilde{\star}} : (D^{\widetilde{\star}} : I^{\widetilde{\star}}))\mathcal{N} = I^{\widetilde{\star}}\mathcal{N} = I\mathcal{N}.$$

Now suppose that $I\mathcal{N}$ is divisorial. Then

$$(D^{\widetilde{\star}} : (D^{\widetilde{\star}} : I^{\widetilde{\star}}))\mathcal{N} = (\mathcal{N} : (\mathcal{N} : I)) = I\mathcal{N},$$

whence

$$(D^{\widetilde{\star}} : (D^{\widetilde{\star}} : I^{\widetilde{\star}})) \subseteq I\mathcal{N} \cap K = I^{\widetilde{\star}}.$$

The "in particular" statement follows from standard considerations. □

Theorem 3.2. *Let $\star$ be a semistar operation on D such that D is a $\star_f$-doublesharp P$\star$MD, and let $P \in \mathrm{QSpec}^{\star_f}(D) \setminus \mathrm{QMax}^{\star_f}(D)$. Then $P^{\star_f}$ is a divisorial ideal of $D^{\star}$.*

Proof. Since $\mathrm{Na}(D, \star)$ is a doublesharp Prüfer domain (Lemma 2.2(5)), $P\mathrm{Na}(D, \star)$ is divisorial by [7, Corollary 4.1.14]. Hence $P^{\star_f}$ is divisorial in $D^{\star}$ by Lemma 3.1. □

Corollary 3.3. *If D is a t-doublesharp PvMD, and P is a non-t-maximal t-prime of D, then P is divisorial.*

Proof. Take $\star = v$ in Theorem 3.2. (More precisely, take $\star$ to be any extension of the star operation v on D to a semistar operation on D, so that $\star_f$ (restricted to D) is the t-operation on D.) Then $P = P^t = P^{\star_f}$ is divisorial by Theorem 3.2. □

Example 3.4. *Let p be a prime integer and let $D := \mathrm{Int}(\mathbb{Z}_{(p)})$. Then D is a two-dimensional Prüfer domain by [2, Lemma VI.1.4 and Proposition V.1.8]. Choose a height 2 maximal ideal M of D, and let P be a height 1 prime ideal of D contained in M. Then $P = q\mathbb{Q}[X] \cap D$ for some irreducible polynomial $q \in \mathbb{Q}[X]$ [2, Proposition V.2.3]. By [2, Theorems VIII.5.3 and VIII.5.15], P is not a divisorial ideal of D. Set $E^{\star} = ED_M$ for $E \in \overline{\boldsymbol{F}}(D)$. Then $\star$ is a finite-type semistar operation on D.*

Clearly, M is the only quasi-$\star$-maximal ideal of D, and, since D_M is a valuation domain, D is a P$\star$MD by Lemma 1.2. Moreover, $\mathrm{Na}(D, \star) = D_M(X)$ *is also a valuation domain and hence a doublesharp Prüfer domain, which yields that D is a $\star_f$-doublesharp P$\star$MD (Lemma 2.2). Finally, since $P = PD_M \cap D = P^\star \cap D$, P is a non-$\star_f$-maximal quasi-$\star_f$-prime of D.* □

In the remainder of the paper, we impose on Prüfer domains (P$\star$MDs) the finite character (finite $\star_f$-character) condition. As we shall see, this allows improved versions of Theorem 2.9 and Corollary 2.10. It also allows a type of unique factorization for (quasi-$\star_f$-)ideals that are simultaneously ($\star_f$-)idempotent and ($\star_f$-)divisorial.

Theorem 3.5. *Let D be a Prüfer domain with finite character, and let I be a nonzero ideal of D. Then:*

(1) *I is idempotent if and only if I is a product of idempotent prime ideals.*
(2) *The following statements are equivalent.*
   (a) *I is idempotent and divisorial.*
   (b) *I is a product of non-maximal idempotent prime ideals.*
   (c) *I is a product of divisorial idempotent prime ideals.*
   (d) *I has a unique representation as the product of incomparable divisorial idempotent primes.*

Proof. (1) Suppose that I is idempotent. By Theorem 2.9, I is the intersection of its minimal primes, each of which is idempotent. Since D has finite character, I is contained in only finitely many maximal ideals, and, since no two distinct minimal primes of I can be contained in a single maximal ideal, I has only finitely many minimal primes and they are comaximal. Hence I is the product of its minimal primes (and each is idempotent). The converse is trivial.

(2) (a) ⇒ (b): Assume that I is idempotent and divisorial. By (1) and its proof, $I = P_1 \cdots P_n = P_1 \cap \cdots \cap P_n$, where the P_i are the minimal primes of I. We claim that each P_i is divisorial. To see this, observe that

$$(P_1)^v P_2 \cdots P_n \subseteq (P_1 \cdots P_n)^v = I^v = I \subseteq P_1.$$

Since the P_i are incomparable, this gives $(P_1)^v \subseteq P_1$, that is, P_1 is divisorial. By symmetry each P_i is divisorial. It is well known that in a Prüfer domain, a maximal ideal cannot be both idempotent and divisorial. Hence the P_i are non-maximal.

(b) ⇒ (c): Since D has finite character, it is a (d)-doublesharp Prüfer domain [13, Theorem 5], whence nonmaximal primes are automatically divisorial by [7, Corollary 4.1.14].

(c) ⇒ (a): Write $I = Q_1 \cdots Q_m$, where each Q_j is a divisorial idempotent prime. Since I is idempotent (by (1)), we may also write $I = P_1 \cdots P_n$, where the P_i are the minimal primes of I. For each i, we have $Q_1 \cdots Q_m = I \subseteq P_i$, from which it follows that $Q_j \subseteq P_i$ for some j. By minimality, we must then have $Q_j = P_i$. Thus each P_i is divisorial, whence $I = P_1 \cap \cdots \cap P_n$ is divisorial.

Finally, we show that (d) follows from the other statements. We use the notation in the proof of (c) $\Rightarrow$ (a). In the expression $I = P_1 \cdots P_n$, the P_i are (divisorial, idempotent, and) incomparable, and it is clear that no P_i can be omitted. To see that this is the only such expression, consider a representation $I = Q_1 \cdots Q_m$, where the Q_i are divisorial, idempotent, and incomparable. Fix a Q_k. Then $P_1 \cdots P_n = I \subseteq Q_k$, and we have $P_i \subseteq Q_k$ for some i. However, as above, $Q_j \subseteq P_i$ for some j, whence, by incomparability, $Q_k = P_i$. The conclusion now follows easily. $\square$

We note that incomparability is necessary for uniqueness above, for example, if D is a valuation domain and $P \subsetneq Q$ are non-maximal (necessarily divisorial) primes, then $P = PQ$.

We close by extending Theorem 3.5 to P$\star$MDs and then to "ordinary" PvMDs. We omit the (by now) straightforward proofs.

Corollary 3.6. *Let $\star$ be a semistar operation on D such that D is a P$\star$MD with finite $\star_f$-character, and let I be a quasi-$\star_f$-ideal of D. Then:*

(1) *I is $\star_f$-idempotent if and only if $I^{\star_f}$ is a $\star_f$-product of $\star_f$-idempotent quasi-$\star_f$-prime ideals in D, that is, $I^{\star_f} = (P_1 \cdots P_n)^{\star_f}$, where the P_i are $\star_f$-idempotent quasi-$\star_f$-primes of D.*
(2) *The following statements are equivalent.*
   (a) *I is $\star_f$-idempotent and $\star_f$-divisorial ($I^{\star_f}$ is divisorial in $D^\star$).*
   (b) *I is a $\star_f$-product of non-quasi-$\star_f$-maximal idempotent quasi-$\star_f$-prime ideals.*
   (c) *I is a $\star_f$-product of $\star_f$-divisorial $\star_f$-idempotent prime ideals.*
   (d) *I has a unique representation as a $\star_f$-product of incomparable $\star_f$-divisorial $\star_f$-idempotent primes.*

Corollary 3.7. *Let D be a PvMD with finite t-character, and let I be a nonzero t-ideal of D. Then:*

(1) *I is t-idempotent if and only if I is a t-product of t-idempotent t-prime ideals in D.*
(2) *The following statements are equivalent.*
   (a) *I is t-idempotent and divisorial.*
   (b) *I is a t-product of non-t-maximal t-idempotent t-primes.*
   (c) *I is a t-product of divisorial t-idempotent t-primes.*
   (d) *I has a unique representation as a t-product of incomparable divisorial t-idempotent t-primes.*

References

1. Arnold, J.: On the ideal theory of the Kronecker function ring and the domain D(X). Can. J. Math. **21**, 558–563 (1969)
2. Cahen, P.-J., Chabert, J.-L.: Integer-Valued Polynomials, vol. 48. Providence (1997) (Amer. Math. Soc. Surv. Monogr.)

3. Fontana, M., Houston, E., Lucas, T.: Toward a classification of prime ideals in Prüfer domains. Forum Math. **22**, 741–766 (2010)
4. Fontana, M., Houston, E., Lucas, T.: Factoring Ideals in Integral Domains. Lectures Notes of U.M.I. Springer, Berlin (2013)
5. Fontana, M., Houston, E., Park, M.H.: Sharpness and semistar operations in Prüfer-like domains. Commun. Algebra **47**, 1478–1489 (2019)
6. Fontana, M., Huckaba, J.A.: Localizing systems and semistar operations. In: Chapman, S.T., Glaz, S. (eds.) Non-Noetherian Commutative Ring Theory. Kluwer Academic Publishers, pp. 169–198 (2000)
7. Fontana, M., Huckaba, J.A., Papick, I.J.: Prüfer Domains. Marcel Dekker Inc., New York (1997)
8. Fontana, M., Jara, P., Santos, E.: Prüfer $\star$-multiplication domains and semistar operations. J. Algebra Appl. **2**, 21–50 (2003)
9. Fontana, M., Loper, K.A.: Nagata rings, Kronecker function rings and related semistar operations. Commun. Algebra **31**, 4775–4805 (2003)
10. Gilmer, R.: Integral domains which are almost Dedekind. Proc. Am. Math. Soc. **15**, 813–818 (1964)
11. Gilmer, R.: Overrings of Prüfer domains. J. Algebra **4**, 331–340 (1966)
12. Gilmer, R.: Multiplicative Ideal Theory. Marcel Dekker, New York (1972)
13. Gilmer, R., Heinzer, W.: Overrings of Prüfer domains, II. J. Algebra **7**, 281–302 (1967)
14. Kang, B.G.: Prüfer v-multiplication domains and the ring $R[X]_{N_v}$. J. Algebra **123**, 151–170 (1989)
15. Okabe, A., Matsuda, R.: Semistar operations on integral domains. Math. J. Toyama Univ. **17**, 1–21 (1994)
16. Zafrullah, M.: Some polynomial characterizations of Prüfer v-multiplication domains. J. Pure Appl. Algebra **32**, 231–237 (1984)

Simultaneous Interpolation and P-adic Approximation by Integer-Valued Polynomials

Sophie Frisch

Abstract Let D be a Dedekind domain with finite residue fields and $\mathcal{F}$ a finite set of maximal ideals of D. Let $r_0, \ldots, r_n$ be distinct elements of D, pairwise incongruent modulo P^{k_P} for each $P \in \mathcal{F}$, and $s_0, \ldots, s_n$ arbitrary elements of D. We show that there is an interpolating P^{k_P}-congruence preserving integer-valued polynomial, that is, $f \in \mathrm{Int}(D) = \{g \in K[x] \mid g(D) \subseteq D\}$ with $f(r_i) = s_i$ for $0 \leq i \leq n$, such that, moreover, the function $f\colon D \to D$ is constant modulo P^{k_P} on each residue class of P^{k_P} for all $P \in \mathcal{F}$.

Keywords Interpolation · Polynomials · Congruence preserving · P-adic approximation · P-adic Lipschitz functions · Lipschitz maps · Integer-valued polynomials · Polynomial functions · Polynomial mappings · Dedekind domains · Commutative rings · Integral domains

Mathematics Subject Classification 2020 Primary 13B25; Secondary 13F20, 13M10, 11C08, 13F05.

1 Introduction

Let D be a Dedekind domain with finite residue fields, K its quotient field, and

$$\mathrm{Int}(D) = \{f \in K[x] \mid f(D) \subseteq D\}$$

the ring of integer-valued polynomials on D.

This research was supported by the Austrian Science Fund FWF grant P30934-N35.

S. Frisch (✉)
Institute of Analysis and Number Theory, Technische Universität Graz,
Kopernikusgasse 24, 8010 Graz, Austria
e-mail: frisch@math.tugraz.at

A. Facchini et al. (eds.), *Advances in Rings, Modules and Factorizations*,
Springer Proceedings in Mathematics & Statistics 321,
https://doi.org/10.1007/978-3-030-43416-8_10

We will show that two different feats that can each be accomplished separately by integer-valued polynomials, namely, interpolation of arbitrary functions on D, and, representation of arbitrary functions on D/P^n, where P^n is a power of a maximal ideal P, can actually be accomplished by one and the same polynomial, simultaneously.

We recall some well-known facts. First, about interpolation by integer-valued polynomials: Newton already used polynomials in Int($\mathbb{Z}$) to interpolate functions on $\mathbb{Z}$, cf. [1]. More generally, when D is a Dedekind domain with finite residue fields, then, given $r_0, \ldots, r_n \in D$ (distinct) and arbitrary $s_0, \ldots, s_n \in D$, we can find $f \in \text{Int}(D)$ with $f(r_i) = s_i$ for $0 \le i \le n$ [3]. If this holds for a domain D, we say that D has the interpolation property. The domains having the interpolation property have been characterized among Noetherian domains and among Prüfer domains [2], and include, as mentioned, all Dedekind domains with finite residue fields.

It turned out that the interpolation property is relevant to the question whether $\text{Int}(D)$ is a Prüfer domain. If D is Prüfer (a necessary condition for $\text{Int}(D)$ to be Prüfer) then $\text{Int}(D)$ is Prüfer if and only if $\text{Int}(D)$ has the interpolation property [2].

Second, about the representation of functions on D/I, where I is an ideal of D: Let $f \in \text{Int}(D)$. We say that f is I-congruence preserving, if, for all $a, b \in D$,

$$a \equiv b \mod I \quad \Longrightarrow \quad f(a) \equiv f(b) \mod I.$$

In that case, f induces a well-defined function on D/I by $f(a + I) = f(a) + I$. Let D be a Dedekind domain with finite residue fields. If I is a power of a maximal ideal of D (and only if I is a power of a maximal ideal), every function on D/I arises from an I-congruence preserving polynomial in $\text{Int}(D)$ in this way. This was shown for $D = \mathbb{Z}$ by Skolem [7] (in the "if" direction) and Rédei and Szele [5, 6] (in the "only if" direction), and later generalized to Dedekind domains [4].

If D is a Dedekind domain with finite residue fields, we will show that, given $r_0, \ldots, r_n \in D$ (distinct) and arbitrary $s_0, \ldots, s_n \in D$, and a finite set of powers P^{k_P} of maximal ideals such that the r_i are pairwise incongruent modulo each P^{k_P}, we can find a polynomial $f \in \text{Int}(D)$ with $f(r_i) = s_i$ for $0 \le i \le n$ and such that

$$a \equiv b \mod P^{k_P} \quad \Longrightarrow \quad f(a) \equiv f(b) \mod P^{k_P}.$$

for each P^{k_P}, cf. Theorem 1.

A note on terminology: if R is any ring and $f \in R[x]$ a polynomial, $f = \sum_k c_k x^k$ induces a function by substitution of elements of R for the variable: $r \mapsto \sum_k c_k r^k$. A function $\varphi\colon R \to R$ thus arising from a polynomial $f \in R[x]$ is called a polynomial function on R.

When R is an infinite domain, then the polynomial f inducing a polynomial function is uniquely determined by its values on an infinite subset of R. Relying on this one-to-one correspondence between polynomials and polynomial functions, in the case where $R = K$ is an infinite field, we will not be as pedantic about the distinction between polynomials and polynomial functions as would be necessary if we were dealing with finite rings or rings with zero-divisors.

In what follows, when we talk about the function associated to an integer-valued polynomial $f \in \mathrm{Int}(D)$, we always mean the function $f: D \to D$ (as opposed to $f: K \to K$).

2 Notation and Definitions

We let $\mathbb{N}$ denote the positive integers (natural numbers) and $\mathbb{N}_0$ the non-negative integers. We use "additive" terminology for Lipschitz functions:

Definition 1. *Let R be a commutative ring, $f: R \to R$ a function, I an ideal of R, and $n \in \mathbb{N}_0$. We say that f is I-adically n-Lipschitz if, for all $m \in \mathbb{N}$ and all $a, b \in R$*

$$a \equiv b \mod I^{m+n} \implies f(a) \equiv f(b) \mod I^m$$

When D is a domain, $g \in \mathrm{Int}(D)$, and I an ideal of D, we will say that g is I-adically n-Lipschitz if the associated function $g: D \to D$ is I-adically n-Lipschitz.

We summarize some elementary consequences of this definition.

Remark 1. Let R be a commutative ring, $f: R \to R$ a function, and I an ideal of R.

1. I-adically n-Lipschitz implies I-adically N-Lipschitz for all $N \geq n$.
2. If $f: R \to R$ is a function induced by a polynomial in $R[x]$ by substitution of the variable, then f is I-adically 0-Lipschitz for all ideals I of R.
3. For fixed I and n, the set of I-adically n-Lipschitz functions on R is closed under addition, subtraction and multiplication and, therefore, forms a subring of the set of all functions R^R.
4. If D is a domain, I an ideal of D and $n \in \mathbb{N}_0$, then the set of $g \in \mathrm{Int}(D)$ that are I-adically n-Lipschitz is a subring of $\mathrm{Int}(D)$.

In what follows, D is always a Dedekind domain with quotient field K, and we always assume $D \neq K$. For such a Dedekind domain, we denote by $\mathrm{Spec}^1(D)$ the set prime ideals of height one, which coincides with the set of maximal ideals of D. For $P \in \mathrm{Spec}^1(D)$, we use v_P to denote the normalized discrete valuation on K associated with P; that is, for $d \in D \setminus \{0\}$, $v_P(d)$ is the maximal exponent v such that $d \in P^v$, and, for an element of $K \setminus \{0\}$ expressed as a fraction a/b with $a, b \in D \setminus \{0\}$, $v_P(a/b) = v_P(a) - v_P(b)$.

Remark 2. Let D a Dedekind domain, $f \in \mathrm{Int}(D)$, and P a maximal ideal of D. If we express f as a fraction $f = g/d$ with $g \in D[x]$ and $d \in D \setminus \{0\}$, we see that f is P-adically $v_P(d)$-Lipschitz. In particular, if $f \in D_P[x]$, then f is P-adically 0-Lipschitz. More generally, if $f \in \mathrm{Int}(D)$ is expressed as a fraction $f = g/d$ with $g \in D_P[x]$ and $d \in D \setminus \{0\}$, then, also, f is P-adically $v_P(d)$-Lipschitz.

Note that $v_P(d)$, in the above remark, is not necessarily the minimal n for which f is P-adically n-Lipschitz (not even if d is relatively prime to the content of g). For instance, when f is a product $f = f_1 \dots f_n$ with $f_i = g_i/d$, $g_i \in D[x]$, then the denominator of f is d^n, but f is P-adically $v_P(d)$-Lipschitz, not just $v_P(d^n)$-Lipschitz, by Remark 1 (3).

We use $\|I\|$ for the norm of an ideal I of D, that is $\|I\| = |D/I|$.

3 P-adic Lipschitz Constants of Interpolating Integer-Valued Polynomials

We recall a Lemma from an earlier paper that we will need for the proof of Lemma 2.

Lemma 1 ([3], **[Lemma 6.1]**). *Let v be a discrete valuation on a field K and R_v its valuation ring. Suppose $g = \sum_{k=0}^{n} d_k x^k$ in $K[x]$ splits over K as*

$$g(x) = d_n(x - b_1)\dots(x - b_m)(x - c_1)\dots(x - c_l),$$

where $v(b_i) < 0$ and $v(c_i) \geq 0$.

Let $\mu = \min_{0 \leq k \leq n} v(d_k)$ and set $g_+(x) = (x - c_1)\dots(x - c_l)$ then, for all $r \in R_v$,

$$v\big(g(r)\big) = \mu + v\big(g_+(r)\big).$$

Definition 2. *For q, m integers with $q > 1$ and $m \geq 0$ define*

$$L(q, m) := \frac{1 - q^m}{1 - q}$$

Lemma 2. *Let D be a Dedekind domain with finite residue fields and a_0, a_1 distinct elements of D. For $P \in \mathrm{Spec}^1(D)$, let $m_P = v_P(a_1 - a_0)$.*

For any finite set $\mathcal{F}$ of maximal ideals of D there exists $f \in \mathrm{Int}(D)$ with $f(a_1) = 0$ and $f(a_0) = 1$, and such that f is P-adically $L(\|P\|, m_P)$-Lipschitz for all $P \in \mathcal{F}$.

Proof. By linear substitution we may assume, w.l.o.g., that $a_0 = 0$. Also, we assume w.l.o.g. that $\mathcal{F}$ contains the set

$$\mathcal{P} = \{P \in \mathrm{Spec}^1(D) \mid a_1 \in P\} = \{P \in \mathrm{Spec}^1(D) \mid m_P > 0\}.$$

The case $\mathcal{P} = \emptyset$ is trivial. Assume $\mathcal{P} \neq \emptyset$. We set $\mathcal{F}_0 = \{P \in \mathcal{F} \mid m_P = 0\}$; such that $\mathcal{F}$ is the disjoint union of $\mathcal{P}$ and $\mathcal{F}_0$.

We will construct a polynomial $g \in K[x]$ with $g(a_1) = 0$, such that for every essential valuation v of D and every $r \in D$, $v\big(g(r)\big) \geq v\big(g(0)\big)$; and then set $f(x) = g(x)/g(0)$.

Let $N = \max_{P \in \mathcal{P}} \|P\|^{m_P}$. Using the Chinese Remainder Theorem modulo P^{m_P+1} for $P \in \mathcal{F}$, we produce a sequence $(b_i)_{i=1}^{N}$ in D with the properties:

1. $b_1 = a_1$
2. For all $P \in \mathcal{F}$, the b_i with $1 \le i \le \|P\|^{m_P}$ form a complete system of residues modulo P^{m_P}.
3. For all $P \in \mathcal{F}$, for all $i > \|P\|^{m_P}$, $b_i \equiv 1$ modulo P^{m_P+1}.

Note that no b_i is in P^{m_P+1} for any $P \in \mathcal{F}$, and, in particular, that no b_i is in P for any $P \in \mathcal{F}_0$.

Let $P \in \mathcal{P}$ and $1 \le k \le m_P$. For any given $r \in D$, the number of b_i with $1 \le i \le \|P\|^{m_P}$ in the residue class $r + P^k$ is the same, namely,

$$\gamma_k(P) := \|P\|^{m_P - k}.$$

Note that, therefore, for all $P \in \mathcal{P}$ and $1 \le k \le m_P$,

$$\forall r \in D \quad \left|\{i \mid v_P(r - b_i) \ge k\}\right| \ge \gamma_k(P),$$

with equality holding for all $r \in D \setminus (1 + P)$ (and, actually, for all $r \in D$ in the case where $\|P\|^{m_P} = N$).

Let $\mathcal{Q} = \{Q \in \mathrm{Spec}^1(D) \setminus \mathcal{P} \mid \exists i \;\; b_i \in Q\}$ and for $Q \in \mathcal{Q}$ let $k_Q = \max_i v_Q(b_i)$. Note that $\mathcal{Q} \cap \mathcal{F} = \emptyset$.

Let $c \in D$ with $v_Q(c) = k_Q + 1$ for all $Q \in \mathcal{Q}$, and $c \equiv 1 \mod P^{m_P+1}$ for all $P \in \mathcal{F}$. Let $\mathcal{Q}' = \{Q \in \mathrm{Spec}^1(D) \mid v_Q(c) > 0\}$. Then $\mathcal{Q} \subseteq \mathcal{Q}'$ and $\mathcal{Q}' \cap \mathcal{F} = \emptyset$.

Let $c_1 = a_1$ and for $1 < i \le N$ let $c_i = c^{-1}b_i$. Then, for every $P \in \mathrm{Spec}^1(D) \setminus \mathcal{Q}'$, and, in particular, for every $P \in \mathcal{F}$, $(c_i)_{i=1}^N$ is a sequence in D_P. Also, for every maximal ideal Q of D that is neither in $\mathcal{Q}'$ nor in $\mathcal{P}$, $v_P(c_i) = 0$ for all i.

We set

$$g(x) = \prod_{i=1}^{N}(x - c_i) = (x - a_1)\prod_{i=2}^{N}(x - c^{-1}b_i)$$

and show that for all essential valuations v of D and all $r \in D$, $v(g(r)) \ge v(g(0))$.

First, assume $P \in \mathcal{P}$. The sequence $(c_i)_{i=1}^N$ enjoys the same properties with respect to PD_P that the sequence $(b_i)_{i=1}^N$ enjoys with respect to P, namely, those c_i with $1 \le i \le \|P\|^{m_P}$ form a complete system of residues modulo $(PD_P)^{m_P}$ and $c_i \equiv 1$ modulo $(PD_P)^{m_P+1}$ for all $i > \|P\|^{m_P}$. Also, no c_i is in P^{m_P+1}.

Consequently, for all $r \in D$, and $1 \le k \le m_P$

$$\left|\{i \mid v_P(r - c_i) \ge k\}\right| = \left|\{i \mid v_P(r - b_i) \ge k\}\right| \ge \gamma_k(P).$$

Let $\gamma_P := \sum_{k=1}^{m_P} \gamma_k(P)$. Then

$$v_P\big(g(r)\big) = \sum_{i=1}^{N} v_P(r - c_i) = \sum_{k=1}^{\infty}\left|\{i \mid v_P(r - c_i) \ge k\}\right| \ge$$

$$\ge \sum_{k=1}^{m_P}\left|\{i \mid v_P(r - c_i) \ge k\}\right| = \sum_{k=1}^{m_P}\left|\{i \mid v_P(r - b_i) \ge k\}\right| \ge \gamma_P,$$

while $v_P(g(0)) =$

$$= \sum_{k=1}^{\infty} \left|\{i \mid v_P(c_i) \geq k\}\right| = \sum_{k=1}^{m_P} \left|\{i \mid v_P(c_i) \geq k\}\right| = \sum_{k=1}^{m_P} \left|\{i \mid v_P(b_i) \geq k\}\right| = \gamma_P.$$

Now consider $Q \in \mathcal{Q}'$. Here $v_Q(c_1) = v_Q(a_1) = 0$ and, for all $i > 1$, $v_Q(c_i) < 0$. Let d_k be the coefficient of x^k in g and $\mu = \min_k v_Q(d_k)$. Using Lemma 1, we see that for all $r \in D$,

$$v_Q\big(g(r)\big) = \mu + v_Q(r - a_1) \geq \mu = \mu + v_Q(a_1) = v_Q\big(g(0)\big).$$

For the remaining essential valuations v of D, $v(c_i) = 0$ for all i, and, therefore, for all $r \in D$, $v\big(g(r)\big) = \sum_i v(r - c_i) \geq 0 = \sum_i v(c_i) = v\big(g(0)\big)$.

Now let $f(x) = g(x)/g(0)$. Then $f(a_1) = 0$, and $f(0) = 1$. Also, $f \in \mathrm{Int}(D)$, because for all $r \in D$ and every essential valuation v of D, $v\big(g(r)\big) \geq v\big(g(0)\big)$ and therefore $v\big(f(r)\big) \geq 0$.

As for the Lipschitz properties: for those $P \in \mathrm{Spec}^1(D)$ for which $v_P(c) = 0$, and, in particular, for all $P \in \mathcal{F}$, g is in $D_P[x]$. f is, therefore, P-adically $v_P(g(0))$-Lipschitz for all $P \in \mathcal{F}$ by Remark 2.

For $P \in \mathcal{F}_0$, $v_P(g(0)) = 0$ and hence f is P-adically 0-Lipschitz for all $P \in \mathcal{F}_0$.

For $P \in \mathcal{P}$,

$$v_P(g(0)) = \gamma_P = \sum_{k=1}^{m_P} \gamma_k(P) = \sum_{k=1}^{m_P} \|P\|^{m_P - k} = \sum_{j=0}^{m_P - 1} \|P\|^j = \frac{1 - \|P\|^{m_P}}{1 - \|P\|}.$$

f is, therefore, P-adically l_P-Lipschitz for all $P \in \mathcal{F}$, for the values of l_P stated in the Lemma. □

Corollary 1. *Let D be a Dedekind domain with finite residue fields, $\mathcal{F}$ a finite set of maximal ideals, and $a_0, \ldots, a_n$ distinct elements of D. For each $P \in \mathcal{F}$, let $m_P \geq \max_{1 \leq i \leq n} v_P(a_i - a_0)$.*

Then there exists $f \in \mathrm{Int}(D)$ with $f(a_i) = 0$ for $1 \leq i \leq n$, and $f(a_0) = 1$, and such that f is P-adically $L(\|P\|, m_P)$-Lipschitz for all $P \in \mathcal{F}$.

Proof. For each $1 \leq i \leq n$ and $P \in \mathcal{F}$, let $m_P(i) = v_P(a_i - a_0)$ and $l_P(i) = L(\|P\|, m_P(i))$. Let $f_i \in \mathrm{Int}(D)$ with $f_i(a_i) = 0$ and $f_i(a_0) = 1$ and such that f_i is P-adically $L(\|P\|, m_P(i))$-Lipschitz for each $P \in \mathcal{F}$. Such an f_i exists by Lemma 2, and it is P-adically $L(\|P\|, m_P)$-Lipschitz, because $m_P(i) \leq m_P$, and $L(q, m)$ is an increasing function in m for fixed q, and l-Lipschitz implies l'-Lipschitz for all for all $l' \geq l$. Now set $f(x) = \prod_{i=1}^{n} f_i(x)$. □

4 Interpolation by Congruence-Preserving Integer-Valued Polynomials

Lemma 3. *Let D be a Dedekind domain with finite residue fields and $r_0, \ldots, r_n$ distinct elements of D.*

Let $\mathcal{F}$ be a finite set of maximal ideals of D. For each $P \in \mathcal{F}$, let $k_P \in \mathbb{N}$ such that the r_i are pairwise incongruent modulo P^{k_P} and $l_P = L(\|P\|, k_P - 1)$ as in Definition 2.

Then there exists $f \in \mathrm{Int}(D)$ such that

1. *$f(r_0) = 1$ and, for $1 \le i \le n$, $f(r_i) = 0$;*
2. *for each $P \in \mathcal{F}$, for every $r \in D \setminus (r_0 + P^{k_P})$, $f(r) \equiv 0 \mod P^{k_P}$;*
3. *for each $P \in \mathcal{F}$, for every $r \in r_0 + P^{k_P + l_P}$, $f(r) \equiv 1 \mod P^{k_P}$.*

Proof. We will first construct a polynomial $f_P \in \mathrm{Int}(D)$ for each $P \in \mathcal{F}$, in several steps. Fix $P \in \mathcal{F}$.

Extend $r_0, \ldots, r_n$ to a complete set of residues $r_0, \ldots, r_{\|P\|^{k_P}-1}$ modulo P^{k_P}, such that for all $i > n$ and all $Q \in \mathcal{F} \setminus \{P\}$, $r_i \equiv r_1$ modulo Q^{k_Q+1}.

Let C be a finite subset of $\prod_{Q \in \mathcal{F}} Q^{k_Q}$ containing a complete system of residues of the residue classes of $P^{k_P+l_P}$ contained in P^{k_P}, and with $0 \in C$.

For each $1 \le i < \|P\|^{k_P}$, and $c \in C$, let f_{ic} a polynomial in $\mathrm{Int}(D)$ with $f_{ic}(r_0) = 1$, $f_{ic}(r_i + c) = 0$, and Q-adically l_Q-Lipschitz for all $Q \in \mathcal{F}$, such as we know to exist by Lemma 2 and its Corollary. Set $f_i = \prod_{c \in C} f_{ic}$. Then $f_i(r_i) = 0$ and $f_i(r_0) = 1$. Also, since $\bigcup_{c \in C} r_i + c + P^{k_P+l_P} = r_i + P^{k_P}$ and $f_i(r_i + c) = 0$ for all $c \in C$, the P-adic Lipschitz property implies that for all $r \in r_i + P^{k_P}$, $f_i(r) \equiv 0$ modulo P^{k_P}. Likewise, the Lipschitz properties of the polynomials f_{ic} imply for all $Q \in \mathcal{F}$ that $f_i(r) \equiv 1$ modulo Q^{k_Q} for all $r \in r_0 + Q^{k_Q+l_Q}$.

Let $f_P = \prod_{i=1}^{\|P\|^{k_P}-1} f_i$. Then f_P satisfies

1. $f_P(r_0) = 1$ and $f_P(r_j) = 0$ for $1 \le j \le n$;
2. $f_P(r) \equiv 0$ modulo P^{k_P} for $r \in D \setminus (r_0 + P^{k_P})$;
3. for all $Q \in \mathcal{F}$, for all $r \in r_0 + Q^{k_Q+l_Q}$, $f_P(r) \equiv 1$ modulo Q^{k_Q}.

Having constructed f_P for each $P \in \mathcal{F}$, we set $f = \prod_{P \in \mathcal{F}} f_P$, and f has the desired properties. □

Theorem 1. *Let D be a Dedekind domain with finite residue fields, $r_0, \ldots, r_n$ distinct elements of D and $s_0, \ldots, s_n$ arbitrary elements of D.*

Let $\mathcal{F}$ be a finite set of maximal ideals of D. For each $P \in \mathcal{F}$ let $k_P \in \mathbb{N}$ such that the r_i are pairwise incongruent modulo P^{k_P}.

Then there exists $f \in \mathrm{Int}(D)$ such that

1. *for $0 \le i \le n$,*
$$f(r_i) = s_i$$

2. *for all $P \in \mathcal{F}$, for all $a, b \in D$,*

$$a \equiv b \mod P^{k_P} \implies f(a) \equiv f(b) \mod P^{k_P}.$$

3. *for all $P \in \mathcal{F}$, for all $r \in D$ with $(r + P^{k_P}) \cap \{r_0, \dots, r_n\} = \emptyset$,*

$$f(r) \equiv 0 \mod P^{k_P}.$$

Proof. It suffices to show, for each index i, the existence of a polynomial $h_i \in \text{Int}(D)$ such that

1. $h_i(r_i) = 1$ and $h_i(r_j) = 0$ for $j \neq i$,
2. for all $P \in \mathcal{F}$, for all $r \in D \setminus (r_i + P^{k_P})$, $h_i(r) \equiv 0 \mod P^{k_P}$, and
3. for all $P \in \mathcal{F}$, for all $r \in r_i + P^{k_P}$, $h_i(r) \equiv 1 \mod P^{k_P}$,

because, then, the polynomial $f = \sum_{i=0}^{n} s_i h_i$ does the job.

W.l.o.g., assume $i = 0$. We construct h_0 with the help of Lemma 3:

For each $Q \in \mathcal{F}$, let $l_Q = L(\|Q\|, k_Q - 1)$.

Let C be a subset of $\prod_{Q \in \mathcal{F}} Q^{k_Q}$ containing, for each $Q \in \mathcal{F}$, a complete system of residues of the residue classes of $Q^{k_Q + l_Q}$ contained in Q^{k_Q}, and with $0 \in C$.

For each $d \in C$, $r_0 + d, r_1, \dots, r_n$ satisfy the premises of Lemma 3. Accordingly, let $f_d \in \text{Int}(D)$ such that

1. $f_d(r_0 + d) = 1$ and, for $1 \le i \le n$, $f_d(r_i) = 0$;
2. for each $P \in \mathcal{F}$, for every $r \in D \setminus (r_0 + d + P^{k_P})$, $f_d(r) \equiv 0 \mod P^{k_P}$;
3. for each $P \in \mathcal{F}$, for every $r \in r_0 + d + P^{k_P + l_P}$, $f_d(r) \equiv 1 \mod P^{k_P}$.

and set $g_d = 1 - f_d$.

Since $r_0 + d + P^{k_P} = r_0 + P^{k_P}$ for all $P \in \mathcal{F}$ and $d \in C$, each g_d satisfies

1. $g_d(r_0 + d) = 0$ and, for $1 \le i \le n$, $g_d(r_i) = 1$;
2. for each $P \in \mathcal{F}$, for every $r \in D \setminus (r_0 + P^{k_P})$, $g_d(r) \equiv 1 \mod P^{k_P}$;
3. for each $P \in \mathcal{F}$, for every $r \in r_0 + d + P^{k_P + l_P}$, $g_d(r) \equiv 0 \mod P^{k_P}$.

Now, set $g = \prod_{d \in C} g_d$.

Considering that, for all $P \in \mathcal{F}$, $\bigcup_{d \in C} r_0 + d + P^{k_P + l_P} = r_0 + P^{k_P}$, we see that the polynomial $g = \prod_{d \in C} g_d$ satisfies

1. $g(r_0) = 0$ and, for $1 \le i \le n$, $g(r_i) = 1$;
2. for each $P \in \mathcal{F}$, for every $r \in D \setminus (r_0 + P^{k_P})$, $g(r) \equiv 1 \mod P^{k_P}$;
3. for each $P \in \mathcal{F}$, for every $r \in r_0 + P^{k_P}$, $g(r) \equiv 0 \mod P^{k_P}$.

Finally, we let $h_0 = 1 - g$. □

Recall that a function $f : D \to D$ satisfying

$$a \equiv b \mod I \implies f(a) \equiv f(b) \mod I,$$

where D is a commutative ring and I an ideal of D, is called *I-congruence preserving*. In this case, f defines a function $\bar{f}_I : D/I \to D/I$ by

$$\bar{f}_I(a + I) = f(a) + I.$$

We call $\bar{f}_I$ the function induced by f on D/I.

We can now sharpen Theorem 1 some more to obtain a completely general form of simultaneous interpolation and P-adic approximation. Given arbitrary arguments and values in D and, for finitely many maximal ideals, a function on the residue class ring modulo a power of the ideal, we can find a polynomial in $\mathrm{Int}(D)$ that interpolates, while simultaneously realizing the given functions on the residue class rings, provided that the requirements are not obviously contradictory.

Theorem 2. *Let D be a Dedekind domain with finite residue fields, $r_0, \ldots, r_n$ distinct elements of D and $s_0, \ldots, s_n$ arbitrary elements of D.*

Let $\mathcal{F}$ be a finite set of maximal ideals of D. For each $P \in \mathcal{F}$ let $k_P \in \mathbb{N}$ a natural number, and

$$\varphi_P : D/P^{k_P} \to D/P^{k_P}$$

a function.

If, for all $P \in \mathcal{F}$ and for all $0 \le i \le n$,

$$s_i \in \varphi_P(r_i + P^{k_P})$$

then there exists $f \in \mathrm{Int}(D)$ such that

1. *for $0 \le i \le n$,*
$$f(r_i) = s_i$$
2. *for all $P \in \mathcal{F}$, for all $a, b \in D$,*
$$a \equiv b \mod P^{k_P} \quad \Longrightarrow \quad f(a) \equiv f(b) \mod P^{k_P}$$
and the function $\bar{f} : D/P^{k_P} \to D/P^{k_P}$ defined by $\bar{f}(a + P^{k_P}) = f(a) + P^{k_P}$ equals φ_P.

Proof. We may, w.l.o.g., assume that for all $P \in \mathcal{F}$ the arguments r_i are pairwise incongruent modulo P^{k_P}. If they are not, we replace each k_P by a possibly larger l_P such that the r_i are incongruent modulo P^{l_P}, and replace each φ_P by a function

$$\psi_P : D/P^{l_P} \to D/P^{l_P}$$

which preserves congruences modulo $P^{k_P} + P^{l_P}$, induces φ_P on D/P^{k_P} and satisfies $\psi_P(r_i + P^{l_P}) = s_i + P^{l_P}$.

Now assume that the r_i are pairwise incongruent modulo P^{k_P}. We apply Theorem 1 to produce $g \in \mathrm{Int}(D)$ such that

1. for $0 \le i \le n$, $g(r_i) = s_i$
2. for all $P \in \mathcal{F}$, g is P^{k_P}-congruence preserving
3. for all $P \in \mathcal{F}$, for all $r \in D$ such that $(r + P^{k_P})$ contains no r_i, we have $g(r) \equiv 0 \mod P^{k_P}$.

Let $\mathcal{F}'$ be the subset of $\mathcal{F}$ consisting of those P for which $r_0, \ldots, r_n$ do not form a complete system of residues modulo P^{k_P}. For all $P \in \mathcal{F} \setminus \mathcal{F}'$, g already induces φ_P on D/P^{k_P}. We now modify g by adding a polynomial $f_Q \in \mathrm{Int}(D)$ for each $Q \in \mathcal{F}'$ to the effect that φ_Q is induced on D/Q^{k_Q}, without affecting the properties 1 and 2 of g and without changing the function induced on D/P^{k_P} for any $P \in \mathcal{F} \setminus \{Q\}$.

Fix $Q \in \mathcal{F}'$. To construct f_Q, first extend $r_0, \ldots, r_n$ to a complete system of residues $r_0, \ldots, r_n, r_{n+1}, \ldots, r_{q-1}$ modulo Q^{k_Q}.

Then, for each i with $n < i < q$, use Theorem 1 to find $h_i \in \mathrm{Int}(D)$ which is Q^{k_Q}-congruence preserving and satisfies $h_i(r_i) = 1$ and $h_i(r_j) = 0$ for $0 \le j < q$ with $j \ne i$.

Also, for $n < i < q$, let $b_i \in \varphi_Q(r_i + Q^{k_Q})$ such that $b_i \equiv 0 \mod P^{k_P}$ for all $P \in \mathcal{F} \setminus \{Q\}$. Then, set

$$f_Q = \sum_{i=n+1}^{q-1} b_i h_i.$$

Having thus defined f_Q for each $Q \in \mathcal{F}'$, finally, set

$$f = g + \sum_{Q \in \mathcal{F}'} f_Q.$$

□

References

1. Cahen, P.J., Chabert, J.L.: Integer-Valued Polynomials. Mathematical Surveys and Monographs, vol. 48. American Mathematical Society, Providence, RI (1997)
2. Cahen, P.J., Chabert, J.L., Frisch, S.: Interpolation domains. J. Algebra **225**(2), 794–803 (2000). http://dx.doi.org/10.1006/jabr.1999.8151
3. Frisch, S.: Interpolation by integer-valued polynomials. J. Algebra **211**(2), 562–577 (1999). http://dx.doi.org/10.1006/jabr.1998.7741
4. Frisch, S.: Polynomial functions on finite commutative rings. In: Dobbs, D.E., Fontana, M., Kabbaj, S.E. (eds.) Advances in Commutative Ring Theory (Fez, 1997). Lecture Notes in Pure and Applied Mathematics, vol. 205, pp. 323–336. Dekker, New York (1999)
5. Rédei, L., Szele, T.: Algebraischzahlentheoretische Betrachtungen über Ringe. I. Acta Math. **79**, 291–320 (1947). https://doi.org/10.1007/BF02404701
6. Rédei, L., Szele, T.: Algebraisch-zahlentheoretische Betrachtungen über Ringe. II. Acta Math. **82**, 209–241 (1950). https://doi.org/10.1007/BF02398278
7. Skolem, T.: Einige Sätze über Polynome. Avh. Norske Vid. Akad. Oslo. I. **1940**(4), 16 (1940)

Length Functions over Prüfer Domains

Gabriele Fusacchia and Luigi Salce

Abstract The first part of this paper is mostly devoted to survey results by Northcott-Reufel on length functions over commutative domains, specifically over valuation domains and 1-dimensional Prüfer domains of finite character. In the second part we present new results leading to the characterization of length functions over arbitrary Prüfer domains of finite character.

Keywords Prüfer domains · Valuation domains · Length functions

2000 Mathematics Subject Classification Primary: 13F05, 13F30, 13G05. Secondary: 13C60

1 Introduction

An intriguing feature in mathematical research arises when papers dating from the same time and dealing with completely different arguments are found, after many years, to have a strict connection, when new researches provide a unifying point of view or unforeseeable relationships. An example of this situation is offered by two papers on different subjects that appeared in 1965.

The first paper by Northcott and Reufel [16], which is the source of the present paper, dealt with suitable generalizations of the notion of the classical composition length for modules over a commutative ring, called *length functions*. They were able

Research supported by the Department of Mathematics "Tullio Levi-Civita" of the University of Padova.

G. Fusacchia · L. Salce (✉)
Dipartimento di Matematica "Tullio Levi-Civita", Università di Padova,
Via Trieste 63, 35121 Padua, Italy
e-mail: salce@math.unipd.it

G. Fusacchia
e-mail: gabfus@gmail.com

A. Facchini et al. (eds.), *Advances in Rings, Modules and Factorizations*,
Springer Proceedings in Mathematics & Statistics 321,
https://doi.org/10.1007/978-3-030-43416-8_11

to characterize all length functions over valuation domains and over 1-dimensional Prüfer domains of finite character. The latter paper, by Adler, Konheim and McAndrew [1], introduced and investigated the notion of topological entropy and, at the very end, gave a sketch definition of *algebraic entropy* for endomorphisms of Abelian groups. At that time it was impossible to realize, or even suspect, a connection between the two papers [1, 16].

After almost half a century, the generalization of the algebraic entropy from Abelian groups (in the meanwhile deeply investigated in [6] and elsewhere) to the more general setting of modules over arbitrary rings presented in [18] used the notion of sub-additive invariant as a crucial tool for the definition of entropy. Furthermore, important theorems for this entropy were proved in [17] when the sub-additive invariant is, indeed, a length function, thus establishing a strict connection between length functions and algebraic entropy.

Length functions have been investigated in the late '60's by Peter Vámos in [20, 21], who obtained their complete classification over Noetherian commutative rings. In recent years the renewed interest in length functions came as a sub-product of the investigation of the algebraic entropy for endomorphisms of modules over arbitrary rings (the second author is indebted to Peter Vámos for addressing him to length functions as a privileged tool in this investigation). However, in the authors' opinion, length functions are a subject which deserve independent interest, especially for people working in commutative algebra.

This opinion motivates the present paper, whose goal is to investigate length functions over Prüfer domains; we must restrict to domains of finite character, since our actual knowledge is confined to this class of domains. Our starting point is the following characterization, due to Northcott and Reufel [16, Theorem 14], of length functions over 1-dimensional Prüfer domains of finite character; note that these domains are, in particular, h-local, i.e., beside finite character, they have the property that every non-zero prime ideal is contained in a unique maximal ideal.

Theorem 1.1. (Northcott-Reufel) *Let R be a 1-dimensional Prüfer domain of finite character and $L : Mod(R) \to \mathbb{R}_{\geq 0} \cup \{\infty\}$ a length function such that $L(R) = \infty$. Then there exists a unique family $\{L^{(P)}\}_{P \in Max(R)}$ of local length functions for R such that $L^{(P)}(R_P) = \infty$, with the property that $L(M) = \sum_{P \in Max(R)} L^{(P)}(M_P)$ for every R-module M. The length functions $L^{(P)}$ coincide with the localizations L_P of L for each $P \in Max(R)$.*

For the notions of local length function and of localization of a length function see the following Section 2, where we provide the general information, mostly borrowed from [16], concerning length functions over commutative rings. The hypothesis that $L(R) = \infty$ in Theorem 1.1 is not restrictive, since length functions satisfying $L(R) < \infty$ are easily classifiable, as we will see in Section 2.

In Section 3 we survey the characterization obtained by Northcott and Reufel in [16] for length functions over valuation domains; it is worth recalling that this classification has been obtained using a different multiplicative approach by Zanardo in [24]. In Section 4 we collect some properties of Prüfer domains used in the sequel of the paper and introduce the notion of their meet-spectrum.

The extension of Theorem 1.1 from 1-dimensional to arbitrary Prüfer domains of finite character, where the structure of the prime spectrum is more complicated, presents problems for representing a length function as sum of its localizations at maximal ideals, as well as for the uniqueness of its representation. These problems are solved in Section 5 where, after an example which makes evident the problems mentioned above and revisiting unpublished notes by the first author [11], we present the most relevant new results leading to the characterization of length functions over semilocal Prüfer domains. These results are based on the ad hoc notions of compatible families of local length functions and of their balanced sum.

Once this step is accomplished, its transfer to general Prüfer domains of finite character follows easily in the final Section 6, where we present the main theorem of the paper (Theorem 6.2). Theorem 1.1, as well as its extension to arbitrary h-local domains, are derived as simple corollaries of the main theorem.

For completeness of information on what is at our knowledge on length functions, we must mention that recently Virili [22] surveyed Vámos' results on length functions over Noetherian commutative rings, retracing complete proofs different in many parts from the original ones. The main goal of Virili's survey was the description of algebraic entropies induced by these length functions. A further investigation by Virili [23] concerns length functions on Grothendieck categories.

Finally, very recently Spirito [19] studied decompositions of length functions over integral domains as sums of length functions constructed from overrings. He found standard representations when the integral domain admits a Jaffard family of flat overrings, and when it is a Prüfer domain such that every ideal has only finitely many minimal primes. His paper has some overlaps with the present paper.

2 Generalities on Length Functions

The starting point which originated the notion of length function is the classical notion of composition length: given a commutative ring R and an R-module M, its *composition length* $l(M)$ is the maximum of the lengths of chains of submodules $0 < M_1 < M_2 < \cdots < M_n = M$ such that each factor M_{i+1}/M_i $(0 \leq i \leq n-1)$ is a simple module, if such a maximum exists, otherwise it is ∞. It is clear that $l(0) = 0$ and that, given a short exact sequence of R-modules $0 \to M' \to M \to M'' \to 0$, $l(M) = l(M') + l(M'')$; furthermore, it is easy to prove that, for each module M, $l(M)$ is the supremum of the composition lengths of its finitely generated submodules.

The idea behind the notion of length function, introduced by Northcott and Reufel [16] in order to generalize the composition length, is to measure the size of modules by associating with each module a "quantity" consisting either of a non-negative real number or ∞. They axiomatized this notion imposing the three conditions satisfied by the composition length recalled above.

Let R be a commutative unitary ring and $Mod(R)$ the category of R-modules. We denote by $\mathbb{R}^*$ the totally ordered set of the non-negative real numbers with the symbol ∞ adjoint, which is strictly bigger than any real number: $\mathbb{R}^* = \mathbb{R}_{\geq 0} \cup \{\infty\}$;

obviously $\infty + \infty = \infty + r = \infty$ for each $r \in \mathbb{R}_{\geq 0}$. A *length function* over $Mod(R)$ (or simply over R) is a map $L : Mod(R) \to \mathbb{R}^*$ which satisfies the following three conditions:

(i) $L(0) = 0$;

(ii) (additivity) $L(M) = L(M') + L(M'')$ if $0 \to M' \to M \to M'' \to 0$ is a short exact sequence in $Mod(R)$;

(iii) (upper continuity) for every $M \in Mod(R)$, $L(M) = sup_X L(X)$, where X ranges over the set of the finitely generated submodules of M.

The first two properties imply that a length function is an invariant, i.e., it assigns the same value to isomorphic modules. Immediate consequences of (ii) are that $M' \leq M$ implies $L(M) = L(M') + L(M/M')$ and that, if $I \leq J$ are ideals of R, then $L(R/J) \leq L(R/I)$.

The set of length functions over R may be endowed with a partial order, by saying the $L \leq L'$ if $L(M) \leq L'(M)$ for every $M \in Mod(R)$.

By a *local length function* for R we mean a length function over $Mod(R_P)$, for some $P \in Max(R)$, the maximal spectrum of R, where R_P denotes, as usual, the localization of R at P. A length function L over R is *discrete* if its finite values form a discrete subset of $\mathbb{R}^*$, necessarily order-isomorphic to $\mathbb{N}$, and it is *faithful* if $L(M) = 0$ implies $M = 0$.

Example 2.1. The most common examples of length functions are:

- the classical *composition length* of a module; it is discrete and faithful;

- if R is an integral domain, the *rank* of a module M, defined as $rk(M) = dim_K(M \otimes_R K)$, where K is the field of fractions of R; it is discrete but not faithful, unless R is a field;

- if $R = \mathbb{Z}$, for every Abelian group G let $L(G) = log|G|$ (obviously setting $L(G) = \infty$ if G is infinite); this is a discrete and faithful length function;

- the *trivial* length functions L_0 and L_∞ always exist and are defined as follows: $L_0(M) = 0$ for all $M \in Mod(R)$, and $L_\infty(M) = \infty$ for all $0 \neq M \in Mod(R)$.

The next result says that every length function is completely determined by its action on cyclic modules.

Proposition 2.2. ([16, Lemma 1]) *Two length functions L, L' on a commutative ring R coincide if and only $L(R/I) = L'(R/I)$ for every ideal I of R.*

Proof. The necessity is trivial. The sufficiency depends on the fact that every finitely generated module is the union of a finite chain of submodules with cyclic quotients of two consecutive submodules, to which we can apply the additivity of L, and on upper continuity. □

A relevant consequence of Proposition 2.2 is that completion properties of quotients of R in the R-topology have no impact on length functions; for instance, a maximal valuation domain and an incomplete non-almost maximal valuation domain with the same value group admit exactly the same length functions, as will be clear in the next section.

The next result shows the relevance of prime ideals in connection with length functions.

Proposition 2.3. *Let P be a prime ideal of the commutative ring R and L a length function over R such that $L(R/P) < \infty$. Then $L(R/I) = 0$ for every ideal I strictly containing P.*

Proof. By additivity, $L(R/P) = L(R/I) + L(I/P)$, hence $L(R/P) \geq L(I/P)$. Choosing $x \in I \setminus P$ provides an embedding $R/P \to I/P$ through the multiplication by x. This gives $L(R/P) \leq L(I/P)$, therefore $L(R/P) = L(I/P)$ and consequently $L(R/I) = 0$. □

Since we will focus on modules over integral domains, the next consequence of Proposition 2.3 is crucial. It tells us that a length function on a domain R taking positive finite value on R coincides with the rank function, up to a positive real multiple.

Corollary 2.4. ([16, Theorem 2]) *Let L be a length function over an integral domain R satisfying $0 < L(R) < \infty$. Then, for every module $M \in Mod(R)$, $L(M) = L(R) \cdot rk(M)$.*

Proof. As $rk(R) = 1$, the equality $L(M) = L(R) \cdot rk(M)$ holds for $M = R$. If $0 < I < R$, then $rk(R/I) = 0$, so, in view of Proposition 2.2, it is enough to prove that $L(R/I) = 0$. But this immediately follows from Proposition 2.3, since 0 is a prime ideal. □

If a length function L satisfies $L(R) = 0$, then obviously it coincides with the trivial length function L_0. So, by Corollary 2.4, the investigation of length functions over an integral domain R may be confined to those L such that $L(R) = \infty$.

Localizations at multiplicatively closed subsets of the commutative ring R and length functions have strict relationships. First, any length function L on R determines a multiplicatively closed set S_L, defined as

$$S_L = \{a \in R : L(R/aR) = 0\}.$$

In fact, it is clear that $1 \in S_L$ and, if $a, b \in S_L$, then $L(R/aR) = 0 = L(R/bR)$, hence from the exact sequence

$$0 \to aR/abR \to R/abR \to R/aR \to 0$$

we deduce that $L(R/abR) = L(aR/abR) + L(R/aR)$. But aR/abR is an epic image of R/bR, hence $L(aR/abR) = 0$, consequently $L(R/abR) = 0$ and $ab \in S_L$.

Recall that the S_L-torsion submodule of an R-module M consists of those elements $x \in M$ such that $ax = 0$ for some $a \in S_L$.

Lemma 2.5. ([16, Lemmas 3 and 4]) *Let L be a length function over the commutative ring R. Let $M \in Mod(R)$ and T its S_L-torsion submodule. Then $L(M) = L(M/T) = L(M_{S_L})$.*

Proof. If R/I is a cyclic S_L-torsion module, then I contains an element $a \in S_L$, so that $L(R/I) \leq L(R/aR) = 0$. It follows that $L(T) = 0$ and, by additivity, $L(M) = L(M/T)$. Since M/T embeds into M_{S_L} with S_L-torsion cokernel, the additivity of L also implies that $L(M/T) = L(M_{S_L})$. □

On the other hand, starting with a multiplicatively closed subset S of R, a length function L over R is said to be *localized at* S if it satisfies $L(M) = L(M_S)$ for every R-module M, that is, if the length of each R-module is the same as the length of its localization at S, so that L can be viewed as a length function on $Mod(R_S)$. The next result tells in particular that a length function L is always localized at S_L.

Proposition 2.6. *Let L be a length function over the commutative ring R, and $S \subseteq T$ two multiplicatively closed subsets of R. Then:*
(1) *if L is localized at T, then it is localized at S;*
(2) *L is localized at S if and only if $S \subseteq S_L$.*

Proof. (1) Since $S \subseteq T$, for every module M we have $M_T = (M_S)_T$. Then $L(M_S) = L((M_S)_T)$, since L is localized at T, therefore $L(M_S) = L(M_T) = L(M)$.

(2) The same argument as in point (1) proves the sufficiency, once we have shown that L is localized at S_L; this follows by Lemma 2.5. For the necessity, if S is not contained in S_L, there exists $s \in S$ such that $L(R/sR) > 0$. On the other hand, $L(R_S/sR_S) = L(0) = 0$, hence L cannot be localized at S. □

We describe some methods to obtain new length functions from pre-assigned ones. We leave as an exercise for the reader to check the details of the straightforward proofs. It is worthwhile to remark that the first example, with the two lemmas and the proposition following it, provide a crucial tool in the development of the paper, notably, in Definition 5.6 and in the proofs of the main Theorems 5.7 and 6.2.

Example 2.7. (Difference) Let $L \geq L'$ be two length functions over $Mod(R)$ and denote by $\mathcal{F}G(R)$ the class of finitely generated R-modules. The function $L - L' : \mathcal{F}G(R) \to \mathbb{R}^*$ defined by setting

$$(L - L')(F) = \begin{cases} L(F) - L'(F) & \text{if } L(F) < \infty \\ \infty & \text{if } L(F) = \infty \end{cases}$$

is additive on short exact sequences of finitely generated modules, as it is easy to check. Extend this function to the whole category $Mod(R)$ by setting, for every R-module M:

$$(L - L')(M) = sup\{(L - L')(F) | F \in \mathcal{F}G(R), F \leq M\}.$$

Then $L - L' : Mod(R) \to \mathbb{R}^*$ is upper continuous by definition, but in general it is not a length function, since it is not additive. As a counter-example, take $R = K(+)V$, where K is a field, V an infinite dimensional K-vector space, and $(+)$ the idealization. Taking $L = dim_K$ it is easy to see that the function $L - L$ as defined above is not additive on the exact sequence $0 \to V \to R \to K \to 0$ (we thank the referee for suggesting us such a counter-example).

We look for properties of L and L' ensuring that $L - L'$ is additive, hence a length function.

Lemma 2.8. *Let L and L' be two length functions over $Mod(R)$ satisfying the inequality $L \geq L'$. If $L(A) < \infty$ for a module A, then $(L - L')(A) = L(A) - L'(A)$.*

Proof. All the involved quantities are finite, so we must equivalently prove that $(L - L')(A) + L'(A) = L(A)$, that is, $sup_{F \leq A}(L - L')(F) + sup_{F' \leq A} L'(F') = sup_{F'' \leq A} L(F'')$, where F, F', F'' are finitely generated modules. If $F'' \leq A$, then $L(F'') = (L - L')(F'') + L'(F'')$, hence $L(A) \leq (L - L')(A) + L'(A)$. Conversely, $(L - L')(F) + L'(F') \leq (L - L')(F + F') + L'(F' + F) = L(F + F')$, hence the converse inequality $(L - L')(A) + L'(A) \leq L(A)$ also holds. □

Lemma 2.9. *Let L and L' be two length functions over $Mod(R)$ satisfying the inequality $L \geq L'$. If $0 \to A \to B \to C \to 0$ is an exact sequence such that B is finitely generated and $L(A) < \infty$, then $L - L'$ is exact on it.*

Proof. If $L(B) < \infty$, then $(L - L')(B) = L(B) - L'(B) = L(A) + L(C) - L'(A) - L'(C) = L(A) - L'(A) + (L - L')(C)$. But $(L - L')(A) = L(A) - L'(A)$, by Lemma 2.8, so the claim follows. If $L(B) = \infty$, then necessarily $L(C) = \infty$, hence $(L - L')(B) = \infty = (L - L')(C)$, thus additivity follows also in this case. □

If in the preceding lemma we drop the hypothesis that $L(A) < \infty$, the claim is no more true, as the counter-example above shows. The next result provides a sufficient condition in order that the difference of two length functions is again a length function.

Proposition 2.10. *Let L and L' be two length functions over $Mod(R)$ satisfying the inequality $L \geq L'$. If for every submodule A of a finitely generated R-module the equality $L(A) = \infty$ implies $(L - L')(A) = \infty$, then $L - L'$ is a length function over $Mod(R)$*

Proof. Since $L - L'$ is upper continuous by definition, it is enough to prove that it is additive. It is immediate to check that $L - L'$ is additive on short exact sequences $0 \to A \to B \to C \to 0$, with A, B, C finitely generated. If it is additive on $0 \to A \to B \to C \to 0$ when only B, and consequently C, are finitely generated, arguing as in the proof of Lemma 3.14 in [22], one can show that $L - L'$ is additive on any type of short exact sequence. So we must check the additivity of $L - L'$ on $0 \to A \to B \to C \to 0$, assuming B finitely generated. If $L(A) < \infty$, then $L - L'$

is additive without any additional hypothesis by Lemma 2.8. If $L(A) = \infty$, then the hypothesis ensures that $(L - L')(A) = \infty$. As $L(B) = \infty = (L - L')(B)$, the additivity of $L - L'$ holds also in this case. □

We continue with four more examples of length functions.

Example 2.11. (Sum) If $\{L_\sigma\}_{\sigma\in\Lambda}$ is a family of length functions over $Mod(R)$, their sum $\sum_\sigma L_\sigma$ is the length function defined by setting, for each $M \in Mod(R)$:

$$\sum_{\sigma\in\Lambda} L_\sigma(M) = sup\{\sum_{\sigma\in F} L_\sigma(M) \ : \ F \text{ finite subset of } \Lambda\}.$$

Example 2.12. (Scalar multiple) Let L be a length function over $Mod(R)$ and λ a positive real number. The scalar multiple $\lambda \cdot L$, defined by setting $(\lambda \cdot L)(M) = \lambda L(M)$ for every R-module M, is a length function as well.

Example 2.13. (Contraction; see [16, Proposition 2]) Let S be a multiplicatively closed subset of R and $H : Mod(R_S) \to \mathbb{R}^*$ a length function. Then $H^c : Mod(R) \to \mathbb{R}^*$, defined by $H^c(M) = H(M_S)$ for each $M \in Mod(R)$, is a length function on $Mod(R)$. Thus H^c is just the composition of H with the localization map $- \otimes_R R_S$.

Example 2.14. (Localization; see [16, Proposition 3]) Let S be a multiplicatively closed subset of R and $L : Mod(R) \to \mathbb{R}^*$ a length function. Then $L_S : Mod(R_S) \to \mathbb{R}^*$ defined by $L_S(M) = L(M)$ for each $M \in Mod(R_S)$ is a length function as well. Thus L_S is just the restriction of L to R_S-modules. When S is the complement set of a prime ideal P, the localization of L at P is simply denoted by L_P.

3 Length Functions over Valuation Domains

In this section we survey the classification obtained by Northcott and Reufel in [16] for length functions over valuation domains. Recall that an integral domain R is a valuation domain if its ideals are linearly ordered by the inclusion. For terminology, notation and basic facts concerning these domains and their modules, we refer to [9, 10]. In particular, we denote by $v : Q(R) \to \Gamma \cup \{\infty\}$ the valuation of R, where $Q(R)$ is the field of fractions of R, Γ is its value group, and $R = \{x \in Q(R) : v(x) \geq 0\}$.

Excluding the trivial length function on $Mod(R)$ which sends all modules to 0, characterized by the property that $L(R) = 0$, the length functions may be distinguished in three different types:

- $0/\infty$-type, (I) and (II)
- Rank-type
- Valuation-type.

These types depend on two subsets of R, which are associated with the length function L:

$$P_\infty = \{a \in R \ : \ L(R/aR) = \infty\} \subseteq P_+ = \{a \in R \ : \ L(R/aR) > 0\}.$$

Note that $L(R) = \infty \Leftrightarrow 0 \in P_\infty \Leftrightarrow P_\infty \neq \emptyset$, and $L(R) > 0 \Leftrightarrow 0 \in P_+ \Leftrightarrow P_+ \neq \emptyset$.

It is easy to check that, if P_∞ and P_+ are non-empty, then they are prime ideals, and that P_+ is the complement of the multiplicatively closed set S_L defined in the previous section.

We have seen that we can exclude from our classification also length functions satisfying $0 < L(R) < \infty$, since, from Corollary 2.4, we know this condition implies that $L = L(R) \cdot rk$. So the classification of length functions will concern only those L such that $L(R) = \infty$, in which case $P_\infty \subseteq P_+$ are prime ideals. The next result shows that idempotency of these prime ideals plays a role in a crucial way.

Lemma 3.1. *If $L(R) = \infty$ and $L(R/P_\infty) < \infty$, then $0 < P_\infty^2 = P_\infty = P_+$.*

Proof. Clearly $0 < P_\infty$. Assume by way of contradiction that $P_\infty^2 < P_\infty$. Take $b \in P_\infty \setminus P_\infty^2$ and consider bR/bP_∞. Then

$$L(bR/P_\infty^2) \leq L(bR/bP_\infty) = L(R/P_\infty) \Rightarrow L(P_\infty/P_\infty^2) = sup_{b \in P_\infty \setminus P_\infty^2} L(bR/P_\infty^2) \leq L(R/P_\infty) < \infty.$$

From the exact sequence:

$$0 \to P_\infty/P_\infty^2 \to R/P_\infty^2 \to R/P_\infty \to 0$$

we deduce that $L(R/P_\infty^2) < \infty$. But this is absurd, since $P_\infty^2 < aR < P_\infty$ implies that $L(R/P_\infty^2) \geq L(R/aR) = \infty$. Thus $P_\infty^2 = P_\infty$. If $a \in R \setminus P_\infty$, then $a^k \in R \setminus P_\infty$ for all positive integers k, hence

$$k \cdot L(R/aR) = L(R/a^k R) \leq L(R/P_\infty) < \infty \Rightarrow L(R/aR) = 0 \Rightarrow a \in R \setminus P_+$$

that is, $P_\infty = P_+$. □

We describe now the three different types of length functions and their relationship with the prime ideals P_∞ and P_+. The first type, which is called "$0/\infty$-*type*", splits in two sub-types, according as we start with an arbitrary or an idempotent prime ideal P and we set $L(R/P) = \infty$ (first sub-type), or $L(R/P) = 0$ (second sub-type).

$0/\infty$ − type (I)

Fixed a prime ideal P of R, for an ideal I of R and a module $M \in Mod(R)$, we set

$$L(R/I) = \begin{cases} 0 & \text{if } P < I \\ \infty & \text{if } I \leq P \end{cases} \qquad L(M) = \begin{cases} 0 & \text{if } PM = 0 \text{ and } M_P = 0 \\ \infty & \text{otherwise} \end{cases}$$

If $P = 0$, $L(M) = 0$ for every torsion module M, and $L(M) = \infty$ for all other modules.

If P is the maximal ideal, $L(M) = \infty$ for every module $M \neq 0$. In each case $L(R) = \infty$.

In this case $P = P_\infty = P_+$ and P can be idempotent or non-idempotent.

$0/\infty$ – type (II)

Fixed an idempotent prime ideal $P = P^2$ of R, for an ideal I of R and a module $M \in Mod(R)$, we set

$$L(R/I) = \begin{cases} 0 & \text{if } P \leq I \\ \infty & \text{if } I < P \end{cases} \qquad L(M) = \begin{cases} 0 & \text{if } PM = 0 \\ \infty & \text{if } PM \neq 0 \end{cases}$$

We must exclude $P = 0$ that gives the trivial function, so we have $L(R) = \infty$.

If P is the maximal ideal, then $L(M) = 0$ for M semisimple, otherwise $L(M) = \infty$.

In this case $P = P_\infty = P_\infty^2 = P_+$. The idempotency of P excludes the possibility that P is maximal and principal (if $P = pR$ is maximal, $L(R/pR) = 0$ and $L(R/p^2R) = \infty$, absurd).

Rank – type

Given an idempotent prime ideal $P = P^2$ of R, and fixed a real number $\lambda > 0$, set:

$$L(R/I) = \begin{cases} 0 & \text{if } P < I \\ \lambda & \text{if } I = P \\ \infty & \text{if } I < P \end{cases} \qquad L(M) = \begin{cases} \lambda \cdot rk_{R/P}(M) & \text{if } PM = 0 \\ \infty & \text{if } PM \neq 0 \end{cases}$$

If $P = 0$, then $L(R) = \lambda$, which implies, by Corollary 2.4, that L coincides with the rank function up to the multiple λ. We have excluded this case from our classification.

Thus in this case also we have $L(R) = \infty$; furthermore $P = P_\infty = P_\infty^2 = P_+$. The idempotency of P excludes the possibility that P is maximal and principal (if $P = pR$ is maximal, $L(R/pR) = \lambda$ and $L(R/p^2R) = \infty$, absurd). Note that a rank-type length function is discrete.

Valuation – type

Given two adjacent prime ideals $P' \subset P$, the valuation domain $S = R_P/P'$ obtained by localizing R at P and factorizing over P' is archimedean, with a valuation:

$$w \,:\, Q(S) \to \mathbb{R} \cup \{\infty\}$$

which is discrete if and only if S is a DVR.

Fixed a real number $\lambda > 0$, we have the induced length function

$$L_w : Mod(S) \to \mathbb{R}^*$$

defined by setting, for every ideal J of S:

$$L_w(S/J) = \lambda \cdot inf_{x \in J} w(x).$$

Note that $L_w(S) = \infty$, and, if S is not a DVR, $0 < L_w(S/J) < \infty$ if and only if $0 < J < PR_P/P'$.

From the length function $L_w : Mod(S) \to \mathbb{R}^*$ we derive a length function $L : Mod(R) \to \mathbb{R}^*$ as follows: given an R-module M, set

$$L(M) = \begin{cases} L_w(M_{PR_P/P'}) & \text{if } P'M = 0 \\ \infty & \text{otherwise} \end{cases}$$

Length functions of this type are the significant ones, since they are related to the valuation of the domain R, and are the only non-discrete length functions provided $S = R_P/P'$ is a non-discrete archimedean domain.

If $P' = 0$, then P is the maximal ideal of R and $w = v$ is the valuation of the archimedean valuation domain R. If furthermore $\lambda = 1$, L is denoted by L_v and is called the *valuation length*. In this case, for every ideal $I \leq R$ we have:

$$L_v(R/I) = inf_{a \in I} v(a).$$

L_v is non-discrete if and only if P is not principal, equivalently, if $P = P^2$.

To sum up, we have seen that if L is a non-trivial length function on $Mod(R)$ and $L(R) < \infty$, then L is essentially the rank function, while every non-discrete length function L satisfying if $L(R) = \infty$ is essentially a valuation length on a non-discrete archimedean factor of a suitable localization of R at a prime ideal. The composition length is of rank-type if the maximal ideal is idempotent, of valuation type otherwise.

As an immediate application of the above results, we conclude this section with the description of the length functions over two simple types of valuation domains, namely, the archimedean ones (i.e., 1-dimensional), and the strongly discrete ones, i.e., without non-zero idempotents prime ideals (see [10, II.8.3] for their characterization).

Corollary 3.2. *Let R be valuation domain. Then*

(1) *if R is 1-dimensional, then the only non-trivial length functions on R are of $0/\infty$-type or, up to a positive real multiple: the composition length and the rank function if R is discrete; the valuation length and the rank function if R is non-discrete;*

(2) *if R is strongly discrete, then the only non-trivial length functions on R are of $0/\infty$-type (I) and of discrete valuation-type.*

4 Properties of Prüfer Domains and Their Meet-Spectrum

We fix the notation and start with some basic results holding for an arbitrary commutative integral domain R. The (prime) spectrum of R is denoted by $Spec(R)$, and the maximal spectrum by $Max(R)$.

If I is an ideal of R, we use the standard notation $\Omega(I)$ to denote the set of the maximal ideals containing I. If S is multiplicatively closed set of R, let $I_{(S)} = I_S \cap R$ be the contraction of the localization of I at S; if S is the complement of a prime ideal P, then we simply write $I_{(P)} = I_P \cap R$.

If Ω is a family of maximal ideals of R, the localization of R at the multiplicatively closed set

$$\bigcap_{P\in\Omega}(R \setminus P) = R \setminus \bigcup_{P\in\Omega} P$$

is denoted by R_Ω; the corresponding localization of an ideal I is denoted by I_Ω and we set $I_{(\Omega)} = I_\Omega \cap R$. If Ω consists of a single maximal ideal P, then we write $I_{(\Omega)} = I_{(P)}$. If $\Omega = \Omega(I)$, then clearly $I_{(\Omega)} = I$.

Lemma 4.1. *Let R be an integral domain and I be a non-zero ideal of R. If $\Omega = \Omega(I)$, then*

$$R/I \cong R_\Omega/I_\Omega.$$

Proof. The composition map

$$R \to R_\Omega \to R_\Omega/I_\Omega$$

has kernel $I_{(\Omega)} = I$, thus R/I embeds into R_Ω/I_Ω. In order to show that this embedding is also epic, we must prove that $R_\Omega = I_\Omega + R$. First we show that $R = I + sR$ for every $s \in \bigcap_{P\in\Omega}(R \setminus P)$. Assume, by way of contradiction, that $I + sR \leq P'$ for some maximal ideal P'. Then $P' \in \Omega$ since it contains I; on the other hand, $P' \notin \Omega$ since it contains s, absurd. As a consequence we have that, if $r/s \in R_\Omega$, since $r = a + sb$ for some $a \in I, b \in R$, then $r/s = a/s + b \in I_\Omega + R$, as desired. □

Lemma 4.2. *Let R be an integral domain, Ω a finite family of maximal ideals of R and I a non-zero ideal of R. The maximal ideals of R_Ω are exactly the extensions of the maximal ideals in Ω. Furthermore, $I_\Omega = \bigcap_{P\in\Omega} I_P$ and $R_\Omega = \bigcap_{P\in\Omega} R_P$.*

Proof. An ideal $P \in \Omega$ clearly survives as a maximal ideal in R_Ω. Conversely, if Q is a maximal ideal in R_Ω, then its contraction $P' = Q \cap R$ is a prime ideal contained in $\bigcup_{P\in\Omega} P$. By prime avoidance, P' is contained in some $P \in \Omega$ and equality must hold, since, as we already noted, P extends to a maximal ideal of R_Ω.

Since every $P \in \Omega$ survives in R_Ω, clearly $I_\Omega \leq \bigcap_{P\in\Omega} I_P$. Conversely, let $x \in \bigcap_{P\in\Omega} I_P$. Then, for each $P \in \Omega$, $(I :_R xR)$ is not contained in P; therefore, since the maximal ideals of R_Ω are the extensions of the ideals in Ω, we obtain that $(I :_R xR)_\Omega = R_\Omega$. Moreover, $(I :_R xR)_\Omega = (I_\Omega :_{R_\Omega} xR_\Omega)$, whence $x \in I_\Omega$. □

We restrict now our consideration to Prüfer domains, which may be defined in several different ways; for our purposes, the right definition is that each localization R_P at a maximal (or non-zero prime) ideal P is a valuation domain. For general notion on Prüfer domains and their modules we refer to the monographs [7, 10]. Since there is an ordered bijection between the prime ideals contained in a fixed prime ideal P of R and the prime ideals of the localization R_P, which form a totally ordered set, $Spec(R)$ is a rooted tree (the ideal 0 is its root) with the order given by the inclusion, according to the following:

Definition 4.3. A tree is a partially ordered set T such that, for every element $t \in T$, the subset $\{t' \in T \mid t' \leq t\}$ is totally ordered.

Remark 4.4. Very often the definition of tree requires the stronger assumption that, for every element $t \in T$, the set $\{t' \in T \mid t' \leq t\}$ is well-ordered. The results by Lewis [15] on the characterization of partially ordered sets as spectra of Prüfer (or Bézout) domains uses Definition 4.3 above.

As usual, we denote by $J(R)$ the Jacobson radical of the integral domain R, i.e., the intersection of its maximal ideals. The next proposition provides an indispensable tool for comparing the length of cyclic modules and of certain localizations of them.

Proposition 4.5. *Let R be a Prüfer domain, P a non-zero prime ideal contained in $J(R)$, and L a length function over R. The following facts hold:*

(1) *every ideal I of R is comparable with P;*

(2) *if $L(R/P) < \infty$, then L is localized at P;*

(3) *if S is a multiplicatively closed set disjoint from P, and $I \leq P$, then $L(R/I) = L(R_S/I_S) = L(R_P/I_P)$.*

Proof. (1) Suppose the ideal I is not contained in P. Then, for each maximal ideal Q, I_Q is not contained in $P_Q \neq R_Q$, and since R_Q is a valuation domain, P_Q is strictly contained in I_Q, so that $I = \bigcap_{Q \in Max(R)} I_Q$ contains $P = \bigcap_{Q \in Max(R)} P_Q$.

(2) If $x \in R \setminus P$, then point (1) implies that $P < xR$, so that $L(R/xR) = 0$, by Proposition 2.3. Then $x \in S_L$, whence S_L contains $R \setminus P$, showing that L is localized at P, by Proposition 2.6.

(3) If L is localized at P, then it is localized at S as well, by Proposition 2.6, and the result is true for every R-module. On the other hand, if L is not localized at P, then point (2) shows that $\infty = L(R/P) \leq L(R/I)$, and the conclusion follows from the existence of canonical inclusions $R/P \leq R_S/P_S \leq R_P/P_P$, and the fact that $L(R/I) \geq L(R/P)$ for every $I \leq P$. □

Given two different non-zero prime ideals P and P' of R, their intersection $P \cap P'$ is never a prime ideal and it contains the product $P \cdot P'$; the prime ideals contained in $P \cap P'$ form a chain, hence their union is the biggest prime ideal contained in $P \cap P'$, which is denoted by $P \wedge P'$. The same holds for the intersection of finitely many prime ideals. With this operation $Spec(R)$ is a meet-semilattice.

The next lemma shows that the localization of R at $P \wedge P'$ coincides with the double localization with respect to P and P'.

Lemma 4.6. *Let $P \neq P'$ be two maximal ideals of the Prüfer domain R. Then $(R_P)_{P'} = (R_{P'})_P = R_{P \wedge P'}$.*

Proof. $(R_P)_{P'} = R_S$, where S is the multiplicatively closed set generated by $(R \setminus P) \cup (R \setminus P') = R \setminus (P \cap P')$. Since $P \wedge P' \leq P \cap P'$, $R \setminus (P \cap P')$ contains S, so $P \wedge P'$ survives in R_S. Moreover, every prime ideal not contained in $P \wedge P'$ cannot be contained in $P \cap P'$, hence it must intersect $R \setminus (P \cap P')$, so that $R_S = \bigcap\{R_J \,:\, J \in Spec(R), JR_S < R_S\} = R_{P \wedge P'}$. □

From now on, R will always denote a Prüfer domain of finite character, i.e., every non-zero element (or ideal) is contained only in finitely many maximal ideals. If R is such a domain and the non-zero prime ideal P is contained in the maximal ideals $M_1, \cdots, M_n$, let

$$\bar{P} = M_1 \wedge \cdots \wedge M_n$$

which is a prime ideal containing P.

Definition 4.7. Let R be a Prüfer domain of finite character. The meet-spectrum of R, denoted by $\wedge Max(R)$, is the subset of $Spec(R)$:

$$\wedge Max(R) = \{P \in Spec(R) \mid 0 \neq P = \bar{P}\}.$$

In other words, the meet-spectrum of R contains those non-zero prime ideals which can be written as the meet of finitely many maximal ideals. Obviously the following inclusions hold:

$$Max(R) \setminus \{(0)\} \subseteq \wedge Max(R) \subseteq Spec(R).$$

The equality $Max(R) = Spec(R)$ occurs if and only if R is 1-dimensional; furthermore, the equality $Max(R) \setminus \{(0)\} = \wedge Max(R)$ occurs if and only if every non-zero prime ideal is contained in a unique maximal ideal, hence, under the hypothesis that R has finite character, if R is h-local.

For Prüfer domains of finite character we have at disposal the following result, due to Brandal [5], on the structure of torsion cyclic modules (see also [10, V.1.2]).

Theorem 4.8. (Brandal [5]) *Let R be a Prüfer domain of finite character and I a non-zero ideal of R. Then:*

(1) *the cyclic module R/I is a direct sum of indecomposable cyclic modules;*

(2) *the cyclic module R/I is indecomposable if and only if, for any partition $\{\Omega_1, \Omega_2\}$ of $\Omega(I)$, there exists $M_i \in \Omega_i$ $(i = 1, 2)$ such that $I \leq M_1 \wedge M_2$.*

We will improve this result using the meet-spectrum of R in the next Theorem 4.10, for which we need the following

Lemma 4.9. *Let I be a non-zero ideal of the Prüfer domain R. If $P \in \Omega(I)$, then $\sqrt{I_{(P)}}$ is the unique minimal prime ideal containing $I_{(P)}$, and $\sqrt{I_{(P)}} \leq P$.*

Proof. Since R_P is a valuation domain, I_P has exactly one minimal prime ideal in R_P, which coincides with $\sqrt{I_P}$. Now we have:

$$\sqrt{I_{(P)}} = (\sqrt{I})_{(P)} = (\sqrt{I})_P \cap R = \sqrt{I_P} \cap R$$

so that $\sqrt{I_{(P)}}$ is a prime ideal contained in P, hence it is the only minimal prime ideal of $I_{(P)}$. □

Given a non-zero ideal $I \leq R$, the meet-spectrum $\wedge Max(R)$ induces a partition on the finite set $\Omega(I)$ as follows. Consider the finite set of the prime ideals of $\wedge Max(R)$ which contain I and are minimal with respect to this containment, denoted by $\wedge Min(I)$. Then clearly $\{\Omega(P) \mid P \in \wedge Min(I)\}$ is a partition of $\Omega(I)$, called the *partition induced by the meet-spectrum*. The following result shows that this partition gives rise to an indecomposable direct decomposition of R/I; notice that item (2) improves Lemma 4.1.

Theorem 4.10. *Let R be a Prüfer domain of finite character and I a non-zero ideal of R. If $\{\Omega_1, \cdots, \Omega_m\}$ is the partition of $\Omega(I)$ induced by the meet-spectrum $\wedge Max(R)$, then:*

(1) $R/I \cong \bigoplus_{1 \leq i \leq m} R/I_{(\Omega_i)}$ *is a direct decomposition with indecomposable direct summands, and*

(2) *for every $i \leq m$ we have the isomorphism $R/I_{(\Omega_i)} \cong R_{\Omega_i}/I_{\Omega_i}$.*

Proof. (1) Consider the canonical map $R \to \bigoplus_{1 \leq i \leq m} R/I_{(\Omega_i)}$, whose kernel is $I = \bigcap_{1 \leq i \leq m} I_{(\Omega_i)}$. This map provides an embedding $\epsilon : R/I \to \bigoplus_{1 \leq i \leq m} R/I_{(\Omega_i)}$. For every index i, the equality $\Omega(\bigcap_{j \neq i} I_{(\Omega_j)}) = \bigcup_{j \neq i} I_{(\Omega_j)}$ holds, which gives $I_{(\Omega_i)} + \bigcap_{j \neq i} I_{(\Omega_j)} = R$. From this equality the surjectivity of the map ϵ follows. Finally, each summand $R/I_{(\Omega_i)}$ is indecomposable since it satisfies condition (2) of Theorem 4.8: if $\Omega_i = \Omega(P)$ with $P \in \wedge Min(I)$ and $P \leq M_1, M_2 \in \Omega(P)$, then $I \leq P \leq M_1 \wedge M_2$.

(2) The proof is similar to that of Lemma 4.1. Let us set $\Omega_i = \Omega$. The composition map $R \to R_\Omega \to R_\Omega/I_\Omega$ has kernel $I_{(\Omega)}$, thus $R/I_{(\Omega)}$ embeds into R_Ω/I_Ω. In order to show that this embedding is also epic, we must prove that $R_\Omega = I_\Omega + R$. First we show that $R = I_{(\Omega)} + sR$ for every $s \in \bigcap_{P \in \Omega}(R \setminus P)$. Assume, by way of contradiction, that $I_{(\Omega)} + sR \leq P'$ for some maximal ideal P'. Then $P' \notin \Omega$ since it contains s, but P' must contain $I_{(P)}$ for some $P \in \Omega$. Thus both P and P' must contain the single minimal prime of $I_{(P)}$. But then $I \subseteq P \wedge P'$, which is impossible by the current hypothesis. The conclusion is now as in the proof of Lemma 4.1. □

Fixed a prime ideal $P \in \wedge Max(R)$, the set

$$T(P) = \{P' \in \wedge Max(R) \mid P \subseteq P'\}$$

is a finite tree with root P, in which the meet of two elements is the same as the meet in $Spec(R)$.

Definition 4.11. Let R be a Prüfer domain of finite character and let $P \in \wedge Max(R)$. Then a child of P is a prime ideal $P' \in T(P)$ different from P which is adjacent to P.

Clearly, every element of the meet-spectrum is either a maximal ideal (with no children), or it is the meet of at least two maximal ideals, hence it has at least two children.

If the ring R is semilocal, so in particular of finite character, then the meet-spectrum of R enjoys nice properties.

Lemma 4.12. *Let R be a semilocal Prüfer domain, then:*

(1) *$\wedge Max(R)$ is a finite set;*

(2) *$\wedge Max(R)$ is a disjoint union of a finite number of rooted trees $T(P)$, ranging P over the set of the minimal elements of $\wedge Max(R)$;*

(3) *$\wedge Max(R)$ is a rooted tree if and only if the largest prime ideal contained in the Jacobson radical is non-zero.*

Proof. (1) and (2) are obvious.

(3) The largest prime ideal contained in the Jacobson radical is the meet of all maximal ideals, so it is the unique minimal element of $\wedge Max(R)$ exactly if it is non-zero. □

We prove now a technical result needed in the proof of the main Theorem 5.7, concerning the meet-spectrum of a semilocal Prüfer domain R in case it is rooted, relating the number of children of the non-maximal elements to the number of the maximal ideals.

For this purpose, we introduce the following notation: let k_P be the number of children of a prime ideal $P \in \wedge Max(R)$, and denote by $\mathcal{C}$ the complement of $Max(R)$ in $\wedge Max(R)$:

$$\mathcal{C} = \wedge Max(R) \setminus Max(R).$$

We exclude the trivial case of R local, so that $\mathcal{C} \neq \emptyset$.

Proposition 4.13. *Let R be a semilocal non-local Prüfer domain with maximal ideals $M_1, \cdots, M_n$ and suppose that the meet-spectrum $\wedge Max(R)$ is a rooted tree. Then the following formula holds:*

$$\sum_{Q \in \mathcal{C}} (k_Q - 1) = n - 1.$$

Proof. Since every element of $\wedge Max(R)$ is a child, except the root, the following equality holds: $\sum_{Q \in \mathcal{C}} k_Q = |\wedge Max(R)| - 1$. Therefore we get:

$$\sum_{Q \in \mathcal{C}} k_Q = |\mathcal{C}| + |Max(R)| - 1 \Rightarrow \sum_{Q \in \mathcal{C}} k_Q - |\mathcal{C}| = |Max(R)| - 1$$

from which the equality $\sum_{Q \in \mathcal{C}} (k_Q - 1) = n - 1$ immediately follows. □

5 Length Functions over Semilocal Prüfer Domains

A semilocal Prüfer domain is trivially of finite character; it is a classical result due to Hinohara in 1962 that it is also a Bézout domain, that is, finitely generated ideals are principal (see [10, III, 5.1]).

In this and in the next section, in order to simplify the notation, we set $\mathcal{M} = Max(R)$ and $\wedge\mathcal{M} = \wedge Max(R)$. Recall that a *local length function for* R is a length function on $Mod(R_P)$ for some $P \in \mathcal{M}$. Since with the symbol L_P we denote the localization at P of a length function $L : Mod(R) \to \mathbb{R}^*$, we will indicate an arbitrary local length function for R with the symbol $L^{(P)}$.

A close inspection to the proof of Theorem 1.1 shows that it is based on the following facts, holding for every ideal $I \neq 0$ of a 1-dimensional Prüfer domain of finite character R: if $\Omega(I) = \{P_1, \cdots, P_n\}$, then

$$R/I \cong R/I_{(P_1)} \oplus \cdots \oplus R/I_{(P_n)}$$

is an indecomposable direct decomposition; and, if $P \in \Omega(I)$, then $R/I_{(P)} \cong R_P/I_P$.

These two facts imply that

$$L(R/I) = \sum_{1 \le i \le n} L(R_{P_i}/I_{P_i}) = \sum_{1 \le i \le n} L_{P_i}(R_{P_i}/I_{P_i});$$

furthermore, if $P \in \mathcal{M}$ and J is a non-zero proper ideal of R_P, then $J = (J \cap R)_P$ (see [9, p. 6, (D)] and $J \cap R$ is $\sqrt{J \cap R}$-primary, therefore $R_P/J \cong R/(J \cap R)$; consequently $L^{(P)}(R_P/J) = L(R/(J \cap R))$, so that $L^{(P)} = L_P$ for every $P \in \mathcal{M}$.

When we try to extend Theorem 1.1 from 1-dimensional to arbitrary Prüfer domains of finite character, the more complex structure of the meet-spectrum $\wedge\mathcal{M}$ creates problems in the decomposition of L as sum of its localizations at maximal ideals and also in the uniqueness of its representation, due to the fact that the action of local length functions can overlap, as the next example shows.

Example 5.1. Let $V = \mathbb{Q}[[X]]$ be the 1-dimensional valuation domain of the rational power series. If $P = X\mathbb{Q}[[X]]$ denotes its maximal ideal, then the residue field V/P is isomorphic to $\mathbb{Q}$. Fixed a prime number $p \in \mathbb{N}$, let $W_p = \mathbb{Z}_p + X\mathbb{Q}[[X]]$ be the subring of V consisting of the power series with constant term in $\mathbb{Z}_p$, the localization at the prime ideal $p\mathbb{Z}$ of the ring of the integers $\mathbb{Z}$. W_p is the pullback of the canonical surjection $V \to \mathbb{Q}$ and the inclusion $\mathbb{Z}_p \to \mathbb{Q}$. Clearly W_p is a 2-dimensional strongly discrete valuation domain, with the same quotient field as V. Its non-zero prime ideals are P and $pW_p = p\mathbb{Z}_p + P$; moreover, $(W_p)_P = V$.

Let now q be a prime number different from p and consider the valuation domain W_q obtained in a similar way as before. The ring $R = W_p \cap W_q$ is a Bézout domain with exactly two maximal ideals (see [10, III, 5.1]). Its spectrum is given by the two maximal ideals pR and qR, the prime ideal $pR \wedge qR = P$, and the null ideal. In particular, $R_{pR} = W_p$, $R_{qR} = W_q$ and $R_P = (W_p)_P = (W_q)_P = V$.

In order to obtain a length function over $Mod(R)$, in view of Example 2.11 one can simply add together two local length functions $L^{(pR)}$ and $L^{(qR)}$ chosen over the two localizations at maximal ideals $R_{pR} = W_p$ and $R_{qR} = W_q$, respectively. Looking at Section 3 and taking care that we don't have non-zero idempotent prime ideals, we have, up to positive multiples, only three cases for a non-trivial length function $L^{(pR)} : W_p \to \mathbb{R}^*$ such that $L^{(pR)}(W_p) = \infty$.

Case 1. $L^{(pR)}$ is of $0/\infty$-type (I), so that $L^{(pR)}(W_p/I) = 0$ or ∞, according as $P < I$ or $I \leq P$;

Case 2. $L^{(pR)}$ is of valuation-type associated with the pair of prime ideals $(P < pW_p)$, so that $L^{(pR)}(W_p/I) = k$ if $I = p^k W_p$ for some $k \geq 0$, and $L^{(pR)}(W_p/I) = \infty$ otherwise;

Case 3. $L^{(pR)}$ is of valuation-type associated with the pair of prime ideals $(0 < P)$, so that $L^{(pR)}(W_p/I) = 0$ if $P \leq I$, and $L^{(pR)}(W_p/I) = k$ if $I = X^k\mathbb{Q}[[X]]$.

We have a similar situation for W_q. Now if we choose $L^{(pR)}$ as in Case 2 above, and we add it to a local length function $L^{(qR)}$ as in Case 1 or Case 3, we always have the same length function $L = L^{(pR)} + L^{(qR)}$; in fact, in both cases we have $L(R/I) = k + 0 = k$, for a certain $k \in \mathbb{N}$, if $P < I$, and $L(R/I) = \infty$ otherwise. So uniqueness is lost.

Let now $L : Mod(R) \to \mathbb{R}^*$ be the length function which is the contraction of the valuation length $H : R_P = V \to \mathbb{R}*$ (see Example 2.13). Then the localizations L_{pR} and L_{qR} both coincide with H, so the equality $L = L_{pR} + L_{qR}$ cannot hold. In fact, $L(R/P) = H(R_P/P) = 1$, while we get

$$L_{pR}(R/P) + L_{qR}(R/P) = L(R_{pR}/P_{pR}) + L(R_{qR}/P_{qR}) = H(R_P/P) + H(R_P/P) = 1 + 1 = 2.$$

So also the decomposition of L as sum of its localizations at maximal ideals is lost.

If we want to express a length function L by means of a sum $\sum_{P\in\mathcal{M}} L^{(P)}$ of local length functions such that each $L^{(P)}$ coincides with the localization of L at P, we must ask that the given family satisfies a certain compatibility condition, according with the following

Definition 5.2. A family of local length functions $\{L^{(P)}\}_{P\in\mathcal{M}}$ is compatible if $L^{(P)}(R_P) = \infty$ for all $P \in \mathcal{M}$ and, for each pair $P, P' \in \mathcal{M}$, $(L^{(P)})_{P\wedge P'} = (L^{(P')})_{P\wedge P'}$.

In fact, the family of the localizations L_P ($P \in \mathcal{M}$) of a length function L such that $L(R) = \infty$ is compatible, since for each pair of maximal ideals P and P', $(L_P)_{P\wedge P'} = (L_{P'})_{P\wedge P'}$, both being equal to $L_{P\wedge P'}$. Note that for a 1-dimensional Prüfer domain the compatibility condition is trivially satisfied.

Another problem is made evident by the final part of Example 5.1, which indicates that $\sum_{P\in\mathcal{M}} L_P$ is strictly larger in general than L, so we must subtract something from it. This problem arises since the meet-spectrum $\wedge\mathcal{M}$ is larger than the maximal spectrum $\mathcal{M}$. Recall that we denoted by $\mathcal{C}$ the complement of $\mathcal{M}$ in $\wedge\mathcal{M}$, that is, $\mathcal{C} = \wedge\mathcal{M} \setminus \mathcal{M}$. If $\{L^{(P)}\}_{P\in\mathcal{M}}$ is a compatible family of local length functions, and if $Q \in \mathcal{C}$, we denote by $L^{(Q)}$ the common localization at $Mod(R_Q)$ of all $L^{(P)}$ such that $P \in \Omega(Q)$. In order to state our main theorem, we introduce the following

Definition 5.3. Given a compatible family $\mathcal{L} = \{L^{(P)}\}_{P \in \mathcal{M}}$ of local length functions, we call $L'_{\mathcal{L}}$ and $L''_{\mathcal{L}}$ the maximal sum and the complement sum of $\mathcal{L}$, respectively, where

$$L'_{\mathcal{L}} = \sum_{P \in \mathcal{M}} L^{(P)} \quad , \quad L''_{\mathcal{L}} = \sum_{Q \in C} (k_Q - 1) L^{(Q)}.$$

Using this definition and the notation in it, we have the following

Lemma 5.4. *If $\mathcal{L} = \{L^{(P)}\}_{P \in \mathcal{M}}$ is a compatible family of local length functions, then its maximal sum $L'_{\mathcal{L}}$ and complement sum $L''_{\mathcal{L}}$ satisfy the inequality $L'_{\mathcal{L}} \geq L''_{\mathcal{L}}$.*

Proof. It is enough to prove that, for every ideal I of R, $L'_{\mathcal{L}}(R/I) \geq L''_{\mathcal{L}}(R/I)$. If $I = 0$, then all the involved length functions assign value ∞ to R, so the above inequality is verified. Assume that $I \neq 0$ and let $\{Q_1, \cdots, Q_m\} = \wedge Min(I)$. The sets $\Omega_i = \Omega(Q_i)$ form a partition of $\Omega(I)$. Denote by $\mathcal{Q}_i$ the subset of $\wedge\mathcal{M}$ consisting of the elements containing Q_i; then $\mathcal{Q}_i$ is a rooted tree with root Q_i, which is order-isomorphic to the meet-spectrum of R_{Ω_i}, and the $\mathcal{Q}_i$ are disjoint sets. Denoting by n_i the cardinality of Ω_i, we have for each i:

$$\sum_{P \in \Omega_i} L^{(P)}(R_P/I_P) = n_i \alpha_i$$

where $\alpha_i = L^{(Q_i)}(R_{Q_i}/I_{Q_i})$, since $L^{(P)}(R_P/I_P) = L^{(Q_i)}(R_{Q_i}/I_{Q_i})$ for all $P \in \mathcal{Q}_i$. Applying Proposition 4.13 to R_{Ω_i} we get:

$$\sum_{Q \in \mathcal{Q}_i \setminus \Omega_i} (k_Q - 1) L^{(Q)}(R_Q/I_Q) = \sum_{Q \in \mathcal{Q}_i \setminus \Omega_i} (k_Q - 1)\alpha_i = (n_i - 1)\alpha_i.$$

We can now conclude that:

$$L'_{\mathcal{L}}(R/I) = \sum_{P \in \mathcal{M}} L^{(P)}(R_P/I_P) = \sum_{P \in \Omega(I)} L^{(P)}(R_P/I_P) = \sum_i n_i \alpha_i$$

$$\geq \sum_i (n_i - 1)\alpha_i = \sum_{Q \in C} (k_Q - 1) L^{(Q)}(R_Q/I_Q) = L''_{\mathcal{L}}(R/I).$$

□

We can now prove the following result.

Proposition 5.5. *Let R be a semilocal Prüfer domain and $\mathcal{L} = \{L^{(P)}\}_{P \in \mathcal{M}}$ a compatible family of local length functions over $Mod(R)$. Let $L'_{\mathcal{L}}$, $L''_{\mathcal{L}}$ be the maximal sum and the complement sum of $\mathcal{L}$, respectively. Then the difference $L'_{\mathcal{L}} - L''_{\mathcal{L}}$ is a length function.*

Proof. In view of Proposition 2.10, it is enough to prove that, if A is a submodule of a finitely generated R-module satisfying $L'_{\mathcal{L}}(A) = \infty$, then $(L'_{\mathcal{L}} - L''_{\mathcal{L}})(A) = \infty$. If there exists a finitely generated submodule F of A such that $L'_{\mathcal{L}}(F) = \infty$, then $(L'_{\mathcal{L}} - L''_{\mathcal{L}})(F) = \infty$, hence $(L'_{\mathcal{L}} - L''_{\mathcal{L}})(A) = \infty$. Suppose on the contrary that all the finitely generated submodules $F \leq A$ satisfy $L'_{\mathcal{L}}(F) < \infty$. These finitely generated modules F are unions of finite chains of submodules with cyclic sections R/I which satisfy also $L'_{\mathcal{L}}(R/I) < \infty$, since $L'_{\mathcal{L}}(F)$ is the sum of the lengths of these sections. Using the notation in the proof of Lemma 5.4 and denoting by N the number of the maximal ideals of R, we have:

$$L'_{\mathcal{L}}(R/I) = \sum_i n_i \alpha_i \leq N \sum_i \alpha_i$$

hence

$$L'_{\mathcal{L}}(R/I) - L''_{\mathcal{L}}(R/I) = \sum_i \alpha_i \geq N^{-1} L'_{\mathcal{L}}(R/I).$$

Therefore $L'_{\mathcal{L}}(F) - L''_{\mathcal{L}}(F)) \geq N^{-1} L'_{\mathcal{L}}(F)$ for all finitely generated modules F. Since $L'_{\mathcal{L}}(F)$ is unbounded, so is $L'_{\mathcal{L}}(F) - L''_{\mathcal{L}}(F))$. □

In view of Proposition 5.5, we can give the following definition.

Definition 5.6. Let $\mathcal{L} = \{L^{(P)}\}_{P \in \mathcal{M}}$ be a compatible family of local length functions for R. The balanced sum of this family is the length function defined as the difference $L_{\mathcal{L}} = L'_{\mathcal{L}} - L''_{\mathcal{L}}$.

We can now state our main theorem, which says that the balanced sums of compatible families of length functions over a semilocal Prüfer domain R represent all the length functions L such that $L(R) = \infty$, and that this representation is unique. Keeping in mind that the involved local length functions are defined over valuation domains, hence covered by the classification illustrated in Section 3, we obtain an unambiguous classification.

Theorem 5.7. *Let R be a semilocal Prüfer domain. Given a length function L over R such that $L(R) = \infty$, there exists a unique compatible family $\{L^{(P)}\}_{P \in \mathcal{M}}$ of local length functions for R such that L equals their balanced sum, and $L^{(P)}$ coincides with the localization L_P for each $P \in \mathcal{M}$.*

Proof. First we prove that L coincides with the balanced sum of its localizations at maximal ideals. So let $\mathcal{L} = \{L_P\}_{P \in \mathcal{M}}$ and $L' = L'_{\mathcal{L}}$, $L'' = L''_{\mathcal{L}}$. Let I be a non-zero ideal of R, $\Omega = \Omega(I)$, $\wedge Min(I) = \{Q_1, \cdots, Q_m\}$ and $\Omega_i = \Omega(I_i)$ for $i = 1, \cdots, m$. By Theorem 4.10 and Lemma 4.1 we get:

$$R/I \cong \bigoplus_{1 \leq i \leq m} R_{\Omega_i}/I_{\Omega_i}.$$

First suppose that $L(R/I)=\infty$. Then, for some index k, $\infty = L(R_{\Omega_k}/I_{\Omega_k}) = L(R_{Q_k}/I_{Q_k}) = \alpha_k$, where the last equality follows from Proposition 4.5 applied to $Mod(R_{\Omega_k})$. Using the notation in the proof of Lemma 5.4, we get:

$$L'(R/I) = \sum_{1\le i\le m} n_i\alpha_i \ge \alpha_k = \infty.$$

Therefore, by definition, $(L'-L'')(R/I)=\infty$. Suppose now that $L(R/I)<\infty$. Then $\alpha_i<\infty$ for all i and, by additivity and by the proof of Lemma 5.4, we get:

$$L(R/I) = \sum_{1\le i\le m} L(R_{\Omega_i}/I_{\Omega_i}) = \sum_{1\le i\le m} \alpha_i$$

$$= \sum_{1\le i\le m} n_i\alpha_i - \sum_{1\le i\le m}(n_i-1)\alpha_i = (L'-L'')(R/I).$$

The case $I=0$ being trivial, since all the involved length functions assume infinite value on the ground ring, we have proved that L coincides on cyclic modules with the balanced sum $L'_{\mathcal{L}} - L''_{\mathcal{L}}$ of its localizations at maximal ideals, so they are equal.

In order to prove uniqueness, given a compatible family $\{\mathcal{L}\} = \{L^{(P)}\}_{P\in\mathcal{M}}$ of local length functions for R such that $L = L'_{\mathcal{L}} - L''_{\mathcal{L}}$, we must prove that, for every $P\in\mathcal{M}, L^{(P)} = L_P$. Let $0\ne I$ be an ideal of R and let $P\in\Omega(I)$. In order to simplify notation, let $W = R_P/I_P$; we may assume, without loss of generality, that $I = I_{(P)}$, so that $\sqrt{I}$ is the only minimal prime ideal containing it. Consider the prime ideal:

$$Q = \bigwedge_{P'\in\Omega(I)} P' = \bigwedge_{P'\in\Omega(\sqrt{I})} P'$$

which is the smallest element of $\mathcal{C}$ containing I. Therefore, if $I\le N\in\mathcal{C}$ and $I\le P'\in\mathcal{M}$, we have:

$$L^{(P)}(W) = L^{(P')}(W_{P'}) = L^{(P\wedge P')}(W_{P\wedge P'}) = L^{(N)}(W_N) = L^{(Q)}(W_Q) = \alpha.$$

Then we get:

$$L'_{\mathcal{L}}(W) = \sum_{P'\in\mathcal{M}} L^{(P')}(W_{P'}) = \sum_{P'\in\Omega(I)} L^{(P')}(W_{P'}) = |\Omega(I)|\cdot\alpha.$$

On the other hand, we get:

$$L''_{\mathcal{L}}(W) = \sum_{N\in\mathcal{C}}(k_N-1)L^{(N)}(W_N) = \sum_{N\in\mathcal{C}}(k_N-1)\alpha = (|\Omega(I)|-1)\cdot\alpha$$

where the last equality follows by Proposition 4.13 applied to the domain $R_{\Omega(I)}$. Now, if $L_P(W) = \infty$, then $\infty = L_{\mathcal{L}}(W) = L'_{\mathcal{L}}(W) = |\Omega(I)| \cdot \alpha$, so $L^{(P)}(W) = \alpha = \infty$. On the other hand, if $L^{(P)}(W) = \alpha < \infty$, then $L_P(W) = L_{\mathcal{L}}(W) = L'_{\mathcal{L}}(W) - L''_{\mathcal{L}}(W) = \alpha = L^{(P)}(W)$.

In conclusion, since $L^{(P)}(R_P) = \infty = L_P(R_P)$, the two length functions L_P and $L^{(P)}$ agree on all cyclic R_P-modules, hence $L_P = L^{(P)}$ is uniquely determined by L. □

6 Length Functions over Prüfer Domains of Finite Character

In this section we extend Theorem 5.7 to Prüfer domains of finite character. This class of rings is of central importance among Prüfer domains; it includes Dedekind domains (the Noetherian case) and semilocal Prüfer domains. An example of Prüfer domain of finite character which is neither Dedekind nor semilocal can be found in [12], while an example of Prüfer (actually, Bézout) domain which fails to be of finite character is the ring of entire functions investigated by Helmer in [13]. For an extensive treatment of these domains and their modules we refer to [10].

In order to underline the importance of these domains, it is worth recalling that Bazzoni proved that their class semigroup is a Clifford semigroup, determining the idempotents, the constituent groups and the bonding homomorphisms (see [2–4, 8] and [10, Chapter III, Sections 2, 3]). Furthermore, in [14] it was proved, solving a conjecture posed by Bazzoni in [2], that a Prüfer domain with the property that every locally finitely generated ideal is finitely generated is, in fact, of finite character.

We will need the following

Lemma 6.1. *Let R be a Prüfer domain of finite character and $\{L^{(P)}\}_{P\in\mathcal{M}}$ a family of compatible local length functions for R. If $\Omega \subseteq \mathcal{M}$, then $\{L^{(P)}\}_{P\in\Omega}$ is a family of compatible local length functions for R_Ω.*

Proof. The proof follows immediately from the observation that every localization R_P at a maximal ideal $P \in \Omega$ coincides with $(R_\Omega)_P$. □

Theorem 5.7 can now be extended *verbatim* to Prüfer domains of finite character, but the proof of its extension requires some additional argument for the reduction to the semilocal case.

Theorem 6.2. *Let R be a Prüfer domain of finite character. Given a length function L over R such that $L(R) = \infty$, there exists a unique compatible family $\{L^{(P)}\}_{P\in\mathcal{M}}$ of local length functions for R such that L equals their balanced sum, and $L^{(P)}$ coincides with the localization L_P for each $P \in \mathcal{M}$.*

Proof. First we show that the balanced sum of a compatible family $\mathcal{L} = \{L^{(P)}\}_{P\in\mathcal{M}}$ of length functions over R is still a well defined length function. Write $L' = L'_{\mathcal{L}}$ and

$L'' = L''_{\mathcal{L}}$. Let I be a non-zero ideal of R and set $\Omega = \Omega(I)$. The elements $Q \in \wedge\mathcal{M}$ containing I are in finite number, an this subset of $\wedge\mathcal{M}$ is order-isomorphic to the meet-spectrum of the semilocal Prüfer domain R_Ω. Denote by E the cyclic module R/I. Then by Lemma 4.1 we have:

$$L'(E) = L'(E_\Omega) = \sum_{P \in \mathcal{M}} L^{(P)}(E_P) = \sum_{P \in \Omega} L^{(P)}(E_P)$$

and similarly

$$L''(E) = L''(E_\Omega) = \sum_{I \subseteq Q \in \mathcal{C}} (k_Q - 1) L^{(Q)}(E_Q).$$

Now, by Lemma 6.1, $\{L^{(P)}\}_{P \in \Omega}$ is a family of compatible length functions for R_Ω, hence we can apply Theorem 5.7 to this semilocal domain, ensuring that $L'(E) \geq L''(E)$. The analogue for $I = 0$ is trivial. We can now extend the proof of Proposition 5.5 to the present situation, taking care that finitely generated torsion R-modules B and their submodules are modules over a suitable localization of R which is semilocal. So we deduce that $L' - L''$ is a length function.

Now we deal with the existence part. We reduce to the semilocal case, by localizing at the finite set of maximal ideals $\Omega = \Omega(I)$. We have that $L(E) = L(E_\Omega) = L_\Omega(E_\Omega)$, and L_Ω is a length function over R_Ω, hence by Theorem 5.7 we have

$$L_\Omega = \sum_{P \in \Omega} (L_\Omega)_P - \sum_{I \subseteq Q \in \mathcal{C}} (k_Q - 1)(L_\Omega)_Q.$$

Clearly, the summands $(L_\Omega)_P$ and $(L_\Omega)_Q$ in the above equation coincide, respectively, with L_P and L_Q, so that:

$$L(E) = (\sum_{P \in \Omega} L_P - \sum_{I \subseteq Q \in \mathcal{C}} (k_Q - 1) L_Q)(E_\Omega)$$

$$= (\sum_{P \in \Omega} L_P - \sum_{I \subseteq Q \in \mathcal{C}} (k_Q - 1) L_Q)(E) = (\sum_{P \in \mathcal{M}} L_P - \sum_{Q \in \mathcal{C}} (k_Q - 1) L_Q)(E).$$

Since $\{L_P\}_{P \in \mathcal{M}}$ is a compatible family, and being the case $I = 0$ trivial, this proves that $L = L_{\{L_P\}}$.

The proof of the uniqueness is identical to the one given in Theorem 5.7 for the semilocal case. □

As an easy consequence of Theorem 6.2, we obtain the characterization of length functions over h-local Prüfer domains, which extends Theorem 1.1.

Corollary 6.3. *Let R be a h-local Prüfer domain and $L : Mod(R) \to \mathbb{R}^*$ a length function such that $L(R) = \infty$. Then there exists a unique family $\{L^{(P)}\}_{P \in \mathcal{M}}$ of local length functions for R such that $L = \sum_{P \in \mathcal{M}} L^{(P)}$, and $L^{(P)}$ coincides with the localization L_P of L for each $P \in \mathcal{M}$.*

Proof. It is enough to recall that for a h-local domain R, $\wedge\mathcal{M} = \mathcal{M}$, hence the balanced sum of $\{L^{(P)}\}_{P\in\mathcal{M}}$ coincides with the usual sum, and the compatibility condition for a family of local length functions is always trivially satisfied. □

As a natural conclusion of this paper we pose the following:

Problem 6.4. *Characterize length functions over arbitrary Prüfer domains.*

Acknowledgements We would like to thank the referee for the numerous remarks that improved the presentation of our paper.

References

1. Adler, R.L., Konheim, A.G., McAndrew, M.H.: Topological entropy. Trans. Am. Math. Soc. **114**, 309–319 (1965)
2. Bazzoni, S.: Class semigroups of Prüfer domains. J. Algebra. **184**, 613–631 (1996)
3. Bazzoni, S.: Idempotents of the class semigroup of a Prüfer domain of finite character, LNPAM, vol. 201, pp. 79-89. Marcel Dekker (1998)
4. Bazzoni, S.: Groups in the class semigroup of a Prüfer domain of finite character. Commun. Algebra **28**(11), 5157–5167 (2000)
5. Brandal, W.: On h-local integral domains. Trans. Am. Math. Soc. **206**, 201–212 (1975)
6. Dikranjan, D., Goldsmith, B., Salce, L., Zanardo, P.: Algebraic entropy of endomorphisms of abelian groups. Trans. Am. Math. Soc. **361**, 3401–3434 (2009)
7. Fontana, M., Huckaba, J.A., Papick, I.J.: Prüfer Domains. Pure Appl. Math. (1997) (Marcel Dekker)
8. Fuchs, L.: On the class semigroups of Prüfer domains. In: Abelian Groups and Modules. Birkhäuser, pp. 319–326 (1999)
9. Fuchs, L., Salce, L.: Modules over Valuation Domains. Marcel Dekker, New York (1985)
10. Fuchs, L., Salce, L.: Modules over Non-Noetherian Domains. Mathematical Surveys and Monographs. Providence, vol. 84 (2001) (Am. Math. Soc.)
11. Fusacchia, G.: Length Functions Over Prüfer Domains of Finite Character, unpublished notes
12. Heinzer, W.: Quotient overrings of integral domains. Mathematika **17**, 139–148 (1970)
13. Helmer, O.: Divisibility properties of integral functions. Duke Math. J. **6**, 351–000 (1940)
14. Holland, W.C., Martinez, J., McGovern, W.W., Tesemmad, M.: Bazzoni's conjecture. J. Algebra **320**(4), 1764–1768 (2008)
15. Lewis, W.J.: The spectrum of a ring as a partially ordered set. J. Algebra **25**, 419–434 (1973)
16. Northcott, D.G., Reufel, M.: A generalization of the concept of length. Q. J. Math. (Oxford) (2), **16**, 297–321 (1965)
17. Salce, L., Vámos, P., Virili, S.: Length functions, multiplicities and algebraic entropy. Forum Math. **25**(2), 255–282 (2013)
18. Salce, L., Zanardo, P.: A general notion of algebraic entropy and the rank-entropy. Forum Math. **21**(4), 579–599 (2009)
19. Spirito, D.: Decomposition and Classification of Length Functions, preprint
20. Vámos, P.: Length functions on modules, Ph.D. thesis, University of Sheffield (1968)
21. Vámos, P.: Additive functions and duality over Noetherian rings. Q. J. Math. (Oxford) (2) **19**, 43–55 (1968)
22. Virili, S.: Length functions and algebraic entropy over commutative Noetherian rings, Internal report, Universitá de Castellon (2012)
23. Virili, S.: Length Functions of Grothendieck Categories with Applications to Infinite Group Representations, preprint (2014). arxiv:1410.8306
24. Zanardo, P.: Multiplicative invariants and additive functions over valuation domains. J. Commut. Algebra. **3**(4), 561–587 (2011)

On Some Arithmetical Properties of Noetherian Domains

Florian Kainrath

Abstract We give a characterization of those noetherian domains, such that for all non zero $a \in R$ there are only (up to associates) finitely many irreducibles, that divide some power of a. In turns out, that a necessary condition for that, is that the integral closure of R is a root extension of R. We also give a description of noetherian domains with this property.

Keywords Quasi finitely · Generated monoid · Root extension · Associated prime

1 Introduction

Let R be a noetherian domain. For a non-zero non-unit $a \in R$ let $f(a) \in \mathbb{N} \cup \{\infty\}$ be the number of essentially different factorizations of a, and $N(a) \in \mathbb{N} \cup \{\infty\}$ the number (up to associates) of irreducibles dividing some power of a. The behaviour of $f(a^n)$ as $n \to \infty$ depends on the number $N(a)$ in the following way:

$$f(a^n) \quad \begin{cases} = An^s + O(n^{s-1}) & \text{if } N(a) < \infty \\ \gg n^r \quad \text{for all integers } r & \text{if } N(a) = \infty, \end{cases}$$

for some $A \in \mathbb{Q}_{>0}$ and some $s \in \mathbb{N}$ (see [6] Theorems 1 and 2). We will give a description of those R, such that $N(a) < \infty$ for all a. As we will see, a necessary condition for that, is that the integral closure $\bar{R}$ of R is a root extension of R, i.e. for all $x \in \bar{R}$ there exists some $m \in \mathbb{N}$, such that $x^m \in R$. We will also give a characterization of noetherian domains having this property. This answers a question, that remained open in the description of tamely inside factorial, noetherian domains, achieved in [1] (Corollary 6).

F. Kainrath (✉)
Institut für Mathematik, Karl-Franzens-Universität Graz,
Heinrichstraße 36, 8010 Graz, Austria
e-mail: florian.kainrath@uni-graz.at

A. Facchini et al. (eds.), *Advances in Rings, Modules and Factorizations*,
Springer Proceedings in Mathematics & Statistics 321,
https://doi.org/10.1007/978-3-030-43416-8_12

2 Quasi Finitely Generated Monoids

In this paper a monoid is commutative, cancellative monoid with unit 1. We let $Q(H)$ be the quotient group of a monoid H, and let $H^\times$ be its group of units. H is called reduced, if $H^\times = \{1\}$. The monoid $H_{\text{red}} = H/H^\times$ is reduced.

Let D be a monoid and $H \subset D$ a submonoid. H is called to be saturated in D, if for all $h_1, h_2 \in H$ we have $h_1 \mid_H h_2 \iff h_1 \mid_D h_2$, or equivalently if $Q(H) \cap D = H$.

For a subset A of H, let $[A]_H$ be the submonoid generated by A in H and let $[[A]]_H$ be the divisor-closed submonoid of H generated by A, i.e. $[[A]]_H$ consists of all $h \in H$ dividing some $a \in [A]_H$. H is called finitely generated, if $H = [A]_H$ for some finite subset $A \subset H$. H is called quasi finitely generated, if H_{red} is finitely generated. Note that this will be so, if and only if every element is a product of irreducibles and H has (up to associates) only finitely many irreducibles. We call H locally finitely generated, if for all $a \in H$ the monoid $[[a]]_H := [[\{a\}]]_H$ is quasi finitely generated.

An extension $H \subset D$ of monoids or rings is called a root extension, if for all $d \in D$, there is some $m \in \mathbb{N}^+$ such that $d^m \in H$. The monoid $\widetilde{H} = \{x \in Q(H) \mid x^m \in H \text{ for some } m \in \mathbb{N}^+\}$ is called the root closure of H. It is contained in the complete integral closure $\widehat{H}$, which consists of all $x \in Q(H)$, such that there exists some $c \in H$ with $cx^n \in H$ for all $n \in \mathbb{N}$. If $H = R\backslash\{0\}$ for some noetherian domain R, then $\widehat{H} = \bar{R}\backslash\{0\}$, where $\bar{R}$ is the integral closure of R.

Finally, if R is a domain and $a \in R\backslash\{0\}$, then let R_a be the ring of fractions $\{x/a^n \mid x \in R \text{ and } n \in \mathbb{N}\}$. Note that $[[a]]_R$ $(= [[a]]_{R\backslash\{0\}}) = R \cap R_a^\times$ and $Q([[a]]_R) = R_a^\times$.

Lemma 1. *Let H be a monoid. Then H is quasi finitely generated if and only if $\widehat{H}$ is quasi finitely generated, $\widetilde{H} = \widehat{H}$ and $\widehat{H}^\times/H^\times$ is finite.*

Proof. We may suppose that H is reduced. Suppose first that H is finitely generated. From Proposition 2.7.11 and Lemma 2.7.12 in [4] we get that $\widetilde{H} = \widehat{H}$ is finitely generated. It remains to show that $\widetilde{H}^\times$ is finite. Clearly it is finitely generated, since $\widetilde{H}$ is so. From the definition of $\widetilde{H}$ and $H \cap \widetilde{H}^\times = H^\times = \{1\}$ (Lemma 5.4 in [3]) we get that $\widetilde{H}^\times$ is a torsion group and hence is finite.

Conversely, assume that $\widehat{H} = \widetilde{H}$ is quasi finitely generated and that $\widetilde{H}^\times = \widehat{H}^\times$ is finite. Then clearly $\widetilde{H}$ is finitely generated, which implies that H is finitely generated by Proposition 6.1 in [3].

Lemma 2. *Let R be a noetherian domain and $a \in R$, $a \neq 0$. Then:*

1. $\widehat{[[a]]_R} = R_a^\times \cap \bar{R}$.
2. *$\widehat{[[a]]_R}$ is a saturated submonoid of $[[a]]_{\bar{R}}$.*
3. *$\widehat{[[a]]_R}$ is quasi finitely generated.*

Proof. (1) and (2) follow from Lemma 3.6 in [5].

(3) follows now from (2) and Propositions 2.7.5.1 and 2.7.8.3 in [4].

Theorem 1. *Let R be a noetherian domain. Then the following are equivalent:*

1. *$R\setminus\{0\}$ is locally finitely generated.*
2. *$R \subset \bar{R}$ is a root extension and for all $a \in R\setminus\{0\}$ the group $(R_a^\times \cap \bar{R}^\times)/R^\times$ is finite.*

If $\bar{R}$ is a finitely generated R-module, then the second condition in (2) is equivalent to the finiteness of the group $\bar{R}^\times/R^\times$.

Proof. We have

$$\bar{R}\setminus\{0\} = \bigcup_{a\in R\setminus\{0\}} R_a^\times \cap \bar{R} = \bigcup_{a\in R\setminus\{0\}} \widehat{[[a]]_R} .$$

Hence the extension $R \subset \bar{R}$ is a root extension if and only if this holds for $[[a]]_R \subset \widehat{[[a]]_R}$ for all $a \in R\setminus\{0\}$. Now the equivalence of 1. and 2. follows from Lemma 1.

Suppose now that $\bar{R}$ is a finitely generated R-module. For all $a \in R\setminus\{0\}$ we have $(R_a^\times \cap \bar{R}^\times)/R^\times \subset \bar{R}^\times/R^\times$. So the finiteness of $\bar{R}^\times/R^\times$ implies the finiteness of $(R_a^\times \cap \bar{R}^\times)/R^\times$ for all $a \in R\setminus\{0\}$. Conversely, let us assume that all the groups $(R_a^\times \cap \bar{R}^\times)/R^\times$ ($a \in R\setminus\{0\}$) are finite. Since $\bar{R}$ is a finitely generated R-module, there exists some $a \in R\setminus\{0\}$ such that $a\bar{R} \subset R$. Then we have $\bar{R}_a = R_a$ and hence $R_a^\times \cap \bar{R}^\times = \bar{R}^\times$, so that $\bar{R}^\times/R^\times$ is finite.

3 Root Extensions of Noetherian Rings

Let R be a (commutative) ring. For a prime ideal p of R we let $k(p)$ be its residue field. If $R \subset S$ is any extension ring of R and if there is only one prime ideal of S lying over p, we denote this prime by p_S. For any R-module M we denote by $\mathrm{Ass}_R(M)$ the set of prime ideals associated to M. If p is any prime ideal of R we denote by R_p the ring of fractions $(R\setminus p)^{-1}R$.

The aim of this section is to prove the following theorem:

Theorem 2. *Let R be a noetherian ring and $R \subset S$ an integral extension of R. Then $R \subset S$ is a root extension if and only if the following holds:*

1. *The induced map* $\mathrm{Spec}(S) \to \mathrm{Spec}(R)$ *is one to one (and hence bijective).*
2. *For any $p \in \mathrm{Ass}_R(S/R)$ we have either $k(p)$ is an algebraic extension of a finite field or* $\mathrm{char}(k(p)) > 0$ *and $k(p_S)$ is purely inseparable over $k(p)$.*

For the proof of this theorem we need a series of lemmas.

Lemma 3. *Let $R \subset S$ be a root extension of rings. Then the induced map* $\mathrm{Spec}(S) \to \mathrm{Spec}(R)$ *is one to one.*

Proof. Let $R \subset S$ be a root extension of rings and let $p \in \mathrm{Spec}(R)$. Choose any $q \in \mathrm{Spec}(S)$ lying over p. Then $q = \{x \in S \mid x^n \in p \text{ for some } n \in \mathbb{N}\}$, so that q is uniquely determined.

Lemma 4. *Let $R \subset S$ be a root extension of rings and let $T \subset R$ be multiplicatively closed and let I be an ideal of S. Then the extensions $T^{-1}R \subset T^{-1}S$ and $R/(I \cap R) \subset S/I$ are root extension, too. In particular for any prime ideal p of R the extension $k(p) \subset k(p_S)$ is a root extension.*

Proof. This is obvious.

Lemma 5. *Let $k \subset k'$ be an algebraic extension of fields. Then it is a root extension if and only if either k is an algebraic extension of a finite field or the extension is purely inseparable.*

Proof. This is well known. See for example Proposition 3.6 in [2].

In the next two Lemmas we consider an extension $R \subset S$ such that R and S are local and noetherian. We denote by m resp. m_S the maximal ideal of R resp. S. We assume further that S is a finitely generated R-module and that $\mathrm{Ass}_R(S/R) = \{m\}$. In particular $R \neq S$. Note that S/R is an R-module of finite length, and hence is annihilated by some power of m. Therefore there is some integer n such that $m^n S \subset R$. Further m_S is the unique prime ideal of S containing mS. Hence $\sqrt{mS} = m_S$. Since S is noetherian, there is some integer k such that $m_S^k \subset mS$. We choose k minimal with this property. Putting these two inclusions together, we obtain $m_S^{kn} \subset R$. Note that $1 + m \subset R^\times$ and $1 + mS \subset 1 + m_S \subset S^\times$ are subgroups and that we have two exact sequences of abelian groups

$$1 \longrightarrow \frac{1+m_S}{1+m} \longrightarrow S^\times/R^\times \longrightarrow k(m_S)^\times/k(m)^\times \longrightarrow 1 \tag{1}$$

$$1 \longrightarrow \frac{1+mS}{1+m} \longrightarrow \frac{1+m_S}{1+m} \longrightarrow \frac{1+m_S}{1+mS} \longrightarrow 1 \quad . \tag{2}$$

Lemma 6. *The group $(1 + m_S)/(1 + m)$ is a torsion group, if and only if $m = m_S$ or* $\mathrm{char}(k(m)) > 0$.

Proof. To begin with let $I \subset J \subset m_S$ be two ideals of S such that $J^2 \subset I$. Then the map $x \mapsto 1 + x$ induces an isomorphism of abelian groups $J/I \to (1 + J)/(1 + I)$. Applying this to $m_S^{l+1} \subset m_S^l$ for $l=1, \ldots, k$, we see that the group $(1+m_S)/(1+m_S^k)$ has a finite filtration, whose quotients are vector spaces over $k(m)$. Hence $(1 + m_S)/(1 + m_S^k)$ is a torsion group if $\mathrm{char}(k(m)) > 0$.

Next consider the group $(1 + m_S)/(1 + mS)$. Suppose that it is a non-trivial torsion group. Then $m_S \neq mS$. Hence $k \geq 2$. Then $(1 + m_S^{k-1})/(1 + m_S^{k-1} \cap mS) \cong (1 + m_S^{k-1})(1 + mS)/(1 + mS)$ is a non-trivial subgroup of $(1 + m_S)/(1 + mS)$. But as we have seen above $(1 + m_S^{k-1})/(1 + m_S^{k-1} \cap mS)$ is isomorphic to $m_S^{k-1}/(m_S^{k-1} \cap mS)$, which is a non-trivial vector space over $k(m)$. Hence $\mathrm{char}(k(m)) > 0$. Conversely suppose that $\mathrm{char}(k(m)) > 0$. Then we already know that $(1 + m_S)/(1 + m_S^k)$ is a torsion group. Since $m_S^k \subset mS$ our group $(1 + m_S)/(1 + mS)$ is a homomorphic image of $(1 + m_S)/(1 + m_S^k)$. Hence it is a torsion group, too. Putting our observations together, we obtain, that $(1 + m_S)/(1 + mS)$ is a torsion group if and only if either $m_S = mS$ or $\mathrm{char}(k(m)) > 0$.

For any integer l let U_l be the image of $1 + m^{l+1}S$ in $(1 + mS)/(1 + m)$. Then the U_l form a filtration of $(1 + mS)/(1 + m)$, whose quotients are again vector spaces over $k(m)$ (see Lemma 2.3 in [7]). Since $m^n S \subset R$ this filtration is finite. So we obtain that $(1 + mS)/(1 + m)$ is a torsion group if and only if either $m = mS$ or $\mathrm{char}(k(m)) > 0$.

The assertion of the Lemma follows now from the exact sequence (2).

Lemma 7. *$R \subset S$ is a root extension, if and only if either $k(m)$ is an algebraic extension of a finite field or* $\mathrm{char}\, k(m) > 0$ *and the extension of fields $k(m) \subset k(m_S)$ is purely inseparable.*

Proof. From $m_S^{kn} \subset R$ and $S = m_S \cup S^\times$ we conclude that $R \subset S$ is a root extension if and only if $S^\times/R^\times$ is a torsion group. Now the exact sequence (1) tells us, that this will be so if and only if $k(m) \subset k(m_S)$ is a root extension and $(1 + m_S)/(1 + m)$ is a torsion group. Since $R \neq S$ we have $k(m) \neq k(m_S)$ if $m = m_S$. Now the assertion of the lemma follows from Lemmas 5 to 6.

Lemma 8. *Let $R \subset S$ be any non-trivial extension of rings and let $T \subset R$ be the set of all non-zero divisors of the R-module S/R. Then $R \subset S$ is a root extension if and only if $T^{-1}R \subset T^{-1}S$ is a root extension.*

Proof. By Lemma 4 we need only show that $R \subset S$ is a root extension if $T^{-1}R \subset T^{-1}S$ is one. So let us assume that $T^{-1}R \subset T^{-1}S$ is a root extension and let $s \in S$. Then there exists some $r \in R$, $t \in T$ and an integer $n \geq 1$ such that $s^n/1 = r/t \in T^{-1}S$. Hence $t'ts^n \in R$ for some $t' \in T$. By definition of T this implies $s^n \in R$.

We are now in the position to prove Theorem 2.

Proof. (*Proof of Theorem* 2) Let $R \subset S$ be as in the statement of the Theorem. Let Σ be the set of all R-subalgebras of S, which are finitely generated R-modules. Then S is a root extension of R if and only if every member of Σ is so. Further we have $\mathrm{Ass}_R(S/R) = \bigcup \mathrm{Ass}_R(S'/R)$ where S' ranges over Σ. We may therefore assume that S is a finitely generated R-module. By Lemma 3 we may further assume that $\mathrm{Spec}(S) \to \mathrm{Spec}(R)$ is one to one. This implies that $\mathrm{Spec}(S) \to \mathrm{Spec}(R')$, $\mathrm{Spec}(R') \to \mathrm{Spec}(R)$ are one to one, too, for any intermediate ring R' of $R \subset S$.

Suppose first that $R \subset S$ is a root extension and let $p \in \mathrm{Ass}_R(S/R)$. From Lemmas 4 to 5 we already know that either $k(p)$ is an algebraic extension of a finite field or $k(p) \subset k(p_S)$ is purely inseparable. So we are left to show that $\mathrm{char}(k(p)) > 0$. Let q be any minimal member of $\mathrm{Ass}_R(S/R)$ contained in p. Then we have an epimorphism $R/q \to R/p$. So if $\mathrm{char}(k(q)) = \mathrm{char}(R/q) > 0$ then $\mathrm{char}(k(p)) = \mathrm{char}(R/p) > 0$, too. Therefore we suppose that p is minimal in $\mathrm{Ass}_R(S/R)$. Then we have $\mathrm{Ass}_{R_p}(S_p/R_p) = \{pR_p\}$. After replacing R and S by R_p and S_p we may assume that R and S are local and p is the maximal ideal of R. Now we are in the situation of Lemma 7, which in particular tells us that $\mathrm{char}(k(p)) > 0$.

Next assume that for all $p \in \mathrm{Ass}_R(S/R)$ either $k(p)$ is an algebraic extension of a finite field or $k(p) \subset k(p_S)$ is purely inseparable. Choose any minimal member p of $\mathrm{Ass}_R(S/R)$ and let $R' = \{x \in S \mid sx \in R \text{ for some } s \in R\backslash p\}$ $(= R_p \cap S)$ be the p-primary component of R in S. Then R' is a subring of S containing R, and we have $\mathrm{Ass}_R(R'/R) = \mathrm{Ass}_R(S/R)\setminus\{p\}$ and $\mathrm{Ass}_R(S/R') = \{p\}$. By Proposition 9. A in [8] we have $\{p\} = \mathrm{Ass}_R(S/R') = \{q \cap R \mid q \in \mathrm{Ass}_{R'}(S/R')\}$. It follows, that $\mathrm{Ass}_{R'}(S/R') = \{p_{R'}\}$. Hence using an induction on the cardinality of $\mathrm{Ass}_R(S/R)$, we reduce to the case that $\mathrm{Ass}_R(S/R) = \{p\}$.

Set $T = R\setminus p$. Then T consists exactly of the non-zero divisors of S/R. By Lemma 8 we may replace $R \subset S$ by $T^{-1}R \subset T^{-1}S$. But now Lemma 7 shows us that $R \subset S$ is a root extension.

References

1. Chapman, S.T., Halter-Koch, F., Krause, U.: Inside factorial monoids and integral domains. J. Algebra **252**, 350–375 (2002)
2. Davis, E.D., Maroscia, P.: Affince curves on which every point is a set-theoretic complete intersection. J. Algebra **87**, 113–135 (1984)
3. Geroldinger, A., Halter-Koch, F.: Arithmetical theory of monoid homomorphisms. Semigroup Forum **48**, 333–362 (1994)
4. Geroldinger, A., Halter-Koch, F.: Non-Unique factorizations. Algebraic, combinatorial and analytic theory. Pure and Applied Mathematics, vol. 278. Chapman & Hall/CRC (2006)
5. Geroldinger, A., Hassler, W.: Arithmetic of Mori domains and monoids. J. Algebra **319**, 3419–3463 (2008)
6. Halter-Koch, F.: On the asymptotic behaviour of the number of distinct factorizations into irreducibles. Ark. Mat. **31**, 297–305 (1993)
7. Kainrath, F.: A note on quotients formed by unit groups of semilocal rings. Houst. J. Math **24**, 613–618 (1998)
8. Matsumura, H.: Commutative Algebra, 2nd ed. Benjamin/Cummings (1989)

Spectral Spaces Versus Distributive Lattices: A Dictionary

Henri Lombardi

Abstract The category of distributive lattices is, in classical mathematics, anti-equivalent to the category of spectral spaces. We give here some examples and a short dictionary for this antiequivalence. We propose a translation of several abstract theorems (in classical mathematics) into constructive ones, even in the case where points of a spectral space have no clear constructive content.

Keywords Distributive lattice · Spectral space · Constructive mathematics · Krull dimension · Zariski lattice · Zariski spectrum · Real spectrum · Valuative spectrum · Lying over · Going up · Going down

Introduction

This paper is written in Bishop's style of constructive mathematics [3, 4, 6, 17, 22]. We give a short dictionary between classical and constructive mathematics w.r.t. properties of spectral spaces and of the associated dual distributive lattices. We give several examples of how this works.

1 Distributive Lattices and Spectral Spaces: Some General Facts

References: [7, 10, 19, 20, 23], [1, Chapter 4] and [17, Chapters XI and XIII].

H. Lombardi (✉)
Laboratoire de Mathématiques de Besançon, Université Bourgogne Franche-Comté, Bourgogne Franche-Comté, France
e-mail: henri.lombardi@univ-fcomte.fr
URL: http://hlombardi.free.fr/

A. Facchini et al. (eds.), *Advances in Rings, Modules and Factorizations*, Springer Proceedings in Mathematics & Statistics 321,
https://doi.org/10.1007/978-3-030-43416-8_13

1.1 The Seminal Paper by Stone

In classical mathematics, a *prime ideal* $\mathfrak{p}$ of a distributive lattice $\mathbf{T} \neq \mathbf{1}$ is an ideal whose complement $\mathfrak{f}$ is a filter (*a prime filter*). The quotient lattice $\mathbf{T}/(\mathfrak{p} = 0, \mathfrak{f} = 1)$ is isomorphic to $\mathbf{2}$. Giving a prime ideal of $\mathbf{T}$ is the same thing as giving a lattice morphism $\mathbf{T} \to \mathbf{2}$. We will write $\theta_{\mathfrak{p}} : \mathbf{T} \to \mathbf{2}$ the morphism corresponding to $\mathfrak{p}$.

If S is a system of generators for a distributive lattice $\mathbf{T}$, a prime ideal $\mathfrak{p}$ of $\mathbf{T}$ is characterised by its trace $\mathfrak{p} \cap S$ (cf. [7]).

The (Zariski) *spectrum of the distributive lattice* $\mathbf{T}$ is the set $\mathsf{Spec}\,\mathbf{T}$ whose elements are prime ideals of $\mathbf{T}$, with the following topology: an open basis is provided by the subsets $\mathfrak{D}_{\mathbf{T}}(a) \stackrel{\text{def}}{=} \{\mathfrak{p} \in \mathsf{Spec}\,\mathbf{T} \mid a \notin \mathfrak{p}\} = \left\{\mathfrak{p} \mid \theta_{\mathfrak{p}}(a) = 1\right\}$. One has

$$\left.\begin{aligned} \mathfrak{D}_{\mathbf{T}}(a \wedge b) &= \mathfrak{D}_{\mathbf{T}}(a) \cap \mathfrak{D}_{\mathbf{T}}(b), \quad \mathfrak{D}_{\mathbf{T}}(0) = \emptyset, \\ \mathfrak{D}_{\mathbf{T}}(a \vee b) &= \mathfrak{D}_{\mathbf{T}}(a) \cup \mathfrak{D}_{\mathbf{T}}(b), \quad \mathfrak{D}_{\mathbf{T}}(1) = \mathsf{Spec}\,\mathbf{T}. \end{aligned}\right\} \tag{1}$$

The complement of $\mathfrak{D}_{\mathbf{T}}(a)$ is a *basic closed set* denoted by $\mathfrak{V}_{\mathbf{T}}(a)$. This notation is extended to $I \subseteq \mathbf{T}$: we let $\mathfrak{V}_{\mathbf{T}}(I) \stackrel{\text{def}}{=} \bigcap_{x \in I} \mathfrak{V}_{\mathbf{T}}(x)$. If $\mathfrak{I}$ is the ideal generated by I, one has $\mathfrak{V}_{\mathbf{T}}(I) = \mathfrak{V}_{\mathbf{T}}(\mathfrak{I})$. The closed set $\mathfrak{V}_{\mathbf{T}}(I)$ is also called *the subvariety of* $\mathsf{Spec}\ \mathbf{T}$ *defined by* I.

The adherence of a point $\mathfrak{p} \in \mathsf{Spec}\ \mathbf{T}$ is provided by all $\mathfrak{q} \supseteq \mathfrak{p}$. Maximal ideals are the closed points of $\mathsf{Spec}\ \mathbf{T}$. The spectrum $\mathsf{Spec}\ \mathbf{T}$ is empty iff $0 =_{\mathbf{T}} 1$.

The spectrum of a distributive lattice is the paradigmatic example of a *spectral space*. Spectral spaces can be characterised as the topological spaces satisfying the following properties:

- the space is quasi-compact,[1]
- every open set is a union of quasi-compact open sets,
- the intersection of two quasi-compact open sets is a quasi-compact open set,
- for two distinct points, there is an open set containing one of them but not the other,
- every irreducible closed set is the adherence of a point.

The quasi-compact open sets then form a distributive lattice, the supremum and the infimum being the union and the intersection, respectively. A continuous map between spectral spaces is said to be *spectral* if the inverse image of every quasi-compact open set is a quasi-compact open set.

Stone's fundamental result [23] can be stated as follows. *The category of distributive lattice is, in classical mathematics, antiequivalent to the category of spectral spaces.*

Here is how this works.

1. *The quasi-compact open sets of* $\mathsf{Spec}\ \mathbf{T}$ *are exactly the* $\mathfrak{D}_{\mathbf{T}}(u)$*'s.*
2. *The map* $u \mapsto \mathfrak{D}_{\mathbf{T}}(u)$ *is well-defined and it is an isomorphism of distributive lattices.*

---

[1]The nowadays standard terminology is quasi-compact, as in Bourbaki and Stacks, rather than compact.

In the other direction, if X is a spectral space we let $\mathsf{Oqc}(X)$ be the distributive lattice formed by its quasi-compact open sets. If $\xi : X \to Y$ is a spectral map, the map

$$\mathsf{Oqc}(\xi) : \mathsf{Oqc}(Y) \to \mathsf{Oqc}(X), \quad U \mapsto \xi^{-1}(U)$$

is a morphism of distributive lattices. This defines Oqc as a contravariant functor.

Johnstone calls *coherent spaces* the spectral spaces [16]. Balbes and Dwinger [1] give them the name *Stone space*. The name *spectral space* is given by Hochster in a famous paper [15] where he proves that all spectral spaces can be obtained as Zariski spectra of commutative rings.

In constructive mathematics, spectral spaces may have no points. So it is necessary to translate the classical stuff about spectral spaces into a constructive rewriting about distributive lattices. It is remarkable that all useful spectral spaces in the literature correspond to simple distributive lattices.

Two other natural spectral topologies can be defined on $\mathsf{Spec}\ \mathbf{T}$ by changing the definition of basic open sets. When one chooses the $\mathfrak{V}(a)$'s as basic open sets, one gets the spectral space corresponding to $\mathbf{T}^{\circ}$ (obtained by reversing the order). When one chooses Boolean combinations of the $\mathfrak{D}(a)$'s as basic open sets one gets the constructible topology (also called the patch topology). This spectral space can be defined as the spectrum of $\mathbb{B}\mathrm{o}(\mathbf{T})$ (the Boolean algebra generated by $\mathbf{T}$).

1.1.1 Spectral Subspaces Versus Quotient Lattices

Theorem 1. (Subspectral spaces) *Let $\mathbf{T}'$ be a quotient lattice of $\mathbf{T}$ and $\pi : \mathbf{T} \to \mathbf{T}'$ the quotient morphism. Let us write $X' = \mathsf{Spec}\,\mathbf{T}'$, $X = \mathsf{Spec}\,\mathbf{T}$ and $\pi^\star : X' \to X$ the dual map of π.*

1. *$\pi^\star$ identifies X' with a* topological subspace *of X. Moreover $\mathsf{Oqc}(X') = \{\, U \cap X' \mid U \in \mathsf{Oqc}(X)\,\}$. We say that X'* is a subspectral space of X.
2. *A subset X' of X is a subspectral space of X if and only if*
   – the induced topology by X on X' is spectral and
   – $\mathsf{Oqc}(X') = \{\, U \cap X' \mid U \in \mathsf{Oqc}(X)\,\}$.
3. *A subset X' of X is a subspectral space if and only if it is closed for the patch topology.*
4. *If Z is an arbitrary subset of $X = \mathsf{Spec}\,\mathbf{T}$, its adherence for the patch topology is given by $X' = \mathsf{Spec}\,\mathbf{T}'$, where $\mathbf{T}'$ is the quotient lattice of $\mathbf{T}$ defined by the following preorder $\preccurlyeq$:*

$$a \preccurlyeq b \iff (\mathfrak{D}_{\mathbf{T}}(a) \cap Z) \subseteq (\mathfrak{D}_{\mathbf{T}}(b) \cap Z). \tag{2}$$

1.1.2 Gluing Distributive Lattices and Spectral Subspaces

Let $(x_1, \dots, x_n)$ be a system of comaximal elements in a commutative ring $\mathbf{A}$. Then the canonical morphism $\mathbf{A} \to \prod_{i \in [\![1..n]\!]} \mathbf{A}[1/x_i]$ identifies $\mathbf{A}$ with a finite subproduct of localisations of $\mathbf{A}$.

Similarly a distributive lattice can be recovered from a finite number of good quotient lattices.

Definition 1. Let $\mathbf{T}$ be a distributive lattice and $(\mathfrak{a}_i)_{i \in [\![1..n]\!]}$ (resp. $(\mathfrak{f}_i)_{i \in [\![1..n]\!]}$) a finite family of ideals (resp. of filters) of $\mathbf{T}$. We say that the ideals $\mathfrak{a}_i$ *cover* $\mathbf{T}$ if $\bigcap_i \mathfrak{a}_i = \{0\}$. Similarly we say that the filters $\mathfrak{f}_i$ *cover* $\mathbf{T}$ if $\bigcap_i \mathfrak{f}_i - \{1\}$.

Let $\mathfrak{b}$ be an ideal of $\mathbf{T}$; we write $x \equiv y \mod \mathfrak{b}$ as meaning $x \equiv y \mod (\mathfrak{b} = 0)$. Let us recall that for $s \in \mathbf{T}$ the quotient $\mathbf{T}/(s = 0)$ is isomorphic to the principal filter $\uparrow s$ (one sees this filter as a distributive lattice with s as 0 element).

Fact 1. *Let $\mathbf{T}$ be a distributive lattice, $(\mathfrak{a}_i)_{i \in [\![1..n]\!]}$ a finite family of principal ideals ($\mathfrak{a}_i = \downarrow s_i$) and $\mathfrak{a} = \bigcap_i \mathfrak{a}_i$.*

1. *If (x_i) is a family in $\mathbf{T}$ s.t. for each i, j one has $x_i \equiv x_j \mod \mathfrak{a}_i \vee \mathfrak{a}_j$, then there exists a unique x modulo $\mathfrak{a}$ satisfying: $x \equiv x_i \mod \mathfrak{a}_i$ ($i \in [\![1..n]\!]$).*
2. *Let us write $\mathbf{T}_i = \mathbf{T}/(\mathfrak{a}_i = 0)$, $\mathbf{T}_{ij} = \mathbf{T}_{ji} = \mathbf{T}/(\mathfrak{a}_i \vee \mathfrak{a}_j = 0)$, $\pi_i : \mathbf{T} \to \mathbf{T}_i$ and $\pi_{ij} : \mathbf{T}_i \to \mathbf{T}_{ij}$ the canonical maps. If the ideals $\mathfrak{a}_i$ cover $\mathbf{T}$, the system $(\mathbf{T}, (\pi_i)_{i \in [\![1..n]\!]})$ is the inverse limit of the diagram*

$$((\mathbf{T}_i)_{1 \le i \le n}, (\mathbf{T}_{ij})_{1 \le i < j \le n}; (\pi_{ij})_{1 \le i \ne j \le n}).$$

3. *The analogous result works with quotients by principal filters.*

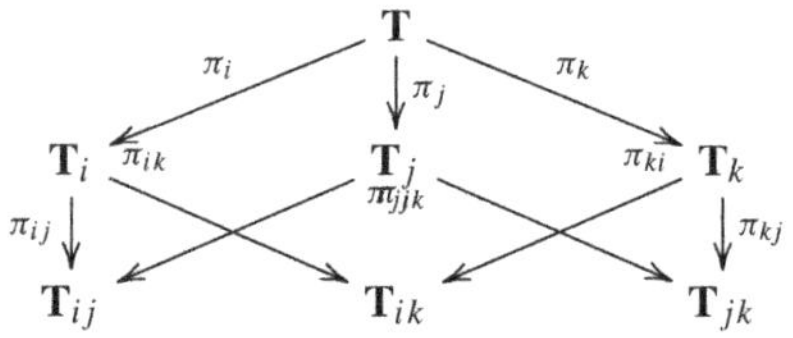

We have also a gluing procedure described in the following proposition.[2]

Proposition 1. (Gluing distributive lattices) *Let I be a finite set and a diagram of distributive lattices*

$$\left((\mathbf{T}_i)_{i \in I}, (\mathbf{T}_{ij})_{i<j \in I}, (\mathbf{T}_{ijk})_{i<j<k \in I}; (\pi_{ij})_{i \ne j}, (\pi_{ijk})_{i<j, j \ne k \ne i}\right)$$

and a family of elements $(s_{ij})_{i \ne j \in I} \in \prod_{i \ne j \in I} \mathbf{T}_i$ satisfying the following properties:

[2] In commutative algebra, a similar procedure works for $\mathbf{A}$-modules [17, XV-4.4]. But in order to glue commutative rings, it is necessary to pass to the category of Grothendieck schemes.

- *the diagram is commutative,*
- *if* $i \neq j$, π_{ij} *is a quotient morphism w.r.t. the ideal* $\downarrow s_{ij}$,
- *if* i, j, k *are distinct,* $\pi_{ij}(s_{ik}) = \pi_{ji}(s_{jk})$ *and* π_{ijk} *is a quotient morphism w.r.t. the ideal* $\downarrow \pi_{ij}(s_{ik})$.

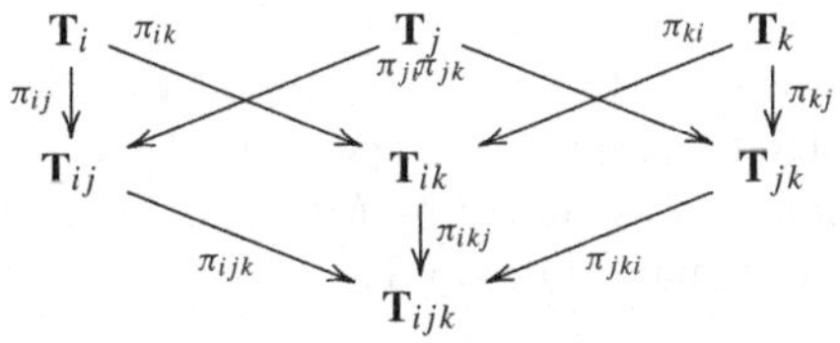

Let $(\mathbf{T}\,;\,(\pi_i)_{i\in I})$ *be the limit of the diagram. Then there exist* s_i*'s in* $\mathbf{T}$ *such that the principal ideals* $\downarrow s_i$ *cover* $\mathbf{T}$ *and the diagram is isomorphic to the one in Fact 1. More precisely each* π_i *is a quotient morphism w.r.t. the ideal* $\downarrow s_i$ *and* $\pi_i(s_j) = s_{ij}$ *for all* $i \neq j$.
The analogous result works with quotients by principal filters.

Remark 1. The reader can translate the previous result in gluing of spectral spaces.

1.1.3 Heitmann Lattice and J-Spectrum

An ideal $\mathfrak{m}$ of a distributive lattice $\mathbf{T}$ is *maximal* when $\mathbf{T}/(\mathfrak{m} = 0) \simeq \mathbf{2}$, i.e. if $1 \notin \mathfrak{m}$ and $\forall x \in \mathbf{T}$ ($x \in \mathfrak{m}$ or $\exists y \in \mathfrak{m}\ x \vee y = 1$).

In classical mathematics we have the following result.

Lemma 1. *The intersection of all maximal ideals containing an ideal* $\mathfrak{J}$ *is called the* Jacobson radical of $\mathfrak{J}$ *and is equal to*

$$\mathrm{J}_{\mathbf{T}}(\mathfrak{J}) \;=\; \{a \in \mathbf{T} \mid \forall x \in \mathbf{T}\ (a \vee x = 1 \Rightarrow \exists z \in \mathfrak{J}\ \ z \vee x = 1)\}\,. \tag{3}$$

We write $\mathrm{J}_{\mathbf{T}}(b)$ *for* $\mathrm{J}_{\mathbf{T}}(\downarrow b)$. *The ideal* $\mathrm{J}_{\mathbf{T}}(0)$ *is the* Jacobson radical of $\mathbf{T}$.

In constructive mathematics, equality (3) is used as definition.

The *Heitmann lattice of* $\mathbf{T}$, denoted by $\mathsf{He}(\mathbf{T})$, is the quotient of $\mathbf{T}$ corresponding to the following preorder $\preccurlyeq_{\mathsf{He}(\mathbf{T})}$:

$$a \preccurlyeq_{\mathsf{He}(\mathbf{T})} b \;\overset{\text{def}}{\Longleftrightarrow}\; \mathrm{J}_{\mathbf{T}}(a) \subseteq \mathrm{J}_{\mathbf{T}}(b) \Longleftrightarrow a \in \mathrm{J}_{\mathbf{T}}(b). \tag{4}$$

Elements of $\mathsf{He}(\mathbf{T})$ can be identified with ideals $\mathrm{J}_{\mathbf{T}}(a)$, via the canonical map

$$\mathbf{T} \longrightarrow \mathsf{He}(\mathbf{T}), \quad a \longmapsto \mathrm{J}_{\mathbf{T}}(a).$$

The next definition follows the remarkable paper by Heitmann [14].

Definition 2. Let $\mathbf{T}$ be a distributive lattice.

1. The *maximal spectrum of* $\mathbf{T}$, denoted by $\mathsf{Max}\,\mathbf{T}$, is the topological subspace of $\mathsf{Spec}\,\mathbf{T}$ provided by the maximal ideals of $\mathbf{T}$.
2. The *j-spectrum of* $\mathbf{T}$, denoted by $\mathsf{jspec}\,\mathbf{T}$, is the topological subspace of $\mathsf{Spec}\,\mathbf{T}$ provided by the primes $\mathfrak{p}$ s.t. $\mathrm{J}_{\mathbf{T}}(\mathfrak{p}) = \mathfrak{p}$, i.e. the prime ideals $\mathfrak{p}$ which are intersections of maximal ideals.
3. The *Heitmann* J-*spectrum of* $\mathbf{T}$, denoted by $\mathsf{Jspec}\,\mathbf{T}$, is the adherence of $\mathsf{Max}\,\mathbf{T}$ in $\mathsf{Spec}\,\mathbf{T}$ for the patch topology. It is a spectral subspace of $\mathsf{Spec}\,\mathbf{T}$.
4. The *minimal spectrum of* $\mathbf{T}$, denoted by $\mathsf{Min}\,\mathbf{T}$, is the topological subspace of $\mathsf{Spec}\,\mathbf{T}$ provided by minimal primes of $\mathbf{T}$.

In general, $\mathsf{Max}\,\mathbf{T}$, $\mathsf{jspec}\,\mathbf{T}$ and $\mathsf{Min}\,\mathbf{T}$ are not spectral spaces.

Theorem 2. $\mathsf{Jspec}\,\mathbf{T}$ *is a spectral subspace of* $\mathsf{Spec}\,\mathbf{T}$ *canonically homeomorphic to* $\mathsf{Spec}\,(\mathsf{He}(\mathbf{T}))$.

1.2 Distributive Lattices and Entailment Relations

A particularly important rule for distributive lattices, known as *cut*, is

$$\big(x \wedge a \leqslant b\big) \;\; \& \;\; \big(a \leqslant x \vee b\big) \quad \Longrightarrow \quad a \leqslant b. \tag{5}$$

For $A \in \mathrm{P}_{\mathrm{fe}}(\mathbf{T})$ (finitely enumerated subsets of $\mathbf{T}$) we write

$$\bigvee A := \bigvee\nolimits_{x \in A} x \qquad \text{and} \qquad \bigwedge A := \bigwedge\nolimits_{x \in A} x.$$

We denote by $A \vdash B$ or $A \vdash_{\mathbf{T}} B$ the relation defined as follows over the set $\mathrm{P}_{\mathrm{fe}}(\mathbf{T})$:

$$A \vdash B \quad \overset{\text{def}}{\Longleftrightarrow} \quad \bigwedge A \leqslant \bigvee B.$$

This relation satisfies the following axioms, in which we write x for $\{x\}$ and A, B for $A \cup B$:

$$\begin{array}{rrcll} & & a \vdash a & & (R)\\ A \vdash B & \Longrightarrow & A, A' \vdash B, B' & & (M)\\ (A, x \vdash B)\ \&\ (A \vdash B, x) & \Longrightarrow & A \vdash B & & (T). \end{array}$$

We say that the relation is *reflexive*, *monotone* and *transitive*. The third rule (transitivity) can be seen as a version of rule (5) and is also called the *cut* rule.

Definition 3. For an arbitrary set S, a relation over $\mathrm{P}_{\mathrm{fe}}(S)$ which is reflexive, monotone and transitive is called an *entailment relation*.

The following theorem is fundamental. It says that the three properties of entailment relations are exactly what is needed for the interpretation in the form of a distributive lattice to be adequate.

Theorem 3. (Fundamental theorem of entailment relations) *[7], [17, XI-5.3], [21, Satz 7] Let S be a set with an entailment relation $\vdash_S$ on* $\mathrm{P_{fe}}(S)$. *We consider the distributive lattice* $\mathbf{T}$ *defined by generators and relations as follows: the generators are the elements of S and the relations are*

$$A \vdash_{\mathbf{T}} B$$

each time that $A \vdash_S B$. Then, for all A, B in $\mathrm{P_{fe}}(S)$, *we have*

$$A \vdash_{\mathbf{T}} B \implies A \vdash_S B.$$

2 Spectral Spaces in Algebra

The usual spectral spaces in algebra are (always?) understood as spectra of distributive lattices associated to coherent theories describing relevant algebraic structures. We describe this general situation and give some examples.

2.1 *Dynamical Algebraic Structures, Distributive Lattices and Spectra*

References: [13, 20]. The paper [13] introduces the general notion of "dynamical theory" and of "dynamical proof". See also the paper [2] which illustrates the usefulness of these notions.

2.1.1 Dynamical Theories and Dynamical Algebraic Structures

Dynamical theories are a version "without logic, purely computational" of *coherent theories* (we say theory for "first order formal theory").

Dynamical theories use only dynamical rules, i.e. deduction rules of the form

$$\Gamma \ \vdash \ \exists \underline{y^1}\, \Delta_1 \ \textsf{or} \ \cdots \ \textsf{or} \ \exists \underline{y^m}\, \Delta_m,$$

where Γ and the Δ_i's are lists of atomic formulae in the language $\mathcal{L}$ of the theory $\mathcal{T} = (\mathcal{L}, \mathcal{A})$.

The computational meaning of "$\exists y\, \Delta$" is "**Introduce** y **such that** Δ". The computational meaning of "U **or** V **or** W" is "open three branches of computations ...".

Axioms (elements of $\mathcal{A}$) are dynamical rules and theorems are valid dynamical rules (validity is described in a simple way and uses only a computational machinery).

A *dynamical algebraic structure* for a dynamical theory $\mathcal{T}$ is given through a presentation (G, R) by generators and relations. Generators are the element of G and they are added to the constants in the language. Relations are the elements of R. They are dynamical rules without free variables and they are added to the axioms of the theory.

A dynamical algebraic structure is intuitively thought of as an incompletely specified algebraic structure. The notion corresponds to lazy evaluation in Computer Algebra.

Purely equational algebraic structures correspond to the case where the only predicate is equality and the axioms are Horn rules.

Dynamical theories whose axioms contain neither **or** nor $\exists$ are called *Horn theories* (algebraic theories in [13]). For example, theories of absolutely flat rings and of pp-rings can be given as Horn theories.

A coherent theory is a *first order geometric theory*. In non-first order geometric theories we accept dynamical rules that use infinite disjunctions at the right of $\vdash$. In this paper we speak only of first order geometric theories.

A fundamental result about dynamical theories says that adding the classical first order logic to a dynamical theory does not change valid rules: first order classical mathematic is conservative over dynamical theories [13, Theorem 1.1].

2.1.2 Distributive Lattices Associated to a Dynamical Algebraic Structure

Let $\mathbf{A} = \big((G, R), \mathcal{T}\big)$ be a dynamical algebraic structure for $\mathcal{T} = (\mathcal{L}, \mathcal{A})$.

• **First example**. If $P(x, y)$ belongs to $\mathcal{L}$ and if Clt is the set of closed terms of $\mathbf{A}$, we get the following entailment relation $\vdash_{\mathbf{A},P}$ for $Clt \times Clt$:

$$\begin{array}{rl} (a_1, b_1), \dots, (a_n, b_n) \vdash_{\mathbf{A},P} (c_1, d_1), \dots, (c_m, d_m) & \overset{\text{def}}{\Longleftrightarrow} \\ P(a_1, b_1), \dots, P(a_n, b_n) \vdash_{\mathbf{A}} P(c_1, d_1) \textbf{ or } \dots \textbf{ or } P(c_m, d_m). & \end{array} \tag{6}$$

Intuitively the distributive lattice $\mathbf{T}$ generated by this entailment relation represents the "truth values" of P in the dynamical algebraic structure $\mathbf{A}$. In fact to give an element $\alpha : \mathbf{T} \to \mathbf{2}$ of $\operatorname{Spec} \mathbf{T}$ amounts to giving the value $\top$ (resp. $\bot$) to $P(a, b)$ when $\alpha(a, b) = 1$ (resp. $\alpha(a, b) = 0$).

• **Second example, the Zariski lattice of a commutative ring**. Let $\mathcal{Al}$ be a dynamical theory of nontrivial local rings, e.g. with signature

$$(\cdot = 0, \mathrm{U}(\cdot)\ ;\ \cdot + \cdot, \cdot \times \cdot, - \cdot, 0, 1\,).$$

This is an extension of the purely equational theory of commutative rings. The predicate $\mathrm{U}(x)$ is defined as meaning the invertibility of x,

- $\mathrm{U}(x) \vdash \exists y\, xy = 1$
- $xy = 1 \vdash \mathrm{U}(x)$

and the axioms of nontrivial local rings are written as

AL $\mathrm{U}(x+y) \vdash \mathrm{U}(x)$ **or** $\mathrm{U}(y)$

- $\mathrm{U}(0) \vdash \bot$

The *Zariski lattice* $\mathsf{Zar}\,\mathbf{A}$ of a commutative ring $\mathbf{A}$ is defined as the distributive lattice generated by the entailment relation $\vdash_{\mathsf{Zar}\,\mathbf{A}}$ for $\mathbf{A}$ defined as

$$\begin{array}{c} a_1, \ldots, a_n \vdash_{\mathsf{Zar}\,\mathbf{A}} c_1, \ldots, c_m \qquad \stackrel{\text{def}}{\Longleftrightarrow} \\ \mathrm{U}(a_1), \ \ldots, \mathrm{U}(a_n) \vdash_{\mathcal{Al}(\mathbf{A})} \mathrm{U}(c_1) \ \textbf{or} \ \ldots \ \textbf{or} \ \mathrm{U}(c_m). \end{array} \tag{7}$$

Here $\mathcal{Al}(\mathbf{A})$ is the dynamical algebraic structure of type $\mathcal{Al}$ over $\mathbf{A}$.
We get the following equivalence (we call it a *formal Nullstellensatz*):

$$a_1, \ldots, a_n \vdash_{\mathsf{Zar}\,\mathbf{A}} c_1, \ldots, c_m \Longleftrightarrow \exists k > 0 \ \ (a_1 \cdots a_n)^k \in \langle c_1, \ldots, c_m \rangle \,.$$

So, $\mathsf{Zar}\,\mathbf{A}$ can be identified with the set of ideals $\mathrm{D}_{\mathbf{A}}(\underline{x}) = \sqrt[\mathbf{A}]{\langle \underline{x} \rangle}$, with $\mathrm{D}_{\mathbf{A}}(\mathfrak{j}_1) \wedge \mathrm{D}_{\mathbf{A}}(\mathfrak{j}_2) = \mathrm{D}_{\mathbf{A}}(\mathfrak{j}_1\mathfrak{j}_2)$ and $\mathrm{D}_{\mathbf{A}}(\mathfrak{j}_1) \vee \mathrm{D}_{\mathbf{A}}(\mathfrak{j}_2) = \mathrm{D}_{\mathbf{A}}(\mathfrak{j}_1 + \mathfrak{j}_2)$.

Now, the usual Zariski spectrum $\mathsf{Spec}\,\mathbf{A}$ is canonically homeomorphic to $\mathsf{Spec}\,(\mathsf{Zar}\,\mathbf{A})$. Indeed, to give a point of $\mathsf{Spec}\,\mathbf{A}$ (a prime ideal) amounts to giving an epimorphism $\mathbf{A} \to \mathbf{B}$ where $\mathbf{B}$ is a local ring, or also, that is the same thing, to giving a minimal model of $\mathcal{Al}(\mathbf{A})$. This corresponds to the intuition of "forcing the ring to be a local ring".

• **More generally.** Let us consider a set S of closed atomic formulae of the dynamical algebraic structure $\mathbf{A} = \big((G, R), \mathcal{T}\big)$. We define a corresponding entailment relation (with the A_i's and B_j's in S):

$$\begin{array}{c} A_1, \ldots, A_n \vdash_{\mathbf{A},S} B_1, \ldots, B_m \qquad \stackrel{\text{def}}{\Longleftrightarrow} \\ A_1, \ \ldots, A_n \vdash_{\mathbf{A}} B_1 \ \textbf{or} \ \ldots \ \textbf{or} \ B_m. \end{array} \tag{8}$$

We may denote by $\mathsf{Zar}(\mathbf{A}, S)$ this distributive lattice.

• **Points of a spectrum and models in classical mathematics.** With a good choice of predicates in the language, to give a point of the spectrum of the corresponding lattice amounts often to giving a minimal model of the dynamical algebraic structure. This is the case when all existence axioms in the theory imply unique existence. The topology of the spectrum is in any case strongly dependent on the choice of predicates.

• **The complete Zarisiki lattice of a dynamical algebraic structure A** is defined by choosing for S the set $Clat(\mathbf{A})$ of all closed atomic formulas of $\mathbf{A}$. When the theory has no existential axioms, this lattice corresponds to the entailment relation for $Clat(\mathbf{A})$ generated by the axioms of $\mathcal{T}$, replacing the variables by arbitrary closed terms of $\mathbf{A}$.

2.2 A Very Simple Case

Let $\mathcal{T}$ be a Horn theory. Any dynamical algebraic structure $\mathbf{A} = ((G, R), \mathcal{T})$ of type $\mathcal{T}$ defines an ordinary algebraic structure $\mathbf{B}$ and there is no significant difference between dynamical algebraic structures and ordinary algebraic structures.

The minimal models of $\mathbf{A}$ are (identified with) the quotient structures $\mathbf{C} = \mathbf{B}/\sim$. If we choose convenient predicates for defining a distributive lattice associated to $\mathbf{B}$, the points of the corresponding spectrum are (identified with) these quotient structures.

For example, in the case of the purely equational theory $\mathcal{T} = \mathcal{M}od_{\mathbf{A}}$ (the theory of modules over a fixed ring $\mathbf{A}$), and choosing the predicate $x = 0$ (or the predicate $x \neq 0$), we get the lattice generated by the following entailment relation for an $\mathbf{A}$-module M:

• $x_1 = 0, \ldots, x_n = 0 \vdash y_1 = 0$ **or** $\ldots$ **or** $y_m = 0,$

or by

• $y_1 \neq 0, \ldots, y_m \neq 0 \vdash x_1 \neq 0$ **or** $\ldots$ **or** $x_n \neq 0,$

which means "*one y_j is in the submodule $\langle x_1, \ldots, x_n \rangle$*" (formal Nullstellensatz for linear algebra).

Here the points of the spectrum are (identified with) submodules of M and a basic open (or a basic closed) set $\mathfrak{D}(a)$ is the set of submodules containing a.

It might be that these kinds of lattices and spectra are too simple to lead to interesting results in algebra.

2.3 The Real Spectrum of a Commutative Ring

The real spectrum $\mathsf{Sper}\,\mathbf{A}$ of a commutative ring corresponds to the intuition of "forcing the ring $\mathbf{A}$ to be an ordered (discrete[3])" field.

A point of $\mathsf{Sper}\,\mathbf{A}$ can be given as an epimorphism $\varphi : \mathbf{A} \to \mathbf{K}$, where $(\mathbf{K}, \mathbf{C})$ is an ordered field.[4] Moreover two such morphisms $\varphi : \mathbf{A} \to \mathbf{K}$ and $\varphi' : \mathbf{A} \to \mathbf{K}'$ define the same point of the spectrum if there exists an isomorphism of ordered fields $\psi : \mathbf{K} \to \mathbf{K}'$ making the suitable diagram commutative.

---

[3] We ask the order relation to be decidable.

[4] $\mathbf{C}$ is the cone of nonnegative elements.

We write "$x \geq 0$" the predicate over **A** corresponding to "$\varphi(x) \geq 0$ in **K**". We get the following axioms:

- $\vdash x^2 \geq 0$
- $x \geq 0,\ y \geq 0 \ \vdash x + y \geq 0$
- $x \geq 0,\ y \geq 0 \ \vdash xy \geq 0$
- $-1 \geq 0 \ \vdash \bot$
- $-xy \geq 0 \ \vdash x \geq 0$ **or** $y \geq 0$

This means that $\{x \in \mathbf{A} \mid x \geq 0\}$ is a prime cone: to give a model of this theory is the same thing as to give a point of Sper **A**.

In order to get the usual topology of Sper **A**, it is necessary to use the opposite predicate $x < 0$. For the sake of comfort, we take $x > 0$. This predicate satisfies the dual axioms to those for $-x \geq 0$:

- $-x^2 > 0 \ \vdash \bot$
- $x + y > 0 \ \vdash x > 0$ **or** $y > 0$
- $xy > 0 \ \vdash x > 0$ **or** $-y > 0$
- $\vdash 1 > 0$
- $x > 0,\ y > 0 \ \vdash xy > 0$

So the *real lattice* of **A**, denoted by Real(**A**), is the distributive lattice generated by the minimal entailment relation for **A** satisfying the following relations (we write $\mathrm{R}(a)$ instead of a):

- $\mathrm{R}(-x^2) \vdash$
- $\mathrm{R}(x + y) \vdash \mathrm{R}(x), \mathrm{R}(y)$
- $\mathrm{R}(xy) \vdash \mathrm{R}(x), \mathrm{R}(-y)$
- $\vdash \mathrm{R}(1)$
- $\mathrm{R}(x), \mathrm{R}(y) \vdash \mathrm{R}(xy)$

So Spec (Real **A**) is isomorphic to Sper **A**, viewed as the set of prime cones of **A**. The spectral topology admits the basis of open sets

$$\mathfrak{R}(a_1, \ldots, a_n) = \left\{ \mathfrak{c} \in \mathsf{Sper}\,\mathbf{A} \mid \&_{i=1}^{n} - a_i \notin \mathfrak{c} \right\}.$$

This approach to the real spectrum was proposed in [7].

An important point is the following *formal Positivstellensatz*.

Theorem 4. (Formal Positivstellensatz for ordered fields) *T.F.A.E.*

1. *We have* $\mathrm{R}(x_1), \ldots, \mathrm{R}(x_k) \vdash \mathrm{R}(a_1), \ldots, \mathrm{R}(a_n)$ *in the lattice* Real **A**.
2. *We have* $x_1 > 0, \ \ldots, x_k > 0 \ \vdash a_1 > 0$ ***or*** $\ldots$ ***or*** $a_n > 0$ *in the theory of ordered fields over* **A**.
3. *We have* $x_1 > 0, \ \ldots, x_k > 0, a_1 \leq 0, \ \ldots, a_n \leq 0 \ \vdash \bot$ *in the theory of ordered fields over* **A**.
4. *We have an equality* $s + p = 0$ *in* **A**, *with* s *in the monoid generated by the* x_i*'s and* p *in the cone generated by the* x_i*'s and the* $-a_j$*'s.*

2.4 Linear Spectrum of a Lattice-Group

The theory of lattice-groups, denoted by $\mathcal{L}gr$, is a purely equational theory over the signature $(\cdot = 0; \cdot + \cdot, -\cdot, \cdot \vee \cdot, 0)$. The following rules express that $\vee$ defines a join semilattice and the compatibility of $\vee$ with $+$:

sdt1 $\vdash x \vee x = x$

sdt2 $\vdash x \vee y = y \vee x$

sdt3 $\vdash (x \vee y) \vee z = x \vee (y \vee z)$

grl $\vdash x + (y \vee z) = (x + y) \ \vee \ (x + z)$

We get the theory $\mathcal{Liog}$ by adding to $\mathcal{L}gr$ the axiom $\vdash x \geq 0$ **or** $-x \geq 0$.

The linear spectrum of an ℓ-group Γ corresponds to the intuition of "forcing the group to be linearly ordered". So a point of this spectrum can be given as a minimal model of the dynamical algebraic structure $\mathcal{Liog}(\Gamma)$, or equivalently by a linearly ordered group G quotient of Γ, or as the kernel H of the canonical morphism $\pi : \Gamma \to G$. This subgroup H is a *prime solid subgroup* of Γ.

The *linear lattice* of Γ, denoted by $\mathsf{Liog}(\Gamma)$, is generated by the entailment relation for Γ defined in the following way:

$$a_1, \ldots, a_n \vdash_{\mathsf{Liog}\,\Gamma} b_1, \ldots, b_m \quad \overset{\text{def}}{\Longleftrightarrow}$$
$$a_1 \geq 0, \ \ldots, a_n \geq 0 \ \vdash_{\mathcal{Liog}\Gamma} \ b_1 \geq 0 \ \textbf{or} \ \ldots \ \textbf{or} \ b_m \geq 0.$$

The spectral space previously defined is (isomorphic to) $\mathsf{Spec}\,(\mathsf{Liog}\,\Gamma)$. We have a *formal Positivstellensatz* for this entailment relation $(m, n \neq 0)$.

$$a_1, \ldots, a_n \vdash_{\mathsf{Liog}(\Gamma)} b_1, \ldots, b_m \iff \exists k > 0 \ (b_1^- \wedge \ldots \wedge b_m^-) \leq k(a_1^- \vee \ldots \vee a_n^-).$$

2.5 Valuative Spectrum of a Commutative Ring

The valuative spectrum $\mathsf{Spev}\,\mathbf{A}$ of a commutative ring corresponds to the intuition of "forcing the ring to be a valued field". A point of this spectrum is given by an epimorphism $\varphi : \mathbf{A} \to \mathbf{K}$ where $(\mathbf{K}, \mathbf{V})$ is a valued field.[5] Moreover two such morphisms $\varphi : \mathbf{A} \to \mathbf{K}$ and $\varphi' : \mathbf{A} \to \mathbf{K}'$ define the same point of the spectrum if there exists an isomorphism of valued fields $\psi : \mathbf{K} \to \mathbf{K}'$ making the suitable diagram commutative.

We denote by $x \mid y$ the predicate over $\mathbf{A} \times \mathbf{A}$ corresponding to "$\varphi(x)$ divides[6] $\varphi(y)$ in $\mathbf{K}$". We get the following axioms:

[5] $\mathbf{V}$ is a valuation ring of $\mathbf{K}$.

[6] That is, $\exists z \in \mathbf{V}\ z\varphi(x) = \varphi(y)$.

- $\vdash 1 \mid 0$
- $\vdash -1 \mid 1$
- $a \mid b \ \vdash ac \mid bc$
- $\vdash a \mid b$ **or** $b \mid a$
- $0 \mid 1 \ \vdash \bot$
- $a \mid b, b \mid c \ \vdash a \mid c$
- $a \mid b, a \mid c \ \vdash a \mid b + c$
- $ax \mid bx \ \vdash a \mid b$ **or** $0 \mid x$

Any predicate $x \mid y$ over $\mathbf{A} \times \mathbf{A}$ satisfying these axioms defines a point in $\mathsf{Spev}\,\mathbf{A}$. So we define the *valuative lattice* of $\mathbf{A}$, denoted by $\mathsf{Val}(\mathbf{A})$ as generated by the minimal entailment relation for $\mathbf{A} \times \mathbf{A}$ satisfying the following relations:

- $\vdash (1, 0)$
- $\vdash (-1, 1)$
- $(a, b) \vdash (ac, bc)$
- $\vdash (a, b), (b, a)$
- $(0, 1) \vdash$
- $(a, b), (b, c) \vdash (a, c)$
- $(a, b), (a, c) \vdash (a, b + c)$
- $(ax, bx) \vdash (a, b), (0, x)$

The two spectral spaces $\mathsf{Spec}\,(\mathsf{Val}\,\mathbf{A})$ and $\mathsf{Spev}\,\mathbf{A}$ can be identified. The spectral topology of $\mathsf{Spec}\,(\mathsf{Val}\,\mathbf{A})$ is generated by the basic open sets $\mathfrak{U}\big((a, b)\big) = \{\, \varphi \in \mathsf{Spev}\,\mathbf{A} \mid \varphi(a) \mid \varphi(b) \,\}$.

We have a formal Valuativstellensatz.

Theorem 5. (Formal Valuativstellensatz for valued fields) *Let* $\mathbf{A}$ *be a commutative ring, t.f.a.e.*

1. *One has* $(a_1, b_1), \ldots, (a_n, b_n) \vdash (c_1, d_1), \ldots, (c_m, d_m)$ *in the lattice* $\mathsf{Val}\,\mathbf{A}$.
2. *Introducing indeterminates* x_i*'s* $(i \in [\![1..n]\!])$ *and* y_j*'s* $(j \in [\![1..m]\!])$ *we have in the ring* $\mathbf{A}[\underline{x}, \underline{y}]$ *an equality*

$$d\big(1 + \sum\nolimits_{j=1}^{m} y_j P_j(\underline{x}, \underline{y})\big) \in \big\langle (x_i a_i - b_i)_{i \in [\![1..n]\!]}, (y_j d_j - c_j)_{j \in [\![1..m]\!]} \big\rangle,$$

where d *is in the monoid generated by the* d_j*'s and the* $P_j(x_1, \ldots, x_n, y_1, \ldots, y_m)$*'s are in* $\mathbb{Z}[\underline{x}, \underline{y}]$.

2.6 Heitmann Lattice and J-Spectrum of a Commutative Ring

In a commutative ring the *Jacobson radical of an ideal* $\mathfrak{J}$ denoted by $\mathrm{J}_{\mathbf{A}}(\mathfrak{J})$ is defined in classical mathematics as the intersection of the maximal ideals containing $\mathfrak{J}$. In constructive mathematics we use the classically equivalent definition

$$\mathrm{J}_{\mathbf{A}}(\mathfrak{J}) \stackrel{\text{def}}{=} \{\, x \in \mathbf{A} \mid \forall y \in \mathbf{A}, \ \ 1 + xy \text{ is invertible modulo } \mathfrak{J} \,\}. \qquad (9)$$

We write $\mathrm{J}_{\mathbf{A}}(x_1, \ldots, x_n)$ for $\mathrm{J}_{\mathbf{A}}(\langle x_1, \ldots, x_n \rangle)$. The ideal $\mathrm{J}_{\mathbf{A}}(0)$ is called *the Jacobson radical of* $\mathbf{A}$.

The *Heitmann lattice of* $\mathbf{A}$ is $\mathsf{He}(\mathsf{Zar}\,\mathbf{A})$, denoted by $\mathsf{Heit}\,\mathbf{A}$; it is a quotient of $\mathsf{Zar}\,\mathbf{A}$. In fact $\mathsf{Heit}\,\mathbf{A}$ can be identified with the set of ideals $\mathrm{J}_{\mathbf{A}}(x_1, \dots, x_n)$, with $\mathrm{J}_{\mathbf{A}}(\mathfrak{j}_1) \wedge \mathrm{J}_{\mathbf{A}}(\mathfrak{j}_2) = \mathrm{J}_{\mathbf{A}}(\mathfrak{j}_1\mathfrak{j}_2)$ and $\mathrm{J}_{\mathbf{A}}(\mathfrak{j}_1) \vee \mathrm{J}_{\mathbf{A}}(\mathfrak{j}_2) = \mathrm{J}_{\mathbf{A}}(\mathfrak{j}_1 + \mathfrak{j}_2)$.

We denote by $\mathsf{Jspec}(\mathbf{A})$ the spectral space $\mathsf{Spec}\,(\mathsf{Heit}\,\mathbf{A})$. In classical mathematics it is the adherence (for the patch topology) of the maximal spectrum in $\mathsf{Spec}\,\mathbf{A}$. We call it the (Heitmann) J-spectrum of $\mathbf{A}$. It is a subspectral space of $\mathsf{Spec}\,\mathbf{A}$. When $\mathbf{A}$ is Noetherian, $\mathsf{Jspec}(\mathbf{A})$ coincides with the subspace $\mathsf{jspec}(\mathbf{A})$ of $\mathsf{Spec}\,\mathbf{A}$ made of the prime ideals which are intersections of maximal ideals.

Remark. $\mathrm{J}_{\mathbf{A}}(x_1, \dots, x_n)$ is a radical ideal but not generally the nilradical of a finitely generated ideal. ∎

3 A Short Dictionary

References: [1, Theorem IV-2.6], [7, 11].

In this section we consider the following context: $f : \mathbf{T} \to \mathbf{T}'$ is a morphism of distributive lattices and $\mathsf{Spec}\,(f)$, denoted by $f^\star : X' = \mathsf{Spec}\,\mathbf{T}' \to X = \mathsf{Spec}\,\mathbf{T}$, is the dual morphism.

3.1 Properties of Morphisms

Theorem 6. ([1, Theorem IV-2.6]) *In classical mathematics we have the following equivalences:*

1. $f^\star$ *is onto* (f *is lying over*) $\iff$ f *is injective* $\iff$ f *is a monomorphism* $\iff$ $f^\star$ *is an epimorphism.*
2. f *is an epimorphism* $\iff$ $f^\star$ *is a monomorphism* $\iff$ $f^\star$ *is injective.*
3. f *is onto*[7] $\iff$ $f^\star$ *is an isomorphism on its image, which is a subspectral space of* X.

There are bijective morphisms of spectral spaces that are not isomorphisms. For example, the morphism $\mathsf{Spec}\,(\mathbb{B}\mathrm{o}(\mathbf{T})) \to \mathsf{Spec}\,\mathbf{T}$ is rarely an isomorphism and the lattice morphism $\mathbf{T} \to \mathbb{B}\mathrm{o}(\mathbf{T})$ is an injective epimorphism which is rarely onto.

Lemma 2. *Let* S *be a system of generators for* $\mathbf{T}$. *The morphism* f *is lying over if and only if for all* $a_1, \dots, a_n, b_1, \dots, b_m \in S$ *we have*

$$f(a_1), \dots, f(a_n) \vdash_{\mathbf{T}'} f(b_1), \dots, f(b_m) \;\Rightarrow\; a_1, \dots, a_n \vdash_{\mathbf{T}} b_1, \dots, b_m.$$

Proposition 2. (Going up vs. lying over) *In classical mathematics t.f.a.e. (see [11]):*

1. *For each prime ideal* $\mathfrak{q}$ *of* $\mathbf{T}'$ *and* $\mathfrak{p} = f^{-1}(\mathfrak{q})$, *the morphism* $f' : \mathbf{T}/(\mathfrak{p} = 0) \to \mathbf{T}'/(\mathfrak{q} = 0)$ *is lying over.*

[7] In other words, f is a quotient morphism.

2. *For each ideal I of $\mathbf{T}'$ and $J := f^{-1}(I)$, the morphism $f_I : \mathbf{T}/(J = 0) \to \mathbf{T}'/(I = 0)$ is lying over.*
3. *For each $y \in \mathbf{T}'$ and $J = f^{-1}(\downarrow y)$, the morphism $f_y : \mathbf{T}/(J = 0) \to \mathbf{T}'/(y = 0)$ is lying over.*
4. *For each $a, c \in \mathbf{T}$ and $y \in \mathbf{T}'$ we have*

$$f(a) \vdash_{\mathbf{T}'} f(c),\, y \implies \exists x \in \mathbf{T} \quad a \vdash_{\mathbf{T}} c,\, x \quad \text{and} \quad f(j) \leq_{\mathbf{T}'} x.$$

Theorem 7. *In classical mathematics we have the following equivalences [11]:*

1. *f is going up $\iff$ for each $a, c \in \mathbf{T}$ and $y \in \mathbf{T}'$ we have*

$$f(a) \leq f(c) \vee y \Rightarrow \exists x \in \mathbf{T}\ (a \leq c \vee x \text{ and } f(x) \leq y).$$

2. *f is going down $\iff$ for each $a, c \in \mathbf{T}$ and $y \in \mathbf{T}'$ we have*

$$f(a) \geq f(c) \wedge y \Rightarrow \exists x \in \mathbf{T}\ (a \geq c \wedge x \text{ and } f(x) \geq y).$$

3. *f has the property of incomparability $\iff$ f is zero-dimensional.*[8]

Theorem 8. *In classical mathematics t.f.a.e.*

1. *$\mathsf{Spec}\,(f)$ is an open map.*
2. *There exists a map $\widetilde{f} : \mathbf{T}' \to \mathbf{T}$ with the following properties:*
   (a) *For $c \in \mathbf{T}$ and $b \in \mathbf{T}'$, one has $b \leq f(c) \Leftrightarrow \widetilde{f}(b) \leq c$.*
   In particular, $b \leq f(\widetilde{f}(b))$ and $\widetilde{f}(b_1 \vee b_2) = \widetilde{f}(b_1) \vee \widetilde{f}(b_2)$.
   (b) *For $a, c \in \mathbf{T}$ and $b \in \mathbf{T}'$, one has $f(a) \wedge b \leq f(c) \Leftrightarrow a \wedge \widetilde{f}(b) \leq c$.*
   (c) *For $a \in \mathbf{T}$ and $b \in \mathbf{T}'$, one has $\widetilde{f}(f(a) \wedge b) = a \wedge \widetilde{f}(b)$.*
   (d) *For $a \in \mathbf{T}$, one has $\widetilde{f}(f(a)) = \widetilde{f}(1) \wedge a$.*
3. *There exists a map $\widetilde{f} : \mathbf{T}' \to \mathbf{T}$ satisfying property 2b.*
4. *For $b \in \mathbf{T}$ the g.l.b. $\bigwedge_{b \leq f(c)} c$ exists, and if we write it $\widetilde{f}(b)$, the property 2b holds.*

For this result in locales' theory see [5, Section 1.6]. We give now a proof for spectral spaces. Implications concerning item *1* need classical mathematics. The other equivalences are constructive.

Lemma 3. *Let $f : A \to A'$ be a nondecreasing map between ordered sets $(A, \leq)$ and $(A', \leq')$ and $b \in A'$. An element $b_1 \in A$ satisfies the equivalence*

$$\forall x \in A \ \ (b \leq' f(x) \iff b_1 \leq x\,)$$

[8] See Theorem 10.

if and only if

- *on the one hand* $b \leq' f(b_1)$,
- *and on the other hand* $b_1 = \bigwedge_{x:b\leq' f(x)} x$.

In particular, if b_1 *exists, it is uniquely determined.*

Proof. If b_1 satisfies the equivalence, one has $b \leq' f(b_1)$ since $b_1 \leq b_1$. If $z \in A$ satisfies the implication $\forall x \in A\ (b \leq' f(x) \Rightarrow z \leq x)$, we get $z \leq b_1$ since $b \leq' f(b_1)$. So when b_1 satisfies the equivalence it is the maximum of $S_b \overset{\text{def}}{=} \bigcap_{b\leq' f(x)} \downarrow x \subseteq A$, i.e. the g.l.b. of $\{x \in A \mid b \leq' f(x)\}$. Conversely, if such a g.l.b. b_1 exists, it satisfies the implication $\forall x \in A\ (b \leq' f(x) \Rightarrow b_1 \leq x)$ since $b_1 \in S_b$. Moreover, if $b \leq' f(b_1)$ we have the converse implication $\forall x \in A\ (b_1 \leq x \Rightarrow b \leq' f(x))$ because if $b_1 \leq x$ then $b \leq' f(b_1) \leq' f(x)$.

Proof of Theorem 8. *3* $\Rightarrow$ *2*. The property *2a* is the particular case of *2b* with $a = 1$. The property *2d* is the particular case of *2c* with $b = 1$. It remains to see that *2b* implies *2c*. Indeed

$$\begin{aligned}\widetilde{f}(f(a) \wedge b) &= \bigwedge\nolimits_{c:f(a)\wedge b\leq f(c)} c \quad \text{(Lemma 3)}\\ &= \bigwedge\nolimits_{c:a\wedge \widetilde{f}(b)\leq c} c \quad \text{(item 2b)}\\ &= a \wedge \widetilde{f}(b)\end{aligned}$$

1 $\Rightarrow$ *3*. We assume the map $f^\star : \mathsf{Spec}\,\mathbf{T}' \to \mathsf{Spec}\,\mathbf{T}$ to be open. If $b \in \mathbf{T}'$, the quasi-compact open set $\mathfrak{D}_{\mathbf{T}'}(b) = \mathfrak{B}$ has as image a quasi-compact open set of $\mathbf{T}$, written as $f^\star(\mathfrak{B}) = \mathfrak{D}_{\mathbf{T}}(\widetilde{b})$ for a unique $\widetilde{b} \in \mathbf{T}$. We write $\widetilde{b} = \widetilde{f}(b)$ and we get a map $\widetilde{f} : \mathbf{T}' \to \mathbf{T}$.

It remains to see that item *2b* is satisfied. For $a, c \in \mathbf{T}$ let us write $\mathfrak{A} = \mathfrak{D}_{\mathbf{T}}(a)$, $\mathfrak{C} = \mathfrak{D}_{\mathbf{T}}(c)$ and $g = f^\star$. We have to prove the equivalence *2b*, written as

$$g^{-1}(\mathfrak{A}) \cap \mathfrak{B} \subseteq g^{-1}(\mathfrak{C}) \iff \mathfrak{A} \cap g(\mathfrak{B}) \subseteq \mathfrak{C}.$$

For the direct implication, we consider an $x \in \mathfrak{B}$ such that $g(x) \in \mathfrak{A}$. We have to show that $g(x) \in \mathfrak{C}$. But $x \in g^{-1}(\mathfrak{A}) \cap \mathfrak{B}$, so $x \in g^{-1}(\mathfrak{C})$, i.e. $g(x) \in \mathfrak{C}$.

For the converse implication, we transform the r.h.s. by g^{-1}. This operation respects inclusion and intersection. We get $g^{-1}(\mathfrak{A}) \cap g^{-1}(g(\mathfrak{B})) \subseteq g^{-1}(\mathfrak{C})$ and we conclude by noticing that $\mathfrak{B} \subseteq g^{-1}(g(\mathfrak{B}))$.

2 $\Rightarrow$ *1*. We show that $f^\star(\mathfrak{D}_{\mathbf{T}'}(b)) = \mathfrak{D}_{\mathbf{T}}(\widetilde{f}(b))$.
First we show $f^\star(\mathfrak{D}_{\mathbf{T}'}(b)) \subseteq \mathfrak{D}_{\mathbf{T}}(\widetilde{f}(b))$. Let $\mathfrak{p}' \in \mathsf{Spec}\,\mathbf{T}'$ with $b \notin \mathfrak{p}'$ and let

$$\mathfrak{p} = f^\star(\mathfrak{p}') = f^{-1}(\mathfrak{p}').$$

If one had $\widetilde{f}(b) \in \mathfrak{p}$ one would have $f(\widetilde{f}(b)) \in f(\mathfrak{p}) \subseteq \mathfrak{p}'$ and since $b \leq f(\widetilde{f}(b))$, $b \in \mathfrak{p}'$. So we have $\mathfrak{p} \in \mathfrak{D}_{\mathbf{T}}(\widetilde{f}(b))$.

For the reverse inclusion, let us consider a $\mathfrak{p} \in \mathfrak{D}_{\mathbf{T}}(\widetilde{f}(b))$. As $\widetilde{f}$ is nondecreasing and respects $\vee$, the inverse image $\mathfrak{q} = \widetilde{f}^{-1}(\mathfrak{p})$ is an ideal.

We have $b \notin \mathfrak{q}$ because if $b \in \mathfrak{q}$ we have $\widetilde{f}(b) \in \widetilde{f}(\widetilde{f}^{-1}(\mathfrak{p})) \subseteq \mathfrak{p}$.
If $y \in \mathfrak{q}$ then $\widetilde{f}(y) = z \in \mathfrak{p}$ so $y \leq f(z)$ for a $z \in \mathfrak{p}$ (item *2a*). Conversely if $y \leq f(z)$ for a $z \in \mathfrak{p}$, then $\widetilde{f}(y) \leq \widetilde{f}(f(z)) \leq z$ (item *2d*), so $\widetilde{f}(y) \in \mathfrak{p}$. So we get

$$\mathfrak{q} = \widetilde{f}^{-1}(\mathfrak{p}) = \left\{ y \in \mathbf{T}' \mid \exists z \in \mathfrak{p} \;\; y \leq f(z) \right\}.$$

So $f^{-1}(\mathfrak{q}) = \{ x \in \mathbf{T} \mid \exists z \in \mathfrak{p} \;\; f(x) \leq f(z) \}$. But $f(x) \leq f(z)$ is equivalent to $x \wedge \widetilde{f}(1) \leq z$ (item 2b with $b = 1$). Moreover $\widetilde{f}(1) \notin \mathfrak{p}$ since $\widetilde{f}(b) \leq \widetilde{f}(1)$ and $\widetilde{f}(b) \notin \mathfrak{p}$. So

$$f^{-1}(\mathfrak{q}) = \left\{ x \in \mathbf{T} \;\middle|\; \exists z \in \mathfrak{p} \;\; x \wedge \widetilde{f}(1) \leq z \right\} = \left\{ x \in \mathbf{T} \;\middle|\; x \wedge \widetilde{f}(1) \in \mathfrak{p} \right\} = \mathfrak{p}.$$

Nevertheless it is possible that $\mathfrak{q}$ be not a prime ideal. In this case let us consider an ideal $\mathfrak{q}'$ which is maximal among those satisfying $f^{-1}(\mathfrak{q}') = \mathfrak{p}$ and $\widetilde{f}(b) \notin \mathfrak{q}'$. We want to show that $\mathfrak{q}'$ is prime. Assume we have y_1 and $y_2 \in \mathbf{T}' \setminus \mathfrak{q}'$ such that $y = y_1 \wedge y_2 \in \mathfrak{q}'$. By maximality there is an element $z_i \in \mathbf{T} \setminus \mathfrak{p}$ such that $f(z_i)$ is in the ideal generated by $\mathfrak{q}'$ and y_i ($i = 1, 2$), i.e. $f(z_i) \leq x_i \vee y_i$ with $x_i \in \mathfrak{q}'$. Taking $z = z_1 \wedge z_2$ (it is in $\mathbf{T} \setminus \mathfrak{p}$) and $x = x_1 \vee x_2$ we get $f(z_i) \leq x \vee y_i$ and $f(z) = f(z_1) \wedge f(z_2) \leq x \vee y_i$, so $f(z) \leq x \vee y \in \mathfrak{q}'$, and finally $z \in f^{-1}(\mathfrak{q}') = \mathfrak{p}$: a contradiction.
4 $\Leftrightarrow$ *3*. Use Lemma 3 by noticing that *2b* implies *2a*.

3.2 Dimension Properties

Theorem 9. (Dimension of spaces, see [12, 18], [17, Chapter XIII]) *In classical mathematics t.f.a.e.*

1. *The spectral space* Spec $(\mathbf{T})$ *is of Krull dimension* $\leq n$ *(with the meaning of chains of primes).*
2. *For each sequence* $(x_0, \ldots, x_n)$ *in* $\mathbf{T}$ *there exists a* complementary sequence $(y_0, \ldots, y_n)$*, which means*

$$\left.\begin{array}{rcl} 1 & \vdash & y_n, x_n \\ y_n, x_n & \vdash & y_{n-1}, x_{n-1} \\ \vdots & \vdots & \vdots \\ y_1, x_1 & \vdash & y_0, x_0 \\ y_0, x_0 & \vdash & 0 \end{array}\right\} \qquad (10)$$

For example, for the dimension $n \leq 2$, the inequalities in (10) correspond to the following diagram in $\mathbf{T}$:

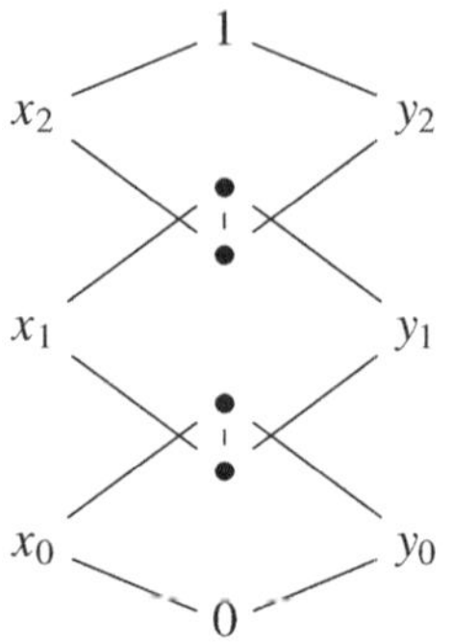

A zero-dimensional distributive lattice is a Boolean algebra.

Theorem 10. (Dimension of morphisms, see [11], [17, Section XIII-7]) *Let* $\mathbf{T} \subseteq \mathbf{T}'$ *and* f *be the inclusion morphism. In classical mathematics t.f.a.e.*

1. *The morphism* $\mathsf{Spec}\,(f) : \mathsf{Spec}\,(\mathbf{T}') \to \mathsf{Spec}\,(\mathbf{T})$ *has Krull dimension* $\leq n$.
2. *For any sequence* $(x_0, \ldots, x_n)$ *in* $\mathbf{T}'$ *there exists an integer* $k \geq 0$ *and elements* $a_1, \ldots, a_k \in \mathbf{T}$ *such that for each partition* (H, H') *of* $\{1, \ldots, k\}$, *there exist* $y_0, \ldots, y_n \in \mathbf{T}'$ *such that*

$$\begin{array}{rcl} \bigwedge_{j\in H'} a_j & \vdash & y_n,\ x_n \\ y_n,\ x_n & \vdash & y_{n-1},\ x_{n-1} \\ \vdots\ \vdots & & \vdots \\ y_1,\ x_1 & \vdash & y_0,\ x_0 \\ y_0,\ x_0 & \vdash & \bigvee_{j\in H} a_j \end{array} \tag{11}$$

For example, for the relative dimension $n \leq 2$, the inequalities in (11) correspond to the following diagram in $\mathbf{T}$, with $u = \bigwedge_{j\in H'} a_j$ and $i = \bigvee_{j\in H} a_j$:

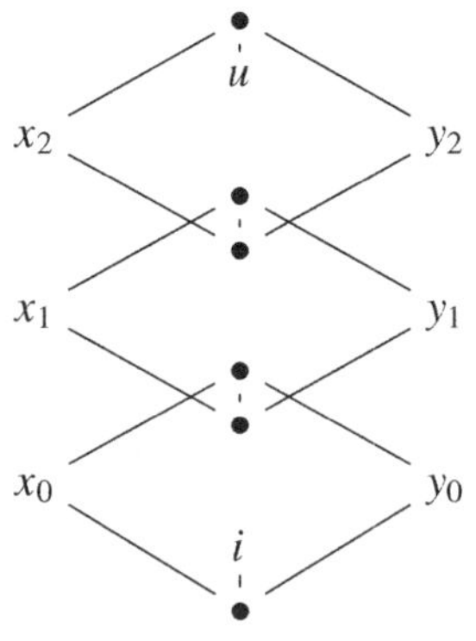

Note that the dimension of the morphism $\mathbf{T} \to \mathbf{T}'$ is less than or equal to the dimension of $\mathbf{T}'$: take $k = 0$ in item 2.

The Krull dimension of a ring **A** and of a morphism $\varphi : \mathbf{A} \to \mathbf{B}$ are those of $\mathsf{Zar}\,\mathbf{A}$ and $\mathsf{Zar}\,\varphi$.

A commutative ring **A** is zero-dimensional when for each $a \in \mathbf{A}$ there exist $n \in \mathbb{N}$ and $x \in \mathbf{A}$ such that $x^n(1 - xa) = 0$. A reduced zero-dimensional ring[9] is a ring in which any element a has a *quasi-inverse* $b = a^\bullet$, i.e. such that $aba = a$ and $bab = b$.

Let $\mathbf{A}^\bullet$ the reduced zero-dimensional ring generated by **A**. Then the Krull dimension of a morphism $\rho : \mathbf{A} \to \mathbf{B}$ equals the Krull dimension of the ring $\mathbf{A}^\bullet \otimes_{\mathbf{A}} \mathbf{B}$.

3.3 Properties of Spaces

The spectral space $\mathsf{Spec}\ \mathbf{T}$ is said to be *normal* if any prime ideal of **T** is contained in a unique maximal ideal.

Theorem 11. *We have the following equivalences:*

1. *The spectral space* $\mathsf{Spec}\,(\mathbf{T})$ *is normal* $\iff$
   for each $x \vee y = 1$ *in* **T** *there exist* a, b *such that* $a \vee x = b \vee y = 1$ *and* $a \wedge b = 0$.
2. *In the spectral space* $\mathsf{Spec}\,(\mathbf{T})$ *each quasi-compact open set is a finite union of irreducible quasi-compact open sets* $\iff$
   the distributive lattice **T** *is constructed from a dynamical algebraic structure where all axioms are Horn rules (e.g. this is the case for purely equational theories).*

4 Some Examples

We give in this section constructive versions of classical theorems. Often, the theorem has exactly the same wording as the classical one. But now, these theorems have a clear computational content, which was impossible when using classical definitions. Sometimes the new theorem is stronger than the previously known classical results (e.g. Theorems 17 or 18 or 19).

4.1 Relative Dimension, Lying Over, Going Up, Going Down

See [11] and [17, Section XIII-9].

Theorem 12. *Let* $\rho : \mathbf{A} \to \mathbf{B}$ *be a morphism of commutative rings or distributive lattices. If* $\mathsf{Kdim}\ \mathbf{A} \le m$ *and* $\mathsf{Kdim}\ \rho \le n$*, then* $\mathsf{Kdim}\ \mathbf{B} \le (m + 1)(n + 1) - 1$.

---

[9] Such a ring is also called absolutely flat or von Neumann regular.

Theorem 13. *If a morphism* $\alpha : \mathbf{A} \to \mathbf{B}$ *of distributive lattices or of commutative rings is lying over and going up (or lying over and going down) one has* $\mathsf{Kdim}(\mathbf{A}) \leq \mathsf{Kdim}(\mathbf{B})$.

Lemma 4. *Let* $\rho : \mathbf{A} \to \mathbf{B}$ *be a morphism of commutative rings. If* $\mathbf{B}$ *is generated by primitively algebraic elements*[10] *over* $\mathbf{A}$, *then* $\mathsf{Kdim}\,\rho \leq 0$ *and so* $\mathsf{Kdim}\,\mathbf{B} \leq \mathsf{Kdim}\,\mathbf{A}$.

Lemma 5. *Let* $\varphi : \mathbf{A} \to \mathbf{B}$ *be a morphism of commutative rings. The morphism* φ *is lying over if and only if for each ideal* $\mathfrak{a}$ *of* $\mathbf{A}$ *and each* $x \in \mathbf{A}$, *one has* $\varphi(x) \in \varphi(\mathfrak{a})\mathbf{B} \Rightarrow x \in \sqrt[\mathbf{A}]{\mathfrak{a}}$.

Lemma 6. *Let* $\varphi : \mathbf{A} \to \mathbf{B}$ *be a morphism of commutative rings. T.F.A.E.*

1. *The morphism* φ *is going up (i.e. the morphism* $\mathsf{Zar}\ \varphi$ *is going up).*
2. *For any ideal* $\mathfrak{b}$ *of* $\mathbf{B}$, *with* $\mathfrak{a} = \varphi^{-1}(\mathfrak{b})$, *the morphism* $\varphi_{\mathfrak{b}} : \mathbf{A}/\mathfrak{a} \to \mathbf{B}/\mathfrak{b}$ *is lying over.*
3. *The same thing with finitely generated ideals* $\mathfrak{b}$.
4. *(In classical mathematics) the same thing with prime ideals.*

Lemma 7. *Let* $\mathbf{A} \subseteq \mathbf{B}$ *be a faithfully flat* $\mathbf{A}$*-algebra. The morphism* $\mathbf{A} \to \mathbf{B}$ *is lying over and going up. So* $\mathsf{Kdim}\ \mathbf{A} \leq \mathsf{Kdim}\ \mathbf{B}$.

Lemma 8. (A classical going up) *Let* $\mathbf{A} \subseteq \mathbf{B}$ *be commutative rings with* $\mathbf{B}$ *integral over* $\mathbf{A}$. *Then the morphism* $\mathbf{A} \to \mathbf{B}$ *is lying over and going up. So* $\mathsf{Kdim}\ \mathbf{A} \leq \mathsf{Kdim}\ \mathbf{B}$.

Lemma 9. *Let* $\varphi : \mathbf{A} \to \mathbf{B}$ *be a morphism of commutative rings. T.F.A.E.*

1. *The morphism* φ *is going down.*
2. *For* $b, a_1, \ldots, a_q \in \mathbf{A}$ *and* $y \in \mathbf{B}$ *such that* $\varphi(b)y \in \sqrt[\mathbf{B}]{\langle \varphi(a_1, \ldots, a_q)\rangle}$, *there exist* $x_1, \ldots, x_p \in \mathbf{A}$ *such that*
$$\langle bx_1, \ldots, bx_p\rangle \subseteq \sqrt[\mathbf{A}]{\langle a_1, \ldots, a_q\rangle} \ \text{ and } \ y \in \sqrt[\mathbf{B}]{\langle \varphi(x_1), \ldots, \varphi(x_p)\rangle}.$$
3. *(In classical mathematics) for each prime ideal* $\mathfrak{p}$ *of* $\mathbf{B}$ *with* $\mathfrak{q} = \varphi^{-1}(\mathfrak{p})$ *the morphism* $\mathbf{A}_{\mathfrak{q}} \to \mathbf{B}_{\mathfrak{p}}$ *is lying over.*

Theorem 14. (Going down) *Let* $\mathbf{A} \subseteq \mathbf{B}$ *be commutative rings. The inclusion morphism* $\mathbf{A} \to \mathbf{B}$ *is going down in the following cases:*

1. $\mathbf{B}$ *is flat over* $\mathbf{A}$.
2. $\mathbf{B}$ *is a domain integral over* $\mathbf{A}$, *and* $\mathbf{A}$ *is integrally closed.*

---

[10] An element of $\mathbf{B}$ is said to be primitively algebraic over $\mathbf{A}$ if it annihilates a polynomial in $\mathbf{A}[X]$ whose coefficients are comaximal.

4.2 Kronecker, Forster–Swan, Serre and Bass Theorems

References: [8, 9] and [17, Chapter XIV].

Theorem 15. (Kronecker–Heitmann theorem, with Krull dimension, without Noetherianity)

1. *Let* $n \geq 0$. *If* $\mathsf{Kdim}\,\mathbf{A} < n$ *and* $b_1, \ldots, b_n \in \mathbf{A}$, *there exist* $x_1, \ldots, x_n$ *such that for all* $a \in \mathbf{A}$, $\mathrm{D}_{\mathbf{A}}(a, b_1, \ldots, b_n) = \mathrm{D}_{\mathbf{A}}(b_1 + ax_1, \ldots, b_n + ax_n)$.
2. *Consequently in a ring with Krull dimension* $\leq n$, *every finitely generated ideal has the same nilradical as an ideal generated by at most* $n + 1$ *elements.*

For a commutative ring $\mathbf{A}$ we define $\mathsf{Jdim}\,\mathbf{A}$ (J-dimension of $\mathbf{A}$) as being $\mathsf{Kdim}(\mathsf{Heit}\,\mathbf{A})$. In classical mathematics it is the dimension of the Heitmann J-spectrum $\mathsf{Jspec}(\mathbf{A})$.

Another dimension, called Heitmann dimension and denoted by $\mathsf{Hdim}(\mathbf{A})$, has been introduced in [8, 9]. One has always $\mathsf{Hdim}(\mathbf{A}) \leq \mathsf{Jdim}(\mathbf{A}) \leq \mathsf{Kdim}(\mathbf{A})$. The following results with Jdim hold also for Hdim.

Definition 4. A ring $\mathbf{A}$ is said to have *stable range (of Bass)* less than or equal to n when unimodular vectors of length $n + 1$ may be shortened in the following meaning:

$$1 \in \langle a, a_1, \ldots, a_n \rangle \implies \exists x_1, \ldots, x_n,\ 1 \in \langle a_1 + x_1 a, \ldots, a_n + x_n a \rangle\,.$$

Theorem 16. (Bass–Heitmann Theorem, without Noetherianity) *Let* $n \geq 0$. *If* $\mathsf{Jdim}\,\mathbf{A} < n$, *then* $\mathbf{A}$ *has* stable range $\leq n$. *In particular each stably free* $\mathbf{A}$*-module of rank* $\geq n$ *is free.*

A matrix is said to be of rank $\geq k$ when the minors of size k are comaximal.

Theorem 17. (Serre's Splitting Off theorem, for Jdim)
Let $k \geq 1$ *and* M *be a projective* $\mathbf{A}$*-module of rank* $\geq k$, *or more generally isomorphic to the image of a matrix of rank* $\geq k$.
Assume that $\mathsf{Jdim}\,\mathbf{A} < k$. *Then* $M \simeq N \oplus \mathbf{A}$ *for a suitable module* N *isomorphic to the image of a matrix of rank* $\geq k - 1$.

Corollary 1. *Let* $\mathbf{A}$ *be a ring such that* $\mathsf{Jdim}\,\mathbf{A} \leq h$ *and* M *be an* $\mathbf{A}$*-module isomorphic to the image of a matrix of rank* $\geq h + s$. *Then* M *has a direct summand which is a free submodule of rang* s. *Precisely, if* M *is the image of a matrix* $F \in \mathbf{A}^{n \times m}$ *of rank* $\geq h + s$, *one has* $M = N \oplus L$ *where* L *is a direct summand that is free of rank* s *in* $\mathbf{A}^n$, *and* N *the image of a matrix of rank* $\geq h$.

In the following theorem we use the notion of finitely generated module *locally generated by* k *elements*. In classical mathematics this means that after localisation at any maximal ideal, M is generated by k elements. A classically equivalent constructive definition is that the k-th Fitting ideal of M is equal to $\langle 1 \rangle$.

Theorem 18. (Forster–Swan *theorem* for Jdim*) If* Jdim **A** $\leq k$ *and if the* **A***-module* $M = \langle y_1, \ldots, y_{k+r+s} \rangle$ *is locally generated by r elements, then it is generated by* $k + r$ *elements: one can compute* $z_1, \ldots, z_{k+r}$ *in* $\langle y_{k+r+1}, \ldots, y_{k+r+s} \rangle$ *such that M is generated by* $(y_1 + z_1, \ldots, y_{k+r} + z_{k+r})$.

Theorem 19. (Bass' cancellation *theorem,* with Jdim*)*
Let M be a finitely generated projective **A***-module of rank* $\geq k$*. If* Jdim **A** $< k$*, then M is cancellative for every finitely generated projective* **A***-module. That is, if Q is finitely generated projective and* $M \oplus Q \simeq N \oplus Q$*, then* $M \simeq N$.

Theorems 17, 18 and 19 were conjectured by Heitmann in [14] (he proved these theorems for the Krull dimension without Noetherianity assumption).

4.3 Other Results Concerning Krull Dimension

In [17] Theorem XII-6.2 gives the following important characterisation. *An integrally closed coherent ring* **A** *of Krull dimension at most* 1 *is a Prüfer domain.* This explains in a constructive way the nowadays classical definition of Dedekind domains as Noetherian, integrally closed domains of Krull dimension 1, and the fact that, from this definition, in classical mathematics, one is able to prove that finitely generated nonzero ideals are invertible.

In [17, Chapter XVI] there is a constructive proof of the Lequain–Simis theorem. This proof uses the Krull dimension.

In [24, Section 2.6] we find the following new result, with a constructive proof. *If* **A** *is a ring of Krull dimension* $\leq d$*, then the stably free modules of rank* $> d$ *over* $\mathbf{A}[X]$ *are free.*

References

1. Balbes, R., Dwinger, P.: Distributive Lattices. University of Missouri Press, Columbia, Mo (1974)
2. Bezem, M., Coquand, T.: Automating coherent logic. In: Logic for Programming, Artificial Intelligence, and Reasoning. Proceedings of the 12th International Conference, LPAR 2005, Montego Bay, Jamaica, 2–6 Dec 2005. pp. 246–260. Springer, Berlin (2005)
3. Bishop, E., Bridges, D.: Constructive analysis. Grundlehren der Mathematischen Wissenschaften [Fundamental Principles of Mathematical Sciences], vol. 279. Springer, Berlin (1985)
4. Bishop, E.: Foundations of Constructive Analysis. McGraw-Hill, New York (1967)
5. Borceux, F.: Handbook of Categorical Algebra, Encyclopedia of Mathematics and its Applications, vol. 3, no. 52. Cambridge University Press, Cambridge (1994). Categories of sheaves
6. Bridges, D., Richman, F.: Varieties of constructive mathematics. London Mathematical Society Lecture Note Series, vol. 97. Cambridge University Press, Cambridge (1987)
7. Cederquist, J., Coquand, T.: Entailment relations and distributive lattices. In: Logic Colloquium '98 (Prague). Lecture Note in Logic, vol. 13, pp. 127–139. Association for Symbolic Logic, Urbana, IL (2000)

8. Coquand, T., Lombardi, H., Quitté, C.: Dimension de Heitmann des treillis distributifs et des anneaux commutatifs. In: Algèbre et théorie des nombres. Années 2003–2006, pp. 57–100. Laboratoire de Mathématiques de Besançon, Besançon, 2006. Revised version in 2017. http://arxiv.org/abs/1712.01958
9. Coquand, T., Lombardi, H., Quitté, C.: Generating non-Noetherian modules constructively. Manuscripta Math. **115**(4), 513–520 (2004)
10. Coquand, T., Lombardi, H.: A logical approach to abstract algebra. Math. Struct. Comput. Sci. **16**(5), 885–900 (2006)
11. Coquand, T., Lombardi, H.: Constructions cachées en algebre abstraite. Dimension de Krull, Going up, Going down. 2001. Updated version. Technical report, Laboratoire de Mathématiques de l'Université de Franche-Comté (2018). http://hlombardi.free.fr/GoingUpDownFrench-2018.pdf
12. Coquand, T., Lombardi, H.: Hidden constructions in abstract algebra: Krull dimension of distributive lattices and commutative rings. In: Commutative Ring Theory and Applications (Fez, 2001). Lecture Notes in Pure and Applied Mathematics, Dekker, New York, vol. 231, pp. 477–499 (2003)
13. Coste, M., Lombardi, H., Roy, M.-F.: Dynamical method in algebra: effective Nullstellensätze. Ann. Pure Appl. Log. **111**(3), 203–256 (2001)
14. Heitmann, R.: Generating non-Noetherian modules efficiently. Mich. Math. J. **31**, 167–180 (1984)
15. Hochster, M.: Prime ideal structure in commutative rings. Trans. Am. Math. Soc. **142**, 43–60 (1969)
16. Johnstone, P.T.: Stone Spaces. Cambridge studies in Advanced Mathematics, vol. 3. Cambridge University Press, Cambridge (1986). Reprint of the 1982 edition
17. Lombardi, H., Quitté, C.: Commutative algebra: constructive methods. Finite projective modules. Algebra and Applications. Springer, Dordrecht (2015). Translated from the French (Calvage & Mounet, 2011, revised and extended by the authors) by Tania K. Roblot
18. Lombardi, H.: Dimension de Krull, Nullstellensätze et évaluation dynamique. Math. Z. **242**(1), 23–46 (2002)
19. Lombardi, H.: Structures algébriques dynamiques, espaces topologiques sans points et programme de Hilbert. Ann. Pure Appl. Log. **137**(1–3), 256–290 (2006)
20. Lombardi, H.: Théories géométriques pour l'algébre constructive (2018). http://hlombardi.free.fr/Theories-geometriques.pdf
21. Lorenzen, P.: Algebraische und logistische Untersuchungen über freie Verbände. J. Symbol. Log. **16**, 81–106 (1951). Translation by Stefan Neuwirth: Algebraic and logistic investigations on free lattices. http://arxiv.org/abs/1710.08138
22. Mines, R., Richman, F., Ruitenburg, W.: A course in constructive algebra. Universitext. Springer, New York (1988)
23. Stone, M.H.: Topological representations of distributive lattices and Brouwerian logics. Cas. Mat. Fys. **67**, 1–25 (1937)
24. Yengui, I.: Constructive commutative algebra: projective modules over polynomial rings and dynamical Gröbner bases. Lecture Notes in Mathematics, vol. 2138. Springer, Cham (2015)

Valuative Marot Rings

Thomas G. Lucas

Abstract For a commutative ring R with total quotient ring $T(R)$, R is said to be valuative if for each nonzero $t \in T(R)$ at least one of the extensions $R \subseteq R[t]$ and $R \subseteq R[(R:t)]$ has no proper intermediate rings. There are weak and strong versions: R is weakly valuative if for each pair $s, t \in T(R)\backslash\{0\}$ such that $st \in R$, at least one of $R \subseteq R[t]$ and $R \subseteq R[s]$ has no proper intermediate rings; R is strongly valuative if for each nonzero $t \in T(R)$ at least one of the extensions $R \subseteq R[(R:(R:t))]$ and $R \subseteq R[(R:t)]$ has no proper intermediate rings. There are weakly valuative rings that are not valuative, and valuative rings that are not strongly valuative. In the special case that R is a Marot ring (each regular ideal can be generated by a set of regular elements), R is weakly valuative if and only if it is strongly valuative. Also a valuative Marot ring has at most three regular maximal ideals.

Keywords Marot ring · Minimal extension · Valuative domain · Valuative ring

Subject Classifications [2100] Primary 13B99 · 13A15; Secondary 13G05 · 13B21

1 Introduction

Throughout this article, R denotes a commutative ring with nonzero identity. Also we use $Z(R)$ to denote the set of zero divisors of R, $T(R)$ to denote the total quotient ring of R, and R' to denote the integral closure of R in $T(R)$. We have $T(R) = \{a/b \mid a, b \in R \text{ with } b \in R\backslash Z(R)\}$. The set $\text{Reg}(R) = R\backslash Z(R)$ denotes the set of regular elements of R and an (fractional) ideal is *regular* if it contains a regular element of R (of $T(R)$). Also, $\text{RMax}(R)$ denotes the set of regular maximal ideals of R. We will concentrate on those R that are *Marot rings*, meaning each regular ideal can

T. G. Lucas (✉)
Department of Mathematics and Statistics, University of North Carolina Charlotte, Charlotte, NC 28223, USA
e-mail: tglucas@uncc.edu

A. Facchini et al. (eds.), *Advances in Rings, Modules and Factorizations*, Springer Proceedings in Mathematics & Statistics 321,
https://doi.org/10.1007/978-3-030-43416-8_14

be generated by a set of regular elements. A related (stronger) condition is that of a *weakly additively regular ring*: R is weakly additively regular if for each pair of elements $f, g \in R$ with f regular, there are elements $s, t \in R$ such that $gs + ft$ is regular and $sR + fR = R$.

An R-module $J \subseteq T(R)$ is an R-fractional ideal if there is a regular element $r \in R$ such that $rJ \subseteq R$. Also, we use the standard notation of $(R : J)$ to denote the set $\{t \in T(R) \mid tJ \subseteq R\}$ and $(R :_R J) = (R : J) \cap R$ (so the same as $\{r \in R \mid rJ \subseteq R\}$). In the case R is a Marot ring and J is an R-fractional ideal that contains a regular element, then J can be generated by regular elements (see Lemma 2.1 below).

As in [13], we say that R is a *weakly valuative* if for each pair of nonzero elements $x, y \in T(R)$ such that $xy \in R$, at least one of $R \subseteq R[x]$ and $R \subseteq R[y]$ has no proper intermediate rings. Also R is *valuative* if for each nonzero $x \in T(R)$ at least one of $R \subseteq R[x]$ and $R \subseteq R[(R : x)]$ has no proper intermediate rings. Finally, R is *strongly valuative* if for each nonzero $x \in T(R)$ at least one of $R \subseteq R[(R : (R : x))]$ and $R \subseteq R[(R : x)]$ has no proper intermediate rings. An alternate (equivalent) definition for weakly valuative is that for each $x \in T(R)$, either $R \subseteq R[x]$ has no proper intermediate rings or for each $y \in (R : x)$, the extension $R \subseteq R[y]$ has no proper intermediate rings. See [3] for a more general notion between comparable pairs of rings $S \subseteq T$.

Clearly a strongly valuative ring is valuative and a valuative ring is weakly valuative, but there are valuative rings that are not strongly valuative and weakly valuative rings that are not valuative [13, Examples 1.5–1.8]. However, if R is integrally closed in $T(R)$, then it is weakly valuative if and only if it is strongly valuative [3, Theorem 2.15]. More recently, it has been shown that if R is a weakly additively regular ring, then it is weakly valuative if and only if it is strongly valuative (see [13, Theorem 3.13] for the case that R is not integrally closed).

A valuative domain has at most three maximal ideals [1, Theorem 2.2] and at most two when it is not integrally closed [1, Theorem 6.2]. Similarly a weakly additively regular weakly valuative ring has at most three regular maximal ideals and at most two when it is not integrally closed [13, Theorem 3.9], but in both cases it may have infinitely many maximal ideals that are not regular. Also an integrally closed valuative Marot ring has at most three regular maximal ideals [13, Theorem 2.10]. Below we show that a weakly valuative Marot ring that is not integrally closed has at most two regular maximal ideals (see Theorem 3.1).

Several results in [1] characterize when an integral domain is valuative. In particular, [1, Theorem 3.7] covers the case that R is integrally closed, [1, Theorems 5.2 and 5.10] is for when R has a unique maximal ideal and is not integrally closed and [1, Theorem 6.2] is for the case R has exactly two maximal ideals and is not integrally closed. In [13], each of these results was extended to weakly additively regular rings, specifically [13, Theorem 3.5], [13, Theorem 3.13], and [13, Theorem 3.17], respectively. Below we give the analogous results for Marot rings: Theorem 2.17 (the same as [13, Theorem 2.10] cited in the previous paragraph), Theorems 3.4 and 3.7.

If $R \subsetneq R[t]$ is a minimal extension for some $t \in T \backslash R$ (i.e., there are no proper intermediate rings), then $M = \sqrt{(R :_R t)}$ is a maximal ideal of R (see, for example, [2]). The maximal ideal M is referred to as the *crucial maximal ideal* of the extension (see, for example, [5, Page 804]). When dealing with a specific ring extension $R \subsetneq T$, if there is an element $s \in T \backslash R$ such that $N = \sqrt{(R :_R s)}$ is a maximal ideal of R, then, as in [3], we refer to N as a *maximal T-radical ideal* of R (no matter whether the extension $R \subsetneq R[s]$ is minimal or not). It is known that if R is weakly valuative, then there are at most three maximal $T(R)$-radical ideals in R [3, Theorem 5.4]. The ring R in [13, Example 1.9] shows that even if a ring is strongly valuative, it can have infinitely many regular maximal ideals.

For a prime ideal P of R, $R_{[P]} = \{t \in T(R) \mid ts \in R$ for some $s \in R \backslash P\}$ and $R_{(P)} = \{t \in T(R) \mid ts \in R$ for some $s \in \mathrm{Reg}(R) \backslash P\}$. The latter is the same ring as R_H where $H = \mathrm{Reg}(R) \backslash P$. If P is not regular, then $\mathrm{Reg}(R) \cap P = \emptyset$ and so $R_{(P)} = T(R)$ in this case. There are different types of extensions of ideals to a ring of the type $R_{[P]}$. For example, one may consider the simple extension $IR_{[P]}$ and the potentially larger extension $[I]R_{[P]} = \{t \in T(R) \mid ts \in I$ for some $s \in R \backslash P\}$ (for example, $[I]R_{[P]} = R_{[P]}$ for each ideal I that is not contained in P, but $IR_{[P]}$ is a proper ideal if $I \subseteq Z(R)$). In contrast, $IR_{(P)} = \{t \in T(R) \mid ts \in I$ for some $s \in \mathrm{Reg}(R) \backslash P\}$. If R is a Marot ring and P is regular, then $R_{[P]} = R_{(P)}$ [14, Proposition 6]. Also in this case, if J is a regular ideal, then $JR_{[P]} = [J]R_{[P]}$, with $JR_{[P]} = R_{[P]}$ if J is not contained in P.

Recall that a ring R together with a prime ideal P is a *valuation pair* of $T(R)$ if for each $t \in T(R) \backslash R$, there is a $p \in P$ such that $tp \in R \backslash P$ (see, for example, [9, Page 25]). Corresponding to a valuation pair (R, P) is a totally ordered additive Abelian group G and a surjective map $\mathrm{v} : T(R) \to G \cup \{\infty\}$ (with $g < \infty$ for all $g \in G$) such that (i) $\mathrm{v}(ab) = \mathrm{v}(a) + \mathrm{v}(b)$, (ii) $\mathrm{v}(a + b) \geq \min\{\mathrm{v}(a), \mathrm{v}(b)\}$, (iii) $\mathrm{v}(1) = 0$ and $\mathrm{v}(0) = \infty$, and (iv) $R = \{t \in T(R) \mid \mathrm{v}(t) \geq 0\}$ and $P = \{t \in T(R) \mid \mathrm{v}(t) > 0\}$. A related weaker notion is that of a *mock valuation ring* meaning that R is such that $T(R) \backslash R$ is multiplicatively closed [3]. While there are mock valuation rings that are not valuation rings, if R is a Marot ring, it is a mock valuation ring if and only if it has a unique regular maximal ideal M and M is such that (R, M) is a valuation pair of $T(R)$ (see [9, Theorem 7.7] and its proof). Also R is a *Prüfer ring* if each finitely generated regular ideal is invertible. This is equivalent to having $(R_{[M]}, [M]R_{[M]})$ a valuation pair for each (regular) maximal ideal M [7, Theorem 13].

2 Preliminary Results

Let P be a prime ideal of a ring R and let S be a proper overring of R. Then we let $S_{(P)} = \{t \in T(R) \mid bt \in S$ for some $b \in \mathrm{Reg}(R) \backslash P\}$ and $S_{[P]} = \{t \in T(R) \mid bt \in S$ for some $b \in R \backslash P\}$. As with $R_{(P)}$ and $R_{[P]}$, if P is not a regular ideal of R, then $S_{(P)} = T(R) = S_{[P]}$.

Lemma 2.1. *Let R be a Marot ring.*

1. *Each regular R-fractional ideal can be generated by a set of regular elements.*
2. *If P is a regular prime ideal and I is a regular ideal, then $IR_{[P]} = [I]R_{[P]}$.*

Proof. Let J be a regular R-fractional ideal. Then there is a regular element $r \in R$ such that $rJ \subseteq R$. Since R is Marot, there is a set of regular elements X in rJ that generates rJ as an (integral) ideal of R. The set $r^{-1}X$ is a set of regular elements in J that generate J as an R-fractional ideal.

Next suppose P is a regular prime ideal of R and let I be a regular ideal. If I is not contained in P, then the Marot property guarantees the existence of a regular element $m \in I\backslash P$. Clearly m is a unit of $R_{[P]}$ and thus $[I]R_{[P]} \subseteq R_{[P]} = mR_{[P]} \subseteq IR_{[P]}$. On the other hand if I is contained in P, we at least know that $IR_{[P]} \subseteq [I]R_{[I]}$. For the reverse containment let $q \in [I]R_{[P]}$. Then there is an element $r \in R\backslash P$ such that $rq \in I$. We also have a regular element $p \in R$ such that $pq \in R$ and a regular element $b \in I$. The ideal $J = rR + pbR$ is a regular ideal contained in $(R :_R q)$ but not in P. Since R is Marot, J is generated by its regular elements, so there is a regular element $x \in J\backslash P$. As $xq \in I$ and x is a unit of $R_{[P]}$, $q \in IR_{[P]}$ as desired. □

Lemma 2.2. *Let M be a regular maximal ideal of a ring R and let S be a ring such that $R \subseteq S \subseteq T(R)$.*

1. *$MR_{[M]}$ is a maximal ideal of $R_{[M]}$, $(R_{[M]})_{[MR_{[M]}]} = R_{[M]}$ and $R_M = (R_{[M]})_{MR_{[M]}}$.*
2. *$(S_{[M]})_{[MR_{[M]}]} = S_{[M]}$ and $S_M = (S_{[M]})_{MR_{[M]}}$.*
3. *If R is a Marot ring, then $S_{[M]} = S_{(M)}$.*

Proof. Let $q \in R_{[M]}\backslash MR_{[M]}$. Then there is an element $r \in R\backslash M$ such that $qr \in R$. If $qr \notin M$, then there are elements $a \in R\backslash M$ and $n \in M$ such that $qra + n = 1$. Clearly both qra and n are in $qR_{[M]} + MR_{[M]}$. Thus $qR_{[M]} + MR_{[M]} = R_{[M]}$. In contrast, if $qr \in M$ for each such $r \in R\backslash M$, then for a given r we have elements $c \in R\backslash M$ and $m \in M$ such that $cr + m = 1$ which puts $q = crq + mq \in MR_{[M]}$, a contradiction. Hence $MR_{[M]}$ is a maximal ideal of $R_{[M]}$.

For (2) and the rest of (1), start with an element $i \in (S_{[M]})_{[MR_{[M]}]}$. Then there is an element $j \in R_{[M]}\backslash MR_{[M]}$ such that $ij \in S_{[M]}$. We have elements $k, m \in R\backslash M$ such that $kj \in R\backslash M$ and $mji \in S$. Also $mkji \in S$ with $mkj \in R\backslash M$. Thus $i \in S_{[M]}$. Hence $(S_{[M]})_{[MR_{[M]}]} = S_{[M]}$ and $(R_{[M]})_{[MR_{[M]}]} = R_{[M]}$.

To see that $S_M = (S_{[M]})_{MR_{[M]}}$, suppose $c/d \in (S_{[M]})_{MR_{[M]}}$ with $c \in S$ and $d \in R_{[M]}\backslash MR_{[M]}$. Then we have elements $x, y \in R\backslash M$ such that $xc \in S$ and $yd \in R$. We also have $xy, xyd \in R\backslash M$ and $xyc \in S$. Hence $c/d = xyc/xyd \in S_M$. Thus $S_M = (S_{[M]})_{MR_{[M]}}$ and $R_M = (R_{[M]})_{MR_{[M]}}$.

For (3), assume R is Marot and let $q \in S_{[M]}$. Then there is an element $t \in R\backslash M$ such that $qt \in S$. We also have a regular element $r \in R$ such that $qr \in R$. Thus the (integral) ideal $(S :_R q)$ is a regular ideal of R that is not contained in M. Since R is Marot, there is a regular element $x \in (S :_R q)\backslash M$. As $qx \in S$, $q \in S_{(M)}$. □

A more general version of the following lemma appears in [3].

Lemma 2.3. (cf. [3, Lemma 2.6]) *Let $R \subsetneq S$ be a minimal extension with corresponding crucial maximal ideal $M \in \mathrm{RMax}(R)$ and $S \subseteq T(R)$. Then $R_{[M]} \subsetneq S_{[M]}$ is also a minimal extension.*

For each $M \in \text{Max}(R)$, $R_{(M)}$ is a regular ring of quotients of R (inside $T(R)$) and thus $R'_{(M)}$ is the integral closure, $(R_{(M)})'$, of $R_{(M)}$ in $T(R)$.

Next we provide a pair of statements equivalent to saying that the integral closure of R is a Prüfer ring. These (and the corollary that follows) will be used to show that the integral closure of a weakly valuative Marot ring is a Prüfer ring.

Theorem 2.4. *Let R be a ring with integral closure $R' \supsetneq R$. Then the following are equivalent:*

1. *R' is a Prüfer ring.*
2. *$R'_{(M)}$ is a Prüfer ring for each $M \in \text{RMax}(R)$.*
3. *For each $M \in \text{RMax}(R)$, the integral closure of $R_{(M)}$ is a Prüfer ring.*

Proof. If R' is a Prüfer ring, then each ring between R' and $T(R)$ is also a Prüfer ring. Hence $R'_{(M)}$ is a Prüfer ring for each $M \in \text{RMax}(R)$. Thus (1) implies (2).

The equivalence of (2) and (3) follows from the fact that $R'_{(M)}$ is the integral closure of $R_{(M)}$ for each $M \in \text{RMax}(R)$.

To complete the proof we show (2) implies (1).

Let N' be a regular maximal ideal of R'. Then $N = N' \cap R$ is a regular maximal ideal of R. The ideal $Q = N'R'_{(N)}$ is a regular maximal ideal of $S = R'_{(N)}$. A proof similar to that used to establish statement (2) in Lemma 2.2 shows that $S_{[Q]} = R'_{[N']}$ and $[Q]S_{[Q]} = [N']R'_{[N']}$.

Since $R'_{(P)}$ is a Prüfer ring for each $P \in \text{RMax}(R)$, $(R'_{[M']}, [M']R'_{[M']})$ is a valuation pair for each $M' \in \text{RMax}(R')$. Therefore R' is a Prüfer ring. □

As a corollary we have the following special case.

Corollary 2.5. *Let R be a ring that is not integrally closed. If there is a unique (regular) maximal ideal M such that the integral closure of $R_{(M)}$ is a Prüfer ring and $(R_{(N)}, NR_{(N)})$ is a Prüfer valuation ring for all other $N \in \text{RMax}(R)$ (if any), then the integral closure of R is a Prüfer ring.*

We will make use of the following lemma in the proofs of Theorems 2.7 and 3.7.

Lemma 2.6. *Let R and S be rings with $R \subsetneq S \subseteq T(R)$. Then $S = \bigcap\{S_{(M)} \mid M \in \text{RMax}(R)\} = \bigcap\{S_{[M]} \mid M \in \text{RMax}(R)\}$.*

Proof. The usual conductor argument shows that $S = \bigcap\{S_{(N)} \mid N \in \text{Max}(R)\} = \bigcap\{S_{[N]} \mid N \in \text{Max}(R)\}$. Since $S_{(N)} = S_{[N]} = T(R)$ for all $N \in \text{Max}(R)\backslash\text{RMax}(R)$, we also have $S = \bigcap\{S_{(M)} \mid M \in \text{RMax}(R)\} = \bigcap\{S_{[M]} \mid M \in \text{RMax}(R)\}$. □

Theorem 2.7. *Let R and S be rings with $R \subsetneq S \subseteq T(R)$.*

1. *The following are equivalent:*
   a. *$R \subsetneq S$ is a minimal extension.*
   b. *There is a regular maximal ideal M of R such that $R_M \subsetneq S_M$ is minimal and $R_N = S_N$ for all other regular maximal ideals $N \in \text{RMax}(R)\backslash\{M\}$ (if any).*

c. *There is a regular maximal ideal M of R such that $R_{[M]} \subsetneq S_{[M]}$ is minimal and $R_{[N]} = S_{[N]}$ for all other regular maximal ideals $N \in \mathrm{RMax}(R)\backslash\{M\}$ (if any).*

2. *If R is a Marot ring, then $R \subsetneq S$ is a minimal extension if and only if there is a regular maximal ideal M of R such that $R_{(M)} \subsetneq S_{(M)}$ is minimal and $R_{(N)} = S_{(N)}$ for all other regular maximal ideals $N \in \mathrm{RMax}(R)\backslash\{M\}$ (if any).*

Proof. To start, if P is a prime ideal of R, then $R_P = S_P$ if and only if $R_{[P]} = S_{[P]}$. To see this first suppose $R_P = S_P$ and let $q \in S_{[P]}$. Then by definition, there is an element $r \in R\backslash P$ such that $qr \subset S$. From the equality $R_P = S_P$ we then have an element $t \in R\backslash P$ such that $qrt \in R$ and it follows that $q \in R_{[P]}$. Similarly, if $R_{[P]} = S_{[P]}$ and $x/y \in S_P$ with $x \in S$ and $y \in R\backslash P$, then there is an element $z \in R\backslash P$ such that $zx \in R$ and it follows that $x/y = zx/zy \in R_P$.

To establish the equivalence of (a) and (b) in (1), we note that from [6, Lemme 1.2 and Théorème 2.2], $R_M \subsetneq S_M$ is minimal for some (necessarily regular) maximal ideal $M \in \mathrm{Max}(R)$ and $R_P = S_P$ for all other prime ideals $P \in \mathrm{Spec}(R)$ when $R \subsetneq S$ is a minimal extension, so certainly $R_N = S_N$ for all regular maximal ideals $N \in \mathrm{RMax}(R)\backslash\{M\}$.

An observation in [2, Pages 1092 and 1093] is that if there is a maximal ideal $N \in \mathrm{Max}(R)$ such that $R_N \subsetneq S_N$ is minimal and $R_P = S_P$ for all $P \in \mathrm{Spec}(R)\backslash\{N\}$, then $R \subsetneq S$ is a minimal extension. We will start with only assuming we have a regular maximal ideal M of R such that $R_M \subsetneq S_M$ is minimal and $R_N = S_N$ for all $N = \mathrm{RMax}(R)\backslash\{M\}$ (if any). From the observation in [2], it suffices to show $R_P = S_P$ for all $P \in \mathrm{Spec}(R)\backslash\{M\}$. First note that if $P \in \mathrm{Spec}(R)$ is not regular, then $R_P = T(R)_P = S_P$ (as each regular element of R is a unit in R_P). For P regular and not contained in M, it is contained in some maximal ideal $N \in \mathrm{RMax}(R)$. As $R_N = S_N$, we also have $R_P = (R_N)_P = (S_N)_P = S_P$. Finally for P regular and (properly) contained in M, the assumption that $R_M \subsetneq S_M$ is a minimal extension yields that $R_P = (R_M)_P = (S_M)_P = S_P$ (from [6, Théorème 2.2]). Hence (a) and (b) are equivalent.

To see that (a) implies (c), we make use of the observation in the first sentence and Lemma 2.6. Suppose $R \subseteq S$ is minimal with corresponding crucial maximal ideal M. Then by Lemma 2.6, $R_{[M]} \subsetneq S_{[M]}$ is minimal. We also have $R_{[N]} = S_{[N]}$ for all $N \in \mathrm{RMax}(R)\backslash\{M\}$.

To complete the proof of (1), we show that (c) implies (a). Suppose we have a regular maximal ideal M of R such that $R_{[M]} \subsetneq S_{[M]}$ is minimal and $R_{[N]} = S_{[N]}$ for all $N \in \mathrm{RMax}(R)\backslash\{M\}$ (if any). From the latter, we have $R_N = S_N$ for all $N \in \mathrm{RMax}(R)\backslash\{M\}$. Let W be a ring such that $R \subsetneq W \subseteq S$. For each $N \in \mathrm{RMax}(R)\backslash\{M\}$, we get $W_{[N]} = R_{[N]}$ as $W_{[P]}$ always lies between $R_{[P]}$ and $S_{[P]}$ for a prime P of R. As M is the only other regular maximal ideal of R (and W properly contains R), $R_{[M]} \subsetneq W_{[M]}$ since $W = \bigcap\{W_{[Q]} \mid Q \in \mathrm{RMax}(R)\}$. We also have $W_{[M]} \subseteq S_{[M]}$ and so $W_{[M]} = S_{[M]}$ (as $R_{[M]} \subsetneq S_{[M]}$ is minimal) and therefore $W = S$. Hence $R \subsetneq S$ is minimal.

Under the assumption that R is Marot, $R_{[Q]} = R_{(Q)}$ and $S_{[Q]} = S_{(Q)}$ for all regular maximal ideals $Q \in \mathrm{RMax}(R)$ (Lemma 2.2). Thus the equivalence in (2) follows from the equivalence of (a) and (c) in (1). $\square$

The strongly valuative ring R in [13, Example 2.19] has infinitely many regular maximal ideals each of which contains the set of regular nonunits of R. Thus $R_{(Q)} = R$ for each regular maximal ideal $Q \in \mathrm{RMax}(R)$. Also the integral closure of R is a minimal extension of R. Hence the equivalence in statement (2) of the previous theorem does not hold in general. However, we do have the following.

Corollary 2.8. *Let R and S be rings such that $R \subsetneq S \subseteq T(R)$. If there is a regular maximal ideal M of R such that $R_{(M)} \subsetneq S_{(M)}$ is a minimal extension and $R_{(N)} = S_{(N)}$ for all other regular maximal ideals N (if any), then $R \subsetneq S$ is a minimal extension.*

Proof. The result follows trivially if M is the only regular maximal ideal of R. Suppose we have a ring W such that $R \subsetneq W \subseteq S$. Then we have $R_{(N)} \subseteq W_{(N)} \subseteq S_{(N)} = R_{(N)}$ for each $N \in \mathrm{RMax}(R)\backslash\{M\}$. Since W properly contains R, it must be that $W_{(M)}$ properly contains $R_{(M)}$ (by way of Lemma 2.6). As $R_{(M)} \subsetneq S_{(M)}$ is minimal and $R_{(M)} \subsetneq W_{(M)} \subseteq S_{(M)}$ we have $W_{(M)} = S_{(M)}$ and so by Lemma 2.6 $W = S$. Therefore $R \subsetneq S$ is a minimal extension. $\square$

A useful concept when dealing with valuative rings is that of a pointwise minimal extension: a pair of rings $S \subseteq T$ is referred to as a *pointwise minimal extension* if for each $t \in T$, the extension $S \subseteq S[t]$ has no proper intermediate rings. For an integral domain that is not integrally closed, a necessary condition for it to be valuative is that its integral closure is a pointwise minimal extension [1, Proposition 4.1].

Corollary 2.9. *Let R be a ring that is not integrally closed and let S be a proper overring that is integral over R. If there is a regular maximal ideal M of R such that $R_{(M)} \subsetneq S_{(M)}$ is a pointwise minimal extension and $R_{(N)} = S_{(N)}$ for all regular maximal ideals $N \in \mathrm{RMax}(R)\backslash\{M\}$ (if any), then $R \subsetneq S$ is a pointwise minimal extension.*

Proof. Assume there is a regular maximal ideal M of R such that $R_{(M)} \subsetneq S_{(M)}$ is a pointwise minimal extension and $R_{(N)} = S_{(N)}$ for all regular maximal ideals $N \in \mathrm{RMax}(R)\backslash\{M\}$. Then for $t \in S\backslash R$, we have $R_{(N)} = R_{(N)}[t] = R[t]_{(N)}$ for all regular maximal ideals N other than M. Thus it must be that $t \in S_{(M)}\backslash R_{(M)}$. Hence $R_{(M)} \subsetneq R_{(M)}[t] = R[t]_{(M)}$ is a minimal integral extension. It follows that $R \subsetneq R[t]$ is minimal and therefore S is a pointwise minimal extension of R. $\square$

We recall several basic results that will be useful in our study of valuative Marot rings.

Lemma 2.10. [cf. [14, Proposition 5]] *If R is a Marot ring and $R \subseteq S \subseteq T(R)$, then S is a Marot ring.*

Lemma 2.11. [13, Lemma 2.2] *Let $S \subsetneq T$ be a ring extension.*

1. *If* $t \in T$ *is such that* $S \subsetneq S[r]$ *is a minimal extension for each* $r \in (S :_T t)\backslash S$*, then there is a unique maximal ideal* M *of* S *such that* $M = \sqrt{(S :_S r)}$ *for each* $r \in (S :_T t)\backslash S$.
2. *If* T *is a pointwise minimal extension of* S*, then there is a unique maximal ideal* M *of* S *such that* $M = \sqrt{(S :_S x)}$ *for each* $x \in T\backslash S$.

Lemma 2.12. [13, Lemma 2.3] *Let* I *be an invertible ideal of a ring* R*. If* $b \in R$ *is a nonunit of* R *such that* $bR + I = R$*, then* $I^{-1} = bI^{-1} + R$.

Lemma 2.13. [13, Lemma 2.4] *Let* I *and* J *be invertible comaximal ideals of a ring* R*. If* $b \in I$ *and* $c \in J$ *are such that* $b + c = 1$*, then* $(R :_R bJ^{-1}) = J$ *and* $(R :_R cI^{-1}) = I$.

Lemma 2.14. [13, Lemma 3.2] *Let* R *be a ring with at least two regular maximal ideals* M *and* N*. If* $R \subsetneq (M : M)$ *and* $b \in N\backslash M$ *and* $c \in M\backslash N$ *are regular elements, then* $R \subsetneq R[b/c]$ *is not a minimal extension of* R.

Theorem 2.15. [13, Theorem 2.5] *Let* R *be a weakly valuative ring. If* $I = bR + rR$ *and* $J = cR + rR$ *are comaximal invertible proper ideals with* $b + c = 1$*, then at least one of* $\sqrt{I}$ *and* $\sqrt{J}$ *is a maximal* $T(R)$*-radical ideal.*

Corollary 2.16. [13, Corollary 2.6] *Let* R *be a weakly valuative ring with exactly three regular maximal ideals* M_1, M_2, *and* M_3*. Also let* $r \in M_1 \cap M_2 \cap M_3$ *be regular and let* $b \in M_1$ *and* $c \in M_2 \cap M_3$ *be such that* $b + c = 1$ *with* $B = bR + rR$ *and* $C = cR + rR$ *invertible. Then* $M_1 = \sqrt{B}$ *and there is an element* $t \in B^{-1}\backslash R$ *such that* $M_1 = \sqrt{(R :_R t)}$ *and* $R \subsetneq R[t]$ *is a minimal extension.*

The next result from [13] gives a characterization of integrally closed Marot rings that are valuative.

Theorem 2.17. [13, Theorem 2.10] *The following are equivalent for an integrally closed Marot ring* R*:*

1. R *is strongly valuative.*
2. R *is valuative.*
3. R *is weakly valuative.*
4. R *is a Prüfer with at most three regular maximal ideals, the set of regular non-maximal prime ideals is linearly ordered under set containment and at most one regular maximal ideal fails to contain each regular nonmaximal prime.*
5. R *is a Prüfer ring with at most three regular maximal ideals and at most one regular maximal ideal fails to contain each regular nonmaximal prime.*

It is convenient to record the following corollary.

Corollary 2.18. *Let* R *be an integrally closed Marot ring with a unique regular maximal ideal. Then* R *is valuative if and only if it is a Prüfer valuation ring.*

Proof. A valuation ring is always valuative. For the converse we have that R is a Marot Prüfer ring with a unique regular maximal ideal M. Since R is Marot, $R = R_{(M)} = R_{[M]}$, and thus (R, M) is a valuation pair of $T(R)$ [7, Theorem 13]. □

The ring R in [13, Example 3.4] (and [11, Example 2.4]) is an integrally closed valuative Marot ring that is a Prüfer ring with exactly two regular maximal ideals but is not weakly additively regular. However, it is rather easy to show that a Marot (Prüfer) valuation ring is weakly additively regular.

With regard to the proof of the next result and Theorem 2.21, we define *weakly additively regular ordered pairs* as follows: for $f, g \in R$ with f regular, if there is a pair $s, t \in R$ such that $gs + ft$ is regular and $fR + sR = R$, we say that (g, f) is a *weakly additively regular ordered pair*. Note that if $g \in fR$, then we have $f = g + f(1 - p)$ where $g = fp$. Hence (g, f) is a weakly additively regular ordered pair whenever f divides g. In particular, (h, u) is a weakly additively ordered pair for each unit u and element $h \in R$.

Theorem 2.19. *Each Marot valuation ring is weakly additively regular.*

Proof. Let V be a Marot ring that is also a valuation ring. Then V has a unique regular maximal ideal M. Let $\mathrm{v} : T(V) \to G \cup \{\infty\}$ be a valuation map corresponding to the valuation pair (V, M) where G is the corresponding value group. Since V is a Marot ring, the regular nonunits of V map onto the positive elements of G [9, Theorem 7.9]. Let $f, g \in V$ with f regular. As noted above, if f divides g in V, then (g, f) is a weakly additively regular ordered pair.

To complete the proof, consider the case that f does not divide g. Then f is not a unit. Since f is regular, $g/f \in T(V)$ with $\mathrm{v}(g/f) = \mathrm{v}(g) - \mathrm{v}(f)$. As we are assuming f does not divide g, it must be that $0 \leq \mathrm{v}(g) < \mathrm{v}(f)$. The ideal $gV + fV$ is regular and thus is generated by regular elements. Moreover, there must be a regular element $h \in gV + fV$ such that $\mathrm{v}(h) = \mathrm{v}(g)$. We have $h = gb + fc$ for some $b, c \in V$. With regard to v, we have $\mathrm{v}(h) = \mathrm{v}(gb + fc) \geq \min\{\mathrm{v}(g) + \mathrm{v}(b), \mathrm{v}(f) + \mathrm{v}(c)\}$. As $\mathrm{v}(h) = \mathrm{v}(g) < \mathrm{v}(f)$, it must be that $\mathrm{v}(b) = 0$. Since M is the only regular maximal ideal and $b \notin M$, we have $bV + fV = V$. Hence we again have that (g, f) is a weakly additively regular ordered pair. Therefore V is weakly additively regular. □

Lemma 2.20. *Let R be a ring with a regular element f that is contained in each regular maximal ideal. Then R is weakly additively regular if and only if (g, k) is a weakly additively regular ordered pair for each $g \in R$ and each $k \in fR \cap \mathrm{Reg}(R)$.*

Proof. Suppose (g, k) is a weakly additively regular ordered pair for each $g \in R$ and each $k \in fR \cap \mathrm{Reg}(R)$. For each $r \in \mathrm{Reg}(R)$, $rf \in fR \cap \mathrm{Reg}(R)$ and thus there are elements $m, n \in R$ (depending on g and rf) such that $gm + rfn$ is regular with $mR + rfR = R$. Since f is contained in each regular maximal ideal, no regular maximal ideal contains m and thus $mR + rR = R$. Hence R is weakly additively regular. The converse is trivial. □

Theorem 2.21. *Let R be a Marot ring with integral closure $R' \supsetneq R$. If R' is a valuation ring and $(R : R')$ is a regular ideal of R, then R is weakly additively regular.*

Proof. Assume R' is a valuation ring and $(R : R')$ is a regular ideal of R. Since R is a Marot ring, so is R'. Hence R' is also a Prüfer ring with unique regular maximal ideal M'. In addition, for the corresponding valuation map v and value group G, $M' = \{r \in T(R) \mid \mathrm{v}(r) > 0\}$ and the regular elements of M' map onto the set of positive elements of G. The ideal $M = M' \cap R$ is the unique regular maximal ideal of R.

Let $I = (R : R')$. Then I is also a regular ideal of R', necessarily proper and thus contained in M'. Let $f' \in I$ and $f \in f'R$ be regular. Also let $k \in R'$ be such that $\mathrm{v}(k) \geq \mathrm{v}(f)$. Since f is regular, $k/f \in T(R)$ with $\mathrm{v}(k/f) \geq 0$. Hence $k/f \in R'$. As I is an ideal of R', $k = f(k/f) \in I$.

Next let $g \in R$. To see that (g, f) is a weakly additively regular ordered pair, first consider the case $\mathrm{v}(g) \geq 2\mathrm{v}(f)$. Then $\mathrm{v}(g/f) \geq \mathrm{v}(f)$ so that $g/f \in I$. In this case f divides g in R and thus (g, f) is a weakly additively regular ordered pair. Next suppose $\mathrm{v}(g) < 2\mathrm{v}(f)$. The ideal $gR + f^2R$ is regular and so is generated by regular elements $f^2, d_1, d_2, \ldots, d_n$. By properties of v, at least one d_i is such that $\mathrm{v}(d_i) = \mathrm{v}(g)$. Without loss of generality we may assume $\mathrm{v}(d_1) = \mathrm{v}(g)$. We also have $d_1 = gs + f^2t = gs + f(ft)$ for some $s, t \in R$. Since $\mathrm{v}(g) < 2\mathrm{v}(f)$, it must be that $\mathrm{v}(s) = 0$ and thus $sR' + fR' = R'$. By integrality we also have $sR + fR = R$. Hence (g, f) is weakly additively regular ordered pair for each $g \in R$ and each regular element $f \in f'R$. Therefore R is weakly additively regular by Lemma 2.20. □

3 Valuative Marot Rings

The first result of this section extends several from [1] that deal with valuative domains that are not integrally closed and some from [13] that deal with weakly additively regular rings that are not integrally closed.

Theorem 3.1. *Let R be a Marot ring that is not integrally closed. If R is weakly valuative, then*

1. *the set of regular nonmaximal primes is linearly ordered (under set containment),*
2. *there is a unique regular maximal ideal M such that $R_{(M)}$ is not a mock valuation ring and for all other regular prime ideals P, $R_{(P)}$ is a Prüfer valuation ring,*
3. *the ideal M is a maximal $T(R)$-radical ideal with $R \subsetneq (M : M)$,*
4. *the ring R has at most two regular maximal ideals,*
5. *if R has (exactly) two regular maximal ideals, then both are maximal $T(R)$-radical ideals, and*
6. *the ideal M contains each regular nonmaximal prime.*

Proof. We establish the six conclusions in the order they appear. The proof that the set of regular nonmaximal prime ideals is linearly ordered is only slightly different than the one for valuative domains. By way of contradiction, suppose P and Q are incomparable regular prime ideals with neither a maximal ideal. Since R is Marot, there is a regular element $r \in P\backslash Q$ and a regular element $s \in Q\backslash P$. Then neither r/s nor s/r is in R. Clearly, $r/s \in (R : s/r)$ and $s/r \in (R : r/s)$. Hence at least one of $R \subsetneq R[r/s]$ and $R \subsetneq R[s/r]$ is a minimal extension. But $P \supseteq (R :_R s/r)$ and $Q \supseteq (R :_R r/s)$ are not maximal ideals. So we have a contradiction by way of [2, Theorems 2.3 and 3.4].

For (2) and (3), [3, Theorem 5.10] yields that R has a unique regular maximal ideal M such that $R_{(M)}$ is not a mock valuation ring. In addition, $R_{(P)}$ is a mock valuation ring for all other regular prime ideals P (if any). As R is Marot, each $R_{(P)}$ is a Prüfer valuation ring with unique regular maximal ideal $PR_{(P)}$ [9, Theorem 7.7]. Hence each $R_{(P)}$ contains R'. We also have an element $t \in R'\backslash R$ such that $R \subsetneq R[t]$ is a minimal extension with $M = (R :_R t)$ and $R[t] \subseteq (M : M)$. Thus M is a maximal $T(R)$-radical ideal and $R \subsetneq (M : M)$.

To see that R has at most two regular maximal ideals, suppose that there are at least two regular maximal ideals N_1 and N_2 other than M. Since R is a Marot ring, there is a regular element $r \in N_1 \cap N_2$ that is not in M. Also there is a regular element $s \in M$ that is not in N_1. By Lemma 2.14, $R \subsetneq R[r/s]$ is not a minimal extension. Thus since R is weakly valuative $R \subsetneq R[s/r]$ is a minimal extension. But both N_1 and N_2 contain $(R :_R s/r)$ and so $R \subsetneq R[s/r]$ cannot be minimum either. Hence R has at most two regular maximal ideals.

For (5), suppose R has a regular maximal ideal N other than M. Since R is a Marot ring there are regular elements $r \in N\backslash M$ and $s \in M\backslash N$. As above, we again have that $R \subsetneq R[r/s]$ is not minimal and thus $R \subsetneq R[s/r]$ is minimal, necessarily with $N = \sqrt{(R :_R s/r)}$. Thus both N and M are maximal $T(R)$-radical ideals in this case.

Finally, if M is the only regular maximal ideal, then it trivially contains all regular nonmaximal primes. So suppose we have a second regular maximal ideal N and, by way of contradiction, also suppose N contains a regular nonmaximal prime P that is not contained in M. Since R is a Marot we can choose a regular element $b \in P\backslash M$ and a regular element $c \in M\backslash P$. We will also have $b \in N$ and thus by Lemma 2.14, $R \subsetneq R[b/c]$ is not minimal. As above we obtain a contradiction since P contains $(R :_R c/b)$ but $R \subsetneq R[c/b]$ must be minimal since R is weakly valuative. Therefore M contains each regular nonmaximal prime. □

Recall from above that for a pair of rings $S \subsetneq T$, T is referred to as a *pointwise minimal extension* of S if for each $t \in T\backslash S$, the extension $S \subsetneq S[t]$ is minimal. We will find this notion to be quite useful in characterizing Marot rings that are valuative. Another useful concept is to extend the definition of the types of valuative rings to individual elements: $x \in T(R)$ is (i) *valuative over* R if at least one of $R \subseteq R[x]$ and $R \subseteq R[(R : x)]$ has no proper intermediate rings, (ii) *weakly valuative over* R if either $R \subseteq R[x]$ has no proper intermediate rings or for each $y \in (R : x)$, $R \subseteq R[y]$ has no proper intermediate rings, and (iii) *strongly valuative over* R is at least one of $R \subseteq R[(R : (R : x))]$ and $R \subseteq R[(R : x)]$ has no proper intermediate rings.

Lemma 3.2. *Let R be a Marot ring that is weakly valuative. If R is not integrally closed, then there is a regular element $t \in R'\backslash R$ such that $R \subsetneq R[t]$ is a minimal extension.*

Proof. If R is not integrally closed, there is an element $b \in R' \backslash R$. The R-fractional ideal $bR + R$ is regular, so it contains a regular element $t \in R' \backslash R$ since R is a Marot ring. Since t is integral over R (and regular), it is contained in $R[t^{-1}]$. That $R \subsetneq R[t]$ is minimal follows from the assumption that R is weakly valuative. □

Theorem 3.3. *Let R be a Marot ring with integral closure $R' \supsetneq R$. If R is weakly valuative, then R' is a Prüfer ring and R and R' have the same number of regular maximal ideals (exactly one, or exactly two).*

Proof. Assume R is weakly valuative. Then it has at most two regular maximal ideals. In the case R has (exactly) two regular maximal ideals, one of these, say M is such that $R_{(M)}$ is not integrally closed and for the other one N, $R_{(N)}$ is a Prüfer valuation ring with a unique regular maximal ideal $NR_{(N)}$. By way of Corollary 2.5, we may conclude that R' is a Prüfer ring by showing the integral closure of $R_{(M)}$ is a Prüfer valuation ring.

The ring $R_{(M)}$ is weakly valuative with unique regular maximal ideal $MR_{(M)}$. Hence we may reduce to the case M which is the unique regular maximal ideal of R.

By Lemma 3.2 there is a regular element $t \in R' \backslash R$ such that $R \subsetneq R[t]$ is a minimal extension. As M is the unique regular maximal ideal of R, we have $M = (R :_R t)$ with $t \in (M : M)$. Thus by [3, Lemma 3.7], there is no $b \in T(R) \backslash R$ such that $R \subsetneq R[b]$ is a closed minimal extension. For units $u, u^{-1} \in T(R) \backslash R$, at least one of $R \subsetneq R[u]$ and $R \subsetneq R[u^{-1}]$ is minimal and thus at least one of u and u^{-1} is in R'. Therefore, since R' is a Marot ring, it is also a Prüfer valuation ring, necessarily with a unique (regular) maximal ideal that lies over M.

For the case that R has exactly two regular maximal ideals N and M as given above, the fact that both $R_{(N)}$ and $R'_{(M)}$ are Prüfer valuation rings implies R' is a Prüfer ring with a unique regular maximal ideal N' that lies over N and a unique regular maximal ideal M' that lies over M. These are the only regular maximal ideals of R' since each regular maximal ideal of R' lies over a regular maximal ideal of R. □

Theorem 3.4. *Let R be a Marot ring that is not integrally closed. If R has a unique regular maximal ideal M, then the following are equivalent:*

1. *R is strongly valuative.*
2. *R is valuative.*
3. *R is weakly valuative.*
4. *The integral closure of R is a Prüfer valuation ring (with a unique regular maximal ideal) and it is a pointwise minimal extension of R.*

Proof. It is clear that (1) implies (2), and (2) implies (3). We next show that (4) implies (3).

Assume R' is a Prüfer valuation ring and a pointwise minimal extension of R. Since R' is a Prüfer valuation ring, it has a unique regular maximal ideal M' [9, Theorem 6.5]. For each $s \in R' \backslash R$, $R \subsetneq R[s]$ is a minimal integral extension necessarily with $M = (R :_R s)$ and $MR[s] = M$. Thus $MR' = M$. So each element of R' is weakly valuative over R.

For $t \in T(R)\backslash R'$, $(R' : t) \subseteq M'$. Thus if $z \in (R : t)$, then $z \in M'$ and so the extension $R \subseteq R[z]$ has no proper intermediate rings. It follows that t is weakly valuative over R and therefore R is weakly valuative.

To complete the proof we show that (3) implies both (1) and (4). By Theorem 3.3, R' is a Prüfer valuation ring with a unique regular maximal ideal M' that lies over M.

First note that since R is not integrally closed, there is an element $m \in \text{Reg}(R')\backslash R$ such that $R \subsetneq R[m]$ is a minimal integral extension (Lemma 3.2). Since M is the only regular maximal ideal, we have $M = (R :_R m)$ and $(M : M) \supseteq R[m]$.

Let $f \in M \cap \text{Reg}(R)$ and let $q \in R'\backslash R$. The R-fractional ideal $qR + fR$ is regular and thus is generated by regular elements. For such a generating set $\{s_1, s_2, \ldots, s_n\}$, we have $s_i M \subseteq M$ for each i. Hence $qM \subseteq M$ and therefore M is an ideal of R'.

By Theorem 2.21, R is weakly additively regular. Therefore from [13, Theorem 3.13], R is strongly valuative and R' is a pointwise minimal extension of R. □

Corollary 3.5. *Let R be a Marot ring with a unique regular maximal ideal M. If R is (weakly, strongly) valuative and not integrally closed, then R is weakly additively regular, R' is a Prüfer valuation ring with a unique maximal ideal M' and either $M' = M$ or there is a regular element $d \in R'$ such that $M' = dR'$ and $M = d^2R'$.*

Proof. That R' is a Prüfer valuation ring with a unique regular maximal ideal M' is from the previous theorem. Also, from the proof of that theorem, M is an ideal of R'. Hence R is weakly additively regular by Theorem 2.21. Thus by [13, Theorem 3.13] we have that either $M = M'$ is the unique regular maximal ideal of both R and R', or there is a regular element $d \in R'$ such that $M = d^2R' \subsetneq M = dR'$. □

Later we give an example of a Marot valuative ring with exactly two regular maximal ideals that are neither weakly additively regular nor integrally closed (Example 3.15).

Lemma 3.6. *Let $t \in T(R)\backslash R$. If $R \subsetneq R[t]$ is a closed minimal extension, then t is strongly valuative over R and $R[t] = R[(R : (R : t))]$.*

Proof. If $R \subsetneq R[t]$ is a closed minimal extension, then $(R :_R t)$ is an invertible M-primary ideal of R where $M = \sqrt{(R :_R t)}$ is a regular maximal ideal of R [2, Theorem 3.4]. Also, we have a pair of elements $a, b \in (R :_R t)$ such that $a + bt = 1$. We claim that $(R : (R : t)) \subseteq R[t]$. Let $u \in (R : (R : t))$. Then $ua, ub \in R$ and so $u = ua + (ub)t \in R[t]$. Therefore $R[(R : (R : t))] = R[t]$ and we have that t is strongly valuative over R. □

For an ideal I of R, $\Psi(I) = \bigcap\{R_{[P]} \mid I \nsubseteq P,\ P \in \text{Spec}(R)\}$ (see, for example, [2, Page 1083]).

Theorem 3.7. *Let R be a Marot ring that is not integrally closed. If R has exactly two regular maximal ideals, then the following are equivalent:*

1. *R is strongly valuative.*
2. *R is valuative.*
3. *R is weakly valuative.*
4. *R' is both a Prüfer ring and a pointwise minimal extension of R, R has a regular maximal ideal N such that $R_{(N)}$ is a Prüfer valuation ring, and for the other regular maximal ideal M, (i) M is a maximal ideal of R', (ii) M contains each regular nonmaximal prime of R and R', and (iii) $R_{(M)}$ is a (weakly) [strongly] valuative ring such that $MR_{(M)}$ is also the maximal ideal of $R'_{(M)}$.*

Proof. In general, a strongly valuative ring is valuative and a valuative ring is weakly valuative. Thus it suffices to show that (3) implies (4), and (4) implies (1).

We start by showing (3) implies (4). Assume R is weakly valuative. Since R has two regular maximal ideals, both are maximal $T(R)$-radical ideals (Theorem 3.1) and thus R' is a pointwise minimal extension of R by [3, Corollary 5.14]. Also, by Theorem 3.3, R' is a Prüfer ring with exactly two regular maximal ideals. By [3, Theorem 5.10] (and the Marot assumption), there is a unique regular maximal ideal M of R such that $R_{(M)}$ is not a mock valuation ring and for the other regular maximal ideal N, $R_{[N]} = R_{(N)}$ is a Prüfer valuation ring with unique regular maximal ideal $NR_{(N)}$. We also have that $R_{(M)}$ is weakly valuative [3, Proposition 5.2]. As $R_{(M)}$ is also a Marot ring, it is strongly valuative and its integral closure, $R'_{(M)}$, is a Prüfer valuation ring that is a pointwise minimal extension of $R_{(M)}$ (Theorem 3.4). In addition, $M = (R :_R t)$ for each $t \in R' \backslash R$ (since R' is a pointwise minimal extension of R). By Corollary 3.5, $MR_{(M)}$ is either the unique regular maximal ideal of $R'_{(M)}$ or $MR_{(M)} = d^2 R'_{(M)} \subseteq dR'_{(M)} = M'R'_{(M)}$ where M' is regular maximal ideal of R' that lies over M. In the first case, $M'R'_{(M)} = MR_{(M)}$ and so $M' = M$, and in the second $M'R'_{(M)} \supsetneq MR_{(M)}$ and so $M' \supsetneq M$. We will show that the second case does not occur.

By Theorem 3.1, the set of regular nonmaximal primes is linearly ordered under set containment and M contains each regular nonmaximal prime.

Since $R_{(P)}$ is a Prüfer valuation ring for each regular prime P other than M, the contraction map from $\mathrm{Spec}(R')$ to $\mathrm{Spec}(R)$ is an order preserving bijective correspondence. It follows that M' contains each regular nonmaximal prime of R'.

Since R is a Marot ring and M and N are the only regular maximal ideals, there is a regular element $s \in N \backslash M$. As M contains each regular nonmaximal prime, $N = \sqrt{sR}$. We also have $N' = \sqrt{sR'}$. Let Q be a nonmaximal prime ideal contained in N and let $m \in Q$. Then the ideal $I = sR + mR$ is such that $IR_{(N)} = sR_{(N)}$ and $IR_{(M)} = sR_{(M)} = R_{(M)}$. Hence s divides m. In fact s^n divides m for each positive integer n. It follows that $\bigcap s^n R \supseteq Q$. Similarly $\bigcap s^n R' \supseteq Q'$ for each nonmaximal prime ideal Q' of R' that is contained in N'. Thus $\bigcap s^n R = P$ and $\bigcap s^n R' = P'$ for some nonmaximal primes P of R and P' of R'.

Consider the extension $R \subsetneq R[1/s]$. We have $N = \sqrt{(R :_R 1/s)}$. For a finitely generated N-primary ideal J, $JR_{(M)} = R_{(M)}$ and $JR_{(N)}$ are both regular principal ideals. Hence J is invertible. By [1, Corollary 3.7], $R \subsetneq R[1/s] = \Psi(N)$ is a closed minimal extension. Since M contains each regular nonmaximal prime, $\Psi(N) = R_{(M)}$. The extension $R \subsetneq R[1/s]$ is the unique closed minimal extension of R since

N and M are the only regular maximal ideals (and M is the crucial maximal ideal of a minimal integral extension).

Finally we will show that M is a common maximal ideal of R and R'. By way of contradiction suppose there is a regular $d \in R'$ such that $M'R'_{(M)} = dR'_{(M)} \supsetneq d^2R'_{(M)} = MR'_{(M)}$. While $sdR'_{(M)} = M'R'_{(M)} = dR'_{(M)}$ implies it is possible to have s divide d in R', we cannot have that each positive power of s divides d in R' as this would put d in P', making P' regular and thus properly contained in M' and hence in M. Thus without loss of generality we may assume d/s is not in R'. Since $d \in M'$ and $s \in R' \backslash M'$, s/d is not in R'. Thus neither $R \subsetneq R[d/s]$ nor $R \subsetneq R[s/d]$ is integral. For the extension $R \subsetneq R[d/s]$, it is clear that this is not $R_{(M)} = \Psi(N)$ as $d \notin R_{(M)}$ (while s is a unit of $R_{(M)}$). Hence $R \subsetneq R[d/s]$ is not a minimal extension.

For the extension $R \subsetneq R[s/d]$, the fact that $d \in M'$ while $s \in R' \backslash M'$ implies s/d is not in $R'_{(M)}$ and so not in $R_{(M)}$. Thus $R[s/d] \neq \Psi(N)$. As $R \subsetneq \Psi(N)$ is the unique closed minimal extension R, $R \subsetneq R[s/d]$ is not a minimal extension, contradicting the assumption that R is weakly valuative. Hence it must be that M is a common maximal ideal of R and R'. We also have that each regular nonmaximal prime of R' is contained in M. Hence (3) implies (4).

To complete the proof we show that (4) implies (1). We assume all of the following: R' is a pointwise minimal extension of R, R has a regular maximal ideal N such that $R_{(N)}$ is a Prüfer valuation ring, and for the other regular maximal ideal M, (i) M is a maximal ideal of R', (ii) M contains each regular nonmaximal prime of R and R', and (iii) $R_{(M)}$ is a weakly valuative ring such that $MR_{(M)}$ is also the maximal ideal of $R'_{(M)}$ which is a Prüfer valuation ring. By Theorem 3.4, $R_{(M)}$ is strongly valuative. Since R is Marot and M contains each regular nonmaximal prime, there is a regular element $s \in N \backslash M$ necessarily with $N = \sqrt{sR}$. Also, for a finitely generated N-primary ideal I, I contains a power of s so I is regular. Since $R_{(N)}$ is a Prüfer valuation ring, $IR_{(N)}$ is invertible. Also $IR_{(M)} = R_{(M)}$. It follows that I is invertible. By [2, Corollary 3.7], $R \subsetneq R[1/s] = \Psi(N)$ is a closed minimal extension of R, and since M contains each regular nonmaximal prime $\Psi(N) = R_{(M)}$. By Lemma 3.6, each element $t \in R_{(M)} \backslash R$ is strongly valuative over R.

We let v be a valuation map corresponding to the valuation pair $(R'_{(M)}, MR'_{(M)})$ and let w be a valuation map corresponding to the valuation pair $(R_{(N)}, NR_{(N)})$. Since R is a Marot ring, both v and w map the units of $T(R)$ onto the elements of the corresponding value group [9, Theorem 7.9].

Let $x \in T(R) \backslash R_{(M)}$. We consider three main cases, the third will have three subcases.

Case 1: x is in neither $R_{(N)}$ nor $R'_{(M)}$.
In this case, $(R_{(N)} : x) \subseteq NR_{(N)}$ and $(R'_{(M)} : x) \subseteq MR'_{(M)} = MR_{(M)}$. Thus if x is in neither $R_{(N)}$ nor $R'_{(M)}$, then $(R : x) \subseteq NR_{(N)} \cap MR_{(M)} = N \cap M \subseteq R$. Hence in this case x is strongly valuative over R.

Case 2: $x \in R_{(N)} \backslash R'_{(M)}$.
We have $(R'_{(M)} : x) \subseteq MR_{(M)}$ which implies $(R : x) \subseteq MR_{(M)}$. It follows that $R \subseteq R[(R : x)]$ has no proper intermediate rings as $R \subsetneq R_{(M)}$ is a closed minimal extension. Therefore x is strongly valuative in this case.

Case 3: $x \in R'_{(M)} \backslash R_{(M)}$.
We have $(R :_R x) \subseteq M$ so there is a regular element $m \in M$ such that $mx \in R$. Since $R_{(M)}$ is a strongly valuative Marot with a single regular maximal ideal, it is weakly additively regular (Corollary 3.5). Thus there is a pair of elements $f, g \in R_{(M)}$ such that $z = fxm + gm^2$ is regular with $fR_{(M)} + m^2R_{(M)} = R_{(M)}$. The element $y' = fx + gm \in xR_{(M)} + mR_{(M)}$ is a unit of $R'_{(M)}$ since $f, x \in R'_{(M)} \backslash MR_{(M)}$. Also, there are elements $c, d \in R_{(M)}$ such that $cf + dm^2 = 1$. Thus $x = x(cf + dm^2) = c(y' - gm) + m(dxm) = cy' + m(dxm - g) \in y'R_{(M)} + mR_{(M)}$. We also have a regular element $q \in R\backslash M$ such that $y = qy' \in R'$. As q is a unit of $R_{(M)}$ and $m/y \in MR_{(M)}$, $xR_{(M)} + mR_{(M)} = yR_{(M)} + mR_{(M)} = yR_{(M)}$. Hence $x/y \in R_{(M)}$. With regard to colons, $(R_{(M)} : x) = (R_{(M)} : y) = (1/y)R_{(M)}$ and $(R_{(M)} : (R_{(M)} : x)) = (R_{(M)} : R_{(M)} : y)) = yR_{(M)}$.

We have three subcases: (i) $\mathrm{w}(x) = \infty$, (ii) $\mathrm{w}(x) < 0$, and (iii) $0 \leq \mathrm{w}(x) < \infty$.

If $\mathrm{w}(x) = \infty$, $(R_{(N)} : x) = T(R)$ and $(R_{(N)} : (R_{(N)} : x)) \subsetneq NR_{(N)}$. We also have $(1/y)R = (R : y) = (R_{(N)} : y) \cap (R_{(M)} : y) \subseteq T(R) \cap (R_{(M)} : x) = (R : x)$. It follows that $(R : (R : x)) \subseteq (R : (R : y)) = yR$. As $y \in R'\backslash R$, $R \subsetneq R[y]$ is a minimal extension and thus $R \subseteq R[(R : (R : x))]$ has no proper intermediate rings.

For the case $\mathrm{w}(x) < 0$, we have $(R_{(N)} : x) \subseteq NR_{(N)}$ and thus $R_{(N)}[(R : x)] = R_{(N)}$. Also $R_{(M)} \subseteq R_{(M)}[(R : x)] \subseteq R_{(M)}[1/y]$ with $R_{(M)} \subsetneq R_{(M)}[1/y]$ minimal. Hence $R \subseteq R[(R : x)]$ has no proper intermediate rings (Corollary 2.8).

Finally we consider the case $0 \leq \mathrm{w}(x) < \infty$. We have a regular element $t \in R_{(N)}$ such that $\mathrm{w}(t) = \mathrm{w}(x)$. Hence $x/t \in R_{(N)}$. We also have $x/y \in R_{(M)}$. Thus there are regular elements $b \in R\backslash N$ and $c \in R\backslash M$ such that $bx/t, cx/y \in R$. It follows that $(b/t)R + (c/y)R \subseteq (R : x)$. Since b is a unit of $R_{(N)}$ and $(R : x) \subseteq (R_{(N)} : x) = (1/t)R_{(N)}$, $(b/t)R_{(N)} + (c/y)_{(N)} \supseteq (R : x)_{(N)}$. Similarly $(R : x) \subseteq (R_{(M)} : x) = (1/y)R_{(M)}$, $(b/t)R_{(M)} + (c/y)_{(M)} \supseteq (R : x)_{(M)}$. Thus $(R : x) = (b/t)R + (c/y)R$. Since $(R : x)$ is a finitely generated fractional R-ideal, we have $(R : (R : x))_{(N)} = (R_{(N)} : (R_{(N)} : x)) \subseteq R_{(N)}$ and $(R : (R : x))_{(M)} = (R_{(M)} : (R_{(M)} : x)) = yR_{(M)}$. Another application of Corollary 2.8 yields that $R \subseteq R[(R : (R : x))]$ has no proper intermediate rings since $R_{(M)} \subsetneq R[(R : (R : x))]_{(M)} = R_{(M)}[y]$ is a minimal extension and $R[(R : (R : x))]_{(N)} = R_{(N)}$.

From all of the cases, we have that x is strongly valuative over R and therefore R is strongly valuative. □

The next corollary provides alternate lists of conditions that are equivalent to those given in (4) of the previous theorem.

Corollary 3.8. *Let R be a Marot ring that has exactly two regular maximal ideals and is not integrally closed. Then the following are equivalent:*

1. *R is (weakly) [strongly] valuative.*
2. *R has a regular maximal ideal N such that $(R_{(N)}, NR_{(N)})$ is a valuation pair and for the other maximal ideal M: M is a maximal ideal of R' that contains each regular nonmaximal prime ideal of R' and $R_{(M)}$ is valuative.*

3. *R' is a Prüfer ring and R has a regular maximal ideal M such that (i) M is a maximal ideal of R', (ii) M contains each regular nonmaximal prime of R and R', and (iii) $R_{(M)}$ is a (weakly) [strongly] valuative ring.*
4. *R' is a pointwise minimal extension of R, R has a regular maximal ideal N such that $R_{(N)}$ is a Prüfer valuation ring, and for the other regular maximal ideal M, (i) M is a maximal ideal of R', (ii) M contains each regular nonmaximal prime of R and R', and (iii) $R_{(M)}$ is a (weakly) [strongly] valuative ring such that $MR_{(M)}$ is also the maximal ideal of $R'_{(M)}$.*

Proof. It is clear that statement (4) of Theorem 3.7 implies statements (2), (3), and (4) here. Also statement (4) in this result clearly implies (2), and taken together (2), (3) and (4) in this result yields statement (4) of Theorem 3.7. For all three of (2), (3), and (4), we have the regular maximal ideal M of R that is also a maximal ideal of R'. Since R has a second regular maximal ideal N, there is a regular maximal ideal N' of R' such that $N' \cap R = N$. Also $R'_{(M)}$ is the integral closure of $R_{(M)}$.

By basic properties of pullbacks, the prime ideals of R that are not contained in M are in one-to-one correspondence with the prime ideals of R' that are not contained in M. Specifically for a prime P' of R' that is not contained in M, $P = P' \cap R$ is the corresponding prime ideal of R and $P' = (M :_{R'} P)$. Since M is a regular ideal of R, MP' is a regular ideal of R whenever P' is a regular ideal of R'. Hence R' has exactly two regular maximal ideals, M and N'. Since R is a Marot ring, there is a regular element $t \in M \backslash N$. So for each $q \in R' \backslash N'$ and $b \in R'$, $tq \in R \backslash N$, $tb \in R$, and $b/q = tb/tq \in R_{(N)}$. It follows that $R_{(N)} = R'_{(N)} = R'_{(N')}$.

To see that (2) implies (3), all we need to show is that R' is a Prüfer ring. Since R is Marot and $R_{(N)}$ is a valuation ring, $(R_{(N)}, NR_{(N)}) = (R'_{(N')}, N'R'_{(N')})$ is a valuation pair and M and N' are the only regular maximal ideals of R'. Next we consider $R'_{(M)}$. We have that $MR_{(M)} = MR'_{(M)}$ is the only regular maximal ideal of $R_{(M)}$, so we simply apply Theorem 3.4 to see that $R'_{(M)}$ is valuation ring (and $R'_{(M)}, MR'_{(M)}$ is a valuation pair). Thus R' is a Prüfer ring and so (2) implies (3).

Finally assume the restrictions in (3). From the arguments above, $R_{(N)}$ is a valuation ring. So to show (3) implies (4), all that remains is to show that R' is a pointwise minimal extension of R. Since $R'_{(M)}$ is the integral closure of $R_{(M)}$, it is also a pointwise minimal extension of $R_{(M)}$ by Theorem 3.4. That R' is a pointwise minimal extension of R now follows from Corollary 2.9 (and the fact that $R_{(N)} = R'_{(N)}$). □

Theorem 3.9. *Let R be a ring whose integral closure, R', is a valuation ring properly contained in $T(R)$. If $R \subsetneq R'$ is a minimal extension, then R is strongly valuative.*

Proof. Since R' is a valuation ring, there is a prime ideal P of R' such that for each $q \in T(R) \backslash R'$, there is an element $p \in P$ such that $pq \in R' \backslash P$. It is also the case that if $f \in R' \backslash P$, then $(R' : f) = R'$. Also note that for $h \in R$, $1 \in (R : h)$ implies $(R : (R : h)) \subseteq R$. Thus h is strongly valuative over R.

Suppose $R \subsetneq R'$ is a minimal extension and let $t \in T(R) \backslash R$. We have two cases to consider, $t \in T(R) \backslash R'$ and $t \in R' \backslash R$. For the former, $(R : t) \subseteq (R' : t) \subseteq R'$. Since $R \subsetneq R'$ is a minimal extension and $(R : t) \subseteq R'$, $R \subseteq R[(R : t)]$ has no proper intermediate rings. Thus, in this case, t is strongly valuative over R.

Next suppose $t \in R'\backslash R$. We split the argument into two subcases: $t \in R'\backslash P$ and $t \in P$. If $t \in R'\backslash P$, then $(R : t) \subseteq (R' : t) = R'$ in which case $R[(R : t)] \subseteq R'$, and thus $R \subseteq R[(R : t)]$ has no proper intermediate rings. Hence t is strongly valuative over R in this case.

Finally suppose $t \in P$. We again have $(R : t) \subseteq (R' : t)$ but now $(R' : t)$ is not contained in R'. If $(R : t) \subseteq R'$, then as above $R \subseteq R[(R : t)]$ has no proper intermediate rings. On the other hand, if $(R : t)$ is not contained in R', then $(R : (R : t)) \subseteq R'$ and so $R \subseteq R[(R : (R : t))]$ $(\subseteq R')$ has no proper intermediate rings. Therefore each element of $T(R)$ is strongly valuative over R and so R is strongly valuative. □

Let D be an integral domain and let $\mathscr{P} = \{P_\alpha\}_{\alpha \in \mathscr{A}}$ be a nonempty set of prime ideals of D. Next let $\mathscr{I} = \mathscr{A} \times \mathbb{N}$ and for each $i = (\alpha, n) \in \mathscr{I}$, let K_i be the quotient field of D/P_α. Form the D-algebra $B = \sum K_i$ and define a ring structure on $D \times B$ by setting $(r, b) + (s, c) = (r + s, b + c)$ and $(r, b)(s, c) = (rs, rc + sb + bc)$. The resulting ring is denoted $D + B$ and is referred to as the ring of the form $A + B$ corresponding to D and $\mathscr{P}$. For $r \in D$, $b \in B$ and $i = (\alpha, n) \in \mathscr{I}$, r_i denotes the image of r in K_i and b_i denotes the ith component of b.

A few of the basic properties of these rings are given in the following theorem.

Theorem 3.10. [10, Theorems 8.3 and 8.4] *Let $\mathscr{P} = \{P_\alpha\}_{\alpha \in \mathscr{A}}$ be a nonempty set of prime ideals of a domain D and let $R = D + B$ be the $A + B$ ring corresponding to D and $\mathscr{P}$.*

1. *An element (r, b) of R is a zero divisor if and only if there is an $i = (\alpha, n) \in \mathscr{I} = \mathscr{A} \times \mathbb{N}$ such that $r_i + b_i = 0$. A necessary condition for $(s, c) \in R$ to be regular is that $s \in D\backslash \bigcup P_\alpha$.*
2. *For each $i \in \mathscr{I}$, the set $M_i = \{(r, b) \in R \mid r_i = -b_i\}$ is both a maximal ideal and a minimal prime ideal of R. All other prime ideals of R are of the form $P + B$ for some prime ideal P of D.*
3. *The total quotient ring of R can be identified with the ring $D_{\mathscr{S}} + B$ where $\mathscr{S} = D\backslash \bigcup\{P_\alpha \mid P_\alpha \in \mathscr{P}\}$.*
4. *If I is an ideal of D such that $I \cap \mathscr{S} \neq \emptyset$, then $IR = I + B$ is a regular ideal of R. Conversely, if J is a regular ideal of R, then $J = I + B = IR$ for some I of D such that $I \cap \mathscr{S} \neq \emptyset$.*
5. *If I is an ideal of D, then IR is an invertible of R if and only if I is an invertible ideal of D and $I \cap \mathscr{S} \neq \emptyset$.*

The following definitions were introduced in [12]. A nonempty set of prime ideals $\mathscr{P} = \{P_\alpha\}_{\mathscr{A}}$ of a domain D is said to be a *weakly additively regular family* if for each $g \in \bigcup P_\alpha$ and $f \in D\backslash \bigcup P_\alpha$, there is a pair of elements $s, t \in D$ such that $gs + ft \in D\backslash \bigcup P_\alpha$ and $sD + fD = D$. Also $\mathscr{P}$ is said to be a *Marot family* if for each ideal $I \not\subseteq \bigcup P_\alpha$, I can be generated by the set $I\backslash \bigcup P_\alpha$.

The next theorem blends two results from [12] concerning special families of prime ideals and rings of the form $A + B$.

Theorem 3.11. (cf. [12, Theorems 3.6 and 3.9]) *Let $R = D + B$ be the ring of the form $A + B$ corresponding to a domain D and a nonempty set of nonzero prime ideals $\mathscr{P} = \{P_\alpha\}_{\alpha \in \mathscr{A}}$ of D corresponding to D and $\mathscr{P}$.*

1. *R is weakly additively regular if and only if $\mathscr{P}$ is a weakly additively regular family.*
2. *R is Marot if and only if $\mathscr{P}$ is a Marot family.*

Another result from [12] gives us the following (stated in a slightly different way).

Theorem 3.12. (cf. [12, Corollary 3.11]) *Let D be a Bezout domain and let $\mathscr{P} = \{P_\alpha\}_{\alpha\in\mathscr{A}}$ be a nonempty set of nonzero prime ideals of D and let $R = D + B$ be the ring of form $A + B$ corresponding to D and $\mathscr{P}$. Then each finitely generated regular ideal of R is principal and thus R is a regular Bezout ring (and a Prüfer ring).*

We will use these characterizations in the proof of the next example.

Example 3.13. Let $D' = \mathbb{Q}[X]$ and let D be the pullback of $\mathbb{Q}$ over the maximal ideal $M = (X^2 + 1)D'$. For $\mathscr{P}' = \text{Max}(D')\backslash\{M\}$, let $R = D + B$ be the ring of form $A + B$ corresponding to D and $\mathscr{P}'$ and let $R' = D' + B$ be the ring of form $A + B$ corresponding to D' and $\mathscr{P}'$.

1. For both R and R', $MR' = MR$ is the only regular prime ideal.
2. R' is both a weakly additively regular valuation ring and a regular Bezout ring.
3. $R \subsetneq R' = R[X]$ is a minimal integral extension and thus R is strongly valuative.
4. R is not a Marot ring.

Proof. The set $\mathscr{P} = \text{Max}(D)\backslash\{M\}$ yields the same module B and thus the same R. For both R' and R, $MR' = M + B = MR$ is the only regular prime ideal. Since D' is a PID, R' is a regular Bezout ring with a unique regular prime ideal. Thus D' is valuation ring and so it is weakly additively regular (Theorem 2.19).

The extension $D \subsetneq D' = D[X]$ is a minimal integral extension (since the same is true for $D/M = \mathbb{Q} \subsetneq \mathbb{Q}[\sqrt{-1}] = D'/M$) and thus $R \subsetneq R' = R[X]$ is a minimal integral extension. Hence R is strongly valuative by Theorem 3.9. We have $D'\backslash \bigcup P_\alpha = \{q(X^2+1)^n \mid n \geq 0,\ q \in \mathbb{Q}\backslash\{0\}\}$. Also $(p, 0)$ is a unit in R for each nonzero $p \in \mathbb{Q}$. It follows that the (proper) regular principal ideals of R have the form $(X^2+1)^m R$ for some $m \geq 1$. Since X is not in D, $X(X^2+1)D$ and $(X^2+1)D$ are incomparable. We also have $X(X^2+1)D \subseteq \bigcup P_\alpha$ and $M = X(X^2+1)D + (X^2+1)D \supsetneq (X^2+1)D$ since $D' = XD + D$ (and $M = (X^2+1)D'$). It follows that MR cannot be generated by regular elements of R. Thus R is a not a Marot ring. □

The last example is the one promised above of a valuative Marot ring with exactly two regular maximal ideals that are neither weakly additively regular nor integrally closed. The next result provides a basis for exhibiting such a ring. Note that the restriction in (a) is used to show the ring R is Marot.

Theorem 3.14. *Let D' be a Dedekind domain with a pair of maximal ideals P and Q where neither is principal, but all three of P^2, Q^2, and $P \cap Q$ are principal. Assume D' also satisfies these additional restrictions.*

(a) *For each nonunit $w \in D'\backslash P$, there is an element $d \in P$ such that $w + d$ is a unit of D'.*

(b) The residue field $L = D'/P$ has a proper subfield K such that $L = K + iK$ for some i.
(c) There are elements $a, b, c \in D$ such that $P \cap Q = aD'$, $P^2 = bD'$, and $Q^2 = cD'$ where D is the pullback of K over P.

Next we let $\mathscr{P} = \{P_\alpha\} = \mathrm{Max}(D')\backslash\{P, Q\}$ and $S' = \bigcup P_\alpha$. Also let $Q_o = D \cap Q$ and $S = D \cap S'$. Finally, let $R = D + B$ and $R' = D' + B$ be the rings of the form $A + B$ corresponding to D and D', respectively, and the set $\mathscr{P}$.

1. *Each P-primary ideal of D can be generated by elements in $P\backslash S'$ and each Q_o-primary ideal of D can be generated by elements in $Q_o\backslash S'$.*
2. *R' is the integral closure of R and it is not weakly additively regular. Also R' is a Prüfer ring with exactly two regular prime ideals.*
3. *R is a valuative Marot ring with exactly two regular maximal ideals and it is not weakly additively regular.*

Proof. Since $L = K + iK$ is an algebraic extension of K, D' is integral over D. The fact that P is a common ideal of D' and D and $P^2R' = aR'$ is a regular ideal with $a \in D$ guarantee's that R' is the integral closure of R. We also have that $D'_P = D'_{D\backslash P}$ is a discrete rank one valuation domains with maximal ideal PD'_P also the maximal ideal of D_P. Hence D_P is a pseudovaluation domain with integral closure D'_P. Since $L = K + iK$, there is an element $j \in D'\backslash P$ such that $j + P = i$. If j is not a unit of D', then there is an element $e \in P$ such that $k = j + e$ is a unit with $k + P = j + P = i$. We have $D' = D + kD$ and $D'_P = D_P + kD_P$.

Since the principal ideal Q^2 is Q-primary, no other prime contains c. Similarly no prime other than P contains b. By checking locally in D' we have that $Q = aD' + cD'$ and $P = aD' + bD'$. Also note that $a, b, c \in S' \cap D$. Thus checking locally in D shows that $Q_0 = aD + cD$. In addition, since $Q_0D_{(Q_0)} = QD'_Q$ and $D_{Q_0} = D'_Q$, Q_0 is locally principal as an ideal of D. Hence Q_0 is an invertible maximal ideal of D. It follows that each Q_0-primary ideal is a power of Q_0 and generated by $\{a, c\}^n$ (the nth power of the set $\{a, c\}$) for some positive n.

With regard to P-primary ideals of D, we make use of [8, Theorem 3.5]. As noted above, $D' = D + kD$ and $D'_P = D_P + kD_P$ for some unit k of D'. Thus $aD'_P = PD'_P = PD_P = aD_P + kaD_P$ with a and ka in $D \cap S'$. Suppose $x \in a^nD'_P\backslash a^{n+1}D'_P$ for some positive integer n. We will show that xD_P contains $P^{n+1}D'_P$.

We have $x = fa^n$ for some unit $f \in D'_P$. We may write $f = g/h$ where $g, h \in D'\backslash P$. Let $q \in a^{n+1}D'_P = P^{n+1}D'_P$. Then $q = ra^{n+1}/s$ for some $r \in D'$ and $s \in D'\backslash P$. Since f is a unit of D'_P, $q = ra^{n+1}/s = x(rf^{-1}a/s)$ with $rf^{-1}a/s \in PD_P$. Thus xD_P contains $P^{n+1}D'_P$.

Let J be a P-primary ideal of D. Then there is a positive integer m such that $b^m \in J$. By [8, Theorem 3.5], we either have $JD_P = P^nD_P$ for some n, or JD_P is a principal ideal of D_P. If $JD_P = P^nD_P$, then we $JD_P = a^nD_P + ka^nD_P$. By checking locally $J = a^nD + ka^nD + b^mD$ with $a^n, ka^n, b^m \in P\backslash S'$.

Next suppose JD_P is a principal ideal of D_P. Then there is a positive integer n and a unit $u \in D'_P$ such that $JD_P = ua^nD_P$. From above, JD_P contains $P^{n+1}D_P$. Since

u is a unit of D'_P, there are elements $y, z \in D' \backslash P$ with $z \in D$ such that $u = y/z$. Since z is a unit of D_P, we may assume $y = u$. Next let $w \in P$ be such that $u + w$ is a unit of D'. Since $wa^n \in P^{n+1}$, $ua^n D_P = (u + w)a^n D_P$. Since $u + w$ is a unit of D', $(u + w)a^n \in P \backslash S'$. Checking locally shows that $J = (u + w)a^n D + b^m D$ (with $(u + w)a^n, b^m \in P \backslash S'$). It follows that R is a Marot ring. Also since R' has exactly two regular maximal ideals with both invertible and neither principal, R' is not weakly additively regular [12, Corollary 3.2]. Thus R is not weakly additively regular.

Finally, since both regular prime ideals of R' are maximal and R' is a Prüfer ring, it is valuative [13, Theorem 2.10] (see Theorem 2.17 above). We also have that $D \subsetneq D'$ is a minimal integral extension and thus $R \subsetneq R'$ is a minimal integral extension. Hence R is valuative by Theorem 3.7. □

For a specific example satisfying the requirements in Theorem 3.14, we adapt the construction used in the proof of [4, Theorem 7].

Example 3.15. Let $E = \mathbb{Z}[\sqrt{10}]$ and $E' = E[z_1, z_2, z_3, \ldots]$. For each pair $f, g \in E'$ where no height one prime contains both, there is a smallest positive integer n such that $f, g \in E[z_1, z_2, \ldots, z_n]$. The polynomials $fz_{n+1} + g$ and $f + gz_{n+1}$ are primes of E' (see, for example, [15, Theorem 29, page 85]). Let $\mathcal{S}$ be the multiplicative set generated by these polynomials. Then $D' = E'_{\mathcal{S}}$ is a Dedekind domain, $P = 2D' + \sqrt{10}D'$ and $Q = 5D' + \sqrt{10}D'$ are invertible maximal ideals such that neither is principal, but $P^2 = 2D'$, $Q^2 = 5D'$, and $P \cap Q = \sqrt{10}D'$. We also have $D'/P = \mathbb{Z}_2(z_1, z_2, z_3, \ldots) = \mathbb{Z}_2(z_1^2, z_2, z_3, \ldots) + z_1\mathbb{Z}_2(z_1^2, z_2, z_3, \ldots)$ so the pullback of $\mathbb{Z}_2(z_1^2, z_2, z_3, \ldots)$ over P is a domain D whose integral closure is $D' = D + z_1 D$. Finally note that for $r \in D' \backslash P$, there is an element $s \in \mathcal{S}$ and an integer m such that $r + 2z_m/s$ is a unit of D'. By Theorem 3.14, R is a valuative Marot ring with exactly two regular maximal ideals that are neither integrally closed nor weakly additively regular.

References

1. Cahen, P.-J., Dobbs, D.E., Lucas, T.G.: Valuative domains. J. Algebra Appl. **9**, 43–72 (2010)
2. Cahen, P.-J., Dobbs, D.E., Lucas, T.G.: Characterizing minimal ring extensions. Rocky Mt. J. Math. **41**, 1081–1125 (2011)
3. Cahen, P.-J., Dobbs, D.E., Lucas, T.G.: Valuative pairs of commutative rings. Houst. J. Math. **41**, 397–431 (2015)
4. Claborn, L.: Every Abelian group is a class group. Pac. J. Math. **18**, 219–222 (1966)
5. Dobbs, D.E., Shapiro, J.: A classification of the minimal overrings of certain commutative rings. J. Algebra **308**, 800–821 (2007)
6. Ferrand, D., Olivier, J.-P.: Homomorphismes minimaux d'anneaux. J. Algebra **16**, 461–471 (1970)
7. Griffin, M.: Prüfer rings with zero divisors. J. Reine Angew. Math. **239**(240), 55–67 (1969)
8. Hedstrom, J., Houston, E.: Pseudo-valuation domains. Pac. J. Math. **75**, 137–147 (1978)
9. Huckaba, J.: Commutative Rings with Zero Divisors. Dekker, New York (1988)

10. Lucas, T.G.: Krull rings, Prüfer v-multiplication rings and the ring of finite fractions. Rocky Mt. J. Math. **35**, 1251–1325 (2005)
11. Lucas, T.G.: Additively regular rings and Marot rings. Palest. J. Math. **5**, 90–99 (2016)
12. Lucas, T.G.: Weakly additively regular rings and special families of prime ideals. Palest. J. Math **7**, 14–31 (2017)
13. Lucas, T.G.: Valuative rings. J. Algebra Appl. **17**(7), 1850123, 29 pp (2018)
14. Portelli, D., Spangher, W.: Krull rings with zero divisors. Commun. Algebra **11**, 1817–1851 (1983)
15. Zariski, O., Samuel, P.: Communications in Algebra, II edn. Springer, Berlin (1960)

Classifying Modules in Add of a Class of Modules with Semilocal Endomorphism Rings

Pavel Příhoda

Abstract We present a dimension theory for modules in $\mathrm{A}dd(\mathcal{C})$, where $\mathcal{C}$ is a class of modules with semilocal endomorphism rings satisfying certain smallness conditions. For example, if $\mathcal{C}$ is the class of all finitely presented modules over a semilocal ring R, then we get cardinal invariants which describe pure projective R-modules up to isomorphism.

Keywords Factor categories · Semilocal endomorphism ring · Direct sum decompositions of modules

Let $\mathcal{C}$ be a class of modules with semilocal endomorphism rings satisfying certain smallness conditions which will be specified later. For example, $\mathcal{C}$ can be a class of countably generated Artinian modules over any ring or a class of finitely presented modules over a semilocal ring. In this note we suggest a dimension theory for modules in $\mathrm{Add}(\mathcal{C})$, that is, a class of functions assigning cardinal invariants to modules of $\mathrm{Add}(\mathcal{C})$ such that two modules of $\mathrm{Add}(\mathcal{C})$ are isomorphic if and only if they have the same invariants. Note that if R is a semilocal ring and $\mathcal{C} = \{R\}$, then $\mathrm{Add}(\mathcal{C})$ is the class of projective R-modules. In [10] we gave the dimension theory for this case, so this paper can be considered as an extension of dimension theory for projective modules over a semilocal ring R to the class of pure projective R-modules. Another example studied so far is the class $\mathcal{C}$ of uniserial R-modules. In [6] the dimension theory for the class of direct summands of serial modules was presented. This theory appeared to be a useful tool for understanding pure projective modules over serial rings. For example, it enabled to characterize chain domains possessing a pure projective module which is not a direct sum of finitely presented modules as those possessing a nontrivial idempotent one-generated ideal (see [11]).

A dimension theory for classes of modules with semilocal endomorphism ring appeared in [4]. In this note we reconsider these dimensions using factor categories. The application of factor categories in the study of direct-sum decompositions is due

P. Příhoda (✉)
Faculty of Mathematics and Physics, Department of Algebra, Charles University, Sokolovská 83, 186 00 Prague, Czech Republic
e-mail: prihoda@karlin.mff.cuni.cz

A. Facchini et al. (eds.), *Advances in Rings, Modules and Factorizations*, Springer Proceedings in Mathematics & Statistics 321,
https://doi.org/10.1007/978-3-030-43416-8_15

to Harada and Sai [8], who considered direct sums of modules with local endomorphism rings.

The paper is organized as follows. The first section is of expository character, we follow mostly [7], where a particular case of modules with finite type was considered. Let us point out that there is a more general treatment of the subject in [3]; however, we decided to present the topic in a way sufficient to understand the dimension theory for modules in $\mathrm{Add}(\mathcal{C})$ presented in the third section of the paper. The second section collects relevant properties of I-small modules introduced in [6].

1 Overview of the Finite Case

Let R be a ring, let $0 \neq M \in \text{Mod-}R$ and let I be an ideal of $\mathrm{End}_R(M)$. Any class of right R-modules $\mathcal{D}$ is always considered as a full subcategory of Mod-R. The ideal $\mathcal{I}$ of $\mathcal{D}$ *associated to* I is the ideal consisting of all morphisms $f\colon X \to Y \in \mathcal{D}$ such that $\beta f \alpha \in I$ for any pair of homomorphisms $\alpha\colon M \to X$ and $\beta\colon Y \to M$. Observe that if M is an object of $\mathcal{D}$, then $I = \mathcal{I} \cap \mathrm{End}_R(M)$. Notice that the ideal $\mathcal{I}$ depends on I and $\mathcal{D}$, so the notation should be, for example, $\mathcal{I}_{I,\mathcal{D}}$ but $\mathcal{D}$ and I will always be specified or obvious from the context.

Let us recall the definition of the Jacobson radical of the category according to Mitchell [9]. Let $\mathcal{D}$ be a full category of Mod-R and set $\mathcal{J}(\mathcal{D}) := \{f \in \mathcal{D}(X, Y) \mid 1_X - gf$ is left invertible in $\mathrm{End}_{\mathcal{D}}(X)$ for every $g \in \mathcal{D}(Y, X)\}$. The original definition in [9] was for small preadditive categories and $\mathcal{J}$ was defined as the intersection of all maximal right ideals of $\mathcal{D}$. However, we just need to know that $\mathcal{J}(\mathcal{D})$ is an ideal of $\mathcal{D}$ and $\mathcal{J}(N, N)$ is $J(\mathrm{End}_R(N))$ for every $N \in \mathcal{D}$. The first statement is proved in [9, Lemma 4.1, Lemma 4.2] and the second one follows directly from the definition of $\mathcal{J}$. Observe that $f \in \mathcal{J}(\mathcal{D})$ if and only if, for every $N \in \mathcal{D}$, the ideal of $\mathcal{D}$ associated to $J(\mathrm{End}_R(N))$ contains f. Note that if M is a module with semilocal endomorphism ring and I is a maximal ideal of $\mathrm{End}_R(M)$, then the ideal of $\mathcal{D}$ associated to $\mathcal{I}$ contains $\mathcal{J}(\mathcal{D})$.

Most of the following results appeared in [3].

Lemma 1.1. *Let M, N be nonzero modules with semilocal endomorphism rings, let I be a maximal ideal of* $\mathrm{End}_R(M)$*, and let $\mathcal{I}$ be the ideal of $\mathcal{D}$ associated to I, where $\mathcal{D}$ is a class of modules containing N. Then either $\mathcal{I} \cap \mathrm{End}_R(N) = \mathrm{End}_R(N)$ or $\mathcal{I} \cap \mathrm{End}_R(N)$ is a maximal ideal of* $\mathrm{End}_R(N)$.

Proof. The arguments from the proof of [7, Lemma 4.4] can be used to check that $J(\mathrm{End}_R(N)) \subseteq J := \mathcal{I} \cap \mathrm{End}_R(N)$. The rest of the proof needs a slight modification: Suppose that $J \neq \mathrm{End}_R(N)$, so we have to prove that $\mathrm{End}_R(N)/J$ is a simple Artinian ring. Let $f, g \in \mathrm{End}_R(N)$ be such that $f + J, g + J$ are nontrivial orthogonal central idempotents of $\mathrm{End}_R(N)/J$. There exist $\alpha, \alpha'\colon M \to N$ and $\beta, \beta'\colon N \to M$ such that $\beta f \alpha \notin I$ and $\beta' g \alpha' \notin I$. Then $(\mathrm{End}_R(M)\beta f \alpha + I)/I$ is a nonzero left ideal and $(\beta' g \alpha' \mathrm{End}_R(M) + I)/I$ is a nonzero right ideal of a simple Artinian ring $\mathrm{End}_R(M)/I$, so $\beta' g \alpha' \mathrm{End}_R(M) \beta f \alpha + I \nsubseteq I$. On the other hand

$g\alpha'\mathrm{End}_R(M)\beta f \subseteq J$ (use $g + J$, $f + J$ are orthogonal and central), and $\beta' J\alpha \subseteq I$, since $J \subseteq \mathcal{I}$. This contradiction proves the lemma. ■

Lemma 1.2. *Let M, N be nonzero modules with semilocal endomorphism rings contained in a class $\mathcal{D} \subseteq$ Mod-R, let I be a maximal ideal of* $\mathrm{End}_R(M)$ *and let $\mathcal{I}$ be the ideal of $\mathcal{D}$ associated to I. Assume $J := \mathcal{I} \cap \mathrm{End}_R(N) \neq \mathrm{End}_R(N)$. Then $\mathcal{I}$ is the ideal of $\mathcal{D}$ associated to J.*

PROOF. Let $f : X \to Y \in \mathcal{D}$ be such that f is not in the ideal of $\mathcal{D}$ associated to J. Then there are $\alpha : N \to X$ and $\beta : Y \to N$ such that $\beta f\alpha \notin J = \mathcal{I} \cap \mathrm{End}_R(N)$. In particular, $f \notin \mathcal{I}$.

Conversely, assume that $f : X \to Y \in \mathcal{D}$ is not in $\mathcal{I}$, i.e., there are $\alpha : M \to X$ and $\beta : Y \to M$ such that $g := \beta f\alpha \notin I$. If g is in the ideal of $\mathcal{D}$ associated to J, then $\mathrm{Hom}_R(M, N)(\mathrm{End}_R(M)g\mathrm{End}_R(M) + I)\mathrm{Hom}_R(N, M) \subseteq J$. It follows that the whole $\mathrm{End}_R(M)$ is in the ideal of $\mathcal{D}$ associated to J. Fix $\varphi : M \to N$, $\psi : N \to M$ such that $\psi\varphi \notin I$. Then $\psi\varphi\mathrm{End}_R(M)\psi\varphi \nsubseteq I$. But $\varphi\mathrm{End}_R(M)\psi \subseteq J \subseteq \mathcal{I}$, which is a contradiction. So g is not in the ideal of $\mathcal{D}$ associated to J and there are homomorphisms $\alpha' : N \to M$ and $\beta' : M \to N$ such that $\beta'\beta f\alpha\alpha' \notin J$. Thus f is not in the ideal of $\mathcal{D}$ associated to J. ■

For a given class $\mathcal{C}$ of modules with semilocal endomorphism rings, let $\mathrm{Spec}(\mathcal{C})$ be the class of ideals in the category $\mathcal{C}$ such that $\mathcal{I} \in \mathrm{Spec}(\mathcal{C})$ if and only if $\mathcal{I}$ is associated to a maximal ideal of $\mathrm{End}_R(M)$ for some nonzero object M of $\mathcal{C}$. For each $M \in \mathcal{C}$, set $V(M) := \{\mathcal{I} \in \mathrm{Spec}(\mathcal{C}) \mid 1_M \notin \mathcal{I}\}$, so that $V(M)$ is the finite set of ideals associated to the maximal ideals of $\mathrm{End}_R(M)$.

Lemma 1.3. *Let $\mathcal{C}$ be a class of modules with semilocal endomorphism rings and let $M, N, M \oplus N \in \mathcal{C}$. Then $V(M) \cup V(N) = V(M \oplus N)$.*

PROOF. If $\mathcal{I} \in V(M)$, then $1_{M\oplus N} \notin \mathcal{I}$, otherwise $1_M = \pi_M 1_{M\oplus N}\iota_M$ would imply $1_M \in \mathcal{I}$. Then $\mathcal{I} \in V(M \oplus N)$. Similarly, $\mathcal{I} \in V(N)$ implies $\mathcal{I} \in V(M \oplus N)$ and $V(M) \cup V(N) \subseteq V(M \oplus N)$ follows.

Conversely, if $\mathcal{I} \in V(M \oplus N)$, then $1_{M\oplus N} = \iota_M 1_M \pi_M + \iota_N 1_N \pi_N$ implies that either $1_M \notin \mathcal{I}$ or $1_N \notin \mathcal{I}$. Consequently, $\mathcal{I} \in V(M)$ or $\mathcal{I} \in V(N)$ and $V(M \oplus N) \subseteq V(M) \cup V(N)$. ■

For each $\mathcal{I} \in \mathrm{Spec}(\mathcal{C})$ we would like to define a 'dimension function' $\dim_{\mathcal{I}}$ assigning a cardinal number to every object of $\mathcal{C}$. These 'functions' should describe modules of $\mathcal{C}$ up to isomorphism, i.e., $X, Y \in \mathcal{C}$ should be isomorphic if and only if $\dim_{\mathcal{I}}(X) = \dim_{\mathcal{I}}(Y)$ for every $\mathcal{I} \in \mathrm{Spec}(\mathcal{C})$.

For any $M, N \in \mathcal{C}$ and $\mathcal{I} \in \mathrm{Spec}(\mathcal{C})$ consider the group $\mathrm{Hom}_R(M, N)/\mathcal{I}(M, N)$. It is zero if $1_M \in \mathcal{I}$. Otherwise $I := \mathcal{I}(M, M)$ is a maximal ideal of $\mathrm{End}_R(M)$ and $\mathrm{End}_R(M)/I$ is a simple Artinian ring. Therefore $\mathrm{Hom}_R(M, N)/\mathcal{I}(M, N)$ has a natural structure of a right module over $\mathrm{End}_R(M)/I$ so we would like to define the $\mathcal{I}$-dimension of N to be the length of this module. Of course, it is necessary to check that the definition is independent of the choice of M. In fact, observe that $\mathrm{Hom}_R(M, N)/\mathcal{I}(M, N) = 0$ if $1_N \in \mathcal{I}$. Otherwise we can check that the composition length of $\mathrm{Hom}_R(M, N)/\mathcal{I}(M, N)$ is just the dual Goldie dimension of the ring $\mathrm{End}_R(N)/\mathcal{I}(N, N)$.

Lemma 1.4. *Let $\mathcal{C}$ be a class of modules with semilocal endomorphism rings, $M, N \in \mathcal{C}$ and let $\mathcal{I} \in \operatorname{Spec}(\mathcal{C})$. Then $\operatorname{Hom}_R(M, N)/\mathcal{I}(M, N) = 0$ if $1_M \in \mathcal{I}$ or $1_N \in \mathcal{I}$. If $1_M \notin \mathcal{I}$ and $1_N \notin \mathcal{I}$, the length of $\operatorname{Hom}_R(M, N)/\mathcal{I}(M, N)$, considered as a right $\operatorname{End}_R(M)/\mathcal{I}(M, M)$-module, is the same as the dual Goldie dimension of $\operatorname{End}_R(N)/\mathcal{I}(N, N)$.*

Proof. The first part of the statement is obvious. Suppose that $1_M, 1_N \notin \mathcal{I}$. Then it follows from Lemma 1.2 that the ideals of $\mathcal{C}$ associated to $\mathcal{I}(M, M)$ and $\mathcal{I}(N, N)$ are both equal to $\mathcal{I}$. Put $I := \mathcal{I}(N, N)$. Let $e_1, \dots, e_k \in \operatorname{End}_R(N)$ be such that $e_1 + I, \dots, e_k + I$ is a complete set of primitive orthogonal idempotents of $\operatorname{End}_R(N)/I$. Then $\operatorname{Hom}_R(M, N)/\mathcal{I}(M, N) = \oplus_{i=1}^k (e_i \operatorname{Hom}_R(M, N) + \mathcal{I}(M, N))/\mathcal{I}(M, N)$. Note that for any $1 \le i \le k$ the module $e_i \operatorname{Hom}_R(M, N) + \mathcal{I}(M, N)/\mathcal{I}(M, N)$ is not zero: Since $e_i \notin \mathcal{I}$ and $\mathcal{I}$ is the ideal of $\mathcal{C}$ associated to $\mathcal{I}(M, M)$, $e_i \operatorname{Hom}_R(M, N) \subseteq \mathcal{I}$ cannot hold.

It remains to prove that for every $1 \le i \le k$ the $\operatorname{End}_R(M)/\mathcal{I}(M, M)$-module $e_i \operatorname{Hom}_R(M, N) + \mathcal{I}(M, N)/\mathcal{I}(M, N)$ is simple. Take $f : M \to N$ such that $e_i f \notin \mathcal{I}$. Then there exists $h : N \to M$ such that $e_i f h \notin \mathcal{I}$ since $\mathcal{I}$ is associated to I. Consequently, $e_i f h t - e_i \in I$ for some $t \in \operatorname{End}_R(N)$. Then $e_i f (htg) - e_i g \in \mathcal{I}(M, N)$ for every $g : M \to N$. ■

Remark 1.5. The previous proof gives a way of finding a decomposition of the $\operatorname{End}_R(M)/\mathcal{I}(M, M)$-module $\operatorname{Hom}_R(M, N)/\mathcal{I}(M, N)$ into a direct sum of simple modules: Fix some complete set of orthogonal primitive idempotents $e_1 + I, \dots, e_k + I$ of the ring $\operatorname{End}_R(N)/I$, where $I = \mathcal{I}(N, N)$. If $f + \mathcal{I}(M, N)$ belongs to $\operatorname{Hom}_R(M, N)/\mathcal{I}(M, N)$, then $(e_1 f + \mathcal{I}(M, N), \dots, e_k f + \mathcal{I}(M, N))$ is the decomposition of $f + \mathcal{I}(M, N)$ in the direct-sum decomposition $\operatorname{Hom}_R(M, N)/\mathcal{I}(M, N) = \oplus_{i=1}^k (e_i \operatorname{Hom}_R(M, N) + \mathcal{I}(M, N))/\mathcal{I}(M, N)$. So $f + \mathcal{I}(M, N) \mapsto e_i f + \mathcal{I}(M, N)$ can be considered as the canonical projection $\pi_i : \operatorname{Hom}_R(M, N)/\mathcal{I}(M, N) \to (e_i \operatorname{Hom}_R(M, N) + \mathcal{I}(M, N))/\mathcal{I}(M, N)$ with respect to this decomposition of $\operatorname{Hom}_R(M, N)/\mathcal{I}(M, N)$. We will use this fact in the proof of Proposition 1.7.

It is also possible to consider $\operatorname{Hom}_R(M, N)/\mathcal{I}(M, N)$ as a left module over $\operatorname{End}_R(N)/\mathcal{I}(N, N)$ and use its length to define the $\mathcal{I}$-dimension of M. Similarly one can show that the length of this left module is the dual Goldie dimension of $\operatorname{End}_R(M)/\mathcal{I}(M, M)$. So these two definitions coincide. Of course, it is not true that the bimodule $\operatorname{Hom}_R(M, N)/\mathcal{I}(M, N)$ has same lengths on both sides. Indeed, take a simple R-module S and set $M := S$, $N := S \oplus S$, let $\mathcal{I}$ be the ideal of Mod-R associated to the zero ideal of $\operatorname{End}_R(M)$, then $\mathcal{I}(M, M), \mathcal{I}(M, N), \mathcal{I}(N, N)$ are all zero, $\operatorname{Hom}_R(M, N)$ is of length two as a right $\operatorname{End}_R(M)$-module but of length one as a left $\operatorname{End}_R(N)$-module.

If $\mathcal{C}$ is a class of modules with semilocal endomorphism rings, $\mathcal{I} \in \operatorname{Spec}(\mathcal{C})$ and $N \in \mathcal{C}$, we define $\dim_{\mathcal{I}}(N)$ to be the dual Goldie dimension of $\operatorname{End}_R(N)/\mathcal{I}(N, N)$ (if $\mathcal{I} \notin V(N)$, then $\dim_{\mathcal{I}}(N) = 0$). As remarked above, there is another interpretation of $\dim_{\mathcal{I}}$. The factor category $\mathcal{C}/\mathcal{I}$ has a fully faithful embedding to a category of finitely generated modules over a division ring (see Proposition 1.7 below). Then $\dim_{\mathcal{I}}(N)$

is the length of the corresponding module. Let us make a remark on uniqueness of such an embedding.

Remark 1.6. Let $\mathcal{C}$ be a class of modules with semilocal endomorphism rings, let M be a nonzero module of $\mathcal{C}$, I a maximal ideal of $\mathrm{End}_R(M)$, and let $\mathcal{I}$ be the ideal of $\mathcal{C}$ associated to I. Suppose that $F: \mathcal{C}/\mathcal{I} \to \text{mod-}K$ and $G: \mathcal{C}/\mathcal{I} \to \text{mod-}K'$ are full and faithful functors into the categories of finitely generated modules over division rings K, K'. Then $K \simeq K'$, since $K \simeq e\mathrm{End}_K(F(M))e$ for every primitive idempotent $e \in \mathrm{End}_K(F(M))$ and $K' \simeq e'\mathrm{End}_{K'}(G(M))e'$ for every primitive idempotent $e' \in \mathrm{End}_{K'}(G(M))$. Moreover, for every $N \in \mathcal{C}$ dimensions (or lengths) $\dim_K(F(N))$ and $\dim_{K'}(G(N))$ both coincide with the dual Goldie dimension of $\mathrm{End}_R(N)/\mathcal{I}(N,N)$. (Observe that $\mathrm{End}_K(F(N)) \simeq \mathrm{End}_{\mathcal{C}/\mathcal{I}}(N)$. The dual Goldie dimension of $\mathrm{End}_K(F(N))$ is $\dim_K(F(N))$.)

Proposition 1.7. *Let $\mathcal{C}$ be a class of modules with semilocal endomorphism rings and let $0 \neq M \in \mathcal{C}$. If $\mathcal{I}$ is the ideal of $\mathcal{C}$ associated to a maximal ideal I of* $\mathrm{End}_R(M)$, *then there exists a division ring K and a full and faithful functor $\mathcal{C}/\mathcal{I} \to$* mod-$K$.

Proof. Let $F: \mathcal{C} \to \mathcal{C}/\mathcal{I}$ be the canonical functor and let $S := \mathrm{End}_R(M)/I$. Consider the functor $G := \mathrm{Hom}_{\mathcal{C}/\mathcal{I}}(F(M), -): \mathcal{C}/\mathcal{I} \to \text{Mod-}S$. In the proof of Lemma 1.4 we have seen that $G(F(N))$ is a finitely generated module, so we have to show that G is full and faithful.

Let $f: X \to Y \in \mathcal{C}$ be such that $F(f) \neq 0$. Then $\beta f \alpha \notin I$ for some $\alpha: M \to X$ and $\beta: Y \to M$. In particular, $f\alpha \notin \mathcal{I}$ and $G(F(f))(F(\alpha)) \neq 0$. So G is faithful.

In order to prove that G is full, consider $N_1, N_2 \in \mathcal{C}$ such that $F(N_1) \neq 0$ and $F(N_2) \neq 0$. Let $I_i := \mathcal{I}(M, N_i)$, $i = 1, 2$. Let $e_1, \dots, e_k \in \mathrm{End}_R(N_1)$ and $f_1, \dots, f_l \in \mathrm{End}_R(N_2)$ be such that $F(e_1), \dots, F(e_k)$ is a complete set of orthogonal primitive idempotents of $\mathrm{End}_{\mathcal{C}/\mathcal{I}}(F(N_1))$ and $F(f_1), \dots, F(f_l)$ is a complete set of orthogonal primitive idempotents of $\mathrm{End}_{\mathcal{C}/\mathcal{I}}(F(N_2))$. By Remark 1.5, $GF(N_1) = \oplus_{j=1}^{k}(e_j\mathrm{Hom}_R(M, N_1) + I_1)/I_1$ and $GF(N_2) = \oplus_{j=1}^{l}(f_j\mathrm{Hom}_R(M, N_2) + I_2)/I_2$.

For every $i \in \{1, \dots, k\}$ there exists $\alpha_i: N_1 \to M$ such that $\alpha_i e_i \notin \mathcal{I}$ and for every $j \in \{1, \dots, l\}$ there exists $\beta_j: M \to N_2$ such that $f_j\beta_j \notin \mathcal{I}$. Since S is simple Artinian and the S-modules $(e_i\mathrm{Hom}_R(M, N_1) + I_1)/I_1$, $(f_j\mathrm{Hom}_R(M, N_2) + I_2)/I_2$ are simple, $\mathrm{Hom}_S((e_i\mathrm{Hom}_R(M, N_1) + I_1)/I_1, (f_j\mathrm{Hom}_R(M, N_2) + I_2)/I_2) = \pi_j^2 \circ GF(\beta_j) \circ \mathrm{End}_S(S) \circ GF(\alpha_i) \circ \iota_i^1$, where $\iota_i^1: (e_i\mathrm{Hom}_R(M, N_1) + I_1)/I_1 \to GF(N_1)$ is the canonical embedding and $\pi_j^2: GF(N_2) \to (f_j\mathrm{Hom}_R(M, N_2) + I_2)/I_2$ is the canonical projection. Observe that $\mathrm{End}_S(S) = G(\mathrm{End}_{\mathcal{C}/\mathcal{I}}F(M))$. Further, let $\iota_i^2: (f_i\mathrm{Hom}_R(M, N_2) + I_2)/I_2 \to GF(N_2)$ be the canonical embedding and $\pi_j^1: GF(N_1) \to (e_j\mathrm{Hom}_R(M, N_1) + I_1)/I_1$ be the canonical projection. As remarked above, $GF(e_i) = \iota_i^1\pi_i^1$ and $GF(f_j) = \iota_j^2\pi_j^2$.

Let $\varphi \in \mathrm{Hom}_S(GF(N_1), GF(N_2))$. Then $\varphi = \sum_{i=1,j=1}^{k,l} G(F(f_j))\varphi G(F(e_i))$ and $G(F(f_j))\varphi G(F(e_i)) = \iota_j^2\pi_j^2\varphi\iota_i^1\pi_i^1$, $\pi_j^2\varphi\iota_i^1 = \pi_j^2 GF(\psi_{j,i})\iota_i^1$ for some $\psi_{j,i}: N_1 \to N_2$. Therefore $\varphi = GF(\sum_{i=1,j=1}^{k,l} f_j\psi_{j,i}e_i)$.

To conclude the proof, observe that mod-S is equivalent to mod-K, where K is the endomorphism ring of the unique simple S-module.

Now let us reformulate several results from [7] to our setting. The next lemma has the same proof as [7, Lemma 4.7, Lemma 4.8].

Lemma 1.8. *Let $\mathcal{C}$ be a class of modules with semilocal endomorphism rings, $M_1, \dots, M_n \in \mathcal{C}$ and $S = \mathrm{End}_R(M_1 \oplus \cdots \oplus M_n)$. The following conditions are equivalent for an endomorphism $f \in S$:*

(i) $f \in J(S)$.
(ii) $\pi_j f \iota_i \in \mathcal{I}(M_i, M_j)$ *for every* $i, j = 1, \dots, n$ *and every* $\mathcal{I} \in \cup_{k=1}^{n} V(M_k)$.
(iii) *For every* $i, j = 1, \dots, n$ *and for every* $\mathcal{P} \in V(M_i) \cap V(M_j)$ *in the category* $\mathcal{C}$, *one has* $\pi_j f \iota_i \in \mathcal{P}$.

Moreover, if A is a direct summand of $\oplus_{k=1}^{n} M_k$, then $g \in J(\mathrm{End}_R(A))$ if and only if g is in every ideal of Mod-R *which is associated to a maximal ideal of* $\mathrm{End}_R(M_i)$ *for some $i = 1, \dots, n$.*

The proof of [7, Theorem 4.10] can be also transferred to our setting.

Proposition 1.9. *Let $\mathcal{C}$ be a class of modules with semilocal endomorphism rings, let M, N be direct summands of a finite direct sum of modules in $\mathcal{C}$. Then $M \simeq N$ if and only if* $\dim_{\mathcal{I}}(M) = \dim_{\mathcal{I}}(N)$ *for every* $\mathcal{I} \in \mathrm{Spec}(\mathrm{add}(\mathcal{C}))$.

PROOF. Note that $V(M) = V(N) = \{\mathcal{I}_1, \dots, \mathcal{I}_k\}$ is a finite set of ideals in $\mathrm{add}(\mathcal{C})$. If $k = 0$, then $M = N = 0$. Assume $k > 0$. For every $1 \le i \le k$ let $I_i = \mathcal{I}_i(M, M)$ and $J_i = \mathcal{I}_i(N, N)$. By Lemma 1.1, $I_1, \dots, I_k$ are the maximal ideals of $\mathrm{End}_R(M)$ and $J_1, \dots, J_k$ are the maximal ideals of $\mathrm{End}_R(N)$. Let $e_i \in \mathrm{End}_R(M)$, $f_i \in \mathrm{End}_R(N)$ be such that $e_i, f_i \in \mathcal{I}_j$ for every $j \ne i$ and $1_M - e_i$, $1_N - f_i \in \mathcal{I}_i$. By Proposition 1.7, for every $1 \le i \le k$ there are homomorphisms $\alpha_i\colon M \to N$ and $\beta_i\colon N \to M$ inducing mutually inverse isomorphisms of M and N in Mod-$R/\mathcal{I}_i$. Consider $\alpha := \sum_{j=1}^{k} \alpha_i e_i\colon M \to N$ and $\beta := \sum_{i=1}^{k} \beta_i f_i$. Then $1_M - \beta\alpha \in I_1 \cap \cdots \cap I_k = J(\mathrm{End}_R(M))$ and $1_N - \alpha\beta \in J_1 \cap \cdots \cap J_k = J(\mathrm{End}_R(N))$. So $\beta\alpha$ and $\alpha\beta$ are isomorphisms, therefore $M \simeq N$. ■

Remark 1.10. Note that, by Lemma 1.3, every ideal of $\mathrm{Spec}(\mathrm{add}(\mathcal{C}))$ is in $V(C)$ for some $C \in \mathcal{C}$. Therefore it is possible to extend dimension functions classifying objects in $\mathcal{C}$ to get a dimension theory for $\mathrm{add}(\mathcal{C})$.

2 *I*-small Modules

Recall that a family f_λ, $\lambda \in \Lambda$, of elements of $\mathrm{Hom}_R(X, Y)$ is called *summable* if, for any $x \in X$, $f_\lambda(x) = 0$, for all but finitely many $\lambda \in \Lambda$. If f_λ, $\lambda \in \Lambda$, is a summable family in $\mathrm{Hom}_R(X, Y)$, then we can define the homomorphism $f := \sum_{\lambda \in \Lambda} f_\lambda$ by $f(x) = \sum_{\lambda \in \Lambda, f_\lambda(x) \ne 0} f_\lambda(x)$.

Let us recall the notion of I-small module: Let M be a module, I an ideal of $\mathrm{End}_R(M)$. We say that M is *I-small* provided for every summable family f_λ, $\lambda \in \Lambda$, of endomorphisms in I the sum $\sum_{\lambda \in \Lambda} f_\lambda$ is in I. If $\mathcal{A}$ is a class of modules closed under arbitrary direct sums and $\mathcal{I}$ is an ideal of $\mathcal{A}$ closed under sums of summable families of morphisms, then the canonical functor $F\colon \mathcal{A} \to \mathcal{A}/\mathcal{I}$ preserves coproducts ([6, Lemma 2.5]). Thus, if we want to define the dimension function $\dim_{\mathcal{I}}$ on objects of $\mathcal{A}$ in such a way that $\dim_{\mathcal{I}}$ is compatible with arbitrary direct sums of objects in $\mathcal{A}$, the condition of $\mathcal{A}$ being $\mathcal{I}$-small introduced below seems to be essential. Let us recall [6, Corollary 2.7].

Proposition 2.1. *Let $\mathcal{A}$ be a full subcategory of* Mod-R *closed under arbitrary direct sums. Suppose that U is a nonzero module and I is an ideal of* $\mathrm{End}_R(U)$ *such that U is I-small. If $\mathcal{I}$ is the ideal of $\mathcal{A}$ associated to I, then the canonical functor $F\colon \mathcal{A} \to \mathcal{A}/\mathcal{I}$ preserves coproducts.*

Let $\mathcal{A}$ be a full subcategory of Mod-R and let $\mathcal{I}$ be an ideal of $\mathcal{A}$. We say that $\mathcal{A}$ is *$\mathcal{I}$-small* if the sum of every summable family of morphisms of $\mathcal{I}$ is in $\mathcal{I}$.

Lemma 2.2. *Let $\mathcal{A}$ be a full subcategory of* Mod-R *containing modules M, N. Let $\mathcal{I}$ be the ideal of $\mathcal{A}$ associated to an ideal I of the ring* $\mathrm{End}_R(M)$. *Then the following are equivalent:*

(i) M is I-small.
(ii) $\mathcal{A}$ is $\mathcal{I}$-small.

In particular, if M is I-small, then N is $\mathcal{I}(N, N)$-small.

PROOF. (i) follows immediately from (ii), let us prove the converse. Take modules $X, Y \in \mathcal{A}$ and a summable family of homomorphisms $f_\lambda\colon X \to Y$, $\lambda \in \Lambda$. Suppose that the sum $f = \sum_{\lambda\in\Lambda} f_\lambda$ is not in $\mathcal{I}$, that is, there are $g\colon M \to X$ and $h\colon Y \to M$ such that $hfg \notin \mathcal{I}$. Observe that $hf_\lambda g$, $\lambda \in \Lambda$ is a summable family in I and that $\sum_{\lambda\in\Lambda} hf_\lambda g = hfg$. So M cannot be I-small in this case.

Recall that an R-module M is *small* if for every family N_λ, $\lambda \in \Lambda$, of R-modules the canonical homomorphism $\alpha\colon \bigoplus_{\lambda\in\Lambda} \mathrm{Hom}_R(M, N_\lambda) \to \mathrm{Hom}_R(M, \bigoplus_{\lambda\in\Lambda} N_\lambda)$ is an isomorphism. Lemma 2.3 explains how to understand the relation between being small and being I-small in our particular case.

Lemma 2.3. *Let $\mathcal{A}$ be a full subcategory of* Mod-R, *let $M, N_\lambda, \lambda \in \Lambda$ be modules of $\mathcal{A}$ such that* $\mathrm{End}_R(M)$ *is semilocal and $\bigoplus_{\lambda\in\Lambda} N_\lambda$ is in $\mathcal{A}$. Let $\mathcal{I}$ be the ideal of $\mathcal{A}$ associated to a maximal ideal I of* $\mathrm{End}_R(M)$. *If M is I-small, then for any $f\colon M \to \bigoplus_{\lambda\in\Lambda} N_\lambda$ there is a finite set $\Lambda_0 \subseteq \Lambda$ such that $\pi_\lambda f \in \mathcal{I}$ for every $\lambda \in \Lambda \setminus \Lambda_0$.*

PROOF. Let $e_1, \dots, e_n \in \mathrm{End}_R(M)$ be such that $e_1 + I, \dots, e_n + I$ is a complete set of primitive orthogonal idempotents of $\mathrm{End}_R(M)/I$. Suppose that the statement is not true for $f\colon M \to \bigoplus_{\lambda\in\Lambda} N_\lambda$. Then there exists $i \in \{1, \dots, n\}$ and pairwise different elements $\lambda_1, \lambda_2, \dots \in \Lambda$ such that $\pi_{\lambda_j} f e_i \notin \mathcal{I}$ for every $j \in \mathbb{N}$. For simplicity suppose that $i = 1$. Then there are $h_j\colon N_{\lambda_j} \to M$ such that $t'_j := e_1 h_j \pi_{\lambda_j} f e_1 \notin I$ for every $j \in \mathbb{N}$. Since $(e_1 + I)\mathrm{End}_R(M)/I(e_1 + I)$ is a division ring, we can suppose that $t_j := e_1 - e_1 h_j \pi_{\lambda_j} f e_1 \in I$ for every $j = 1, 2, \dots$. Consider a countable family $t_1, t_2 - t_1, t_3 - t_2, \dots$. This is a summable family of homomorphisms in I but the sum of this family is $e_1 \notin I$, which is impossible if M is I-small. ∎

Corollary 2.4. *Let $\mathcal{A}$ be a full subcategory of* Mod-R *closed under (arbitrary) direct sums. Let M be a module of $\mathcal{A}$ such that* $\mathrm{End}_R(M)$ *is semilocal and let I be a maximal ideal of* $\mathrm{End}_R(M)$. *If M is I-small, then the functor*

$$\mathrm{Hom}_R(M,-)/\mathcal{I}(M,-)\colon \mathcal{A} \to \text{Mod-End}_R(M)/I$$

commutes with direct sums.

Lemma 2.5. *Let M be a nonzero module with semilocal endomorphism ring. Then M is $J(\mathrm{End}_R(M))$-small if and only if M is I-small for every maximal ideal I of* $\mathrm{End}_R(M)$.

PROOF. Suppose that M is I-small for every maximal ideal $I \subseteq \mathrm{End}_R(M)$. Take any summable family f_λ, $\lambda \in \Lambda$ in $J(\mathrm{End}_R(M))$. Then $f := \sum_{\lambda\in\Lambda} f_\lambda \in I$ for every maximal ideal $I \subseteq \mathrm{End}_R(M)$. Therefore also $f \in J(\mathrm{End}_R(M))$.

Conversely, suppose M is $J(\mathrm{End}_R(M))$-small and there is a maximal ideal $I' \subseteq \mathrm{End}_R(M)$ such that M is not I'-small. Then there exists a summable family f_λ, $\lambda \in \Lambda$ in I' such that $f := \sum_{\lambda\in\Lambda} f_\lambda$ is not in I'. Let $g \in \mathrm{End}_R(M)$ be such that $g \in I$ for every maximal ideal $I' \neq I \subseteq \mathrm{End}_R(M)$ and $1_M - g \in I'$. Then gf_λ, $\lambda \in \Lambda$ is a summable family in $J(\mathrm{End}_R(M))$ with the sum $g\sum_{\lambda\in\Lambda} f_\lambda \notin I'$. This is a contradiction. ∎

Example 2.6. (i) If M is a finitely generated module, then M is I-small for every ideal $I \subseteq \mathrm{End}_R(M)$. This is obvious since every summable family in $\mathrm{End}_R(M)$ has only finitely many nonzero members. In particular, a Noetherian module of finite dual Goldie dimension has semilocal endomorphism ring (cf. [5, Theorem 4.3(b)]) and is I-small for every maximal ideal $I \subseteq \mathrm{End}_R(M)$. Similarly, if M is a finitely presented module over a semilocal ring R, then, by [2, Theorem 3.3], $\mathrm{End}_R(M)$ is semilocal and M is I-small for every maximal ideal I of $\mathrm{End}_R(M)$.
(ii) Let M be an R-module such that $\mathrm{End}_R(M)$ is local. Then M is $J(\mathrm{End}_R(M))$-small by [6, beginning of Section 6].
(iii) Let M be an Artinian R-module with socle S. It is not true in general that $\mathrm{End}_R(M)$ is local if M is indecomposable (see [5, Example 8.19]). However, by [1, Corollary 6], $\mathrm{End}_R(M)$ is semilocal. Note that $J_0 := \{f \in \mathrm{End}_R(M) \mid f(S) = 0\} \subseteq J(\mathrm{End}_R(M))$. Any summable family of $\mathrm{End}_R(M)$ has almost all its members in J_0. In particular, M is $J(\mathrm{End}_R(M))$-small. By Lemma 2.5, M is I-small for every maximal ideal $I \subseteq \mathrm{End}_R(M)$.

In the next section, we give a dimension theory for modules in $\mathrm{Add}(\mathcal{C})$, where $\mathcal{C}$ is either a class of Noetherian modules of finite dual Goldie dimension or a class of (unfortunately) countably generated Artinian modules or a class of finitely presented modules if the ring is semilocal.

3 The Infinite Case

We would like to extend Proposition 1.9 to infinite direct sums of modules with semilocal endomorphism rings. Of course, it is necessary to extend the definition of $\dim_{\mathcal{I}}$ to infinite direct sums. There is an obvious way: Let M be a module with semilocal endomorphism ring, I a maximal ideal of $\mathrm{End}_R(M)$, and let $\mathcal{I}$ be the ideal of the category Mod-R associated to I. Then $\mathrm{Hom}_R(M, N)/\mathcal{I}(M, N)$ has a canonical structure of a module over a simple Artinian ring $\mathrm{End}_R(M)/I$. Let S be the simple right $\mathrm{End}_R(M)/I$-module. For $N \in$ Mod-R we would like to define $\dim_{\mathcal{I}}(N) = \kappa$, if $\mathrm{Hom}_R(M, N)/\mathcal{I}(M, N) \simeq S^{(\kappa)}$. Of course, it is necessary to check that such a definition is independent of the choice of M and I.

Proposition 3.1. *Let $\mathcal{C}$ be a class of modules with semilocal endomorphism rings, $M, N \in \mathcal{C}$ and $\mathcal{I}' \in V(M) \cap V(N)$. Let $\mathcal{I}$ be the ideal of* Mod-R *associated to $\mathcal{I}'(M, M)$. Then for an arbitrary module X, the length of* $\mathrm{End}_R(M)/\mathcal{I}(M, M)$-*module* $\mathrm{Hom}_R(M, X)/\mathcal{I}(M, X)$ *is the same as the length of* $\mathrm{End}_R(N)/\mathcal{I}(N, N)$-*module* $\mathrm{Hom}_R(N, X)/\mathcal{I}(N, X)$ *(since there are no assumptions on X these lengths can be infinite).*

PROOF. Let $I := \mathcal{I}(M, M)$, $J := \mathcal{I}(N, N)$, $S := \mathrm{End}_R(M)/I$ and $T := \mathrm{End}_R(N)/J$. Assume that M and N are isomorphic in Mod-$R/\mathcal{I}$. Let $\varphi\colon M \to N$, $\psi\colon N \to M$ be homomorphisms such that $1_M - \psi\varphi, 1_N - \varphi\psi \in \mathcal{I}$ and let us denote $\overline{\varphi}, \overline{\psi}$ the corresponding morphisms in Mod-$R/\mathcal{I}$. In this case S and T are isomorphic via $\theta\colon S \to T$, $\theta(s) = \overline{\varphi} s \overline{\psi}$. Observe that the homomorphism $\mathrm{Hom}_R(\psi, X)$ induces a group isomorphism $\Phi\colon \mathrm{Hom}_R(M, X)/\mathcal{I}(M, X) \to \mathrm{Hom}_R(N, X)/\mathcal{I}(N, X)$ given by $\Phi(f + \mathcal{I}(M, X)) := f\psi + \mathcal{I}(N, X)$. Moreover, $\Phi((f + \mathcal{I}(M, X))s) = \Phi((f + \mathcal{I}(M, X))\theta(s)$ for every $s \in S$, hence a decomposition of $\mathrm{Hom}_R(M, X)/\mathcal{I}(M, X)$ into a direct sum of simple S-modules is mapped by Φ to a direct-sum decomposition of $\mathrm{Hom}_R(N, X)/\mathcal{I}(N, X)$ into simple T-modules. Therefore the lengths of these modules are the same.

In general $M^{\mathrm{codim}(T)} \simeq N^{\mathrm{codim}(S)}$ in Mod-$R/\mathcal{I}$ by Lemma 1.4 and Proposition 1.7, therefore we can restrict to the case $N = M^m$ for some $m \in \mathbb{N}$. Let $\iota_i\colon M \to N$, $\pi_i\colon N \to M$, $i = 1, \dots, m$ be the canonical embeddings and projections and let $\overline{\iota_i}, \overline{\pi_i}, i = 1, \dots, m$ be the corresponding morphisms in Mod-$R/\mathcal{I}$. Consider a homomorphism $\tau\colon S \to T$ given by $\tau(s) := \sum_{i=1}^m \overline{\iota_i} s \overline{\pi_i}$. Then any right T-module can be considered as an S-module via τ. Let S_0 be a minimal right ideal of S, then it is possible to verify that $T_0 = \{\sum_{j=1}^m \overline{\iota_1} s_j \overline{\pi_j} \mid s_1, \dots, s_m \in S_0\}$ is a minimal right ideal of T and $T_0 \simeq S_0^m$ as S-modules. Further, it is easy to verify that $\Psi(f + \mathcal{I}(N, X)) := (f\iota_1 + \mathcal{I}(M, X), \dots, f\iota_m + \mathcal{I}(M, X))$ gives an S-module isomorphism $\Psi\colon \mathrm{Hom}_R(N, X)/\mathcal{I}(N, X) \to \mathrm{Hom}_R(M, X)/\mathcal{I}(M, X)^m$. Therefore m times the length of the S-module $\mathrm{Hom}_R(M, X)/\mathcal{I}(M, X)$ is the same as m times the length of the T-module $\mathrm{Hom}_R(N, X)/\mathcal{I}(N, X)$. ■

The previous proposition allows us to define the $\mathcal{I}$-dimension for an arbitrary module. Let $\mathcal{D}$ be a class of modules, let $M \in \mathcal{D}$ be a nonzero module with semilocal endomorphism ring and let $\mathcal{I}$ be the ideal of $\mathcal{D}$ associated to a maximal ideal of $I \subseteq$

$\mathrm{End}_R(M)$. For every $X \in \mathcal{D}$, we define $\dim_{\mathcal{I}}(X)$ to be the length of the $\mathrm{End}_R(M)/I$-module $\mathrm{Hom}_R(M, X)/\mathcal{I}(M, X)$.

Let us recall [6, Remark 2.9], which provides a useful criterion to establish whether a morphism between two direct sums belongs to an ideal.

Remark 3.2. Let $\mathcal{I}$ be an ideal of a category $\mathcal{C} \subseteq \text{Mod-}R$ such that $\mathcal{C}$ is $\mathcal{I}$-small and let $M_i, i \in I, N_j, j \in J$ be two families of modules from $\mathcal{C}$ such that $\oplus_{i\in I} M_i, \oplus_{j\in J} N_j \in \mathcal{C}$. Then $f \in \mathrm{Hom}_R(\oplus_{i\in I} M_i, \oplus_{j\in J} N_j)$ belongs to $\mathcal{I}$ if and only if $\pi_j f \iota_i \in \mathcal{I}$ for every $i \in I$, $j \in J$, where $\iota_i: M_i \to \oplus_{k\in I} M_k$ is the canonical embedding and $\pi_j: \oplus_{k\in J} N_k \to N_j$ is the canonical projection.

Proposition 3.3. *Let $\mathcal{C}$ be a class of modules with semilocal endomorphism rings, and $M \in \mathcal{C}$ be nonzero. Let I be a maximal ideal of* $\mathrm{End}_R(M)$ *and $\mathcal{I}$ be the ideal of* $\mathrm{Add}(\mathcal{C})$ *associated to I. If M is I-small, then the functor* $\mathrm{Hom}_R(M, -)$ *induces a full and faithful functor G:* $\mathrm{Add}(\mathcal{C})/\mathcal{I} \to \text{Mod-}\mathrm{End}_R(M)/I$.

Proof. Let $F: \mathrm{Add}(\mathcal{C}) \to \mathrm{Add}(\mathcal{C})/\mathcal{I}$ be the canonical functor and let $G := \mathrm{Hom}_{\mathrm{Add}(\mathcal{C})/\mathcal{I}}(M, -)$. As remarked above F and GF preserve direct sums. Let $\mathcal{D}$ be the full subcategory of $\mathrm{Add}(\mathcal{C})$ whose objects are arbitrary direct sums of modules of $\mathcal{C}$. First let us check that the restriction GF to $\mathcal{D}$ is a full functor on $\mathcal{D}$. Recall that by Proposition 1.7, the functor GF gives an onto map $\varphi_{C_1,C_2}: \mathcal{D}(C_1, C_2) \to \mathrm{Hom}_{\mathrm{End}_R(M)/I}(GF(C_1), GF(C_2))$ for every $C_1, C_2 \in \mathcal{C}$. Consider $D_1 := \oplus_{j\in J_1} C_j$, $D_2 := \oplus_{j\in J_2} C'_j \in \mathcal{D}$ and let $\iota_{j_1}: C_{j_1} \to D_1$, $j_1 \in J_1$, $\pi_{j_2}: D_2 \to C'_{j_2}$, $j_2 \in J_2$ be the corresponding embeddings and projections. Let $f \in \mathrm{Hom}_{\mathrm{End}_R(M)/I}(GF(D_1), GF(D_2))$. For $j_1 \in J_1$, $j_2 \in J_2$ let f_{j_2,j_1} be such that $f_{j_2,j_1} \in \varphi^{-1}_{C_{j_1},C'_{j_2}}(GF(\pi_{j_2}) \circ f \circ GF(\iota_{j_1}))$. Moreover, put $f_{j_2,j_1} = 0$ if $GF(\pi_{j_2}) \circ f \circ GF(\iota_{j_1}) = 0$. By Lemma 1.4, $GF(C_{j_1})$ is finitely generated, therefore for any $j_1 \in J_1$ there are only finitely many $j_2 \in J_2$ such that $f_{j_2,j_1} \neq 0$. Therefore there exists $f': D_1 \to D_2$ such that $\pi_{j_2} f' \iota_{j_1} = f_{j_2,j_1}$. Then it is easily seen that $GF(f') = f$.

Now we can prove that GF is full on $\mathrm{Add}(\mathcal{C})$. Let $A_1, A_2 \in \mathrm{Add}(\mathcal{C})$. Then there are $D_1, D_2 \in \mathcal{D}$ such that A_i is a direct summand of D_i, $i = 1, 2$. Let $\pi_i: D_i \to A_i$, $\iota_i: A_i \to D_i$, $i = 1, 2$ be the corresponding projections and embeddings. Let $f: GF(A_1) \to GF(A_2)$. From the previous part of the proof there exists $f': D_1 \to D_2$ such that $GF(f') = GF(\iota_2) f GF(\pi_1)$. Then $f'' := \pi_2 f' \iota_1$ satisfies $GF(f'') = f$.

We have proved that G is full. It remains to prove that G is faithful. Let $A_1, A_2 \in \mathrm{Add}(\mathcal{C})$ and $f: A_1 \to A_2$ be such that $GF(f) = 0$. We have to prove that $f \in \mathcal{I}$. Suppose the contrary. Then there are $\alpha: M \to A_1$ and $\beta: A_2 \to M$ such that $\beta f \alpha \notin I$, hence $\mathrm{Hom}_{\mathrm{Add}(\mathcal{C})/\mathcal{I}}(M, f)(\alpha) \neq 0$ which is not possible as $\mathrm{Hom}_{\mathrm{Add}(\mathcal{C})/\mathcal{I}}(M, f) = 0$. ■

Any module over a simple Artinian ring is determined up to isomorphism by its length. Therefore

Corollary 3.4. *Let $\mathcal{C}$ be a class of modules with semilocal endomorphism rings, let $M \in \mathcal{C}$ be nonzero, and let I be a maximal ideal of* $\mathrm{End}_R(M)$ *such that M is I-small. If $\mathcal{I}$ is an ideal of* $\mathrm{Add}(\mathcal{C})$ *associated to I and $X, Y \in \mathrm{Add}(\mathcal{C})$, then $\dim_{\mathcal{I}}(X) = \dim_{\mathcal{I}}(Y)$ if and only if X and Y are isomorphic in $\mathrm{Add}(\mathcal{C})/\mathcal{I}$.*

Lemma 3.5. *Let $\mathcal{C}$ be a class of modules with semilocal endomorphism rings. Consider a nonzero module $M \in \mathcal{C}$, a maximal ideal $I \subseteq \mathrm{End}_R(M)$ such that M is I-small and an ideal $\mathcal{I}$ of $\mathrm{Add}(\mathcal{C})$ associated to I. Moreover let $M_i, i \in \mathbb{N}$, be a countable family of modules in $\mathcal{C}$. If A, B are direct summands of $\oplus_{i\in\mathbb{N}} M_i$ such that $\dim_{\mathcal{I}}(A) = \dim_{\mathcal{I}}(B)$, then there are $f, g \in \mathrm{End}_R(\oplus_{i\in\mathbb{N}} M_i)$ satisfying the following conditions:*

(i) For any $i, j \in \mathcal{C}$ $\pi_j f \iota_i \neq 0$ or $\pi_j g \iota_i \neq 0$ implies $\mathcal{I} \cap \mathcal{C} \in V(M_i) \cap V(M_j)$.
(ii) $1_A - \pi_A g \iota_B \pi_B f \iota_A$, $1_B - \pi_B f \iota_A \pi_A g \iota_B \in \mathcal{I}$.
(iii) If $\mathcal{I} \cap \mathcal{C} \neq \mathcal{L} \in \mathrm{Spec}(\mathcal{C})$ such that $\mathcal{C}$ is $\mathcal{L}$-small, then $\pi_j f \iota_i, \pi_j g \iota_i \in \mathcal{L}$ for every $i, j \in \mathbb{N}$.

Proof. Let $F\colon \mathrm{Add}(\mathcal{C}) \to \mathrm{Add}(\mathcal{C})/\mathcal{I}$ be the canonical functor. By Corollary 3.4, $F(A) \simeq F(B)$. We can proceed as in [6, Lemma 3.3]: Suppose that $M = \oplus_{i\in\mathbb{N}} M_i = A \oplus A' = B \oplus B'$ and let $\iota_A\colon A \to M$, $\iota_B\colon B \to M$, $\pi_A\colon M \to A$, $\pi_B\colon M \to B$ be the canonical embeddings and projections. Let $f_0\colon F(A) \to F(B)$ and $g_0\colon F(B) \to F(A)$ be mutually inverse isomorphisms and let $f_0', g_0' \in \mathrm{End}_{\mathrm{Add}(\mathcal{C})/\mathcal{I}}(F(M))$ be such that $F(\pi_B) f_0' F(\iota_A) = f_0$, $F(\pi_A) g_0' F(\iota_B) = g_0$. Let $f', g' \in \mathrm{End}_R(M)$ be such that $F(f') = f_0'$, $F(g') = g_0'$. Because of Remark 3.2, we may choose f', g' satisfying $\pi_j f' \iota_i = 0$, $\pi_j g' \iota_i = 0$ for every $i, j \in \mathbb{N}$ such that $\mathcal{I}' := \mathcal{I} \cap \mathcal{C} \notin V(M_i) \cap V(M_j)$.

Observe that for every $i \in \mathbb{N}$ there exists $h_i \in \mathrm{End}_R(M_i)$ such that $1_{M_i} - h_i \in \mathcal{I}'$ and $h_i \in \mathcal{L}$ for every $\mathcal{I}' \neq \mathcal{L} \in \mathrm{Spec}(\mathcal{C})$. Indeed, fix $i \in \mathbb{N}$. If $\mathcal{I}' \notin V(M_i)$, put $h_i = 0$. If $V(M_i) = \{\mathcal{I}', \mathcal{L}_1, \dots, \mathcal{L}_n\}$, then $\mathcal{I}(M_i, M_i), \mathcal{L}_1(M_i, M_i), \dots, \mathcal{L}_n(M_i, M_i)$ are the maximal ideals of $\mathrm{End}_R(M_i)$. Hence, the Chinese Reminder Theorem gives $h_i \in \mathrm{End}_R(M_i)$ such that $1_{M_i} - h_i \in \mathcal{I}'$ and $h_i \in \cap_{i=1}^n \mathcal{L}_i$. Of course, $h_i \in \mathcal{L}$ for every $\mathcal{L} \notin V(M_i)$.

Let $h := \oplus_{i\in\mathbb{N}} h_i \in \mathrm{End}_R(M)$, $f := hf'$, $g := hg'$. Notice $F(f) = F(f')$, $F(g) = F(g')$, therefore (ii) holds. The choice of f', g' guarantees that (i) is satisfied too. Finally, if $\mathcal{L} \neq \mathcal{I}'$ is an ideal of $\mathrm{Spec}(\mathcal{C})$, then every h_i is in $\mathcal{L}$ which implies (iii). ■

Let $\mathcal{C}$ be a full subcategory of Mod-R and let $\mathcal{I}$ be an ideal of $\mathcal{C}$. We say that the ideal $\mathcal{J}$ of $\mathrm{Add}(\mathcal{C})$ is *associated* to $\mathcal{I}$ if $\mathcal{J}$ is the largest ideal of $\mathrm{Add}(\mathcal{C})$ such that $\mathcal{C} \cap \mathcal{J} = \mathcal{I}$. Observe that if $M \in \mathcal{C}$, I is an ideal of $\mathrm{End}_R(M)$, $\mathcal{I}$ is the ideal of $\mathcal{C}$ associated to I, and $\mathcal{J}$ is the ideal of $\mathrm{Add}(\mathcal{C})$ associated to $\mathcal{I}$, then $\mathcal{J}$ is the ideal of $\mathrm{Add}(\mathcal{C})$ associated to I.

We have already recalled that a module M is *small* if for every family of modules N_λ, $\lambda \in \Lambda$, and every homomorphism $f\colon M \to \oplus_{\lambda\in\Lambda} N_\lambda$ there exists a finite set $\Lambda_0 \subseteq \Lambda$ such that $f(M) \subseteq \oplus_{\lambda\in\Lambda_0} N_\lambda$. A module is said to be *σ-small* if it is a countable union of small modules (observe that every countably generated module is σ-small).

Proposition 3.6. *Let $\mathcal{C}$ be a class of modules with semilocal endomorphism rings, let M_i, $i \in \mathbb{N}$ be a countable family of objects in $\mathcal{C}$ such that for every $i \in \mathbb{N}$ the module M_i is σ-small and $J(\mathrm{End}_R(M_i))$-small. If A, B are direct summands of $\oplus_{i\in\mathbb{N}} M_i$ such that $\dim_{\mathcal{I}}(A) = \dim_{\mathcal{I}}(B)$ for every ideal $\mathcal{I}$ of $\mathrm{Add}(\mathcal{C})$ associated to an ideal of $\mathrm{Spec}(\mathcal{C})$, then $A \simeq B$.*

PROOF. Let $A \oplus A_1 = M = B \oplus B_1$. Using the standard Eilenberg's trick (cf. the proof of [6, Lemma 7.2]) we can suppose $A_1 \simeq B_1$. Let us denote $\iota_A \colon A \to M$, $\iota_B \colon B \to M$, $\pi_A \colon M \to A$, and $\pi_B \colon M \to B$ the canonical embeddings and projections. Let $\mathcal{S}$ be the set of ideals in $\mathrm{Add}(\mathcal{C})$ such that $\mathcal{I} \in \mathcal{S}$ if and only if $\mathcal{I}$ is associated to an ideal from $\cup_{i\in\mathbb{N}} V(M_i)$.

Step 1: There are $f, g \in \mathrm{End}_R(M)$ such that $1_A - \pi_A g \iota_B \pi_B f \iota_A$, $1_B - \pi_B f \iota_A \pi_A g \iota_B \in \mathcal{I}$ for every $\mathcal{I} \in \mathcal{S}$: For every $\mathcal{I} \in \mathcal{S}$ take $f_{\mathcal{I}}, g_{\mathcal{I}}$ satisfying the conditions (i)–(iii) of Lemma 3.5. Observe that $f_{\mathcal{I}}, \mathcal{I} \in \mathcal{S}$ and $g_{\mathcal{I}}, \mathcal{I} \in \mathcal{S}$ are summable families of homomorphisms (in fact, condition (i) of Lemma 3.5 says that $f_{\mathcal{I}}(M_i) \neq 0$ only if $\mathcal{I}$ is associated to an ideal from $V(M_i)$ and similarly for $g_{\mathcal{I}}$). Put $f := \sum_{\mathcal{I}\in\mathcal{S}} f_{\mathcal{I}}$, $g := \sum_{\mathcal{I}\in\mathcal{S}} g_{\mathcal{I}}$. It is easy to verify $f - f_{\mathcal{I}} \in \mathcal{I}$, $g - g_{\mathcal{I}} \in \mathcal{I}$ for every $\mathcal{I} \in \mathcal{S}$: Note that $f - f_{\mathcal{I}} = \sum_{\mathcal{I}\neq\mathcal{L}\in\mathcal{S}} f_{\mathcal{L}}$, every $f_{\mathcal{L}}$ on the right hand side is in $\mathcal{I}$. By Lemma 2.2, $\mathrm{Add}(\mathcal{C})$ is $\mathcal{I}$-small, hence also $f - f_{\mathcal{I}} \in \mathcal{I}$. Lemma 3.5(ii) gives $1_A - \pi_A g \iota_B \pi_B f \iota_A \in \mathcal{I}$ and $1_B - \pi_B f \iota_A \pi_A g \iota_B \in \mathcal{I}$ for every $\mathcal{I} \in \mathcal{S}$.

Step 2: Assume $f\colon A \to B$ and $g\colon B \to A$ satisfy $1_A - gf, 1_B - fg \in \mathcal{I}$ for every $\mathcal{I} \in \mathcal{S}$. For a small module A' contained in A there exists $g'\colon B \to A$ such that $1_A - g'f, 1_B - fg' \in \mathcal{I}$ for every $\mathcal{I} \in \mathcal{S}$ and $g'f(a) = a$ for any $a \in A'$. Similarly, for a small module B' contained in B there exists $f'\colon A \to B$ such that $1_A - gf', 1_B - f'g \in \mathcal{I}$ for every $\mathcal{I} \in \mathcal{S}$ and $f'g(b) = b$ for any $b \in B'$.

We prove the first statement only. Let $f_1 \colon A_1 \to B_1$, $g_1 \colon B_1 \to A_1$ be mutually inverse isomorphisms and let $f_0 := f \oplus f_1$, $g_0 := g \oplus g_1$. Observe that $1_M - g_0 f_0 \in \mathcal{I}$ for every $\mathcal{I} \in \mathcal{S}$. Take $n \in \mathbb{N}$ such that A' and $g_0 f_0(A')$ are both submodules of $\oplus_{i=1}^n M_i$ (such an n exists since A' is small). Let $\iota\colon \oplus_{i=1}^n M_i \to M$ be the canonical embedding and let $\pi\colon M \to \oplus_{i=1}^n M_i$ be the canonical projection. By Proposition 1.9, $\pi(1_M - g_0 f_0)\iota \in J(\mathrm{End}_R(\oplus_{i=1}^n M_i))$, therefore $\pi g_0 f_0 \iota$ is invertible. Let $h \in \mathrm{End}_R(\oplus_{i=1}^n M_i)$ be its inverse and define $h_0 := h \oplus 1_{\oplus_{i\geq n+1} M_i} \in \mathrm{End}_R(M)$. Observe that $1_M - h_0 \in \mathcal{I}$ for every $\mathcal{I} \in \mathcal{S}$. Put $g' := \pi_A h_0 \iota_A g$. Since $1_M - h_0 \in \mathcal{I}$ for every $\mathcal{I} \in \mathcal{S}$, $g - g' \in \mathcal{I}$ and hence $1_A - g'f, 1_B - fg' \in \mathcal{I}$ for every $\mathcal{I} \in \mathcal{S}$. Let $a \in A'$. Then $g'f(a) = \pi_A h_0 \iota_A g f(a) = \pi_A(h_0\iota)\pi g_0 f_0 \iota\pi(a) = \pi_A(\iota h)(\pi g_0 f_0 \iota)\pi(a) = \pi_A \iota\pi(a) = a$.

Step 3: Since M_i's are σ-small modules there are small modules A'_i, $i \in \mathbb{N}$ contained in A and small modules B'_i, $i \in \mathbb{N}$ contained in B such that $A = \sum_{i\in\mathbb{N}} A'_i$ and $B = \sum_{i\in\mathbb{N}} B'_i$. Now we define inductively small modules A_i, $i \in \mathbb{N}$ contained in A, small modules B_i, $i \in \mathbb{N}$ contained in B, homomorphisms $f_i \colon A \to B$, $i \in \mathbb{N}$, $g_i \colon B \to A$ such that

(i) $g_i f_i(a) = a$ for every $a \in A_i$,
(ii) $f_{i+1} g_i(b) = b$ for every $b \in B_i$,
(iii) $f_i(A_i) \subseteq B_i$, $g_i(B_i) \subseteq A_{i+1}$,

(iv) $1_A - g_i f_i, 1_B - f_i g_i, 1_A - g_i f_{i+1}, 1_B - f_{i+1} g_i \in \mathcal{I}$ for every $\mathcal{I} \in \mathcal{S}$ and every $i \in \mathbb{N}$, and
(v) $\sum_{i\in\mathbb{N}} A_i = A, \sum_{i\in\mathbb{N}} B_i = B$.

Put $A_1 := A_1'$, and let $f_1 \colon A \to B$, $g_1 \colon B \to A$ be such that $g_1 f_1(a) = a$, for every $a \in A_1$ and $1_A - g_1 f_1, 1_B - f_1 g_1 \in \mathcal{I}$ for every $\mathcal{I} \in \mathcal{S}$. The previous steps of the proof imply the existence of these homomorphisms.

Suppose that $f_1, \dots, f_n, g_1, \dots, g_n, A_1, \dots, A_n, B_1, \dots, B_{n-1}$ have been found. Define $B_n := B_n' + f_n(A_n)$, then define $f_{n+1} \colon A \to B$ such that $1_A - g_n f_{n+1}, 1_B - f_{n+1} g_n \in \mathcal{I}$ for every $\mathcal{I} \in \mathcal{S}$ and $f_{n+1} g_n(b) = b$ for every $b \in B_n$. Put $A_{n+1} := A_{n+1}' + g_n(B_n)$ and another application of Step 2 gives $g_{n+1} \colon B \to A$ such that $g_{n+1} f_{n+1}(a) = a$ for every $a \in A_{n+1}$ and $1_A - g_{n+1} f_{n+1}, 1_B - f_{n+1} g_{n+1} \in \mathcal{I}$ for every $\mathcal{I} \in \mathcal{S}$.

It is easily seen that $f_n|_{A_n} = f_{n+1}|_{A_n}, g_n|_{B_n} = g_{n+1}|_{B_n}$. Therefore we can define $f \colon A \to B$ and $g \colon B \to A$ by $f(a) = f_n(a)$ if $a \in A_n$ and $g(b) = g_n(b)$ if $b \in B_n$. Then f and g are mutually inverse isomorphisms. ∎

Theorem 3.7. *Let $\mathcal{C}$ be a class of σ-small modules with semilocal endomorphism rings such that every module $M \in \mathcal{C}$ is $J(\mathrm{End}_R(M))$-small. Two modules $A, B \in \mathrm{Add}(\mathcal{C})$ are isomorphic if and only if $\dim_{\mathcal{I}}(A) = \dim_{\mathcal{I}}(B)$ for every ideal of $\mathrm{Add}(\mathcal{C})$ associated an element of $\mathcal{I} \in \mathrm{Spec}(\mathcal{C})$.*

Proof. We proceed as in [6, Theorem 7.4]. Apply [5, Theorem 2.47] to the class of σ-small modules to deduce that A has a decomposition $A = \oplus_{x\in X} A_x$, such that, for every $x \in X$, the module A_x is isomorphic to a direct summand of a countable direct sum of modules from $\mathcal{C}$. Similarly let $B = \oplus_{y\in Y} B_y$ be a decomposition of B into direct summands of countable direct sums of modules from $\mathcal{C}$. The additivity of $\dim_{\mathcal{I}}$ gives $\sum_{x\in X} \dim_{\mathcal{I}}(A_x) = \sum_{y\in Y} \dim_{\mathcal{I}}(B_y)$ for every $\mathcal{I} \subseteq \mathrm{Add}(\mathcal{C})$ associated to an ideal of $\mathrm{Spec}(\mathcal{C})$, where the sums indicate the cardinalities of disjoint unions of $\dim_{\mathcal{I}}(A_x)$, $x \in X$, and $\dim_{\mathcal{I}}(B_y)$, $y \in Y$. Observe that there is a set $\mathcal{C}' \subseteq \mathcal{C}$ such that $A, B \in \mathrm{Add}(\mathcal{C}')$, let $\mathcal{S}$ be the set of ideals associated to ideals from $\cup_{M\in\mathcal{C}'} V(M)$ and notice that $\dim_{\mathcal{I}}(A) = \dim_{\mathcal{I}}(B) = 0$ if $\mathcal{I} \notin \mathcal{S}$. For any $\mathcal{I} \in \mathcal{S}$ let $E_{\mathcal{I}}$ be a fixed set of cardinality $\dim_{\mathcal{I}}(A)$ (hence $E_{\mathcal{I}}$ is empty if $\dim_{\mathcal{I}}(A) = 0$) and let $p_{\mathcal{I}} \colon E_{\mathcal{I}} \to X$, $q_{\mathcal{I}} \colon E_{\mathcal{I}} \to Y$ be such that $|p_{\mathcal{I}}^{-1}(x)| = \dim_{\mathcal{I}}(A_x)$ for every $x \in X$ and $|q_{\mathcal{I}}^{-1}(y)| = \dim_{\mathcal{I}}(B_y)$ for every $y \in Y$. Let us consider the unoriented bipartite graph G on the set $X \dot\cup Y$ with multiple edges whose set of edges is $\{\{p_{\mathcal{I}}(e), q_{\mathcal{I}}(e)\} \mid e \in E_{\mathcal{I}}, \mathcal{I} \in \mathcal{S}\}$.

We claim that degree of any vertex in the graph G is at most countable: Fix $x \in X$. The module A_x is isomorphic to a direct summand of $\oplus_{i\in\mathbb{N}} M_i$ for some $M_i \in \mathcal{C}$. Therefore $\dim_{\mathcal{I}}(A_x) = 0$ whenever $\mathcal{I} \notin \mathcal{S}'$, where $\mathcal{S}'$ is the set of ideals of $\mathrm{Add}(\mathcal{C})$ associated to the ideals from $\cup_{i\in\mathbb{N}} V(M_i)$. Since $V(M_i)$ are finite subsets of $\mathrm{Spec}(\mathcal{C})$, $\mathcal{S}'$ is countable. Moreover $\dim_{\mathcal{I}}(A_x) \le \dim_{\mathcal{I}}(\oplus_{i\in\mathbb{N}} M_i) \le \aleph_0$. Therefore the set of edges incident to x is $T = \{\{x, q(e)\} \mid e \in p_{\mathcal{I}}^{-1}(x), \mathcal{I} \in \mathcal{S}'\}$. Obviously $|T| \le \aleph_0$ which proves the claim (the number of edges incident to $y \in Y$ can be expressed similarly).

Let κ be an ordinal of cardinality at least $|X|$. As in the proof of [6, Theorem 7.4] we construct families of sets $X_\lambda \subseteq X$, $\lambda \in \kappa$, $Y_\lambda \subseteq Y$, $\lambda \in \kappa$ such that $\{X_{\lambda+1} \setminus X_\lambda \mid \lambda <$

$\kappa\}$ is a partition of X, $\{Y_{\lambda+1} \setminus Y_\lambda \mid \lambda < \kappa\}$ is a partition of Y. Moreover, for every $\lambda < \kappa$ the sets $X_{\lambda+1} \setminus X_\lambda$ and $Y_{\lambda+1} \setminus Y_\lambda$ are at most countable and $\dim_{\mathcal{I}}(\bigoplus_{x \in X_{\lambda+1} \setminus X_\lambda} A_x) = \dim_{\mathcal{I}}(\bigoplus_{y \in Y_{\lambda+1} \setminus Y_\lambda} B_y)$ for every $\mathcal{I} \in \mathcal{S}$. The construction is literally the same, so we do not repeat it here. Then we conclude by Proposition 3.6. ■

Acknowledgements The author was partially supported by grant GA ČR P201/12/G028. The author is indebted to the referee for valuable comments.

References

1. Camps, R., Dicks, W.: On semilocal rings. Israel J. Math. **81**(1–2), 203–211 (1993)
2. Facchini, A., Herbera, D.: Local morphisms and modules with a semilocal endomorphism ring. Algebr. Represent. Theor **9**(4), 403–422 (2006)
3. Facchini, A., Perone, M.: Maximal ideals in preadditive categories and semilocal categories. J. Algebra Appl. **10**(1), 1–27 (2011)
4. Facchini, A., Wiegand, R.: Direct sum decompositions of modules, semilocal endomorphism rings, and Krull monoids. J. Algebra **256**, 280–307 (2002)
5. Facchini, A.: Module Theory. Endomorphism Rings and Direct Sum Decompositions in Some Classes of Modules, Progress in Math, vol. 167. Birkhäuser Verlag, Basel (1998)
6. Facchini, A., Příhoda, P.: Factor categories and infinite direct sums. Int. Electron. J. Algebra **5**, 135–168 (2009)
7. Facchini, A., Příhoda, P.: Endomorphism rings with finitely many maximal right ideals. Comm. Algebra **39**(9), 3317–3338 (2011)
8. Harada, M., Sai, Y.: On categories of indecomposable modules I. Osaka J. Math. **7**, 323–344 (1970)
9. Mitchell, B.: Rings with several objects. Adv. Math. **8**, 1–161 (1972)
10. Příhoda, P.: Projective modules are determined by their radical factors. J. Pure Appl. Algebra **210**(3), 827–835 (2007)
11. Příhoda, P., Puninski, G.: Pure projective modules over chain domains with Krull dimension. J. Algebra **459**, 189–212 (2016)

The Multiplicative Ideal Theory of Leavitt Path Algebras of Directed Graphs—A Survey

Kulumani M. Rangaswamy

Abstract Let L be the Leavitt path algebra of an arbitrary directed graph E over a field K. This survey article describes how this highly non-commutative ring L shares a number of the characterizing properties of a Dedekind domain or a Prüfer domain expressed in terms of their ideal lattices. Special types of ideals such as the prime, the primary, the irreducible, and the radical ideals of L are described in terms of the graphical properties of E. The existence and the uniqueness of the factorization of a non-zero ideal of L as an irredundant product of prime or primary or irreducible ideals are established. Such factorization always exists for every ideal in L if the graph E is finite or if L is two-sided Artinian or two-sided Noetherian. In all these factorizations, the graded ideals of L seem to play an important role. Necessary and sufficient conditions are given under which L is a generalized ZPI ring, that is, when every ideal of L is a product of prime ideals. Intersections of various special types of ideals are investigated and an analogue of Krull's theorem on the intersection of powers of an ideal in L is established.

Keywords Leavitt path algebras · Multiplicative ideal theory · Factorization of ideals

1 Introduction

Leavitt path algebras of directed graphs are algebraic analogues of graph C*-algebras and, ever since they were introduced in 2004, have become an active area of research [1]. Every Leavitt path algebra $L := L_K(E)$ of a directed graph E over a field K is equipped with three mutually compatible structures: L is an associative K-algebra, L is a $\mathbb{Z}$-graded algebra, and L is an algebra with an involution $*$. Further, L possesses a large supply of idempotents, but it is highly non-commutative. Indeed, in most of the cases, the center of this K-algebra is trivial, being just the field K. In spite

K. M. Rangaswamy (✉)
Department of Mathematics, University of Colorado, Colorado Springs 80918, USA
e-mail: kmranga@gmail.com

A. Facchini et al. (eds.), *Advances in Rings, Modules and Factorizations*,
Springer Proceedings in Mathematics & Statistics 321,
https://doi.org/10.1007/978-3-030-43416-8_16

of this, it is somewhat intriguing and certainly interesting that the ideals of such a non-commutative algebra L exhibit the behavior of the ideals of a Prüfer domain and sometimes that of a Dedekind domain, thus making the multiplicative ideal theory of these algebras L worth investigating. The purpose of this survey is to give a detailed account of some of these properties of L and the resulting factorizations of its ideals. To start with, the ideal multiplication in L is commutative: $AB = BA$ for any two ideals A, B of L. As we shall see, the Prüfer-domain-like properties of L lead to satisfactory factorizations of ideals of L as products of prime, primary, or irreducible ideals. The graded ideals of L seem to possess interesting properties such as coinciding with their own radical, being realizable as Leavitt path algebras of suitable graphs, possessing local units and many others. They play an important role in the factorization of non-graded ideals of L. As noted in ([1], Theorem 2.8.10 and in [19]), the two-sided ideal structure of L can be described completely in terms of the hereditary saturated subsets and breaking vertices and cycles without exits in the graph E and irreducible polynomials in $K[x, x^{-1}]$, and the association preserves the lattice structures. This fact facilitates the description of various factorization properties of the two-sided ideals in L.

This paper is organized as follows. After the preliminaries, Section 3 describes the various properties of the graded ideals of L which are foundational to the study of non-graded ideals and in the factorization of ideals in L. In Section 4, L is shown to be an arithmetical ring, that is, its ideal lattice is distributive and, as a consequence, the Chinese Remainder Theorem holds in L. In addition, L is shown to be a multiplication ring. The ideal version of the number-theoretic theorem $\gcd(m, n) \cdot \text{lcm}(m, n) = mn$ for positive integers m, n holds in L, namely, for any two ideals M, N in L, $(M \cap N)(M + N) = MN$, again a characterizing property of Prüfer domains. In the next section, the prime, the primary, the irreducible, and the radical ideals of L are described in terms of the graph properties of E. It is interesting to note that for a graded ideal I of L the first three of these properties coincide and that I is always a radical ideal. In Section 6, we consider the existence and the uniqueness of factorizations of a non-zero ideal I as a product of prime, primary, or irreducible ideals of L. It is shown that if E is a finite graph or more generally, if L is two-sided Noetherian or Artinian, then every ideal of L is a product of prime ideals. This leads to a complete characterization of L as a generalized ZPI ring, that is, a ring in which every ideal of L is a product of prime ideals. Finally, an analogue of the Krull's theorem on powers of an ideal is proved for Leavitt path algebras. The results of this paper indicate the potential for successful utilization of the ideas and results from the ideal theory of commutative rings in the deeper study of the ideal theory of Leavitt path algebras (of course using different techniques, as L is non-commutative, and using the graphical properties of E and the nature of the graded ideals of L).

2 Preliminaries

For the general notation, terminology and results in Leavitt path algebras, we refer to [1, 18, 22] and for those in graded rings, we refer to [14, 17]. We refer to [8–13, 16] for results in commutative rings. Below we give an outline of some of the needed basic concepts and results.

A (directed) graph $E = (E^0, E^1, r, s)$ consists of two sets E^0 and E^1 together with maps $r, s : E^1 \to E^0$. The elements of E^0 are called *vertices* and the elements of E^1 *edges*. For each $e \in E^1$, say,

$$\bullet_{s(e)} \xrightarrow{e} \bullet_{r(e)}$$

$s(e)$ is called the **source** of e and $r(e)$ the **range** of e. If $\overset{u}{\bullet} \xrightarrow{e} \overset{v}{\bullet}$ is an edge, then $\overset{u}{\bullet} \xleftarrow{e^*} \overset{v}{\bullet}$ denotes the **ghost edge** e^* with $s(e^*) = v$ and $r(e^*) = u$.

A vertex v is called a **sink** if it emits no edges and a vertex v is called a **regular vertex** if it emits a non-empty finite set of edges. An **infinite emitter** is a vertex which emits infinitely many edges.

A **path** μ of length n is a sequences of edges $\mu = e_1 \dots e_n$ where $r(e_i) = s(e_{i+1})$ for all $i = 1, \cdots, n-1$. $|\mu|$ denotes the length of μ. The path $\mu = e_1 \dots e_n$ in E is **closed** if $r(e_n) = s(e_1)$, in which case μ is said to be *based at the vertex* $s(e_1)$. A closed path μ as above is called **simple** provided it does not pass through its base more than once, i.e., $s(e_i) \neq s(e_1)$ for all $i = 2, \dots, n$. The closed path μ is called a **cycle** if it does not pass through any of its vertices twice, that is, if $s(e_i) \neq s(e_j)$ for every $i \neq j$.

An *exit* for a path $\mu = e_1 \dots e_n$ is an edge e such that $s(e) = s(e_i)$ for some i and $e \neq e_i$.

If there is a path from vertex u to a vertex v, we write $u \geq v$. A subset D of vertices is said to be **downward directed** if for any $u, v \in D$, there exists a $w \in D$ such that $u \geq w$ and $v \geq w$. A subset H of E^0 is called **hereditary** if, whenever $v \in H$ and $w \in E^0$ satisfy $v \geq w$, then $w \in H$. A hereditary set is **saturated** if, for any regular vertex v, $r(s^{-1}(v)) \subseteq H$ implies $v \in H$.

Definition 1. *Given an arbitrary graph E and a field K, the Leavitt path algebra $L_K(E)$ is defined to be the K-algebra generated by a set $\{v : v \in E^0\}$ of pair-wise orthogonal idempotents, together with a set of variables $\{e, e^* : e \in E^1\}$ which satisfy the following conditions:*

(1) $s(e)e = e = er(e)$ for all $e \in E^1$.

(2) $r(e)e^* = e^* = e^*s(e)$ for all $e \in E^1$.

(3) (The "CK-1 relations") For all $e, f \in E^1$, $e^*e = r(e)$ and $e^*f = 0$ if $e \neq f$.

(4) (The "CK-2 relations") For every regular vertex $v \in E^0$,

$$v = \sum_{e \in E^1, s(e) = v} ee^*.$$

Note that L need not have an identity. Indeed, L will have the identity 1 exactly when the vertex set E^0 is finite and in that case $1 = \sum_{v \in E^0} v$. However, L possesses **local units**, namely, given any finite set of elements $a_1, \ldots, a_n \in L$, there is an idempotent u such that $ua_i = a_i = a_i u$ for all $i = 1, \ldots, n$. Every element $a \in L := L_K(E)$ can be written as $a = \sum_{i=1}^{n} k_i \alpha_i \beta_i^*$ where α_i, β_i are paths and $k_i \in K$. Here $r(\alpha_i) = s(\beta_i^*) = r(\beta_i)$. From this, it is easy to see that $L = \bigoplus_{u \in E^0} Lu$.

Many well-known examples of rings occur as Leavitt path algebras.

Example 1. The Leavitt path algebra of the straight line graph E :

$$\bullet_{v_1} \xrightarrow{e_1} \bullet_{v_2} \xrightarrow{e_2} \cdots \xrightarrow{e_{n-1}} \bullet_{v_n}$$

is isomorphic to the matrix ring $M_n(K)$.

(Indeed, if $p_1 = e_1 \cdots e_{n-1}$, $p_2 = e_2 \ldots e_{n-1}$, ..., $p_{n-1} = e_{n-1}$, $p_n = v_n$, then $\{\epsilon_{ij} = p_i p_j^* : 1 \leq i, j \leq n\}$ is a set of matrix units, that is, $\epsilon_{ii}^2 = \epsilon_{ii}$ and $\epsilon_{ij}\epsilon_{jk} = \epsilon_{ik}$. Then $\epsilon_{ij} \longmapsto E_{ij}$ induces the isomorphism, where E_{ij} is the $n \times n$ matrix with 1 at (i, j) position and 0 everywhere else.)

Example 2. If E is the graph with a single vertex and a single loop

then $L_K(E) \cong K[x, x^{-1}]$, the Laurent polynomial ring, induced by the map $v \mapsto 1$, $x \mapsto x$, $x^* \mapsto x^{-1}$.

The defining relations of a Leavitt path algebra $L_K(E)$ show that it is a non-commutative ring. Indeed if e is an edge in E, say, $\overset{u}{\bullet} \overset{e}{\longrightarrow} \overset{v}{\bullet}$ where $u \neq v$, then by defining relation (1), $ue = e$, but $eu = evu = e(vu) = 0$. The following proposition describes when $L_K(E)$ becomes a commutative ring.

Proposition 1. *Let E be a connected graph. Then the Leavitt path algebra $L_K(E)$ is commutative if and only if either E consists of just a single vertex $\{\bullet\}$ or E is the graph with a single vertex and a single loop as in Example 2. In this case $L_K(E) \cong K$ or $K[x, x^{-1}]$.*

Every Leavitt path algebra $L_K(E)$ is a $\mathbb{Z}$ **-graded algebra**, namely, $L_K(E) = \bigoplus_{n \in \mathbb{Z}} L_n$ induced by defining, for all $v \in E^0$ and $e \in E^1$, $\deg(v) = 0$, $\deg(e) = 1$, $\deg(e^*) = -1$. Here the L_n, called **homogeneous components**, are abelian subgroups satisfying $L_m L_n \subseteq L_{m+n}$ for all $m, n \in \mathbb{Z}$. Further, for each $n \in \mathbb{Z}$, the subgroup L_n is given by

$$L_n = \{\sum k_i \alpha_i \beta_i^* \in L : |\alpha_i| - |\beta_i| = n\}.$$

An ideal I of $L_K(E)$ is said to be a **graded ideal** if $I = \bigoplus_{n \in \mathbb{Z}} (I \cap L_n)$. If I is a non-graded ideal, then $\bigoplus_{n \in \mathbb{Z}} (I \cap L_n)$ is the largest graded ideal contained in I and is called the **graded part** of I, denoted by $gr(A)$.

We will also be using the fact that the Jacobson radical (and in particular, the prime/Baer radical) of $L_K(E)$ is always zero (see [1]).

Let Λ be an arbitrary non-empty (possibly, infinite) index set. For any ring R, we denote by $M_\Lambda(R)$ the ring of matrices over R whose entries are indexed by $\Lambda \times \Lambda$ and whose entries, except for possibly a finite number, are all zero. It follows from the works in [4] that $M_\Lambda(R)$ is Morita equivalent to R.

Throughout this paper L will denote the Leavitt path algebra $L_K(E)$ of an arbitrary directed graph E over a field K.

3 Graded Ideals of a Leavitt Path Algebra

In this section, we shall describe some of the salient properties of the graded ideals of a Leavitt path algebra L. As we shall see in a later section, these properties impact the factorization of ideals of L. Every ideal of L, whether graded or not, is shown to possess an orthogonal set of generators. As a consequence, we get the interesting property that every finitely generated ideal of L is a principal ideal. It is interesting to note that if I is a graded ideal of L, then both I and L/I can be realized as Leavitt path algebras of suitable graphs.

Suppose H is a hereditary saturated subset of vertices. A **breaking vertex** of H is an infinite emitter $w \in E^0 \backslash H$ with the property that $0 < |s^{-1}(w) \cap r^{-1}(E^0 \backslash H)| < \infty$. The set of all breaking vertices of H is denoted by B_H. For any $v \in B_H$, v^H denotes the element $v - \sum_{s(e)=v, r(e) \notin H} ee^*$. The following theorem of Tomforde describes graded ideals of L by means of their generators.

Theorem 1. *([22]) Suppose H is a hereditary saturated set of vertices and S is a subset of B_H. Then the ideal $I(H, S)$ generated by the set of idempotents $H \cup \{v^H : v \in S\}$ is a graded ideal of L, and conversely every graded ideal I of L is of the form $I(H, S)$ where $H = I \cap E^0$ and $S = \{u \in B_H : u^H \in I\}$.*

Given a pair (H, S) where H is a hereditary saturated set of vertices in the graph E and S is a subset of B_H, one could construct the **Quotient graph** $E \backslash (H, S)$ given by $(E \backslash (H, S))^0 = E^0 \backslash H \cup \{u' : u \in B_H \backslash S\}$, $(E \backslash (H, S))^1 = \{e \in E^1 : r(e) \notin H\} \cup \{e' : e \in E^1$ with $r(e) \in B_H \backslash S\}$ and r, s are extended to $(E \backslash (H, S))^0$ by setting $s(e') = s(e)$ and $r(e') = r(e)'$.

The next theorem describes a generating set Y for a not necessarily graded non-zero ideal of L. This set Y is actually an orthogonal set of generators.

Theorem 2. *([19]) Let E be an arbitrary graph and let I be an arbitrary non-zero ideal of $L = L_K(E)$ with $H = I \cap E^0$ and $S = \{u \in B_H : u^H \in I\}$. Then I is generated by the set*

$$Y = H \cup \{v^H : v \in S\} \cup \{f_t(c_t) : t \in T\},$$

where T is some index set (which may be empty), for each $t \in T$, c_t is a cycle without exits in $E \backslash (H, S)$, no v in S is on any cycle c_t, and $f_t(x) \in K[x]$ is a polynomial with a non-zero constant term and is of the smallest degree such that $f_t(c_t) \in I$. Any two elements $x \neq y$ in Y are orthogonal, that is, $xy = 0 = yx$.

If I is a finitely generated ideal, then the orthogonal set Y of generators mentioned in the above theorem can be shown to be finite and, in that case, the single element $a = \sum_{y \in Y} y$ will be a generator for the ideal I. Consequently, we obtain the following interesting result.

Theorem 3. *([19]) Every finitely generated ideal in a Leavitt path algebra is a principal ideal, i.e., of the form LaL for some $a \in L$.*

Remark 1. In [3], the above theorem has been extended by showing that every finitely generated one-sided ideal of L is a principal ideal, that is, L is a Bêzout ring.

An important property of graded ideals is the following.

Theorem 4. *([21]) Every graded ideal $I(H, S)$ of L can be realized as a Leavitt path algebra $L_K(F)$ of some graph F and further the corresponding quotient ring $L/I(H, S)$ is also a Leavitt path algebra, being isomorphic to the Leavitt path algebra $L_K(E \backslash (H, S))$ of the quotient graph $E \backslash (H, S)$.*

Since Leavitt path algebras possess local units, we conclude that the graded ideals I of L possess local units. Using this, we obtain some interesting properties of graded ideals.

Proposition 2. *([20]) (i) Let A be a graded ideal of L. Then*
(a) for any ideal B of L, $AB = A \cap B$, $BA = B \cap A$ and, in particular, $A^2 = A$;
(b) $AB = BA$ for all ideals B;
(c) If $A = A_1 \cdots A_m$ is a product of ideals, then $A = \bigcap_{i=1}^{m} gr(A_i) = \prod_{i=1}^{m} gr(A_i)$.

Similarly, if $A = A_1 \cap \cdots \cap A_m$ is an intersection of ideals A_i, then $A = \bigcap_{i=1}^{m} gr(A_i) = \prod_{i=1}^{m} gr(A_i)$.

(ii) If $A_1, \ldots, A_m$ are graded ideals of L, then $\prod_{i=1}^{m} A_i = \bigcap_{i=1}^{m} A_i$.

Proof. We shall point out the easy proof of (i)(a). We need only to prove $A \cap B \subseteq AB$. Let $x \in A \cap B$. Since the graded ideal A has local units, there is an idempotent $u \in A$ such that $ua = a = au$. Clearly then $a = ua \in AB$. So $A \cap B = AB$. Similarly, $B \cap A = BA$. Hence $AB = BA$. In particular, $A^2 = A \cap A = A$.

A natural question is when every ideal of L will be a graded ideal. This can happen when E satisfies the following graph property.

Definition 2. *A graph E satisfies* **Condition (K)** *if whenever a vertex v lies on a simple closed path α, v also lies on another simple closed path β distinct from α.*

Here is a simple graph satisfying Condition (K), where every vertex satisfies the required property.

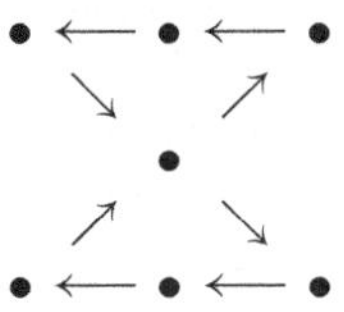

Theorem 5. *([18, 22]) The following conditions are equivalent for $L := L_K(E)$:*
(a) Every ideal of L is graded;
(b) Every prime ideal of L is graded;
(c) The graph E satisfies Condition (K).

4 The Lattice of Ideals of a Leavitt Path Algebra

This section describes how the ideals of a Leavitt path algebra L share lattice-theoretic properties and module-theoretic properties of the ideals of a Dedekind domain or a Prüfer domain. We start with noting that, in this non-commutative ring L, the multiplication of ideals is commutative. Moreover, L is left/right hereditary, that is, every left/right or two-sided ideal of L is projective as a left or a right ideal. The ideal lattice of L is distributive and multiplicative. It is also shown how many of the characterizing properties of a Prüfer domain stated in terms of its ideals hold in L.

Using a deep theorem of George Bergman, Ara and Goodearl proved the following result that every Leavitt path algebra is a left/right hereditary ring, a property shared by Dedekind domains.

Theorem 6. *(Theorem 3.7, [5]) Every ideal (including any one-sided ideal) of a Leavitt path algebra L is projective as a left/right L-module.*

In Section 3, we noted that if A is a graded ideal of L, then $AB = BA$ for any ideal B of L. What happens if A is not a graded ideal? With an analysis of the "non-graded parts" of A and B, it was shown in [1, 20] that even though L is, in general, non-commutative, the multiplication of its ideals is commutative as noted next.

Theorem 7. *([1, 20]) For any two arbitrary ideals A, B of a Leavitt path algebra L, $AB = BA$.*

The next result shows that every Leavitt path algebra L is an arithmetical ring, that is, the ideal lattice of L is distributive, a property that characterizes Prüfer domains.

Theorem 8. *([20]) For any three ideals A, B, C of the Leavitt path algebra L, we have*

$$A \cap (B + C) = (A \cap B) + (A \cap C).$$

Remark 2. A well-known result in commutative rings (see, e.g., Theorem 18, Chapter V, [23]) states that if the ideal lattice of a commutative ring R is distributive (such as when R is a Dedekind domain), then the Chinese Remainder Theorem holds in R: This means that the simultaneous congruences $x \equiv x_i \pmod{A_i}$ $(i = 1, \cdots, n)$ where the A_i are ideals and the elements $x_i \in R$, admits a solution for x in R provided the compatibility condition $x_i + x_j \equiv 0 \pmod{A_i + A_j}$ holds for all $i \neq j$. The proof of this theorem does not require R to be commutative and nor does it require the existence of a multiplicative identity in R. So, as a consequence of Theorem 8, one can show that the Chinese Remainder Theorem holds in Leavitt path algebras. (Thus Leavitt path algebras satisfy another property of Dedekind domains.)

We next use Theorem 8 to show that every Leavitt path algebra is a multiplication ring, a useful property in the multiplicative ideal theory of Leavitt path algebras.

Theorem 9. *([20]) The Leavitt path algebra $L = L_K(E)$ of an arbitrary graph E is a multiplication ring, that is, for any two ideals A, B of L with $A \subseteq B$, there is an ideal C of L, such that $A = BC = CB$. Moreover, if A is a prime ideal, then $AB = A = BA$.*

A well-known property of a Dedekind domain R is that if there are only finitely many prime ideals in R, then R is a principal ideal domain (see Theorem 16, Chapter V in [23]). Interestingly, as the next theorem shows, a Leavitt path algebra possesses this property.

Theorem 10. *([6]) Let $L := L_K(E)$ be the Leavitt path algebra of an arbitrary graph E. If L has only a finite number of prime ideals, then every ideal of L is a principal ideal, i.e., of the form LaL for some $a \in L$.*

Recently, it was shown (see [7]) that the ideals of a Leavitt path algebra satisfy two more characterizing properties of Prüfer domains.

Theorem 11. *([7]) Let A, B, C be any three ideals of a Leavitt path algebra L. Then*
(i) $A(B \cap C) = AB \cap AC$;
(ii) $(A \cap B)(A + B) = AB$.

Note that the statement (ii) in the preceding theorem is the ideal version of a well-known theorem in elementary number theory that, for any two positive integers a, b, $\gcd(a, b) \cdot \mathrm{lcm}(a, b) = ab$.

However not all characterizing properties of a Prüfer domain hold in a Leavitt path algebra. For instance, a domain R is a Prüfer domain if and only if finitely generated ideals of R are cancellative, that is, if A is a non-zero finitely generated ideal, then for any two ideals B, C of R, $AB = AC$ implies $B = C$. This property may not hold in a Leavitt path algebra as the next example shows.

Example 3. Consider the graph E

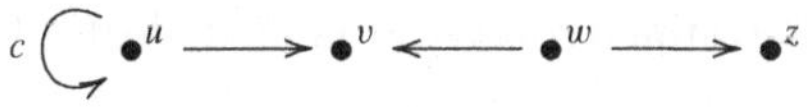

Here $H = \{v\}$ is a hereditary saturated subset. Let $A = \langle H \rangle$, the ideal generated by H. Clearly the cycle c has no exits in $E \backslash H$. Let B be the non-graded ideal $A + \langle p(c) \rangle$, where $p(x) = 1 + x \in K[x]$. Clearly $gr(B) = A$. Since A is a graded ideal, we apply Proposition 2 (a), to conclude that $AB = A \cap B = A = A^2 = AA$. But $A \neq B$.

5 Prime, Radical, Primary, and Irreducible Ideals of a Leavitt Path Algebra

In this section, we describe special types of ideals in L such as the prime, the irreducible, the primary, and the radical (= semiprime) ideals using graphical properties. While these concepts are independent for ideals in a commutative ring, we show that the first three properties of ideals coincide for graded ideals in the Leavitt path algebra L. We also show that a non-graded ideal I of L is irreducible if and only if I is a primary ideal if and only if $I = P^n$, a power of a prime ideal P. This is useful in the factorization of ideals in the next section. We also characterize the radical ideals of L. It may be some interest to note that every graded ideal of L is a radical ideal.

The following description of prime ideals of L was given in [18].

Theorem 12. *(Theorem 3.2, [18]) An ideal P of $L := L_K(E)$ with $P \cap E^0 = H$ is a prime ideal if and only if P satisfies one of the following properties:*

(i) $P = I(H, B_H)$ and $E^0 \backslash H$ is downward directed;

(ii) $P = I(H, B_H \backslash \{u\})$, $v \geq u$ for all $v \in E^0 \backslash H$ and the vertex u' that corresponds to u in $E \backslash (H, B_H \backslash \{u\})$ is a sink;

(iii) P is a non-graded ideal of the form $P = I(H, B_H) + \langle p(c) \rangle$, where c is a cycle without exits based at a vertex u in $E \backslash (H, B_H)$, $v \geq u$ for all $v \in E^0 \backslash H$ and $p(x)$ is an irreducible polynomial in $K[x, x^{-1}]$ such that $p(c) \in P$.

Recall, an ideal I of a ring R is called an **irreducible ideal** if, for ideals A, B of R, $I = A \cap B$ implies that either $I = A$ or $I = B$. Given an ideal I, the **radical of the ideal** I, denoted by $Rad(I)$ or $\sqrt{I}$, is the intersection of all prime ideals containing I. A useful property is that if $a \in Rad(I)$, then $a^n \in I$ for some integer $n \geq 0$. (The proof of this property is given in Theorem 10.7 of [15] for non-commutative rings with identity, but the proof also works for rings without identity but with local units.)

If $Rad(I) = I$ for an ideal I, then I is called a **radical ideal** or a **semiprime ideal**. An ideal I of R is said to be a **primary ideal** if, for any two ideals A, B, if $AB \subseteq I$ and $A \nsubseteq I$, then $B \subseteq Rad(I)$.

Remark 3. We note in passing that for any graded ideal I of L, say $I = I(H, S)$, $Rad(I) = I$. Because, $Rad(I)/I$ is a nil ideal in L/I and L/I, being isomorphic to the Leavitt path algebra $L_K(E\backslash(H, S))$, has no non-zero nil ideals.

We now point out an interesting property of graded ideals of L.

Theorem 13. *([20]) Suppose I is a graded ideal of L. Then the following are equivalent:*
(i) I is a primary ideal;
(ii) I is a prime ideal;
(iii) I is an irreducible ideal.

The next theorem extends the above result to arbitrary ideals of L.

Theorem 14. *([20]) Suppose I is a non-graded ideal of L. Then the following are equivalent:*
(i) I is a primary ideal;
(ii) $I = P^n$, a power of a prime ideal P for some $n \geq 1$;
(iii) I is an irreducible ideal.

The final result of this section describes the radical (also known as semiprime) ideals of L.

Theorem 15. *([2]) Let A be an arbitrary ideal of L with $A \cap E^0 = H$ and $S = \{v \in B_H : v^H \in A\}$. Then the following properties are equivalent:*
(i) A is a radical ideal of L;
(ii) $A = I(H, S) + \sum_{i \in Y} \langle f_i(c_i) \rangle$, where Y is an index set which may be empty, for each $i \in Y$, c_i is a cycle without exits based at a vertex v_i in $E\backslash(H, S)$ and $f_i(x)$ is a polynomial with its constant term non-zero which is a product of distinct irreducible polynomials in $K[x, x^{-1}]$.

6 Factorization of Ideals in L

As noted in the introduction, ideals in an arithmetical ring admit interesting representations as products of special types of ideals ([10–12]). In this section, we explore the existence and the uniqueness of factorizations of an arbitrary ideal in a Leavitt path algebra L as a product of prime ideals and as a product of irreducible/primary ideals. The prime factorization of graded ideals of L seems to influence that of the non-graded ideals in L. Indeed, an ideal I is a product of prime ideals in L if and only its graded part $gr(I)$ has the same property and, moreover, $I/gr(I)$ is finitely generated with a generating set of cardinality no more than the number of distinct prime ideals in an irredundant factorization of $gr(I)$. It is interesting to note that if I is a graded ideal and if $I = P_1 \cdots P_n$ is an irredundant product of prime ideals, then necessarily each of the ideals P_j must be graded. We also show that I is an intersection of irreducible ideals if and only if I is an intersection of prime ideals. If L is the Leavitt path algebra of a finite graph or, more generally, if L is two-sided Noetherian or two-sided Artinian, then every ideal of L is shown to be a product of prime ideals. We also give necessary and sufficient conditions under which every non-zero ideal of L is a product of prime ideals, that is, when L is a generalized ZPI ring. We end this section by proving for L an analogue of the Krull's theorem on the intersection of powers of an ideal.

We begin with the following useful proposition.

Proposition 3. *([20]) Suppose I is a non-graded ideal of L. If $gr(I)$ is a prime ideal, then I is a product of prime ideals.*

Using this, we obtain the following main factorization theorem.

Theorem 16. *([20]) Let E be an arbitrary graph. For a non-graded ideal I of $L := L_K(E)$, the following are equivalent:*

(i) I is a product of prime ideals;

(ii) I is a product of primary ideals;

(iii) I is a product of irreducible ideals;

(iv) $gr(I)$ is a product of (graded) prime ideals;

(v) $gr(I) = P_1 \cap \cdots \cap P_m$ is an irredundant intersection of m graded prime ideals P_j and $I/gr(I)$ is generated by at most m elements and is of the form $I/gr(I) = \bigoplus_{r=1}^{k} \langle f_r(c_r) \rangle$ where $k \leq m$ and, for each $r = 1 \cdots k$, c_r is a cycle without exits in $E^0 \backslash I$ and $f_r(x) \in K[x]$ is a polynomial with non-zero constant term of smallest degree such that $f_r(c_r) \in I$.

As a consequence of Theorem 16, we obtain a number of corollaries.

Corollary 1. *([20]) Let E be a finite graph, or more generally, let E^0 be finite. Then every non-zero ideal of $L = L_K(E)$ is a product of prime ideals.*

Using a minimal or maximal argument, the above corollary can be extended to the case when the ideals of L satisfy the DCC or ACC as noted below.

Corollary 2. *([20]) Suppose L is two-sided Artinian or two-sided Noetherian. Then every non-zero ideal of L is a product of prime ideals.*

We now give the necessary and sufficient conditions under which L is a generalized ZPI ring, that is, when every ideal of L is a product of prime ideals.

Theorem 17. *([20]) Let E be an arbitrary graph and let $L := L_K(E)$. Then every proper ideal of L is a product of prime ideals if and only if every homomorphic image of L is either a prime ring or contains only finitely many minimal prime ideals.*

The next theorem states that an irredundant factorization of an ideal A as a product of prime ideals in L is unique up to a permutation of the factors. It also points out the interesting fact that if A is a graded ideal, then every factor in this irredundant factorization must also be a graded ideal.

Recall that $A = P_1 \cdots P_n$ is an **irredundant product** of the ideals P_i, if A is not the product of a proper subset of the set $\{P_1, \ldots, P_n\}$.

Theorem 18. *([6]) (a) Suppose A is an arbitrary ideal of L and $A = P_1 \cdots P_m = Q_1 \cdots Q_n$ are two representations of A as irredundant products of prime ideals $P_i$and Q_j. Then $m = n$ and $\{P_1, \ldots, P_m\} = \{Q_1, \ldots, Q_n\}$;*

(b) If A is a graded ideal of L and if $A = P_1 \cdots P_m$ is an irredundant product of prime ideals P_j, then the ideals are all graded and $A = P_1 \cap \cdots \cap P_m$.

From Proposition 2(c) and the equivalence of conditions (i) and (iv) of Theorem 16, we derive following proposition.

Proposition 4. *If an ideal I of L is an intersection of finitely many prime ideals, then I is a product of (finitely many) prime ideals.*

But a product of prime ideals in L need not be an intersection of prime ideals as the next example shows.

Example 4. If E is the graph with a single vertex and a single loop

then $L_K(E) \cong K[x, x^{-1}]$, the Laurent polynomial ring, induced by the map $v \mapsto 1$, $x \mapsto x$, $x^* \mapsto x^{-1}$. So it is enough to find a ideal A in $K[x, x^{-1}]$ with the desired property. Consider the prime ideal $A = \langle p(x) \rangle$ in $K[x, x^{-1}]$, where $p(x)$ is an irreducible polynomial. We claim that $B = A^2$ is not an intersection of prime ideals in $K[x, x^{-1}]$. Suppose, on the contrary, $B = \bigcap_{\lambda \in \Lambda} M_\lambda$ where Λ is some (finite or infinite) index set and each M_λ is a (non-zero) prime ideal of $K[x, x^{-1}]$ and hence a maximal ideal of the principal ideal domain $K[x, x^{-1}]$. Now there is a homomorphism $\phi : R \longrightarrow \prod_{\lambda \in \Lambda} R/M_\lambda$ given by $r \mapsto (\ldots, r + M_\lambda, \ldots)$ with $\ker(\phi) = B$. Then $\bar{A} = \phi(A) \cong A/B \neq 0$ satisfies $(\bar{A})^2 = 0$ and this is impossible since $\prod_{\lambda \in \Lambda} R/M_\lambda$, being a direct product of fields, does not contain any non-zero nilpotent ideals.

The next proposition is new and gives necessary and sufficient conditions under which a product of prime ideals in a Leavitt path algebra is also an intersection of prime ideals. This happens exactly when every ideal of L is a radical ideal.

Proposition 5. *Let E be an arbitrary graph and let $L := L_K(E)$. Then the following properties are equivalent:*

(i) Every product of prime ideals in L is an intersection of prime ideals;
(ii) The graph E satisfies Condition (K);
(iii) Every ideal of L is a radical ideal;
(iv) Every ideal of L is a graded ideal.

Proof. Assume (i). Assume, by way of contradiction, that the graph E does not satisfy Condition (K). Then, for some admissible pair (H, S), the quotient graph $E\backslash(H, S)$ does not satisfy Condition (L) (see [1]) and thus there is a cycle c without exits in $E\backslash(H, S)$. By [1, Lemma 2.7.1], the ideal M of $L_K(E\backslash(H, S))$ generated by $\{c^0\}$ is isomorphic to the matrix ring $M_\Lambda(K[x, x^{-1}])$ where Λ is some index set. Then [7, Proposition 1] and Example 4 imply that, for any prime ideal P of M, P^2 is not an intersection of prime ideals of M. Since the graded ideal M is a ring with local units ([1, Corollary 2.5.23]), every ideal (prime ideal) of M is an ideal (prime ideal) of $L_K(E\backslash(H, S))$ and, for any prime ideal Q of $L_K(E\backslash(H, S))$, $M \cap Q$ is a prime ideal of M. Consequently, P^2 cannot be an intersection of prime ideals of $L_K(E\backslash(H, S))$. This is a contradiction, since $L_K(E\backslash(H, S))$, being isomorphic to the quotient ring $L/I(H, S)$, satisfies (i). Consequently, the graph E must satisfy Condition (K), thus proving (ii).

Assume (ii). By [1, Proposition 2.9.9], every ideal of L is graded. On the other hand if $I = I(H, S)$ is a graded ideal, then L/I is isomorphic to the Leavitt path algebra $L_K(E\backslash(H, S)$ and since the prime radical (the intersection of all prime ideals of $L_K(E\backslash(H, S))$ is zero, I is the intersection of all the prime ideals containing I and hence is a radical ideal. This proves (iii).

Assume (iii). We claim that every ideal of L must be a graded ideal. Suppose, by way of contradiction, there is a non-graded ideal I in L, say, $I = I(H, S) + \sum_{i \in Y}\langle f_i(c_i)\rangle$, where Y is an index set and, for each $i \in Y$, $f_i(x) \in K[x]$ and c_i is a cycle without exits in $E\backslash(H, S)$. Now for a fixed $i \in Y$ and an irreducible polynomial $p(x) \in K[x, x^{-1}]$, $P = I(H, S) + \langle p(c_i)\rangle$ is a prime ideal and $\bar{P} = P/I(H, S) = \langle p(c_i)\rangle \subsetneqq M = \langle\{c_i^0\}\rangle$. As noted in the proof of (i) $\Longrightarrow$ (ii), $\bar{P}^2$ is not a radical ideal of $L/I(H, S)$ and hence P^2 is not a radical ideal in L, a contradiction. Hence every ideal of L is a graded ideal. This proves (iv).

Now (iv) $\Longrightarrow$ (i), by Proposition 2(c).

We end this section by considering the powers of an ideal in L. From Proposition 2, it is clear that if A is a graded ideal of L, then $A = A^2$ and so $A = A^n$ for all $n \geq 1$. What happens if A is a non-graded ideal? The next proposition implies that, for such an A, $A \neq A^n$ for any $n > 1$.

Proposition 6. *([6]) If A is a non-graded ideal in L, then $\bigcap_{n=1}^{\infty} A^n$ is a graded ideal, being equal to $gr(A)$.*

As a corollary, we obtain

Corollary 3. *An ideal A of L is a graded ideal if and only if $A = A^n$ for all $n \geq 1$.*

W. Krull showed that if A is an ideal of a commutative Noetherian ring with identity 1, then $\bigcap_{n=1}^{\infty} A^n = 0$ if and only if $1 - x$ is not a zero divisor for all $x \in A$ (see Theorem 12, Section 7 in [23]). As a consequence of Proposition 6, we obtain an analogue of Krull's theorem for Leavitt path algebras.

Corollary 4. *([6]) Let A be an arbitrary ideal of L. Then $\bigcap_{n=1}^{\infty} A^n = 0$ if and only if A contains no vertices of the graph E.*

Acknowledgements My thanks to Gene Abrams for carefully reading this article, making corrections, and offering suggestions.

References

1. Abrams, G., Ara, P., Siles Molina, M.: Leavitt path algebras. Lecture Notes in Mathematics, vol. 2191. Springer (2017). ISBN 978-1-4471-7344-1
2. Abrams, G., Greenfeld, B., Mesyan, Z., Rangaswamy, K.M.: Chains of semiprime and prime ideals in Leavitt path algebras. Contemp. Math. **715**, 1–16 (2018). https://doi.org/10.1090/conm/715/14400
3. Abrams, G., Mantese F., Tonolo, A.: Leavitt path algebras are Bézout. Israel J. Math. (To appear)
4. Ánh, P.N., Márki, L.: Morita equivalence of rings without identity. Tsukuba J. Math. **11**, 1–16 (1987). https://doi.org/10.21099/tkbjm/1496160500
5. Ara, P., Goodearl, K.: Leavitt path algebras of separated graphs. J. Reine Angew. Math. **669**, 165–224 (2012). https://doi.org/10.1515/CRELLE.2011.146
6. Esin, S., Kanuni, M., Rangaswamy, K.M.: On intersections of two-sided ideals of Leavitt path algebras. J. Pure Appl. Algebra **221**, 632–644 (2017). https://doi.org/10.1016/j.jpaa.2016.07.009
7. Esin, S., Kanuni, M., Koç, A., Radler, K., Rangaswamy, K.M.: On Prüfer-like Properties of Leavitt Path Algebras (To appear)
8. Fontana, M., Houston, E., Lucas, T.: Factoring Ideals in Integral Domains. Springer (2013)
9. Fuchs, L., Mosteig, F.: Ideal theory in Prüfer domains—an unconventional approach. J. Algebra **252**, 411–430 (2002). https://doi.org/10.1016/S0021-8693(02)00040-6
10. Fuchs, L., Heinzer, W., Olberding, B.: Maximal prime divisors in arithmetical rings, Rings, modules, algebras and abelian groups. Lecture Notes in Pure and Applied Mathematics, vol. 236. Dekker, New York (2004)
11. Fuchs, L., Heinzer, W., Olberding, B.: Commutative ideal theory without finiteness conditions—primal ideals. Trans. Am. Math. Soc. **357**, 2771–2798 (2005). https://doi.org/10.1090/S0002-9947-04-03583-4
12. Fuchs, L., Heinzer, W., Olberding, B.: Commutative ideal theory without finiteness conditions—completely irreducible ideals. Trans. Am. Math. Soc. **358**, 3113–3131 (2006). https://doi.org/10.1090/S0002-9947-06-03815-3
13. Halter-Koch, F.: Ideal Systems, an Introduction to Multiplicative Ideal Theory. Marcel Dekker (1998)
14. Hazrat, R.: Graded rings and Graded Grothendieck Groups. LMS Lecture Notes Series, vol. 435. Cambridge University Press, Cambridge (2016)

15. Lam, T.Y.: Lectures on Modules and Rings, Grad. Texts in Mathematics, vol. 189. Springer (1999)
16. Larsen, M.D., McCarthy, P.J.: Multiplicative Theory of Ideals. Academic Press, New York, London (1971)
17. Nastasescu, C., van Oystaeyen, F.: Graded Ring Theory. North-Holland, Amsterdam (1982)
18. Rangaswamy, K.M.: Theory of prime ideals of Leavitt path algebras over arbitrary graphs. J. Algebra **375**, 73–90 (2013). https://doi.org/10.1016/j.jalgebra.2012.11.004
19. Rangaswamy, K.M.: On generators of two-sided ideals of Leavitt path algebras over arbitrary graphs. Communications in Algebra **42**, 2859–2868 (2014). https://doi.org/10.1080/00927872.2013.765008
20. Rangaswamy, K.M.: The multiplicative ideal theory of Leavitt path algebras. J. Algebra **487**, 173–199 (2017). https://doi.org/10.1016/j.jalgebra.2017.05.031
21. Ruiz, E., Tomforde, M.: Ideals in graph algebras. Algebras Repres. Theory **17**, 849–861 (2014). https://doi.org/10.1007/s10468-013-9421-3
22. Tomforde, M.: Uniqueness theorem and ideal structure of Leavitt path algebras. J. Algebra **318**, 270–299 (2007). https://doi.org/10.1016/j.jalgebra.2007.01.031
23. Zariski, O., Samuel, P.: Commutative Algebra, vol. 1. Van Nostrand Company (1969)

When Two Principal Star Operations Are the Same

Dario Spirito

Abstract We study when two fractional ideals of the same integral domain generate the same star operation.

Keywords Star operations · Principal star operations · m-canonical ideals

2010 Mathematics Subject Classification 13G05 · 13A15

1 Introduction

Throughout the paper, R will denote an integral domain with quotient field K and $\mathcal{F}(R)$ will be the set of *fractional ideals* of R, that is, the set of R-submodules I of K such that $xI \subseteq R$ for some $x \in K \setminus \{0\}$.

A *star operation* on R is a map $\star : \mathcal{F}(R) \longrightarrow \mathcal{F}(R)$ such that, for every $I, J \in \mathcal{F}(R)$ and every $x \in K$:

- $I \subseteq I^\star$;
- if $I \subseteq J$, then $I^\star \subseteq J^\star$;
- $(I^\star)^\star = I^\star$;
- $(xI)^\star = x \cdot I^\star$;
- $R^\star = R$.

The usual examples of star operations are the identity (usually denoted by d), the v-*operation* (or *divisorial closure*) $J \mapsto J^v := (R : (R : J))$, the t- and the w-operation (which are defined from v) and the star operations $I \mapsto \bigcap_{T \in \Delta} IT$, where Δ is a set of overrings of R intersecting to R. While these examples are the easiest to work with, they usually cover only a rather small part of the set of star operations.

D. Spirito (✉)
Dipartimento di Matematica e Fisica, Università degli Studi "Roma Tre", Roma, Italy
e-mail: spirito@mat.uniroma3.it

A. Facchini et al. (eds.), *Advances in Rings, Modules and Factorizations*, Springer Proceedings in Mathematics & Statistics 321,
https://doi.org/10.1007/978-3-030-43416-8_17

A much more general construction is given in [9, Proposition 3.2]: if $(I : I) = R$, then the map $J \mapsto (I : (I : J))$ is a star operation. This construction is much more flexible than the more "classical" ones, and allows to construct a much higher number of star operations (see, e.g., [10, Proposition 2.1(1)] or [11, Theorem 2.1] for its use to construct an infinite family of star operations, or [14, 15] for constructions in the case of numerical semigroups). In this paper, we slightly generalize this construction (removing the condition $(I : I) = R$), associating to each ideal I a star operation $v(I)$ (which we call the star operation *generated* by I); we study under which conditions two ideals I and J generate the same star operation and, in particular, we are interested in understanding when this happens only for isomorphic ideals.

The structure of the paper is as follows: in Section 3 we give some general properties of principal star operations; in Section 4, we generalize some results of [9] from m-canonical ideals to general ideals; in Section 5 we study the effect of localizations on principal star operations; in Section 6 we study operations generated by ideals whose v-closure is R (and, in particular, what happens when R is a unique factorization domain); in Section 7 we study the Noetherian case, reaching a necessary and sufficient condition for $v(I) = v(J)$ under the assumption $(I : I) = (J : J) = R$.

2 Background

By an *ideal* of R we shall always mean a fractional ideal of R, reserving the term *integral ideal* for those contained in R.

Let $\star$ be a star operation on R. An ideal I of R is $\star$-*closed* if $I = I^\star$; the set of $\star$-closed ideals is denoted by $\mathcal{F}^\star(R)$. When $\star = v$ is the divisorial closure, the elements of $\mathcal{F}^v(R)$ are called *divisorial ideals*.

Let $\mathrm{Star}(R)$ be the set of star operations on R. Then, $\mathrm{Star}(R)$ has a natural order structure, where $\star_1 \leq \star_2$ if and only if $I^{\star_1} \subseteq I^{\star_2}$ for every $I \in \mathcal{F}(R)$, or equivalently if $\mathcal{F}^{\star_1}(R) \supseteq \mathcal{F}^{\star_2}(R)$. Under this order, $\mathrm{Star}(R)$ is a complete lattice whose minimum is the identity and whose maximum is the v-operation.

A star operation is said to be *of finite type* if it is determined by its action on finitely generated ideals, or equivalently if

$$I^\star = \bigcup\{J^\star \mid J \subseteq I \text{ is finitely generated}\}$$

for every $I \in \mathcal{F}(R)$. A star operation is *spectral* if there is a subset $\Delta \subseteq \mathrm{Spec}(D)$ such that

$$I^\star = \bigcap\{IR_P \mid P \in \Delta\}$$

for every $I \in \mathcal{F}(R)$.

If $\star$ is a star operation of R, a prime ideal P is a $\star$-*prime* if it is $\star$-closed; the set of the $\star$-primes, denoted by $\mathrm{Spec}^\star(R)$, is called the $\star$-*spectrum*. A $\star$-*maximal ideal* of R is an ideal maximal among the set of proper ideals of R that are $\star$-closed;

their set is denoted by $\mathrm{Max}^\star(R)$. Any $\star$-maximal ideal is prime; however, $\star$-maximal ideals need not exist. If $\star$ is a star operation of finite type, then every $\star$-closed proper integral ideal is contained in some $\star$-maximal ideal; furthermore, for every $\star$-closed ideal I we have $I = \bigcap\{IR_P \mid P \in \mathrm{Spec}^\star(R)\}$.

3 Principal Star Operations

Definition 3.1. Let R be an integral domain. For every $I \in \mathcal{F}(R)$, the *star operation generated by* I, denoted by $v(I)$, is the supremum of all the star operations $\star$ on R such that I is $\star$-closed. If $\star = v(I)$ for some ideal I, we say that $\star$ is a *principal* star operation. We denote by $\mathrm{Princ}(R)$ the set of principal star operations of R.

We can give a more explicit representation of $v(I)$.

Proposition 3.2. *For every fractional ideal J, we have*

$$J^{v(I)} = J^v \cap (I : (I : J)) = J^v \cap \bigcap_{\alpha \in (I:J)\setminus\{0\}} \alpha^{-1} I. \tag{1}$$

Furthermore, if $(I : I) = R$ then $J^{v(I)} = (I : (I : J))$.

Proof. The fact that the two maps $J \mapsto J^v \cap (I : (I : J))$ and $J \mapsto J^v \cap \bigcap_{\alpha \in (I:J)\setminus\{0\}} \alpha^{-1} I$ give star operations and coincide follows in the same way as [9, Lemma 3.1 and Proposition 3.2]. The second representation clearly implies that they close I; furthermore, if I is closed then J^v and each $\alpha^{-1}I$ are closed, and thus the two representations of (1) give exactly $v(I)$.

The "furthermore" statement follows again from [9, Lemma 3.1 and Proposition 3.2]. □

In the paper [9] that introduced the map $J \mapsto (I : (I : J))$ when $(I : I) = R$, an ideal I was said to be *m-canonical* if $J = (I : (I : J))$ for every ideal J. This is equivalent to saying that $(I : I) = R$ and that $v(I)$ is the identity.

The definition of $v(I)$ can be extended to semistar operations, as in [13, Example 1.8(2)]; such construction was called the *divisorial closure with respect to* I in [4]. The terminology "generated" is justified by the following Proposition 3.3.

Proposition 3.3. *Let $\star$ be a star operation on R. Then, $\star = \inf\{v(I) \mid I \in \mathcal{F}^\star(R)\}$.*

Proof. Let $\sharp := \inf\{v(I) \mid I \in \mathcal{F}^\star(R)\}$. By definition, $\star \leq v(I)$ for every $I \in \mathcal{F}^\star(R)$, and thus $\star \leq \sharp$. Conversely, let J be a $\star$-ideal; then, $\sharp \leq v(J)$ and thus J is $\sharp$-closed. It follows that $\star \geq \sharp$, and thus $\star = \sharp$. □

Our main interest in this paper is to understand when two ideals generate the same star operation. The first cases are quite easy.

Lemma 3.4. *Let I be a fractional ideal of R. Then, the following hold.*

(a) *$v(I) = v$ if and only if I is divisorial.*
(b) *If $(I : I) = R$, then $v(I) = d$ if and only if I is m-canonical.*
(c) *For every $a \in K$, $a \neq 0$, we have $v(I) = v(aI)$.*
(d) *If L is an invertible ideal of R, then $v(I) = v(IL)$.*

Proof. The only non-trivial part is the last point. If L is invertible, then

$$I^{v(IL)}L \subseteq (I^{v(IL)}L)^{v(IL)} = (IL)^{v(IL)} = IL$$

and thus $I^{v(IL)} \subseteq IL(R : L) = I$, i.e., I is $v(IL)$-closed; it follows that $v(I) \geq v(IL)$. Symmetrically, we have $v(IL) \geq v(IL(R : L)) = v(I)$, and thus $v(I) = v(IL)$. □

We note that if $J = IL$ for some invertible ideal L, then I and J are locally isomorphic. However, the latter condition is neither necessary nor sufficient for I and J to generate the same star operation, even excluding divisorial ideals. For example, if R is an almost Dedekind domain that is not Dedekind, then all ideals are locally isomorphic but not all are divisorial, and two nondivisorial maximal ideals generate different star operations (if $M \neq N$ are two such ideals, then $(M : N) = M$ and so $N^{v(M)} = N^v \cap (M : (M : N)) = R$). For an example of non-locally isomorphic ideals generating the same star operation see Example 7.10.

The following necessary condition has been proved in [14, Lemma 3.7] when I and J are fractional ideals of a numerical semigroup; the proof of the integral domain case (which was also stated later in the same paper) can be obtained in exactly the same way.

Proposition 3.5. *Let R be an integral domain and I, J be nondivisorial ideals of R. If $v(I) = v(J)$ then*

$$I = I^v \cap \bigcap_{\gamma \in (I:J)(J:I)\setminus\{0\}} (\gamma^{-1}I).$$

4 Local Rings

As the construction of the principal star operation $v(I)$ generalizes the definition of m-canonical ideal, we expect that I is in some way "m-canonical for $v(I)$". Pursuing this strategy, we obtain the following generalization of [9, Lemma 2.2(e)].

Lemma 4.1. *Let I be an ideal of a domain R such that $(I : I) = R$. Let $\{J_\alpha \mid \alpha \in A\}$ be $v(I)$-ideals such that $\bigcap_{\alpha \in A} J_\alpha \neq (0)$. Then,*

$$\left(I : \bigcap_{\alpha \in A} J_\alpha\right) = \left(\sum_{\alpha \in A}(I : J_\alpha)\right)^{v(I)}.$$

Proof. Let $J := \sum_{\alpha\in A}(I : J_\alpha)$. Since $(I : I) = R$, we have $L^{v(I)} = (I : (I : L))$ for every ideal L; therefore,

$$(I : J) = \left(I : \sum_{\alpha\in A}(I : J_\alpha)\right) = \bigcap_{\alpha\in A}(I : (I : J_\alpha)) = \bigcap_{\alpha\in A} J_\alpha^{v(I)} = \bigcap_{\alpha\in A} J_\alpha$$

and thus

$$J^{v(I)} = (I : (I : J)) = \left(I : \bigcap_{\alpha\in A} J_\alpha\right),$$

as claimed. □

The following definition abstracts a property proved, for m-canonical ideals of local domains, in [9, Lemma 4.1].

Definition 4.2. Let $\star$ be a star operation on R. We say that an ideal I of R is *strongly $\star$-irreducible* if $I = I^\star \neq \bigcap\{J \in \mathcal{F}^\star(R) \mid I \subsetneq J\}$.

Lemma 4.3. *Let R be a domain and I be a nondivisorial ideal of R. If I is strongly $v(I)$-irreducible and $v(I) = v(J)$, then $I = uJ$ for some $u \in K$.*

Proof. Suppose $v(I) = v(J)$. Then

$$I = I^{v(J)} = I^v \cap \bigcap_{\alpha\in(J:I)\setminus\{0\}} \alpha^{-1}J.$$

Both I^v and each $\alpha^{-1}J$ are $v(I)$-ideals; hence, either $I = I^v$ (which is impossible since I is not divisorial) or $I = \alpha^{-1}J$ for some $\alpha \in K$. □

Lemma 4.4. *Suppose (R, M) is a local ring and $R = (I : I)$. If M is $v(I)$-closed, then I is strongly $v(I)$-irreducible.*

Proof. Let $\{J_\alpha\}$ be a family of $v(I)$-ideals such that $I = \bigcap J_\alpha$. Then,

$$R = (I : I) = \left(I : \bigcap_\alpha J_\alpha\right) = \left(\sum_\alpha (I : J_\alpha)\right)^{v(I)}$$

by Lemma 4.1.

Hence $(I : J_\alpha) \subseteq R$ for every α; suppose $I \subsetneq J_\alpha$ for all α. Then, $1 \notin (I : J_\alpha)$ and thus $(I : J_\alpha) \subseteq M$; therefore, $\sum(I : J_\alpha) \subseteq M$ and, since M is $v(I)$-closed, also $\left(\sum_\alpha(I : J_\alpha)\right)^{v(I)} \subseteq M$, a contradiction. Therefore, we must have $J_\alpha = I$ for some α, and I is strongly $v(I)$-irreducible. □

As a consequence of the previous two lemmas, we have a very general result for local rings.

Proposition 4.5. *Let (R, M) be a local domain and I a nondivisorial ideal of R such that $(I : I) = R$. If $M = M^{v(I)}$ (in particular, if M is divisorial), then $v(I) = v(J)$ for some ideal J if and only if $I = uJ$ for some $u \in K$.*

Proof. By Lemma 4.4, I is strongly $v(I)$-irreducible; by Lemma 4.3 it follows that $I = uJ$. □

Corollary 4.6. *Let (R, M) be a local domain, and I and J two nondivisorial ideals of R. If R is completely integrally closed and M is divisorial, then $v(I) = v(J)$ if and only if $I = uJ$ for some $u \in K$.*

Proof. Since R is completely integrally closed, $(L : L) = R$ for all ideals L; furthermore, since M is divisorial $M^{v(L)} = M$ for every L. The claim follows from Proposition 4.5. □

One problem of the previous results is the hypothesis $(I : I) = R$. In the following proposition we eliminate it at the price of forcing more properties on R.

Proposition 4.7. *Let (R, M) be a local ring, and let $T := (M : M)$. Let I, J be ideals of R, properly contained between R and T, such that $v(I) = v(J)$.*

(a) If $(I : I), (J : J) \subseteq T$, then $(I : I) = (J : J)$.
(b) Suppose also that $(I : I) =: A$ is local with divisorial maximal ideal, and that I and J are not divisorial over A. Then, there is a $u \in K$ such that $I = uJ$.

Proof. If M is principal, $T = R$ and the statement is vacuous. Suppose thus M is not principal: then, we also have $T = (R : M)$. We first claim that $L^v = T$ for every ideal L properly contained between R and T. Indeed, the containment $R \subsetneq L$ implies that $(R : L) \subsetneq R$ and thus, since R is local, $(R : L) \subseteq M$ and $L^v \supseteq T \supsetneq L$; hence, $L^v = T$.

(a) Let $T_1 := (I : I)$ and $T_2 := (J : J)$, and define $\star_i$ as the star operation $L^{\star_i} := L^v \cap LT_i$. Since T contains T_1 and T_2, it is both a T_1- and a T_2-ideal. We claim that $L \neq R$ is $\star_i$-closed if and only if it is a T_i-ideal: the "if" part is obvious, while if $L = L^v \cap LT_i$ then $L^v = T$ is a T_i-ideal and thus L is intersection of two T_i-ideals.

If $v(I) = v(J)$, then I is $\star$-closed if and only if J is $\star$-closed; therefore, since I is $\star_1$-closed and J is $\star_2$-closed, both I and J are T_1- and T_2-ideals. But $(I : I)$ (respectively, $(J : J)$) is the maximal overring of R in which I (respectively, J) is an ideal; thus $(I : I) = (J : J)$.

(b) Consider the star operation generated by I on A, i.e., $v_A(I) : L \mapsto (A : (A : L)) \cap (I : (I : L))$ for every $L \in \mathcal{F}(A)$. By the first paragraph of the proof, applied on the A-ideals, we have $(A : (A : L)) = T$ for all ideals L of A properly contained between A and T; in particular, this happens for J (since $R \subset J$ implies $A = AR \subseteq AJ = J$, and $A \neq J$ since J is not divisorial), and thus $J^{v_A(I)} = J^{v(I)} = J$. Symmetrically, $I^{v_A(J)} = I$; hence, $v_A(I) = v_A(J)$. By Proposition 4.5, applied to A, we have $I = uJ$ for some $u \in K$, as claimed. □

Recall that a *pseudo-valuation domain* (PVD) is a local domain (R, M) such that M is the maximal ideal of a valuation overring of R (called the valuation domain *associated* to R) [8].

Corollary 4.8. *Let (R, M) be a pseudo-valuation domain with associated valuation ring V, and suppose that the field extension $R/M \subseteq V/M$ is algebraic. Let I, J be nondivisorial ideals of R. Then, $v(I) = v(J)$ if and only if $I = uJ$ for some $u \in K$.*

Proof. By [12, Proposition 2.2(5)], there are $a, b \in K$ such that $a^{-1}I$ and $b^{-1}J$ are properly contained between R and $V = (M : M)$. Furthermore, since $R/M \subseteq V/M$ is algebraic, every ring between R and V is the pullback of some intermediate field, and in particular it is itself a PVD with maximal ideal M. The claim follows from Proposition 4.7. □

5 Localizations

Let $\star$ be a star operation on R and T a flat overring of R. Then, $\star$ is said to be *extendable* to T if the map

$$\begin{aligned} \star_T : \mathcal{F}(T) &\longrightarrow \mathcal{F}(T) \\ IT &\longmapsto I^\star T \end{aligned}$$

is well-defined; when this happens, $\star_T$ is called the *extension* of $\star$ to T and is a star operation on T [16, Definition 3.1]. In general, not all star operations are extendable, although finite-type operations are (see [10, Proposition 2.4] and [16, Proposition 3.3(d)]).

We would like to have an equality $v(I)_T = v(IT)$, where the latter is considered as a star operation on T. In general, this is false, both because $v(I)$ may not be extendable and because the extension $v(I)_T$ may not be equal to $v(IT)$.

For example, let V be a valuation domain and suppose that its maximal ideal M is principal. Let P be a prime ideal of V. Then, the only star operation on V is the identity, and thus $v(I) = d$ for all ideals I; in particular, $v(I)$ is extendable to V_P and the extension $v(I)_{V_P}$ is the identity on V_P. Suppose now that $P = PV_P$ is not principal as an ideal of V_P. Then, V_P has two star operations (the identity and the v-operation) and if $a \in K \setminus \{0\}$ then aV_P generates the v-operation. Hence, the extension of $v(aV) \in \mathrm{Star}(V)$ to V_P is different from $v(aV_P) \in \mathrm{Star}(V_P)$.

In the Noetherian case, however, everything works.

Proposition 5.1. *If R is Noetherian, then $v(I)_T = v(IT)$ for every flat overring T of R.*

Proof. By definition, $J^{v(I)} = (R : (R : J)) \cap (I : (I : J))$; multiplication by a flat overring commutes with finite intersections, and since every ideal is finitely generated, the colon localizes, and thus

$$\begin{aligned} J^{v(I)}T &= (R:(R:J))T \cap (I:(I:J))T = \\ &= (T:(T:JT)) \cap (IT:(IT:JT)) = \\ &= (JT)^{v_T} \cap (IT:(IT:JT)) = (JT)^{v(IT)}, \end{aligned}$$

i.e., $v(I)_T = v(IT)$. □

Another case where localization works well is for Jaffard families. If R is an integral domain with quotient field K, a *Jaffard family* of R is a set Θ of flat overrings of R such that [6, Section 6.3.1]:

- Θ is locally finite;
- $I = \prod\{IT \cap R \mid T \in \Theta, IT \neq T\}$ for every integral ideal I;
- $(IT_1 \cap R) + (IT_2 \cap R) = R$ for every integral ideal I and every $T_1 \neq T_2$ in Θ.

Jaffard families can be used to factorize the set of star operations of a domain R into a direct product of sets of star operations.

Theorem 5.2. *Let R be an integral domain and let Θ be a Jaffard family on R. Then, every star operation on R is extendable to every $T \in \Theta$, and the map*

$$\begin{aligned} \lambda_\Theta \colon \mathrm{Star}(R) &\longrightarrow \prod_{T\in\Theta} \mathrm{Star}(T) \\ \star &\longmapsto (\star_T)_{T\in\Theta} \end{aligned}$$

is an order-preserving order-isomorphism.

Proof. It is a part of [16, Theorem 5.4]. □

For principal star operations, the previous result must be modified using, instead of the direct product, a "direct sum"-like construction. Given a family Θ of overrings, we set

$$\bigoplus_{T\in\Theta} \mathrm{Princ}(T) := \left\{ (\star^{(T)}) \in \prod_{T\in\Theta} \mathrm{Princ}(T) \mid \star^{(T)} \neq v^{(T)} \text{ for only finitely many } T \right\}.$$

Using this terminology, we have the following.

Proposition 5.3. *Let R be an integral domain and Θ be a Jaffard family on R. For every ideal I of R and every $T \in \Theta$, we have $v(I)_T = v(IT)$; furthermore, the map*

$$\begin{aligned} \Upsilon \colon \mathrm{Princ}(R) &\longrightarrow \bigoplus_{T\in\Theta} \mathrm{Princ}(T) \\ v(I) &\longmapsto (v(IT))_{T\in\Theta} \end{aligned}$$

is a well-defined order-isomorphism.

Proof. By Theorem 5.2 $v(I)$ is extendable to any $T \in \Theta$; furthermore, by [16, Lemma 5.3], we have $(J : L)T = (JT : LT)$ for every pair of fractional ideals J, L of R. Using the same calculation of Proposition 5.1 we get $v(I)_T = v(IT)$.

In particular, it follows that the map Υ is just the restriction of the localization map λ_Θ to $\mathrm{Princ}(R)$; since λ_Θ is an isomorphism (by Theorem 5.2), we have only to show that the image of Υ is the direct sum $\bigoplus_{T\in\Theta} \mathrm{Princ}(T)$.

Since $IT = T$ for all but a finite number of T (by definition of a Jaffard family), we have $v(IT) = v(T) = v^{(T)}$ for all but a finite number of T. In particular, the image of Υ lies inside the direct sum.

Suppose, conversely, that $(v(J_T))_{T\in\Theta} \in \bigoplus_{T\in\Theta} \mathrm{Princ}(T)$. We can suppose that $J_T \subseteq T$ for every T, and that $J_T = T$ if $v(J_T) = v^{(T)}$. Define thus $I := \bigcap_{T\in\Theta} J_T$: then, I is nonzero (since $J_T \neq T$ for only a finite number of T) and $IT = J_T$ for every T [16, Lemma 5.2]. Therefore, $v(I)_T = v(IT) = v(J_T)$, and the image of Υ is exactly $\bigoplus_{T\in\Theta} \mathrm{Princ}(T)$. □

Proposition 5.3 can be interpreted as a way to factorize principal star operations.

Corollary 5.4. *Let R be an integral domain and Θ be a Jaffard family on R. Let I be an integral ideal of R. Then, there are $T_1, \dots, T_n \in \Theta$ such that $v(I) = v(IT_1 \cap R) \wedge \cdots \wedge v(IT_n \cap R)$.*

Proof. Since $I \subseteq R$, we have $IT = T$ for all but finitely many $T \in \Theta$; let $T_1, \dots, T_n$ be the exceptions. The claim follows from Proposition 5.3. □

Recall that an integral domain is said to be *h-local* if every ideal is contained in a finite number of maximal ideals and every prime ideal is contained in only one maximal ideal.

Corollary 5.5. *Let R be an h-local Prüfer domain, and let $\mathcal{M}$ be the set of non-divisorial maximal ideals of R. Then, there is a bijective correspondence between* $\mathrm{Princ}(R)$ *and the set $\mathcal{P}_{\mathrm{fin}}(\mathcal{M})$ of finite subsets of $\mathcal{M}$. Furthermore, $\mathcal{M}$ is finite if and only if every star operation is principal.*

Proof. Since R is h-local, $\{R_M \mid M \in \mathrm{Max}(R)\}$ is a Jaffard family of R, and thus by Proposition 5.3 there is a bijective correspondence Υ between $\mathrm{Princ}(R)$ and $\bigoplus_{M\in\mathrm{Max}(R)} \mathrm{Princ}(R_M)$. If $M \notin \mathcal{M}$, then MR_M is principal and thus $\mathrm{Star}(R_M) = \mathrm{Princ}(R_M) = \{d = v\}$; hence, Υ restricts to a bijection Υ' between $\mathrm{Princ}(R)$ and $\bigoplus_{M\in\mathcal{M}} \mathrm{Princ}(R_M)$. Since R_M is a valuation domain, each $\mathrm{Princ}(R_M)$ is composed by two elements (the identity and the v-operation). Thus, we can construct a bijection Υ_1 from the direct sum to $\mathcal{P}_{\mathrm{fin}}(\mathcal{M})$ by associating to $\star := (\star^{(M)})$ the finite set $\Upsilon_1(\star) := \{M \in \mathcal{M} \mid \star^{(M)} \neq v\}$. The composition $\Upsilon_1 \circ \Upsilon'$ is a bijection from $\mathrm{Princ}(R)$ to $\mathcal{P}_{\mathrm{fin}}(\mathcal{M})$.

The last claim follows immediately. □

A factorization property similar to Corollary 5.4 can be proved for ideals having a primary decomposition with no embedded primes.

Proposition 5.6. *Let $Q_1, \ldots, Q_n$ be primary ideals, let $P_i := \mathrm{rad}(Q_i)$ for all i and let $I := Q_1 \cap \cdots \cap Q_n$. If the P_i are pairwise incomparable, then $v(I) = v(Q_1) \wedge \cdots \wedge v(Q_n)$.*

Proof. For every i, the ideal Q_i is $v(Q_i)$-closed, and thus I is $(v(Q_1) \wedge \cdots \wedge v(Q_n))$-closed; hence, $v(I) \geq v(Q_1) \wedge \cdots \wedge v(Q_n)$. To prove the converse, we need to show that each Q_i is $v(I)$-closed.

Without loss of generality, let $i = 1$, and define $\widehat{Q} := Q_2 \cap \cdots \cap Q_n$; we claim that $Q_1 = (I :_R \widehat{Q})$. Since $Q_1\widehat{Q} \subseteq Q_1 \cap \widehat{Q} = I$, clearly $Q_1 \subseteq (I :_R \widehat{Q})$. Conversely, let $x \in (I :_R \widehat{Q})$. Since the radicals of the Q_i are pairwise incomparable, $Q_i \nsubseteq P_1$ for every $i > 1$, and so $\widehat{Q} \nsubseteq P_1$; therefore, there is a $q \in \widehat{Q} \setminus P_1$. Then, $xq \in I$, and in particular $xq \in Q_1$. If $x \notin Q_1$, then since Q_1 is primary we would have $q^t \in Q_1$ for some $t \in \mathbb{N}$; however, this would imply $q \in \mathrm{rad}(Q_1) = P_1$, against the choice of q. Thus, $Q_1 \subseteq (I :_R \widehat{Q})$ and so $Q_1 = (I :_R \widehat{Q})$.

By definition, I is $v(I)$-closed; hence, also $(I :_R \widehat{Q})$ is $v(I)$-closed. It follows that Q_1 is $v(I)$-closed, and thus that each Q_i is $v(I)$-closed, i.e., $v(I) \leq v(Q_1) \wedge \cdots \wedge v(Q_n)$. The claim is proved. □

6 v-Trivial Ideals

In this section, we analyze principal operations generated by v-trivial ideals.

Definition 6.1. An ideal I of a domain R is *v-trivial* if $I^v = R$.

Lemma 6.2. *If I is v-trivial, then $(I : I) = R$.*

Proof. If $I^v = R$, then $(R : I) = R$, and thus $(I : I) \subseteq (R : I) = R$. □

Definition 6.3. A star operation $\star$ is *semifinite* (or *quasi-spectral*) if every $\star$-closed ideal $I \subsetneq R$ is contained in a $\star$-prime ideal.

All finite type and all spectral operations are semifinite; on the other hand, if V is a valuation domain with maximal ideal that is branched but not finitely generated, the v-operation on V is not semifinite. The class of semifinite operations is closed by taking infima, but not by taking suprema (see [5, Example 4.5]).

Lemma 6.4. *Let R be an integral domain, and let I, J be v-trivial ideals of R.*

(a) If $J \subsetneq I$, then $J^{v(I)} = I$, and in particular $v(I) \neq v(J)$.

Suppose v is semifinite on R.

(b) $I \cap J$ is v-trivial.
(c) $I \subseteq J^{v(I)}$.
(d) If $I \neq J$, then $v(I) \neq v(J)$.

Proof. (a) Since I is v-trivial, by Lemma 6.2 and Proposition 3.2 we have $J^{v(I)} = (I : (I : J))$.
However, $R \subseteq (I : J) \subseteq (R : J) = R$ (using the v-triviality of J) and thus $J^{v(I)} = (I : R) = I$, as claimed. In particular, $J = J^{v(J)} \neq J^{v(I)}$ and so $v(I) \neq v(J)$.

(b) If $(I \cap J)^v \neq R$, then by semifiniteness there is a prime ideal P such that $I \cap J \subseteq P = P^v$. However, this would imply $I \subseteq P$ or $J \subseteq P$, against the hypothesis that I and J are v-trivial.

(c) Since $J \subseteq J^{v(I)}$, it follows that $J^{v(I)}$ is v-trivial, and by the previous point so is $J^{v(I)} \cap I$. If $I \nsubseteq J^{v(I)}$, it would follow that $J^{v(I)} \cap I \subsetneq I$, but $J^{v(I)} \cap I$ is $v(I)$-closed, against (a). Hence $I \subseteq J^{v(I)}$.

(d) If both I and J are $v(I)$-closed, then so is $I \cap J$; by (b), $(I \cap J)^v = R$. The claim follows applying (a) to $I \cap J$ and I (or J). □

Corollary 6.5. *Let R be a domain such that v is semifinite. Let I, J be ideals of R such that I^v and J^v are invertible; then, $v(I) = v(J)$ if and only if $I = LJ$ for some invertible ideal L.*

Proof. By invertibility, we have

$$R = I^v(R : I^v) = (I^v(R : I^v))^v = (I(R : I^v))^v;$$

since $I \subseteq I(R : I^v) \subseteq R$, the ideal $I(R : I^v)$ is v-trivial. Analogously, $R = (J(R : J^v))^v$ and $J(R : J^v)$ is v-trivial. Hence, by Lemma 6.4(d) $I(R : I^v) = J(R : J^v)$; thus, $I = I^v(R : J^v)J$, and $L := I^v(R : J^v)$ is invertible. □

We denote by $h(I)$ the height of the integral ideal I.

Corollary 6.6. *Let R be a unique factorization domain. Then,*

(a) for every principal star operation $\star \neq v$ there is a proper ideal I such that $h(I) > 1$ and $\star = v(I)$;
(b) if I, J are fractional ideals of R, $v(I) = v(J)$ if and only if $I = uJ$ for some $u \in K$.

Proof. Let $\star = v(I)$ for some ideal I. By [7, Corollary 44.5], every v-closed ideal of R is principal; hence, let $I^v = pR$. Then, $(p^{-1}I)^v = R$, i.e., $p^{-1}I$ is v-trivial. In particular, $\star = v(I) = v(p^{-1}I)$, and $p^{-1}I$ is a proper ideal of R with $h(p^{-1}I) > 1$ (since all prime ideals of height 1 are v-closed).

Suppose that we also have $\star = v(J)$. With the same reasoning of the previous paragraph, $q^{-1}J$ is v-trivial for some q; thus $v(p^{-1}I) = v(I) = v(J) = v(q^{-1}J)$. Applying Lemma 6.4 (d) to $p^{-1}I$ and $q^{-1}J$ we get $p^{-1}I = q^{-1}J$, i.e., $I = (pq^{-1})J$. □

For star operations generated by v-trivial prime ideals, we can also determine the set of closed ideals.

Proposition 6.7. *Let R be a domain such that v is semifinite and such that I^v is invertible for every ideal I, and let $P \in \mathrm{Spec}(R)$. Then $\mathcal{F}^{v(P)}(R) = \mathcal{F}^v(R) \cup \{LP \mid L$ is an invertible ideal$\}$. In particular, $v(P)$ is a maximal element of $\mathrm{Princ}(R) \setminus \{v\}$.*

Proof. Let I be a nondivisorial ideal; multiplying by an invertible ideal L, we can suppose $I^v = R$. If $I \subseteq P$, by Lemma 6.4 (a) $I^{v(P)} = P$, and thus $I \neq I^{v(P)}$ unless $I = P$; suppose $I \nsubseteq P$. Then $(P : I) = P$: we have $(P : I) \subseteq (R : I) = R$, and thus if $xI \subseteq P$ then $x \in P$. Therefore, $I^{v(P)} = I^v \cap (P : (P : I)) = R \cap (P : P) = R \neq I$.

For the "in particular" claim, note that if $v(I) \geq v(P)$ then I should be $\star$-closed; by the previous part of the proof, this means that either I is divisorial (and so $v(I) = v$) or $I = LP$ for some invertible L (and thus $v(I) = v(P)$ by Lemma 3.4(d)). □

Corollary 6.8. *Let R be a unique factorization domain, and let $P \in \mathrm{Spec}(R)$. Then, $\mathcal{F}^{v(P)}(R) = \mathcal{F}^v(R) \cup \{aP \mid a \in K\}$.*

We have seen in Proposition 3.3 that all star operations can be "generated" by principal star operations; we can use v-trivial ideals to show that in many cases we need infinitely many of them.

Proposition 6.9. *Let R be a domain such that v is semifinite, and let $I_1, \dots, I_n$ be v-trivial ideals; let $\star := v(I_1) \wedge \cdots \wedge v(I_n)$. Then, the ideal $I_1 \cap \cdots \cap I_n$ is the minimal v-trivial ideal that is $\star$-closed.*

Proof. Let $J := I_1 \cap \cdots \cap I_n$. By Lemma 6.4 (b), J is v-trivial. Clearly J is $\star$-closed. Suppose L is v-trivial; then, applying Lemma 6.4(c),

$$L^\star = L^{v(I_1) \wedge \cdots \wedge v(I_n)} \supseteq I_1 \cap \cdots \cap I_n = J.$$

Therefore, J is the minimum among v-trivial $\star$-closed ideals. □

Corollary 6.10. *Let R be a unique factorization domain, and let $\star \in \mathrm{Star}(R)$ be such that $\star \neq v$. If $\bigcap\{J \in \mathcal{F}^\star(R) \mid J^v = R\} = (0)$, then $\star$ is not the infimum of a finite family of principal star operations.*

Proof. Since R is a UFD, the v-operation is semifinite, and every principal star operation can be generated by a v-trivial ideal. If $\star$ were to be finitely generated, say $\star = v(I_1) \wedge \cdots \wedge v(I_n)$, then $J := I_1 \cap \cdots \cap I_n$ would be the minimal v-trivial $\star$-closed ideal; however, by hypothesis, there must be a v-trivial $\star$-closed ideal J' not containing J, and thus $\star$ cannot be finitely generated. □

Proposition 6.11. *Let R be a domain, and let Δ be a set of overrings whose intersection is R. Let $\star$ be the star operation $I \mapsto \bigcap\{IT \mid T \in \Delta\}$. Suppose that*

(1) v is semifinite;
(2) every v-trivial ideal contains a finitely generated v-trivial ideal;
(3) there is a v-trivial $\star$-closed ideal.

Then, $\star$ is not the infimum of a finite family of principal star operations.

Proof. By substituting an overring $T \in \Delta$ with $\{T_M \mid M \in \mathrm{Max}(T)\}$, we can suppose without loss of generality that each member of Δ is local.

If $\star$ were finitely generated, by Proposition 6.9 there would be a minimal v-trivial $\star$-closed ideal, say J. By hypothesis, there is finitely generated v-trivial ideal $I \subseteq J$; since $I^\star = J$, by [1, Theorem 2], we have $IT = JT$ for every $T \in \Delta$.

Since $I^\star \neq R$, there must be an $S \in \Delta$ such that $IS \neq S$; by Nakayama's lemma, $I^2S = (IS)^2 \subsetneq IS$, and so $(I^2)^\star \subseteq I^2S \cap R \subsetneq I$. In particular, $(I^2)^\star$ is a v-trivial $\star$-closed ideal, against the definition of I. Thus, $\star$ is not finitely generated. □

The first two hypothesis hold, for example, for unique factorization domains of dimension $d > 1$; the third one holds, for example, in the following cases:

- $\star$ is a spectral star operation of finite type different from the w-operation (see [2, 17]);
- if R is integrally closed and (at least) one maximal ideal is not divisorial, and $\star$ is the b-operation/integral closure;
- if R is a UFD, all star operations coming from overrings, except the v-operation.

7 Noetherian Domains

In this section, we study in more detail the case of Noetherian domains; in particular, we shall give in Theorem 7.8 a necessary and sufficient condition on when $v(I) = v(J)$, under the assumption that $(I : I) = R = (J : J)$. We first state a case that is already settled, even without this hypothesis.

Proposition 7.1. *[14, Proposition 5.4] Let (R, M) be a local Noetherian integral domain of dimension 1 such that its integral closure V is a discrete valuation domain that is finite over R; suppose also that the induced map of residue fields $R/M \subseteq V/M_V$ is an isomorphism. Then, $v(I) = v(J)$ if and only if $I = uJ$ for some $u \in K$, $u \neq 0$.*

We denote by $\mathrm{Ass}(I)$ the set of associated primes of I.

Proposition 7.2. *Let R be a domain and I an ideal of R. Then, $\mathrm{Spec}^{v(I)}(R) \supseteq \mathrm{Spec}^{v}(R) \cup \mathrm{Ass}(I)$, and if R is Noetherian the two sets are equal.*

Proof. If $P \in \mathrm{Ass}(I)$, then $P = (I :_R x) = x^{-1}I \cap R$ for some $x \in R$, and thus it is $v(I)$-closed; if $P \in \mathrm{Spec}^v(R)$ then $P = P^v$ and thus $P = P^{v(I)}$.

Conversely, suppose R is Noetherian and $P = P^{v(I)}$. Then $P = P^v \cap (I : (I : P)) = P^v \cap (I : J)$, where $J = (I : P)$; let $J = j_1R + \cdots + j_nR$. We have

$$\begin{aligned} P = P^v \cap (I : J) = P^v \cap R \cap (I : J) = P^v \cap (I :_R J) = \\ = P^v \cap (I :_R j_1R + \cdots + j_nR) = P^v \cap \textstyle\bigcap_{i=1}^n (I :_R j_iR), \end{aligned}$$

and, since P is prime, this implies that $P^v = P$ or $(I :_R j_iR) = P$ for some i. In the latter case, since $j_i \in K$, $j_i = a/b$ for some $a, b \in R$; hence $(I :_R j_iR) = (I : ab^{-1}R) \cap R = (bI :_R aR)$, and thus P is associated to bI. There is an exact sequence

$$0 \longrightarrow \frac{bR}{bI} \longrightarrow \frac{R}{bI} \longrightarrow \frac{R}{bR} \longrightarrow 0$$

and, since R is a domain, $bR/bI \simeq R/I$ and thus $\mathrm{Ass}(bI) \subseteq \mathrm{Ass}(I) \cup \mathrm{Ass}(bR)$ [3, Chapter IV, Proposition 3]; therefore, P is associated to I or it is divisorial (since an associated prime of a divisorial ideal—in this case, bR—is divisorial). □

Remark 7.3. Note that, if $P^v = R$, then $(I : P) \subseteq (R : P) = R$, and thus $j_i \in R$; in this case, $b = 1$ and the last part of the proof can be greatly simplified.

The following is a slight improvement of Proposition 6.7. We denote by $X^1(R)$ the set of height-1 prime ideals of R.

Corollary 7.4. *Let R be an integrally closed Noetherian domain. Then, the maximal elements of* $\mathrm{Princ}(R) \setminus \{v\}$ *are the $v(P)$, as P ranges in* $\mathrm{Spec}(R) \setminus X^1(R)$.

Proof. Since R is integrally closed, the divisorial prime ideals of R are the height 1 primes. In particular, if P is a prime ideal of height > 1, then $v(P)$ is maximal by Proposition 6.7.

Conversely, suppose $v(I)$ is maximal in $\mathrm{Princ}(R) \setminus \{v\}$. If all associated primes of I have height 1, then $I = \bigcap_{P \in X^1(R)} IR_P$, and so I is divisorial, against $v(I) \neq v$. Hence, there is a $P \in \mathrm{Ass}(I) \setminus X^1(R)$; by Proposition 7.2, $P \in \mathrm{Spec}^{v(I)}(R)$, and thus $v(I) \leq v(P)$. As $v(I)$ is maximal, it follows that $v(I) = v(P)$. The claim is proved. □

Corollary 7.5. *Let R be a Noetherian unique factorization domain. Then, $v(I)$ is a maximal element of* $\mathrm{Princ}(R) \setminus \{v\}$ *if and only if $I = uP$ for some prime ideal $P \in \mathrm{Spec}(R) \setminus X^1(R)$ and some $u \in K$.*

Proof. It is enough to join Corollary 7.4 (the maximal elements are the $v(P)$) with Corollary 6.6 ($v(I) = v(P)$ if and only if $I = uP$). □

Proposition 7.2 allows to determine, in the Noetherian case, all the spectra of the principal star operations. We need a lemma.

Lemma 7.6. *Let $\star_1, \ldots, \star_n \in$* $\mathrm{Star}(R)$, *and let $\star := \star_1 \wedge \cdots \wedge \star_n$. Then,* $\mathrm{Spec}^\star(R) = \bigcup_i \mathrm{Spec}^{\star_i}(R)$.

Proof. If $P = P^{\star_i}$ for some i then $P^\star \subseteq P^{\star_i} = P$ and thus $P = P^\star$. Conversely, if $P = P^\star$ then $P = P^{\star_1} \cap \cdots \cap P^{\star_n}$; since P is prime, it follows that $P = P^{\star_i}$ for some i. The claim is proved. □

Proposition 7.7. *Let R be a Noetherian domain, and let $\Delta \subseteq \mathrm{Spec}(R)$. Then, the following are equivalent:*

(i) $\Delta = \mathrm{Spec}^{v(I)}(R)$ *for some ideal I;*
(ii) $\Delta = \mathrm{Spec}^\star(R)$ *for some $\star = v(I_1) \wedge \cdots \wedge v(I_n)$;*
(iii) $\Delta = \mathrm{Spec}^v(R) \cup \Delta'$*, for some finite set Δ'.*

Proof. (i) $\Longrightarrow$ (ii) is obvious, while (ii) $\Longrightarrow$ (iii) follows from Lemma 7.6.

If (iii) holds, then by [18, Chapter 4, Theorem 21] there is an ideal I whose set of associated primes is Δ'. By Proposition 7.2, $\mathrm{Spec}^{v(I)}(R) = \mathrm{Spec}^v(R) \cup \Delta' = \Delta$, and so (i) holds. □

We now characterize when two nondivisorial ideals with $(I : I) = (J : J) = R$ generate the same star operation.

Theorem 7.8. *Let R be a Noetherian domain, and let I, J be nondivisorial ideals such that $(I : I) = (J : J) = R$. Then, $v(I) = v(J)$ if and only if $\mathrm{Ass}(I) \cup \mathrm{Spec}^v(R) = \mathrm{Ass}(J) \cup \mathrm{Spec}^v(R)$ and, for every $P \in \mathrm{Ass}(I) \cup \mathrm{Spec}^v(R)$, there is an $a_P \in K$ such that $IR_P = a_P JR_P$.*

Proof. Suppose the two conditions hold. By Proposition 7.2, $\mathrm{Ass}(I) \cup \mathrm{Spec}^v(R) = \mathrm{Spec}^{v(I)}(R)$, and thus $\mathrm{Spec}^{v(I)}(R) = \mathrm{Spec}^{v(J)}(R) =: \Delta$. For every ideal L, using Proposition 5.1 we have

$$L^{v(I)} = \bigcap_{P\in\Delta} L^{v(I)}R_P = \bigcap_{P\in\Delta} (LR_P)^{v(I)_{R_P}} = \bigcap_{P\in\Delta} (LR_P)^{v(IR_P)}.$$

Since IR_P and JR_P are isomorphic, $(LR_P)^{v(IR_P)} = (LR_P)^{v(JR_P)}$; it follows that $v(I) = v(J)$.

Conversely, suppose $v(I) = v(J) =: \star$. Then, $\mathrm{Spec}^\star(R)$ is equal to both $\mathrm{Ass}(I) \cup \mathrm{Spec}^v(R)$ and $\mathrm{Ass}(J) \cup \mathrm{Spec}^v(R)$, which thus are equal. Note also that $(I : I) = R$ implies that $R_P = (I : I)R_P = (IR_P : IR_P)$ for every prime ideal P.

Let now $P \in \mathrm{Spec}^\star(R)$. Since $v(I) = v(J)$, clearly $v(I)_{R_P} = v(J)_{R_P}$, which by Proposition 5.1 implies that $v(IR_P) = v(JR_P)$. However, PR_P is $v(IR_P)$-closed because P is $v(I)$-closed; it follows, by Proposition 4.5, that $IR_P = a_P JR_P$ for some $a_P \in K$, as claimed. □

Corollary 7.9. *Let R be an integrally closed Noetherian domain, and let I, J be nondivisorial ideals. Then, $v(I) = v(J)$ if and only if $\mathrm{Ass}(I) \cup X^1(R) = \mathrm{Ass}(J) \cup X^1(R)$ and for every $P \in \mathrm{Ass}(I)$ there is an $a_P \in R_P$ such that $IR_P = a_P JR_P$.*

Proof. Since R is integrally closed and Noetherian, we have $(I : I) = R$ for every ideal I; furthermore, the divisorial primes are the height 1 primes, and for any such P the localizations IR_P and JR_P are isomorphic since R_P is a DVR. The claim now follows from Theorem 7.8. □

Example 7.10. Let R be a Noetherian integrally closed domain, and suppose that R_M is not a UFD for some maximal ideal M. Let P be an height 1 prime contained in M such that PR_M is not principal, and let Q be a prime ideal of height bigger than 1 such that $P + Q = R$ (in particular, $Q \nsubseteq M$). We claim that $v(PQ) = v(Q)$ but PQ and Q are not locally isomorphic.

In fact, since they are coprime, $PQ = P \cap Q$, and thus $\mathrm{Ass}(PQ) = \{P, Q\}$ while $\mathrm{Ass}(Q) = \{Q\}$; moreover, $P \nsubseteq Q$ and thus $PQR_Q = QPR_Q = QR_Q$. Since $P \in X^1(R)$, by Corollary 7.9 it follows that $v(PQ) = v(Q)$. However, $QR_M = R_M$ is principal, while $PQR_M = PR_M$, by hypothesis, is not; therefore, Q and PQ are not locally isomorphic. In particular, there cannot be an invertible ideal L such that $Q = LPQ$, because LR_M would be principal and thus Q and PQ would be locally isomorphic.

References

1. Anderson, D.D.: Star-operations induced by overrings. Comm. Algebra **16**(12), 2535–2553 (1988)
2. Anderson, D.D., Cook, S.J.: Two star-operations and their induced lattices. Comm. Algebra **28**(5), 2461–2475 (2000)
3. Bourbaki, N.: Commutative Algebra. Chapters 1–7. Elements of Mathematics (Berlin). Springer, Berlin, 1989. Translated from the French, Reprint of the 1972 edition
4. Elliott, J.: Semistar operations on Dedekind domains. Comm. Algebra **43**(1), 236–248 (2015)
5. Finocchiaro, C.A., Fontana, M., Spirito, D.: Spectral spaces of semistar operations. J. Pure Appl. Algebra **220**(8), 2897–2913 (2016)
6. Fontana, M., Houston, E., Lucas, T.: Factoring Ideals in Integral Domains. Lecture Notes of the Unione Matematica Italiana, vol. 14. Springer, Heidelberg; UMI, Bologna (2013)
7. Gilmer, R.: Multiplicative Ideal Theory. Marcel Dekker Inc., New York, 1972. Pure and Applied Mathematics, No. 12
8. Hedstrom, J.R., Houston, E.G.: Pseudo-valuation domains. Pacific J. Math. **75**(1), 137–147 (1978)
9. Heinzer, W.J., Huckaba, J.A., Papick, I.J.: m-canonical ideals in integral domains. Comm. Algebra **26**(9), 3021–3043 (1998)
10. Houston, E.G., Mimouni, A., Park, M.H.: Integral domains which admit at most two star operations. Comm. Algebra **39**(5), 1907–1921 (2011)
11. Houston, E.G., Mimouni, A., Park, M.H.: Noetherian domains which admit only finitely many star operations. J. Algebra **366**, 78–93 (2012)
12. Houston, E.G., Mimouni, A., Park, M.H.: Integrally closed domains with only finitely many star operations. Comm. Algebra **42**(12), 5264–5286 (2014)
13. Picozza, G.: Star operations on overrings and semistar operations. Comm. Algebra **33**(6), 2051–2073 (2005)
14. Spirito, D.: Star Operations on Numerical Semigroups. Comm. Algebra **43**(7), 2943–2963 (2015)

15. Spirito, D.: Star operations on numerical semigroups: the multiplicity 3 case. Semigroup Forum **91**(2), 476–494 (2015)
16. Spirito, D.: Jaffard families and localizations of star operations. J. Commut. Algebra **11**(2), 265–300 (2019)
17. Wang, F., McCasland, R.L.: On w-modules over strong Mori domains. Comm. Algebra **25**(4), 1285–1306 (1997)
18. Zariski, O., Samuel, P.: Commutative algebra. Vol. 1. Springer, New York, Heidelberg, Berlin, 1975. With the cooperation of I. S. Cohen, Corrected reprinting of the: edition, p. 28. Graduate Texts in Mathematics, No (1958)

Tilting Modules and Tilting Torsion Pairs

Filtrations Induced by Tilting Modules

Francesco Mattiello, Sergio Pavon and Alberto Tonolo

Abstract Tilting modules, generalising the notion of progenerator, furnish equivalences between pieces of module categories. This paper is dedicated to study how much these pieces say about the whole category. We will survey the existing results in the literature, introducing also some new insights.

Keywords Tilting modules · Torsion pairs · t-structures · t-tree

1 Introduction

In 1958, Morita characterised equivalences between the entire categories of left (or right) modules over two rings. Let A be an arbitrary associative ring with $1 \neq 0$. A left A-module ${}_AP$ is a *progenerator* if it is projective, finitely generated and generates the category A-Mod of left A-modules. Set $B := \mathrm{End}({}_AP)$, the covariant functor $\mathrm{Hom}_A(P, ?)$ gives an equivalence between A-Mod and B-Mod; moreover, any equivalence between modules categories is of this type.

The notion of tilting module has been axiomatised in 1979 by Brenner and Butler [BB], generalising that of progenerator for modules of projective dimension 1. The various forms of generalisations to higher projective dimensions considered until today continue to follow their approach.

F. Mattiello · S. Pavon
Dipartimento di Matematica "Tullio Levi-Civita",
Università degli studi di Padova, Padua, Italy
e-mail: fran.mattiello@gmail.com
e-mail: sergio.pavon@math.unipd.it

A. Tonolo (✉)
Dipartimento di Scienze Statistiche, Università degli studi di Padova, Padua, Italy
e-mail: alberto.tonolo@unipd.it

A. Facchini et al. (eds.), *Advances in Rings, Modules and Factorizations*,
Springer Proceedings in Mathematics & Statistics 321,
https://doi.org/10.1007/978-3-030-43416-8_18

A tilting module T of projective dimension n naturally gives rise to $n+1$ corresponding classes of modules in A-Mod and B-Mod, the Miyashita classes, with $n+1$ equivalences between them. These classes are

$$\mathrm{KE}_e(T) = \{M \in A\text{-Mod} : \mathrm{Ext}^i_A(T, M) = 0 \ \forall i \neq e\}$$

$$\mathrm{KT}^e(T) = \{N \in B\text{-Mod} : \mathrm{Tor}^B_i(T, M) = 0 \ \forall i \neq e\}, \quad e = 0, 1, ..., n$$

and the $n+1$ equivalences are

$$KE_e(T) \underset{\mathrm{Tor}^B_e(T,?)}{\overset{\mathrm{Ext}^e_A(T,?)}{\rightleftarrows}} KT^e(T)\ , \quad e = 0, 1, ..., n.$$

In the $n = 0$ case (progenerator), there is only one class on each side, and so every module is subject to the equivalence of categories (that of Morita); for $n = 1$, on each side, the two Miyashita classes form torsion pairs, so every module in both A-Mod and B-Mod can be decomposed in terms of modules in the Miyashita classes: precisely every module admits a composition series of length 2 with composition factors in the Miyashita classes.

For $n > 1$, the Miyashita classes fail to decompose every module; the way to recover a similar decomposition is the subject of this paper.

In Section 2, we define classical n-tilting modules and Miyashita classes; we show that they give a torsion pair for $n = 1$, and hence they can be used to decompose every module; we give an example showing that a similar decomposition does not exist for $n > 1$, and characterise those modules which can be decomposed.

In Section 3, we present some previous attempts to recover the decomposition for $n > 1$ as well, by extending the Miyashita classes, due to Jensen, Madsen, Su [11] and to Lo [13]. A useful tool in our analysis will be a characterisation of modules in $\bigcap_{i>e} \mathrm{Ker}\,\mathrm{Ext}^i_A(T, ?)$, $0 \leq e \leq n$ (see Lemma 1), which generalises the characterisation of modules in $\bigcap_{i>0} \mathrm{Ker}\,\mathrm{Ext}^i_A(T, ?)$ given by Bazzoni [3, Lemma 3.2]. These extensions deform in an irreversible way the Miyashita classes, weakening their role.

In Section 4, we recall some introductory notions about the derived category of an abelian category and about t-structures.

In Section 5, we drop the finiteness assumptions on the tilting modules, recalling the definition of non classical n-tilting modules [2]. In this setting, we recall the definition of the t-structure associated to such a module; we then study its interaction with the natural t-structure of the derived category.

In Section 6, we exploit the results of Section 5 to construct in the derived category the t-tree of a module with respect to a tilting module. This procedure, discovered in the classical tilting case by Fiorot, the first and the third author in [8], solves satisfactorily the decomposition problem for $n > 1$: the classes used for the decomposition intersect the module category exactly in the Miyashita classes. As a result of the work of the previous section, we prove that this construction can be reproduced also in the non classical case.

Throughout the paper, the concrete case considered in Example 1 introduced in Section 2 will be used to illustrate the various attempts to solve the decomposition problem (see Examples 2, 3).

2 Classical n-tilting Modules

In 1986, Miyashita [14] and Cline, Parshall and Scott [7] gave similar definitions of a *tilting module of projective dimension n*.

Definition 1 (Miyashita [14]) A left A-module T is a classical n-tilting module, for some integer $n \geq 0$, if

p_n) T has a finitely generated projective resolution of length n, i.e. a projective resolution

$$0 \longrightarrow P_n \longrightarrow \cdots \longrightarrow P_0 \longrightarrow T \longrightarrow 0$$

with the P_i finitely generated;

e_n) T is rigid, i.e. $\operatorname{Ext}^i_A(T, T) = 0$ for every $0 < i \leq n$;

g_n) the ring A admits a coresolution of length n

$$0 \longrightarrow A \longrightarrow T_0 \longrightarrow \cdots \longrightarrow T_n \longrightarrow 0$$

with the T_i finitely generated direct summands of arbitrary coproducts of copies of T.

In the case when $n = 0$, p_0) says that the module is a finitely generated projective, e_0) is empty and g_0) says that it is a generator: this is then the definition of a progenerator module. As such, a classical 0-tilting module T induces a Morita equivalence of categories of modules, as follows. Let $B = \operatorname{End}_A(T)$ be its ring of endomorphisms, which acts on the right on T, and consider the category B-Mod of left B-modules. There are functors

$$\begin{aligned} \operatorname{Hom}_A(T, ?) : &\quad A\text{-Mod} \to B\text{-Mod} \\ T \otimes_B ? : &\quad B\text{-Mod} \to A\text{-Mod} \end{aligned}$$

which are category equivalences, with the unit and counit morphisms being those of the adjunction. This is the motivating example for the definition of tilting modules, along with the next case.

In the case when $n = 1$, we find what was originally (see Brenner and Butler [6]) defined as a *tilting module*; we will give a brief and incomplete overview of what is known about them.

Let T be a classical 1-tilting left A-module, and let as before $B = \mathrm{End}_A(T)$ be its ring of endomorphisms. In this case, the pair $(\mathrm{Hom}_A(T, ?), T \otimes_B ?)$ does not induce an equivalence of A-Mod and B-Mod anymore; however, a little less can be proved, as follows.

Define the following pairs of full subcategories of A-Mod and B-Mod, respectively,

$$\begin{aligned} KE_0(T) &= \left\{X \in A\text{-Mod} : \ \mathrm{Ext}^1_A(T, X) = 0\right\} \\ KE_1(T) &= \{X \in A\text{-Mod} : \ \mathrm{Hom}_A(T, X) = 0\} \\ KT^0(T) &= \left\{Y \in B\text{-Mod} : \ \mathrm{Tor}^B_1(T, Y) = 0\right\} \\ KT^1(T) &= \{Y \in B\text{-Mod} : \ T \otimes_B Y = 0\} . \end{aligned}$$

Then we have the following results.

Theorem 1 (Brenner and Butler [6]) *Let A be a ring, T a classical 1-tilting left A-module, $B = \mathrm{End}_A(T)$.*

i) The pairs $(KE_0(T), KE_1(T))$ and $(KT^1(T), KT^0(T))$ defined above are torsion pairs, respectively, in A-Mod and B-Mod.
ii) There are equivalences of (sub)categories

$$KE_0(T) \underset{T \otimes_B ?}{\overset{\mathrm{Hom}_A(T,?)}{\rightleftarrows}} KT^0(T)$$

$$KE_1(T) \underset{\mathrm{Tor}^B_1(T,?)}{\overset{\mathrm{Ext}^1_A(T,?)}{\rightleftarrows}} KT^1(T) .$$

This theorem shows that the 1-tilting case is slightly more complex than the 0-tilting one. Instead of having an equivalence of the whole categories A-Mod and B-Mod, we have two pairs of equivalent subcategories, giving a functorial decomposition of every module in its torsion and torsion free parts.

For an arbitrary $n \geq 0$, following Miyashita, we find that every classical n-tilting module T gives rise to two sets of $n + 1$ full subcategories of A-Mod and B-Mod, respectively, defined as follows for $e = 0, \ldots, n$:

$$\begin{aligned} KE_e(T) &= \left\{X \in A\text{-Mod} : \ \mathrm{Ext}^i_A(T, X) = 0 \text{ for every } i \neq e\right\} \ \subset A\text{-Mod} \\ KT^e(T) &= \left\{Y \in B\text{-Mod} : \ \mathrm{Tor}^B_i(T, Y) = 0 \text{ for every } i \neq e\right\} \ \subset B\text{-Mod} \end{aligned}$$

where conventionally $\mathrm{Ext}^0_A(T, X) = \mathrm{Hom}_A(T, X)$ and $\mathrm{Tor}^B_0(T, Y) = T \otimes_B Y$. As a generalisation of point (ii) of Theorem 1, we may state the following result.

Theorem 2 (Miyashita [14, Theorem 1.16]) *In the setting above, there are equivalences of (sub)categories, for every $e = 0, \ldots, n$:*

$$KE_e(T) \underset{\mathrm{Tor}^B_e(T,?)}{\overset{\mathrm{Ext}^e_A(T,?)}{\rightleftarrows}} KT^e(T)\ .$$

For $n \geq 2$, however, the Miyashita classes do not provide a decomposition of every module, as it used to happen for $n = 1$. This is proved by the existence of simple modules (which can have only a trivial decomposition in the module category) not belonging to any class.

Example 1 ([20, Example 2.1]) Let k be an algebraically closed field. Let A be the k-algebra associated to the quiver $1 \xrightarrow{a} 2 \xrightarrow{b} 3$ with the relation $b \circ a = 0$. The indecomposable projectives are $\substack{1\\2}, \substack{2\\3}, 3$, while the indecomposable injectives are $1, \substack{1\\2}, \substack{2\\3}$. It follows that the module $T = \substack{2\\3} \oplus \substack{1\\2} \oplus 1$ is a classical 2-tilting module: a p_2) resolution is

$$P^\bullet \to T \to 0: \qquad 0 \to 0 \oplus 0 \oplus 3 \to 0 \oplus 0 \oplus \substack{2\\3} \to \substack{2\\3} \oplus \substack{1\\2} \oplus \substack{1\\2} \to \substack{2\\3} \oplus \substack{1\\2} \oplus 1 \to 0\,;$$

T is a direct sum of injectives, so it is rigid; lastly, $A = 3 \oplus \substack{2\\3} \oplus \substack{1\\2}$ and so a g_2) co-resolution can be easily found. We shall show that the simple module 2 does not belong to any of the Miyashita classes.

In order to compute the $\mathrm{Ext}^i_A(T, 2)$, we apply the contravariant functor $\mathrm{Hom}_A(?, 2)$ to the resolution $P^\bullet$, obtaining

$$0 \to \mathrm{Hom}_A(\substack{2\\3} \oplus \substack{1\\2} \oplus \substack{1\\2}, 2) \to \mathrm{Hom}_A(0 \oplus 0 \oplus \substack{2\\3}, 2) \to \mathrm{Hom}_A(0 \oplus 0 \oplus 3, 2) \to 0$$

which is isomorphic to

$$0 \longrightarrow \mathrm{Hom}_A(\substack{2\\3}, 2) \xrightarrow{0} \mathrm{Hom}_A(\substack{2\\3}, 2) \xrightarrow{0} 0 \longrightarrow 0\,.$$

Hence, $\mathrm{Hom}_A(T, 2) \simeq \mathrm{Ext}^1_A(T, 2) \simeq \mathrm{Hom}_A(\substack{2\\3}, 2) \neq 0$ as abelian groups.

Indeed, those modules for which the $KE_i(T)$ (resp. the $KT^i(T)$) provide a decomposition can be characterised in the following way.

Definition 2 A left A-module M (resp. a left B-module N) is *sequentially static* (resp. *costatic*) if for every $i \neq j \geq 0$,

$$\mathrm{Tor}^B_i(T, \mathrm{Ext}^j_A(T, M)) = 0 \quad (\text{resp. } \mathrm{Ext}^i_B(T, \mathrm{Tor}^A_j(T, N)) = 0).$$

Notice that for an A-module M (resp. a B-module N) to be sequentially static (resp. costatic) means that for every $e = 0, \ldots, n$, we have that $\mathrm{Ext}^e_A(T, M)$ belongs to KT^e (resp. $\mathrm{Tor}^B_e(T, N)$ belongs to KE_e).

Proposition 1 *([20, Theorem 2.3]) A left A-module M is sequentially static if and only if there exists a filtration*

$$M = M_n \geq M_{n-1} \geq M_{n-2} \geq \cdots \geq M_0 \geq M_{-1} = 0$$

such that for every $i = 0, \dots, n$, the quotient M_i/M_{i-1} belongs to $KE_i(T)$. In this case, for every such i, we have that $M_i/M_{i-1} \simeq \mathrm{Tor}_i^B(T, \mathrm{Ext}_A^i(T, M))$.

Dually, a left B-module M is sequentially costatic if and only if there exists a filtration

$$N = N_{-1} \geq N_0 \geq N_1 \geq \cdots \geq N_{n-1} \geq N_n = 0$$

such that for every $i = 0, \dots, n$, the quotient N_{i-1}/N_i belongs to $KT^i(T)$. In this case, for every such i, we have that $N_{i-1}/N_i \simeq \mathrm{Ext}_A^i(T, \mathrm{Tor}_i^B(T, N))$.

Remark 1 In Example 1, the module 2 was not sequentially static. Let us check that

$$\mathrm{Tor}_2^B(T, \mathrm{Hom}_A(T, 2)) \neq 0.$$

The ring $B = \mathrm{End}_A(T)$ (with multiplication the composition left to right) is the k-algebra associated to the quiver $4 \xrightarrow{c} 5 \xrightarrow{d} 6$ with the relation $d \circ c = 0$. In detail, the idempotents are the endomorphisms of T induced by the identities of its direct summands, e_4 of 1, e_5 of $\begin{smallmatrix}1\\2\end{smallmatrix}$ and e_6 of $\begin{smallmatrix}2\\3\end{smallmatrix}$, respectively; and c and d are the endomorphisms of T induced by the morphisms $\begin{smallmatrix}1\\2\end{smallmatrix} \to 1$ and $\begin{smallmatrix}2\\3\end{smallmatrix} \to \begin{smallmatrix}1\\2\end{smallmatrix}$, respectively.

In order to compute the right B-module structure of T, we notice first that as a k-vector space T is generated by five elements: $x \in \begin{smallmatrix}2\\3\end{smallmatrix} \setminus 3$ and $y = bx \in 3$, $v \in \begin{smallmatrix}1\\2\end{smallmatrix} \setminus 2$ and $w = av \in 2$, and $z \in 1$. If we look at how B acts on the right on these elements, we see that T as a right B-module is isomorphic to $\begin{smallmatrix}5\\4\end{smallmatrix} \oplus \begin{smallmatrix}6\\5\end{smallmatrix} \oplus 6 = \begin{smallmatrix}v\\z\end{smallmatrix} \oplus \begin{smallmatrix}x\\w\end{smallmatrix} \oplus y$.

To compute $\mathrm{Ext}_A^1(T, 2)$, we consider the injective coresolution of 2 in A-Mod

$$0 \longrightarrow 2 \longrightarrow \begin{smallmatrix}1\\2\end{smallmatrix} \longrightarrow 1 \longrightarrow 0$$

and compute $\mathrm{coker}\left[\mathrm{Hom}_A(T, \begin{smallmatrix}1\\2\end{smallmatrix}) \to \mathrm{Hom}_A(T, 1)\right]$ as left B-modules.

The left B-module $\mathrm{Hom}_A\left(T, \begin{smallmatrix}1\\2\end{smallmatrix}\right)$ is generated as a k-vector space by (the morphisms induced on T by) two morphisms $\begin{smallmatrix}2\\3\end{smallmatrix} \to \begin{smallmatrix}1\\2\end{smallmatrix}$ and $\begin{smallmatrix}1\\2\end{smallmatrix} \to \begin{smallmatrix}1\\2\end{smallmatrix}$. When we look at how B acts on the left on these elements, we see that the module is isomorphic to ${}_B\left(\begin{smallmatrix}5\\6\end{smallmatrix}\right)$. Similarly, it can be seen that $\mathrm{Hom}_A(T, 1)$ as a left B-module is isomorphic to $\begin{smallmatrix}4\\5\end{smallmatrix}$, hence the cokernel we are interested in is the simple 4. To compute $\mathrm{Tor}_2^B(T, 4)$, we now consider the presentation

$$0 \longrightarrow 5 \longrightarrow \begin{smallmatrix}4\\5\end{smallmatrix} \longrightarrow 4 \longrightarrow 0$$

where ${}^{4}_{5}$ is a projective left B-module. It can be easily seen that $\mathrm{Tor}_2^B(T, 4) \simeq \mathrm{Tor}_1^B(T, 5)$. Take the injective coresolution of ${}_B 5$

$$0 \longrightarrow 6 \longrightarrow {}^{5}_{6} \longrightarrow 5 \longrightarrow 0\,;$$

similarly to what we did to compute $\mathrm{Ext}_A^1(T, 2)$, we can compute $\mathrm{Tor}_1^B(T, 5)$ as the kernel of $T \otimes_B 6 \to T \otimes_B {}^{5}_{6}$ as a morphism of left A-modules.

If we call t a generator of 6, with the previous notation for the generators of T_B, as a k-vector space $T \otimes_B 6$ is generated by $v \otimes t, z \otimes t, x \otimes t, w \otimes t, y \otimes t$. Since however $e_6 t = t$, the only generators of these which are not zero are $x \otimes t = xe_6 \otimes t$ and $y \otimes t = ye_6 \otimes t$. If we look at the action of A on the left of these elements, we deduce that $T \otimes_B 6$ is isomorphic to ${}^{2}_{3}$ as a left A-module. Similarly, $T \otimes_B {}^{5}_{6}$ turns out to be isomorphic to 2, so in the end

$$\mathrm{Tor}_2^B(T, \mathrm{Hom}_A(T, 2)) \simeq 3 \neq 0.$$

3 First Attempts to Recover the Decomposition

In order to recover a decomposition of every module induced by a classical n-tilting module, different strategies have been proposed.

In [11], Jensen, Madsen and Su suggested a solution for the $n = 2$ case by enlarging the subcategories KE_0, KE_1, KE_2 in the following way. Let $\mathcal{K}_0$ be the full subcategory of cokernels of monomorphisms from objects in KE_2 to objects in KE_0; let $\mathcal{K}_1$ be KE_1; let $\mathcal{K}_2$ be the full subcategory of kernels of epimorphisms from objects in KE_2 to objects in KE_0:

$$\begin{aligned}
\mathcal{K}_0 &= \left\{ \operatorname{coker} f : X_2 \overset{f}{\hookrightarrow} X_0, \quad X_2 \in KE_2, X_0 \in KE_0 \right\} \\
\mathcal{K}_1 &= KE_1 \\
\mathcal{K}_2 &= \left\{ \ker g : X_2 \overset{g}{\twoheadrightarrow} X_0, \quad X_2 \in KE_2, X_0 \in KE_0 \right\}.
\end{aligned}$$

By considering the morphisms $f : 0 \hookrightarrow X_0$ and $g : X_2 \twoheadrightarrow 0$, we can see that $KE_i \subset \mathcal{K}_i$ for every $i = 0, 1, 2$, so this is indeed an enlargement.

Now, for $i = 0, 1, 2$, let $\mathcal{E}_i$ be the extension closure of $\mathcal{K}_i$, i.e. the smallest subcategory containing $\mathcal{K}_i$ and closed under extensions.

Proposition 2 *([11, Corollary 15, Theorem 19, Lemma 24]) For any left A-module X, there exists a unique filtration*

$$0 = X_0 \subseteq X_1 \subseteq X_2 \subseteq X_3 = X$$

with the quotients $X_{i+1}/X_i \in \mathcal{E}_i$ for every $i = 0, 1, 2$. Moreover, such a filtration is functorial.

Example 2 Let us apply this construction to find a decomposition of the simple module 2 considered in the Example 1. In a way similar to that used to study the $\mathrm{Ext}^i_A(T, 2), i = 0, 1, 2$, we may prove that $\substack{2\\3}$ belongs to $K E_0$ and 3 belongs to $K E_2$. Then, 2 belongs to $\mathcal{K}_0 \subseteq \mathcal{E}_0$, being the cokernel of the monomorphism $3 \to \substack{2\\3}$. Therefore, the trivial filtration $0 \leq 2$ has its only filtration factor in the new class $\mathcal{E}_0$.

In [13], Lo generalised this filtration to the $n > 2$ case as well. After giving a different proof of Proposition 2, he introduced the following subcategories. For a class of objects $\mathcal{S}$, denote by $[\mathcal{S}]$ the extension closure of the full subcategory of quotients of objects of $\mathcal{S}$:

$$[\mathcal{S}] = \langle \{X : \exists (S \twoheadrightarrow X) \text{ for some } S \in \mathcal{S}\} \rangle_{\text{ext}}.$$

This subcategory is closed under quotients ([13, Lemma 5.1]). Then set, for $i = 0, \ldots, n$:

$$\mathcal{T}_i = \left[\mathrm{Ker}\,\mathrm{Ext}^i_A(T, ?) \cap \cdots \cap \mathrm{Ker}\,\mathrm{Ext}^n_A(T, ?)\right]$$
$$\mathcal{F}_i = \mathrm{Ker}\,\mathrm{Hom}_A(\mathcal{T}_i, ?) = \{X : \mathrm{Hom}_A(\mathcal{T}_i, X) = 0\}$$

with our usual convention that $\mathrm{Ext}^0_A = \mathrm{Hom}_A$. Define also $\mathcal{T}_{n+1} = A$-Mod and $\mathcal{F}_{n+1} = 0$.

This provides pairs $(\mathcal{T}_i, \mathcal{F}_i)$ of full subcategories, which are torsion pairs since the $\mathcal{T}_i$'s are closed under extensions and quotients (see Polishchuk [16]). The following easy proposition can then be applied to these torsion pairs.

Proposition 3 *([13, Theorem 5.3]) Let $(\mathcal{T}_i, \mathcal{F}_i)$ be torsion pairs in A-Mod, for $i = 0, \ldots, n + 1$, such that*

$$0 = \mathcal{T}_0 \subseteq \mathcal{T}_1 \subseteq \cdots \subseteq \mathcal{T}_{n+1} = A\text{-}Mod.$$

Then for every left A-module X, there exists a functorial filtration

$$0 = X_0 \subseteq X_1 \subseteq \cdots \subseteq X_{n+1} = X$$

such that $X_i \in \mathcal{T}_i$ for $i = 0, \ldots, n + 1$ and $X_i/X_{i-1} \in \mathcal{T}_i \cap \mathcal{F}_{i-1}$ for $i=1, \ldots, n + 1$. Moreover, the $\mathcal{T}_i \cap \mathcal{F}_{i-1}$ have pairwise trivial intersection.

We now prove that the subcategories $\mathcal{T}_i \cap \mathcal{F}_{i-1}$ introduced by Lo are indeed enlargements of the Miyashita classes using the following generalisation of [3, Lemma 3.2], which we find of independent interest.

Lemma 1 *Let X be a module belonging to $\bigcap_{i>e} \operatorname{Ker} \operatorname{Ext}_A^i(T, X)$ for some $0 \leq e \leq n$. Then, there exists a sequence of direct summands of coproducts of copies of T,*

$$\cdots \longrightarrow T_{-1} \xrightarrow{d_{-1}} T_0 \xrightarrow{d_0} \cdots \longrightarrow T_e \longrightarrow 0$$

which is exactly everywhere except for degree 0, and having $\ker d_0 / \operatorname{im} d_{-1} \simeq X$. In particular, for $e = n$, $\bigcap_{i>n} \operatorname{Ker} \operatorname{Ext}_A^i(T, X) = A$-Mod and hence X may be any module.

Proof Set $T^{\perp_\infty} := \bigcap_{i>0} \operatorname{Ker} \operatorname{Ext}^i(T, ?)$ and, for a family of modules $\mathcal{S}$, ${}^\perp\mathcal{S} := \operatorname{Ker} \operatorname{Ext}^1(?, \mathcal{S})$. It is well known (see [9], after Definition 5.1.1) that the pair of subcategories $({}^\perp(T^{\perp_\infty}), T^{\perp_\infty})$ is a complete hereditary cotorsion pair. This means (see [9, Lemma 2.2.6]) that X (as any other module) admits a special ${}^\perp(T^{\perp_\infty})$-precover

$$0 \longrightarrow J \longrightarrow K \longrightarrow X \longrightarrow 0\,.$$

In particular, J belongs to $({}^\perp(T^{\perp_\infty}))^\perp$, which equals $T^{\perp_\infty}$ by definition of cotorsion pair. Now we can apply [3, Lemma 3.2] to J and [9, Proposition 5.1.9] to K in order to construct a sequence of direct summands of coproducts of copies of T

$$\begin{array}{ccccccccccc}
\cdots \longrightarrow & T_{-2} & \longrightarrow & T_{-1} & \overset{d_{-1}}{\dashrightarrow} & T_0 & \xrightarrow{d_0} & \cdots \longrightarrow & T_n & \longrightarrow 0 & \qquad (*) \\
 & & & \downarrow & & \uparrow & & & & & \\
 & & & J & \hookrightarrow & K & \overset{\pi}{\twoheadrightarrow} & X & & &
\end{array}$$

By construction, the first row is a sequence which is exact everywhere except for degree 0, where $\ker d_0 / \operatorname{im} d_{-1} \simeq K/J \simeq X$.

This concludes the proof for the case where $e = n$. Otherwise, it can be easily proved that since by hypothesis $\operatorname{Ext}_A^i(T, X) = 0$ for $i > e$, then for these indices $T_i = 0$: let us show it for $i = n$, then the other cases follow similarly. First, notice that since $\operatorname{Ext}_A^j(T, J) = 0$, one gets $\operatorname{Ext}_A^j(T, K) = \operatorname{Ext}_A^j(T, X)$ for every $j > 0$. Then, if we call $K_j = \ker d_j$ for $j \geq 0$, we have

$$\operatorname{Ext}_A^1(T, K_{n-1}) \cong \operatorname{Ext}_A^n(T, K_0) = \operatorname{Ext}_A^n(T, X) = 0;$$

applying the functor $\operatorname{Hom}_A(T, ?)$ to the short exact sequence

$$0 \to K_{n-1} \to T_{n-1} \to K_n = T_n \to 0$$

we get that $\operatorname{Hom}_A(T, T_{n-1}) \to \operatorname{Hom}_A(T, T_n)$ is an epimorphism and hence all morphisms $T \to T_n$ factorise through T_{n-1}. Using the universal property of the coproduct of which T_n is a direct summand, it is easy to prove that this implies that

$0 \to K_{n-1} \to T_{n-1} \to T_n \to 0$ splits. Thus K_{n-1} is a direct summand of a coproduct of copies of T. Therefore, we may truncate the sequence $(*)$ as

$$\cdots \longrightarrow T_{n-3} \longrightarrow T_{n-2} \longrightarrow K_{n-1} \longrightarrow 0\ .$$

Notice that this lemma generalises [3, Lemma 3.2], which is the case where $e = 0$.

Remark 2 We shall prove that $KE_e \subseteq \mathcal{T}_{e+1} \cap \mathcal{F}_e$ for every $e = 0, \ldots, n$. Indeed, it is obvious that $KE_e \subseteq \mathcal{T}_{e+1}$. To see that any $M \in KE_e$ belongs to $\mathcal{F}_e$ as well, we will proceed in subsequent steps.

First, we prove that for every $X \in \bigcap_{i>e-1} \operatorname{Ker} \operatorname{Ext}^i_A(T, ?) \subseteq \mathcal{T}_e$, there are no non zero morphisms $X \to M$. Indeed, if $e = 0$, then $X = 0$; if $e > 0$, consider the sequence

$$T^\bullet := \quad \cdots \longrightarrow T_{-1} \xrightarrow{d_{-1}} T_0 \xrightarrow{d_0} \cdots \longrightarrow T_{e-1} \longrightarrow 0$$

given by Lemma 1 applied to X. Set $K_j = \ker d_j$ for $j \geq 0$ (and so $K_0 = K$), applying the functor $\operatorname{Hom}(-, M)$ to the epimorphism $K_0 \to X$, one gets

$$\operatorname{Hom}_A(X, M) \hookrightarrow \operatorname{Hom}_A(K_0, M) \cong \operatorname{Ext}^1_A(K_1, M) \cong \cdots$$

$$\cdots \cong \operatorname{Ext}^{e-1}_A(K_{e-1}, M) = \operatorname{Ext}^{e-1}_A(T_{e-1}, M) = 0,$$

and hence $\operatorname{Hom}_A(X, M) = 0$.

Now, if X' is the epimorphic image of some $X \in \bigcap_{i>e-1} \operatorname{Ker} \operatorname{Ext}^i_A(T, ?)$, we have $\operatorname{Hom}_A(X', M) \hookrightarrow \operatorname{Hom}_A(X, M) = 0$ so $\operatorname{Hom}_A(X', M) = 0$ as well. Lastly, if X'' is an extension of such epimorphic images, we still find that $\operatorname{Hom}_A(X'', M) = 0$.

This proves the claim that M has no non zero morphisms from objects of $\mathcal{T}_e$, and therefore it belongs to $\mathcal{F}_e$.

The last result of [13] is the proof that for $n = 2$, the filtration procedure of Proposition 3 reduces to that provided by Jensen, Madsen, and Su.

It should be noted that these results, while providing a way to generalise the decomposition of every module found in the $n = 1$ case, do so by introducing enlargements of the Miyashita classes KE_i which are not very natural, at the point that the connection to the tilting object they originate from seems a bit weak.

The rest of the article is devoted to the description of an alternative approach to this enlarging strategy, introduced in [8], which takes place in the derived category $\mathcal{D}(A)$ of A-Mod. In the following section, we recall some basic facts about derived categories and t-structures.

4 Introducing Derived Categories and t-structures

Given an abelian category $\mathcal{A}$, one may construct its derived category $\mathcal{D}(\mathcal{A})$ defining objects and morphisms in the following way. As objects, one takes the cochain complexes with terms in $\mathcal{A}$:

$$\cdots \longrightarrow X^n \xrightarrow{d_X^n} X^{n+1} \xrightarrow{d_X^{n+1}} X^{n+2} \longrightarrow \cdots$$

In order to define morphisms, one first takes the quotient of morphisms of complexes modulo those satisfying the *nullohomotopy* condition; the category having these equivalence classes as morphisms is called the *homotopy category*. The step from this to the derived category is performed by an argument of *localisation*; in this way, morphisms of complexes which induce isomorphisms on the cohomologies get an inverse in the derived category.

The category $\mathcal{D}(\mathcal{A})$ so obtained is not abelian anymore, but it is a *triangulated category*. This means that it is equipped with the following structure. First, there is an autoequivalence, whose action on the complex $X^\bullet$ is denoted as $X^\bullet[1]$ and is defined as follows:

$$(X^\bullet[1])^n = X^{n+1} \qquad d^n_{X[1]} = -d_X^{n+1}.$$

This functor is called the *suspension functor*; its natural definition on chain morphisms induces a good definition on morphisms in $\mathcal{D}(\mathcal{A})$. We will sometimes denote this functor also as Σ; its inverse as Σ^{-1} or $?[-1]$; their powers as Σ^i or $?[i]$ for $i \in \mathbb{Z}$.

Given this autoequivalence, one calls *triangles* the diagrams of the form

$$X^\bullet \xrightarrow{u} Y^\bullet \xrightarrow{v} Z^\bullet \xrightarrow{w} X[1]$$

such that $v \circ u = 0 = w \circ v$; in $\mathcal{D}(\mathcal{A})$, a particular role is played by the triangles isomorphic (as diagrams) to those of the form

$$X^\bullet \xrightarrow{f} Y^\bullet \longrightarrow \operatorname{Cone} f \longrightarrow X[1]$$

where Cone f is defined as the complex having terms $(\operatorname{Cone} f)^i = X^{i+1} \oplus Y^i$ and differentials $d^i_{\operatorname{Cone} f} = \begin{bmatrix} -d_X^{i+1} & 0 \\ f^{i+1} & d_Y^i \end{bmatrix}$. These triangles are called *distinguished triangles* and are the analogous of short exact sequences in abelian categories.

In a triangulated category, hence also in $\mathcal{D}(\mathcal{A})$, products and coproducts of distinguished triangles, when they exist, are distinguished (see [15, Proposition 1.2.1, and its dual]). In particular, if $\mathcal{A}$ has arbitrary products or coproducts, $\mathcal{D}(\mathcal{A})$ has them as well: they are constructed degree-wise using those of $\mathcal{A}$.

Once we have set our context, we now define the main object which we will work with.

Definition 3 Let $\mathcal{S} = (\mathcal{S}^{\leq 0}, \mathcal{S}^{\geq 0})$ be a pair of full, strict (i.e. closed under isomorphisms) subcategories of $\mathcal{D}(\mathcal{A})$, and denote $\mathcal{S}^{\leq i} = \mathcal{S}^{\leq 0}[-i]$ and $\mathcal{S}^{\geq i} = \mathcal{S}^{\geq 0}[-i]$, for every $i \in \mathbb{Z}$.

The pair $\mathcal{S}$ is a *t-structure* if it satisfies the following properties:

T1) $\mathcal{S}^{\leq 0} \subseteq \mathcal{S}^{\leq 1}$ and $\mathcal{S}^{\geq 0} \supseteq \mathcal{S}^{\geq 1}$;
T2) $\mathrm{Hom}_{\mathcal{D}(\mathcal{A})}(\mathcal{S}^{\leq 0}, \mathcal{S}^{\geq 1}) = 0$;
T3) For any complex $X^\bullet$ in $\mathcal{D}(\mathcal{A})$, there exist complexes $A^\bullet \in \mathcal{S}^{\leq 0}$ and $B^\bullet \in \mathcal{S}^{\geq 1}$ and morphisms such that

$$A^\bullet \longrightarrow X^\bullet \longrightarrow B^\bullet \longrightarrow A^\bullet[1]$$

is a distinguished triangle in $\mathcal{D}(\mathcal{A})$. This is called an *approximating triangle* of $X^\bullet$.

In this case, $\mathcal{S}^{\leq 0}$ is called an *aisle*, $\mathcal{S}^{\geq 0}$ a *coaisle*. The t-structure $\mathcal{S}$ is called *non degenerate* if $\bigcap_{i\in\mathbb{Z}} \mathcal{S}^{\leq i} = 0$ (or equivalently $\bigcap_{i\in\mathbb{Z}} \mathcal{S}^{\geq i} = 0$). The full subcategory $\mathcal{H}_\mathcal{S} = \mathcal{S}^{\leq 0} \cap \mathcal{S}^{\geq 0}$ is called the *heart* of $\mathcal{S}$.

This definition immediately resembles that of a torsion pair in an abelian category. As it holds for torsion pairs, the approximating triangle of a complex with respect to a t-structure is functorial, as we are going to state.

Given a t-structure $\mathcal{S}$ in $\mathcal{D}(\mathcal{A})$, it can be proved that the embeddings of subcategories $\mathcal{S}^{\leq 0} \subseteq \mathcal{D}(\mathcal{A})$ and $\mathcal{S}^{\geq 0} \subseteq \mathcal{D}(\mathcal{A})$ have a right adjoint $\sigma^{\leq 0} : \mathcal{D}(\mathcal{A}) \to \mathcal{S}^{\leq 0}$ and a left adjoint $\sigma^{\geq 0} : \mathcal{D}(\mathcal{A}) \to \mathcal{S}^{\geq 0}$, respectively.

For $i \in \mathbb{Z}$, let us write $\sigma^{\leq i} = \Sigma^{-i} \circ \sigma^{\leq 0} \circ \Sigma^i : \mathcal{D}(\mathcal{A}) \to \mathcal{S}^{\leq i}$ and similarly $\sigma^{\geq i} = \Sigma^{-i} \circ \sigma^{\geq 0} \circ \Sigma^i : \mathcal{D}(\mathcal{A}) \to \mathcal{S}^{\geq i}$; $\sigma^{\leq i}$ and $\sigma^{\geq i}$ will be called, respectively, the left and the right *truncation functors at i* with respect to $\mathcal{S}$, for $i \in \mathbb{Z}$.

It can be proved that for every $X^\bullet$ in $\mathcal{D}(\mathcal{A})$, the approximation triangle for $X^\bullet$ provided by the definition of the t-structure $\mathcal{S}$ is precisely (isomorphic to):

$$\sigma^{\leq 0}(X^\bullet) \longrightarrow X^\bullet \longrightarrow \sigma^{\geq 1}(X^\bullet) \longrightarrow (\sigma^{\leq 0}(X^\bullet))[1]\ .$$

The truncation functors of $\mathcal{S}$ can be used to define the *i-th cohomology* with respect to $\mathcal{S}$. It can be proved that for every $i, j \in \mathbb{Z}$, there is a canonical natural isomorphism $\sigma^{\leq i}\sigma^{\geq j} \simeq \sigma^{\geq j}\sigma^{\leq i}$. Then, for every $i \in \mathbb{Z}$, the functor $H^i_\mathcal{S} = \Sigma^i\sigma^{\leq i}\sigma^{\geq i} \simeq \Sigma^i\sigma^{\geq i}\sigma^{\leq i} : \mathcal{D}(\mathcal{A}) \to \mathcal{H}_\mathcal{S}$ is called the i-th *cohomology functor* with respect to the t-structure $\mathcal{S}$ (or simply $\mathcal{S}$*-cohomology*).

We introduce now the first t-structure in $\mathcal{D}(\mathcal{A})$ we are going to use.

Definition 4 The *natural* t-structure $\mathcal{D}$ of $\mathcal{D}(\mathcal{A})$ has aisle and coaisle:

$$\mathcal{D}^{\leq 0} = \{X^\bullet \in \mathcal{D}(\mathcal{A}) :\ H^i(X^\bullet) = 0 \text{ for every } i > 0\}$$
$$\mathcal{D}^{\geq 0} = \{X^\bullet \in \mathcal{D}(\mathcal{A}) :\ H^i(X^\bullet) = 0 \text{ for every } i < 0\}\ .$$

Notice that by construction, the ith $\mathcal{D}$-cohomology of $X^\bullet$ is a complex having zero cohomology everywhere except for degree 0, where it has $H^i(X^\bullet)$, the usual i-th cohomology of $X^\bullet$: i.e., $H^i_{\mathcal{D}}(X^\bullet) = H^i(X^\bullet)[0]$.

The original proof that this is indeed a t-structure can be found in [5].

We now state the following fundamental theorem about t-structures. One may read it with our example $\mathcal{D}$ in mind.

Theorem 3 *Let $\mathcal{S}$ be a non degenerate t-structure in $\mathcal{D}(\mathcal{A})$. Then*

1. *The heart $\mathcal{H}_\mathcal{S}$ is an abelian category; moreover, a short sequence*

$$0 \longrightarrow X^\bullet \longrightarrow Y^\bullet \longrightarrow Z^\bullet \longrightarrow 0$$

in $\mathcal{H}_\mathcal{S}$ is exact if and only if there exists a morphism $Z \to X[1]$ in $\mathcal{D}(\mathcal{A})$ such that the triangle

$$X^\bullet \longrightarrow Y^\bullet \longrightarrow Z^\bullet \longrightarrow X^\bullet[1]$$

is distinguished.

2. *Given any distinguished triangle*

$$X^\bullet \longrightarrow Y^\bullet \longrightarrow Z^\bullet \longrightarrow X^\bullet[1]$$

in $\mathcal{D}(\mathcal{A})$, there is a long exact sequence in $\mathcal{H}_\mathcal{S}$

$$\cdots \longrightarrow H^{i-1}_\mathcal{S} Z^\bullet \longrightarrow H^i_\mathcal{S} X^\bullet \longrightarrow H^i_\mathcal{S} Y^\bullet \longrightarrow H^i_\mathcal{S} Z^\bullet \longrightarrow H^{i+1}_\mathcal{S} \longrightarrow \cdots$$

As can be easily seen, the heart $\mathcal{H}_\mathcal{D}$ of the natural t-structure of $\mathcal{D}(\mathcal{A})$ is (equivalent to) $\mathcal{A}$ itself via the embedding $\mathcal{A} \to \mathcal{D}(\mathcal{A})$ defined by

$$X \mapsto X[0] = (\cdots \longrightarrow 0 \longrightarrow X \longrightarrow 0 \longrightarrow \cdots)$$

whose quasi-inverse is H^0, the usual 0th-cohomology functor.

As it happens for torsion pairs, the aisle or the coaisle of a t-structure is sufficient to characterise the whole t-structure. Indeed, we give the following lemma by Keller and Vossieck [12].

Lemma 2 *Let $\mathcal{R} = (\mathcal{R}^{\leq 0}, \mathcal{R}^{\geq 0})$ be a t-structure in $\mathcal{D}(\mathcal{A})$. Then*

$$\mathcal{R}^{\leq 0} = \left\{X^\bullet \in \mathcal{D}(\mathcal{A}) : \operatorname{Hom}_{\mathcal{D}(\mathcal{A})}(X^\bullet, Y^\bullet) = 0 \textit{ for all } Y^\bullet \in \mathcal{R}^{\geq 1}\right\}$$
$$\mathcal{R}^{\geq 0} = \left\{Y^\bullet \in \mathcal{D}(\mathcal{A}) : \operatorname{Hom}_{\mathcal{D}(\mathcal{A})}(X^\bullet, Y^\bullet) = 0 \textit{ for all } X^\bullet \in \mathcal{R}^{\leq -1}\right\}.$$

We now give the following proposition, which gives a very useful way to construct t-structures in the derived category of a Grothendieck category $\mathcal{G}$.

Proposition 4 *([1, Lemma 3.1, Theorem 3.4]) Let $\mathcal{G}$ be a Grothendieck category. Given any complex E in $\mathcal{D}(\mathcal{G})$, let $\mathcal{U}$ be the smallest* cocomplete pre-aisle *contain-*

ing E, that is, the smallest full, strict subcategory of $\mathcal{D}(\mathcal{G})$ closed under positive shifts, extensions and coproducts. Then, $\mathcal{U}$ is an aisle and the corresponding coaisle is

$$\begin{aligned}\mathcal{U}^{\perp} &= \left\{Y^{\bullet} \in \mathcal{D}(\mathcal{G}) : \operatorname{Hom}_{\mathcal{D}(\mathcal{G})}(X^{\bullet}, Y^{\bullet}) = 0 \text{ for every } X^{\bullet} \in \mathcal{U}[1]\right\} \\ &= \left\{Y^{\bullet} \in \mathcal{D}(\mathcal{G}) : \operatorname{Hom}_{\mathcal{D}(\mathcal{G})}(E, Y^{\bullet}[i]) = 0 \text{ for every } i < 0\right\}.\end{aligned}$$

Remark 3 As a first application of this proposition, it is easy to see that if $\mathcal{G}$ has a projective generator E, then the natural t-structure of $\mathcal{D}(\mathcal{G})$ will be that generated by E (in the sense of the proposition). This will be the case when we will consider $\mathcal{G} = A$-Mod, with $E = A$.

Remark 4 In the case where the object E is in fact a module, that is, a complex concentrated in degree zero, we shall give a characterisation of the aisle $\mathcal{U}$ generated by E.

First, $\mathcal{U}$ contains E; and it is closed under positive shifts, hence it contains $E[i]$ for every $i > 0$. $\mathcal{U}$ is closed under arbitrary coproducts; let then $J = \cup_{i>0} J_i$ be a set of indices, and let $E_j = E[i]$ for every $j \in J_i$. Then the coproduct $\coprod_{j \in J} E_j = \coprod_{i>0} E^{(J_i)}[i]$ belongs to $\mathcal{U}$ as well. If $\mathcal{V}$ is the full subcategory of all objects isomorphic to these coproducts, this means that $\mathcal{V} \subseteq \mathcal{U}$. Since $\mathcal{U}$ is also closed under extensions, if we call $\mathcal{V}'$ the extension closure of $\mathcal{V}$, we have $\mathcal{V}' \subseteq \mathcal{U}$ as well. Moreover, since coproducts of distinguished triangles are distinguished, from the fact that $\mathcal{V}$ is closed under arbitrary coproducts follows easily that $\mathcal{V}'$ is as well. Hence, $\mathcal{V}'$ is a cocomplete pre-aisle, and by definition $\mathcal{U} \subseteq \mathcal{V}'$.

In conclusion, objects of $\mathcal{U}$ are isomorphic to complexes having zero terms in positive degrees and coproducts of E in nonpositive degrees.

5 n-Tilting Objects and Associated t-structures

In the following, we are going to work with a generalisation of classical n-tilting modules, introduced by Angeleri Hügel and Coelho [2]; the equivalent definition we give is more oriented towards the derived category $\mathcal{D}(A)$ of A-Mod, which will be our setting.

Definition 5 A left A-module T is (non necessarily classical) *n-tilting* if it satisfies the following properties:

P_n) T has projective dimensions at most n, i.e. there exists an exact sequence

$$0 \longrightarrow P_n \longrightarrow \cdots \longrightarrow P_1 \longrightarrow P_0 \longrightarrow T \longrightarrow 0$$

in A-Mod with the P_i projectives;

E_n) T is rigid, i.e. $\operatorname{Ext}^i_A(T, T^{(\Lambda)}) = 0$ for every index $0 < i \leq n$ and set Λ;

G_n) T is a *generator* in $\mathcal{D}(A)$, meaning that if for a complex $X^{\bullet}$ we have $\operatorname{Hom}_{\mathcal{D}(A)}(T, X[i]) = 0$ for every $i \in \mathbb{Z}$, then $X^{\bullet} = 0$ in $\mathcal{D}(A)$.

Notice that a classical n-tilting module is indeed n-tilting: in particular, p_n) implies P_n), p_n) and e_n) imply E_n) (see the Stacks Project [19, Proposition 15.72.3] and g_n) implies G_n) (see Positselski and Stovicek [17, Corollary 2.6]).

The discussion about t-structures in the previous section is justified by the following construction. Let T be a n-tilting left A-module and consider the pair $\mathcal{T} = (\mathcal{T}^{\leq 0}, \mathcal{T}^{\geq 0})$ of subcategories of $\mathcal{D}(A)$

$$\mathcal{T}^{\leq 0} = \left\{X^\bullet \in \mathcal{D}(A) : \operatorname{Hom}_{\mathcal{D}(A)}(T, X^\bullet[i]) = 0 \text{ for every } i > 0\right\}$$
$$\mathcal{T}^{\geq 0} = \left\{X^\bullet \in \mathcal{D}(A) : \operatorname{Hom}_{\mathcal{D}(A)}(T, X^\bullet[i]) = 0 \text{ for every } i < 0\right\}.$$

Remark 5 This is the t-structure generated by T in the sense of Proposition 4, as proved in [4, Lemma 3.4], which in turn follows [18, Lemma 4.4]. We provide here another proof.

Let $\mathcal{G} = (\mathcal{G}^{\leq 0}, \mathcal{G}^{\geq 0})$ be the generated t-structure. We have $\mathcal{T}^{\geq 0} = \mathcal{G}^{\geq 0}$. For the aisle, notice that $\mathcal{T}^{\leq 0}$ contains T by (E_n); and it is clearly closed under positive shifts, hence it contains any $T[i]$ for $i > 0$. Now, we show that it is closed under arbitrary coproducts of such complexes $T[i]$. Let $J = \cup_{i>0} J_i$ be a set, let $T_j = T[i]$ for every $j \in J_i$, and consider the coproduct $\coprod_{j\in J} T_j = \coprod_{i>0} T^{(J_i)}[i]$. Notice that since by (P_n) T has projective dimension n, we have

$$\operatorname{Hom}_{\mathcal{D}(A)}\left(T, \coprod_{j\in J} T_j\right) = \operatorname{Hom}_{\mathcal{D}(A)}\left(T, \coprod_{i>0} T^{(J_i)}[i]\right) = \operatorname{Hom}_{\mathcal{D}(A)}\left(T, \coprod_{1\leq i\leq n} T^{(J_i)}[i]\right).$$

Now, since $\mathcal{D}(A)$ is an additive category, this is itself isomorphic to

$$\operatorname{Hom}_{\mathcal{D}(A)}\left(T, \prod_{1\leq i\leq n} T^{(J_i)}[i]\right) \simeq \prod_{1\leq i\leq n} \operatorname{Hom}_{\mathcal{D}(A)}\left(T, T^{(J_i)}[i]\right) = 0$$

which is zero by property (E_n). Lastly, $\mathcal{T}^{\leq 0}$ is clearly closed under extensions, and so by Remark 4, it contains $\mathcal{G}^{\leq 0}$.

For the inclusion $\mathcal{T}^{\leq 0} \subseteq \mathcal{G}^{\leq 0}$, take an object $X^\bullet \in \mathcal{T}^{\leq 0}$ and consider its approximation triangle with respect to $\mathcal{G}$,

$$A^\bullet \longrightarrow X^\bullet \longrightarrow B^\bullet \xrightarrow{+1} .$$

We have $A^\bullet \in \mathcal{G}^{\leq 0} \subseteq \mathcal{T}^{\leq 0}$; and since $\mathcal{T}^{\leq 0}$ is clearly closed under cones, $B^\bullet \in \mathcal{T}^{\leq 0}$ as well. So in the end $B^\bullet \in \mathcal{T}^{\leq 0} \cap \mathcal{G}^{\geq 1} = \mathcal{T}^{\leq 0} \cap \mathcal{T}^{\geq 1}$ which is 0 by G_3).

As a side note, observe that if T is classical n-tilting, it induces a triangulated equivalence $R\operatorname{Hom}_A(T, ?) : \mathcal{D}(A) \to \mathcal{D}(B)$ (see [7]); then, by the fact that

$$\operatorname{Hom}_{\mathcal{D}(A)}(T, X^\bullet[i]) = H^i R\operatorname{Hom}_A(T, X^\bullet)$$

we may recognise in $\mathcal{T} := (\mathcal{T}^{\leq 0}, \mathcal{T}^{\geq 0})$ the "pullback" of the natural t-structure of $\mathcal{D}(B)$ along $R\operatorname{Hom}_A(T, ?)$.

Remark 6 Without further study, the t-structure $\mathcal{T}$ can be immediately used to review some previous results.

First, we can greatly simplify the proof of Remark 2. In the notation used there, to prove that there are no non zero morphisms $X \to M$, for X in $\bigcap_{i>e-1} \operatorname{Ker}\operatorname{Ext}^i_A(T, ?)$ and M in KE_e, we may just recognise that we have $X = X[0] \in \mathcal{T}^{\leq e-1}$ and $M \in \mathcal{T}^{\geq e}$, and use axiom (T2) of t-structures.

Second, we may read our Lemma 1 under a different light: given the characterisation of objects in $\mathcal{T}^{\leq 0}$ as in Remark 4, the lemma can be seen to be the equality

$$\bigcap_{i>e} \operatorname{Ker}\operatorname{Ext}^i_A(T, ?) = A\text{-Mod} \cap \mathcal{T}^{\leq e}.$$

In the following, T will be a n-tilting module; $\mathcal{T}$ will be the associated t-structure, as defined above. The solution that we are going to give to our decomposition problem originates from the interaction of the t-structure $\mathcal{T}$ with the natural t-structure $\mathcal{D}$ of $\mathcal{D}(A)$ (see Definition 4). First, we make an easy observation.

Proposition 5 *The following inclusions of aisles and coaisles hold:*

$$\mathcal{D}^{\leq -n} \subseteq \mathcal{T}^{\leq 0} \subseteq \mathcal{D}^{\leq 0} \quad \text{and} \quad \mathcal{D}^{\geq 0} \subseteq \mathcal{T}^{\geq 0} \subseteq \mathcal{D}^{\geq -n}.$$

Proof Some of the inclusions are easy to prove: if $X^\bullet \in \mathcal{D}^{\leq -n}$, then for every $i > 0$ we will have $X^\bullet[i] \in \mathcal{D}^{\leq -n-i} \subseteq \mathcal{D}^{\leq -n-1}$, hence $\operatorname{Hom}_{\mathcal{D}(A)}(T, X^\bullet[i]) = 0$ since T has projective dimension n. On the other hand, if $X^\bullet \in \mathcal{D}^{\geq 0}$, then for every $i < 0$ we will have $X^\bullet[i] \in \mathcal{D}^{\geq 0-i} \subseteq \mathcal{D}^{\geq 1}$, hence again $\operatorname{Hom}_{\mathcal{D}(A)}(T, X^\bullet[i]) = 0$ since $T \in \mathcal{D}^{\leq 0}$. The other two inclusions can be easily proved from these using Lemma 2.

Remark 7 With Proposition 5, we are ready to notice an important fact, which will be key later. Take a module X in KE_e, for some $e = 0, \ldots, n$; in particular, being a module, it belongs to $A\text{-Mod} \simeq \mathcal{H}_\mathcal{D} \subseteq \mathcal{D}^{\geq 0} \subseteq \mathcal{T}^{\geq 0}$. Moreover, by definition, for every $i = 0, \ldots, e-1$, we have $0 = \operatorname{Ext}^i_A(T, X) \simeq \operatorname{Hom}_{\mathcal{D}(A)}(T, X[i])$, hence X belongs in fact to $\mathcal{T}^{\geq e}$. Lastly, again by definition, for every $i > e$, we have $0 = \operatorname{Ext}^i_A(T, X) \simeq \operatorname{Hom}_{\mathcal{D}(A)}(T, X[i])$, hence X belongs to $\mathcal{T}^{\leq e}$ as well.

This proves that, after identifying $A\text{-Mod} \simeq \mathcal{H}_\mathcal{D}$, for every $e = 0, \ldots, n$ the e-th Miyashita class is

$$KE_e = A\text{-Mod} \cap \mathcal{H}_\mathcal{T}[-e].$$

Let us now look at Proposition 5 in the $n = 1$ case. Its proof suggests that we may focus on the inclusions between the aisles (those between the coaisles being their "dual" in the sense of Lemma 2). If T is a 1-tilting module, we will then have

$$\mathcal{D}^{\leq -1} \subseteq \mathcal{T}^{\leq 0} \subseteq \mathcal{D}^{\leq 0}. \tag{$*$}$$

In other words, complexes in $\mathcal{T}^{\leq 0}$ are allowed to have any cohomology (with respect to $\mathcal{D}$, which means the usual complex cohomology H^i) in degrees ≤ -1 and some kind of cohomology in degree 0, while they must have 0 cohomology in higher degrees.

Remark 8 In this situation, with T a 1-tilting module, we may try to characterise $H^0(X^\bullet)$ for $X^\bullet \in \mathcal{T}^{\leq 0}$. Notice that $X^\bullet$ sits in the approximation triangle with respect to $\mathcal{D}$

$$\delta^{\leq -1}(X^\bullet) \longrightarrow X^\bullet \longrightarrow H^0(X^\bullet)[0] \xrightarrow{+1}$$

where

$H^0(X^\bullet)[0] = H^0_{\mathcal{D}}(X^\bullet) = \delta^{\geq 0}\delta^{\leq 0}(X^\bullet) \simeq \delta^{\geq 0}(X^\bullet)$ since $X^\bullet \in \mathcal{D}^{\leq 0}$. If we apply the homological functor $\mathrm{Hom}_{\mathcal{D}(A)}(T, ?)$ to it, we get the long exact sequence of abelian groups

$$\cdots \to \mathrm{Hom}_{\mathcal{D}(A)}(T, X^\bullet[1]) \to \mathrm{Hom}_{\mathcal{D}(A)}(T, H^0(X^\bullet)[1]) \to \mathrm{Hom}_{\mathcal{D}(A)}(T, \delta^{\leq -1}(X^\bullet)[2]) \to \cdots$$

The last term is 0 because $\delta^{\leq -1}(X^\bullet) \in \mathcal{D}^{\leq -1} \subseteq \mathcal{T}^{\leq 0}$; similarly, the first is 0 because $X^\bullet \in \mathcal{T}^{\leq 0}$. This means that

$$\mathrm{Ext}^1_A(T, H^0(X^\bullet)) \simeq \mathrm{Hom}_{\mathcal{D}(A)}(T, H^0(X^\bullet)[1]) = 0$$

as well, i.e. that $H^0(X^\bullet) \in KE_0$. □

The inclusions (∗) are precisely the hypothesis of the following proposition by Polishchuk.

Proposition 6 *([16, Lemma 1.1.2]) Let $\mathcal{R}, \mathcal{S}$ be two t-structures in a triangulated category $\mathcal{C}$ such that*

$$\mathcal{R}^{\leq -1} \subseteq \mathcal{S}^{\leq 0} \subseteq \mathcal{R}^{\leq 0} \qquad \textit{(or equivalently } \mathcal{R}^{\geq 0} \subseteq \mathcal{S}^{\geq 0} \subseteq \mathcal{R}^{\geq -1}\textit{).}$$

Then the classes:

$$\mathcal{X} = \mathcal{H}_{\mathcal{R}} \cap \mathcal{S}^{\leq 0} = \mathcal{R}^{\geq 0} \cap \mathcal{S}^{\leq 0}, \qquad \mathcal{Y} = \mathcal{H}_{\mathcal{R}} \cap \mathcal{S}^{\geq 1} = \mathcal{R}^{\leq 0} \cap \mathcal{S}^{\geq 1}$$

form a torsion pair $(\mathcal{X}, \mathcal{Y})$ in $\mathcal{H}_{\mathcal{R}}$. $\mathcal{S}$ can be reconstructed from $\mathcal{R}$ and $(\mathcal{X}, \mathcal{Y})$ as

$$\mathcal{S}^{\leq 0} = \{X^\bullet \in \mathcal{R}^{\leq 0} : H^0_{\mathcal{R}}(X^\bullet) \in \mathcal{X}\}$$
$$\mathcal{S}^{\geq 0} = \{X^\bullet \in \mathcal{R}^{\geq -1} : H^{-1}_{\mathcal{R}}(X^\bullet) \in \mathcal{Y}\}.$$

This procedure to recover $\mathcal{S}$ is called *tilting* of the t-structure $\mathcal{R}$ with respect to the torsion pair $(\mathcal{X}, \mathcal{Y})$ in $\mathcal{H}_{\mathcal{R}}$. It was introduced by Happel et al. [10], and it is a central tool in the construction we are going to present.

Remark 9 It can be proved without too much effort that in our case the torsion pair $(\mathcal{X}, \mathcal{Y})$ in $\mathcal{H}_{\mathcal{D}} \simeq A$-Mod so identified is exactly the pair (KE_0, KE_1) induced by the 1-tilting module T; this confirms Remark 8.

We would like to use a procedure analogous to the Happel-Reiten-Smalø tilting of Proposition 6 in order to link $\mathcal{D}$ and $\mathcal{T}$ in the $n > 1$ case. Notice that if we repeat this tilting operation n times, the first and last of the produced t-structures will be related by the inclusions of Proposition 5. Indeed, let $\mathcal{R}_0, \ldots, \mathcal{R}_n$ be t-structures such that $\mathcal{R}_i$ is obtained by tilting $\mathcal{R}_{i-1}$ with respect to some torsion pair on $\mathcal{H}_{\mathcal{R}_{i-1}}$, for every $i = 1, \ldots, n$. Then, we have by construction

$$\mathcal{R}_0^{\leq -1} \subseteq \mathcal{R}_1^{\leq 0} \subseteq \mathcal{R}_0^{\leq 0} \quad \text{and} \quad \mathcal{R}_1^{\leq -1} \subseteq \mathcal{R}_2^{\leq 0} \subseteq \mathcal{R}_1^{\leq 0}$$

which combined give

$$\mathcal{R}_0^{\leq -2} \subseteq \mathcal{R}_2^{\leq 0} \subseteq \mathcal{R}_0^{\leq 0}.$$

One can then clearly prove by induction that

$$\mathcal{R}_0^{\leq -n} \subseteq \mathcal{R}_n^{\leq 0} \subseteq \mathcal{R}_0^{\leq 0}.$$

If T is an n-tilting A-module, we shall show that the associated t-structure $\mathcal{T}$ in $\mathcal{D}(A)$ can indeed be constructed from the natural t-structure $\mathcal{D}$ with this iterated procedure. To do so, we are going to construct the "intermediate" t-structures produced after each tilting.

For $i = 0, \ldots, n$, consider the strict full subcategories $\mathcal{D}_i^{\geq} = \mathcal{D}^{\geq -i} \cap \mathcal{T}^{\geq 0}$ (notice that we are working with the coaisles). We have as wanted that

$$\mathcal{D}^{\geq 0} = \mathcal{D}_0^{\geq} \subseteq \mathcal{D}_1^{\geq} \subseteq \cdots \subseteq \mathcal{D}_n^{\geq} = \mathcal{T}^{\geq 0}$$

and $\mathcal{D}_{i-1}^{\geq} \subseteq \mathcal{D}_i^{\geq} \subseteq \mathcal{D}_{i-1}^{\geq}[1]$ for $i = 1, \ldots, n$. The only thing needed to proceed with an iterated application of Proposition 6 is to prove that these $\mathcal{D}_i^{\geq}$ are indeed the coaisles of some t-structures, for $i = 1, \ldots, n-1$.

Lemma 3 *The $\mathcal{D}_i^{\geq} = \mathcal{D}^{\geq -i} \cap \mathcal{T}^{\geq 0}$ are coaisles of t-structures.*

Proof As we noticed before (see Remark 3 and the definition of $\mathcal{T}$), we have

$$\mathcal{D}^{\geq -i} = \left\{ Y^\bullet \in \mathcal{D}(A) : \ \mathrm{Hom}_{\mathcal{D}(A)}(A[i], Y^\bullet[j]) = 0 \text{ for every } j < 0 \right\}$$
$$\mathcal{T}^{\geq 0} = \left\{ Y^\bullet \in \mathcal{D}(A) : \ \mathrm{Hom}_{\mathcal{D}(A)}(T, Y^\bullet[j]) = 0 \text{ for every } j < 0 \right\}.$$

Hence, we have

$$\mathcal{D}^{\geq -i} \cap \mathcal{T}^{\geq 0} = \left\{ Y^\bullet \in \mathcal{D}(A) : \ \mathrm{Hom}_{\mathcal{D}(A)}(T \oplus A[i]), Y^\bullet[j]) = 0 \text{ for every } j < 0 \right\}$$

which is the coaisle of the t-structure generated by $T \oplus A[i]$ in the sense of Proposition 4.

This concludes our previous discussion, making sure that $\mathcal{T}$ can be constructed from $\mathcal{D}$ with (at most) n applications of the procedure of tilting a t-structure with respect to a torsion pair on its heart.

6 The t-tree

We are now going to exploit this fact to solve our decomposition problem.

First, we characterise the torsion pairs involved. According to Proposition 6, at the i-th step, the t-structure $\mathcal{D}_i$ (having coaisle $\mathcal{D}_i^{\geq 0} = \mathcal{D}_i^{\geq} = \mathcal{D}^{\geq -i} \cap \mathcal{T}^{\geq 0}$) is tilted with respect to the torsion pair $(\mathcal{X}_i, \mathcal{Y}_i) = (\mathcal{D}_i^{\geq 0} \cap \mathcal{D}_{i+1}^{\leq 0}, \mathcal{D}_i^{\leq 0} \cap \mathcal{D}_{i+1}^{\geq 1})$ in the heart $\mathcal{H}_i$ of $\mathcal{D}_i$, $i = 0, \dots, n-1$, thus producing the t-structure $\mathcal{D}_{i+1}$.

Theorem 4 *Let T be a n-tilting left A-module. We can associate to each left A-module X a tree (we call it the t-tree of X with respect to the t-structure induced by the tilting module T)*

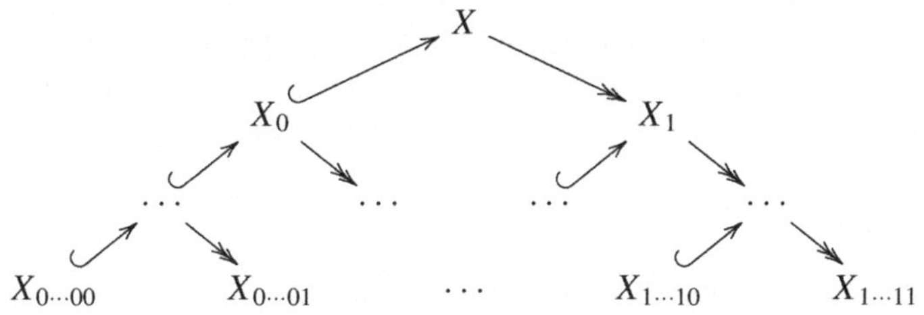

with $n+1$ rows, where

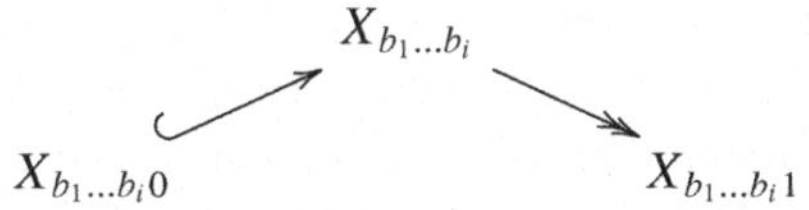

is the short exact sequence obtained decomposing $X_{b_1 \dots b_i}$ with respect to the torsion pair $(\mathcal{X}_i[-(b_1 + \dots + b_i)], \mathcal{Y}_i[-(b_1 + \dots + b_i)])$ in $\mathcal{H}_i[-(b_1 + \dots + b_i)]$.

Proof We may regard the left A-module X as a complex concentrated in degree 0, $X[0]$ in the heart $\mathcal{H}_\mathcal{D} = \mathcal{H}_0$. The first torsion pair $(\mathcal{X}_0, \mathcal{Y}_0)$ provides then a decomposition

$$X_0 \hookrightarrow X \twoheadrightarrow X_1 \quad \text{in } \mathcal{H}_0$$

with $X_0 \in \mathcal{X}_0$, $X_1 \in \mathcal{Y}_0$. Notice that by construction $\mathcal{X}_0 \subseteq \mathcal{H}_1$ and $\mathcal{Y}_0 \subseteq \mathcal{H}_1[-1]$ (see Proposition 6); this means that we can use $(\mathcal{X}_1, \mathcal{Y}_1)$ and $(\mathcal{X}_1[-1], \mathcal{Y}_1[-1])$ to further decompose X_0 and X_1, respectively, obtaining

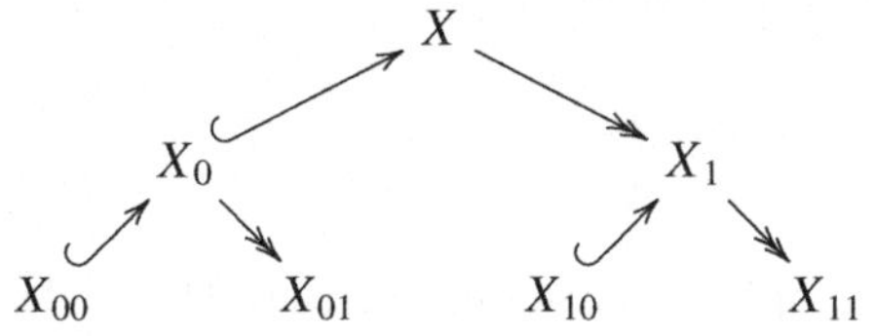

with the exact sequences in the respective abelian categories:

$$\begin{array}{ccccc} & & X & & \\ & \nearrow & & \searrow & \\ X_0 & & & & X_1 \end{array} \qquad \text{in } \mathcal{H}_0$$

$$\begin{array}{ccc} & X_0 & \\ \nearrow & & \searrow \\ X_{00} & & X_{01} \end{array} \quad \text{in } \mathcal{H}_1 \qquad \begin{array}{ccc} & X_1 & \\ \nearrow & & \searrow \\ X_{10} & & X_{11} \end{array} \quad \text{in } \mathcal{H}_1[-1]$$

Now, notice again that since $\mathcal{X}_1 \subseteq \mathcal{H}_2$ and $\mathcal{Y}_1 \subseteq \mathcal{H}_2[-1]$, we have that $X_{00} \in \mathcal{H}_2$, $X_{01}, X_{10} \in \mathcal{H}_2[-(0+1)] = \mathcal{H}_2[-(1+0)]$ and $X_{11} \in \mathcal{H}_2[-(1+1)]$.

Inductively, by decomposing each $X_{b_1 \ldots b_i}$ with respect to the torsion pair $(\mathcal{X}_i[-(b_1 + \cdots + b_i)], \mathcal{Y}_i[-(b_1 + \cdots + b_i)])$ in $\mathcal{H}_i[-(b_1 + \cdots + b_i)]$ we obtain objects $X_{b_1 \ldots b_i 0} \in \mathcal{H}_{i+1}[-(b_1 + \cdots + b_i)]$ and $X_{b_1 \ldots b_i 1} \in \mathcal{H}_{i+1}[-(b_1 + \cdots + b_i + 1)]$.

After n steps, we obtain the complete diagram.

We claim that Theorem 4 solves our decomposition problem. Indeed, by construction each object $X_{b_1 \cdots b_n}$ in the last row (called a *t-leaf*) belongs to $\mathcal{H}_n[-(b_1 + \cdots + b_n)] = \mathcal{H}_{\mathcal{T}}[-(b_1 + \cdots + b_n)]$: as noted in Remark 7, these shifted hearts are extensions of the Miyashita classes: $KE_{b_1+\cdots+b_n} = A\text{-Mod} \cap \mathcal{H}_{\mathcal{T}}[-(b_1 + \cdots + b_n)]$. Moreover, these shifted hearts are obtained by adding only non-module objects (i.e., objects of $\mathcal{D}(A)$ outside of $\mathcal{H}_{\mathcal{D}}$) to the corresponding Miyashita class; for this reason, they are less artificial than other enlargments, and instead shed a new light on the Miyashita classes. The latter can indeed be regarded as the piece of the shifted hearts of $\mathcal{T}$ visible in the category of modules.

Example 3 We recall one last time the situation considered in Example 1 to show an application of the construction of the t-tree; we will do it for the simple module 2 again.

First, a computation shows that the indecomposable complexes in $\mathcal{D}(A)$ are (shifts of)

$$\left\{ 1\,,\ 2\,,\ 3\,,\ {}^{1}_{2}\,,\ {}^{2}_{3}\,,\ {}^{2}_{3} \rightarrow {}^{1}_{2} \right\}.$$

Since we know that $\mathcal{D}^{\geq 0} \subseteq \mathcal{T}^{\geq 0}$, any bounded below complex will belong to $\mathcal{T}^{\geq 0}$, up to shifiting it enough to the right. We can then check for each of the indecomposable complexes what is their leftmost shift which still belongs to $\mathcal{T}^{\geq 0}$; with an easy computation, the following is the result:

$$\mathcal{T}^{\geq 0} = \left\langle 1\,,\ 2\,,\ 3\,[2],\ {}^{1}_{2}\,,\ {}^{2}_{3}\,,\ {}^{2}_{3} \rightarrow \overset{\bullet}{{}^{1}_{2}} \right\rangle$$

where the dot over a complex indicates its degree 0. The angle brackets will be used to denote the closure under direct sums and negative shifts.

Following the construction, we can compute the intermediate coaisles:

$$\mathcal{D}_{\bar{0}}^{\geq 0} = \left\langle 1, 2, 3, \begin{smallmatrix}1\\2\end{smallmatrix}, \begin{smallmatrix}2\\3\end{smallmatrix}, \overset{\bullet}{\begin{smallmatrix}2\\3\end{smallmatrix}} \to \begin{smallmatrix}1\\2\end{smallmatrix} \right\rangle = \mathcal{D}^{\geq 0}$$

$$\mathcal{D}_{\bar{1}}^{\geq 0} = \left\langle 1, 2, 3[1], \begin{smallmatrix}1\\2\end{smallmatrix}, \begin{smallmatrix}2\\3\end{smallmatrix}, \begin{smallmatrix}2\\3\end{smallmatrix} \to \overset{\bullet}{\begin{smallmatrix}1\\2\end{smallmatrix}} \right\rangle$$

$$\mathcal{D}_{\bar{2}}^{\geq 0} = \left\langle 1, 2, 3[2], \begin{smallmatrix}1\\2\end{smallmatrix}, \begin{smallmatrix}2\\3\end{smallmatrix}, \begin{smallmatrix}2\\3\end{smallmatrix} \to \overset{\bullet}{\begin{smallmatrix}1\\2\end{smallmatrix}} \right\rangle = \mathcal{T}^{\geq 0}.$$

Now we compute the hearts of the respective t-structures: to do this, we use Lemma 2. An object X of $\mathcal{D}_{\bar{i}}^{\geq 0}$ will belong to $\mathcal{H}_i$ if and only if $\operatorname{Hom}_{\mathcal{D}(A)}(X, Y) = 0$ for every $Y \in \mathcal{D}_{\bar{i}}^{\geq 1} = \mathcal{D}_{\bar{i}}^{\geq 0}[-1]$. In particular, it is easy to see that we must look for objects of the heart only among the "leftmost shifts" we have listed. The resulting computation gives (only indecomposable objects are listed)

$$\mathcal{H}_0 = \left\{ 1, 2, 3, \begin{smallmatrix}1\\2\end{smallmatrix}, \begin{smallmatrix}2\\3\end{smallmatrix} \right\} = \mathcal{H}_{\mathcal{D}} = A\text{-Mod}$$

$$\mathcal{H}_1 = \left\{ 1, 2, 3[1], \begin{smallmatrix}1\\2\end{smallmatrix}, \begin{smallmatrix}2\\3\end{smallmatrix}, \begin{smallmatrix}2\\3\end{smallmatrix} \to \overset{\bullet}{\begin{smallmatrix}1\\2\end{smallmatrix}} \right\}$$

$$\mathcal{H}_2 = \left\{ 1, 3[2], \begin{smallmatrix}1\\2\end{smallmatrix}, \begin{smallmatrix}2\\3\end{smallmatrix}, \begin{smallmatrix}2\\3\end{smallmatrix} \to \overset{\bullet}{\begin{smallmatrix}1\\2\end{smallmatrix}} \right\} = \mathcal{H}_{\mathcal{T}}.$$

Notice that neither 2 nor its shifts belong to $\mathcal{H}_{\mathcal{T}}$, which means exactly that it does not belong to any Miyashita class.

Lastly, we can compute the torsion pairs $(\mathcal{X}_i, \mathcal{Y}_i)$ in $\mathcal{H}_i$, for $i = 0, 1$. We have

$$\mathcal{X}_0 = \mathcal{H}_0 \cap \mathcal{H}_1 = \left\{ 1, 2, \begin{smallmatrix}1\\2\end{smallmatrix}, \begin{smallmatrix}2\\3\end{smallmatrix} \right\}, \qquad \mathcal{Y}_0 = \mathcal{H}_0 \cap \mathcal{H}_1[-1] = \{3\}$$

$$\mathcal{X}_1 = \mathcal{H}_1 \cap \mathcal{H}_2 = \left\{ 1, \begin{smallmatrix}1\\2\end{smallmatrix}, \begin{smallmatrix}2\\3\end{smallmatrix}, \begin{smallmatrix}2\\3\end{smallmatrix} \to \overset{\bullet}{\begin{smallmatrix}1\\2\end{smallmatrix}} \right\}, \qquad \mathcal{Y}_1 = \mathcal{H}_1 \cap \mathcal{H}_2[-1] = \{3[1]\}.$$

The t-tree for the module 2 is then

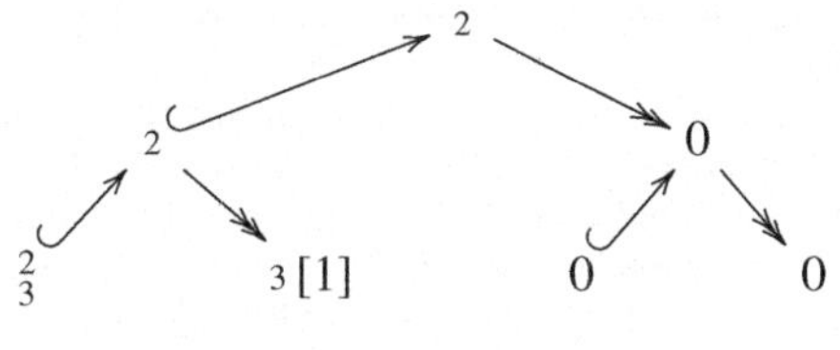

where the bottom left exact sequence is that associated to the distinguished triangle

$$\begin{smallmatrix}2\\3\end{smallmatrix} \longrightarrow 2 \longrightarrow 3[1] \xrightarrow{+1} .$$

Notice that this triangle can be shifted to become $3 \longrightarrow {}^{2}_{3} \longrightarrow 2 \xrightarrow{+1}$, which can be read as a short exact sequence of modules. This says that 2 is realised as the cokernel of the monomorphism $3 \to {}^{2}_{3}$, which is what was found following Jensen et al. in Example 2.

We conclude with a remark about the construction presented above, giving a possible direction for future developments.

Remark 10 Notice that while our motivation comes from an n-tilting A-module, the construction of the t-tree only relies on the existence of some t-structures $\mathcal{D}_i$, for $i = 0, \dots, n$, having the property that $\mathcal{D}_i^{\geq 0} \subseteq \mathcal{D}_{i+1}^{\geq 0} \subseteq \mathcal{D}_i^{\geq -1}$. Therefore, it can be replicated in an arbitrary triangulated category $\mathcal{C}$, given such t-structures: to any object in the heart $\mathcal{H}_0$ of $\mathcal{D}_0$, it is possible to associate a tree-like diagram with $n + 1$ rows having leaves in the shifts $\mathcal{H}_n[-e]$ of the heart $\mathcal{H}_n$ of $\mathcal{D}_n$, for $e = 0, \dots, n$. It is of natural interest to investigate other situations in which such t-structures may appear.

References

1. Alonso Tarrío, L., Jeremías López, A., Souto Salorio, M.J.: Construction of t-structures and equivalences of derived categories. Trans. Amer. Math. Soc. **355**(6), 2523–2543 (2003). https://doi.org/10.1090/S0002-9947-03-03261-6
2. Angeleri Hügel, L., Coelho, F.U.: Infinitely generated tilting modules of finite projective dimension. Forum Math. **13**(2), 239–250 (2001). https://doi.org/10.1515/form.2001.006
3. Bazzoni, S.: A characterization of n-cotilting and n-tilting modules. J. Algebra **273**(1), 359–372 (2004). https://doi.org/10.1016/S0021-8693(03)00432-0
4. Bazzoni, S.: The t-structure induced by an n-tilting module. Trans. Amer. Math. Soc. **371**(9), 6309–6340 (2019). https://doi.org/10.1090/tran/7488
5. Beĭlinson, A.A., Bernstein, J., Deligne, P.: Faisceaux pervers. In: Analysis and Topology on Singular Spaces, I (Luminy, 1981). In: *Astérisque*, vol. 100, pp. 5–171. Soc. Math. France, Paris (1982)
6. Brenner, S., Butler, M.C.R.: Generalizations of the Bernstein-Gel'fand-Ponomarev reflection functors. In: Representation theory, II (Proceedings of the Second International Conference, Carleton University, Ottawa, ON, 1979). Mathematics, vol. 832, pp. 103–169. Springer, Berlin and New York (1980)
7. Cline, E., Parshall, B., Scott, L.: Derived categories and Morita theory. J. Algebra **104**(2), 397–409 (1986). https://doi.org/10.1016/0021-8693(86)90224-3
8. Fiorot, L., Mattiello, F., Tonolo, A.: A classification theorem for t-structures. J. Algebra **465**, 214–258 (2016). https://doi.org/10.1016/j.jalgebra.2016.07.008. URL http://www.sciencedirect.com/science/article/pii/S0021869316301843
9. Göbel, R., Trlifaj, J.: Approximations and endomorphism algebras of modules. In: De Gruyter Expositions in Mathematics, vol. 41. Walter de Gruyter GmbH & Co. KG, Berlin (2006). https://doi.org/10.1515/9783110199727
10. Happel, D., Reiten, I., Smalø, S.O.: Tilting in abelian categories and quasitilted algebras. Mem. Amer. Math. Soc. **120**(575), viii+ 88 (1996). https://doi.org/10.1090/memo/0575
11. Jensen, B.T., Madsen, D.O., Su, X.: Filtrations in abelian categories with a tilting object of homological dimension two. J. Algebra Appl. **12**(2), 1250149, 15 (2013). https://doi.org/10.1142/S0219498812501496

12. Keller, B., Vossieck, D.: Aisles in derived categories. Bull. Soc. Math. Belg. Sér. A **40**(2), 239–253 (1988). Deuxième Contact Franco-Belge en Algèbre, Faulx-les-Tombes (1987)
13. Lo, J.: Torsion pairs and filtrations in abelian categories with tilting objects. J. Algebra Appl. **14**(8), 1550121, 16 (2015). https://doi.org/10.1142/S0219498815501212
14. Miyashita, Y.: Tilting modules of finite projective dimension. Math. Z. **193**(1), 113–146 (1986). https://doi.org/10.1007/BF01163359
15. Neeman, A.: Triangulated categories. In: Annals of Mathematics Studies, vol. 148. Princeton University Press, Princeton, NJ (2001). https://doi.org/10.1515/9781400837212
16. Polishchuk, A.: Constant families of t-structures on derived categories of coherent sheaves. Mosc. Math. J. **7**(1), 109–134, 167 (2007)
17. Positselski L., Stovicek J.: The tilting-cotilting correspondence, arXiv:1710.02230. To appear in International Mathematics Research Notices (2017). https://doi.org/10.1093/imrn/rnz116
18. Stovicek, J.: Derived equivalences induced by big cotilting modules. Adv. Math. **263** (2013). https://doi.org/10.1016/j.aim.2014.06.007
19. The Stacks project authors: The stacks project (2018). https://stacks.math.columbia.edu
20. Tonolo, A.: Tilting modules of finite projective dimension: sequentially static and costatic modules. J. Algebra Appl. **1**(3), 295–305 (2002). https://doi.org/10.1142/S0219498802000197

The manufacturer's authorised representative in the EU is Springer
Nature Customer Service Centre GmbH, Europaplatz 3, 69115 Heidelberg,
Germany. If you have any concerns regarding our products, please
contact ProductSafety@springernature.com

Printed and bound by CPI Group (UK) Ltd, Croydon, CR0 4YY
30/09/2026
02225510-0013

Surgical Decision Making in Geriatrics

Rifat Latifi
Editor

Surgical Decision Making in Geriatrics

A Comprehensive Multidisciplinary Approach

Springer

Editor
Rifat Latifi
Department of Surgery
New York Medical College
School of Medicine and
Westchester Medical Center Health Network
Valhalla, NY
USA

ISBN 978-3-030-47962-6 ISBN 978-3-030-47963-3 (eBook)
https://doi.org/10.1007/978-3-030-47963-3

This Springer imprint is published by the registered company Springer Nature Switzerland AG
The registered company address is: Gewerbestrasse 11, 6330 Cham, Switzerland

Prologue: COVID-19 Pandemic and the Elderly, Disparity

When we conceived this book, none of us could imagine that before the book goes to the print, the world would be in chaos from a disease that would affect all of us, but particularly will be deadly for the elderly and disparity population. None of us thought that the entire world would change as it did. Over the past six months, not just the world has changed and with that the hospitals and practice surgery have also changed. For months we did not perform any elective surgery; we only performed emergency surgery. Everyone has a mask on, everyone who works in the hospitals and patient alike. But as bad as it was, we learned a lot during these past months, and chances are we will continue to learn more about the disease that has spread throughout the world and that affects every organ.

With each day passing we learned new things. Who is dying, but even more importantly, why patients are dying? How do we protect each other, ourselves, our families, our patients, and hospital staff? It is very difficult to keep up with new information. One paragraph written today on your paper on COVID-19 or information you read is challenged next day by new data or by lack of data. The first report on deaths from COVID-19 that came from China was alarming with death rates of 11.1–14.6% of those infected. However, as New York become the hottest zone in the world, the study from New York City reported higher mortality rates. We have learned that elderly, obese patients, those with hypertension, lung diseases, chronic kidney disease, malignancy, myocardial infarction, cerebral infraction, and arrhythmia, had worse outcomes. Those who died from COVID-19 had significantly increased white blood cell (WBC) count and decreased lymphocyte and platelet counts. On rounds in the ICU now we talk about biomarkers of inflammation, interleukins 6 (IL-6) and 10 (IL-10), serum ferritin, and neutrophil-to-lymphocyte ratio (NLR). One thing that we knew, but it became even more apparent from this pandemic, is the disparity and inequality of healthcare. It has become evident that blacks, Native Americans, and patients from Hispanic communities are dying at a much higher rate. In other words, poor people and elderly in nursing homes are dying in disturbing numbers worldwide, particularly in Europe and USA. Vascular complications of COVID-19 (thrombosis both of venous and arterial tree) are severe including causing major abdominal catastrophes (Figures). In order to prevent these complications, patients are placed on anticoagulants, which in critically ill patients can cause major bleeding such as bleeding from ulcers or other intestinal bleeding. Stopping anticoagulants causes stroke, entering this way in a vicious cycle, that often ends fatally.

Due to severely depressed mental status in severely ill COVID-19 patients, often neither stroke nor abdominal catastrophe is recognized in a timely manner and may have severe consequences. As we were only performing emergency and cancer surgery, there is a fear that many patients stayed at home and did not seek help. Are patients with gallbladder disease, major hernias, reflux disease, and cardiovascular problems dying at home because they are afraid of calling 911 and go to the hospital? Is this the new post-COVID-19 surgical world order where we will continue to operate only when things get bad? Is this new surgical world order of healthcare: only emergency surgery. Hope not. Will today's "elective" surgery become emergency one day, or perhaps patients will select to suffer or simply die at home? What will the outcomes of all "emergency surgery only" patients be? More questions than answers.

Valhalla, NY, USA Rifat Latifi

Preface: The Last Sunset and Dying Alone

The Last Sunset

The times have changed. What was considered old a few decades ago now is not old at all. In fact, despite the fact that World Health Organization has defined the elderly as those above 65 years of age, there are significant changes in demographics of world population, and we researchers of the subject are struggling with definitions. Should the elderly be defined chronologically (e.g., those above the age of 85) or based on frailty index and physiology? So, I have to admit that even the title of this book, *Surgical Decision Making in Geriatrics,* is unclear and may raise an eyebrow of those who are totally functional but have passed their 70s long ago. A good number of faculty in my department are there already. Moreover, it is not uncommon for us surgeons to perform complex surgeries in patients in their 80s and 90s.

As we age, our bodies and our minds change, our needs change, our goals and objectives change. Aging is beautiful though. Recently, a well-known surgeon said while receiving a lifetime achievement award, "I am very happy to be with you, but frankly nowadays, I am very happy be anywhere." It is a beautiful thing to be able to walk without major help, to feed yourself, to love and be loved, to put clothes on and bathe and take care of your own hygiene, to go out to dinner with family, friends, or your spouse or partner, to watch your own grandkids or other kids grow. Simply talking to friends is a beautiful thing. Going to the movies or the park is a wonder. It is all simple stuff maybe – stuff that we take for granted when we are young.

When my paternal grandfather died at age 54, the kids in our village of Kllodernice, Kosova, told my uncle that "an o*ld man has died*," not knowing that he was the son of the dead man. I never met my grandfather, but everyone has told me that I look like him. When I celebrated my 54th birthday in Tucson, AZ, thousands of miles from the village I grew up, I was relieved that at least I had passed that mark. My grandfather died at home – probably from tuberculosis or lung cancer or both. And he knew that he was dying, like many patients do. I was told that the night before his death, he finished praying and looked at the sunset, saying "This was the last sunset I will see." He died early in the morning before the next sunrise. There was no doctor at the bedside, no nurse, no intravenous fluid, no test, no surgery, no nursing home, no rehab center. The nearest place to get a chest x-ray was days away by horse carriage. He died surrounded by 9 out of 10 of his kids, his brothers, his cousins, and his friends. The entire village and many people from surrounding villages came to his funeral to pay their respects.

By the time my grandmother died, I had just finished medical school and was attending her at home. She was in her 90s and died surrounded by near 100 kids (10 of whom she birthed and raised), grandkids, sons and daughters-in-law, cousins, and friends. I saw her take the last breath. It was a peaceful death. It was a beautiful death and it took place in her own large bedroom.

Many decades later, my father was 90 years old living in Prishtina, Kosova. One day, my daughter Kalterina, who at the time was living in Prishtina, called to tell me that my father was not doing well, and the surgeon would like to take him to the operating room for a left ruptured iliac artery aneurism.

I had to make a difficult decision. I was told that he has been anuric for almost 9 hours. Because he was 90 years old, the prospect of him making it out of hospital functional was very grim. He had lived a full life and was functional till the last day of his life. Now it was time for the end. I told the vascular surgeon not to operate but instead to find him a quiet room and control his pain while his family gathered to say goodbye.

Three hours later, he died peacefully after waving goodbye to all around him. He died surrounded by my two sisters and their husbands, my brother, his grandkids, cousins, friends, and many others. He was a soldier in three wars, fought the bad guys of the time, nearly died a few times, but lived to be old. This was a beautiful death.

My mother passed has few years earlier. She was 82 years old. She died at home, but she had doctors at her bedside, providing care in the last 3 days of her life. When she died, I was in a surgical mission in Tagbilaran, Philippines, caring for others. She died peacefully, I was told, at her home being cared by my two sisters, their husbands, and many grandkids. I went to her funeral in Prishtine, but for me, my mother died in Philippines, not in Prishtine.

Lonely Death

My grandparents died at their homes, not a bustling and noisy hospital ward or intensive care unit. They grew old and truly enjoyed their lives. They spent their years with plenty of family and social interaction, enjoying seeing kids and grandkids and friends. They enjoyed life as it is meant to be. This is how we are supposed to live and die – independent and with dignity, fully aware of our age and life and participating actively in life.

But, unfortunately, this is not the case all the time. With the modernization of our lives, the family fabric and family supporting infrastructure has changed dramatically. Kids move out, parents live alone, and they grow old alone. And they die alone.

One of the most difficult things that I had to do as trauma and general surgeon in Tucson, AZ, was watch patients die alone without family around. Sometimes their family was living on the other side of the country and the patient was alone in Tucson, trying to avoid the harsh winters of the East Coast. Often, when finally, I would get a hold of someone on the phone to give the bad news or the good news, the conversation was strange. Almost non-human. The kids often would not even know that their parents were in Tucson. They may not have seen them for years.

Once, I performed a complex abdominal wall reconstruction on an elderly, pleasant woman, a known medical illustrator, for disruption of the abdominal wall from a seatbelt injury. She survived two surgeries and was doing great but her son, a priest, insisted on us stopping everything and extubating her. "She suffered enough," was what he kept saying, but it felt like he was in a rush to return to his church in Connecticut. I did not agree with his opinion and his decision and refused to extubate her prematurely. At his insistence, while I was out of town, one of my partners extubated her. To everyone's surprise, she lived and recovered and went to rehab. I asked her to be an illustrator of my book on abdominal wall reconstruction. She replied softly "I will give it the best shot." I never forgot her and often wonder if she eventually died alone.

About 2 years ago, my youngest daughter Lulejeta and her dance team Pulse from Katonah, New York, were performing at the nursing home in our town. The nursing home is a beautiful complex of many buildings on the hill overseeing the reservoir, and I pass by it daily. I went to see the show and support my daughter. Watching the interaction of support staff with the residents and seeing how the staff were treating the patients like they were in kindergarten, I became acutely depressed and told my wife I had to leave. It was depressing to see how these old people were being treated like kids. Old people are not kids. I made my wife promise me that if and when I grow old, no one will take me to a nursing home. No one; I will not go. Recently, my friend's mother was in the IC for few days; an old lady with serious problems. She was visibly upset when nurses were calling her mother "honey," "sweetheart," "baby." As a patient is Mr. and Ms. or Mss., and not honey or baby or sweetheart.

Do We Have to Die from Surgery?

So much has changed since my grandfather died. We live longer but we also require more resources, more medical care, and more extensive surgical care. We have the most sophisticated medical and surgical advances to postpone death and have the ability to perform the most complex surgical procedures to sustain life. As surgeons, we can operate in any cavity, any organ, any part of the body. We can repair, remove, or replace an organ. That is not a question anymore.

The fundamental question that we, as surgeons, and for that matter healthcare providers, must ask ourselves is, what is the mission and the goal of the treatment that we are proposing to our elderly patient? My personal answer to this is: restore the function, relieve the suffering, and improve the quality of life. We must also ask at what cost? This book will explore this fundamental question. One of my dearest friends, Tom, underwent radical cystectomy and prostatectomy with creation of a neobladder. I had significant reservations and suggested much less radical surgery. He had significant complications postoperatively. And once he recovered the pathology report came back. "Suspicious for invasion…" The recommended treatment was chemoradiation therapy. I pleaded with him not to undergo either one. He did. He did not survive. At his funeral, his grandson read from Ecclesiastes: "There is time for everything, and a season for every activity under the heavens: A time to be born and a time to die…." The truth is that we should not offer therapies with significant complications for any kind of suspicions to frail 76-year-old men. We should not.

About this Book

While there are other books in the field of geriatric surgery, this is a unique book that deals with surgical decision-making and other elements in caring for the elderly. Just like the majority of healthcare institutions around the country and the world, the fastest growing population that we care for at Westchester Medical Center Health Network, Valhalla, New York, is the geriatric population. More than 35% of all surgical procedures performed occurs in this population.

As it is evident from all chapters in this book, the fastest growing segment of the US population is over 65 years of age. Similar growths are seen and projected to continue globally. This demographic shift in the population has serious implications in many aspects of life, but from the surgery standpoint, the elderly will undergo increasingly frequent major surgeries and other interventional procedures to maintain quality of life and physical and social independence. At the same time, the cost of health services will increase and the elderly will become the major consumers of healthcare and hospital resources. Deciding what surgical approach to take for an elderly patient is not an easy task. Should we perform definitive surgery in the elderly at the time of presentation or should we try a minimalistic approach initially and give time for the patient to recuperate and then perform the definitive surgery? This book will explore surgical decision-making in this population, particularly, in all aspect of surgery. The hardest decision that we surgeons have to make is whether everyone who has a potential surgical problem actually needs surgery? Does an 80-year-old lady with serious comorbidities and massive hernia for about 20 years need an operation? The answer is complex and must be searched for with care.

The idea for this book was born during the preparation for the Third Retreat of Department of Surgery at Westchester Medical Center and New York Medical College which took place in October 2018, where the focus was a multidisciplinary team approach to caring for the elderly patient where every aspect of geriatric surgery was discussed by each surgical specialist and those who are part of the geriatric healthcare team including emergency medicine, cardiology, anesthesia, and others. I want to thank all the authors for their time and devotion first to their patients and to this book. I hope that this book will serve those who make difficult decisions for their geriatric patients. To my team of research fellows, coordinators, and research scientists and to the Springer editorial team: thank you. Publishing these books truly takes more than a village. This time, it took an entire county – Westchester County.

Valhalla, NY, USA
November 2019

Rifat Latifi, MD, FACS

Contents

Contributors

Husham Abdelrahman, MD Department of Surgery, Trauma Surgery, Hamad General Hospital, Doha, Qatar

Kaveh Alizadeh, MD, MSc, FACS Westchester Medical Center, Valhalla, NY, USA

Patrice L. Anderson, MD Department of Surgery, New York Medical College, School of Medicine and Westchester Medical Center, Valhalla, NY, USA

Gabriela Andries, MD Cardiology Division, Department of Medicine, Westchester Medical Center and New York Medical College, Valhalla, NY, USA

Clara Angeles, MD Department of Surgery, Westchester Medical Center, Valhalla, NY, USA

Wilbert S. Aronow, MD, FACC, FAHA Cardiology Division, Department of Medicine, Westchester Medical Center and New York Medical College, Valhalla, NY, USA

Mohammad Asim, PhD Department of Surgery, Trauma Surgery, Hamad General Hospital, Doha, Qatar

Ammar Athar, MD Cardiology Division, Department of Medicine, Westchester Medical Center and New York Medical College, Valhalla, NY, USA

Sateesh C. Babu, MD Section of Vascular Surgery, Department of Surgery, Westchester Medical Center, Valhalla, NY, USA

Samuel Barasch, MD Department of Pathology, Westchester Medical Center, New York Medical College, Valhalla, NY, USA

Saleha Batool, MD Department of Hematology and Oncology, New York Medical College and Westchester Medical Center, Valhalla, NY, USA

Ira Bedzow, MA, PhD Biomedical Ethics and Humanities Program, New York Medical College, Valhalla, NY, USA

Aspen Center for Social Values, Aspen, Colorado (CO), USA

Roberto Bergamaschi, MD, PhD, FRCS, FASCRS, FACS Department of Surgery, Westchester Medical Center and New York Medical College, School of Medicine, Valhalla, NY, USA

Vincent Finbarr Blood, MD Department of Surgery, Westchester Medical Center, Valhalla, NY, USA

Christian A. Bowers, MD Department of Neurosurgery, Westchester Medical Center, New York Medical College, Valhalla, NY, USA

Susan Boyle, MSN APN AGNP Department of Emergency Medicine, Morristown Medical Center, Morristown, NJ, USA

Francis X. Carroll, MD, FACS, FCCP Division of Thoracic Surgery, New York Medical College, School of Medicine and Westchester Medical Center, Valhalla, NY, USA

Maria Castaldi, MD, FACS Department of Surgery, Westchester Medical Center, Valhalla, NY, USA

Department of Surgery, New York Medical College, School of Medicine, WMCHealthcare Network, Valhalla, NY, USA

Vasu Chirumamilla, MD Department of Surgery, Westchester Medical Center, Valhalla, NY, USA

James K. Choi, MD Department of Surgery, New York Medical College, School of Medicine and Westchester Medical Center, Valhalla, NY, USA

Muhammad S. Choudhury, MD, FACS Department of Urology, New York Medical College, Valhalla, NY, USA

Jorge Con, MD, FACS Department of Surgery, New York Medical College, School of Medicine and Westchester Medical Center, Valhalla, NY, USA

Miles Dale, MD Department of Surgery, Westchester Medical Center, Valhalla, NY, USA

Gizelka David-West, MD, FACOG Department of Obstetrics and Gynecology, Division of Gynecologic Oncology, New York Medical College/Westchester Medical Center, Valhalla, NY, USA

Kartik Dhaduk, MD Cardiology Division, Department of Medicine, Westchester Medical Center and New York, Valhalla, NY, USA

Department of Medicine, New York Medical College and Westchester Medical Center, Valhalla, NY, USA

Thomas Diflo, MD, FACS Department of Surgery, New York Medical College, Westchester Medical Center, Valhalla, NY, USA

Xiang Da Dong, MD, FACS Surgical Oncology, Department of Surgery, Westchester Medical Center and New York Medical College, Valhalla, NY, USA

Ayman El-Menyar, MD, FRCP Department of Surgery, Trauma Surgery, Hamad General Hospital, Doha, Qatar

Clinical Medicine, Weill Cornell Medical College, Doha, Qatar

Jason Elyaguov, MD Department of Urology, New York Medical College, Valhalla, NY, USA

Ashraf Elzanie, MD Westchester Medical Center, Valhalla, NY, USA

James M. Feeney, MD, FACS Associate Professor of Surgery, Department of Surgery, Westchester Medical Center, and New York Medical College, School of Medicine, Valhalla, NY, USA

William H. Frishman, MD Cardiology Division, Department of Medicine, Westchester Medical Center and New York, Medical College, Valhalla, NY, USA

Mahir Gachabayov, MD, PhD Department of Surgery, Westchester Medical Center and New York Medical College, School of Medicine, Valhalla, NY, USA

Geena George, MPH Department of Surgery, Clinical Research Unit, Westchester Medical Center, Valhalla, NY, USA

Shekhar Gogna, MD Department of Surgery, Westchester Medical Center, Valhalla, NY, USA

Joshua B. Goldberg, MD Department of Surgery, Division of Cradiothoracic Surgery, Westchester Medical Center and New York Medical College, Valhalla, NY, USA

Nikhil Gopal, MD Department of Urology, New York Medical College, Valhalla, NY, USA

Cara Grimes, MD, FACOG Department of Obstetrics and Gynecology, Advanced Urogynecology and Female Pelvic Medicine and Reconstructive Surgery, New York Medical College/Westchester Medical Center, Valhalla, NY, USA

Ansab Haider, MD Department of General Surgery, Westchester Medical Center, Valhalla, NY, USA

Firas Jafri, MD Department of Medicine, New York Medical College and Westchester Medical Center, Valhalla, NY, USA

Faisal Jehan, MD Department of Surgery, Westchester Medical Center and New York, Medical College, Valhalla, NY, USA

Ashutosh Kaul, MD Department of Surgery, Westchester Medical Center, Valhalla, NY, USA

Sarvesh Kaul Department of Surgery, Westchester Medical Center, Valhalla, NY, USA

Colgate University, Hamilton, NY, USA

Jaclyn Klimczak, MD Department of Otolaryngology Head and Neck Surgery, Icahn School of Medicine at Mount Sinai, New York, NY, USA

Donna C. Koo New York Medical College, Valhalla, NY, USA

Harry Kopolovich, MD MBA FAAEM FAEMS Department of Emergency Medicine, Richmond University Medical Center, Staten Island, NY, USA

Department of Emergency Medicine, Albert Einstein College of Medicine, Bronx, NY, USA

Danny Lascano, MD Department of Surgery, Westchester Medical Center, Valhalla, NY, USA

Igor Laskowski, MD, PhD Section of Vascular Surgery, Department of Surgery, Westchester Medical Center, Valhalla, NY, USA

Rifat Latifi, MD, FACS, FICS Department of Surgery, New York Medical College, School of Medicine and Westchester Medical Center Health Network, Valhalla, NY, USA

Robert G. Lerner, MD Department of Hematology and Oncology, New York Medical College and Westchester Medical Center, Valhalla, NY, USA

Alessandra Marino, MHS, PA-C Department of Obstetrics and Gynecology, Division of Gynecologic Oncology, New York Medical College/Westchester Medical Center, Valhalla, NY, USA

Romeo Mateo, MD Section of Vascular Surgery, Department of Surgery, Westchester Medical Center, Valhalla, NY, USA

Jaqueline Maxwell, MS, RD, CNSC Clinical Nutrition, Westchester Medical Center, Valhalla, NY, USA

Matthew K. McIntyre, BA School of Medicine, New York Medical College, Valhalla, NY, USA

Ivan T. Miller, MD, FACEP Department of Emergency Medicine, Westchester Medical Center, Valhalla, NY, USA

Department of Emergency Medicine, New York Medical College, Valhalla, NY, USA

Augustine L. Moscatello, MD, FACS Department of Otolaryngology/Head and Neck Surgery, New York Medical College, Valhalla, NY, USA

Department of Otolaryngology/Head and Neck Surgery, Westchester Medical Center, Valhalla, NY, USA

Seigo Nishida, MD, PhD, FACS Department of Surgery, Transplant Section, Westchester Medical Center and New York Medical College, Valhalla, NY, USA

Julia Pazniokas, MS New York Medical College, Valhalla, NY, USA

John L. Phillips, MD, FACS Department of Urology, New York Medical College, Valhalla, NY, USA

Anthony J. Policastro, MD, FACS Trauma and Critical Care, Department of Surgery, Westchester Medical Center and New York Medical College, Valhalla, NY, USA

Kartik Prabhakaran, MD, FACS Section of Trauma and Acute Care Surgery, Department of Surgery, Westchester Medical Center, Valhalla, NY, USA

Department of Surgery, New York Medical College, Valhalla, NY, USA

Tana Pradhan, DO, FACOG Department of Obstetrics and Gynecology, Division of Gynecologic Oncology, New York Medical College/Westchester Medical Center, Valhalla, NY, USA

J. K. Rasamny, MD Department of Otolaryngology, Head and Neck Surgery, ENT Faculty Practice, Ardsley, NY, USA

Aditya Safaya, MD Section of Vascular Surgery, Department of Surgery, Westchester Medical Center, Valhalla, NY, USA

David J. Samson, MS Department of Surgery, Westchester Medical Center and New York Medical College, Valhalla, NY, USA

Anar Shah, MD MBA Geriatric Emergency Department, Morristown Medical Center, Morristown, NJ, USA

Department of Emergency Medicine, Sidney Kimmel Medical College, Thomas Jefferson University, Philadelphia, PA, USA

Abbas Smiley, MD, PhD Department of Surgery, Westchester Medical Center and New York Medical College, School of Medicine, Valhalla, NY, USA

Sarah C. Smith, MD Division of Cardiothoracic Anesthesia, Department of Anesthesiology, Westchester Medical Center, New York Medical College, Valhalla, NY, USA

Anitha Srinivasan, MD, MPH, FACS Metropolitan Hospital Breast Health Center, Department of Surgery, New York Medical College, School of Medicine, Valhalla, NY, USA

Metropolitan Hospital, New York, NY, USA

Noam Stadlan, MD Department of Neurosurgery, NorthShore University Healthcare System, Evanston, IL, USA

Division of Neurosurgery, Pritzker School of Medicine at the University of Chicago, Evanston, IL, USA

Brianna Theriault, PhD New York Medical College, Valhalla, NY, USA

Stephanie Twomey, PA-C Department of Obstetrics and Gynecology, Division of Gynecologic Oncology, New York Medical College/Westchester Medical Center, Valhalla, NY, USA

Gregory Veillette, MD, FACS Section of Hepatobiliary and Pancreatic Surgery, Department of Surgery, Westchester Medical Center, and New York Medical College, School of Medicine, Valhalla, NY, USA

Tracey L. Weigel, MD, FACS, FACCP Division of Thoracic Surgery, New York Medical College, School of Medicine and Westchester Medical Center, Valhalla, NY, USA

Dan E. Wiener, MD, FACEP Department of Emergency Medicine, Morristown Medical Center, Morristown, NJ, USA

Department of Emergency Medicine, Sidney Kimmel Medical College, Thomas Jefferson University, Philadelphia, PA, USA

Part I

General Considerations for the Geriatric Surgical Patient

Decision-Making in Geriatric Surgery: More Surgery or Less Surgery?

1

Rifat Latifi, Shekhar Gogna, and Geena George

Introduction

The NIH report on aging predicts as the world population ages, the elderly will be the fastest-growing subset of the population, as fertility rates have fallen to low levels in most parts of the world and life expectancy is increasing. When the global population reached 7 billion in 2012, 562 million (8.0%) were aged 65 and over. Post World War II baby boomers in the United States and Europe recently joining the older ranks and with the accelerated growth of older populations in Asia and Latin America, the next 10 years will witness an increase of about 236 million people aged 65 and older throughout the world. Thereafter, from 2025 to 2050, the older population is projected to be twice the size to 1.6 billion globally, whereas the total population will grow by just 34 percent over the same period [1–3].

Only in the United States it is expected that by the year 2050, there will be over 80 million adults older than 65 years, representing more than one-fifth of the population. As the longevity of geriatric population will increase, it will have more independent and active lifestyles. Hence, the burden of surgical disease and injuries in this population is expected to increase. Importantly, this population accounts for 23% of all trauma admissions, and trauma is the fifth leading cause of death in the elderly [4]. Because of the high prevalence of multiple comorbidities in the elderly, there is an increased likelihood of death or severe disability following trauma. Up to one-third of all patients presenting with an Injury Severity Score (ISS) greater than 15 can be expected to have in-hospital mortality. Furthermore, elderly patients experience higher economic and societal costs, following trauma. Falls are the leading cause of trauma in the elderly. Approximately one-third of geriatric adults are at risk for falls each year. With an average hospital cost of $18,000 per fall and further costs associated with long-term nursing care following trauma, the economic implications of all trauma to the elderly are astonishing (the United States, 2012). Looking forward, the social and economic implications of the expected increase in geriatric trauma cannot be overlooked, and clinicians must continue to strive toward a more standardized and evidence-based approach to the diagnosis and treatment of these patients. (See Chap. 10).

With dramatic changes in our population, there is a growing need for surgeons, surgery

R. Latifi (✉)
Department of Surgery, New York Medical College, School of Medicine and Westchester Medical Center Health Network, Valhalla, NY, USA
e-mail: rifat.latifi@wmchealth.org

S. Gogna
Department of Surgery, Westchester Medical Center, Valhalla, NY, USA

G. George
Department of Surgery, Clinical Research Unit, Westchester Medical Center, Valhalla, NY, USA

R. Latifi (ed.), *Surgical Decision Making in Geriatrics*,
https://doi.org/10.1007/978-3-030-47963-3_1

residents, medical students, and nurses across all disciplines (medicine, anesthesia, gynecology, orthopedics, etc.). They will need to learn the basic surgical and other major decision-making in managing our oldest population. While there are other books in the field, this is a unique book that will deal with surgical decision-making and other elements in caring for the elderly. During perioperative period, the elderly often require a different level of care than younger patients. Many geriatric patients have multiple chronic illnesses other than the one for which surgery is required and therefore are prone to developing postoperative complications, functional decline, loss of independence, and other untoward outcomes. To provide optimal care for the older surgical patient, a thorough assessment of the individual's health status and a plan of care that identifies and addresses deficits during the perioperative period are essential. While this assessment is possible for elective surgery, many times in emergency situation, there is no time for such preparation, and often the elderly undergo major emergency surgery under suboptimal conditions.

The American College of Surgeons (ACS) has partnered with the American Geriatrics Society (AGS) and in collaboration with John A. Hartford Foundation has developed guidelines for the optimal surgical care of older adults. The first part of these guidelines, the American College of Surgeons National Surgical Quality Improvement Program (NSQIP)/American Geriatrics Society (AGS) Best Practices Guidelines: Optimal Preoperative Assessment of the Geriatric Surgical Patient, was published in 2012. This resource defined nine assessment categories: cognitive/behavioral disorders, cardiac evaluation, pulmonary evaluation, functional/performance status, frailty, nutritional status, medication management, patient counseling, and preoperative testing [3].

Surgical Decision-Making in the Elderly

How we surgeons make decisions under what can be inauspicious conditions is still a matter of debate and not well understood overall even though millions of surgical procedures are performed daily worldwide [5]. It is not easy to decide what surgical approaches to take in the elderly when there are multiple other options available. The surgical diseases once approached radically are now commonly treated with the "minimalistic approach." So the question to be answered is: Should we perform "more surgery" (that is taking care of the problem entirely in one sitting, as soon as clinically and physiologically possible) or perform a "less invasive" procedure and come fight another day? There is no simple answer for this question yet. For example, would you consider performing cholecystectomy to remove the infected gallbladder versus tube cholecystostomy in the elderly?

Intraoperative surgical decision-making (SDM) in the elderly should not be different from the SDM in the younger patient. However, one has to not only keep in mind the ability to execute the procedure itself but to remember the long-term goals of the surgical procedure, the implications on the physiology, and overall outcomes of the patient.

For the most part, SDM has been described as "intuition" or "gut-level" responses. Many factors affect the decision-making process of the surgeons before and during operations. These factors are the physiology of the patient, the balance and implications of doing or not doing a procedure, the goals of surgical care, and the patients' and their families' wishes. To name a few, other factors that affect SDM are the physiologic state of the surgeon, the harmony of teamwork, and the surgeon's ability to adapt quickly to a changing environment. The favorable outcomes in the postoperative elderly are obtained by incorporating evidence-based medicine (EBM), SDM, and the joint decision-making with the patient and family (Fig. 1.1). Yet, the question remains: How to perform an evaluation of the surgical decision and gain a better understanding of a seemingly gut-level process, which helps surgeons combat the external factors experienced before and during surgery?

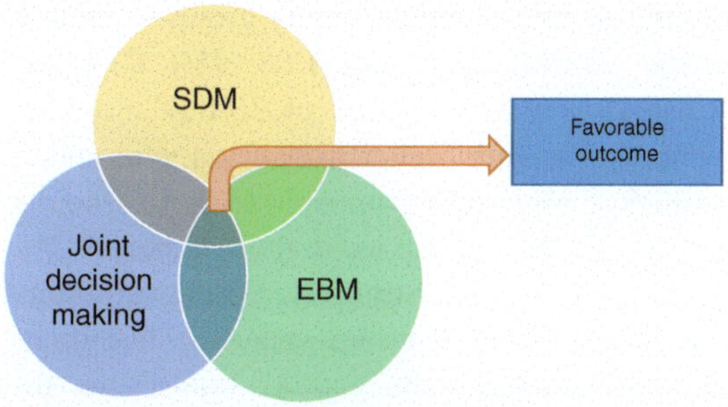

Fig. 1.1 The favorable outcome in elderly is achieved by amalgamating SDM, EBM, and joint decision-making. SDM Surgical decision-making, EBM Evidence based medicine

Surgical Decisions in Emergency Surgery in the Elderly: The Tale of Four Common Clinical Problems

Complex surgical procedures, particularly in the elderly, carry significant risks and potential complications. These complications can occur despite the most conscientious preoperative preparations. Unforeseen surprising events may occur during the operation (less likely), immediately in the postoperative period (most likely), and after a long time after the surgery itself. The complications, such as cognitive decline, and simply "giving up" are often thought to be due to anesthesia (see Chap. 7), delirium, or a slew of other factors. Aging promotes the physiological changes in the protoplasm and functioning of the heart, lungs, kidneys, and liver. So, deciding to operate in the elderly is not an easy one and requires the surgeon, patient, and family to understand the gravity of the procedure. Despite enormous significance and implications that SDM has on surgical outcomes, this topic and particularly its role in the elderly have received minimal attention in the literature. Only in recent years, both scientific and public media [3, 6] have taken up reporting surgical outcomes of the geriatric population. Subsequently, there are only a few studies that investigate how these decisions are made, although the decision-making process (DMP) is of great importance both for training and patient safety purposes [5]. According to the paper from, American Association for the Surgery of Trauma (AAST), most common emergency surgery in elderly occurs in the following clinical conditions: hepatobiliary, hernia, colorectal cancer, bowel obstruction [7]. In this chapter we will further dissect the role of SDM in these four clinical problems.

Acute Cholecystitis

Hepatobiliary emergency surgery is common among the elderly. While much progress has been made in this clinical discipline, there is a wide difference on how acute cholecystitis is managed from country to country, institution to institution, or even among individual surgeons within the same group. The question is: How do we surgeons decide which procedure to perform on an elderly patient with acute cholecystitis as compared to the non-elderly? Is there a difference in SDM? Let's assume that you have seen an 89-year-old patient, living independently who undergoes TEVAR and has known gallstones. He has a remote history of laparotomy for the perforated gastric ulcer but otherwise is in good shape. Postoperatively the patient develops acute cholecystitis. What would you do? One surgeon may perform cholecystectomy, perhaps even an open one, and the other one will perform a percutaneous tube cholecystostomy (PTC). Either approach has become an "acceptable" treatment. So it comes to a surgeon's SDM. I think we should take the gallbladder out. Remove the infected gallbladder and be done with it. Although I would start with laparoscopic approach with a very low threshold to convert into an open cholecystectomy.

Percutaneous tube cholecystostomy (PTC) could also be an acceptable treatment. Although it was intended for acute acalculous cholecystitis (AAC) [8] and not for those who have stones, nowadays it is has become a common practice. What is wrong with PCT? Well for starters, it commits the patient to many more procedures, morbidity, and eventually an operation that usually is not an easier one. While recent studies on the use of cholecystostomy tube for AAC reported that drainage can be achieved in up to 90%, PCT has a high rate of tube dysfunction requiring frequent re-intervention, such as tube exchange or replacement at an average of 2 per patient (range 1–10) [9, 10]. In another study of 288 patients with ACC, undergoing PCT, PCT dysfunction occurred in 132 patients (46%), with 80 patients (28%) requiring re-intervention, while 7% developed procedure-related complications [11]. Interval cholecystectomy reduced the risk of recurrent biliary events from 21% to 7% (p = 0.002). Cholecystectomy was performed laparoscopically in 45% of patients receiving an interval operation vs. 22% of those undergoing urgent surgery for PCT failure/dysfunction or recurrent biliary event (p = 0.03).

While the use of PCT in AAC is “more” acceptable, however, there is a new trend of the use of PTC in acute calculous cholecystitis. It is very commonly seen that a sick patient, with a gallbladder full of stones and subsequent sepsis, undergoes PTC drainage. Sometime ago an elderly gentleman on his way to Tucson felt sick on the plane. He was removed from the flight and underwent emergent PTC in one of the local hospitals in Dallas. He eventually improved after weeks in the ICU and came to see me. His gallbladder was filled with 54 stones. The question that I discussed with residents that day was: Is this what we do nowadays? (Fig. 1.2).

Recent data in elderly patients with grade II acute cholecystitis demonstrated the worst outcomes that were seen in those who underwent cholecystectomy tube placement. The in-hospital mortality was similar between the two groups

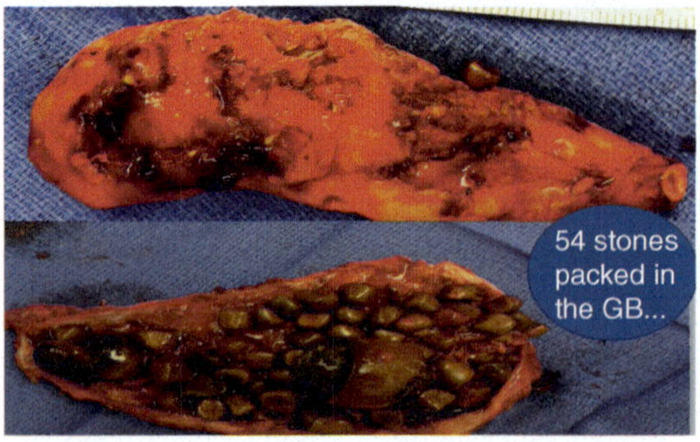

Fig. 1.2 Post-PTC cholecystectomy specimen showing 54 stones in gentleman with acute calculous cholecystitis

(24.0% in the cholecystostomy group vs. 22.6% in the control group; OR 1.08, 95% CI 0.86–1.35). However, the odds of 30-day mortality were significantly higher in patients who underwent PCT placement compared to those who did not (38.9% PCT group vs. in those versus 32.7% in no PCT group). Similarly, 90-day mortality was 46.7% in PCT group versus 39.6% in no PCT group. Also, the authors report that the 2-year survival rates in patients in PCT group were significantly shorter compared to those who did not (35% vs 41%, $p < 0.0059$) [12].

Recently, I performed an open cholecystectomy in a patient with a severely contracted gallbladder that was treated with a cholecystostomy tube. This tube was making her life miserable. While I did not see her previously, I am pretty sure that there was no good indication for draining the gallbladder. What needed to be done correctly actually happened months later. She did very well postoperatively and is back at work. Just the fact that you wait for a later day to perform a cholecystectomy will not make the procedure any better or easier. It may even make the situation worse. I remember a case from a few years ago when a surgeon performed a cholecystectomy 4 months after the PCT placement. The fibrosis that set in caused severe contraction of the gallbladder (GB), and major injury to the common bile duct (CBD) ensued. The good thing was that it was recognized intraoperatively (due to inadvertent injury to the right hepatic artery) causing the surgeon to convert to an open procedure and prompt hepaticojejunostomy. Most of the time, inflamed gallbladders are easier to remove surgically compared to a fibrotic, contracted GB.

In acute gallstone cholecystitis, patients should have a cholecystectomy. If there are no gallstones and the patient improves, then there is no reason to perform a cholecystectomy. This decision is assuming that the tube cholangiogram demonstrates patent cystic duct along with clinical signs of improvement. This can be done either laparoscopically or with the open technique. In sick patients with difficult gallbladders, partial or subtotal cholecystectomy is a viable and safe procedure [13].

A recent paper has proposed to discard the term partial cholecystectomy and substitute it with "fenestrated" and "reconstituting" types based on how you deal with the cystic duct and or the lower end of the gallbladder [14]. I agree with the concept of subtotal cholecystectomy, as long as great care is given to remove any stones in the remnant of the gallbladder, as the patient may return with symptoms of cholecystitis in the future. The patient must understand what type of procedure was done, so in the case of recurrence of symptoms, the patients knew what to expect. I have performed a completion cholecystectomy in a patient, and one has to remind the patient what the other surgeon did, but particularly why they did a subtotal cholecystectomy. In elderly critically ill patients, this is a very good option to remove the gallbladder while not risking major injury to the common bile duct.

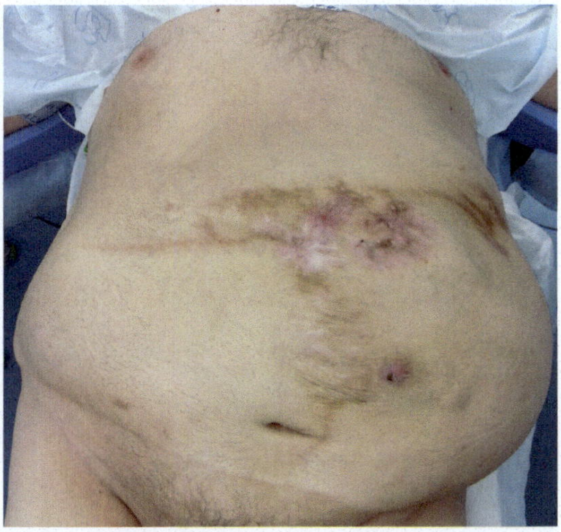

Fig. 1.3 An elderly gentleman with chronic small bowel obstruction with concomitant large "reducible" hernia and associated with loss of abdominal wall domain

Bowel Obstruction: Operate Early or Wait?

Another very common surgical dilemma in current surgical practice is partial or complete small bowel obstruction (SBO). While the second scenario (complete bowel obstruction) is less controversial, in partial recurrent SBO, SDM tends to be more difficult. The SDM gets even more complex when there is a concomitant large "reducible" hernia and often with loss of abdominal wall domain (Fig. 1.3). Add morbid obesity and things become even more complicated. It is not uncommon that these patients are in and out of the hospital being managed with nil per Os (NPO) for weeks, nasogastric tube (NGT) decompression, and Foley catheter and severe deconditioning ensues. Malnutrition, already present at admission, gets worse because the patients are "too old to have an operation" or "still do not have peritonitis" or because "let's give conservative" approach a few more days. Eventually these patients deteriorate and develop complete bowel obstruction, requiring an emergency surgery or even a damage control approach (Fig. 1.4), with intestinal diversion. They will have to return to the operating room, have another hernia, or may require major abdominal wall reconstruction under suboptimal conditions (Fig. 1.5). Hence, the vicious cycle goes on, and should be stopped.

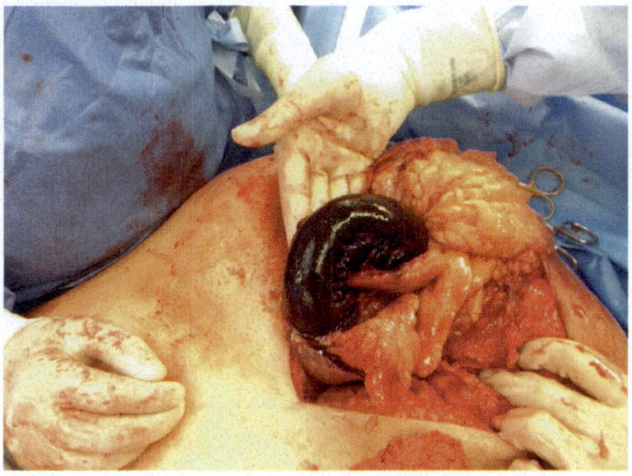

Fig. 1.4 An elderly patient undergoing damage control surgery for bowel ischemia due to failure of "conservative approach"

While many have adopted the use of gastrografin as a first line of "treatment challenge," this tactic may work only in those who have an early episode of partial SBO, mostly due to electrolyte imbalance and not true adhesive SBO. Use of water-soluble contrast medium (gastrografin) does not decrease the need for operative intervention nor the duration of hospital stay in uncomplicated acute adhesive small bowel obstruction.

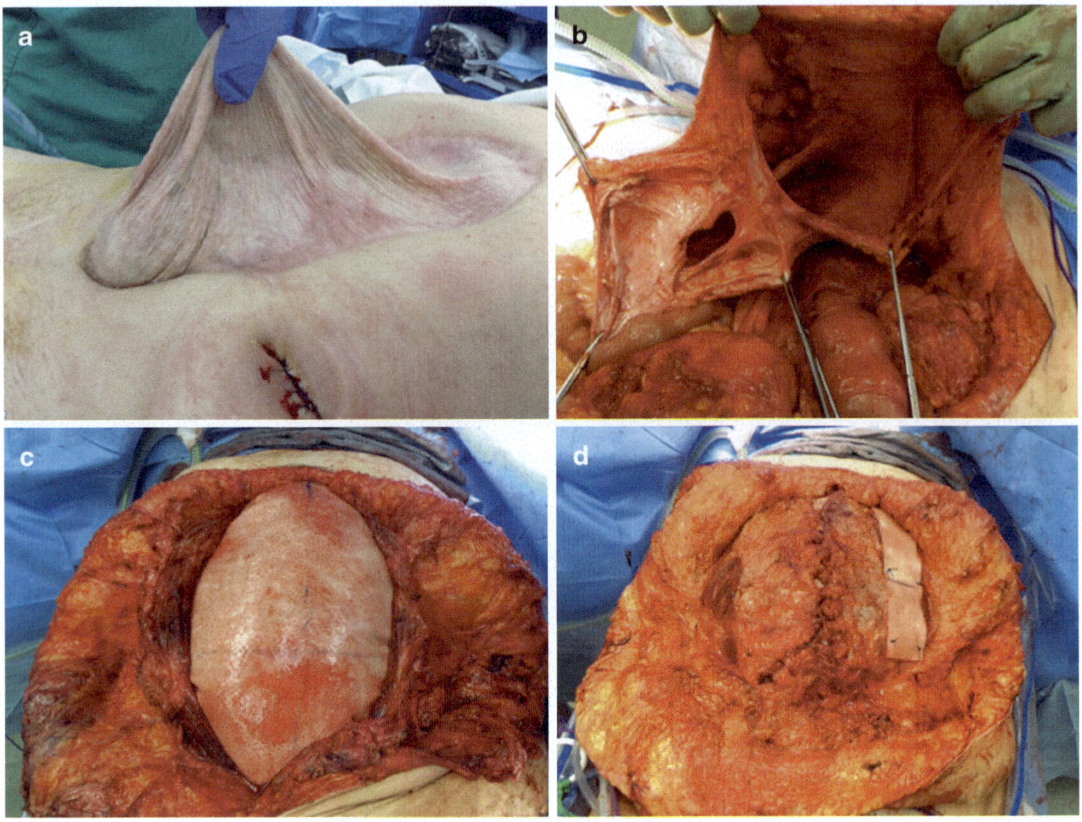

Fig. 1.5 (**a**) Patient being operated for complex abdominal wall hernia with loss of domain; (**b**) complex anatomy due to fibrosis and high burden of hernia sacs; (**c**, **d**) surgical decision-making to determine type of complex abdominal wall reconstruction

A multicenter, randomized, clinical trial (Adhesive Small Bowel Obstruction Study) and systematic review evaluated the association between oral gastrografin and the need for operative intervention. In this trial, the rate of operative intervention in gastrografin vs. saline solution arms was 24% and 20%, and the bowel resection rate was 8% and 4%, respectively. Moreover, age was the only identified potential risk factor for the failure of nonoperative management. The evidence thus suggests that gastrografin challenge is of no benefit in patients with adhesive small bowel obstruction [15].

On the other side, studies have concluded that the use of water-soluble contrast medium as a predictive test for nonoperative resolution of adhesive small bowel obstruction may reduce the need for operation and appears to shorten the hospital stay for those who do not require surgery [16].

What Are the Data on Small Bowel Obstruction in Geriatrics?

Small bowel obstruction (SBO) is one of the most frequent emergencies in general surgery, commonly affecting elderly patients. Morbidity and mortality from small bowel obstruction in the elderly are high [17]. Up to 50% of emergency laparotomies are due to small bowel obstruction, and 10–12% of these are seen in the elderly [18, 19]. The elderly have higher mortality with emergency [20]. These measures translate into the fact that we should not deprive major surgery in this group of patients as not offering the surgery would probably lead to worse outcomes. Although recent advances in diagnostic modalities have made it easier to diagnose the SBO, nonetheless the surgical treatments need better prioritizing, and one should be cognizant that the elderly may have a major catastrophe in the abdomen, without demonstrating signs of peritonitis, normal lactate, and simply just looking sick.

Recently, I operated on a cachectic 82-year-old lady with visible loops of intestines under thinned out skin in the lower abdomen that had gangrene of more than 80 cm of the small bowel, but preoperatively she was sitting up in her bed with no complaints whatsoever. So, just because they do not demonstrate the usual clinical signs and symptoms or biochemical indicators of small bowel obstruction does not mean that the elderly do not need an operation. The real art and science of being a surgeon lie in determining when and which operation will enhance the outcomes. To achieve that level of surgical decision-making takes time and years of practice, and, more importantly, it requires a sincere and honest interest in surgical discipline.

Diverticulitis

Diverticular disease (DD) of the colon accounts for more than 300,000 hospitalizations and nearly $2.4 billion in direct healthcare costs each year in the United States [21, 22]. The prevalence of diverticulitis is age-dependent. It is estimated to be approximately 5% in those under the age of 40, with this prevalence increasing to 65% in those above 65 years of age [23].

Patients with contained perforation (Hinchey Stages I and II) are managed with intravenously administered antibiotics with or without percutaneous drainage of the abscess, depending on abscess size. Those with free perforation resulting in either purulent or fecal peritonitis (Hinchey Stages III and IV) require surgery [24].

The standard of care for patients with acute diverticulitis who fail conservative treatment has also changed over time. Since the 1980s, a two-stage procedure, sigmoid resection plus colostomy followed by colostomy takedown, has become widely accepted as the surgical standard of care for acute diverticulitis. Over the past 2 decades, however, a significant amount of research and opinion has advocated for a shift in the surgical approach to patients with acute diverticulitis. The use of a Hartmann's procedure for patients with the mild or moderate disease has been questioned, with the idea that many (if not most) patients are better served by a single-stage procedure (resection, primary anastomosis). In a systematic review, the morbidity and mortality of primary anastomosis were found to be similar to that of a Hartmann's procedure for patients with even the most severe acute disease [25]. In another study based on ACS-NSQIP database, 1314 patients showed that partial colectomy with end colostomy and closure of distal segment (Hartman's procedure), colectomy with primary anastomosis (PA), and colectomy with PA with proximal diversion (PAPD) had similar outcomes. Resection and PA can be performed safely in acute diverticulitis with no difference in postoperative morbidity or 30-day mortality when compared with the Hartmann's procedure [26]. One of the most controversial aspects of diverticulitis in recent years is the role of laparoscopic lavage (LL), particularly in Hinchey III diverticulitis. There are mixed results; in the SCANDIV trial, patients with the suspected perforated diverticular disease were randomized to undergo either LL or colonic resection. The LL does not reduce serious complication rates, and patients had higher reoperation rates [27]. Another trial from Belgium showed similar results. The primary outcome was a composite endpoint of major morbidity and mortality within 12 months. Recruitment terminated early after an interim analysis of results demonstrated poorer outcomes in the LL group [28].

A contrast-enhanced CT scan is typically the examination of choice for patients with suspected diverticulitis [29]. Clinical examination and CT scan in conjunction should be used to decide the subsequent treatment. A large abscess found on initial CT scan may prompt early percutaneous catheter drainage (PCD), and the drainage serves as the bridge to surgery because most surgeons regard PCD as a temporary procedure and not a definitive treatment [30]. Failure to respond within 48–72 hours is an indication to proceed to surgery. Interestingly, the severity of diverticulitis at the time of the first CT scan predicts not only an increased risk of failure of medical therapy on index admission but also a high risk of secondary complications after initial nonoperative management [31].

When we consider the type of surgical approach in the elderly, a recent Cochrane analysis from 2017 comparing laparoscopic versus open resection for sigmoid diverticulitis showed that there is no evidence to support or refute the safety and effectiveness of laparoscopic surgery versus open surgical resection for treatment of patients with acute diverticular disease [32].

Appendicitis

Only 5–10% of all cases of appendicitis are diagnosed in patients older than 65 years, but they have mortality rates 5–8 times higher than younger patients [33, 34]. The classic triad of appendicitis, anorexia, fever, and right lower quadrant pain, is seen only in 20% of the elderly at presentation [35]. They present later in the course of their illness, with 85% presenting after 24 hours of pain. This delay in diagnosis leads to higher rates of perforation, i.e., 72% in the elderly as opposed to 20–30% in younger populations [36, 37]. Early surgical consultation should be obtained even in the absence of clear radiographic signs of acute appendicitis.

Several randomized controlled trials (RCTs) and meta-analyses have suggested that nonoperative management (NOM) for acute appendicitis is a viable alternative to the long-standing practice of immediately proceeding with an appendectomy on the diagnosis. Limitations of the current studies assessing NOM for acute appendicitis include (1) selection bias, (2) influence of comorbidity on decision-making for operation, and (3) exclusion of potential high-risk patients [38, 39]. The use of NOM in the elderly is not established to date.

The other important issue about surgical treatment is to use interval appendectomy (IA) in the elderly as an alternative. However, recent evidence has shed doubt on the necessity of this procedure. The IA should not be performed on a routine basis. However, neoplasia must be actively ruled out, particularly in the older age group [40]. Appendectomy in elderly patients has a low rate of complications similar to younger patients and should be offered early [41]. Laparoscopic appendectomy can be safely performed in the elderly with acute appendicitis or with a complicated one [42].

Complex Hernia Repair in the Elderly

Although a separate chapter is dedicated to this topic (see Chap. 16), we believe that the patient with a large complex incisional hernia should be repaired, unless they are at prohibitive risk for perioperative complication.

Intraoperative Surgical Decision-Making

The question is: How do we surgeons make intraoperative decisions under what can be inauspicious conditions? Some describe these decisions as "intuition," "gut-level," or "gray hair effect." On whatever factor the decision is made, we surgeons have difficulty in describing exactly how we came to the specific decision and why we did what we did during surgery. Many factors affect our decision-making before and during operations. This is probably more important during the surgery itself. These factors are the physiologic state of the surgeon, the harmony of teamwork, external factors at work such as scheduling, and the surgeon's ability to adapt quickly to a changing environment, to name only a few. So, perhaps while we may understand how we made the decision, the question remains: How to perform an evaluation of the surgical decision and gain a better understanding of a seemingly gut-level process, which helps surgeons combat the external factors experienced before and during surgery?

When a patient is dying from bleeding that we cannot control when irreversible metabolic shock does not respond to anything that we do, when new problems emerge unexpectedly, and when things go alarmingly wrong in such dire moments during a carefully planned operation, how do we decide what to do next? Many surgeons make such decisions based on "a gut feeling" or "intuition" or the "gray hair effect," among other techniques. In this chapter, we review theoretical as well as objective data that we as surgeons use to make intraoperative decisions. Most of the many theories and hypotheses in the literature have been created by individuals who are not surgeons. But, our collective firsthand experience as surgeons points to a combination of factors contributing to our intraoperative decision-making process, including education, clinical expertise, mentoring, and the creativity and excellence that come with long practice and with surgical strict discipline.

Concomitant Medical Problems Should Not Stop Surgical Care

A 78-year-old male, who is an active farmer, presented with stable metastatic melanoma to the ribs and lungs after undergoing immunotherapy for a few years and subsequently developed myelofibrosis and splenomegaly requiring almost monthly blood transfusions. The risk of getting injured at the farm due to frequent drop in his hemoglobin was concerning. The patient told me in the office, "I do not feel good when my hemoglobin goes under 6.5," and one could almost see the outline of enlarged spleen on his left abdomen.

After careful planning, including blood transfusion and selective splenic artery embolization, I performed an open splenectomy (Fig. 1.6). I started with a long midline incision and extended to the left subcostal incision. A month later the patient reported, "I baled 20 bales of hay." Much of planning went into the DMP. The decision came down to: If we do not take his spleen, he will bleed to death (should he hurt himself) and continue to suffer. Neither one was a good option.

Lesson: Otherwise healthy-looking people with the surgical problem and medical issues should be operated. His frailty score was 0. The only issue was malignancy which was not part of the modified frailty index (mFi).

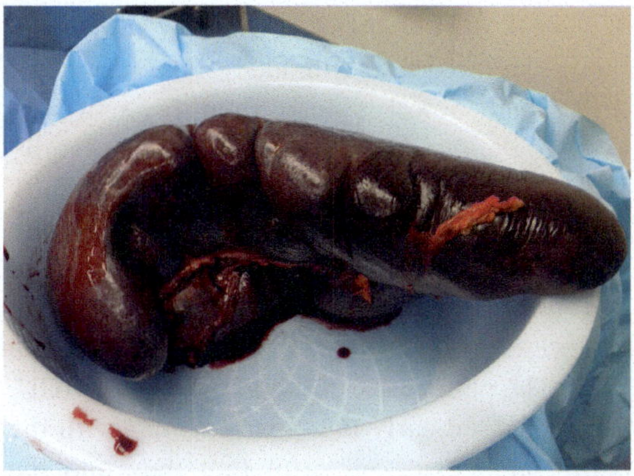

Fig. 1.6 Massive splenomegaly in an elderly gentleman due to secondary myelofibrosis after chemotherapy for metastatic melanoma

Final Thoughts: The Art of an Exploratory Laparotomy?

In the days before the CT scan and other modern imaging technologies, physical examination was truly a clinical art, and an exploratory laparotomy was commonly practiced to make the definitive diagnosis. Recently, in a very difficult situation, a colleague of mine asked me to see a patient that was caring in the ICU for the last few months and who now has taken a wrong turn, septic but without a clear source, requiring vasopressors to maintain blood pressure, intubated, pale-looking, and critically ill. The *succus* was coming out of the incisional wound VAC. The team has asked many questions: Is there a dead gut? Is there just a small hole in the bowel, the so called enetero-atmospheric fistula, or this is a reflection of some real catastrophe? Should the surgical team go back to the operating room and make the diagnosis and decide which direction should go? The patient in is his 70s, on continuous veno-venous hemodialysis (CVVH). One can make an argument to do nothing. I saw the patient briefly, I discussed with the family, and I suggested to them and my colleagues: you have two options – do nothing and terminate the surgical care, or go back and make the diagnosis, which will give you a clear direction of care. If this was my decision to make, I would go back to the OR, I told them. If there is something small like the dead gallbladder, take it out, but if the small bowel is dead, then this will help you and the family to stop the care. The family agreed to do just that. The team and the family did all they could. Intraoperatively, the entire small and large bowel was leathery looking, with no life into them. Two spontaneous fistulas were being drained by the incisional VAC. The patient was brought back to the ICU, and he passed early the next morning with the family at bedside, grateful that they and the team of surgeons and nurses were so helpful in the decision-making process and had done everything possible.

Conclusions

Surgical decision-making has implications for surgeons, patients, and their family members. The SDM is not merely a "gut feeling"; it is the amalgamation of the vast experience, learned evidence-based concepts, and their practical application over the years. The elderly do present with atypical signs and symptoms of pathology, and high index of suspicion with early decision-making is the key. The geriatric population will be the major shareholders in the healthcare industry; the composite synthesis of SDM, current evidence-based medicine, and joint decision-making among multidisciplinary teams, patients, and their families will enhance the outcomes.

References

1. Ortman JM, Velkoff VA, Hogan H. An aging nation: the older population in the United States. Hyattsville: US Census Bureau; 2014. Report Accessed 4/30/2019.
2. Harris A, Sharma A. Estimating the future health and aged care expenditure in Australia with changes in morbidity [Internet]. PloS One. Public Library of Science; 2018 [cited 2019 Jun 25]. Available from: https://www.ncbi.nlm.nih.gov/pmc/articles/PMC6084938/.
3. Mohanty S, Rosenthal RA, Russell MM, Neuman MD, Ko CY, Esnaola NF. Optimal perioperative management of the geriatric patient: a best practices guideline from the American College of Surgeons NSQIP and the American Geriatrics Society. J Am Coll Surg. 2016;222(5):930–47.
4. Bonne S, Schuerer DJ. Trauma in the older adult: epidemiology and evolving geriatric trauma principles. Clin Geriatr Med. 2013;29(1):137–50.
5. Latifi R. Surgical decision making. 1st ed. Cham: Springer; 2016.
6. Paula S. The elderly are getting complex surgeries. often it doesn't end well. New York Times [newspaper on the Internet]. 2019 Jun 07 [cited 2019 Jun 25]; Health: [about 3 p.]. Available from: https://www.nytimes.com/2019/06/07/health/elderly-surgery-complications.html.
7. Schuster KM, Holena DN, Salim A, Savage S, Crandall M. American Association for the Surgery of Trauma emergency general surgery guideline summaries 2018: acute appendicitis, acute cholecystitis, acute diverticulitis, acute pancreatitis, and small bowel obstruction. Trauma Surg Acute Care Open. 2019;4(1):e000281. Published 2019 Mar 27. https://doi.org/10.1136/tsaco-2018-000281.
8. Soria Aledo V, Galindo Iniguez L, Flores Funes D, Carrasco Prats M, Aguayo Albasini JL. Is cholecystectomy the treatment of choice for acute acalculous cholecystitis? A systematic review of the literature. Rev Esp Enferm Dig. 2017;109(10):708–18.
9. Fleming CA, Ismail M, Kavanagh RG, Heneghan HM, Prichard RS, Geoghegan J, et al. Clinical and survival outcomes using percutaneous cholecystostomy tube alone or subsequent interval cholecystectomy to treat acute cholecystitis. J Gastrointest Surg. 2020;24:627–32.
10. Simorov A, Ranade A, Parcells J, Shaligram A, Shostrom V, Boilesen E, et al. Emergent cholecystostomy is superior to open cholecystectomy in extremely ill patients with acalculous cholecystitis: a large multicenter outcome study. Am J Surg. 2013;206(6):935–40; discussion 940–1.
11. Alvino DML, Fong ZV, McCarthy CJ, Velmahos G, Lillemoe KD, Mueller PR, et al. Long-term outcomes following percutaneous cholecystostomy tube placement for treatment of acute calculous cholecystitis. J Gastrointest Surg. 2017;21(5):761–9.
12. Dimou FM, Adhikari D, Mehta HB, Riall TS. Outcomes in older patients with grade III cholecystitis and cholecystostomy tube placement: a propensity score analysis. J Am Coll Surg. 2017;224(4):502–511.e1.
13. Bodvall B, Overgaard B. Cystic duct remnant after cholecystectomy: incidence studied by cholegraphy in 500 cases, and significance in 103 reoperations. Ann Surg. 1966;163:382e390.
14. Strasberg SM, Pucci MJ, Brunt LM, Deziel DJ. Subtotal cholecystectomy-"fenestrating" vs "reconstituting" subtypes and the prevention of bile duct injury: definition of the optimal procedure in difficult operative conditions. J Am Coll Surg. 2016;222:89–96.
15. Scotte M, Mauvais F, Bubenheim M, Cosse C, Suaud L, Savoye-Collet C, et al. Use of water-soluble contrast medium (gastrografin) does not decrease the need for operative intervention nor the duration of hospital

stay in uncomplicated acute adhesive small bowel obstruction? A multicenter, randomized, clinical trial (Adhesive Small Bowel Obstruction Study) and systematic review. Surgery. 2017;161(5):1315–25.
16. Abbas SM, Bissett IP, Parry BR. A meta-analysis of oral water-soluble contrast agent in the management of adhesive small bowel obstruction. Br J Surg. 2007 Apr;94(4):404–11.
17. Ozturk E, van Iersel M, Stommel MM, Schoon Y, Ten Broek RR, van Goor H. Small bowel obstruction in the elderly: a plea for comprehensive acute geriatric care. World J Emerg Surg. 2018;13:48-018-0208-z. eCollection 2018.
18. Spangler R, Van Pham T, Khoujah D, Martinez JP. Abdominal emergencies in the geriatric patient. Int J Emerg Med. 2014;7:43-014-0043-2. eCollection 2014.
19. Bugliosi TF, Meloy TD, Vukov LF. Acute abdominal pain in the elderly. Ann Emerg Med. 1990;19(12):1383–6.
20. Arenal JJ, Bengoechea-Beeby M. Mortality associated with emergency abdominal surgery in the elderly. Can J Surg. 2003;46(2):111–6.
21. Kozak LJ, DeFrances CJ, Hall MJ. National hospital discharge survey: 2004 annual summary with a detailed diagnosis and procedure data. Vital Health Stat. 2006;13(162):1–209.
22. Sandler RS, Everhart JE, Donowitz M, et al. The burden of selected digestive diseases in the United States. Gastroenterology. 2002;122(5):1500–11.
23. Painter NS, Burkitt DP. Diverticular disease of the colon: a deficiency disease of Western civilization. BMJ. 1971;ii:450–4.
24. Ballian N, Rajamanickam V, Harms BA, et al. Predictors of mortality after emergent surgery for acute colonic diverticulitis: analysis of National Surgical Quality Improvement Project data. J Trauma Acute Care Surg. 2013;74(2):611–6.
25. Salem L, Flum DR. Primary anastomosis or Hartmann's procedure for patients with diverticular peritonitis? A systematic review. Dis Colon Rectum. 2004;47:1953–64.
26. Tadlock M, Karamanos E, Skiada D, et al. Emergency surgery for acute diverticulitis: which operation? A National Surgical Quality Improvement Program study. J Trauma Acute Care Surg. 2013;74:1385–91. https://doi.org/10.1097/TA.0b013e3182924a82.
27. Schultz JK, Yaqub S, Wallon C, et al. Laparoscopic lavage vs primary resection for acute perforated diverticulitis: the SCANDIV randomized clinical trial. JAMA. 2015;314:1364–75.
28. Swank HA, Vermeulen J, Lange JF, et al. The ladies trial: laparoscopic peritoneal lavage or resection for purulent peritonitis and Hartmann's procedure or resection with primary anastomosis for purulent or fecal peritonitis in perforated diverticulitis (NTR2037). BMC Surg. 2010;10:29.
29. Ambrosetti P, Jenny A, Becker C, Terrier TF, Morel P. Acute left colonic diverticulitis—compared performance of computed tomography and water-soluble contrast enema: a prospective evaluation of 420 patients. Dis Colon Rectum. 2000;43:1363–7.
30. Liu CK, Hsu HH, Cheng SM. Colonic diverticulitis in the elderly. Int J Gerontol. 2009;3:9–12.
31. Chautems RC, Ambrosetti P, Ludwig A, et al. Long-term follow-up after the first acute episode of sigmoid diverticulitis: is surgery mandatory? A prospective study of 118 patients. Dis Colon Rectum. 2002;45:962–6.
32. Abraha I, Binda GA, Montedori A, et al. Laparoscopic versus open resection for sigmoid diverticulitis. Cochrane Database Syst Rev. 2017;11:CD009277.
33. Gupta H, Dupuy D. Abdominal emergencies: has anything changed? Surg Clin North Am. 1997;77:1245–64.
34. Young YR, Chiu TF, Chen JC, et al. Acute appendicitis in the octogenarians and beyond: a comparison with younger geriatric patients. Am J Med Sci. 2007;334(4):255–9.
35. Horattas MC, Guyton DP, Wu D. A reappraisal of appendicitis in the elderly. Am J Surg. 1990;160(3):291–3.
36. Elangovan S. Clinical and laboratory findings in acute appendicitis in the elderly. J Am Board Fam Parct. 1996;9(2):75–8.
37. Omari AH, Khammash MR, Qasaimeh GR, et al. Acute appendicitis in the elderly: risk factors for perforation. World J Emerg Surg. 2014;9(1):6.
38. Vons C, et al. Amoxicillin plus clavulanic acid versus appendicectomy for treatment of acute uncomplicated appendicitis: an open-label, non-inferiority, randomized controlled trial. Lancet. 2011;377:1573–9.
39. Salminen P, et al. Antibiotic therapy vs appendectomy for treatment of uncomplicated acute appendicitis: the APPAC randomized clinical trial. JAMA. 2015;313:2340–8.
40. Al-Kurd A, Mizrahi I, Siam B, et al. Outcomes of interval appendectomy in comparison with appendectomy for acute appendicitis. J Surg Res. 2018;225:90–4.
41. Renteria O, Shahid Z, Huerta S. Outcomes of appendectomy in elderly veteran patients. Surgery. 2018;164:460–5. https://doi.org/10.1016/j.surg.2018.04.027.
42. Yeh CC, Wu SC, Liao CC, et al. Laparoscopic appendectomy for acute appendicitis is more favorable for patients with comorbidities, the elderly, and those with complicated appendicitis: a nationwide population-based study. Surg Endosc. 2011;25:2932–42. https://doi.org/10.1007/s00464-011-1645-x.

2 Is There Room for Common Sense in Surgical Decision-Making?

Ira Bedzow and Noam Stadlan

Introduction

The topic of this chapter arose from a conversation that one of the authors (IB) had with the editor of this book. In the conversation, the editor expressed the difficulty for physicians to be, at times, critical of some of the choices that they make regarding which medical intervention they either suggested or acquiesced to implement. After discussing the influences of hospital policies, healthcare law, insurance, clinical specialization, and patient autonomy on medical and surgical decision-making, the editor of this book asked, or rather exclaimed: "Is there room for common sense?" While this author (IB) appreciated the sentiment and understood the source of frustration while discussing the issue, he later had trouble communicating the difficulty of the issue and its importance in written form. Of course, what influences decision-making and how shared decision-making should occur between patients and physicians have produced a large body of literature. Yet, descriptions of the internal components, processes of thinking, and external pressures that affect what options are possible and plausible do not seem to convey accurately the meaning of the exclamation. The editor's point of "common sense" conveys that there must be something intangible or non-reductionistic that separates decisions that follow certain protocols and those that are correct for the situation at hand, even when they seem to diverge from the literature regarding the processes of thinking and the expected norms proposed for general cases.

We understand the question, "Is there room for common sense in surgical decision-making?" to mean something very specific. In this instance, "common sense" does not mean something akin to "folk wisdom" or an unreflective knowledge that has no basis in specialized training or deliberation. Such a definition would hardly fit the needs of shared decision-making or providing care in a healthcare setting, which necessitates specialization, training, and an ability to recognize and account for changing circumstances and the needs of various stakeholders over time. Nor does this connotation fit with the question presented in the title of this chapter. Rather, for the sake of this chapter, we take the term "common sense" to mean something akin to "prudence" or "practical reasoning." Prudence, or practical rea-

I. Bedzow (✉)
Biomedical Ethics and Humanities Program, New York Medical College, Valhalla, NY, USA

Aspen Center for Social Values, Aspen, Colorado (CO), USA
e-mail: ira_bedzow@nymc.edu

N. Stadlan
Department of Neurosurgery, NorthShore University Healthcare System, Evanston, IL, USA

Division of Neurosurgery, Pritzker School of Medicine at the University of Chicago, Evanston, IL, USA

R. Latifi (ed.), *Surgical Decision Making in Geriatrics*, https://doi.org/10.1007/978-3-030-47963-3_2

son, is the intellectual skill of knowing what one wants to achieve in a given situation as well as knowing the means through which one can achieve it. It is typically gained through experience, which implies it is gained over time and practice. Yet the individual skills that give rise to prudence are not simply mechanical; they serve as a foundation to allow the prudent individual to consider more than what may narrowly be in primary focus and to act on more than the skills that their training hones.

The relationship between "common sense" and "prudence" is evident in common parlance, even if the philosophical understandings of the two terms are distinct and even contradictory at times. The reason for choosing to define "common sense" as prudence is because it is a faculty that elucidates the editor's question in two very important ways. First, prudence and its influence on decision-making are very difficult to describe operationally; it, therefore, is very difficult to explain how to utilize it when considering the various factors and influences on decision-making. Second, prudence is very difficult to inculcate explicitly; it, therefore, seems as if there is no explicit room for it in medical training. In this chapter, we will give a description of prudence and explain how it can direct decision-making effectively through a case example. We will then conclude with a brief vision for how it can be incorporated as part of medical training.

Prudence

Prudence is traditionally conceived of as the mother of all the virtues, which gives it a moral or spiritual connotation, since virtue ethics and moral theology have historically been the only fields to speak of virtue. However, prudence has a very practical meaning at its root, and, as stated above, it should be defined as the ability to discern what one wants to achieve in a given situation as well as knowing the means through which one can achieve it. In this sense, prudence as a virtue aligns with the view of virtue as a capability without a specific moral or religious connotation. Nevertheless, when prudence is applied to clinical or surgical decision-making, the goals of professional care necessarily entail professional or ethical values. As such, prudential decision-making would not simply be technical know-how; it would encompass the professional and personal values of the various stakeholders who are affected by the decision. Moreover, those professional and personal values of the various stakeholders are what provides the foundation for prudence to arise over and above one's technical skills, since they form part of the motivation to achieve one's goals when the simple rote of acting competently does not fulfill what one wants to accomplish.

Prudence emerges from the utilization of the various components of decision-making in such a way that the total can be greater than the sum of its parts. Those components can be categorized broadly as recognizing reasons and responding to reasons, though these two broader categories include subcategories in themselves.

Reasons can be explanatory or normative. In other words, reasons can describe a situation, or they can impose expectations on what should be done in a situation given the goals and values of which the normative reasons are a consequence. Explanatory reasons describe a given state of affairs. As one of the authors (IB) defines elsewhere, explanatory reasons are facts that describe relationships between other facts in the world. Their existence as facts does not depend on whether they are perceived and understood by a person, yet the person must perceive and understand them in order to employ them as reasons. Normative reasons, on the other hand, consist of facts that describe relationships between other facts in the world in such a way that the person who recognizes that fact may be motivated to respond in a particular way because of them. Normative reasons create duties for the person to act even if he or she does not recognize them as such, yet they only become motivation for acting when the person recognizes them as applying to him or her. Again, their existence as facts does not depend on whether they are perceived and understood by a person, yet the person must perceive and understand them in order to be motivated by them as reasons for acting [1].

For example, suppose that *the reason* why a person has pain shooting down his leg is due to *the fact* that he has a compressed nerve in his spine. This *fact* is an *explanatory reason*. It describes a state of affairs that gives rise to the person's situation. Yet it does not obligate or motivate anyone to do anything. Now, assume *the fact* that the person wants to relieve his back pain. This may motivate the person to act so as to relieve his pain, but it does not obligate or motivate anyone to assist him in doing so.

Now consider a member of the medical profession who is dedicated to curing and alleviating suffering of patients; the fact of his membership and the dedication that the profession demands is a *normative fact* (consisting of professional values) – i.e., that the person is dedicated to curing and alleviating suffering – that may impose an *obligation* on him or her to consider the additional facts that the person has back pain/a compressed nerve and that he or she has the ability to alleviate the person's pain. Anyone who does not have the ability to relieve that pain would not have an obligation to do so, since all three facts are part of the normative reason to treat. (While "ought" implies "can," "cannot" implies "ought not" or the lack of an obligation). Yet, even if a medical professional does have the ability, the *normative fact* may nevertheless still not impose an *obligation* if the person does not become the medical professional's patient. This is a legal fact that serves as an additional component to create a normative reason in this case.

Recognizing the *fact* that the person is in pain and the *fact* that one is a member of the medical profession, as well as the other facts of the situation that entail an obligation to treat, will *motivate* the person to engage in a patient-physician relationship and seek to treat the person, if he or she is able to do so. In order words, the *explanatory and normative facts* of the case which obligate medical professionals in general become the source of *motivation* for the particular physician who recognizes them as *reasons* to act for the sake of the patient. Yet how the physician acts to best alleviate that pain depends on more than simply recognizing that he or she is motivated to do so; it also depends on many specific details of the case, including the different treatment alternatives that could be applied, the probabilities of success for each of the different alternatives, the particularities of the unique patient that may influence which alternative is best, as well as the resources at the disposal of the physician. Moreover, all of these factors are not static, and all of them can influence how the physician determines the best way to respond to the reason to treat. The prudent physician must continually re-evaluate his or her recognition of the state of affairs (explanatory facts) of the case so that he or she can respond in the most optimal way to the normative reason that motivates him or her to treat the patient.

Recognizing reasons for action includes the following components: attention, perception, and orientation. By attention, we mean the active self-urging to sustain focus. The importance of attention relates to the fact that clinicians in general, and surgeons in particular, must continually assess the situation in which they are acting. Surgery entails many complex and interconnected systems, both physiologically and within the dynamics of the clinical team. If one does not attend to the situation at hand, one may miss many of the cues needed for prudential decision-making. As a person becomes more skilled at maintaining attention, he or she can use less cognitive energy to attend to the same area of focus, thereby becoming more able to attend to a larger scope. This broadens the person's receptivity to recognize more explanatory facts so that they can be incorporated into better responses. For example, an inexperienced surgeon may focus entirely on a particular compressed nerve and miss other physiological aberrations which a more experienced surgeon might be able to perceive because of his or her greater and broader attention span. The more experienced surgeon's attention would also allow him or her to be better prepared for unexpected changes, either in the patient or in the broader situational environment.

By perception, we do not mean simply recognition of what objectively occurs. This is because any given situation may allow for multiple descriptions which bear competing or even conflicting claims. Moreover, emphasizing different details will highlight different considerations for how to relate to a particular scenario. Perception is the ability to understand nuanced differences between one situation and another and between the needs of one person and another. It is a skill of recognizing the particular while maintaining a grasp of commonalities. For example, consider two patients with a compressed nerve. It may be possible that a physician perceives that one patient is suitable for surgery, while the other is more suitable for physical therapy. It may also be the case that one physician recommends surgery, while another recommends physical therapy for the same patient. When perception is conceived in this way, it is heavily tied to orientation. Orientation is the lens through which perception is interpreted. It provides the epistemological frame that helps a person see different claims and prioritize which perspectival interpretation is most appropriate for the goals of care. It is tied not only to the facts that one perceives but also with what one may expect to see and with one's experience responding to similar situations in the past. In this way, prudence provides a person with the ability to understand how framing and other biases may influence perception by making salient certain factors while dismissing others. This is because prudence incorporates more than simply scientific knowledge or technical skill; it is tied to a social or professional orientation that provides a framework through which one can prioritize conflicting values so that one's decision aligns with one's overall goal.

After a prudential person has evaluated the situation and recognized the various reasons for acting, responding to reasons in action includes assessing the various ways to respond and choosing the alternative that best fits the goals of the situation and one's own ability to implement the chosen alternative. In assessing the various ways to respond, the prudential person does more than simply apply moral maxims or decision-making rules to the situation at hand. Prudential decision-making includes and carries over the experience of previous decisions that the person has made and continuously compares the current case with previous cases through use of analogical reasoning. Analogical reasoning is different than deductive reasoning. In deductive reasoning, the governing rule is given first, and the person derives particularities from the rule. When it comes to clinical decision-making, those rules consist of "best practices," hospital policies, or professional codes. In analogical reasoning, the reasoning itself helps a person to identify the principle for action and its relevance to the situation, since the person compares the situation at hand to similar experiences, accounting for both the similarities encountered and the differences that must be addressed. Through analogical reasoning, a person will not simply apply "best practices" because that is what is recommended. He or she will recognize that best practice recommendations apply in general to most cases but may not apply to every case, given the circumstances of the case. There may be other factors that override the applicability of a standard practice. For example, consider the guideline that one does not transfuse a patient who appears to be hemodynamically stable if the patient's hemoglobin level is more than 7. If that patient has compression of their spinal cord or other areas of tissue at risk if oxygen levels fall, a prudent physician may transfuse an apparently stable patient even if the patient has a hemoglobin level greater than 7. It is for this reason that experience is such an integral part of acquiring the virtue of prudence. Experience not only refines one's intellectual skills, but it also provides a bank of information so that decisions have a deeper source of knowledge that can inform them.

While previous examples have been more technical in nature, in medicine especially, where every situation necessitates communicating and responding to explanatory and normative facts that are social and value-laden, prudence demands honing interpersonal skills as much as deliberative and physical skills. Therefore, choosing the best alternative includes knowledge of what one is able to accomplish, given dispositional and situational considerations. Dispositional considerations include a person's tendency to act in certain ways, such as whether one is introverted or extroverted or whether one communicates better face-to-face or in written form, as well as his or her emotional and motivational affect. These latter dispositional traits influence more than just what one can accomplish; they also influence how one recognizes reasons as well. For example,

Walter Mischel and Yuichi Shoda have argued that features of current situations as well as how those situations compare to previous experiences activate cognitive and affective reactions, which influence how one categorizes and reacts to current cases. Moreover, previous experience need not be actual. How a person has thought, planned, fantasized, and imagined a situation, as well as the emotional states that they invoke, also influence recognition and response to reasons in a current case. "Thus," they write, "what constitutes a situation in part depends on the perceiver's constructs and subjective maps, that is, on the acquired meaning of situational features for that person, rather than being defined exclusively by the observing scientist" [2]. Thus, it is not only actual experience which can be utilized to develop prudence. Even vicarious experience gained through reviewing other professionals' cases, as well as imagining how one may act in a hypothetical case, can serve to develop prudence. Because prudence incorporates lessons from experience as well as hones skills of perception and deliberation, a prudent person may respond to a situation by seeing commonalities between the current case and previous cases, whereas a novice or an outside observer may not.

In a given situation, the prudent person will thus choose a particular response and choose to perform it in a particular way, accounting for what the situation demands and the person's abilities, social role, and disposition. Prudence can thus be described as the combination of a ready disposition to respond to situations in a certain way and an intellectual ability to discern the best way to respond *for that person*. Factors that influence how a prudent person responds to a situation include not only what one would want to do generally but also the various stakeholders and what would be most persuasive to them in a given situation, what consequences the person foresees for himself or herself as well as for the other stakeholders, the systemic or organizational limitations or support the person might encounter when implementing his or her decision, and whether interpersonal relationships which the choice affects seem to support or disallow the desired choice.

The emergent quality of prudence appears in understanding that acting efficaciously entails more than sensory perception, intellectual cognition, and affective states, it also demands that these components of reasoning work in tandem and in relation to each other. A person cannot simply deliberate on a given situation and each potentially appropriate response as if he or she were an outside observer, since what he or she affirms as reasons to act are based on his or her relationship to the situation, i.e., his or her perception and receptivity to see alternative interpretations and responses, his or her goals, responsibilities, and capabilities. Moreover, in many interpersonal situations, there will be reasons to act in contradictory ways. Prudence allows a person to prioritize the values that different reasons engender to apprehend which reason has the greatest normative force.

Case Example

The following case example comes from the experience of one of the authors of this chapter (NS). The vast majority of spinal surgery consists of removing structures such as herniated discs which are compressing neural structures such as nerves or the spinal cord. The surgeon dissects and removes tissue until he or she identifies the neural structure that is being compressed. Since the removal of tissue is done with cutting instruments, one of the most challenging parts of the surgery is accurately identifying where the neural structures are and where they are not. Obviously much greater care needs to be taken near the neural structures. The method to understand where the neural structures are located is to use specific parts of the vertebrae (bone) as landmarks. The neural structures are almost always found in a specific relation to the landmark, but unusual anatomic variations occur.

During the dissection, on the way to identifying landmarks, a relative novice may pay more attention to every detail, not having the experience to know, or, perhaps, more accurately, to know with confidence, where the safe areas are and where unanticipated damage may be inflicted on a nerve. On the other hand, the experienced surgeon, confident of where he or she is, and having done the procedure numerous times, may need to expend less cognitive energy to attend to the details that he or she considers routine and also to dissections in areas where he or she is confident no danger lies. The input of vision and touch have to be mapped to the surgeon's mind's view and understanding of where the instruments are located in relation to the crucial parts of the patient's anatomy. The perception is then filtered through the surgeon's orientation. The experienced surgeon may be better oriented to understand the visual and tactile data, and the result is (hopefully) a more accurate map in the surgeon's mind of where the instruments are and where the important structures lie.

The spinal nerves are almost always found in a specific relationship to a part of the vertebra called the pedicle. The medial part of the pedicle marks the lateral borders of the thecal sac. In the lumbar spine, this sac contains the spinal nerves that are exiting from the spinal cord. Nerves exit from the sac at every level of the spine, similar to branches from a tree. In the lower (lumbar) spine, the nerve gracefully passes the pedicle and turns laterally to exit the spinal canal directly under (inferior to) the pedicle. Therefore, the thecal sac can only be present medial to the pedicles, and the nerves exiting are usually only directly under the pedicle. This means that the space directly above the pedicle, where herniated discs are most likely to occur, is free of nerves and therefore a safe place to dissect. The boundaries of the safe area are usually bordered by an imaginary line extending superiorly from the medial border of the pedicle and extending upward until the nerve above (exiting under the pedicle above) is encountered. The nerve root above is the superior border of the safe zone and is also the lateral border of the safe zone, because the nerve root angles inferiorly as it travels laterally. Therefore, identifying the pedicle provides the surgeon with confidence of where the nerves are and are not and where a safe dissection can be done and cannot be done.

However, anatomic variants are known to occur, the most frequent one known as a conjoined nerve root. In this case, the more proximal (higher up) nerve root extends much lower than usual, sometimes over the disc space, and sometimes it can be mistaken for a disc that is out of place. Mistakenly incising the nerve, thinking that it is the disc, can result in severe nerve damage and significant unnecessary pain, weakness, and/or numbness. Dealing with the aberrant nerve requires not only attention to the anatomy but perceiving it in the context of the known information regarding conjoined nerve roots and then dealing with it in an acceptable fashion. A knowledgeable surgeon will have a significant advantage over one who either lacks the knowledge or experience to properly identify landmarks and account for unanticipated variations.

While the most frequent goal of surgery is to remove the pressure from the nerve by removing the piece of disc that is compressing the nerve, sometimes that goal may also incorporate an additional decision to fuse vertebrae together. A fusion is done if the joint is causing significant pain or if there is concern that the vertebrae will move too much in relationship to each other, perhaps causing compression of nerves or becoming painful in the future. Fusion sometimes is the primary way to achieve the goal of the surgery, and it is also sometimes contemplated as something to be done in order to keep the patient from needing an additional surgery. In other words, sometimes there is a choice of doing a larger surgery initially or a smaller surgery initially. The expectation is that the pain and risk of an initial larger surgery will have the benefit of decreasing or eliminating the need for surgery in the future. With the smaller surgery, the benefit of less pain and less risk is offset by a larger risk of requiring another surgery in the future. Frequently there is no surgical consensus, and the patient's hierarchy of values has the most influence in which surgery is chosen.

As an example, elderly patients frequently develop pain from lumbar stenosis. This is where the gradual growth of arthritic joints and disc bulging result in compression of the adjacent nerve roots. Patients can develop severe pain with walking, frequently somewhat alleviated with leaning over shopping carts or walkers. While surgery is the most successful treatment, sometimes non-surgical treatment can be beneficial. Many patients, in addition to having compression of the nerve roots (lumbar stenosis), also have slippage of the vertebrae (spondylolisthesis). The slippage is not dangerous in and of itself. However, surgery to enlarge the spinal canal and relieve the stenosis can destabilize an already somewhat unstable segment. To complicate matters further, patients frequently have not only the classic buttock and leg pain of stenosis but also back pain. That back pain sometimes can be caused by the stenosis but can also be caused by slippage of the vertebrae and the resultant stress on the joints. Frequently there is uncertainty as to whether one, the other, or both are the source of the back pain.

There are two major surgical options for such a situation: decompression (removing the offending joints, ligament, and disc and relieving the pressure on the nerves) only and decompression with a fusion. Decompression has the advantage of being a smaller procedure and can reliably relieve the buttock and leg pain. It may relieve the back pain as well, but it may not. It also may lead to the need for another surgery in the future due to further slippage of the vertebrae. The other option is to decompress the stenosis and also place devices and/or bone graft to fuse the vertebrae together. This has the benefit of not only relieving the buttock and leg pain but having the best chance of relieving the back pain as well. In addition, it is the procedure that carries the smallest chance of requiring another surgery.

When surgery is contemplated for these patients, there can be a number of goals. Usually the primary goal is relief of the stenosis symptoms – the debilitating buttock and leg pain. However, if the back pain is severe, the patient may want to make relief of the back pain another primary goal. The patient also could want to have the smallest surgery possible or may prioritize the likelihood of never having to undergo another surgery. All of these are reasonable goals that serve as potential reasons to provide different alternative responses, each of varying degrees of priority. The prudent surgeon not only has as much data on outcomes as possible to discuss with the patient but works with the patient to decide on the goals of surgery. Only then can the best surgery for the patient be planned.

One of the more common methods of accomplishing a fusion is to remove most of the disc and place a plastic support in the space along with bone graft and chemicals designed to "trick" the body into growing a bony bridge from one vertebra to the other across the disc space. This is termed posterior lumbar interbody fusion (PLIF). In order to implant the device, the surgeon needs to retract (gently move) the nerves out of the way in order to have space to place the device. Retracting nerve roots in the lumbar spine is a very common maneuver. However, too much retraction can result in damage to the nerve. Not enough retraction can mean having to place a smaller and perhaps inadequate device.

Returning to our example above, a conjoined nerve root sometimes cannot be easily retracted, and therefore there can be technical difficulty in finding enough space to place the PLIF device. In this situation the surgeon needs to assess the risk of nerve damage with the amount of retraction needed to place the device, his or her technical skill in performing the required dissections/mobilization, whether achieving the fusion is part of the integral goal of surgery, and what other options he or she may have to fuse or stabilize the spine (there would be options of making the surgery larger, placing screws or other types of devices, etc.). Other factors that may have an effect include previous experience (both successful and unsuccessful), training, whether aborting this part of the procedure would affect the surgeon's self-esteem or image to the or staff or peers, concerns that the procedure may not be as described in the informed consent, or that the surgery does not comply with the patient's stated values. Other issues that may come into play would be time and convenience. A long dissection or using alternate methods of fusion could be time consuming and affect the rest of the schedule for the day.

The prudent surgeon, having identified the problem and potential solutions, attempts to find the solution that provides maximal benefit to the patient in the context of what the patient wants to accomplish while at the same time minimizing risk. In some situations, that might mean an extended dissection to allow safe retraction and placement of the device that the surgeon and patient agreed was the best option. In other situations, it may mean aborting that part of the procedure and explaining to the patient the rationale for the change in plan. Some surgeons by their nature are more likely to accept more risk on behalf of the patient in order to accomplish the goal, while others by nature are more risk-averse. Prudent surgeons should recognize their tendencies and do their best to make sure that they are doing their best for the patient, not just for themselves.

How Can Prudence Be Taught

Once one recognizes that prudence is a capability that consists of various skills of recognizing and responding to reasons, then training to hone those skills becomes easier to implement. Of course, experience will always be essential to develop practical skills, whether they be intellectual or physical skills, yet certain exercises can focus on the various components of prudence independently and holistically to make the acquisition of prudence more effective so that it need not take as much experience to acquire it. The analogy would be to how coaches give guided training to an athlete to build his or her technique and strength rather than simply having the athlete improve through playing without any guidance or practice drills. In a similar way, guiding younger surgeons to perceive complications differently and verbalize the ramifications of the different descriptions would expand their ability to frame circumstances – and thus decisions – differently when they are acting on their own. Similarly, thinking through many and various ways to approach a complication and examining the benefits and detriments of each alternative – and not just describing what occurred and what should occur – would expand surgeon's capabilities of evaluating different responses to reasons. While personal experience will always be the ideal practice to hone one's skills, relying on the vicarious experience through engaging others in analysis of case studies and observation can allow people to see options that they may not have seen before and to assess different components of how decisions are made and in which ways one can best respond to reasons.

References

1. Bedzow I. Maimonides for moderns: a statement of contemporary Jewish philosophy. Cham: Springer; 2016.
2. Mischel W, Shoda Y. A cognitive-affective system theory of personality: reconceptualizing situations, dispositions, dynamics, and invariance in personality structure. Psychol Rev. 1995;102(2):246–68.

3 Emergency Medicine Management of the Elderly

Ivan T. Miller, Susan Boyle, Harry Kopolovich, Anar Shah, and Dan E. Wiener

Introduction: A Brief History of Emergency Medicine

Emergency medicine is a clinical field that consists of the knowledge required for the unscheduled diagnosis and management of acute illness including injury. It is positioned at the intersection of prehospital and inhospital care and represents a critical entry point for the management and disposition of patients in our healthcare system. Emergency medicine, as a field evolved in the 1960s during a time in which a rising number of patients sought immediate unscheduled care for emergency conditions. Emergency departments of this era were staffed by community physicians of varying expertise, and, eventually, the need for specialized training became evident. In 1979, the American Board of Medical Specialties recognized emergency medicine as a distinct medical specialty in the United States. Today, emergency medicine is a thriving academic specialty that welcomes over 2000 new residents each year. Fellowships and subspecialty training in emergency medicine span critical care, toxicology, pediatrics, sports medicine, disaster, prehospital medicine, and ultrasound, among others. The emergency department is a key partner for surgical specialties, as many patients requiring surgical interventions will receive assessments and stabilization in the ED, including some who will be referred from the ED to the outpatient setting for definitive surgical management.

According to the United States Census Bureau, the population age 65 and over increased

I. T. Miller (✉)
Department of Emergency Medicine, Westchester Medical Center, Valhalla, NY, USA

Department of Emergency Medicine, New York Medical College, Valhalla, NY, USA
e-mail: Ivan.Miller@wmchealth.org

S. Boyle
Department of Emergency Medicine, Morristown Medical Center, Morristown, NJ, USA

H. Kopolovich
Department of Emergency Medicine, Richmond University Medical Center, Staten Island, NY, USA

Department of Emergency Medicine, Albert Einstein College of Medicine, Bronx, NY, USA

A. Shah
Geriatric Emergency Department, Morristown Medical Center, Morristown, NJ, USA

Department of Emergency Medicine, Sidney Kimmel Medical College, Thomas Jefferson University, Philadelphia, PA, USA

D. E. Wiener
Department of Emergency Medicine, Morristown Medical Center, Morristown, NJ, USA

Department of Emergency Medicine, Sidney Kimmel Medical College, Thomas Jefferson University, Philadelphia, PA, USA

R. Latifi (ed.), *Surgical Decision Making in Geriatrics*,
https://doi.org/10.1007/978-3-030-47963-3_3

from 37.2 million in 2006 to 49.2 million in 2016. The Administration on Aging, a division of the US Department of Health and Human Services, projects this group to almost double to 98 million by 2060. In 2015, 7.1 million people age 65 and over had at least one overnight hospitalization during the year. In 2016, older Americans spent over 13% of their total expenditures on health [1]. The number of patients over the age 65 presenting to emergency departments (EDs) has been steadily increasing. The 2009 National Hospital Ambulatory Care Survey showed 52,200 ED visits per year for every 100,000 patients over age 65 [2]. For this age group, the ED visit rate per 100,000 was consistently higher than that for other age groups for the 10 years between 2006 to 2015, and rising overall, reaching 56,803 in 2015 [3] (Fig. 3.1). In 2016, there were 145,591,000 ED visits in the United States, with those age 65 years and over representing 23,108,000 or 15.9%.

According to the Census Bureau's projections, the elderly populations will more than double between 2017 and the year 2050, to 80 million. By that year, as many as 1 in 5 Americans could be elderly. Most of this growth will have occurred between 2010 and 2030 when the "baby boom" generation enters their elderly years. During that period, the number of elderly will grow by an average of 2.8 percent annually.

Geriatric Emergency Medicine: An Overview

As the geriatric population has continued to increase in size, hospitals around the world have designed models of care that make the ED friendlier for geriatric patients. Numerous studies and scholarly articles focusing on the emergency care of the elderly have been published. Multiple papers have been published exploring many aspects of geriatric emergency medicine from predictors of emergency department use through triage, disposition, and beyond.

Emergency medicine triage is the process of rapidly sorting patients by clinical priority. It is typically performed by a nurse following an algorithm that includes vital signs, chief complaint, brief history of present illness, past medical history, and medications. The emergency severity index (ESI) has become standard in US emer-

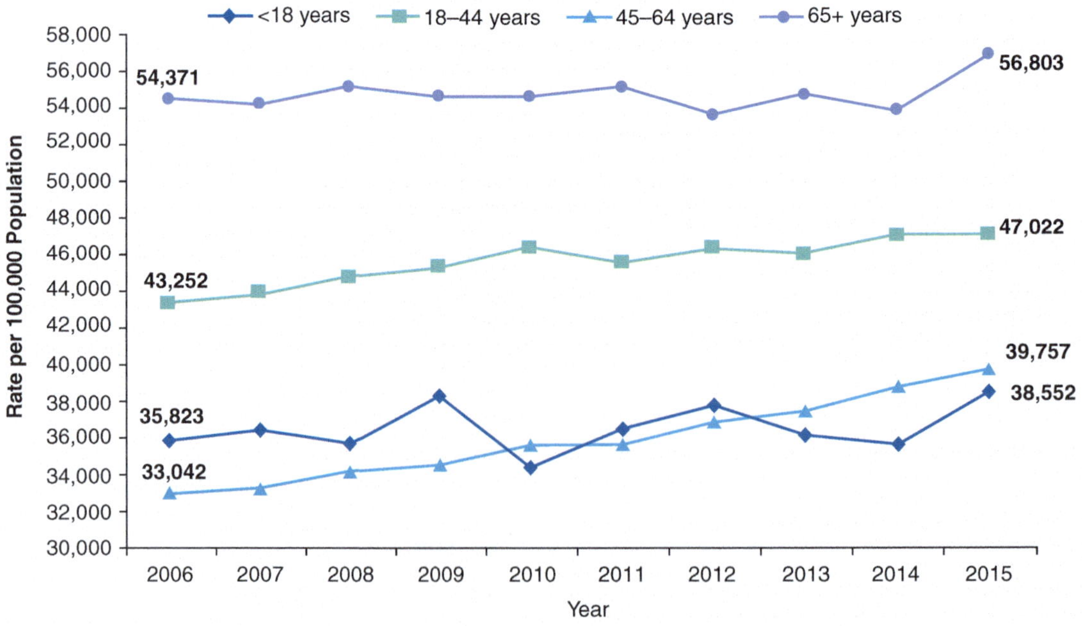

Fig. 3.1 Rate of ED visits, per 100,000 population by age group, 2006–2015. (From Sun et al. [3])

gency department triage since the American College of Emergency Medicine and the Emergency Nurse Association endorsed it in 2003 and 2010 [4]. Studies of its use in the geriatric population are mixed. On the one hand, the ESI score did predict likelihood of hospitalization, length of stay and 1 year outcomes [5]. On the other, ESI score identified fewer than half of elderly patients receiving an immediate lifesaving intervention [6]. This finding is not surprising considering the difficulty some elderly patients have in describing their chief complaint. Given the increased prevalence of acute surgical and medical emergencies in the elderly population, triage protocols that place this population in the waiting room may be high risk. Various efforts to reduce undertriage have been studied. Across the spectrum of outcome studies, ESI level 3 has been recognized as the most heterogeneous. In many crowded emergency departments, ESI level 3 patients may wait hours prior to being seen by a provider. One approach that has been implemented as a QI project at a geriatric emergency department is to forbid ESI level 3 in elder patients. This approach forces a triage nurse to choose level 2 which requires immediate placement in a treatment room or level 4 which is reserved for patients who do not need a workup. Without the option of ESI level 3, the choice is either immediate bed placement or minimal workup. This change would likely reduce undertriage at the expense of overtriage. Outcome studies are needed to assess this age-based alteration to the standard ESI-based triage protocols.

Frailty

Frailty is a phenomenon estimated to occur in approximately 25% of those over the age of 65 and in over 50% of those over the age of 85, but it is not an inevitable consequence of normal aging [7]. There is no one single definition of frailty, but most geriatricians describe it as a syndrome that combines a cumulative loss of physiologic reserve and function with an increased vulnerability to stressors [7, 8]. The consequence of this decline is the characteristic presentation of frailty: weakness, fatigue, reduced mobility and loss of muscle mass, poor balance and gait instability, susceptibility to urinary tract infections, anorexia, weight loss, and some loss of cognitive function, with increased susceptibility to extrinsic stressors. Frailty is now recognized as a risk factor for poor outcomes in healthcare and especially postsurgical interventions. A meta-analysis showed that frailty is associated with higher postoperative inhospital and 1-year mortality, longer length of hospital stay, and increased need for residential post-acute care [8]. While there is widespread agreement on the negative effect that frailty has on outcomes, there remains a lack of consensus on an exact causal mechanism or method of measurement.

Pathophysiology of Frailty

While normal aging is thought to arise from an accumulation of molecular and cellular damage, despite the efforts of a complex repair and maintenance network, frail patients have an increased presence of inflammatory markers and epigenetic changes. A range of inflammatory cytokines have been independently associated with frailty including interleukin-6 (IL-6), C-reactive protein (CRP), tumor necrosis factor-α (TNFα), and CXC chemokine ligand-10 (CXCL-10), a potent pro-inflammatory mediator [7]. Advanced glycation end products (AGEs) may also cause widespread cellular damage through the upregulation of inflammation. This leads to an abnormal, low-grade inflammatory response that is hyperresponsive to stimuli and that persists for prolonged periods following removal of the initial inflammatory stimulus.

Frailty Models

There are two principal overlapping and statistically convergent models of frailty: the frailty phenotype by Fried and the cumulative deficit model, which was proposed by Rockwood [7]. Fried's model is based on five criteria: weakened grip strength, self-reported exhaustion, slow gait speed, unintentional weight loss, and low physical activity, where people with three out of the five factors are classed as frail and those with two of the factors are classed as pre-frail. The cumulative deficit model proposes that it is a cumulative accumulation of deficits (symptoms, signs, and laboratory tests) that eventually tips the balance. It uses a frailty index based on the presence or absence of 92 parameters, the higher the number of deficits, the greater the frailty.

Measurements of Frailty

A variety of tools have been used to detect and measure frailty including measurement of gait speed, the timed up and go test (TUG), hand grip strength, pulmonary function tests, and many questionnaires, but the gold standard is the comprehensive geriatric assessment, usually conducted by a multidisciplinary team.

Frailty is a risk factor for death and disability following a range of surgical treatments including hip fracture, aortic valve replacement and likely plays a pivotal role in the older patient's potential for recovery [9, 10, 11]. While national guidelines recommend a preoperative evaluation of frailty and other risk factors, the lack of consensus surrounding frailty assessment tools is a major reason why frailty is often not measured in clinical practice.

In a meta-analysis of a wide range of surgical interventions, frailty was associated with a higher inhospital mortality rate, a higher 1-year mortality rate, a longer hospital stay, and a higher rate of discharge to post-acute residential nursing care [8]. Similar data exists in emergency department measures of frailty. In Dresden et al. the clinical frailty score (CFS) correlated with admission or transfer to a skilled nursing facility (SNF) [12]. The implications for emergency medicine are that frailty must be considered along with other risk factors and comorbidities. Another attempt to measure premorbid risk is the Identification of Seniors at Risk (ISAR) score, which is a six question self-reported screening tool [13]. ISAR has been evaluated in multiple studies in emergency medicine and in other surgical contexts. The results are mixed. Tavares et al. found that ISAR could predict return to the ED within 30 and 180 days but not admission during the repeat visit [14]. Suffoletto at al found that ISAR was not able to predict seniors who would die or return to the emergency department within 30 days of discharge [15]. A meta-analysis of 10 studies using ISAR (8860 patients) concluded that ISAR is not suitable "alone for identifying seniors at risk for adverse outcomes in the ED" [16].

Delirium and Dementia

One dimension of frailty that has been studied in more detail is dementia, especially in its early phases [17, 18]. Dementia may prevent a patient from providing an accurate history which is crucial to guiding the emergency department workup. The absence of a reliable history along with the increased prevalence of multiple morbid conditions leads to increasing use of radiological imaging in the emergency department with advancing patient age [19]. Especially without collateral information about baseline status, delirium, and dementia may be difficult to distinguish in the emergency department setting [20, 21]. Efforts to screen for both conditions have been combined in tools such as the 4AT and 6-CIT. O'Sullivan et al. found that 6-CIT and 4AT accurately exclude delirium and dementia in older ED attendees. 6-CIT does not require collateral history but has lower positive predictive value (PPV) for delirium [22]. Much of the literature on screening for dementia and delirium has been produced by clinicians who provide geriatric assessment upon clinical consultation to the ED. In a meta-analysis, Carpenter et al. found variable diagnostic accuracy of screening tests for dementia. In their assessment, the AMT-4 appeared to be the most accurate positive predictor of dementia, and the Brief Alzheimer's Screen is the best negative predictor of dementia [23].

Delirium or acute encephalopathy is a neuropsychiatric disorder with an acute onset and a fluctuating course, characterized by disturbances in consciousness, memory, cognition, perception, and behavior [24]. It occurs in hyperactive, hypo-active, or mixed forms, is more common in those with preexisting dementia, and is associated with increased mortality, hospital length of stay, functional disability, rates of admission to long-term care institutions, and healthcare costs [24, 25]. Frailty and dementia are risk factors for delirium and acute delirium in the emergency department and are significant predictors for bad outcomes [11]. Delirium results from the interaction of multiple predisposing factors together with precipitating stressors. The most common causes of delirium seen in the emergency department are infection, drugs/medications, CNS disorders, metabolic disturbances, pain, and hypoxia [7]. Guidelines encourage the need to screen all older patients for the presence of delirium, although there is no consensus on the best tool. Many, such as the b-CAM and CAM-ICU, are a variation of the Confusion Assessment Method (CAM) [26]. The CAM diagnostic algorithm is based on the four cardinal features of delirium: (1) acute onset and fluctuating course, (2) inattention, (3) disorganized thinking, and (4) altered level of consciousness. A brief two-part screening tool – the Delirium Triage Screen (DTS) can be used as an initial step to rule out delirium, and those who have a positive DTS are then screened using the bCAM, which has higher specificity. A diagnosis of delirium according to the CAM requires the presence of features 1, 2, and either 3 or 4. Management of delirium focuses on prevention, symptom management, and the need to consider and treat all possible reversible causes.

Similar to the task of evaluating abdominal pain without an accurate history is the challenge presented by nonspecific complaints in general. Every emergency physician learns to approach complaints such as generalized weakness ("weak and dizzy") in elderly patients with caution. Dementia or even delirium may cause an elderly patient with a suggestive history to describe the reason for their visit to the ED in nonspecific terms. Nemec et al. found that in a cohort of patient's median age of 82 who presented to an ED with nonspecific complaints, a serious condition was diagnosed within 30 days in 59%, and the 30-day mortality was 6% [27]. Interestingly, a secondary analysis of this cohort by Liu et al. found that the subgroup with recent falls had similar outcomes to those without falls suggesting to the authors that falls in the elderly might best be considered "under the broader entity of nonspecific complaints [28]."

In the authors' collective experience, tools such as AMT-4, 4AT, 6-CIT, CFS, and ISAR are not commonly deployed in emergency departments at this time. Nevertheless, the emergency physician's perception of frailty, dementia, and other risk factors including delirium and nonspecific complaints found in this body of work should influence the extent of diagnostic workup and choice of disposition. Further study is needed before a standardized scale can be recommended for routine use in the emergency medicine setting. In other words, the older the patient, the more likely they will receive tests in the emergency department due to the pretest probability of significant findings and to the challenges in obtaining an accurate history. In this context, we now turn to the emergency department evaluation of abdominal pain in the elderly.

Abdominal Pain in the Elderly

Abdominal pain is the chief complaint of 3–4% of elderly patients presenting to the ED [29]. Of these patients, 51% will be admitted [30], 24–34% will require surgery [31, 32] and mortality rates are reported as 6–35% [33]. All these rates are significantly higher than those of younger patients [34]. The diagnostic approach to the elderly patient is challenging as their symptoms are less severe, nonspecific, and often atypical. A retrospective review of 131 patients >65 all with culture proven intra-abdominal infections found that geriatric patients presented more frequently without the usual symptoms of abdominal pathology including an absence of abdominal pain, nausea, vomiting, diarrhea, and fever [35]. Symptoms were present for a longer period before presentation [31–33, 35, 36]. Taking an accurate history can be challenging as it is estimated 10–20% of persons older than 65 have significant degrees of memory deficit, disorientation, or decline in intellectual performance [37].

Classic findings on physical examination may be absent as the multisystem changes associated with aging lead to altered physiologic responses. Atrophy of the abdominal wall musculature can diminish rebound and guarding and changes in peripheral nerve function lead to later and subtler presentation of pain [38]. Older patients may not be able to mount a fever and those with an intra-abdominal infection are four times more likely than a younger patient is to present with hypothermia [35]. In one study 30% of patients greater than 80 years old in the ED with an intra-abdominal abnormality requiring surgery developed neither fever nor leukocytosis [39]. In older patients with peritonitis, only 34% manifest guarding or rebound tenderness [40]. Medications commonly taken by elderly patients, including beta-blockers, steroids, NSAIDs, and opiates, make it less likely that they will manifest fever or tachycardia [38]. Many elderly patients present later in the course of their disease as they fear losing independence, hospitalization, and death or may lack health insurance, transportation, or a reliable caregiver for their spouse or pet [41].

A 2005 prospective, multicenter, observational study of 360 patients ≥60 presenting with atraumatic abdominal pain found that patients had the following diagnosis: nonspecific (14.8%), UTI (8.6%), bowel obstruction (8%), gastroenteritis (7.4%), gallbladder disease (5.5%), diverticulitis (6.2%), constipation (5.9%), cancer/mass (5.6%), PUD/GERD (4.5%), pancreatitis (3.9%), and urinary retention(3.6%) [42]. A retrospective review of ED records and operative reports of 117 patients ≥80 years old presenting with acute abdominal pain requiring operation found that cholecystitis, hernia, obstruction, peptic ulcer disease, ischemia, and perforation were the most common diagnosis [39]. A second retrospective cohort study of patients with intra-abdominal infections compared the frequency of diagnosis in those ≤65 with younger patients and found the following incidence of disease [35] (Table 3.1).

Table 3.1 Common causes of abdominal pain

Disease	Incidence in those ≤65	Incidence in those ≥65
Appendicitis	28	61
Diverticulitis	28	6
Cholangitis	12	8
Cholecystitis	12	2
Intra-abdominal abscess	9	14

Cooper et al. [35]

In addition to the changing incidence of common diagnosis, the risks for certain diseases such as ruptured AAA mesenteric ischemia, colon cancer, and atypical presentations of MI significantly increase in patient is older than 50 years. Given the challenges of making an accurate and timely diagnosis in an elderly patient presenting with abdominal pain, CT is often the diagnostic modality of choice. CT is performed in 37% to 59% of older patents presenting to the ED with abdominal pain and leads to a diagnosis in 67–75% of cases [19, 34, 42]. Given the challenges of atypical presentations and the significant morbidity and mortality of elderly patients presenting to the ED with abdominal pain, clinicians need to maintain a large differential diagnosis focus on potentially immediate life-threatening conditions.

Biliary Tract Disease

The incidence of gallstones increases with age and in postmortem studies, gallstones were found in approximately 50% of patients >60 years [43]. The increases in the prevalence of gallstones, lithogenicity of bile, percentage of pigmented stones, and CBD diameter all contribute the greater incidence and severity of biliary disease in elderly patients [44]. The incidence of complications ranges from 7 to 20% overall and rises to 40–60% in patients over age 70. These complications include acute cholecystitis, empyema, perforation, abscess, jaundice, cholangitis, and pancreatitis [45]. Gallstone pancreatitis increases with age accounts for 50% of acute pancreatitis adults >56 years old compared to 31% in those younger <21 years old [46]. A retrospective review of 88 male patients older than age 60 with biliary tract disease reported a mortality rate of 6.7–6.8% [45, 47].

As with other abdominal emergencies, the prevalence of classic signs and symptoms may be absent. In a retrospective review of patients over age 65, pain was in the epigastric area or RUQ in 84% of patients and in the back or flank in 36%, and 5% had no pain. The prevalence of other symptoms was noted to be nausea (57%), emesis (38%), jaundice (8%), and fever (44%) [48]. An elevated WBC was found in 58%–65% of patients, and 13% had no fever, and all tests were normal [47, 48]. Bilirubin and alkaline phosphatase were elevated in 59% of patients, and SGOT and SGPT were elevated in 62% [47]. Evaluation by ultrasound is the test of choice when there is a high index of suspicion of biliary tract disease. However, it may be less helpful than in younger patients given the increased prevalence of acalculous cholecystitis [37].

Appendicitis

Appendicitis in elderly is characterized by atypical and late presentations resulting in higher perforation rates and increases in morbidity and mortality compared to younger patients [49]. Atypical presentations and perforation are attributable to depressed T-cell function and increased vascular sclerosis [50]. The classic presentation of appendicitis (RLQ pain, fever >37.6C, WBC >10,000, nausea, and vomiting) is seen in 10–30% of elderly patients [49–52]. On physical examination findings include RLQ tenderness (91%), fever (71%), rebound tenderness (54%), pain and guarding (40%), Rovsing's sign (16%), abdominal distention (7%), and abdominal mass on palpation (5%) [53]. A study done in 2011 found the sensitivity of CT scanning with contrast in the elderly population to be 100% sensitive with a specificity of 99.1% [54]. While non-contrast CTs have a sensitivity of 92.7% and a specificity of 96.1% [55], ultrasound has a sensitivity of 76% and a specificity of 95% [56].

Elderly patients have increased risks of significant morbidity and mortality, ranging as high as 70% versus approximately 1% in the general population [52]. Perforation rates are 50–70% in the elderly compared to 20–30% in the general population [52, 57]. A study published in 2002 noted a decrease in perforation over the previous 20 years from 72% to 51% which was attributed to the increased use of CT [52]. The overall mortality rate for elderly with appendicitis has been reported to approach 15% [54].

Diverticulitis

Diverticula form as patients age with an incidence of 50% in those over 70 and 66% in those over 80 [37]. Approximately 80% of patients with diverticulosis are asymptomatic [37]. The most common complication of diverticulosis is diverticulitis, which occurs in 10–30% of patients [58, 59]. Uncomplicated diverticulitis is defined as a microperforation that results in peridiverticulitis and/or phlegmon. In complicated diverticulitis, this progresses to free perforation, fistula or abscess formation, sepsis, or obstruction [60].

Diverticulitis is characterized by LLQ pain and tenderness, fever, and leukocytosis. Peritoneal signs may be present if the diverticula have perforated. 30% of the geriatric patients with acute diverticulitis abdominal tenderness is absent on physical examination [37]. CT is the test of choice to confirm diverticulitis as it reliably detects the location and extent of the inflammation [61]. Following recovery from the first episode of simple, uncomplicated diverticulitis surgery is seldom indicated as only 20–30% of patients that will have recurrent episodes.

Rectal bleeding is associated with diverticulosis in up to 30% of patients. The bleeding is painless [58] and can be massive. Angiodysplasia of the right colon must be considered in the differential diagnosis [62]. In an actively bleeding patient, angiography is the test of choice, as it is both specific and sensitive, provided the rate of bleeding is sufficient (0.5–1.0 ml/min.) [61]. Angiography also provides the option to treat with either infusion of vasospastic substances or selective embolization [61]. In patients with slower bleeding rates, nuclear scanning techniques using Technetium-99m sulfur colloid and Technetium-tagged RBCs may be useful [63]. When the bleeding site cannot be identified with these modalities, colonoscopy has been successfully used for electrocoagulation of arteriovenous malformations [64]. Patients may require massive transfusion. This should be done cautiously as elderly patients with systolic or diastolic dysfunction can experience fluid overload.

Peptic Ulcer Disease

Several physiologic changes associated with aging increase the incidence of PUD in elderly. Gastric emptying becomes delayed leading to an increased prevalence of atrophic gastritis and PUD [65, 66]. The incidence of colonization with *Helicobacter pylori* increases, and approximately 50–70% of elderly ulcer patients are *H. pylori* positive. This pathogen has been associated with an increased incidence of PUD [67]. In one study of Medicare patients, 26% tested positive for *H. pylori*, while recent use of NSAIDs was documented in 82% of patients [67]. The odds ratio for bleeding ulcers in elderly patients is 4.9 for NSAIDs and 1.8 in patients with *H. pylori*. When both of these risk factors are present, it increases to 6.1 [68].

The incidence of gastric ulcers increases from 13% to 45% in elderly patients compared to younger patients with a concurrent decrease in the incidence of duodenal ulcers from 84% to 51% [52]. Symptoms of PUD are usually absent or vague and poorly localized. Approximately 30–35% of patients older than 60 with confirmed PUD have no abdominal pain compared to 6.9% of younger patients [69, 70]. These atypical presentations are associated with an increased incidence of complications which occur in 50% of patients older than 70. Complications include hemorrhage, perforation, penetration, and obstruction [71]. When hemorrhage occurs in the elderly, they have more severe bleeding and require more transfusions [72]. Endoscopy may be considered to evaluate for the indicators that predict recurrent bleeding and identify findings that may require endoscopic intervention [71]. Perforations in the elderly may be difficult to diagnose because of lack of symptoms, lack of physical signs, and lack of history of PUD [71]. The mortality from perforation has been reported as 9% for duodenal ulcers and 13% gastric ulcers [71].

Treatment should include lifestyle changes such as discontinuing NSAIDS, ETOH, tobacco, and caffeine. Medications including antacids, sucralfate, h-2 receptor antagonists, proton-pump inhibitors, misoprostol, and antibiotics for *H. pylori* have been shown to be effective. Proton-pump inhibitor infusion for acute bleeding does not lower mortality but does decrease the incidence of finding a lesion that requires intervention when endoscopy is performed [73]. If endoscopy does not control the bleeding, interventional radiology can attempt angiography with arterial embolization. Surgery presently is reserved for individuals with ulcers that are refractory to medication or are associated with complications [73].

Small Bowel Obstruction

Small bowel obstruction (SBO) can be caused by adhesions from previous surgeries (50–70%), incarcerated hernias (15–30%), and neoplasms (15%). Gallstone ileus is the cause of up to 25% of cases among older adults, in contrast to 2% in younger patients [74]. Patients with SBO present with diffuse abdominal pain, distension, vomiting, and constipation. In the elderly the symptoms of SBO are often subtler than in younger patients [75]. The treatment is conservative consisting of NG decompression, NPO, and IV fluids. Surgery should be considered if the patient does not improve in 24–48 hours [37].

Large Bowel Obstructions

Large bowel obstructions (LBO) occur more frequently in elderly patients than in the general population. The most common causes are malignancy and diverticulosis [37]. Colonic volvulus can also cause obstruction. Sigmoid volvulus is the most common type (75–80%) followed by cecal volvulus [76].

The symptoms of sigmoid volvulus include crampy abdominal pain, vomiting, obstipation, small liquid stools, and abdominal distension [74]. Plain films demonstrate a dilated sigmoid loop and air/fluid levels in 67% of cases [77] but if not present CT is diagnostic [78]. A sigmoid volvulus can often be decompressed with a rectal tube, sigmoidoscopy, or barium enema. Sigmoid or cecal volvulus has a high risk for perforation and should be decompressed urgently [37]. After nonoperative reduction, recurrence rates and mortality rates are as high as 90% and 35%, respectively. Therefore, elective resection is recommended to avoid recurrences [78].

Patients with cecal volvulus present with episodic cramping abdominal pain, nausea, vomiting, and obstipation. Recent surgery, adhesions, congenital bands, chronic, and mental disorders constipation are all risk factors [78, 79]. On physical examination, patients will have diffusely distended abdomens, and they can progress to exhibit rebound tenderness if peritonitis or bowel ischemia develops. Complications include bowel ischemia, necrosis, or perforation. There is a 25% incidence of gangrenous bowel at the time of diagnosis [72]. The classic finding on plain film is a "coffee bean sign" in the LUQ, but this is present only in 17% of patients [80]. On CT scan a "whirl sign" (twisting of the mesentery around the ileocolic vessels) is pathognomonic for cecal volvulus, CT may also demonstrate signs of bowel obstruction (a massively dilated cecum with associated small bowel dilation) or signs of colonic or small bowel ischemia (mural thickening or mesenteric edema) [81, 82]. The treatment of cecal volvulus requires surgical repair [78].

Colonic pseudo-obstruction or Ogilvie syndrome is a functional obstruction of the GI tract, which occurs more commonly in the elderly debilitated patient, particularly those who are institutionalized or have a prolonged hospital course [80]. Patients present with gradually worsening abdominal distension and abdominal pain (83%), nausea (63%), and emesis (57%) [83]. Passage of stool is absent in 50% of patients [84]. Physical exam findings will include massive abdominal distension [82]. Treatment involves discontinuing narcotics, sedatives and anticholinergics, and making the patient NPO while providing IV hydration and electrolyte replacement. A rectal tube can help decompress an air-filled sigmoid colon [82]. About 75% of patients with Ogilvie syndrome will resolve with conservative therapy within 8 days [84]. If symptoms do not resolve after a few weeks, mechanical obstruction should be considered.

Vascular Problems in Elderly

Both acute mesenteric ischemia (AMI) and abdominal aortic aneurysm (AAA) should be considered in elderly patients presenting with abdominal pain. These vascular emergencies require early identification and treatment. AAA is found in 5% to 10% of men aged 65 to 79 years [85]. Risk factors for AAA include tobacco use, HTN, PVD, and a family history of AAA [86, 87]. The classic triad of ruptured AAA (hypotension, flank pain, pulsatile mass) is present in only 25–50% of patients [40]. Patients may also present with deceased blood flow to the lower extremities manifesting as numbness and coldness [88]. In those patients presenting to the hospital with rupture, the mortality is 80% [85].

AMI most often involves embolism or thrombosis of the superior mesenteric artery. Patients older than 60 with a history of recent myocardial infarction, CHF, arterial emboli, or atrial fibrillation are at increased risk. Classically, patients present with severe, poorly localized pain that is out of proportion to physical findings [89]. One third will have nausea, vomiting, or diarrhea. On physical examination there is abdominal distension, and patients may develop peritoneal signs and shock [40]. Laboratory abnormalities consistent with ischemic bowel include leukocytosis and an elevated lactate, which result in an anion gap metabolic acidosis. Plain films are diagnostic in only 30% of patients (focal edema or pneumatosis intestinalis) [90]. CT angiography should be done as soon as possible in patients with suspected AMI. CTA has a sensitivity of 93% and a specificity of 100% [91]. Survival is dependent on early recognition with mortality rates of 50% when the diagnosis is made within 24 hours increasing to

<70% when there is a delay [92]. The mainstays of therapy are fluid resuscitation, anticoagulation, superior mesenteric artery infusion of vasodilators, and emergency laparotomy [89].

Emergency Management of Elderly Trauma Patient

Customarily, trauma models have mirrored the centers for Medicare services definition of elderly as those aged 65 and above. This definition was validated in the largest trial evaluating the relationship between age and trauma mortality, the MTOS study, which demonstrated that patients older than 65 years of age had elevated mortality across matched Injury Severity Scales (ISS), mechanisms of trauma, and body region injured [93]. However, restricting patients to intensity of initial evaluation based solely on chronological age has severe limitations for the potential improvements in morbidity and mortality of patients. In fact, physiologic changes that impact trauma-associated morbidity and mortality begin to occur at age 40 [94]. Currently, the American College of Surgeons recommends that emergency medical services transport patients older than 55 years of age to a designated trauma center regardless of apparent injury severity [95]. Furthermore, these same trauma activation criteria recommend universal activation of the trauma team of all patients older than 75 who present regardless of the mechanism of injury or physiologic status [96].

The elderly are not just old adults. In fact, age may be an overly simplistic measure to understand outcomes in geriatric patients. A more robust methodology may be the degree of frailty [97]. Frailty is composed of both the loss of physiologic reserve and increased incidence and severity of the comorbid disease. Biologically, this translates to musculoskeletal, neuroendocrine, and immunologic deficits that synergistically contribute to a state of muscular weakness and overall physical impairment [98]. Frailty is an independent predictor of postoperative mortality, complications, and hospital duration [99]. Frail patients not only have short-term worse outcomes but long-term ones as well. For example, frail elderly patients have decreased functional ability 1 year after a fall and increased mortality that persists 3 years after the event [100]. Chronological age is, therefore, one of many other components that must be assessed to determine the correlation between pre-injury state and long-term outcome. As described above, multiple factors with particular attention to frailty interplay to determine complications, consumption of hospital resources, and readmission rates.

Falls

Although the causes of injury in elderly persons are many, falls are the most common mechanism in this population subset. More deaths occur from falls in patients aged 85 or older than deaths from motor vehicle crashes in the 18- to 19-year age group [101]. Coexistent medical conditions that collectively contribute to the pre-injury frailty of the individual result in a myriad of medical complications. These inevitably result in poor outcome and ultimate demise.

Ground level falls tend to prevail in this age group, whereas falls from height are less prevalent. Lack of adequate protective mechanisms exacerbate the injury patterns seen in low-energy falls. Similar height falls tend to cause considerably less injury in younger cohorts likely due to the ability to brace for the impact on an outstretched hand. Older patients with impaired reflexes or slowed cognition are less likely to have such a protective mechanism, often resulting in more severe and combined injuries. Elderly persons are predisposed to fall for a myriad of reasons (Table 3.2).

Table 3.2 Factors predisposing to falls in the elderly person

Impaired sensation and proprioception	Degenerative joint disease
Visual disturbances	Dementia
Unsteady gait and balance	Predisposition to syncope and near-syncope
Neuromuscular disorder	Stroke
Weakness of musculature	Polypharmacy

When evaluating the elderly patient who has fallen, the emergency physician must evaluate not just the injuries that result from the fall but also acute illness that may have caused the injury. The latter may have a far greater ramification on the patient's post-convalescent period than the injury itself. Over 80% of patients evaluated after accidental fall are found to be on medications implicated in contributing to their fall [102]. The presence of four or more medications is directly correlated with increased risk for falls [102].

Initial Evaluation

Expert panels as well as the American College of Surgeons Committee on Trauma (ACS COT) field triage algorithm have recommended transport of the elderly to a trauma center even in the absence of high-energy injuries [103, 104]. Once at the trauma center, an elderly patient with a fall from standing may not meet the criteria for full trauma team activation. However, patients in this age group may have more serious injuries than younger patients with higher-energy mechanisms. Without trauma activation, a patient such as this may have to wait while patients with higher triage acuity are evaluated. Some centers have created an automatic activation for elderly patients. This approach is resource intensive as many of these patients do not have significant injury. One of our trauma centers has established a "level 3" activation which is limited to emergency department staff. A level 3 patient is immediately placed in a treatment area and fully exposed by a smaller team including emergency department physician, nurse, and patient care tech. Decisions about c spine immobilization, analgesia, and imaging are made rapidly, and the full trauma team can be consulted when indicated. Level 3 activation can be helpful for all patients whose mechanism of injury is below the typical threshold for level 1 or 2 activation with an emphasis on older patients at elevated risk for injury for the reasons described above.

Several unique considerations are relevant during the trauma primary and secondary survey of the elderly trauma patient. Physicians should be aware that because of unique physiologic and anatomic changes that occur in this group as well as the complicating role of medications and prosthetic devices, many of the traditional hallmarks of clinical instability may be initially masked. Attention to vital signs is particularly important in the elderly. Although abnormal vital signs certainly warrant additional investigation, normal prehospital or initial vital signs should not impede a thorough investigation. Normal blood pressure in an older patient with a history of chronic hypertension may be abnormal for that patient and an indication of serious injury. Vital sign trends are critical in the elderly trauma patient. The emergency department team should track and trend vital signs, reporting any change during the emergency department course.

The remainder of the physical examination should proceed with emphasis on the disability and exposure portion of the primary survey. Poor nutrition, loss of lean muscle mass, microvascular changes, and blunted hypothalamic function increase the elderly trauma patient's risk for hypothermia and sores [105]. All efforts should be made to quickly clear the cervical spine when applicable and remove the patient from the hard backboard if it has been applied. Several orthopedic and wound management studies have demonstrated that in the elderly, the pathologic process of pressure ulcers begins early in the hospital course [106]. Pressure sores result in increased hospital length of stay, as well as preventable patient morbidity and mortality. Action thus taken in the trauma bay can thus positively impact the patient in these domains.

Computed tomography scanning has been shown to help reduce mortality in the severely injured elderly patient [107]. Maintaining a low threshold for axial imaging is a mainstay of emergency medicine evaluation in this cohort. Even in the absence of high-injury mechanisms, the routine usage of CT scans can result in the earlier detection of injury that would otherwise prove difficult to diagnose. Earlier management can ultimately decrease ultimate resource utilization, more rapid mobilization, and decreased fasting period, both of which are important for the prevention of delirium and other morbidities.

Despite a relatively low-injury rate, the elderly patient has the highest trauma-related mortality of any age group [101]. Elderly patients that have been injured do not have the physiologic reserve of their younger counterparts. They are less tolerant in delays of diagnosis and definitive management of their injuries. Elderly trauma patients have reported inhospital complication rates of 33% compared with 19% for younger patients [108]. Cardiovascular events (23%) and pneumonia (22%) are the most common and most clinically significant delayed complications. Prevention begins in the emergency department through aggressive evaluation coupled with timely and thorough management.

Elderly trauma patients have increased mortality for given ISS, RTS, and GCS as compared to their younger cohorts. However, as with younger patients, shock and occult hypoperfusion reliably predict mortality in the elderly trauma patient [109]. Two retrospective studies demonstrate that an elderly blunt trauma patient with a systolic blood pressure <90 mmHg was associated with mortality rates of 82% to 100% [110]. Unfortunately, preexisting conditions can mask the diagnosis as well as complicate resuscitative efforts. Specifically, heart disease and chronic kidney insufficiency can complicate the resuscitative effort, resulting in volume overload and a worsening of the clinical picture. Further compounding the resuscitative effort is the challenge of assessing the response to resuscitation. A combination of radiologic, laboratory, and clinical parameters may reflect the effectiveness of the resuscitative effort. Initial blood gas and lactic acid should be routinely measured in elderly trauma patients. Both lactic acid and base deficit correlate with systemic hypoperfusion and shock. Initial values sampled upon emergency department presentation correlate with ICU length of stay (LOS), hospital LOS, ISS, and overall mortality [111].

Vital signs can remain remarkably normal in the elderly trauma patient, especially early in their evaluation and management. This is despite the occult shock that can result in precipitous decompensation. Careful attention to stability prior to allowing a patient to leave the resuscitation area is crucial. Explanations for why the vital signs do not respond in a predictable fashion as compared to a younger cohort are multiple. Elderly patients may have significant reductions in coronary blood flow even in the absence of atherosclerotic disease which may cause demand ischemia in the setting of early hemorrhagic shock. Couple this to the exaggerated response that is seen to hypothermia, acidosis, and hypoxia in the elderly. Furthermore, as the myocardium ages, it becomes less responsive to circulating catecholamines because of beta-receptor insensitivity [112].

Under periods of stress, the circulating catecholamines are less able to produce an increase in heart rate and therefore less able to increase cardiac output to effectively compensate for blood loss. Furthermore, the elderly trauma patient has a higher likelihood of being on medications such as beta-blockers and calcium channel blockers that limit their maximal achievable heart rate below what would be expected at a given age. Because of this, emphasis on tachycardia in this cohort may curtail identification of occult shock early in the course of evaluation.

Specific Sites of Injury

In a study of prehospital data of trauma patients over age 70 presenting to the emergency department, the majority of injuries were due to fall (60.7%), followed by motor vehicle accidents (21.5%) with a combination of pedestrian struck by automobiles and penetrating injuries making up the most of the remainder [113]. Low-level falls, such as those from standing corresponds to the largest proportion of injuries in the elderly. Complications resulting from falls are the leading cause of death from injury in men and women over the age of 65. Injuries sustained by the geriatric trauma patient tend to be more significant than their younger counterparts. Specific sites of injuries tend to include head and face, pelvis, and lower extremities [111].

Head injury is a serious contributor to traumatic death in the geriatric population. Elderly patients experience a higher mortality rate as well as poorer overall functional outcomes as a result of their injuries. Traumatic brain injury (TBI) and intracranial hemorrhage (IH) are common injuries in the elderly trauma patient. Age is an independent predictor of mortality and disability in patients who have moderate to severe head trauma [114]. Compared with younger patients with equivalent or lower injury severity, the elderly have longer hospital stays, increased ICU usage, lower rates of functional recovery, and significantly higher mortality [115]. Patients older than 65 years have mortality rates two-to-five times of those younger patients with matched intracranial injuries [116].

Multiple factors contribute to the increase in morbidity and mortality in the elderly head injury patient. Brain weight decreases as a result of aging by10% between the ages of 30–70. As a result of this cerebral atrophy, bridging veins traverse a greater distance and are at a higher likelihood to rupture due to shearing forces. Cerebral atrophy also increases the amount of intracranial free space resulting in subtle symptoms and potentially delaying diagnosis of intracranial hemorrhage. In addition, the prevalence of antiplatelet and anticoagulant agents in this age group is higher, which dramatically increases morbidity associated with the traumatically brain injury [117, 118].

Blunt thoracic trauma is responsible for 25% of all trauma deaths in the United States. Two-thirds of these patients have rib fractures, and up to 35% is affected by pulmonary complications [119]. The elderly patient is more susceptible to rib fractures from a minor injury due to loss of bone density. In addition, Bulger and colleagues study of 277 patients older than 65 years of age who had rib fractures, mortality increased 19% and risk for pneumonia increased 27% for each rib fractured [119]. The elderly patient who has rib fractures is more likely to present with hypotension and has a significantly higher risk for sternal fracture. Aggressive pain management, hemodynamic monitoring, and early resuscitation are of particular importance. Control of fracture-associated pain decreases splinting and atelectasis and may limit subsequent pulmonary sequelae.

Pelvic fractures in the elderly are a distinct clinical entity. In the younger population, a pelvic fracture requires a significant amount of kinetic energy. In the elderly, however, low-energy trauma such as those from falls from standing height is the most common mechanism of injury. This is followed closely by motor vehicle collisions.

Fracture patterns are similar between younger and older trauma patients. The pubic rami are most commonly fractured (56%) followed by the acetabulum (19%) and ischium (11%) [107]. More than 50% of elderly patients with a pelvic fracture have two or more pelvic fractures. The mechanisms and clinical sequelae differ significantly in the elderly. An elderly patient with a pelvic fracture is less likely to suffer more severe vascular and solid organ injuries and yet have higher rates of mortality than younger patients with pelvic fractures. In addition, the elderly have higher rates of hemorrhage, blood and blood product transfusion, need for interventional embolization, and admission to the ICU [120, 121].

Elderly patients undergoing radiography of the cervical spine after trauma are twice as likely as their younger counterparts to have fractures diagnosed in this region. In particular, elderly patients who fall from standing height are at significant risk to injure the region between the occiput and C2 as well as to the spine in general. Given the high incidence of injuries to the atlantoaxial complex in cases of isolated head trauma, a common strategy is to pursue computed tomography of the cervical spine in any situation where an elderly patient is going to be undergoing imaging of the head for trauma.

Due to factors associated with aging such as the development of osteophytes in the cervical spine and other advanced degenerative changes, the elderly are at higher risk for development of spinal cord injuries without concomitant fractures. Specifically, the distinct clinical entities of Brown-Sequard and central cord syndrome must be carefully evaluated clinically in any senior who sustains a fall. Mechanisms for this phenomenon have been attributed to narrowing of the spinal canal, making the spinal cord more susceptible to compression in instances that result in hyperextension such as that which occurs in a fall [122].

ED Analgesia

Inadequate treatment of pain termed oligoanalgesia has become a recognized phenomenon that occurs in acute care settings. Specifically, recent work done by Quattromani and colleagues [123] has demonstrated that the elder trauma patient tends to wait on average 30 minutes longer to analgesic administration, with the average time being 92 minutes from the time of arrival to analgesic to be administered. In this study, there was no difference in morphine equivalence by age group. However, the significant delay in time-to-analgesia was interpreted by the authors of this study as being representative of oligoanalgesia. Their findings were recapitulated in the study by Jones and colleagues, where age was determined to be an independent risk factor for inadequate analgesia in the setting of acute fractures [124].

When treating pain in older adults, it is important to incorporate patient preference and goals of care. Pain treatment in older adults requires balancing the relief of symptoms with the unwanted side effects of medications. Age-related changes in metabolism can cause varying levels of increased sensitivity to opioids in the elderly. This can result in altered mental status and respiratory depression [125]. Cautious management by starting slow and titrating to the desired level of analgesia with frequent reassessment and careful simultaneous attention to patient and vital signs is paramount in the elderly. Opioid medications impair balance and make subsequent falls even more likely.

Antiplatelet/Anticoagulant

The average elder traumatic patient arrives into the emergency department taking four or more prescribed medications. Aspirin and other platelet inhibitors are commonly used for primary and secondary prevention of atherosclerotic disease.

The last decade has seen a dramatic increase in the number of older individuals on dual antiplatelet agents or oral anticoagulants (Table 3.3). Therapeutic indications are multifactorial and are outside the scope of this chapter. However, elderly trauma patients on one or more of these agents require a special mention. Bleeding in these patients is common, and so the liberal use of imaging and in some cases reversal of iatrogenic coagulopathy is indicated.

Table 3.3 Anticoagulants and potential reversal agents

Anticoagulant agent	Mechanism of action	Potential reversal agent
Warfarin	Vitamin K antagonist	FFP, PCC, vitamin K
Dabigatran	Direct thrombin inhibitor	Idarucizumab
Rivaroxaban	Factor XA direct inhibitor	PCC
Apixaban	Factor XA direct inhibitor	PCC
Enoxaparin	Antithrombin III inhibitor	Protamine
Heparin	Antithrombin III inhibitor	Protamine
Aspirin	Antiplatelet agent	Platelets
Clopidogrel	Antiplatelet agent	Platelets

Current trauma guidelines call for the uniform assessment of all patients with particular attention to the international normalized ratio (INR), prothrombin time (PT), and partial thromboplastin time (PTT). Furthermore, guidelines recommend rapid reversal of all abnormalities when indicated. However, normalization of PT or INR, which reflects the effects of warfarin but not the newer oral anticoagulants, does not always correlate with decreased bleeding. However, what has emerged is that traumatic intracranial hemorrhage when present is more profound in those individuals on antiplatelet or anticoagulants [126, 127].

Disposition

Despite the high incidence of falls, very little is known about the long-term outcome of geriatric patients who have fallen, especially if their initial workup is negative. The observed readmission rate of 5% to 10% is higher than the younger trauma patient [128]. Hall and colleagues determined readmissions were strongly associated with 6-month mortality, and home living status at the time of the fall was a predictor of 30-day readmission [129]. This is in direct contrast to other studies that demonstrated that elderly trauma patients discharged to a rehabilitation facility are more likely to be readmitted than those who are discharged to home [130]. From this conflicting set of data, what is clear is that the elderly are a heterogeneous population with a variety of comorbidities and physical limitations. Factors that should be evaluated prior to discharging the elderly trauma patient including a thorough evaluation of predisposing factors to recurrent accidental injury including polypharmacy, home situation, and its associated barriers to mobility, assessment of pain, and social support network.

The Geriatric ED

The introduction of emergency departments specializing in the care of the elderly over the past decade has increased the focus on geriatric-centered safety measures, coordination of both inpatient and outpatient services, and comprehensive care to assist patients and their families, all concurrent with typical emergency care.

Geriatric patients can pose unique challenges to medical and surgical providers. Patients with advanced age are often sicker than the average younger patient. In 2016, 14.7% of patients aged 65 or older were triaged as ESI level 1 (immediate) or level 2 (emergent), highlighting the acuity of elderly patients who present to the ED [131]. Older patients often have health histories that are more complex with aging, with the potential for multiple comorbidities, long medication lists, and nonspecific/vague complaints that lack classic features of emergent pathology. Furthermore, older patients with cognitive impairment may be unable to provide a reliable and complete history. Care of the geriatric patient therefore requires a high index of clinical suspicion and careful coordination with caregivers, family members, friends, and outpatient providers. The Geriatric ED structure supports this by tasking case managers, social workers, patient advocates, and pharmacists to partner with physician and nursing teams to evaluate clinical cases comprehensively with a patient-centered approach. This allows for safer care to be delivered in the acute care setting and can serve to streamline the care of admitted patients, as well as those patients who are discharged to the community or rehabilitation/longtime care centers.

Accreditation of Geriatric Emergency Departments

Geriatric emergency departments were first established in the United States in 2008. In 2014, The American College of Emergency Physicians, The American Geriatrics Society, Emergency Nurses Association, and the Society for Academic Emergency Medicine created a consensus guideline, with recommendations for the accreditation of Geriatric EDs [132]. The Geriatric ED guidelines are similar to recommendations made by the British Geriatrics Society in *The Silver Book: Quality Care for Older People with Urgent or Emergency Care Needs* [133]. ACEP recognized the positive impact that accredited trauma centers has had on trauma care nationally, and in 2018 ACEP launched their own accreditation process for Geriatric EDs. The ACEP Geriatric Emergency Department Accreditation awards Levels I, II, and III certification to departments that meet certain benchmarks (https://www.acep.org/globalassets/sites/geda/documnets/geda-criteria-final_1.17.2019.pdf) [134]. Currently, over 50 departments hold formal accreditation across the United States, with many more to join in the near future.

The goal of geriatric accreditation is to improve adherence to the geriatric guidelines and catalyze adoption of best geriatric-focused practices. The accreditation process provides an opportunity for emergency departments to demonstrate their commitment to these standards and their investment in optimizing the care of the older patient. Accreditation is a means of assuring that GEDs meet a set of standardized, measurable criteria, provide screening for geriatric syndromes, and provide additional resources to support safe discharge and transitions of care.

The geriatric guidelines provide a template and guidance in regards to critical issues that affect this population such as delirium, frailty, cognitive impairment, and falls; however, each GED is challenged to build upon these by producing evidence-based policies, procedures, and protocols, or adapt existing policies to meet the needs of their specific geriatric population.

Critical to the model of the Geriatric ED is a multidisciplinary team inclusive of a Medical Director, Nurse Manager, Emergency Medicine physicians, Staff Nurses, Case Managers, Social Workers, Mid-Level Providers, Occupational/ Physical Therapists, and Pharmacists. Successful Geriatric EDs also ensure that these stakeholders receive geriatric-specific education and seamlessly integrate geriatrics into the ongoing quality improvement (QI) programs. Geriatric-focused QI may focus on deaths, return visits to the ED, readmissions, fall prevention, catheter use and preventing associated urinary tract infection, polypharmacy and medication reconciliation, restraints, and delirium assessment, screening for dementia, and other diseases of the elderly such as abdominal aortic aneurysms.

The clinical team can deliver care in a separate ED when available or when the general emergency department treatment areas are made more "geriatric friendly." Geriatric patients can have unique challenges to vision, mobility, behavior, incontinence, and memory; these can be supported with physical modifications and integration of geriatric supplies. The physical part may include features like chairs and stretchers that are more easily reclined/transferred from, with fabrics that are soft and moisture proof. Equipment such as bedside commodes, non-slip fall mats, natural and soft lighting, and enhanced signage can aid the care of these patients by making the clinical environment both safer and more comfortable.

Another key component of the Geriatric ED model is a formalized follow-up and transition of care system that can facilitate communication with family, providers, and long-term care facilities. The overarching goal is to ensure that at the point of entry into the hospital, geriatric patients receive focused care in an environment that attends to their unique needs. This framework of care facilitates formalized assessments by a team that approaches the patient comprehensively, allowing for overall care both inpatient and outpatient to be coordinated, safe, and more patient-centered.

Palliative Care

The World Health Organization (WHO) defines palliative care as "an approach that improves the **quality of life** of patients and their families facing the problem associated with life-threatening illness, through the **prevention and relief of suffering** by means of early identification and impeccable assessment and treatment of pain and other problems, physical, psycho-social and spiritual." [135]. Palliative care is a specialized medical care for people living with a serious illness and focuses on improving the quality of life for both patients and their families (CAPC). It is appropriate for patients of any age and at any stage in a serious illness and can be provided in conjunction with curative treatments. While timely palliative care consultations have been shown to reduce hospital length of stay [136], days in intensive care [137], and improve patient and family satisfaction [138], there are large gaps in the availability and provision of palliative care in the outpatient environment. Elderly patients with life-limiting diseases who present to the emergency department may be the subject of aggressive life-saving measures that are not in alignment with their goals of care, and it is important that measures are taken both to screen patients for unmet palliative care needs, to initiate palliative care earlier in the patient's hospitalization, to provide better symptom management, and to establish the patient's goals and healthcare preferences [139, 140]. In 2013, ACEP included the following statement in their Choosing Wisely campaign: "Don't delay engaging available palliative and hospice care services in the Emergency Department for patients likely to benefit." [141]. Geriatric patients undergoing emergency surgery for illness or trauma often have higher morbidity and mortality outcomes than those undergoing elective surgery [142], and this is where physicians can utilize effective communication with the patient to share the outcomes and possible burdens of surgical interventions, with a focus on their quality, and not just quantity of life.

Unfortunately, palliative care consultations for surgical patients are less commonly requested than those with a medical diagnosis [143], and often only as the patient is approaching end of life, when the patient might be more appropriately referred to hospice.

In the United States, hospice and palliative care (HPM) fellowship training and board certification is growing, and since the American Board of Emergency Medicine held its first HPM board certification exam in 2008, it is an increasingly common pathway following emergency medicine residency training.

References

1. 2017 Profile of Older Americans. Administration for Community Living|ACL Web site. https://acl.gov/sites/default/files/Aging%20and%20Disability%20in%20America/2017OlderAmericansProfile.pdf. Published 2017. Updated 2019. Accessed August 06, 2019.
2. Morris JN, Howard EP, Steel K, Schreiber R, Fries BE, Lipsitz LA, Goldman B. Predicting risk of hospital and emergency department use for home care elderly persons through a secondary analysis of cross-National Data. [in eng], BMC health Serv res. 2014;519 https://doi.org/10.1186/s12913-014-0519-z.
3. Sun R (AHRQ), Karaca Z (AHRQ), Wong HS (AHRQ). Trends in Hospital Emergency Department Visits by Age and Payer, 2006–2015. HCUP Statistical Brief #238. 2018. Agency for Healthcare Research and Quality, Rockville, MD. https://www.hcup-us.ahrq.gov/reports/statbriefs/sb238-Emergency-Department-Age-Payer-2006-2015.pdf/
4. https://www.acep.org/patient-care/policy-statements/triage-scale-standardization/
5. Baumann, M. R., and T. D. Strout. "Triage of Geriatric patients in the emergency department: validity and survival with the emergency severity index. [in eng]. Ann Emerg Med 49, no. 2 (Feb 2007): 234–240. https://doi.org/10.1016/j.annemergmed.2006.04.011.
6. Platts-Mills TF, Travers D, Biese K, McCall B, Kizer S, LaMantia M, Busby-Whitehead J, Cairns CB. Accuracy of the emergency severity index triage instrument for identifying elder emergency department patients receiving an immediate life-saving intervention. [in eng]. Acad Emerg Med. 2010;17(3):238–43. https://doi.org/10.1111/j.1553-2712.2010.00670.x.
7. Clegg A, Young J, Iliffe S, Rikkert MO, Rockwood K. Frailty in elderly people. Lancet. 2013;381(9868):752–62.
8. Oakland K, Nadler R, Cresswell L, Jackson D, Coughlin PA. Systematic review and meta-analysis of the association between frailty and outcome in surgical patients. The Annals of The Royal College of Surgeons of England. 2016;98(2):80–5.
9. Afilalo J, Lauck S, Kim DH, Lefèvre T, Piazza N, Lachapelle K, Martucci G, Lamy A, Labinaz M, Peterson MD, Arora RC. Frailty in older adults undergoing aortic valve replacement: the FRAILTY-AVR study. J Am Coll Cardiol. 2017;70(6):689–700.
10. van de Ree CLP, Landers MJF, Kruithof N, et al. Effect of frailty on quality of life in elderly patients after hip fracture: a longitudinal study. BMJ Open. 2019;9:e025941. https://doi.org/10.1136/bmjopen-2018-025941.
11. Rizk P, Morris W, Oladeji P, Huo M. Review of post-operative delirium in geriatric patients undergoing hip surgery. Geriatric orthopaedic surgery & rehabilitation. 2016;7(2):100–5.
12. Dresden SM, Platts-Mills TF, Kandasamy D, Walden L, Betz ME. Patient versus physician perceptions of frailty: a comparison of clinical frailty scale scores of older adults in the emergency department. Acad Emerg Med. 2019;00:1–4.
13. ISAR questionnaire: https://www.researchgate.net/figure/Six-items-of-ISAR-screening-tool_fig1_24023264; https://www.researchgate.net/profile/Fabio_Salvi/publication/24023264/figure/fig 1/AS:336610982940672@1457265720159/Six-items-of-ISAR-screening-tool.png" alt="-Six items of ISAR screening tool."/></a>
14. Tavares JPA, Sá-Couto P, Boltz M, Capezuti E. Identification of seniors at risk (ISAR) in the emergency room: a prospective study. Int Emerg Nurs. 2017;35:19–24. https://doi.org/10.1016/j.ienj.2017.05.008. Epub 2017 Jun 24.
15. Suffoletto B, Miller T, Shah R, Callaway C, Yealy DM. Predicting older adults who return to the hospital or die within 30 days of emergency department care using the ISAR tool: subjective versus objective risk factors. Emerg Med J. 2016;33(1):4–9. https://doi.org/10.1136/emermed-2014-203936. Epub 2015 May 18.
16. Yao JL, Fang J, Lou QQ, Anderson RM. eCollection 2015 a systematic review of the identification of seniors at risk (ISAR) tool for the prediction of adverse outcome in elderly patients seen in the emergency department. Int J Clin Exp Med. 2015;8(4):4778–86.
17. Clevenger CK, Chu TA, Yang Z, Hepburn KW. Clinical care of persons with dementia in the emergency department: a review of the literature and agenda for research. J Am Geriatr Soc. 2012;60:1742–8.
18. Carpenter CR. Evidence-based emergency medicine/rational clinical examination abstract. Does

this patient have dementia? Ann Emerg Med. 2008;52:554–6.
19. Hustey FM, Meldon SW, Banet GA, Gerson LW, Blanda M, Lewis LM. The use of abdominal computed tomography in older ED patients with acute abdominal pain. Am J Emerg Med. 2005;23(3):259–65. PubMed [citation] PMID: 15915395).
20. Han JH, Zimmerman EE, Cutler N, Schnelle J, Morandi A, Dittus RS, et al. Delirium in older emergency department patients: recognition, risk factors, and psychomotor subtypes. Acad Emerg Med. 2009;16:193–200.
21. Kakuma R, Du Fort GG, Arsenault L, Perrault A, Platt RW, Monette J, et al. Delirium in older emergency department patients discharged home: effect on survival. J Am Geriatr Soc. 2003;51:443–50.
22. O'Sullivan D, Brady N, Manning E, O'Shea E, O'Grady S, O'Regan N, Timmons S. Validation of the 6-item cognitive impairment test and the 4AT test for combined delirium and dementia screening in older emergency department attendees. Age Ageing. 2018;47(1):61–8. https://doi.org/10.1093/ageing/afx149.
23. Carpenter CR, Banerjee J, Keyes D, et al. Accuracy of dementia screening instruments in emergency medicine: a diagnostic meta-analysis. Acad Emerg Med. 2019;26(2):226–45.
24. Fong TG, Tulebaev SR, Inouye SK. Delirium in elderly adults: diagnosis, prevention and treatment. Nat Rev Neurol. 2009;5(4):210.
25. Hshieh TT, Yang T, Gartaganis SL, Yue J, Inouye SK. Hospital elder life program: systematic review and meta-analysis of effectiveness. Am J Geriatr Psychiatry. 2018;26(10):1015–33.
26. Inouye SK, van Dyck CH, Alessi CA, Balkin S, Siegal AP, Horwitz RI. Clarifying confusion: the confusion assessment method: a new method for detection of delirium. Ann Intern Med. 1990;113(12):941–8.
27. Nemec M, Koller MT, Nickel CH, Maile S, Winterhalder C, Karrer C, Laifer G, Bingisser R. Patients presenting to the emergency department with non-specific complaints: the Basel non-specific complaints (BANC) study. Acad Emerg Med. 2010;17(3):284–92. https://doi.org/10.1111/j.1553-2712.2009.00658.x.
28. Liu SW, Sri-On J, Tirrell GP, Nickel C, Bingisser R. Serious conditions for ED elderly fall patients: a secondary analysis of the Basel non-specific complaints study. Am J Emerg Med. 2016;34(8):1394–9. https://doi.org/10.1016/j.ajem.2016.04.007. Epub 2016 Apr 9.
29. Wofford JL, Schwartz E, Timerding BL, Folmar S, Ellis SD, Messick CH. Emergency department utilization by the elderly: analysis of the National Hospital Ambulatory Medical Care Survey. Acad Emerg Med. 1996;3(7):694–9. PubMed [citation] PMID: 8816186.
30. Brewer BJ, Golden GT, Hitch DC, Rudolf LE, Wangensteen SL. Abdominal pain. An analysis of 1,000 consecutive cases in a university hospital emergency room. Am J Surg. 1976;131(2):219–23. PubMed [citation] PMID: 1251963.
31. Fenyö G. Diagnostic problems of acute abdominal diseases in the aged. Acta Chir Scand. 1974;140(5):396–405. No abstract available. PubMed [citation].
32. Fenyö G. Acute abdominal disease in the elderly: experience from two series in Stockholm. Am J Surg. 1982;143(6):751–4. PubMed [citation] PMID: 7091511.
33. van Geloven AA, Biesheuvel TH, Luitse JS, Hoitsma HF, Obertop H. Hospital admissions of patients aged over 80 with acute abdominal complaints. Eur J Surg. 2000;166(11):866–71. PubMed [citation].
34. Lewis LM, Banet GA, Blanda M, Hustey FM, Meldon SW, Gerson LW. Etiology and clinical course of abdominal pain in senior patients: a prospective, multicenter study. J Gerontol A Biol Sci Med Sci. 2005;60(8):1071–6. PubMed [citation] PMID: 16127115.
35. Cooper GS, Shlaes DM, Salata RA. Intraabdominal infection: differences in presentation and outcome between younger patients and the elderly. Clin Infect Dis. 1994;19(1):146–8. PubMed [citation] PMID: 7948517.
36. de Dombal FT. Acute abdominal pain in the elderly. J Clin Gastroenterol. 1994;19(4):331–5. Review. PubMed [citation].
37. Bassuk EL, Minden S, Apsler R. Geriatric emergencies: psychiatric or medical? Am J Psychiatry. 1983;140(5):539–42. PubMed [citation].
38. Leuthauser A, McVane B. Abdominal pain in the Geriatric patient. Emerg Med Clin North Am. 2016;34(2):363–75. https://doi.org/10.1016/j.emc.2015.12.009. Epub 2016. Review. PubMed [citation].
39. Potts FE 4th, Vukov LF. Utility of fever and leukocytosis in acute surgical abdomens in octogenarians and beyond. J Gerontol A Biol Sci Med Sci. 1999;54(2):M55–8. PubMed [citation].
40. Wroblewski M, Mikulowski P. Peritonitis in geriatric inpatients. Age Ageing. 1991;20(2):90–4. PubMed [citation].
41. Lyon C, Clark DC. Diagnosis of acute abdominal pain in older patients. Am Fam Physician. 2006;74(9):1537–44.
42. Lewis LM, Klippel AP, Bavolek RA, Ross LM, Scherer TM, Banet GA. Quantifying the usefulness of CT in evaluating seniors with abdominal pain. Eur J Radiol. 2007;61(2):290–6. Epub 2006 Nov 7. PubMed [citation] PMID: 17085004.
43. Ibach JR Jr, Hume HA, Erb WH. Cholecystectomy in the aged. Surg Gynecol Obstet. 1968;126(3):523–8. No abstract available. PubMed [citation] PMID: 5637311.
44. Ross SO, Forsmark CE. Pancreatic and biliary disorders in the elderly. Gastroenterol Clin North Am. 2001;30(2):531–45, x. Review PubMed [citation] PMID: 11432304.

45. Arnold DJ. 28,621 cholecystectomies in Ohio. Results of a survey in Ohio hospitals by the Gallbladder Survey Committee, Ohio Chapter, American College of Surgeons. Am J Surg. 1970;119(6):714–7. No abstract available. PubMed [citation] PMID: 5445995.
46. Nesvaderani M, Eslick GD, Vagg D, Faraj S, Cox MR. Epidemiology, aetiology and outcomes of acute pancreatitis: A retrospective cohort study. Int J Surg. 2015;23(Pt A):68–74. https://doi.org/10.1016/j.ijsu.2015.07.701. Epub 2015 Sep 16. PubMed [citation] PMID: 26384834.
47. Morrow DJ, Thompson J, Wilson SE. Acute cholecystitis in the elderly: a surgical emergency. Arch Surg. 1978;113(10):1149–52. PubMed [citation] PMID: 708235.
48. Parker LJ, Vukov LF, Wollan PC. Emergency department evaluation of geriatric patients with acute cholecystitis. Acad Emerg Med. 1997;4(1):51–5. PubMed [citation] PMID: 9110012.
49. Paranjape C, Dalia S, Pan J, Horattas M. Appendicitis in the elderly: a change in the laparoscopic era. Surg Endosc. 2007;21(5):777–81. Epub 2007 Feb 7. PubMed. [citation] PMID: 17285390.
50. Lunca S, Bouras G, Romedea NS. Acute appendicitis in the elderly patient: diagnostic problems, prognostic factors and outcomes. Rom J Gastroenterol. 2004;13(4):299–303. PubMed [citation] PMID: 15624027.
51. Hui TT, Major KM, Avital I, Hiatt JR, Margulies DR. Outcome of elderly patients with appendicitis: effect of computed tomography and laparoscopy. Arch Surg. 2002;137(9):995–8; discussion 999-1000. PubMed [citation] PMID: 12215147.
52. Storm-Dickerson TL, Horattas MC. What have we learned over the past 20 years about appendicitis in the elderly? Am J Surg. 2003;185(3):198–201. PubMed [citation] PMID: 12620555.
53. Rub R, Margel D, Soffer D, Kluger Y. Appendicitis in the elderly: what has changed? Isr Med Assoc J. 2000;2(3):220–3. PubMed [citation] PMID: 10774271.
54. Pooler BD, Lawrence EM, Pickhardt PJ. MDCT for suspected appendicitis in the elderly: diagnostic performance and patient outcome. Emerg Radiol. 2012;19(1):27–33. https://doi.org/10.1007/s10140-011-1002-3. Epub 2011 Dec 1. PubMed. [citation] PMID: 22131057.
55. Hlibczuk V, Dattaro JA, Jin Z, Falzon L, Brown MD. Diagnostic accuracy of noncontrast computed tomography for appendicitis in adults: a systematic review. Ann Emerg Med. 2010;55(1):51–9. https://doi.org/10.1016/j.annemergmed.2009.06.509. Epub 2009. Review. PubMed [citation] PMID: 19733421.
56. van Randen A, Laméris W, van Es HW, van Heesewijk HP, van Ramshorst B, Ten Hove W, Bouma WH, van Leeuwen MS, van Keulen EM, Bossuyt PM, Stoker J, Boermeester MA, OPTIMA Study Group. A comparison of the accuracy of ultrasound and computed tomography in common diagnoses causing acute abdominal pain. Eur Radiol. 2011;21(7):1535–45. https://doi.org/10.1007/s00330-011-2087-5. Epub 2011 Mar 2. PubMed [citation] PMID: 21365197, PMCID: PMC3101356.
57. Horattas MC, Guyton DP, Wu D. A reappraisal of appendicitis in the elderly. Am J Surg. 1990;160(3):291–3. PubMed [citation] PMID: 2393058.
58. Young-Fadok TM, Roberts PL, Spencer MP, Wolff BG. Colonic diverticular disease. Curr Probl Surg. 2000;37(7):457–514. Review. No abstract available. PubMed [citation] PMID: 10932672.
59. Farrell RJ, Farrell JJ, Morrin MM. Diverticular disease in the elderly. Gastroenterol Clin North Am. 2001;30(2):475–96. Review. PubMed [citation] PMID: 11432301.
60. Roberts P, Abel M, Rosen L, Cirocco W, Fleshman J, Leff E, Levien D, Pritchard T, Wexner S, Hicks T, et al. Practice parameters for sigmoid diverticulitis. The Standards Task Force American Society of Colon and Rectal Surgeons. Dis Colon Rectum. 1995;38(2):125–32. PubMed [citation] PMID: 7851165.
61. Place RJ, Simmang CL. Diverticular disease. Best Pract Res Clin Gastroenterol. 2002;16(1):135–48. Review. PubMed [citation] PMID: 11977933.
62. Welch CE, Athanasoulis CA, Galdabini JJ. Hemorrhage from the large bowel with special reference to angiodysplasia and diverticular disease. World J Surg. 1978;2(1):73–83. No abstract available. PubMed [citation] PMID: 307312.
63. Winzelberg GG, Froelich JW, McKusick KA, Waltman AC, Greenfield AJ, Athanasoulis CA, Strauss HW. Radionuclide localization of lower gastrointestinal hemorrhage. Radiology. 1981;139(2):465–9. PubMed [citation] PMID: 6971455.
64. Brandt LJ, Boley SJ. The role of colonoscopy in the diagnosis and management of lower intestinal bleeding. Scand J Gastroenterol Suppl. 1984;102:61–70. PubMed [citation] PMID: 6332370.
65. Russell RM. Changes in gastrointestinal function attributed to aging. Am J Clin Nutr. 1992;55(6 Suppl):1203S–7S. https://doi.org/10.1093/ajcn/55.6.1203S. Review. PubMed [citation] PMID: 1590257.
66. Fischer J, Johnson MA. Low body weight and weight loss in the aged. J Am Diet Assoc. 1990 Dec;90(12):1697-706. Review PubMed [citation].
67. Pilotto A, Salles N. Helicobacter pylori infection in geriatrics. Helicobacter. 2002;7 Suppl 1:56–62. Review. PubMed [citation].
68. Huang JQ, Sridhar S, Hunt RH. Role of Helicobacter pylori infection and non-steroidal anti-inflammatory drugs in peptic-ulcer disease: a meta-analysis. Lancet. 2002 359(9300):14-22. PubMed [citation].
69. Hilton D, Iman N, Burke GJ, Moore A, O'Mara G, Signorini D, Lyons D, Banerjee AK, Clinch D. Absence of abdominal pain in older persons with endoscopic ulcers: a prospective study. Am

J Gastroenterol. 2001 Feb;96(2):380-4. PubMed [citation].
70. Clinch D, Banerjee AK, Ostick G. Absence of abdominal pain in elderly patients with peptic ulcer. Age Ageing. 1984 Mar;13(2):120-3. PubMed [citation].
71. Borum ML. Peptic-ulcer disease in the elderly. Clin Geriatr Med. 1999 Aug;15(3):457-71. Review. PubMed [citation].
72. Permutt RP, Cello JP. Duodenal ulcer disease in the hospitalized elderly patient. Dig Dis Sci. 1982 Jan;27(1):1-6. PubMed [citation].
73. Sreedharan A, Martin J, Leontiadis GI, Dorward S, Howden CW, Forman D, Moayyedi P. Proton pump inhibitor treatment initiated prior to endoscopic diagnosis in upper gastrointestinal bleeding. Cochrane Database Syst Rev. 2010 Jul 7;(7):CD005415. doi: https://doi.org/10.1002/14651858.CD005415.pub3. Review. PubMed [citation].
74. Kauvar DR. The geriatric acute abdomen. Clin Geriatr Med. 1993 Aug;9(3):547-58. Review. PubMed [citation].
75. Bugliosi TF, Meloy TD, Vukov LF. Acute abdominal pain in the elderly. Ann Emerg Med. 1990 Dec;19(12):1383-6. PubMed [citation].
76. Avots-Avotins KV, Waugh DE. Colon volvulus and the geriatric patient. Surg Clin North Am. Apr;62(2):249-60. PubMed [citation]. 1982;
77. Atamanalp SS. Sigmoid volvulus: diagnosis in 938 patients over 45.5 years. Tech Coloproctol. 2013 Aug;17(4):419–24. doi: https://doi.org/10.1007/s10151-012-0953-z. Epub 2012 Dec 6. PubMed [citation].
78. Kapadia MR. Volvulus of the Small Bowel and Colon. Clin Colon Rectal Surg. 2017 Feb;30(1):40–45. doi: https://doi.org/10.1055/s-0036-1593428. Review. PubMed [citation] PMID: 28144211, PMCID: PMC5179272.
79. Ballantyne GH, Brandner MD, Beart RW Jr, Ilstrup DM. Volvulus of the colon. Incidence and mortality. Ann Surg. 1985 Jul;202(1):83–92. PubMed [citation] PMID: 4015215, PMCID: PMC 1250842.
80. Rabinovici R, Simansky DA, Kaplan O, Mavor E, Manny J. Cecal volvulus. Dis Colon Rectum. 1990 Sep;33(9):765-9. Review. PubMed [citation].
81. Delabrousse E, Sarliève P, Sailley N, Aubry S, Kastler BA. Cecal volvulus: CT findings and correlation with pathophysiology. Emerg Radiol. 2007 Nov;14(6):411–5. Epub 2007 Jul 6. PubMed [citation].
82. Batke M, Cappell MS. Adynamic ileus and acute colonic pseudo-obstruction. Med Clin North Am. 2008 May;92(3):649–70, ix. doi: https://doi.org/10.1016/j.mcna.2008.01.002. Review. PubMed [citation].
83. Vanek VW, Al-Salti M. Acute pseudo-obstruction of the colon (Ogilvie's syndrome). An analysis of 400 cases. Dis Colon Rectum. 1986 Mar;29(3):203-10. PubMed [citation].
84. Haj M, Haj M, Rockey DC. Ogilvie's syndrome: management and outcomes. Medicine (Baltimore). 2018 Jul;97(27):e11187. doi: https://doi.org/10.1097/MD.0000000000011187. PubMed [citation] PMID: 29979381, PMCID: PMC6076157.
85. Cosford PA, Leng GC. Screening for abdominal aortic aneurysm. Cochrane Database Syst Rev. 2007 18;(2):CD002945. Review. PubMed [citation].
86. Lederle FA, Johnson GR, Wilson SE, Chute EP, Littooy FN, Bandyk D, Krupski WC, Barone GW, Acher CW, Ballard DJ. Prevalence and associations of abdominal aortic aneurysm detected through screening. Aneurysm Detection and Management (ADAM) Veterans Affairs Cooperative Study Group. Ann Intern Med. 1997 15;126(6):441-9. PubMed [citation].
87. Wilmink TB, Quick CR, Day NE. The association between cigarette smoking and abdominal aortic aneurysms. J Vasc Surg. 1999 Dec;30(6):1099-105. PubMed [citation].
88. Magidson PD, Martinez JP. Abdominal Pain in the Geriatric Patient. Emerg Med Clin North Am. 2016 Aug;34(3):559–74. doi: https://doi.org/10.1016/j.emc.2016.04.008. Review. PubMed [citation].
89. Oldenburg WA, Lau LL, Rodenberg TJ, Edmonds HJ, Burger CD. Acute mesenteric ischemia: a clinical review. Arch Intern Med. 2004;164(10):1054–62. Review. PubMed [citation].
90. Smerud MJ, Johnson CD, Stephens DH. Diagnosis of bowel infarction: a comparison of plain films and CT scans in 23 cases. AJR Am J Roentgenol. 1990;154(1):99–103. PubMed [citation].
91. Bala M, Kashuk J, Moore EE, Kluger Y, Biffl W, Gomes CA, Ben-Ishay O, Rubinstein C, Balogh ZJ, Civil I, Coccolini F, Leppaniemi A, Peitzman A, Ansaloni L, Sugrue M, Sartelli M, Di Saverio S, Fraga GP, Catena F. Acute mesenteric ischemia: guidelines of the world Society of Emergency Surgery. World J Emerg Surg. 2017;12:38. https://doi.org/10.1186/s13017-017-0150-5. eCollection 2017. Review. PubMed [citation] PMID: 28794797, PMCID: PMC5545843.
92. Boley SJ, Feinstein FR, Sammartano R, Brandt LJ, Sprayregen S. New concepts in the management of emboli of the superior mesenteric artery. Surg Gynecol Obstet. 1981;153(4):561–9. PubMed [citation].
93. Champion HR, Copes SW, Buyer D, et al. Major trauma in geriatric patients. Am J Public Health. 1989;79(9):1278–82.
94. Aschkenasy MT, Rothenhaus TC. Trauma and falls in the elderly. Emerg Med Clin N Am. 2006;24:413–32.
95. American College of Surgeons. Advanced trauma life support for doctors. American College of Surgeons Committee on trauma. 7th ed. Chicago (IL): American College of Surgeons; 2004.
96. Demetriades D, Sava J, Alo K, et al. Old age as a criterion for trauma team activation. J Trauma. 2001;51:754–6.

97. McGwin G Jr, Nunn AM, Mann JC, et al. Reassesment of the tri-modal mortality distribution in the presence of a regional trauma system. J Trauma. 2009;66(2):526–30.
98. Bortz WM 2nd. A conceptual framework of frailty: a review. J Gerontol A Biol Sci Med Sci. 2002;57(5):M283–8.
99. Orouji Jokar T, Ibraheem K, Rhee P, et al. Emergency general surgery specific frailty index: a validation study. J Trauma Acute Care Surg. 2016;222(5):805–13.
100. Tan MP, Kamaruzzaman SB, Zakaria MI, et al. Ten-year mortality in older patients attending the emergency department after a fall. Geriatr Gerontol Int. 2016;16(1):111–7.
101. Broos P. Multiple trauma in elderly patients: factors influencing outcome. Injury. 1993;24:365.
102. Nordell E, Jarnlo GB, Jetsen C, et al. Accidental falls and related fractures in 65-74 year olds: a retrospective study of 332 patients. Acta Orthop Scand. 2000;71:175–9.
103. Sasser SM, Hunt RC, Faul M, et al. Guidelines for field triage of injured patients: recommendations of the national expert panel on field triage, 2011. MMWR Recomm Rep. 2012;61(RR-1): 1–20.
104. Meldon SW, Reilly M, Drew B, Mancusco C, Fallon W. Trauma in the very elderly: a community-based study of outcomes at trauma and non-trauma centers. Acad Emerg Med. 2000;7(10):1166.
105. Morris JA, MacKenzie EJ, Edelstein SL. The effects of preexisting conditions on mortality in trauma patients. JAMA. 1990;263(14):1942–6.
106. Baumgarten M, Margolis DJ, Localio AR, et al. Pressure ulcers among elderly patients in the hospital stay. Gerontol A Biol Ned Sci. 2006;61(7):749–54.
107. Callaway DW, Wolfe R. Geriatric trauma. Emerg Med Clin N Am. 2007;25:837–60.
108. Knudson MM, Lieberman J, Morris J, et al. Mortality factors in geriatric blunt trauma patients. Arch Surg. 1994;129(4):448–53.
109. van Aalst JA, Morris JA Jr, Yates HK, et al. Severely injured geriatric patients return to independently living: a study of factors influencing function and independence. J Trauma. 1991;31(8):1096–101.
110. Schwab CW, Kauder DR. Trauma in the geriatric patient. Arch Surgery. 1992;127:701.
111. McNelis J, Marini CP, Jurkiewicz A, et al. Prolonged lactate clearance is associated with increased mortality in the surgical intensive care unit. Am J Surg. 2001;182(5):481–5.
112. Asckhenasy MT, Rothenhaus TC. Trauma and falls in the elderly. Emerg Med Clin N Am. 2006;24:413–32.
113. Masson F, Thicoipe M, Mokni T, et al. Aquitaine group for severe brain injury study. Epidemiology of traumatic comas: a prospective population – based study. Brain Inj. 2003;17(4):279–93.
114. Susman M, DiRosso SM, Sullivan T, et al. Et al. traumatic brain injury in the elderly: increased mortality and worse functional outcome at discharge despite lower injury severity. J Trauma. 2002;53:219–24.
115. Cohen DB, Rinker C, Wilberger JE. Traumatic brain injury in the anticoagulated patients. J Trauma. 2002;53:219–24.
116. Ohm C, Mina A, Howells G, et al. Effect of antiplatelet agents on outcome for elderly patients with traumatic intracranial hemorrhage. J Trauma. 2005;58:518–22.
117. Franko J, Kish KJ, O'Connell BG, et al. Advanced age and preinjury warfarin anticoagulation increase the risk of mortality after head trauma. J Trauma. 2006;61:107–10.
118. Ziegler DW, Agarwal NN. The morbidity and mortality of rib fractures. J Trauma. 1994;37:975–9.
119. Bulger EM, Arneson MA, Mock CN, et al. Rib fractures in the elderly. J Trauma. 2000;48(6):1040–7.
120. Blackmore CC, Cummings P, Jurkovich GJ, et al. Predicting major hemorrhage in patients with pelvic fracture. J Trauma. 2006;61(2):346–52.
121. Kimbrell BJ, Velhamos GC, Chan LS, et al. Angiographic embolization for pelvic fractures in older patients. Arch Surg. 2004;139(7):728–32.
122. Brown CV, Antevil JL, Sise MJ, et al. Spiral computed tomography for the diagnosis of cervical, thoracic, and lumbar spine fractures: its time has come. J Trauma. 2005;58(5):890–5.
123. Quattromani E, Normansell D, Storkan M, et al. Olioganalgesia in blunt geriatric trauma. The Journal of Em Med. 2014;48(6):653–9.
124. Jones JS, Johnsons K, McNinch M. Age as a risk factor for inadequate emergency department analgesia. Am Jour of Em Med. 1996;14(2):157–60.
125. Turnheim K. Drug therapy in the elderly. Exp gerontology. 2004;39:1731–8.
126. Boltz MM, Podany AB, Hollenbeck CS, et al. Injuries and outcomes associated with traumatic falls in the elderly population on oral anticoagulant therapy. Injru. 2015, 46(9):1765–71.
127. Smith K, Weeks S. The impact of pre-injury anticoagulation therapy in the older adult patient experiencing a traumatic brain injury: a systematic review. JBI Libr Syst Rev. 2012;10(58):4610–21.
128. Petrey LB, Weddle RJ, Richardson B, et al. Trauma patient readmissions: why do they come back for more? J Trauma Acute Care Surg. 2015;79(5):717–24.
129. Hall C, Essler S, Dandashi J, et al. Impact of fraility and anticoagulation status on readmission and mortality rates following falls in patients over 80. Proc Bayl Univ Med Cent. 2019;32(2):181–6.
130. Strosberg DS, Housley BC, Vasquez D, et al. Discharge destinations and readmission rates in older trauma patients. J Surg Res. 2017;207:27–32.
131. https://www.cdc.gov/nchs/fastats/emergency-department.htm
132. American GS, Geriatric ED. American College of Emergency Physicians, emergency nurses association. Geriatric emergency department guidelines. Ann Emerg Med. 2014;63(5):e7.

133. Banerjee J, Conroy S, O'Leary V. The silver book: quality care for older people with urgent and emergency care needs. British Geriatric Society. 2012:1–02.
134. Adapted from https://www.acep.org/globalassets/sites/geda/documnets/geda-criteria-final_1.17.2019.pdf
135. (World Health Organization. Palliative care 2002. Available online: www.who.int/cancer/palliative/definition/en/)
136. Wu FM, Newman JM, Lasher A, Brody AA. Effects of initiating palliative care consultation in the emergency department on inpatient length of stay. J Palliat Med. 2013;16(11):1362–7.
137. Khandelwal N, Kross EK, Engelberg RA, Coe NB, Long AC, Curtis JR. Estimating the effect ofpalliative care interventions and advance care planning on ICU utilizaation: a systematic review. Crit Care Med. 2015;43(5):1002–11.
138. George N, Phillips E, Zaurova M, Song C, Lamba S, Grudzen C. Palliative care screening and assessment in the emergency department: a systematic review. J Pain Symptom Manag. 2016;51(1):108–119.e102. https://doi.org/10.1016/j.jpainsymman.2015.07.017.
139. Lamba S, DeSandre PL, Todd KH, Bryant EN, Chan GK, Grudzen CR, et al. Integration of palliative care into emergency medicine: the improving palliative Care in Emergency Medicine (IPAL-EM) collaboration. J Emerg Med. 2014;46(2):264–70.
140. Grudzen CR, et al. Palliative care needs of seriously ill, older adults presenting to the emergency department. Acad Emerg Med. 2010:1253–7.
141. http://www.choosingwisely.org/nations-emergency-physicians-announce-list-of-test-and-procedures-to-question-as-part-of-choosing-wisely-campaign/
142. Ballou JH, Brasel KJ. Palliative care and Geriatric surgery. Clin Geriatr Med. 2019;35(1):35–44.
143. Rivet EB, et al. Characteristics of palliative care consultation at an academic level one trauma center. Am J Surg. 2017;214(4):657–60.

Cardiovascular Comorbidities in the Elderly Undergoing Surgery

4

Kartik Dhaduk, Ammar Athar, Gabriela Andries, Wilbert S. Aronow, and William H. Frishman

Introduction

Advancing age is associated with a reduction in the physiologic reserve of all organ systems, even in the absence of any underlying pathology. With aging there is also an associated increased prevalence of multiple comorbidities and chronic diseases as well as polypharmacy. These factors contribute to the increased likelihood of developing complications during the perioperative and postoperative periods [1–3]. This issue is increasingly important as the elderly population continues to grow. Studies have estimated that approximately 53% of all surgical procedures are performed on patients over the age of 65 [4]. In the United States alone, the elderly population (above 65 years old) has tripled over the last 50 years. It is estimated that by 2030 there will be approximately 72 million individuals >65 years of age which represents almost 20% of the US population [5]. Unfortunately, although the aging population is rising, they remain poorly represented in clinical trials which mostly include patients between the ages of 18 and 64 years. Inadequate evidence and knowledge regarding responses of geriatric patients pose a unique challenge for physicians when evaluating this population prior to major operative procedures [6].

Cardiac disease is undoubtedly the most common comorbid condition in the elderly. Up to 80% of patients over 80 years old have identifiable cardiovascular diseases, which account for 40% of mortality. The age-related cardiovascular physiologic changes involve structural remodeling such as increased vascular intimal thickness and stiffness, less compliance from increased left ventricular wall thickness, modified contractile behavior, interstitial fibrosis, increased left atrial stiffness, altered regulation of the vascular tone, and a reduced responsiveness to beta-adrenergic modulation [1]. These physiologic changes subject the elderly to a substantially higher risk of adverse cardiovascular events related to the stress of surgery, blood loss, potential fluid and electrolyte shifts, drug-drug interactions, and the effect of anesthetic and analgesic agents during the perioperative period [4]. A preoperative evaluation of geriatric patients must be based on a thorough understanding of the patient's physical and functional status to allow the implementation of an appropriate perioperative care plan which aims to reduce risk and delay in recovery [5]. In this chapter, we will review common cardiovascular comorbidities and the pre- and perioperative

K. Dhaduk
Cardiology Division, Department of Medicine, Westchester Medical Center and New York, Valhalla, NY, USA

A. Athar · G. Andries · W. S. Aronow
W. H. Frishman (✉)
Cardiology Division, Department of Medicine, Westchester Medical Center and New York Medical College, Valhalla, NY, USA
e-mail: William.Frishman@wmchealth.org

R. Latifi (ed.), *Surgical Decision Making in Geriatrics*,
https://doi.org/10.1007/978-3-030-47963-3_4

Table 4.1 Perioperative management of cardiovascular medications

Beta-blockers	Angiotensin converting enzyme inhibitors/ angiotensin receptor blockers (ACEI/ARB)	Calcium channel blockers	Diuretics
Continue beta-blockers for patients already taking them (class I)	It is reasonable to continue therapy perioperatively within patients with stable left ventricular heart failure (class IIa)	Insufficient evidence to recommend for or against the use of these agents perioperatively	Continue if needed for management of volume overload
It is reasonable to initiate beta-blockers >1 day prior to surgery for patients considered to be intermediate-high preoperative risk or for those that have three or more cardiovascular risk factors as per the Revised Cardiac Risk Index (class IIb)	It is reasonable to hold therapy 1 day prior to surgery in the absence of heart failure or unstable hypertension (class IIa)	Calcium channel blockers have been associated with postoperative bleeding presumably due to inhibition of platelet aggregation	Consider dose reduction in the presence of hypovolemia, hypotension, or electrolyte abnormalities
Statins	Nitrates	Alpha-2 agonists	Antiarrhythmics
Continue for patients already on treatment (class I)	Prophylactic intravenous nitroglycerin is not effective in reducing myocardial infarction in patients undergoing noncardiac surgery (class III)	Prophylactic use not recommended in patients undergoing noncardiac surgery (class III)	Amiodarone: continue for stable, monomorphic ventricular tachycardia
Preoperative initiation is reasonable in patients undergoing vascular surgery (class IIa)		Avoid abrupt discontinuation as it may result in rebound hypertension, headache, agitation, and tremor	Continued for sustained polymorphic ventricular tachycardia if no long QT syndrome
Preoperative initiation may be considered in patients with clinical conditions who are undergoing elevated risk procedures (class IIb)		Isoproterenol: reasonable to continue for torsades de pointes with recurrent pauses	

strategies that may help to modify surgical morbidity and mortality. Specific guidelines regarding particular cardiovascular medications are summarized in Tables 4.1 and 4.2 [7–13].

Systemic Hypertension

Background

Hypertension (HTN) is known to increase the risk of cardiovascular events, bleeding, cerebrovascular accidents, and mortality in the peri- and postoperative period [14]. It predisposes patients to a variety of systemic changes such as cardiovascular remodeling resulting in congestive heart failure, chronically elevated blood pressure causing renal impairment, and coronary and cerebrovascular occlusive disease increasing the risk for ischemic events. Existing hypertension is considered by most medical practitioners when determining whether or not to proceed with noncardiac surgery and is often the culprit for delaying procedures. In 2001, a meta-analysis of 30 studies found that 25% of patients undergoing major noncardiac surgery have perioperative hypertension and that this increases their risk of cardiovascular complications by 35%. In addition, induction of anesthesia can result in sympathetic inhibition and loss of the baroreceptor reflex, contributing to intraoperative blood pressure lability, which may increase the risk of myocardial ischemia [15]. While patients with untreated

Table 4.2 Antiplatelet and anticoagulation recommendations

Antiplatelet therapy (aspirin, clopidogrel, prasugrel, ticagrelor, cangrelor)		
Patients without coronary stents	*Patients with coronary stents*	
In patients who have received coronary stents and must undergo surgical procedures that require the discontinuation of P2Y12-inhibitors, it is recommended that aspirin be continued if possible and the P2Y12-inhibitor be restarted as soon as possible after surgery (class I)	Elective noncardiac surgery should be delayed 30 days after bare metal stent (BMS) and 6 months after drug-eluting stent (DES) placement (class I)	
In patients with no prior history of stenting and undergoing nonemergency/nonurgent noncardiac surgery, it may be reasonable to continue aspirin when the risk of potential cardiac event outweighs risk of bleeding (class IIb)	Elective noncardiac surgery after DES implantation in patients taking P2Y2-inhibitor may be considered after 3 months if risk of delay of surgery > risk of stent thrombosis (class IIb)	
	Elective noncardiac surgery should not be performed within 30 days after BMS implantation or within 3 months after DES implantation in patients in whom dual antiplatelet therapy will need to be discontinued perioperatively (class III)	
Anticoagulation		
Nonvalvular atrial fibrillation	*Bridging therapy*	*Prosthetic heart valves*
Direct-acting oral anticoagulants	*Bridging indicated*	*Unfractionated heparin* (UFH)
Interrupt therapy for all procedures for a duration based on the estimated creatinine clearance	Low thrombotic risk (risk <5%/year, CHA2DS2-VASc score ≤4 and no prior history of ischemic stroke, transient ischemic attack (TIA) or venous thromboembolism (VTE))	No substantiated specific recommendation currently exists regarding anticoagulation for prosthetic heart valves in the perioperative period
	Prior cerebrovascular accident (stroke, TIA) with no bleed risk	Available evidence seems to favor the use of UFH with further clinical trials ongoing
Vitamin K antagonist (warfarin)	High thrombotic risk (risk >10%/year, CHA2DS2-VASc score of >7, stoke, TIA, or VTE within last 3 months) consider bridging	*Mechanical mitral valve*: recommend perioperative anticoagulation bridging regardless of any risk factors for thromboembolism
Do not interrupt in patients undergoing procedures with no or low bleeding risk and absence of patient-related factors that increase the risk of bleeding	*Bridging NOT indicated*: moderate thrombotic risk with increased risk of bleeding	*Aortic valve*: recommend bridging if one of more additional risk factors (such as atrial fibrillation, previous thromboembolism, left ventricular dysfunction, hypercoagulable condition, or an older generation prosthetic aortic valve) are present
Interrupt therapy for procedures with intermediate or high bleed risk or uncertain bleed risk but with presence of patient-related factors that increase the risk of bleeding	No history of prior stroke, TIA, or VTE with no bleed risk	

hypertension have been found to have an exaggerated responses to the induction of anesthesia, patients with well-controlled preexisting hypertension were shown to respond similarly to those without hypertension [16] seemingly favoring treatment of perioperative hypertension. In regard to the geriatric population, Williamson et al. recommend targeting a perioperative blood pressure of less than 130/80 mmHg prior to surgery.

Perioperative Guidelines

Despite such reports of increased risk, existing guidelines for the management of perioperative hypertension more or less agree that mild to moderate preexisting hypertension is not a major contributor to increased intraoperative cardiovascular mortality. The American College of Cardiology/American Heart Association (ACC/AHA) Guidelines list uncontrolled hypertension as a minor perioperative risk factor [17] as most patients are not at an increased risk of intraoperative complications in the absence of severe hypertension, abnormal renal function or electrolyte abnormalities [18]. The European Society of Cardiology/European Society of Anesthesiology (ESC/ESA) guidelines recommend patients with new hypertension be screened for evidence of end organ damage in the perioperative period. Their recommendation is supported by the fact that end organ damage as evidenced by congestive heart failure or renal insufficiency significantly increases the probability of adverse cardiac outcomes [17]. In terms of postponement of surgery, the American and European guidelines seem to be in agreement for delaying elective surgery in the presence of severe hypertension as defined by systolic blood pressure >180 mmHg or diastolic blood pressure >110 mmHg [7, 8]. Isolated systolic hypertension has been known to produce a 40% increase in perioperative cardiovascular events [19], while diastolic pressure greater than 110 mmHg in the perioperative period has been associated with myocardial ischemia, infarction, dysrhythmias, neurologic complications, and renal failure [20]. Patients with diastolic blood pressure less than 110 mmHg were not found to be at increased operative risk [21]. See Table 4.3 for current recommendations [7–9].

Table 4.3 Hypertension recommendations

Mild-moderate (systolic blood pressure <180 or diastolic blood pressure <110)	Reasonable to proceed with surgery without initiating antihypertensive medication
	Treatment for hypertension may be started postoperatively after recovery
Severe (systolic blood pressure >180 or diastolic blood pressure >110)	Elective/urgent surgery: delay surgery until hemodynamic stability is achieved with medications
	Emergent surgery: proceed with surgery as the risks of postponement outweigh the benefit of treatment

Beta-Blockers

Although beta-blockers have been known to reduce the risk of intraoperative myocardial ischemia in untreated hypertensive patients [22], their initiation in the immediate perioperative period has been questioned by a higher rate of mortality (3.1% vs. 2.3%) and stroke (1% vs. 0.5%) [17]. The beneficial effects of beta-blockers likely stem from their ability to delay cardiovascular remodeling, thus delaying the development of significant perioperative risk factors such as advanced heart failure and progressively worse aortic stenosis. Both the ACC/AHA and ESC/ESA take a similar position recommending against initiation of a beta-blocker in the perioperative period [23] based on the evidence presented in the POISE trial [24]. Both the American and European guidelines also agree that beta-blockers should be continued perioperatively in patients already taking the medication. This is further supported by the finding that beta-blocker withdrawal in patients with underlying coronary artery disease can result in heightened angina, myocardial infarction, or sudden death. In patients already receiving beta-blockers, stopping treatment perioperatively was associated with a significant increase in postoperative mortality (50% vs. 1.5%) [25].

In patients with intermediate or high preoperative risk (Revised Cardiac Risk Index (RCRI) >3), the ACC/AHA recommends initiation of beta-blockers preoperatively at least 24 hours prior to surgery similarly to the Canadian Cardiovascular Society (CCS) Guidelines [9] and to the ESC/ESA guidelines which advise against perioperative initiation of beta-blockers in patients undergoing low-risk surgery [8].

Angiotensin Converting Enzyme (ACE) Inhibitors/Angiotensin Receptor Blockers (ARBs)

Currently available data suggest a lack of conclusive evidence to support absolute continuation or discontinuation of ACE inhibitors or ARBs in the perioperative period. The American, Canadian, and European guidelines appear to have reached a consensus that it is reasonable to hold ACE inhibitors and ARBs 1 day prior to surgery in the absence of heart failure or inadequately treated hypertension. One study of 150 vascular surgery patients found that the incidence of hypotension during anesthesia induction was significantly lower in patients who stopped taking captopril or enalapril on the evening before surgery compared to those who took the medication on the morning of surgery [10]. A higher incidence of severe hypotension has also been reported in patients who undergo general anesthesia taking an ARB. The European guidelines, however, defer slightly in their recommendations by advising continuation of ACEI/ARB under close monitoring in stable patients with left ventricular systolic heart failure [14].

Calcium Channel Blockers

Cardiovascular society guidelines are unanimously unable to recommend the use of calcium channel blockers perioperatively due to a lack of substantial evidence. Although preoperative use of diltiazem has been shown to reduce the risk of death and/or myocardial infarction [26], large-scale trials to demonstrate the efficacy of verapamil and dihydropyridines are lacking. In addition, calcium channel blockers have been associated with an increased incidence of postoperative bleeding presumably due to an inhibition of platelet aggregation [27].

Clonidine

The ACC/AHA, ESC/ESA, and CCS guidelines unanimously advise against the use of alpha-agonists perioperatively largely due to the findings of the POISE-2 trial which demonstrated an increased rate of nonfatal cardiac arrest and lack of reduction in death with perioperative clonidine use [11].

Diuretics

Though evidence in the current literature is limited, the ESC/ESA guidelines suggest that perioperative use of diuretics up to the day of surgery is reasonable in the management of volume overload. However, dose reduction and/or discontinuation of diuretics should be considered in the presence of hypovolemia, hypotension, or electrolyte disturbances as the use of diuretics in these conditions may increase the risk of hypokalemia, anesthesia-related complications, and cardiac arrhythmias [8].

Heart Failure

Background

Heart failure (HF), a cardiovascular syndrome usually defined by the presence of dyspnea, edema, fatigue, and/or decreased ventricular function [7], is playing an increasingly common role in preoperative risk assessment. With advances in available medical therapy and the development of more effective assistive devices, many adults who may have previously succumbed to the complications of HF are now progressing to an advanced age [28]. In fact, heart failure (HF) has been found to be present in as much as 20% of older adults undergoing surgery in the United States [29] and has been identified as a major risk factor for adverse cardiac events following noncardiac surgery [30]. Elderly patients with HF undergoing surgery have also been found to be at higher risk of hospital readmission and postoperative mortality as compared to those with coronary artery disease undergoing similar procedures [29]. It is thus prudent to identify which patients warrant further investigation when considering HF as an independent risk factor for those undergoing noncardiac surgery.

Risk Assessment

The clinical assessment of left ventricular (LV) function in patients with new or existing HF is paramount in the perioperative period and may significantly impact the postoperative outcome. Patients with newly diagnosed heart failure with reduced LV function are known to be at higher risk than those with existing HF [31] with an increase in postoperative risk with lower left ventricular ejection fraction (LVEF) [32, 33]. It is also important to identify certain clinical features such as hypotension, persistent tachycardia, worsening renal function, hyponatremia, history of multiple implantable cardioverter defibrillator (ICD) shocks, or recent hospitalizations as these are associated with worse short-term survival in patients with reduced LVEF [34]. Existing guidelines appear to support the role of perioperative LV assessment for patients with suspected HF or clinical evidence of worsening HF [7, 8]; however, the American and Canadian guidelines recommend against routine assessment of LV function for patients without heart failure [7, 9]. Although the ACC/AHA guidelines do not support the measurement of natriuretic peptides in perioperative LV assessment, the European and Canadian guidelines stated that measurement of natriuretic peptides is an acceptable method of LV assessment with the CCS guidelines specifically highlighting patients 65 years of age or older as appropriate candidates [9]. This recommendation is supported by a study in 2013 that found brain natriuretic peptide (BNP) or N-terminal prohormone BNP (NT-proBNP) levels may be helpful in identifying patients at higher risk for adverse outcomes when the diagnosis of HF is unclear [35].

Perioperative Medical Management of HF

The ESC/ESA guidelines are unique in that they advocate patients with existing HF undergoing intermediate or high-risk surgery be therapeutically optimized with medical management and that patients with newly diagnosed HF postpone surgery for at least 3 months to allow time for initiation and optimization of medical therapy [8]. This recommendation was corroborated and expanded upon by studies that support delaying elective surgery for at least 1 month for patients with decompensated heart failure with reduced ejection fraction (HFrEF) or new HF with preserved ejection fraction (HFpEF) or delaying urgent surgery by a couple of days to allow for medical optimization [31, 34]. See Table 4.4 for a summary of current recommendations [7–9].

Beta-Blockers

The ACC/AHA and ESC/ESA guidelines agree that patients including those with existing heart failure who are already receiving beta-blockers should continue them in the perioperative period [7, 8]. However, beta-blockers should not be initiated perioperatively as the beneficial effects of

Table 4.4 Heart failure recommendations

Risk assessment	Asymptomatic/HF not suspected: recommend against evaluation of LV function
	Symptomatic/suspected, existing or worsening HF: recommend assessment of LV function with echocardiography or if not feasible, BNP/NT-proBNP level
Perioperative management	Low-risk surgery: reasonable to proceed with surgery
	Intermediate-/high-risk surgery: recommend optimization of HF with medical therapy prior to surgery
	Newly diagnosed HF: recommend postponing nonemergent surgery for at least 3 months to allow for medical optimization
	Decompensated HF: recommend postponing nonemergent surgery for at least 1 month to allow for medical optimization

therapy can take weeks or months to manifest whereas the potential effects of intraoperative hypotension would be immediate.

ACE Inhibitors/ARBs

While all existing guidelines consider it reasonable to hold ACE/ARB in the perioperative period in order to reduce the risk of intraoperative hypotension, the European guidelines state that ACE inhibitors or ARBs may be given safely up until the night prior to the day of surgery [8].

Mineralocorticoid Receptor Antagonist

There is currently a lack of substantial evidence to support the use of these medications in the perioperative period, and we therefore advise deferring initiation of mineralocorticoid receptor antagonists until after surgery.

Digoxin

The role of digoxin is not well defined in the perioperative period, but it is reasonable to consider as an adjunct to other therapies in patients with HFrEF with persistent symptoms in the absence of renal failure.

Diuretics

These agents are recommended for HF patients with signs and symptoms of volume overload.

Arrhythmias

Background

Arrhythmias and conduction disorders are an increasingly common finding among the elderly population. The presence of symptomatic arrhythmias may indicate an increased risk of mortality and should be considered in the preoperative evaluation of patients undergoing noncardiac surgery. High-grade AV blocks, symptomatic ventricular arrhythmias, and supraventricular arrhythmias with uncontrolled ventricular rate are known to be major predictors of an increased perioperative risk [36]. Aberrant rhythms such as atrial fibrillation and ventricular tachycardia often signifies underlying heart disease and therefore merits evaluation in the perioperative period [8] no differently than they would in a patient not undergoing surgery. During preoperative risk stratification, the emphasis lies in identification and correction of reversible causes such as electrolyte abnormalities, acid-base disorders, or decompensated heart failure [28] as the use of antiarrhythmic drugs is not recommended in the absence of hemodynamic instability.

However, in the case of asymptomatic patients, evidence is not in favor of further evaluation during the perioperative period as patients with intraventricular conduction delays with or without the presence of bundle branch blocks were rarely found to progress to complete AV block during this time [37]. Furthermore, one study utilizing continuous electrocardiogram (EKG) found that ventricular arrhythmias including couplets and non-sustained ventricular tachycardia (NSVT) were not associated with an increase in adverse cardiac events after noncardiac surgery [38].

Perioperative Guidelines

The ACC/AHA guidelines are limited in their recommendations regarding the management of arrhythmias and conduction abnormalities in the perioperative period citing a lack of substantial evidence to support or oppose specific treatment modalities. They instead focus on the management of patients with cardiovascular implantable electronic devices (CIEDs) and advocate the early involvement of a designated clinical team to oversee the function of the CIED perioperatively and if need be during the procedure [7].

The ESC/ESA guidelines offer more specific recommendations based on the nature of the involved arrhythmia. In the presence of supraventricular tachycardia (SVT), the European guidelines advise continuation of oral antiarrhythmic drugs prior to surgery as well as electrical cardioversion if the patient is hemodynamically unstable [8]. For patients with bradyarrhythmias, there is no difference in the indications for temporary pacemakers during the perioperative period when compared to criteria for placement of permanent pacemakers. The ESC/ESA also advocate the presence of external defibrillation equipment in patients whose ICDs were deactivated preoperatively. In regard to ventricular tachycardia (VT), the European guidelines recommend continuation of any oral antiarrhythmic drugs especially in patients with sustained VT [8].

Atrial Fibrillation/Flutter

Paroxysmal atrial fibrillation (AF) can be seen in approximately 10% of elderly patients in the community with incidence of chronic AF doubling with each decade of life after 60 years of age [28]. Both the American and European guidelines currently advocate the preoperative treatment of atrial fibrillation if the patient experiences hemodynamic instability. Beta-blockers have been shown to facilitate the conversion of AF to sinus rhythm after noncardiac surgery [39] and are thus the treatment of choice along with non-dihydropyridine calcium channel blockers [40]. Amiodarone is the preferred agent in patients with heart failure as digoxin has been found to be ineffective during surgery. In addition, preoperative catheter ablation may be reasonable in patients with Wolff-Parkinson-White syndrome and preexcitation [8] based on the urgency and risk category of the planned procedure.

Recently revised ACC guidelines (2017) regarding perioperative bridging for patients with atrial fibrillation take into consideration both the type of anticoagulation the patient is already receiving as well as the future bleeding risk and thrombotic risk [12]. Bleeding risk is defined by the presence of at least one of the following factors: platelet abnormalities including aspirin use, major bleed or intracranial hemorrhage <3 months, supratherapeutic INR, or bleeding as the result of previous bridging. Thrombotic risk is assessed using CHADs-VASc with a score 1–4 corresponding to low risk, 5–6 moderate risk, and >7 high risk. Patients on a direct oral anticoagulants (DOAC) or patients on warfarin with CHADs-VASc <5 do not require bridging, and anticoagulation can simply be held for the procedure. Patients receiving warfarin with CHADs-VASc >7 without increased bleeding risk should receive periprocedural bridging with UFH or low molecular weight heparin (LMWH). Patients on warfarin with CHADs-VASc >5 and at least one increased bleeding risk factor are subject to individualized clinical judgment when determining the appropriateness of bridging anticoagulation.

Bradyarrhythmias

Patients with high-grade conduction abnormalities and symptomatic bradycardia may require perioperative temporary pacemaker placement [28] as unexpected complete atrioventricular block may increase operative risk [41] if necessary pacing equipment is not readily available during the procedure. Although prophylactic pacing prior to noncardiac surgery is not always indicated, preoperative temporary pacing may be beneficial in patients with prior symptomatic asystolic episodes [8]. We recommend having transvenous or transcutaneous pacing equipment on standby during the procedure in the event of intraoperative complete heart block.

Supraventricular Tachycardia (SVT)

According to the ESC/ESA guidelines, beta-blockers, calcium channel blockers, or amiodarone can be safely used as prophylactic treatment where needed in the perioperative management of SVT [8].

Ventricular Arrhythmias

Although considered general risk factors of development of arrhythmias, premature ventricular contractions and NSVT have not been shown to increase the risk of nonfatal MI or cardiac death in the perioperative period [38, 42]. Current evidence does not indicate a worse prognosis with these rhythm aberrancies [8].

Perioperative management of sustained monomorphic VT consists of intravenous amiodarone for hemodynamically stable patients [43] and electrical cardioversion in the presence of hemodynamic compromise [8].

For patients with hemodynamically stable, sustained polymorphic VT, amiodarone is an acceptable treatment option in the absence of the long QT syndrome [43]. Current evidence supports the use of magnesium sulfate for patients with the long QT syndrome and torsades de pointes [44]. In patients with torsades de pointes and concurrent sinus bradycardia, the use of beta-blockade with temporary pacing is advised with isoproterenol reserved for use in patients with recurrent, pause-dependent torsades de pointes without congenital long QT syndrome [43].

Cardiac Implantable Electronic Device (CIED)

Pre-procedural planning and evaluation of implantable cardiac devices are crucial steps during the perioperative period to prevent adverse cardiac events related to conduction abnormalities and device malfunction as the use of electrocautery during surgery may interfere with device function. The management of CIEDs must be individualized based on the exact type of device and is ideally conducted by a multidisciplinary team consisting of electrophysiologists and device manufacturer technicians. The Heart Rhythm Society (HRS) recommends pre-procedure device interrogation within 6 months of surgery for an ICD, 12 months for a PPM, and 3–6 months for a cardiac resynchronization (CRT) device [45] in order to ensure appropriate device functioning prior to surgery. This assessment should be performed perioperatively in order to confirm appropriate device programming and function based on the patient's underlying rhythm [45–52]. In the event of an emergent surgery, a magnet may be placed over the device for reprogramming in the presence of a trained cardiologist who can interpret EKG recordings to evaluate the device backup mode and function. Pacemakers should be switched to a non-sensing mode with pacemaker interrogation following surgery. If the use of electrocautery is planned during surgery, the ICD should be turned off intraoperatively and switched back on during post-op recovery [8]. See Table 4.5 for a summary of current recommendations [7–9, 43, 45].

Coronary Artery Disease (CAD)

The stress related to noncardiac surgery has many physiological effects on the body with regard to the cardiovascular system resulting in tachycardia, high blood pressure, and an increased free fatty acid concentration. These changes in turn increase myocardial oxygen demand and may precipitate ischemic heart disease [53, 54].

Patients with risk factors for CAD, known CAD or poor exercise capacity, need to be evaluated for perioperative cardiovascular risks. The role of the 12-lead EKG is limited but reasonable in patients with known CAD [7]. The perioperative assessment of left ventricular function with echocardiography may have a role in heart failure patients but not necessarily in CAD assessment [7]. Stress tests such as an exercise EKG, dobutamine echocardiography, or treadmill test are dynamic tests that have a valuable role in assessing a patient's ability to tolerate external stress. Stress tests can evaluate the ischemic threshold of the myocardium and localize CAD with regard to the amount of threatened myocardium with an excellent negative predictive value (95–100%) although the positive predictive value is limited (20–30%) [55]. The ACC/AHA 2014 guidelines recommend the use of noninvasive stress tests for patients with elevated cardiovascular risk and unknown or poor exercise capacity if it will change management. It is reasonable to forgo further exercise testing in patients with an elevated risk if the patient reports good (≥10 metabolic equivalent of task (METs)) or moderate to good functional capacity (≥4 METs to 10 METs). Relatively newer cardiovascular diagnostic modalities such as coronary calcium scores, CT angiography, and cardiac MRI continue to expand

Table 4.5. Arrhythmia recommendations

Atrial fibrillation (AF)	*Elective/urgent surgery*
	Proceed with surgery if hemodynamically stable
	Preoperative medical treatment indicated if hemodynamic stability is compromised
	Consider catheter ablation if Wolff-Parkinson-White
	Emergent surgery
	Consider electrical cardioversion if hemodynamically unstable
Anticoagulation	*Patients taking DOAC*
	Hold therapy. No indication for bridging anticoagulation
	Patients taking warfarin
	If CHADs-VASc <5, hold warfarin, no need for bridging
	If CHADs-VASc >7, hold warfarin, begin bridging anticoagulation with LMWH or UFH
Bradyarrhythmias	*Asymptomatic*: prophylactic pacing not indicated
	Symptomatic asystolic episodes: recommend transvenous/transcutaneous pacing equipment on standby intraoperatively
Supraventricular tachycardia	Treatment required only for hemodynamic instability
Ventricular arrhythmias	*Premature Ventricular Contractions (PVCs)/NSVT*: preoperative treatment not required
	Sustained monomorphic VT
	If hemodynamically stable, treat with IV amiodarone
	If hemodynamically unstable, electrical cardioversion
	Sustained polymorphic VT
	If no long QT syndrome, treat with IV amiodarone
	If suspected long QT syndrome, treat with IV magnesium
	Torsades de pointes (TdP)
	If concurrent sinus bradycardia, recommend temporary pacing
	If TdP only with recurrent pauses, can use isoproterenol
Cardiac implantable electronic devices	Recommend pre-procedure interrogation of device by a cardiologist to confirm proper settings and function
	Permanent pacemaker: expert interrogation within 12 months of surgery
	Device should be switched to non-sensing mode intraoperatively and interrogated postoperatively
	Implantable cardiac defibrillator
	Expert interrogation within 6 months of surgery
	If electrocautery use is planned, device should be turned off intraoperatively and switched on during recovery
	Cardiac resynchronization therapy: expert interrogation within 3–6 months of surgery

our ability to monitor different stages of ischemic heart disease [56]. The widespread use of these modalities still requires further study to establish efficacy and cost-effectiveness.

The evaluation of exercise capacity is vital in the perioperative assessment of elderly patients in order to determine whether further investigation is warranted. Metabolic equivalents (METs) is an objective method to describe the functional capacity or exercise tolerance where one MET is equivalent to a resting or basal oxygen consumption of a 40-year-old 70 kg man [57]. For example, daily living activities such as dressing, eating, and toilet use are equivalent to one to two METs, and sports activities such as swimming, singles tennis, and football require more than ten METs. The number of METs associated with the normal physiological stress level of most noncardiac surgeries with general anesthesia is four to five METs and can be correlated to a patients' capacity to walk four blocks or climb two flights of stairs without stopping due to limiting symptoms. Studies have shown patients with METs less than 4–5 are at increased perioperative cardiovascular risk [7]. One study with 600 participants reported association of major perioperative cardiovascular events with poor functional capacity (<4 METs) [58]. The objective assessment of exercise capacity is crucial as a study with 5939 patients found the subjective assessment of exercise capacity has minimal to poor correlation in prediction of cardiovascular complications and mortality [59]. The Duke Activity Status Index is a standardized 12-questionnaire-based index for functional status assessment, and studies have demonstrated that it correlates well with gold standard measures of functional capacity [60, 61].

According to the major perioperative cardiology guidelines [7, 8], there is no role for routine coronary revascularization prior to noncardiac surgery to reduce perioperative cardiac events [62]. If the perioperative ischemic work-up findings are significant and suggestive revascularization per existing guidelines, then coronary revascularization is recommended before planned major noncardiac surgery. Studies have shown reduced risk with intervention and better outcomes with coronary artery bypass graft (CABG) than percutaneous coronary intervention (PCI) [63]. For patients on dual antiplatelet therapy, it is recommended to delay noncardiac surgery for 2 weeks after balloon angioplasty, at least 30 days after bare metal stent (BMS) placement and 6 months after drug eluting stent (DES) placement. The risk of stent thrombosis is less with newer generation drug eluting stents. The updated guidelines for perioperative dual antiplatelet therapy (DAPT) have a class I recommendation to proceed with surgery if it is greater than 6 months post-DES implantation and class IIb recommendation to operate if 3–6 months have passed since DES implantation if the risk of delaying surgery is higher than the risk of stent thrombosis. Asymptomatic patients who have undergone CABG in the past 6 years can undergo elective or nonurgent noncardiac surgery without further angiographic evaluation except for patients with a high perioperative risk.

Several studies have reported on the prognostic use of postoperative elevation of troponin and BNP in predicting higher short-term mortality [64–66]. A prospective Vascular Events In Noncardiac Surgery Patients Cohort Evaluation (VISION) trial with 15,000 patients reported that age >75 years and troponin ≥0.04 are independent risk factors of 30-day mortality [65]. ACC/AHA guideline [7] for perioperative management recommends checking troponin and EKG in the setting of probable myocardial ischemia or infarction, but doing so in high-risk patients regardless of symptomatology is of uncertain value. Routine measurement of troponin levels in asymptomatic patients is not useful. Perioperative medication management is also crucial for better outcomes. A major trial for the use of perioperative aspirin (POISE 2 trial) [67] showed no benefit in outcomes including death or nonfatal MI (7.0% vs. 7.1%, $p = 0.92$) even in subgroup analysis, regardless of Revised Cardiac Risk Index (RCRI) score or aspirin use prior to randomization. However, it does increase the risk of major bleeding (4.6% vs. 3.8%, $p = 0.04$). A case-controlled study including 2816 patients undergoing major vascular surgery reported a lower adjusted odds ratio 0.22 (95% confidence interval 0.10–0.47) for perioperative mortality among statin users as compared with nonusers [68]. A study including 1163 patients of a predominantly elderly population showed a significant protective effect of statin use (odds ratio 0.52, $p = 0.001$) on various cardiovascular complications [69]. Another study with 577 elderly patients undergoing noncardiac vascular surgery reported significantly reduced perioperative MI or mortality rates in statin users compared to patients not on a statin (11% vs. 27%, $p < 0.0001$) [70]. The major guidelines [7, 8] recommend continuous perioperative use of statins in clinically indicated cases and initiation of statin therapy before vascular surgery. A propensity-matched study using perioperative beta-blockers reported a reduced mortality rate among high-risk patients (RCRI >2) undergoing major noncardiac surgery [71]. A randomized control study across 23 countries

(POISE trial) [72], with nearly 50% of the population age >70 years, showed beneficial effect of metoprolol use in MI (4.2% vs. 5.7%, $p = 0.0017$) within 30 days postoperatively. The same study also showed a higher incidence of stroke in the metoprolol group than with a placebo (1% vs. 0.5%, $p = 0.0053$) and higher mortality (3.1% vs. 2.3%, $p = 0.0317$). Thus, the utility of perioperative beta-blocker therapy remains questionable, but it is recommended not to withhold perioperative beta-blockers in the absence of bradycardia or hypotension. Patients with known ischemic heart disease who are at high risk for perioperative MI or those with ≥3 RCRI risk factors may benefit from starting a beta-blocker perioperatively. However, it would be reasonable to initiate beta-blocker therapy more than 1 week prior to surgery to determine safety and tolerability [7]. See Table 4.6 for a summary of current recommendations [7–9].

Dyslipidemia

Hypercholesterolemia is associated with various cardiovascular diseases. High total serum cholesterol, high low-density lipoproteins (LDL), or low high-density lipoproteins (HDL) have been reported as a strong risk factor for atherosclerosis and coronary events [73–75]. Aronow and colleagues reported the impact of HDL level on new coronary events in elderly patients as decrease of every 10 mg/dl of serum HDL cholesterol would increase the event rates by 1.7 times in men (at 40-month follow-up) and 1.95 times in women (at 4-year follow-up) [73].

Erosions and rupture of the coronary plaques are the primary precipitants of acute coronary events in the operative and nonoperative setting [76]. Statins have anti-inflammatory and plaque-stabilizing properties in addition to cholesterol-lowering property. A 6-month follow-up study

Table 4.6 Coronary artery disease recommendations

Risk assessment	*Exercise capacity*
	Poor functional capacity (<4 METs) is associated with major perioperative cardiovascular events
	Objective assessment of exercise capacity is crucial; subjective assessment of exercise capacity has minimal to poor correlation in prediction of cardiovascular complications and mortality
	12-lead EKG: reasonable in patients with known CAD
	Echocardiography: no role in preoperative CAD assessment
	Stress test: recommend in patients with elevated cardiovascular risk and unknown or poor exercise capacity if it will change management
	It is reasonable to forgo further exercise testing evaluation in patients with elevated risk if the patient has good (≥10 METs) or moderate to good functional capacity (≥4 METs to 10 METS)
	Coronary calcium scores, CT angiography, cardiac MRI: further randomized controlled trials are needed to evaluate the role in perioperative evaluation
Perioperative management	*Role of pre-op angiography and interventions*
	If perioperative ischemic work-up warrants revascularization per existing guidelines, then coronary revascularization is recommended
	No role for routine pre-op coronary revascularization
	Asymptomatic patients who have undergone coronary bypass graft (CABG) in the past 6 years can undergo elective or nonurgent, noncardiac surgery without further angiographic evaluation except for patients with high perioperative risk
	Checking biomarkers: routine measurement of troponin levels in asymptomatic patients is not useful
	Low-intermediate-risk patients: recommend checking troponin and EKG in the setting of probable myocardial ischemia or infarction
	High-risk patients: checking troponin and EKG regardless of symptomatology is of uncertain evidence

(with >66% patients with age >65 years) on postoperative cardiovascular outcomes after vascular surgery comparing placebo group to patients on atorvastatin showed three times higher incidence rate of cardiac events in placebo group (26% vs. 8%, $p = 0.018$) where statins were started on average of 30 days prior to surgery [77]. The major studies on fluvastatin [78, 79] also reported the cardiovascular benefit of statin in the perioperative period of noncardiac surgery. A study comparing 577 elderly patients (mean age 74 years) undergoing noncardiac vascular surgery treated with and without statins reported significant positive impact of statin on lowering perioperative MI rate (6% vs. 14%, $p = 0.001$) and 2-year mortality rate (6% vs. 16%, $p = 0.0002$), showing patients using statins had 57% lesser odds of perioperative MI or death at 2-year follow-up after controlling for other variables [68]. While studies have shown positive results of statin in the perioperative period, a prospective study comparing postoperative cardiac outcomes after continuing or discontinuing the chronic statin therapy after major vascular surgery (infrarenal aortic surgery) was done which showed that the discontinuation of postoperative statin (>4 days) was an independent predictor of postoperative cardiac myonecrosis (OR 2.9, 95% confidence interval 1.6–5.5) [80].

Current perioperative guidelines from ACC/AHA recommend continuing the statin therapy in patients currently taking it as a class I recommendation. It is reasonable to begin the perioperative statin in patients undergoing the vascular surgery, and perioperative initiation of statins may be considered if a patient has clinical indications according to the guideline and undergoing high-risk surgeries. During the perioperative period, statins with the longer half-life (e.g., atorvastatin) or extended release formulations (e.g., lovastatin) are preferred as a bridge immediately post-surgery when oral intake is not feasible [7, 8]. The main concern for the statin use is a side effect profile which includes muscle damage. Due to multiple factors perioperatively such as renal function impairment after major surgery and anesthetic drug use that may increase the risk of statin-induced myopathy, early initiation of the statin is recommended to detect and manage the potential side effects. If the statin-naive patient is undergoing vascular intervention, the statin should ideally be started at least 2 weeks prior to the procedure for a maximal plaque-stabilizing effect and should be continued at least 1-month post-intervention [7, 8]. See Table 4.7 for a summary of current recommendations regarding dyslipidemia and peripheral arterial disease [7–9].

Table 4.7 Dyslipidemia and peripheral arterial disease

Dyslipidemia recommendations	It is reasonable to begin statins perioperatively in patients undergoing vascular surgery
	Perioperative initiation of statins may be considered if a patient has clinical indications according to established guidelines and is undergoing high-risk surgery
	If a statin-naïve patient is undergoing vascular intervention, statins should ideally be started at least 2 weeks prior to the procedure for a maximal plaque-stabilizing effect *and* should be continued at least 1 month post-intervention
Peripheral arterial disease recommendations	Conservative and medical management of modifiable risk factors that affect perioperative morbidity and mortality is strongly recommended for better outcomes

Peripheral Arterial Disease (PAD)

Peripheral arterial disease (defined as an ankle-brachial ratio of less than 0.9 or previously vascularized with surgery or percutaneous transluminal angioplasty) is one of the common manifestations of atherosclerosis. Patients with PAD have a higher risk of concomitant atherosclerotic disease in other arteries including coronaries and cerebral vasculature [81], and therefore, these patients are at elevated risk of all-cause mortality, cardiovascular mortality, and mortality from coronary artery disease [82, 83]. Considering this, it is appropriate to evaluate the presence of ischemic heart disease in these patients from the medical history, a routine clinical exam and tests, but there is no evidence supporting routine investigations with an exercise or imaging test to detect asymptomatic atherosclerosis, unless the patient has more than two of the following risk factors: ischemic heart disease, heart failure, stroke or transient ischemic attack, renal dysfunction (serum creatinine >2 mg/dl or creatinine clearance <60 ml/min/1.73 m^3), and diabetes mellitus requiring insulin treatment [8, 84].

Not only the atherosclerotic cardiac disease but also the presence of atherosclerotic disease in other vessels has an impact on perioperative complications. A study showed a 2–3 times elevated risk in patients with two or more atherosclerotic diseases than only one [85]. An elderly population study [86] reported the prevalence of modifiable risk factors for symptomatic PAD as age (odds ratio, men = 1.052, women 1.025), cigarette smoking (odds ratio, men = 2.552, women 4.634), hypertension (odds ratio, men = 2.196, women 2.777), diabetes mellitus (odds ratio, men = 6.054, women 3.594), serum HDL cholesterol (odds ratio, men = 0.948, women 0.965), and serum LDL cholesterol (odds ratio, men and women = 1.019). Conservative and medical management of modifiable risk factors, which affects the perioperative morbidity and mortality, is strongly recommended in the guideline for better outcomes [84].

Valvular Heart Disease

Background

Valvular heart lesions are common in elderly people. Significant valvular stenosis or insufficiencies are of important consideration because it can lead to an unfavorable hemodynamic state in the perioperative period. Anticoagulation management in valvular heart disease, especially in prosthetic heart valves, needs to be managed properly considering the risks of surgery-related bleeding events versus a prothrombotic valvular state and consequences of thromboembolism.

Aortic Stenosis (AS)

AS is most common valvular heart disease in elderly people and usually results from degenerative calcification of the aortic valve. A 2013 meta-analysis study reported AS prevalence rate was 12.4% in the elderly (age more than 75 years) with a severe AS prevalence rate of 3.4% [87]. A study by Kertai reported higher perioperative cardiovascular complications (MI and mortality) from severe aortic stenosis compared to moderate aortic stenosis (31% vs. 11%, $p = 0.04$) [88]. This shows a substantial rate of disease burden in the general elderly population which requires medical attention before major surgeries.

Advances in the anesthetic approach and surgical approach and in perioperative care have had a major impact on declining cardiac risk in patients with significant AS undergoing noncardiac surgery. A recent tertiary center study [89] with a propensity score analysis showed that patients undergoing nonemergent noncardiac surgery with moderate AS (aortic valve area: 1.0 cm^2 to 1.5 cm^2) or severe AS (aortic valve area <1.0 cm^2) compared to patients without AS had a 30-day mortality rate of 2.1% vs. 1.0% ($p = 0.036$) and a postoperative myocardial infarction rate of 3.0% vs. 1.1% ($p = 0.001$). The worst primary outcomes (defined as a composite of 30-day mortality and postoperative MI) were reported in moderate AS (4.4% vs. 1.7%; $p = 0.002$) and severe AS (5.7% versus 2.7%; $p = 0.02$) compared to patients without AS, and the predictors were high-risk surgery, symptomatic AS, coexisting mitral regurgitation, and preexisting CAD.

Transthoracic echocardiography (TTE) is a standard method of assessment for the aortic valve. The perioperative management ACC/AHA guideline recommends preoperative echocardiography in suspected moderate or higher degree of AS if no prior echocardiogram is available within 1 year or if there is a significant clinical status change since the last evaluation (level of evidence: C), and in patients who meet standard indication of valvular intervention based on symptoms and severity of the valvular problem, an intervention before elective noncardiac surgery is effective in reducing perioperative risk. In those patients who are high-risk candidates for surgical AVR (aortic valve replacement), studies have demonstrated that TAVR (transcatheter aortic valve replacement) is a good alternative option [90]. Recent studies have reported on favorable outcome of TAVR in intermediate- and low-risk patients as well [91, 92]. If the patient is not a candidate for valve replacement, percutaneous balloon aortic valvuloplasty may be considered as a bridging procedure during a need for urgent noncardiac surgery with hemodynamic instability [1]. A study in patients aged >75 years with asymptomatic severe AS [93] reported intraoperative hypotension is frequent and requires aggressive management, but intermediate- to low-risk noncardiac surgery is relatively safe in a controlled setting. ACC/AHA guideline reports emergency noncardiac surgery may occur in patients with significant valvular disease, but the risk can be minimized by making an accurate diagnosis of the type and severity of valvular diseases, the appropriate choice of anesthetic agents, and a high level of both perioperative and postoperative patient care in the ICU setting.

Mitral Stenosis (MS)

Patients with severe MS are at an increased risk of perioperative complications and should be managed similarly to patients with AS. The main perioperative concerns for the MS patient are hemodynamic alterations, pulmonary edema from volume status changes, and the occurrence of arrhythmias such as atrial fibrillation.

The guidelines on perioperative MS management recommend valvular interventions before elective noncardiac surgery if the patient meets an indication for intervention. The noncardiac surgery may be considered without intervention if it's emergency or the valve anatomy is not favorable for percutaneous mitral balloon commissurotomy but with invasive cardiac hemodynamics monitoring and optimization [7]. The European guidelines on perioperative valvular management [94] recommend no indication of any mitral valve interventions if patient is asymptomatic with systolic pulmonary artery pressure (SPAP) <50 mmHg or nonsignificant mitral stenosis (valve area <1.5 cm^2). If a SPAP is >50 mmHg in an asymptomatic patient or the noncardiac surgery-related risk is high, a patient may benefit from intervention (percutaneous mitral commissurotomy or open surgical repair) prior to major surgery.

Prosthetic Valves

Patients with mechanical heart valves have a similar risk of perioperative morbidity and mortality compared to patients with native cardiac valves, reported by one study [95], as a risk of bleeding (18.6% vs. 14.2%, $p = 0.989$), thromboembolism (3.6% vs. 2%, $p = 0.989$), and mortality at 3 months (1.4% vs. 1.3%, $p = 0.825$). The guidelines also suggest that patients with prosthetic valves can undergo noncardiac surgery without any additional risk when there is no evidence of valve or ventricular dysfunction [7, 8]. The main clinical problem for prosthetic heart valve patients remains with anticoagulation management when a brief interruption of therapy is required to achieve adequate hemostasis. With major surgery the risk of perioperative bleeding and thrombosis should be weighed in individual cases. Overall, mechanical valves are comparatively more thrombogenic than bioprosthetic valves, and caged ball valves are more thrombogenic than tilting disk valves. Currently, the anticoagulation in prosthetic heart valves is an uncertain topic with multiple ongoing clinical trials, but the evidence is more toward the use of intravenous UFH perioperatively [7, 8].

The latest guidelines regarding the anticoagulation recommend that perioperative anticoagulation bridging is appropriate with a mechanical mitral valve regardless of any risk factors for thromboembolism and for the aortic valve if one or more of the following risk factors such as AF, previous thromboembolism, LV dysfunction, hypercoagulable condition, or an older generation prosthetic aortic valve are present [7]. Patients receiving an oral vitamin K antagonist (VKA) such as warfarin may require bridging with unfractionated heparin or low molecular weight heparin, starting once the INR goes subtherapeutic, until 4 hours before surgery, and then restarting warfarin based on hemostatic status of the patient but at least 12 hours post-surgery. VKA should be restarted at a preoperative maintenance dose plus a 50% booster dose for two consecutive days followed by a preoperative maintenance dose. Heparin should be continued until the therapeutic INR level from VKA is achieved [7, 8]. When rapid correction of anticoagulation is required, fresh frozen plasma and prothrombin complex concentrate (PCC) are the choices. Vitamin K does not have any immediate effects, but it delays the INR level to quickly reach the therapeutic level again. Factor Xa inhibitors (rivaroxaban and apixaban) and direct thrombin inhibitors (dabigatran) did not have any reversal agent until the recent FDA approval of andexanet alfa and idarucizumab, but their use in various clinical settings, safety and efficacy have yet to be investigated [96, 97]. See Table 4.8 for a summary of current recommendations [7–9].

The other important perioperative point to consider is antibiotic prophylaxis as prosthetic heart valves are high risk for infective endocarditis. The antibiotic approach and regimen for prophylaxis during major surgery should be considered as per standard guidelines [98].

Table 4.8 Valvular heart disease recommendations

Aortic stenosis	*Pre-op echocardiography (Echo)*: may be considered in suspected moderate to severe AS if no prior echo is available within 1 year or if there is a significant clinical status change since the last evaluation (level of evidence: C)
	Interventions: recommended if symptoms and severity of stenosis meet criteria for valve replacement or repair
	Emergent surgery: the risk can be minimized by high-level perioperative care and post-op care in an intensive care unit
Mitral stenosis	*Interventions*: recommended if symptoms and severity of stenosis meet criteria for valve replacement or repair
	Systolic pulmonary artery pressure (SPAP) <50 mm Hg: no indication of mitral valve intervention if nonsignificant mitral stenosis (valve area <1.5 cm^2)
	Systolic pulmonary artery pressure (SPAP) >50 mm Hg: if patient is asymptomatic or surgery related risk is high, patient may benefit from an intervention (percutaneous mitral commissurotomy or open surgical repair) prior to major surgery
	Emergent surgery: can be considered without valvular intervention *but* with close perioperative care
Prosthetic heart valves	Mechanical valves are comparatively more thrombogenic than bioprosthetic valves
	Caged ball valves are more thrombogenic than tilting disk valves

Conclusion

The aging process brings physiological, anatomical, and cognitive changes, with resultant loss of ability to cope with operative stress. In addition, elderly patients often have multiple cardiac comorbidities, complex medication regimens, functional limitation, cognitive impairment, and frailty. All these factors may play a role in increasing the likelihood of postoperative cardiac and noncardiac complications. As the world population is becoming older, geriatric surgery has become more frequent and requires physicians to become more familiar with special issues that are unique to this group. Preoperative cardiac risk assessment involves a multidisciplinary approach and should focus on the urgency of the procedure, patient's baseline functional capacity, and cardiac risk profile. It is also critical that physicians carefully weigh the risk and benefit of surgery, value of additional cardiac testing, and patient's long-term goals.

References

1. Colloca G, Santoro M, Gambassi G. Age-related physiologic changes and perioperative management of elderly patients. Surg Oncol. 2010;19(3):124–30.
2. Baquero GA, Rich MW. Perioperative care in older adults. J Geriatr Cardiol. 2015;12(5):465–9.

3. Partridge JS, Collingridge G, Gordon AL, Martin FC, Harari D, Dhesi JK. Where are we in perioperative medicine for older surgical patients? A UK survey of geriatric medicine delivered services in surgery. Age Ageing. 2014;43(5):721–4.
4. Yang R, Wolfson M, Lewis MC. Unique aspects of the elderly surgical population: an anesthesiologist's perspective. Geriatr Orthop Surg Rehabil. 2011;2(2):56–64.
5. White PF, White LM, Monk T, Jakobsson J, Raeder J, Mulroy MF, et al. Perioperative care for the older outpatient undergoing ambulatory surgery. Anesth Analg. 2012;114(6):1190–215.
6. Shenoy P, Harugeri A. Elderly patients' participation in clinical trials. Perspect Clin Res. 2015;6(4):184–9.
7. Fleisher LA, Fleischmann KE, Auerbach AD, Barnason SA, Beckman JA, Bozkurt B, et al. 2014 ACC/AHA guideline on perioperative cardiovascular evaluation and management of patients undergoing noncardiac surgery: executive summary: a report of the American College of Cardiology/American Heart Association Task Force on practice guidelines. Developed in collaboration with the American College of Surgeons, American society of Anesthesiologists, American Society of Echocardiography, American Society of Nuclear Cardiology, Heart Rhythm Society, Society for Cardiovascular Angiography and Interventions, Society of Cardiovascular Anesthesiologists, and Society of Vascular Medicine Endorsed by the Society of Hospital Medicine. J Nucl Cardiol. 2015;22(1):162–215.
8. Kristensen SD, Knuuti J, Saraste A, Anker S, Botker HE, Hert SD, et al. 2014 ESC/ESA guidelines on non-cardiac surgery: cardiovascular assessment and management: The Joint Task Force on non-cardiac surgery: cardiovascular assessment and management of the European Society of Cardiology (ESC) and the European Society of Anaesthesiology (ESA). Eur Heart J. 2014;35(35):2383–431.
9. Duceppe E, Parlow J, MacDonald P, Lyons K, McMullen M, Srinathan S, et al. Canadian cardiovascular society guidelines on perioperative cardiac risk assessment and management for patients who undergo noncardiac surgery. Can J Cardiol. 2017;33:17.
10. Coriat P, Richer C, Douraki T, et al. Influence of chronic angiotensin-converting enzyme inhibition on anesthetic induction. Anesthesiology. 1994;81(2):299–307.
11. Cegarra-Sanmartin V, Paniagua-Iglesias P, Popova E, de Nadal CM, Alonso-Coello P, Plou P, et al. Perioperative acetylsalicylic acid and clonidine in noncardiac surgery patients (POISE-2 trial). Rev Esp Anestesiol Reanim. 2015;62(5):270–4.
12. Doherty JU, Gluckman TJ, Hucker WJ, Januzzi JL Jr, Ortel TL, Saxonhouse SJ, et al. 2017 ACC expert consensus decision pathway for periprocedural management of anticoagulation in patients with nonvalvular atrial fibrillation: a report of the American College of Cardiology Clinical Expert Consensus Document Task Force. J Am Coll Cardiol. 2017;69(7):871–98.
13. Levine GN, Bates ER, Bittl JA, Brindis RG, Fihn SD, Fleisher LA, et al. 2016 ACC/AHA guideline focused update on duration of dual antiplatelet therapy in patients with coronary artery disease: a report of the American College of Cardiology/American Heart Association Task Force on Clinical Practice Guidelines: An Update of the 2011 ACCF/AHA/SCAI Guideline for Percutaneous Coronary Intervention, 2011ACCF/AHA Guideline for Coronary Artery Bypass Graft Surgery, 2012, ACC/AHA/ACP/AATS/PCNA/SCAI/STS Guideline for the Diagnosis and Management of Patients With Stable Ischemic Heart Disease, 2013 ACCF/AHA Guideline for the Management of ST-Elevation Myocardial Infarction, 2014 AHA/ACC Guideline for the Management of Patients With Non-ST-Elevation Acute Coronary Syndromes, and 2014 ACC/AHA Guideline on Perioperative Cardiovascular Evaluation and Management of Patients Undergoing Noncardiac Surgery. Circulation. 2016;134(10):e123–55.
14. Aronow WS. Management of hypertension in patients undergoing surgery. Ann Transl Med. 2017;5(10):227.
15. Prys-Roberts C. Anaesthesia and hypertension. Br J Anaesth. 1971;43(2):122–37.
16. Foex P, Meloche R, Prys-Roberts C. Studies of anaesthesia in relation to hypertension. 3. Pulmonary gas exchange during spontaneous ventilation. Br J Anaesth. 1971;43(7):644.
17. Eagle KA, Berger PB, Calkins H, Chaitman BR, Ewy GA, Fleischmann KE, et al. ACC/AHA guideline update for perioperative cardiovascular evaluation for noncardiac surgery–executive summary. a report of the American College of Cardiology/American Heart Association Task Force on Practice Guidelines (Committee to Update the 1996 Guidelines on Perioperative Cardiovascular Evaluation for Noncardiac Surgery). Anesth Analg. 2002;94(5):1052–64.
18. Sellevold OF, Raeder J, Stenseth R. Undiagnosed phaeochromocytoma in the perioperative period. Case reports. Acta Anaesthesiol Scand. 1985;29(5):474–9.
19. Aronson S, Boisvert D, Lapp W. Isolated systolic hypertension is associated with adverse outcomes from coronary artery bypass grafting surgery. Anesth Analg. 2002;94:1079.
20. Wolfsthal SD. Is blood pressure control necessary before surgery? Med Clin North Am. 1993;77:349.
21. Goldman L, Caldera DL. Risks of general anesthesia and elective operation in the hypertensive patient. Anesthesiology. 1979;50(4):285–92.
22. Stone JG, Foëx P, Sear JW, et al. Myocardial ischemia in untreated hypertensive patients: effect of a single small oral dose of a beta-adrenergic blocking agent. Anesthesiology. 1988;68:495.
23. Fleisher LA. Preoperative evaluation of the patient with hypertension. J Am Med Assoc. 2002;287(2043):6.
24. POISE Study Group, Devereaux PJ, Yang H, Yusuf S, Guyatt G, Leslie K, et al. Effects of extended-release metoprolol succinate in patients undergoing

non-cardiac surgery (POISE trial): a randomised controlled trial. Lancet. 2008;371(9627):1839–47.
25. Shammash JB, Trost JC, Gold JM, Berlin JA, Golden MA, Kimmel SE. Perioperative beta-blocker withdrawal and mortality in vascular surgical patients. Am Heart J. 2001;141(1):148–53.
26. Wijeysundera DN, Beattie WS. Calcium channel blockers for reducing cardiac morbidity after noncardiac surgery: a meta-analysis. Anesth Analg. 2003;97(3):634–41.
27. Zuccala G, Pahor M, Landi F, et al. Use of calcium antagonists and need for perioperative transfusion in older patients with hip fracture: observational study. BMJ. 1997;314:643.
28. Aronow WS, Fleg JL, Rich MW, editors. Tresch and Aronow's cardiovascular disease in the elderly. 6th ed. Boca Raton: Taylor & Francis; 2019.
29. Hammill BG, Curtis LH, Bennett-Guerrero E, O'Connor CM, Jollis JG, Schulman KA, et al. Impact of heart failure on patients undergoing major noncardiac surgery. Anesthesiology. 2008;108(4):559–67.
30. Patel AY, Eagle KA, Vaishnava P. Cardiac risk of noncardiac surgery. J Am Coll Cardiol. 2015;66:2140.
31. Upshaw J, Kiernan MS. Preoperative cardiac risk assessment for noncardiac surgery in patients with heart failure. Curr Heart Fail Rep. 2013;10:147.
32. Lerman BJ, Popat RA, Assimes TL, Heidenreich PA, Wren SM. Association of left ventricular ejection fraction and symptoms with mortality after elective noncardiac surgery among patients with heart failure. JAMA. 2019;321(6):572–9.
33. Healy KO, Waksmonski CA, Altman RK. Perioperative outcome and long-term mortality for heart failure patients undergoing intermediate- and high-risk noncardiac surgery: impact of left ventricular ejection fraction. Congestive Heart Fail. 2010;16:45–9.
34. Meyer TE, Kiernan MS, McManus DD, Shih J. Decision-making under uncertainty in advanced heart failure. Curr Heart Fail Rep. 2014;11(2):188–96.
35. Farzi S, Stojakovic T, Marko T, Sankin C, Rehak P, Gumpert R, et al. Role of N-terminal pro B-type natriuretic peptide in identifying patients at high risk for adverse outcome after emergent non-cardiac surgery. Br J Anaesth. 2013;110(4):554–60.
36. American College of Cardiology Foundation/American Heart Association Task Force on Practice Guidelines, American Society of Echocardiography, American Society of Nuclear Cardiology, Heart Rhythm Society, Society of Cardiovascular Anesthesiologists, Society for Cardiovascular Angiography and Interventions, et al. 2009 ACCF/AHA focused update on perioperative beta blockade incorporated into the ACC/AHA 2007 guidelines on perioperative cardiovascular evaluation and care for noncardiac surgery. J Am Coll Cardiol. 2009;54(22):e13–e118.
37. Pastore JO, Yurchak PM, Janis KM, Murphy JD, Zir LM. The risk of advanced heart block in surgical patients with right bundle branch block and left axis deviation. Circulation. 1978;57(4):677–80.
38. O'Kelly B, Browner WS, Massie B, Tubau J, Ngo L, Mangano DT. Ventricular arrhythmias in patients undergoing noncardiac surgery. The Study of Perioperative Ischemia Research Group. JAMA. 1992;268(2):217–21.
39. Balser JR, Martinez EA, Winters BD, Perdue PW, Clarke AW, Huang W, et al. Beta-adrenergic blockade accelerates conversion of postoperative supraventricular tachyarrhythmias. Anesthesiology. 1998;89(5):1052–9.
40. Camm AJ, Lip GY, De Caterina R, Savelieva I, Atar D, Hohnloser SH, et al. 2012 focused update of the ESC guidelines for the management of atrial fibrillation: an update of the 2010 ESC guidelines for the management of atrial fibrillation–developed with the special contribution of the European Heart Rhythm Association. Europace. 2012;14(10):1385–413.
41. Gregoratos G, Abrams J, Epstein AE, Freedman RA, Hayes DL, Hlatky MA, et al. ACC/AHA/NASPE 2002 guideline update for implantation of cardiac pacemakers and antiarrhythmia devices: summary article. A report of the American College of Cardiology/American Heart Association Task Force on Practice Guidelines (ACC/AHA/NASPE Committee to Update the 1998 Pacemaker Guidelines). J Cardiovasc Electrophysiol. 2002;13(11):1183–99.
42. Mahla E, Rotman B, Rehak P, Atlee JL, Gombotz H, Berger J, et al. Perioperative ventricular dysrhythmias in patients with structural heart disease undergoing noncardiac surgery. Anesth Analg. 1998;86(1):16–21.
43. Zipes DP, Camm AJ, Borggrefe M, Buxton AE, Chaitman B, Fromer M, et al. ACC/AHA/ESC 2006 guidelines for management of patients with ventricular arrhythmias and the prevention of sudden cardiac death: a report of the American College of Cardiology/American Heart Association Task Force and the European Society of Cardiology Committee for Practice Guidelines (writing committee to develop Guidelines for Management of Patients With Ventricular Arrhythmias and the Prevention of Sudden Cardiac Death): developed in collaboration with the European Heart Rhythm Association and the Heart Rhythm Society. Circulation. 2006;114(10):e385–484.
44. Drew BJ, Ackerman MJ, Funk M, Gibler WB, Kligfield P, Menon V, et al. Prevention of torsade de pointes in hospital settings: a scientific statement from the American Heart Association and the American College of Cardiology Foundation. J Am Coll Cardiol. 2010;55(9):934–47.
45. Wilkoff BL, Auricchio A, Brugada J, Cowie M, Ellenbogen KA, Gillis AM, et al. HRS/EHRA expert consensus on the monitoring of cardiovascular implantable electronic devices (CIEDs): description of techniques, indications, personnel, frequency and ethical considerations. Heart Rhythm. 2008;5(6):907–25.

46. Schulman PM, Rozner MA. Case report: use caution when applying magnets to pacemakers or defibrillators for surgery. Anesth Analg. 2013;117(2):422–7.
47. Rooke GA, Lombaard SA, Van Norman GA, Dziersk J, Natrajan KM, Larson LW, et al. Initial experience of an anesthesiology-based service for perioperative management of pacemakers and implantable cardioverter defibrillators. Anesthesiology. 2015;123(5):1024–32.
48. American Society of Anesthesiologists. Practice advisory for the perioperative management of patients with cardiac implantable electronic devices: pacemakers and implantable cardioverter-defibrillators: an updated report by the American Society of Anesthesiologists task force on perioperative management of patients with cardiac implantable electronic devices. Anesthesiology. 2011;114(2):247–61.
49. Crossley GH, Poole JE, Rozner MA, Asirvatham SJ, Cheng A, Chung MK, et al. The Heart Rhythm Society (HRS)/American Society of Anesthesiologists (ASA) Expert Consensus Statement on the perioperative management of patients with implantable defibrillators, pacemakers and arrhythmia monitors: facilities and patient management this document was developed as a joint project with the American Society of Anesthesiologists (ASA), and in collaboration with the American Heart Association (AHA), and the Society of Thoracic Surgeons (STS). Heart Rhythm. 2011;8(7):1114–54.
50. Healey JS, Merchant R, Simpson C, Tang T, Beardsall M, Tung S, et al. Society position statement: Canadian Cardiovascular Society/Canadian Anesthesiologists' Society/Canadian Heart Rhythm Society joint position statement on the perioperative management of patients with implanted pacemakers, defibrillators, and neurostimulating devices. Can J Anaesth. 2012;59(4):394–407.
51. Rooke GA, Bowdle TA. Perioperative management of pacemakers and implantable cardioverter defibrillators: it's not just about the magnet. Anesth Analg. 2013;117(2):292–4.
52. Cima RR, Brown MJ, Hebl JR, Moore R, Rogers JC, Kollengode A, et al. Use of lean and six sigma methodology to improve operating room efficiency in a high-volume tertiary-care academic medical center. J Am Coll Surg. 2011;213(1):83–92; discussion 93–4.
53. Sametz W, Metzler H, Gries M, et al. Perioperative catecholamine changes in cardiac risk patients. Eur J Clin Invest. 1999;29(7):582–7.
54. Devereaux PJ, Goldman L, Cook DJ, Gilbert K, Leslie K, Guyatt GH. Perioperative cardiac events in patients undergoing noncardiac surgery: a review of the magnitude of the problem, the pathophysiology of the events and methods to estimate and communicate risk. CMAJ. 2005;173(6):627–34.
55. Mangano DT, Goldman L. Preoperative assessment of patients with known or suspected coronary disease. N Engl J Med. 1995;333(26):1750–6.
56. Dweck MR, Williams MC, Moss AJ, Newby DE, Fayad ZA. Computed tomography and cardiac magnetic resonance in ischemic heart disease. J Am Coll Cardiol. 2016;68(20):2201–16.
57. Jette M, Sidney K, Blumchen G. Metabolic equivalents (METS) in exercise testing, exercise prescription, and evaluation of functional capacity. Clin Cardiol. 1990;13(8):555–65.
58. Reilly DF, McNeely MJ, Doerner D, et al. Self-reported exercise tolerance and the risk of serious perioperative complications. Arch Intern Med. 1999;159(18):2185–92.
59. Wiklund RA, Stein HD, Rosenbaum SH. Activities of daily living and cardiovascular complications following elective, noncardiac surgery. Yale J Biol Med. 2001;74(2):75–87.
60. Hlatky MA, Boineau RE, Higginbotham MB, et al. A brief self-administered questionnaire to determine functional capacity (the Duke Activity Status Index). Am J Cardiol. 1989;64(10):651–4.
61. Wijeysundera DN, Pearse RM, Shulman MA, et al. Assessment of functional capacity before major noncardiac surgery: an international. prospective cohort study. Lancet. 2018;391(10140):2631–40.
62. Wong EY, Lawrence HP, Wong DT. The effects of prophylactic coronary revascularization or medical management on patient outcomes after noncardiac surgery–a meta-analysis. Can J Anaesth. 2007;54(9):705–17.
63. Livhits M, Gibbons MM, de Virgilio C, O'Connell JB, Leonardi MJ, Ko CY, et al. Coronary revascularization after myocardial infarction can reduce risks of noncardiac surgery. J Am Coll Surg. 2011;212(6):1018–26.
64. Bittl JA, Baber U, Bradley SM, Wijeysundera DN. Duration of dual antiplatelet therapy: a systematic review for the 2016 ACC/AHA guideline focused update on duration of dual antiplatelet therapy in patients with coronary artery disease: a report of the American College of Cardiology/American Heart Association Task Force on Clinical Practice Guidelines. J Am Coll Cardiol. 2016;68(10):1116–39.
65. Botto F, Alonso-Coello P, Chan MT, Villar JC, Xavier D, Srinathan S, et al. Myocardial injury after noncardiac surgery: a large, international, prospective cohort study establishing diagnostic criteria, characteristics, predictors, and 30-day outcomes. Anesthesiology. 2014;120(3):564–78.
66. Levy M, Heels-Ansdell D, Hiralal R, Bhandari M, Guyatt G, Yusuf S, et al. Prognostic value of troponin and creatine kinase muscle and brain isoenzyme measurement after noncardiac surgery: a systematic review and meta-analysis. Anesthesiology. 2011;114(4):796–806.
67. Devereaux PJ, Mrkobrada M, Sessler DI, et al. Aspirin in patients undergoing noncardiac surgery. N Engl J Med. 2014;370(16):1494–503.
68. Poldermans D, Bax JJ, Kertai MD, et al. Statins are associated with a reduced incidence of perioperative mortality in patients undergoing major noncardiac vascular surgery. Circulation. 2003;107(14):1848–51.
69. Katsimaglis G, Danias PG, Danias PG, et al. Statins decrease perioperative cardiac complications in

patients undergoing noncardiac vascular surgery. J Am Coll Cardiol. 2004;45(3):336–42.
70. Desai H, Aronow WS, Gandhi K, et al. Incidence of perioperative myocardial infarction and of 2-year mortality in patients undergoing noncardiac vascular surgery treated with and without statins. Chest. 2016;136(4):46S.
71. Lindenauer PK, Pekow P, Wang K, Mamidi DK, Gutierrez B, Benjamin EM. Perioperative beta-blocker therapy and mortality after major noncardiac surgery. N Engl J Med. 2005;353(4):349–61.
72. Devereaux PJ, Yang H, Yusuf S, et al. Effects of extended-release metoprolol succinate in patients undergoing non-cardiac surgery (POISE trial): a randomised controlled trial. Lancet. 2008;371(9627):1839–47.
73. Aronow WS, Ahn C. Risk factors for new coronary events in a large cohort of very elderly patients with and without coronary artery disease. Am J Cardiol. 1996;77(10):864–6.
74. Aronow WS, Ahn C. Correlation of serum lipids with the presence or absence of coronary artery disease in 1793 men and women aged ≥62 years. Am J Cardiol. 1994;73(9):702–3.
75. Aronow WS, Frishman WH. Management of hypercholesterolemia in older persons for the prevention of cardiovascular disease. Cardiol Rev. 2010;18(3):132–40.
76. Libby P, Pasterkamp G, Crea F, Jang IK. Reassessing the mechanisms of acute coronary syndromes. Circ Res. 2019;124(1):150–60.
77. Durazzo AE, Machado FS, Ikeoka DT, De Bernoche C, Monachini MC, Puech-Leao P, et al. Reduction in cardiovascular events after vascular surgery with atorvastatin: a randomized trial. J Vasc Surg. 2004;39(5):967,75; discussion 975–6.
78. Schouten O, Boersma E, Hoeks SE, et al. Fluvastatin and perioperative events in patients undergoing vascular surgery. N Engl J Med. 2009;361(10):980–9.
79. Dunkelgrun M, Boersma E, Schouten O, Koopman-van Gemert AW, van Poorten F, Bax JJ, et al. Bisoprolol and fluvastatin for the reduction of perioperative cardiac mortality and myocardial infarction in intermediate-risk patients undergoing noncardiovascular surgery: a randomized controlled trial (DECREASE-IV). Ann Surg. 2009;249(6):921–6.
80. Le Manach Y, Godet G, Coriat P, Martinon C, Bertrand M, Fleron MH, et al. The impact of postoperative discontinuation or continuation of chronic statin therapy on cardiac outcome after major vascular surgery. Anesth Analg. 2007;104(6):1081.
81. Poredos P, Jug B. The prevalence of peripheral arterial disease in high risk subjects and coronary or cerebrovascular patients. Angiology. 2007;58(3):309–15.
82. Aronow WS, Ahmed MI, Ekundayo OJ, Allman RM, Ahmed A. A propensity-matched study of the association of peripheral arterial disease with cardiovascular outcomes in community-dwelling older adults. Am J Cardiol. 2009;103(1):130–5.
83. Criqui MH, Ninomiya JK, Wingard DL, Ji M, Fronek A. Progression of peripheral arterial disease predicts cardiovascular disease morbidity and mortality. J Am Coll Cardiol. 2008;52(21):1736–42.
84. Gerhard-Herman MD, Gornik HL, Barrett C, et al. 2016 AHA/ACC guideline on the management of patients with lower extremity peripheral artery disease. J Am Coll Cardiol. 2017;135(12):e726–79.
85. van Kuijk JP, Flu WJ, Welten GMJM, et al. Long-term prognosis of patients with peripheral arterial disease with or without polyvascular atherosclerotic disease. Eur Heart J. 2010;31(8):992–9.
86. Ness J, Aronow WS, Ahn C. Risk factors for symptomatic peripheral arterial disease in older persons in an academic hospital-based geriatrics practice. J Am Geriatr Soc. 2000;48(3):312–4.
87. Osnabrugge RLJ, Mylotte D, Head SJ, et al. Aortic stenosis in the elderly: disease prevalence and number of candidates for transcatheter aortic valve replacement: a meta-analysis and modeling study. J Am Coll Cardiol. 2013;62(11):1002–12.
88. Kertai MD, Bountioukos M, Boersma E, et al. Aortic stenosis: an underestimated risk factor for perioperative complications in patients undergoing noncardiac surgery. Am J Med. 2004;116(1):8–13.
89. Agarwal S, Rajamanickam A, Bajaj NS, et al. Impact of aortic stenosis on postoperative outcomes after noncardiac surgeries. Circ Cardiovasc Qual Outcomes. 2013;6(2):193–200.
90. Smith CR, Leon MB, Mack MJ, et al. Transcatheter versus surgical aortic-valve replacement in high-risk patients. N Engl J Med. 2011;364:2187–98.
91. Baron SJ, Wang K, House JA, et al. Cost-effectiveness of transcatheter versus surgical aortic valve replacement in patients with severe aortic stenosis at intermediate risk. Circulation. 2019;139:877–88.
92. Mack MJ, Leon MB, Thourani VH, et al. Transcatheter aortic-valve replacement with a balloon-expandable valve in low-risk patients. N Engl J Med. 2019;380:1695–705.
93. Calleja AM, Dommaraju S, Gaddam R, Cha S, Khandheria BK, Chaliki HP. Cardiac risk in patients aged >75 years with asymptomatic, severe aortic stenosis undergoing noncardiac surgery. Am J Cardiol. 2010;105(8):1159–63.
94. Guarracino F, Baldassarri R, Priebe HJ. Revised ESC/ESA guidelines on non-cardiac surgery: cardiovascular assessment and management. Implications for preoperative clinical evaluation. Minerva Anestesiol. 2015;81(2):226–33.
95. Biteker M, Tekkesin AI, Can MM, Dayan A, Ilhan E, Turkmen FM. Outcome of noncardiac and nonvascular surgery in patients with mechanical heart valves. Am J Cardiol. 2012;110(4):562–7.
96. Abuan I, Wong KH, Bolinske B, Hale KS. Andexanet Alfa: a recombinant modified human factor Xa protein for drug reversal of rivaroxaban and Apixaban. J Pharm Technol. 2019;35(3):119.

97. Pollack CV Jr, Reilly PA, Eikelboom J, Glund S, Verhamme P, Bernstein RA, et al. Idarucizumab for dabigatran reversal. N Engl J Med. 2015;373(6):511–20.
98. Habib G, Hoen B, Tornos P, et al. Guidelines on the prevention, diagnosis, and treatment of infective endocarditis (new version 2009): the Task Force on the Prevention, Diagnosis, and Treatment of Infective Endocarditis of the European Society of Cardiology (ESC). Endorsed by the European Society of Clinical Microbiology and Infectious Diseases (ESCMID) and the International Society of Chemotherapy (ISC) for Infection and Cancer. Eur Heart J. 2009;30(19):2369–413.

Reversal of Oral Anticoagulants in the Elderly

5

Firas Jafri, Saleha Batool, Kartik Dhaduk, and Robert G. Lerner

Introduction

The geriatric population is increasing in the United States and throughout the world. There are many factors that contribute to making aging itself a hypercoagulable state (Table 5.1) [1–9]. This hypercoagulable state results in a high prevalence of venous thromboembolic disease. Additionally the rising prevalence of cardiac disease, particularly atrial fibrillation (AF) with its risk of stroke, has resulted in a great need for anticoagulation treatment. For many years the only available oral anticoagulants were vitamin K antagonists (VKAs), which in the United States is predominantly Warfarin.

Venous thromboembolism (VTE) is widely prevalent with an increase with age. According to the report by American Heart Association, roughly the incidence of VTE each year in United States is roughly about 375,000 to 425,000 new cases each year [9]. A study which looked at the prevalence of VTE and future trend shows that the incidence of VTE increases as the age of the patient progresses [10]. A 25-year population-based study reveals the annual reported incidence of venous thromboembolism ranging widely from 43.7 to 145.0 per 100,000 people [11].

Table 5.1 Aging is a hypercoagulable state

Fibrinogen level increases [1]
D-dimer increases [2]
F1+2 increase [3]
Factor V, VII, VIII, IX levels increase [1]
Von Willebrand factor increases [1]
ADAMTS13 decreases [4]
Factor VIII and IX activation peptides increase [5]
Factor VIIa increases [5]
Thrombomodulin decreases [6]
PAI-1 increases [7]
Homocysteine increases [8]
Endothelial dysfunction increases [6]

The many shortcomings of Warfarin anticoagulation as well as advances in biochemistry and the understanding of the thrombotic pathways have led to the development of small molecules that directly target specific molecular entities in the coagulation mechanism. These agents collectively have been called new oral anticoagulants (NOACs), direct oral anticoagulants (DOACs), or targeted oral anticoagulants (TOACs). Several have been approved by the FDA and introduced to the market.

Although the NOACs have been widely accepted as having many advantages over Warfarin, there has been a hesitancy to use them because of a lack of a way to reverse the anticoagulant effect, i.e., an antidote. Warfarin has had well-known reversal mechanisms including vitamin K, fresh frozen plasma (FFP), and coagulation

F. Jafri (✉) · K. Dhaduk
Department of Medicine, New York Medical College and Westchester Medical Center, Valhalla, NY, USA

S. Batool · R. G. Lerner
Department of Hematology and Oncology, New York Medical College and Westchester Medical Center, Valhalla, NY, USA

R. Latifi (ed.), *Surgical Decision Making in Geriatrics*,
https://doi.org/10.1007/978-3-030-47963-3_5

factor concentrates, but only recently have reversal agents for the NOACs been available. Accordingly we will review and contrast VKAs and NOACs and their reversal agents. Figure 5.1 shows the coagulation cascade and point of action of VKAs and NOACs.

History of Drug Development

The history of the VKAs is especially interesting. The events that led to the discovery of Warfarin started with a hay shortage due to a drought in the northern plains. This led to the importation

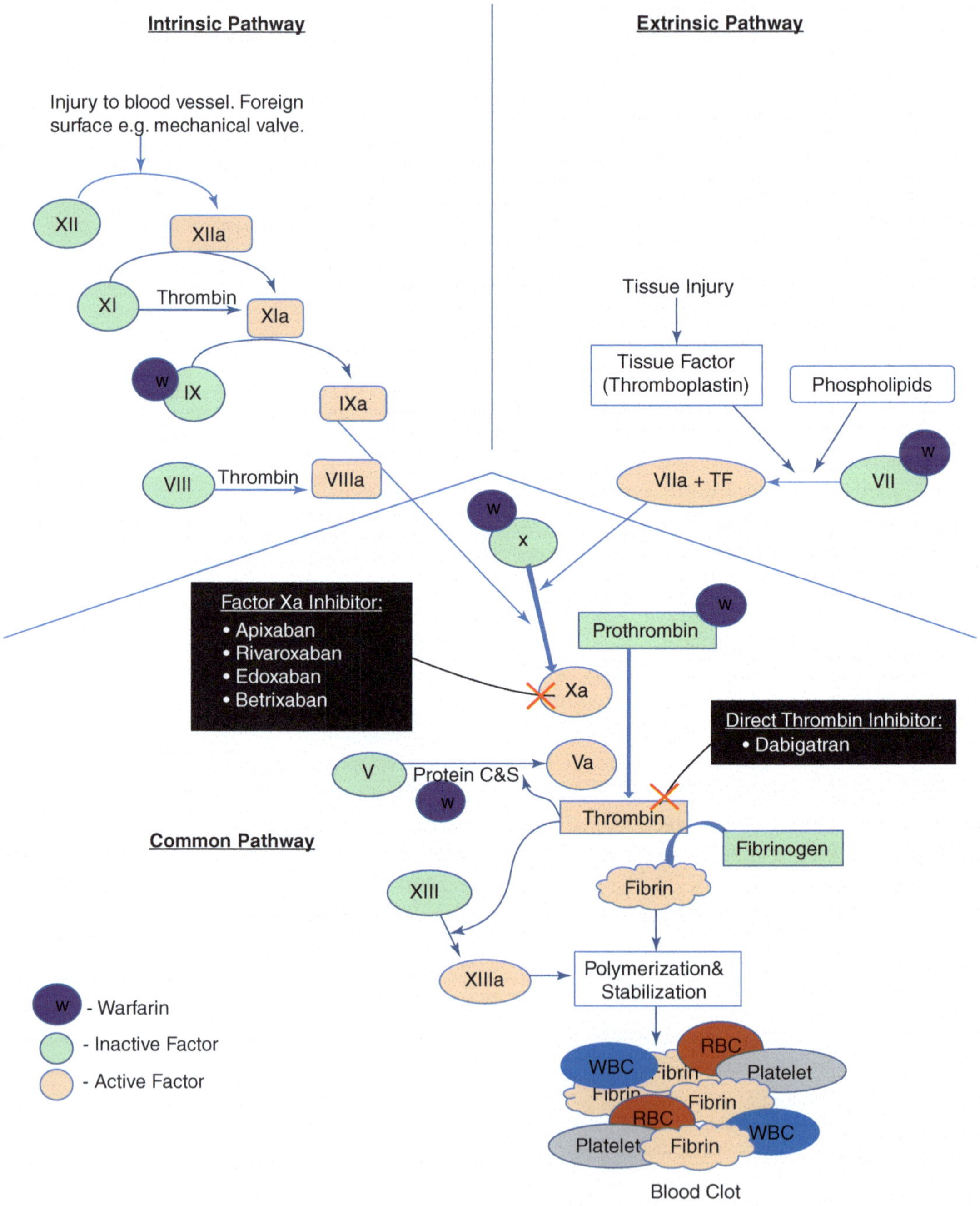

Fig. 5.1 The coagulation cascade and point of action of VKAs and NOACs

of sweet clover plants from Europe for cattle feed. Cattle fed spoiled sweet clover developed a hemorrhagic disease called sweet clover disease described by Schofield in 1922. Roderick found that these cattle were deficient in prothrombin in 1929. One snowy morning in February 1933, Ed Carlson, a farmer from Deer Park, Wisconsin, came into Karl Paul Link's laboratory carrying a milk can full of blood that refused to coagulate. Outside he had a small heap of spoiled sweet clover hay and a dead heifer freezing in the back of his truck [12]. Dr. Link finally synthesized dicumarol in 1940. The work was funded by the Wisconsin Alumni Research Foundation and led to the product named Warfarin. The Nobel Prize in Medicine in 1943 was awarded to Drs. Dam and Doisy for the discovery of vitamin K and its chemical nature. Link had postulated that dicumarol might be useful as a rodenticide and in 1948 Warfarin was launched as a rat poison. The first clinical studies of Warfarin were in 1955, and President Eisenhower was treated for his heart attack with Warfarin in the same year [13].

The development of the NOACs is much more a routine story of modern biochemistry and pharmacology. Factor Xa (FXa) is a critical serine protease situated at the confluence of the intrinsic and extrinsic pathways of the blood coagulation cascade. FXa catalyzes the conversion of prothrombin to thrombin via the prothrombinase complex. Its singular role in thrombin generation, coupled with its potentiating effects on clot formation, renders it an attractive target for therapeutic intervention [14]. The ability to synthesize molecules targeted to the active site of FXa resulted in the anti-Xa NOACs.

Leeches have been used medicinally for 2500 years. Direct thrombin inhibitors were initially derived from leeches and subsequently synthesized. Dabigatran (Pradaxa®) was synthesized and introduced to the market as an orally active direct anticoagulant directed against thrombin rather than FXa.

There have been many studies comparing Warfarin use to the NOACs for anticoagulation studying safety, efficacy, and bleeding risks. A recent review in the Journal of the American geriatric society provided a meta-analysis of Rivaroxaban, Apixaban, and Dabigatran compared with Warfarin use. They evaluated the bleeding risk, stroke risk, and thrombosis risk. They concluded that in elderly adults enrolled in randomized trials, bleeding with NOACs was not different from that with VKAs [15].

Warfarin Use and Reversal

According to the FDA-approved package label, Coumadin® is a vitamin K antagonist indicated for prophylaxis and treatment of venous thrombosis and its extension, pulmonary embolism (PE), prophylaxis, and treatment of thromboembolic complications associated with AF and/or cardiac valve replacement and reduction in the risk of death, recurrent myocardial infarction (MI), and thromboembolic events such as stroke or systemic embolization (SE) after MI [16].

There are many issues with Warfarin use in addition to the risk of hemorrhage (Table 5.2). According to the package label although there are no overall differences in effectiveness or safety observed between geriatric and younger patients, patients 60 years or older appear to exhibit greater than expected INR response to the anticoagulant effects of Warfarin [16].

Reversal of Warfarin anticoagulation is often required because of excessive elevation of the INR, bleeding, or clinical circumstances such as preparation for surgery. Reversal of anticoagulation does

Table 5.2 Risk of Warfarin use

The dosing is variable and must be individualized [16]
Initial and maintenance doses are different [16]
Continuous monitoring is required to achieve optimal anticoagulation [16]
Tissue necrosis can occur especially in individuals lacking protein C either congenitally or acquired, with limb ischemia in heparin-induced thrombocytopenia or heparin-induced thrombosis thrombocytopenia syndrome [16]
Calciphylaxis may occur especially in dialysis patients [16]
There are numerous drug-drug, drug-food, and botanical interactions [16]
Use in pregnancy and with liver disease has limitations [16]

leave the patient with the previously present thrombotic risk. Reversal may be accomplished by discontinuing Warfarin or administration of oral or parenteral vitamin K1. One study suggested that even with markedly elevated INRs in the absence of bleeding results were similar.

Rapid reversal can be accomplished using prothrombin complex concentrate (PCC), FFP, or activated factor VII treatment.

Management of Warfarin in the perioperative period is complicated and depends on clinical evaluation of the level of risk of thrombosis and bleeding in the individual patient and the planned surgery. Many different guidelines have been published, and a convenient compendium of these guidelines which uses an algorithm to guide dosing is available as an online app [look for MAPPP (management of anticoagulation in the periprocedural period) in your app store].

Dabigatran Use and Reversal

Dabigatran (US brand name Pradaxa®) is a direct thrombin inhibitor approved in 2010 by the US Food and Drug Administration (FDA). It undergoes 80% renal elimination and must be used cautiously in renal impairment. The RE-LY trial (Dabigatran vs. Warfarin) [17] randomized 18,113 patients to Dabigatran 150 mg (D150) twice daily, Dabigatran 110 mg (D110) twice daily, or open-label Warfarin. Majority of the participants (82% percent of patients) were aged >65 years, and 40% were aged >75 years. The rates of stroke were 1.53% per year for D110, 1.11% per year for D150, and 1.69% per year for Warfarin [18]. Both doses of Dabigatran were non-inferior to Warfarin, and the 150 mg dose of Dabigatran was superior to Warfarin. Both doses of Dabigatran were associated with a lower risk for intracranial hemorrhage (ICH) when compared to Warfarin, across all age groups. However, in patients >75 years of age, both doses of Dabigatran were associated with a higher risk for extracranial bleeding, particularly gastrointestinal (GI) bleeding [19]. Because of increased risk of stroke and major bleeding with D110, D150 was approved by FDA, even though the rate of major bleeding was higher with D150.

Renal function (CrCl) and age were found to be the two most important determinants of the plasma concentration of Dabigatran, with a 68% higher trough concentration in patients >75 years of age as compared to those <65 years of age. In addition, low body weight and female sex were associated with higher plasma concentrations of Dabigatran [20]. According to the FDA-approved package label even though the risk of stroke and bleeding increases with age, the risk and benefit profile are favorable in all age groups.

RE-COVER trial randomized 2500 patients with acute VTE to either Dabigatran 150 mg orally twice daily or Warfarin. At approximately 6 months of follow-up, there were similar rates of the primary outcome of VTE or VTE-related death in each arm (2.5–2.7%). There was no difference in major bleeding risk (1.6–1.9%), but the risk of any bleeding was reduced by 26% with Dabigatran (16% vs. 22%) [21].

According to the FDA-approved package label, Pradaxa® (Dabigatran etexilate mesylate) capsules are indicated:

- To reduce the risk of stroke and SE in patients with non-valvular AF
- For the treatment of deep venous thrombosis (DVT) and PE in patients who have been treated with a parenteral anticoagulant for 5–10 days to reduce the risk of recurrence of DVT and PE in patients who have been previously treated for the prophylaxis of DVT and PE in patients who have undergone hip replacement surgery

There is a black box warning about the risk of thrombotic events with premature discontinuation of the drug and the risk of spinal/epidural hematoma.

The spinal/epidural hematoma problem arose with all the NOACs when routine preoperative lab results failed to indicate that a patient was anticoagulated.

A specific reversal agent (Idarucizumab) is available and marketed as Praxbind®.

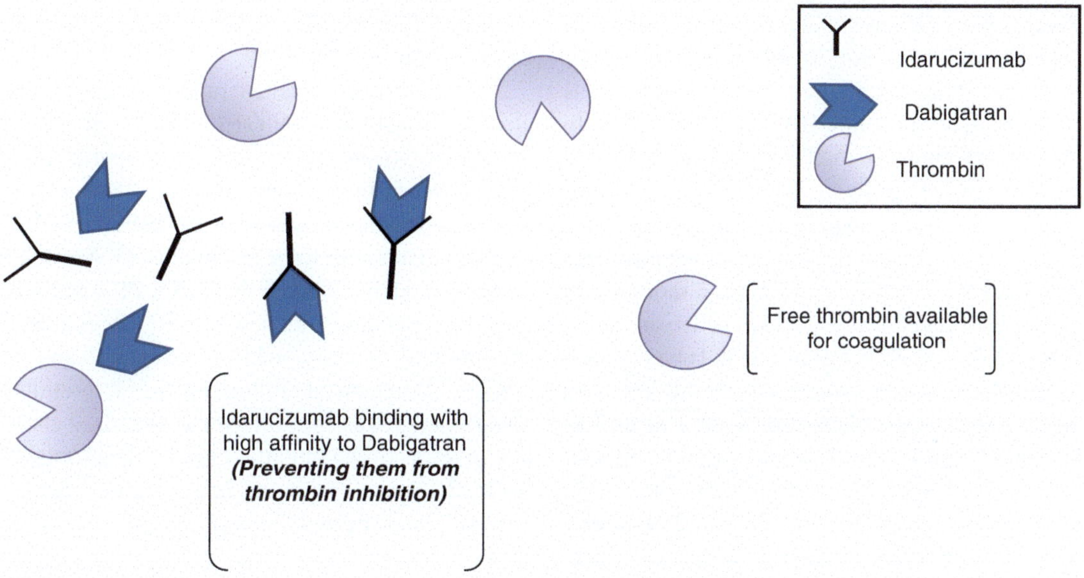

Fig. 5.2 The mechanism of action of Idarucizumab, which binds with Dabigatran molecule making more thrombin available for coagulation, is demonstrated

Figure 5.2 demonstrates the mechanism of action of Idarucizumab which binds with Dabigatran molecule making more thrombin available for coagulation.

According to the package label, Praxbind® is a humanized monoclonal antibody fragment (Fab) indicated in patients treated with Pradaxa® [22] when reversal of the anticoagulant effects of Dabigatran is needed:

- For emergency surgery/urgent procedures [22]
- In life-threatening or uncontrolled bleeding [22]

This indication is approved under accelerated approval based on a reduction in unbound Dabigatran and normalization of coagulation parameters in healthy volunteers. Continued approval for this indication may be contingent upon the results of an ongoing cohort case series study. The subsequent cohort study REVERSE-AD trial concluded that in emergency situations, Idarucizumab rapidly, durably, and safely reversed the anticoagulant effect of Dabigatran [23].

The recommended dose is 5 g given as two separate vials of 2.5 g with limited data to support additional dosing.

Rivaroxaban and Apixaban Use and Reversal

Rivaroxaban (brand name Xarelto®) is a direct FXa inhibitor predominantly metabolized by the liver, with one-third of the drug undergoing renal elimination. It has a long half-life, once-daily dosing, and reliable anticoagulation without a need for laboratory monitoring. As per the package insert, it has been approved by FDA for treatment of DVT and PE; for prophylaxis of DVT in patients undergoing hip or knee replacement surgery, to reduce the risk of stroke and SE in patients with non-valvular AF; and for the reduction in the risk of recurrence of DVT and/or PE in patients at continued risk for recurrent DVT and/or PE after completion of initial treatment lasting at least 6 months and in combination with aspirin, to reduce the risk of major cardiovascular events (cardiovascular (CV) death, MI and stroke) in patients with chronic coronary artery disease (CAD) or peripheral artery disease (PAD) [24].

ROCKET AF (Rivaroxaban compared with Warfarin for prevention of stroke and embolism trial in AF) [25]. This trial randomized 14,264 patients with AF and a $CHADS_2$ score of ≥ 2 to

Rivaroxaban (20 mg daily or 15 mg daily in patients with a CrCl of 30–49 ml per minute) or Warfarin. The mean age of the patients was 73 years, with a mean $CHADS_2$ score of 3.5. In the primary analysis, the Rivaroxaban group had a lower rate of stroke (1.7% per year) as compared to Warfarin group (2.2% per year). Rates of major bleeding were similar for Rivaroxaban and Warfarin (14.9% vs. 14.5%), although rate of major GI bleed was higher in the Rivaroxaban group (3.2% vs. 2.2%). Rates of ICH were lower in the Rivaroxaban group than in the Warfarin group (0.5% vs. 0.7% per year). At a mean follow-up of 2 years, Rivaroxaban was non-inferior to Warfarin for stroke or SE, without increasing the bleeding rates [25]. A study of 6229 elderly patients (>75 years of age) as a part of a pre-specified secondary analysis of the ROCKET AF trial found no difference in the primary efficacy outcome (stroke and SE) between Warfarin therapy and Rivaroxaban. Similarly, there were no differences in major bleeding events between the Warfarin and Rivaroxaban groups. However, the incidence of GI bleeding was higher in patients treated with Rivaroxaban compared to Warfarin [26].

RECORD 1 trial conducted in 2008 compared the efficacy and safety of Rivaroxaban, an oral direct inhibitor of FXa. 4541 patients were randomized to receive either 10 mg of oral Rivaroxaban once daily, beginning after surgery, or 40 mg of Enoxaparin subcutaneously once daily, beginning the evening before surgery, plus a placebo tablet or injection. The primary efficacy outcome (DVT, PE, or death) occurred in 18 of 1595 patients (1.1%) in the Rivaroxaban group and in 58 of 1558 patients (3.7%) in the Enoxaparin group. Major VTE occurred in 4 of 1686 patients (0.2%) in the Rivaroxaban group and in 33 of 1678 patients (2.0%) in the Enoxaparin group. Major bleeding occurred in 6 of 2209 patients (0.3%) in the Rivaroxaban group and in 2 of 2224 patients (0.1%) in the Enoxaparin group. A once-daily, 10-mg oral dose of Rivaroxaban was significantly more effective for extended thromboprophylaxis than a once-daily, 40-mg subcutaneous dose of Enoxaparin in patients undergoing elective total hip arthroplasty [27].

In 2010 an open-labelled, randomized trial was conducted, namely, EINSTEIN-DVT trial, a trial compared Rivaroxaban with Enoxaparin plus VKAs (Warfarin or Acenocumarol). Rivaroxaban showed non-inferior efficacy with respect to DVT in comparison with Enoxaparin plus VKAs [28].

Published in 2012, EINSTEIN-PE randomized 4832 patients with acute PE to Rivaroxaban or standard therapy with Enoxaparin and a VKAs. At a mean follow-up of 7 months, Rivaroxaban was non-inferior to standard therapy in terms of the rate of recurrent symptomatic VTE (2.1% vs. 1.8%) and had a similar risk of clinically significant bleeding (10.3% vs. 11.4%) [29].

Apixaban (brand name Eliquis®) is a direct FXa inhibitor approved by the FDA in 2012. As per the package insert, it can be used for stroke prevention in AF, for prophylaxis of DVT in patients undergoing hip or knee replacement surgery, and in 2014 for treatment of DVT and PE [30].

The ARISTOTLE trial [31] randomized 18,201 patients with AF and 1 additional risk factor for stroke to Apixaban or dose-adjusted Warfarin. The median age of the population was 70 years with 31% of the population >75 years of age. The mean $CHADS_2$ score was 2.1, and the mean time in therapeutic range of Warfarin cohort was 66%. The primary outcome of stroke or SE occurred in 1.27% patients per year in the Apixaban group versus 1.60% per year in the Warfarin group. The rate of ICH was 0.33% per year in the Apixaban group and 0.80% per year in the Warfarin group. The clinical benefit of Apixaban over Warfarin was consistent across all patients irrespective of their $CHADS_2$, CHA_2DS_2-VASc, and HAS-BLED scores. In addition, Apixaban decreased all-cause mortality when compared to Warfarin [31].

The Apixaban Versus Acetylsalicylic Acid to Prevent Stroke in AF Patients Who Have Failed or Are Unsuitable for VKAs Treatment (AVERROES) trial [32] showed greater efficacy for Apixaban in preventing stroke or SE events and similar rates of bleeding compared to Aspirin, making this a good option in elderly patients considered unsuitable for Warfarin therapy.

Apixaban was given 5 mg twice daily in the ARISTOTLE study, although 2.5 mg twice daily was used in a subset of patients with two or more of the following criteria: ≥80 years, body weight ≤60 kg, or a serum creatinine level of ≥1.5 mg/dl. The superiority of Apixaban relative to Warfarin for preventing stroke was consistent, irrespective of the degree of renal impairment [31].

Both Rivaroxaban and Apixaban have an FDA-approved reversal agent marketed as ANDEXXA® (coagulation FXa (recombinant), inactivated-zhzo).

Figure 5.3 demonstrates the mechanism of action of Andexanet Alfa. The molecule of Andexanet Alfa is similar in size as that of FXa. Andexanet Alfa binds with FXa inhibitors making it unavailable for it to bind and inhibit FXa.

ANDEXXA®, coagulation FXa (recombinant), inactivated-zhzo is a recombinant modified human FXa protein indicated for patients treated with Rivaroxaban and Apixaban, when reversal of anticoagulation is needed due to life-threatening or uncontrolled bleeding [33]. This indication is approved under accelerated approval based on the change from baseline in anti-FXa activity in healthy volunteers. There have been two cohort studies ANNEXA-A and ANNEXA-R. On the basis of these two studies in healthy volunteers in which Andexanet Alfa reversed the anticoagulant effects of Apixaban and Rivaroxaban, the FDA granted accelerated approval to Andexanet Alfa in May 2018 for the treatment of life-threatening or uncontrolled bleeding in patients receiving either of these two FXa inhibitors [34]. An improvement in hemostasis has not been established. As with Praxbind®

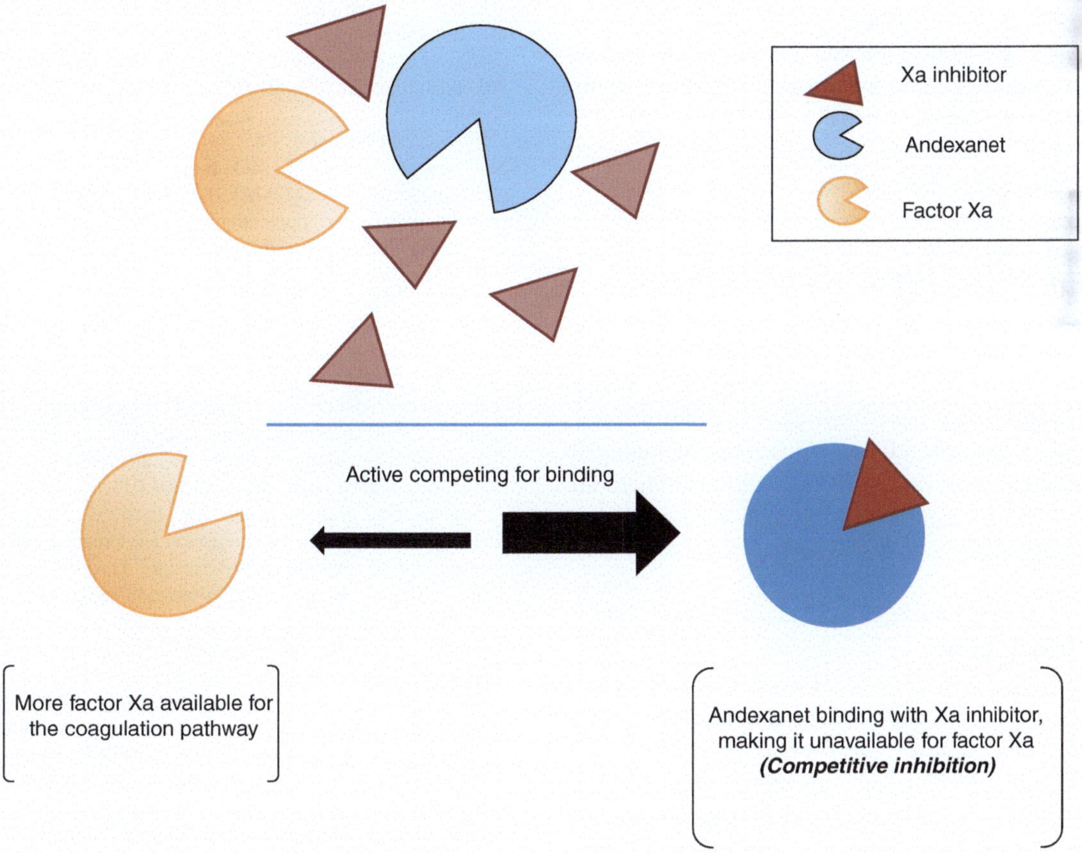

Fig. 5.3 The mechanism of action of Andexanet Alfa is demonstrated

continued approval for this indication may be contingent upon the results of studies to demonstrate an improvement in hemostasis in patients. ANDEXXA® has not been shown to be effective for, and is not indicated for, the treatment of bleeding related to any FXa inhibitors other than Apixaban and Rivaroxaban [33].

Coagulation FXa (recombinant), inactivated-zhzo should be able to reverse the anticoagulant effect of other FXa inhibitors, but there are no published clinical trials and it is not indicated.

Another small molecule, Ciraparantag, has been shown to reverse FXa inhibitors in vitro, but there is no application for FDA approval at the time of this writing.

There are other FXa inhibitors like Edoxaban (Savaysa®) which is a direct inhibitor of activated Factor X with a rapid onset of action. It is administered orally once daily and has proven antithrombotic efficacy [35]. In 2015 Edoxaban was approved by FDA for prevention of thromboembolic events in patients with valvular AF [36]. Another FDA-approved FXa inhibitor for extended VTE prophylaxis after discharge from a medical illness is betrixaban (Bevyxxa®) [37]. However there are no specific reversal agents approved for these agents.

Prothrombin Complex Concentrates

Prothrombin complex concentrates have been widely used as a reversal agent for VKAs. However they have been used to control bleeding in multiple clinical situations, including in attempting to reverse FXa inhibitors. There are individual reports but no convincing clinical information of their efficacy.

Conclusion

There have been multiple randomized clinical trials demonstrating safety and efficacy of NOACs in multiple clinical indications with notable exception of mechanical heart valves. Depending upon the indication and patient population, NOACs may be no different, safer, or more effective than Warfarin. The decision to select a NOAC or a more traditional medicine such as Warfarin should involve a discussion with patient reviewing the risk and benefits or both approaches. Warfarin has always had multiple specific reversal agents. With the introduction of specific reversal agents for Rivaroxaban, Apixaban, and Dabigatran, a major cause of hesitancy to their use has been removed.

References

1. Favaloro EJ, Franchini M, Lippi G. Aging hemostasis: changes to laboratory markers of hemostasis as we age – a narrative review. Semin Thromb Hemost. 2014;40(6):621–33. https://doi.org/10.1055/s-0034-1384631.
2. Gómez-Jabalera E, Bellmunt Montoya S, Fuentes-Camps E, Escudero Rodríguez JR. Age-adjusted D-dimer for the diagnosis of deep vein thrombosis. Phlebology. 2018;33(7):458–63. https://doi.org/10.1177/0268355517718762.
3. Cadroy Y, Pierrejean D, Fontan B, Sié P, Boneu B. Influence of aging on the activity of the hemostatic system: prothrombin fragment 1+2, thrombin-antithrombin III complexes and D-dimers in 80 healthy subjects with age ranging from 20 to 94 years. Nouv Rev Fr Hematol. 1992;34(1):43–6.
4. Kokame K, Sakata T, Kokubo Y, Miyata T. von Willebrand factor-to-ADAMTS13 ratio increases with age in a Japanese population. J Thromb Haemost. 2011;9(7):1426–8. https://doi.org/10.1111/j.1538-7836.2011.04333.x.
5. Mari D, Mannucci PM, Coppola R, Bottasso B, Bauer KA, Rosenberg RD. Hypercoagulability in centenarians: the paradox of successful aging. Blood. 1995;85(11):31449.
6. Sepúlveda C, Palomo I, Fuentes E. Mechanisms of endothelial dysfunction during aging: predisposition to thrombosis. Mech Ageing Dev. 2017;164:91–9. https://doi.org/10.1016/j.mad.2017.04.011.
7. Vaughan DE, Rai R, Khan SS, Eren M, Ghosh AK. Plasminogen activator inhibitor-1 is a marker and a mediator of senescence. Arterioscler Thromb Vasc Biol. 2017;37(8):1446–52. https://doi.org/10.1161/ATVBAHA.117.309451.
8. Bakulski KM, Park SK, Weisskopf MG, Tucker KL, Sparrow D, Spiro A 3rd, Vokonas PS, Nie LH, Hu H, Weuve J. Lead exposure, B vitamins, and plasma homocysteine in men 55 years of age and older: the VA normative aging study. Environ Health Perspect. 2014;122(10):1066–74. https://doi.org/10.1289/ehp.1306931.
9. Heart disease and stroke statistics-2017 update: a report from the American Heart Association. (PMID: 28122885 PMCID:pmc5408160).

10. Deitelzweig SB, Johnson BH, Lin J, Schulman KL. Prevalence of clinical venous thromboembolism in the USA: current trends and future projections. Am J Hematol. 2011;86(2):217–20.
11. Silverstein MD, Heit JA, Mohr DN, et al. Trends in the incidence of deep vein thrombosis and pulmonary embolism a 25-year population-based study. Arch Intern Med. 1998;158(6):585–93. https://doi.org/10.1001/archinte.158.6.585.
12. Mike Scully (Thrombosis Research Institute, London, UK) The Biochemist — February 2002. ©2002 The Biochemical Society.
13. Norn S, Permin H, Kruse E, et al. On the history of vitamin K, dicoumarol and Warfarin. Dan Medicinhist Arbog. 2014;42:99–119.
14. Guertin KR, Choi YM. The discovery of the Factor Xa inhibitor otamixaban: from lead identification to clinical development. Curr Med Chem. 2007;14(23):2471–81.
15. Partha Sardar MD, Saurav Chatterjee MD, Shobhana Chaudhari MD, et al. New oral anticoagulants in elderly adults: evidence from a meta-analysis of randomized trials. J Am Geriatr Soc. 2014;62(5):857–64. https://doi.org/10.1111/jgs.12799.
16. www.accessdata.fda.gov/drugsatfda_docs/label/2011/009218s107lbl.pdf.
17. Connolly SJ, Ezekowitz MD, et al. Dabigatran vs Warfarin in patients with atrial fibrillation. N Engl J Med. 2009;361:1139–51. https://doi.org/10.1056/NEJMoa0905561.
18. www.accessdata.fda.gov/drugsatfda_docs/label/2011/022512s007lbl.pdf.
19. Eikelboom JW, Wallentin L, Connolly SJ, Ezekowitz M, Healey JS, et al. Risk of bleeding with 2 doses of Dabigatran compared with Warfarin in older and younger patients with atrial fibrillation: an analysis of the randomized evaluation of long-term anticoagulant therapy (RE-LY) trial. Circulation. 2011;123:2363–72.
20. Reilly PA, Lehr T. The effect of dabigatran plasma concentrations and patient characteristics on the frequency of ischemic stroke and major bleeding in atrial fibrillation patients: The RE-LY trial (randomized evaluation of long-term anticoagulation therapy). J Am Coll Cardiol. 2014;63:321–8.
21. Schulman S, et al. Dabigatran versus Warfarin in the treatment of acute venous thromboembolism. N Engl J Med. 2009;361(24):2342–52.
22. www.accessdata.fda.gov/drugsatfda_docs/label/2015/761025lbl.pdf.
23. Pollack CV, Reilly PA, van Ryn J, et al. Idarucizumab for dabigatran reversal — full cohort analysis. N Engl J Med. 2017;77:431–41.
24. www.accessdata.fda.gov/drugsatfda_docs/label/2016/202439s017lbl.pdf.
25. Patel MR, Mahaffey KW, Garg J. Rivaroxaban versus Warfarin in nonvalvular atrial fibrillation. N Engl J Med. 2011;365:883–91.
26. Halperin J, Hankey G, Wojdyla DM, et al. Efficacy of rivaroxaban to prevent stroke in atrial fibrillation maintained in elderly. Circulation. 2014;130(2):138–46.
27. Eriksson BI, Borris LC, Friedman RJ. Rivaroxaban versus enoxaparin for thromboprophylaxis after hip arthroplasty. N Engl J Med. 2008;358:2765–75. https://doi.org/10.1056/NEJMoa0800374.
28. Bauersachs R, Berkowitz SD, Brenner B, et al. Oral rivaroxaban for symptomatic venous thromboembolism. N Engl J Med. 2010;363(26):2499–510. https://doi.org/10.1056/NEJMoa1007903.
29. Büller HR, et al. Oral rivaroxaban for the treatment of symptomatic pulmonary embolism. N Engl J Med. 2012;366(14):1287–97.
30. www.accessdata.fda.gov/drugsatfda_docs/label/2012/202155s000lbl.pdf.
31. Connolly SJ, Eikelboom J, Joyner C, Diener HC, Hart R, et al. Apixaban in patients with atrial fibrillation. N Engl J Med. 2011;364:806–17.
32. Lopes RD, Al-Khatib SM, Wallentin L, Yang H, Ansell J, et al. Efficacy and safety of apixaban compared with Warfarin according to patient risk of stroke and of bleeding in atrial fibrillation: a secondary analysis of a randomised controlled trial. Lancet. 2012;380:1749–58.
33. www.fda.gov/files/vaccines%2C%20blood%20%26%20biologics/published/Package-Insert%2D%2D-ANDEXXA.pdf.
34. Siegal DM, Curnutte JT, Connolly SJ, et al. Andexanet Alfa for the reversal of factor Xa inhibitor activity. N Engl J Med. 2015;373:2413–24. https://doi.org/10.1056/NEJMoa1510991.
35. Büller HR, Décousus H, Grosso MA. Edoxaban versus Warfarin for the treatment of symptomatic venous thromboembolism. N Engl J Med. 2013;369:1406–15. https://doi.org/10.1056/NEJMoa1306638.
36. www.accessdata.fda.gov/drugsatfda_docs/label/2015/206316lbl.pdf.
37. www.accessdata.fda.gov/drugsatfda_docs/label/2017/208383s000lbl.pdf.

6 Frailty Assessment as Measurement of Physiologic Reserves in the Elderly

Kartik Prabhakaran, David J. Samson, and Rifat Latifi

The number of elderly persons in the world population has seen a continued increase in the twenty-first century that is expected to continue [1, 2]. As the population continues to age, an understanding of the physiologic consequences of aging and resultant effects on risk upon patient-centered outcomes has become very important. Advanced age is associated with high risk of developing disease processes and leads to both pathologic and natural age-related organ dysfunction, which in turn lead to a higher risk of morbidity and mortality [3]. Concomitant with the overall increase in elderly, there is an increase in the demand for surgery among the aging population [4]. The physiologic decline associated with aging is increased with adverse postoperative outcomes [5, 6]. Yet, within the elderly population, there is substantial heterogeneity among surgical outcomes [7], owing to heterogeneity among the elderly population with respect to physiologic reserve, burden of comorbid conditions, and functional independence. This heterogeneity among the elderly has led to the development of a concept termed "frailty" which refers to both intrinsic and extrinsic factors that contribute to a state of vulnerability to health-related stressors [3, 8]. Understanding the distinction between chronologic age and frailty is predicated on the understanding that not all elderly persons are frail [9] and that frailty itself is a distinct disease process worthy of measurement and study [10].

Although frailty is widely recognized and accepted, there continues to be significant debate on the most appropriate method for its assessment. Whereas most frailty researchers agree on the necessity of a multidimensional approach, there is significant variability among the specific variables used in assessment, inclusion/exclusion criteria, clinical applicability, and validation across the available instruments of assessment [11].

This chapter provides an overview of frailty indices and measurements of physiologic reserves in the elderly by discussing the physiologic changes associated with aging, frailty as a measure of biological age, the various tools and indices available for the assessment of frailty, and the application of frailty to clinical practice (specifically in emergency general surgery).

K. Prabhakaran (✉)
Section of Trauma and Acute Care Surgery, Department of Surgery, Westchester Medical Center, Valhalla, NY, USA

Department of Surgery, New York Medical College, Valhalla, NY, USA
e-mail: Kartik.Prabhakaran@wmchealth.org

D. J. Samson
Department of Surgery, Westchester Medical Center and New York Medical College, Valhalla, NY, USA

R. Latifi
Department of Surgery, New York Medical College, School of Medicine and Westchester Medical Center Health Network, Valhalla, NY, USA

R. Latifi (ed.), *Surgical Decision Making in Geriatrics*,
https://doi.org/10.1007/978-3-030-47963-3_6

Biologic Basis for Frailty

Aging is accompanied by decline and deterioration of organ systems on a functional level which leads to an overall decrease in physiologic reserve and the ability to withstand stress and physiologic insults [12]. This functional decline on an organ system level has its origins in the process of aging on a cellular level and is characterized by loss of homeostasis and a breakdown in regulatory pathways and molecular structures [13]. Normal mechanisms of cell maintenance such as DNA repair, clearance of defective mediators of cell function, and defense against pathogens are integral to cellular and organ system homeostasis. However, these mechanisms begin to fail with advanced aging. Instead, these homeostatic mechanisms are often replaced by DNA damage, cell cycle dysregulation and/or senescence, oxidative damage, and aberrant enzymatic degradation [14, 15]. This cellular dysregulation goes hand in hand with endocrine dysfunction as the normal production and regulation of essential hormones diminish and are replaced by hormonal upregulation of cellular aging and death in the form of apoptosis and senescence [16].

The Impact of Aging on Organ Function

The cellular changes associated with aging translate into significant organ-specific alterations that affect overall physiology. For example, with aging, the cardiopulmonary system is affected over time by a progressive loss of cardiac myocytes, which in turn leads to increased myocyte volume across both ventricles [17]. The concomitant decrease in arterial compliance leads to increased afterload, impaired diastolic filling, and left ventricular dysfunction [18]. When combined with increasing prevalence of underlying cardiovascular disease in the elderly, these "normal" physiologic changes associated with aging portend a state of decreased cardiovascular reserve. The aging process also results in progressive compromise of pulmonary function. When combined with changes in pulmonary mechanics, gas exchange, and respiratory muscle strength, the aging process results in a significant reduction in normal gas exchange as measured by pulmonary function tests [19, 20]. Gradual compromise of the mucociliary reflex and diminishing ventilator response to hypercapnia impair the elderly patient's ability to clear respiratory secretions and respond to physiologic stressors and insults [21].

Kidney function and nutrition are also progressively impaired during the normal aging process. Kidneys undergo an overall loss of volume and function of the renal cortex, primarily in the form of glomerulosclerosis and senescence of the renal tubule; and these processes are further accelerated by comorbid conditions such as hypertension, diabetes, and atherosclerosis [22]. In the face of physiologic stress (e.g., disease, injury, or surgery [23]), these alterations in renal physiology cause a significant reduction in renal physiology reserve which places the elderly patient at significant risk of acute kidney injury and renal failure. With respect to nutritional status, the prevalence of malnutrition in the elderly population is well known. In addition to decreased appetite and oral intake, the gastrointestinal system undergoes significant atrophy and deterioration as a consequence of the normal aging process [24]. Neuromuscular atrophy and autonomic dysregulation associated with aging produce significant changes, in the form of discordant muscular contractions along the gastrointestinal tract which gives rise to impaired swallowing and peristalsis. Aging of the gastrointestinal system is also accompanied by declining secretory and absorptive capacity of both the stomach and small intestine mediated by decreased production of gastric acid and pepsin, as well as villous atrophy of the small bowel. Taken together, these changes have a profound impact on the elderly patient's nutritional status which comprises a principal defense against physiologic stress.

The cognitive decline associated with aging is a well-known phenomenon. Apart from the increased prevalence of neurologic disease processes such as dementia and neurodegenerative diseases in the elderly population [25, 26], the normal aging brain is associated with cortical atrophy and impairment of cerebral blood flow [27]. These changes not only place the elderly patient at increased risk for cerebrovascular accidents but also contribute to profound impairments in visual, auditory, and sensory function which give rise to potentially detrimental effects upon quality of life, self-perception, functional independence, and fall risk. Moreover, increased sensitivity to analgesia and anesthesia, as well as increased risk for delirium, places the hospitalized elderly patient at a significantly higher risk for morbidity and mortality [28].

Frailty as a Measure of Biological and Physiologic Reserve

The heterogeneity among the elderly population with respect to both the prevalence of comorbid conditions, functional status, and age-related physiologic decline has given rise to the concept of frailty. Despite variations in measurement, interpretation, and application, frailty is conceptually understood to represent an entity, related to but distinct from chronologic age, that correlates with increased vulnerability to disease [29]. Given the discordance between chronologic age and frailty, it is important to note that not all elderly persons are frail. Studies have shown that approximately 20% of community-dwelling persons over the age of 50 years are frail, with higher percentages of frailty observed among patients in clinical settings [9, 30]. While there is no universally applicable or accepted definition of frailty, it is important to understand that frailty is a dynamic and multidimensional entity that exists across multiple domains including disability, comorbidity, functionality, and psychosocial integrity [31, 32]. Despite inherent genetic predispositions toward the development of comorbid conditions which are included in most definitions of frailty, it is important to recognize frailty as a conglomerate of symptoms or "syndrome" that is not merely a phenotype but acquired during various stages of aging based on the accumulation or risk factors that span the domains of lifestyle, socioeconomic status, psychological well-being, nutrition, disability, and comorbid conditions [33]. Furthermore, it is important to note that variables such as gender, geographic/demographic effects, and educational and cognitive function can affect the onset, severity, and prevalence of frailty.

Measurement of Frailty: Tools and Indices

Measurement of frailty should be informed by an established definition of this condition. A consensus definition of frailty does not currently exist. This may explain the many dozens of instruments available to assess frailty. Frequently described elements of definitions are weakness, challenges to mobility, weight loss, progressive decline in system function, decreased physiologic reserve across multiple organ systems, decreased resistance or increased susceptibility/vulnerability to stressors, and increased risk of adverse events such as surgical complications, disability, decline, and death [34, 35].

The mere fact that today we have 13 scales of frailty (Table 6.1) means that none of them is "the best" and validated in a randomized multi-institutional long-term study. The table includes details about the component domains included in each instrument along with ranges of possible scores and threshold values for what should be considered frailty or different levels of frailty. Specific domains encompassed by different instruments include activities of daily living (ADLs), instrumental ADLs (IADLs), cognition, comorbidities/disease history, demographics, energy, history of falls, mobility, musculoskeletal signs, neurologic signs, nutritional status, physical activity, polypharmacy, sensory function, social relations/support, strength, and weight loss.

From a theoretical perspective, the first notable instrument for measuring frailty was introduced in 2001 by Fried et al. [8]. Expressing their perspective as a phenotype model of frailty, these authors targeted a wasting syndrome. They stated that declines in lean body mass, strength, fitness, balance, ability to ambulate, and level of activity combine into a frailty cycle marked by low

Table 6.1 Measures of frailty frequently found in the literature

Index	Items	Domains	Range of scores, thresholds	Reference(s)
1. Fried Frailty Phenotype Cardiovascular Health Study (CHS)	5	Weight loss, low physical activity, exhaustion, slowness, weakness	0–5 present; frailty ≥3 items; pre-frail 1–2 items; robust = none	[8, 36–41]
2. Canadian Study of Health and Aging (CHSA) Accumulated Deficits Frailty Index and modifications	5–70	Presence and severity of current diseases, ability in the activities of daily living, and physical and neurological signs from the clinical examinations	Proportion present: (0.0–1.0); frailty = score ≥ 0.25	[37, 42–51]
3. CSHA Clinical Frailty Scale	1	Global clinical impression	7–9 levels 1 very fit, 2 well, 3 managing well, 4 vulnerable, 5 mildly frail, 6 moderately frail, 7 severely frail, 8 very severely frail, 9 terminally frail	[43, 52]
4. FRAIL Scale	5	Fatigue, resistance, ambulation, illness, and loss of weight	0–5; frailty ≥3 items; pre-frail 1–2 items; robust = 0 items	[56, 57]
5. Study of Osteoporotic Fractures (SOF) Frailty Index	3	Weight loss, exhaustion, unable to rise from chair 5 times	0–3; frailty ≥2 items; pre-frail = 1 item; robust = 0 items	[38, 58]
6. Edmonton Frailty Scale	8	Cognition, general health status, functional independence, social support, medication use, nutrition, mood, continence, functional performance	0–17, frailty ≥7	[59]
7. Program of Research to Integrate the Services for the Maintenance of Autonomy (PRISMA-7)	7	Age, sex, activities of daily living, social support, physical mobility	0–7; frailty ≥3	[61]
8. Sherbrooke Postal Questionnaire	6	Living situation, number of medications, physical mobility, vision, hearing, memory	0–6; frailty ≥2	[60]
9. Vulnerable Elders Survey	13	Age, physical activities, functional activities, self-rated health	0–10; frailty ≥3	[62]
10. Strawbridge Functional Domains Model	16	Physical functioning, nutritive functioning, cognitive functioning, sensory problems	0–4; frailty ≥2	[63]
11. Comprehensive Assessment of Frailty	8	Weakness, self-reported exhaustion, slow gait speed, low physical activity, standing balance, rising from chair 3 times, picking up pen from floor, putting on and removing jacket, supplemented by selected lab tests	0–35; not frail = 1–10; moderately frail = 11–25; severely frail = 26–36	[64, 65]
12. Groningen Frailty Indicator	25	Mobility, physical fitness, vision, hearing, nourishment, morbidity, cognition, psychosocial	0–15; moderately to severely frail ≥4	[66, 67]
13. Tilburg Frailty Indicator	15	Sex, age, marital status, country of birth, highest educational level, monthly household income, diseases/chronic disorders, recent health/loss events, living situation	0–15; frail ≥5	[68]

energy and physiologic reserve. They analyzed data from the prospective Cardiovascular Health Study (CHS). Among five criteria comprising the CHS Fried Frailty Phenotype instrument, the first was unintentional weight loss (shrinking), defined as >10 pounds lost in prior year. Weakness was characterized by grip strength in the lowest quintile by gender and body mass index. Poor endurance was given by self-report of exhaustion. Slowness of ambulation was noted if performance was in the slowest quintile for gender and height. Low activity was measured by kilocalories/week: males <383 Kcals/week and females <270 Kcals/week. The range of possible scores is 0–5. The threshold for frailty was ≥3 criteria present. Intermediate or pre-frail was defined as 1 or 2 criteria present. Patients who lack any of the criteria would be classified as robust. Other authors have since adapted the frailty phenotype instrument, using the same domains but different specific criteria [36–41].

The second main theoretical approach to frailty is the deficit accumulation model put forward by Rockwood and Mitnitski, investigators who participated in the Canadian Study of Health and Aging (CSHA) [42]. The underlying concept of this model is simply that frailty develops as the number of clinical deficits rises. The original CSHA Frailty Index included 70 items in the domains of the presence/severity of current diseases, impaired ADLs, and physical and neurological signs from clinical examination. The Frailty Index is scored in terms of proportion of items present, so the range is between 0.0 and 1.0. An individual is considered frail if the proportion is ≥0.25.

Rating all 70 items on the full Frailty Index can be quite time-consuming; thus many adaptations of the Frailty Index have appeared that used fewer numbers of items, including 51 [43], 50 [44], 48 [45], 47 [37], 44 [46], 40 [47], 38 [48], 11 [49, 50], and 5 [51]. One widely used adaptation of the Frailty Index is the 11-factor modified frailty index (mFI), developed by Farhat, Velanovich, and associates [49, 50]. Items included in this instrument are diabetes mellitus, congestive heart failure, hypertension requiring medication, transient ischemic attack or resolved cerebrovascular accident, dependent functional status, myocardial infarction, peripheral vascular disease or rest pain, cerebrovascular accident with residual neurological deficit, chronic obstructive pulmonary disease (COPD) or pneumonia, impaired sensorium and either prior percutaneous coronary intervention, prior cardiac surgery, or angina. Specific definitions for all of these items are given in an appendix of the 2013 Velanovich article [49]. A proposed threshold for frailty using the mFI is ≥3 items present. Subramaniam et al. studied a five-factor version of the mFI [51] that included dependent functional status, diabetes, COPD, congestive heart failure, and hypertension requiring medication. These investigators found that the 5-factor mFI and the 11-factor mFI performed similarly in predicting mortality and postoperative complications using the American College of Surgeons (ACS) National Surgical Quality Improvement Program (NSQIP) dataset.

In 2005, Rockwood et al. proposed a simpler alternative to the 70-item CSHA Frailty Index [43] that relies on a clinician's global judgment of the patient's frailty status. In its original form, the CSHA Clinical Frailty Scale separates individuals into seven levels: very fit; well; well, with treated comorbid disease; apparently vulnerable; mildly frail; moderately frail; and severely frail. A revision of the Clinical Frailty Scale has nine levels: very fit; well; managing well; vulnerable; mildly frail; moderately frail; severely frail; very severely frail; and terminally ill [52].

A third theoretical framework concerning frailty has not yet been converted into an assessment instrument. The World Health Organization (WHO) World Report on Ageing and Health from 2015 [53] identifies a construct called intrinsic capacity. While most frailty measurement instruments focus on deficits and weaknesses, this framework focuses more positively on functions that are present. It makes use of the International Classification of Functioning, Disability, and Health framework. Intrinsic capacity is defined as "…the composite of all the physical and mental capacities that an individual can draw upon at any point in their life…" and is composed of these domains: cognition, locomotion, psychological states, vitality, and sensory functions [54]. Preparatory work to construct an instrument based on this approach is ongoing [55].

Among less widely published assessment instruments, the FRAIL Scale name is an acronym of the included domains: fatigue, resistance, ambulation, illness, and loss of weight [56, 57]. The scale was produced by the International Academy Nutrition and Aging Task Force on Frailty Assessment of Older People in Clinical Practice. Fatigue is rated by self-report. Resistance is defined as the ability to climb one flight of stairs. Ambulation is the ability to walk one block. The threshold for illnesses is >5. Loss of weight means >5%. Scores range from 0 to 5; frailty is ≥3 items present, pre-frail is 1–2 items, and robust is 0 item.

The Study of Osteoporotic Fractures Frailty was described by Ensrud et al. [39, 58]. It was developed within the conduct of this prospective cohort study. It includes three items: weight loss (>5%); exhaustion (unable to rise from a chair 5 times without using arms); and self-report reduced energy level. Ranging from 0 to 3, frail is ≥2 items, pre-frail is 1 item, and robust is 0 item.

The Edmonton Frail Scale entails eight items on these domains: cognition, general health status, functional independence, social support, medication use, nutrition, mood, and continence and functional performance [59]. Most items could be scored as either 0 or 1 point, while others can be scored up to 2 points. The range of possible scores was 17 and frailty was ≥7 points.

Raîche et al. developed two instruments at Sherbrooke University in Quebec, Canada. The first was the Sherbrooke Postal Questionnaire [60], encompassing six items addressing living alone; taking ≥3 medications daily; using a cane, walker, or wheelchair; impaired visual acuity; impaired hearing acuity; and memory problems. Frailty is ≥2 items present. The Program of Research to Integrating the Services for the Maintenance of Autonomy (PRISMA-7) [61] instrument was developed for assessing community-dwelling adults aged ≥75. The seven items in this tool were age ≥85, male sex, health problems limiting ADLs, need for regular assistance, being homebound, social support, and need for a cane, walker, or wheelchair. Frailty is judged if ≥3 items are present.

The Vulnerable Elders Survey (VES-13) [62] is a 13-item instrument developed in 2001 using data from the Medicare Current Beneficiary Survey. Domains include age, physical activities, functional activities, and self-rated health. Score can range from 0 to 10 and frailty is defined by scores ≥3.

The Strawbridge Functional Domains Model instrument [63] was developed in 1994 for the prospective Alameda County Study. It included 16 items in four domains: physical functioning (sudden loss of balance, weakness in arms, weakness in legs, dizziness/faintness upon standing quickly); nutritive function (loss of appetite, unexplained weight loss); cognitive functioning (attentional difficulty, word finding difficulty, difficulty remembering things, and misplacing things); and sensory problems (reading a newspaper, recognizing a friend across the street, reading signs at night, hearing over the phone, hearing a normal conversation, hearing a conversation in a noisy room). Problems in ≥2 domains represented frailty.

The Comprehensive Assessment of Frailty [64, 65] combines an adapted version of the Fried Frailty Phenotype and other information. Additional data is gathered for body mass index (BMI, with gender-specific thresholds), serum albumin, forced expiratory volume in 1 second (FEV1), creatinine, and a modified physical performance test. The last part evaluates standing static balance, chair rise, putting on and removing a jacket, picking up a pen from the floor, and turning 360 degrees. The maximum score is 35. Scores ≤10 are not frail, 11–25 are moderately frail, and 26–35 are severely frail.

The Groningen Frailty Indicator [66, 67] was created in the Netherlands for evaluation of community-dwelling elderly. Fifteen items belong to eight domains: mobility, physical fitness, vision, hearing, nourishment, morbidity, cognition (perception), and psychosocial. Scores range between 0 and 15. Scores ≥4 are considered moderately to severely frail.

The Tilburg Frailty Indicator [68] comes from Denmark and has undergone thorough evaluation of reliability and validity. Unscored items include demographics (sex, age, marital status, country of birth, education, and monthly household income), overall assessment of healthy lifestyle, ≥2 diseases and/or chronic disorders, personal losses in past year, and satisfaction with home living environment. Fifteen-scored items addressed physical, psychological, and social domains. Score ranges from 0 to 15 and the frailty threshold is ≥5.

Application of Frailty Measurements to Surgical Practice

While frailty score may have academic importance and has become a subject of intense literature debate and writings, the fact remains that only in select elective surgical procedure this may be modified.

The modifications and improvement of frailty may be important on the four questions:

1. Whether to perform surgery at all?
2. When to perform surgery, if truly necessary?
3. Is there a need for any preoperative intervention?
4. How can we modify and improve postoperative outcomes?

   Regarding the first class, forgoing surgery may in some instances be the better choice for the most frail. Abundant evidence has accumulated to show that frailty measures predict adverse outcomes such as surgical complications and postoperative mortality [45, 49, 69–81]. Shared decision-making between patients, caregivers, and surgeons should carefully weigh the potential benefits and harms of both operative and nonoperative options. Consultation with palliative care specialists may be desirable.

If decision-makers (patient, family, and the surgeon and other members of the team) have selected the surgical option, choosing the optimal time could be crucial to achieving good outcomes. Assuming that it is possible to delay surgery, time could be well spent assuring that comorbid conditions are appropriately managed. Furthermore, surgeons are increasingly pursuing prehabilitation of surgical candidates. Prehabilitation can be broadly defined as any intervention intended to bolster the physiologic reserve of frail patients and mitigate risks of adverse postoperative outcomes [81]. Interventions may include exercise, improvements to nutrition, and inspiratory muscle training.

Preoperative and postoperative actions would also be essential to optimizing surgical outcomes for frail individuals. Prior to surgery, anesthesiologists and surgeons must apply meticulous care to assessment, monitoring, and management of major organ system functioning, hemodynamics, laboratory test results, and mental status. Following surgery, frail patients may require prolonged stays in intensive care units. Discharge to rehabilitation programs and skilled nursing facilities may be necessary.

References

1. United States Census Bureau. https://www.census.gov/.
2. Population ageing and sustainable development. August 2014.
3. Fried LP, Ferrucci L, Darer J, Williamson JD, Anderson G. Untangling the concepts of disability, frailty, and comorbidity: implications for improved targeting and care. J Gerontol A Biol Sci Med Sci. 2004;59(3):255–63. https://doi.org/10.1093/gerona/59.3.m255.
4. Etzioni DA, Liu JH, Maggard MA, Ko CY. The aging population and its impact on the surgery workforce. Ann Surg. 2003;238(2):170–7. https://doi.org/10.1097/01.SLA.0000081085.98792.3d.
5. Turrentine FE, Wang H, Simpson VB, Jones RS. Surgical risk factors, morbidity, and mortality in elderly patients. J Am Coll Surg. 2006;203(6):865–77. https://doi.org/10.1016/j.jamcollsurg.2006.08.026.
6. Hamel MB, Henderson WG, Khuri SF, Daley J. Surgical outcomes for patients aged 80 and older: morbidity and mortality from major noncardiac surgery. J Am Geriatr Soc. 2005;53(3):424–9. https://doi.org/10.1111/j.1532-5415.2005.53159.x.
7. Oresanya LB, Lyons WL, Finlayson E. Preoperative assessment of the older patient: a narrative review. JAMA.2014;311(20):2110–20.https://doi.org/10.1001/jama.2014.4573.
8. Fried LP, Tangen CM, Walston J, et al. Frailty in older adults: evidence for a phenotype. J Gerontol A Biol Sci Med Sci. 2001;56(3):M146–56.

9. Collard RM, Boter H, Schoevers RA, Oude Voshaar RC. Prevalence of frailty in community-dwelling older persons: a systematic review. J Am Geriatr Soc. 2012;60(8):1487–92. https://doi.org/10.1111/j.1532-5415.2012.04054.x.
10. Theou O, Rockwood K. Should frailty status always be considered when treating the elderly patient? Aging Health. 2012;8(3):261–71.
11. Sourial N, Bergman H, Karunananthan S, et al. Implementing frailty into clinical practice: a cautionary tale. J Gerontol A Biol Sci Med Sci. 2013;68(12):1505–11. https://doi.org/10.1093/gerona/glt053.
12. Varadhan R, Seplaki CL, Xue QL, Bandeen-Roche K, Fried LP. Stimulus-response paradigm for characterizing the loss of resilience in homeostatic regulation associated with frailty. Mech Ageing Dev. 2008;129(11):666–70. https://doi.org/10.1016/j.mad.2008.09.013.
13. Holliday R. Understanding Ageing, vol. 30. New York: Cambridge University Press; 1995.
14. Wohlgemuth SE, Seo AY, Marzetti E, Lees HA, Leeuwenburgh C. Skeletal muscle autophagy and apoptosis during aging: effects of calorie restriction and life-long exercise. Exp Gerontol. 2010;45(2):138–48. https://doi.org/10.1016/j.exger.2009.11.002.
15. Ramasamy R, Vannucci SJ, Yan SSD, Herold K, Yan SF, Schmidt AM. Advanced glycation end products and RAGE: a common thread in aging, diabetes, neurodegeneration, and inflammation. Glycobiology. 2005;15(7):16R–28R. https://doi.org/10.1093/glycob/cwi053.
16. Hertoghe T. The "multiple hormone deficiency" theory of aging: is human senescence caused mainly by multiple hormone deficiencies? Ann N Y Acad Sci. 2005;1057:448–65. https://doi.org/10.1196/annals.1322.035.
17. Olivetti G, Melissari M, Capasso JM, Anversa P. Cardiomyopathy of the aging human heart. Myocyte loss and reactive cellular hypertrophy. Circ Res. 1991;68(6):1560–8. https://doi.org/10.1161/01.res.68.6.1560.
18. Fleg JL, O'Connor F, Gerstenblith G, et al. Impact of age on the cardiovascular response to dynamic upright exercise in healthy men and women. J Appl Physiol. 1995;78(3):890–900. https://doi.org/10.1152/jappl.1995.78.3.890.
19. Knudson RJ, Lebowitz MD, Holberg CJ, Burrows B. Changes in the normal maximal expiratory flow-volume curve with growth and aging. Am Rev Respir Dis. 1983;127(6):725–34. https://doi.org/10.1164/arrd.1983.127.6.725.
20. Kronenberg RS, Drage CW. Attenuation of the ventilatory and heart rate responses to hypoxia and hypercapnia with aging in normal men. J Clin Invest. 1973;52(8):1812–9. https://doi.org/10.1172/JCI107363.
21. Brenner BM, Meyer TW, Hostetter TH. Dietary protein intake and the progressive nature of kidney disease: the role of hemodynamically mediated glomerular injury in the pathogenesis of progressive glomerular sclerosis in aging, renal ablation, and intrinsic renal disease. N Engl J Med. 1982;307(11):652–9. https://doi.org/10.1056/NEJM198209093071104.
22. Dastur DK. Cerebral blood flow and metabolism in normal human aging, pathological aging, and senile dementia. J Cereb Blood Flow Metab. 1985;5(1):1–9. https://doi.org/10.1038/jcbfm.1985.1.
23. Jin F, Chung F. Minimizing perioperative adverse events in the elderly. Br J Anaesth. 2001;87(4):608–24. https://doi.org/10.1093/bja/87.4.608.
24. Clegg A, Young J, Iliffe S, Rikkert MO, Rockwood K. Frailty in elderly people. Lancet. 2013;381(9868):752–62. https://doi.org/10.1016/S0140-6736(12)62167-9.
25. Mitnitski A, Song X, Skoog I, et al. Relative fitness and frailty of elderly men and women in developed countries and their relationship with mortality. J Am Geriatr Soc. 2005;53(12):2184–9. https://doi.org/10.1111/j.1532-5415.2005.00506.x.
26. Gobbens RJ, Luijkx KG, Wijnen-Sponselee MT, Schols JM. Toward a conceptual definition of frail community dwelling older people. Nurs Outlook. 2010;58(2):76–86. https://doi.org/10.1016/j.outlook.2009.09.005.
27. Hogan DB, MacKnight C, Bergman H. Steering Committee, Canadian Initiative on Frailty and Aging. Models, definitions, and criteria of frailty. Aging Clin Exp Res. 2003;15(3 Suppl):1–29.
28. Heuberger RA. The frailty syndrome: a comprehensive review. J Nutr Gerontol Geriatr. 2011;30(4):315–68. https://doi.org/10.1080/21551197.2011.623931.
29. Warnier RMJ, van Rossum E, van Velthuijsen E, Mulder WJ, Schols JMGA, Kempen GIJM. Validity, reliability and feasibility of tools to identify frail older patients in inpatient hospital care: a systematic review. J Nutr Health Aging. 2016;20(2):218–30. https://doi.org/10.1007/s12603-015-0567-z.
30. de Vries NM, Staal JB, van Ravensberg CD, Hobbelen JSM, Olde Rikkert MGM. Nijhuis-van der Sanden MWG. Outcome instruments to measure frailty: a systematic review. Ageing Res Rev. 2011;10(1):104–14. https://doi.org/10.1016/j.arr.2010.09.001.
31. Hogan DB, Freiheit EA, Strain LA, et al. Comparing frailty measures in their ability to predict adverse outcome among older residents of assisted living. BMC Geriatr. 2012;12:56. https://doi.org/10.1186/1471-2318-12-56.
32. Woo J, Leung J, Morley JE. Comparison of frailty indicators based on clinical phenotype and the multiple deficit approach in predicting mortality and physical limitation. J Am Geriatr Soc. 2012;60(8):1478–86. https://doi.org/10.1111/j.1532-5415.2012.04074.x.
33. Kristjansson SR, Rønning B, Hurria A, et al. A comparison of two pre-operative frailty measures in older surgical cancer patients. J Geriatr Oncol. 2012;3(1):1–7.
34. Ensrud KE, Ewing SK, Cawthon PM, et al. A comparison of frailty indexes for the prediction of falls, disability, fractures, and mortality in older men.

J Am Geriatr Soc. 2009;57(3):492–8. https://doi.org/10.1111/j.1532-5415.2009.02137.x.
35. Gruenewald TL, Seeman TE, Karlamangla AS, Sarkisian CA. Allostatic load and frailty in older adults. J Am Geriatr Soc. 2009;57(9):1525–31. https://doi.org/10.1111/j.1532-5415.2009.02389.x.
36. Bandeen-Roche K, Xue Q-L, Ferrucci L, et al. Phenotype of frailty: characterization in the women's health and aging studies. J Gerontol A Biol Sci Med Sci. 2006;61(3):262–6.
37. Rockwood K, Song X, MacKnight C, et al. A global clinical measure of fitness and frailty in elderly people. CMAJ. 2005;173(5):489–95. https://doi.org/10.1503/cmaj.050051.
38. Rockwood K, Mitnitski A, Song X, Steen B, Skoog I. Long-term risks of death and institutionalization of elderly people in relation to deficit accumulation at age 70. J Am Geriatr Soc. 2006;54(6):975–9. https://doi.org/10.1111/j.1532-5415.2006.00738.x.
39. Joseph B, Pandit V, Rhee P, et al. Predicting hospital discharge disposition in geriatric trauma patients: is frailty the answer? J Trauma Acute Care Surg. 2014;76(1):196–200. https://doi.org/10.1097/TA.0b013e3182a833ac.
40. Kulminski AM, Ukraintseva SV, Kulminskaya IV, Arbeev KG, Land K, Yashin AI. Cumulative deficits better characterize susceptibility to death in elderly people than phenotypic frailty: lessons from the Cardiovascular Health Study. J Am Geriatr Soc. 2008;56(5):898–903. https://doi.org/10.1111/j.1532-5415.2008.01656.x.
41. Hastings SN, Purser JL, Johnson KS, Sloane RJ, Whitson HE. Frailty predicts some but not all adverse outcomes in older adults discharged from the emergency department. J Am Geriatr Soc. 2008;56(9):1651–7. https://doi.org/10.1111/j.1532-5415.2008.01840.x.
42. Collerton J, Martin-Ruiz C, Davies K, et al. Frailty and the role of inflammation, immunosenescence and cellular ageing in the very old: cross-sectional findings from the Newcastle 85+ Study. Mech Ageing Dev. 2012;133(6):456–66. https://doi.org/10.1016/j.mad.2012.05.005.
43. Cigolle CT, Ofstedal MB, Tian Z, Blaum CS. Comparing models of frailty: the Health and Retirement Study. J Am Geriatr Soc. 2009;57(5):830–9. https://doi.org/10.1111/j.1532-5415.2009.02225.x.
44. Velanovich V, Antoine H, Swartz A, Peters D, Rubinfeld I. Accumulating deficits model of frailty and postoperative mortality and morbidity: its application to a national database. J Surg Res. 2013;183(1):104–10. https://doi.org/10.1016/j.jss.2013.01.021.
45. Farhat JS, Velanovich V, Falvo AJ, et al. Are the frail destined to fail? Frailty index as predictor of surgical morbidity and mortality in the elderly. J Trauma Acute Care Surg. 2012;72(6):1526–1530; discussion 1530-1531. https://doi.org/10.1097/TA.0b013e3182542fab.
46. Subramaniam S, Aalberg JJ, Soriano RP, Divino CM. New 5-factor modified frailty index using American College of Surgeons NSQIP Data. J Am Coll Surg. 2018;226(2):173–181.e8. https://doi.org/10.1016/j.jamcollsurg.2017.11.005.
47. Clinical Frailty Scale. 2008. https://www.dal.ca/sites/gmr/our-tools/clinical-frailty-scale.html. Accessed 17 July 2019.
48. World Report on Ageing and Health. 2015. https://www.who.int/ageing/events/world-report-2015-launch/en/. Accessed 17 July 2019.
49. Araujo de Carvalho I, Martin FC, Cesari M, Sumi Y, Thiyagarajan JA, Beard J. Operationalising the concept of intrinsic capacity in clinical settings. March 2017. https://www.who.int/ageing/health-systems/clinical-consortium/CCHA2017-backgroundpaper-1.pdf. Accessed 17 July 2019.
50. Cesari M, Araujo de Carvalho I, Amuthavalli Thiyagarajan J, et al. Evidence for the domains supporting the construct of intrinsic capacity. J Gerontol A Biol Sci Med Sci. 2018;73(12):1653–60. https://doi.org/10.1093/gerona/gly011.
51. Abellan van Kan G, Rolland Y, Bergman H, Morley JE, Kritchevsky SB, Vellas B. The I.A.N.A Task Force on frailty assessment of older people in clinical practice. J Nutr Health Aging. 2008;12(1):29–37.
52. Abellan van Kan G, Rolland YM, Morley JE, Vellas B. Frailty: toward a clinical definition. J Am Med Dir Assoc. 2008;9(2):71–2. https://doi.org/10.1016/j.jamda.2007.11.005.
53. Ensrud KE, Ewing SK, Taylor BC, et al. Comparison of 2 frailty indexes for prediction of falls, disability, fractures, and death in older women. Arch Intern Med. 2008;168(4):382–9. https://doi.org/10.1001/archinternmed.2007.113.
54. Rolfson DB, Majumdar SR, Tsuyuki RT, Tahir A, Rockwood K. Validity and reliability of the Edmonton Frail Scale. Age Ageing. 2006;35(5):526–9. https://doi.org/10.1093/ageing/afl041.
55. Hébert R, Bravo G, Korner-Bitensky N, Voyer L. Predictive validity of a postal questionnaire for screening community-dwelling elderly individuals at risk of functional decline. Age Ageing. 1996;25(2):159–67. https://doi.org/10.1093/ageing/25.2.159.
56. Raîche M, Hébert R, Dubois M-F. PRISMA-7: a case-finding tool to identify older adults with moderate to severe disabilities. Arch Gerontol Geriatr. 2008;47(1):9–18. https://doi.org/10.1016/j.archger.2007.06.004.
57. Min LC, Elliott MN, Wenger NS, Saliba D. Higher vulnerable elders survey scores predict death and functional decline in vulnerable older people. J Am Geriatr Soc. 2006;54(3):507–11. https://doi.org/10.1111/j.1532-5415.2005.00615.x.
58. Strawbridge WJ, Shema SJ, Balfour JL, Higby HR, Kaplan GA. Antecedents of frailty over three decades in an older cohort. J Gerontol B Psychol Sci Soc Sci. 1998;53(1):S9–16. https://doi.org/10.1093/geronb/53b.1.s9.
59. Sündermann S, Dademasch A, Rastan A, et al. One-year follow-up of patients undergoing elective cardiac surgery assessed with the Comprehensive Assessment

of Frailty test and its simplified form. Interact Cardiovasc Thorac Surg. 2011;13(2):119–123; discussion 123. https://doi.org/10.1510/icvts.2010.251884.
60. Sündermann S, Dademasch A, Praetorius J, et al. Comprehensive assessment of frailty for elderly high-risk patients undergoing cardiac surgery. Eur J Cardiothorac Surg. 2011;39(1):33–7. https://doi.org/10.1016/j.ejcts.2010.04.013.
61. Steverink N, Slaets J, Schuurmanns H, van Lis M. Measuring frailty: developing and testing the GFI (Groningen Frailty Indicator). Gerontologist. 2001;41:236–7.
62. Schuurmans H, Steverink N, Lindenberg S, Frieswijk N, Slaets JPJ. Old or frail: what tells us more? J Gerontol A Biol Sci Med Sci. 2004;59(9):M962–5. https://doi.org/10.1093/gerona/59.9.m962.
63. Gobbens RJJ, van Assen MALM, Luijkx KG, Wijnen-Sponselee MT, Schols JMGA. The Tilburg Frailty Indicator: psychometric properties. J Am Med Dir Assoc. 2010;11(5):344–55. https://doi.org/10.1016/j.jamda.2009.11.003.
64. Akyar S, Armenia SJ, Ratnani P, Merchant AM. The impact of frailty on postoperative cardiopulmonary complications in the emergency general surgery population. Surg J (N Y). 2018;4(2):e66–77. https://doi.org/10.1055/s-0038-1655756.
65. Eamer G, Al-Amoodi MJH, Holroyd-Leduc J, Rolfson DB, Warkentin LM, Khadaroo RG. Review of risk assessment tools to predict morbidity and mortality in elderly surgical patients. Am J Surg. 2018;216(3):585–94. https://doi.org/10.1016/j.amjsurg.2018.04.006.
66. Seib CD, Rochefort H, Chomsky-Higgins K, et al. Association of patient frailty with increased morbidity after common ambulatory general surgery operations. JAMA Surg. 2018;153(2):160–8. https://doi.org/10.1001/jamasurg.2017.4007.
67. Congiusta DV, Palvannan P, Merchant AM. The impact of frailty on morbidity and mortality following open emergent colectomies. Biomed Res Int. 2017;2017:5126452. https://doi.org/10.1155/2017/5126452.
68. Wahl TS, Graham LA, Hawn MT, et al. Association of the modified frailty index with 30-day surgical readmission. JAMA Surg. 2017;152(8):749–57. https://doi.org/10.1001/jamasurg.2017.1025.
69. Mogal H, Vermilion SA, Dodson R, et al. Modified frailty index predicts morbidity and mortality after pancreaticoduodenectomy. Ann Surg Oncol. 2017;24(6):1714–21. https://doi.org/10.1245/s10434-016-5715-0.
70. Wen Y, Jabir MA, Dosokey EMG, et al. Using modified frailty index to predict safe discharge within 48 hours of ileostomy closure. Dis Colon Rectum. 2017;60(1):76–80. https://doi.org/10.1097/DCR.0000000000000722.
71. Mosquera C, Spaniolas K, Fitzgerald TL. Impact of frailty on approach to colonic resection: laparoscopy vs open surgery. World J Gastroenterol. 2016;22(43):9544–53. https://doi.org/10.3748/wjg.v22.i43.9544.
72. Augustin T, Burstein MD, Schneider EB, et al. Frailty predicts risk of life-threatening complications and mortality after pancreatic resections. Surgery. 2016;160(4):987–96. https://doi.org/10.1016/j.surg.2016.07.010.
73. Joseph B, Zangbar B, Pandit V, et al. Emergency general surgery in the elderly: too old or too frail? J Am Coll Surg. 2016;222(5):805–13. https://doi.org/10.1016/j.jamcollsurg.2016.01.063.
74. Ehlert BA, Najafian A, Orion KC, Malas MB, Black JH, Abularrage CJ. Validation of a modified Frailty Index to predict mortality in vascular surgery patients. J Vasc Surg. 2016;63(6):1595–1601.e2. https://doi.org/10.1016/j.jvs.2015.12.023.
75. Louwers L, Schnickel G, Rubinfeld I. Use of a simplified frailty index to predict Clavien 4 complications and mortality after hepatectomy: analysis of the National Surgical Quality Improvement Project database. Am J Surg. 2016;211(6):1071–6. https://doi.org/10.1016/j.amjsurg.2015.09.015.
76. Arya S, Kim SI, Duwayri Y, et al. Frailty increases the risk of 30-day mortality, morbidity, and failure to rescue after elective abdominal aortic aneurysm repair independent of age and comorbidities. J Vasc Surg. 2015;61(2):324–31. https://doi.org/10.1016/j.jvs.2014.08.115.
77. Keller DS, Bankwitz B, Nobel T, Delaney CP. Using frailty to predict who will fail early discharge after laparoscopic colorectal surgery with an established recovery pathway. Dis Colon Rectum. 2014;57(3):337–42. https://doi.org/10.1097/01.dcr.0000442661.76345.f5.
78. Karam J, Tsiouris A, Shepard A, Velanovich V, Rubinfeld I. Simplified frailty index to predict adverse outcomes and mortality in vascular surgery patients. Ann Vasc Surg. 2013;27(7):904–8. https://doi.org/10.1016/j.avsg.2012.09.015.
79. Obeid NM, Azuh O, Reddy S, et al. Predictors of critical care-related complications in colectomy patients using the National Surgical Quality Improvement Program: exploring frailty and aggressive laparoscopic approaches. J Trauma Acute Care Surg. 2012;72(4):878–83. https://doi.org/10.1097/TA.0b013e31824d0f70.
80. Makary MA, Segev DL, Pronovost PJ, et al. Frailty as a predictor of surgical outcomes in older patients. J Am Coll Surg. 2010;210(6):901–8. https://doi.org/10.1016/j.jamcollsurg.2010.01.028.
81. Milder DA, Pillinger NL, Kam PCA. The role of prehabilitation in frail surgical patients: a systematic review. Acta Anaesthesiol Scand. 2018;62(10):1356–66. https://doi.org/10.1111/aas.13239.

Part II

System Based Specialized Surgical Problems in Geriatric Surgical Patients

Anesthetic Concerns in the Elderly

7

Sarah C. Smith

The aging body is more susceptible to illness and less resilient in the face of disease, such that geriatric patients consume a disproportionate share of healthcare resources. Of the more than 100 million in- and out-patient procedures performed annually in the USA, 32–35% are for those aged 65 and older [1]. As many of these procedures require anesthesia management, anesthesiologists must be familiar with the physiological effects of aging and how these changes influence the response to surgery and anesthesia.

As patients age, there are fundamental changes in body composition and organ function that alter the pharmacokinetics and pharmacodynamics of drugs, including those used during the administration of anesthesia. The accumulation of comorbid conditions that is typical of advancing age also alters responses to drugs administered in the perioperative period. Not only are elderly persons more sensitive to sedative hypnotics, but they also have a higher incidence of certain side effects, including perioperative neurocognitive dysfunction (PND).

While these general principles are widely known, applying them to individual patients is particularly challenging. What is known of changes in pharmacokinetics and pharmacodynamics in the older population is generally based on limited studies in otherwise healthy persons between the ages of 60 and 80. Extrapolating this knowledge to the increasing numbers of octogenarians, nonagenarians, and even centenarians who present for surgery and anesthesia is problematic. Older persons are also more likely to be taking a number of medications on a routine basis compared to their younger counterparts. The anesthesiologist must also consider how each of these drugs may interact with medications administered within the perioperative period [2].

Aging-Related Changes in Organ System Function: Impact on Anesthetic Management

Central Nervous System (CNS)

Aging is strongly associated with a decrease in cognitive function, characterized by impairments in memory [3], reaction time [4, 5], and creativity [6, 7]. While previously acquired knowledge and skills often remain intact in "normal" aging, even these abilities can be lost in age-related pathological states such as Alzheimer's disease (AD), other forms of dementia, or as a result of cerebrovascular accidents (CVA). The magnitude of cognitive decline increases steadily with age, such that among nonagenarians and centenarians only one third have cognitive functioning that could be

S. C. Smith (✉)
Division of Cardiothoracic Anesthesia, Department of Anesthesiology, Westchester Medical Center, New York Medical College, Valhalla, NY, USA
e-mail: Sarah.Smith@wmchealth.org

R. Latifi (ed.), *Surgical Decision Making in Geriatrics*,
https://doi.org/10.1007/978-3-030-47963-3_7

classified as normal while another third are moderately impaired and the remaining third are profoundly demented [8].

Age-related cognitive decline parallels changes in the CNS at the gross anatomical, cellular, and neurochemical level. A recent MRI study of healthy adults showed that whole brain volume decreases 0.32% annually after age 20, with an acceleration of volume loss observed after age 70 [9]. Some of this decrease in volume is due to neuronal loss that occurs throughout adulthood, such that by age 90, even cognitively normal individuals have only 90% of the neurons they had as a young adult. The loss of neurons is not uniform, with the hippocampus being largely spared while the neocortex, basal forebrain nuclei, and brainstem monoaminergic systems are more significantly affected [8].

Alterations in neuronal morphology that occur with advancing age may be even more important functionally than overall neuronal loss. For example, the dendritic arbors of pyramidal neurons in the hippocampus and prefrontal cortex regress with age, with losses of dendritic length and number being associated with loss of neuronal connectivity and cognitive functioning [10]. Neurogenesis decreases with age, as do neuroprotective mechanisms, leaving the aging brain highly susceptible to oxidative stress and other sources of cellular damage. These changes result in the aged brain being more vulnerable to injury and less able to adapt to pathological states through neuronal plasticity.

Levels of several neurotransmitters including acetylcholine, dopamine, serotonin, and glutamate (Glu) decrease with age, resulting in corresponding changes in receptors for these chemicals [11]. Proton magnetic resonance spectroscopy (H^1-MRS) allows for the measurement of certain neurometabolites to be measured noninvasively in humans, and several such studies have shown significant changes associated with aging. Levels of Glu, the primary excitatory neurotransmitter in the brain, have been shown to fall with aging, as do levels of n-acetyl aspartate (NAA), a marker for neuronal metabolic efficiency and functioning. The aging brain also has increased levels of choline (Cho), an indicator of neuroinflammation. Although changes in NAA, Glu, and Cho are associated with normal aging, the changes are more pronounced in patients who have lower scores on neurocognitive testing [12].

The neurocognitive changes associated with aging parallel a generalized increase in sensitivity to anesthetics and analgesics. The minimum alveolar concentration (MAC) is defined as the percentage of volatile anesthetic at 1 atmosphere required to produce immobility in 50% of a population in the presence of surgical stimuli. After peaking at 6 months, MAC decreases steadily with age, such that patients in their eighth and ninth decades of life require only approximately two thirds the level of volatile agent as a patient in young adulthood [13]. Changes in GABA receptor function result in an increased sensitivity to benzodiazepines in geriatric patients, while other agents including propofol, etomidate, and opioids are also noted to be more potent. A decrease in cerebrospinal fluid (CSF) volume and an increase in dural permeability result in geriatric patients requiring lower volumes of local anesthetics for neuraxial anesthesia, while a decrease in peripheral nerve myelination also decreases the dose required for peripheral nerve blockade [2].

Cardiovascular

The incidence of cardiac pathologies including ischemic and valvular heart disease increases with age, but even if these conditions are absent, there are changes in the cardiovascular system that affect anesthetic management. Aging is associated with a generalized loss of compliance in the tissues of the heart and vascular system as elastin production declines leading to its replacement by stiffer collagen fibers. There are several clinical manifestations of this loss of compliance. Because stroke volume is preserved, but the aorta is less compliant, systolic blood pressure (SBP) is generally higher in older adults. The pressure wave generated by ventricular contraction travels more quickly through the stiffened arterial system such that a reflective pressure wave may return to the heart before ejection is complete

[14]. This increase in afterload over time may contribute to the development of left ventricular hypertrophy (LVH) and subsequent diastolic dysfunction, which may progress to diastolic heart failure. The incidence of diastolic heart failure has been increasing in recent decades such that it is now more prevalent than systolic heart failure. Importantly, both diastolic dysfunction and diastolic heart failure are associated with an increased risk of adverse cardiac events following non-cardiac surgery as well as prolonged lengths of stay (LOS) [15].

Although resting sympathetic tone is high in geriatric patients, the response to beta-stimulation is diminished, such that maintenance of cardiac output (CO) is more preload dependent. Preload stability is dependent upon the function of the venous system to serve as a reservoir for most of the body's blood volume. As the venous system also becomes less compliant with age, this stabilizing factor on preload is also diminished. The result is significant blood pressure lability in geriatric patients under anesthesia, with precipitous hypotension often occurring due to the vasodilating effects of induction agents and potential rebound hypertension following treatment with vasoconstrictors. Goals for anesthetic management for these elderly patients should include careful maintenance of preload, as too little will not allow sufficient ventricular filling due to noncompliance of the myocardium, while too much may raise left atrial pressures and cause pulmonary edema. Elderly patients are more reliant upon the atrial kick for ventricular filling, so maintenance of sinus rhythm and prompt treatment of tachycardia are also important [14].

Pulmonary

Lung function peaks in the third decade of life and steadily declines thereafter due to changes in the lung parenchyma, chest wall, and respiratory muscles. The lung parenchyma becomes more compliant with age, resulting in a loss of structural support that helps keep small airways open, such that atelectasis develops more easily, particularly in the perioperative setting due to mechanical ventilation, pharmacologic paralysis, and surgical positioning. There is an enlargement of airspaces, such that the alveolar surface area declines, leading to ventilation-perfusion (VQ) mismatching, decreased diffusing lung capacity for carbon monoxide (DLCO), and impaired oxygenation [16]. The dead space ventilation increases due to this expansion of airspaces as well, such that geriatric patients must maintain a higher resting minute ventilation than younger persons to maintain normocarbia [17].

Aging is associated with a loss of intervertebral disk space, leading to kyphosis and a chest cavity that is shorter and more barrel-shaped than in young adulthood. This anatomical change is mechanically disadvantageous, resulting in increased work of breathing. Simultaneously, diaphragmatic and intercostal muscle strength decreases, such that it is more challenging for geriatric patients to increase minute ventilation in response to stress or exercise [17]. Together these changes result in a chest wall that is less compliant with age, contributing to deleterious changes in lung mechanics. The total lung capacity (TLC) is largely preserved, but a rise in functional residual capacity (FRC) and residual volume (RV) results in a corresponding decline in vital capacity (VC). The forced vital capacity (FVC) and forced expiratory volume in 1 second (FEV1) decrease with age, while the ratio FEV1/FVC is also lower due to an increase in airway resistance [16].

Geriatric patients also have a diminished respiratory drive in response to hypoxia and hypercarbia and a higher incidence of central sleep apnea. Decreased mucociliary function [18], diminished cough reflexes, and weakness of respiratory musculature frequently impair the ability of geriatric patients to clear mucus and debris from the airway, increasing the risk of perioperative aspiration [17]. Predictably, age is an independent risk factor for a variety of postoperative pulmonary complications (PPC) [19, 20], including unplanned reintubation [21] and pneumonia [22].

A number of strategies can be employed in the perioperative period to minimize the risk of PPCs in elderly patients, including avoidance of general anesthesia and utilization of lung-protective ventilation. During mechanical ventilation, low tidal volumes (6–8 mL/kg), avoidance of unnecessarily high levels of inspired oxygen (FiO_2), and the use of recruitment maneuvers may decrease the risk of PPCs. The ideal level of positive end-expiratory pressure (PEEP) during general anesthesia is unclear as studies have shown both low (<2 cm H_2O) and high (>10 cm H_2O) levels of PEEP to be deleterious. Reversal of nondepolarizing neuromuscular blockade should be complete in geriatric patients as residual weakness in the recovery period can also increase the risk of PPCs. Adequate analgesia, chest physiotherapy, incentive spirometry, and early mobilization are also strategies that can be implemented in the immediate postoperative period [23].

Gastrointestinal

Liver mass decreases 20–40% from young adulthood to old age with a corresponding decline in hepatic blood flow. Bile acid secretion declines with age, while biliary cholesterol increases, contributing to an increased risk of gallstones and gallbladder disease in the elderly. There is also evidence that the regenerative capacity of the liver decreases in old age, particularly in the presence of significant hepatic disease. It may be presumed, therefore, that geriatric patients are at a higher risk of hepatic dysfunction in the setting of pathophysiologic stress such as shock. Human and nonhuman primate studies of the effects of aging on Phase I metabolism have yielded mixed results, as there is much interindividual variability in the activity of the involved hepatic enzymes regardless of age. However, the increased risk of adverse drug events in elderly patients suggests there may be an overall decline in hepatic drug metabolism associated with aging [24]. Additionally, mortality due to liver disease increases markedly after age 45 before beginning to decline again after age 85. The decrease in liver disease mortality in those in their ninth and tenth decades of life likely reflects the presence of underlying protective physiologic factors, suggesting that the absence of significant liver disease is essentially a prerequisite for achieving extremely old age [25].

Both esophageal motility and gastric emptying decline with age, perhaps contributing to the higher risk of perioperative aspiration that is observed in geriatric patients [14]. The American Society of Anesthesiologists (ASA) perioperative fasting guidelines make no specific recommendations for the elderly, however [26]. Atrophic gastritis is associated with aging and may contribute to hypocalcemia in this population. Pancreatic function, however, remains largely intact in advanced age [14].

Renal

While it was once thought that aging itself resulted in a significant decline in renal function, more recent evidence suggests that when confounders such as hypertension, diabetes, and heart failure are accounted for, renal function remains relatively normal in advanced age. Although renal mass and the number of glomeruli decrease approximately 20–30% by age 80, clearance of drugs remains in the low-normal range when compared to younger adults. The clearance of waste products and maintenance of fluid homeostasis also remain largely intact [27].

Nonetheless, all causes of renal dysfunction increase in frequency with advancing age, such that a large portion of the geriatric population has renal pathology, although not as a result of aging *per se*. Because the serum creatinine commonly obtained as part of preoperative laboratory studies is a poor indicator of renal function, many geriatric patients presenting for surgery may have occult renal dysfunction and will have decreased clearance of drugs common in anesthetic practice, particularly certain neuromuscular blocking agents. Changes in the renin-angiotensin-aldosterone system, a diminished sense of thirst, and a decreased ability to thermoregulate in the setting of high temperatures make geriatric patients susceptible to dehydration. Hypovolemia in the preoperative fasting state may, therefore, be more pronounced in older patients, and the anesthesiologist must be prepared to administer necessary intravenous volume expansion. The aging kidney may also be more susceptible to acute insults, including intraoperative hypotension. Careful titration of vasopressors and volume expansion during anesthesia are particularly important to maintain blood pressure within a range consistent with baseline to minimize the risk of perioperative acute kidney injury (AKI) [14].

Endocrine and Thermoregulatory Systems

The incidence of Type 2 diabetes increases with age, but even those geriatric patients without such a diagnosis have altered responses to serum glucose levels and insulin compared to their younger counterparts. Aging is associated with a decrease in insulin secretion as well as an increase in insulin resistance, primarily in skeletal muscle, whereas the hepatic response to insulin remains relatively intact. This change may parallel the sarcopenia, or loss of muscle mass, that is common in old age [28]. Perioperative hyperglycemia is associated with a variety of adverse outcomes, including infection, poor wound healing, and perioperative neurocognitive dysfunction, so glucose monitoring should be strongly considered in elderly patients undergoing major surgery. The precise range in which to maintain perioperative blood glucose levels is unclear, however, intensive glucose management (maintenance of serum glucose at 80–100 mg/dL) is associated with an increased risk of stroke and death. Subsequently, there is a great deal of variability between the recommendations of different specialty organizations regarding when perioperative insulin therapy should be initiated. However, most recommend treatment for glucose levels over 180 mg/dL [29].

Aging is associated with a number of endocrine disturbances that appear to contribute to sarcopenia, the progressive loss of muscle mass that occurs in older people contributing to generalized weakness, increased fall risk, and susceptibility to musculoskeletal injury. Sex hormones decrease in both men and women, with testosterone decline being more significantly linked to sarcopenia in men. Unfortunately, studies investigating testosterone supplementation in older men have shown an increased risk of cardiovascular events. Dehydroepiandrosterone (DHEA), a precursor for estrogen and androgens in skeletal muscle, also steadily falls with age. Decreased vitamin D levels, declines in insulin-like growth factor (IGF) and growth hormone (GH), as well as perturbations of the hypothalamic pituitary adrenal (HPA) axis resulting in elevated cortisol levels are also thought to contribute to sarcopenia [30].

Although thyroid function is not affected by aging itself, disorders resulting in hypothyroidism become progressively more common with age, particularly among women, for whom hypothyroidism is ten times more common than in men. Hypothyroidism in the elderly is underdiagnosed, as several of its symptoms, such as fatigue, weight gain, and cold intolerance, are also common in normal aging. Postoperatively hypothyroid patients have an increased risk of atrial fibrillation, particularly after cardiovascular procedures. Severe hypothyroidism could manifest in the postoperative period as poor wound healing, infection, or even myxedema coma. While severe hypothyroidism should be corrected before elective surgery, there is a paucity of evidence that correction of mild to moderate hypothyroidism improves surgical outcomes [31].

The elderly are more prone than younger adults to both hypo- and hyperthermia due to diminished thermoregulatory mechanisms. A variety of systems are engaged to maintain constant body temperature. Thermosensory afferents send signals regarding body and skin temperature to the CNS which then activates compensatory mechanisms to maintain temperature within a narrow range. Increased temperature results in vasodilation at the periphery such that heat can be discharged to the surroundings. Lower temperature conversely results in vasoconstriction and increased production of norepinephrine as well as increased thermogenesis from shivering. Aging is associated with a loss of sensory function such that temperature information is not relayed to the CNS as effectively as in younger adults. Further, thermogenesis is limited in the elderly, while vasoconstrictive and vasodilatory responses to temperature changes are also blunted [32].

General, neuraxial, and regional anesthesia impairs all aspects of the thermoregulatory system, and the effects are more pronounced in the elderly [33–35]. Because operating rooms are kept relatively cold and patients are at least partially exposed during surgery, this generally manifests as intraoperative and postoperative hypothermia, which is associated with prolonged anesthesia recovery, cardiac morbidity, and impaired drug metabolism. Perioperative hypothermia also promotes infection, impedes wound healing, and may result in coagulopathy which can exacerbate surgical bleeding. Intraoperative temperature monitoring is, therefore, of particular importance in geriatric surgical patients, and multiple warming methods should be considered, including heating of intravenous fluids, hot air or water blankets, and avoidance of unnecessary skin exposure [33].

Pharmacology of Anesthetics and Analgesics in the Geriatric Population

Geriatric patients are generally more sensitive to anesthetics and analgesics and display prolonged reactions to these drugs when compared with younger adults. Although there are some modest changes in pharmacokinetics associated with aging, most of this alteration in the response to sedative hypnotics is due pharmacodynamics effects. As mentioned in the preceding section, renal and hepatic clearance of drugs declines only minimally in the geriatric population in the absence of pathology. Phase I and Phase II metabolism in the liver is for the most part unchanged in geriatric patients. Protein binding of drugs may also decrease in advanced age, but clinically this seems to be of limited significance [27].

Aging is associated with an increased percentage of adipose tissue and lower total body water, resulting in a higher volume of distribution (V_d) and subsequent duration of action for lipophilic drugs [2]. Propofol is highly lipophilic and is noted to have an exaggerated sedative effect for a given dose in the elderly. Further, because of the increased V_d, the time to reach a steady state during a propofol infusion and the time required for recovery are also increased. Because of the cardiovascular changes associated with aging that have been previously discussed, hypotension may be profound after propofol induction in elderly patients. A 25–50% reduction in induction dose for propofol is recommended in the geriatric population, particularly in the setting of cardiac disease or frailty. Although etomidate provides a more hemodynamically stable induction than propofol, the dosing of this drug should also be decreased.

Long-acting opiates such as morphine and hydromorphone should be used with caution in the elderly, particularly in the setting of end-organ dysfunction. The pharmacokinetics of the shorter-acting fentanyl and remifentanil are largely unchanged in advanced age, but changes in pharmacodynamics result in these drugs being notably more potent. By the ninth decade of life, a 50% reduction in dosing of these drugs is recommended compared to young adults [36].

Prolonged duration of action of neuromuscular blockade (NMB) in the geriatric population is of particular concern. Residual NMB in the postoperative recovery period may result in hypoxia and hypercarbia, as well as increase the risk for reintubation, aspiration, and pneumonia. The duration of action of the aminosteroid NMBs, vecuronium and rocuronium, is prolonged by approximately 50% in the geriatric patients due to changes in pharmacokinetics as well as aging-related changes at the neuromuscular junction. By comparison, the benzylisoquinoline NMBs, atracurium and cisatracurium, are eliminated by Hoffman elimination, and their duration of action does not appear to be prolonged in the elderly [37].

As previously described, for volatile anesthetics MAC peaks at age 6 months and then declines throughout life, likely due to changes in GABA receptor function. After age 40 MAC decreases approximately 6% per decade, particularly during coadministration of other sedatives, including benzodiazepines, opiates, and nitrous oxide. Despite this well-known phenomenon and the fact that many modern anesthesia machines can be programmed to calculate age-adjusted MAC values automatically, there is evidence that elderly patients are routinely overdosed with volatile agents during general anesthesia [38]. Similar overdosing of fentanyl and propofol during induction of anesthesia has also been demonstrated [39]. Excess doses of anesthetics and analgesics contribute to a variety of postoperative complications, including PND, which will be discussed in detail in a following section of this chapter. It is important for anesthesiologists to be aware of the increased sensitivity among elderly patients to a variety of drugs used in the perioperative setting and to make dosing adjustments accordingly.

Preoperative Optimization and Assessment

Geriatric patients scheduled for elective surgery should undergo a comprehensive assessment addressing not only comorbidities but functional status, mental health, and social circumstances. Structured preoperative assessment allows for risk stratification such that patients and their families can be appropriately counseled regarding probable surgical outcomes and the likelihood of complications. This is also an opportunity for older patients to designate an appropriate surrogate and ideally develop a written advance directive that reflects their individual values and preferences. Establishing an advance directive in anticipation of a scheduled surgery and hospitalization can ensure that goals of care are understood by the patient, family, and perioperative medical team, minimizing the risk of psychosocial stress and conflict that may arise in the setting of an adverse outcome or complication [40].

The American Society for Enhanced Recovery and Perioperative Quality Initiative (ASER/PQI) has issued a joint consensus statement recommending nutritional screening for all elective surgery patients, with special recommendations for those in the geriatric age group. Initial screening includes an assessment of weight and inquiring about recent eating habits and unintended weight loss. For those over age 65, all-cause mortality begins to rise below a BMI of 24 kg/m^2 and doubles at 22 kg/m^2 for women and 20 kg/m^2 for men. For this reason, the ASER/PQI joint statement recommends that elderly adults scheduled for elective surgery with a BMI of less than 20 kg/m^2 should be referred to nutritional counseling, while the trigger for those under age 65 is 18.5 kg/m^2. Although neither specific nor sensitive for malnutrition, the serum albumin level is a simple and inexpensive indicator, with a value less than 3 g/dL being of concern. Patients reporting an unintended weight loss of greater than 10% in 6 months or those who report consuming less than half their normal intake in the preceding week are also considered high risk and should be referred for nutritional counseling.

Preoperative nutritional optimization for high-risk individuals should include nutritional supplementation for 7 days prior to surgery and avoidance of unnecessary prolonged fasts. More than increasing caloric intake, preoperative nutritional supplementation should emphasize increased protein consumption as this promotes wound healing, decreases the risk of pressure ulcers, and bolsters immune function. Arginine supplementation can also be considered, as it is required for proper T-cell function and the synthesis of endothelial nitric oxide. Omega-3s, as found in fish oil, and antioxidants are also sometimes used during the nutritional optimization period to further support the immune system. If a patient is unable to consume nutritional supplementation orally, enteral or parenteral supplementation should be considered with the guidance of a dietician [41]. Adults at normal risk for aspiration according to current ASA guidelines can consume a light meal within 6 hours of surgery and may continue to consume clear liquids until 2 hours before surgery [26]. Additionally, a preoperative drink containing greater than 45 g carbohydrates is recommended for all adults undergoing major surgery [40]. Oral, enteral, or parenteral nutrition should be reinitiated as soon as possible postoperatively, and supplementation in the post-discharge period may be necessary for elderly adults following major surgery [41].

Preoperative geriatric assessment should include an assessment of frailty, an age-related decline in strength, functional status, and physiologic reserve associated with adverse postoperative outcomes. Frailty can be assessed by a number of different scoring systems, incorporating quantitative measurements, such as grip strength or walking speed, as well as qualitative metrics such as independence in activities of daily living (ADL), mood, and degree of social support [40]. Derived from the Canadian Study of Health and Aging, the seven-point Clinical Frailty Scale is a simple global assessment of frailty that can be completed easily in the preoperative clinic. Using this scale, a patient who is robust, active, and energetic would receive a score of 1, while one who is somewhat dependent on others in ADLs would receive a score of 5. A score of 7 would indicate severe frailty, characterized by complete dependency or terminal illness [42]. The Edmonton Frail Scale (EFS) involves both subjective and objective assessments across 17 different domains, including cognition, mood, medication use, and functional independence. Although somewhat more complex than the Clinical Frailty Scale, EFS can generally be completed within 5 minutes and is, therefore, also appropriate for use by non-geriatricians in the preoperative clinic setting [43].

Regardless of the particular scale or scoring system utilized, frailty is strongly associated with mortality, complications, and prolonged LOS among geriatric surgical patients [44]. While it is not clear that frailty is a process that can be meaningfully reversed in the preoperative period, it does provide useful risk stratification and may guide surgical and anesthesia-related decision-making. For example, frail geriatric patients may be better candidates for less invasive forms of surgery and other treatment, while the anesthetic management of these patients should include measures such as close attention to maintenance of normothermia and strategies to minimize the risk of postoperative delirium [40].

Implementation of comprehensive perioperative optimization programs for geriatric patients have been demonstrated to decrease complications, shorten LOS, and be cost saving. The Duke Perioperative Optimization of Senior Health program involves a preoperative visit with a multidisciplinary team for elective surgery patients over age 85 or 65 in the presence of multiple comorbidities, cognitive dysfunction, unintended weight loss, significant hearing or visual deficit, or polypharmacy. During this assessment, patients are assessed and optimized across multiple domains, including mobility, nutrition, pain, cognition, and advanced care planning. Postoperatively, these patients are followed daily by a geriatric consult team. A cohort study of this program demonstrated that the intervention cohort had a lower risk of any complication and shock. Average LOS was shorter by 2 days, while 7- and 30-day readmission rates were also reduced. These patients returned home to self-care more frequently than those who did not participate in the program [45]. The University of Michigan Surgical Home incorporates several preoperative interventions including a home-based walking program, smoking cessation, incentive spirometry, and education regarding nutrition, stress management, and advanced care planning. Although not specifically targeted at the geriatric population, a cohort study of the Michigan Surgical Home in which the average patient was 60 years old demonstrated a LOS decrease of 2 days and an average total cost savings of nearly $7000 per patient [46].

Perioperative Complications in the Elderly

Stroke

Perioperative ischemic stroke is a rare, but potentially devastating complication that occurs with increasing frequency in advanced age. Strokes can occur intraoperatively or in the immediate recovery period, but most frequently strike over 24 hours after surgical intervention [47]. One retrospective study of over 350,000 patients undergoing non-cardiac, non-vascular, and non-neurological surgery found that while the incidence of perioperative stroke was only 0.2/10,000 for those under age 45, for those over 75, the rate rose to 15.7/10,000 [48]. The risk is substantially higher for cardiac and vascular procedures. A prospective study of over 16,000 open heart surgery patients found an overall risk of 4.6% with the highest incidence observed following double valve surgery (9.7%) [49]. In addition to age, there are several other modifiable and nonmodifiable risk factors for perioperative stroke, including renal failure, diabetes, hypertension, and female sex [49, 50]. Those who have experienced a myocardial infarction (MI) in the preceding 6 months and those with a history of prior stroke are also at increased risk [51]. For the open heart population, prolonged periods of cardiopulmonary bypass (CPB) and high transfusion rates are also factors [49].

While it is possible that better blood pressure and glycemic control in the preoperative period may decrease stroke risk, there is no evidence to support a specific strategy to address this. Additionally, it is unclear whether avoidance of general anesthesia (GA) and the utilization of neuraxial or regional techniques offer any advantage. There is evidence that allowing blood pressure to deviate more than 20–30% from baseline intraoperatively probably increases risk, so careful attention to hemodynamics in the elderly is important to minimize stroke risk [47]. Among non-cardiac and non-vascular procedures, orthopedic surgery is associated with a higher risk of stroke, perhaps because deliberate hypotension is frequently utilized in these procedures to minimize bleeding and improve surgical exposure [48]. Avoidance of hypo- and hyperglycemia is also warranted both for the prevention and management of perioperative stroke. Monitoring of EEG activity or cerebral oximetry should be strongly considered during high-risk procedures, including open heart surgery or major vascular procedures.

Anesthesiologists should maintain a high index of suspicion for stroke in elderly patients displaying neurological changes during or after surgery. This is challenging, however, because residual effects of sedative hypnotics or delirium can either mask or be confused with an acute stroke. Suspected strokes should be investigated quickly by a multidisciplinary stroke team, referred to by many institutions as "stroke codes." Rapid evaluation with imaging should be pursued and treatment with thrombolytics considered. The risk-benefit evaluation of thrombolytic therapy and other anticoagulants in the immediate postoperative period is challenging, however, because of the risk of postsurgical bleeding [47].

Cardiovascular Events

The risk of perioperative cardiovascular complications, including acute myocardial infarction (MI), acute heart failure, arrhythmia, and cardiovascular mortality, increases in a nonlinear fashion with age [52]. A Danish study of over 300,000 non-cardiac elective surgery patients showed that risk for major adverse cardiovascular events (MACE), defined as non-fatal MI, stroke, or cardiovascular mortality, dramatically increased with age. For example, in the 40–50-year-old age group, the risk of MACE and death was 0.06% and 0.07%, respectively, while in the 80–90-year-old age group, this had increased to 1.67% and 1.95%, respectively. Among those above 90 years old, the risk of MACE was 3% and risk of death was 4%. This corresponded to an odds ratio (OR) of 1.87 per decade for MACE. These authors emphasized that while the absolute risk remained relatively low in even those over age 90, other series had found higher risk levels, perhaps because in the Danish health system, high-risk elderly patients were being effectively excluded from surgery [53]. Another study of over 11,000 Canadian men undergoing prostatectomy for cancer revealed similar results. These authors identified an OR of 1.54 for perioperative cardiovascular complications among those 60–69 years old when compared to those under age 60, while the OR increased to 3.33 for those in the 70–79-year-old age bracket [54].

Among cardiac surgical patients, a similar nonlinear relationship has been observed with age for postoperative complications and mortality. However, improvements in surgical techniques and intensive care have resulted in more patients in extreme old age undergoing these procedures. A study of octogenarians undergoing open cardiac surgery showed an in-hospital mortality rate of 3.9%, with emergent surgery being associated with a higher risk of death. However, these authors noted that 91% were discharged home in good clinical condition [55]. A series of 42 nonagenarian cardiac surgery patients demonstrated very high rates of complications, including respiratory failure and infection, but a 30-day survival rate of 95%. At the time of publication, 81% of the patients in the series were still alive an average of 2.53 years after surgery, with one patient surviving for over 7 years [56].

Therefore, anesthesiologists can anticipate encountering more patients in the ninth and tenth decades of life presenting for elective surgery, even for cardiac procedures. Prevention of cardiovascular complications in this age group begins with preoperative optimization, including management of heart failure by existing guidelines and continuation of beta-blockade if applicable. As the elderly tend to have higher blood pressures than their younger counterparts, management should be tailored to an individual's baseline to prevent a variety of complications that may result from hypotension, including myocardial ischemia. Elderly patients should be closely monitored in the perioperative period for evidence of ischemia, heart failure, and arrhythmia given their increased risk [40]. Commonly used cardiac risk indices, including the Revised Cardiac Risk Index and Gupta MI and Cardiac Arrest calculator, have been shown to be inaccurate in the elderly and underestimate risk. A recently derived and validated Geriatric-Sensitive Perioperative Cardiac Risk Index based on National Surgical Quality Improvement Program (NSQIP) data was found to outperform these other two models in those over age 65 and may provide better risk stratification for geriatric patients [52].

Perioperative Neurocognitive Disorders

The PND include delirium and postoperative cognitive dysfunction (POCD), both of which occur with increasing frequency in the elderly. Delirium is characterized by altered or fluctuating levels of consciousness, disorientation, and disordered thinking and may be either hyper- or hypo-active. Both forms of delirium are associated with prolonged LOS and poor surgical outcomes, while the acute incidence of delirium also increases the risk of POCD [57]. A definitive diagnosis of POCD is challenging in most clinical scenarios as it requires a documented change in neurocognitive testing before and after a surgical intervention. Patients rarely undergo this type of testing outside of the research setting. Nonetheless, a decline in cognitive functioning is often recognized by the families of elderly patients after surgery and is a primary concern of elderly patients undergoing surgery.

Of the 16 million Americans over age 60 who undergo surgery annually, 10–40% will experience PND, such that the personal and financial costs of this problem are quite vast [58].

In addition to being associated with poor outcomes, there is also some concern that POCD is associated with an increased risk of dementia or acceleration of otherwise normal cognitive decline. The clinical data on this issue is quite mixed, with some studies showing that older adults undergoing surgery and anesthesia have no increased risk of dementia or cognitive dysfunction 1–5 years post-surgery [59, 60] while others indicate that there is indeed an increased risk [61, 62]. Certain elderly patients may experience an improvement in cognitive functioning following surgery, as has been reported following coronary artery bypass grafting (CABG), presumably due to improvements in cerebral perfusion and overall levels of functioning due to coronary revascularization [63].

Among the nonmodifiable risk factors for PND are longer and more invasive surgical procedures, a prior diagnosis of stroke or dementia, and a lower level of educational attainment. Other factors that can potentially be optimized pre- or postoperatively include smoking, diabetes, metabolic syndrome, and postoperative renal or pulmonary complications [63].

There is growing evidence that the risk of PND can be decreased by minimizing the exposure to anesthetics and analgesics. The CODA trial investigated this possibility by randomizing non-cardiac surgery patients over age 60 either to have anesthetic depth titrated to bispectral index

(BIS) or to have the BIS values blinded to the anesthesia provider. Titration of anesthesia to a BIS of 40–60 resulted in lower cumulative doses of propofol and average concentrations of volatile agent, while rates of delirium and POCD at 3 months were significantly reduced [64]. A recent Cochrane meta-analysis of several trials, including CODA, confirmed this finding [65].

The ASA Perioperative Neurotoxicity Working Group has published several recommendations for the prevention and management of PND. The first of these is that elderly patients should be informed of the risk of PND such that they can plan to make cognitively taxing decisions before undergoing surgery and engage with family and friends, as effective social support is a mitigating factor in the development of PND. The Working Group also recommends that geriatric surgical patients undergo a brief cognitive screening as part of the preoperative evaluation, particularly as patients with preexisting neurocognitive dysfunction are at increased risk for PND. Finally, these authors recommend using age-adjusted MAC values to titrate anesthetic depth in older adults, maintaining cerebral perfusion by limiting intraoperative hypotension, and utilizing EEG-based monitoring (such as BIS) when available [58].

Limiting exposure to postoperative opiates may also be effective in reducing PND in older adults, as suggested by the recent DEXACET trial. This randomized placebo-controlled trial included patients over age 60 (median age 69) undergoing on-pump CABG with or without valve replacement. Patients were randomized postoperatively to receive either placebo or standing intravenous acetaminophen every 6 hours for the first 48 hours in conjunction with propofol or dexmedetomidine sedation. Although there was no difference between the propofol and dexmedetomidine groups, patients receiving intravenous acetaminophen had less breakthrough pain and required less morphine analgesia postoperatively, resulting in a lower risk for delirium (10% vs. 28%), shorter duration of delirium (1 vs. 2 days), and reduced length of ICU stay (29.5 vs. 46.7 hours) [66].

Conclusion

As the US and global populations continue to age, the number of geriatric patients, including those in extreme old age, will present with increasing frequency for both elective and emergent procedures requiring anesthesia management. The anesthesiologist must be aware of the normal processes of physiologic aging and how these changes adversely impact different organ systems. Older patients also typically have higher rates of comorbidities than their younger counterparts, and anesthesia management plans should be individualized to address the specific risk factors of each patient. While pharmacokinetic changes attributable specifically to advancing age are relatively modest, the frequency of renal and hepatic dysfunction is high in the geriatric population such that response to drugs used during anesthetic management may be exaggerated and difficult to predict. It is very well established that sedative hypnotics, including anesthetic agents and opiates, are more potent in the geriatric population, and careful titration of these drugs is essential. Residual effects of anesthesia or excess sedation from analgesics and anxiolytics increase the risk of a variety of complications. Particularly noteworthy is the increasing recognition that relative overdose of anesthetics and analgesics in the elderly contributes to the development of PND, a costly complication of surgery and anesthesia that may contribute to permanent cognitive decline. The adoption of perioperative optimization programs for geriatric patients can likely mitigate this and other complications.

References

1. Hall MJ, DeFrances CJ, Williams SN, Golosinskiy A, Schwartzman A. National hospital discharge survey: 2007 summary. Natl Health Stat Report. 2010;29:1–20, 24.
2. Rana MV, Bonasera LK, Bordelon GJ. Pharmacologic considerations of anesthetic agents in geriatric patients. Anesthesiol Clin. 2017;35(2):259–71.
3. Nyberg L, Lovden M, Riklund K, Lindenberger U, Backman L. Memory aging and brain maintenance. Trends Cogn Sci. 2012;16(5):292–305.

4. Gorus E, De Raedt R, Lambert M, Lemper JC, Mets T. Reaction times and performance variability in normal aging, mild cognitive impairment, and Alzheimer's disease. J Geriatr Psychiatry Neurol. 2008;21(3):204–18.
5. Svetina M. The reaction times of drivers aged 20 to 80 during a divided attention driving. Traffic Inj Prev. 2016;17(8):810–4.
6. Abra J. Changes in creativity with age: data, explanations, and further predictions. Int J Aging Hum Dev. 1989;28(2):105–26.
7. Mazzucchi A, Sinforiani E, Boller F. Artistic creativity, artistic production, and aging. Prog Brain Res. 2013;204:45–69.
8. Drachman DA. Aging of the brain, entropy, and Alzheimer disease. Neurology. 2006;67(8):1340–52.
9. Scahill RI, Frost C, Jenkins R, Whitwell JL, Rossor MN, Fox NC. A longitudinal study of brain volume changes in normal aging using serial registered magnetic resonance imaging. Arch Neurol. 2003;60(7):989–94.
10. Morrison JH, Baxter MG. The ageing cortical synapse: hallmarks and implications for cognitive decline. Nat Rev Neurosci. 2012;13(4):240–50.
11. Brown EN, Purdon PL. The aging brain and anesthesia. Curr Opin Anaesthesiol. 2013;26(4):414–9.
12. Cleeland C, Pipingas A, Scholey A, White D. Neurochemical changes in the aging brain: a systematic review. Neurosci Biobehav Rev. 2019;98:306–19.
13. Mapleson WW. Effect of age on MAC in humans: a meta-analysis. Br J Anaesth. 1996;76(2):179–85.
14. Alvis BD, Hughes CG. Physiology considerations in geriatric patients. Anesthesiol Clin. 2015;33(3):447–56.
15. Singh A, Mehta Y. Heart failure with preserved ejection fraction (HFpEF): implications for the anesthesiologists. J Anaesthesiol Clin Pharmacol. 2018;34(2):161–5.
16. Sprung J, Gajic O, Warner DO. Review article: age related alterations in respiratory function – anesthetic considerations. Can J Anaesth. 2006;53(12):1244–57.
17. Ramly E, Kaafarani HM, Velmahos GC. The effect of aging on pulmonary function: implications for monitoring and support of the surgical and trauma patient. Surg Clin North Am. 2015;95(1):53–69.
18. Svartengren M, Falk R, Philipson K. Long-term clearance from small airways decreases with age. Eur Respir J. 2005;26(4):609–15.
19. Mazo V, Sabate S, Canet J, Gallart L, de Abreu MG, Belda J, et al. Prospective external validation of a predictive score for postoperative pulmonary complications. Anesthesiology. 2014;121(2):219–31.
20. McAlister FA, Bertsch K, Man J, Bradley J, Jacka M. Incidence of and risk factors for pulmonary complications after nonthoracic surgery. Am J Respir Crit Care Med. 2005;171(5):514–7.
21. Hua M, Brady JE, Li G. A scoring system to predict unplanned intubation in patients having undergone major surgical procedures. Anesth Analg. 2012;115(1):88–94.
22. Gupta H, Gupta PK, Schuller D, Fang X, Miller WJ, Modrykamien A, et al. Development and validation of a risk calculator for predicting postoperative pneumonia. Mayo Clin Proc. 2013;88(11):1241–9.
23. Miskovic A, Lumb AB. Postoperative pulmonary complications. Br J Anaesth. 2017;118(3):317–34.
24. Schmucker DL. Age-related changes in liver structure and function: implications for disease? Exp Gerontol. 2005;40(8–9):650–9.
25. Regev A, Schiff ER. Liver disease in the elderly. Gastroenterol Clin N Am. 2001;30(2):547–63, x–xi.
26. Practice guidelines for preoperative fasting and the use of pharmacologic agents to reduce the risk of pulmonary aspiration: application to healthy patients undergoing elective procedures: an updated report by the American Society of Anesthesiologists task force on preoperative fasting and the use of pharmacologic agents to reduce the risk of pulmonary aspiration. Anesthesiology. 2017;126(3):376–93.
27. Klotz U. Pharmacokinetics and drug metabolism in the elderly. Drug Metab Rev. 2009;41(2):67–76.
28. Scheen AJ. Diabetes mellitus in the elderly: insulin resistance and/or impaired insulin secretion? Diabetes Metab. 2005;31(2):5S27–34.
29. Duggan EW, Carlson K, Umpierrez GE. Perioperative hyperglycemia management: an update. Anesthesiology. 2017;126(3):547–60.
30. Vitale G, Cesari M, Mari D. Aging of the endocrine system and its potential impact on sarcopenia. Eur J Intern Med. 2016;35:10–5.
31. Vacante M, Biondi A, Basile F, Ciuni R, Luca S, Di Saverio S, et al. Hypothyroidism as a Predictor of Surgical Outcomes in the Elderly. Front Endocrinol (Lausanne). 2019;10:258.
32. Szekely M, Garai J. Thermoregulation and age. Handb Clin Neurol. 2018;156:377–95.
33. Insler SR, Sessler DI. Perioperative thermoregulation and temperature monitoring. Anesthesiol Clin. 2006;24(4):823–37.
34. Ozaki M, Sessler DI, Matsukawa T, Ozaki K, Atarashi K, Negishi C, et al. The threshold for thermoregulatory vasoconstriction during nitrous oxide/sevoflurane anesthesia is reduced in the elderly. Anesth Analg. 1997;84(5):1029–33.
35. Vassilieff N, Rosencher N, Sessler DI, Conseiller C. Shivering threshold during spinal anesthesia is reduced in elderly patients. Anesthesiology. 1995;83(6):1162–6.
36. Akhtar S. Pharmacological considerations in the elderly. Curr Opin Anaesthesiol. 2018;31(1):11–8.
37. Stankiewicz-Rudnicki M. Neuromuscular blockade in the elderly. Anaesthesiol Intensive Ther. 2016;48(4):257–60.

38. Van Cleve WC, Nair BG, Rooke GA. Associations between age and dosing of volatile anesthetics in 2 academic hospitals. Anesth Analg. 2015;121(3):645–51.
39. Akhtar S, Heng J, Dai F, Schonberger RB, Burg MM. A retrospective observational study of anesthetic induction dosing practices in female elderly surgical patients: are we overdosing older patients? Drugs Aging. 2016;33(10):737–46.
40. Chan SP, Ip KY, Irwin MG. Peri-operative optimisation of elderly and frail patients: a narrative review. Anaesthesia. 2019;74(Suppl 1):80–9.
41. Wischmeyer PE, Carli F, Evans DC, Guilbert S, Kozar R, Pryor A, et al. American society for enhanced recovery and perioperative quality initiative joint consensus statement on nutrition screening and therapy within a surgical enhanced recovery pathway. Anesth Analg. 2018;126(6):1883–95.
42. Rockwood K, Song X, MacKnight C, Bergman H, Hogan DB, McDowell I, et al. A global clinical measure of fitness and frailty in elderly people. CMAJ. 2005;173(5):489–95.
43. Rolfson DB, Majumdar SR, Tsuyuki RT, Tahir A, Rockwood K. Validity and reliability of the Edmonton Frail Scale. Age Ageing. 2006;35(5):526–9.
44. Lin HS, Watts JN, Peel NM, Hubbard RE. Frailty and post-operative outcomes in older surgical patients: a systematic review. BMC Geriatr. 2016;16(1):157.
45. McDonald SR, Heflin MT, Whitson HE, Dalton TO, Lidsky ME, Liu P, et al. Association of integrated care coordination with postsurgical outcomes in high-risk older adults: the perioperative optimization of senior health (POSH) initiative. JAMA Surg. 2018;153(5):454–62.
46. Englesbe MJ, Grenda DR, Sullivan JA, Derstine BA, Kenney BN, Sheetz KH, et al. The Michigan Surgical Home and Optimization Program is a scalable model to improve care and reduce costs. Surgery. 2017;161(6):1659–66.
47. Vlisides P, Mashour GA. Perioperative stroke. Can J Anaesth. 2016;63(2):193–204.
48. Wang H, Li SL, Bai J, Wang DX. Perioperative acute ischemic stroke increases mortality after noncardiac, nonvascular, and non-neurologic surgery: a retrospective case series. J Cardiothorac Vasc Anesth. 2019;33(8):2231–6.
49. Bucerius J, Gummert JF, Borger MA, Walther T, Doll N, Onnasch JF, et al. Stroke after cardiac surgery: a risk factor analysis of 16,184 consecutive adult patients. Ann Thorac Surg. 2003;75(2):472–8.
50. Sharifpour M, Moore LE, Shanks AM, Didier TJ, Kheterpal S, Mashour GA. Incidence, predictors, and outcomes of perioperative stroke in noncarotid major vascular surgery. Anesth Analg. 2013;116(2):424–34.
51. Mashour GA, Shanks AM, Kheterpal S. Perioperative stroke and associated mortality after noncardiac, nonneurologic surgery. Anesthesiology. 2011;114(6):1289–96.
52. Alrezk R, Jackson N, Al Rezk M, Elashoff R, Weintraub N, Elashoff D, et al. Derivation and validation of a geriatric-sensitive perioperative cardiac risk index. J Am Heart Assoc. 2017;6(11):e006648.
53. Hansen PW, Gislason GH, Jorgensen ME, Kober L, Jensen PF, Torp-Pedersen C, et al. Influence of age on perioperative major adverse cardiovascular events and mortality risks in elective non-cardiac surgery. Eur J Intern Med. 2016;35:55–9.
54. Alibhai SM, Leach M, Tomlinson G, Krahn MD, Fleshner N, Holowaty E, et al. 30-day mortality and major complications after radical prostatectomy: influence of age and comorbidity. J Natl Cancer Inst. 2005;97(20):1525–32.
55. Scandroglio AM, Finco G, Pieri M, Ascari R, Calabro MG, Taddeo D, et al. Cardiac surgery in 260 octogenarians: a case series. BMC Anesthesiol. 2015;15:15.
56. Bacchetta MD, Ko W, Girardi LN, Mack CA, Krieger KH, Isom OW, et al. Outcomes of cardiac surgery in nonagenarians: a 10-year experience. Ann Thorac Surg. 2003;75(4):1215–20.
57. Hshieh TT, Inouye SK, Oh ES. Delirium in the elderly. Psychiatr Clin North Am. 2018;41(1):1–17.
58. Berger M, Schenning KJ, Brown CH, Deiner SG, Whittington RA, Eckenhoff RG, et al. Best practices for postoperative brain health: recommendations from the fifth international perioperative neurotoxicity working group. Anesth Analg. 2018;127(6):1406–13.
59. Avidan MS, Searleman AC, Storandt M, Barnett K, Vannucci A, Saager L, et al. Long-term cognitive decline in older subjects was not attributable to noncardiac surgery or major illness. Anesthesiology. 2009;111(5):964–70.
60. Moller JT, Cluitmans P, Rasmussen LS, Houx P, Rasmussen H, Canet J, et al. Long-term postoperative cognitive dysfunction in the elderly ISPOCD1 study. ISPOCD investigators. International Study of Post-Operative Cognitive Dysfunction. Lancet. 1998;351(9106):857–61.
61. Chen CW, Lin CC, Chen KB, Kuo YC, Li CY, Chung CJ. Increased risk of dementia in people with previous exposure to general anesthesia: a nationwide population-based case-control study. Alzheimers Dement. 2014;10(2):196–204.
62. Chen PL, Yang CW, Tseng YK, Sun WZ, Wang JL, Wang SJ, et al. Risk of dementia after anaesthesia and surgery. Br J Psychiatry. 2014;204(3):188–93.
63. Berger M, Nadler JW, Browndyke J, Terrando N, Ponnusamy V, Cohen HJ, et al. Postoperative cognitive dysfunction: minding the gaps in our knowledge of a common postoperative complication in the elderly. Anesthesiol Clin. 2015;33(3):517–50.
64. Chan MT, Cheng BC, Lee TM, Gin T, CODA Trial Group. BIS-guided anesthesia decreases postoperative delirium and cognitive decline. J Neurosurg Anesthesiol. 2013;25(1):33–42.

65. Punjasawadwong Y, Chau-In W, Laopaiboon M, Punjasawadwong S, Pin-On P. Processed electroencephalogram and evoked potential techniques for amelioration of postoperative delirium and cognitive dysfunction following non-cardiac and non-neurosurgical procedures in adults. Cochrane Database Syst Rev. 2018;5:CD011283.
66. Subramaniam B, Shankar P, Shaefi S, Mueller A, O'Gara B, Banner-Goodspeed V, et al. Effect of intravenous acetaminophen vs placebo combined with propofol or dexmedetomidine on postoperative delirium among older patients following cardiac surgery: the DEXACET randomized clinical trial. JAMA. 2019;321(7):686–96.

Nutrition Support in Elderly Patients Undergoing Surgery

8

Shekhar Gogna, Jaqueline Maxwell, Anthony J. Policastro, and Rifat Latifi

"How we nourish and treat our elderly population during next decade or two will influence greatly how we define our character individually and as a society culture and nation. Nutrition support involves more than putting food on the table, putting an enteral tube feed into the gastrointestinal tract or putting a parenteral feeding catheter into a central vein." – Stanley J Dudrick, MD, FACS

Advanced age is an independent predictor of poor nutritional status and postoperative mortality in surgical patients [1]. The etiology of malnutrition in the elderly is multifactorial and encompasses low nutrient intake, reduced appetite, chronic disease, multiple medication, and/or psychological condition [2]. Malnutrition is associated with higher rates of perioperative complication such as impaired wound healing, longer length of hospital stay, and increased morbidity and mortality [3, 4]. The nutritional interventions alter this risk when applied appropriately. In this chapter we will describe the systematic approach to assess, intervene, and analyze the nutritional status in the elderly undergoing major surgery. We begin with expounding the prevalence and pathogenesis of malnutrition in the elderly undergoing surgery. The second part defines the malnutrition in the elderly and the methods to screen and diagnose malnutrition in the elderly. This is followed by a description of the nutrition intervention plan for elderly surgical patients. The nutrition intervention plan includes the nutritional counseling, mode of nutritional support, and the timing of administration of nutrition in pre- and postoperative period. The final section of this chapter will highlight the areas of current and future research on malnutrition in the elderly undergoing surgery.

Prevalence and Pathogenesis of Malnutrition in the Elderly Undergoing Surgery

More than one-third of all inpatient surgical procedures in the USA were performed on patients aged 65 years and older in 2007, and this number is expected to double by 2020 [5]. The prevalence of malnutrition in non-institutionalized elderly in the USA is 1–15%. This proportion increases dramatically to 25–60% for patients in geriatric care facilities and 35–65% in hospitalized elderly [6]. Malnutrition in the elderly is prevalent in pre-hospital, hospital, and post-hospital periods,

S. Gogna
Department of Surgery, Westchester Medical Center, Valhalla, NY, USA

J. Maxwell
Clinical Nutrition, Westchester Medical Center, Valhalla, NY, USA

A. J. Policastro
Trauma and Critical Care, Department of Surgery, Westchester Medical Center and New York Medical College, Valhalla, NY, USA

R. Latifi (✉)
Department of Surgery, New York Medical College, School of Medicine and Westchester Medical Center Health Network, Valhalla, NY, USA
e-mail: Rifat.Latifi@wmchealth.org

R. Latifi (ed.), *Surgical Decision Making in Geriatrics*,
https://doi.org/10.1007/978-3-030-47963-3_8

with gradual deterioration in nutritional stores as patients traverse through each of these phases. Aging promotes malnutrition, which in turn is an increased risk for hospitalization in the elderly [7]. Inadequate nutritional intake continues in hospital and after discharge [8]. Despite being frequent, malnutrition remains under-diagnosed and untreated and has been correctly referred to as the "skeleton in the hospital closet" [9]. Studies have shown that there was a threefold increase in hospital costs when patients were malnourished. Nationally the annual cost of disease-associated malnutrition is over $15.5 billion. The elderly bear a disproportionate share of this cost on both the state and national levels [10].

The pathogenesis of malnutrition is multi-pronged and is depicted in Fig. 8.1. Malnutrition is often due to a combination of the following factors: inadequate food intake; food choices that lead to dietary deficiencies; and systemic illness that causes increased nutrient requirements, increased nutrient loss, poor nutrient absorption, or a combination of these factors [11]. Physiologic aging promotes loss of cells in the myenteric plexus and decreased gastric emptying, possibly associated with reduced nitric oxide concentrations [12]. This neuronal dysfunction promotes increased transcription of IL-6, TNF-α, and glucagon-like peptide which promotes anorexia, loss of muscle mass, and sarcopenia [12, 13]. Malnutrition resulting from this promotes further immune dysregulation, hence precipitating vicious cycle.

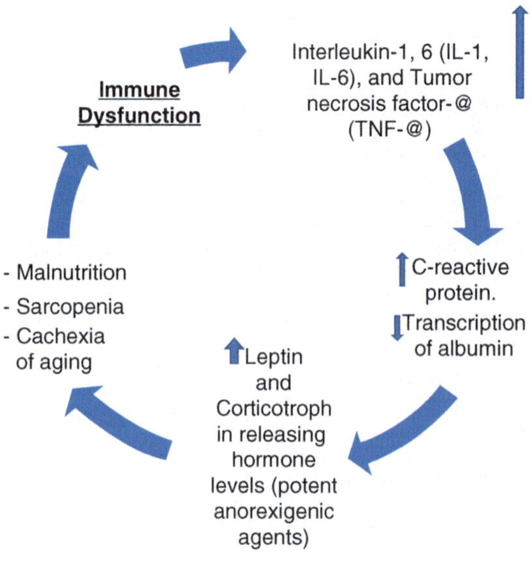

Fig. 8.1 Pathophysiology of malnutrition

Definition and Detection of Malnutrition in Elderly

Disagreement exists among health professionals regarding the use of terminologies such as malnutrition, starvation, cachexia, and sarcopenia. To address this issue, the representatives of the four of the prominent global parenteral and enteral nutrition societies, European Society for Clinical Nutrition and Metabolism (ESPEN), American Society for Parenteral and Enteral Nutrition (ASPEN), The Parenteral and Enteral Nutrition Society of Asia (PENSA), and Latin American Congress of Clinical Nutrition, Nutrition Therapy, and Metabolism (FELANPE), met at ESPEN Congress in Copenhagen, Denmark (September 2016) to create a consensus statement on diagnosing malnutrition [14]. The ESPEN suggested a grading approach that could encompass various types of signs, symptoms, and etiologies to support diagnosis. ASPEN emphasized weight loss as a key indicator for malnutrition, while FELANPE suggested that the anticipated consensus approach needs to prioritize a diagnostic methodology that is available for everybody since resources differ globally. Finally, and PENSA highlighted that BMI varies by ethnicity/ race and that sarcopenia/muscle mass evaluation is important for the diagnosis of malnutrition. Malnutrition can be defined as "a state resulting from lack of intake or uptake of nutrition that leads to altered body composition (decreased fat free mass) and body cell mass leading to diminished physical and mental function and impaired clinical outcome from disease" [15]. The global consensus and uniform guidelines are an ongoing work, but the identification of two or more of the following six characteristics is recommended for diagnosis for malnutrition: (1) insufficient energy intake, (2) weight loss, (3) loss of muscle mass, (4) loss of subcutaneous fat, (5) localized or generalized fluid accumulation, and (6) diminished functional status as measured by handgrip strength [16].

We believe that improved knowledge and awareness of classifications of malnutrition are crucial for optimal nutritional treatment in order to minimize complications after surgery in the elderly. Table 8.1 depicts the various subclasses of malnutrition with or without inflammation and malnutrition/undernutrition without the disease [17]. In order to intervene appropriately to optimize nutritional status in elderly patients, malnutrition must first be identified correctly. The first step in the identification process is a thorough history and physical examination. It should be kept in mind that due to physiological changes induced by the aging process, interpretation of standard nutritional assessment parameters should be adjusted for the geriatric population. This is especially important for anthropometric and biochemical tests. Medical and surgical history should focus on recent unexplained weight loss, depression, dementia, other neurologic diseases, chronic infections, malignancy, and end-organ disease or failure, and the malnutrition can very well be a presentation of a serious disease or condition. It is important to keep in mind that depression and malignancy are the top two reasons for malnutrition in the elderly [18].

Table 8.1 Sub-classification of clinical nutrition concepts, i.e., nutrition disorders and nutrition-related conditions. [ESPEN committee consensus, 80% agreement]

Malnutrition
Disease-related malnutrition (DRM) with inflammation
Chronic DRM with inflammation; synonym, cachexia
A cancer cachexia and other disease-specific forms of cachexia
Acute disease- or injury-related malnutrition
DRM without inflammation
Hunger-related malnutrition
Socioeconomic- or psychologic-related malnutrition
Sarcopenia
Frailty
Overnutrition
Overweight
Obesity
Sarcopenic obesity
Central obesity
Micronutrient abnormalities
Deficiency
Excess
Refeeding syndrome

Adapted from Cederholm et al. [17], with permission

Nutritional Screening: Definition and Description

Both ASPEN and ESPEN guidelines have a consensus that the next key step in the evaluation of nutritional status is malnutrition risk screening to identify "the risk" status using any validated screening tool (Table 8.2). The real reason behind the development of these multidimensional screening and scoring tools is the limited information provided by anthropometric or biochemical tests. ESPEN recognizes the following risk screening tools to be used in the hospital and elderly care and community settings: Nutritional Risk Screening 2002 (NRS-2002), Mini Nutritional Assessment-Short Form (MNA-SF), and Malnutrition Universal Screening Tool (MUST) [19].

Nutritional Risk Screening-2002 (NRS-2002) scoring system utilizes the three markers of nutrition, BMI score, weight loss score, and acute illness score [20]. The Mini Nutritional Assessment (MNA) was designed specifically as an assessment tool for the elderly. The distinct advantage of MNA is that all screening methods detect undernutrition in elderly patients, but for the frail elderly, the MNA screening is more sensitive because it evaluates both the physical and mental aspects that impact nutrition in geriatric patients. The MNA-SF is a quick and easy questionnaire to use and has been shown to predict adverse outcomes and mortality and predicts the risk of developing undernutrition at an early stage [21]. The Malnutrition Universal Screening Tool (MUST), though not developed specifically for geriatric patients, is another assessment tool available for adult patients of any age group and in any care setting. The MUST consists of three components – BMI score, weight loss score, and acute illness score – which are each given a numerical value ranging from 0 to 2 [22].

Subjective Global Assessment (SGA) introduced by Detsky in 1987 includes functional

Table 8.2 Survey of existing approaches used in screening and assessment of malnutrition and cachexia

	NRS-2002[a]	MNA-SF[a,b]	MUST[a]	ESPEN 2015[a]	ASPEN/AND[a]	SGA[a]	Evans 2008[c]	PEW 2008[d]	Fearon 2011[c]
Etiologies									
Reduced food intake	X	X	X	X	X	X		X	X
Disease burden/inflammation	X	X	X	X	X	X	X	X	X
Symptoms									
Anorexia		X				X	X		
Weakness									
Signs/phenotype									
Weight loss	X	X	X	X	X	X	X	X	X
Body mass index	X	X	X	X			X	X	X
Lean/fat free/muscle mass		X		X	X	X	X	X	X
Fat mass					X	X	X		
Fluid retention/ascites					X	X			
Muscle function; e.g., grip strength					X	X	X		
Biochemistry						X	X		

Adapted from Cederholm et al. [17], with permission

NRS-2002 Nutritional Risk Screening-2002, *MNA-SF* ¼ Mini Nutritional Assessment-Short Form, *MUST* Malnutrition Universal Screening Tool, *ESPEN* European Society for Clinical Nutrition and Metabolism, *ASPEN* American Society of Parenteral and Enteral Nutrition, *AND* Academy of Nutrition and Dietetics, *SGA* Subjective Global Assessment, *PEW* Protein Energy Wasting

[a]Malnutrition approach

[b]Adapted for older adults

[c]Cachexia approach

[d]Adapted for chronic kidney disease

capacity and the clinician's overall impression of the patient's status in which they are designated as "normal," "mildly malnourished," or "significantly malnourished." This screening test includes a history of weight loss, poor dietary intake, and loss of subcutaneous tissue and muscle wasting on physical exam [23]. However, the important limitation of SGA is that a physician has to be experienced to be able to use this tool in order to diagnose the elderly with malnutrition with higher sensitivity.

Cachexia is often used as a synonym for malnutrition in the elderly on day-to-day basis in clinical practice. To clarify for the healthcare providers, the consensus panel developed a set of diagnostic criteria to make a definitive diagnosis of cachexia. The key component is at least a 5% loss of edema-free body weight during the previous 12 months or less. The time frame may be disease-specific and is likely to be shorter in cancer (3–6 months) and longer in the chronic kidney or heart failure or COPD (12 months). The synonymous use of cachexia with malnutrition should be avoided because it suggests that the disease is mainly associated with nutritional problems and will be resolved by adequate nutrition which is not true; hence the utility of screening for cachexia is still far from practical use in surgical practice [24, 25].

Biochemical Tests

Serum proteins synthesized by the liver have been used as markers of nutrition including albumin, prealbumin, transferrin, total lymphocyte count, retinol-binding proteins, and thyroxine-binding globulin [26]. Serum albumin is the most commonly used marker since it can predict mortality in older people. However, albumin with a long half-life of 18 days apart from the nutritional state is also affected by inflammation, infection, cirrhosis, fluid status, and steroid intake, thus limiting its utility in acute settings.

Aging itself is associated with a modest decline in serum albumin levels, with a decrease of 0.8 g/L per decade after the age of 60 [27]. Prealbumin [also known as transthyretin (TTR)] has been used as a surrogate marker for protein malnutrition since the 1970s. A consensus statement regarding the use of prealbumin from 1995 states that a level between 50 and 109 mg/L is indicative of the significant risk of malnutrition and a level less than 50 mg/L was an indicator of poor prognosis. However similar to albumin the serum values of prealbumin also change quickly in acute inflammatory phase [28]. Transferrin is more sensitive marker of early protein-energy malnutrition but is also affected by a number of other conditions including but not limited to pregnancy, iron deficiency, hypoxia, chronic infection, and hepatic disease [29]. Total lymphocyte count (TLC) has also been suggested as a marker of malnutrition with TLC of less than 1500/mm^3. This value was found to be associated with a fourfold increase in mortality [30]. Similarly, insulin-like growth factors (IGF) and fibronectin have been used as a measure of malnutrition, but similar to abovementioned biochemical markers, their use has not been validated internationally. To date, there is no single biochemical marker of malnutrition as a screening test. The main value of biochemical markers is in a detailed assessment and monitoring [31]. They can be used to support the presence of systemic inflammatory response and further contribute to the identification of the etiologic basis for the diagnosis of malnutrition proposed by ESPEN (Table 8.1).

Body Composition and Anthropometric Tests

There have been rapid advances in the field of nutritional support in recent years; the updated guidelines from all the major societies (ASPEN and ESPEN) use anthropometric tests and body composition in some form or the other in defining and stratifying patients. As shown in Tables 8.1 and 8.2, the etiology-based classification and screening tools all use anthropometric tests. The recent approach is to use a combination of phenotypic and etiologic criteria for malnutrition diagnosis, and anthropometric tests are included under phenotypic tests [14, 32]. The anthropometric tests utilized in the assessment are weight loss, body mass index (BMI), and assessment of muscle mass (mid-brachial, mid-calf) and muscle function by using grip strength exercise. The data on the correlation between the rate of weight loss and poor surgical outcomes is very robust, and Global Leadership Initiative on Malnutrition (GLIM) meeting at the ESPEN Congress 2016 strongly considers non-intentional weight loss as one of the robust criteria of diagnosis of the malnutrition [33]. The utility of low BMI as one of the criteria has met with mixed skepticism. The proportion of the overweight population in the USA is comparatively higher than Europe and Asia, so low BMI as criteria of malnutrition in the USA is the area of controversy and future research. Measurement of muscle mass has strong evidence of utility as an indicator of malnutrition. The recommended radiologic methods to measure it are the use of dual-energy absorptiometry, bioelectrical impedance, ultrasound, computed tomography, or magnetic resonance imaging, but these methods are still not widely available. Physical examination or anthropometric measures of the calf or arm muscle circumference are therefore included as alternative measures [34]. Importantly, the reference standards for muscle mass may warrant an adjustment for age, race, and sex.

Studies have shown that functional measurement with handgrip strength and quadriceps strength is also related to mortality while muscle mass was not [35]. The concepts that underlie the utility of function as a surrogate for nutrition status is that muscle structure changes with malnutrition leading to loss of contractile elements, increased muscle fatigue, and altered contraction patterns [36]. Nevertheless, the assessment of muscle function using grip strength is recommended as a supportive measure in the GLIM consensus. In situations where muscle mass cannot be readily assessed, the muscle strength, e.g., handgrip strength, is an appropriate supporting proxy.

The underlying aim is to screen the elderly at the risk of malnutrition using the validated screening tool in the elderly population such as MNA-SF than to diagnose them by using above-mentioned tests "in collaboration" to utilize the information to improve nutritional status pre- or postoperatively.

Nutrition Intervention Plan

The specific nutritional requirements of older patients are difficult to quantify exactly because of the physiologic diversity and heterogeneity of this population along with the high prevalence of the chronic disease. The plan to utilize the nutritional intervention starts with estimating the nutritional requirements of the elderly after they have been positively screened and diagnosed with the help of tests mentioned above. After estimating the need for requirements, we will discuss the preferred and employed mode of nutrition delivery in elderly who are waiting for surgery or might have already faced it.

Estimating the Need for Nutritional Requirements in Elderly

The first step in estimating the need for nutritional requirement begins with calculating the nutrient balance of the patient. Nutrient balance is a gradient between nutrient intake and loss of nutrients [37]. Nutrient intake is estimated by obtaining a dietary history and then calculating the calorie and protein intake, whereas nutrient losses are estimated by calculating the basal energy expenditure (BEE) and then adding any abnormal nutrient losses, such as those from external fistula output, diarrhea, or proteinuria. BEE can be estimated by applying the Harris-Benedict equations and adding a correction factor depending on the degree of metabolic stress in the patient [38].

$$\begin{aligned}\text{Men } 66.47 + 13.75 &(\text{weight in kg}) \\ + 5.0 &(\text{Height in cm}) \\ - 6.76 &(\text{Age in years})\end{aligned}$$

$$\begin{aligned}\text{Women } 665.1 + 9.56 &(\text{weight in kg}) \\ + 1.85 &(\text{Height in cm}) \\ - 4.68 &(\text{Age in years})\end{aligned}$$

BEE is sufficient to initiate nutritional therapy in the elderly [39]. The primary nutrient substrate required by the surgical patient, especially the elderly patient, is protein, which plays an essential role in the metabolic response to stress. The rationale behind this concept is that amino acids are ordinarily mobilized from endogenous muscle to provide precursors for hepatic gluconeogenesis, and this step does not function efficiently in elderly; hence the provision of protein is necessary in the times of surgical intervention which is a major stress factor for elderly. The protein requirement is estimated by nitrogen loss in the urine by measuring 24-hour urinary urea nitrogen (UUN). The accurate formula accounts for urinary nitrogen excreted in forms other than urea (20%) together with fecal and cutaneous losses (2 g/d):

$$\begin{aligned}\text{Total nitrogen losses} = {} & 24\,\text{h UUN}(\text{g}/\text{d}) \\ & + 0.20 \times 24\,\text{h UUN} \\ & + 2\,\text{g}/\text{d}\end{aligned}$$

The protein requirement in elderly patients is slightly higher to the amount of 1–1.2 g/kg body weight, because of their age-related impaired capacity for protein synthesis. They require around 20–30 kcal/kg/day of energy, depending on their activity, and also energy expenditure decreases with age due to the decline in body cell mass. Many elderly people also suffer from fluid and specific micronutrient deficiencies, which should be corrected by supplementation [40].

Baseline fluid requirements of elderly patients without renal and/or cardiac insufficiency are usually estimated at 25 mL/kg/d. Fluid requirements may be increased with vomiting, diarrhea, enterocutaneous fistula, polyuria, or excess perspiration; fluid intake may require restriction in chronic renal insufficiency, pulmonary insufficiency, or congestive heart failure. The risk of both dehydration and fluid overload is high in the elderly [41].

Hospitalized elderly have an increased risk of developing vitamin D deficiency due to the lack of sun exposure. This can further lead to depression, cognitive changes, and increased fracture risk. The recommended daily allowance (RDA) of vitamin D in this patient population is 800–1000 IU. Calcium deficiency is very commonly seen in geriatric patients. To reduce the risk of osteoporosis, the RDA for calcium is 1200 or 1500 mg/dl. Vitamin B12 and folate are obtained adequately by diet alone, but in some patients, oral supplementation may be required. Postoperatively, elderly are at higher risk for developing a vitamin B12 deficiency due to impact of general anesthesia (nitrous oxide), which irreversibly binds and oxidizes the cobalt atom, thus inactivating B12 [42]. Some other well-known reasons are decreased intake, decreased absorption, hypochlorhydria, atrophic gastritis, and long-term use of acid-reducing medications. Vitamin B12 deficiency is often missed in the elderly. Symptoms of vitamin B12 deficiency such as depression, cognitive decline, mild to moderate dementia, irritability, tremors, confusion, and neuropathies are commonly seen in the elderly and often assumed to be symptoms of aging rather than a consequence of a nutritional deficiency and therefore are not further investigated [43].

Elderly patients that show signs of neurological deterioration after an operation should be further evaluated for vitamin B12 deficiency. The RDA for vitamin B12 is 1000 mg/day orally. Doubling of the daily multivitamin dose can be safely done while nutritional support is ongoing and until the normal nutritional status is achieved [43].

The Mode and Timing of Nutritional Delivery – What, When, and Why?

Elderly undergo surgery for various reasons, and depending on the acuity, it can be emergent or elective. In the case of emergent settings, "life first" dictum holds the priority followed by an evidence-based nutritional intervention which plays a huge role in improving the quality and longevity of life. Elective settings allow the time to screen, diagnose, and intervene preoperatively and to plan for postoperative nutritional intervention.

Nutritional intervention/therapy is defined according to ESPEN as:

Nutrition therapy is the provision of nutrition or nutrients either orally (regular diet, therapeutic diet, e.g. fortified food, oral nutritional supplements) or via enteral nutrition (EN) or parenteral nutrition (PN) to prevent or treat malnutrition. "Medical nutrition therapy" is a term that encompasses oral nutritional supplements, enteral tube feeding (enteral nutrition) and parenteral nutrition. [17, 44]

Whenever possible oral intake is the preferred and recommended route for nutritional intervention in all age groups. In a recent Cochrane analysis of 33 studies analyzed, twenty-nine showed that oral nutritional therapy (ONT) led to an increase in energy and nutrient intake (level of evidence Ia). We do understand that there are inherent practical limitations of ONT such as poor compliance, low palatability, the inability of family support to arrange for food, severe dysphagia, and side effects such as nausea and diarrhea [45]. Postoperative ileus due to abdominal surgery may inhibit early oral food intake. Major open gastrointestinal surgery, excess fluid administration, opioids for pain management, and surgical complications are some of the commonest causes of ileus limiting oral intake. A wholehearted effort should be made to avoid them as far as possible.

One of the major issues that warrant the attention is the mythical concept of preoperative fasting in the elderly. The patients are often placed on restrictive diets based on their comorbidities or made NPO for tests and interventions. We all are familiar with the overbookings of the operating room and the increased workload that leads to frequent NPO orders in patients, further increasing the risk of malnutrition. The American College of Surgeons National Surgical Quality Improvement Program (NSQIP)/American Geriatrics Society (AGS) best practices guidelines published in 2012 did address this concern [46]. In elderly undergoing non-emergent surgical procedures, fasting from the intake of clear liquids for at least 2 hours, light meal and/or milk for 6 hours, and additional fasting (8 hours or more) may be required depending on the amount and type of food ingested such as fried, fatty foods, or meat that may prolong gastric emptying before elective procedures requiring general anesthesia or regional anesthesia or recommending sedation/analgesia. However, caution should be addressed in the elderly with certain comorbidity such as diabetes or a hiatal hernia where the timing of preoperative fasting has to be modified.

Early initiation of feeds whenever feasible is recommended to enhance the outcomes. More independence to choose from the menu, friendly atmosphere, ad lib oral intake, early mobilization, and assisted feeding are some of the common useful steps that can be instituted to promote oral intake in the elderly population.

Enteral nutrition is started if the patient is not expected to eat for more than 7 days after surgery or if they cannot maintain more than 60% of the recommended oral intake for 10 days or more [47]. Enteral nutrition is provided with the help of tube feeds. As shown in Fig. 8.2, adequate access can be achieved in most patients with minimal intervention, with fewer patients requiring more complex approaches. Once determined to be an appropriate candidate for the initiation of enteral support, the naso- or orogastric tube provides temporary conduits for delivery that can be used immediately following clinical or radiological confirmation of appropriate placement. The pliable naso-enteric tubes can be placed with endoscopic assistance or placed "blindly" with the assistance of either promotility agents or various signaling devices (commonly known as @@ Coatrack/Kao tubes). As with all minor and major procedures, the complications and risk associated with placement include esophageal perforation, increased incidence of sinusitis, and unintentional placement into the bronchopulmonary tree [48].

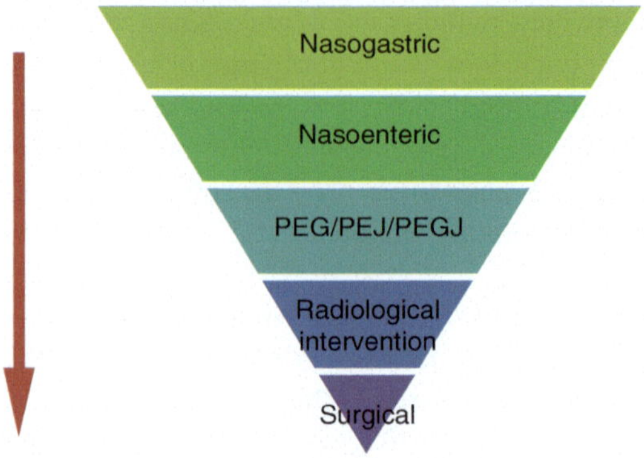

Fig. 8.2 Increased clinical complexity demands a more invasive surgical gut access (SGA) for nutrition; however, the subset of patients decreases with increased complexity of SGA

Greater than 4 weeks of enteral access requires the placement of surgical gut access (SGA) by the percutaneous, endoscopic, or surgical route. Percutaneous endoscopic gastrostomy (PEG), PEG with the jejunal extension (PEGJ), direct percutaneous endoscopic jejunostomy (DPEJ), and surgical gastrostomy/jejunostomy are the available options. The surgical steps can be found in any standard surgical textbook or standard review articles [49]. Studies have shown that after supplemental tube feeding, total energy and nutrient intake was markedly improved [50, 51]. The enteral feedings via percutaneous endoscopic gastrostomy (PEG) are tolerated much better as compared to nasogastric (NG) tube feeding in the elderly [46]. The practical questions regarding the timing of the intervention need to be addressed and applied uniformly in the hospitals. Both ESPEN and ASPEN guidelines recommend that elective operative interventions be postponed for enteral nutrition in patients with elevated nutritional risk. These patients should receive nutritional intervention sooner than later

[41]. Nutritional intervention after emergency surgery in elderly should be started as soon as feasible because the early nutritional support reduces the disease severity, diminishes complications, and decreases the intensive care unit (ICU) length of stay [52, 53].

The issue of tube feeds (TFs) in the elderly is inherently associated with ethical and humane considerations. This is important to mention that TFs are not a replacement of oral feeding as elderly should get an equal chance of having oral feeds because they can maintain their nutritional needs with oral nutrition and assisted feeding so the geriatric care programs should inculcate this concept. Additionally, the patient's expressed wishes and goals of care should be discussed with them and their families while considering SGA.

The formula feeds can be tailor-made in elderly depending on the disease process. Patients with normal digesting ability should receive polymeric feedings. Patients with malabsorption syndromes require elemental or predigested feedings. Fiber-supplemented feedings are contraindicated in hemodynamic unstable patients with dysmotility [54]. Antioxidant vitamins and trace minerals are recommended in both enteral and parenteral regimens. Immune-modulating enteral formulations have been shown to be beneficial in surgical critically ill patients. These formulations contain arginine, glutamine, nucleic acids, omega-3 fatty acids, and vitamin antioxidants [55].

The enteral route should always be preferred except for the following contraindications: intestinal obstructions or ileus, severe shock, intestinal ischemia, high-output fistula, and severe intestinal hemorrhage.

Apart from these indications, if the energy and nutrient requirements cannot be met by oral and enteral intake alone (<50% of caloric requirement) for more than 7 days, a combination of enteral and parenteral nutrition (PN) is recommended (grade of recommendation, good clinical practice, ASPEN). This idea of combining the two modes of nutrition therapy has 100% consensus among the ESPEN committee members [56]. Combined nutrition is not necessary if the expected time period of PN is <4 days. If the expected PN period is expected to last between 4 and 7 days, nutrition can be hypocaloric with 2 g carbohydrate and 1 g amino acids/kg body weight administered via a peripheral catheter, and if it is likely to last more than 7–10 days [57], it is recommended that a central venous catheter should be inserted. The long-term parenteral nutrition is delivered via appropriate devices such as port and Broviac or Hickman catheter. However, in the elderly population, routine postoperative PN use is associated with higher complication rates because of higher rates of insulin resistance leading to hyperglycemia and cardiac and renal dysfunction [58]. The potential solution is to increase the lipid content of PN formula and keep the calories between 18 and 20 kcals/kg and a glucose infusion rate at 3.0–3.5 mg/kg/minute in the elderly instead of carbohydrates.

Refeeding Syndrome

During the initiation of any nutritional intervention in an undernourished elderly patient, it is mandatory to pay attention to the risk of refeeding syndrome. In this syndrome, phosphate can drop precipitously by introducing glucose rapidly, and electrolyte shifts also result in lower serum levels of potassium and magnesium. All of these changes occurring with rapid refeeding can invoke neurologic symptoms. Therefore, malnourished elderly require a gradual intake of nutrients (especially glucose) with strict monitoring of serum electrolyte levels [59]. Thiamine should be provided when refeeding syndrome is suspected or anticipated. Thiamine is an important co-enzyme in several metabolic enzymes such as pyruvate dehydrogenase and α-ketoglutarate dehydrogenase. When reintroducing carbohydrates in undernourished patients, the demand for thiamine increases [60].

Areas of Future Research: Metabolomics in Nutrition

The area of nutrition is a complex one due to the underlying interaction between numerous internal and external factors. Thus, decoding this complexity requires dedicated research strategies and exploring the future ventures pertinent to it.

Efforts are underway to support evidence-based nutritional research and achieve effective diet-based disease prevention. In this context, the global metabolite analysis, or "metabolomics," is becoming an appealing research tool for the scientists. Metabolomics is defined as the screening of small-molecule metabolites present in samples of plants, animals, or microorganisms. By comparing metabolome profiles (metabolic phenotypes or "metabotypes"), the patterns of variations between different groups can be determined and labeled as healthy versus diseased [61]. By discerning the molecular composition of food and individual's nutrition and health status, metabolomics will provide valuable information to physicians in terms of diagnosis and diet counseling. Metabolomics will identify the individual variations in dietary requirements classifying individuals into specific groups based on their "metabotype." This strategy could lead to the development of "personalized nutrition," in which diet is attuned to the nutritional needs of individual patients [62, 63]. The introduction of metabolomics in the field of nutrition is bound to open the doors of revolution in the field of nutrition science in the near future.

Summary

The elderly population in the world is rising, and this group is at high risk for malnutrition and hence poor outcomes after surgical intervention. Prompt recognition is therefore required in order to initiate timely interventions. Multimodal assessments using screening tools, serum, and radiological testing are utilized to assess the severity of malnutrition. Nutritional assessment should be comprehensive and continue at regular intervals during hospitalization. The prominent societies have published their evidence-based guidelines which are available to help clinicians; the common theme is that early feeding should be instituted whenever possible. The combination of oral and enteral feed (tube feeds) is acceptable when needed. The use of parenteral nutrition can bridge the crises until the definite form of gut access is available. Along with the carbohydrates and proteins, the micronutrient supplementation holds an important place in nutritional intervention for the elderly. Further research in basic and clinical areas is indicated and necessary to define more nutritional and metabolic changes that occur with aging and to define the indications and nutritional support techniques for optimal management of these changes. The meticulously designed controlled study protocols specifically for geriatric populations with subgroup analysis on frail elderly are the need of the hour.

References

1. Forster S, Gariballa S. Age as a determinant of nutritional status: a cross sectional study. Nutr J. 2005;4:28.
2. Drewnowski A, Shultz JM. Impact of aging on eating behaviors, food choices, nutrition, and health status. J Nutr Health Aging. 2001;5(2):75–9.
3. Braunschweig C, Gomez S, Sheean PM. Impact of declines in nutritional status on outcomes in adult patients hospitalized for more than 7 days. J Am Diet Assoc. 2000;100:1316–22.
4. Allison SP. Malnutrition, disease and outcome. Nutrition. 2000;16:590–3.
5. Hall MJ, DeFrances CJ, Williams SN, Golosinskiy A, Schwartzman A. National hospital discharge survey: 2007 summary. Natl Health Stat Report. 2010;29:1–20.
6. Gil-Montoya JA, Ponce G, Sánchez Lara I, Barrios R, Llodra JC, Bravo M. Association of the oral health impact profile with malnutrition risk in Spanish elders. Arch Gerontol Geriatr. 2013;57:398–402.
7. Mowe M, Bohmer T, Kindt E. Reduced nutritional status in an elderly population (.70 y) is probable before disease and possibly contributes to the development of disease. Am J Clin Nutr. 1994;59:317.
8. Hiesmayr M, Schindler K, Pernicka E, et al. Decreased food intake is a risk factor for mortality in hospitalized patients: the Nutrition Day Survey 2006. Clin Nutr. 2009;28(5):484–91.
9. Guenter P, Jensen G, Patel V, et al. Addressing disease-related malnutrition in hospitalized patients: a call for a national goal. Jt Comm J Qual Saf. 2015;41(10):469–73.

10. Goates S, Du K, Braunschweig CA, Arensberg MB. Economic burden of disease-associated malnutrition at the state level. PLoS One. 2016;11:e0161833.
11. Jeong SJ. Nutritional approach to failure to thrive. Korean J Pediatr. 2011;54(7):277–81. https://doi.org/10.3345/kjp.2011.54.7.277.
12. Morley JE. Anorexia of aging: physiologic and pathologic. Am J Clin Nutr. 1997;66:760–73.
13. Hanna JS. Sarcopenia and critical illness: a deadly combination in the elderly. JPEN J Parenter Enteral Nutr. 2015;39:273–81.
14. Cederholm T, Jensen GL. To create a consensus on malnutrition diagnostic criteria: a report from the Global Leadership Initiative on Malnutrition (GLIM) meeting at the ESPEN Congress 2016. Clin Nutr. 2017;36:7–10.
15. Sobotka L, editor. Basics in clinical nutrition. 4th ed. Prague: Galen; 2012.
16. White JV, Guenter P, Jensen G, et al. Consensus statement of the Academy of Nutrition and Dietetics/American Society for Parenteral and Enteral Nutrition: characteristic recommended for the identification and documentation of adult malnutrition (undernutrition). J Acad Nutr Diet. 2012;112(5):730–8.
17. Cederholm T, Bosaeus I, Barazzoni R, et al. Diagnostic criteria for malnutrition – an ESPEN consensus statement. Clin Nutr. 2015;34(3):335–40.
18. Cederholm T, Morley J. Ageing: biology and nutrition. Curr Opin Clin Nutr Metab Care. 2013;16(1):1–2.
19. Guaitoli PR, Jansma EP, de Vet HC. Nutrition screening tools: does one size fit all? A systematic review of screening tools for the hospital setting. Clin Nutr. 2014;33:39–58.
20. Kondrup J, et al. ESPEN guidelines for nutrition screening 2002. Clin Nutr. 2003;22(4):415–21.
21. Kaiser MJ, et al. Validation of the Mini Nutritional Assessment short-form (MNA-SF): a practical tool for identification of nutritional status. J Nutr Health Aging. 2009;13(9):782–8.
22. Stratton RJ, Hackston A, Longmore D, et al. Malnutrition in hospital outpatients and inpatients: prevalence, concurrent validity and ease of use of the 'malnutrition universal screening tool' ('MUST') for adults. Br J Nutr. 2004;92:799–808.
23. Detsky AS, et al. What is subjective global assessment of nutritional status? J Parenter Enter Nutr. 1987;11(1):8–13.
24. Evans WJ, Morley JE, Argiles J, et al. Cachexia: a new definition. Clin Nutr. 2008;27:793–9.
25. Fearon K, Strasser F, Anker SD, et al. Definition and classification of cancer cachexia: an international consensus. Lancet Oncol. 2011;12:489–95.
26. Alpers DH, Klein S. Approach to the patient requiring nutritional supplementation. In: Yamada T, editor. Textbook of gastroenterology. 4th ed. Baltimore: Lipponcott and Wilkins; 2003.
27. Chan YC, Suzuki M, Yamamoto S. A comparison of anthropometry, biochemical variables and plasma amino acids among centenarians, elderly and young subjects. J Am Coll Nutr. 1999;18(4):358–65.
28. Group PiNCC. Measurement of visceral protein status in assessing protein and energy malnutrition: standard of care. Prealbumin in Nutritional Care Consensus Group. Nutrition. 1995;11(2):169–71.
29. Jeejeebhoy KN, Baker JP, Wolman S, et al. Critical evaluation of the role of clinical assessment and body composition studies in patients with malnutrition and after total parenteral nutrition. Am J Clin Nutr. 1982;35:1117–27.
30. Seltzer MH, et al. Instant nutritional assessment in the intensive care unit. JPEN J Parenter Enteral Nutr. 1981;5(1):70–2.
31. Ahmed T, Haboubi N. Assessment and management of nutrition in older people and its importance to health. Clin Interv Aging. 2010;5:207–16.
32. Jensen GL, Mirtallo J, Compher C, International Consensus Guideline Committee, et al. Adult starvation and disease related malnutrition: a proposal for etiology-based diagnosis in the clinical practice setting from the International Consensus Guideline Committee. J Parenter Enter Nutr. 2010;34:156e9. Clin Nutr 2010;29:151–3.
33. Bosch X, Monclús E, Escoda O, et al. Unintentional weight loss: clinical characteristics and outcomes in a prospective cohort of 2677 patients. PLoS One. 2017;12(4):e0175125. https://doi.org/10.1371/journal.pone.0175125.
34. Cruz-Jentoft AJ, Baeyens JP, Bauer JM, et al. European working group on sarcopenia in older people. Sarcopenia: European consensus on definition and diagnosis: report of the European working group on sarcopenia in older people. Age Ageing. 2010;39:412–23.
35. Newman AB, et al. Strength, but not muscle mass, is associated with mortality in the health, aging and body composition study cohort. J Gerontol A BiolSci Med Sci. 2006;61(1):72–7.
36. Norman K, et al. Hand grip strength: outcome predictor and marker of nutritional status. Clin Nutr. 2001;30(2):135–42.
37. Dudrick SJ. Nutrition management of geriatric surgical patients. Surg Clin North Am. 2011;91:877–96.
38. Harris J, Benedict F. A biometric study of basal metabolism in man: publication #279. Washington, DC: Carnegie Institution; 1919.
39. Rolandelli RH, Ullrich JR. Nutritional support in the frail elderly surgical patient. Surg Clin North Am. 1994;74(1):79–92.
40. Gersovitz M, Motil K, Munro HN, et al. Human protein requirements: assessment of the adequacy of the current recommended dietary allowance for dietary protein in elderly men and women. Am J Clin Nutr. 1982;35:6–14.
41. Randall HT. Fluid, electrolyte, and fluid base balance. Surg Clin North Am. 1976;56:1019–58.
42. Sanders RD, Weimann J, Maze M. Biologic effects of nitrous oxide: a mechanistic and toxicologic review. Anesthesiology. 2008;109:707–22.
43. Stover PJ. Vitamin B_{12} and older adults. Curr Opin Clin Nutr Metab Care. 2010;13:24–7.

44. Cederholm T, Barrazoni R, Austin P, et al. ESPEN guidelines on definitions and terminology of clinical nutrition. Clin Nutr. 2017;36:49–64.
45. Milne AC, Potter J, Avenell A. Protein and energy supplementation in elderly people at risk from malnutrition. Cochrane Database Syst Rev. 2005;(1):CD003288. https://doi.org/10.1002/14651858.
46. Mohanty S, Rosenthal RA, Russell MM, et al. Optimal perioperative management of the geriatric patient: a best practices guideline from the American College of Surgeons NSQIP and the American Geriatrics Society. J Am Coll Surg. 2016;222:930–47.
47. Volkert D, Berner YN, Berry E, et al. ESPEN guidelines on enteral nutrition: geriatrics. Clin Nutr. 2006;25(2):330–60. https://doi.org/10.1016/j.clnu.2006.01.012.
48. Miller KR, McClave SA, Kiraly LN, et al. A tutorial on enteral access in adult patients in the hospitalized setting. JPEN J Parenter Enteral Nutr. 2014;38(3):282–95.
49. Tuna QM, Latifi R, El-Menyar A, et al. Gastrointestinal tract access for enteral nutrition in critically ill and trauma patients: indications, techniques, and complications. Eur J Trauma Emerg Surg. 2013;39:235–42. https://doi.org/10.1007/s00068-013-0274-6.
50. Hartgrink HH, Wille J, Konig P, Hermans J, Breslau PJ. Pressure sores and tube feeding in patients with a fracture of the hip: a randomized clinical trial. Clin Nutr. 1998;17:287–92.
51. Sullivan DH, Nelson CL, Bopp MM, et al. Nightly enteral nutrition support of elderly hip fracture patients: a phase I trial. J Am Coll Nutr. 1998;17:155–61.
52. Latifi R. Nutritional therapy in critically ill and injured patients. Surg Clin North Am. 2011;91:579–93.
53. Afifi I, Elazzazy S, Abdulrahman Y, Latifi R. Nutrition therapy for critically ill and injured patients. Eur J Trauma Emerg Surg. 2013;39(3):203–13.
54. McClave SA, Martindale RG, Vanek VW, et al. Guidelines for the provision and assessment of nutrition support therapy in the adult critically ill patient: Society of Critical Care Medicine (SCCM) and American Society for Parenteral and Enteral Nutrition (A.S.P.E.N.). JPEN J Parenter Enteral Nutr. 2009;33(3):277–316.
55. Joseph B, Wynne JL, Dudrick SJ, Latifi R. Nutrition in trauma and critically Ill patients. Eur J Trauma. 2010;36:25. https://doi.org/10.1007/s00068-010-9213-y.
56. Weimann A, Braga M, Carli F, Higashiguchi T, Hubner M, Klek S, et al. ESPEN guideline: clinical nutrition in surgery. Clin Nutr. 2017;36(3):623–50.
57. Latifi R, Uraneus S. Feed the gut, feed early and with the right stuff, but do not forget total parenteral nutrition. Eur J Trauma Emerg Surg. 2013;39:201–2.
58. Sobotka L, et al. ESPEN guidelines on parenteral nutrition: geriatrics. Clin Nutr. 2009;28(4):461–6.
59. Crook MA, Hally V, Panteli V. The importance of the refeeding syndrome. Nutrition. 2001;17:632–7.
60. Kraft MD, Btaiche IF, Sacks GS. Review of the refeeding syndrome. Nutr Clin Pract. 2005;20:625–33.
61. Nicholson JK, Holmes E, Kinross JM, Darzi AW, Takats Z, Lindon JC. Metabolic phenotyping in clinical and surgical environments. Nature. 2012;491:384–92.
62. German JB, Watkins SM, Fay LB. Metabolomics in practice: emerging knowledge to guide future dietetic advice toward individualized health. J Am Diet Assoc. 2005;105:1425–32.
63. German JB, Zivkovic AM, Dallas DC, Smilowitz JT. Nutrigenomics and personalized diets: what will they mean for food? Annu Rev Food Sci Technol. 2011;2:97–123.

Neurosurgery in the Elderly

9

Julia Pazniokas, Brianna Theriault, and Christian A. Bowers

The conversation surrounding neurosurgery in elderly patients has drastically changed in the past 30 years. It was only 30 years ago that an innovative groundbreaking geriatric neurosurgery review determined that the current age limit of 65 was arbitrary and must be reconsidered [1]. In contrast, now there are reports of successful neurosurgical interventions in patients as old as 103 years of age, and neurosurgery in the geriatric population is common. Recent literature in nearly every neurosurgical subspecialty concludes that advanced age alone should not preclude a patient from surgical consideration. In fact, a growing body of literature suggests that comorbidity status, or frailty, as measured typically by the modified frailty index (mFI), is significantly more predictive of outcomes and should be emphasized over age in the surgical decision-making process since not all elderly patients are the same.

The combination of a continuously increasing life expectancy, advancements in neurosurgical technologies, and improved neurocritical care and neurosurgical postoperative care have all led to a rapidly growing pool of potential candidates for neurosurgical procedures. The American population of people >60 years old increased by 25% between 2002 and 2010 [2] and is projected to increase to 90 million by 2050 [3]. This particular predicted demographic growth is extremely relevant to neurosurgery as the peak incidence for many operative neurosurgical conditions occurs after age 60. Elderly patients represent an increasingly larger proportion of patients presenting to hospitals and requesting neurosurgical consultation [4].

In particular, elderly spine surgery is constantly increasing annually. Between 1990 and 2004, there was 28-fold increase in anterior discectomy and fusion surgeries performed on patients aged 65 years and above [5]. Intracranial tumor diagnoses are also more common in the elderly, with a reported peak incidence at age 75 [6]. Also, the elderly have a relative risk of brain cancer of 3.18 when compared to young adults [7]. Likewise, vascular pathology increases constantly as age increases. The incidence of subarachnoid hemorrhage increases fivefold between the fourth to six decades and the eighth decade, from 15 per 100,000 per year to 78 per 100,000 per year [8].

The historical dogma is that elderly patients have increased mortality rates and increased complication rates following neurosurgical interventions. However, an abundance of recent data demonstrates that many elderly patients undergo neurosurgery with good outcomes and low mortality rates. Unfortunately, the traditional

J. Pazniokas · B. Theriault
New York Medical College, Valhalla, NY, USA

C. A. Bowers (✉)
Department of Neurosurgery, Westchester Medical Center, New York Medical College, Valhalla, NY, USA
e-mail: Christian.Bowers@wmchealth.org

R. Latifi (ed.), *Surgical Decision Making in Geriatrics*,
https://doi.org/10.1007/978-3-030-47963-3_9

belief that elderly patients are unfit for neurosurgical interventions has contributed to decreased neurosurgical access. The elderly have been shown to have decreased rates of transfer to neurosurgical units and experience longer delays for urgent diagnostic CT scans upon emergency department presentation [9]. Elderly patients with glioblastoma are also less likely to receive multimodal treatment and are more likely to receive diagnosis without histopathologic verification [10].

There is a vast body of subspecialty neurosurgical literature reporting outcomes in elderly patients. Although elderly patients traditionally demonstrated higher mortality rates and complication rates, both have decreased significantly over time [4].

Spine

The disease burden caused by spinal pathology in elderly patients is associated with more disability than CHF, COPD, or diabetes [11]. This disability is secondary to various manifestations of degenerative spinal pathology, including osteoporotic fractures, lumbar disc herniation, lumbar stenosis, and spondylolisthesis. The exponential explosion in the incidence, and treatment costs, of spine surgery in the Medicare population over the past decades have led many to question the benefit and medical necessity of spine surgery in elderly patients. Other reports suggest that elderly patients undergo unwarranted prolonged trials of conservative treatment without any evidence of its efficacy due only to their advanced age [12]. Prolonged medical management in one small series of elderly patients with disc herniation, stenosis, or spondylolisthesis demonstrated no significant improvement in any patient-centered outcome [13]. While a trial of medical management is absolutely indicated for elderly patients with low back pain, age alone should not prevent patients from being considered for beneficial surgical interventions. This message, along with multiple examples of the good outcomes achievable in elderly patients, is one of the key tenets provided by this chapter.

Degenerative spine conditions are extremely common in the elderly and increase with age. Operations that are simple low-risk procedures in young, healthy patients often require more nuanced surgical decision-making in older patients with more comorbidities. As with any surgery, elderly neurosurgical patients carry some increased risk for adverse outcomes associated with anesthetic complications, including memory and cognition deficits. The long average operative time of complicated neurosurgical procedures makes this anesthetic risk even more concerning. Elderly degenerative spine patients are also more likely to have multilevel disease compared to their younger counterparts which complicates the diagnostic and the treatment considerations. For example, elderly patients may require multilevel fusions, which require longer anesthesia time and have increased risk, while also being more predisposed to complications (e.g., osteoporotic bone, thinner dura mater, etc.) [12].

Even after deciding to recommend surgical treatment for spinal pathology in an elderly patient, there is frequently no consensus on which specific procedure to perform. From 1980 to 2000, spinal stenosis surgery was the fastest-growing type of lumbar surgery performed in the USA [12]. In 2007, over 37,000 patients >65 in the Medicare Provider Analysis and Review database underwent surgery for spinal stenosis, with an associated total hospital bill of $1.65 billion [13]. RCTs have demonstrated patients with severe symptoms from spinal stenosis benefit more from decompression without fusion than from nonsurgical options [14]. Nevertheless, rates of decompression surgery or simple fusion in elderly patients declined between 2002 and 2007, while rates of complex fusions increased over the same period [14]. While major complications occurred in only 3.1% of total patients and the mortality rate was 0.4% within 30 days of discharge, "life-threatening complications" increased with age, complex fusions, and increasing comorbidities [13]. This analysis demonstrates that many elderly patients can safely undergo and benefit from degenerative spine surgery, but neurosurgeons must be more selective in which patients require complex fusions.

Osteoporotic spinal fractures are another major cause of disability and pain in the senior population. They are present in approximately one-quarter of women older than the age of 70 years and up to one-third of those older than the age of 80 years [15]. A significant portion of these fractures are refractory to non-operative treatment, leading many elderly patients to opt for surgical treatment with vertebroplasty and kyphoplasty. A systematic review of 69 clinical trials demonstrated surgery had a significant clinical benefit with ~90% rate of pain relief and significant decrease in disability postoperatively [16]. Despite this evidence supporting the efficacy of vertebroplasty and kyphoplasty, it is important to note that more invasive surgery may be required to address severe underlying pathology such as sagittal imbalance and to achieve long-term efficacy. However, many older patients with comorbidities may not be able to tolerate more invasive surgery.

It is important to recognize that surgical interventions in the elderly are often geared toward improving quality of life and functional status, rather than curing disease. The vast majority of studies on geriatric spinal surgery analyze complication and mortality rates, rather than functional outcomes. Studies have shown that elderly patients may even experience better improvement in health-related quality of life scores after surgical decompression of the lumbar spine when compared to their younger counterparts [17], even in the setting of increased rates of adverse events. Still, clear evidence-based guidelines are lacking for determining which elderly patients will benefit from lumbar spine surgery and how surgically aggressive surgeons should be with this population.

Intracranial Neuro-oncology

Intracranial tumors increase in incidence with age and peak around age 75 [18]. Nevertheless, elderly patients have not historically always been considered for surgical resection. Furthermore, despite elderly patients often being thought of as having worse outcomes after craniotomy for tumor resection, the evidence demonstrating poor outcomes after geriatric intracranial surgery is lacking, but many treatment guidelines are based off studies in which patients age >70 are excluded or underrepresented [19]. The most obvious example of differing treatment patterns between young and old patients is with the most common primary brain cancer – glioblastoma multiforme (GBM). The standard of care for patients with GBM is rapidly evolving in all age groups. While older patients were not necessarily offered resection in the past, more recent literature shows that they too may benefit from surgical interventions. Elderly patients with more benign tumors, including pituitary adenomas, meningioma, and vestibular schwannomas, frequently tolerate surgery nearly as well, if not as well, as their younger counterparts.

GBM

The belief that elderly patients with aggressive brain tumors (GBM, brain metastases, etc.) are poor surgical candidates arises from the fact that age is a poor prognostic marker for these pathologies. The overall median survival for GBM patients in 2015 was 15 months, but reported median survival for patients >65 was 4 months [20]. Despite the median age at diagnosis of GBM being 64 years, patients age >65 are statistically less likely to be offered multimodal treatment. Standard treatment in non-elderly patients involves surgical resection followed by radiotherapy with concurrent chemotherapy. Surgical resection in elderly patients, on the other hand, is not always offered or performed. This is despite the fact that many studies suggest that elderly patients benefit from maximal safe surgical resection, just like normal non-geriatric adult patients, as significant tumor resection prolongs patient survival due to cytoreduction [20].

Benign Tumors: Meningioma, Pituitary Adenoma, and Vestibular Schwannoma

The increasing availability and frequency of neuroimaging has contributed to an increase in the detection of incidental benign masses. In particular, elderly patients are more likely to undergo neuroimaging for a variety of reasons (working up cognitive decline, after a fall or car accident, etc.). Meningiomas are the most common benign intracranial tumor, and their incidence in the elderly has continued to increase. Surgical resection of meningiomas is done electively, due to their slow growth and insidious symptom onset. The natural history of meningioma varies, but studies of incidental meningiomas have reported that the majority exhibited no growth [21]. Additionally, tumors in elderly patients may actually grow slower than in younger patients [22]. A 2016 review of the literature on meningioma in the elderly generally reported few differences between elderly and non-elderly patients undergoing meningioma resection, but a large 2004 study demonstrated higher in-hospital mortality, adverse outcomes, and permanent deficits in elderly meningioma resection patients [23]. Elderly patients have higher frequencies of WHO grade II and III atypical meningiomas. Patients over 80 years old had higher rates of 1- and 5-year mortality, but not at 3 months after surgery. Many studies identify that ASA score, or other risk identification scales, predicts mortality better than age [24].

Optimal timing of meningioma resection is particularly important in elderly patients. Surgical decision-making is simple for small tumors that are amenable to stereotactic radiosurgery or for tumors causing significant symptoms or significant mass effect. Conversely, when elderly patients are diagnosed with large tumors causing few or no symptoms, surgeons must balance the complication risk associated with aggressive surgery with the possibility that delaying treatment may increase the likelihood of suboptimal outcomes in the future. While it has been demonstrated that elderly patients undergoing meningioma resection have favorable survival and clinical outcomes [24], the risk factors associated with good or bad outcomes have yet to be delineated, and therefore clear guidelines are still lacking on how best to manage geriatric patients presenting with symptomatic meningiomas. Incidental meningiomas should generally be followed with observation due to the shortened life expectancy.

There has been a large increase in the detection of pituitary adenomas in the elderly due to the increased neuroimaging. Most will remain asymptomatic and will not require treatment, but age should not preclude surgery in symptomatic patients. Elderly patients are more likely to have nonfunctioning PAs compared to patients <60 years old [25], and visual disturbances are the most common indication for surgery. These symptoms improve in elderly patients at similar rates to the general population, but elderly patients are at higher risk of postoperative complications including pituitary hemorrhage, hypertension, and SIADH. Mortality rates do not appear to increase with age [25].

There is a smaller body of literature on vestibular schwannoma resection in the elderly, likely due to the smaller number of patients. Studies have suggested a complication rate as high as 60% in patients age >65, while others report no difference in outcome between older and younger patients [26]. A 2016 study by Bowers et al. demonstrated that patients >65 had equivalent facial nerve outcomes, hearing preservation, and complication rates when compared to patients <65 [27].

As has been repeatedly demonstrated across the neurosurgical literature, good preoperative clinical status, frequently measured by frailty, is a better predictor of outcomes in elderly patients with benign intracranial tumors than age alone, as an emerging body of literature is supporting.

Traumatic Brain Injury

Traumatic brain injury is a major cause of morbidity and mortality in patients of all ages worldwide. While many young patients sustain TBIs after motor vehicle accidents, the majority of TBIs in the elderly are caused from falls around the home. Since interventions to reduce traffic accidents have been implemented, the incidence of TBI is decreasing in young patients. However, there has been no change in TBI incidence in the elderly. Additionally, as the elderly population increases in number, more TBIs will occur. In general, although TBI outcomes are improving, increasing age is still a significant risk factor for poor outcome. Elderly patients' increased preexisting medical comorbidities, i.e., their frailty, lead to increased postoperative complications and difficult hospital courses. Additionally, elderly patients are more likely to be on antiplatelet or anticoagulation medications, thereby increasing their risk for a severe intracerebral bleed after trauma.

Several studies have considered how age affects outcome after TBI. Mosenthal et al. found that mortality progressively increased with each decade after age 50 and was doubled in patients >65 compared to younger patients. This increase in mortality occurred at all injury severity levels. Additionally, elderly patients were more likely to have poor functional outcome compared to young patients. Multivariate logistic regression analysis demonstrates age is an independent risk factor for increased mortality, even after controlling for complications and comorbidities. Similarly, Kuhne et al. saw that mortality steadily increased with age starting at age 56. This increase was independent of injury severity [28]. Furthermore, Stochetti et al. found that patients older than 59 years had six times the probability of having unfavorable outcomes after TBI [29]. The odds ratio was equally high for patients with poor health status before TBI. This finding was corroborated by Susman et al., who saw that despite similar GCS at admission to young patients and lower injury severity scores, elderly patients tended to have more unfavorable outcomes [30].

A series published by Lau et al. compared patients >80 years old to a younger patient cohort [31]. They found that these older patients tended to have longer hospital stays and more postoperative complications. Unexpectedly, they found that there were no differences in 30-day mortality or rates in return to baseline functional status. This suggests that elderly patients may require longer postoperative medical care and increased complication rates, but they can also recover back to their baseline. Interestingly, one study found that the mortality rate of elderly patients beyond 6 months post-injury was comparable to rates for the general population. The amazing conclusion that "6-month survivors could expect to have a normal lifespan" is encouraging given how bleak this elderly TBI literature looked in the past [32].

Conclusions

While the published data and subjective evidence show that increased age is a predictor for mortality and worse outcome after TBI, it likely is not the only factor. Elderly patients tend to sustain more severe injuries despite an often less severe mechanism of injury. Their increased likelihood of comorbidities also predisposes them to postoperative complications. These complications may be survivable in a young patient but can be devastating in an elderly patient. Taken together, the data suggests that careful assessment of risk must be conducted on a patient-by-patient basis to determine if early surgical intervention or medical management is appropriate following TBI.

Subdural Hematoma

Acute Subdural Hematoma

Acute subdural hematoma is a common neurosurgical condition, occurring in up to 29% of all cases of TBI [33]. While young patients often suffer aSDH after high-velocity trauma, older patients may suffer SDH after low-velocity injuries, such as falls. This difference is due to age-related brain atrophy in the elderly, which causes increased tension on the bridging subdural veins leading to shearing injuries, most commonly after falls. Elderly aSDH patients do not commonly present with primary brain injury and swelling, like that seen in younger aSDH patients due to the differences in force of the injuries causing the aSDH [34]. The force from a ground-level fall, a common cause for aSDH in elderly patients, is the opposite of the high-speed mechanism that most commonly causes aSDH in younger patients.

Older data suggests that acute subdural hematoma (aSDH) in the elderly is nearly a distinct entity compared to that in younger patients, with elderly patients suffering mortality rates as high as 74–88% [35, 36]. Other studies demonstrate that surgical intervention in elderly aSDH patients results in few functional survivors [37]. Although few studies have been published in the past decade, more recent data suggests that mortality rates in elderly aSDH patients have improved and are similar to younger patients (35–50%) [35]. In 2016 Raj et al. studied patients age >75 and demonstrated an overall 1-year mortality rate of 50% [38]. Nonsurvivors had worse GCS score on presentation, but there were no significant differences in those on antithrombotic medication or with INR >1.5. Still, there was a considerable absolute difference in mortality between INR <1.5 and >1.5 groups, which would likely be significant in a larger study. Premorbid functional dependence at time of injury was also predictive of mortality [37]. This study demonstrates the importance of premorbid status and comorbidities in stratifying elderly patients with TBI.

Chronic Subdural Hematoma

Chronic subdural hematoma is a distinct neurosurgical entity from acute SDH and is much more common in the older patient population. Large studies have demonstrated that older age is an independent risk factor for morbidity and mortality after surgical drainage of cSDH [39]. Other studies have suggested that clinical exam on presentation is a more pertinent predictor of outcomes. There is no consensus on outcome prognostication in elderly patients with cSDH and which elderly patients are likely to suffer complications.

Multiple studies have been published on the outcomes of the oldest elderly patients, nonagenarians, but the conclusions are unclear and contradictory. Stippler et al. concluded that patients >90 years of age have poor outcomes after suffering chronic subdural hematomas, regardless of whether they undergo surgical drainage or conservative treatment [40]. Dobran et al. also concluded that nonagenarians were at higher risk of postoperative complications compared to patients <80 years of age, but that average functional recovery, at 1 month and 6 months postop, were equivalent in the two groups [41]. Lee et al. also demonstrated that cSDH patients aged 90+ benefited from surgical drainage compared to conservative treatment with better mortality rates, improvement in neurological function, and discharge disposition [42]. They also note that conservative management is actually less likely to be successful in the very elderly, because their brain atrophy means that only large cSDH will produce symptoms in the first place.

It is important to note that the most common surgical intervention for cSDH, burr hole craniostomy, is relatively noninvasive. An alternative procedure, twist-drill craniostomy, can even be done at bedside without any general anesthesia. Even patients of this advanced age should be considered for surgical intervention given that nonagenarians do better with evacuation compared to conservative treatment [42].

Aneurysm

The literature regarding cerebral aneurysm management in elderly patients primarily focuses on comparing outcomes between endovascular coiling and open neurosurgical aneurysm clipping. While rates of good functional outcomes vary between studies, generally, there is a consensus by most clinicians that endovascular techniques are less invasive and may be a safer and more effective treatment alternative to clipping in elderly aneurysm patients [43–45]. This is particularly true in elderly patients with multiple comorbidities, who do better with less invasive interventions across all surgical disciplines. Gonzalez et al. saw that after endovascular coiling of unruptured or ruptured aneurysms, 65% of elderly patients had a good functional outcome (mRS ≤ 2) [46].

Unruptured Aneurysms

Although there is still controversy on which asymptomatic unruptured aneurysms should receive therapeutic intervention, especially in elderly patients where their shortened life expectancy significantly reduces the cumulative aneurysm rupture risk, treatment outcomes have been reported in multiple studies. Brijiniki et al. observed that while increasing age was associated with greater mortality and worse discharge location after any treatment for unruptured aneurysms, endovascularly coiled patients fared better than surgically clipped patients [43]. These patients had decreased length of stay, lower rates of discharge to a long-term care facility, and decreased mortality rates. The authors noted that this difference was more pronounced in patients over 80 years old. Barker et al. also found that endovascular coiling for unruptured aneurysms was associated with reduced mortality compared to surgical clipping [47]. Gonzalez et al. saw that 91% of patients aged 70 years and older with unruptured aneurysms had good functional outcome (modified Rankin score ≥2) after endovascular embolization [46]. Finally, Cai et al. found that 91% of patients over age 70 with symptomatic unruptured aneurysm treated endovascularly achieved excellent outcomes (modified Rankin score of 0–1) [48].

Aneurysmal SAH (Ruptured Aneurysms)

Patients with ruptured aneurysmal subarachnoid hemorrhage have worse outcomes than patients with unruptured aneurysm [46]. However, coiling ruptured aneurysms still produces good outcomes in the elderly population, despite their advanced age. In a series by Zhang et al., the efficacy of endovascular coiling versus clipping was compared in 198 Chinese patients over age 60 with ruptured aneurysms [45]. While there was no difference in complication rates between techniques, coil embolization was associated with decreased length of stay and decreased procedural time. Additionally, coiling was associated with a favorable outcome, defined by a Glasgow Outcome Scale (GOS) score of 4–5, in 88.2% of patients. Similarly, a good or excellent outcome (GOS 4–5) was found in 59% of patients aged 65–80 with SAH secondary to aneurysm rupture treated in a series by Lubicz et al. [44] and 48% of patients aged 65 and above in a series published by Sedat et al. [49]. The International Subarachnoid Aneurysm Trial (ISAT) found that endovascular coiling was associated with better outcomes than surgical clipping in the general population. Of note, a substudy of the ISAT on elderly patients found that, to optimize outcomes, endovascular coiling should be the treatment of choice for ruptured ICA and PCA aneurysms while surgical treatment should be considered for patients with MCA aneurysms.

Aneurysm Treatment and Considerations for Elderly Patients

There are specific considerations to treating aneurysms in the elderly population. First, intracranial aneurysms are more common in an aging population as the prevalence in some series is doubled in elderly patients [45, 50]. Also, the Hunt and Hess (HH) grades tend to be higher in elderly patients [49], while the vasospasm risk is lower [51]. Sedat et al. noted that the rate of high grade aneurysmal SAH (Hunt-Hess 4–5) was 27.3% in young patients but 36.6% in patients over age 64 [49]. These considerations highlight the need to determine the best approach to managing ruptured aneurysms in this population.

The technical endovascular challenges unique to elderly patients are that the increased tortuosity and stenosis of the elderly vasculature may lead to navigation difficulties during the procedure and make complete embolization more challenging. Additionally, due to the atheromatous nature of vessels in the elderly, there is an increased risk of thromboembolic events during embolization [49]. In a series published by Sedat et al., elderly patients had an increased rate of thromboembolic events compared to younger patients (13% vs. 4.2%) [49]. These significant increases in complication rate warrant extra consideration for this patient population.

Conclusions

As the population continues to age and with the increase in the frequency, and the technological advances, in neuroimaging, aneurysm incidence will continue to rise. While the superiority of endovascular coiling versus microsurgical clipping remains controversial for all adult patients, particularly in the very young, there is an increasingly large and growing body of literature to support the use of endovascular treatment in elderly patients with unruptured or ruptured aneurysms. Treatment that is noninvasive and associated with less morbidity and mortality is likely best in these patients who may have multiple comorbidities.

Frailty

The contradictory evidence on the management of elderly neurosurgery patients across subspecialties may be due to the heterogeneous nature of the 65+-year-old population. A constantly growing body of literature suggests that age alone is not the best method of stratifying risk. This concept is well illustrated by Farhat et al.'s statement that "surgeons routinely use phrases such as a 'young 85-year-old' or the converse '50 going on 90'" [52]. In place of age, a patient's likelihood to tolerate surgery can be better defined by his or her level of frailty.

Broadly defined as a "a biologic syndrome of decreased reserve and resistance to stressors, resulting from cumulative declines across multiple physiologic systems" [53], frailty is a measure of biological age, rather than chronological. Although frailty increases with age [53] and is higher in the elderly surgical population than the nonsurgical population [54], up to 75% of people over the age of 85 are not frail [55].

The medical literature has long demonstrated frailty's effect on outcomes in the elderly population, usually defining frailty by "some combination of history, physical examination, and determination of physical capability, such as walking speed and grip strength" [56]. Surgical fields have also analyzed the effect of frailty on outcomes and complications, and since 2015 many neurosurgical studies have been published on the topic.

Between 2015 and mid-2019, 25 studies have been published on frailty in neurosurgical subspecialties including spinal deformity, degenerative spine, spinal tumor, intracranial hemorrhage, and intracranial tumor (Table 9.1). Although frailty was initially described as a way to discern the "biological age" specifically of the elderly population, the vast majority of these studies include all adult patients, so that frailty acts more like a traditional comorbidity index than a measure of physiologic reserve.

There are a variety of different frailty indices made up of a combination of functional status and comorbidities. The modified frailty index (mFI), an 11-factor index extrapolated from the Canadian Study of Health and Aging, is the most commonly used index. The neurosurgical literature consists of eight other frailty indices, and many are specific to pathology, such as the Adult Spinal Deformity-Frailty Index, Cervical Deformity-FI, and the Metastatic Spinal Tumor-FI. Each of these indices was created specifically to be predictive for patients undergoing procedures specific to these pathologies. The 9 indices combined include over 40 factors, but most commonly include functional status, congestive heart failure, hypertension, anemia, chronic kidney disease, and COPD or recent pneumonia. The use of so many different frailty indices can be confusing initially. However, increasing frailty will affect outcomes differently for neurosurgery patients depending on the extent of the surgical procedure. Frailty will play a much smaller role in a patient undergoing a bedside burr hole chronic subdural hematoma evacuation, or a muscle biopsy procedure, when compared to an all-day spinal deformity correction surgery. Studies comparing the predictive value of the mFI or the simplified 5i-mFI to the more complicated pathology-specific FIs are lacking. Still, the mFI has demonstrated predictive value across a range of neurosurgical procedures including kyphoplasty [57], lumbar [58] and cervical fusion [59], spinal deformity surgery [60], oncologic neurosurgery [61], GBM

Table 9.1 25 studies published on frailty in neurosurgery by mid-2019

Author	Year	Category of procedure	Sample size	Data source	Age
Cranial					
Tomlinson	2017	Mixed	27,098	NSQIP	All
Cloney	2015	Glioblastoma	243	One hospital (US)	**>65**
Youngerman	2017	Tumor	9149	NSQIP	>18
Imaoka	2018	Spontaneous ICH	156	2 hospitals (Japan)	All
Shimizu	2018	Chronic subdural hematoma	211	One hospital (Japan)	**>65**
Spine					
Ali	2016	Mixed spinal surgery	18,294	NSQIP	>18
Deformity					
Leven	2016	Spinal deformity	1001	NSQIP (2005–2012)	>18
Miller	2018	Spinal deformity	266	ESSG	>18
Miller	2018	Cervical deformity	61	ISSG	>18
Miller	2017	Spinal deformity	417	ISSG	>18
Miller	2018	Spinal deformity	267	SCOLI-RISK-1 database	>18
Reid	2018	Spinal deformity	332	Multicenter	>18
Yagi	2018	Spinal deformity and degenerative spine**	481	2 high volume spine centers	**>50**
Degenerative					
Charest-Morin	2018	Degenerative lumbar spine disease	102	One hospital (Canada)	**>65**
Flexman	2016	Degenerative spine disease	52,671	NSQIP (2006–2012)	All
Kessler	2018	Thoracolumbar spinal fracture	303	NSQIP (2007–2012)	>18
Leven	2017	Lumbar fusion	6094	NSQIP (2005–2012)	>18
Phan	2017	Anterior lumbar interbody fusion	3920	NSQIP (2010–2014)	>18
Rothrock	2018	Spine – degenerative	87	One hospital (US)	**>65**
Segal	2018	Kyphoplasty vertebral augmentation	2399	NSQIP	>18
Shin	2017	Cervical spinal fusion	6965	NSQIP (2005–2012)	>18
Weaver	2019	Elective posterior lumbar fusion	23,516	NSQIP	All
Tumor					
Ahmed	2017	Primary spinal tumor	1589	NIS (2002–2011); NSQIP*	All
De la Garza Ramos	2016	Metastatic spinal tumor	4583	NIS (2002–2011); NSQIP*	>18
Lakomkin	2018	Spinal tumor surgery	2170	NSQIP (2008–2014)	All

ICH intracerebral hemorrhage, *NIS* nationwide inpatient sample, *NSQIP* national surgical quality improvement project
*NSQIP used for external validation
**Adult spinal deformity, degenerative lumbar
Bold values: When the study is specifically focusing on older adults (>65 or one study was >50).

resection [62], and surgery for intracranial hemorrhage [63].

The lack of consensus in the neurosurgical literature is also reflected in the cutoffs used to define "non-frail," "pre-frail," "frail," or "severely frail" (Table 9.2). To allow for comparison, each cutoff can be converted to a score between 0 and 1 by dividing the number by the total number of factors in the index. Using the mFI, for example, a mFI of 3 out of a possible total of 11 is equivalent to a frailty score of 0.27. Comparing definitions of frailty between studies shows vast variability. As Fig. 9.1 demonstrates, a score of about 0.3 would be considered non-frail, pre-frail, frail,

Table 9.2 Literature published between 2016 and mid-2019 on frailty in neurosurgery demonstrating association between frailty and complications across many neurosurgical procedures

Author	Category of procedure	Frailty measure	Frailty cutoffs	Outcomes associated with frailty	Predictive value of frailty in comparison to other risk factors
Cranial					
Tomlinson	Mixed	Novel Pre-op Frailty Scale	Non-frail: <0.19 Frail: ≥0.19	↑Mortality ↑Complications	Frailty predictive across age stratifications
Cloney	Glioblastoma	mFI	Least frail: 0 Moderately frail: 0.09–0.18 Frailest: I≥0.27	↑Mortality ↑Complications ↑Length of stay	Predictive value independent of age, KPS, CCI
Youngerman	Tumor	mFI	No frailty: 0 Low: 0.01–0.19 Intermediate: 0.2–0.29 High frailty: ≥0.3	↑Mortality ↑Complications ↑Length of stay ↑Discharge to higher care	mFI better predictor than age *ASA equally predictive as mFI*
Imaoka	Spontaneous ICH	mFI	Non-frail: <0.18 Frail: ≥0.18	↑Mortality ↑Poor outcome	ICH score + mFI best predictor
Shimizu	Chronic subdural hematoma	Clinical Frailty Score	Non-frail: ≤4 out of 9 Frail: ≥5 out of 9	↑Discharge to higher care	*Worse predictor of poor outcome than age*
Spine					
Ali	Mixed spinal surgery	mFI	Low vs. high mFI: 0 vs. 0.27	↑Complications ↑CD-G4 complications ↑Mortality	
Deformity					
Leven	Spinal deformity	mFI	Frail: ≥0.18 used for multivariable analysis	↑Complications ↑Reoperation	mFI better predictor than age >60 of reoperation but not complications
Miller	Cervical deformity	CD-FI	Not frail: <0.2 Frail: 0.2–0.4 Severely frail: >0.4	↑Complications	
$Miller_1$	Spinal deformity	ASD-FI	Non-frail: <0.3 Frail: 0.3–0.5 Severely frail: >0.5	↑Complications ↑Length of stay ↑Reoperation ↑Poor outcome	Increase in complications independent of age
$Miller_2$	Spinal deformity	ASD-FI	Non-frail: <0.3 Frail: 0.3–0.5 Severely frail: >0.5	↑Complications ↑Reoperation ↑Length of stay ↑Poor outcome	Significant on multivariate analysis (did not include age)
$Miller_3$	Spinal deformity	ASD-FI	Not frail: <0.3 Frail: 0.3–0.5 Severely frail: >0.5	↑Complications ↑LOS	
Reid	Spinal deformity	ASD-FI	Non-frail: <0.3 Frail: 0.3–0.5 Severely frail: >0.5	*More clinical benefit in frail* vs. *significantly frail OR non-frail*	

Table 9.2 (continued)

Author	Category of procedure	Frailty measure	Frailty cutoffs	Outcomes associated with frailty	Predictive value of frailty in comparison to other risk factors
Yagi	Spinal deformity and degenerative spine	mFI	No frail: 0 Pre-frail: >0, < 0.21 Frail: >0.21	Deformity: ↑Poor Outcomes Degenerative*: No significant association*	
Degenerative					
Charest-Morin	Degenerative lumbar spine disease	mFI	Not frail: 0 Pre-frail: >0 and <0.21 Frail: ≥0.21	↑Mortality	
Flexman	Degenerative spine disease	mFI	Not frail: 0 Pre-frail: 0–0.21 Frail: ≥0.21	↑Mortality ↑Complications ↑Length of stay ↑Discharge disposition	Predictive on multivariate analysis
Kessler	Thoracolumbar spinal fracture	mFI	Non-frail: ≤0.27 Frail: ≥0.27	↑Complications	
Leven	Lumbar fusion	mFI	Stepwise analysis Frail: ≥0.36 used for multivariate analysis	↑Complications ↑Readmission ↑Reoperation ↑Length of stay	mFI only better predictor of reoperation
Phan	Anterior lumbar interbody fusion	mFI	Low vs. high mFI: 0 vs. 0.27	↑Complications	
Rothrock	Degenerative spine disease	FRAIL scale	0: robust 1–2: pre-frail 3–5: frail	***No significant difference in return to cognitive function***	
Segal	Kyphoplasty vertebral augmentation	5i – mFI	1) mFI-5 = 0 2) mFI-5 = 1 3) mFI-5 ≥ 2	↑Complications ↑Readmission ↑Length of stay ↑Discharge disposition	Better predictor of complications than age
Shin	Cervical spinal fusion	mFI	ACDF group: 0 vs. ≥0.27 PCF group: 0 vs. ≥0.36	Both groups: ↑Mortality ↑Complications ↑CDG4 complications	ACDF: mFI better predictor of CDG4 complications than age>75, ASA>3 ------------- *PCF: Age>75 and ASA>3 better predictors than mFI*
Weaver	Elective posterior lumbar fusion	mFI-5	1. mFI-5 = 0 2. mFI-5 = 1 3. mFI-5 ≥ 2	↑Complications ↑Readmission ↑Discharge disposition	Predictive value independent of age and ASA
Tumor					
Ahmed	Primary spinal tumor	Spinal Frailty Index	Not frail: 0 Mildly frail: 0.11 Moderately frail: 0.22 Severely frail: ≥0.33	↑Complications ↑Length of stay	

(continued)

Table 9.2 (continued)

Author	Category of procedure	Frailty measure	Frailty cutoffs	Outcomes associated with frailty	Predictive value of frailty in comparison to other risk factors
De la Garza Ramos	Metastatic spinal tumor	Metastatic Spinal Tumor-FI	Not frail: 0 Mildly frail: 0.11 Moderately frail: 0.22 Severely frail: ≥0.33	↑Mortality ↑Complications ↑Length of stay	
Lakomkin	Spinal tumor surgery	mFI,	Stepwise analysis	↑Mortality ↑Length of stay	*High CCI better predictor than mFI*

mFI 11-factor modified frailty index
ASD-FI adult spinal deformity frailty index
5i-mFI 5-factor mFI
mCD-FI modified cervical deformity frailty index
∗Poor outcome from bleed but associated with less deterioration after surgery
Wide variation in cutoffs used to demonstrate frailty in these studies. Many different frailty indices used

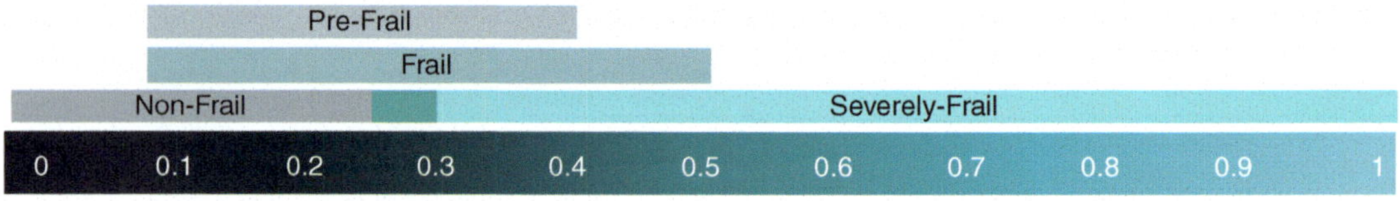

Fig. 9.1 Frailty cutoffs

or significantly frail by different studies. Even within one study analyzing different approaches to cervical fusion, different cutoffs were used to analyze frailty for the anterior vs. posterior approaches [58].

While it must be noted that the studies define frailty using different indices and cutoffs, the vast majority demonstrate an association between frailty and complication rates, including infection, reoperation, pneumonia, UTI, DVT, PE, and sepsis. Most importantly, 13 studies demonstrated that various frailty measurements are better predictors of outcomes and complication rates than age alone. This highlights that age alone must not be used as a treatment cutoff and why further study is needed across the spectrum of neurosurgical pathologies.

Case Reports

We present two recent cases that demonstrate that outcomes and complications are better predicted by comorbidities than by age (Table 9.3).

Case Illustrations

Patient A (Fig. 9.2) is a 76-year-old woman who presented to an outside hospital after syncopizing and was found to have a left temporal mass. Her past medical history was significant only for hyperlipidemia and remote seizure disorder (mFI of 0, non-frail or robust). Upon transfer she underwent left-sided craniotomy and temporal lobectomy for resection of an intracranial mass, which was later revealed to be glioblastoma. Postoperatively the patient had mild word-finding difficulties and lower extremity weakness, both of which improved significantly by discharge on POD6. She was discharged to acute rehab without any significant postoperative complications.

Patient B (Fig. 9.3) is a 65-year-old woman who was found to have a large right cerebellopontine angle mass on imaging after a fall. At the time she also complained of 1 year of gait instability and right-sided hearing loss. Her PMH was significant for HTN on medication, DM2, CHF, morbid obesity, previous TIAs, and a large abdominal hernia that had caused bowel ischemia status post

Table 9.3 Case illustrations of complications and outcomes in neurosurgery patients

	Patient A	Patient B
Pathology	Intracranial temporal mass – GBM	Intracranial CP angle mass – acoustic neuroma
Age	76	65
Presentation	Syncope	1-year gait instability and hearing loss, mass found on imaging after fall
Comorbidities	Seizures, HLD (mFI = 0)	HTN, DM2, CHF, TIA, abdominal hernia with bowel ischemia requiring ostomy formation (mFI = 4)
mFI	0.00	0.36
Medications at time of admission	Acidophilus, atorvastatin, levetiracetam, metoprolol succinate	Metformin, lisinopril, terbinafine, ventolin prn
BMI	20.9	46.0
Surgery	Left craniotomy and temporal lobectomy for mass resection	Left retrosigmoid craniotomy for resection of schwannoma with placement of lumbar drain
Outcome	Recovered well. At discharge had mild word-finding difficulties. Post-op mild weakness improved before discharge. Moving all extremities with full strength at discharge	Required intubation overnight postoperatively. Recovered well but had 3/6 right-sided facial paralysis Discharged to acute rehab on POD 11 but was readmitted 2 weeks later with wound dehiscence requiring reoperation
Complications	None	Readmission and reoperation for surgical site infection
Length of stay	6 days	11 days, 24 days after readmission
Discharge disposition	Acute rehab	Acute rehab, readmitted and then discharged to subacute rehab

Age is often not predictive of complications or outcomes in neurosurgery patients, such as with these two patients

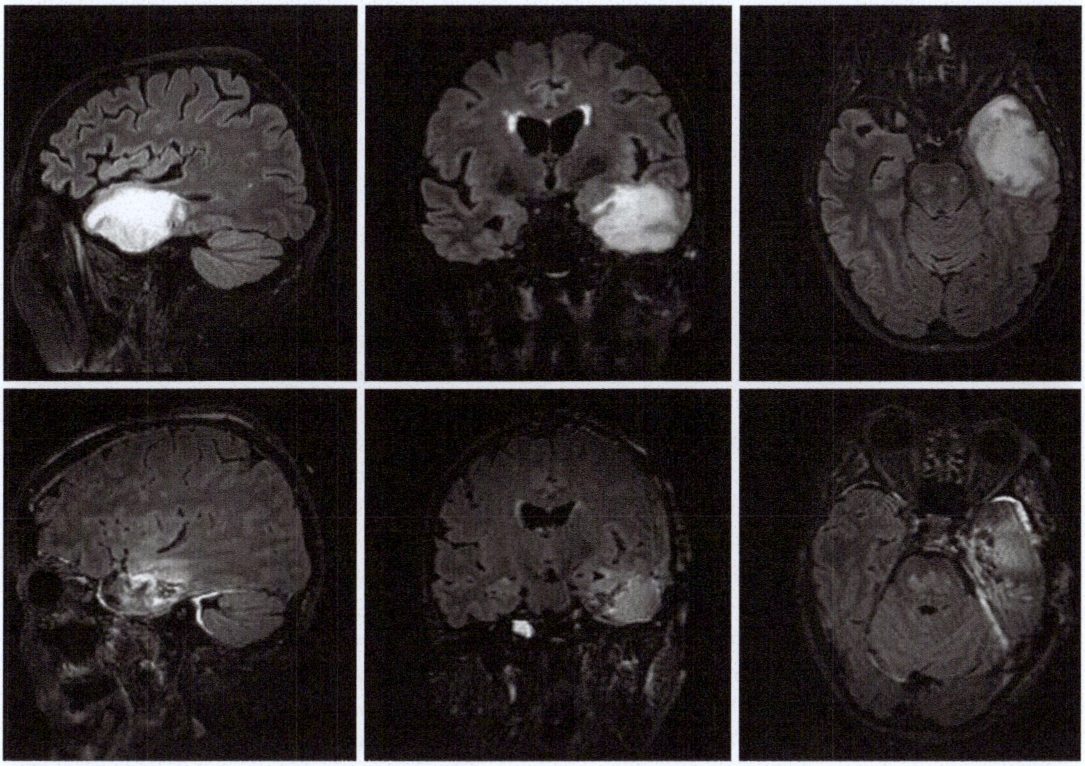

Fig. 9.2 Patient A, a 76-year-old woman. Preoperative (left row) and postoperative imaging (middle row) demonstrating resection of a large temporal GBM

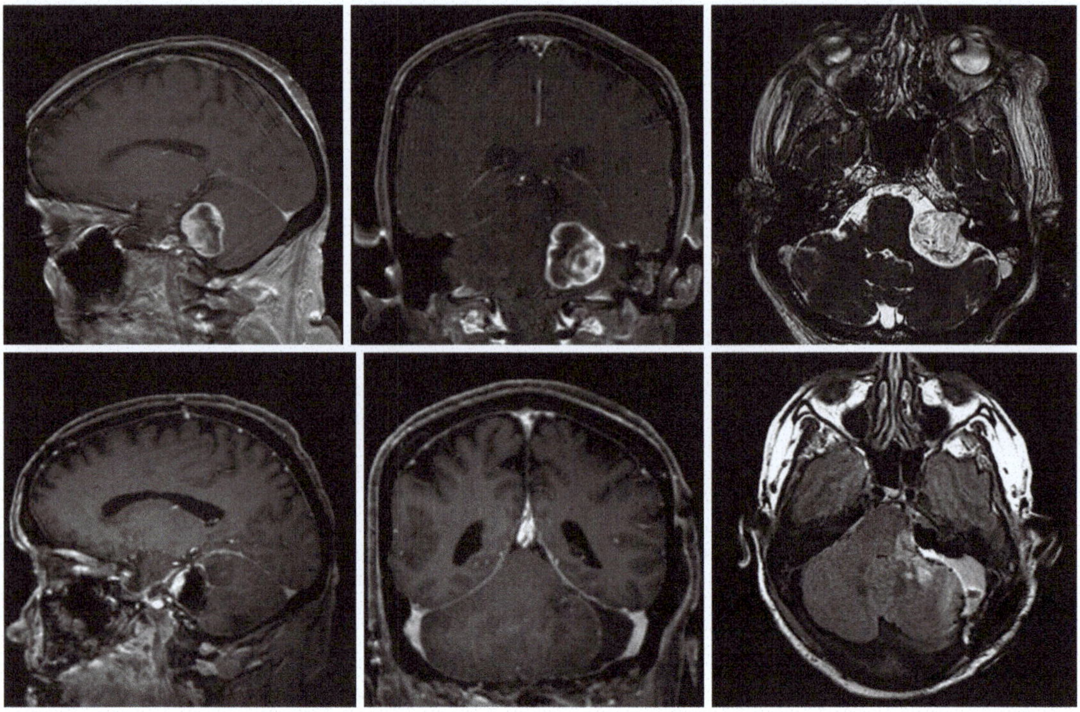

Fig. 9.3 Patient B, a 65-year-old woman. Preoperative (left row) and postoperative imaging (middle row) demonstrating near total resection of a vestibular schwannoma at the cerebellar-pontine angle

permanent ostomy diversion (mFI of 4, severe frailty). She was admitted for elective craniotomy and tumor resection. Her surgery went well with good resection of the mass. She remained intubated overnight and had transient 3/6 facial paralysis postoperatively, which eventually returned to normal. On postoperative day 11, she was discharged to acute rehab. Two weeks later she was readmitted with fevers, leukocytosis, and purulent discharge from her surgical site. She underwent reoperation for wound exploration and washout. She was treated with prolonged courses of vancomycin and aztreonam and was discharged to subacute rehab 24 days later.

Centenarian Case Reports

There are three case reports in the literature of centenarians undergoing neurosurgical procedures with good outcomes: a 101-year-old with an intraspinal meningioma, a 102-year-old with an acute subdural hematoma, and a 103-year-old with a thoracic compression fracture. These reports support our assertion that there should not be any absolute limit for neurosurgical intervention based on age alone.

101-Year-Old: Resection of Intraspinal Meningioma

In 2008, Cavanaugh et al. [64] published their case of a 101-year-old woman with an intraspinal meningioma found after presenting with leg spasms, bilateral leg weakness, stocking-type sensory loss distal to the knee bilaterally, and loss of ankle jerk reflexes. She had no other focal neurologic deficits. Imaging demonstrated an enhancing mass anterolateral to the spinal cord spanning from C7 to T1 that was compressing the spinal cord into a crescent shape. Surgery was offered due to the high risk of developing paraplegia without treatment. She then underwent laminectomy and resection of intradural mass that was adherent to the spinal cord. Her postoperative course was uneventful with recovery of strength in bilateral legs and no new neurologic deficits. On postoperative day 4, MRI confirmed no residual tumor and re-expansion of the compressed spinal cord.

At the time of presentation, this patient was living independently, and her only medical illness was mild congestive heart failure, which was well-compensated on medication (mFI = 1). Her low preoperative frailty likely contributed to her good outcome despite undergoing an invasive procedure.

103-Year-Old: Vertebroplasty for Thoracic Compression Fracture

In 2012 Kale et al. [65] published a case out of Turkey of a 103-year-old woman who successfully underwent vertebroplasty for a thoracic compression fracture. She presented with severe back pain refractory to conservative treatment 2 months after falling on her back. She had no focal neurologic deficits and exam only significant for tenderness to palpation at mid-thoracic spine. CT and MRI demonstrated T7 compression fracture. She underwent cement vertebroplasty with continuous fluoroscopic guidance under general anesthesia and was discharged on postoperative day 2 with significant improvement in her pain. Preoperative Visual Analogue Scale score was 9 and had decreased to 3 immediately postoperatively. On postoperative day 10, her VAS score was 1, and at 9-month follow-up, the patient reported no complications and required no medications for pain control.

Importantly, this patient was reportedly completely healthy preoperatively (mFI = 0). She was living independently without help and was taking no medications. Her injury was limited to one vertebral body and was not associated with any neurologic deficits. The procedure she underwent, percutaneous transpedicular vertebroplasty, can be performed with low complication rates in most patients. If any surgeon was presented this pathology unaware of the patient's age, they would offer treatment, but if the same set of clinicians were first presented the case knowing the age, then very few would offer treatment.

102-Year-Old: 2× Craniotomy and Evacuation of Subdural Hematoma

In 2007, Vyas et al. [66] presented the case of a 102-year-old woman who underwent two craniotomies for evacuation of two distinct acute subdural hemorrhages and who recovered all premorbid neurologic function and continued to do well through 7-year follow-up to the age of 110. This patient presented with mild drowsiness and right-sided hemiparesis after a fall. Head CT demonstrated acute SDH 2 cm in thickness. After being observed her neurologic condition deteriorated and her family opted for surgical intervention. She underwent left frontotemporal-parietal craniotomy and evacuation of subdural hematoma and was discharged to a rehabilitation after 11 days in the hospital and to home 2 weeks later. She recovered all premorbid neurologic function. Six months later, at the age of 103, she presented again after a fall with a GCS of 11. Head CT again demonstrated a large acute SDH, and the family again opted for surgery. After right-sided craniotomy and evacuation of SDH, the patient recovered to her baseline state of health and neurologically continued to do well through 7 years of follow-up.

This patient was in good premorbid condition relative to her age. Her past medical history at the time of presentation was significant only for impaired vision, difficulty walking, recurrent UTIs, and mild memory problems (mFI = 1). Her frailty level would likely be mild or prefail, depending on which scale was used to evaluate. This level of frailty is lower than would be expected when compared to her peers and even patients decades younger. Nonsurgical treatment would have likely resulted in severe neurological deficits or mortality. This patient continued to have good quality of life and no prolonged effects of surgery through 7 years of follow-up. There should be no neurosurgical treatment cutoff based solely on age.

Conclusions

The older neurosurgical literature regarding outcomes in elderly patients was extremely heterogeneous, and the results were conflicting. However, the baseline patient characteristics (frailty, comorbidities, etc.) were frequently not used to differentiate the patients, and clinicians were not as aggressive in treating these older patients sometimes. Furthermore, elderly patients from previous eras may not be representative of today's elderly patients as the baseline health of older patients continues to improve each decade. A Danish study analyzed how two cohorts of nonagenarians born 10 years apart scored in activities of daily living and cognitive tests [67]. The cohort born later scored significantly higher, suggesting that today's elderly population may not be equivalent to the elderly that were studied in the past.

Future prospective research needs to be performed that separates elderly neurosurgical patients by their baseline frailty or comorbidity status and by their specific pathology. This way "apples" can be compared to "apples" instead of "oranges" like the older literature when all patients >65 were lumped into one "elderly" category irrespective of their vastly different states of health that would lead to obvious and dramatic differences in outcomes with any potential treatments. Elderly non-frail patients can do great with certain neurosurgical procedures, while extremely frail elderly patients will often do poorly with significant invasive surgeries. The literature is replete with examples of non-frail elderly patients achieving excellent outcomes, and further research should continue trying to identify and then predict which patients will do well for each neurosurgical pathology subtype.

References

1. Dujovny M, Charbel F, Berman SK, Diaz FG, Malik G, Ausman JI. Geriatric neurosurgery. Surg Neurol. 1987;28(1):10–6.
2. Administration on aging. 2010 census data on aging [Internet]. 2011. Available from: http://www.aoa.gov/aoaroot/aging_statistics/Census_Population/census2010/Index.aspx.
3. U.S. Census Bureau. Projections of the population by selected age groups and sex for the United States: 2010 to 2050 [Internet]. 2008. Available from: http://www.census.gov/population/www/projections/summarytables.html
4. Chibbaro S, Di Rocco F, Makiese O, Mirone G, Marsella M, Lukaszewicz AC, et al. Neurosurgery and elderly: analysis through the years. Neurosurg Rev. 2010;34(2):229–34.
5. Nilasena DS, Vaughn RJ, Mori M, Lyon JL. Surgical trends in the treatment of diseases of the lumbar spine in Utah's Medicare population, 1984 to 1990. Med Care. 1995;33(6):585–97.
6. Johannesen TB, Angell-Andersen E, Tretli S, Langmark F, Lote K. Trends in incidence of brain and central nervous system tumors in Norway, 1970-1999. Neuroepidemiology. 2004;23(3):101–9.
7. Deorah S, Lynch CF, Sibenaller ZA, Ryken TC. Trends in brain cancer incidence and survival in the United States: Surveillance, Epidemiology, and End Results Program, 1973 to 2001. Neurosurg Focus. 2006;20(4):E1.
8. Sacco RL, Wolf PA, Bharucha NE, Meeks SL, Kannel WB, Charette LJ, et al. Subarachnoid and intracerebral hemorrhage: natural history, prognosis, and precursive factors in the Framingham Study. Neurology. 1984;34(7):847–54.
9. Munro PT. Effect of patients' age on management of acute intracranial haematoma: prospective national study. BMJ. 2002;325(7371):1001.
10. Gulati S, Jakola AS, Johannesen TB, Solheim O. Survival and treatment patterns of glioblastoma in the elderly: a population-based study. World Neurosurg. 2012;78(5):518–26.
11. Cheng JS. Elderly issues: our society getting older. Congress Quarterly. 2013.
12. McGirt MJ. Prolonged medical management of lumbar spine disorders in the elderly: questioning its value, an argument for surgery. Congress Quarterly. 2013.
13. Deyo RA, Mirza SK, Martin BI, Kreuter W, Goodman DC, Jarvik JG. Trends, major medical complications, and charges associated with surgery for lumbar spinal stenosis in older adults. JAMA. 2010;303(13):1259–65.
14. Weinstein JN, Tosteson TD, Lurie JD, Tosteson ANA, Blood E, Hanscom B, et al. Surgical versus nonsurgical therapy for lumbar spinal stenosis. N Engl J Med. 2008;358(8):794–810.
15. Kanter AS, Asthagiri AR, Shaffrey CI. Aging spine: challenges and emerging techniques. Clin Neurosurg. 2007;54:10–8.
16. Hulme PA, Krebs J, Ferguson SJ, Berlemann U. Vertebroplasty and kyphoplasty: a systematic review of 69 clinical studies. Spine. 2006;31(17):1983–2001.
17. Glassman SD, Carreon LY, Dimar JR, Campbell MJ, Puno RM, Johnson JR. Clinical outcomes in older

patients after posterolateral lumbar fusion. Spine J. 2007;7(5):547–51.
18. Pirracchio R, Resche-Rigon M, Bresson D, Basta B, Welschbillig S, Heyer L, et al. One-year outcome after neurosurgery for intracranial tumor in elderly patients. J Neurosurg Anesthesiol. 2010;22(4):342–6.
19. Hutchins LF, Unger JM, Crowley JJ, Coltman CA, Albain KS. Underrepresentation of patients 65 years of age or older in cancer-treatment trials. N Engl J Med. 1999;341(27):2061–7.
20. Jordan JT, Gerstner ER, Batchelor TT, Cahill DP, Plotkin SR. Glioblastoma care in the elderly. Cancer. 2016;122(2):189–97.
21. Yano S, Kuratsu J. Kumamoto brain tumor research group. Indications for surgery in patients with asymptomatic meningiomas based on an extensive experience. J Neurosurg. 2006;105(4):538–43.
22. Chamoun R, Krisht KM, Couldwell WT. Incidental meningiomas. Neurosurg Focus. 2011;31(6):E19.
23. Bateman BT, Pile-Spellman J, Gutin PH, Berman MF. Meningioma resection in the elderly: nationwide inpatient sample, 1998–2002. Neurosurgery. 2005;57(5):866–72; discussion 866–872.
24. Ikawa F, Kinoshita Y, Takeda M, Saito T, Yamaguchi S, Yamasaki F, et al. Review of current evidence regarding surgery in elderly patients with meningioma. Neurol Med Chir (Tokyo). 2017;57(10):521–33.
25. Gondim JA, Almeida JP, de Albuquerque LAF, Gomes E, Schops M, Mota JI. Endoscopic endonasal transsphenoidal surgery in elderly patients with pituitary adenomas. J Neurosurg. 2015;123(1):31–8.
26. Glasscock ME, Pappas DG, Manolidis S, Von Doersten PG, Jackson CG, Storper IS. Management of acoustic neuroma in the elderly population. Am J Otol. 1997;18(2):236–41; discussion 241–242.
27. Bowers CA, Gurgel RK, Brimley C, Hawryluk GWJ, Taggart M, Braden S, et al. Surgical treatment of vestibular schwannoma: does age matter? World Neurosurg. 2016;96:58–65.
28. Kuhne CA, Ruchholtz S, Kaiser GM, Nast-Kolb D. Working group on multiple trauma of the German Society of Trauma. Mortality in severely injured elderly trauma patients–when does age become a risk factor? World J Surg. 2005;29(11):1476–82.
29. Stocchetti N, Paternò R, Citerio G, Beretta L, Colombo A. Traumatic brain injury in an aging population. J Neurotrauma. 2012;29(6):1119–25.
30. Susman M, DiRusso SM, Sullivan T, Risucci D, Nealon P, Cuff S, et al. Traumatic brain injury in the elderly: increased mortality and worse functional outcome at discharge despite lower injury severity. J Trauma. 2002;53(2):219–23; discussion 223–224.
31. Lau D, El-Sayed AM, Ziewacz JE, Jayachandran P, Huq FS, Zamora-Berridi GJ, et al. Postoperative outcomes following closed head injury and craniotomy for evacuation of hematoma in patients older than 80 years. J Neurosurg. 2012;116(1):234–45.
32. Flaada JT, Leibson CL, Mandrekar JN, Diehl N, Perkins PK, Brown AW, et al. Relative risk of mortality after traumatic brain injury: a population-based study of the role of age and injury severity. J Neurotrauma. 2007;24(3):435–45.
33. Langlois JA, Rutland-Brown W, Wald MM. The epidemiology and impact of traumatic brain injury: a brief overview. J Head Trauma Rehabil. 2006;21(5):375–8.
34. Evans JA, Bailey M, Vail A, Tyrrell PJ, Parry-Jones AR, Patel HC. A simple tool to identify elderly patients with a surgically important acute subdural haematoma. Injury. 2015;46(1):76–9.
35. Howard MA, Gross AS, Dacey RG, Winn HR. Acute subdural hematomas: an age-dependent clinical entity. J Neurosurg. 1989;71(6):858–63.
36. Taussky P, Hidalgo ET, Landolt H, Fandino J. Age and salvageability: analysis of outcome of patients older than 65 years undergoing craniotomy for acute traumatic subdural hematoma. World Neurosurg. 2012;78(3–4):306–11.
37. Cagetti B, Cossu M, Pau A, Rivano C, Viale G. The outcome from acute subdural and epidural intracranial haematomas in very elderly patients. Br J Neurosurg. 1992;6(3):227–31.
38. Raj R, Mikkonen ED, Kivisaari R, Skrifvars MB, Korja M, Siironen J. Mortality in elderly patients operated for an acute subdural hematoma: a surgical case series. World Neurosurg. 2016;88:592–7.
39. Pilitsis J, Atwater B, Warden D, Deck G, Carroll J, Smith J, et al. Outcomes in octogenarians with subdural hematomas. Clin Neurol Neurosurg. 2013;115(8):1429–32.
40. Stippler M, Ramirez P, Berti A, Macindoe C, Villalobos N, Murray-Krezan C. Chronic subdural hematoma patients aged 90 years and older. Neurol Res. 2013;35(3):243–6.
41. Dobran M, Marini A, Nasi D, Liverotti V, Benigni R, Costanza MD, et al. Clinical outcome of patients over 90 years of age treated for chronic subdural hematoma. J Korean Neurosurg Soc. 2019; https://doi.org/10.3340/jkns.2018.0011.
42. Lee L, Ker J, Ng HY, Munusamy T, King NKK, Kumar D, et al. Outcomes of chronic subdural hematoma drainage in nonagenarians and centenarians: a multicenter study. J Neurosurg. 2016;124(2):546–51.
43. Brinjikji W, Rabinstein AA, Lanzino G, Kallmes DF, Cloft HJ. Effect of age on outcomes of treatment of unruptured cerebral aneurysms: a study of the National Inpatient Sample 2001–2008. Stroke. 2011;42(5):1320–4.
44. Lubicz B, Leclerc X, Gauvrit J-Y, Lejeune J-P, Pruvo J-P. Endovascular treatment of ruptured intracranial aneurysms in elderly people. AJNR Am J Neuroradiol. 2004;25(4):592–5.
45. Zhang Q-R, Zhang X, Wu Q, Shi J-X, Wang H-D, Hang C-H, et al. The impact of microsurgical clipping and endovascular coiling on the outcome of cerebral aneurysms in patients over 60 years of age. J Clin Neurosci. 2012;19(8):1115–8.
46. Gonzalez NR, Dusick JR, Duckwiler G, Tateshima S, Jahan R, Martin NA, et al. Endovascular coiling of intracranial aneurysms in elderly patients:

report of 205 treated aneurysms. Neurosurgery. 2010;66(4):714–20; discussion 720–721.
47. Barker FG, Amin-Hanjani S, Butler WE, Hoh BL, Rabinov JD, Pryor JC, et al. Age-dependent differences in short-term outcome after surgical or endovascular treatment of unruptured intracranial aneurysms in the United States, 1996-2000. Neurosurgery. 2004;54(1):18–28; discussion 28–30.
48. Cai Y, Spelle L, Wang H, Piotin M, Mounayer C, Vanzin JR, et al. Endovascular treatment of intracranial aneurysms in the elderly: single-center experience in 63 consecutive patients. Neurosurgery. 2005;57(6):1096–102; discussion 1096–1102.
49. Sedat J, Dib M, Lonjon M, Litrico S, Von Langsdorf D, Fontaine D, et al. Endovascular treatment of ruptured intracranial aneurysms in patients aged 65 years and older: follow-up of 52 patients after 1 year. Stroke. 2002;33(11):2620–5.
50. Winn HR, Jane JA, Taylor J, Kaiser D, Britz GW. Prevalence of asymptomatic incidental aneurysms: review of 4568 arteriograms. J Neurosurg. 2002;96(1):43–9.
51. Wachter D, Hans F, Kreitschmann-Andermahr I, Rohde V. Lower incidence of transcranial Doppler and symptomatic vasospasm after aneurysmal subarachnoid hemorrhage and aneurysm clipping in the elderly patient? Neurosurgery. 2011;69(2):261–6; discussion 266–267.
52. Farhat JS, Velanovich V, Falvo AJ, Horst HM, Swartz A, Patton JH, et al. Are the frail destined to fail? Frailty index as predictor of surgical morbidity and mortality in the elderly. J Trauma Acute Care Surg. 2012;72(6):1526–30; discussion 1530–1531.
53. Fried LP, Tangen CM, Walston J, Newman AB, Hirsch C, Gottdiener J, et al. Frailty in older adults: evidence for a phenotype. J Gerontol A Biol Sci Med Sci. 2001;56(3):M146–56.
54. Iqbal J, Denvir M, Gunn J. Frailty assessment in elderly people. Lancet. 2013;381(9882):1985–6.
55. Song X, Mitnitski A, Rockwood K. Prevalence and 10-year outcomes of frailty in older adults in relation to deficit accumulation. J Am Geriatr Soc. 2010;58(4):681–7.
56. Velanovich V, Antoine H, Swartz A, Peters D, Rubinfeld I. Accumulating deficits model of frailty and postoperative mortality and morbidity: its application to a national database. J Surg Res. 2013;183(1):104–10.
57. Segal DN, Wilson JM, Staley C, Michael KW. The 5-item modified frailty index is predictive of 30-day postoperative complications in patients undergoing kyphoplasty vertebral augmentation. World Neurosurg. 2018;116:e225–31.
58. Shin JI, Kothari P, Phan K, Kim JS, Leven D, Lee NJ, et al. Frailty index as a predictor of adverse postoperative outcomes in patients undergoing cervical spinal fusion. Spine. 2017;42(5):304–10.
59. Flexman AM, Charest-Morin R, Stobart L, Street J, Ryerson CJ. Frailty and postoperative outcomes in patients undergoing surgery for degenerative spine disease. Spine J. 2016;16(11):1315–23.
60. Miller EK, Neuman BJ, Jain A, Daniels AH, Ailon T, Sciubba DM, et al. An assessment of frailty as a tool for risk stratification in adult spinal deformity surgery. Neurosurg Focus. 2017;43(6):E3.
61. Youngerman BE, Neugut AI, Yang J, Hershman DL, Wright JD, Bruce JN. The modified frailty index and 30-day adverse events in oncologic neurosurgery. J Neuro-Oncol. 2018;136(1):197–206.
62. Cloney M, D'Amico R, Lebovic J, Nazarian M, Zacharia BE, Sisti MB, et al. Frailty in geriatric glioblastoma patients: a predictor of operative morbidity and outcome. World Neurosurg. 2016;89:362–7.
63. Imaoka Y, Kawano T, Hashiguchi A, Fujimoto K, Yamamoto K, Nishi T, et al. Modified frailty index predicts postoperative outcomes of spontaneous intracerebral hemorrhage. Clin Neurol Neurosurg. 2018;175:137–43.
64. Cavanaugh DA, Jawahar A, Lee JA, Wilkinson K, Kerr EJ, Nunley PD. Intraspinal meningioma in a 101-year old: should age determine the aggressiveness of intervention? Surg Neurol. 2008;69(2):130–4.
65. Kale A, Aydoğmuş E, Gun EG, Açıkgoz B. The oldest alive vertebroplasty patient with thoracic compression fracture. J Clin Case Rep. 2015;5(619):2.
66. Vyas NA, Chicoine MR. Extended survival after evacuation of subdural hematoma in a 102-year-old patient: case report and review of the literature. Surg Neurol. 2007;67(3):314–6; discussion 317.
67. Christensen K, Thinggaard M, Oksuzyan A, Steenstrup T, Andersen-Ranberg K, Jeune B, et al. Physical and cognitive functioning of people older than 90 years: a comparison of two Danish cohorts born 10 years apart. Lancet. 2013;382(9903):1507–13.

10 Caring for the Geriatrics Trauma Patient: The Challenges and the Opportunities

Abbas Smiley and Rifat Latifi

It is expected that by 2050, almost 40% of trauma hospital admissions will be older adults [1]. Well-established and well-coordinated efforts in proper triage, preoperative, operative, and postoperative care of elderly trauma patients are crucial to decreasing preventable morbidity and mortality. While prevention is the most important component to reducing trauma in the elderly, this chapter will focus on general principles of management of trauma in elderly based on the physiology and mechanism of injury [2]. Additionally, creating the infrastructure for caring for elderly trauma patients such as geriatric consult service has proven critical. Preoperative considerations, such as time to surgery and nutritional status of the patient, are described in this chapter. Implementation of standardized perioperative assessment regimens has been shown to reduce poor outcomes and is discussed in this chapter as well. Consideration of anesthetic agents as part of the strategy to optimize care for elderly trauma patients is very important. Palliative care while in the ICU has received increased attention and acceptance in recent years since many elderly patients prefer this form of care over life-extending therapies. Finally, implementation of a discharge process that effectively addresses follow-up care solutions is the final component to proper care of elderly geriatric patients.

A. Smiley
Department of Surgery, Westchester Medical Center and New York Medical College, School of Medicine, Valhalla, NY, USA

R. Latifi (✉)
Department of Surgery, New York Medical College, School of Medicine and Westchester Medical Center Health Network, Valhalla, NY, USA
e-mail: Rifat.Latifi@wmchealth.org

The Burden of Trauma in Elderly

Currently, trauma is the fifth leading cause of death in people aged ≥65 years [3, 4]. Within the next 15 years, elderly trauma will consume about $34 billion dollars annually [5]. Falls are recognized as the leading cause of trauma-related mortality and morbidity in elderly [6], which occurs in about 30–35% of elders annually [7–9]. Road traffic accidents comprised about one-fourth of elderly trauma [10], and its average fatality rate is about 9% [10]. Elders have 16% higher chance of road traffic accidents compared to those 25- to 64-year-old drivers [11]. The burden of elderly trauma on the healthcare system and national resources is enormous [12–16]. By 2050, almost 40% of trauma admissions will be older adults [1].

Trends and Risk Factors

In the United States, trauma is the leading cause of death for people up to the age of 45 [17] and fifth leading cause of death in people aged

R. Latifi (ed.), *Surgical Decision Making in Geriatrics*,
https://doi.org/10.1007/978-3-030-47963-3_10

≥65 years as mentioned above. After age 65, the characteristics of trauma change dramatically. While elderly people are still involved in motor vehicle accidents, falls are the leading cause of injury. Outcomes from injuries such as in hip fractures can be devastating, because they further reduce the quality of life for a patient who may already be frail. As people get older, the risk of falling increases because of frailty, weakness, chronic disorders and long-term use of medications, and cognitive, vision, and hearing problems. Moreover, living in the homes where they grew up (stairs and sleeping room on the second or third floor) and not modified for their age poses a major problem that should be addressed.

Several other factors for higher incidence of trauma among elders and their higher rates of morbidity and mortality exist; however, the mere aging process is the strongest one [18]. Constant declines in function of different organs predispose the elderly to various injuries even by low-energy level trauma. Osteoporosis, muscle atrophy, decreased subcutaneous tissue, osteoarthritis, renal, respiratory, and cardiovascular dysfunction, neurologic deficits, polypharmacy, and chronic medications are all parts of the aging process especially when combined with inappropriate lifestyle [1, 16]. History of fall is number one predictor for the increased risk of falls [19]. Females fall more, but males have worse outcome [20].

Other major mechanisms of injuries are road traffic accidents, elder abuse, and blunt/penetrating trauma [21]. Elder drivers with chronic disorders such as obesity, depression, or anxiety have higher risk of road traffic accidents [22, 23] and also higher mortality and morbidity following trauma [24]. Age distribution of pedestrian accidents in all age groups below 60 years usually follows the age distribution of the whole population. But the probability of pedestrian accidents in elders is double to triple of that in younger population [25, 26], and the risk of mortality is also 2–5 times higher [27, 28]. Pedestrian accidents are the worst mechanism of elderly trauma in terms of final outcome [29].

Elderly people constitute significant proportion of drivers on the road and have distinctive risk patterns as drivers. Chronic disorders, take multiple medications, fatigue, and sleep insufficiency are the main factors leading to reduced concentration and lower performance in elders [25, 30, 31]. In a long-term study, males had fewer falls and more road traffic accidents, whereas females had higher risk of more severe injuries. The latter could be explained by more frequent and more severe osteoporosis in elderly women [32, 33]. Elder men had significantly higher chance of being run over [34]. Marriage and education were protective factors against road traffic accidents and falls [32, 35]. Road traffic accidents were more frequent during weekends, mornings, and evening [36]. Fatal accidents were three times greater at night [37]. Driving in the country roads and rural areas had higher risk of road traffic accidents than in highways [32, 33, 38]. Road traffic accidents associated with elders occurred more frequently at intersections [25]. Alcohol was also associated with higher chance of road traffic accidents and falls [32].

Primary Prevention

Continuous weakening in various cellular functions in aging limits the response capacity to trauma. Demineralization and osteoporosis lead to weak bone which predisposes to insufficiency fractures and more severe fractures compared to the young population by minor trauma. Older adults are further predisposed to severe fractures by poorer mobility due to degenerative joint disease and stiffness of the spinal column [39]. Fall-related all extremity fractures, pelvic fractures and its major subtype, lateral compression injuries, and the related complications such as bleeding are more common in older adults than in younger ones. Traffic-associated limb injuries are also more common in elderly [40–42]. Poorer mobility also delays healing and increases the chance of venous thromboembolism [43]. Injuries in elderly victims are more severe and fatal [44–46] especially in those ≥75 years [29, 47]. So, age is the strongest predictor of mortality and morbidity in elderly trauma [29]; the older the victim, the higher the risk of mortality and morbidity. This age-related pattern was observed for both motorcyclists and pedestrians [48]. Measures that can potentially reduce or prevent harm include maintaining or strengthening the functions of musculoskeletal, cardiovascular, respiratory, renal, gastrointestinal, and neurological systems as well as ensuring proper levels of vitamin D, proper nutrition, physical activity, and enough sleep because they all prevent osteoporosis. Osteoporosis is often underdiagnosed. Screening for osteoporosis is a must, and education plays a key role [49].

Smart homes combined with remote health system monitoring can provide important health information and preventive measures for elderly population. Many required medical services could also be delivered to elderly patient in the comfort of their homes. Smart cloths, smart cameras, smart watches, and the associated health apps are the major components of smart homes [50]. More than 95% of falls could be accurately detected and promptly managed through smart homes and remote health system monitoring [51]. Smart homes provide the faster medical response, which can significantly reduce the mortality rate. In fact, the mortality rate of elderly trauma can be reduced significantly by just 10-minute faster medical response [52].

Secondary Prevention, Timely Diagnosis, and Proper Care

The aims of secondary prevention are timely and correct diagnosis of trauma and prompt surgical/non-surgical treatment. Early and correct diagnosis of the severity of elderly trauma patient is difficult, and, consequently, proper treatment could be delayed or incomplete. Lack of ability to demonstrate tachycardia due to beta-blocker medications, or simply inability to mount response to pain, could be detrimental in delaying the diagnosing major injuries. Incidence of fever in elderly patients is lower, and they are more often hyperglycemic and azotemic [53]. Decreases in renal, respiratory, and cardiovascular function further limit the response to trauma in hemodynamically compromised patients [1, 16]. These factors are especially important in internal bleeding following pelvic fracture, chest trauma, rib fractures, abdominal trauma, epidural/subdural hematoma, and cervical spinal cord injuries [54–56]. Other comorbid conditions may be hidden at the time of presentation. No awareness of comorbid conditions such as heart failure by healthcare providers may further complicate the management of patients [57].

Age >60 years, by itself, is a significant risk factor for bleeding [58]. Neurologic deficits such as reduced pain sensation, visual/hearing loss, and cognitive declines add more limits to the picture of hemodynamic response to trauma. Occult presentations of intracranial hemorrhage are more common in older adults [59]. Moreover, older adults take long-term medications such as anticoagulants, antihypertensives, antipsychotics, and corticosteroids. This polypharmacy may further complicate trauma care in aged population and change or limit their response to administered treatments. A clear examples are beta-blockers which decrease the proper response of sympathetic nervous system to trauma [60–62]. Therefore, the outcome in elderly trauma is multifactorial and more complex than the outcome of trauma in other age groups.

Given all the abovementioned issues, triage in elderly trauma to identify high-risk victims is difficult, but one should err on over-triage. Under-triage significantly increases the mortality in elderly trauma [41, 63]. Neither traditional severity scores such as the injury severity score and revised trauma score nor alternative outcome scores such as geriatric trauma outcome score and shock indices provide optimum measurement of triage in elderly trauma [5, 64–67]. Similarly, although CT scan is the primary imaging measure in geriatric trauma [41], the utility of repeated head CT scan in traumatic brain injuries has not proven useful. Repeated head CT scan is mainly valuable in unexaminable victims or those with neurologic deterioration [68]. Elder victims with head injury have higher risk of mortality specially within the first-day post-trauma. Age and Glasgow Coma Scale are the most important prognostic factors in these patients [69].

Given the complexity of elderly trauma, there are five keys that should be applied in practice in order to help with the management of geriatric trauma.

1. *Age alone as an indicator for trauma activation*. Many centers use vital signs and trauma mechanism as a venue for trauma team activation, whereas none of them could be a good indicator of severity of trauma. Many old patients with trauma have normal vital signs [70]. The mortality rate of elderly patients with mild, moderate, and severe trauma (injury severity score <15, 15–29, and >29, respectively) is more than triple of that in younger population [71]. Similarly, the mortality rate of elderly patients with mild, moderate, and severe traumatic brain injury (Glasgow Coma Scale <9, 9–12, and >12, respectively) is sextuple, triple, and double of that in younger population, respectively [72]. Health outcomes and cost-effectiveness will be improved if we treat trauma patients aged ≥70-year old at trauma centers, irrespective of their vital signs or trauma mechanism [73]. Then, a fully active and experienced trauma team will evaluate their frailty. Consequently, mortality will decrease dramatically compared to the non-trauma centers [74].
2. *Identification of frailty*: Evidence shows that identification of frailty and quantifying the frailty score are an important factor in trauma management in elders [47, 55, 75–77]. Frailty indices have been discussed extensively in Chap. 7, but one that has shown promising results in elderly trauma is Trauma-Specific Frailty Index. Its validity and reliability have been superior to the abovementioned scores. Frailty indices may also be able to predict the poor outcomes in elderly trauma [47, 55, 75–77]. Proper frailty estimate leads also the healthcare professional to decide whether the elder should be managed in a trauma center or not. In an interesting study conducted at American College Level I trauma center, starting on a specific date, all elderly patients aged ≥70 years were managed by the highest-level trauma activation system upon ER admission regardless of injury mechanism (933 patients). Patients admitted and managed for the last 2 years before the specific date were considered as the control group (1271 patients). The intervention group had 60% higher chance of staying in ER for 2 hours and shorter and also had 30% less mortality compared to the control group [78]. We know the importance of time in elderly trauma management. For example, fat embolism following long bone fractures may happen within the first 72 hours after long bone fractures. Definitive fixation of fractures within the first 48 hours after trauma is associated with uncomplicated hospital course [79]. Referral to Level I trauma centers or centers that treat a high proportion of older adults can significantly reduce the mortality rate in elderly trauma [5, 28, 64, 80]. In a recent cohort study, the mortality rates of elderly trauma patients admitted to trauma centers were compared with those admitted to non-trauma centers in seven counties in Oregon and Washington in 2011 and followed up for a year. More than 80% of all deaths occurred in patients managed by non-trauma centers [74]. Improving the qualities of medical and surgical care and using more advanced technologies in hospitals will enhance the efficiency of secondary prevention. The higher the quality of trauma center, the lower the years of lost life and the years lived with disability. Overall, the proper frailty assessment and management in a trauma center is the first important key in elderly trauma.
3. *Identification and evaluation of coagulation profile*: The third key in elderly trauma management is to know the coagulation profile of the patient at the time of injury and whether patient was taking anticoagulants before trauma or not [81–86]. The coagulation profile should be assessed in all elderly trauma victims especially those who were taking anticoagulation medications before injury. Since pre-trauma anticoagulants and antiplatelets significantly increase the rate of mortality following elderly trauma [81], quick reversal of anticoagulation is essential in preventing the progress of traumatic brain

injury and further internal bleeding [63]. Shock in elderly trauma is an ominous sign, and hypotension increases the odds of in-hospital death, especially in brain injuries [87]. Elderly trauma victims with bleeding, hypotension, or shock should be treated aggressively and monitored invasively [88–90]. If the elders were receiving warfarin before trauma, INR-correcting treatments should be started in less than 2 hours, and their INR should be fully corrected (<1.6 of normal) within 4 hours using fresh frozen plasma and intravenous vitamin K [81–86]. Also, patients on anticoagulants with suspected head injuries when there is headache, nausea and vomiting, reduced Glasgow Coma Scale, or external trauma should have CT scan to assess their possible head injury. Bleeding is also part of the picture in most patients with pelvic fracture. MRI can identify 99% of insufficiency fractures in pelvis and proximal femur [91] whereas CT scan better identifies periprosthetic fractures [92]. The amount of bleeding, severity of fracture, mental status change, and mechanical ventilation can predict the mortality of pelvic fractures in elderly [42, 93, 94]. Sarcopenia and osteopenia can also predict 1-year mortality in elderly trauma patients [95]. In those with Abbreviated Injury Scale (AIS) less than 3, masseter sarcopenia and brain atrophy may accurately predict 1-year mortality [96]. Finally, in elderly trauma with Glasgow Coma Scale<8 and no improvement within 3 days of injury, more aggressive treatments are better to be limited [78].

4. *ICU monitoring and management*: Should every elder trauma patient be transferred to ICU for close monitoring and management? The answer to this question provides the fourth key in elderly trauma care management. In a systematic review conducted by Eastern Association for the Surgery of Trauma (EAST) which covered 90 related references, it was revealed that the elderly patients with AIS ≥3 in one body system or a base deficit of −6 mEq/L or less must be managed at ICU [78]. Correction of acid-base balance (pH to more than 7.25) within the first 8 hours after trauma is associated with fewer pulmonary complications [79]. Compared to pH, initial lactate could be a stronger predictor of pneumonia [79]. Also, blunt thoracic trauma in elderly patients is a reasonable indication for admission to ICU because it will significantly decrease complications, length of stay, and mortality rate [97].

Persistent and systematic monitoring of ICU patients, in terms of vital signs and physical exam, is crucial. Other adjuncts of monitoring include urinary output, central venous pressure, stroke volume variation, ECG, RASS score, CAM score, CBC, BMP, magnesium, phosphate and ionized calcium, and, in some patients, EEG. Patients with special signs and symptoms such as fever, cough, diarrhea, and rash need extra measures of monitoring such as LFT, CXR, and *Clostridium difficile* testing. Systematic assessment of drug interactions before prescribing any new medication is extremely important [98–101].

Several studies have evaluated the advantages and disadvantages of early vs. late tracheostomy in trauma patients admitted to ICU. Although few studies have shown no significant difference between them, most investigations have been in favor of early tracheostomy. It seems early tracheostomy improves the health outcomes in terms of nosocomial pneumonia, duration of mechanical ventilation, ICU length of stay, and hospital length of stay [102–122]. No study has shown a significant difference in terms of mortality rate between early and late tracheostomy in elderly trauma patients admitted to ICU.

The factors associated with poorer outcomes in victims of road traffic accidents admitted to ICU were old age, development of secondary systemic problems such as respiratory, circulatory, hepatic, and metabolic impairments, Glasgow Coma Scale less than 8, and subdural hematoma [123, 124]. Severe traumatic brain injury (Glasgow Coma Scale less than 9) in elderly patients is accompanied with more than 80% mortality rate or long-term placement facilities.

5. *Geriatric emergency service*: The fourth key in optimal management of elderly trauma patients includes participation of geriatric practitioners in their care as soon as the patient is admitted. This protocol may consist of standardized perioperative assessment regimens by both surgeons and geriatricians, expeditious surgical treatment, and continued surgical/geriatric care postoperatively, which will result in reductions in lengths of stay, ICU admissions, and hospital costs per patient [125–130]. In an interesting interventional study on older adults with hip fracture, 210 patients were assigned into two groups, orthogeriatric co-management group vs. orthopedic care with geriatric consultation group. In orthogeriatric co-management group, both orthopedic surgeons and geriatricians shared the responsibility of patient management in an integrated model of care. Patients in orthogeriatric co-management group showed significantly lower 1-year mortality, shorter hospital length of stay, and higher probability of having surgery within 2 days after admission [131]. Delay of the surgical intervention occurs due to arrhythmia, severe anemia, electrolyte disturbance, coagulopathy, uncontrolled hyperglycemia, uncontrolled heart failure, pneumonia, and sepsis [132]. Similar findings were reported by other centers from different countries using orthogeriatric co-management by conducting retrospective cohort studies [125–130].

Integration of Other Services in the Healthcare System

Emergency medical services: The faster the victim transports to trauma center, the better the prognosis of the victims. The closer the trauma care center, the lower the rates of mortality and morbidity. Transportation improvement and accessibility of more ambulance services for elder trauma victims are critical steps in improving trauma care [80, 123]. Ten-minute faster medical response can reduce the mortality rate by one-third in elderly trauma [52].

Pharmacy services: Development of more efficient and safer medications can save victims lives. For example, it has been estimated that if all trauma victims with significant bleeding around the world receive tranexamic acid on time, at least 100,000 premature deaths will be prevented annually [133]. This could be more lifesaving for elderly who has limited hemodynamic response. On the other hand, polypharmacy significantly increases the risk of falls, trauma, and fractures [134, 135]. Cardiovascular, psychotropic, and nonsteroidal anti-inflammatory drugs increase the risk of falls through dizziness, confusion, sedation, orthostatic hypotension, and sleep disturbance [135]. A population-based study showed more than 8 times increase in the risk of hip fracture in older adults who used ≥10 medications vs. those who used ≤1 medication. Elders aged ≥85 years who took ≥10 medications had 23 times higher risk of hip fracture compared to 65- to 74-year-old elders who took ≤1 medication [136]. Then, careful monitoring of medications and their side effects is an essential element of preventive measures in elderly trauma.

Rehabilitation services: Inpatient and outpatient rehabilitation facilities play important roles in recovery of elderly patients after trauma and retrieval of the pre-trauma quality of life. It is important to compare the following indices in every patient before and after injury in order to prepare the best rehabilitation plan: functional independence, vision, and memory status. It is further discussed in detail below.

Preoperative Considerations

Optimization of preoperative abnormalities is essential as long as it is possible. Elderly trauma patients have significantly more nutritional deficits than the younger patients after operations, and prolonged hospitalization worsens the nutritional deficits [137]. Measurements of initial and follow-up prealbumin level may provide important prognostic information about the risk of malnutrition and inflammation and the response to treatment in critically sick elderly trauma patients. Correction of prealbumin within 4 days after admission lowers injury severity score, shortens length of stay and ventilator days, and improves the survival [138]. Preparing a proper diet which covers essential nutrients for recovery has a vital role. Nutritional support should be initiated, if possible, in malnourished patient. Nutritional requirements of elderly population are further discussed in other chapters of this book (see Chap. 8 on nutritional support).

The sensitivity of elders to anesthetic medications is higher due to less functional reserve and limited renal, respiratory, and cardiovascular function response to stress (see Chap. 7 on anesthesia). Serum albumin level as the main protein binding to acidic drugs decreases in elderly. Total body water and hepatorenal metabolism decrease which leads to higher serum concentrations of medications. Total body fat increases which provide the higher volume of drug distribution and consequently, the longer drug action. Overall, elders need less medications, and they experience prolonged effects of medications.

Comorbid disorders and conditions such as diabetes mellitus, cardiovascular disease, depression, malnutrition, immobility, dehydration, and cognitive dysfunctions are more common in elders and associated with higher perioperative morbidity and poorer outcomes. Faster recovery and avoiding functional decline are essential objectives in elderly patients [139].

Postoperative Care

A recent study showed that 86% of patients with orthopedic trauma had sleep disturbance, and in about 55%, the sleep disturbance was very severe, similar to the level observed in major depression [140]. Similar findings of sleep disturbance have been observed in eye trauma in elderly [141]. Sleep disturbance not only postpones recovery, it also intensifies depression [142]. In order to reduce sleep disturbance, special attention should be paid to body pain and mental health since they are the main predictors of sleep quality in patients with orthopedic trauma [140, 143]. Depression also increases the morbidity and mortality and reduces the efficacy of rehabilitation [144]. Depression is not only common after trauma [145], it is also under-detected and under-treated in elderly population [146]. Counseling has a very positive impact on quality of life in elderly trauma patients [143].

Sleep insufficiency and depression are both predisposing factors for development of postoperative delirium [147–149]. Up to a third of elderly patients with trauma experience delirium postoperatively [150]. Delirium increases the mortality rate, the length of hospital stay, cognitive impairment at the time of discharge, and the probability of discharging to rehabilitation institutes or nursing home [150, 151]. In addition to age, other associated risk factors are length of surgery, American Society of Anesthesiologists classification >2 level, opioid use, and history of alcohol excess. Precipitating factors are polypharmacy, infections (meningitis, pneumonia, and UTI) systemic illnesses (hepatic failure, malignancy, pulmonary disease, CHF, and anemia), metabolic derangements (volume depletion, diabetes, serum albumin level, and vitamin B_{12} deficiency), and electrolyte disturbances (hyponatremia and hypercalcemia). Multicomponent interventions are necessary to tackle postoperative delirium in older adults [152–155]. Several systematic reviews have shown that improving sleep hygiene through measures such as bright light exposure, earplugs, and melatonin has huge positive effects on preventing delirium, whether postoperatively or in ICU [149, 156–160]. Other preventive measures of postoperative delirium are adequate pain management, dietary modification, exercise training, and hearing and visual aid [161]. Several medications have also shown promising effects such as dexmedetomidine, analgosedatives, rivastigmine, olanzapine, risperidone, and droperidol [162–165].

On the other hand, prolonged immobility triggers serious complications. Pressure ulcers, urinary tract infection, pneumonia, and venous thromboembolism are among the most important ones. Immobility-induced hypercalcemia and the associated complications could potentiate a worse outcome [166]. Orthostatic hypotension after prolonged immobility and the resultant falls and secondary trauma are other concerning risks [166]. Overall, functional mobility and the ability to conduct daily living activities are among the main predictors of long-term outcome [145]. Therefore, early mobility and early involvement of physical therapy are crucial for almost all cases [63]. Before the implementation of stroke center certification program, elderly trauma patients were more likely to discharge to skilled nursing facilities than to inpatient rehabilitation facilities. But, after the implementation of stroke center certification program, the trend has shifted toward more discharge to rehabilitation facilities rather than to nursing facilities. One of the eight domains of this program is focused on assessing and monitoring the mobility and the rehabilitation after surgeries [167].

Follow-up bone mineral density is a must, but it has not been always practiced properly [168]. In some instances, such as hip fracture, the risk of second hip fracture will persist for more than 10 years among the survivors. This means secondary prevention must be continued beyond the early postoperative care [169]. Dedicated fracture prevention service and clinic will reduce further fracture 2 years after the first incident [170, 171]. Elderly patients are more susceptible to postoperative complications such as stroke, low cardiac output, respiratory failure, and renal failure [172].

On the other hand, elderly patients with severe traumatic brain injury who are not improving in 72 hours and not responding to treatment should have limited aggressive treatments. Many elderly patients near the end of life select palliative intervention vs. life-extending therapies. Goals of end-of-life care and decision-making should be further discussed with families [173–175]. End-of-life care has been discussed in other chapters of this book. Palliative care is an important topic for elderly trauma but sometimes is not paid enough attention. In a recent study, 118 trauma surgeons were compared with 244 palliative care specialists. Trauma surgeons were less familiar with goals of end-of-life care, less likely reported high-quality training in perfuming conversations, and showed less interest in conducting related conversations. But they rated themselves better than the palliative care specialists in performing goals of care discussions with patients [174].

Trauma care in the elderly is in its primary stages of development. Specialized research projects dedicated to trauma and trauma care in the elderly are needed to provide evidence-based medicine on the table.

Post-Traumatic Quality of Life and Tertiary Prevention

The goal of tertiary prevention is to prevent or reduce post-injury and postoperative complications and progress the quality of life after discharge from hospital. Resuming the pre-trauma quality of life among elderly trauma patients might be difficult. Quality of life is significantly reduced after the trauma and even 3 months later [176]. Higher percentage of elderly trauma patients are eventually discharged to rehabilitation care or assisted living facilities compared to younger people [145, 167]. Age, severity of injury, number of comorbidities, fracture, length of stay in ICU, having Medicare, and ambulation status were significant predictors of discharging to rehabilitation care or assisted living facilities [177]. Furthermore, longer length of stay and discharge to rehabilitation care and assisted living facilities were the main predictors for hospital readmissions in elderly trauma [178, 179]. A recent meta-analysis showed that comprehensive geriatric care significantly improves the percentage of older adults who regained the same level of walking and daily living activities as before fracture [180]. Comorbid conditions and frailty play major roles in long-term quality of life [181]. For example, diabetic patients who experience traumatic brain injury have more than 3 times higher risk of dementia [182]. Multidimensional interventions in different aspects of lifestyle such as sleep, exercise, nutrition, recreation, and cognition are necessary to improve their quality of life [183]. Advanced follow-up studies are needed to reveal the pattern of quality of life and the areas that need more attention in rehabilitation in elderly trauma.

Summary

Trauma is the fifth leading cause of death in elders, and fall is the leading cause of trauma in this population. Osteoporosis, musculoskeletal weakness, osteoarthritis, renal, respiratory and cardiovascular dysfunction, neurologic deficits, chronic medications, and polypharmacy are all parts of the aging process. Prevention by vitamin D, proper nutrition, physical activity, and enough sleep is essential. Smart homes and remote healthcare monitoring carry further preventive and therapeutic benefits. When trauma occurred, the fast transport of elderly victim to the trauma center and proper triage become crucial. Employment of the following five keys will upgrade the patient management:

1. Age 70 is an indication for activation of trauma team.
2. Application of Trauma-Specific Frailty Index and admission in a trauma center to recognize the severity of frailty improve the health outcomes.
3. Assessment of the current coagulation profile, rapid reversal of anticoagulation effects, and aggressive management of shock are essential.
4. Transfer to ICU in case of AIS $\geq$3, blunt thoracic trauma, or a base deficit of −6 mEq/L or less is critical. Persistent and systematic monitoring of ICU patient is extremely important. Early tracheostomy shortens nosocomial pneumonia, duration of mechanical ventilation, and hospital/ICU length of stay.
5. Inclusion of geriatric practitioners to consult as soon as the patient is admitted is beneficial.

Minimizing anesthesia, medications, and increasing the mobility as much as possible following injury are most important factors. Faster recovery and avoiding functional decline are essential objectives in elderly operations. Resuming the pre-trauma quality of life among elderly trauma patients can be difficult. Higher percentage of victims of elderly trauma patients will be eventually discharged to rehabilitation care or assisted living facilities if early mobility and rehabilitation therapy are not implemented. Rehabilitation and physical therapy are critical parts of postoperative management, even in the ICU settings. Also, early involvements of counseling to treat sleep and depression are crucial. Multidimensional interventions in different aspects of lifestyle such as social life, nutrition, recreation, and cognition are necessary for long-term management. Many elderly patients near the end of life select palliative intervention vs. life-extending therapies. Goals of end-of-life care and decision-making should be further discussed with patients and families.

References

1. Banks SE, Lewis MC. Trauma in the elderly: considerations for anesthetic management. Anesthesiol Clin. 2013;31(1):127–39. https://doi.org/10.1016/j.anclin.2012.11.004.
2. Cubitt M, Downie E, Shakerian R, Lange PW, Cole E. Timing and methods of frailty assessments in geriatric trauma patients: a systematic review. Injury. 2019; https://doi.org/10.1016/j.injury.2019.07.026.
3. Siram SM, Sonaike V, Bolorunduro OB, et al. Does the pattern of injury in elderly pedestrian trauma mirror that of the younger pedestrian? J Surg Res. 2011;167(1):14–8.
4. Stevenson J. When the trauma patient is elderly. J Perianesth Nurs. 2004;19(6):392–400.
5. Reske-Nielsen C, Medzon R. Geriatric Trauma. Emerg Med Clin North Am. 2016;34(3):483–500. https://doi.org/10.1016/j.emc.2016.04.004.
6. Ambrose AF, Paul G, Hausdorff JM. Risk factors for falls among older adults: a review of the literature. Maturitas. 2013;75(1):51–61. https://doi.org/10.1016/j.maturitas.2013.02.009.
7. Yang C-S, Chen SC-C, Yang Y-C, Huang L-C, Guo H-R, Yang H-Y. Epidemiology and patterns of facial fractures due to road traffic accidents in Taiwan: a 15-year retrospective study. Traffic Inj Prev. 2017;18(7):724–9. https://doi.org/10.1080/15389588.2017.1309650.
8. Liu-Ambrose T, Davis JC, Hsu CL, et al. Action seniors! - secondary falls prevention in community-dwelling senior fallers: study protocol for a randomized controlled trial. Trials. 2015;16:144. https://doi.org/10.1186/s13063-015-0648-7.

9. Sherrington C, Fairhall NJ, Wallbank GK, et al. Exercise for preventing falls in older people living in the community. Cochrane Database Syst Rev. 2019;1:CD012424. https://doi.org/10.1002/14651858.CD012424.pub2.
10. Azami-Aghdash S, Aghaei MH, Sadeghi-Bazarghani H. Epidemiology of Road Traffic Injuries among Elderly People; A Systematic Review and Meta-Analysis. Bull Emerg Trauma. 2018;6(4):279–91. https://doi.org/10.29252/beat-060403.
11. Loughran DS, Seabury SA, Zakaras L. What Risks Do Older Drivers Pose to Traffic Safety? https://www.rand.org/pubs/research_briefs/RB9272.html. Published 2007. Accessed 20 Aug 2019.
12. Chan LYL, Moran JL, Clarke C, Martin J, Solomon PJ. Mortality and cost outcomes of elderly trauma patients admitted to intensive care and the general wards of an Australian tertiary referral hospital. Anaesth Intensive Care. 2009;37(5):773–83.
13. Mackenzie EJ, Morris JJ, Smith GS, Fahey M. Acute hospital costs of trauma in the United States: implications for regionalized systems of care. J Trauma. 1990;30(9):1096–101.
14. Ross N, Timberlake GA, Rubino LJ, Kerstein MD. High cost of trauma care in the elderly. South Med J. 1989;82(7):857–9.
15. Young JS, Cephas GA, Blow O. Outcome and cost of trauma among the elderly: a real-life model of a single-payer reimbursement system. J Trauma Acute Care Surg. 1998;45(4):800–4.
16. Bonne S, Schuerer DJE. Trauma in the older adult: epidemiology and evolving geriatric trauma principles. Clin Geriatr Med. 2013;29(1):137–50. https://doi.org/10.1016/j.cger.2012.10.008.
17. Rhee P, Joseph B, Pandit V, et al. Increasing trauma deaths in the United States. Ann Surg. 2014;260(1):13–21. https://doi.org/10.1097/SLA.0000000000000600.
18. Carpenter CR, Avidan MS, Wildes T, Stark S, Fowler SA, Lo AX. Predicting geriatric falls following an episode of emergency department care: a systematic review. Acad Emerg Med. 2014;21(10):1069–82. https://doi.org/10.1111/acem.12488.
19. Jin J. Prevention of falls in Older adults. JAMA. 2018;319(16):1734. https://doi.org/10.1001/jama.2018.4396.
20. El-Menyar A, Tilley E, Al-Thani H, Latifi R. Females fall more from heights but males survive less among a geriatric population: insights from an American level 1 trauma center. BMC Geriatr. 2019;19(1):238. https://doi.org/10.1186/s12877-019-1252-6.
21. Labib N, Nouh T, Winocour S, et al. Severely injured geriatric population: morbidity, mortality, and risk factors. J Trauma. 2011;71(6):1908–14. https://doi.org/10.1097/TA.0b013e31820989ed.
22. Alavi SS, Mohammadi MR, Souri H, Mohammadi Kalhori S, Jannatifard F, Sepahbodi G. Personality, driving behavior and mental disorders factors as predictors of road traffic accidents based on logistic regression. Iran J Med Sci. 2017;42(1):24–31.
23. Joseph B, Hadeed S, Haider AA, et al. Obesity and trauma mortality: sizing up the risks in motor vehicle crashes. Obes Res Clin Pract. 2017;11(1):72–8. https://doi.org/10.1016/j.orcp.2016.03.003.
24. Joseph B, Zangbar B, Haider AA, et al. Hips don't lie: waist-to-hip ratio in trauma patients. J Trauma Acute Care Surg. 2015;79(6):1055–61. https://doi.org/10.1097/TA.0000000000000876.
25. Abou-Raya S, ElMeguid LA. Road traffic accidents and the elderly. Geriatr Gerontol Int. 2009;9(3):290–7. https://doi.org/10.1111/j.1447-0594.2009.00535.x.
26. Abdulrazzaq H, Zarour A, El-Menyar A, et al. Pedestrians: the daily underestimated victims on the road. Int J Inj Control Saf Promot. 2013;20(4):374–9. https://doi.org/10.1080/17457300.2012.748811.
27. Keller JM, Sciadini MF, Sinclair E, O'Toole RV. Geriatric trauma: demographics, injuries, and mortality. J Orthop Trauma. 2012;26(9):e161–5. https://doi.org/10.1097/BOT.0b013e3182324460.
28. Zafar SN, Obirieze A, Schneider EB, et al. Outcomes of trauma care at centers treating a higher proportion of older patients: the case for geriatric trauma centers. J Trauma Acute Care Surg. 2015;78(4):852–9. https://doi.org/10.1097/TA.0000000000000557.
29. Ang BH, Chen WS, Lee SWH. Global burden of road traffic accidents in older adults: a systematic review and meta-regression analysis. Arch Gerontol Geriatr. 2017;72:32–8. https://doi.org/10.1016/j.archger.2017.05.004.
30. Bilban M. Road traffic accidents caused by elderly drivers. Coll Antropol. 1997;21(2):573–83.
31. Orriols L, Philip P, Moore N, et al. Benzodiazepine-like hypnotics and the associated risk of road traffic accidents. Clin Pharmacol Ther. 2011;89(4):595–601. https://doi.org/10.1038/clpt.2011.3.
32. Raina P, Sohel N, Oremus M, et al. Assessing global risk factors for non-fatal injuries from road traffic accidents and falls in adults aged 35-70 years in 17 countries: a cross-sectional analysis of the prospective urban rural epidemiological (PURE) study. Inj Prev. 2016;22(2):92–8. https://doi.org/10.1136/injuryprev-2014-041476.
33. Ernstberger A, Joeris A, Daigl M, et al. Decrease of morbidity in road traffic accidents in a high income country - an analysis of 24,405 accidents in a 21 year period. Injury. 2015;46(Suppl 4):S135–43. https://doi.org/10.1016/S0020-1383(15)30033-4.
34. Fernandes CM, Boing AC. Pedestrian mortality in road traffic accidents in Brazil: time trend analysis, 1996-2015. Epidemiol Serv Saude. 2019;28(1):e2018079. https://doi.org/10.5123/S1679-49742019000100021.
35. Spoerri A, Egger M, von Elm E. Swiss National Cohort Study. Mortality from road traffic accidents in Switzerland: longitudinal and spatial analyses. Accid Anal Prev. 2011;43(1):40–8. https://doi.org/10.1016/j.aap.2010.06.009.
36. Mascarenhas MDM, Souto RMCV, Malta DC, Silva MMA. da, Lima CM de, Montenegro

M de MS. characteristics of motorcyclists involved in road traffic accidents attended at public urgent and emergency services. Cien Saude Colet. 2016;21(12):3661–71. https://doi.org/10.1590/1413-812320152112.24332016.
37. Growing Older, and Adjusting to the Dark - The New York Times. https://www.nytimes.com/2007/03/13/health/13brody.html?mtrref=undefined&gwh=9D85479C8AB096D8063051ACB18C039D&gwt=pay&assetType=REGIWALL. Accessed 20 2019.
38. Xie S-H, Wu Y-S, Liu X-J, et al. Mortality from road traffic accidents in a rapidly urbanizing Chinese city: a 20-year analysis in Shenzhen, 1994-2013. Traffic Inj Prev. 2016;17(1):39–43. https://doi.org/10.1080/15389588.2015.1035370.
39. Touger M, Gennis P, Nathanson N, et al. Validity of a decision rule to reduce cervical spine radiography in elderly patients with blunt trauma. Ann Emerg Med. 2002;40(3):287–93. https://doi.org/10.1067/mem.2002.125708.
40. Mahdian M, Fazel MR, Sehat M, Khosravi G, Mohammadzadeh M. Epidemiological profile of extremity fractures and dislocations in road traffic accidents in Kashan, Iran: a glance at the related disabilities. Arch Bone Jt Surg. 2017;5(3):186–92.
41. Sadro CT, Sandstrom CK, Verma N, Gunn ML. Geriatric trauma: a Radiologist's guide to imaging trauma patients aged 65 years and Older. Radiographics. 2015;35(4):1263–85. https://doi.org/10.1148/rg.2015140130.
42. Henry SM, Pollak AN, Jones AL, Boswell S, Scalea TM. Pelvic fracture in geriatric patients: a distinct clinical entity. J Trauma. 2002;53(1):15–20. https://doi.org/10.1097/00005373-200207000-00004.
43. Knudson MM, Ikossi DG, Khaw L, Morabito D, Speetzen LS. Thromboembolism after trauma: an analysis of 1602 episodes from the American College of Surgeons National Trauma Data Bank. Ann Surg. 2004;240(3):490–6.; discussion 496-498. https://doi.org/10.1097/01.sla.0000137138.40116.6c.
44. Aharonson-Daniel L, Avitzour M, Barell V. Road traffic accidents, circumstances, diagnoses and severity of injury--data from the Israeli National Trauma Registry. Harefuah. 2001;140(10):919–22. 991
45. Kourouma K, Delamou A, Lamah L, et al. Frequency, characteristics and hospital outcomes of road traffic accidents and their victims in Guinea: a three-year retrospective study from 2015 to 2017. BMC Public Health. 2019;19(1):1022. https://doi.org/10.1186/s12889-019-7341-9.
46. Valent F, Schiava F, Savonitto C, Gallo T, Brusaferro S, Barbone F. Risk factors for fatal road traffic accidents in Udine. Italy Accid Anal Prev. 2002;34(1):71–84.
47. Hashmi A, Ibrahim-Zada I, Rhee P, et al. Predictors of mortality in geriatric trauma patients: a systematic review and meta-analysis. J Trauma Acute Care Surg. 2014;76(3):894–901. https://doi.org/10.1097/TA.0b013e3182ab0763.
48. Etehad H, Yousefzadeh-Chabok S, Davoudi-Kiakalaye A, Moghadam Dehnadi A, Hemati H, Mohtasham-Amiri Z. Impact of road traffic accidents on the elderly. Arch Gerontol Geriatr. 2015;61(3):489–93. https://doi.org/10.1016/j.archger.2015.08.008.
49. van der Vet PCR, Kusen JQ, Rohner-Spengler M, et al. Secondary prevention of minor trauma fractures: the effects of a tailored intervention-an observational study. Arch Osteoporos. 2019;14(1):44. https://doi.org/10.1007/s11657-019-0595-0.
50. Guan K, Shao M, Wu S. A remote health monitoring system for the elderly based on smart home gateway. J Healthc Eng. 2017;2017:5843504. https://doi.org/10.1155/2017/5843504.
51. Yu M, Rhuma A, Naqvi SM, Wang L, Chambers J. A posture recognition based fall detection system for monitoring an elderly person in a smart home environment. IEEE Trans Inf Technol Biomed. 2012;16(6):1274–86. https://doi.org/10.1109/TITB.2012.2214786.
52. Sánchez-Mangas R, García-Ferrrer A, de Juan A, Arroyo AM. The probability of death in road traffic accidents. How important is a quick medical response? Accid Anal Prev. 2010;42(4):1048–56. https://doi.org/10.1016/j.aap.2009.12.012.
53. Frankenfield D, Cooney RN, Smith JS, Rowe WA. Age-related differences in the metabolic response to injury. J Trauma. 2000;48(1):49–56; discussion 56-57. https://doi.org/10.1097/00005373-200001000-00009.
54. Elmistekawy EM, Hammad AAM. Isolated rib fractures in geriatric patients. Ann Thorac Med. 2007;2(4):166–8. https://doi.org/10.4103/1817-1737.36552.
55. Joseph B, Pandit V, Rhee P, et al. Predicting hospital discharge disposition in geriatric trauma patients: is frailty the answer? J Trauma Acute Care Surg. 2014;76(1):196–200. https://doi.org/10.1097/TA.0b013e3182a833ac.
56. Stawicki SP, Grossman MD, Hoey BA, Miller DL, Reed JF III. Rib fractures in the elderly: a marker of injury severity. J Am Geriatr Soc. 2004;52(5):805–8.
57. Bergman Schieman K, Pattison K, Early C. Caring for trauma patients with coexisting heart failure. J Trauma Nurs. 2017;24(5):312–6. https://doi.org/10.1097/JTN.0000000000000316.
58. Samaras N, Chevalley T, Samaras D, Gold G. Older patients in the emergency department: a review. Ann Emerg Med. 2010;56(3):261–9. https://doi.org/10.1016/j.annemergmed.2010.04.015.
59. Rathlev NK, Medzon R, Lowery D, et al. Intracranial pathology in elders with blunt head trauma. Acad Emerg Med. 2006;13(3):302–7. https://doi.org/10.1197/j.aem.2005.10.015.
60. Pal R, Ghosh A, Kumar R, et al. Public health crisis of road traffic accidents in India: risk factor assess-

ment and recommendations on prevention on the behalf of the academy of family physicians of India. J Family Med Prim Care. 2019;8(3):775–83. https://doi.org/10.4103/jfmpc.jfmpc_214_18.
61. McCoy GF, Johnston RA, Duthie RB. Injury to the elderly in road traffic accidents. J Trauma. 1989;29(4):494–7. https://doi.org/10.1097/00005373-198904000-00013.
62. Atkins RM, Turner WH, Duthie RB, Wilde BR. Injuries to pedestrians in road traffic accidents. BMJ. 1988;297(6661):1431–4. https://doi.org/10.1136/bmj.297.6661.1431.
63. Adams SD, Holcomb JB. Geriatric trauma. Curr Opin Crit Care. 2015;21(6):520–6. https://doi.org/10.1097/MCC.0000000000000246.
64. Kozar RA, Arbabi S, Stein DM, et al. Injury in the aged: geriatric trauma care at the crossroads. J Trauma Acute Care Surg. 2015;78(6):1197–209. https://doi.org/10.1097/TA.0000000000000656.
65. Brown JB, Gestring ML, Forsythe RM, et al. Systolic blood pressure criteria in the National Trauma Triage Protocol for geriatric trauma: 110 is the new 90. J Trauma Acute Care Surg. 2015;78(2):352–9. https://doi.org/10.1097/TA.0000000000000523.
66. Barea-Mendoza JA, Chico-Fernández M, Sánchez-Casado M, et al. Predicting survival in geriatric trauma patients: a comparison between the TRISS methodology and the geriatric trauma outcome score. Cir Esp. 2018;96(6):357–62. https://doi.org/10.1016/j.ciresp.2018.02.014.
67. Ahl R, Phelan HA, Dogan S, Cao Y, Cook AC, Mohseni S. Predicting in-hospital and 1-year mortality in geriatric trauma patients using geriatric trauma outcome score. J Am Coll Surg. 2017;224(3):264–9. https://doi.org/10.1016/j.jamcollsurg.2016.12.011.
68. Haider AA, Rhee P, Orouji T, et al. A second look at the utility of serial routine repeat computed tomographic scans in patients with traumatic brain injury. Am J Surg. 2015;210(6):1088–93; discussion 1093-1094. https://doi.org/10.1016/j.amjsurg.2015.07.004.
69. El-Menyar A, Abdelrahman H, Al-Hassani A, et al. Clinical presentation and time-based mortality in patients with chest injuries associated with road traffic accidents. Arch Trauma Res. 2016;5(1):e31888. https://doi.org/10.5812/atr.31888.
70. Demetriades D, Sava J, Alo K, et al. Old age as a criterion for trauma team activation. J Trauma Acute Care Surg. 2001;51(4):754–7.
71. Taylor MD, Tracy JK, Meyer W, Pasquale M, Napolitano LM. Trauma in the elderly: intensive care unit resource use and outcome. J Trauma Acute Care Surg. 2002;53(3):407–14.
72. Leblanc J, de Guise E, Gosselin N, Feyz M. Comparison of functional outcome following acute care in young, middle-aged and elderly patients with traumatic brain injury. Brain Inj. 2006;20(8):779–90.
73. Hammer PM, Storey AC, Bell T, et al. Improving geriatric trauma outcomes: a small step toward a big problem. J Trauma Acute Care Surg. 2016;81(1):162–7. https://doi.org/10.1097/TA.0000000000001063.
74. Newgard CD, Caughey A, McConnell KJ, et al. Comparison of Injured Older Adults Included in vs Excluded From Trauma Registries With 1-Year Follow-up. JAMA Surg. 2019:e192279. https://doi.org/10.1001/jamasurg.2019.2279.
75. Joseph B, Pandit V, Zangbar B, et al. Validating trauma-specific frailty index for geriatric trauma patients: a prospective analysis. J Am Coll Surg. 2014, 219;(1):10–17.e1. https://doi.org/10.1016/j.jamcollsurg.2014.03.020.
76. Joseph B, Phelan H, Hassan A, et al. The impact of frailty on failure-to-rescue in geriatric trauma patients: a prospective study. J Trauma Acute Care Surg. 2016;81(6):1150–5. https://doi.org/10.1097/TA.0000000000001250.
77. Joseph B, Pandit V, Zangbar B, et al. Superiority of frailty over age in predicting outcomes among geriatric trauma patients: a prospective analysis. JAMA Surg. 2014;149(8):766–72. https://doi.org/10.1001/jamasurg.2014.296.
78. Calland JF, Ingraham AM, Martin N, et al. Evaluation and management of geriatric trauma: an eastern Association for the Surgery of trauma practice management guideline. J Trauma Acute Care Surg. 2012;73(5 Suppl 4):S345–50. https://doi.org/10.1097/TA.0b013e318270191f.
79. Vallier HA, Wang X, Moore TA, Wilber JH, Como JJ. Timing of orthopaedic surgery in multiple trauma patients: development of a protocol for early appropriate care. J Orthop Trauma. 2013;27(10):543–51. https://doi.org/10.1097/BOT.0b013e31829efda1.
80. Castillo-Manzano JI, Castro-Nuño M, Fageda X. Can health public expenditure reduce the tragic consequences of road traffic accidents? The EU-27 experience. Eur J Health Econ. 2014;15(6):645–52. https://doi.org/10.1007/s10198-013-0512-1.
81. Peck KA, Calvo RY, Schechter MS, et al. The impact of preinjury anticoagulants and prescription antiplatelet agents on outcomes in older patients with traumatic brain injury. J Trauma Acute Care Surg. 2014;76(2):431–6. https://doi.org/10.1097/TA.0000000000000107.
82. Coimbra R, Hoyt DB, Anjaria DJ, Potenza BM, Fortlage D, Hollingsworth-Fridlund P. Reversal of anticoagulation in trauma: a north-American survey on clinical practices among trauma surgeons. J Trauma. 2005;59(2):375–82. https://doi.org/10.1097/01.ta.0000174728.46883.a4.
83. Pieracci FM, Eachempati SR, Shou J, Hydo LJ, Barie PS. Degree of anticoagulation, but not warfarin use itself, predicts adverse outcomes after traumatic brain injury in elderly trauma patients. J Trauma Acute Care Surg. 2007;63(3):525–30.
84. Fortuna GR, Mueller EW, James LE, Shutter LA, Butler KL. The impact of preinjury antiplatelet and anticoagulant pharmacotherapy on outcomes in elderly patients with hemorrhagic brain injury. Surgery. 2008;144(4):598–605.

85. McMillian WD, Rogers FB. Management of prehospital antiplatelet and anticoagulant therapy in traumatic head injury: a review. J Trauma. 2009;66(3):942–50. https://doi.org/10.1097/TA.0b013e3181978e7b.
86. Ivascu FA, Howells GA, Junn FS, Bair HA, Bendick PJ, Janczyk RJ. Rapid warfarin reversal in anticoagulated patients with traumatic intracranial hemorrhage reduces hemorrhage progression and mortality. J Trauma. 2005;59(5):1131–7. discussion 1137-1139. https://doi.org/10.1097/01.ta.0000189067.16368.83.
87. Dams-O'Connor K, Cuthbert JP, Whyte J, Corrigan JD, Faul M, Harrison-Felix C. Traumatic brain injury among older adults at level I and II trauma centers. J Neurotrauma. 2013;30(24):2001–13. https://doi.org/10.1089/neu.2013.3047.
88. Haider AA, Azim A, Rhee P, et al. Substituting systolic blood pressure with shock index in the National Trauma Triage Protocol. J Trauma Acute Care Surg. 2016;81(6):1136–41. https://doi.org/10.1097/TA.0000000000001205.
89. Kim SY, Hong KJ, Shin SD, et al. Validation of the shock index, modified shock index, and age shock index for predicting mortality of geriatric trauma patients in emergency departments. J Korean Med Sci. 2016;31(12):2026–32. https://doi.org/10.3346/jkms.2016.31.12.2026.
90. McNab A, Burns B, Bhullar I, Chesire D, Kerwin A. An analysis of shock index as a correlate for outcomes in trauma by age group. Surgery. 2013;154(2):384–7. https://doi.org/10.1016/j.surg.2013.05.007.
91. Cabarrus MC, Ambekar A, Lu Y, Link TM. MRI and CT of insufficiency fractures of the pelvis and the proximal femur. AJR Am J Roentgenol. 2008;191(4):995–1001. https://doi.org/10.2214/AJR.07.3714.
92. Brady OH, Garbuz DS, Masri BA, Duncan CP. The treatment of periprosthetic fractures of the femur using cortical onlay allograft struts. Orthop Clin North Am. 1999;30(2):249–57.
93. Wang H, Phillips JL, Robinson RD, et al. Predictors of mortality among initially stable adult pelvic trauma patients in the US: data analysis from the National Trauma Data Bank. Injury. 2015;46(11):2113–7. https://doi.org/10.1016/j.injury.2015.08.039.
94. Roth T, Kammerlander C, Gosch M, Luger TJ, Blauth M. Outcome in geriatric fracture patients and how it can be improved. Osteoporos Int. 2010;21(Suppl 4):S615–9. https://doi.org/10.1007/s00198-010-1401-4.
95. Kaplan SJ, Pham TN, Arbabi S, et al. Association of Radiologic Indicators of frailty with 1-year mortality in Older trauma patients: opportunistic screening for sarcopenia and osteopenia. JAMA Surg. 2017;152(2):e164604. https://doi.org/10.1001/jamasurg.2016.4604.
96. Tanabe C, Reed MJ, Pham TN, Penn K, Bentov I, Kaplan SJ. Association of Brain Atrophy and Masseter Sarcopenia with 1-year mortality in Older trauma patients. JAMA Surg. 2019; https://doi.org/10.1001/jamasurg.2019.0988.
97. Pyke OJ, Rubano JA, Vosswinkel JA, McCormack JE, Huang EC, Jawa RS. Admission of elderly blunt thoracic trauma patients directly to the intensive care unit improves outcomes. J Surg Res. 2017;219:334–40. https://doi.org/10.1016/j.jss.2017.06.054.
98. Ely EW, Stephens RK, Jackson JC, et al. Current opinions regarding the importance, diagnosis, and management of delirium in the intensive care unit: a survey of 912 healthcare professionals. Crit Care Med. 2004;32(1):106–12. https://doi.org/10.1097/01.CCM.0000098033.94737.84.
99. Ely EW, Truman B, Shintani A, et al. Monitoring sedation status over time in ICU patients: reliability and validity of the Richmond agitation-sedation scale (RASS). JAMA. 2003;289(22):2983–91. https://doi.org/10.1001/jama.289.22.2983.
100. Aberg T, Hentschel J. Improved total quality by monitoring of a cardiothoracic unit. Medical, administrative and economic data followed for 9 years. Interact Cardiovasc Thorac Surg. 2004;3(1):33–40. https://doi.org/10.1016/S1569-9293(03)00218-4.
101. Brown CVR, Shoemaker WC, Wo CCJ, Chan L, Demetriades D. Is noninvasive hemodynamic monitoring appropriate for the elderly critically injured patient? J Trauma. 2005;58(1):102–7. https://doi.org/10.1097/01.ta.0000105990.05074.4a.
102. Affronti A, Casali F, Eusebi P, et al. Early versus late tracheostomy in cardiac surgical patients: a 12-year single center experience. J Cardiothorac Vasc Anesth. 2019;33(1):82–90. https://doi.org/10.1053/j.jvca.2018.05.041.
103. Ahmed N, Kuo Y-H. Early versus late tracheostomy in patients with severe traumatic head injury. Surg Infect. 2007;8(3):343–7. https://doi.org/10.1089/sur.2006.065.
104. Bickenbach J, Fries M, Offermanns V, et al. Impact of early vs. late tracheostomy on weaning: a retrospective analysis. Minerva Anesthesiol. 2011;77(12):1176–83.
105. Gandía-Martínez F, Martínez-Gil I, Andaluz-Ojeda D, Bobillo de Lamo F, Parra-Morais L, Díez-Gutiérrez F. Analysis of early tracheostomy and its impact on development of pneumonia, use of resources and mortality in neurocritically ill patients. Neurocirugia (Astur). 2010;21(3):211–21.
106. Hyde GA, Savage SA, Zarzaur BL, et al. Early tracheostomy in trauma patients saves time and money. Injury. 2015;46(1):110–4. https://doi.org/10.1016/j.injury.2014.08.049.
107. Jeon Y-T, Hwang J-W, Lim Y-J, Lee S-Y, Woo K-I, Park H-P. Effect of tracheostomy timing on clinical outcome in neurosurgical patients: early versus late tracheostomy. J Neurosurg Anesthesiol. 2014;26(1):22–6. https://doi.org/10.1097/ANA.0b013e31829770a0.
108. Kang BH, Cho J, Lee JC-J, Jung K. Early versus late tracheostomy in trauma patients: a propensity-matched cohort study of 5 years' data at a single

institution in Korea. World J Surg. 2018;42(6):1742–7. https://doi.org/10.1007/s00268-018-4474-4.
109. Koch T, Hecker B, Hecker A, et al. Early tracheostomy decreases ventilation time but has no impact on mortality of intensive care patients: a randomized study. Langenbeck's Arch Surg. 2012;397(6):1001–8. https://doi.org/10.1007/s00423-011-0873-9.
110. Lozano CP, Chen KA, Marks JA, Jenoff JS, Cohen MJ, Weinstein MS. Safety of early tracheostomy in trauma patients after anterior cervical fusion. J Trauma Acute Care Surg. 2018;85(4):741–6. https://doi.org/10.1097/TA.0000000000002045.
111. Miyoshi R, Chen-Yoshikawa TF, Hamaji M, et al. Effect of early tracheostomy on clinical outcomes in critically ill lung transplant recipients. Gen Thorac Cardiovasc Surg. 2018;66(9):529–36. https://doi.org/10.1007/s11748-018-0949-3.
112. Möller MG, Slaikeu JD, Bonelli P, Davis AT, Hoogeboom JE, Bonnell BW. Early tracheostomy versus late tracheostomy in the surgical intensive care unit. Am J Surg. 2005;189(3):293–6. https://doi.org/10.1016/j.amjsurg.2005.01.002.
113. Pinheiro B. Do V, Tostes R de O, Brum CI, Carvalho EV, Pinto SPS, Oliveira JCA de. Early versus late tracheostomy in patients with acute severe brain injury. J Bras Pneumol. 2010;36(1):84–91. https://doi.org/10.1590/s1806-37132010000100014.
114. Puentes W, Jerath A, Djaiani G, Cabrerizo Sanchez R, Wąsowicz M. Early versus late tracheostomy in cardiovascular intensive care patients. Anaesthesiol Intensive Ther. 2016;48(2):89–94. https://doi.org/10.5603/AIT.a2016.0016.
115. Sugerman HJ, Wolfe L, Pasquale MD, et al. Multicenter, randomized, prospective trial of early tracheostomy. J Trauma. 1997;43(5):741–7. https://doi.org/10.1097/00005373-199711000-00002.
116. Young D, Harrison DA, Cuthbertson BH, Rowan K. TracMan collaborators. Effect of early vs late tracheostomy placement on survival in patients receiving mechanical ventilation: the TracMan randomized trial. JAMA. 2013;309(20):2121–9. https://doi.org/10.1001/jama.2013.5154.
117. Zagli G, Linden M, Spina R, et al. Early tracheostomy in intensive care unit: a retrospective study of 506 cases of video-guided Ciaglia blue rhino tracheostomies. J Trauma. 2010;68(2):367–72. https://doi.org/10.1097/TA.0b013e3181a601b3.
118. Wang F, Wu Y, Bo L, et al. The timing of tracheotomy in critically ill patients undergoing mechanical ventilation: a systematic review and meta-analysis of randomized controlled trials. Chest. 2011;140(6):1456–65. https://doi.org/10.1378/chest.11-2024.
119. Diaz-Prieto A, Mateu A, Gorriz M, et al. A randomized clinical trial for the timing of tracheotomy in critically ill patients: factors precluding inclusion in a single center study. Crit Care. 2014;18(5):585. https://doi.org/10.1186/s13054-014-0585-y.
120. Correia IAM, Sousa V, Pinto LM, Barros E. Impact of early elective tracheotomy in critically ill patients. Braz J Otorhinolaryngol. 2014;80(5):428–34. https://doi.org/10.1016/j.bjorl.2014.07.008.
121. Terragni PP, Antonelli M, Fumagalli R, et al. Early vs late tracheotomy for prevention of pneumonia in mechanically ventilated adult ICU patients: a randomized controlled trial. JAMA. 2010;303(15):1483–9. https://doi.org/10.1001/jama.2010.447.
122. Schneider GT, Christensen N, Doerr TD. Early tracheotomy in elderly patients results in less ventilator-associated pneumonia. Otolaryngol Head Neck Surg. 2009;140(2):250–5. https://doi.org/10.1016/j.otohns.2008.11.006.
123. Lee YY, Fang E, Weng Y, Ganapathy S. Road traffic accidents in children: the "what", "how" and "why". Singapore Med J. 2018;59(4):210–6. https://doi.org/10.11622/smedj.2017114.
124. Chelly H, Bahloul M, Ammar R, et al. Clinical characteristics and prognosis of traumatic head injury following road traffic accidents admitted in ICU "analysis of 694 cases". Eur J Trauma Emerg Surg. 2019;45(2):245–53. https://doi.org/10.1007/s00068-017-0885-4.
125. Buecking B, Timmesfeld N, Riem S, et al. Early orthogeriatric treatment of trauma in the elderly: a systematic review and metaanalysis. Dtsch Arztebl Int. 2013;110(15):255.
126. Stenqvist C, Madsen CM, Riis T, et al. Orthogeriatric service reduces mortality in patients with hip fracture. Geriatr Orthop Surg Rehabil. 2016;7(2):67–73. https://doi.org/10.1177/2151458515625296.
127. Henderson CY, Shanahan E, Butler A, et al. Dedicated orthogeriatric service reduces hip fracture mortality. Ir J Med Sci. 2017;186(1):179–84. https://doi.org/10.1007/s11845-016-1453-3.
128. Pioli G, Bendini C, Pignedoli P, Giusti A, Marsh D. Orthogeriatric co-management - managing frailty as well as fragility. Injury. 2018;49(8):1398–402. https://doi.org/10.1016/j.injury.2018.04.014.
129. Della Rocca GJ, Moylan KC, Crist BD, Volgas DA, Stannard JP, Mehr DR. Comanagement of geriatric patients with hip fractures: a retrospective, controlled, cohort study. Geriatr Orthop Surg Rehabil. 2013;4(1):10–5. https://doi.org/10.1177/2151458513495238.
130. Kusen JQ, Schafroth B, Poblete B, et al. The implementation of a geriatric fracture Centre for hip fractures to reduce mortality and morbidity: an observational study. Arch Orthop Trauma Surg. 2019; https://doi.org/10.1007/s00402-019-03229-0.
131. Baroni M, Serra R, Boccardi V, et al. The orthogeriatric comanagement improves clinical outcomes of hip fracture in older adults. Osteoporos Int. 2019;30(4):907–16. https://doi.org/10.1007/s00198-019-04858-2.
132. Colquhoun AD, Zuelzer W, Butterworth JF. Improving the management of hip fractures in the elderly: a role for the perioperative surgical home?

Anesthesiology. 2014;121(6):1144–6. https://doi.org/10.1097/ALN.0000000000000479.
133. Roberts I. The truth about road traffic accidents. Br J Surg. 2012;99(Suppl 1):8–9. https://doi.org/10.1002/bjs.7786.
134. Frazier SC. Health outcomes and polypharmacy in elderly individuals: an integrated literature review. J Gerontol Nurs. 2005;31(9):4–11. https://doi.org/10.3928/0098-9134-20050901-04.
135. Glab KL, Wooding FGG, Tuiskula KA. Medication-related falls in the elderly: mechanisms and prevention strategies. Consult Pharm. 2014;29(6):413–7. https://doi.org/10.4140/TCP.n.2014.413.
136. Lai S-W, Liao K-F, Liao C-C, Muo C-H, Liu C-S, Sung F-C. Polypharmacy correlates with increased risk for hip fracture in the elderly: a population-based study. Medicine (Baltimore). 2010;89(5):295–9. https://doi.org/10.1097/MD.0b013e3181f15efc.
137. Wade CE, Kozar RA, Dyer CB, Bulger EM, Mourtzakis M, Heyland DK. Evaluation of nutrition deficits in adult and elderly trauma patients. JPEN J Parenter Enteral Nutr. 2015;39(4):449–55. https://doi.org/10.1177/0148607114523450.
138. Akella K, Ibrahim S, Chendrasekhar A. Initial prealbumin level and ability to normalize prealbumin are highly predictive of increased mortality and morbidity in elderly trauma patients. Inter J Sci Res. 2019;8(5). http://worldwidejournals.net/index.php/IJSR/article/view/44. Accessed 19 Sep 2019.
139. Kanonidou Z, Karystianou G. Anesthesia for the elderly. Hippokratia. 2007;11(4):175–7.
140. Swann MC, Batty M, Hu G, Mitchell T, Box H, Starr A. Sleep disturbance in Orthopaedic trauma patients. J Orthop Trauma. 2018;32(10):500–4. https://doi.org/10.1097/BOT.0000000000001276.
141. Moreira SHN, Kasahara N. Patterns of ocular trauma among the elderly in a south-American urban area and the association between eye traumas with sleep disorders. J Emerg Trauma Shock. 2017;10(3):111–5. https://doi.org/10.4103/JETS.JETS_90_16.
142. Smiley A, Wolter S, Nissan D. Mechanisms of Association of Sleep and Metabolic Syndrome. In. 2019.
143. Gambatesa M, D'Ambrosio A, D'Antini D, et al. Counseling, quality of life, and acute postoperative pain in elderly patients with hip fracture. J Multidiscip Healthc. 2013;6:335–46. https://doi.org/10.2147/JMDH.S48240.
144. Billig N, Ahmed SW, Kenmore PI. Approaches to senior care #1. Hip fracture, depression, and cognitive impairment: a follow-up study. Orthop Rev. 1988;17(3):315–20.
145. Devore S, Parli SE, Oyler DR, Bernard A. Comprehensive geriatric assessment for trauma: operationalizing the trauma quality improvement program directive. J Trauma Nurs. 2016;23(6):337–42. https://doi.org/10.1097/JTN.0000000000000244.
146. Blumstein T, Benyamini Y, Shmotkin D, Lerner-Geva L. Gender differences in the prevalence and correlates of psychotropic medication use among older adults in Israel. Isr J Psychiatry Relat Sci. 2014;51(2):118–25.
147. Thorsteinsdóttir SA, Sveinsdóttir H, Snædal J. Delirium after open cardiac surgery:systematic review of prevalence, risk factors and consequences. Laeknabladid. 2015;101(6):305–11. https://doi.org/10.17992/lbl.2015.06.31.
148. Aitken SJ, Blyth FM, Naganathan V. Incidence, prognostic factors and impact of postoperative delirium after major vascular surgery: a meta-analysis and systematic review. Vasc Med. 2017;22(5):387–97. https://doi.org/10.1177/1358863X17721639.
149. Lu Y, Li Y-W, Wang L, et al. Promoting sleep and circadian health may prevent postoperative delirium: a systematic review and meta-analysis of randomized clinical trials. Sleep Med Rev. 2019;48:101207. https://doi.org/10.1016/j.smrv.2019.08.001.
150. Salluh JIF, Wang H, Schneider EB, et al. Outcome of delirium in critically ill patients: systematic review and meta-analysis. BMJ. 2015;350:h2538. https://doi.org/10.1136/bmj.h2538.
151. Maagaard M, Barbateskovic M, Perner A, Jakobsen JC, Wetterslev J. Dexmedetomidine for the prevention of delirium in critically ill patients - a protocol for a systematic review. Acta Anaesthesiol Scand. 2019;63(4):540–8. https://doi.org/10.1111/aas.13313.
152. Zhu C, Wang B, Yin J, et al. Risk factors for postoperative delirium after spinal surgery: a systematic review and meta-analysis. Aging Clin Exp Res. 2019; https://doi.org/10.1007/s40520-019-01319-y.
153. Scholz AFM, Oldroyd C, McCarthy K, Quinn TJ, Hewitt J. Systematic review and meta-analysis of risk factors for postoperative delirium among older patients undergoing gastrointestinal surgery. Br J Surg. 2016;103(2):e21–8. https://doi.org/10.1002/bjs.10062.
154. Oberai T, Laver K, Crotty M, Killington M, Jaarsma R. Effectiveness of multicomponent interventions on incidence of delirium in hospitalized older patients with hip fracture: a systematic review. Int Psychogeriatr. 2018;30(4):481–92. https://doi.org/10.1017/S1041610217002782.
155. Trogrlić Z, van der Jagt M, Bakker J, et al. A systematic review of implementation strategies for assessment, prevention, and management of ICU delirium and their effect on clinical outcomes. Crit Care. 2015;19:157. https://doi.org/10.1186/s13054-015-0886-9.
156. Fadayomi AB, Ibala R, Bilotta F, Westover MB, Akeju O. A systematic review and meta-analysis examining the impact of sleep disturbance on postoperative delirium. Crit Care Med. 2018;46(12):e1204–12. https://doi.org/10.1097/CCM.0000000000003400.
157. Flannery AH, Oyler DR, Weinhouse GL. The impact of interventions to improve sleep on delirium in the ICU: a systematic review and research framework. Crit Care Med. 2016;44(12):2231–40. https://doi.org/10.1097/CCM.0000000000001952.

158. Zhang Q, Gao F, Zhang S, Sun W, Li Z. Prophylactic use of exogenous melatonin and melatonin receptor agonists to improve sleep and delirium in the intensive care units: a systematic review and meta-analysis of randomized controlled trials. Sleep Breath. 2019; https://doi.org/10.1007/s11325-019-01831-5.
159. Litton E, Carnegie V, Elliott R, Webb SAR. The efficacy of earplugs as a sleep hygiene strategy for reducing delirium in the ICU: a systematic review and meta-analysis. Crit Care Med. 2016;44(5):992–9. https://doi.org/10.1097/CCM.0000000000001557.
160. Groves RL. Increasing light exposure for the prevention of delirium: a systematic review. Dimens Crit Care Nurs. 2019;38(2):96–107. https://doi.org/10.1097/DCC.0000000000000343.
161. Janssen T, Alberts A, Hooft L, Mattace-Raso F, Mosk C, van der Laan L. Prevention of postoperative delirium in elderly patients planned for elective surgery: systematic review and meta-analysis. Clin Interv Aging. 2019;14:1095–117. https://doi.org/10.2147/CIA.S201323.
162. Liu Y, Li X-J, Liang Y, Kang Y. Pharmacological prevention of postoperative delirium: a systematic review and meta-analysis of randomized controlled trials. Evid Based Complement Alternat Med. 2019;2019:9607129. https://doi.org/10.1155/2019/9607129.
163. Wang C-T, Mao Y, Zhao L, Ma B. The impact of analgosedation on mortality and delirium in critically ill patients: a systematic review and meta-analysis. Intensive Crit Care Nurs. 2019;54:7–14. https://doi.org/10.1016/j.iccn.2019.06.004.
164. Sun R, Wang S, Li S, Yang C, Zhao Y, Luo A. Effects of dexmedetomidine on delirium and mortality during sedation in ICU patients: a systematic review and meta-analysis protocol. BMJ Open. 2019;9(4):e025850. https://doi.org/10.1136/bmjopen-2018-025850.
165. Cerveira CCT, Pupo CC, dos Santos SDS, Santos JEM. Delirium in the elderly: a systematic review of pharmacological and non-pharmacological treatments. Dement Neuropsychol. 2017;11(3):270–5. https://doi.org/10.1590/1980-57642016dn11-030009.
166. Mobily PR, Kelley LS. Iatrogenesis in the elderly: factors of immobility. J Gerontol Nurs. 1991;17(9):5–9. https://doi.org/10.3928/0098-9134-19910901-04.
167. Ayoung-Chee PR, Rivara FP, Weiser T, Maier RV, Arbabi S. Beyond the hospital doors: improving long-term outcomes for elderly trauma patients. J Trauma Acute Care Surg. 2015;78(4):837–43. https://doi.org/10.1097/TA.0000000000000567.
168. Baba T, Hagino H, Nonomiya H, et al. Inadequate management for secondary fracture prevention in patients with distal radius fracture by trauma surgeons. Osteoporos Int. 2015;26(7):1959–63. https://doi.org/10.1007/s00198-015-3103-4.
169. Sobolev B, Sheehan KJ, Kuramoto L, Guy P. Risk of second hip fracture persists for years after initial trauma. Bone. 2015;75:72–6. https://doi.org/10.1016/j.bone.2015.02.003.
170. Van der Kallen J, Giles M, Cooper K, et al. A fracture prevention service reduces further fractures two years after incident minimal trauma fracture. Int J Rheum Dis. 2014;17(2):195–203. https://doi.org/10.1111/1756-185X.12101.
171. Bell K, Strand H, Inder WJ. Effect of a dedicated osteoporosis health professional on screening and treatment in outpatients presenting with acute low trauma non-hip fracture: a systematic review. Arch Osteoporos. 2014;9:167. https://doi.org/10.1007/s11657-013-0167-7.
172. Albert A, Ujvari Z, Mauser M, Ennker J. Cardiac surgery in the elderly: perioperative care and operative strategies. Dtsch Med Wochenschr. 2008;133(46):2393–402.; quiz 2403-2406. https://doi.org/10.1055/s-0028-1100931.
173. Lilley EJ, Williams KJ, Schneider EB, et al. Intensity of treatment, end-of-life care, and mortality for older patients with severe traumatic brain injury. J Trauma Acute Care Surg. 2016;80(6):998–1004. https://doi.org/10.1097/TA.0000000000001028.
174. Cunningham HB, Scielzo SA, Nakonezny PA, et al. Trauma surgeon and palliative care physician attitudes regarding goals-of-care delineation for injured geriatric patients. Am J Hosp Palliat Care. 2019;36(8):669–74. https://doi.org/10.1177/1049909118823182.
175. O'Connell K, Maier R. Palliative care in the trauma ICU. Curr Opin Crit Care. 2016;22(6):584–90. https://doi.org/10.1097/MCC.0000000000000357.
176. Aitken LM, Davey TM, Ambrose J, Connelly LB, Swanson C, Bellamy N. Health outcomes of adults 3 months after injury. Injury. 2007;38(1):19–26. https://doi.org/10.1016/j.injury.2006.05.020.
177. James MK, Robitsek RJ, Saghir SM, Gentile PA, Ramos M, Perez F. Clinical and non-clinical factors that predict discharge disposition after a fall. Injury. 2018;49(5):975–82. https://doi.org/10.1016/j.injury.2018.02.014.
178. Crijns TJ, Caton T, Teunis T, et al. Longer length of stay increases 1-year readmission rate in patients undergoing hip fracture surgery. Arch Bone Jt Surg. 2018;6(6):492–500.
179. Strosberg DS, Housley BC, Vazquez D, Rushing A, Steinberg S, Jones C. Discharge destination and readmission rates in older trauma patients. J Surg Res. 2017;207:27–32. https://doi.org/10.1016/j.jss.2016.07.015.

180. Wang H, Li C, Zhang Y, et al. The influence of inpatient comprehensive geriatric care on elderly patients with hip fractures: a meta-analysis of randomized controlled trials. Int J Clin Exp Med. 2015;8(11):19815–30.
181. Abdulle AE, de Koning ME, van der Horn HJ, et al. Early predictors for long-term functional outcome after mild traumatic brain injury in frail elderly patients. J Head Trauma Rehabil. 2018;33(6):E59–67. https://doi.org/10.1097/HTR.0000000000000368.
182. Gilsanz P, Albers K, Beeri MS, Karter AJ, Quesenberry CP, Whitmer RA. Traumatic brain injury associated with dementia risk among people with type 1 diabetes. Neurology. 2018;91(17):e1611–8. https://doi.org/10.1212/WNL.0000000000006391.
183. Rangel EL. Future directions of geriatric trauma care: function and quality of life beyond survival. JAMA Surg. 2017;152(2):e164642. https://doi.org/10.1001/jamasurg.2016.4642.

ENT Surgery in the Elderly

11

Jaclyn Klimczak, Augustine L. Moscatello, and J. K. Rasamny

The geriatric age group is defined as individuals over the age of 65 and is one of the United States' fastest expanding populations. With the continuous expansion of medical technology, we have seen an increase in life expectancy occurring in congruence with the decline in birth rates. If the current trend continues, the expected census of people over the age of 65 by 2050 will be in excess of 83 million [1]. The change in demographic trends in the United States has an immense impact on the field of medicine and, more specifically, the practice of otolaryngology. Older individuals not only harbor acute and chronic medical problems specific to the geriatric population, but even their common problems need to be treated differently than their younger counterparts. Medical and surgical care for the geriatric population is also more complex because of the higher rate of comorbidities that must be considered. The aging process is burdened with deficits specific to the field of ENT that include but are not limited to hearing impairment, balance impairment, an increased risk of head and neck cancer, swallowing disorders, and voice concerns. These problems in turn contribute to a lack of independence and often social isolation as quality of life becomes compromised due to functional disability. It has been found that up to one-third of all health-related visits are related to complaints managed by otolaryngologists [2]. In response to the demands of this newly expanding population, it is imperative to understand the otolaryngologic disorders specific to this aging population and with that how to deliver an individualized treatment plan specific to their needs.

J. Klimczak
Department of Otolaryngology Head and Neck Surgery, Icahn School of Medicine at Mount Sinai, New York, NY, USA

A. L. Moscatello (✉)
Department of Otolaryngology/Head and Neck Surgery, New York Medical College, Valhalla, NY, USA

Department of Otolaryngology/Head and Neck Surgery, Westchester Medical Center, Valhalla, NY, USA
e-mail: Augustine.Moscatello@wmchealth.org

J. K. Rasamny
Department of Otolaryngology, Head and Neck Surgery, ENT Faculty Practice, Ardsley, NY, USA

Head and Neck Surgery in the Geriatric Population

Thyroid and Parathyroid Disease

Recently, the prevalence of both benign and malignant tumors of the thyroid has increased in concordance with increased utilization of diagnostic imaging. Statistics estimate about 50% of patients over the age of 65 years demonstrate nodules on ultrasound examination and about 90% of women present with thyroid nodules after

R. Latifi (ed.), *Surgical Decision Making in Geriatrics*,
https://doi.org/10.1007/978-3-030-47963-3_11

the age of 60 years, and 60% of men after the age of 80 years [3, 4]. Although the American Thyroid Association (ATA) continuously update their management guidelines on thyroid disorders, there are none that are specific to the geriatric population. However, the staging system of papillary thyroid cancer is stratified by age less than or over 55 to account for the less aggressive nature of these tumors in younger individuals [5]. That carries over to the literature on thyroid disease and surgery in the geriatric population with there being a paucity in literature regarding specific management.

In general, it has been noted that increasing age has been an independent factor associated with postoperative morbidity and mortality [6]. Thyroid and parathyroid surgery is considered an intermediate-risk surgical procedure, and data is conflicting regarding its risk in the geriatric population [7]. Although smaller single institution–based studies have demonstrated similar outcomes among patients regardless of age, larger population–based research has shown that the elderly are at a higher risk for longer hospitalization stays, higher readmission rates, and major systemic complications following thyroid and parathyroid surgery [3, 8–12]. Tuggle et al. reviewed readmission rates in geriatric patients using the Surveillance, Epidemiology, and End Results (SEER)–Medicare-linked database to identify patients 65 years with thyroid cancer who underwent thyroidectomy over a 5-year period. Rehospitalization rates were found to be at 8% in their population and associated with increased comorbidity, advanced stage, the number of lymph nodes, increased length of stay, and small hospital size. Unplanned readmission cost was greater than $5000 for a mean length of stay of 3.5 days and significantly associated with death at 1 year [8]. Readmission rates after thyroidectomy and/or parathyroidectomy are found to be lower in larger hospitals with postoperative discharge follow-ups to outpatient providers and among high-volume surgeons who perform greater than 30 thyroidectomies per year [8, 13].

The elderly are at increased risk of malignancy and harbor more aggressive thyroid tumors [4, 10, 11, 14]. Thyroidectomy is used to treat both benign and malignant disease. With increasing incidence due to better diagnostic tools, diagnosis and treatment become more important. Risks and benefits need to be carefully reviewed despite the increased safety profile of anesthesia and surgical procedures. For thyroid and parathyroid surgery, injury to the recurrent laryngeal nerve (RLN) is one of the most serious complications and intraoperative monitoring has become mainstay to avoid postoperative morbidity. In a patient population where nodules tend to be larger and more aggressive, and comorbidities are more prevalent, injury to the RLN are increased [15]. Calcified vessels and scar tissue can often be disorienting; therefore, intraoperative nerve monitoring is of great value when performing thyroid or parathyroid surgery in elderly patients. Postoperative RLN injury which can lead to symptoms of hoarseness, difficulty breathing, and dysphagia is often more difficult to overcome in the geriatric population and can lead to devastating morbidity and prolonged hospitalizations. As such, the weight of the benefits of surgery has to be carefully weighed against the risks.

Individual risk–benefit analyses must be performed and are even more important in the geriatric patient. Current guidelines stress the importance of surveillance in nodular thyroid disease and recommend against fine-needle aspiration (FNA) unless they meet criteria for size and suspicious characteristics on ultrasound. Those that do meet criteria and are found to be benign on FNA should also undergo routine imaging surveillance in 6–24 months depending on sonogram characteristics [5]. Studies have shown that although the incidence of papillary thyroid carcinoma increases with time, the incidence of nonpapillary thyroid carcinoma remains relatively constant, and increase in incidence of thyroid cancer during the last decades without concomitant rise in mortality may reflect the growing detection of indolent forms of thyroid cancer [16]. In a prospective study stratified by age that looked at disease progression in patients with papillary thyroid microcarcinoma, it was found that less than 2% of patients greater than 60 years of age had progression to clinical disease. These findings suggest that observation may play a large role in older patients with low-risk papillary microcarcinoma, thereby minimizing the risks of surgical interventions in the geriatric patient [17].

Parathyroid surgery has become more streamlined and effectively a safer procedure over the years due to the improvement in preoperative localization studies. Primary hyperparathyroidism has a peak incidence between 55 and 70 years of age and is associated with being a direct cause of osteoporosis, constipation, and mental fogging. In geriatric patients, complications from primary hyperparathyroidism such as osteoporosis and increased risk of fracture are devastating and appropriate treatment of the disease by surgical removal is imperative [18]. Because parathyroid surgery has become a fast, outpatient procedure with immediate cure and benefit, it is recommended even in older individuals after appropriate preoperative clearance.

With the increase in the prevalence of thyroid and parathyroid disease, there has been an increase in surgical procedures as a means of first-line treatment. In the geriatric population, surgical thyroid and parathyroidectomy can be safely and appropriately performed after careful risk–benefit assessments and preoperative counseling has been performed with the ultimate benefit of improving or prolonging each patient's life.

Head and Neck Oncology

Head and neck cancers represent 6% of all cancer diagnoses worldwide and account for 650,000 new cases and 350,000 deaths yearly [19]. That number accounts for 52,000 new cases diagnosed in the United States each year with the median age being 60 years old at the time of diagnosis. Head and neck cancers encompass a broad group of epithelial malignancies that can arise from the nasal cavity, ears, oral cavity, pharynx, larynx, and paranasal sinuses. Squamous cell carcinoma represents the most common histological presentation with tobacco and alcohol use and prior infection with the human papillomavirus (HPV) being the greatest risk factors [20]. As the population increases and lifespans continue to lengthen, head and neck cancer in the geriatric population will continue to grow. Current estimates suggest that 20–25% of patients with head neck cancers are over the age of 65 with over 50% of new diagnoses occurring in this age group [21–23]. Despite being a predominantly male-dominated disease with ratios ranging from 8:1 to 15:1, the proportion narrows with age as more females become diagnosed in patients over 65 years of age [24–26].

Management of head and neck cancers requires a multimodality treatment regimen targeted by surgery, radiotherapy, systemic chemotherapy, or a combination of the three depending on tumor stage, location, regional disease, distant disease, and intent of treatment. Approximately two-thirds of patients with head and neck cancer present with locoregionally advanced disease commonly involving the cervical lymph nodes (stage III or IV), where distant metastatic spread at initial presentation is found in about 10% of patients [27, 28]. Treatment of head and neck cancer is complex with an extensive perioperative period and long-term close follow-up appointments with often multiple physicians involved as part of the treatment team. It does not follow a one-size-fits-all model. Its careful diagnosis and management in the geriatric population is necessarily much more specific to the individual needs of the patient.

Although the biological age is more telling in the geriatric patient than the chronological age, with age there are particular physiologic organ-specific changes as well as a decline in the integrity of certain anatomical functions. Functional ability and comorbidities are important to assess as complex medical histories are more common in elderly patients. Elderly patients with poor physiologic reserve and complex medical problems are vulnerable to an increased burden of symptoms and have reduced tolerance for medical interventions, thereby necessitating a comprehensive geriatric assessment as the first and essential step prior to selecting the most appropriate treatment options [20]. Age-specific problems include functional decline, impaired cognition, decreased psychological well-being, reduced nutritional status, and the absence of social support. All of these should be vigilantly evaluated and perform an important role in determining the appropriate treatment. Presenting a comprehensive plan for the patient that illustrates its fluidity as the treatment process is implemented is crucial in the geriatric. Treatment of head and neck cancer is often a long and arduous path toward cure and prolongment of lifespan, so it is imperative that understanding the risks and potential complications that can arise throughout the course is thoroughly discussed with the patient. Informed decision by the patient should include knowledge of all the potential complications that can arise from each of the treatment options that can have adverse effects on their quality of life. These should be frequently revisited throughout their treatment course to reevaluate goals of care.

Early-state head and neck cancer is typically treated with a single-modality regimen including either surgical resection or radiotherapy and is determined based on the anatomical site. Prognostic factors well studied in head and neck cancer patients are disease specific such as tumor size, staging, lymph node involvement, margin status, perineural or vascular invasion, and extra lymphatic extension. Age as a prognostic factor is less understood but gains more attention in recent years as the population has increased. A Surveillance, Epidemiology and End Results (SEER) Program in 9386 subjects found comorbidities to be an independent predictor of survival for patients older than 65 years with head and neck cancer [28]. Multiple prospective and retrospective studies in head and neck cancer survival outcomes have not shown any differences when stratified by chronological age [25, 28, 29]. Over 2500 cases of squamous cell carcinoma of the glottic larynx, oral tongue, and tonsil were extracted from the SEERS program over a 10-year period and stratified by chronological age. Overall survival and disease-specific survival were compared among younger and elderly patients and found no statistically or practically significant differences [30]. Sarini et al. assessed outcomes in patients 74 years of age and older compared to younger patients undergoing head and neck cancer treatment and found no differences when it came to tolerance to treatments and survival outcomes [24]. Although older patients with head and neck cancer are more likely to harbor comorbidities, the chronological age in this heterogeneous geriatric population is not an independent predictor of response to therapy or survival. Therefore, the biological age becomes a crucial parameter that may have a large impact on prognosis when considering treatment plans. Those patients who are more frail or harbor multiple comorbidities which increase with age may be unable to tolerate treatment or its toxicities, putting them at higher risk for receiving suboptimal care and less aggressive treatment options needed to achieve complete remedy of their disease [27].

As head and neck oncology has evolved, so have treatments, which have led toward the development of less-invasive modalities such as transoral laser and robotic surgery, endoscopic surgery, and intensity-modulated radiotherapy which are better tolerated while minimizing the side effect profile. The use of transoral robotic surgery for early-stage oropharyngeal carcinoma offers a function sparing surgery that is less intrusive than prior open techniques. With complete resection and lymph node dissection, these newer methods allow delivery of lower doses of adjuvant radiation therapy, sparing critical structures such as the pharyngeal constrictors, brachial plexus, and carotid artery from high-dose radiation damage and functional impairment, disabilities that can be devastating to overcome in an elderly patient [31, 32]. Due to the expanding evidence in the literature that geriatric patients have similar head and neck cancer treatment outcomes when compared to control matched younger individuals, primary treatment plans for these patients do not need to stray from the gold standard to compensate for age.

However, because comorbidities and symptom incidence, prevalence, and severity are known to increase with age, it is imperative to be meticulous in the overall management of elderly head and neck cancer patients [22, 33, 34]. Developments in oncologic practices have put forth standardized quality of life, symptom assessment, and comorbidity tools to best determine prognosis and response to treatment modalities which should be incorporated into initial pretreatment assessments [33–35]. One of the most important assessments that need to be done at the initial evaluation is the determination of frailty in the elderly. Elderly patients once stratified based on their biological age or level of frailty can be appropriately intervened prior to treatment to improve outcomes. Head and neck patients are even more at risk for functional declines because the location of their tumors impact many important functions of daily living such as swallowing, speaking, hearing, and breathing. Dysfunction in any of these can lead to dysphagia, malnutrition, aspiration, infection, and even death. Due to the nature of these tumors, patients who are functionally viable and independent can decompensate as they lose these basic functions that greatly impact quality of life. This may only be exacerbated with treatment modalities as surgical resections violate this functional anatomy. Alternatively, radiation therapy is not without side effects. In patients that may have difficulty swallowing post-surgical intervention or radiotherapy treatment for instance may benefit with early intervention enteral feeding to prevent malnutrition throughout their cancer therapy. Comprehensive geriatric assessment's (CGA) is a multifactorial measurement that can be used in the geriatric population to determine one's frailty by measuring their functional status, medical conditions, cognition, mental health, social support, nutritional status, and polypharmacy. The evaluation takes into account functional status through the means of completing activities of daily living and objective measurements of functional status, such as the 'Get up and Go' test [36]. Promptly creating a plan, using preoperative assessment tools, and mapping out the proposed long-term treatment plan in patients can identify those at risk for complications.

Primary surgical treatment is a well-established modality for the treatment of most head and neck cancers but is not without its complications. Many studies in the literature that have looked at patients with comorbidities have found they are in fact at a higher risk for postoperative complications [37, 38]. In head and neck cancer and reconstructive surgeries where the patient undergoes anesthesia for a prolonged period of time, these comorbidities become more important in the preoperative evaluation. Pedicle and free flap reconstructions have improved immensely over the years with success rates ranging from 95.4 to 100% for free flaps and 87.7 to 100% for pedicle flaps [39–41]. These statistics have not been shown to vary based on elderly patients, but concurrent comorbidities regardless of age have been shown to play a large role [42, 43]. Zhang et al. analyzed the treatment outcomes including the type of flap, the length of the intervention, and the outcomes of over 600 elderly head and neck cancer patients who underwent surgical treatment and reconstruction with either a pedicle flap or a free flap. They found outcomes in surgical intervention and reconstruction with a pedicle or free flap tissue were not significantly affected by age. This has been consistently shown in the literature to suggest head and neck cancer surgeries with reconstruction can be performed safely in patients regardless their age, but comorbidities need to be preoperatively assessed and monitored [39, 44].

Radiotherapy is a means of treatment for head and neck cancers that are either early stage and not suitable for surgery, inoperable, locally advanced, or used for adjuvant treatment after surgical resection. Treatment can be difficult for the cancer patient because of the time commitment dedicated to completing its course. The average fractionations are 5 times a week for 7 weeks, necessitating the need for a reliable support system and consistent care. Furthermore, radiotherapy is known to induce acute morbidity due to the xerostomia and fibrosis induced by this therapy [45]. The responsibility of transportation to the appointments, dedication of time, and post-radiation complications all can be very overwhelming for any patient. In the geriatric population, this can lead to a loss of independence and depression. The use of concurrent chemotherapy, complications of ototoxicity, and nephrotoxicity can acutely impact the patient's quality of life and lead to negative physical, emotional, and social outcomes [46]. Close monitoring and frequent reevaluations by the head and neck surgeon and oncologist in this patient population are important to maintain goal-oriented expectations and be attentive to make adjustments as needed for intolerable morbidities that may arise.

Geriatric patients should be presented the best treatment plan for head and neck cancer regardless of age. Surgical treatment which can be disfiguring and long-term morbidity from radiation therapy can have considerable effects on their quality of life. Pretreatment evaluations with geriatric assessments to determine biological age and fragility are useful in initial pretreatment planning to unmask any foreseeable morbidities and identify patients who can or cannot tolerate standard of care treatment interventions.

Laryngology in the Geriatric Population

Aging Voice

Physiologic differences exist between not only males and females, but also exist among different age populations [47–49]. The prevalence of voice disorders in the United States is about 6.6 to 7.6% with nearly $300 million annually related to dysphonia-related disorders. In the elderly population of patients over the age of 65, hoarseness and dysphonia are common complaints that draw patients to see an otolaryngologist and are widely due to natural physiologic changes of senescence that take place with age. As we age, pitch and loudness diminish, and speech slows down with more pauses for air. The geriatric voice without aberrant pathologies is found to have [50]:

- Increased pitch in men
- Decreased pitch in women
- Reduced vocal endurance (pauses for breaths, fatigability increased)
- Reduced vocal volume
- Decreased vocal projection
- Vocal tremor

The structure and function of the geriatric larynx deteriorate with age as the laryngeal framework weakens and ossification takes place of the cricoid, thyroid, and arytenoid cartilages, leading to a more rigid structure and limiting the adduction ability of the inferior pharyngeal constrictors [51]. The composition of the extracellular matrix (ECM) of the vocal folds, laryngeal muscles, cartilage, and nerves all continue to change with age and lead to functional changes such as dysphonia and hoarseness. Structurally, a normal vocal fold is made up of five layers: the epithelium, the superficial, intermediate and deep layer of the lamina propria and the vocalis muscle. The three layers of the lamina propria have different viscoelastic properties and thereby differ in the composition of the ECM [52]. The ratio of hyaluronic acid to collagen is decreased as collagen fibrils increase and elastin fibers thin in the superficial layer of the lamina propria. This compromises the viscoelasticity and vibratory properties of the folds [53, 54].

Vocal fold bowing and atrophy are common anatomical changes that take place with age as connective tissue is lost in the superficial and intermediate lamina propria, making the ventricle and vocal process more prominent, sulci to develop, and atrophy to ensue. This is commonly seen as a glottic gap on flexible laryngoscopy examinations [55, 56]. Vocal fold bowing and thinning of the vocal process is more common in men than women and increases the pitch and fundamental frequency. Women, especially with the lack of hormonal signaling after menopause, experience an edematous thickening to the vocal cords that change the viscoelasticity of the mucosa and stiffening the vocal cord, thereby deepening the voice and decreasing the fundamental frequency [57]. Muscle mass in both men and women decreases and is replaced by fibrofatty tissue; incoordination of muscle movement and timing becomes more prevalent and mucus membranes lose their lubricating properties [58]. Presbyphonia is categorized into synonymous disorders encompassed by terminology that includes vocal cord atrophy, vocal cord bowing, vocal fold thinning, vocal fold edema, or posterior glottic gap.

Voice changes in the elderly patient are a common complaint that can be a significant handicap in an elderly patient's quality of life. It leads to impairment in communication and can have detrimental social effects, leading to depression and isolation. Consequently, early recognition and intervention are crucial to maintaining quality of life. The causes of geriatric vocal dysfunction represent a broad diagnosis that can be divided into these subsets [59, 60]:

- Anatomical changes (vocal fold atrophy, glottic gap, bowing)
- Neurologic conditions (e.g., Parkinson's disease, cerebral vascular accident, tremor, spasmodic dysphonia)
- Impaired conditioning (e.g., decreased lung elasticity and compliance, reduced physical fitness)
- Malignant lesions (laryngeal carcinoma)
- Benign lesions (e.g., epithelial lesions, Reinke's edema, cysts, granulomas)
- Inflammatory disorders (e.g., reflux laryngitis)
- Infectious disorders (e.g., laryngitis)
- Vocal fold paralysis
- Autoimmune (e.g., Amyloidosis, Sarcoidosis)

In-office examination can be performed with simple flexible laryngoscopy to rule out pathological conditions in geriatric patients presenting with voice disorders. Complaints related to prebysphonia include a weak or breathy voice that easily fatigues. Patients may complain of pain as their phonation becomes stressed due to the application of hyperfunctinoal compensatory mechanisms [52]. It is important to understand normal age-related changes in the geriatric population to provide earlier intervention and avoid inaccurate diagnosis and unnecessary costly workups. In relatively benign examinations on flexible laryngoscopy, the first step is to obtain comprehensive information on the vocal folds themselves through videolaryngostroboscopy. This allows high-power interpretation of glottic closure and the mucosal wave. Acoustic parameters are used to analyze and measure the jitter, shimmer, noise to harmonic ratio (NHR), and fundamental frequency [61]. Once the cause can be attributed to organic anatomical age-related changes, treatment with goal-oriented voice therapy can be initiated. Voice therapy and its success are correlated directly with the effort the patient puts into the treatment, which includes removal of aberrant hyper-functional compensatory mechanisms and learning adaptive strategies that allow patients to voice appropriately with the anatomical changes [52, 62]. These adaptive strategies are focused on improving vocal cord adduction to improve phonatory duration and strength and reduce fatigability. Vocal therapy requires multiple visits before patients experience improvement. It may be difficult to organize a reliable treatment plan in some patients.

Geriatric patients who do not respond to conservative therapy can be considered for surgical management. Injection laryngoplasty or thyroplasty should only be considered in a particular subset of patients with anatomical characteristics that would benefit from intervention. Patients with glottic insufficiency and gaps are excellent candidates for injection laryngoplasty. In geriatric patients this is a favorable intervention because it can be performed during an in-office visit under local anesthesia [63]. Injection laryngoplasty does not recover laryngeal function, but only provides additional vocal fold bulk to minimize the glottic thinning and subsequent glottic gap, and allow for more complete adduction. There are many different substances used for injection ranging from temporary to permanent materials that are injected into the paraglottic space deep to the lamina propria to cause medialization of the vocal fold [64]. Often used to treat recurrent laryngeal nerve paralysis by improving glottic incompetence, injection laryngoplasty has gained popularity for improvement in voice for elderly patients with anatomic changes amenable to this procedure.

Permanent materials have fallen out of favor for injection laryngoplasty because of increased risk of complications if not injected accurately. However, the aging larynx that benefits from this temporary medialization may be an excellent candidate for a more permanent result that can be achieved with thyroplasty augmentation. In this procedure the neck is open while the patient is responsive to expose the thyroid cartilage and carve a window out of its framework, exposing the paraglottic space. There implant materials can be advanced through the paraglottic space until an optimal voice is changed through medialization, while avoiding glottic compromise [54]. In a select subset of elderly patients, thyroplasty procedures should be considered, but not before determining clinical benefit from prior injection laryngoplasty and completing a thorough workup to ensure a preoperative risk benefit.

Swallowing in the Geriatric Patient

Dysphagia is defined as difficulty in moving food through the oral cavity, pharynx, and esophagus that can be structural or functional in nature [65]. The ability to eat and drink is a basic human need and pleasure that when impaired can have significant physical, psychosocial, and financial implications. Eating not only provides a means of nutrition but interpersonal gatherings and rituals are typically centered around feasts and meals [66, 67]. Dysphagia can have devastating health-related complications such as dehydration, malnutrition, and aspiration that have significant effects on morbidity and mortality. In addition, it can also weigh on the patient's mental health, leading to decreased social interaction and depression. It has been noted in the literature that more than 40% of patients over the age of 60 experience symptoms of dysphagia [68]. Swallowing mechanisms become affected as we age due to normal physiologic changes. Presbyphagia refers to the natural changes in the swallowing mechanism in otherwise healthy adults that occur as we age and is associated more with biological age rather than chronologic. Dysphagia is associated with many age-related comorbidities and becomes more common in patients with multiple medical problems.

Swallowing is a complex function that can be affected at any point in one of the three phases: oral, pharyngeal, and esophageal (Table 11.1).

Alterations in the oral phase that take place lead to difficulties preparing the bolus and lifting it toward the pharynx. Bolus preparation in large part is successfully completed by retained dentition, providing the necessary force of mastication. Age is associated with alveolar bone loss and a decline in dental hygiene, leading to problems with eating and swallowing. Studies have shown that older patients who were edentulous were more likely to have laryngeal penetration than those with complete dentition [69]. The tongue is another crucial component of the oral cavity that aids in the propulsion of the food bolus into the pharynx. With normal aging, it has been found that the generation of lingual pressure needed for propulsion decreases in congruence with a decline in overall tongue strength contributing to prebysphagia [70]. The muscles of mastication that move in a coordinated fashion are known to deteriorate with age similar to the

Table 11.1 Causes of dysphagia at different levels of the swallowing phase

	Age-related changes	Effect
Oral phase	Dental compromise Decreased masticatory muscle strength Decreased tongue strength and pressure Sensory changes Xerostomia	Poor bolus preparation Poor propulsion toward pharynx
Pharyngeal phase	Delayed swallowing reflex Reduced pharyngeal muscle strength Reduced suprahyoid muscle strength Dilated upper esophageal sphincter opening	Increased pharyngeal swallow time Reduced hyoid elevation Aspiration Pooling in the pharynx
Esophageal phase	Dysfunction in upper/ lower esophageal sphincter Decreased peristalsis Uncoordinated contractions Dilation/ constriction	Laryngo-esophageal reflux Esophagitis Gastroesophageal reflux Bolus retention in the esophagus

atrophy and strength decrease seen in other aging muscle groups [71]. Salivation is stimulated with the onset of mastication and produced by the submandibular, parotid, sublingual, and minor salivary glands. It serves many important functions in the physiological homeostasis of the oral cavity that includes the digestion of carbohydrates, maintaining dental mineralization, providing immune support, and maintaining a lubricated environment to facilitate in the acts of swallowing and speaking. Xerostomia is common in the elderly population, not because of a global age-related decline in salivary production, but can be caused by several possible etiologies such as medical comorbidities (diabetes, Sjogren syndrome, rheumatoid arthritis), medication side effects, and even prior radiation to the head and neck region [72, 73]. Lastly, sensory changes and diminished tactile sensation especially in the oral cavity lead to a delayed response in swallowing. Since swallowing is a coordinated process, the diminished response can hinder the natural swallow pattern. Sensory, salivary, and muscular changes in the elderly are not inherent to the oral cavity but occur throughout the pharynx and esophagus as well.

Pharyngeal abnormalities that occur as a result of aging are of more clinical relevance. This is due to a risk of aspiration when protective mechanisms to secure the glottis are lost due to a decline in the pharyngeal strength and coordination. Both aspiration and penetration are adverse results of a misdirected bolus and swallowing dysfunction that can lead to significant morbidity and mortality [74]. Nasogastric tubes are commonly placed in head and neck patients during times when swallowing is compromised, or the patient is required to be kept nil per os (translated: nothing through the mouth) (NPO) for a prolonged period of time such as with the healing phase after a free flap microvascular reconstruction. One study found that swallowing function and liquid penetration were more pronounced in patients over 70 years of age where nasogastric tubes were placed. This suggests that elderly patients are unable to compensate or overcome even slight modifications to the swallowing process under stressful situations [74]. Otolaryngology patients are commonly faced with these changes because their aerodigestive tract is at the forefront of surgical procedures, and otolaryngologists commonly introduce foreign bodies such as nasogastric tubes and tracheostomy tubes. Consideration and early intervention should be taken in elderly patients that will undergo any form of alteration even if minor to prevent complications. These problems can lead to prolonged hospital stays and increased costs as the functional ability to obtain nutrition is compromised.

Lastly the esophageal phase of the swallow mechanism becomes less coordinated and fluid as we age, leading to decreased peristalsis and risk of stasis of the bolus while making its way to the lower digestive tract. Furthermore, the upper and lower esophageal sphincter that aids in being a gateway for the bolus from the pharynx to the stomach can become less functional, leading to reflux of food and acid and incomplete passage. Although these esophageal sphincter functions decrease with age, it is the decrease in the extent of its ability to open that varies in older individuals and can manifest into symptoms of dysphagia and frailty [75].

Diverticulum

Zenker's diverticulum is a pulsion diverticulum that occurs at Killian's triangle between the cricopharyngeus muscle and inferior aspect of the inferior pharyngeal constrictor. Typically, a disorder that is inherent to middle-aged and elderly patients with a predisposition for men in their seventh and eighth decade, studies have suggested that its pathology is due to reduced compliance and muscular contractility of the cricopharyngeus (contributing to the upper esophageal sphincter) [76, 77]. Histological findings from muscle biopsies of the cricopharyngeaus suggest progressive denervation may contribute. Symptoms of intermittent dysphagia, regurgitation, and hoarseness are common in patients with esophageal diverticulum; and diagnosis is based on clinical suspicion of radiographic swallowing studies. Treatment of a

Zenker's diverticulum has evolved over the years from a transcervical open repair with cricopharyngeal myotomy to endoscopic repair with laser or stapling and subsequent cricopharyngeal myotomy [78]. These less-invasive techniques have gained popularity as a safe and effective treatment for a largely geriatric-centered patient population who have notoriously more comorbidities that put them at a risk for surgery. Asymptomatic lesions that are less than 2 centimeters can be monitored, while larger diverticulum or those that are symptomatic should undergo surgical resection and cricopharyngeal myotomy [79]. Howell et al. performed a meta-analysis and systematic review of over 800 patients undergoing an open vs endoscopic approach to Zenker's diverticulectomy and found no difference in patient-reported improved symptoms of dysphagia [80]. Endoscopic approaches, however, have been found to have statistically significantly shorter operating times and hospital stays, and time to resume oral feedings [81, 82]. In patients unfit for surgery, this less-invasive approach may be the most suitable. However multiple factors including the size of the diverticulum, the exposure of the diverticulum, surgeon ability, and the patient's tolerance to undergo surgery and withstand its complications are all important considerations that must be evaluated.

The differential diagnosis of dysphagia in the elderly patient is broad with causes ranging from neurologic, autoimmune, infectious, traumatic, neoplastic, anatomic, medication-induced and iatrogenic. In the otolaryngology patient, more common causes of dysphagia are related to head and neck tumors, radiation therapy, esophageal diverticulum, strictures/webs, and cricopharyngeal bars [83]. Elderly patients should be thoroughly screened for swallowing disorders as these conditions are more common in their population. Complications from dysphagia can be devastating if not diagnosed and treated in a timely manner. It has been found that over 60% geriatric patients with oropharyngeal dysphagia are initially misdiagnosed and over 66% do not receive timely treatment once a diagnosis has been achieved [84]. A multidisciplinary team of nurses, therapists, and physicians all play critical roles in diagnosing and treating dysphagia in the elderly. First and foremost, a thorough history and physical exam can promptly identify the cause of dysphagia. Screening questionnaires such as the Gugging Swallowing Screen (GUSS), the repetitive saliva-swallowing test (RSST), and the 3-oz water swallowing test [85, 86] have shown promise in being early intervention tools to diagnose and treat dysphagia. Palli et al. trained nurses in a stoke unit to perform dysphagia screening using the GUSS and found dysphagia was significantly reduced in the intervention group and had lower rates of pneumonia and a reduced length of hospital stay [87]. More comprehensive studies to diagnose dysphagia include those that quantify swallowing dysfunction such as the barium swallow radiography, videofluoroscopic swallowing study (VFSS), fiberoptic endoscopic evaluation of swallowing (FEES), upper endoscopy, and esophageal manometry. Depending on the underlying pathology, treatment can vary from compensatory management to surgical intervention.

Otologic Problems in the Geriatric Population

Hearing Loss

One of the most common reasons elderly patients seek out an otolaryngologist is for hearing and otologic-based complaints. A large retrospective review concluded the most common chief complaints in patients who visited an otolaryngologist over the age of 65 and among 418 patients were ear and hearing disorders (24.2%). Age-related hearing loss also known as presbycusis is a common phenomenon that is estimated to impact 25–30% in the United States between the ages of 65 and 74, and nearly 80% of individuals over the age of 75 [88]. It carries a large economic burden with about $3 billion annually spent on its diagnosis and treatment in the United States [89]. Hearing impairment is a serious health threat in the geriatric population as it contributes to impaired balance, deterioration in mobility, an increased risk of falls and traumatic injuries, and cognitive impairment. The functional disability that ensues as a consequence of hearing loss can therefore lead to depression, anxiety, loss of independence, social isolation, and an overall impaired quality of life [90, 91]. Lin et al. conducted a cross-sectional analysis of the association of hearing loss with self-reported falls in patients in the United States participating in the National Health and Nutritional Examination Survey (NHANES) and found hearing loss to be independently associated with an increased risk of falls [90].

Presbycusis is the most common form of hearing loss in the geriatric population. It is classically diagnosed by its pattern of bilateral symmetric high frequency sensorineural hearing loss with difficulty in central auditory processing and speech discrimination. In its earliest stages, patients notice a slow decline in their ability to understand high-frequency speech which is most pronounced in noisy environments [92]. As it progresses, the ability to localize and detect sounds becomes increasingly more difficulty, creating situations where failure to hear warning signals is exceedingly dangerous. Advanced hearing loss affects the lower frequencies, making speech detection and understanding difficult [91, 93]. The cause of presbycusis is multifactorial with studies citing intrinsic causes to include genetic disorders, metabolic disease, and systemic disease such as hypertension and diabetes. Extrinsic factors include ototoxic medications, loud noise exposure, and diet [94]. Historically, presbycusis was categorized into four different categories based on audiometric tests and temporal bone pathology to include sensory, neural, strial (metabolic), and cochlear conductive. Two additional categories, mixed and indeterminate, were added over 20 years after [95, 96].

Sensory presbycusis affects the organ of Corti in the cochlea that contains the outer hair cells vulnerable to intrinsic and extrinsic factors. Damage begins at the basal turn of the cochlea which is responsible for high-frequency hearing in this subset of presbycusis. It has been found that sensory presbycusis demonstrates a pattern, much like noise induced, with outer hair cell loss and steeply sloping high frequency sensorineural hearing loss on audiograms. Age itself is not a direct cause of outer hair cell loss, but more so than environmental factors [93].

In *Neural presbycusis* there is a loss of 50% (or more than 35,500) spiral ganglion auditory nerve fibers seen on histopathology. Audiograms show a gradual downward sloping pure tone threshold toward high-frequency with diminished speech discrimination out of proportion to pure tone thresholds [95, 96].

Strial (metabolic) presbycusis is characterized by atrophy (30% or more) of the stria vascularis in all turns of the cochlea. The stria vascularis provides the inner ear blood supply and balances the chemical and bioelectric potential across the cochlea. Atrophy of strial tissue causes a decrease in endolymphatic sac potential through potassium recycling [97]. Audiograms in patients with strial presbycusis have diminished pure tone thresholds and hearing loss across all frequencies with the preservation of speech discrimination [96].

Cochlear conductive presbycusis is believed to be caused by an age-related degenerative thickening and stiffening of the basilar membrane but has not yet been verified. Audiograms show a low-frequency sensorineural hearing loss with unaffected speech recognition [95, 96].

Mixed presbyscusis is a combination of the above forms of presbyscusis. It is characterized by a loss of outer hair cells at the basal turn of the cochlea, stria vascularis atrophy, and a severe loss of cochlear neurons. Audiograms show a down-slope high-frequency hearing loss.

Indeterminate presbycusis represents about 25% of cases and shows none of the patterns seen as described earlier. There are no visualized abnormalities in cochlear tissue under microscopy or correlation between audiometric or pathologic patterns [94].

There is currently no diagnostic test to predict which individuals will have hearing loss as a result of presbycusis. Furthermore, there is not a single treatment to reverse its effects. Screening geriatric patients as a routine part of their physical exam is important in recognizing and providing early intervention for hearing loss. Studies have found an independent association between hearing loss and cognitive decline [98–100]. Lin et al. completed a prospective study of over 600 patients undergoing audiometric testing who were dementia free at its initiation and found over an 11-year period a strong correlation between hearing loss and incident all-cause dementia. Patients with mild-to–severe hearing loss have a 2- to five-fold increased risk of developing dementia which increased linearly with the severity of hearing loss [99]. Neuroimaging is found to be congruent with these findings where temporal lobe and overall brain atrophy is more prevalent in patients with hearing impairment [101, 102]. Current treatment modalities for presbycusis fall under hearing amplification devices and cochlear implantation which should be introduced early in their disease process to prevent morbidity and mortality that have considerable socioeconomic impacts on patients.

Hearing aids can provide patients with substantial benefits, yet it is estimated that less than 20% of patients with hearing loss use hearing aids [103]. Limited resources, the absence of education of their benefit and usage, and expense are factors that affect a patient's reasoning for doubting this therapy modality. Patients who have severe to profound hearing loss without any benefit from traditional hearing aid amplification are strong candidates for cochlear implantation. Studies have shown that early intervention in these patients can be shown to improve cognitive skills and positively influence their social interactions and quality of life [104, 105].

When considering a cochlear implantation in an elderly individual, not only should their operative fitness and perceived benefit be taken into account, but more specifically, the integrity and thickness of the scalp. Cochlear implant survival relies on a substantial and healthy scalp to protect the implant from extrusion or exposure. The older individual is more at risk for extrusion because of the predictable senescent thinning of the scalp with the loss of collagen, which should be taken into consideration when operating on these patients [75]. Close surveillance with appropriate wound care is critical to ensure the implant remains in a good position. In individuals with significant comorbidities or at risk of general anesthesia, it has been found that cochlear implantation under local anesthesia and sedation via a suprameatal approach is a well-tolerated and safe alternative [104, 106]. Hearing aids and cochlear implants are both excellent treatment modalities for qualified elderly patients with hearing loss and should be introduced early in their treatment to maintain and prolong their quality of life. Age should not be a primary determining factor in surgical hearing aid placement as the benefits in the majority of patients far exceed the risks.

Vertigo/Balance

The physiological process of aging is associated with a steady decline in vestibular function. With a decline in muscle mass, strength, and agility, the range of motion becomes more limited and reaction time is delayed. Due to the progressive loss of peripheral and central vestibular system nerve cells, dizziness, imbalance, gait disturbances, and vertigo are commonly seen complaints in the elderly [107, 108]. Agrawal et al. found vestibular dysfunction in the United States affects 18% of adults aged 40 to 49, 49% of persons over 60, and 80% in individuals over the age of 80 [109]. As with hearing loss, an impaired vestibular system can also lead to reduced quality of life as mobility and independence become compromised. With the significantly increased risk of falls in patients with vestibular dysfunction, morbidity, mortality, and health care costs are enormous in this population. Therefore, the importance of screening is imperative to promptly diagnose and treat these patients to prevent potentially disastrous consequences.

The vestibular system is comprised of three semicircular canals (superior, posterior, and horizontal) and two otolith organs (utricle and saccule) housed within the inner ear. These structures transduce information to the central nervous system in regard to rotational and linear head movement along with spatial orientation relative to gravity. The integration of this tightly controlled system through a central pathway from the inner ear structures produces the necessary information to the cortex and controls the vestibulo-ocular reflex and vestibulospinal reflex which maintains gaze and postural stability [109, 110] Peripheral vestibular disorders warranting otolaryngology visits in the elderly typically include benign paroxysmal positional vertigo (BPPV) and Meniere's with vestibular migraines. Vestibular neuritis is less specific to this age population.

The most common vestibular disorder in the elderly population is BPPV, which is caused by the accumulation of dislodged otoconia in the endolymphtic duct of the semicircular canals [111]. Otoconia are calcium carbonate–based biocrystals found linked to the dense gelatinous maculae layer of the utricle and saccule by filaments and serve the purpose of sensing linear acceleration and gravity. As we age otoconia undergo a decrease in number, volume, and change in shape, facilitating crystal dislocation to the ampulla and causing disruption balance and equilibrium [111, 112]. Theories for degeneration of otoconia have been proposed, but no treatment targeting their compromise currently exists. BPPV can be easily diagnosed in the clinic with a thorough history and physical exam. Patients typically complain of short attacks of rotary vertigo that occur when changing position or moving their head without associated hearing loss. Diagnosis can be confirmed in the office with maneuvers such as such as the Dix Hallpike for the posterior semicircular canal which is affected in most cases (>90%) [113]. Although BPPV typically resolves spontaneously, treatment can also be completed with certain maneuvers such as the Epley, Semont, or Gans [114–116]. Care must be taken in elderly patients to consider their range of motion and neck extension as it does require some manipulation. Vestibular rehabilitation therapy for refractory cases may also prove to be useful in its treatment.

Meniere's disease is generally a disorder of the middle-age population but found to reported in up to 15% of patients over the age of 65 [117]. Typical symptoms include vertigo, aural fullness, hearing loss, and tinnitus with a unilateral presentation. Vertigo episodes last longer and are more variable in Meniere's disease, lasting from minutes to a few hours at a time. The condition is thought to be caused by an excess of endolymph in the semicircular canals, and no one therapy exists as a cure for the disease [118]. Patients often respond to a combination of a low-salt diet and diuretics medications with more invasive therapies including intratympanic injections of gentamicin and corticosteroids to cause ototoxicity, thereby alleviating the vestibular symptoms [117].

Surgery is rarely a means of treating BPPV-specific vertigo and a path when all other therapies have been exhausted. Newer techniques include those that occlude the posterior semicircular canal with argon lasers, or performing vestibular neurectomy [119]. Meniere's patients, because of the severity in some cases of their symptoms, can also undergo vestibular neurectomy, endolymphatic sac shunting, or labyrinthectomy [120, 121]. In elderly patients, the risk of surgery for a primarily medically managed disorder may propose to outweigh the benefits. In addition, complications such as hearing loss and incomplete resolution of symptoms may not be superior to medical management. Compensatory mechanisms in the elderly are often slowed or lacking, so surgical management may cause more complications if patient selection is not precise. All these factors need to be taken into consideration in determining the best patient population, especially in older individuals prior to surgical management for vertigo.

Conclusion

The elderly population over 65 years of age is becoming one of the most rapidly growing segments of the population and, with the rise in life expectancy in part because of medical advances, will continue to increase. The American Society of Geriatric Otolaryngology was established in 2006 recognizing the increase in demand in geriatric focused medicine with the goal to provide knowledge to practitioners on disorders of the ears, nose and throat in this population. While some otolaryngologic diseases are specific to the geriatric patient, management of even common disorders should take into consideration the needs of this subset of patients. Over one third of patients that visit an otolaryngologist are found to be over the age of 65, and care should be taken to recognize the specifics in treating and managing their presentation otolaryngologic diseases.

References

1. Centers for Disease Control and Prevention. The State of Aging and Health in America 2013. Atlanta, Ga.: Centers for Disease Control and Prevention, U.S. Department of Health and Human Services; 2013. https://www.cdc.gov/aging/pdf/state-aging-health-in-america-2013.pdf. Accessed 2 April 2019.
2. Eibling D, Kost K. The emerging field of geriatric otolaryngology. Otolaryngol Clin N Am. 2018;51(4):847–52. https://doi.org/10.1016/j.otc.2018.03.011.
3. Mekel M, Stephen AE, Gaz RD, Perry ZH, Hodin RA, Parangi S. Thyroid surgery in octogenarians is associated with higher complication rates. Surgery. 2009;146(5):913–21. https://doi.org/10.1016/j.surg.2009.05.004.
4. Seybt MW, Khichi S, Terris DJ. Geriatric thyroidectomy: safety of thyroid surgery in an aging population. Arch Otolaryngol Neck Surg. 2009;135(10):1041. https://doi.org/10.1001/archoto.2009.138.
5. Haugen BR, Alexander EK, Bible KC, et al. 2015 American Thyroid Association management guidelines for adult patients with thyroid nodules and differentiated thyroid Cancer: the American Thyroid Association guidelines task force on thyroid nodules and differentiated thyroid Cancer. Thyroid. 2016;26(1):1–133. https://doi.org/10.1089/thy.2015.0020.
6. Hanover Operative N. Mortality with elective surgery in older adults. Eff Clin Pract. 2001;4:172–7.
7. Fleisher L. Cardiac risk stratification for noncardiac surgery. Cleve Clin J Med. 2009;76:S9.
8. Tuggle CT, Park LS, Roman S, Udelsman R, Sosa JA. Rehospitalization among elderly patients with thyroid Cancer after thyroidectomy are prevalent and costly. Ann Surg Oncol. 2010;17(11):2816–23. https://doi.org/10.1245/s10434-010-1144-7.
9. Grogan RH, Mitmaker EJ, Hwang J, et al. A population-based prospective cohort study of complications after thyroidectomy in the elderly. J Clin Endocrinol Metab. 2012;97(5):1645–53. https://doi.org/10.1210/jc.2011-1162.
10. Passler C. Thyroid surgery in the geriatric patient. Arch Surg. 2002;137(11):1243. https://doi.org/10.1001/archsurg.137.11.1243.
11. Tartaglia F, Russo G, Sgueglia M, et al. Total thyroidectomy in geriatric patients: a retrospective study. Int J Surg. 2014;12:S33–6. https://doi.org/10.1016/j.ijsu.2014.08.386.
12. Sahli ZT, Ansari G, Gurakar M, et al. Thyroidectomy in older adults: an American College of Surgeons National Surgical Quality Improvement Program study of outcomes. J Surg Res. 2018;229:20–7. https://doi.org/10.1016/j.jss.2018.03.057.
13. Sosa JA, Mehta PJ, Wang TS, Boudourakis L, Roman SA. A population-based study of outcomes from thyroidectomy in aging Americans: at what cost? J Am Coll Surg. 2008;206(6):1097–105. https://doi.org/10.1016/j.jamcollsurg.2007.11.023.
14. Ríos A, Rodríguez JM, Galindo PJ, Canteras M, Parrilla P. Surgical treatment for multinodular goitres in geriatric patients. Langenbeck's Arch Surg. 2005;390(3):236–42. https://doi.org/10.1007/s00423-004-0521-8.
15. Kai H, Xixia L, Miaoyun L, et al. Intraoperative nerve monitoring reduces recurrent laryngeal nerve injury in geriatric patients undergoing thyroid surgery. Acta Otolaryngol (Stockh). 2017;137(12):1275–80. https://doi.org/10.1080/00016489.2017.1354397.
16. Jegerlehner S, Bulliard J-L, Aujesky D, et al. Overdiagnosis and overtreatment of thyroid cancer: a population-based temporal trend study. Tordjman KM, ed. PLoS One. 2017;12(6):e0179387. https://doi.org/10.1371/journal.pone.0179387.
17. Ito Y, Miyauchi A, Kihara M, Higashiyama T, Kobayashi K, Miya A. Patient age is significantly related to the progression of papillary microcarcinoma of the thyroid under observation. Thyroid. 2014;24(1):27–34. https://doi.org/10.1089/thy.2013.0367.
18. Oltmann SC, Rajaei MH, Sippel RS, Chen H, Schneider DF. Primary hyperparathyroidism across the ages: presentation and outcomes. J Surg Res. 2014;190(1):185–90. https://doi.org/10.1016/j.jss.2014.04.010.
19. Jemal A, Bray F, Center MM, Ferlay J, Ward E, Forman D. Global cancer statistics. CA Cancer J Clin. 2011;61(2):69–90. https://doi.org/10.3322/caac.20107.
20. Santana-Davila R. Head and neck Cancer in elderly patients: what to do when data are limited? J Oncol Pract. 2018;14(9):549–50. https://doi.org/10.1200/JOP.18.00480.
21. Syrigos KN, Karachalios D, Karapanagiotou EM, Nutting CM, Manolopoulos L, Harrington KJ. Head and neck cancer in the elderly: An overview on the treatment modalities. Cancer Treat Rev. 2009;35(3):237–45. https://doi.org/10.1016/j.ctrv.2008.11.002.
22. Kruse AL, Bredell M, Luebbers HT, Grätz KW. Head and neck Cancer in the elderly: a retrospective study over 10 years (1999 - 2008). Head Neck Oncol. 2010, 2(1) https://doi.org/10.1186/1758-3284-2-25.
23. Huang SH, O'Sullivan B, Waldron J, et al. Patterns of Care in Elderly Head-and-Neck Cancer Radiation Oncology Patients: a single-Center cohort study. Int J Radiat Oncol. 2011;79(1):46–51. https://doi.org/10.1016/j.ijrobp.2009.10.052.

24. Sarini J, Fournier C, Lefebvre J-L, Bonafos G, Van JT, Coche-Dequéant B. Head and neck squamous cell carcinoma in elderly patients: a long-term retrospective review of 273 cases. Arch Otolaryngol Neck Surg. 2001;127(9):1089. https://doi.org/10.1001/archotol.127.9.1089.
25. Lusinchi A, Bourhis J, Wibault P, Le Ridant AM, Eschwege F. Radiation therapy for head and neck cancers in the elderly. Int J Radiat Oncol. 1990;18(4):819–23. https://doi.org/10.1016/0360-3016(90)90403-7.
26. Koch WM, Patel H, Brennan J, Boyle JO, Sidransky D. Squamous cell carcinoma of the head and neck in the elderly. Arch Otolaryngol - Head Neck Surg. 1995;121(3):262–5. https://doi.org/10.1001/archotol.1995.01890030006001.
27. Siddiqui F, Gwede CK. Head and neck Cancer in the elderly population. Semin Radiat Oncol. 2012;22(4):321–33. https://doi.org/10.1016/j.semradonc.2012.05.009.
28. Argiris A, Eng C. Epidemiology, staging, and screening of head and neck cancer. Cancer Treat Res. 2003;114:15–60.
29. Airoldi M, Cortesina G, Giordano C, et al. Postoperative adjuvant Chemoradiotherapy in older patients with head and neck Cancer. Arch Otolaryngol Neck Surg. 2004;130(2):161. https://doi.org/10.1001/archotol.130.2.161.
30. Bhattacharyya N. A matched survival analysis for squamous cell carcinoma of the head and neck in the elderly. Laryngoscope. 2003;113(2):368–72. https://doi.org/10.1097/00005537-200302000-00030.
31. Hurtuk A, Agrawal A, Old M, Teknos TN, Ozer E. Outcomes of Transoral robotic surgery: a preliminary clinical experience. Otolaryngol Neck Surg. 2011;145(2):248–53. https://doi.org/10.1177/0194599811402172.
32. Bhayani MK, Holsinger FC, Lai SY. A shifting paradigm for patients with head and neck cancer: transoral robotic surgery (TORS). Oncol Williston Park N. 2010;24(11):1010–5.
33. Paleri V, Wight RG, Silver CE, et al. Comorbidity in head and neck cancer: a critical appraisal and recommendations for practice. Oral Oncol. 2010;46(10):712–9. https://doi.org/10.1016/j.oraloncology.2010.07.008.
34. Bond SM, Dietrich MS, Murphy BA. Association of age and symptom burden in patients with head and neck cancer. ORL-Head Neck Nurs Off J Soc Otorhinolaryngol Head-Neck Nurses. 2011;29(1):8–14.
35. Murphy BA, Ridner S, Wells N, Dietrich M. Quality of life research in head and neck cancer: a review of the current state of the science. Crit Rev Oncol Hematol. 2007;62(3):251–67. https://doi.org/10.1016/j.critrevonc.2006.07.005.
36. Extermann M, Hurria A. Comprehensive geriatric assessment for older patients with Cancer. J Clin Oncol. 2007;25(14):1824–31. https://doi.org/10.1200/JCO.2007.10.6559.
37. Nesic VS, Petrovic ZM, Sipetic SB, Jesic SD, Soldatovic IA, Kastratovic DA. Comparison of the adult comorbidity evaluation 27 and the Charlson comorbidity indices in patients with laryngeal squamous cell carcinoma. J Laryngol Otol. 2012;126(5):516–24. https://doi.org/10.1017/S0022215112000254.
38. Bøje CR, Dalton SO, Grønborg TK, et al. The impact of comorbidity on outcome in 12 623 Danish head and neck Cancer patients: a population based study from the DAHANCA database. Acta Oncol. 2013;52(2):285–93. https://doi.org/10.3109/0284186X.2012.742964.
39. Zhao X, Zhang Y, Fu S, Zhang C, Li M, Wu Y. Clinical application of a Pedicled forearm flap in the reconstruction after Oral Cancer resection. J Craniofac Surg. 2017;28(3):e222–5. https://doi.org/10.1097/SCS.0000000000003406.
40. Spanio di Spilimbergo S, Nordera P, Mardini S, et al. Pedicled temporalis muscle flap for craniofacial reconstruction: a 35-year clinical experience with 366 flaps. Plast Reconstr Surg. 2017;139(2):468e–76e. https://doi.org/10.1097/PRS.0000000000003011.
41. Rahpeyma A, Khajehahmadi S, Sedigh HS. Facial artery Myomucosal flap, Pedicled solely on the facial artery: experimental design study on survival. J Craniofac Surg. 2016;27(7):e614–5. https://doi.org/10.1097/SCS.0000000000002947.
42. Piccirillo JF. Prognostic importance of comorbidity in a hospital-based Cancer registry. JAMA. 2004;291(20):2441. https://doi.org/10.1001/jama.291.20.2441.
43. Peters TTA, Langendijk JA, Plaat BEC, et al. Co-morbidity and treatment outcomes of elderly pharyngeal cancer patients: a matched control study. Oral Oncol. 2011;47(12):1159–64. https://doi.org/10.1016/j.oraloncology.2011.08.004.
44. Grammatica A, Piazza C, Paderno A, Taglietti V, Marengoni A, Nicolai P. Free flaps in head and neck reconstruction after oncologic surgery: expected outcomes in the elderly. Otolaryngol Neck Surg. 2015;152(5):796–802. https://doi.org/10.1177/0194599815576905.
45. Nguyen NP, Sallah S, Karlsson U, Antoine JE. Combined chemotherapy and radiation therapy for head and neck malignancies: quality of life issues. Cancer. 2002;94(4):1131–41. https://doi.org/10.1002/cncr.10257.
46. Pottel L, Lycke M, Boterberg T, et al. Serial comprehensive geriatric assessment in elderly head and neck cancer patients undergoing curative radiotherapy identifies evolution of multidimensional health problems and is indicative of quality of life: geriatric assessment in elderly head and neck cancer. Eur J Cancer Care (Engl). 2014;23(3):401–12. https://doi.org/10.1111/ecc.12179.
47. Roy N, Merrill RM, Gray SD, Smith EM. Voice disorders in the general population: prevalence, risk factors, and occupational impact. Laryngoscope.

2005;115(11):1988–95. https://doi.org/10.1097/01.mlg.0000179174.32345.41.
48. Bhattacharyya N. The prevalence of voice problems among adults in the United States. Laryngoscope. 2014;124(10):2359–62. https://doi.org/10.1002/lary.24740.
49. Cohen SM, Kim J, Roy N, Asche C, Courey M. Direct health care costs of laryngeal diseases and disorders. Laryngoscope. 2012;122(7):1582–8. https://doi.org/10.1002/lary.23189.
50. Çiyiltepe M, Şenkal ÖA. The ageing voice and voice therapy in geriatrics. Aging Clin Exp Res. 2017;29(3):403–10. https://doi.org/10.1007/s40520-016-0597-3.
51. Kandogan T, Seifert E. Influence of aging and sex on voice parameters in patients with unilateral vocal cord paralysis. Laryngoscope. 2005;115(4):655–60. https://doi.org/10.1097/01.mlg.0000161344.23128.9d.
52. Hagen P, Lyons GD, Nuss DW. Dysphonia in the elderly: diagnosis and management of age-related voice changes. South Med J. 1996;89(2):204–7.
53. Ohno T, Hirano S, Rousseau B. Age-associated changes in the expression and deposition of vocal fold collagen and Hyaluronan. Ann Otol Rhinol Laryngol. 2009;118(10):735–41. https://doi.org/10.1177/000348940911801009.
54. Johns MM, Arviso LC, Ramadan F. Challenges and opportunities in the Management of the Aging Voice. Otolaryngol Neck Surg. 2011;145(1):1–6. https://doi.org/10.1177/0194599811404640.
55. Honjo I, Isshiki N. Laryngoscopic and voice characteristics of aged persons. Arch Otolaryngol - Head Neck Surg. 1980;106(3):149–50. https://doi.org/10.1001/archotol.1980.00790270013003.
56. Pontes P, Brasolotto A, Behlau M. Glottic characteristics and voice complaint in the elderly. J Voice. 2005;19(1):84–94. https://doi.org/10.1016/j.jvoice.2004.09.002.
57. Lundy DS, Silva C, Casiano RR, Lu FL, Xue JW. Cause of hoarseness in elderly patients. Otolaryngol Neck Surg. 1998;118(4):481–5. https://doi.org/10.1177/019459989811800409.
58. Roy N, Gray SD, Simon M, Dove H, Corbin-Lewis K, Stemple JC. An evaluation of the effects of two treatment approaches for teachers with voice disorders: a prospective randomized clinical trial. J Speech Lang Hear Res. 2001;44(2):286–96. https://doi.org/10.1044/1092-4388(2001/023).
59. Woo P, Casper J, Colton R, Brewer D. Dysphonia in the aging: physiology versus disease. Laryngoscope. 1992;102(2):139–44. https://doi.org/10.1288/00005537-199202000-00007.
60. Ward PH, Colton R, McConnell F, Malmgren L, Kashima H, Woodson G. Aging of the voice and swallowing. Otolaryngol Neck Surg. 1989;100(4):283–6. https://doi.org/10.1177/019459988910000405.
61. Hartl DM, Hans S, Vaissière J, Riquet M, Brasnu DF. Objective voice quality analysis before and after onset of unilateral vocal fold paralysis. J Voice. 2001;15(3):351–61. https://doi.org/10.1016/S0892-1997(01)00037-6.
62. Rapoport SK, Menier J, Grant N. Voice changes in the elderly. Otolaryngol Clin N Am. 2018;51(4):759–68. https://doi.org/10.1016/j.otc.2018.03.012.
63. Choi YH, Kim SH, Kim IG, Lee JH, Kwon SK. Injectable basic fibroblast growth factor-loaded alginate/hyaluronic acid hydrogel for rejuvenation of geriatric larynx. Acta Biomater. 2019;89:104–14. https://doi.org/10.1016/j.actbio.2019.03.005.
64. Seino Y, Allen JE. Treatment of aging vocal folds: surgical approaches. Curr Opin Otolaryngol Head Neck Surg. 2014;22(6):466–71. https://doi.org/10.1097/MOO.0000000000000099.
65. Ney DM, Weiss JM, Kind AJH, Robbins J. Senescent swallowing: impact, strategies, and interventions. Nutr Clin Pract. 2009;24(3):395–413. https://doi.org/10.1177/0884533609332005.
66. Patel DA, Krishnaswami S, Steger E, et al. Economic and survival burden of dysphagia among inpatients in the United States. Dis Esophagus. 2018;31(1) https://doi.org/10.1093/dote/dox131.
67. Baijens LW, Clavé P, Cras P, et al. European Society for Swallowing Disorders – European Union geriatric medicine society white paper: oropharyngeal dysphagia as a geriatric syndrome. Clin Interv Aging. 2016;11:1403–28. https://doi.org/10.2147/CIA.S107750.
68. Rofes L, Arreola V, Almirall J, et al. Diagnosis and Management of Oropharyngeal Dysphagia and its Nutritional and respiratory complications in the elderly. Gastroenterol Res Pract. 2011;2011:1–13. https://doi.org/10.1155/2011/818979.
69. Yoshikawa M, Yoshida M, Nagasaki T, Tanimoto K, Tsuga K, Akagawa Y. Effects of tooth loss and denture wear on tongue-tip motion in elderly dentulous and edentulous people. J Oral Rehabil. 2008;35(12):882–8. https://doi.org/10.1111/j.1365-2842.2008.01882.x.
70. Nicosia MA, Hind JA, Roecker EB, et al. Age effects on the temporal evolution of isometric and swallowing pressure. J Gerontol A Biol Sci Med Sci. 2000;55(11):M634–40.
71. Kikutani T, Tamura F, Nishiwaki K, et al. Oral motor function and masticatory performance in the community-dwelling elderly. Odontology. 2009;97(1):38–42. https://doi.org/10.1007/s10266-008-0094-z.
72. Ouanounou A. Xerostomia in the geriatric patient: causes, Oral manifestations, and treatment. Compend Contin Educ Dent Jamesburg NJ 1995. 2016;37(5):306–11;quiz312.
73. Anil S, Vellappally S, Hashem M, Preethanath RS, Patil S, Samaranayake LP. Xerostomia in geriatric patients: a burgeoning global concern. J Investig Clin Dent. 2016;7(1):5–12. https://doi.org/10.1111/jicd.12120.
74. Robbins J, Coyle J, Rosenbek J, Roecker E, Wood J. Differentiation of Normal and abnormal airway protection during swallowing using the penetration–

aspiration scale. Dysphagia. 1999;14(4):228–32. https://doi.org/10.1007/PL00009610.
75. Logemann JA, Pauloski BR, Rademaker AW, Colangelo LA, Kahrilas PJ, Smith CH. Temporal and biomechanical characteristics of oropharyngeal swallow in younger and older men. J Speech Lang Hear Res JSLHR. 2000;43(5):1264–74.
76. Achem SR, Devault KR. Dysphagia in aging. J Clin Gastroenterol. 2005;39(5):357–71.
77. Bizzotto A, Iacopini F, Landi R, Costamagna G. Zenker's diverticulum: exploring treatment options. Acta Otorhinolaryngol Ital Organo Uff Della Soc Ital Otorinolaringol E Chir Cerv-facc. 2013;33(4):219–29.
78. Bock JM, Van Daele DJ, Gupta N, Blumin JH. Management of Zenker's diverticulum in the endoscopic age: current practice patterns. Ann Otol Rhinol Laryngol. 2011;120(12):796–806. https://doi.org/10.1177/000348941112001205.
79. Nesheiwat Z, Antunes C. Zenker Diverticulum. In: StatPearls. Treasure Island (FL): StatPearls Publishing; 2019. http://www.ncbi.nlm.nih.gov/books/NBK499996/. Accessed 8 May 2019.
80. Howell RJ, Giliberto JP, Harmon J, et al. Open versus endoscopic surgery of Zenker's diverticula: a systematic review and meta-analysis. Dysphagia. 2019; https://doi.org/10.1007/s00455-019-09994-9.
81. Smith SR, Genden EM, Urken ML. Endoscopic stapling technique for the treatment of Zenker diverticulum vs standard open-neck technique: a direct comparison and charge analysis. Arch Otolaryngol Head Neck Surg. 2002;128(2):141–4.
82. Philippsen LP, Weisberger EC, Whiteman TS, Schmidt JL. Endoscopic stapled diverticulotomy: treatment of choice for Zenker's diverticulum. Laryngoscope. 2000;110(8):1283–6. https://doi.org/10.1097/00005537-200008000-00011.
83. Aslam M, Vaezi MF. Dysphagia in the elderly. Gastroenterol Hepatol. 2013;9(12):784–95.
84. Ekberg O, Hamdy S, Woisard V, Wuttge-Hannig A, Ortega P. Social and psychological burden of dysphagia: its impact on diagnosis and treatment. Dysphagia. 2002;17(2):139–46. https://doi.org/10.1007/s00455-001-0113-5.
85. DePippo KL, Holas MA, Reding MJ. Validation of the 3-oz water swallow test for aspiration following stroke. Arch Neurol. 1992;49(12):1259–61.
86. Oguchi K, Saitoh E, Baba M, Kusudo S, Tanaka T, Onogi K. The repetitive saliva swallowing test (RSST) as a screening test of functional dysphagia. (2). Validity of RSST. Jpn J Rehabil Med. 2000;37(6):383–8. https://doi.org/10.2490/jjrm1963.37.383.
87. Palli C, Fandler S, Doppelhofer K, et al. Early dysphagia screening by trained nurses reduces pneumonia rate in stroke patients: a clinical intervention study. Stroke. 2017;48(9):2583–5. https://doi.org/10.1161/STROKEAHA.117.018157.
88. Składzień J, Skalska A. Challenges of modern medicine - geriatric otolaryngology or the advantages of cooperation between an otolaryngologist and a geriatrician. Otolaryngol Pol. 2018;72(6):1–5. https://doi.org/10.5604/01.3001.0012.7218.
89. Foley DM, Frick KD, Lin FR. Association between hearing loss and healthcare expenditures in older adults. J Am Geriatr Soc. 2014;62(6):1188–9. https://doi.org/10.1111/jgs.12864.
90. Lin FR. Hearing loss and falls among older adults in the United States. Arch Intern Med. 2012;172(4):369. https://doi.org/10.1001/archinternmed.2011.728.
91. Liljas AEM, Wannamethee SG, Whincup PH, et al. Socio-demographic characteristics, lifestyle factors and burden of morbidity associated with self-reported hearing and vision impairments in older British community-dwelling men: a cross-sectional study. J Public Health. 2016;38(2):e21–8. https://doi.org/10.1093/pubmed/fdv095.
92. Liu X, Yan D. Ageing and hearing loss. J Pathol. 2007;211(2):188–97. https://doi.org/10.1002/path.2102.
93. Gates GA, Mills JH. Presbycusis. Lancet. 2005;366(9491):1111–20. https://doi.org/10.1016/S0140-6736(05)67423-5.
94. Lee K-Y. Pathophysiology of age-related hearing loss (peripheral and central). Korean J Audiol. 2013;17(2):45. https://doi.org/10.7874/kja.2013.17.2.45.
95. Merchant SN, Nadol JB Jr. Schuknecht's pathology of the ear. 3rd ed. People's Medical Pub. House-USA: Shelton, CT; 2010.
96. Schuknecht HF, Gacek MR. Cochlear pathology in Presbycusis. Ann Otol Rhinol Laryngol. 1993;102(1_suppl):1–16. https://doi.org/10.1177/00034894931020S101.
97. Mills DM, Schmiedt RA. Metabolic Presbycusis: differential changes in auditory brainstem and Otoacoustic emission responses with chronic furosemide application in the gerbil. JARO - J Assoc Res Otolaryngol. 2004;5(1):1–10. https://doi.org/10.1007/s10162-003-4004-3.
98. Gallacher J, Ilubaera V, Ben-Shlomo Y, et al. Auditory threshold, phonologic demand, and incident dementia. Neurology. 2012;79(15):1583–90. https://doi.org/10.1212/WNL.0b013e31826e263d.
99. Lin FR, Metter EJ, O'Brien RJ, Resnick SM, Zonderman AB, Ferrucci L. Hearing loss and incident dementia. Arch Neurol. 2011;68(2) https://doi.org/10.1001/archneurol.2010.362.
100. Nirmalasari O, Mamo SK, Nieman CL, et al. Age-related hearing loss in older adults with cognitive impairment. Int Psychogeriatr. 2017;29(1):115–21. https://doi.org/10.1017/S1041610216001459.
101. Lin FR, Ferrucci L, An Y, et al. Association of hearing impairment with brain volume changes in older adults. NeuroImage. 2014;90:84–92. https://doi.org/10.1016/j.neuroimage.2013.12.059.
102. Peelle JE, Troiani V, Grossman M, Wingfield A. Hearing loss in older adults affects neural systems supporting speech comprehension. J Neurosci.

2011;31(35):12638–43. https://doi.org/10.1523/JNEUROSCI.2559-11.2011.

103. Chien W, Lin FR. Prevalence of hearing aid use among older adults in the United States. Arch Intern Med. 2012;172(3):292–3. https://doi.org/10.1001/archinternmed.2011.1408.
104. Mosnier I, Bebear J-P, Marx M, et al. Improvement of cognitive function after Cochlear implantation in elderly patients. JAMA Otolaryngol Neck Surg. 2015;141(5):442. https://doi.org/10.1001/jamaoto.2015.129.
105. Voelter C, Götze L, Dazert S, Falkenstein M, Thomas JP. Can cochlear implantation improve neurocognition in the aging population? Clin Interv Aging. 2018;13:701–12. https://doi.org/10.2147/CIA.S160517.
106. Kecskeméti N, Szőnyi M, Küstel M, Gáborján A, Tamás L, Répássy G. Cochlear implantation under local anesthesia: a possible alternative for elderly patients. Eur Arch Otorhinolaryngol. 2019; https://doi.org/10.1007/s00405-019-05407-7.
107. Jahn K, Kressig RW, Bridenbaugh SA, Brandt T, Schniepp R. Dizziness and unstable gait in Old age. Dtsch Aerzteblatt Online. 2015; https://doi.org/10.3238/arztebl.2015.0387.
108. Ji L, Zhai S. Aging and the peripheral vestibular system. J Otol. 2018;13(4):138–40. https://doi.org/10.1016/j.joto.2018.11.006.
109. Agrawal Y, Carey JP, Della Santina CC, Schubert MC, Minor LB. Disorders of balance and vestibular function in US adults: data from the National Health and nutrition examination survey, 2001-2004. Arch Intern Med. 2009;169(10):938. https://doi.org/10.1001/archinternmed.2009.66.
110. Minor LB. Gentamicin-induced bilateral vestibular Hypofunction. JAMA. 1998;279(7):541. https://doi.org/10.1001/jama.279.7.541.
111. Lundberg YW, Zhao X, Yamoah EN. Assembly of the otoconia complex to the macular sensory epithelium of the vestibule. Brain Res. 2006;1091(1):47–57. https://doi.org/10.1016/j.brainres.2006.02.083.
112. Thalmann R, Ignatova E, Kachar B, Ornitz DM, Thalmann I. Development and maintenance of otoconia: biochemical considerations. Ann N Y Acad Sci. 2001;942:162–78.
113. Iwasaki S, Yamasoba T. Dizziness and imbalance in the elderly: age-related decline in the vestibular system. Aging Dis. 2015;6(1):38. https://doi.org/10.14336/AD.2014.0128.
114. Epley JM. The Canalith repositioning procedure: for treatment of benign paroxysmal positional Vertigo. Otolaryngol Neck Surg. 1992;107(3):399–404. https://doi.org/10.1177/019459989210700310.
115. Semont A, Freyss G, Vitte E. Curing the BPPV with a liberatory maneuver. Adv Otorhinolaryngol. 1988;42:290–3.
116. Iwasaki S, Chihara Y, Ushio M, Ochi A, Murofushi T, Yamasoba T. Effect of the canalith repositioning procedure on subjective visual horizontal in patients with posterior canal benign paroxysmal positional vertigo. Acta Otolaryngol (Stockh). 2011;131(1):41–5. https://doi.org/10.3109/00016489.2010.514008.
117. Ballester M, Liard P, Vibert D, Häusler R. Menière's disease in the elderly. Otol Neurotol Off Publ Am Otol Soc Am Neurotol Soc Eur Acad Otol Neurotol. 2002;23(1):73–8.
118. Ekvall Hansson E, Magnusson M. Vestibular asymmetry predicts falls among elderly patients with multi- sensory dizziness. BMC Geriatr. 2013;13(1) https://doi.org/10.1186/1471-2318-13-77.
119. Lin S-Z, Fan J-P, Sun A-H, Guan J, Liu H-B, Zhu Q-B. Efficacy of laser occlusion of posterior semicircular canal for benign paroxysmal positional vertigo: case report. J Laryngol Otol. 2010;124(5) https://doi.org/10.1017/S0022215109991381.
120. Canzi P, Manfrin M, Perotti M, et al. Translabyrinthine vestibular neurectomy and simultaneous cochlear implant for Ménière's disease. Acta Neurochir. 2017;159(1):123–30. https://doi.org/10.1007/s00701-016-2996-9.
121. Brosel S, Strupp M. The vestibular system and ageing. In: Birbrair A, editor. Pericyte biology - novel concepts, vol. 1109. Cham: Springer; 2019. p. 195–225. https://doi.org/10.1007/978-981-13-3681-2_8.

Thoracic Surgery in the Elderly

12

Donna C. Koo, Francis X. Carroll,
and Tracey L. Weigel

General thoracic surgery, i.e., non-cardiac thoracic surgery, is a broad field including relatively minor procedures such as tube thoracostomy, diagnostic video-assisted thoracoscopic surgery (VATS), and mediastinoscopy, as well as major operations primarily for resection of malignant or benign thoracic tumors. Operations such as lobectomy, esophagectomy, and thymectomy traditionally required large, painful incisions in the chest or upper abdomen. At high-volume academic centers, these surgeries are now being carried out by minimally invasive technique with VATS or robotic assistance (RATS) with improved patient outcomes (Fig. 12.1) [1, 2]. These procedures are still associated with significant morbidity and mortality. As the general population ages, so too does our patient population. As surgery, anesthesia, and perioperative care become less invasive and more advanced, major surgeries are increasingly considered in the care of elder patients. This has appropriately become an area of increased interest in research, such that the special challenges related to the surgical treatment of elder patients can be better understood and addressed. It is now recognized that there is a difference between chronologic age and biologic age, perhaps best understood and assessed by various frailty measurements. Similarly, with newer research and insight, it is recognized that in states of advanced age, physiology is altered. This may lead directly to enhanced care by targeting specific pathways. This chapter will attempt to address the preoperative assessment of older candidates for thoracic surgery, with a view toward optimizing patient selection for major thoracic operations and introducing the concept of prehabilitation when possible in preparation for major thoracic operations.

Perioperative Risk Assessment for the Geriatric Patient

With advancements in medicine and an aging population, physicians and surgeons are increasingly caring for a greater number of older patients [3]. By one study, greater than 50% of surgeries are now performed in patients older than 65 years. Although chronologic age is often associated with increased perioperative risk, increased age is only one variable in perioperative complications in the geriatric population [4]. Other causes of increased operative risk in elderly

D. C. Koo
New York Medical College, Valhalla, NY, USA

F. X. Carroll · T. L. Weigel (✉)
Division of Thoracic Surgery, New York Medical College, School of Medicine and Westchester Medical Center, Valhalla, NY, USA
e-mail: Tracey.Weigel@wmchealth.org

R. Latifi (ed.), *Surgical Decision Making in Geriatrics*,
https://doi.org/10.1007/978-3-030-47963-3_12

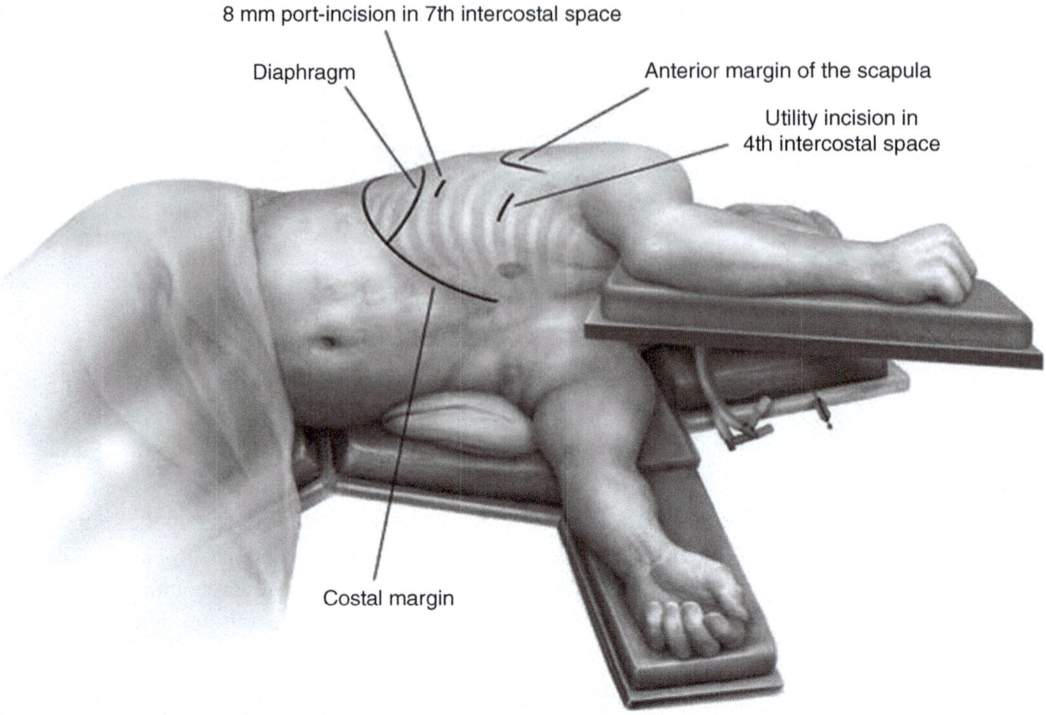

Fig. 12.1 Video-assisted thoracoscopic surgery (VATS). Minimally invasive techniques such as video-assisted thoracoscopic surgery (VATS) employed at high-volume academic centers minimize the need for large, painful incisions during procedures such as lobectomies that were traditionally performed as open procedures. VATS are typically associated with improved patient outcomes. (From Veronesi et al. [1], with permission)

patients include higher number of comorbidities, incidence of geriatric syndromes, higher levels of disability, and increased frailty.

Perioperative Risk Factors for the Geriatric Patient

Comorbidity

Comorbidity is defined as the concurrent presence of more than one medical condition in the same individual. While only 35.3% of adults ages 65–79 in the USA have medical comorbidities, they are present in 70.2% of adults older than 80 years [5]. Geriatric patients typically have higher number of comorbidities when compared to their younger counterparts [4]. Such comorbidities include higher rates of anemia, cardiac disease, COPD, heart failure, diabetes and its complications, peripheral vascular disease, and renal failure. Chronic comorbidities have been shown to increase perioperative complications in a variety of operations including orthopedic, urologic, and thoracic procedures [6–9].

Disability

Another area of operative risk for geriatric patients is associated disability, which is more prevalent in older populations. Disability is failure of independence in performing acts of daily living such as eating, bathing, dressing, transferring objects, toileting, and continence. Acts of daily living (ADL) include fundamental tasks in self-care, while instrumental acts of daily living (IADL) include additional tasks of household management [5]. As adults age, they become increasingly unable to live independently and rely on caregivers to manage daily tasks. A 2008 study by the SIOG surgical task force demonstrated that the likelihood of complications is increased by 50% when patients have dependency in IADL, abnormal performance status, or impaired scores by other various measures [10]. In fact, the presence of dependent ADL was the best predictor of extended length of stay in the hospital after surgery. Such studies have identified disability as a better predictor of postsurgical outcomes compared to presence of comorbidities. Unfortunately, up to 20–30% of adults living in communities have impairments in ADL or IADL [5].

Geriatric Syndromes

Other sources of increased operative risk in geriatric patients are geriatric syndromes [4]. These syndromes include preoperative cognitive impairment and preoperative malnutrition. Cognitive impairment is associated with postoperative delirium, while malnutrition can be associated with an increase in surgical complications and can be seen even in "healthy" geriatric patients. Malnutrition in geriatric patients confers a high risk of mortality with malnourished patients suffering from a mortality rate 55% greater than their adequately nourished peers [11]. Studies estimate that up to 60% of geriatric hospitalized patients are malnourished [11].

Frailty

Decreased physiologic reserve also contributes to increased perioperative risk. In particular, cardiorespiratory reserve can be affected by older patients who demonstrate difficulty walking on flat ground and up a flight of stairs. Across all age groups, lower anaerobic exercise threshold is predictive for adverse postoperative outcomes. The concept of "frailty" in recent literature is defined as a state of decreased physiological reserve and resiliency associated with age that causes patients to have decreased tolerance for handling stressors such as illness or surgery [12]. Increased frailty correlates with morbidity and mortality risk postoperatively in geriatric patients, and higher scores on various frailty scales are correlated with increased length of stay and increased functional decline.

Tools for Preoperative Assessment

Screening for factors contributing to increased perioperative risk in the geriatric patient can help identify areas of optimization prior to surgery. To date, several preoperative assessments exist for assessing such risk, including the Comprehensive Geriatric Assessment (CGA), Abbreviated CGA (aCGA), Preoperative Assessment in Elderly Cancer Patients (PACE), and various preoperative assessments using frailty, disability, and comorbidity [10, 13–16] (Table 12.1). Together, geriatric assessment markers for frailty, disability, and comorbidity have a high sensitivity and specificity when predicting 6-month mortality rates [15]. The use of preoperative assessments, such as CGA and aCGA in elderly patients, has been shown to improve surgical outcomes [4].

Table 12.1 Comprehensive Geriatric Assessment – domains of health

Physical medical conditions	Comorbid conditions and disease severity Medication list Nutritional status Past medical history
Mental health	Cognitive status Mood Fears and anxieties
Functional status	Gait, mobility, and balance Activities of daily living Instrumental acts of daily living Patient's priorities for independent living and functional status
Social health	Social network Support Socioeconomic status
Environment	Housing and facilities Ability to use technology Availability of transport Accessibility to local resources

Adapted from Welsh et al. [16], with permission
The Comprehensive Geriatric Assessment (CGA) describes domains of health that physicians should evaluate in geriatric patients [16]. This and other such scales can be used to assess surgical risk preoperatively

Comorbidity

The burden of comorbidity in geriatric patients can be assessed using various scales. One such scale is the Charlson Comorbidity Index (CCI) that contains 19 categories of comorbid conditions. These conditions are assigned weight values based on their independent 1-year mortality rates [15]. Other measures of comorbidity include American Society of Anesthesiologists (ASA) score and polypharmacy [15].

Disability

Disability and dependence can be measured using scales such as the Katz Index of Independence in Activities of Daily Living [15, 17]. In this scale, patients are given 1 point for ability to maintain independence in each of the six domains (eating, bathing, dressing, transferring objects, toileting, and continence). Higher scores are associated with higher levels of independence and improved postsurgical outcomes (Table 12.2).

Geriatric Syndromes

Preoperative cognitive impairment and malnutrition can be assessed by a variety of scales. Cognitive impairment can be measured using traditional tests such as the Montreal Cognitive Assessment (MoCA) or the Mini-Mental State Examination (MMSE). Clinical malnutrition can be assessed using scales such as the MNA-SF (Mini-Nutritional Assessment-Short Form) or using serum albumin levels [18].

Frailty

To date, frailty can be measured using two main scales. The Frailty Index (FI) is a comprehensive geriatric assessment that scores patients based on many factors including cognitive status, emotional well-being, motivation, communication, strength, mobility, balance, elimination, nutrition, ADL, IADL, quality of sleep, social support, and presence of comorbidities [19, 20] (Table 12.3). Patients with a FI score <0.10 are considered non-frail, while patients whose FI scores are between 0.1 and 0.2 are considered vulnerable. Patients with FI scores between 0.21 and 0.45 are considered frail, and FI scores >0.45 are considered the frailest. Although most of these variables are self-reported, this scale can be time-consuming due to potential cognitive deficits present in the elderly [12]. Despite the time investment, however, the Frailty Index identifies specific areas that can be optimized preoperatively due to its comprehensive nature. Several scales have attempted to improve on the Frailty Index, such as Edmonton Frail Scale, Modified Frailty Index (mFI), and Clinical Frail Scale.

Another main assessment used to determine frailty, the Frailty Phenotype, focuses more on the clinical manifestations of frailty. In this scale, points are assigned to various domains including unintentional weight loss >10 pounds in 1 year, self-reported exhaustion, weakness as determined by grip strength, slow walking speed

Table 12.2 Katz Index of Independence in Activities of Daily Living

Activities	Independence	Dependence
	(1 Point) Performed *without* supervision, direction, or assistance	(0 Points) Performed *with* supervision, direction, or assistance
Eating	Gets food from plate into mouth without help. Food may be prepared by another person	Needs partial or total help with feeding or requires parenteral feeding
Bathing	Bathes self completely or needs help bathing only a single part of the body (back, genital area, disabled extremity)	Needs help with bathing more than one part of the body and getting in or out of bathing area or requires total bathing
Dressing	Gets clothes from storage area and puts on clothes and outer garments with fasteners. May have help tying shoes	Needs help dressing self or needs to be completely dressed
Transferring	Moves in and out of bed or chair unassisted. May use mechanical transfer aids	Needs help moving from bed to chair or requires a complete transfer
Toileting	Goes to toilet, gets on and off, arranges clothes, and cleans genital area without help	Needs help transferring to the toilet and cleaning self or uses a bedpan or commode
Continence	Maintains complete control over urination and defecation	Is partially or totally incontinent of bladder or bowel

From Katz et al. [17], with permission

The Katz Index of Independence in Activities of Daily Living is a scale in which patients are given 1 point for ability to maintain independence in each of the six domains (eating, bathing, dressing, transferring objects, toileting, and continence). Higher scores are associated with higher levels of independence and improved postsurgical outcomes [17]

Table 12.3 Frailty Index

Variables			Deficit points		
Needs help bathing		Yes = 1	No = 0		
Needs help dressing		Yes = 1	No = 0		
Needs help transferring from chair		Yes = 1	No = 0		
Needs help walking around house		Yes = 1	No = 0		
Needs help eating		Yes = 1	No = 0		
Needs help grooming		Yes = 1	No = 0		
Needs help toileting		Yes = 1	No = 0		
Needs help using stairs		Yes = 1	No = 0		
Needs help lifting 10 lbs		Yes = 1	No = 0		
Needs help shopping		Yes = 1	No = 0		
Needs help with housework		Yes = 1	No = 0		
Needs help with meal preparation		Yes = 1	No = 0		
Needs help managing medications		Yes = 1	No = 0		
Needs help with finances		Yes = 1	No = 0		
Lost >10 lbs in the last year		Yes = 1	No = 0		
Self-rating of health	Poor = 1	Fair = 0.75	Good = 0.5	Very good = 0.25	Excellent = 0
How health has changed in last year	Worse = 1	Same = 0	Better = 0		
Stayed in bed > half the day in last month due to health reasons		Yes = 1	No = 0		
Cut down on usual activity in last month		Yes = 1	No = 0		
Walks outside		<3 days = 1	≤3 days = 0		
Feels everything requires a lot of effort		Most of the time = 1	Sometimes = 0.5	Rarely = 0	
Feels depressed		Most of the time = 1	Sometimes = 0.5	Rarely = 0	
Feels happy		Rarely = 1	Sometimes = 0.5	Most of the time = 0	
Feels lonely		Most of the time = 1	Sometimes = 0.5	Rarely = 0	
Has trouble getting going		Most of the time = 1	Sometimes = 0.5	Rarely = 0	
Has high blood pressure		Yes = 1	Suspect = 0.5	No = 0	
History of heart attack		Yes = 1	Suspect = 0.5	No = 0	
Has CHF		Yes = 1	Suspect = 0.5	No = 0	
History of stroke		Yes = 1	Suspect = 0.5	No = 0	
Has cancer		Yes = 1	Suspect = 0.5	No = 0	
Has diabetes		Yes = 1	Suspect = 0.5	No = 0	
Has arthritis		Yes = 1	Suspect = 0.5	No = 0	
Has chronic lung disease		Yes = 1	Suspect = 0.5	No = 0	
MMSE score	<10 = 1	11–17 = 0.75	18–20 = 0.5	20–24 = 0.25	>24 = 0
Peak flow		*Men:* ≤340 = 1	*Women:* ≤310 = 1		
Shoulder strength		*Men:* ≤12 = 1	*Women:* ≤9 = 1		
BMI		*Men:* <18.5 = 1 >30 = 1	*Women:* >25 and <30 = 0.5		

(continued)

Table 12.3 (continued)

Variables			Deficit points		
Grip strength	*Men:* BMI ≤24: BMI 24.1–28: BMI >28:	GS ≤29 = 1 GS ≤30 = 1 GS ≤32 = 1	*Women:* BMI ≤23: BMI 23.1–26: BMI 16.1–29: BMI >29	GS ≤17 = 1 GS ≤17.3 = 1 GS ≤18 = 1 GS ≤21 = 1	
Usual walking pace		>16 s = 1	<16 s = 0		
Rapid walking pace		>10 s = 1	<10 s = 0		

From Searle et al. [20], with permission.
The Frailty Index (FI) is a geriatric assessment that scores patients based on many the presence of deficits in many domains including cognitive status, emotional well-being, motivation, communication, strength, mobility, balance, elimination, nutrition, ADL, IADL, quality of sleep, social support, and presence of comorbidities [19, 20]. The total number of points is tallied and divided by 40 (the number of variables in the index). Patients with a FI score <0.10 are considered non-frail, while patients whose FI scores are between 0.1 and 0.2 are considered vulnerable. Patients with FI scores between 0.21 and 0.45 are considered frail, and FI scores >0.45 are considered most frail

(measured in seconds needed to walk 15 feet), and low physical activity (as measured by the Minnesota Leisure Time Activity Questionnaire). Patients with deficits in more than three domains are considered frail, while patients with deficits in 1–2 domains are considered pre-frail.

Perioperative Risk in Thoracic Surgery

Common postoperative complications for thoracic surgery include atrial fibrillation, allogenic blood transfusion, pneumonia, reintubation, delirium, acute renal failure, ARDS, sepsis, cerebral vascular accident, pulmonary embolism myocardial infarction, bronchopleural fistula, and empyema [21]. It is estimated that the risk of postoperative complications in thoracic non-cardiac procedures can be as high as 27%. Patients who require reintubation or develop pneumonia and delirium postoperatively have a higher risk of death 3–18 months after surgery, while patients who develop sepsis have higher risk of death after 18 months. Patients who require allogenic blood transfusions have higher risk of death in both time frames.

Fitness for surgery can be determined via functional capacity tests to stratify surgical risk in thoracic surgery candidates. The Shuttle Walk Test (SWT) can be used as an initial screening tool that does not require cardiopulmonary testing [22]. Patients walk at an incrementally increasing pace between two cones placed 10 m apart. A pre-recorded "beep" signals the allotted time that patients have to walk 10 m at each speed. Patients who are able to walk more than 400 m were found to have lower rates of complications after thoracic surgery. Other simple tests, such as stair climbing, can also be used. In general, patients with maximal oxygen consumption less than 15 mL/kg or with an anaerobic exercise threshold less than 10–11 mL/kg/min were defined to have poor functional status and at higher risk for postsurgical complications [21].

Additionally, preoperative pulmonary function tests can be used to determine appropriateness of thoracic surgery. In particular, predicted postoperative FEV1% (ppoFEV1%) can be used to predict pulmonary complications.

$$\text{ppoFEV1\%} = \text{preoperative FEV1\%} \cdot 1 - \frac{\text{\%functional lung tissue removed}}{100}$$

In general, a ppoFEV1% >40% is considered low-risk for respiratory complications, while a ppoFEV1% < 30% is considered high-risk [23]. A nuclear quantitative perfusion scan should be performed to determine the fraction of function that comes from each lobe to attempt to quantify post-op predicted FEV1 and FVC, i.e., after the affected lobe is removed.

Thoracic Surgery in Relation to the Geriatric Patient

The issue of perioperative risk in geriatrics is an especially relevant problem for thoracic surgery. Thoracic surgery patients are typically older and have comorbidities pertaining to pulmonary disease [24]. This causes them to have decreased performance on screening tests traditionally used for surgical risk stratification (shuttle walk and stair climbing). However, despite increased surgical risk, elderly patients with early-stage lung cancer still benefit from surgical treatment due to poor survival rates in unresected lung cancer [25].

Tools for Preoperative Assessment of Geriatric Patients in Thoracic Surgery

Several screening tools exist for determining perioperative risk for geriatric patients undergoing thoracic surgery. One such tool is the Thoracic Onco Geriatric Assessment (TOGA) which screens patients for functional status, nutrition, and mood via a combination of other scales [26, 27]. Functional status is assessed through independence with ADLs and IADLs [17, 28], while nutrition is assessed using the NSI Nutrition Health Checklist [29]; mood is assessed using the Geriatric Depression Scale [30]. The TOGA is quick to administer and allows physicians to identify areas of risk prior to surgery, as well as determine the effect of surgery on a patient's independence (Table 12.4).

Another tool used to predict postoperative complications for patients older than 75 years undergoing video-assisted thoracoscopic surgery (VATS) is the Simplified Comorbidity Scale (SCS) [31]. The SCS scores patients based on whether or not they have various comorbidities, including smoking status, diabetes, renal insufficiency, respiratory distress, cardiovascular disease, cancer, and alcoholism. Higher SCS scores are associated with more frequent postoperative complications.

Table 12.4 Thoracic Onco Geriatric Assessment (TOGA)

Screening test	Assesses	# of items	Estimated administration time
Activities of Daily Living (ADLs) [26]	Functional status	6	2–5 minutes
Instrumental Activities of Daily Living (IADLs) [27]	Functional status	8	2–5 minutes
NSI Nutrition Health Checklist [28]	Nutrition	10	3–6 minutes
Geriatric Depression Scale [29]	Mood	15	3–6 minutes

The Thoracic Onco Geriatric Assessment (TOGA) is a preoperative screening tool for thoracic general surgery that uses various pre-existing screening tools to assess functional status, nutrition, and mood. It is quick to administer and allows surgeons to identify areas of risk prior to surgery, as well as determine the effect of surgery on a patient's independence [26, 27]

Improving Surgical Outcomes for the Geriatric Patient

Ultimately, the goal of assessing perioperative risk in the geriatric patient is to determine how we can best mitigate such risks and prevent adverse outcomes. Several perioperative domains of intervention have been used with some success to improve prognosis for geriatric patients: shared decision-making, prehabilitation, and interdisciplinary geriatric co-management [12].

Shared Decision-Making

Shared decision-making refers to the practice of having in-depth conversations with patients and their families about goals of care [32]. These conversations allow physicians to determine decision-making capacity and allow patients to have realistic expectations regarding the outcomes of surgery and help them to make informed decisions prior to surgery [33]. Topics of discussion include personal goals, treatment preferences, as well as the importance of specific outcomes, such as the possibility of loss of independence. Additionally, it is important to ensure that patients undergoing surgery should have advance directives and a designated health care proxy in place prior to surgery. If applicable, palliative care should be consulted prior to surgery for high-risk geriatric patients with a particularly poor prognosis.

Prehabilitation

Prehabilitation and preoperative exercise therapy have been shown to improve postoperative outcomes and decrease postoperative complications and length of stay in geriatric patients undergoing surgery, especially thoracic surgery [34]. Prehabilitation, or the optimization of geriatric syndromes, is especially helpful for patients with high frailty scores. Interventions include exercise, nutritional support, and psychological interventions lasting greater than 4 weeks pre-op [12]. Poor exercise capacity and muscle weakness are predictors of poor surgical outcomes and high complication rates [35]; thus, prehabilitation to improve exercise tolerance and muscle strength can optimize patients for surgery and improve surgical outcomes. This is especially true in general thoracic surgery, where preoperative exercise capacity can be used to predict the length of stay after thoracic cancer surgeries [36].

Prehabilitation regimens can take place in prehabilitation centers or in patients' homes [37]. Patients receiving prehabilitation care at formalized centers are required to attend various supervised exercise sessions weekly for a minimum of 4 weeks. Exercises can include aerobic exercises at target heart rates, high-intensity interval training, and stretching [38]. Home-based exercise training programs can similarly include aerobic activities such as walking and cycling and/or muscle training exercises [35]. Home-based prehabilitation programs have been shown to be equally as effective as center-based prehabilitation programs [37]. Improved surgical outcomes after home-based prehabilitation have been noted for patients undergoing lung resection, as well as coronary artery bypass graft (CABG) and valve surgeries [35, 39]. Additionally, home-based regimens are more easily accessible to patients who are unable to travel or stay in care centers due to physical condition or other financial factors, thus increasing prehabilitation compliance.

Recent research has shown promising results in the use of biomarkers to predict the efficacy of certain components of prehabilitation. One molecule, surfactant protein-D (SP-D), has been proven to be a sensitive predictor [40]. SP-D, a lung-derived biomarker for inflammatory lung disease, was shown to decrease significantly after a week of preoperative pulmonary rehab for patients with lung cancer undergoing thoracic surgery. Lower serum levels of SP-D were also correlated with lower rates of postoperative pulmonary complications.

Other strategies used in prehabilitation programs include branched-chain amino acid and herbal medicine supplementation in addition to exercise, which have also shown improvement in postoperative complication rates after lobectomy [41, 42].

Interdisciplinary Geriatric Co-management

Perioperative outcomes improve when interdisciplinary teams work together toward preventing postoperative morbidity and mortality [43]. The Acute Care for Elders Program (ACE) is an inpatient interdisciplinary system that creates a team of a nurse coordinator, geriatrician, nurses, physical and occupational therapists, dietician, pharmacist, and social workers who communicate with the surgical and anesthesia teams to improve surgical outcomes. Currently, the Society of Perioperative Assessment and Quality Improvement (SPAQI) recommends creating interdisciplinary teams for frail patients undergoing high-risk (aortic and vascular) or intermediate (intraperitoneal/intrathoracic, head/neck, orthopedic, prostate, carotid) procedures [12].

General Perioperative Management Techniques for Reducing Surgical Risk

Perioperative management techniques for improving postoperative complications for general surgical patients hold true for geriatric thoracic surgical patients as well. Guidelines such as the Enhanced Recovery After Surgery (ERAS) and Enhanced Recovery After Thoracic Surgery (ERATS) recommend VTE prophylaxis and antibiotic prophylaxis in thoracic surgery patients to prevent the risk of pulmonary embolism and reduce the risk of postoperative infection [24]. Other techniques include early removal of chest tubes and Foley catheters, and discharging patients early, so they can sleep without the constant interruptions of the hospital. In particular, early removal of chest tubes after VATS has been shown to lower length of hospital stays, costs, and patient morbidities [44].

Postoperative Pain Control

Other recommendations by the ERATS include using multimodal non-opioid analgesics for postoperative pain control to circumvent the respiratory depressant effects of opioids [24]. The decision to use opioid and other systemic analgesics should be made cautiously in geriatric patients in particular [45]. Stress and pain reduction techniques such as guided imagery can also be useful techniques to reduce perioperative pain and anxiety. Guided imagery techniques direct patients to visualize and focus on calming scenes, such as walking along a beach, to invoke feelings of relaxation. Guided imagery has been associated with lower levels of distress in lung cancer patients and can also benefit caregivers [46]. Although guided imagery is typically associated with high levels of patient acceptance and satisfaction, they have not been shown to consistently improve pain and reduce postoperative analgesic requirements in the elderly [47]. Other studies argue that guided imagery techniques remain a cost-effective method of reducing perioperative anxiety [48] and can be a powerful non-pharmacologic tool that can be employed by patients to control pain [49].

Other non-narcotic pharmacologic regimens for postoperative pain control include usage of intraoperative intercostal nerve (ICN) blocks with liposomal bupivacaine (Exparel®) after thoracotomy [50]. Such techniques allow patients to have more long-term continuous pain relief (up to 48–72 hours) and decrease the need for additional needlesticks and IV drips. ICNs have also been associated with lower opioid usage in the first 24 hours after VATS [51]. They have also been associated with a significant decrease in pulmonary complications compared to patients receiving epidural analgesia [50]. This benefit may be due to the increased mobilization of patients postoperatively with sufficient analgesia, which has been shown to reduce postoperative complications [2].

Fluid Management

Fluid management provides another potential area for reducing postoperative risk. Although thoracic surgery patients are traditionally volume-restricted, recent ERAS/ERATS guidelines have advocated for preventing preoperative dehydration and hypoglycemia by allowing patients to drink clear liquids up to 2 hours prior to surgery [24]. Ultimately, the goal of fluid therapy is to maintain euvolemia intraoperatively and to resume enteral hydration as soon as possible during the postoperative period.

Incentive Spirometry

Finally, early use of incentive spirometry is thought to prevent atelectasis and improve lung re-expansion following major thoracic surgery [52] and can be used to assess lung function after surgery and reduce postoperative complications [53]. Despite the widespread use of incentive spirometry postoperatively, however, the literature does not indicate a significant reduction of length of stay or significantly improved pulmonary function for patients who use incentive spirometry compared to other postoperative breathing exercises and airway clearance techniques [54, 55]. However, preoperative instruction and practice with an incentive spirometer can optimize preoperative volumes and serve as a personal benchmark and realistic objectively quantifiable goal in the postoperative period.

Conclusion

All medical and surgical specialties are striving to serve our growing elder population better by studying and understanding of the alterations in physiology which occur in advanced age. In reference to thoracic surgery in general, rehabilitation with pulmonary rehab programs, medical optimization, and nutritional supplementation are long-standing and not new. Emphasis is now added to the elder population as their needs and disabilities are recognized as greater, with added perioperative risk due to their potential geriatric syndromes.

More accurate risk assessment in the elder or frail may also at times redirect therapies. Currently, medically inoperable patients with early-stage lung cancer may be offered stereotactic radiation therapy with good results in nonsurgical candidates. Sub-lobar resections, such as segmentectomy or wedge resection, may be offered to high-risk, elderly thoracic patients with subcentimeter lesions, instead of standard lobectomy. For patients with stage IIIa disease, who may be reasonably offered either induction chemotherapy followed by surgery or definitive chemotherapy/radiation therapy without surgery (as suggested by NCCN guidelines), more detailed preoperative assessment and shared decision-making may better inform our recommendations and patient choices. In instances when risk factors such as frailty, debilitation, and malnutrition can be favorably modified, patients may not need to be excluded from curative surgical options.

Researchers are continuing to investigate new measures of assessing and stratifying general thoracic surgical risk in older patients. Sarcopenia, which reflects wasting, malnutrition, debilitation, and frailty, is one potential direction. A recent study proposes a Psoas Index, calculated by measuring psoas muscle diameter on CT and dividing by BMI, as a risk assessment tool for patients considered for transcatheter aortic valve replacement (TAVR) versus surgical aortic valve replacement (SAVR) [56]. Another study suggests a link between sarcopenia and transforming growth factor beta, (TGF-B), satellite cell function, and impaired muscle repair [57]. It is further suggested that this pathway may be favorably affected by administration of angiotensin receptor antagonist. Ongoing research in the area of risk assessment and particularly risk modification by pharmacology, rehabilitation, nutritional support, or other means is vital, in particular as it pertains to the growing elder population.

References

1. Veronesi G, Novellis P, Bottoni E, Alloisio M. Robotic lobectomy: right upper lobectomy. Oper Tech Thorac Cardiovasc Surg. 2017;21:249–68.
2. Batchelor TJP, Ljunggvist O. A surgical perspective of ERAS guidelines in thoracic surgery. Curr Opin Anaesthesiol. 2019;32(1):17–22.
3. Etzoni DA, Kiu JH, Maggard MA, Ko CY. The aging population and its impact on the surgery workforce. Ann Surg. 2003;238(2):170–7.
4. Gordon AL, Evans BJ, Dhesi J. The physician's role in perioperative management of older patients undergoing surgery. Clin Med (Lond). 2017;17(4):357–9.
5. Fried LP, Ferrucci L, Darer J, Williamson JD, Anderson G. Untangling the concepts of disability, frailty, and comorbidity: implications for improved targeting and care. J Gerontol A Biol Sci Med Sci. 2004;59(3):255–63.
6. Everhart JS, Altneu E, Calhoun JH. Medical comorbidities are independent preoperative risk factors for surgical infection after total joint arthroplasty. Clin Orthop Relat Res. 2013;471(10):3112–9.
7. Schoenfeld AJ, Carey PA, Cleveland AW, Bader JO, Bono CM. Patient factors, comorbidities, and surgical characteristics that increase mortality and complication risk after spinal arthrodesis: A prognostic study based on 5,887 patients. Spine J. 2013;13(10):1171–9.
8. Townsend NT, Robinson TN. Surgical risk and comorbidity in older urologic patients. Clin Geriatr Med. 2015;31(4):591–601.
9. Irons JF, Martinez G. Complex, high-risk thoracic surgery – does risk always outweigh the benefit or can we manage it safely? Video-assist Thorac Surg. 2017;2:63–74.
10. Audisio RA, Pope D, Ramesh HS, Gennari R, van Leeuwen BL, West C, et al. Shall we operate? Preoperative assessment in elderly cancer patients (PACE) can help. A SIOG surgical task force prospective study. Crit Rev Oncol Hematol. 2008;65(2):156–63.
11. Orlandoni P, Venturini C, Peladic NJ, Costantini A, di Rosa M, Cola C, et al. Malnutrition upon hospital admission in geriatric patients: why assess it? Front Nutr. 2017;4:50.
12. Alvarez-Nebreda ML, Bentov N, Urman RD, Setia S, Huang JC, Pfeifer K, et al. Recommendations for preoperative management of frailty from the Society for

Perioperative Assessment and Quality Improvement (SPAQI). J Clin Anesth. 2018;47:33–42.
13. Rubenstein LZ, Stuck AE, Siu AL, Wieland D. Impacts of geriatric evaluation and management programs on defined outcomes: overview of the evidence. J Amer Geriatr Soc. 1991;39(9 Pt 2):8–16; discussion 17–8.
14. Overcash JA, Beckstead J, Moody L, Extermann M, Cobb S. The abbreviated comprehensive geriatric assessment (aCHA) for use in the older cancer patient as a prescreen: scoring and interpretation. Crit Rev Oncol Hematol. 2006;59(3):205–10.
15. Robinson TN, Eiseman B, Wallacce JI, Church SD, McFann KK, Pfister SM, et al. Redefining geriatric preoperative assessment using frailty, disability, and co-morbidity. Ann Surg. 2009;250(3):449–55.
16. Welsh TJ, Gordon AL, Gladman JR. Comprehensive geriatric assessment – a guide for the non-specialist. Int J Clin Pract. 2014;68(3):290–3.
17. Katz S, Downs TD, Cash HR, Grotz RC. Progress in development of the index of ADL. Gerontologist. 1970;10(1):20–30.
18. Sanchez-Rodriguez D, Annweiler C, Ronquillo-Moreno N, Tortosa-Rodriguez A, Guillen-Sola A, Vazquez-Ibar O, et al. Clinical application of the basic definition of malnutrition proposed by the European Society for Clinical Nutrition and Metabolism (ESPEN): comparison with classical tools in geriatric care. Arch Gerontol Geriatr. 2018;76:210–4.
19. Rockwood K, Mitnitski A. Frailty in relation to the accumulation of deficits. J Gerontol. 2007;52(7):722–7.
20. Searle SD, Mitnitski A, Gahbauer EA, Gill TM, Rockwood K. A standard procedure for creating a frailty index. BMC Geriatr. 2008;8:24.
21. Pedoto A, Perrino AC. Delayed recovery following thoracic surgery: persistent issues and potential interventions. Curr Opin Anaesthesiol. 2019;32(1):3–9.
22. Fennelly J, Potter L, Pompili C, Brunelli A. Performance in the shuttle walk test is associated with cardiopulmonary complications after lung resections. J Thorac Dis. 2017;9(3):789–95.
23. Nakahara K, Ohno K, Hashimoto J, Miyoshi S, Maeda H, Matsumara A. Prediction of postoperative respiratory failure in patients undergoing lung resection for lung cancer. Ann Thor Surg. 1988;46(5):549–52.
24. Teeter EG, Kolarczyk LM, Popescu WM. Examination of the enhanced recovery guidelines in thoracic surgery. Curr Opin Anaesthesiol. 2019;32(1):10–6.
25. Castillo M. An overview of perioperative considerations in elderly patients for thoracic surgery: demographics, risk/benefit, and resource planning. Curr Opin Anaesthesiol. 2018;31(1):1–5.
26. Kothari A, Philips S, Bretl T, Block K, Weigel T. Components of geriatric assessments predict thoracic surgery outcomes. J Surg Res. 2011;166(1):5–13.
27. Weigel T, Kothari A, Bretl T, Block K, Munoz-del-Rio A, LoConte NK. Use of a thoracic oncogeriatric assessment (TOGA) to predict surgical outcomes. J Clin Oncol. 2010;28(15):e19528.
28. Sonn U, Frandin K, Grimby G. Instrumental activities of daily living related to impairments and functional limitations in 70-year-olds and changes between 70 and 76 years of age. Scand J Rehabil Med. 1995;27(2):119–28.
29. de Groot LC, Beck AM, Schroll M, van Staveren WA. Evaluating the DETERMINE Your Nutritional Health Checklist and the Mini Nutritional Assessment as tools to identify nutritional problems in elderly Europeans. Eur J Clin Nutr. 1998;52(12):877–83.
30. Yesavage JA, Brink TL, Rose TL, Lum O, Huang V, Adey M, et al. Development and validation of a geriatric depression screening scale: a preliminary report. J Psychiatr Res. 1982-1983;17(1):37–49.
31. Haruki T, Yurugi Y, Wakahara M, Matsuoka Y, Miwa K, Araki K, et al. Simplified comorbidity score for elderly patients undergoing thoracoscopic surgery for lung cancer. Surg Today. 2017;47(6):718–25.
32. Mohanty S, Rosenthal RA, Russell MM, Neuman MD, Ko CY, Esnaola NF. Optimal preoperative management of the geriatric patient: a best practices guideline from the American College of Surgeons NSQIP and the American Geriatrics Society. J Am Coll Surg. 2016;222(5):930–47.
33. Chow WB, Rosentha RA, Merkow RP, Ko CY, Esnaola NF. Optimal preoperative assessment of the geriatric surgical patient: a best practices guideline from the American College of Surgeons National Surgical Quality Improvement Program and the American Geriatrics Society. J Am Coll Surg. 2012;215(4):453–66.
34. Pouwels S, Fiddelaers J, Teijink JA, Woorst JF, Siebenga J, Smeenk FW. Preoperative exercise therapy in lung surgery patients: a systematic review. Respir Med. 2015;109(12):1495–504.
35. Coats V, Maltais F, Simard S, Frechette E, Tremblay L, Riberio F, et al. Feasibility and effectiveness of a home-based exercise training program before lung resection surgery. Can Respir J. 2013;20(2):e10–6.
36. Weinstein H, Bates AT, Spaltro BE, Thaler HT, Steingart RM. Influence of preoperative exercise capacity on length of stay after thoracic cancer surgery. Ann Thorac Surg. 2007;84(1):197–202.
37. Anderson L, Sharp GA, Norton RJ, Dalal H, Dean SG, Jolly K, et al. Home-based versus centre-based cardiac rehabilitation. Cochrane Database Syst Rev. 2017;6:CD007130.
38. Stammers AN, Kehler DS, Afilalo J, Avery LJ, Bagshaw SM, Grocott HP, et al. Protocol for the PREHAB study-pre-operative rehabilitation for reduction of hospitalization after coronary bypass and valvular surgery: a randomised controlled trial. BMJ Open. 2015;5(3):e007250.
39. Waite I, Deshpande R, Baghai M, Massey T, Wendler O, Greenwood S. Home-based preoperative rehabilitation (prehab) to improve physical function and reduce hospital length of stay for frail patients undergoing coronary artery bypass graft and valve surgery. J Cardiothorac Surg. 2017;12(1):91–8.

40. Huang J, Lai Y, Gao K, Wang Y, Du H, Su J, et al. Surfactant protein-D: a sensitive predictor for efficiency of preoperative pulmonary rehabilitation. Int J Surg. 2017;41:136–42.
41. Hararda H, Takahama M, Nakao J, Yamashita Y, Taniyama K. Comprehensive preoperative pulmonary prehabilitation including intensive nutritional support reduces the postoperative morbidity rate of sarcopenia-related patients with lung cancer. Phys Med Rehabil. 2018;2(1):1–9.
42. Boujibar F, Bonnevie T, Debeaumont D, Bubenheim M, Cuvellier A, Peillon C, et al. Impact of prehabilitation on morbidity and mortality after pulmonary lobectomy by minimally invasive surgery: a cohort study. J Thorac Dis. 2018;10(4):2240–8.
43. Amrock LG, Deiner S. Perioperative frailty. Int Anesthesiol Clin. 2014;52(4):26–41.
44. Gottgens KW, Siebenga J, Belgers EH, van Huijstee PJ, Bollen EC. Early removal of chest tube after complete video-assisted thoracoscopic lobectomies. Eur J Cardiothorac Surg. 2011;39(4):575–8.
45. Luchting B, Azad SC. Pain therapy for the elderly patient: is opioid-free an option? Curr Opin Anaesthesiol. 2019;32(1):86–91.
46. Winger JG, Rand KL, Hanna N, Jalal SI, Einhorn LH, Birdas TJ, et al. Coping skills practice and symptom change: a secondary analysis of a pilot telephone symptom management intervention for lung cancer patients and their family caregivers. J Pain Symptom Manag. 2018;55(5):1341–9.
47. Haase O, Schwenk W, Hermann C, Muller JM. Guided imagery and relaxation in conventional colorectal resections: a randomized, controlled, partially blinded trial. Dis Colon Rectum. 2005;48(10):1955–63.
48. Renzi C, Peticca L, Pescatori M. The use of relaxation techniques in the perioperative management of proctological patients: preliminary results. Int J Color Dis. 2000;15(5–6):313–6.
49. Kwekkeboom KL. Pain management strategies used by patients with breast and gynecologic cancer with postoperative pain. Cancer Nurs. 2001;24(5):378–86.
50. Khalil KG, Boutrous ML, Irani AD, Miller CC 3rd, Pawelek TR, Estrera AL, et al. Operative intercostal nerve blocks with long-acting bupivacaine liposome for pain control after thoracotomy. Ann Thorac Surg. 2015;100(6):2013–8.
51. Kelley TM Jr, Bailey DW, Sparks P, Rice R, Caddell E, Currier H, Gallo D. Intercostal nerve blockade with Exparel® results in lower opioid usage during the first 24 hours after video-assisted thoracoscopic surgery. Am Surg. 2018;84(9):1433–8.
52. Agostini P, Singh S. Incentive spirometry following thoracic surgery: what should we be doing? Physiotherapy. 2009;95(2):76–82.
53. Agostini P, Calvert R, Subramanian H, Naidu B. Is incentive spirometry effective following thoracic surgery? Interact Cardiovasc Thorac Surg. 2008;792:297–300.
54. Agostini P, Naidu B, Cieslik H, Steyn R, Rajesh PB, Bishay E, et al. Effectiveness of incentive spirometry in patients following thoracotomy and lung resection including those at high risk for developing pulmonary complications. Thorax. 2013;68(6):580–5.
55. Tyson AF, Kendig CE, Mabedi C, Cairns BA, Charles AG. The effect of incentive spirometry on postoperative pulmonary function following laparotomy: a randomized clinical trial. JAMA Surg. 2015;150(3):229–36.
56. Kim HJ. You cannot assess until you measure it: using psoas index for measuring frailty. J Thor Dis. 2018;10(12):6412–3.
57. Burks TN, Andres-Mateos E, Marx R, Mejias R, Van Erp C, Simmers JL, et al. Losartan restores skeletal muscle remodeling and protects against disuse atrophy in sarcopenia. Sci Transl Med. 2011;3(82):82ra37.

13 Cardiac Surgery in the Elderly

Joshua B. Goldberg

The United States Census Bureau estimates by 2030, 20% of the United States will be older than 65, accounting for over 78 million people [1]. Cardiovascular disease is a leading cause of morbidity and mortality among all adult patients. Independent of age, gender, or ethnicity, heart disease accounts for greater than 20% of deaths in the United States [2]. With advancing age, heart disease becomes far and away the leading cause of death, accounting for over 25% of all deaths. Furthermore, the proportion of deaths attributed to heart disease increases with age. According to the CDC, among people aged 45–64, heart disease accounts for 20.8% of deaths, 25.1% of deaths among patients greater than 65, and 28.6 of deaths among patients greater than 85 (Fig. 13.1). Furthermore, the incidence and prevalence of cardiovascular disease is only expected to increase as the population continues to age.

Cardiac surgery is a surgical subspecialty that focuses on the surgical management of diseases of the heart and great vessels. Given the aging population and the incidence, prevalence, and associated mortality of cardiovascualr disease among the patients of advancing age, cardiac surgery is a field largely focused on the care of the elderly population. For instance, the average age of patients undergoing coronary artery bypass grafting or aortic valve replacement, two of the most commonly performed cardiac surgical procedures, is 74 years [2, 3]. Furthermore, recent advances in the management of valvular heart disease has resulted in surgeries being routinely performed on octogenarians and nonagenarians. This chapter will focus on the surgical management of coronary artery disease, valvular disease, and aortic pathology in the elderly population. Each section will consist of a general overview of the disease process, its specific focus on older patients, procedural outcomes, and quality of life in the elderly.

Coronary Artery Disease in the Elderly

Coronary artery disease (CAD) is defined as flow limiting obstruction to one or more coronary artery(ies). CAD is the most common form of heart disease in the general population. It is present in 12% of the general population and 20% of people over 65 [4, 5].

CAD is not only common, but it is a source of considerable morbidity and mortality. CAD represents the leading cause of death among men and women of advancing age. Obstructive coronary artery disease (CAD) is defined as a greater

J. B. Goldberg (✉)
Department of Surgery, Division of Cradiothoracic Surgery, Westchester Medical Center and New York Medical College, Valhalla, NY, USA
e-mail: Joshua.Goldberg@wmchealth.org

R. Latifi (ed.), *Surgical Decision Making in Geriatrics*,
https://doi.org/10.1007/978-3-030-47963-3_13

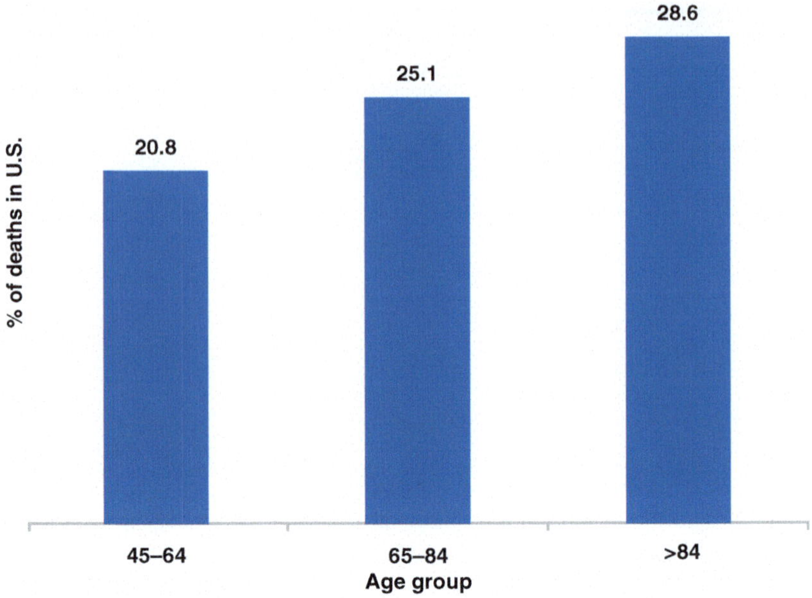

Fig. 13.1 Percentage of deaths in the United States due to heart disease increases with age. (From Centers for Disease Control and Prevention (CDC) [4])

than 50% obstruction of one or more coronary arteries. Obstructions are due to the accumulation of atherosclerotic plaques composed primarily of cholesterol and calcium that accumulate over many years. An atherosclerotic plaque can grow in size, resulting in limitations of downstream myocardial perfusion causing angina or myocardial infarction. Alternatively, atherosclerotic plaques can rupture showering atherosclerotic debris downstream while triggering platelet and coagulation factor activation which may also result in angina and myocardial infarction.

There are numerous genetic, comorbidity, and lifestyle components that contribute to the development of CAD, including diabetes, tobacco use, hypercholesterolemia, obesity, and diet. However, age has been thought to be the number-one associated condition linked with CAD as the deleterious effects of various chronic conditions and atherosclerosis accumulate over time. The incidence and prevalence of CAD increases with advancing age. Approximately 6% of patients between 45 and 65 are diagnosed with CAD, 14% between 65 and 75, and 24% greater than 75 (Fig. 13.2) [5]. The natural history of CAD culminates in decreased myocardial perfusion, resulting in angina or myocardial infarction. As expected, the incidence of MI and deaths from CAD increase with age (Figs. 13.1 and 13.3). There is an annual incidence of MI in patients <60 of approximately 3% which increases to 17% in patients over 80 [5].

There are 3 main treatment modalities for coronary artery disease: medical management, percutaneous therapies, and coronary artery bypass grafting. Medical management primarily consists of cholesterol-lowering therapies with lifestyle modification and medications such as statins, antihypertensives, and other preventive therapies. Percutaneous therapies and coronary artery bypass grafting are utilized when coronary artery disease progresses to a symptomatic level, resulting in angina or myocardial infarction. Percutaneous interventions (PCI) include: angioplasty or stent placement via a femoral or radial artery percutaneous puncture. In the setting of acute ST elevation MI, PCI is a preferred intervention as long as the patient's anatomy is sufficient. In addition, PCI is a preferred modality in patients with single or double vessel coronary artery disease with suitable anatomy for stent placement.

In the setting of multivessel coronary artery disease, left main coronary artery disease or left main equivalent coronary artery disease, diabetes, patients with heart failure

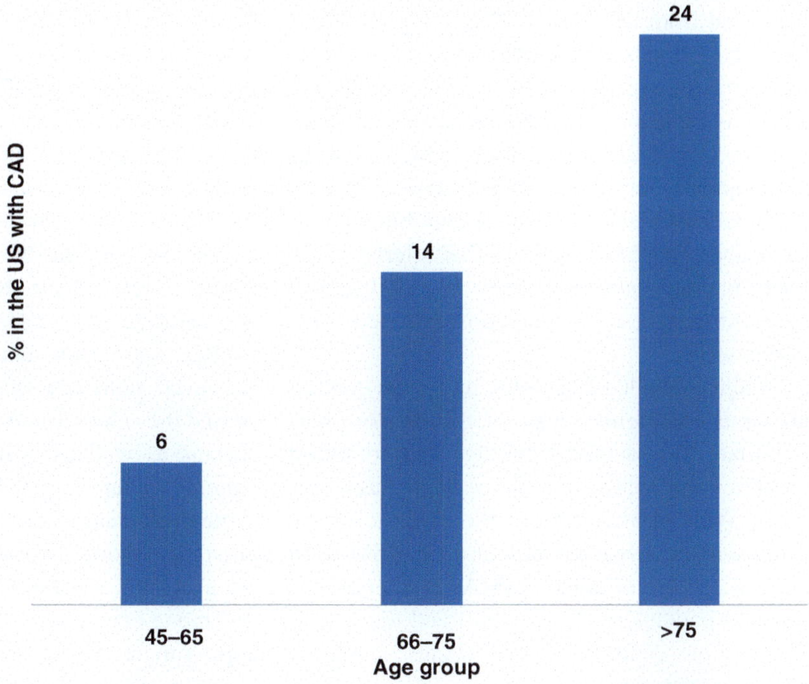

Fig. 13.2 Prevalence of CAD in the United States increases with age. (From Centers for Disease Control and Prevention (CDC) [4])

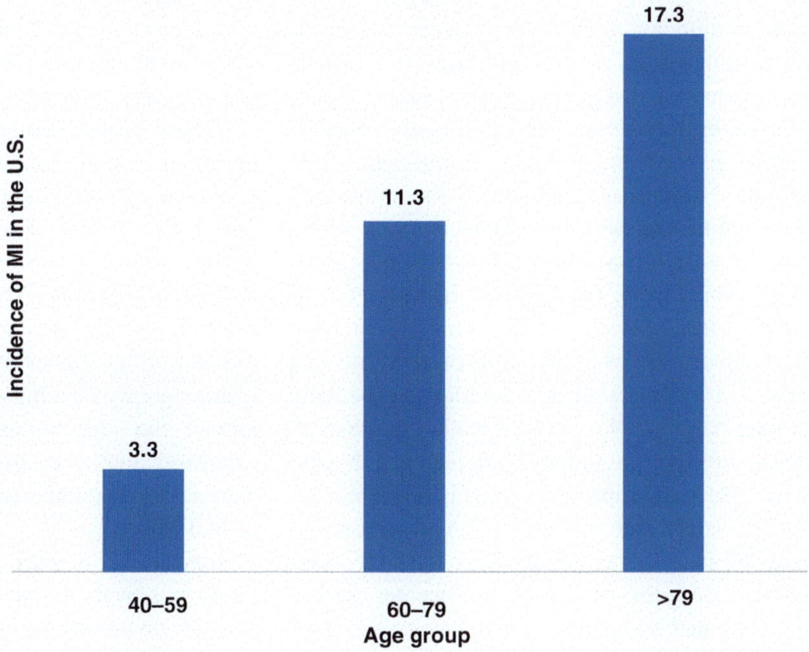

Fig. 13.3 Increasing prevalence of MI in United States with advancing age. (From Centers for Disease Control and Prevention (CDC) [4])

or reduced ejection fraction, coronary artery bypass grafting (CABG) is the recommended treatment modality as it has been shown to improve survival and freedom from future coronary events [2, 6].

Traditionally, CABG consists of a median sternotomy and bypassing significant coronary lesions utilizing a variety of arterial and venous conduits while on cardiopulmonary bypass. The left internal mammary artery (LIMA) is typically harvested from the chest wall, leaving its origin from the left subclavian artery intact. Due to its unique histologic and biochemical properties, the LIMA is typically anastomosed to the left anterior descending artery, the artery which perfuses the major portion of the left ventricle as well as the interventricular septum. A combination of saphenous vein grafts and/or other arterial grafts (right internal mammary artery or radial artery) are used to bypass the remaining diseased coronary arteries.

While a detailed discussion of bypass conduit choice is outside the scope of this chapter, it is important to understand that age and comorbid conditions influence the choice of bypass conduits. With rare exception, the disease of the LAD will be bypassed utilizing the LIMA. In general, the use of bilateral IMAs is reserved for younger patients (<65 years old) and those with few chronic conditions (no diabetes, smoking, immunosuppression, obesity) as bilateral IMA use is associated with sternal healing and infectious complications in older patients as well as those with significant comorbidities. It is important to mention that in the current era most saphenous veins are harvested in a minimally invasive, endoscopic technique that minimizes early mobility limitations compared with traditional, open saphenous vein harvest techniques which entailed an incision along the length of the saphenous vein. Lastly, radial artery harvest is typically performed with an open technique (incision from elbow to wrist) in most centers and can result in early mobility limitations, especially among patients who may be walker dependent. Thus, in older or polymorbid patients, bypass with LIMA and saphenous veins is preferred.

The safety and efficacy of CABG among all patients, independent of age, has been well established. Decades of large, multicenter series, national and international databases, and randomized controlled trials have concluded that, in the general population, CABG is a low morbidity and mortality procedure with an expected mortality of <1% [2, 6]. In comparison to PCI, multiple large randomized trials have concluded that CABG is superior to PCI in terms of survival as well as freedom from future coronary events. Long-term follow-up has demonstrated that CABG is cost effective with superior quality of life compared with PCI. The superiority of CABG has held true even with advances in stent technology as well as medical management.

The safety and efficacy of CABG observed in the general population also holds true with advancing age. In the not so distant past, a patient's age greater than 70 and definitely 80 was considered a relative contraindication for CABG and open heart surgery in general. However, numerous publications have demonstrated the safety of CABG performed in octogenarians as well as nonagenarians. While older patients undergoing CABG tend to have more acute presentations and more comorbidities than younger cohorts, short- and long-term survival are excellent. Overall CABG-associated mortality among octogenarians is around 2–4% with an approximate 5-year survival of 76%. Patients between 80 and 85 undergoing isolated CABG have a median survival of 7.4 years while patients ≥85 have a median survival of 5.4 years [7]. Thus, contemporary data concludes that CABG is a safe option for well-selected older patients.

When treating patients at extremes of age optimizing quality of life is often the prevailing goal of the patient and physician over increasing long-term survival. Exertional chest pain, dyspnea, and/or heart failure associated with CAD can be debilitating. Thus, in some patients of extreme age the goal of surgery may be symptomatic relief rather than long-term survival. Contemporary data has demonstrated that CABG among the elderly and debilitated results in improved quality of life and decrease in future cardiac events compared to medical management or PCI [8].

Innovations in cardiac surgery technique will likely positively impact the elderly population. Hybrid revascularization is a concept that is growing in popularity especially among higher risk surgical cohorts. Hybrid revascularization entails robotic LIMA dissection and LIMA to LAD anastomosis through a small right thoracotomy incision without the need for cardiopulmonary bypass with PCI revascularization of the remaining diseased vessels. This revascularization strategy takes advantage of the long-term patency of the LIMA to LAD anastomosis without the morbidity and mortality associated with a sternotomy and cardiopulmonary bypass. Preliminary data suggests the safety and efficacy of this strategy especially in higher risk and frail patients.

Thus, CABG can increase the longevity and quality of life of older patients including octogenarians and nonagenarians. As with other surgical procedures, patient selection is critical to a successful procedure and postoperative course. Age, in and of itself, should not deter a patient, or a patient's provider from the consideration of CABG in the setting of significant coronary disease.

Valvular Heart Disease

Valvular heart disease is a heterogenous assortment of independent and at times interdependent disease processes resulting from dysfunction of the aortic, mitral, pulmonic, and/or tricuspid valves. By far, aortic and mitral valvular diseases are the two most common valvular pathologies and will be the focus of this chapter [9]. While valvular heart disease is less prevalent than CAD (present in approximately 3% of the general population), it represents approximately 20% of open-heart surgical procedures. As with CAD, the incidence and prevalence of valvular heart disease increases with advancing age.

Aortic Valve Disease

The aortic valve represents the most common valve requiring surgery. The disease of the aortic valve can present in two, not mutually exclusive, forms: stenosis or regurgitation. Aortic insufficiency (AI) is the least common. It can be caused by a number of pathophysiologic processes including leaflet dysfunction, aneurysmal disease, connective tissue disorders, endocarditis, or aortic dissection. Primary AI is less common in older patients compared with younger patients due to a link with connective tissue disorders which tend to present at younger ages. There is a grading system based primarily on echocardiographic-derived hemodynamic parameters combined with clinical features which ranges from mild to severe. Surgery is indicated in the setting of severe symptomatic AI [10]. Most patients are able to tolerate gradual increasing severity with minimal symptoms until the AI becomes severe. While medical management can temporize the effects of AI, ultimately surgery with valve repair or replacement is the only definitive treatment.

Aortic stenosis is the most common valve pathology affecting the patients of advancing age. Classically there have been two categories of aortic stenoisis: congenital, which is usually associated with a bicuspid aortic valve and presents in the fifth to sixth decade of life, and degenerative aortic stenosis which results from calcific degeneration of the valve with aging. The vast majority of AS is secondary to a slowly progressing degenerative process of a trileaflet valve. Because of its increasing incidence and prevalence in progressively older patients, it was classically named "senile aortic stenosis." As with aortic insufficiency, there is a universally accepted AS severity grading system based primarily on echocardiographic or angiographic hemodynamic parameters and symptoms ranging from mild to critical [10].

Regardless of the etiology, the hemodynamic and mechanical effects are the same. As the valve becomes progressively stenotic, so does the LV afterload and work required by the LV to maintain perfusion pressure. Aortic stenosis progresses slowly over years if not decades. As such, the heart does an excellent job compensating for the increased afterload with increasing left ventricular hypertrophy. During this stage patients are usually asymptomatic. However, the heart reaches a "tipping point" when the AS is severe and the heart is no longer able to compensate; patients become symptomatic with evidence of diastolic heart failure which presents as increasing dyspnea on exertion, syncope, or angina. Classic studies that laid the foundation for the surgical management of aortic valve disease demonstrated that severe symptomatic AS is ominous without treatment as up to 50% will die with 1–3 years without valve replacement [11] (Fig. 13.4).

which was defined as a predicted risk of operative mortality of at least 8%. As of 2017 there were nearly 43,000 TAVRs performed in the United States and nearly 50% were performed in patients ≥85 years of age [25]. The subanalysis of TAVRs performed in nonagenarians has demonstrated its safety and efficacy with a 30-day mortality of 4% in a population with a mean age of 93 [26].

Mitral Valve Disease

Mitral regurgitation (MR) is the most common indication for mitral valve surgery. There are two types of MR: functional MR and degenerative MR. Degenerative MR is the least common and is caused by failure of one or more components of the valvular apparatus (valve leaflets and chordae tendineae) resulting in regurgitation. Functional MR is the most common cause of MR in the elderly. In functional MR the valvular apparatus is normal but becomes regurgitant due to annular dilation or leaflet restriction. The most common cause of functional MR is ventricular dilation from ischemic heart disease. As with aortic valve disease, MR is graded from mild to severe largely based on echocardiographic-derived hemodynamic and functional patterns [6]. Progression of MR is typically gradual, allowing for myocardial compensation. As MR becomes severe and myocardial compensatory mechanisms become exhausted, patients develop heart failure symptoms. Severe symptomatic MR is an indication for surgical repair or replacement. Mitral valve repair is preferred over replacement if a functional and lasting result can be achieved as tissue prosthetic valves are subject to limited functional lifespan and mechanical valves require lifelong anticoagulation.

Significant MR is common with aging. An estimated 10% of the patients greater than 75 have significant mitral regurgitation [27]. However, the vast majority of elderly patients who may benefit from surgery to treat their MR are denied surgery due to age and other risk factors [28]. Mitral surgery can be performed safely in older patients as demonstrated in a study investigating over 14,000 procedures which demonstrated a mortality rate of less than 3% among older patients. Furthermore, the long-term survival was equivalent to that of the general population matched for age and gender [29] (Fig. 13.6). Wide and varying outcomes have been reported among octogenarians undergoing mitral surgery with mortality rates ranging from 2% to 25%. The vast majority of reported morbidity and mortality is associated with preoperative comorbidities; and many elderly, higher-risk patients are referred for surgery later in the disease process. Modern series report superior survival of octogenarians after mitral surgery compared to predicted risk of mortality based on risk assessment models [30]. The analysis of mitral repair among octogenarians demonstrates excellent 30-day survival >97%. Replacement survival was less impres-

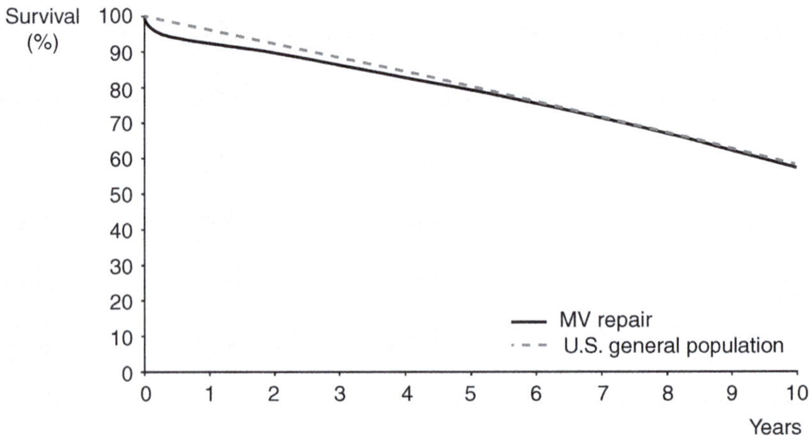

Fig. 13.6 Ten-year survival after mitral valve repair in patients aged 65 years or more. The Kaplan-Meier 10-year survival after valve repair (solid line) is equivalent to the age and sex-matched US population (dashed line). MV mitral valve. (From Ghoreishi et al. [44], with permission)

sive at 86%. However, there are important differences between the repair and replacement groups that have an impact on survival and confound survival analysis. Replacement patients tend to have more comorbid conditions, including a significant number with ischemic heart disease and significant number requiring concomitant procedures, such as CABG [31]. Elderly patients, including octogenarians experience improvement and freedom for heart failure as well as overall quality of life after mitral valve surgery. While mitral surgery among octogenarians is associated with adequate survival and improvement in cardiac-related symptoms, approximately 50% of patients >80 require some sort of assisted living at 1 year. Thus, mitral surgery can be performed safely and effectively in the elderly; patients need to be well selected for optimum outcomes.

There are multiple surgical approaches to the mitral valve that have an important impact on elderly patients. The traditional, and most common, surgical approach is via a median sternotomy. A minimally invasive approach through a right thoracotomy or robotically are increasingly being used. Data on right thoracotomy or robotic mitral surgery reports that it is associated with fewer postoperative mobility limitations and shorter lengths of stay with improved quality of life which may have a particular importance with elderly patients [32].

Just as TAVR has revolutionized aortic valve surgery for elderly and frail patients, Transcatheter Mitral Valve Repair (TMVR) is revolutionizing the treatment of MR. TMVR is an endovascular procedure during which a clip is advanced from the femoral vein into the right atrium and then left atrium through a transeptal puncture. Under TEE and fluoroscopic guidance, the clip is then passed through the mitral valve and the mitral leaflets are approximated. The concept is that the clip will improve leaflet coaptation and, therefore, reduce the MR.

TMVR has shown improvements in survival and quality of life in patients who are at high risk for surgical mitral valve repair or replacement. The EVEREST study randomized patients between surgery and clip for degenerative MR and concluded that, while the clip was less effective at reducing the MR than open surgery, it was associated with lower morbidity with equivalent mortality [33]. Subsequent long-term follow-up data demonstrates equivalent survival between the clip and surgery cohorts, with the clip cohort more likely to need subsequent mitral valve surgery. The COAPT trial investigated the high-risk, functional MR population who are traditionally denied surgery due to risk. The COAPT trial randomized high-risk patients to the clipping procedure or optimal medical therapy and discovered that the clip significantly improved survival as well as quality of life as measured by a decrease in admissions for heart failure. While the data is clear that the clip is not as effective at eliminating all of the MR compared to surgery, it is has less associated morbidity with equivalent risk-adjusted survival even among high-risk surgical patients, approximately 50–80% of whom are denied surgery [34, 35]. Other published series have demonstrated safety and efficacy of the transcatheter mitral repair with low procedural morbidity and mortality rates less than 4% and significant MR reduction in over 90% of patients [36–38].

Transcatheter mitral repair results in a significant improvement in patients' quality of life in the short and long term. While in the short term patients do not have to undergo the challenges of recovering from surgery, at 1 year the majority of patients have improvements from heart failure symptoms and improvements in functional status. This translates to fewer admissions for heart failure exacerbations which is a common problem among patients with severe MR. Furthermore, given the fact that TMVR is approved for patients with high and prohibitive predicted risk of mortality with mitral surgery patients, the vast majority of the patients undergoing the procedure are elderly with multiple comorbidities with median age in the mid-80s [37].

While mitral valve surgery can be performed safely in well-selected older patients, TMVR has changed the treatment paradigm of MR among high-risk patients including patients of advanced age.

Ascending Aortic Surgery

There are two main reasons to operate on the ascending aorta: aneurysm or acute aortic syndrome. Ascending aortic aneurysms are often associated with some form of connective tissue disorder and are asymptomatic until it reaches extremes of size and interacts with surrounding structures or develops acute aortic syndrome. Most aneurysms are detected incidentally during imaging for other reasons. Surgical indications for ascending aneurysms are based on size and growth rate with the intent of preventing dissection or rupture which is more likely to occur with increasing size. Acute aortic syndrome is defined as either aortic dissection or rupture and represents a life-threatening emergency. Aortic dissection has a 50% mortality rate at 48 hours without surgical repair [39]. Elective ascending aortic surgery is safe in the general population with a risk of mortality less than 2%. While the risk of morbidity and mortality of 8–10% is elevated among older patients, it more reflects the risk burden of comorbidities such as CAD and renal dysfunction rather than age alone [40].

Acute aortic syndrome is a lethal diagnosis and in the setting of ascending aorta pathology can only be definitively treated with open surgery. Two separate disease processes comprise acute aortic syndrome: aortic dissection and aortic rupture. Type A aortic dissections involve the ascending aorta and is associated with a 50% mortality within 48 hours of onset and 90% at 1 month without surgical repair [41]. The morbidity and mortality of surgical repair of acute type A AD is high and is largely associated with the preoperative state and comorbidities of the patient [42]. For instance, dissection patients who present to the operating room with evidence of coronary, cerebral, mesenteric, or extremity malperfusion have a significantly higher mortality (30–45%) versus those who do not (6–14%) [43].

Surgery has not been traditionally offered to elderly patients who present with acute type A aortic dissections given the high mortality associated with repair. However, with the aging of the population and increased experience and success with cardiac surgery in the elderly population, there is increasing experience with Type A aortic dissection repair in older patients. Current published experiences have reported mortality range among octogenarians ranging from 8–37% with mortality largely being associated with malperfusion syndromes and preoperative comorbidities. As with other areas of cardiac surgery, success is largely dependent on patient selection when considering aortic dissection repair in the elderly.

Conclusion

Cardiovascular disease is common in the elderly, and many may require surgery for definitive treatment. Current data suggests that age should not be a contraindication for surgical intervention on coronary artery disease, valvular heart disease, or ascending aortic disease as well-selected older patients, even those of very advanced age (90s), have good outcomes after surgery.

References

1. United States Census Bureau. Older people projected to outnumber children for first time in U.S. History. 2018. Available from: https://census.gov/newsroom/press-releases/2018/cb18-41-population-projections.html.
2. Weintraub WS, et al. Comparative effectiveness of revascularization strategies. N Engl J Med. 2012;366(16):1467–76.
3. Ando T, et al. Is transcatheter aortic valve replacement better than surgical aortic valve replacement in patients with chronic obstructive pulmonary disease? A nationwide inpatient sample analysis. J Am Heart Assoc. 2018;7(7):e008408.
4. Centers for Disease Control and Prevention (CDC). Prevalance of coronary heart disease – United States, 2006–2010. 2011. Available from: https://www.cdc.gov/mmwr/preview/mmwrhtml/mm6040a1.htm.
5. U.S Department of Health and Human Services, Centers for Disease Control and Prevention. Summary health statistics: National Health Interview Survey, 2017. 2017 [cited 2019; Available from: https://ftp.cdc.gov/pub/Health_Statistics/NCHS/NHIS/SHS/2017_SHS_Table_A-1.pdf.
6. Patel MR, et al. ACC/AATS/AHA/ASE/ASNC/SCAI/SCCT/STS 2016 appropriate use criteria for coronary revascularization in patients with acute

coronary syndromes: a report of the American College of Cardiology appropriate use criteria task force, American Association for Thoracic Surgery, American Heart Association, American Society of Echocardiography, American Society of Nuclear Cardiology, Society for Cardiovascular Angiography and Interventions, Society of Cardiovascular Computed Tomography, and the Society of Thoracic Surgeons. J Am Coll Cardiol. 2017;69(5):570–91.
7. Likosky DS, et al. Long-term survival of the very elderly undergoing coronary artery bypass grafting. Ann Thorac Surg. 2008;85(4):1233–7.
8. Yamaji K, et al. Effects of age and sex on clinical outcomes after percutaneous coronary intervention relative to coronary artery bypass grafting in patients with triple-vessel coronary artery disease. Circulation. 2016;133(19):1878–91.
9. Iung B, Vahanian A. Epidemiology of valvular heart disease in the adult. Nat Rev Cardiol. 2011;8(3):162–72.
10. Nishimura RA, et al. 2017 AHA/ACC focused update of the 2014 AHA/ACC guideline for the management of patients with valvular heart disease: a report of the American College of Cardiology/American Heart Association task force on clinical practice guidelines. Circulation. 2017;135(25):e1159–95.
11. Ross J Jr, Braunwald E. Aortic stenosis. Circulation. 1968;38(1 Suppl):61–7.
12. Lester SJ, et al. The natural history and rate of progression of aortic stenosis. Chest. 1998;113(4):1109–14.
13. Johnston LE, et al. Outcomes for low-risk surgical aortic valve replacement: a benchmark for aortic valve technology. Ann Thorac Surg. 2017;104(4):1282–8.
14. Melby SJ, et al. Aortic valve replacement in octogenarians: risk factors for early and late mortality. Ann Thorac Surg. 2007;83(5):1651–6; discussion 1656–7.
15. Thourani VH, et al. Long-term outcomes after isolated aortic valve replacement in octogenarians: a modern perspective. Ann Thorac Surg. 2008;86(5):1458–64; discussion 1464–5.
16. Barreto-Filho JA, et al. Trends in aortic valve replacement for elderly patients in the United States, 1999–2011. JAMA. 2013;310(19):2078–85.
17. Mack MC, et al. Outcomes of treatment of nonagenarians with severe aortic stenosis. Ann Thorac Surg. 2015;100(1):74–80.
18. Elgendy IY, et al. In-hospital outcomes of transcatheter versus surgical aortic valve replacement for nonagenarians. Catheter Cardiovasc Interv. 2019;93(5):989–95.
19. Jansen Klomp WW, et al. Survival and quality of life after surgical aortic valve replacement in octogenarians. J Cardiothorac Surg. 2016;11:38.
20. Leon MB, et al. Transcatheter aortic-valve implantation for aortic stenosis in patients who cannot undergo surgery. N Engl J Med. 2010;363(17):1597–607.
21. Smith CR, et al. Transcatheter versus surgical aortic-valve replacement in high-risk patients. N Engl J Med. 2011;364(23):2187–98.
22. Leon MB, et al. Transcatheter or surgical aortic-valve replacement in intermediate-risk patients. N Engl J Med. 2016;374(17):1609–20.
23. Mack MJ, et al. Transcatheter aortic-valve replacement with a balloon-expandable valve in low-risk patients. N Engl J Med. 2019;380(18):1695–705.
24. Popma JJ, et al. Transcatheter aortic-valve replacement with a self-expanding valve in low-risk patients. N Engl J Med. 2019;380(18):1706–15.
25. Carroll JD, et al. Procedural experience for transcatheter aortic valve replacement and relation to outcomes: the STS/ACC TVT registry. J Am Coll Cardiol. 2017;70(1):29–41.
26. Weintraub WS. TAVR in nonagenarians: pushing the boundaries. J Am Coll Cardiol. 2016;67(12):1396–8.
27. Nkomo VT, et al. Burden of valvular heart diseases: a population-based study. Lancet. 2006;368(9540):1005–11.
28. Mirabel M, et al. What are the characteristics of patients with severe, symptomatic, mitral regurgitation who are denied surgery? Eur Heart J. 2007;28(11):1358–65.
29. Badhwar V, et al. Longitudinal outcome of isolated mitral repair in older patients: results from 14,604 procedures performed from 1991 to 2007. Ann Thorac Surg. 2012;94(6):1870–7; discussion 1877–9.
30. Seeburger J, et al. Mitral valve surgical procedures in the elderly. Ann Thorac Surg. 2012;94(6):1999–2003.
31. Nloga J, et al. Mitral valve surgery in octogenarians: should we fight for repair? A survival and quality-of-life assessment. Eur J Cardiothorac Surg. 2011;39(6):875–80.
32. Seeburger J, et al. Minimally invasive mitral valve surgery in octogenarians-a brief report. Ann Cardiothorac Surg. 2013;2(6):765–7.
33. Feldman T, et al. Percutaneous repair or surgery for mitral regurgitation. N Engl J Med. 2011;364(15):1395–406.
34. Stone GW, et al. Transcatheter mitral-valve repair in patients with heart failure. N Engl J Med. 2018;379(24):2307–18.
35. Kortlandt F, et al. Survival after MitraClip treatment compared to surgical and conservative treatment for high-surgical-risk patients with mitral regurgitation. Circ Cardiovasc Interv. 2018;11(6):e005985.
36. Sorajja P, et al. Initial experience with commercial transcatheter mitral valve repair in the United States. J Am Coll Cardiol. 2016;67(10):1129–40.
37. Lim DS, et al. Improved functional status and quality of life in prohibitive surgical risk patients with degenerative mitral regurgitation after transcatheter mitral valve repair. J Am Coll Cardiol. 2014;64(2): 182–92.
38. Nickenig G, et al. Percutaneous mitral valve edge-to-edge repair: in-hospital results and 1-year follow-up of 628 patients of the 2011–2012 Pilot European Sentinel Registry. J Am Coll Cardiol. 2014;64(9):875–84.
39. Hiratzka LF, et al. 2010 ACCF/AHA/AATS/ACR/ASA/SCA/SCAI/SIR/STS/SVM guidelines for the

diagnosis and management of patients with Thoracic Aortic Disease: a report of the American College of Cardiology Foundation/American Heart Association task force on practice guidelines, American Association for Thoracic Surgery, American College of Radiology, American Stroke Association, Society of Cardiovascular Anesthesiologists, Society for Cardiovascular Angiography and Interventions, Society of Interventional Radiology, Society of Thoracic Surgeons, and Society for Vascular Medicine. Circulation. 2010;121(13):e266–369.

40. Peterss S, et al. Elective surgery for ascending aortic aneurysm in the elderly: should there be an age cut-off? Eur J Cardiothorac Surg. 2017;51(5):965–70.
41. Erbel R, et al. 2014 ESC guidelines on the diagnosis and treatment of aortic diseases: document covering acute and chronic aortic diseases of the thoracic and abdominal aorta of the adult. The task force for the diagnosis and treatment of aortic diseases of the European Society of Cardiology (ESC). Eur Heart J. 2014;35(41):2873–926.
42. D'Agostino RS, et al. The Society of Thoracic Surgeons adult cardiac surgery database: 2016 update on outcomes and quality. Ann Thorac Surg. 2016;101(1):24–32.
43. Geirsson A, et al. Significance of malperfusion syndromes prior to contemporary surgical repair for acute type A dissection: outcomes and need for additional revascularizations. Eur J Cardiothorac Surg. 2007;32(2):255–62.
44. Ghoreishi M, et al. Mitral valve surgery in elderly patients with mitral regurgitation: repair or replacement with tissue valve? Curr Opin Cardiol. 2013;28(2):164–9.

Surgical Decisions on Breast Cancer in the Elderly

14

Maria Castaldi

A leaf in the fall susceptible, like fallen leaf, to be dragged to extinction by the smallest wind blow. –Mimnermus

Aging and Immune Response

Breast cancer is the most common cancer diagnosed in women [1, 2], and age is a major risk factor for breast cancer. Based on Surveillance, Epidemiology, and End Results (SEER) data that provides age-specific rates of breast cancer in the United States [3], the median age of diagnosis for women is 62 [3]. Over 40% of all breast cancers diagnosed are in women aged 65 years or older. It is anticipated that by the year 2030, approximately 20% of the population will be aged over 65 years, predicting a greater proportion of older women diagnosed with early breast cancer in the near future with an average of three co-morbidities [4]. This increasing number of comorbidities translates to reduced life expectancy. Healthy elderly patients diagnosed with breast cancer may have 11.6 years of remaining life expectancy. Life expectancy in the elderly is reduced to 9.4 when comorbidities that may lead to organ failure are present and further reduced to 7.3 years in a frail patient. A large-scale European study shows co-morbidities to be independent prognostic factors in breast cancer patients aged 50–79 years [4]. Older women with newly diagnosed breast cancer and the presence of three geriatric domain deficits in sociodemographic, clinical, physical function, or psychosocial measures experienced a near twofold higher breast cancer–specific death rate at 5 and 10 years [5]. Further, optimal treatment for older breast cancer patients is also not well established [6]. The treatment approach for elderly patients requires more than chronological age and includes comorbidities, social and economic circumstances, and life expectancy. As the majority of treatment recommendations are based on retrospective analyses and the extrapolation of study results from younger patients, resultant under treatment and poor survival outcomes in the older populations exists [6]. Undertreatment strongly decreases prognosis and survival of breast cancer in elderly women [7]. Large population databases from Europe show that even when patients over 70 years of age have early-stage tumors and favorable histologies, they still have lower disease-free survival than women diagnosed at less than 70 years of age [8]. Such a trend is possibly attributed to the fact that these patients are less likely to receive standard treatments of sur-

M. Castaldi (✉)
Department of Surgery, New York Medical College, School of Medicine, WMCHealthcare Network, Valhalla, NY, USA

Department of Surgery, Westchester Medical Center, Valhalla, NY, USA
e-mail: Maria.Castaldi@wmchealth.org

R. Latifi (ed.), *Surgical Decision Making in Geriatrics*,
https://doi.org/10.1007/978-3-030-47963-3_14

gery, radiation and adjuvant therapy. There are a number of factors that contribute toward differential treatment. The majority of geriatric recommendations are based on analyses that utilized younger patients with study results being extrapolated for older patients [9–11]. As such, although there is research on various topics pertaining to breast cancer and treatment, often these results are not generalizable to breast cancer in the elderly. Other factors that contribute toward differences in treatment reflect that geriatric patients frequently present with multiple comorbidities and treatment may be perceived as less tolerated and thus less effective [12] due to poor performance status, limited social support, and lower life expectancy [9, 12]. Although clinical tools such as functional status aid in the identification of vulnerable patients from healthy patients and may avoid suboptimal treatments for fit, elderly patients, the biology of aging must be incorporated into treatment decision algorithms.

Inflammatory responses can be altered in the elderly as a consequence of the biology of aging. Aging is associated with chronic subclinical levels of systemic inflammation called inflammaging. Inflammaging is characterized by a state of chronic low-grade sterile inflammation and increased pro-inflammatory cytokines [13]. The most marked changes that occur with aging in the adaptive immune system determine the state of immunosenescence, the age-related dysfunction of the immune system. Immunosenescence and inflammaging involve changes in both the innate and adaptive immune system, which include the changes in function of a variety of immune cells aging of the immune system [14]. Several mechanisms involved in aging and cancer development lead us to believe that aging predisposes patients to cancer. Pathways of genomic instability, telomere attrition, epigenetic changes, loss of proteostasis, decreased nutrient sensing, altered metabolism, cellular senescence, and stem cell function link aging to cancer [15]. These changes of the immune system and indicators of senescence and progressive deterioration of the immune system may have clinical significance. Immunosenescence and inflammaging may predict the development of chronic diseases mostly considered age-related, such as cancer, and frailty [16]. Not all age-related immune changes however are detrimental, and cancer immunotherapies could have an advantageous response.

Frailty from an evolutionary point of view is the depletion of the physiological, biological, and molecular reserves of the aging organs, leading to less-efficient responses to stresses. However, interventions are not withheld in elderly patients based on opinions that aging-related immune changes are detrimental. If we consider immune changes related to gaining as an adaption or remodeling, interventions may be unpredictable. Interventions do depend on the level of inflammation, on interactions of the innate immune system with other systems, and on the appropriate inflammatory state of the individual. Treatment thus should be adapted for the elderly and frailty and should be different than that for young people [16].

For women younger than 40, the risk of developing breast cancer is 1 in 51, whereas the risk for women over the age of 70 is 1 in 15 [17]. The increased risk of developing cancer could be due to two biological processes: the gradual accumulation of DNA damage and the progressive decline of the human body to defend against tumor growth [4]. As both of these biologic processes are natural by-products of aging, it is clear that the incidence of cancer will increase with age. The incidence of breast cancer increases exponentially until menopause, and continues to rise after menopause, only at a slower rate [18]. While early-onset breast cancer is largely inherited, late-onset breast cancer is the result of growth of susceptible epithelium that fails to age normally. These cells are exposed to persistent growth stimuli from secreted products from senescent fibroblasts, although the degree of the influence is not known. Understanding the biology of breast cancer and its age dependency is crucial when one prognosticates breast cancer. Clinically, late-onset breast cancer grows more slowly than early-onset breast cancer and is less aggressive even when controlled for hormone receptors and growth factor receptors [18].

Older patients, especially those with cancer, are at risk of limited physiologic reserve of multiple organ systems and the ability to withstand stress. Geriatric assessment (GA) is currently the best way to assess the level of fitness in oncogeriatric patients in order to plan adequate therapeutic strategy. It allows detection of geriatric deconditioning in order to adapt treatment regimens accordingly. Several frailty models have been developed in general geriatrics (see Frailty Assessment as Measurement of Physiologic Reserves in the Elderly). GA alone is not sufficient to predict outcome and tolerance to treatment regimens for breast cancer [17]. In frail women with early-stage breast cancer, the predictors of 3-year overall survival included cognitive or functional impairment, limitations in instrumental activities of daily living, and the American Society of Anesthesiology grade [19]. Further, functional decline in the months preceding therapy resulted in an increased risk of death [20]. A combination of geriatric multimorbidities predicts overall survival in older patients with cancer. From a cellular stand point, frailty markers could be predictive and may even be used in the place of the extensive GA that is usually suggested.

Brouwers et al. studied the relationship of potential aging and frailty biomarkers with age and clinical frailty [21]. Plasma levels of interleukin-6 (IL-6) and other biomarkers were studied in non-metastatic young and old breast cancer patients. IL-6 levels correlated with chronological age in both groups and with clinical frailty in the older breast cancer group. Such a relationship was not found with other biomarkers such as telomere length, insulin-like growth factor (IGF-1), and Monocyte Chemoattractant Protein-1 (MCP-1), meaning there is no age correlate. This study indicated that plasma IL-6 levels should be further explored as a frailty biomarker in cancer patients [14]. Additionally, this group integrated GA in a single semi-continuous score and introduced a new instrument to document clinical frailty. The relationship between frailty biomarkers and clinical frailty should be sought.

Gene expression databases can identify genes regulated in aging and potential frailty biomarkers. Forty-four markers in aging pathways were evaluated for frailty [22]. Investigators grouped the biomarkers into seven categories (inflammation, mitochondria and apoptosis, calcium homeostasis, fibrosis, neuromuscular junctions and neurons, cytoskeleton and hormones, and other) and labeled them based on priority. They identified nineteen markers with high priority score, twenty-two markers with medium priority, and three markers with low priority. A panel of biomarkers should thus be utilized for assessing frailty rather than a single marker. They proposed a core panel of frailty biomarkers consisting of several signaling proteins, which includes IL6 [23].

In a large Polish Study, the PolSenior Study [23], the role of serum levels of two pro-inflammatory factors, interleukin-6 (IL-6) and C-reactive protein (CRP), was investigated in physical and cognitive performance to predict mortality in the successfully aging elderly. IL-6 and CRP levels were statistically lower in the successfully aging group rather than in the other participants in this study. The investigators found that higher IL-6 and CRP levels were statistically significant and associated with poor physical performance and cognitive performance. They noted the same association when they adjusted for age and other variables. Moreover, patients with lower concentrations of IL-6 and CRP survived longer [23]. IL-6 thus has been shown to be a promising marker in predicting frailty. Time aspect of aging however may not always correlate with clinical frailty or the possible functional consequences of the ageing process. Further, it is not clear if presumed clinically important biomarkes merely reflect chronological age, or rather the presence of clinical frailty [21]. Additionally, the oncogeriatric field is a specific subset and extrapolation of general findings from geriatric research and still requires validation. Conversely, the lack of correlation with frailty status at the time of diagnosis does not necessarily mean that these markers have no value in guiding treatment choices.

At the present time, research on biomarkers of frailty and ability to predict treatment toxicity may influence clinical decision making on the most appropriate management of geriatric cancer patients. Biomarker profiles could be incorporated into current geriatric assessment tools to enhance their ability in assessing the risk of adverse events in elderly cancer patients. Furthermore, identifying these biomarkers can elucidate the pathophysiology and contribute to better goal-directed therapies [24]. Elucidating the role of biomarkers in pathophysiology of cancer may detect unknown geriatric problems and allow adapting treatment regimens accordingly. Due to the importance of frailty as an underlying factor, frailty models, beyond the GA, must be utilized to guide treatment decision and tolerance to cancer treatments [21, 25].

Surgery

Although elderly women represent a large portion of breast cancer diagnoses, these women frequently experience undertreatment compared to their younger counterparts. Elderly cancer patients may experience differential survival due to consistent undertreatment [6].

Axillary Assessment

Axillary staging can have significant effects on decisions regarding the need for adjuvant treatment. The ACOSOG Z0011 study [26] and 10-year follow-up outcomes on disease-free survival and axillary events support the elimination of omitting axillary dissections in patients with early-stage breast cancer [8]. Sentinel lymph node biopsy is unlikely to be beneficial when the tumor is small, hormone receptor positive, and clinically negative axilla [27]. One study found that although omission of axillary staging in elderly patients may suggest an increased risk of regional recurrence, it leads to no difference in overall or breast cancer–specific mortality [28]. Other studies also support omitting axillary staging when elderly women have low-risk breast cancers [29, 30]. Guidelines from the Society of Surgical Oncology also suggest that routine use of sentinel lymph node biopsy would be avoided in a subset of elderly populations [31]. Some consider omitting sentinel lymph node biopsy in clinically lymph node negative, hormone positive invasive breast cancer in patients over the age of 70 [32]. In a study of 193,728 patients from the National Cancer Database, 79.8% had regional lymph node surgery, of which 15% were found to have positive lymph nodes. After multivariate analysis, luminal positivity in the cancer subtype was an independent factor positively correlated with the need for chemotherapy, hormone therapy, and/or radiation therapy [32]. Thus information obtained in the tumor alone may be sufficient to predict behavior and guide therapy regardless of lymph node status. Although pathological information from the axillary sentinel lymph node is important in directing the use of adjuvant therapy, oncologic surgeons generally perform axillary lymph node analysis for axillary staging and survival predictions rather than to guide treatment. Axillary lymph node manipulation in the form of sentinel lymph node excision or axillary dissection is associated with 7–20% rate of lymphedema and surrounding sensory and range of motion derangements. Two randomized controlled trials, the ALMANAC and NSABP B-32, demonstrated increased morbidity for patients who received sentinel lymph node biopsy [33, 34]. Added morbidity for elderly patients may have greater consequences than in younger counterparts, as many elderly patients already have preexisting comobordities [28] and do worse with extended surgery.

The involvement of axillary lymph nodes with cancer metastases strongly prognosticates early-stage breast cancer. In light of being minimally invasive, and highly sensitive, positive sentinel lymph node biopsy has not been shown to increase breast cancer–specific or all-cause mortality in women over 65 years of age. Surveillance, Epidemiology, and End Results (SEER) data research has not shown any 5-year all-cause survival benefit for performing axillary lymph node dissection in patients with positive sentinel lymph node. Additionally, SEER data research did not show a statistically significant difference in five-year breast cancer–specific survival (94.6% vs. 91.6%) [35] Hence, consideration could be given to omitting axillary lymph node analysis in a subset of elderly patients.

Breast Assessment

Surgery to the breast is the standard of care for any age for local control unless life expectancy is less than 2–3 years. In octogenarians with early-stage breast cancer, surgery improves breast cancer–specific survival and has a low postoperative mortality rate. However, this was not observed in 70-year-old patients and older with locally advanced disease defined by large tumor size and, or axillary metastases where there is no cancer-specific survival benefit [27]. In the elderly breast cancer patient, concerning breast conservation surgery and complete breast removal, studies have demonstrated variable use of breast conservation in the elderly. One study utilizing the SEER database demonstrated that approximately 66.55% of women over the age of 70 have improved cancer-specific survival compared with those who received mastectomy [36]. However, other studies have demonstrated how the addition of radiation with breast conservation in older populations does not improve survival and may lend to omission of radiation therapy when breast conservation surgery is performed [37–39]. Mastectomy is the preferred option for women with increased tumor size, wanting to omit radiotherapy or when patients prefer this option to others [40]. The clinical decision to undergo mastectomy using SEER data was based on disease-specific factors such as large tumor size, higher stage disease, and clinically positive lymph nodes [36].

Hormone receptor positive tumors in a patient that is not a candidate for surgery because of frailty with significant comorbidity, primary endocrine therapy is the preferred therapeutic option to halt progression and downsize tumor burden.

Frailty increases the risk of surgical complications. Based on phenotypic frailty criteria categorized as pre-frail (2–3 criteria) and frail (4–5 criteria), there is a step-wise increased risk of postoperative complications within 30 days. There is also longer length of stay to an institution among all, cancer and non-cancer, pre-frail and frail elective surgical patients. This phenotypic frailty is an additive to the increased risk estimation from popular surgical risk scores [41, 42].

There is a high incidence of ductal carcinoma in situ (DCIS) in patients at the age over 80 [43]. It has been shown that surgery does not improve overall survival benefit, but can be beneficial when there is a high-grade ductal carcinoma in situ as this most likely indicates disease aggressiveness. As surgical removal of DCIS is currently the standard of care, treatment with endocrine therapy and no surgery may be an option in low-risk DCIS [44]. The conclusions based on lead time between the development of DCIS and appearance of invasive breast cancer allow for active surveillance only in lieu of surgery in the elderly with comorbidities. For low-grade, small, nonpalpable in situ ductal carcinomas, there may be no benefit to surgery or radiation and a lesser approach such as active surveillance and endocrine therapy alone may suffice as treatment [44].

Adjuvant Therapy

Conventional treatment for breast cancer includes surgery (either breast conservation or total breast removal), radiotherapy, and chemotherapy. Surgery has been well established for local control. The supporting data on efficacy of chemotherapy in comparison to surgery is less robust [45]. Morbidity is largely due to subclinical metastatic disease at diagnosis. A recent 2019 study found that relative survival for elderly patients between the ages of 65–75 years increased over the last few decades. This increase in tandem with the increased use of systematic treatment in this same group. The authors reported that the in patients aged 65–75 years, the survival gain can most likely be attributed to the increased use of chemotherapy [46]. The authors report in another study that in the Netherlands only about 10% of patients with Stage III breast cancer over the age of 70 receive chemotherapy. This is compared to the 35.2% that receive chemotherapy in Belgium. This trend is important because survival outcomes are better in Belgium [47]. As such, it is clear that systemic therapy plays an important role in increased survival. However, the limitations of conventional breast cancer therapy are the low therapeutic index of a drug, side effects, therapy resistance, and heterogeneity of the drug concentration within the tissue. In the elderly, the availability of the drug in the bloodstream as well as pharmacokinetics is subject to the physiologic derangements that occur with aging.

When selecting a therapeutic regimen for an elderly patient, the physiological changes in the body of elder patients may increase the likelihood of toxicities. Aging causes a decrease in the total amount of water in the body, which leads to a decrease in the volume of distribution for water-soluble medications. Reciprocally, total body fat increases, which changes the distribution of fat-soluble medications. Apart from these physiologic changes, normal tissues are more susceptible to harmful effects of medications. Chemotherapy regimens in elderly patients lean toward being less aggressive, or may be excluded from the therapeutic protocol with resultant lower survival. One study found that regardless of a risk category (low, intermediate, high), patients ≥65 years were less likely to receive adjuvant chemotherapy [48]. These findings were consistent even when the population included fit older women. In this particular study, while elderly patients were more likely to experience toxicity, this did not correspond with increased mortality [49]. A systematic review conducted in 2011 also found an association between the increasing number of comorbidities and the decreased use of chemotherapy [50]. This trend is concerning for a number of reasons. First, there are studies that demonstrate that comorbidities in elderly breast cancer patients do not impact their disease-specific survival [51]. Second, some studies suggest that there is increased mortality for elderly patients with comorbidities undergoing chemotherapy. However, it is not elucidated whether this increased mortality is due to toxicity from treatment, or due to undertreatment. Modifications to standard treatments due to the patients being elderly could potentially explain the difference in survival. For example, withholding certain adjuvant therapies, or decreasing the dose of a chemotherapy regimen [52]. An alternative to excluding chemotherapy is enhancing therapy with prophylactic growth factors to offset toxicity, thus making it possible for elderly patients to benefit from full doses of the chemotherapy [53]. As toxicity must be managed, this should not result in undertreatment.

In the past few years, many new approaches have been developed for targeting drugs to breast cancer cells in ways that do not harm normal cells. Further, delivery of combination therapy approaches have been found to improve antiproliferative and anticlonogenetic effects on breast cancer cells due to their drug delivery systems. Nanotechnology could be further developed in elderly populations, and improved drug delivery systems can markedly improve drug absorption rates which are significant concerns in the elderly. Recent works with drug delivery systems in elderly groups are needed with targeted delivery systems customized to elderly and frail patients that could result in increased release, loading, and permeation of drugs tailored to elderly populations.

Treatment of breast cancer in the elder subpopulation is further challenged when elderly patients are subdivided based on their age. Extreme age (age >80) is often a barrier to participating in the clinical trials. Data about how to approach patients who age >80 years comes from the generalizations of the data obtained from studying younger patients. These patients are subject to overtreatment or undertreatment because treatment of this age category is adapted from the treatment of other age categories [27]. In the United States, the incidence of breast cancer in women older than age of 80 is approximately 400 cases in 100,000 women. Age alone is not a determining factor for tailoring the treatment. There are many healthy patients at the age of 80 or older that have life expectancy of 10–15 years and can tolerate treatment [27]. There are a number of online tools such as ePrognosis that help to estimate life expectancy and customize therapeutic plans. Research using SEER data has shown that most breast cancer characteristics in octogenarians are similar to younger patients [54]. Chatzidac et al. reviewed tumor characteristics in patients aged 80 and older. Data revealed 72% of the tumors were estrogen receptor positive and 56% progesterone receptor positive [55]. A high percentage of estrogen receptor positive tumors in the elderly treated with endocrine treatment may prolong survival even when the patient is not a candidate for surgery or chemotherapy. Most concerning in elderly patients is overall compliance to endocrine therapy. As it is well tolerated, completion of 5-year treatments at least should be encouraged [27].

Radiation therapy after breast conservation surgery is the standard of care. However, there is a high rate of non-compliance in octogenarians [56]. The main benefit of radiation therapy reduces the risk of local recurrence [27]. In women over the age of 70 with early-stage, estrogen receptor (ER) positive tumors [23] at 10.5 years, there was an increased rate of local recurrence with endocrine therapy alone compared to patients that had radiation therapy in addition to endocrine therapy treatment. However, a lack of radiation therapy did not change distant recurrence or overall survival. There was also a higher rate of morbidity in patients that received radiation [57]. The PRIME II trial showed the same lower recurrence rate with radiation therapy in patients over 65 years old [39].

However, Martelli et al., in a 15-year non-randomized trial with stage-one breast cancer, showed no statistical difference in mortality and distant metastasis [58]. Radiation therapy lends to problems with adherence, especially in the elderly. This could be due to any number of factors such as a lack of transportation or decreased mobility [2]. However, the inconsistent sessions could also decrease the benefits to these patients, while subjecting them to the negative side effects. Thus, in octogenarians when the tumor is early stage and hormone receptor positive, radiation therapy may be safely omitted [27]. As stated previously sentinel lymph node biopsy in this group may be omitted as well, as there is no survival benefit.

Web-based tools allow the calculation of 5- and 10-year survival and disease recurrence benefits of adjuvant therapies. However, these models do not predict survival and disease recurrence well in patients 80 years of age or older, and they might overestimate the benefit of adjuvant therapies [27].

Systemic treatment, adjuvant therapy, is typically reviewed with patients and their families during the geriatric assessment and discussion about life expectancy. Adjuvant therapy includes endocrine therapy, chemotherapy, and radiation. Edocrine therapy is a good option for patients who are frail with limited life expectancy, or those who are unwilling to undergo surgery [2]. The optimum duration of endocrine therapy in extreme ages is also uncertain. Studies have shown that severe side effects are most common after 2 years and as such a shorter duration of therapy could be wiser but overall effects of shorter time courses have not been proven [27]. While considering adjuvant therapy in patients aged 80 or older, although biological age is more important than chronological age. Side effects often overweigh the benefits in the elderly [27].

Breast Cancer Screening

Two main factors in the treatment of breast cancer are breast screening and adjuvant therapy after surgery as a mainstay for local control. However, the efficiency of breast cancer screening in patients who age >70 years is not certain as there is of insufficient data to assess benefits in this age group [59–62]. Currently the American Cancer Society (ACS) [63] and the US preventive Services Trask Froce (USPSTF) [64] offer guidelines for screening mammography in the elderly. Both the ACS and the USPSTF recommend regular screening for women between the ages of 65–74 years. The ACS recommending annual testing starting at age 40 [63] and the USPSTF recommending biennial screening from age 50 to 74 [64]. Beyond the age of 74, the ACS and the USPSTF vary in their recommendations. The ACS calls for the decision to screen being individualized [65], whereas the USPSTF notes that a patient over the age of 74 who does elect to undergo screening should do so biennially. However, both organization lack sufficient evidence-based guidelines due to the sparsity of randomized trials and observational studies utilizing women above the age of 74 [66].

Physiologic and precision breast imaging has markedly enhanced anatomic imaging. Precision imaging is used over standard imaging modalities the same way targeted adjuvant therapies are applied to tumor characteristics. Precision imaging with advanced software analysis of breast MRI (magnetic resonance imaging) and the like, novel molecular tracers, and intravenous contrast mammography could have the same potential benefits in being able to characterize cancers in a way that histology may not.

Precision imaging with MRI and physiologic imaging with contrast-enhanced spectral mammography may have significant benefit in elderly frail women with breast cancer.

Advanced software tools for texture analysis of tumors on contrast-enhanced MRI identify characteristics that are not seen by the eye alone and define tumor phenotypes and heterogeneity not appreciated on core biopsy. Spectral analysis of an MRI image allows the entire tumor to be assessed, rather than a core needle biopsy which is limited by sampling. Radiogenomics [67] can look at imaging analysis with underlying gene expression patterns. Radionomic phenotypes of tumor heterogeneity [68] is additive to radionomic software to obtain additional tumor information. These phenotypes predict tumor behavior patterns and allow morphology to be studied. Genotyping and phenotyping based on imaging characteristics is interesting in that the most heterogeneous phenotypes can be used to predict tumor behavior. Identifying heterogeneous natures of certain tumors might inform decisions on therapy or predict which tumors may fail to respond to treatment. Breast cancer tumor analysis with precision imaging modalities may potentially guide treatment based on MRI characteristics by texture imaging.

Contrast-Enhanced Spectral Mammography (CESM)

CESM may be comparable to MRI in the assessment of residual breast cancer following neoadjuvant therapy [69]. Prior work shows MRI is superior to clinical exam, mammography, and ultrasound and that there is very good correlation when looking for residual disease and tumor response after neoadjuvant chemotherapy in terms of positive and negative predictive values. Contrast-enhanced spectral mammography may be the same if not superior to breast MRI. IV contrast mammography is mammography with contrast infused. CESM is much easier and cheaper and could be preferable in the elderly. Studies have suggested for known breast cancer, CESM is extremely accurate in depicting tumor size, and in following tumor response to treatment. Regarding assessments on overall residual disease and complete response determination, MRI and CESM are comparable and performance is equal. Thus, response to therapy and determination for surgery if at all needed could be predicted by these advanced imaging modalities. A prospective study found that CESM may underestimate tumor size but specificity and sensitivity were adequate. CESM could be used to potentially replace MRI and even surgery [70]. A change in strategy based in imaging patterns could mean that a frail patient may avoid surgery as tumor burden is downstaged with neoadjuvant therapy.

Summary

Breast cancer in the elderly is common. Individual multidisciplinary and detailed discussions regarding the care and consensus of surgery, radiation, and adjuvant therapy in the elderly breast cancer must be performed. The treatment of breast cancer in the elderly and frail populations should encompass overall biological age as well as social and economic implications of treatment. The adoption of any practice changes in the treatment of the elderly with breast cancer requires consideration of multiple sources of information, but most importantly, more than chronologic age.

References

1. Ferlay J, Soerjomataram I, Dikshit R, et al. Cancer incidence and mortality worldwide: sources, methods and major patterns in GLOBOCAN 2012. Int J Cancer. 2015;136(5):E359–86.
2. Gosain R, Pollock Y, Jain D. Age-related disparity: breast cancer in the elderly. Curr Oncol Rep. 2016;18(11):69-016-0551-8.
3. Howlader N, Noone AM, Krapcho M, Miller D, Brest A, Yu M, Ruhl J, Tatalovich Z, Mariotto A, Lewis DR, Chen HS, Feuer EJ, Cronin KA (eds). SEER cancer statistics review, 1975–2016, National Cancer Institute. Bethesda. https://seer.cancer.gov/csr/1975_2016/, based on November 2018 SEER data submission, posted to the SEER web site, April 2019. Accessed on 7 May 2019.
4. Turner N, Zafarana E, Sanna G, Mottino G, Biganzoli L. The best treatment for older patients with breast cancer. EJC Suppl. 2013;11(2):299–300. https://doi.org/10.1016/j.ejcsup.2013.07.056.
5. Clough-Gorr KM, Stuck AE, Thwin SS, Silliman RA. Older breast cancer survivors: geriatric assessment domains are associated with poor tolerance of treatment adverse effects and predict mortality over 7 years of follow-up. J Clin Oncol. 2010;28(3):380–6.
6. Bouchardy C, Rapiti E, Fioretta G, et al. Undertreatment strongly decreases prognosis of breast cancer in elderly women. J Clin Oncol. 2003;21(19):3580–7.
7. Citron ML, Berry DA, Cirrincione C, et al. Randomized trial of dose-dense versus conventionally scheduled and sequential versus concurrent combination chemotherapy as postoperative adjuvant treatment of node-positive primary breast cancer: first report of intergroup trial C9741/Cancer and leukemia group B trial 9741. J Clin Oncol. 2003;21(8):1431–9.
8. Giuliano AE, Ballman KV, McCall L, et al. Effect of axillary dissection vs no axillary dissection on 10-year overall survival among women with invasive breast cancer and sentinel node metastasis: the ACOSOG Z0011 (alliance) randomized clinical trial. JAMA. 2017;318(10):918–26.
9. Tesarova P. Breast cancer in the elderly-should it be treated differently? Rep Pract Oncol Radiother. 2012;18(1):26–33.
10. Inwald EC, Ortmann O, Koller M, et al. Screening-relevant age threshold of 70 years and older is a stronger determinant for the choice of adjuvant treatment in breast cancer patients than tumor biology. Breast Cancer Res Treat. 2017;163(1):119–30.
11. Xiao W, et al. A genomic storm in critically injured humans. J Exp Med. 2011;208(13):2581–90.
12. Weng NP. Aging of the immune system: how much can the adaptive immune system adapt? Immunity. 2006;24(5):495–9.
13. Aunan JR, Cho WC, Soreide K. The biology of aging and cancer: a brief overview of shared and divergent molecular hallmarks. Aging Dis. 2017;8(5):628–42.
14. Fulop T, Dupuis G, Witkowski JM, Larbi A. The role of immunosenescence in the development of age-related diseases. Rev Investig Clin. 2016;68(2): 84–91.
15. American Cancer Society. Ancer facts & figures 2019. American Cancer Society, Atlnat available from https://www.cancer.org/content/dam/cancer-org/research/cancer-facts-and-statistics/annual-cancer-facts-and-figures/2019/cancer-facts-and-figures-2019.pdf; 2019. Accessed 9.18.2019.
16. Benz CC. Impact of aging on the biology of breast cancer. Crit Rev Oncol Hematol. 2008;66(1):65–74.
17. Stotter A, Reed MW, Gray LJ, Moore N, Robinson TG. Comprehensive geriatric assessment and predicted 3-year survival in treatment planning for frail patients with early breast cancer. Br J Surg. 2015;102(5):525–33; discussion 533.
18. Wildes TM, Ruwe AP, Fournier C, et al. Geriatric assessment is associated with completion of chemotherapy, toxicity, and survival in older adults with cancer. J Geriatr Oncol. 2013;4(3):227–34.
19. Brouwers B, Dalmasso B, Hatse S, et al. Biological ageing and frailty markers in breast cancer patients. Aging (Albany NY). 2015;7(5):319–33. https://doi.org/10.18632/aging.100745.
20. Cardoso AL, Fernandes A, Aguilar-Pimentel JA, de Angelis MH, Guedes JR, Brito MA, Ortolano S, Pani G, Athanasopoulou S, Gonos ES, Schosserer M, Grillari J, Peterson P, Tuna BG, Dogan S, Meyer A, van Os R, Trendelenburg AU. Towards frailty biomarkers: candidates from genes and pathways regulated in aging and age-related diseases. Ageing Res Rev. 2018;47:214–77.
21. Puzianowska-Kuźnicka M, Owczarz M, Wieczorowska-Tobis K, Nadrowski P, Chudek J, Slusarczyk P, Skalska A, Jonas M, Franek E, Mossakowska M. Interleukin-6 andC-reactive protein, successful aging, and mortality: the PolSenior study. Immun Ageing. 2016;13:21.
22. Hubbard JM, Jatoi A. Incorporating biomarkers of frailty and senescence incancer therapeutic trials. J Gerontol A Biol Sci Med Sci. 2015;70(6): 722–8.
23. Hazeldine J, Lord JM, Hampson P. Immunesenescence and inflammaging: Acontributory factor in the poor outcome of the geriatric trauma patient. Ageing Res Rev. 2015;24(Pt B):349–57.
24. Van Leeuwen BL, Rosenkranz KM, Feng LL, et al. The effect of under-treatment of breast cancer in

women 80 years of age and older. Crit Rev Oncol Hematol. 2011;79(3):315–20.
25. Bergman L, Kluck HM, van Leeuwen FE, et al. The influence of age on treatment choice and survival of elderly breast cancer patients in South-Eastern Netherlands: a population-based study. Eur J Cancer. 1992;28A(8–9):1475–80.
26. Caudle AS, Hunt KK, Tucker SL, et al. American College of Surgeons Oncology Group (ACOSOG) Z0011: impact on surgeon practice patterns. Ann Surg Oncol. 2012;19(10):3144–51.
27. Glaser R, Marinopoulos S, Dimitrakakis C. Breast cancer treatment in women over the age of 80: a tailored approach. Maturitas. 2018;110:29–32.
28. Liang S, Hallet J, Simpson JS, Tricco AC, Scheer AS. Omission of axillary staging in elderly patients with early stage breast cancer impacts regional control but not survival: a systematic review and meta-analysis. J Geriatr Oncol. 2017;8(2):140–7.
29. Blackhall V, Bugelli M, Abbott N, Mullen R. The role of sentinel lymph node biopsy in planning adjuvant therapy for elderly women with low-risk breast cancer. Breast J. 2020;26(4):716–20.
30. Poodt IGM, Schipper RJ, Vugts G, et al. The rationale for and long-term outcome of incomplete axillary staging in elderly women with primary breast cancer. Eur J Surg Oncol. 2018;44(11):1714–9.
31. Society of Surgical Oncology Position Statement. Sentinel Node biopsy in clinically node negative women. Choosing Wisely Campaign. 2016. http://www.choosingwisely.org/clinician-lists/sso-sentinel-node-biopsy-in-node-negative-women-70-and-over/. Accessed 30 Sept 2019.
32. Chagpar AB, Hatzis C, Pusztai L, DiGiovanna MP, Moran M, Mougalian S, Sanft T, Evans S, Hofstatter E, Wilson LD, Lannin DR. Association of LN evaluation with survival in women aged 70 years or older with clinically node-negative hormone receptor positive breast cancer. Ann Surg Oncol. 2017;24(10):3073–81.
33. Mansel RE, Fallowfield L, Kissin M, et al. Randomized multicenter trial of sentinel node biopsy versus standard axillary treatment in operable breast cancer: the ALMANAC trial. J Natl Cancer Inst. 2006;98(9):599–609.
34. Krag DN, Anderson SJ, Julian TB, et al. Sentinel-lymph-node resection compared with conventional axillary-lymph-node dissection in clinically node-negative patients with breast cancer: overall survival findings from the NSABP B-32 randomised phase 3 trial. Lancet Oncol. 2010;11(10):927–33.
35. Chatterjee A, Serniak N, Czerniecki BJ. Sentinel lymph node biopsy in breast cancer: a work in progress. Cancer J. 2015;21(1):7–10.
36. Wallwiener CW, Hartkopf AD, Grabe E, et al. Adjuvant chemotherapy in elderly patients with primary breast cancer: are women >/=65 undertreated? J Cancer Res Clin Oncol. 2016;142(8):1847–53.
37. Muss HB, Berry DA, Cirrincione C, et al. Toxicity of older and younger patients treated with adjuvant chemotherapy for node-positive breast cancer: the cancer and leukemia group B experience. J Clin Oncol. 2007;25(24):3699–704.
38. Lee L, Cheung WY, Atkinson E, Krzyzanowska MK. Impact of comorbidity on chemotherapy use and outcomes in solid tumors: a systematic review. J Clin Oncol. 2011;29(1):106–17.
39. Klepin HD, Pitcher BN, Ballman KV, et al. Comorbidity, chemotherapy toxicity, and outcomes among older women receiving adjuvant chemotherapy for breast cancer on a clinical trial: CALGB 49907 and CALGB 361004 (alliance). J Oncol Pract. 2014;10(5):e285–92.
40. Garg P, Rana F, Gupta R, Buzaianu EM, Guthrie TH. Predictors of toxicity and toxicity profile of adjuvant chemotherapy in elderly breast cancer patients. Breast J. 2009;15(4):404–8.
41. Eagle KA, Brundage BH, Chaitman BR, et al. Guidelines for perioperative cardiovascular evaluation for noncardiac surgery. Report of the American College of Cardiology/American Heart Association task force on practice guidelines (Committee on Perioperative Cardiovascular Evaluation for Noncardiac Surgery). J Am Coll Cardiol. 1996;27(4):910–48.
42. Lee TH, Marcantonio ER, Mangione CM, et al. Derivation and prospective validation of a simple index for prediction of cardiac risk of major noncardiac surgery. Circulation. 1999;100(10):1043–9.
43. Virnig BA, Wang SY, Shamilyan T, Kane RL, Tuttle TM. Ductal carcinoma in situ: risk factors and impact of screening. J Natl Cancer Inst Monogr. 2010;2010(41):113–6. https://doi.org/10.1093/jncimonographs/lgq024.
44. Hwang ES, Hyslop T, Lynch T, et al. The COMET (comparison of operative versus monitoring and endocrine therapy) trial: a phase III randomised controlled clinical trial for low-risk ductal carcinoma in situ (DCIS). BMJ Open. 2019;9(3):e026797-2018-026797.
45. Du X, Goodwin JS. Patterns of use of chemotherapy for breast cancer in older women: findings from medicare claims data. J Clin Oncol. 2001;19(5):1455–61.
46. de Glas N, Bastiaannet E, de Boer A, Siesling S, Liefers GJ, Portielje J. Improved survival of older patients with advanced breast cancer due to an increase in systemic treatments: a population-based study. Breast Cancer Res Treat. 2019;178(1):141–9.
47. Derks MGM, Bastiaannet E, Kiderlen M, et al. Variation in treatment and survival of older patients with non-metastatic breast cancer in five european countries: a population-based cohort study from the EURECCA Breast Cancer Group. Br J Cancer. 2018;119(1):121–9.
48. Kunkler IH, Williams LJ, Jack WJ, Cameron DA, Dixon JM, PRIME II investigators. Breast-conserving surgery with or without irradiation in women aged 65 years or older with early breast cancer (PRIME II): a randomised controlled trial. Lancet Oncol. 2015;16(3):266–73.

49. Feigelson HS, James TA, Single RM, et al. Factors associated with the frequency of initial total mastectomy: results of a multi-institutional study. J Am Coll Surg. 2013;216(5):966–75.
50. Huisingh-Scheetz M, Walston J. How should older adults with cancer be evaluated for frailty? J Geriatr Oncol. 2017;8(1):8–15. https://doi.org/10.1016/j.jgo.2016.06.003.
51. Ward EP, Weiss A, Blair SL. Incidence and treatments of DCIS in octogenarians: grade matters. Breast Cancer Res Treat. 2017;165:403–9.
52. Strader LA, Helmer SD, Yates CL, Tenofsky PL. Octogenarians: noncompliance with breast cancer treatment recommendations. Am Surg. 2014;80:1119–23.
53. Hughes KS, Schnaper LA, Bellon JR, et al. Lumpectomy plus tamoxifen with or without irradiation in women age 70 years or older with early breast cancer: long-term follow-up of CALGB 9343. J Clin Oncol. 2013;31(19):2382–7.
54. Schonberg MA, Marcantonio ER, Li D, Silliman RA, Ngo L, McCarthy EP. Breast cancer among the oldest old: tumor characteristics, treatment choices, and survival. J Clin Oncol. 2010;28(12):2038–45.
55. Chatzidaki P, Mellos C, Briese V, Mylonas I. Does primary breast cancer in older women (>/=80 years) have unfavorable histological characteristics? Arch Gynecol Obstet. 2011;284(3):705–12.
56. Mogal HD, Clark C, Dodson R, Fino NF, Howard-McNatt M. Outcomes after mastectomy and lumpectomy in elderly patients with early-stage breast cancer. Ann Surg Oncol. 2017;24(1):100–7.
57. Hughes KS, Schnaper LA, Berry D, et al. Lumpectomy plus tamoxifen with or without irradiation in women 70 years of age or older with early breast cancer. N Engl J Med. 2004;351(10):971–7.
58. Martelli G, Boracchi P, Guzzetti E, et al. Omission of radiotherapy in elderly patients with early breast cancer: 15-year results of a prospective non-randomised trial. Eur J Cancer. 2015;51:1358–64.
59. Vacek PM, Skelly JM. A prospective study of the use and effects of screening mammography in women aged 70 and older. J Am Geriatr Soc. 2015;63(1):1–7.
60. Humphrey LL, Helfand M, Chan BK, Woolf SH. Breast cancer screening: a summary of the evidence for the U.S. Preventive Services Task Force. Ann Intern Med. 2002;137(5 Part 1):347–60.
61. Nystrom L, Andersson I, Bjurstam N, Frisell J, Nordenskjold B, Rutqvist LE. Long-term effects of mammography screening: updated overview of the swedish randomised trials. Lancet. 2002;359(9310):909–19.
62. Final Recommendation Statement: breast cancer: screening. U.S. Preventive Services Task Force. May 2019. https://www.uspreventiveservicestaskforce.org/Page/Document/RecommendationStatementFinal/breast-cancer-screening1.
63. Smith RA, Brooks D, Cokkinides V, Saslow D, Brawley OW. Cancer screening in the United States, 2013: a review of current American Cancer Society guidelines, current issues in cancer screening, and new guidance on cervical cancer screening and lung cancer screening. CA Cancer J Clin. 2013;63(2):88–105.
64. US Preventive Services Task Force. Screening for breast cancer: U.S Preventive Services Task Force recommendation statement. Ann Intern Med. 2009;151(10):716–26. [PubMed: 19920272].
65. Walter LC, Schonberg MA. Screening mammography in older women: a review. JAMA. 2014;311(13):1336–47.
66. Sanderson M, Levine RS, Fadden MK, et al. Mammography screening among the elderly: a research challenge. Am J Med. 2015;128(12):1362.e7–14.
67. Pinker K, Chin J, Melsaether AN, Morris EA, Moy L. Precision medicine and radiogenomics in breast cancer: new approaches toward diagnosis and treatment. Radiology. 2018;287(3):732–47.
68. Chitalia R, Rowland J, McDonald E, et al. Radiomic phenotypes of tumor heterogeneity from pre-operative DCE-MRI independently predict breast cancer recurrence after 10-year follow-up from primary invasive diagnosis. 2017 San Antonio Breast Cancer Symposium;78(4):PD2–10.
69. Patel BK, Hilal T, Covington M, et al. Contrast-enhanced spectral mammography is comparable to MRI in the assessment of residual breast cancer following neoadjuvant systemic therapy. Ann Surg Oncol. 2018;25(5):1350–6.
70. Whitaker R, Guindalini E, Abe H, et al. Breast cancer surveillance in high-risk women with dynamic contrast-enhanced magnetic resonance imaging every 6 months: results from a single institution study. Am Assoc Cancer Res. 2018;78(4):4-02-10.

Emergency General Surgery in the Elderly

15

Jorge Con, Shekhar Gogna, and Rifat Latifi

Definition of Emergency General Surgery

The term "acute care surgery" (ACS) seems to have been created following the realization that trauma surgeons at both academic and community hospitals were increasingly responsible for overall emergency surgical care that was extended way beyond trauma [1, 2]. The term "ACS" included three components: trauma, surgical critical care, and emergency general surgery [3]. The common characteristic shared between these three components is that an urgent treatment is required either in the form of surgery, the recovery from surgery or injury, or requiring surgical expertise.

The urgency in which hemorrhage control is needed in trauma parallels the urgency of certain non-traumatic conditions that care encompasses what constitutes EGS. Common examples include conditions such as viscus perforation, small and large intestinal obstruction, diverticulitis, fistulas, non-traumatic intra-abdominal bleeding, bowel infarction, intra-abdominal sepsis of any cause, and abdominal wall defects, particularly incarcerated and strangulated hernias, any acute pancreatico-hepato-biliary tract diseases but mainly cholecystitis, and acute pancreatitis. These conditions fall within the confines of the real of general surgery as defined by the American Board of Surgery [4] and as such are an integral part of surgical training. Furthermore, in 2013, in order to provide a common nomenclature for research on EGS, the American Association for the Surgery of Trauma (AAST) Committee on Severity Assessment and Patient Outcomes provides a list of ICD–9 diagnosis codes that further defines the scope of EGS based on the clinical practices of the time [5].

Outcomes of Emergency General Surgery in the Elderly

Elderly patients carry a higher risk of undergoing a major surgical procedure. A NSQIP study looking at a total of 7696 surgical procedures showed that surgical morbidity increased in a linear fashion across age groups [6]. This included postoperative wound occurrences, respiratory occurrences, renal occurrences, and cardiovascular occurrences. Morbidity for those patients

J. Con
Department of Surgery, New York Medical College, School of Medicine and Westchester Medical Center, Valhalla, NY, USA

S. Gogna
Department of Surgery, Westchester Medical Center, Valhalla, NY, USA

R. Latifi (✉)
Department of Surgery, New York Medical College, School of Medicine and Westchester Medical Center Health Network, Valhalla, NY, USA
e-mail: Rifat_Latifi@nymc.edu

R. Latifi (ed.), *Surgical Decision Making in Geriatrics*,
https://doi.org/10.1007/978-3-030-47963-3_15

older than 80 years of age reached 51% versus 28%. Similarly, increasing age was associated with increasing mortality, with those greater than 80 years of age having a mortality of 7% versus 2.3%. Similarly, the odds ratio for mortality for undergoing emergency operations increased from OR = 3.0 in the 18–59 years age group, to OR = 5.2 in the 60–79 years age group, to OR = 11.4 in the 80 and above years old age group. Similar findings have been reported in low- to middle-income countries [7].

A study using the Healthcare Cost and Utilization Project – Nationwide Inpatient Sample comparing patient's ≥80 years of age vs. <80 years of age, looked at risk adjusted outcome measures found that although overall risk-adjusted odds of mortality were higher (OR1.67), the older group had lower risk–adjusted odds of morbidity, shorter length of stay, and lower total hospital costs [8]. Although the overall mortality was higher, the unexpected finding of these improved outcomes suggests that perhaps EGS in the elderly may be safer than we thought.

Demographics of Emergency General Surgery

Of all ACS patients, EGS patients tend to be older, more likely to be female, and more likely to be on Medicare or Medicaid [5]. Not only does the need for emergency surgery increase with age, but so does the rate of complications and mortality [9]. Many of the conditions requiring EGS involve re-operative surgery, which is inherently more complex. In another scenario, a non-emergent condition, such as a reducible hernia, may have become an emergent one, such as an incarcerated one, because of the patient's lack of health care access. In our experience, performing EGS in the elderly is particularly challenging because of the higher likelihood of encountering re-operative surgery, the possibility of a delayed or advanced stage presentation, sometimes due to the patient's lack of access to healthcare.

A study utilizing the National Surgical Quality Improvement Program (NSQIP) database published in 2012 describes the most common EGS operations performed in 35,334 patients older than 60, in which partial colon resection and small bowel resection topped the list [10] (Table 15.1). Another study the same year describes the conditions affecting 94 patients of age 80 or over who underwent emergency surgery for acute abdominal conditions [11] (Fig. 15.1). The most frequent surgical indication was acute cholecystitis (24.5%), yet only 3 of 24 cholecystectomies were performed laparoscopically, while the rest were done via a laparotomy. This series also supported the finding that intestinal resection was a more common procedure overall than cholecystectomy and appendectomy.

Table 15.1 Most common EGS operations in the elderly

Procedure type	Percentage
Colectomy	22.1%
Small bowel resection	10.1%
Laparoscopic appendectomy	9.5%
Exploratory laparotomy	5.9%
Cholecystectomy	6.4%
Appendectomy	5.4%
Lysis of adhesions	4.7%
Gastric perforation repair	2.5%
Ventral hernia repair	1.9%

From Farhat et al. [10] with permission

Chronological Age vs. Frailty

With demographic changes resulting in an increasing elderly population, and with the need of emergency surgery increasing with age, understanding the effects of physiological decline from age on surgical outcomes becomes important [1, 2]. Studies have shown that the odds of morbidity and mortality following emergency surgery increases with age [6, 7, 12]. The prevalence of comorbidities such as diabetes, hypertension, and coronary artery disease also increases with age, so does the need to be on medication such as antiplatelet agents and anticoagulation. Although it may not necessarily correlate with chronological age itself, other factors possibly affecting outcomes in the elderly are the nutritional status and activity level. Many factors associated with age affect the increased rate of morbidity and mortality in elderly patients undergoing EGS.

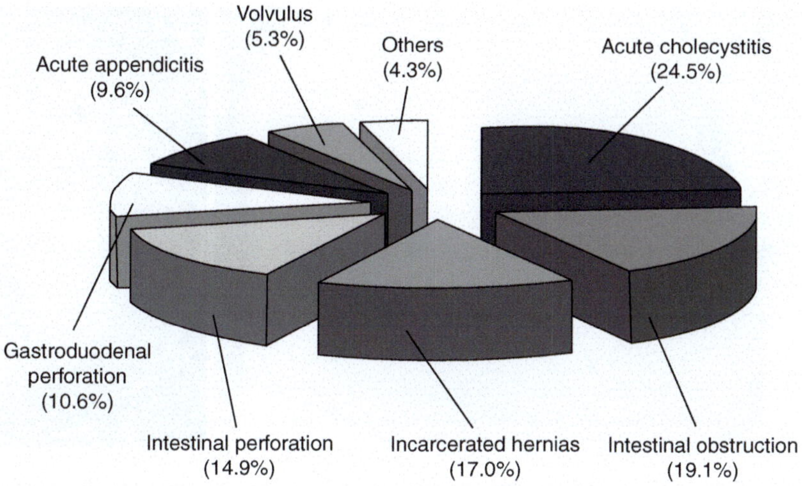

Fig. 15.1 Distribution of surgical conditions requiring emergency surgery in the elderly. (From Fukuda et al. [11] with permission)

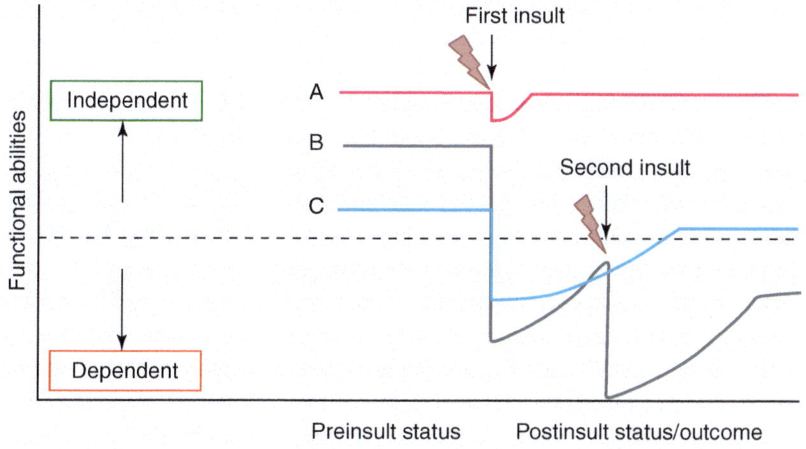

Fig. 15.2 Two-hit model in the frail elderly undergoing emergent general surgery. (From Desserud et al. [2] with permission)

Frailty syndrome results from physiological decline and in an increased vulnerability to adverse health outcomes. Frail patients are less able to adapt to acute illnesses or to recover their pre-illness level of function. A major insult (something requiring an emergency laparotomy) followed by a second insult (a complication) may result in the patient never regaining the pre-injury level of function or independence [2] (Fig. 15.2). The term "frailty" is usually employed to try to encompass several of these factors that are distinct from chronological age and that together are better predictors of outcome than chronological age alone.

Predictors of Outcome in the Elderly

Numerous frailty models have been proposed [13], yet the exact definition of frailty has remained evasive. Although the idea of frailty is that it is distinct from chronological age, age itself has been studied as a predictor. In a study by Marco in 1998, geriatric emergency patients aged 65 and older who came in with a complaint of abdominal pain were enrolled in the longitudinal case series, and it was found that surgical intervention was required for 22.1% of those patients. Advanced age and other more ominous signs such as hypotension, radiographic abnor-

4. SCORE Curriculum Outline for General Surgery. https://www.surgicalcore.org/public/curriculum. Accessed 30 May 2019.
5. Shafi S, Aboutanos MB, Agarwal S Jr, et al. Emergency general surgery: definition and estimated burden of disease. J Trauma Acute Care Surg. 2013;74(4):1092–7.
6. Turrentine FE, Wang H, Simpson VB, Jones RS. Surgical risk factors, morbidity, and mortality in elderly patients. J Am Coll Surg. 2006;203(6):865–77.
7. Shah AA, Haider AH, Riviello R, et al. Geriatric emergency general surgery: survival and outcomes in a low-middle income country. Surgery. 2015;158(2):562–9.
8. Shah AA, Zafar SN, Kodadek LM, et al. Never giving up: outcomes and presentation of emergency general surgery in geriatric octogenarian and nonagenarian patients. Am J Surg. 2016;212(2):211–20.
9. Deiner S, Westlake B, Dutton RP. Patterns of surgical care and complications in elderly adults. J Am Geriatr Soc. 2014;62(5):829–35.
10. Farhat JS, Velanovich V, Falvo AJ, et al. Are the frail destined to fail? Frailty index as predictor of surgical morbidity and mortality in the elderly. J Trauma Acute Care Surg. 2012;72(6):1526–31.
11. Fukuda N, Wada J, Niki M, Sugiyama Y, Mushiake H. Factors predicting mortality in emergency abdominal surgery in the elderly. World J Emerg Surg WJES. 2012;7:12. https://doi.org/10.1186/1749-7922-7-12.
12. Hashmi A, Ibrahim-Zada I, Rhee P, et al. Predictors of mortality in geriatric trauma patients: a systematic review and meta-analysis. J Trauma Acute Care Surg. 2014;76(3):894–901.
13. McDonald VS, Thompson KA, Lewis PR, Sise CB, Sise MJ, Shackford SR. Frailty in trauma: a systematic review of the surgical literature for clinical assessment tools. J Trauma Acute Care Surg. 2016;80(5):824–34.
14. Marco CA, Schoenfeld CN, Keyl PM, Menkes ED, Doehring MC. Abdominal pain in geriatric emergency patients: variables associated with adverse outcomes. Acad Emerg Med. 1998;5(12):1163–8.
15. Arenal JJ, Bengoechea-Beeby M. Mortality associated with emergency abdominal surgery in the elderly. Can J Surg. 2003;46(2):111–6.
16. Merani S, Payne J, Padwal RS, Hudson D, Widder SL, Khadaroo RG. Predictors of in-hospital mortality and complications in very elderly patients undergoing emergency surgery. World J Emerg Surg. 2014;9(1):43.
17. Du Y, Karvellas CJ, Baracos V, Williams DC, Khadaroo RG, Group ESA. Sarcopenia is a predictor of outcomes in very elderly patients undergoing emergency surgery. Surgery. 2014;156(3):521–7.
18. Jones TS, Dunn CL, Wu DS, Cleveland JC, Kile D, Robinson TN. Relationship between asking an older adult about falls and surgical outcomes. JAMA Surg. 2013;148(12):1132–8.
19. Lee JS-J, He K, Harbaugh CM, et al. Frailty, core muscle size, and mortality in patients undergoing open abdominal aortic aneurysm repair. J Vasc Surg. 2011;53(4):912–7.
20. Robinson TN, Wu DS, Sauaia A, et al. Slower walking speed forecasts increased postoperative morbidity and 1-year mortality across surgical specialties. Ann Surg. 2013;258(4):582–8.; discussion 588-590. https://doi.org/10.1097/SLA.0b013e3182a4e96c.
21. Kristjansson S, Benedicte R, Hurria A, et al. A comparison of two pre-operative frailty measures in older surgical cancer patients. J Geriatr Oncol. 2012;3(1).
22. Rix TE, Bates T. Pre-operative risk scores for the prediction of outcome in elderly people who require emergency surgery. World J Emerg Surg. 2007;2(1):16.
23. Samy AK, Murray G, MacBain G. Glasgow aneurysm score. Cardiovasc Surg. 1994;2(1):41–4.
24. Hardman DTA, Fisher CM, Patel MI, et al. Ruptured abdominal aortic aneurysms: who should be offered surgery? J Vasc Surg. 1996;23(1):123–9.
25. Boey J, Choi SK, Poon A, Alagaratnam TT. Risk stratification in perforated duodenal ulcers. A prospective validation of predictive factors. Ann Surg. 1987;205(1):22.
26. Altaca G, Sayek I, Onat D, Cakmakci M, Kamiloğlu S. Risk factors in perforated peptic ulcer disease: comparison of a new score system with the Mannheim peritonitis index. Eur J Surg Acta Chir. 1992;158(4):217–21.
27. Kenig J, Zychiewicz B, Olszewska U, Barczynski M, Nowak W. Six screening instruments for frailty in older patients qualified for emergency abdominal surgery. Arch Gerontol Geriatr. 2015;61(3):437–42.
28. Joseph B, Zangbar B, Pandit V, et al. Emergency general surgery in the elderly: too old or too frail? J Am Coll Surg. 2016;222(5):805–13.

Inguinal Hernia Repair in the Elderly

16

Shekhar Gogna, James K. Choi, and Rifat Latifi

In 2050, the population aged 65 and over in the United States is projected to be 83.7 million, double its estimated number of 43.1 million in 2012 [1]. This demographic shift in the population will have its implications in terms of a higher proportion of the elderly undergoing surgery or other interventional procedures. Inguinal hernia repair is one of the most common surgical treatments performed worldwide [2]. Aging promotes physiological and pathological changes in the elderly that leads to increased prevalence of inguinal hernia (IH) in them as compared to the younger population. Decreased collagen synthesis, weak abdominal musculature, and increased intraabdominal pressures as a result of chronic prostate disorders or pelvic floor weakening are few of the common causes [3]. Surgical intervention is offered for inguinal pain and discomfort interfering with the quality of life (QOL) and to prevent emergency surgery in case of incarceration and/or strangulation, which is associated with very high rates of morbidity and mortality [4, 5]. The elderly population has a higher burden of comorbidities, and they tend to present late to seek medical care. Some studies have shown that 40% of inguinal and femoral hernia repairs in patients above 65 years of age are performed for incarceration or bowel obstruction [3, 6].

S. Gogna
Department of Surgery, Westchester Medical Center, Valhalla, NY, USA

J. K. Choi
Department of Surgery, New York Medical College, School of Medicine and Westchester Medical Center, Valhalla, NY, USA

R. Latifi (✉)
Department of Surgery, New York Medical College, School of Medicine and Westchester Medical Center Health Network, Valhalla, NY, USA
e-mail: Rifat_Latifi@nymc.edu

Inguinal Hernia in the Elderly Presents Differently from the Non-elderly

While considering the acuity of presentation, it is noteworthy that the elderly population is more likely to present with complicated inguinal hernia (incarceration or strangulation) as compared to young patients [7]. The overall risk of incarceration and strangulation approaches is up to 40% [3].

Studies have shown that time from diagnosis to inguinal hernia repair is usually higher in the elderly as compared to young patients [8]. This disparity is an active area of future research and intervention. The role of physicians practicing in the community is of paramount importance in early detection and post-operative care of the elderly undergoing IH repair [9].

Anatomically inguinal hernia is divided into two common types, indirect inguinal hernia when

R. Latifi (ed.), *Surgical Decision Making in Geriatrics*,
https://doi.org/10.1007/978-3-030-47963-3_16

the hernial contents travel through the inguinal canal and exit via the superficial inguinal ring. Protrusion of the hernial contents through the weakened floor of the inguinal canal is known as direct inguinal hernia. An indirect hernia is more common in the young and direct hernia in the elderly [10].

The elderly generally present with larger indirect inguinal hernia than young adults. In the case of large indirect hernia, an acquired defect of the transversalis fascia is always present. Direct inguinal hernias are more often bilateral, and its occurrence is closely related to age [11]. These large hernias may contain large or small bowel (please refer to Figs. 1.3 and 1.5 in Chap. 1, as well as Fig. 16.1) and often are bilateral (Fig. 16.2), and are combined with umbilical hernia (Fig. 16.2).

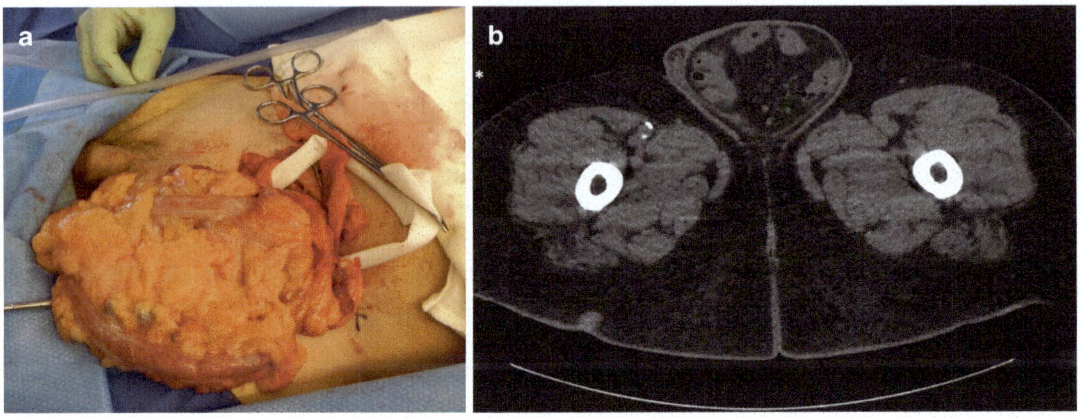

Fig. 16.1 (**a**) 73-year-old male with left incarcerated inguinal hernia containing sigmoid colon. (**b**) CT scan of the same patient with incarcerated sigmoid colon in scrotum (∗)

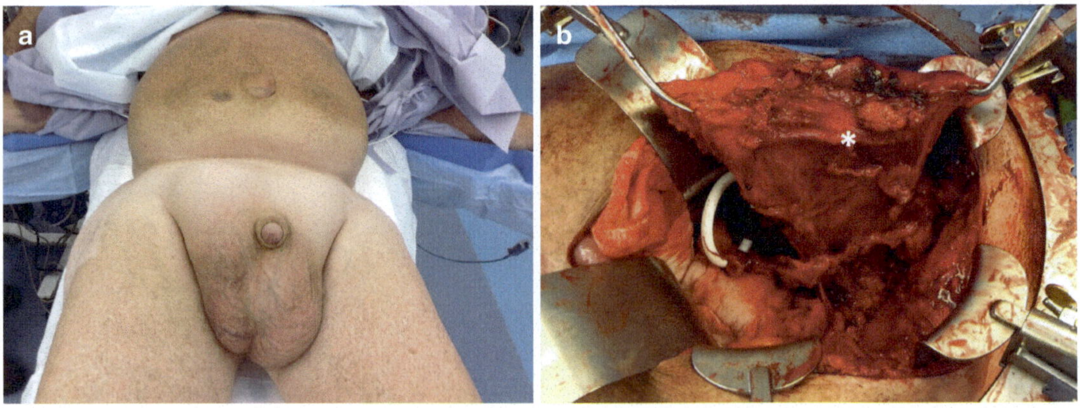

Fig. 16.2 (**a**) 61-year-old male with bilateral inguinal hernias and concurrent umbilical hernia. (**b**) Herniated bladder of the same patient with the left ureter (dilated) as it inserts into the herniated bladder (∗). (**c**) CT scan of the same patient illustrating bilateral inguinal hernias (∗)

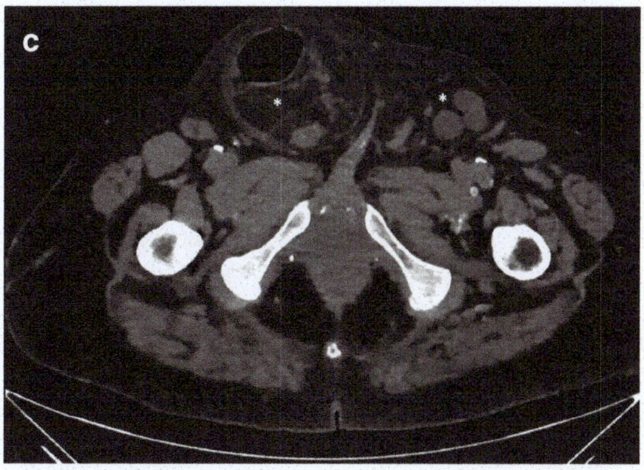

Fig. 16.2 (Continued)

A very practical issue that we surgeons face routinely in our day-to-day surgical practice of IH repair is the presence of a combined hernia in the elderly (Fig. 16.3). The pathogenesis of combined hernia stems from the fact that it is a progressive disease caused by chronic compressive structural damage due to long-term degenerative changes and hence more common in the elderly [12]. The simultaneous presence of hernial contents in different anatomic compartments of the inguinal area may lead to increased chances of early recurrence and morbidity if the surgical

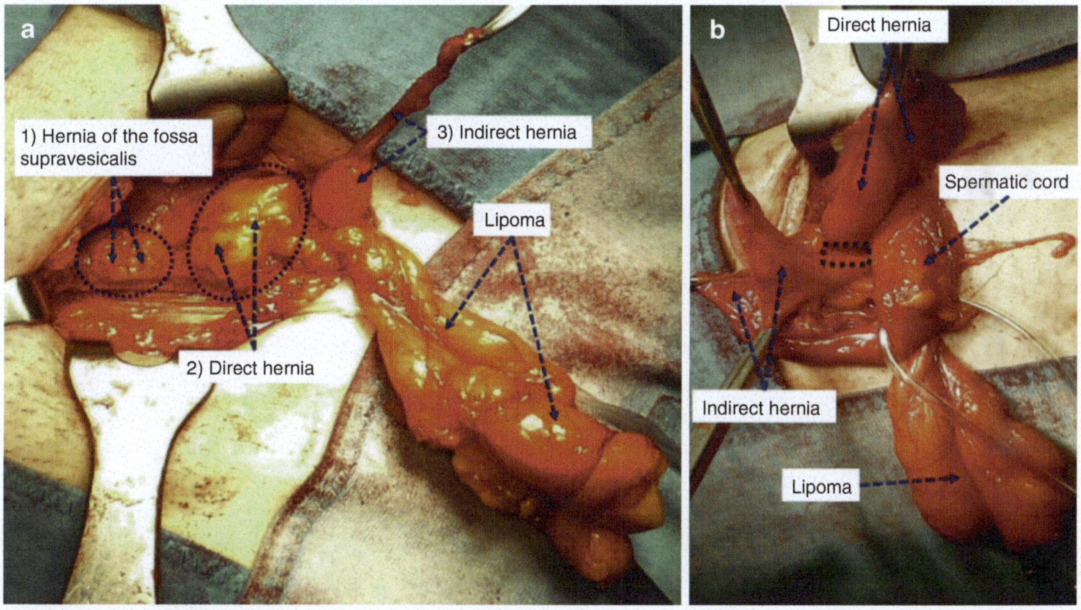

Fig. 16.3 (**a**) Tricomponent multiple ipsilateral herniae composed by a hernia of the fossa supravesicalis, a direct hernia and indirect hernia with a lipoma. (**b**) Double ipsilateral inguinal hernia composed by one direct and one indirect hernia (with opened sac). The medial aspect of the internal ring and the small portion of the back wall containing the inferior epigastric vessels (blue rectangle) divide the two protrusions. (From Amato et al. [12]. CC BY NC ND, license number: 4587921038730)

exposure is compromised due to faulty technique or lack of awareness.

Choice of Anesthesia and Setting of Inguinal Surgery

While there is an entire chapter dedicated to anesthesia in the elderly, in this chapter we briefly describe some of the techniques used for inguinal hernia surgery. Aging is linked to adverse postoperative outcomes affecting various organ systems and deterioration of cognitive function in elderly surgical patients [13]. The solution to prevent adverse postoperative outcomes is to have better integrated perioperative care. The aim of perioperative care is to improve the likelihood of very elderly surgical patients returning to their same pre-morbid status.

While considering the type of anesthesia, European Hernia Society (EHS) recommends that elective open (anterior) inguinal hernia surgery can be safely and effectively performed under local anesthesia (LA) in the elderly (Grade A recommendation, level of evidence Ib) [14]. All elderly patients should have long-acting local anesthetic infiltration preoperatively for better postoperative pain control. General anesthesia (GA) with short-acting agents along with local infiltration with long-acting LA anesthesia is a valid alternative to surgery under LA. The obvious indications for GA in the elderly are incarceration, obstruction, or strangulation.

In a three-arm multicenter randomized trial by Nordin et al., 616 patients at ten hospitals were randomly assigned to local, regional, or general anesthesia in patients undergoing groin hernia surgery. Patients in LA arm had substantial advantages in terms of shorter duration of admission, less postoperative pain, and fewer micturition difficulties [15]. Inguinal hernia repair with local anesthesia is quite safe and results in a good success rate in elderly patients despite a higher rate of comorbidity [16].

Multimodal analgesia combines different drugs with the aim of reducing doses and minimizing side effects of analgesics. Figure 16.4 depicts an effective way to control postoperative pain. All

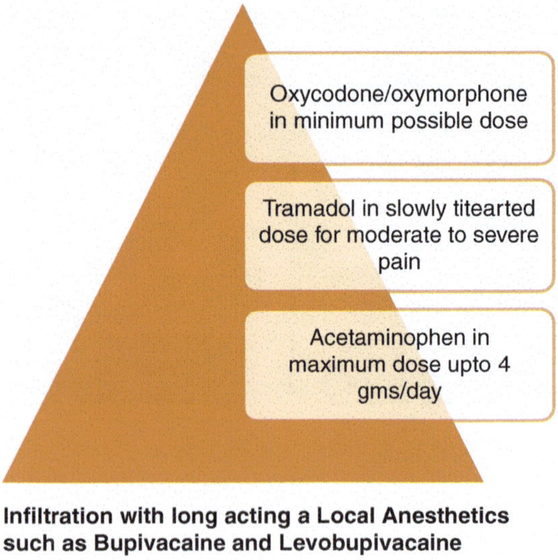

Fig. 16.4 Multimodal analgesia with escalation to more potent drugs

patients get long-acting LA and oral Tylenol to begin with. The slowly escalating doses of tramadol followed by opioids (in minimum possible concentration) are added. In a study by Seib et al., on the association between frailty in the elderly and outcomes after ambulatory surgery, the two important factors associated with decreased odds of complications were the use of local anesthesia and monitored anesthesia care [17]. With the shifting demographics of the aging population, the number of elderly patients requiring surgical procedures is increasing [18]. This has been a strong driving force in tilting the favor toward the ambulatory surgery centers performing the higher number of groin hernia repairs. Ambulatory surgery improves the quality of care and life with low morbidity [19]. A randomized control trial (RCT) comparing ambulatory care vs. inpatient care in elderly patients (excluding ASA IV and unstable ASA III) undergoing open inguinal hernia repair (Lichtenstein or repair with Proline hernia system) under local anesthesia showed no significant differences between both groups in the first 2 weeks postoperatively. Patients in the LA group had a high satisfaction rate and no readmissions [20]. Surgeons have pushed the boundaries little further; elective inguinal hernia repair in the elderly with significant comorbidities under LA has a good outcome [21].

Both open tension-free repairs and endoscopic techniques can be safely performed at daycare centers. The published series showed that other surgical and anesthesiologic techniques can also be effectively used as day surgery [14].

Is Surgical Technique Any Different in the Elderly?

Table 16.1 shows the various surgical techniques employed based on patient and surgeon's preference. The inherent steps in a particular surgical technique essentially remain the same in the elderly as well as the non-elderly. However, surgical intricacies might differ. As pointed earlier, the readers must keep in mind that "hernia is a progressive disease, which always continues to evolve" so the elderly have more propensity to present with a combined hernia, which has both direct and indirect components; this simply means that anatomy of the inguinal canal is distorted in the elderly [12]. Surgeons must be aware of the burden of comorbidity and frailty in elderly patients. The goal of IH surgery is a quick functional recovery after the operation using the "tension-free" technique and whenever possible under local anesthesia [22]. The mesh repair seems to be more prudent and strongly advocated in the elderly in elective cases than tissue repairs (Grade A recommendation) [14].

Table 16.1 Surgical technique of inguinal hernia repair

Tension-free prosthetic repairs	Technique
Anterior repair	Lichtenstein repair and its modifications
	Patch and plug repairs
	Double-layer devices (Proline Hernia system)
Posterior repair	(a) Open techniques via inguinal incision
	(b) Stoppa's repair
Laparoscopic repairs	Transabdominal preperitoneal (TAPP)
Endoscopic repairs	Total extraperitoneal (TEP)
Tissue-suture repairs	Bassini's repair
	Shouldice technique
	Desarda's technique
	Marcy repair

As mentioned in Table 16.1, anterior repair, also known as pre-muscular repair, strengthens the posterior wall of the inguinal canal. The primary goal of the posterior repair, also known as preperitoneal repair is to strengthen the entire myopectineal orifice [23]. Open anterior techniques are well established in the elderly; however, posterior repair necessitates general anesthesia (which might be unsuitable for the elderly with cardiopulmonary disorders).

The laparoscopic approach is safe in carefully selected elderly patients. In a retrospective analysis on 3203 cases (3847 hernias) by Zirui et al., who underwent LIHR, there were no significant differences in the recurrence rate and overall complication rate between the two arms ($P > 0.05$) [24]. The other two retrospective studies comparing laparoscopic approach vs. open approach in octogenarians concluded that laparoscopic inguinal hernia repair can be performed as a safe alternative to open repair with comparable rates of morbidity and mortality [25, 26]. In a prospective study by Vigneswaran et al. which aimed to analyze patient-centered outcomes for open and laparoscopic hernia repairs in the elderly concluded that laparoscopic inguinal hernia repair is safe and effective in the elderly with no major morbidities or mortalities. Although they are at greater risk for postoperative seroma, urinary retention, and octogenarians are at greater risk for readmission [27]. In essence, laparoscopic inguinal hernia repair can be safely performed in the elderly, but when it comes to safety, "open repair under local anesthesia" is still considered the gold standard in elective settings with which all techniques are compared. Emerging literature suggests that robotic inguinal hernia repairs are an option [28] and may be performed safely in the elderly [29]; however, the cost is still very prohibitive for most countries around the world.

It is well established that emergency hernia repair rates for incarceration, obstruction, and strangulation increase exponentially with the age in patients once they cross more than 50 years of their life [5]. Males predominate among the patients up to 75 years of age, while females prevail in the later age after 75 [30]. An emergency operation carries a substantial mortality risk. In the largest prospective study published in Sweden, the mortality was 7%, and it increased seven-fold after emergency operations and 20-fold if bowel resection was undertaken [31]. The mortality in the elderly population after emergency hernia repair is even higher. The surgical principles for the management of acute presentation in the elderly remain the same, but the delay in treatment has higher mortality and poor outcomes in the elderly as compared to a younger cohort.

Current Guidelines of Inguinal Hernia Management in the Elderly and Conclusions

There are no dedicated guidelines available for IH management in the elderly. There are three prominent guidelines on IH issued by hernia societies: the European Hernia Society guidelines (EHS), HerniaSurge Group (international guidelines for groin hernia management published by American Hernia Society, 2018), and International Endo Hernia Society guidelines (IEHS, published in 2011) covering laparo-endoscopic groin hernia repair [14, 32, 33]. The composite recommendations are presented here.

Indications for Treatment

Minimally symptomatic or asymptomatic inguinal hernia in men can be managed by the watchful waiting strategy (Grade A). It is recommended that symptomatic inguinal hernias be treated surgically (Grade D). The strangulated hernias should be operated on urgently (Grade D). In patients with a femoral hernia, early surgery should be performed, even if the symptoms are vague or absent. For recurrent IHs, use the opposite approach (e.g., for recurrence after anterior repair use a posterior technique, and vice versa is recommended [14].

Type of Anesthesia and Setting of Surgery

Most of the open inguinal hernia repairs can be done safely under local anesthesia at the daycare surgery center. Most of the laparo-endoscopic hernia repairs can also be safely performed at daycare centers (Level 2B, grade B).

The use of spinal anesthesia, especially, or long-acting anesthetic agents, should be avoided. General anesthesia with short-acting agents and with local infiltration anesthesia for prolonged pain control is strongly recommended (Grade B recommendation) [14].

Prophylactic Antibiotics

In open surgery, they are not recommended in low-risk patients. They are also not recommended in laparo-endoscopic surgery. In the presence of recurrence, advanced age, immunosuppressive conditions, expected long operating times and use of drains, antibiotic prophylaxis should be considered (Grade C recommendation).

References

1. Ortman JM, Velkoff VA, Hogan H. An aging nation: the older population in the United States. Hyattsville: US Census Bureau; 2014. Report Accessed 4/30/2019.
2. Devlin HB. Trends in hernia surgery in the land of Astley Cooper. In: Soper NJ, editor. Problems in general surgery, vol. 12. Philadelphia: Lippincott-Raven; 1995. p. 85–92.
3. Gianetta E, De Cian F, Cuneo S, et al. Hernia repair in elderly patients. Br J Surg. 1997;84:983–5.
4. Schumpelick V, Treutner KH, Arlt G. Inguinal hernia repair in adults. Lancet. 1994;344:375–9.

5. Primatesta P, Goldacre MJ. Inguinal hernia repair: incidence of elective and emergency surgery, readmission and mortality. Int J Epidemiol. 1996;25:835–9.
6. Nano M. Technique for inguinal hernia repair in the elderly patient. Am J Surg. 1983;146(3):373–5.
7. Rutkow IM, Robbins AW. Classification systems and groin hernias. Surg Clin North Am. 1998;78:1117–27.
8. Dunne JR, Malone DL, Tracy JK, Napolitano LM. Abdominal wall hernias: risk factors for infection and resource utilization. J Surg Res. 2003;111:78–84.
9. McIntosh A, Hutchinson A, Roberts A, et al. Evidence-based management of groin hernia in primary care—a systematic review. Fam Pract. 2000;17(5):442–7.
10. Kulacoglu H. Current options in inguinal hernia repair in adult patients. Hippokratia. 2011;15(3):223–31.
11. Hubner M, Schafer M, Raiss H, et al. A tailored approach for the treatment of indirect inguinal hernia in adults-an old problem revisited. Langenbecks Arch Surg. 2011;396:187–92. https://doi.org/10.1007/s00423-010-0635-0.
12. Amato G, Agrusa A, Rodolico V, et al. Combined inguinal hernia in the elderly. Portraying the progression of hernia disease. Int J Surg. 2016;33:20–9.
13. Jin F, Chung F. Minimizing perioperative adverse events in the elderly. Br J Anaesth. 2001;87(4):608–24.
14. Simons MP, Aufenacker T, Bay-Nielsen M, et al. European Hernia Society guidelines on the treatment of inguinal hernia in adult patients. Hernia. 2009;13:343–403. https://doi.org/10.1007/s10029-009-0529-7.
15. Amato B, Compagna R, Della Corte GA, et al. Feasibility of inguinal hernioplasty under local anesthesia in elderly patients. BMC Surg. 2012;12(Suppl 1):S2. https://doi.org/10.1186/1471-2482-12-S1-S2.
16. Nordin P, Zetterström H, Gunnarsson U, Nilsson E. Local, regional, or general anesthesia in groin hernia repair: multicenter randomized trial. Lancet. 2003;362:853–8.
17. Seib CD, Rochefort H, Chomsky-Higgins K, et al. Association of patient frailty with increased morbidity after common ambulatory general surgery operations. JAMA Surg. 2017;153(2):160–8. https://doi.org/10.1001/jamasurg.2017.4007.
18. Etzioni DA, Liu JH, Maggard MA, et al. The aging population and its impact on the surgery workforce. Ann Surg. 2003;238:170–7.
19. Gogna S, Latifi R. Ambulatory surgery services: changing the paradigm of surgical practice. In: Latifi R, editor. The modern hospital. Cham: Springer; 2019.
20. Mattila K, Vironen J, Eklund A, et al. Randomized clinical trial comparing ambulatory and inpatient care after inguinal hernia repair in patients aged 65 years or older. Am J Surg. 2011;201(2):179–85. https://doi.org/10.1016/j.amjsurg.2010.04.024.
21. Kurzer M, Karkk A, Hussain ST. Day-case inguinal hernia repair in the elderly: a surgical priority. Hernia. 2009;13:131–6.
22. Garavello A, Manfroni S, Antonellis D. Inguinal hernia in the elderly. Indications, techniques, results. Minerva Chir. 2004;59(3):271–6.
23. Li J, Wang X, Feng X, et al. Comparison of open and laparoscopic preperitoneal repair of groin hernia. Surg Endosc. 2013;27:4702–10.
24. He Z, Hao X, Li J, et al. Laparoscopic inguinal hernia repair in elderly patients: single center experience in 12 years. Ann Laparosc Endosc Surg. 2017;2(1):88. https://doi.org/10.21037/3930.
25. Hope WW, Bools L, Menon A, et al. Comparing laparoscopic and open inguinal hernia repair in octogenarians. Hernia. 2013;17:719–22.
26. Hernandez-Rosa J, Lo CC, Choi JJ, et al. Laparoscopic versus open inguinal hernia repair in octogenarians. Hernia. 2011;15:655–8.
27. Vigneswaran Y, Gitelis M, Lapin B, et al. Elderly and octogenarian cohort: comparable outcomes with nonelderly cohort after open or laparoscopic inguinal hernia repairs. Surgery. 2015;158:1137–43; discussion 1143–34.
28. Peters BS, Armijo PR, Krause C, et al. Review of emerging surgical robotic technology. Surg Endosc. 2018;32:1636. https://doi.org/10.1007/s00464-018-6079-2.
29. Buchs NC, Addeo P, Bianco FM, et al. Safety of robotic general surgery in elderly patients. J Robotic Surg. 2010;4:91. https://doi.org/10.1007/s11701-010-0191-1.
30. Nilsson H, Stylianidis G, Haapamäki M, et al. Mortality after groin hernia surgery. Ann Surg. 2007;245(4):656–60.
31. Bay-Nielsen M, Kehlet H, Strand L, et al., Danish Hernia Database Collaboration. Quality assessment of 26,304 herniorrhaphies in Denmark: a prospective nationwide study. Lancet. 2001;358: 1124–8.
32. Simons MP, Smietanski M, Bonjer HJ, Bittner R, Miserez M, Aufenacker TJ. HerniaSurge Group International guidelines for groin hernia management. Hernia. 2018;22:1–165.
33. Bittner R, Arregui ME, Bisgaard T, et al. Guidelines for laparoscopic (TAPP) and endoscopic (TEP) treatment of inguinal hernia [International Endohernia Society (IEHS)]. Surg Endosc. 2011;25(9):2773–843. https://doi.org/10.1007/s00464-011-1799-6.

Abdominal Wall Reconstruction in the Elderly: Techniques, Outcomes, and Pitfalls

17

Rifat Latifi and Ansab Haider

Ventral hernias are defined as non-inguinal and non-hiatal defects in the fascia of the abdominal wall. Based on the etiology of the defect, they are either congenital or acquired. The vast majority of these hernias are acquired and result from previous abdominal surgeries or trauma. Incisional hernia repair is the most common complication after abdominal surgery. It is a consequence of failure of the abdominal wall fascia to heal after a previous surgery. Evidence-based guidelines have emerged to suggest optimal closure techniques for the fascia. However, despite the improvements in the surgical techniques, the incidence of abdominal wall hernias remains high anywhere from 2% to 50% [1–3]. Several factors contribute to this high incidence of incisional hernia which includes patient-related factors, native disease process, and the surgical technique. As a result of this high prevalence, ventral abdominal wall hernia repairs are one of the most commonly performed operations in the United States, and nearly 350,000 of these surgeries are performed every year with an estimated associated cost of $3.2 billion [4].

There has been a rapid growth in the aging population not only in the United States but in the entire world. Every single day 10,000 people turn 65 in the United States [5]. This is mostly a result of improved screening strategies, early diagnosis, and advancements in medical technology. The elderly population has several comorbidities associated with them, and they are well known to be associated with increased morbidity and mortality after surgery. Previously, the presence of these comorbidities was considered prohibitive for elective surgery. With the remarkable improvements in healthcare technology and outcomes and the advent of minimally invasive surgical techniques, many complex abdominal operations can now be safely performed in the elderly population which was previously not possible [6, 7]. This however poses a new set of challenges as well as opportunities for general surgeons. With increasing number of elderly patients undergoing major abdominal surgeries, the incidence of incisional hernia is expected to increase as well. Prevalence of incisional hernia is higher in the elderly compared to their young counter parts, and this is mostly due to the age-related weakness in collagen [8]. This weakness in collagen also puts these patients at higher risk for poor postoperative wound healing, fascial dehiscence, as well as recurrence of the hernia. The incidence of cardiac events and postoperative respiratory failure is also higher in the elderly population. Similarly, poor tolerance to opioids

R. Latifi (✉)
Department of Surgery, New York Medical College, School of Medicine and Westchester Medical Center Health Network, Valhalla, NY, USA
e-mail: Rifat_Latifi@nymc.edu

A. Haider
Department of General Surgery, Westchester Medical Center, Valhalla, NY, USA

R. Latifi (ed.), *Surgical Decision Making in Geriatrics*,
https://doi.org/10.1007/978-3-030-47963-3_17

makes the postoperative pain control significantly more challenging in the elderly population and puts these patients with large hernia repairs at risk for postoperative respiratory failure and pneumonias.

Many of these abdominal wall hernias repairs are complex and may involve multiple fascial defects, presence of extensive underlying scarring to the bowel, associated stomas, concurrent malignant process, prior irradiation, previous hernia repair, malnutrition, and/or wound infection. With increasing geriatric trauma population and increasing number of trauma patients undergoing major abdominal procedures and the expanding utility of damage control surgery, the need for complex abdominal wall reconstruction in geriatric population for these complex defects is also expected to increase. The objective in the management of patients undergoing complex abdominal wall reconstruction is to restore the continuity of the gastrointestinal tract and the integrity of the abdominal wall by utilizing both native tissue approximation as well as reinforcement with biologic or synthetic mesh.

Anatomy of the Abdominal Wall

The boundaries of the abdomen are formed by the xiphoid process and the costal margins superiorly. Inferiorly the abdomen is defined by the pubic bones and the iliac bones. The integrity of the abdominal contents is maintained by the abdominal musculature and the associated ligaments and fasciae. Understanding these components is integral to abdominal wall reconstruction as weakness or defect in one or more of these components results in the ventral hernias. From superficial to deep, the abdominal wall is composed of skin, subcutaneous tissue, muscles and fasciae, extraperitoneal fat, and peritoneum. The subcutaneous tissue is composed of two layers. The superficial, more fatty layer is called the Camper's fascia, and the deeper more fibrous and membranous layer is called the Scarpa's fascia. The abdominal wall is comprised of a pair of strap muscles in the midline called the rectus abdominis which is attached to the xiphoid process and 5th–7th costal cartilages. Inferiorly it is attached to the pubic symphysis and the pubic crest. Laterally the abdominal wall is formed by the external oblique, internal oblique, and transversus abdominis muscles on each side. Fibrous aponeurosis of each of these muscles contributes to formation of the rectus sheath also known as the rectus fascia which is primarily responsible for the integrity of the abdominal wall. External oblique is most superficial of the abdominal wall muscles; its aponeurosis passes anterior to the rectus abdominis muscle and forms the anterior rectus sheath [9]. The internal oblique aponeurosis divides at the lateral edge of the rectus abdominis and passes anterior and posterior to the rectus muscle above the arcuate line forming the anterior and posterior rectus fasciae but passes only anterior to the rectus muscle below the arcuate line. Similarly, the aponeurosis of the transversus abdominis muscle contributes to both anterior and posterior rectus fasciae above the arcuate line but only to anterior rectus fascia below the arcuate line. Therefore, the abdomen is devoid of a posterior rectus sheath below the arcuate line, and rectus muscle lie directly on top of the fascia tranversalis and peritoneum [10]. Fascia transversalis is a thin layer of fibrous tissue that lines the inner surface of the transversus abdominis muscle. Beneath it lies the preperitoneal fat and the peritoneum. The abdominal wall receives its blood supply primarily by the superior and inferior epigastric arteries which run deep to the rectus abdominis muscle and anastomose which each other at the level of the umbilicus. Several perforating branches arise from both these vessels and supply both the superficial and deeper musculocutaneous tissues of the abdominal wall (Fig. 17.1). Laterally, the neurovascular bundle runs transversally between the internal oblique and the transversus abdominis layers of the abdominal wall. Therefore, vertical incisions that transect these nerves result in sensory deficits medial to the level of incision.

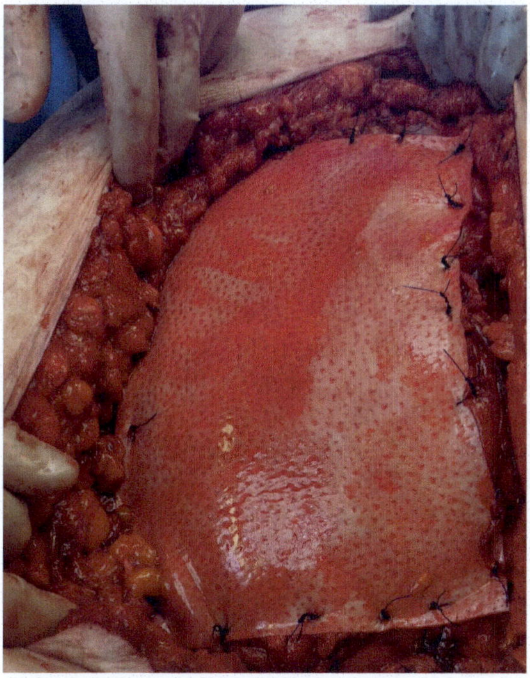

Fig. 17.1 Onlay biologic mesh re-enforcement

Preoperative Consideration and Risk Stratification

When geriatric patients present for elective repair of ventral hernia, it is imperative that a thorough evaluation of the patient with a detailed history and physical examination be performed [11]. In our experience, when these patients present to us for evaluation, it is not infrequent that they have had multiple previous attempts at repair of their ventral hernias. The risk of recurrent hernia increases with each subsequent repair as the abdominal wall strength decreased with each attempted repair. During the physical examination an attempt to should be made to estimate the size of the fascial defect. Often patients will have multiple fascial defects sometimes termed as Swiss cheese fascia which will require more extensive surgery and dissection to bring the fascial edges together without tension and a larger mesh for adequate reinforcement. Overall body habitus is an important factor in determining the risk of future recurrence. Patients with BMI >30 kg/m^2 have significantly higher risk for future recurrence. Similarly, obese patients have a higher risk for perioperative complications during hernia repair and prolonged hospital and ICU length of stay [12]. Due to the safety of bariatric surgery even in the elderly population, some centers will routinely offer preoperative bariatric surgery for patients who meet criteria for bariatric surgery prior to their abdominal wall reconstruction. Once the bariatric surgery has taken place and patients have lost weight, they will undergo a second operation to repair their ventral hernia. This two-staged repair has shown promise in reducing the perioperative morbidity and overall risk of recurrence of hernia [13, 14]. However, these results have not been replicated in the geriatric population, and therefore it is not our routine practice to offer preoperative bariatric surgery to these obese geriatric patients who require abdominal wall reconstruction for ventral hernias. It is imperative to assess the preoperative nutritional status in these geriatric patients using albumin, prealbumin, ferritin, and transferrin levels [15]. In elective patients, preoperative nutritional counseling and intervention can be performed to ameliorate the complications associated with poor nutritional status. While in the acute setting, no interventions can take place in terms of nutritional optimization, however, nutritional evaluation can still provide useful information during counseling patients and their families prior to surgery. Similarly, we routinely counsel our patients who are active smokers to quit smoking for 6 weeks as their risk of perioperative respiratory and wound complications as well as overall risk for recurrence is significantly higher compared to their non-smoking counterparts. In addition to the standard preoperative laboratory workup, we also check HbA1c levels in diabetic patients. Patients with HbA1c >7.0% have significantly higher wound related complications after ventral hernia repair [16].

Special attention should be paid in assessing the presence of preoperative respiratory, cardiac, or hepatic dysfunction. Patients with preexisting respiratory compromise such as those with severe asthma, COPD, or restrictive lung disease who have large ventral are at very high risk for postoperative respiratory failure, ventilator dependence, and need for tracheostomy. When the abdominal domain is restored and the large amount of abdominal viscera are returned back to the abdominal cavity, they have significant increase in intra-abdominal pressure which limits the movement of their diaphragm which combined with their preexisting respiratory dysfunction, and postoperative pain contributes to difficulty weaning them off the ventilator. Patients with preexisting cardiac disease should be evaluated by their cardiologist to determine the need for preoperative stress test, echocardiography, or invasive angiography to determine the safety to undergo general anesthesia. Cirrhosis from chronic alcohol use or hepatitis is one of the major causes of postoperative morbidity and mortality. These patients may look deceptively well prior to surgery; however they remain at extremely high risk for postoperative liver failure which is the major cause of mortality after major abdominal wall reconstruction. The candidacy of such patient to undergo a major abdominal surgery such as abdominal wall reconstruction should be discussed in detail through a combined team approach that should involve the surgeon, hepatologist, anesthesiologist, and the intensivist responsible for postoperative management of these patients. For imaging, CT scan is the most common and useful imaging modality which is used both to diagnose ventral hernias and for planning operative management. It provides reliable information about the number and size of defects and their relation relative to surrounding bony landmarks.

Multiple scoring systems have been developed to predict postoperative morbidity associated with ventral hernia repairs. ACS NSQIP calculator is a useful tool that provides relative risk for different complications and mortality. In our experience, ACS NSQIP calculator is a very useful instrument in the elective patient population but significantly underestimates the relative risk in emergent patient population. The likely explanation is the lack of ability of the NSQIP calculator to take into account hemodynamic parameters when assessing the relative risk of postoperative complications. Frailty plays an important role in determining the postoperative outcomes of geriatric patients as it provides information about the functional age of the patient. At our center, we use the modified frailty index which is an 11-point scoring system that assesses the functional status of the patient [17]. We will use this preoperatively on all of our elective geriatric ventral hernia patients. At this point, the presence of frail status does not directly impact the candidacy of patient to undergo a major abdominal wall reconstruction; however it helps us stratify patients who are at high risk for postoperative complications and helps guide informed decision-making. Joseph et al. have also proposed a frailty index derived from the Rockwood frailty index that is specific for emergency general surgery and may be more applicable to geriatric patients requiring emergency surgery for ventral hernia compared to the modified frailty index [18]. The score is based on 15 variables and uses a cutoff of 0.325 for frail status. The frailty index has been shown to reliably predict postoperative complications and mortality after emergency general surgery procedures.

Principles of Ventral Hernia Repair in the Elderly

The principles that guide the management of ventral hernias in the elderly are primarily the same as those of general population with few additional considerations. The surgical management of these hernias in the elderly has gained recent interest due to the increasing incidence seen in the elderly as well the associated morbidity and mortality when these patients presents with complications associated with hernias, similar to complications seen in inguinal hernia [8]. These patients have a higher risk of strangulation and incarceration and are much more likely to require bowel resection compared to their younger counterparts [19]. This results in higher ICU and hospital lengths of stays and increased associated morbidity and mortality.

All ventral hernias which are strangulated, obstructed, or acutely incarcerated require operative repair regardless of their size. Our practice is to repair these hernias on an urgent or emergent basis during the same hospital admission. When these patients present to us in an acute setting, they are evaluated with routine laboratory workup, and if there is no evidence of peritonitis, imaging with a CT scan of the abdomen and pelvis is performed. Elderly patients who present with acutely symptomatic ventral hernias are immediately resuscitated in the ED or the ICU and undergo operative repair ASAP depending on the acuity of presentation. While the overall management of emergent ventral hernia repairs is much more challenging, the decision-making regarding the need for elective ventral hernia repairs in the elderly population who are not acutely symptomatic is much more challenging for the majority of surgeons. In these patients the decision to perform elective ventral hernia repair depends on the size of the hernia, associated symptoms, quality of life, frailty, obesity, nutritional status, presence of fistulas if any, overall health, and patient preference. If the hernia is truly small and asymptomatic, these hernias can be observed non-operatively with an expected annual risk of complications of around 2.6%. However, it must be kept in mind that over the course of their lifetime, the cumulative risk of complications is much higher. It must also be noted that if hernia-related complications were to occur, the morbidity and mortality associated with an emergent hernia repair are significantly higher than that of elective repair [20]. Therefore, unless truly contraindicated, ventral hernia repairs should be repaired electively whenever possible to avoid life-threatening complications and to provide relief of symptoms. In our practice, we do not use any absolute cutoff of age as a contraindication for elective hernia repair, and only those who have truly prohibitive chances of mortality are not offered the surgery. Several studies have shown that elective ventral hernia repairs can be performed safely even in the elderly morbid population [6].

Open, laparoscopic, and robotic approaches have been utilized in the repair of abdominal wall hernias [21, 22]. However, the tenants that guide these approaches remain the same. The most important is to achieve a tension-free repair of the fascial defect with reinforcement of the native abdominal wall with a suitable mesh with at least an overlap of 3–5 cm in all directions from the fascial edge. The choice of mesh use in these situations can be challenging particularly in patients with contaminated fields, and this will be discussed in detail later on in this chapter. Primary repair of fascial defects alone is rarely, if ever, utilized due to the reported high incidence of recurrence. Primary repair results in fascial closure with very high tension which leads to recurrence rates often greater than 50% recurrence rates. This has been proven in multiple randomized clinical trials comparing primary and mesh repair for ventral hernias [23]. Laparoscopic and robotic repair of ventral hernias with mesh has gained recent popularity, and several studies have shown them to be at least equal or superior to open in terms of pain control and postoperative morbidity [21]. I have topped performing laparoscopic ventral or incisional hernia for the last few years, partly due to poor results and partly due to complex re-operative practice. However, these approaches are frequently not suitable for large complex abdominal wall defects and are much more technically challenging compared to open approach. Robotic repair offers the advantage of allowing technically easier primary fascial closure as opposed to the laparoscopic repair; however, currently there is no evidence to suggest that robotic approach has resulted in superior repair compared to laparoscopic approach at the expense of higher costs [24].

Choice of Mesh

As discussed earlier, nearly all ventral hernias will require placement of mesh due to its proven efficacy in reducing the rate of ventral hernia compared to primary suture repair. The next section will focus on the type of mesh and different placement techniques of these meshes. Broadly speaking meshes can be classified into two categories, i.e., synthetic and biologic mesh. The use of each kind of mesh depends on individual clinical situation as well as surgeon preference. Synthetic mesh has traditionally been used for abdominal wall reconstruction and hernia repairs in clean fields with low risk for infection. These include polypropylene mesh, polyester mesh, and ePTFE mesh. It is needless to say that the risk of infection is significantly higher with the use of synthetic mesh. The other major disadvantage with synthetic mesh is that once they are infected, they will require surgical removal to allow adequate control of infection. This however can be particularly challenging due to the extensive adhesion and scar formation associated with their use and can lead to further complications such as bowel injury and fistulas. Biologic mesh on the other hand is derived from human or animal (porcine, bovine) sources. Biologic meshes are nonimmunogenic as they have been stripped off all their cellular components and are primarily composed of acellular matrix. This acellular matrix acts as a scaffold and promotes native tissue to form collagen scar around it, and it is eventually replaced. Biologic mesh like any synthetic mesh can be infected; however the major advantage is that it does not require to be explanted and will therefore avoid the major morbidity associated with infected mesh removal and its complications. All hernias associated with fistulas and stomas are contaminated by default and should preferably be repaired with biologic mesh where its use has become the standard of care [25–27]. The biggest challenge with the use of biologic mesh is its prohibitive high cost. There is no data to suggest that one type of biologic mesh is superior to the other in terms of reducing the risk of recurrence and their use if primarily driven by surgeon preference. We compared the outcomes of human-derived and porcine-derived acellular dermal matrix and found a significantly higher hernia recurrence rate of 22.5% for patients with porcine-derived mesh compared with a 2.9% recurrence rate for patients with human-derived mesh placement with a mean follow-up time of 16 months. However, there was no difference in the rates of infections, reoperations, or mesh explantation between the two groups. Our experience has been mostly limited to Alloderm™ (human-derived) and Strattice™ (porcine-derived) and the difference in the recurrence rates seen in our study may be due to a selection bias. For the last 10 years, we have used exclusively Strattice™, and currently are the evaluation phase of both retrospective and prospective studies [28].

Mesh Placement Technique

Several techniques exist for the placement of mesh during repair of ventral hernias. During ventral hernia repair, the mesh can be placed in four different locations each with its own pros and cons. The mesh can be placed superficial to the fascia (onlay), in the retrorectus space (sublay or extraperitoneal underlay), or in situations where the fascial edges cannot be approximated at all the mesh can be used as a bridge.

Onlay Mesh Placement

In this technique, the mesh is placed above the primary fascial closure to provide reinforcement (Fig. 17.1). The hernia sac is first dissected and freed from the edges of the fascial defect and then either inverted or excised. After that the hernia defect is repaired primarily and the fascia is closed with non-absorbable continuous or interrupted sutures. The mesh is then placed over the anterior rectus sheath and is anchored to it. We prefer fixing the mesh to the fascia using non-absorbable sutures, either interrupted or continuous. We then place and use three to four large, closed-suction drains (19 French) under the subcutaneous tissue until the output has significantly slowed down (<25 cc/24) to avoid any seroma formation. The biggest advantage of the onlay technique is the ease with which it can be performed [29]. This technique often does not involve extensive intraperitoneal manipulation as long as the hernia sac can be free from the fascial edges and therefore avoids intraabdominal adhesion formation. It also prevents contact between the mesh and the underlying abdominal viscera and decreases the likelihood of fistula formation. Despite these advantages, this technique is associated with a high recurrence rate compared to other techniques [30]. This technique is also associated with higher rate of ischemic necrosis of skin flaps due to the creation of large skin flaps in patients with large hernia defects [31]. For these reasons, this technique is used infrequently and we reserve its use in patients in which the creation of retrorectus planes is impossible.

Sublay or Retrorectus Mesh Placement

This is the most widely used technique for abdominal wall reconstruction in our practice (Figs. 17.2, 17.3, and 17.4). Technically it is a more challenging repair compared to the onlay technique. The sublay technique, also known as

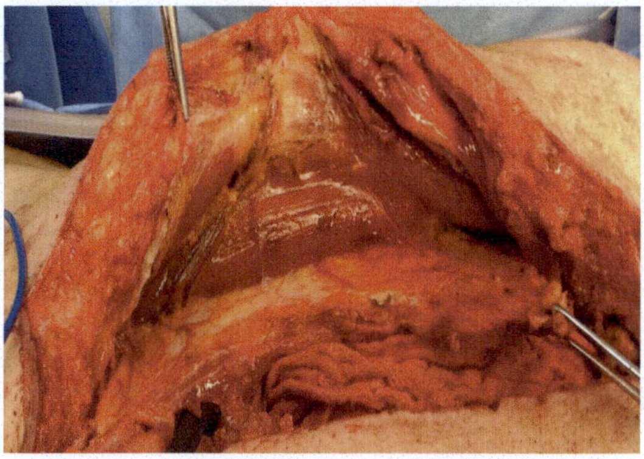

Fig. 17.2 Dissection of posterior rectus sheet

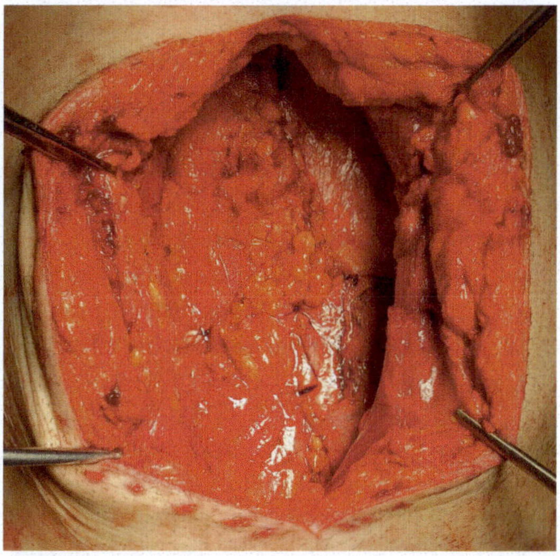

Fig. 17.3 Closed posterior rectus sheet

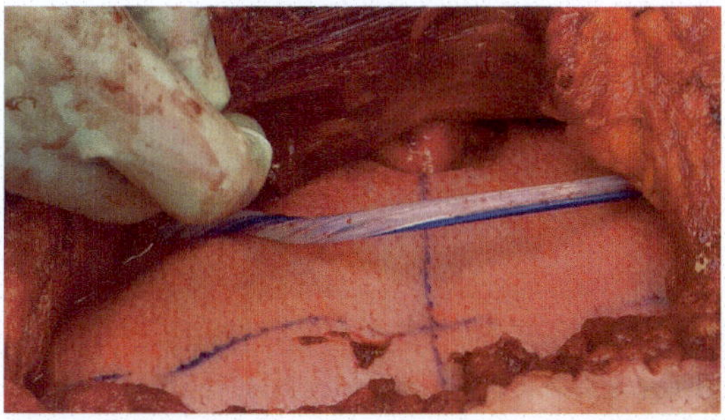

Fig. 17.4 Biologic mesh placed sublay and fashioned around the colostomy

the extraperitoneal underlay or the retrorectus technique, is a more recently described technique. This technique was first described by Stoppa in 1989 and therefore is also known as Rives-Stoppa technique [32, 33]. In this technique the mesh is placed retromuscularly (behind rectus muscle) and preperitoneally, after closure of the posterior rectus sheath and the peritoneum (Fig. 17.2). This is followed by primary closure of the anterior fascia. When performing this repair, the hernia sac is dissected down to the margins of the fascia. The hernia sac is then opened, and adhesiolysis is performed from ligament of Treitz to rectosigmoid junction. Previously placed synthetic mesh is explanted. After this a plane is created behind the rectus muscle using a combination of blunt dissection and electrocautery starting approximately 1 cm laterally to the linea alba. Dissection should be continued sufficiently cranially up to the level of the xiphoid process. Similarly, dissection is carried down inferiorly as far as possible to allow at least a 5 cm of overlap from the fascial edge of the defect. Once dissected, the posterior rectus sheath and the peritoneum are closed with continuous sutures, and the mesh is placed in the retromuscular space. If synthetic mesh is used, it should be ensured with certainty that no direct contact occurs between the mesh and the bowel to avoid the development of adhesions, erosions, and fistulas [32]. Whenever possible, an overlap of 5 cm is essential in all directions. To fix the mesh, we place multiple u-shaped sutures that are brought out transfascially through multiple small 5 mm stab incisions in the the skin which are tied down at the end. This step is critical as it must be insured that the mesh lays completely flat on top of the posterior rectus sheath and fascia which is made possible by placing the transfacial sutures as far laterally as possible. Finally, the anterior rectus sheath is closed using continuous slowly absorbable sutures ensuring no undue tension exists. If there is tension while closing the anterior fascia, additional mobilization using anterior component separation should be considered.

The retrorectus technique reduces the amount of soft tissue dissection and protects the mesh from environmental exposure due to native tissue coverage, which reduces the chances of mesh infection following a superficial surgical site infection. There is no direct contact between the mesh and the viscera which reduces adhesion formation. This technique is also associated with lower seroma formation rates due to the lack of need for large subcutaneous flap formation. Since the mesh is not placed in the subcutaneous plane, this reduces the likelihood of seroma formation [26]. Some recent studies have also demonstrated that sublay placement of mesh is associated with lower odds of recurrence and surgical site infection compared with onlay and underlay [31].

Intraperitoneal Underlay

The intraperitoneal underlay technique was first described in 1981 by McCarthy et al. [34]. When this technique was introduced, polypropylene mesh was used intraperitoneally; however, the intraperitoneal use of polypropylene mesh caused adhesions to the bowel and was, therefore, abandoned. This was then replaced by the use of a laminar polytetrafluoroethylene (ePTFE) mesh or bilayer composite prosthesis (PTFE and polypropylene) in order to avoid adhesions with the underlying bowel [35]. This technique can also be performed laparoscopically with the same principles of repair. The key component of this technique is to free up the internal aspect of the abdominal wall along its entirety. The mesh is placed intraperitoneally and secured into position with multiple transfacial or tacking sutures sufficiently close to one another in order to prevent the intra-abdominal contents from sliding and herniating between the mesh and the abdominal wall. The advantage of the underlay mesh is that the mesh is placed under abdominal wall and the intra-abdominal pressure presses the mesh against the wall which helps with incorporation. In contrast, with an onlay mesh placement, increase in intra-abdominal pressure forces the mesh away from the defect [36]. It also minimizes the risk of bowel injury during placement because of direct visualization of the intraperitoneal contents. However, it increases the risk of adhesions to the underlying bowel particularly if materials such as polypropylene are used. Studies that have compared the outcomes between onlay and underlay mesh placement and found underlay mesh placement had a lower risk for recurrence [0.59 (0.069–1.504)] and surgical site infection (SSI) [0.878 (0.291–1.985)] compared to onlay [31].

Bridge Mesh Placement

This is the least desired form of repair and is used when primary fascial closure cannot be achieved despite other attempts at mobilizing the components of abdominal wall [37, 38]. When the fascial defect cannot be approximated, an interposition mesh is placed to bridge the fascial defect. The mesh is sutured to the fascial edges and this type of abdominal closure is associated with high recurrence rates [38]. This may also be used where component separation cannot be performed due to prior surgeries or any other reason. In our experience, we prefer the use of biologic mesh. When inlay repair has been performed to bridge the fascia, all attempts should be made to cover the mesh with native skin and subcutaneous tissue. If no tissue coverage is possible, then a wound vacuum-assisted closure (VAC) may be employed to promote granulation that can later be used as the bed to perform skin graft for final closure [4, 11].

Summary

Complex abdominal wall reconstruction in elderly is possible and should be done in a timely fashion, when the presence of incisional hernia or complex abdominal wall defect is diagnosed. These hernias have significant risk of strangulation and require emergency surgery, and unless there are prohibitive risks of death, or when patient is deemed unable to survive the procedure, it should be offered to the patient.

References

1. Anthony T, Bergen PC, Kim LT, Henderson M, Fahey T, Rege RV, Turnage RH. Factors affecting recurrence following incisional herniorrhaphy. World J Surg. 2000;24(1):95–100;discussion 1.
2. Read RC, Yoder G. Recent trends in the management of incisional herniation. Arch Surg (Chicago, Ill: 1960). 1989;124(4):485–8.
3. van der Linden FT, van Vroonhoven TJ. Long-term results after surgical correction of incisional hernia. Neth J Surg. 1988;40(5):127–9.
4. Latifi R. Practical approaches to definitive reconstruction of complex abdominal wall defects. World J Surg. 2016;40(4):836–48.
5. Stone AA, Schwartz JE, Broderick JE, Deaton A. A snapshot of the age distribution of psychological well-being in the United States. Proc Natl Acad Sci. 2010;107(22):9985–90.
6. Neupane R, Fayezizadeh M, Majumder A, Novitsky YW. Is old age a contraindication to elective ventral hernia repair? Surg Endosc. 2017;31(11):4425–30.
7. Caglià P, Tracia A, Borzì L, Amodeo L, Tracia L, Veroux M, Amodeo C. Incisional hernia in the elderly: risk factors and clinical considerations. Int J Surg. 2014;12:S164–S9.
8. Işıl RG, Demir U, Kaya C, Bostancı Ö, İdiz UO, Işıl CT, Demircioğlu MK, Mihmanlı M. Approach to inguinal hernia in high-risk geriatric patients: should it be elective or emergent? Turk J Trauma Emerg Surg. 2017;23(2):122–7.
9. Rizk NN. A new description of the anterior abdominal wall in man and mammals. J Anat. 1980;131(Pt 3):373.
10. Flament JB. [Functional anatomy of the abdominal wall]. Der Chirurg. Zeitschrift fur alle Gebiete der operativen Medizen. 2006;77(5):401–7.
11. Patel NG, Ratanshi I, Buchel EW. The best of abdominal wall reconstruction. Plast Reconstr Surg. 2018;141(1):113e–36e.
12. Giordano SA, Garvey PB, Baumann DP, Liu J, Butler CE. The impact of body mass index on abdominal wall reconstruction outcomes: a comparative study. Plast Reconstr Surg. 2017;139(5):1234.
13. Newcomb W, Polhill J, Chen A, Kuwada T, Gersin K, Getz S, Kercher K, Heniford B. Staged hernia repair preceded by gastric bypass for the treatment of morbidly obese patients with complex ventral hernias. Hernia. 2008;12(5):465.
14. Borbély Y, Zerkowski J, Altmeier J, Eschenburg A, Kröll D, Nett P. Complex hernias with loss of domain in morbidly obese patients: role of laparoscopic sleeve gastrectomy in a multi-step approach. Surg Obes Relat Dis. 2017;13(5):768–73.
15. Martindale RG, Deveney CW. Preoperative risk reduction: strategies to optimize outcomes. Surg Clin. 2013;93(5):1041–55.
16. Huntington C, Gamble J, Blair L, Cox T, Prasad T, Lincourt A, Augenstein V, Heniford BT. Quantification of the effect of diabetes mellitus on ventral hernia repair: results from two national registries. Am Surg. 2016;82(8):661–71.
17. Farhat JS, Velanovich V, Falvo AJ, Horst HM, Swartz A, Patton JH Jr, Rubinfeld IS. Are the frail destined to fail? Frailty index as predictor of surgical morbidity and mortality in the elderly. J Trauma Acute Care Surg. 2012;72(6):1526–31.
18. Jokar TO, Ibraheem K, Rhee P, Kulavatunyou N, Haider A, Phelan HA, Fain M, Mohler MJ, Joseph B. Emergency general surgery specific frailty index: a validation study. J Trauma Acute Care Surg. 2016;81(2):254–60.
19. Rutkow IM, Robbins AW. Classification systems and groin hernias. Surg Clin N Am. 1998;78(6):1117–27.
20. Helgstrand F, Rosenberg J, Kehlet H, Bisgaard T. Outcomes after emergency versus elective ventral hernia repair: a prospective nationwide study. World J Surg. 2013;37(10):2273–9.
21. Forbes S, Eskicioglu C, McLeod R, Okrainec A. Meta-analysis of randomized controlled trials comparing open and laparoscopic ventral and incisional hernia repair with mesh. Br J Surg. 2009;96(8):851–8.
22. Allison N, Tieu K, Snyder B, Pigazzi A, Wilson E. Technical feasibility of robot-assisted ventral hernia repair. World J Surg. 2012;36(2):447–52.
23. Nguyen MT, Berger RL, Hicks SC, Davila JA, Li LT, Kao LS, Liang MK. Comparison of outcomes of synthetic mesh vs suture repair of elective primary ventral herniorrhaphy: a systematic review and meta-analysis. JAMA Surg. 2014;149(5):415–21.
24. Warren JA, Cobb WS, Ewing JA, Carbonell AM. Standard laparoscopic versus robotic retromuscular ventral hernia repair. Surg Endosc. 2017;31(1):324–32.
25. Itani KM, Rosen M, Vargo D, Awad SS, Denoto G 3rd, Butler CE. Prospective study of single-stage repair of contaminated hernias using a biologic porcine tissue matrix: the RICH study. Surgery. 2012;152(3):498–505.
26. Patton JH Jr, Berry S, Kralovich KA. Use of human acellular dermal matrix in complex and contaminated

abdominal wall reconstructions. Am J Surg. 2007;193(3):360–3; discussion 3.
27. Latifi R, Gustafson M. Abdominal wall reconstruction in patients with enterocutaneous fistulas. Eur J Trauma Emerg Surg. 2011;37(3):241–50.
28. Latifi R, Samson D, Haider A, Azim A, Iftikhar H, Joseph B, Tilley E, Con J, Gashi S, El-Menyar A. Risk-adjusted adverse outcomes in complex abdominal wall hernia repair with biologic mesh: a case series of 140 patients. Int J Surg. 2017;43:26–32.
29. Klinge U, Conze J, Krones CJ, Schumpelick V. Incisional hernia: open techniques. World J Surg. 2005;29(8):1066–72.
30. Misiakos EP, Patapis P, Zavras N, Tzanetis P, Machairas A. Current trends in laparoscopic ventral hernia repair. JSLS: J Soc Laparoendosc Surg. 2015;19(3):e2015.00048.
31. Holihan JL, Nguyen DH, Nguyen MT, Mo J, Kao LS, Liang MK. Mesh location in open ventral hernia repair: a systematic review and network meta-analysis. World J Surg. 2016;40(1):89–99.
32. Stoppa RE. The treatment of complicated groin and incisional hernias. World J Surg. 1989;13(5):545–54.
33. Heartsill L, Richards M, Arfai N, Lee A, Bingener-Casey J, Schwesinger WH, Sirinek KR. Open Rives-Stoppa ventral hernia repair made simple and successful but not for everyone. Hernia. 2005;9(2):162–6.
34. McCarthy JD, Twiest MW. Intraperitoneal polypropylene mesh support of incisional herniorraphy. Am J Surg. 1981;142(6):707–11.
35. Misiakos EP, Machairas A, Patapis P, Liakakos T. Laparoscopic ventral hernia repair: pros and cons compared with open hernia repair. JSLS: J Soc Laparoendosc Surg. 2008;12(2):117–25.
36. Breuing K, Butler CE, Ferzoco S, Franz M, Hultman CS, Kilbridge JF, Rosen M, Silverman RP, Vargo D, Ventral Hernia Working Group. Incisional ventral hernias: review of the literature and recommendations regarding the grading and technique of repair. Surgery. 2010;148(3):544–58.
37. Cobb WS, Kercher KW, Heniford BT. Laparoscopic repair of incisional hernias. Surg Clin North Am. 2005;85(1):91–103, ix.
38. Perrone JM, Soper NJ, Eagon JC, Klingensmith ME, Aft RL, Frisella MM, Brunt LM. Perioperative outcomes and complications of laparoscopic ventral hernia repair. Surgery. 2005;138(4):708–15; discussion 15–6.

Bariatric Surgery in the Elderly

18

Vasu Chirumamilla, Miles Dale, Sarvesh Kaul, and Ashutosh Kaul

Obesity is a growing global epidemic characterized by the excessive accumulation and storage of fat in the body. Excess fat leads to severe health consequences, including both a significant increase and earlier onset of obesity-associated comorbidities and a decrease in both length and quality of life [1, 2].

Morbid obesity is an independent risk factor for cardiovascular disorders, including coronary artery disease, congestive heart failure, hypertension, and atherosclerosis. Obesity is also a major cause of metabolic syndrome, especially type 2 diabetes mellitus. From an epidemiological perspective, almost 90% of type 2 diabetes cases can be prevented by avoiding obesity [2]. The probability of acquiring musculoskeletal disorders including osteoarthritis, mobility problems, musculoskeletal pains, and gout are all increased in the morbidly obese. Moreover, obstructive sleep apnea and other respiratory conditions are also a common problem. Obesity is also associated with an increased risk of several types of cancer (i.e., breast, colon, gallbladder, pancreas, renal, bladder, uterine, cervical, and prostate cancers). Obese subjects have an approximately 1.5–3.5-fold increased risk of developing these cancers compared with normal-weight subjects. It has been estimated that between 15% and 45% of these cancers can be attributed to being overweight (BMI 25.0–29.9 kg/m^2) and obese (30–40 kg/m^2) in Europe [3]. The World Health Organization states that obesity is the most important known avoidable cause of cancer after tobacco. Obesity can also have harmful psychosocial effects, notably a perception that thin and underweight people are hardworking and attractive while obese people are lazy. This can lead to unfair treatment, lower wages, and depression among the overweight and obese. Overall, obesity significantly lowers both quality and length of life.

Body mass index, or BMI, is a simple index of weight-for-height that is commonly used to classify overweight and obesity in adults. BMI is calculated by dividing a person's weight in kilograms by the square of height in meters. BMI levels are used to classify obesity as follows: <18.5 kg/m^2, underweight; between 18.5 and <25 kg/m^2, normal; 25 to <30 kg/m^2, overweight; between 30 and 40 kg/m^2, obese; and >40 kg/m^2, severely obese. Presently about 64.5% of the US population is overweight (BMI >25 kg/m^2), 30.5% of

V. Chirumamilla · M. Dale · A. Kaul (✉)
Department of Surgery,
Westchester Medical Center, Valhalla, NY, USA

S. Kaul
Department of Surgery, Westchester Medical Center, Valhalla, NY, USA

Colgate University, Hamilton, NY, USA

R. Latifi (ed.), *Surgical Decision Making in Geriatrics*,
https://doi.org/10.1007/978-3-030-47963-3_18

the US population is obese BMI >30 kg/m^2), and 4.7% of the Americans are characterized as severely obese (BMI >40 kg/m^2) [4, 5]. The prevalence of obesity tends to increase with age and peaks in the 45–64 year range [5].

Most developed countries characterize a person as elderly when their chronological age is equal to or greater than 65 years. The United Nations, however, characterizes persons that are 60 years and above as elderly, while in the African population, a person that is 55 years or older is considered elderly [6].

The incidence of weight-related comorbidities continues to increase in the elderly population [5]. Obesity decreases the quality of life and diminishes the life expectancy of patients [2, 7]. Life expectancy in the USA dropped in 2016 and 2017, partly due to the continued obesity epidemic [8]. An increase in obesity contributed to about 186,000 additional deaths in 2011 alone [8].

Obesity is not solely a middle-age disease; geriatric obesity rates are also at an all-time high. 35% of people older than 60 are obese [9]. Aging is associated with a decrease in total energy expenditure, which leads to fat mass reaching maximum levels at 60–70 years of age [10]. Diabetes, metabolic syndrome, hypertension, arthritis, cardiovascular disease, and urinary incontinence are associated with obesity in advancing age. Obesity is a risk factor for chronic diseases and premature mortality, but the extent of these associations among the elderly is under debate [11]. In a pooled meta-analysis, it was found that the mortality rate was highest among malnourished subjects; it then decreased with increasing BMI but rose again in BMI over 30 kg/m^2 [11] . The lowest mortality is correlated with an average BMI of between 23.5 and 27.5 kg/m^2, [11]. A strong correlation between waist circumference and mortality risk in the elderly was confirmed in multiple studies [11]. Moderately high BMI was also found to be a protective factor against acute events that occur frequently in geriatric age groups [11, 12].

Obesity Treatment Options

In the morbidly obese, conservative measures, including diet, exercise, and medical treatments, usually fail to produce an effective sustained weight loss. On an average these measures only lead to a 2–3% weight loss [1]. Bariatric surgery has been shown to be superior to current medical therapies with regard to the induction and maintenance of weight loss and improvement in weight-associated comorbidities in the morbidly obese [13]. Bariatric surgery is the only consistent method for morbidly obese patients to lose weight in the long term according to the National Institutes of Health Consensus Development Conference in 1991 [14].

Elderly patients are a unique population with more comorbidities than their younger counterparts; surgery in these patients is often associated with negative outcomes. Over 70% of people over 65 years have associated comorbidities such as diabetes, hypertension, or hyperlipidemia [15]. As the population of the elderly grows and surgical techniques improve, the previous age restrictions of surgical procedures have become less defined. Bariatric surgery in elderly patients can greatly improve their health, but its repercussions need to be carefully considered [16]. A surgeon's decision process in the expanding requirement of surgical procedures for the elderly has to overcome the boundary of limitation based on age [17]. Therefore, selecting the right surgical candidate remains an integral part of the decision-making process in surgery, and age should no longer be the main component of the process.

Bariatric procedures in the elderly are not unique to the negative bias against geriatric surgery. Historically, bariatric surgical treatment of obesity in the elderly has been controversial based on past studies reporting increased surgical risk and a lack of data demonstrating bariatric surgery's long-term benefit. Several studies revealed better comorbidity, mortality, and weight loss outcomes in younger bariatric surgery patients compared to older subjects [8, 18]. Factors leading to decreased weight loss in elderly patients include sedentary lifestyle, sarcopenia, a higher calorie intake, poor fitness, chronic inflammation, hormonal changes, age-related decreased energy requirements, and lower lipolytic capacity, especially after sympathetic stimulation [19].

Several surgeons have explored utilizing bariatric surgery in the elderly population. The National Institutes of Health in 2006 and the French National Authority for Health in 2009 recognized the possibility of using bariatric surgery in patients older than 60. Despite its benefits, bariatric surgery is not performed at the same rate in the elderly compared to younger patients. As of 2015, only 10% of bariatric surgery is performed in patients >60 years of age [20], despite the fact that over 35% of this population was obese [9]. NSQIP data show that the proportion of bariatric procedures in patients >65 years of age significantly increased from 1.9% in 2005 to 4.8% in 2009 [21]. Irrespective of this progress, a hesitation regarding performing bariatric surgery in the elderly still exists.

Bariatric Surgery in the Elderly

Between 1987 and 2030, the total US population is projected to increase by 26% from 252 to 317 million, while the population of people 65 and older is expected to increase by more than 100%, from the present 12% of the total population to nearly 21% of the total population (67 million) [22]. Old age is not immune to obesity [23]. In the USA, the prevalence of obesity in the elderly ranges from 42.5% in women aged 60–69 years to 19.5% in those aged 80 years or older. The prevalence of obesity is 38.1% in men aged 60–69 years and 9.6% for those men aged 80 years or older [11]. The increase in BMI over the period of 1988–2011 is estimated to have reduced life expectancy at age 40 by 0.9 years in 2011 (95% confidence interval 0.7–1.1 years); it is also expected to have accounted for 186,000 excess deaths that year. Older age (>60) had been considered a relative contraindication for bariatric surgery due to increased complication risk until recently, but multiple studies have studied the outcomes of bariatric surgery in aged patients. Bariatric surgery in the elderly is not as efficient as in younger patients, but it is still superior to conservative medical treatment considering the resolution rates of comorbidities. In some initial reports, including those by Flum et al. and Livingston et al., an increased risk of complications and deaths were seen in patients over 65 years [24, 25]. In these studies, many patients had an open surgical or a laparoscopic procedure performed at the beginning of the RYGB learning curve, the procedures were performed 10–15 years ago, and the patients were probably frail (under the Medicaid scheme) [1]. Recent studies have highlighted the safety of bariatric surgery in this age demographic [1], and several case-control studies have shown mortality is equivalent across age groups when adjusted for other variables [9]. Qin et al. retrospectively reviewed the ACS-NSQIP registry procedures from 2005 to 2012 and found that advanced age was significantly associated with increased overall medical complications, but not surgical complications in laparoscopic sleeve gastrectomy. Their findings suggested that the risk conferred by advancing age in laparoscopic sleeve gastrectomy is predominantly for medical morbidity. Thus, they advocated for improved perioperative management of medical complications. They felt that laparoscopic sleeve gastrectomy may be the preferable option to RYGB for elderly patients, as neither procedure is riskier with regard to 30-day morbidity, while LSG has been shown to be safer with regard to long-term reoperation and readmission risk [9].

In a cohort study of 9,044 bariatric surgical procedures, there were no deaths among the 451 older group patients, while two deaths occurred in the younger group [26]. However, older patients lost less excess weight than younger patients (72.44% versus 86.11%, respectively). Additionally, older patients presented higher rates of complications (8.42% versus 5.59%), comorbidities (77.60% versus 55.45%), and revisions (1.33% versus 0.77%). There was no statistical difference in hospital stay between the older group and control group (2.27 versus 2.23, respectively). When performing a Clavien-Dindo classification of complications, the review demonstrated significant differences in class 3B and 4A and no differences in other classes. Diabetes, fatty liver, and sleep apnea improved or remitted in >90% of patients in both groups, while hypertension and hyperlipidemia improved by >80%. Hyperuricemia and ischemic heart disease were improved or resolved in >70% of the patients [26].

In a review of patients above the age of 60 years in Metabolic and Bariatric Surgery Accreditation and Quality Improvement Program database, laparoscopic sleeve gastrectomy was compared Roux-en-Y gastric bypass in 3371 patients [27]. They discovered that LRYGB and LSG procedures in patients aged 60 years or older are relatively safe in the short term with an acceptable complication rate and low mortality. However, LRYGB was more challenging and was associated with significantly increased rates of leakage events (1.01% versus 0.47%, $p = 0.011$), 30-day reoperation (2.49% versus 0.89%, $p < 0.001$), 30-day readmission rate (6.08% versus 3.74%, $p < 0.001$), longer operative times (122 versus 84 min., $P < 0.001$), and a longer hospital stay [27].

When data were analyzed from the NSQIP database between 2010 and 2011 of patients over the age of 65 years, 1005 patients were found to have undergone SG and RYGB. Diabetes was more frequent in the RYGB group (43.2% versus 55.6%, $P = .004$) but the 30-day post-op mortality (0.6% versus 0.6%, OR 1.1, 95% CI .11–9.49), serious morbidity (5.2% versus 5.6%, OR .91, 95% CI .42–0.96) and overall morbidity (9% versus 9.1%, OR 1.0, 95% CI .55–1.81) were similar. In elderly patients undergoing laparoscopic bariatric surgery, they found that SG was not found to be associated with significantly different 30-day outcomes compared to RYGB. Both procedures were followed by acceptably low morbidity and mortality [21]. Improvement of intraoperative surgical management and optimization of perioperative care have decreased the complication rate [1].

With increased experience, the operative times for BS have decreased, and they compare favorably to operative times seen in elderly patients undergoing cardiac, orthopedic, or oncologic surgical procedures [28]. As life expectancy continues to increase on average, it is imperative to redefine previously held age restrictions on procedures that have been shown to improve people's health [7, 18].

The main goals of bariatric surgery include facilitating weight loss, alleviating or reversing comorbidities, and improving both the quality and length of life. The drive to push the envelope with bariatric surgery for the elderly is not aimed only for losing weight but also at reducing the use of medications for their comorbidities. The percentage of weight loss after BS is usually less in the elderly compared to younger patients. In a study comparing different procedures, weight loss (% of initial weight) was lower after a LAGB than after RYGB or SG. After LAGB, weight loss (%) did not differ between patients above 60 years and those aged <40 (difference 1.7 ± 1.5%, $P = .26$). Weight loss (%) was lower in patients aged >60 years when compared to those <40 years after both RYGB (−5.6 ± 1.7%, $P = .001$) and SG (−7.0 ± 2.6%, $P = .01$) [1].

Despite losing less weight compared to younger patients, significant comorbidity alleviation can still be achieved, as only a 5–10% decrease in weight helps in improving obesity-related comorbidities [1]. This beneficial impact on comorbidities elimination can be seen with a reduction in medication requirement and a decrease in overall medical care expenditure [1].

Which Bariatric Procedure to Choose

There is no standard protocol of preoperative evaluation for bariatric surgery in the elderly [29]. The standard general assessment may be insufficient for a population with specific needs and concerns, like that of older individuals [29]. Batsis and Dolkart proposed an individualized approach to older candidates and bariatric surgery that encompasses and contemplates factors that are not as prominent in younger individuals. These authors propose that older candidates should be evaluated by an extended multidisciplinary team that would include a geriatrician and a social worker. This proposed preoperative assessment must incorporate five specific topics for this population: (1) evaluation of functional status; (2) assessment of frailty; (3) cognitive assessment; (4) identification of depression; and (5) social support and discharge planning [30]. Their proposal is based on the fact that worse surgical outcomes are mostly related to impaired functional status, presence of frailty, delirium that occurs in previously cognitive impaired individuals, presence of depression, and other psychiatric conditions that may be underestimated prior to surgery, as well as the unavailability of caretakers and ideal housing conditions for these individuals after surgery [29]. The evolution of surgical techniques, especially the development of minimally invasive approaches, constitutes an important factor in the improvement of surgical outcomes over time, along with cautious and careful patient selection and preoperative assessment [29].

The four most common bariatric surgery procedures being done in the USA presently are sleeve gastrectomy, RYBG, AGB, and DS. Generally, the mortality and the 1-year excess weight loss after surgery progresses as follows (from the least to the most): LAGB< LSG< RYGB< DS (i.e., both perioperative mortality and weight loss after 1 year are least with LAGB and most with DS).

The number of BS done in the elderly has shown an increase from about 1.9% in 2005 to 4.8% in 2009 [31]. Between 2008 and 2012, the use of laparoscopic sleeve gastrectomy increased from 0.9% to 36.3%, while the use of laparoscopic adjustable gastric banding (LAGB) and RYGB both decreased from 23.8% to 4.1% and 66.8% to 56.4%, respectively [21, 32]. In the last few years, the number of SG has continued to increase, and multiple reviews have shown that both LSG and LRYGB can be safely performed in elderly both with low morbidity and mortality.

A MBSAQIP database analysis [26] of perioperative outcomes comparing laparoscopic sleeve gastrectomy with LRYGB in patients aged 60 years or older demonstrated that laparoscopic Roux-en-Y gastric bypass and laparoscopic sleeve gastrectomy are relatively safe in the short term, with an acceptable complication rate and low mortality. However, LRYGB was associated with significantly increased rates of leakage events, 30-day reoperation, 30-day readmission, longer operative time, and longer hospital stay, as compared to LSG. LRYGBP showed a trend toward higher perioperative mortality than LSG. In the short term, sleeve gastrectomy was associated with significantly fewer 30-day complications than gastric bypass [26].

Another study compared LSG with LRYGB and discovered that both procedures are followed by acceptably low morbidity and mortality [21]. While diabetes was more frequent in the RYGB group (43.2% versus 55.6%, $P = .004$), the 30-day mortality (0.6% versus 0.6%, OR 1.1, 95% CI .11–9.49), while serious morbidity (5.2% versus 5.6%, OR .91, 95% CI .42–0.96), and overall morbidity (9% versus 9.1%, OR 1.0, 95% CI .55–1.81) were similar. The authors concluded that in elderly patients undergoing laparoscopic bariatric surgery, SG was not associated with significantly different 30-day outcomes compared to RYGB. Both procedures are followed by acceptably low morbidity and mortality [21].

Results of Bariatric Surgery in the Elderly

Older patients may have more comorbidities but demonstrate an insignificant increase in mortality in relation to having bariatric surgery [7]. A meta-analysis concluded that mortality has ranged from 0.5% to 2.0% (30 days) and 2.8% (90 days) for gastric bypass patients [23]. When the age group was increased to a range of 65–90 years old, the mortality increased to 3.2% [24]. Laparoscopic Roux-en-Y gastric bypass can be done to achieve weight loss in the elderly population [5]. The same operations were offered regardless of age (LGB, LGS, and LRYGB – two converted to open). The elderly have more awareness of the negative consequences of their comorbidities and thus are more motivated for lifestyle changes that will facilitate a better quality of life [6]. Older age or Medicare status should not be part of the decision-making process for bariatric surgery [6]. In a study of 43,378 patients, of which 1994 were >65 years old, there was no difference in the complications after bariatric surgery, but the length of stay was longer in older patients than their younger counterparts [31, 33]. Another study of 1339 patients over 60 years old undergoing BS also supported the view of Dorman et al. [31, 33].

Insurance Coverage of Bariatric Surgery

Insurance companies have recognized the cost-benefit of bariatric surgery and approve BS for patients meeting their criteria. Analysis of outcomes of BS show a strong return of investment up to 5 years after surgery and find that cost savings start accruing by the third postoperative month [34, 35]. The short-term return on investment associated with bariatric surgery is consistent with a decrease in multiple comorbid conditions, including diabetes, coronary artery disease, hypertension, and sleep apnea [34]. The cost reductions in these diseases take into account prescription drug usage, hospital stays, and physician visits. Type II diabetes is greatly improved by bariatric surgery, and the estimated annual costs of managing a diabetes patient (US$13,243) are five times more than a patient without diabetes (US$2560) [36].

The majority of the geriatric population in USA has some form of Medicare or Medicaid coverage. After a review, Medicare authorized approval of bariatric surgery for older patients in 2006 [33]. However, the reimbursement amount was found to be inadequate in a study, and the low reimbursement amount was felt to lead to unequal access to care [14] by threatening the ability of patients to find a bariatric surgeon willing to perform surgery on Medicare patients. Combinations of the increased comorbidities of obesity accelerate senescence and increase the cost of acute and chronic rehabilitation and palliative care in the elderly [11].

Future of Bariatric Surgery in Elderly

At the moment, there is not a proven successful lifestyle change or medical therapy which leads to sustained weight loss to result in a decrease of comorbidities. Improving quality of life is an important goal of any treatment especially as people age. Obesity has been shown to negatively affect quality of life, and bariatric surgery can offer the obese elderly population improvement not only in their health and activities of daily living, thereby enhancing the quality of life. Continued research on bariatric surgery in the elderly is necessary to increase the strength of this recommendation.

The upper age limit for recommending bariatric surgery in the elderly is yet to be clearly established. Most programs stratify the patient's risk and expected benefits of bariatric surgery based on preoperative assessment and then make a decision on the suitability of surgical intervention. A geriatrician can assist in the surgical decision-making process by performing a comprehensive preoperative assessment including a detailed patient's history, comorbidities, and cardiopulmonary evaluation. It is also important to evaluate the level of recovery support required postoperatively. An ideal bariatric surgical preoperative approach would thus involve a multidisciplinary team consisting of a geriatrician, cardiologist, pulmonologist, psychiatrist, nutritionist, and social worker. This approach would help evaluate a patient's functional status, frailty index, social support, and cognitive state, and rule out psychological disorders.

Conclusion

BS provides an effective and long-lasting treatment option for morbidly obese elderly patients with significant improvement in weight- and obesity-related comorbidities. Bariatric surgery is safe in the elderly, with slightly increased risks when compared to younger populations. The decision on the type of the procedure to be recommended depends on the individual needs, the experience of the surgical team, and the health profile of the patient.

References

1. Ritz P, Topart P, Benchetrit S, Tuyeras G, Lepage B, Mouiel J, et al. Benefits and risks of bariatric surgery in patients aged more than 60 years. Surg Obes Relat Dis. 2014. http://dx.doi.org/10.1016/j.soard.2013.12.012.
2. Kaul A, Sharma J. Impact of bariatric surgery on comorbidities. Surg Clin North Am. 2011;91(6):1295–312, ix.
3. Pischon T, Nöthlings U, Boeing H. Obesity and cancer. Proc Nutr Soc. 2008;67(2):128–45.
4. Flegal KM, Carroll MD, Ogden CL, Johnson CL. Prevalence and trends in obesity among US adults, 1999–2000. JAMA. 2002;288(14):1723–7.
5. Trieu HT, Gonzalvo JP, Szomstein S, Rosenthal R. Safety and outcomes of laparoscopic gastric bypass surgery in patients 60 years of age and older. Surg Obes Relat Dis. 2007;3(3):383–6.
6. Keren D, Matter I, Rainis T. Sleeve gastrectomy in different age groups: a comparative study of 5-year outcomes. Obes Surg. 2016;26(2):289–95.
7. Flegal KM, Kit BK, Orpana H, Graubard BI. Association of all-cause mortality with overweight and obesity using standard body mass index categories: a systematic review and meta-analysis. JAMA. 2013;309(1):71–82.
8. Scozzari G, Passera R, Benvenga R, Toppino M, Morino M. Age as a long-term prognostic factor in bariatric surgery. Ann Surg. 2012;256(5):724–8; discussion 8–9.
9. Qin C, Luo B, Aggarwal A, De Oliveira G, Kim JY. Advanced age as an independent predictor of perioperative risk after laparoscopic sleeve gastrectomy (LSG). Obes Surg. 2015;25(3):406–12.
10. Amarya S, et al. Health consequences of obesity in the elderly. J Clin Gerontol Geriatr. 2014;5:63–7.
11. Donini LM, Savina C, Gennaro E, De Felice MR, Rosano A, Pandolfo MM, et al. A systematic review of the literature concerning the relationship between obesity and mortality in the elderly. J Nutr Health Aging. 2012;16(1):89–98.
12. Galanos AN, Pieper CF, Cornoni-Huntley JC, Bales CW, Fillenbaum GG. Nutrition and function: is there a relationship between body mass index and the functional capabilities of community-dwelling elderly? J Am Geriatr Soc. 1994;42(4):368–73.
13. Colquitt JL, Pickett K, Loveman E, Frampton GK. Surgery for weight loss in adults. Cochrane Database Syst Rev. 2014;(8):CD003641.
14. Madan AK, Powelson JE, Tichansky DS. Cost analysis of laparoscopic gastric bypass practice using current Medicare reimbursement and practice costs. Surg Obes Relat Dis. 2008;4(2):131–6.
15. Susmallian S, Raziel A, Barnea R, Paran H. Bariatric surgery in older adults: should there be an age limit? Medicine. 2019;98(3):e13824.
16. Pajecki D, Santo MA, Joaquim HD, Morita F, Riccioppo D, de Cleva R, et al. Bariatric surgery in the elderly: results of a mean follow-up of five years. Arq Bras Cir Dig: ABCD (Braz Arch Dig Surg). 2015;28(Suppl 1):15–8.
17. Neuman MD, Bosk CL. The redefinition of aging in American surgery. Milbank Q. 2013;91(2):288–315.
18. Contreras JE, Santander C, Court I, Bravo J. Correlation between age and weight loss after bariatric surgery. Obes Surg. 2013;23(8):1286–9.
19. Giordano S, Victorzon M. Bariatric surgery in elderly patients: a systematic review. Clin Interv Aging. 2015;10:1627–35.
20. Gebhart A, Young MT, Nguyen NT. Bariatric surgery in the elderly: 2009–2013. Surg Obes Relat Dis. 2015;11(2):393–8.
21. Spaniolas K, Trus TL, Adrales GL, Quigley MT, Pories WJ, Laycock WS. Early morbidity and mortality of laparoscopic sleeve gastrectomy and gastric bypass in the elderly: a NSQIP analysis. Surg Obes Relat Dis. 2014;10(4):584–8.
22. Thereaux J, Poitou C, Barsamian C, Oppert JM, Czernichow S, Bouillot JL. Midterm outcomes of gastric bypass for elderly (aged ≥60 yr) patients: a comparative study. Surg Obes Relat Dis. 2015;11(4):836–41.
23. Preston SH, Vierboom YC, Stokes A. The role of obesity in exceptionally slow US mortality improvement. Proc Natl Acad Sci U S A. 2018;115(5):957–61.
24. Flum DR, Salem L, Elrod JA, Dellinger EP, Cheadle A, Chan L. Early mortality among Medicare beneficiaries undergoing bariatric surgical procedures. JAMA. 2005;294(15):1903–8.
25. Livingston EH, Langert J. The impact of age and Medicare status on bariatric surgical outcomes. Arch Surg. 2006;141(11):1115–20; discussion 21.
26. Susmallian S, Barnea R, Weiss Y, Raziel A. Outcome of bariatric surgery in older patients. Surg Obes Relat Dis. 2018;14(11):1705–13.
27. Janik MR, Mustafa RR, Rogula TG, Alhaj Saleh A, Abbas M, Khaitan L. Safety of laparoscopic sleeve gastrectomy and Roux-en-Y gastric bypass in elderly patients – analysis of the MBSAQIP. Surg Obes Relat Dis. 2018;14(9):1276–82.

28. Willkomm CM, Fisher TL, Barnes GS, Kennedy CI, Kuhn JA. Surgical weight loss >65 years old: is it worth the risk? Surg Obes Relat Dis. 2010;6(5):491–6.
29. Cazzo E, Gestic MA, Utrini MP, Chaim FDM, Callejas-Neto F, Pareja JC, et al. Bariatric surgery in the elderly: a narrative review. Rev Assoc Med Bras (1992). 2017;63(9):787–92.
30. Batsis JA, Dolkart KM. Evaluation of older adults with obesity for bariatric surgery: geriatricians' perspective. J Clin Gerontol Geriatr. 2015;6:45–53.
31. Dorman RB, Abraham AA, Al-Refaie WB, Parsons HM, Ikramuddin S, Habermann EB. Bariatric surgery outcomes in the elderly: an ACS NSQIP study. J Gastrointest Surg. 2012;16(1):35–44; discussion.
32. Nguyen NT, Nguyen B, Gebhart A, Hohmann S. Changes in the makeup of bariatric surgery: a national increase in use of laparoscopic sleeve gastrectomy. J Am Coll Surg. 2013;216(2):252–7.
33. Marihart CL, Brunt AR, Geraci AA. Older adults fighting obesity with bariatric surgery: benefits, side effects, and outcomes. SAGE Open Med. 2014;2:2050312114530917.
34. Cremieux PY, Buchwald H, Shikora SA, Ghosh A, Yang HE, Buessing M. A study on the economic impact of bariatric surgery. Am J Manag Care. 2008;14(9):589–96.
35. Bradley DW, Sharma BK. Centers of excellence in bariatric surgery: design, implementation, and one-year outcomes. Surg Obes Relat Dis. 2006;2(5):513–7.
36. Klein S, Ghosh A, Cremieux PY, Eapen S, McGavock TJ. Economic impact of the clinical benefits of bariatric surgery in diabetes patients with BMI ≥35 kg/m^2. Obesity (Silver Spring). 2011;19(3):581–7.

Vascular Surgery in the Elderly

19

Sateesh C. Babu, Aditya Safaya, Romeo Mateo, and Igor Laskowski

People have always fantasized about youth, health, fitness, and beauty that is everlasting and are frightened at the prospects of getting old and frail. The most extreme example of this is the Chinese Emperor Quin who ordered his personal physician to invent the "elixir of life" that would make him immortal. It was not until the late 1960s and 1970s did physicians and surgeons realize that there is more to the aging patient than graying of hair and wrinkling of the skin. Vascular disease is the disease of the elderly; hence, most patients fall into the category of geriatric population. The physiology of aging, significance of frailty, and its impact on treatment and outcome are dealt with in other chapters of this book. Hence this section will describe the most frequently encountered vascular conditions in the elderly and will outline current guidelines for management. These include the diseases involving the aorta, peripheral arterial disease (PAD), extracranial cerebrovascular disease, visceral artery disease, venous thromboembolic disease, amputations, and a brief review and update on dialysis access. Vascular surgeons in reality are vascular disease specialists. Thus all aspects of diagnosis and treatment – surgical, interventional, and nonsurgical options of management – will be described.

S. C. Babu (✉) · A. Safaya · R. Mateo · I. Laskowski
Section of Vascular Surgery, Department of Surgery, Westchester Medical Center, Valhalla, NY, USA
e-mail: Sateesh.babu@wmchealth.org

Vascular disease management is complex. It is rendered even more difficult by its prevalence in the aged and presence of multiple coexisting other system disease processes. In order to achieve a successful outcome, the care has to be comprehensive involving multi-team approach. The old system of a "single physician" delivering total care of a patient is not feasible in this special cohort of patients. The team caring for the elderly patient with vascular disease should, in addition to vascular surgeon, include cardiologist, infectious disease specialist, podiatrist, endocrinologist, nutritionist, physiatrist, and orthotic specialist. Other specialists from hematology, plastic surgery, and neurology may also be called upon to render their expertise as needed.

Scope of the Problem

Age-related physiological changes combined with frailty make addressing management strategies in the elderly a unique problem. This is further compounded by factors like sedentary activity, subclinical depression, polypharmacy, and changes in cognitive function.

Frailty is fairly common in elderly vascular patients, though not all elderly are frail. As described in previous chapters, frailty is a condition characterized by loss of biological reserves across multiple organ systems together with failure of homeostatic mechanisms and vulnerability to physiological

R. Latifi (ed.), *Surgical Decision Making in Geriatrics*,
https://doi.org/10.1007/978-3-030-47963-3_19

decompensation after minor stressor events. In other words, even though normal aging is characterized by normal physiological decline, with frailty this decline appears to be accelerated [1]. Frailty is seen in 10% of the patients 65 years old or older, and the incidence increases to 25% in patients aging 85 and older. The incidence of arterial diseases increases with age, owing to age-related changes in the vascular bed. PAD is uncommon before the age of 50 years. The incidence rates rise sharply around 20% after the age of 80 years [2].

Diseases Involving the Aorta

Aneurysms and Aortic Dissections

Aortic pathologies in the elderly usually present as acute, chronic, or acute on chronic conditions. Aortic dissections, aneurismal diseases, and occlusive diseases form the majority of the pathology. The newly defined category of acute aortic syndrome includes (a) acute aortic dissection, (b) penetrating aortic ulcer, (c) intra-mural hematoma, and (d) symptomatic thoracic aortic aneurysms. Acute aortic syndrome should be considered in differential diagnosis especially in an elderly patient presenting with signs of chest pain, hypotension, or neurological manifestations mainly syncope [3].

In contrast to aortic diseases in younger population where etiology would include connective tissue disorders such as Marfan's syndrome and/or Ehlers-Danlos syndrome, atherosclerosis is the most common etiology for aortic diseases in the elderly.

Aortic Dissection

In the elderly, symptoms of syncope, heart failure, and stroke are more common presenting features of acute aortic dissection than mere chest pain [3]. Stanford University classification of aortic dissection to type A and type B is based on the site of intimal tear, type A those involving the ascending aorta and type B those not involving the ascending aorta.

Acute aortic dissection is the most common catastrophe of human aorta with mortalities of 2% per hour if undiagnosed and untreated which is even higher in octogenarians [4]. Medical management is the standard for type B aortic dissection with the goal of keeping blood pressure normal and reduction of the force of pulse. Indications for surgical treatment are persistent pain despite medical management, presence of pseudoaneurysm, rupture or impending rupture, hemothorax, and visceral malperfusion. However, endovascular treatment has now become the treatment of choice in these situations. **T**horacic **e**ndo-**v**ascular **a**ortic **r**epair (TEVAR) is minimally invasive, effective, and superior to surgery in terms of in-hospital mortality, incidence of paraplegia, and long-term survival [5]. On the other hand, standard of care for type A dissection is still open surgery. This consists of replacement of ascending aorta with or without replacement of aortic valve and replantation of coronary arteries. Replacement of the entire aortic arch and reconstruction of great vessels may be required in those cases where the disease process involves the arch. Versatile stent grafts and sophisticated delivery systems are being developed to offer endovascular modality even for type A dissections. This will have significant impact on the ability to offer treatment for even the very elderly and frail in whom open surgery may be attendant with prohibitive risk [6, 7].

Aortic Aneurysms

Aneurysms are generally a disease of the elderly with incidence of 5% in people 65 years and 12.5% in men above 75 years of age [3]. The risk of rupture is directly associated with the size of the aneurysm.

The current guidelines recommend treatment at size of 5.5 cm in males and 5 cm in female patients. Aneurysms 4 cm or less can be followed annually, and sizes from 4 to 5.5 cm are to be followed every 6 months. In addition to these general indications, all symptomatic aneurysms and aneurysms with rapid growth (>1 cm/year) are indication for treatment [8].

The US Preventive Services Task Force favors screening for abdominal aortic aneurysms (AAA) in all men aged 65 and above with history of smoking, family history of aneurysms, or other coexisting risk factors like coronary artery disease (CAD), hyperlipidemia (HLD), hypertension (HTN), and obesity [9]. Ultrasonography is a simple noninvasive test with high sensitivity and specificity for detecting and following AAA. CT angiogram (CTA) is considered for planning treatment.

In 2000 the FDA approved the use of endovascular devices for the management of abdominal aortic aneurysms. It is now the current standard of care for this disease. Around 80% or more of patients with abdominal aneurysms are treated with this minimally invasive technique – **e**ndo**v**ascular **a**neurysm **r**epair (EVAR) [10]. The overall mortality of EVAR is 1.6%. This clearly has enabled high-risk elderly patients, who otherwise would have been denied treatment, get repaired. The incidence of late complications related to EVAR is around 8–10% [11]. Endoleak is the term used to describe entry of blood into the aneurysm sac after EVAR. This can be from some of the branches like lumbar, inferior mesenteric, or middle sacral artery. When not associated with enlarging sac, these type II endoleaks do not require intervention. When the leak into the sac is from inadequate seal zones (type I) or separation of graft limbs (type III), these require treatment. Graft migration and occlusion of one or the other limbs of stent graft are some of the delayed complications, which require prompt recognition and intervention. All patients who are treated with EVAR have to be followed lifelong to identify adverse events. Aortic duplex scan is noninvasive, safe, and repeatable making it the main imaging test of choice to follow patients after EVAR, generally at 6-month intervals. CTA is done if the duplex indicates increase in size of the sac or other findings indicating problems.

Aorto-iliac Occlusive Disease

Aorto-iliac occlusive disease due to atherosclerosis presents with symptoms of calf, thigh, or gluteal claudication and/or erectile dysfunction in men. Stenotic and sometimes even occlusive aorto-iliac segment lends itself to interventions like angioplasty and stenting with excellent results approaching the results of surgical procedures like aortofemoral bypass which had been the gold standard for treating aorto-iliac occlusive disease, before the era of interventions. In fact, open procedures for aorto-iliac diseases are now uncommon. Even in patients who are not amenable to endovascular approaches, extra-anatomic bypass (axillo-femoral or femoro-femoral bypass) is the preferred procedure in the elderly and frail owing to less systemic implications. Extra-anatomic bypass procedures may not have the same long-term patency of direct aortic reconstructions, but have the advantage of being less invasive and less stress on the elderly patient's cardiovascular hemodynamics [12].

Peripheral Vascular Disease

Incidence of chronic occlusive diseases of lower extremity arteries, commonly referred to as PAD (peripheral arterial disease), increases with age affecting nearly 25% of patients 80 years and above [3]. High incidence of chronic coexisting illnesses like diabetes, chronic kidney disease, chronic obstructive pulmonary diseases, coronary artery disease, etc. in this age group renders management of PAD an onerous task.

Intermittent claudication which is the classic presentation of PAD is often not the primary presenting symptom in elderly patients. Diseases such as hip and knee arthritis, degenerative diseases of the spine, frailty, COPD, and CAD compromise the functional status of the patient making them sedentary. Hence it is not uncommon for this group of patients to present with **c**hronic **l**imb-**t**hreatening **i**schemia (CLTI) – rest pain, non-healing ulcer, and/or gangrene as the initial presentation. Thus, medical management that focuses on lifestyle modifications, pharmacological agents, and exercise therapy which is the initial line of treatment for most patients with intermittent claudication may play little or no role in this group. Surgical or endovascular interventions may be the first line of treatment for these patients.

Historically open direct surgical reconstructions are known to have better long-term patency. Intuitively lesser invasive endovascular procedures may seem more suitable for this cohort of patient owing to less operative stress. Due to dearth of conclusive evidence regarding which approach is better, this particular field of research has become of important interest for a number of nationally and internationally funded studies [3].

Owing to all the factors mentioned above, these patients with CLTI have long hospital stays, frequent readmissions, and discharge to facilities other than their own home. Treatment should not only be directed to patients' physical illness, but their social, economic, and emotional components may also need to be addressed.

As mentioned in the beginning, an interdisciplinary team approach is vital for managing these patients with the goal of patient-centered care.

Use of alternative vascular conduits such as cadaveric vein grafts and prosthetic grafts may help in early management of symptoms in a subset of patients where achieving long-term patency with traditional autologous vein grafts may not be feasible or may add to increased morbidity.

This thought process or the new mind-set aims at achieving short-term goals leading to early return of patients to basal functional status. This is a departure from the traditional practice of focusing on long-term graft patency which in the elderly and frail comes at the cost of increased perioperative morbidity, long hospital stays, and failure to return home.

Advances in technology have made it possible for more distal revascularization procedures such as tibial and pedal artery interventions and reconstruction. This has expanded the scope of limb salvage in recent years.

Acute Limb Ischemia

Acute extremity ischemia in the elderly has the same etiology as in general population, the cause being embolism, thrombosis, and trauma. Embolus may originate from the heart (cardio-arterial) or from proximal arterial segment (arterio-arterial embolization). Patients who present with ischemic toes and palpable pedal pulses (blue toe syndrome, trash foot) should be worked up for the source of arterio-arterial embolization. Popliteal aneurysms and less commonly iliac and aortic aneurysms may be the cause. Severely atherosclerotic and ulcerated plaques in either thoracic or abdominal aorta (shaggy aorta) also can cause this syndrome. Acute aortic dissection can manifest with extremity ischemia, which should be considered in differential diagnosis of acute limb ischemia.

Recently, there is increased occurrence of acute limb ischemia caused by low-flow state. Patients who suffer cardiogenic shock are now being resuscitated and maintained on mechanical circulatory assist devices like left ventricular assist device (LVAD), intra-aortic balloon pump (IABP), and extracorporeal membrane oxygenator (ECMO). These critically ill patients are maintained on vasopressors and other vasoactive agents to maintain blood pressure. The combination of shock, vasopressors, and placement of large caliber cannulas in femoral and/or axillary arteries can cause limb ischemia. This "low-flow" state-induced limb ischemia has poor prognosis for salvage of limb and life [13]. Attempts should be made to treat the primary condition that caused circulatory collapse. Frequent discussions between multiple specialists – cardiology, cardiothoracic surgery, critical care specialist, and vascular surgery – are essential in managing this new brand of super ill patients.

Amputation

The incidence of lower extremity amputations is increasing owing to several factors – increasing aging population, increase incidence of vascular diseases in the elderly, and associated comorbidities like diabetes and kidney diseases. Significant innovations in open surgical and endovascular limb salvage techniques have reduced the amputation rates in the United States by 25% as per recent data. Despite this improvement, there is still need for amputations particularly in the elderly [14].

Indications for primary amputations are life-threatening infections, gangrene involving the foot or a part of the leg, and non-reconstructible pattern of vascular disease in patients with CLTI.

Failure of surgical or endovascular revascularization procedures is the more frequent etiology of amputations in elderly patients with CLTI. These secondary amputations are generally done at a higher anatomic level, thus diminishing the potential for rehabilitation. In select patients, decision to amputation is not reflection of failure of treatment but a primary palliative procedure.

Cerebrovascular Disease

The risk of stroke and transient ischemic attacks (TIA) increases exponentially with age [15]. The burden of extracranial cerebrovascular disease (CVD) in the elderly is quite established. Carotid artery disease is the most common cause for ischemic strokes. Elderly patient should be screened for presence of extracranial vascular disease, particularly when they have other risk factors like hypertension, diabetes, smoking history, and heart disease.

Screening as well as follow-up of carotid artery diseases is simple and convenient by means of ultrasound duplex scan. The Intersocietal Commission for Accreditation of Vascular Laboratories (ICAVL) has created reporting standards to classify the extent of carotid disease as 0–15%, 16–49%, 50–79%, and 80–99% stenosis and total occlusion. No additional imaging modality is needed for diagnosis and follow-up. MRA and CTA of the head and neck are required when the treatment is planned.

While there is universal agreement in treatment of symptomatic patients with carotid disease, there is controversy in the management of asymptomatic hemodynamically significant carotid stenosis [16].

Advances in medical management have significantly improved, since the days of NASCET (North American Symptomatic Carotid Endarterectomy Trial) and ACAS (Asymptomatic Carotid Atherosclerosis) trials which demonstrated superiority of surgical management.

Currently, the best medical therapy (BMT) for asymptomatic carotid disease includes antihypertensive and antiplatelet agent and statins. As per recent data, most patients with asymptomatic carotid stenosis can be best managed with intensive medical therapy [17].

However, there is a subset of patients with critical carotid stenosis (80–99% stenosis category) who are asymptomatic but at risk of stroke who would benefit from carotid endarterectomy or stenting. The challenge is to identify this select group of patients.

There is growing interest in characterizing the morphology of the plaque on imaging studies and not just the degree of stenosis. A smooth and fibrous plaque is likely to have a benign prognosis, whereas a soft plaque with signs of intraplaque hemorrhage, degeneration, and ulceration indicates a more ominous pathology with potential for neurologic events [18].

We routinely perform carotid endarterectomy under regional anesthesia with neuro-monitoring modalities (continuous EEG and SSEP) and selective shunting with the goal of reducing the risk of general anesthesia and intra- or perioperative complications. In addition to the conduct of the surgery itself, the immediate perioperative care is equally important. Creating a carotid care pathway model that standardizes post-op care and involves the recovery room and ICU nursing staff in decision-making is important. This protocol-driven care by nurses and doctors facilitates a more efficient and safe patient outcome.

Carotid Stenting

The seminal NIH-sponsored CREST trial has shown that over a 10-year follow-up, there was no significant difference between patients who underwent stenting and those who underwent carotid endarterectomy with respect to periprocedural stroke, myocardial infarction or death, and subsequent ipsilateral stroke. The rate of post-procedural ipsilateral stroke also did not differ between groups [19]. It should be noted that the criteria to participate in such a trial were extremely rigid and highly selective.

TCAR (Trans-Carotid Artery Revascularization)

As most neurological events during stenting are the results of embolization, a new concept has been developed. This entails deliberately reversing the flow in the carotid artery during the time of stent placement. This technique called trans-carotid artery revascularization (TCAR) is becoming the technique of choice for stenting carotids [20].

Other Extra-Cranial Vascular Diseases

Carotid dissection, extracranial aneurysms, and stenotic and occlusive diseases of the vertebral-basilar circulation are less common than the carotid disease. The clinical entity that requires special attention in the elderly population is subclavian steal syndrome. Significant stenosis at the origin or proximal subclavian artery results in decreased pressure beyond the stenosis causing a physiological reversal of flow in the ipsilateral vertebral artery. This condition called subclavian steal syndrome may be an incidental finding on imaging studies with no symptoms in majority of patients. However in the small group of symptomatic patients, symptoms include diplopia, loss of balance, and drop attacks (falling without losing consciousness). Symptoms may be triggered by exercising the affected upper extremity in some. When symptoms do occur, it is frequently misinterpreted as consequences of old age. It is good practice to measure blood pressure in both upper extremities in elderly patients. A difference in values of more than 20 mmHg is considered significant. Diagnosis can be confirmed with other imaging modalities like ultrasound duplex, MRA, and/or CTA. Once the diagnosis is established, and if patients have symptoms, treatment in most patients is by endovascular approach. Subclavian artery angioplasty and stenting via ipsilateral brachial artery approach is safe with good immediate and long-term cure. Carotid-subclavian bypass performed by neck incision is the surgical option when stenting is not feasible.

Mesenteric Vascular Disease

Chronic Mesenteric Ischemia (CMI)

The classic picture of chronic mesenteric ischemia is the clinical triad of postprandial abdominal pain (abdominal angina), weight loss, and steatorrhea. In most patients two of the three intestinal arteries have to be occluded to cause symptoms. However a small number of patients may get mesenteric ischemic syndrome when only one of the three visceral arteries is affected. We have termed this "territorial mesenteric ischemia" [21]. This occurs when there are incomplete or insufficient collaterals to compensate for the stenosed or occluded visceral artery. In these cases, symptoms will be from the territory supplied by the affected visceral artery. Thus stenosis or occlusion of celiac artery may present with gastric and hepatobiliary symptoms, that of SMA with intestinal symptoms and that of IMA presenting with ischemic events of left colon. Owing to the natural process of weight loss and decreasing appetite in the aging population, recognition and diagnosis of CMI may be missed or delayed. This disease may be effectively screened and diagnosed using duplex ultrasound imaging techniques. CTA/MRA and conventional visceral angiography will confirm the diagnosis of CMI. Lateral (sagittal) views are needed to assess the visceral arteries since most atherosclerotic plaques occur at the origin of these arteries as they take off from the aorta. Correction of SMA disease with or without treatment of celiac artery to restore intestinal perfusion is the treatment in most patients.

Angioplasty/stenting of mesenteric arteries has become the first therapeutic option in CMI [22]. Surgical bypasses to the vessels are reserved only for those patients where endovascular management is not feasible. SMA and celiac arteries are the usual targets for revascularization. In select patients who have occlusion of all three visceral arteries, angioplasty/stenting of the IMA as a single vessel treatment can be safely performed in patients with CMI, with satisfactory short- and long-term results [23].

Acute Mesenteric Ischemic Syndromes

Acute mesenteric ischemic (AMI) syndromes should be considered in any elderly patient that comes in with an acute abdomen. Delay or failure in diagnosis can result in high rates of mortality [24]. AMI may be caused by acute mesenteric artery occlusion by embolus (AMOE), acute mesenteric artery occlusion by thrombosis (AMOT), nonocclusive acute mesenteric ischemia (NOAMI), and acute mesenteric venous thrombosis (AMVT).

Acute Mesenteric Ischemia due to Embolus

Mesenteric embolic occlusion is the commonest cause of acute mesenteric ischemia. Among the visceral arteries, the superior mesenteric artery is the most common site for embolic occlusion. Generally, the embolus lodges just beyond the origin of the middle colic artery, which is clearly distinct from acute superior mesenteric artery thrombosis that occurs secondary to pre-occlusive stenosis at the origin of SMA. Clinically, embolic occlusion tends to spare the left colon. The most common source for the emboli is the heart with causes being atrial fibrillation, myocardial infarction, valvular heart disease, ventricular aneurysm, and left atrial thrombus. Severe atherosclerotic aortic disease of the descending aorta with heavy plaque burden and ulcerated disease – the so-called shaggy aorta syndrome – has also been a known source for the embolus. In any patient with AMOE, if a cardiac source is ruled out, we recommend imaging of the entire thoracoabdominal aorta to look for the source (arterio-arterial embolization). Treatment options will be discussed together with management of acute thrombotic occlusion of mesenteric vessels.

Acute Mesenteric Ischemia due to Thrombosis

As mentioned earlier, atheromatous plaque disease generally occurs at the orifice of the mesenteric arteries, leading to stenosis. Again, SMA is most frequently involved. Thrombosis, superimposed on a pre-existing critical stenosis, is the most typical finding. Thrombosis of severely stenosed SMA may be triggered by a variety of low-flow states like CHF, cardiogenic shock, dehydration, or any clinical state causing prolonged hypotension. The ischemic segment involved would be the entire small and large intestine that may lead to devastating consequences.

Nonocclusive Acute Mesenteric Ischemia (NOAMI)

In the late 1960s, it was recognized that mesenteric ischemia can occur without any major mechanical obstruction to the mesenteric flow. The classic picture was an elderly patient presenting with a clinical picture of mesenteric ischemia in the setting of severe heart failure, arrhythmias, or severe volume depletion. This "low-flow" state-induced ischemia of the bowel is termed as nonocclusive mesenteric ischemia (NOAMI) which accounts for 20–30% of all cases of acute mesenteric ischemia. This condition was brought to the attention of physicians by Scott Boley [25]. Mortality of NOAMI is high. Treatment should be directed to correcting the primary condition that caused the low-flow state [26].

Acute Mesenteric Venous Thrombosis

Acute interruption of venous return from the intestines due to thrombosis of the superior mesenteric vein and/or portal vein will present with clinical symptoms which may be indistinguishable from acute mesenteric arterial occlusion. The common etiologies for this condition comprise of either primary or secondary hypercoagulable states. Primary hypercoagulable state may occur due to polycythemia vera, leukemias, myelodysplastic syndromes, and inherited thrombophilias (protein C and S deficiency, prothrombin gene mutation, hyperhomocysteinemia, etc.). Secondary hypercoagulable state can be brought on by cancer, contraceptive pills, etc. Advance liver cirrhosis causing portal venous hypertension and direct extrinsic compression by periportal neoplasms are other causes.

Management of Acute Mesenteric Ischemic Syndromes

High index of suspicion, early diagnosis, and prompt intervention remain the hallmark tenets for successful management of this condition. Delay or misdiagnosis almost always results in mortality [24]. Once the diagnosis is confirmed, systemic anticoagulation should be initiated as the patient is being prepared for surgery. Surgical embolectomy is the treatment of choice for an embolic occlusion. Promptly done, this results in good outcome. In mesenteric thrombosis, the treatment is generally more complex owing to the need for correction of the stenotic or occlusive lesion in addition to thrombectomy. Unlike thrombosis of peripheral vascular bed, thrombolytic therapy for mesenteric thrombosis is generally not recommended because of the urgency to restore flow to the ischemic bowel.

Angioplasty and stenting of the mesenteric vessels can be performed retrograde through the superior mesenteric artery at the time of exploration or through a traditional endovascular brachial approach [22, 26]. Inability to perform angioplasty or stenting will need an urgent bypass to the superior mesenteric artery. We recommend iliac artery as an inflow for the bypass in order to avoid clamping of the aorta in these critically ill patients, who may not tolerate the consequences of aortic clamping [21].

For NOAMI, treatment should be aimed at correcting the low-flow state while the systemic anticoagulation is initiated. In select patients, mesenteric angiography and selective catheterization of the superior mesenteric artery for infusion of papaverine to dilate the mesenteric vascular bed may be needed [26]. Mesenteric venous thrombosis has a more insidious clinical course. Management mostly comprises of systemic anticoagulation, with fluid resuscitation and serial clinical examination and trending biochemical markers such as white count and serum lactate levels. For unprovoked or idiopathic mesenteric venous thrombosis, a hematologic workup is mandatory to determine the need for lifelong anticoagulation. The indication for surgery in these patients is reserved for clinical deterioration despite adequate treatment and/or frank peritonitis or other symptoms of bowel infarction.

Venous Thromboembolism (VTE)

VTE in the elderly patients has the same etiological factors as in the general population (Virchow's triad of stasis, endothelial damage, and hypercoagulability). Advance age and frailty resulting in a sedentary lifestyle render this subset of population at high risk for VTE [27].

When VTE occurs in an elderly without any triggering causative factors, it is imperative to rule out malignancy. This would require further imaging studies of the chest and abdomen to rule out a primary source. Pancreatic, renal, and lung cancers are among the most common neoplasms presenting with VTE.

Upper extremity venous thrombosis, including the central venous occlusive syndromes, is also frequent in this category of patients. Presence of central venous access catheters such as those for dialysis and chemotherapy is now the most common cause for upper extremity and neck vein deep venous thrombosis in these patients [28]. Effort thrombosis (Paget-Schroetter syndrome), which is upper extremity DVT occurring in young patients with thoracic outlet syndrome, is uncommon in the elderly.

Anticoagulation is the standard treatment for VTE. Elderly patients have been found to be more sensitive to anticoagulants. This along with a higher incidence of fall in the elderly calls for close follow-up and modification of the dose and duration of the anticoagulation treatment [27].

Surgical and/or endovascular therapy is reserved for limb-threatening conditions such as phlegmasia (acute iliofemoral thrombosis). Catheter-directed thrombolysis (CDT) and pharmaco-mechanical thrombectomy (PMT) have replaced surgery in most patients.

Indications and use of inferior vena cava filters (IVC filters) are the same for general population, namely: (1) contraindication to anticoagulation and (2) progression of VTE despite adequate anticoagulation. While there is general recommendation to remove these filters once the risk VTE has passed, owing to the risk of additional procedures, this may not be applicable in elderly patients with multiple co-morbidities and limited lifespan.

Hemodialysis Access

The number of patients above the age 65 years requiring hemodialysis access for end-stage renal disease is increasing. As reported by the European Renal Association-European Dialysis and Transplant Association (ERA-EDTA), approximately 25% of patients undergoing dialysis for end-stage renal disease (ESRD) are above the age of 65 years [29]. The initial recommendation by the National Kidney Foundation to follow the strategy of "fistula first" may not be realistic when considered in the subgroup of elderly patients. High incidence of comorbid conditions like diabetes, congestive heart failure, and vascular disease of the upper extremities influences the choice and location of placement of dialysis access procedures.

Although AVF is superior to arteriovenous grafts (AVG) across all age groups, the fistula failure rate in patients above 65–70 years is nearly double than when compared to those younger [29]. All of these issues underscore the need for careful selection. Whether fistula or graft is better, and in very select cases, just the tunneled catheter may be the access of choice – need to be individualized. It would be prudent to choose AVG in an elderly patient with limited longevity. The goal is to minimize the stress of repetitive interventions in the elderly as each hospitalization increases the risk of other complications like hospital-acquired pneumonia or *Clostridium difficile* infections, delirium, and falls.

Close communication between the nephrologist and vascular surgeon is important throughout the course of ESRD patients. The time to maturation from the creation of AVF is about 12 weeks. Patients should be referred at least 3–4 months in advance when the nephrologist anticipates any one progressing from chronic kidney disease to ESRD. Arterial and vein mapping using duplex ultrasound to predict the successful AVF placement is routine. Once placed, duplex surveillance of AVF or AVG at regular intervals will help identify impending failures so that prompt intervention can improve the patency of the access.

The failure rate of a radio-cephalic fistula is higher in the elderly population when compared to their younger counterparts; hence, choosing a more proximal anatomic site (e.g., near the antecubital fossa) must be strongly considered [30]. To facilitate an open but poorly maturing fistula, balloon-assisted maturation (BAM) can be tried in some cases.

Central venous catheters in patients for HD present with their own set of problems and serious complications such as central vein stenosis/occlusions, superior vena cava syndrome, infections and need for repeat interventions. Frequent presence of pacemakers/defibrillators and central venous access catheters (e.g., for oncological conditions) may further limit the options available for placement of HD access. Tunneled catheter may be the only choice in some select patients with limited life expectancy and who have no veins or exhausted all access sites. Novel interventions such as minimally invasive endovascular AV access are soon becoming a reality. However, their long-term efficacy is yet to be determined.

Decision-Making in the Elderly

Decision-making in the elderly has to be based on a realistic outcome, as opposed to the ideal outcome. This comprises not only of symptom- and pathology-guided treatment but also takes into consideration the patients' expected longevity, functional status, established goals of care, and expectations. For example, although we are aware that a lower extremity revascularization bypass is superior in terms of long-term patency, for an elderly patient, an endovascular modality for the same lesion may be preferable to facilitate minimal hospital stay and early mobilization. Aggarwal et al., in their analysis of nationwide trends of hospital admissions and outcomes among CLI patients, had shown that endovascular procedures are associated with reduced hospital stay, major amputation, and mortality [31].

The other important surgical aspect of decision-making comes from the understanding of age-related physiological decline in homeostatic mechanisms. Involving the anesthesiologist in the process of preoperative assessment of the patient becomes an essential component of decision-making in this population. Recognizing the importance of limiting the use of general anesthesia for vascular procedures has led to a paradigm shift in performing long and complex procedures under regional anesthesia.

One also needs to understand that while initial medical management such as lifestyle modifications, diet changes, exercise, high-intensity statin therapy, and antihypertensive and antiplatelet medications may form the basis of treatment for all PAD patients, this may not be a practical solution for an octogenarian. Expecting compliance for lifestyle changes, regular exercises, and diet modifications from an elderly is unlikely to produce satisfactory results.

Conclusion

Scientific evidence has clearly shown us that the geriatric population is at risk for adverse surgical outcomes due to the prevalence of chronic diseases such as hypertension, hyperlipidemia, diabetes, coronary artery disease, and chronic lung and kidney disease. These long-standing chronic medical conditions, together with years of exposure to environmental stressors, poor diet, nicotine, alcohol, polypharmacy etc., are strong risk factors for developing vascular pathologies requiring medical and surgical intervention.

The incidence of arterial diseases increases with age. Vascular surgery as a branch of surgery deals with treating and attending to this unique age group of the population. It is important for the treating physician to understand the concepts of physiological changes in the vasculature in an elderly, the different disease processes that arise as a result of them, and the various modalities available for their management. The decision to perform a vascular procedure in the elderly does not depend on knowing merely what surgeries to perform. Understanding the patient's social background, anticipated quality of life after intervention, and presence or absence of family support forms the other half of the toolkit required to provide not only the best available treatment for the disease but the best-suited management for that patient. This patient-centered approach with patient participating in the overall treatment plan is the new paradigm. Finally, it should be acknowledged that the wide range of therapeutic spectrum – medical, endovascular, and surgical – is merely an intervention for an organic structural lesion. Our understanding of the process of atherogenesis and the formation and progression of a plaque still remains elusive. Advanced research directed toward preventing this process of atherosclerosis would be the most ideal solution.

References

1. Clegg A, Hassan-Smith Z. Frailty and the endocrine system. Lancet Diabetes Endocrinol. 2018;6:743.
2. Criqui MH, Aboyans V. Epidemiology of peripheral artery disease. Circ Res. 2015;116(9):1509–26.
3. Miller AP, Huff CM, Roubin GS. Vascular disease in the older adult. J Geriatr Cardiol JGC. 2016;13(9):727.
4. Mehta RH, O'Gara PT, Bossone E, Nienaber CA, Myrmel T, Cooper JV, Smith DE, Armstrong WF, Isselbacher EM, Pape LA, Eagle KA. Acute type A aortic dissection in the elderly: clinical characteristics, management, and outcomes in the current era. J Am Coll Cardiol. 2002;40(4):685–92.
5. Lou X, Chen EP, Duwayri YM, Veeraswamy RK, Jordan WD Jr, Zehner CA, Leshnower BG. The impact of thoracic endovascular aortic repair on long-term survival in type B aortic dissection. Ann Thorac Surg. 2018;105(1):31–8.
6. Nienaber CA, Sakalihasan N, Clough RE, Aboukoura M, Mancuso E, Yeh JS, Defraigne JO, Cheshire N, Rosendahl UP, Quarto C, Pepper J. Thoracic endovascular aortic repair (TEVAR) in proximal (type A) aortic dissection: ready for a broader application? J Thorac Cardiovasc Surg. 2017;153(2):S3–11.
7. Leshnower BG, Veeraswamy RK, Duwayri YM, Chen EP. The "TEVAR-first" approach to DeBakey I aortic dissection with mesenteric malperfusion. Ann Thorac Surg. 2014;97(2):693–6.
8. Chaikof EL, Dalman RL, Eskandari MK, Jackson BM, Lee WA, Mansour MA, Mastracci TM, Mell M, Murad MH, Nguyen LL, Oderich GS. The Society for Vascular Surgery practice guidelines on the care of patients with an abdominal aortic aneurysm. J Vasc Surg. 2018;67(1):2–77.
9. LeFevre ML. Screening for abdominal aortic aneurysm: US Preventive Services Task Force recommendation statement. Ann Intern Med. 2014;161(4):281–90.
10. Dua A, Kuy S, Lee CJ, Upchurch GR Jr, Desai SS. Epidemiology of aortic aneurysm repair in the United States from 2000 to 2010. J Vasc Surg. 2014;59(6):1512–7.
11. Schermerhorn ML, Buck DB, O'Malley AJ, Curran T, McCallum JC, Darling J, Landon BE. Long-term outcomes of abdominal aortic aneurysm in the Medicare population. N Engl J Med. 2015;373(4):328–38.
12. Burrell MJ, Wheeler JR, Gregory RT, Synder SO Jr, Gayle RG, Mason MS. Axillofemoral bypass: a ten-year review. Ann Surg. 1982;195(6):796.
13. Babu SC, Shah PM, Clauss RH. Unappreciated causes of ischemia in the leg. Am J Surg. 1982;144(2):225–7.
14. Mozaffarian D, Benjamin EJ, Go AS, Arnett DK, Blaha MJ, Cushman M, Das SR, De Ferranti S, Després JP, Fullerton HJ, Howard VJ. Heart disease and stroke statistics-2016 update a report from the American Heart Association. Circulation. 2016;133(4):e38–48.

15. Savji N, Rockman CB, Skolnick AH, Guo Y, Adelman MA, Riles T, Berger JS. Association between advanced age and vascular disease in different arterial territories: a population database of over 3.6 million subjects. J Am Coll Cardiol. 2013;61(16):1736–43.
16. North American Symptomatic Carotid Endarterectomy Trial Collaborators*. Beneficial effect of carotid endarterectomy in symptomatic patients with high-grade carotid stenosis. N Engl J Med. 1991;325(7):445–53.
17. Spence JD, Song H, Cheng G. Appropriate management of asymptomatic carotid stenosis. Stroke Vasc Neurol. 2016;1(2):64–71.
18. Wendorff C, Wendorff H, Pelisek J, Tsantilas P, Zimmermann A, Zernecke A, Kuehnl A, Eckstein HH. Carotid plaque morphology is significantly associated with sex, age, and history of neurological symptoms. Stroke. 2015;46(11):3213–9.
19. Brott TG, Hobson RW, Howard G, Roubin GS, Clark WM, Brooks W, Mackey A, Hill MD, Leimgruber PP, Sheffet AJ, Howard VJ. Stenting versus endarterectomy for treatment of carotid-artery stenosis. N Engl J Med. 2010;363(1):11–23.
20. Kashyap VS, King AH, Foteh MI, Janko M, Jim J, Motaganahalli RL, Apple JM, Bose S, Kumins NH. A multi-institutional analysis of transcarotid artery revascularization compared to carotid endarterectomy. J Vasc Surg. 2019;70:123.
21. Babu SC, Shah PM. Celiac territory ischemic syndrome in visceral artery occlusion. Am J Surg. 1993;166(2):227–30.
22. Loffroy R, Steinmetz E, Guiu B, Molin V, Kretz B, Gagnaire A, Bouchot O, Cercueil JP, Brenot R, Krausé D. Role for endovascular therapy in chronic mesenteric ischemia. Can J Gastroenterol Hepatol. 2009;23(5):365–73.
23. Wohlauer M, Kobeiter H, Desgranges P, Becquemin JP, Cochennec F. Inferior mesenteric artery stenting as a novel treatment for chronic mesenteric ischemia in patients with an occluded superior mesenteric artery and celiac trunk. EJVES Extra. 2014;27(3):e21–3.
24. Kassahun WT, Schulz T, Richter O, Hauss J. Unchanged high mortality rates from acute occlusive intestinal ischemia: six year review. Langenbeck's Arch Surg. 2008;393(2):163–71.
25. Boley SJ, Brandt LJ, Sammartano RJ. History of mesenteric ischemia: the evolution of a diagnosis and management. Surg Clin N Am. 1997;77(2): 275–88.
26. Trompeter M, Brazda T, Remy CT, Vestring T, Reimer P. Non-occlusive mesenteric ischemia: etiology, diagnosis, and interventional therapy. Eur Radiol. 2002 May 1;12(5):1179–87.
27. Tritschler T, Aujesky D. Venous thromboembolism in the elderly: a narrative review. Thromb Res. 2017;155:140–7.
28. Joffe HV, Kucher N, Tapson VF. Upper-extremity deep vein thrombosis: a prospective registry of 592 patients. J Vasc Surg. 2005;41(2):371.
29. Lomonte C, Forneris G, Gallieni M, Tazza L, Meola M, Lodi M, Senatore M, Morale W, Spina M, Napoli M, Bonucchi D. The vascular access in the elderly: a position statement of the Vascular Access Working Group of the Italian Society of Nephrology. J Nephrol. 2016;29(2):175–84.
30. Lok CE, Allon M, Moist L, Oliver MJ, Shah H, Zimmerman D. Risk equation determining unsuccessful cannulation events and failure to maturation in arteriovenous fistulas (REDUCE FTM I). J Am Soc Nephrol. 2006;17(11):3204–12.
31. Agarwal S, Sud K, Shishehbor MH. Nationwide trends of hospital admission and outcomes among critical limb ischemia patients: from 2003–2011. J Am Coll Cardiol. 2016;67(16):1901–13.

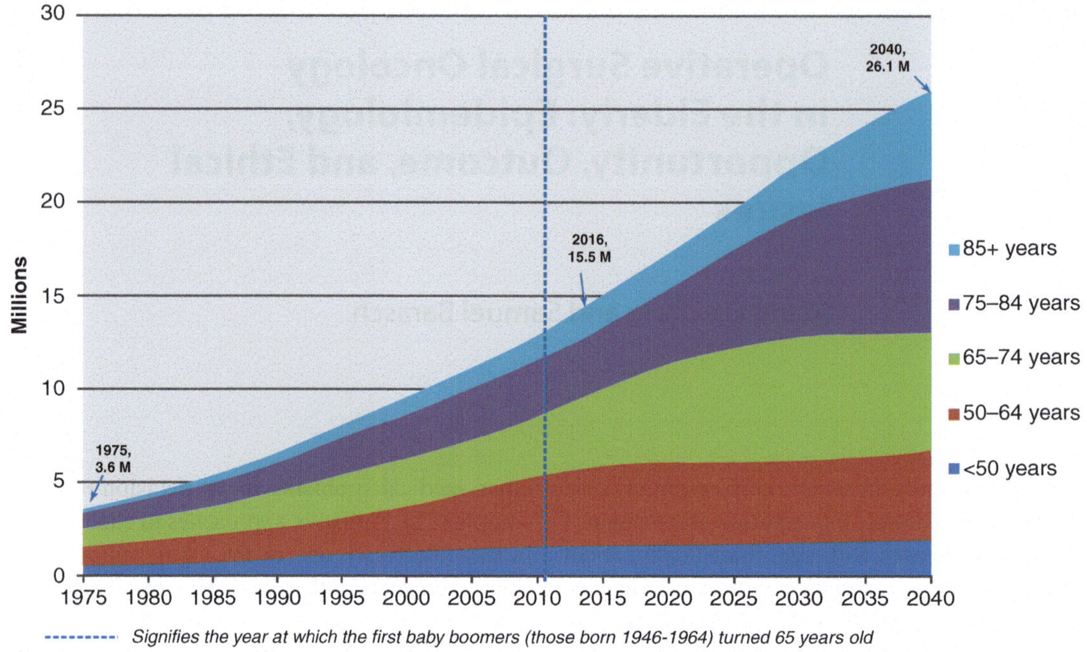

Fig. 20.1 Estimated cancer prevalence by age in the US population from 1975 (216 M) to 2040 (380 M). (From Bluethman et al. [4], with permission)

cost of cancer care in 2020 is going to be around $174 billion, second only to heart disease at $199 billion [7, 8].

On average, estimates of comorbid conditions in the elderly diagnosed with cancer are around five comorbid conditions [4]. The incidence of comorbidities and frailty is even higher in the oldest adults (aged 85 and greater) [9–12]. With advances in treatment and early detection, elderly patients can often be cured and placed in remission. But, the act of treatment poses two additional problems. Firstly, the treatment of malignancies often leads to cancer-related toxicities that will persist after completion of treatment. Secondly, the effective treatments for cancer will surely lead to long-term survivors that will require additional resources for management including posttreatment surveillance and management of complications from treatment. It is estimated that 64 percent of patients become 5-year survivors of cancer and 40 percent become 10-year survivors [4]. This will surely add to the burden of treatment when considering the elderly patient.

Although the treatment options as outlined by national guidelines for elderly patients with cancer do not differ significantly from their younger cohorts, there are differences because of the challenges from the decrease in the physiological reserve in these patients [7, 8]. The studies on management of cancer were often done on younger patients with less medical comorbidities and declining health parameters. It is also well-known that older patients have differences in tumor biology, alterations in their DNA mismatch repair systems, increases in microsatellite instability, and higher likelihood of advanced disease on presentation [13]. However, little adjustments have traditionally been made on the workup and management of elderly patients with cancer solely based on age.

Surgical Oncology Geriatric Risk Assessment

Surgeon attitude toward management of elderly patients differs widely. Almost all surgeons view a patient's chronological age as relative and physiological age as more representative of their true performance status [14–16]. However, assessment of physiological age preoperatively is much more varied in terms of consistency and undertaking. In order to address the attitudes of surgeons toward this problem, the Surgical Task Force at SIOG (International Society of Geriatric Oncology) sent a web-based survey to members of the ESSO (European Society of Surgical Oncology) and SSO (Society of Surgical Oncology) to gather their response and management of geriatric patients [16]. As expected, almost all surgeons (>90 percent) offer surgery to patients regardless of age [16]. Only 48 percent consider mandatory preoperative frailty assessment. Collaboration with geriatricians to assess risk and optimize surgical outcome is also low at 36.3 percent [16]. In terms of cognitive status, the differences are even more pronounced. Only one in two would not offer surgery regardless of cognitive status bearing in mind that meaningful survival will be impacted [16]. As much as one in three would offer surgery regardless of cognitive status due to patient or family requests or wishes. There seems to be a significant lack of data-driven prospective studies to address onco-geriatric surgery in terms of its risks, benefits, and approach.

In light of the risks in treating elderly patients due to their medical comorbidities and declining health status, several societies have created guidelines in terms of medication adjustments and management recommendations [17]. Perhaps the most obvious field in need of stricter guidelines is surgery as the effects of frailty associated with older patients can lead to detrimental surgical outcomes [18–20]. Therefore, frailty assessment and medical optimization is recommended for surgical planning for the geriatric patient population.

Surgical oncology patients benefit from frailty assessment preoperatively. One of the easiest means to assess surgical risk is a chronological age of the patient [21, 22]. However, although it is the simplest means to document surgical risk, it is also the least accurate as it lacks any consideration in the medical comorbidities of the patient, the physiologic status of the patient, the type of surgical intervention involved, or the type of anesthesia required. As a result, multiple nomograms, indexes, and assessment tools were developed to assess the operative risk of the patient in relation to the perceived risk of the surgical intervention [23–25]. Each of the assessment tools aims to balance certain aspect of the patient performance status with respect to the operation involved, with each having particular strengths and weaknesses.

One of the most comprehensive assessments supported by the American Geriatric Society is the Comprehensive Geriatric Assessment (CGA) [16]. This assessment takes into account the physical, mental, social, economic, functional, and environmental aspects of the elderly. However, the assessment is impractical for surgical risk assessment due to the details required. It also involves different aspects of the patient well-being that is not directly related to surgical outcome. Only 6.4 percent of surgeons have reportedly used this assessment prior to surgery [16].

Several other useful tools that are easier to manage are supported by the American College of Surgeons and the American Society of Anesthesiologists (ASA). An easier and often useful tool, as endorsed by the American College of Surgeons (ACS), is the modified Frailty Index (mFI) [24]. The mFI combines 11 variables linked to a patient's medical comorbidities and generates a risk for the patient using data derived from the National Surgical Quality Improvement Program (NSQIP) database [22, 24] (Table 20.1). The mFI is predictive of postoperative complications as well as morbidity and mortality associ-

Table 20.1 The modified Frailty Index

History of diabetes
History of congestive heart failure
HTN requiring medication
History of transient ischemic attack or cerebral vascular accident
Non-independent functional status
History of myocardial infarction
History of either peripheral vascular disease or rest pain
History of cerebral vascular accident with neurologic deficit
History of COPD or pneumonia
History of percutaneous coronary intervention, stent placement, or angina
History of impaired sensorium

From Velanovich et al. [24], with permission

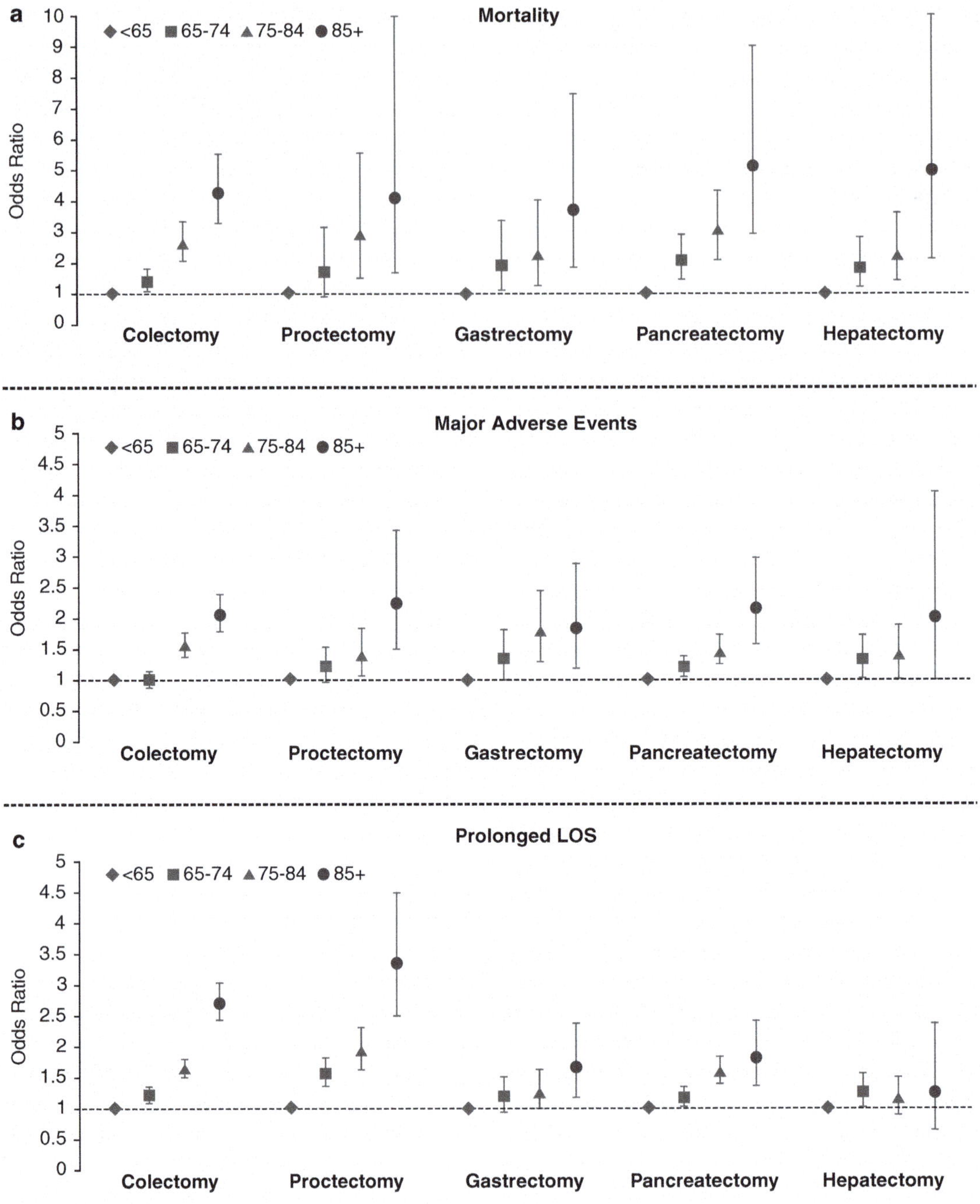

Fig. 20.2 (**a**–**c**) Age-adjusted mortality rates after major GI cancer surgery. (From Yeo [9], with permission)

oncology patients, individual decision-making capacity can be compromised by both the disease and by comorbid conditions. Patients who have increased reliance on family or social networks of caregivers due to comorbid medical conditions or cognitive disability may have reduced autonomy in medical decision-making. Because the principle of autonomy can interdict between non-malfeasance and beneficence, a living will or clearly defined health-care-power-of-attorney can continue to allow autonomy when balancing non-malfeasance with beneficence. In situations where a patient's predetermined preferences are known, autonomy can be preserved and weighted more greatly in balancing ethical concerns.

Lack of clarity about patient wishes could play a role in providing futile end of life care in close to 40 percent of situations in which futile care is provided [45]. Futile care which could increase complications and decrease quality of life raises ethical concerns of non-malfeasance (i.e., "First Do No Harm"). Perceptions of patient suffering by the family and the clinical patient care team can differ; this difference influences the perception of suffering and the interpretation of medical care as futile [44]. Doctor, patient, and hospital/healthcare system factors can all contribute to situations in which futile care is provided. For physicians, a focus on curative treatment during training and inexperience with death and dying are leading contributors to the provision of futile end of life care. Patient or family requests as well as a possible uncertain prognosis drive family decision to pursue care which may, in retrospect, be perceived as futile. Finally, medical systems which include specialization, hierarchy, and a focus on acute care can influence a trend to ineffective treatment in the end of life setting [45].

While avoiding futile care can be an example of non-malfeasance, it is important in the ethical principal of justice as well. Justice includes appropriate distribution of healthcare resources and allocation to individuals based upon need as well as appropriately within the medical system. It is unjust to deny patients medical care based solely upon their age. Patients who perform well on a CGA should be treated as autonomous individuals and have the right to consider all options for their oncologic treatment – including surgery. While advanced stage cancer is generally discussed and understood in the geriatric/palliative patient, full determination of stage may require surgical staging for a complete clinical assessment of the disease in the appropriate geriatric patient. Comprehensive geriatric assessment allows for evaluation of an individual's comorbidities and decision-making capacity. This aids in just allocation of healthcare resources for the individual and healthcare system [46, 47].

Beneficence, service of the greatest good, combines a disease-oriented approach with acknowledgment of patient goals for care, in the light of a CGA, which includes assessment of patient comorbid diseases and medications. This allows the greatest good for the patient while preserving the greatest good for the medical system, and the patient's family. Beneficence has been overplayed in the twentieth century and has trespassed on paternalism. Thus, the principle of beneficence, in current medical ethics decision-making, is balanced by autonomy. For patients with manageable comorbidities including a good mental and psychological state, functional completion of activities of daily life, participation in instrumental activities of daily life, appropriate nutritional status, and good management of poly-pharmacy, the greatest good is preserved by allowing patients and families curative surgical oncologic care.

For example, in contrast to oncologic considerations, current guidelines for joint replacement do not include patient age and focus on provision of quality of life, relief of pain, and maintenance of joint mobility for the appropriate patient. While a specific roadmap is not available to outline definitive treatment for geriatric oncology patients, the CGA can provide a framework for ensuring that patients, families, doctors, and nurses can have all the information needed to make decisions about geriatric surgical oncology. This information allows the clinical team to balance ethical principles of medical care provision. In appropriate patients, surgical oncologic care can preserve quality of life for geriatric patients.

Conclusions

Management of geriatric oncology patients are complex and require multimodality treatment and development of strategies to minimize treatment side effects while optimizing care. Although several studies have confirmed feasibility of major surgical approaches to complex diseases, short-term outcomes need to be balanced with long-term quality of life. Limited data is available at this time in terms of patient outcome in the elderly.

References

1. National Cancer Institute. Cancer prevalence and cost of care projections external. Accessed 30 May 2019.
2. Howlader N, Noone AM, Krapcho M, et al. SEER cancer statistics review, 1975–2016, National Cancer Institute. Bethesda, MD, https://seer.cancer.gov/csr/1975_2016/, based on November 2018 SEER data submission, posted to the SEER web site, April 2019.
3. Soto-Perez-de-Celis E, Li D, Yuan Y, et al. Functional versus chronological age: geriatric assessments to guide decision making in older patients with cancer. Lancet Oncol. 2018;19:e305–16. https://doi.org/10.1016/S1470-2045(18)30348-6.
4. Bluethmann SM, Mariotto AB, Rowland JH. Anticipating the 'Silver Tsunami': prevalence trajectories and co-morbidity burden among older cancer survivors in the United States. Cancer Epidemiol Biomark Prev. 2016;25:1029–36. https://doi.org/10.1158/1055-9965.
5. Bentrem DJ, Cohen ME, Hynes DM, et al. Identification of specific quality improvement opportunities for the elderly undergoing gastrointestinal surgery. Arch Surg. 2009;144:1013–20.
6. Miller KD, Siegel RL, Lin CC, et al. Cancer treatment and survivorship statistics, 2016. CA Cancer J Clin. 2016;66:271–89.
7. Mariotto AB, Wang Z, Klabunde C, et al. Life tables adjusted for comorbidity more accurately estimate noncancer survival for recently diagnosed cancer patients. J Clin Epidemiol. 2013;66:1376–85. https://doi.org/10.1016/j.jclinepi.2013.07.002.
8. Mariotto AB, Yabroff KR, Shao Y, et al. Projections of the cost of cancer care in the United States: 2010-2020. J Natl Cancer Inst. 2011;103:117–28. https://doi.org/10.1093/jnci/djq495.
9. Yeo HL, O'Mahoney PRA, Lachs M, et al. Surgical oncology outcomes in the aging US population. J Surg Res. 2016;205:11–8.
10. Finlayson E, Fan Z, Birkmeyer JD. Outcomes in octogenarians undergoing high risk cancer operation: a national study. J Am Coll Surg. 2007;205(6):729–34.
11. Al-Refaie WB, Parsons HM, Henderson WG, et al. Major cancer surgery in the elderly: results from the American College of Surgeons national surgical quality improvement program. Ann Surg. 2010;251:311–8.
12. Van Gestel YR, Lemmens VE, de Hingh IH, et al. Influence of comorbidity and age on 1-, 2-, and 3-month postoperative mortality rates in gastrointestinal cancer patients. Ann Surg Oncol. 2013;20:371–80.
13. Neri S, Gardini A, Facchini A, et al. Mismatch repair system and aging: microsatellite instability in peripheral blood cells from differently aged participants. J Gerontol. 2005;60:285–92. https://doi.org/10.1093/gerona/60.3.285.
14. Loh KP, Soto-Perez-de-Celis E, Hsu T, et al. What every oncologist should know about geriatric assessment for older patients with cancer: Young International Society of Geriatric Oncology position paper. J Oncol Pract. 2018;14:85–94. https://doi.org/10.1200/JOP.2017.026435.
15. Fagard K, Leonard S, Deschodt M, et al. The impact of frailty on postoperative outcomes in individuals aged 65 and over undergoing elective surgery for colorectal cancer: a systematic review. J Geriat Oncol. 2016;7:479–91.
16. Ghignone F, van Leeuwen BL, Montroni I, et al. The assessment and management of older cancer patients: a SIOG surgical task force survey on surgeons' attitudes. EJSO. 2016;42:297–302.
17. Janssen-Heijnen ML, Maas HA, Houterman S. Comorbidity in older surgical cancer patients: influence on patient care and outcome. Eur J Cancer. 2007;43:2179–93.
18. Turrentine FE, Wang H, Simpson VB, et al. Surgical risk factors, morbidity, and mortality in elderly patients. J Am Coll Surg. 2006;203:865–77.
19. Hamel MB, Hendrson WG, Khuri SF, et al. Surgical outcomes for patients aged 80 and older: morbidity and mortality from major noncardiac surgery. J Am Geriatr Soc. 2005;53:424–9.
20. Gajdos C, Kile D, Hawn MT, et al. Advancing age and 30-day adverse outcomes after nonemergent general surgeries. J Am Geriatr. 2013;61:1608–14.
21. Maekawa Y, Sugimoto K, Yamasaki M, et al. Comprehensive geriatric assessment is a useful predictive tool for postoperative delirium after gastrointestinal surgery in old-old adults. Geriatr Gerontol Int. 2016;16:1036–42. https://doi.org/10.1111/ggi.12587.
22. Griebling TL. Re: geriatric assessment in surgical oncology: a systematic review. J Urol. 2016;195:1079–80. https://doi.org/10.1016/j.juro.2015.12.060.
23. Charlson ME, Pompei P, Ales KL, et al. A new method of classifying prognostic comorbidity in longitudinal studies: development and validation. J Chronic Dis. 1987;40:373–83.
24. Velanovich V, Antoine H, Swartz A, et al. Accumulating deficits model of frailty and postoperative mortality and morbidity: its application to a national database. J Surg Res. 2013;183:103–10.
25. Dripps RD. New classification of physical status. Anesthesiology. 1963;24:111.
26. Scott S, Lund JN, Gold S, et al. An evaluation of POSSUM and P-POSSUM scoring in predicting post-operative mortality in a level 1 critical care setting. BMC Anesthesiol. 2014;14:104. https://doi.org/10.1186/1471-2253-14-104.
27. Ceccarelli G, Andolfi E, Biancafarina A, et al. Robot-assisted surgery in elderly and very elderly population: our experience in oncologic and general surgery with literature review. Aging Clin Exp Res. 2017;29:55–63. https://doi.org/10.1007/s40520-016-0676-5.

28. Weber DM. Laparoscopic surgery – an excellent approach in elderly patients. Arch Surg. 2003;138:1083–8.
29. Caglia P, Tracia A, Buffone A, et al. Physiopathology and clinical considerations of laparoscopic surgery in the elderly. Int J Surg. 2016;33:S97–S102.
30. Amelung FJ, Borstlap WAA, Consten ECJ, et al. Propensity score-matched analysis of oncological outcome between stent as bridge to surgery and emergency resection in patients with malignant left-sided colonic obstruction. Br J Surg. 2019; https://doi.org/10.1002/bjs.11172.
31. Chesney T, Acuna SA. Do elderly patients have the most to gain from laparoscopic surgery? Ann Med Surg. 2015;4:321–3.
32. Bellio G, Troian M, Pasquali A, et al. Outcomes of laparoscopic surgery for pT3/pT4 colorectal cancer in young vs. old patients. Minerva Chir. 2019; https://doi.org/10.23736/S0026-4733.19.07895-7.
33. Giuliani A, Ceccarelli G, Rocca A, et al. The role of laparoscopic distal pancreatectomy in elderly patients. Minerva Chir. 2018;73:179–87. https://doi.org/10.23736/S0026-4733.18.07594-6.
34. Ciria R, Gomez-Luque I, Ocana S, et al. A systematic review and meta-analysis comparing the short- and long-term outcomes for laparoscopic and open liver resections for hepatocellular carcinoma: updated results from the European guidelines meeting on laparoscopic liver surgery, Southampton, UK, 2017. Ann Surg Oncol. 2017;26:252–63. https://doi.org/10.1245/s10434-018-6926-3.
35. Chen K, Pan Y, Maher H, et al. Laparoscopic hepatectomy for elderly patients – major findings based on a systematic review and meta-analysis. Medicine. 2018;97(30):1–12. https://doi.org/10.1097/MD.0000000000011703.
36. Bandoh T, Isoyama T, Toyoshima H. Total gastrectomy for gastric cancer in the elderly. Surgery. 1991;109(2):136–42.
37. Bourcier K, Dinart D, Le Cesne A, et al. Outcome of patients with soft-tissue sarcomas: an age-specific conditional survival analysis. Oncologist. 2019;24(7):3559–e564.
38. Fong Y, Blumgart LH, Forner JG, et al. Pancreatic or liver resection for malignancy is safe and effective for the elderly. Ann Surg. 1995;222:426–37.
39. Poon RT, Law SY, Chu KM, et al. Esophagectomy for carcinoma of the esophagus in the elderly: results of current surgical management. Ann Surg. 1998;227:357–64.
40. Ellis FH Jr, Williamson WA, Heatley GJ. Cancer of the esophagus and cardia: does age influence treatment selection and surgical outcomes? J Am Coll Surg. 1998;187:345–51.
41. Frakes JM, Strom T, Springett GM, et al. Resected pancreatic cancer outcomes in the elderly. J Geriatr Oncol. 2015;6:127–32.
42. Kishida N, Hibi T, Itano O, et al. Validation of hepatectomy for elderly patients with hepatocellular carcinoma. Ann Surg Oncol. 2015;22:3094–101.
43. Makary MA, Winter JM, Cameron JL, et al. Pancreaticoduodenectomy in the very elderly. J Gastrointest Surg. 2006;10:347–56.
44. Daher M. Ethical issues in the geriatric patient with advanced cancer 'living to the end'. Ann Oncol. 2013;23:vii55–vii58. Supplement 7.
45. Willmott L, White B, Gallois C, et al. Reasons doctors provide futile treatment at the end of life: a qualitative study. J Med Ethics. 2016;42:496–503.
46. Chiabi P, Magne N, Breton S, et al. Influence of geriatric consultation with comprehensive geriatric assessment on final therapeutic decision in elderly cancer patients. Crit Rev Oncol/Hem. 2011;79:302–7.
47. Terret C, Zulian G, Droz JP. Statements on the interdependence between the oncologist and the geriatrician in geriatric oncology. Crit Rev Oncol/Hem. 2004;52:127–33.

Nonetheless, the authors emphasized the fact that being frail is not a contraindication for surgical approach [9].

Geriatric Assessment

Geriatric assessment is a systematic assessment of an elderly patient encompassing a multitude of domains such as comorbidities, nutrition, medications, functional status, cognitive abilities, and signs of depression [8]. Geriatric assessment holds three major aims: estimation of patient's physiological reserves and capacity; decision-making support; and evaluation for potentially reversible conditions such as depression, malnourishment, and harmful polypharmacy [10]. In fact, International Society for Geriatric Oncology recommended geriatric assessment in all oncologic patients more than 70 years of age. Likewise, personalized management of elderly patients with rectal cancer including geriatric assessment and preoperative evaluation by a geriatrician was supported by expert recommendations of the European Society of Surgical Oncology, European Society of Coloproctology, International Society of Geriatric Oncology, and American College of Surgeons Commission on Cancer [11]. Geriatric assessment should result in an estimation of the patient's frailty and fitness for standard of care. Due to the changes in treatment strategy and emergence of individualized approach, the value of geriatric assessment has been evaluated in several studies with controversial findings. A recent study found significantly reduced delirium rates in patients undergoing geriatric assessment as compared with usual care (11% vs. 29%; $p < 0.001$) [12]. Similarly, other studies reported that geriatric assessment may serve as a tool to predict postoperative morbidity [13] and mortality [14]. On the other hand, a recent randomized trial compared colorectal cancer patients undergoing surgery with geriatric assessment to the standard of care and found no difference in outcomes [15]. Similarly, Puts et al. concluded in a systematic review of 13 prospective studies that evidence supporting the role of geriatric assessment in the prediction of adverse outcomes is insufficient [16]. Further clinical trials as well as quality control studies evaluating the impact of geriatric assessment are mandated.

Enhanced Recovery in the Elderly

Enhanced recovery after surgery (ERAS) has become the gold standard of care for surgical patients across a range of specialties. Developed by Kehlet in 1995 [17], the majority of data for its use and application currently relate to colorectal surgery.

ERAS consists of a range of techniques such as preoperative carbohydrate loading, small incision surgery, reduced use of opioid analgesia, early postoperative mobilization, and early oral feeding to achieve improved recovery. Currently, there is no standardization of the various ERAS components, and they have remained unchanged in their application to elderly patients undergoing colorectal surgery. Elderly patients undergoing colorectal surgery are at higher risk compared with their younger counterparts for postoperative complications, such as pulmonary infection, urinary tract infection, heart failure, and delirium. A recent systematic review found that ERAS reduces the occurrence of these complications in older patients [18]. It remains questionable whether elderly patients undergoing colorectal surgery have the ability to adhere to ERAS protocols to the same extent as younger adults. Moreover, it is not known which components of ERAS protocols contribute to superior outcomes in elderly patients who differ physiologically from younger adults undergoing colorectal resection. Great emphasis has been placed on length of stay when evaluating the effectiveness of ERAS. For younger patients, this seems reasonable as discharge home and return to baseline functioning is likely. However, there is limited evidence as to whether the same can be said for elderly patients. It is unclear what proportion of elderly patients are discharged directly back to their original environment and which need to first undertake a further interim placement or rehabilitation before they return home. Moreover, one study found that elderly patients, who met the criteria for safe discharge, remained in hospital for a further 3–5 days after this date [19]. This may be due to limited availability of social support in the community, patient anxieties, and an unwillingness of surgeons to discharge their elderly patients in as timely a fashion as they may be inclined to with the younger population.

More recently, functional status is gaining ground as a key outcome measure. Relying on length of stay as an outcome measure for older patients after major surgery is disingenuous. A recent study went beyond the assessment of length of stay when qualifying the effectiveness of the ERAS protocol for older patients; it compared pre- and postoperative functional status in individuals undergoing major colorectal resections and found that for 83% of patients, postoperative function was at least as good as preoperative function [20]. Other centers have addressed the concept of prehabilitation with the aim of maintaining or improving functional status in elderly patients postoperatively. One study compared prehabilitation, based on resistance activities, nutritional counselling, and relaxation exercises, with rehabilitation in patients undergoing colorectal resection. Patients who had undergone prehabilitation had a significant improvement in functional exercise capacity and returned to their functional baseline postoperatively [21]. Further evaluation of elderly patient outcomes beyond discharge is needed. Both disease absence and preserved functional ability following major surgery should be considered in order to maintain physical and social independence for elderly patients.

Colon Cancer in the Elderly

The incidence of colon cancer increases with age, but the progression also varies by anatomic site, population, and gender. The mean age at diagnosis for men was 63 years, and for women, 62 years [22]. In women, colon cancer ranks third in the United States in number of cancer deaths. Adult males have a slight excess of cancer of the descending and transverse colon. There is a controversy as to whether the incidence of right colon cancer is the same for both genders or whether women have more right-sided colon cancer [23]. Generally, there are no differences with respect to presentation, location, TNM stage, and prognosis in comparison with younger adults. All studies confirm that there are no statistically significant differences in age-corrected survival curves. However, emergency operations are associated with a higher morbidity and mortality in elderly patients. Colon cancer in the elderly may present with the simple change in bowel habits, lower GI bleed (the second most common symptom), discharge of mucus mixed with stool and blood, abdominal pain, palpable abdominal mass, unintentional weight loss, symptoms of "appendicitis," history of inguinal hernia of short duration, colocutaneous fistula, and intussusception. Colonoscopy and CT scan are the main elements of the workup. Colonoscopy with biopsy should be performed first in case of change of bowel habit, chronic lower GI bleed, mucus discharge, and colocutaneous fistula. CT scan is the preferred first test in case of abdominal pain, palpable abdominal mass, unintentional weight loss, symptoms of "appendicitis," history of inguinal hernia of short duration, and intussusception. Unlike rectal cancer, the preoperative staging of colon cancer by CT scan can generally only rule out stage IV disease, namely, liver and/or lung metastases. Occasionally, when enlarged lymph nodes are detectable on CT scan, stage III disease can be suspected. Cardiac, pulmonary, and medical clearance are essential in order to risk stratify elderly patients. A multicenter prospective cohort study from Norway [5] recently described in-hospital and post-discharge complications in patients over 65 years of age concluding that most complications were medical and related to preexisting comorbidities. Nonetheless, there is no justification for avoiding necessary surgery on the mere basis of chronological age. There is a general trend in longer survival for colon cancer which is applicable to the elderly as well. Owing to improved survival, elderly colon cancer patients more frequently develop lung metastases over time. A recent study reported the 3-year survival rate in elderly patients to be 50% [24]. A recent NSQIP study assessed risk factors for readmission in geriatric patients undergoing colon cancer surgery with stratification by age ranges. The authors found chemotherapy to be a risk factor in 65–74 and 75–84 age groups and sarcopenia in >85 group [6]. A Dutch cancer registry study on 33,000 patients found significant decreased use of chemotherapy and laparoscopy in patients older than 75 years [25]. Regarding laparoscopy for colon cancer in the elderly, there is a never-ending controversy regarding the association between pneumoperitoneum and lower respiratory compliance, increased airway pressure, enhanced venous stasis, reduced portal venous pressure, and impaired cardiac function. A Cochrane meta-analysis of 5 RCTs published in 2013 found no difference in complication rates between pneumoperitoneum and abdominal wall lift. A meta-analysis comparing laparoscopy to laparotomy for colon cancer in the elderly [26] reported better 3-year survival rates after laparoscopy, although the 5-year survival rates did not differ. Heterogeneity was high in both outcomes; hence, no robust conclusions may be drawn from this meta-analysis.

In elderly patients who are not candidates for resection due to severe comorbidities, palliation may be achieved with one of the following options: argon plasma coagulation, endoscopic stenting, palliative resection and colostomy, and endoscopic submucosal dissection (ESD). In case of bleeding, palliation can be achieved endoscopically with argon plasma coagulation delivered through a colonoscope. In case of obstruction, stenting should be the first choice provided expertise is available. However, stenting should be considered as bridge to resection to be performed within a week following patient stabilization. Resection may have to be performed with palliative intent, which means no lymphadenectomy, no anastomosis, but colostomy. Nonetheless, there will always be a few cases where stenting will become the definitive therapy. ESD is an option in patients with early stages of colon cancer (T1) when the patient is not fit for surgery. Although colonic ESD (especially right colon) may be technically challenging due to narrow tubular lumen as compared to stomach, in experienced hands ESD can provide low margin involvement rates as well as cancer recurrence rates. Having minimally invasive palliation methods, the question whether we should operate on the elderly with colon cancer will always exist. A recent SEER database study compared survival rates in patients who underwent surgery to those who did not undergo surgery stratified by age ranges and cancer stages. The authors found that in all ages and all cancer stages, survival following surgery was significantly better than survival following no surgery. The only category in which survival was comparable was patients older than 94 with stage III cancers.

Rectal Cancer in the Elderly

In high incidence countries, 25% of intestinal cancers develop in the rectum with a slightly higher prevalence in men (30%: 21%). Over the past 40 years, several advances have been made in the management of rectal cancer, one of which is certainly the understanding that it is very much a team effort. A range of experts is required to treat patients with rectal cancer: enterostomal therapists, radiologists, interventional radiologists, pathologists, radiation oncologists, medical oncologists, and colorectal surgeons. Most of the time patients are asymptomatic, and occasionally the diagnosis is established on a CT scan ordered for an unrelated reason. The symptoms of rectal cancer include bleeding per rectum, mucous discharge, and incomplete evacuation. Weight loss and pain are rather symptoms of advanced disease. Although surgical technique is very relevant in the treatment of rectal cancer, its role and goals may drastically change when it comes to frail patients. There are a number of options that can be considered in the management of elderly patients with rectal cancer such as local excision and palliative procedures. Prior to making a decision, local and distant staging must be established by means of MRI and CT scan, respectively. PET scan is traditionally added as rectal cancer (unlike colon cancer) can be associated to ectopic metastatic lymph nodes such as paraesophageal, axillary, etc. Digital examination is reliable in assessing fixity of the tumor in the pelvis, whereas rigid proctoscopy is essential in establishing the level of the tumor. If necessary both might be performed under light sedation. A flexible sigmoidoscope should not be preferred for measuring although it is better tolerated by patients. In order to ascertain accurate measurements, the scope must be seen exiting the anal verge rather than emerging from the contour of the buttocks. Rigid (rather than flexible) scopes are also reliable in terms of determining whether the tumor is located on the anterior or posterior wall of the rectum. In the former, the depth of anterior invasion has to be carefully assessed as the female or male organs might be involved. The treatment options appropriate for frail elderly patients with rectal cancer may include transanal excision, trans-sphincteric excision, trans-coccygeal excision, laser photocoagulation, electrocoagulation, cryosurgical destruction, endocavitary irradiation, and others.

The first approach to the management of rectal cancer was transanal, and it predated the era of laparotomy [27]. Over a century passed before a transanal, this time endoscopic approach would be grabbing the headlines [28]. This endoscopic approach consisted of a rigid platform, which required a long learning curve and a capital budget purchase. Both factors most likely limited its popularization in the United States. American surgeons developed and implemented the concept of soft platform named transanal minimally invasive surgery (TAMIS) [29]. The procedure is performed by inserting a disposable single port in the anal canal with the patient prone or supine depending on the location of the lesion in the rectum. A pneumorectum is created insufflating CO_2, and laparoscopic instruments are inserted. Notwithstanding the opportunities for better visualization, simpler technique, and less expensive instrumentation, TAMIS is not flawless. Its limitations include restricted working angles within the confined space of the rectal lumen and external torque in attempt to compensate for the lack of instruments' articulation. It is not surprising that the abovementioned limitations of laparoscopic instruments created an opportunity for a robotic transanal approach. One of the goals of a robotic soft platform is to facilitate the suture closure of an anterior rectal wall defect [30]. Clearly, the 360-degree range of motion of the robotic instruments in the rectal lumen is the key advantage. The reassignment of the robotic arms from left to right and vice versa after instrument crossing and the 3D high-definition visualization allow for instrument manipulation similar to open surgery. The only available clinical data consist of a multicenter study including 16 patients only and reporting involved margins in 2 of 16 cases (13%) [31]. Hence, transanal excision of rectal cancer (regardless of the platform being rigid or soft) should be restricted to elderly patients with prohibitive comorbidities precluding curative resection. A trans-sphincteric approach does not seem to be the best choice particularly in elderly women. In fact, this procedure entails complete division of the posterior internal sphincter and levator ani muscles [32]. Although this approach offers access to only low-lying tumors, it does allow for segmental resection with transanal anastomosis as it is the case with the trans-sacral approach [33]. In addition to transanal, trans-sphincteric, and trans-sacral excision, there are a number of other local procedures that may be considered for palliation of bleeding, obstruction, and pain. Although electrocoagulation is rarely practiced today, it was advocated as early as in 1913 for palliation in frail patients with rectal cancer [34]. Electrocoagulation may be considered when the tumor is exophytic, confined to the rectal wall, not circumferential, when the patient with known distant metastases can have bleeding palliated, or when the patient refuses or cannot manage a colostomy due to blindness. The most common complication is fever for which antibiotics are recommended pre- and postoperatively for at least 24 hours. Pelvic sepsis may occur without recognizable rectal perforation. There are at least two more ablative methods for palliative endoscopic treatment of malignant strictures in the rectum: laser photocoagulation and cryosurgical destruction. The neodymium:yttrium-aluminum-garnet (Nd:YAG) laser is unique in that it is applicable to tumor located above and below the peritoneal reflection. Although laser therapy can be performed as outpatient, it often requires hospitalization. Concerns have been expressed about the laser power settings which allow to enhance the hemostatic or the vaporizing effects. In essence, excess cavitation must be avoided.

Laser therapy may be combined with the implantation of self-expanding metal stents, which however tend to migrate distally toward the anal canal and cause pain. Cryotherapy may also be used to palliate bleeding or obstruction with a 8% risk for perforation thank to its reduction of tumor bulk [35]. Endocavitary irradiation has been employed in the palliative treatment of rectal cancer for over 50 years [36]. The procedure requires a device to be inserted through a large diameter rigid proctoscope. The unit develops a 10–20 Gy/minute radiation output with 50 kV x-rays to be applied to a 3 cm surface with absorption limited to a 2 cm depth.

The combined use of endoscopic laser and radiation is viewed by many as a safe alternative to stenting in frail patients with obstructed rectal cancer. However, endoscopic transanal resection (ETAR) with a urologic resectoscope has been advocated mostly in Europe for frail elderly patients with obstructed rectal cancer [37]. The potential metabolic implications of absorbing glycine-irrigating fluid during ETAR have been directly correlated to operating time [38].

Complicated Diverticular Disease of the Colon in the Elderly

Diverticular disease resulted in 312,000 admissions and 1.5 million days of inpatient care per year in the United States [39] at a cost of over 2.6 billion dollars [40]. Recent studies have found the prevalence of colonic diverticula to be up to 50% in patients older than 60 years [41]. Hospital admission rates due to diverticulitis have increased by 26% with an average age decreased from 64.6 to 61.8 years [42]. The incidence of diverticular disease remains stable in patients aged 65 to 74 years but is decreasing in patients 75 years of age or older [43]. As opposed to the striking male predominance in younger adults, diverticulitis had a slight female prevalence in elderly subjects. A recent population-based study reported that up to 12% of patients with acute diverticulitis may develop complications at some point of time with no significant change over time [44]. The most common complication is microperforation with abscess and/or extraluminal air accounting for 69% of the cases. Dissimilarly, free perforation with peritonitis is not common in elderly patients with previous episodes of diverticulitis. Chronic intramuscular fibrosis of the colon wall can develop in up to 15% of patients leading to stricture. Similarly, the rate of fistula formation has been reported around 14% [45]. Several classifications of diverticular disease and its complications have been proposed by different authors. The first clinical classification of acute diverticulitis was proposed in 1963 [46]. In 1978, Hinchey et al. published a 4-stage classification [47] later modified by several authors [48]. A recently published review of six guidelines has concluded that clinical assessment of acute diverticulitis is insufficiently precise, hence agreed on the need for imaging, namely, CT scan [49]. In fact, several CT-based classifications have been proposed [50–52]. Moreover, the DICA (Diverticular Inflammation and Complication Assessment) classification based on endoscopy was recently proposed focusing on four findings: diverticulosis extension, number of diverticula, presence of inflammation, and complications [53].

Microperforation with Abscess and/or Extraluminal Air

It has been estimated that 10–20% of the individuals who develop acute diverticulitis will develop an abscess [54]. The fact that immediate operative management would most likely entail a colostomy constituted a robust argument in favor of initial nonoperative management, which gradually gained acceptance approximately 15 years ago [55]. Nonoperative management typically includes nil per os, IV fluids, IV antibiotics, and CT-guided percutaneous drainage. Inpatient total parenteral nutrition (TPN) may be considered in case of persistent ileus, and patients may receive outpatient TPN in case of low albumin and/or weight loss. Although the long-term outcomes following nonoperative management remain unclear in terms of rates and severity of disease recurrence, it seems the logical approach in elderly patients. A meta-analysis of 22 retrospective studies amounting to 1051 patients with acute diverticulitis complicated by abscess concluded that nonoperative management can lead to recurrent symptoms with high probability for resection [56]. Patients with acute diverticulitis may only exhibit at initial CT scan extraluminal air typically pericolic or retroperitoneal. Similar to acute diverticulitis with abscess, the practice of urgent resection in presence of extraluminal air has been gradually replaced by nonoperative management [57]. Although extraluminal air may also occur as isolated finding, it may be followed by the development of an abscess. A recent RCT reported that a repeat CT scan with rectal contrast on hospital day 3 may show interval development of an abscess ranging up to 13 cm in diameter in patients with extraluminal air and no detectable abscess on initial CT scan [58]. The data of this RCT are not in line with the American Society of Colon and Rectal Surgeons (ASCRS) clinical practice guidelines claiming that extraluminal gas alone is not to be viewed as complicated disease unless the patient is immunocompromised [59]. Patients who recovered after nonoperative management of an acute episode of diverticulitis with microperforation are typically recommended to undergo colonoscopy at a several-week interval. This practice should be tailored to the individual case when it comes to elderly patients. The practice of considering elective resection following resolution of an acute episode of diverticulitis with microperforation treated with nonoperative management has been supported by the ASCRS clinical practice guidelines [59] but recently challenged by an RCT [58] and certainly not to be recommended to frail patients. Unfortunately, not all patients with acute diverticulitis complicated by microperforation will recover with nonoperative management. Retrospective reports have suggested that patients with larger size abscesses particularly if located in the pelvis, as well as renal failure or immunosuppression, are unlikely to recover with nonoperative management. A recent RCT found that the percentage of failed nonoperative management was as high as 14% and such patients had an average of 19 locules of pericolic air on initial CT [58].

Perforation with Peritonitis

Free perforation with peritonitis is a life-threatening complication of diverticulitis. Although uncommon in elderly patients, perforated diverticulitis with peritonitis has been reported to account for more than half of emergency operations [60]. The prevalence of perforated diverticulitis has increased in developed countries from 2.4/100000 in 1986 to 3.8/100000 in 2000 [61]. The question of whether perforated diverticulitis with peritonitis should be treated by resection or non-resectional surgery has been addressed by one review published in the 1980s [62] and two randomized-controlled trials in the 1990s [63, 64]. Mortality rates following primary resection and non-resectional surgery were reported to be 12% and 28% ($p < 0.05$), respectively [62]. Regretfully, the two RCTs were heavily underpowered, used different outcome measures (postoperative death and postoperative peritonitis), and drew opposite conclusions. When it comes to elderly and/or frail patients with peritonitis, it is imperative to provide the patient and the patient's family with realistic expectations about the high mortality rates following resection with end colostomy and discuss alternative options such as (but not limited to) comfort care, etc.

Fistulae

Fistula formation is a complication affecting 2% of the patients with a prevalence in the elderly with long history with diverticular disease. Starting from the less frequent ones, diverticular fistulae may present as colovaginal, coloenteric, colouterine, colocutaneous, or colovesical [65]. Although the management of diverticular fistulae is in principle surgical since fistulae generally do not close spontaneously, the presence of a fistula is rarely an indication for urgent surgery. Hence, it is unlikely that frail and/or elderly patients should be benefitting for surgery.

Colovesical fistula is the most common form of colonic fistula in men, and more than half of them result from complicated diverticular disease. A clinical diagnosis is based on pathognomonic signs including pneumaturia, fecaluria, and recurrent urinary tract infections with mixed organisms. Colonoscopy and cystoscopy have a limited role in diagnosis because of poor sensitivity but are useful for clarifying anatomy and excluding strictures and malignancy. CT-cystogram is the preferred diagnostic imaging modality owing to its over 90% sensitivity. Fistulae may be demonstrated by gas or contrast in the bladder, or local colonic and bladder wall thickening. The surgical management of colovesical fistulae due to diverticular disease generally consists of colon resection with anastomosis or colostomy. The most sensible procedure to consider (if any) in frail patients is a transverse colostomy. Most colovaginal fistulas originate from diverticular disease [66], and it is uncommon for colovaginal fistula to occur in the absence of hysterectomy. The diagnosis of colovaginal fistula is largely based on clinical presentation and imaging. CT and contrast enema have demonstrated greater sensitivity for diagnosing colovaginal fistula and identifying etiology, while endoscopy has the advantage of biopsies and excluding malignancy. The surgical management of colovaginal fistula involves either a multi-stage repair or primary resection, both unsuitable for frail patients, who may consider a colostomy. A colouterine fistula secondary to diverticulitis was first described in 1929, and about 27 cases have been reported in the literature since [67]. The clinical presentation can be with or without vaginal discharge. Diagnostic hysteroscopy is the first-choice diagnostic tool for investigation of any abnormal vaginal discharge such as blood or stool because it enables direct vision and biopsy of the lesions of the lower genital tract quickly and at low cost. In terms of surgical management, colon resection with or without en-bloc hysterectomy is not indicated with frail patients, who may be offered a colostomy. Colocutaneous fistula is a rare complication of complicated diverticulitis. Mainstay of surgical therapy for colocutaneous fistula is primary resection with anastomosis or colostomy, similar to other forms of diverticular-related fistulas. However, in frail patients a colostomy can be considered, and the cutaneous fistula opening can be debrided and packed either with gauze or wound vacuum dressing applied.

Obstruction

Although the current incidence of large bowel obstruction due to diverticulitis is unknown, there is reason to believe that its incidence has increased in the past four decades. The main cause for the increased rates seems to be the implementation of nonoperative management. Recurrent episodes of diverticulitis, sometimes subclinical, can initiate progressive fibrosis and stricturing of the colonic wall without persisting inflammation. Ultimately complete obstruction can develop. Strictures may present in an insidious manner with non-specific symptoms being typical or less often with acute obstruction. Intestinal obstruction due to diverticular disease may also occur with and/or without colonic stricture. In the latter case, a small bowel obstruction can result from adhesion of a loop of the small bowel to an inflamed sigmoid colon. This presentation usually improves as inflammation subsides with treatment. A colon stricture can be identified at CT scan and/or colonoscopy. The challenge is to distinguish between a diverticular stricture and a stenosing neoplasm, which is particularly true in older patients. Although it may not always be possible to advance the instrument into the sigmoid colon owing to angulation, narrowing, and tortuosity of the bowel, brush cytology should be performed. Colonic stenting may have a role as definitive treatment in frail patients with limited life expectancy. However, in most cases stenting is viewed as bridge to resection (within 7 days) and carries a 38% complication rate (such as re-obstruction, migration, and perforation) [68]. Performing a loop transverse colostomy through a trephine incision has been the preferred treatment option for decades particularly in patients who are not candidates for resection. Nonetheless, the challenge of a so-called blind colostomy remains the uncertainty of the diagnosis, especially in emergent situations. This statement is supported by the delayed diagnosis of cecal ischemia in patients undergoing blind colostomy for acute large bowel obstruction [69].

Ulcerative Colitis in the Elderly

Samuel Wilks is credited with coining the term ulcerative colitis (UC) in England [70]. The condition was known under the same term to surgeons in the Union Army during the Civil War [71]. Regretfully, to this day our understanding of the pathogenesis of UC still remains obscure despite extensive research on the possible role of genetics, autoimmunity, infection, diet, oxidative metabolism, stool, smoking, psychological factors, and appendectomy. Evaluation of the epidemiology is rendered difficult by the plethora of diarrheal states that present with symptoms not unlike those of nonspecific UC. Most of the data have been accumulated in the Western countries although UC is not a reportable condition in the United States. Although UC is most commonly affecting individuals younger than 30 years, recent population-based cohort studies have reported an increased prevalence in the elderly population [72]. The main concern in elderly UC patients is, however, the relationship to colorectal dysplasia and carcinoma. Older studies had already made known the bimodal distribution of UC with the highest incidence of colon cancer in the seventh decade [73]. Furthermore, the development of UC in the older population has become a source of a difficult differential diagnosis. It is not uncommon for several elderly patients with UC to be misdiagnosed for having ischemic colitis. Colonoscopic biopsies should be obtained from a mucosa that appears macroscopically uninvolved in order to distinguish UC from Crohn's colitis. In fact, granulomas may be found underlying ulcerated mucosa in UC patients. In fact, in the elderly UC may have a sudden and fulminating onset. Although a detailed description of the medical management of UC is beyond the scope of this chapter, a critical appraisal of the common errors in the cornucopia of available medications is briefly enumerated: the irritable bowel components of UC (bloating, gas, fullness) are not in and of themselves indications to steroids; undertreating with aminosalicylates should be avoided, and elderly patients should be encouraged to use enemas or suppositories to maintain remission; steroids have no role in maintaining remission; under-treating with antimetabolites commonly occurs in delaying introduction, underdosing, and early discontinuation; misusing Infliximab by giving to patients with bowel obstruction and/or internal fistulas; misusing Cyclosporine by giving it in fulminant or hemorrhagic disease. Unfortunately, gastroenterologists generally speaking tend to accept as criterion of success their ability to keep the patients from surgery [74]. While subtotal colectomy with end-ileostomy and mucous fistula remains the standard in emergency, the question remains whether there is any role for elective restorative proctocolectomy in the elderly. Unlike for Crohn's disease, there is no role for segmental colectomy in UC. Recent population-based data on 1749 elderly-onset subjects with UC have reported that such patients are more likely to undergo surgery with no significant difference in mortality by chronological age (when compared to young patients) [72]. The International Pouch Database encompassing 101 variables has suggested that elective restorative proctocolectomy may be performed in selected elderly patients, albeit with higher complications rates and longer length of stay [75]. Finally, advanced age does not seem to be a contraindication to ileostomy. Large volume centers have found that although lifestyle did not appear to be altered in the elderly than in younger individuals, elderly patients experienced more appliance management difficulties [76].

Crohn's Disease in the Elderly

The US data published in 2019 indicate that the prevalence of Crohn's disease (CD) is rising in the elderly population, a fact related to its increased incidence but also to improved life expectancies [77]. Presentation, natural history, and treatment of CD differ between young adult-onset patients who progress to elderly age and elderly-onset patients. Presentation at elderly age can also make diagnosis challenging due to pre-existing comorbidities that mimic the symptoms of CD or UC. Identifying patients at high risk for progression or aggressive disease is particularly important as elderly patients may respond differently to medical and surgical treatment, and may be at higher risk for adverse effects. Despite newer agents being approved for CD, the data regarding efficacy and safety in the elderly are currently limited. Balancing symptom management with risks of medical and surgical therapy is an ongoing challenge and requires special consideration in the elderly populations. The 2019 Scandinavian registry report found that elderly-onset UC patients are at increased risk of death when compared to reference population [78]. Moreover, the increased mortality remained also after the introduction of biologic therapies [78]. A US multiregional data set confirmed the trends reported in similar countries and also confirmed lower rates of perianal CD in the elderly [79]. In contrast to UC patients, a patient with CD may have obvious findings on physical examination. Abdominal findings are common in cases of regional enteritis and may include a mass felt in the right iliac fossa or a large mesenteric abscess can be palpated. Patients with Crohn's colitis most commonly have normal findings on physical examination. Inspection of the perianal skin and digital rectal exam may raise the suspicion of CD. Increasing evidence supports the statement that CD of the intestine is actually a systemic disease rather than one limited to the small and large bowel. Hence, extraintestinal symptoms are not uncommon and may include polyarteritis nodosa, erythema nodosum, psoriasis, arthritis, spondylitis, polymyositis, uveitis, and others. Endoscopy, MRI-enterography, and capsule endoscopy are part of the workup. Similar to the case of elderly UC patients, the relationship to dysplasia and carcinoma is a concern in elderly patients with CD. It is very difficult to assess the true incidence of CD and carcinoma of the small intestine, as the latter is such a rare condition. Moreover, the distribution of carcinoma of the small intestine in CD patients differs from its counterpart in patients without CD. The most common (70%) location to develop a carcinoma in CD patients is the ileum, whereas patients without CD most commonly develop tumors in the jejunum [80]. Furthermore, the likelihood of developing a carcinoma is increased by intestinal by-pass surgery [81]. The mortality rate is as high as 80% in CD patient who develop cancer in the excluded bowel. In the large intestine, CD with cancer is most commonly found in the right colon, as opposed to the more common left location in the general population. With regard to dysplasia, both radiology and endoscopy are of little benefit in establishing the diagnosis. Unfortunately, with the exception of the very distal ileum, the small intestine does not lend itself to endoscopy with biopsy. Although a detailed description of the medical management of CD is beyond the scope of this chapter, it is useful to provide a critical appraisal of the safety of drugs used in the treatment of CD. There are, in fact, concerns about the safety and adverse events associated with biologic agents. A review from Australia examined safety data on thiopurines, methotrexate, anti-TNFα agents, vedolizumab and ustekinumab in elderly patients focusing on trial and real-world data. The review found that the data are in support of concerns about an elevated risk of serious infections, skin cancer, and lymphoma particularly with thiopurines and anti-TNFα agents. The review also concluded that long-term data are required to identify risks with extended use of vedolizumab and ustekinumab [82]. In addition, smoking is an additive risk factor for relapse in CD, whereas nutrition is essential to decrease complication rates. Although the concept of short-term intravenous hyperalimentation has advantages, needed emergency surgery such as for free perforation of the small intestine or colon should not be deferred. Dissimilarly, lower GI bleeding very rarely will require surgery. Regardless of the choice of operation, it must be absolutely clear that surgery for Crohn's disease is palliative, not curative. This concept is particularly relevant when it comes to elderly patients with CD. Unlike for UC, segmental colectomy may become indicated in Crohn's colitis although endoscopic stenting can be an option in frail patients. The presence of abscess or fistula should prompt an assessment as to whether parenteral nutrition and interventional radiology may be indicated in frail patients.

Furthermore, in elderly patients the risk of carcinoma of the ileum must be kept in mind. This point is particularly relevant given the fact that achieving uninvolved resection margins and dissecting lymph nodes are not part of the standard of care in surgery for CD.

Pelvic Floor Disorders in the Elderly

Full-Thickness Rectal Prolapse in the Elderly

More than 150 years have elapsed since the first report on external full-thickness rectal prolapse surgery (FTRP) appeared in the literature [83]. Although there are a number of abdominal procedures for the treatment of external FTRP, which differ technically as to whether sigmoid resection is added to rectal mobilization and rectopexy, none of such procedures are actually appropriate in elderly patients. In fact, any abdominal operation for external FTRP will require general anesthesia with endotracheal intubation, whereas perineal procedures can be performed under epidural or spinal anesthesia. In elderly patients the most frequent primary complaint is referable to the prolapse itself. Three quarter of patients report a lump in the perineum. Problems with bowel regulation and incontinence are also common presenting features. Almost one half of our patients had a history of constipation. Major bleeding is rare unless the prolapse is massive or irreducible. The patient may however complain of small amounts of blood and mucous soiling the underwear.

Fecal incontinence is frequently associated with prolapse. Indeed, among the one quarter of patients who do not report a lump, many will complain of fecal incontinence as the main presenting problem, and rectal prolapse should be suspected whenever this symptom presents, especially in the elderly. Incontinence becomes more severe as the protrusion increases in degree. Dilatation of the canal by the mass results in further relaxation of the sphincter muscles and further prolapse. Protrusion may occur when lifting or coughing and not necessarily solely on defecation. Initially the prolapse may retract spontaneously but manual reduction eventually becomes necessary, and ultimately external FTRP may protrude for most of the time. Infrequently, the prolapse may become incarcerated or even strangulated resulting in a surgical emergency.

The patient with rectal prolapse may present with a myriad of different symptoms that range from constipation and straining to fecal incontinence. Therefore, it is of utmost importance that all patients undergo a complete preoperative workup before surgery. Upon physical examination, inspection may reveal an external rectal prolapse, especially during straining. However, external FTRP must be differentiated from mucosal prolapse. The mucosal prolapse can be differentiated from the full-thickness presentation because of the radially oriented grooves, while the FTRP has concentric grooves. In order to measure the prolapse adequately, the patient is asked to position himself in a squatting position. The patient is then asked to increase straining and the prolapse enlarges and lengthens. While the patient is straining, the distance from the perianal skin to the top of the prolapse is measured. Digital rectal examination may also add valuable information by detecting anal pathology and assessing sphincter tone and squeeze pressures. This information is important and aids the surgeon to choose the appropriate procedure for each individual patient.

Although not a standard test ordered for the evaluation of a patient with rectal prolapse, MRI defecography provides the surgeon with valuable anatomic and functional information for pelvic floor abnormalities. In addition, the results may indicate the presence of sigmoidocele or enterocele.

Colorectal transit time (also known sitz-marker study) provides essential information when confronted with a patient with concomitant constipation. Among the different methods for establishing the colorectal transit time, we recommend the method described by Gore et al. [84]. The patient receives six numbered day-packs. The five first packs contain 10 rings and the sixth contains 10 rings and 20 cylinders. Each day, at the same time, a pack is ingested. A plain abdominal radiography is taken on day 7. Rings are counted and the transit time is measured by the hospital's radiology department protocol. Colonoscopy must be performed in order to rule out any mucosal abnormality, especially in patients with a prior diagnosis of diverticulitis, inflammatory bowel disease, or cancer. Anal manometry is important in evaluating the patient with a longstanding history of rectal prolapse and incontinence. Patients with pudendal nerve damage, from either obstetric trauma, diabetes, or neoplasms, must also undergo manometric evaluation prior to surgery.

There are, in essence, two perineal procedures that should be considered in elderly patients with external FTRP. Such procedures are referred to with the names of the surgeons who described them initially: Altemeier's procedure and Delorme's procedure. Altemeier's is the procedure of choice in case of external FTRP exceeding 5 cm length, but it is also indicated in the rare circumstance of strangulated unreducible external FTRP where gangrene may or may not occur. Moreover, a levator-plasty and a colonic J-pouch may be considered in case of associated severe fecal incontinence. Delorme's procedure should be considered in case of external FTRP of less than 5 cm in length. Because the peritoneal cavity is not entered, Delorme's procedure is the procedure of choice in case of cirrhotic patients with ascites. Dissimilarly, it should be kept in mind that the peritoneal cavity is always entered during Altemeier's procedure.

In Altemeier's procedure, a circumferential incision is made through all layers of the outer bowel wall 1 cm proximal to the mucocutaneous junction. When the circumferential incision is completed, clamps are reapplied to the distal edge of rectum, and the prolapse is delivered as a single loop of externalized bowel. With a deep Douglas pouch, it is straightforward to enter the peritoneal cavity by incising the peritoneum anteriorly. The redundant colon is delivered through the defect. The peritoneum is repaired using a continuous suture to obliterate the sac. A plication of the levator ani muscles is carried out. The redundant intestine is transected. The anastomosis is fashioned with an interrupted long-term absorbable suture technique.

In Delorme's procedure, a circumferential incision is made 1 cm proximal to the dentate line, similar to that for Altemeier's procedure. The bowel is not transected, however. A submucosal dissection with a mucosal stripping is carried out to the apex of the protruding bowel. Dissection may be facilitated by infiltrating the submucosa with saline solution. The redundant mucosa is excised, and the denuded muscularis propria is plicated longitudinally, collapsing the bowel like an accordion. The edges of the mucosa are then sutured.

Postoperatively, the usual considerations are addressed. Early activity and incentive spirometry are encouraged. Pain is initially controlled with an epidural catheter with local anesthetic. Diet is advanced with the return of bowel function, and the pain medication is transitioned to oral formulations. The Foley catheter can be removed on the first postoperative day unless other comorbidities are present. Upon discharge, the patient is instructed to avoid heavy lifting. Dietary goals should be addressed. Avoidance of constipation or overly loose stool should be discussed. The patient should be seen in the office within 1–2 weeks of discharge. Continued follow-up will assist the surgeon in his or her evaluation of the success of the repair.

Incontinence

Fecal incontinence is very common in the elderly population and can cause morbidity and decrease quality of life. Many patients feel embarrassed to discuss the condition with a physician. Incontinence can result from intestinal disease-causing irritability and urgency of defecation due to inflammation or loss of capacity of the rectum or neorectum or from a defective anal sphincter mechanism. The management of incontinence associated with intestinal diseases is part of the treatment of those conditions.

Previously fecal incontinence was regarded as a mechanical disorder, whereas it has become increasingly clear that there is a large functional component. Behavioral therapies have been developed and their value is now established. Importantly, surgical correction is reserved for patients who have a mechanical disruption of the sphincter mechanism. Its use has therefore declined having been largely replaced by neuromodulation including stimulation of sacral, posterior tibial, or pudendal nerves. Artificial sphincters either biological in the form of graciloplasty (stimulated or non-stimulated) or prosthetic are not indicated in elderly patients. Anal procedures including the injection of bulking agents into the anal canal lining and so-called radio frequency energy delivery (SECCA) to promote scarring of the anal canal have been developed, particularly for patients with passive incontinence due to failure of the internal sphincter.

Clinical assessment combined with anal ultrasound to establish the anatomy of the sphincter and physiological testing by manometry will result in satisfactory decision-taking in most circumstances. Anal ultrasound is now routine in the assessment of continence. Some of the physiological parameters thought to be important in the past, for example, electromyography and pudendal nerve terminal motor latency, are no longer regarded to be useful in clinical practice.

The outcome for the patient is now expressed as much by quality of life as by the frequency of incontinent episodes. The important symptom of urgency, which may be the only manifestation of a continence disturbance, can now be quantified enabling improved assessment. The development of continence scoring systems has also improved our ability to assess the effectiveness of treatment.

The development of scoring systems has been an important feature of practice. In 1993, Jorge and Wexner proposed a continence grading scale which is now termed the Cleveland Clinic Incontinence Score that has now come to be used by many investigators [85].

Attempts have been made to compare functional estimations of anal sphincter capacity in the elderly population as compared to younger adults. Some studies have concluded that elderly patients have lower anal pressures, required lower rectal volumes to inhibit anal sphincter tone, and have increased rectal pressures as measured by balloon implying lower compliance [86]. Some authors suggested that internal anal sphincter dysfunction may be the important factor [87]. Others suggested that, especially in women, the pudendal and somatic pelvic nerves are injured when there is perineal descent on straining as occurs with aging [88].

Patients with loss of sphincter function through injury, whether surgical, obstetric, or trauma, are the most amenable to reconstructive surgery, whereas those with incontinence due to a general disease or a locally diffuse weakness of the pelvic are generally poor candidates for an attempt at repair. For these, conservative treatment in the form of dietary advice, antidiarrheal medication, and behavioral therapy such as biofeedback should be the initial approach along with treatment of any underlying disease. Some of these patients will be candidates for neuromodulation if conservative management is not successful.

The management strategy of fecal incontinence depends on the results of assessment. The key to decision-taking depends on whether the sphincter is intact or not. This is determined on clinical examination by inspection, palpation, and anal ultrasound. A patient with no sphincter defect will not respond to surgical repair. Conservative management should be tried, and if that fails, the options include neuromodulation, anal canal injection, SECCA, or irrigation.

A patient with a sphincter defect will benefit from surgical repair if the defect is large. If the sphincter defect is small, the patient may be suitable for neuromodulation as long as the muscle has contractility. A failed repair is not a contraindication to a second attempt at repair provided that the sphincter muscle is not atrophic. Irrigation is also an option when the above treatments fail.

In general, conservative treatment should be tried first unless a cloacal deformity with severe incontinence is diagnosed, a condition that requires immediate repair is indicated. Conservative management should be abandoned when the patient feels with full knowledge of the options and disadvantages that another treatment is necessary. When this situation is the case, the treatment options depend on whether the sphincter is intact or not.

Treatment options are traditionally classified in non-invasive and invasive. The former includes perineal exercises, biofeedback, electrical stimulation, anal plug, and anterograde or retrograde colonic irrigation. Unlike anterograde, retrograde irrigation is easy to perform and has very low morbidity. The available evidence strongly indicates that it can have a success rate of 50% or more. It may be sufficiently successful to avoid a colostomy when all other treatments have been tried and failed. In current practice, this option is probably underused. Invasive treatment options include injection of bulking agents, radiofrequency energy (SECCA procedure), neuromodulation, sacral nerve modulation, posterior tibial nerve modulation, pudendal nerve stimulation, and sphincter repair.

Two different forms of sphincter dysfunction (diffuse weakness and localized sphincter trauma) have been traditionally treated with sphincter repair. Patients present with an acute injury caused by direct lacerating or blunt trauma often associated with other injuries. There may be fecal contamination of the injured field, and management will involve a decision whether or not to perform a defunctioning stoma along with debridement of the wound removing non-viable tissue and foreign material. Defunctioning stoma may be necessary. After recovery of the patient, any pelvic floor damage can be repaired electively. It is recommended that reconstruction should not be performed for at least 3 months to allow the acute inflammation to settle. Although earlier publications demonstrated improvement in continence after delayed sphincter repair in 80% of patients [89], more recent reports showed that an initially good result was followed by deterioration with time [90].

Performing a colostomy in a patient with fecal incontinence is thought generally to be an admission of failure, but it should not be regarded as such. Elderly individuals should not be submitted to the morbidity of a sphincter-saving operation. The likelihood of success in elderly patients who have neurologic deficit, or who have severe bowel function problems, is limited. Fecal diversion is virtually often the optimal choice for elderly patients confined to a nursing home or to a convalescent facility unless they are blind. Irrespective of the method of treatment, more frequently than in any other colorectal disease, elderly patients with fecal incontinence need to be partners in the decision-making process.

Constipation

Elderly patients who present with constipation may have a mechanical reason, or the condition may be functional. In the latter case, the constipation could be due to slow transit through the intestine or to a difficulty in evacuating the rectum. Not infrequently both factors can present in elderly patients. Moreover, delayed intestinal transit can be present in bowel of normal or dilated caliber. In the latter circumstance, the condition is usually chronic, but it can also be acute such as in pseudo-obstruction (Ogilvie's syndrome) [91]. Normal caliber constipation may also be caused by factors such as advanced age, institutionalization and immobility, drugs, metabolic disorders (hypothyroidism, hypercalcemia), depression, or neurological disease. Obstructed defecation should be viewed as mechanical and functional, although overlap is not uncommon. In the last 10 years there has been a marked move away from surgery toward behavioral treatment (biofeedback and neuromodulation) for patients with functional constipation due to delayed transit. In the United States, the cost of over-the-counter laxatives is in excess of 400 million dollars, and the outpatient medical care of elderly females with constipation is twice as costly as those without [92]. The prevalence in the community of the so-called functional constipation has been reported to be between 20% and 30% [93]. Constipation is more common in adults older than 60 years of age (23%), and in those who were inactive [94]. Painful anal disease may also result in constipation through the reluctance of the patient to have bowel movements leading to impaction in the rectum. Non-mechanical causes of constipation include conditions resulting in reduced transit. The causes of the various forms are unknown, but delayed transit may involve the entire intestine including gastric function or be confined to the colon [95]. The differential diagnosis can be made by radio-opaque markers or isotope scintigrams, which can measure gastric emptying as well as oro-cecal and colonic transit times. Obstructed defecation is a common cause of constipation. Functional obstructed defecation may be caused by failure of the pelvic floor adequately to relax as in anismus including the solitary rectal ulcer syndrome or it may be due to laxity of the pelvic floor associated with perineal descent or even rectocele [96]. The diagnosis is based on the Rome III criteria [97], and its severity can be assessed with validated constipation scores [98]. Immobility plays a major role in constipation. In fact, there is evidence that physical activity is associated with a higher stool frequency [99]. In a survey of elderly patients, the overall incidence of constipation was no different than in younger adults but it was related to immobility and/or depression [100]. Bedridden elderly patients are more likely to have constipation than those who are able to walk and even more so than patients who walk several hundred yards a day [101]. Constipation is often due to fecal loading in the rectum. Impaction is commonly seen in institutions where immobility and constipating drugs are both factors delaying transit and evacuation. Exercise improves bowel function in normal sedentary men and women [102], but not in patients who are already severely constipated [103]. There is evidence from institutional studies that exercise, adequate hydration, and fiber intake can reduce the need for laxatives [104]. There is a large literature regarding the questions whether laxatives can affect the structure and/or function of the intestine. With respect to the former, there is poor evidence that laxatives can cause intestinal nerve and smooth muscle damage. The effect of laxatives on the enteric nerves and smooth muscle was reported in the 1960s [105]. In fact, such elderly patients had severe constipation due to the primary disease itself rather than to secondary damage by laxatives. Moreover, ultrastructural changes seen in elderly patients on long-term laxatives can also be found in amyloid, diabetic neuropathy, and inflammatory bowel disease [106]. More recent studies have demonstrated reduced numbers of Cajal cells and enteric neurons in patients with severe colonic inertia [107, 108]. However, such changes may be due to the disease and not to the laxatives [109]. Laxatives can certainly cause electrolyte inbalance when taken in high doses. A review of the literature showed 70 publications including 240 patients with diarrhea caused by hidden laxative administration. The laxative in question was phenolphthalein and the metabolic disturbance was hypokalemia [110]. Certainly, laxatives can lead to melanosis coli if used over time. This condition is due to the deposition of pigment due to staining by anthroquinones of cell debris from colocytes ingested by macrophages in the submucosa [111]. This condition is of no clinical significance and disappears within months after stopping the laxatives.

Drug-induced constipation may be caused by opioids, anticholinergics, antidepressants, iron, bismuth, antiparkinsonians, aluminum-containing antacids, antihypertensives, and anticonvulsants. In fact, both symptoms and radiologic findings of colonic dilatation may completely resolve following discontinuation of the medications [112]. Low levels of thyroid hormones have been regarded as a cause of constipation although there is a low prevalence of hypothyroidism in a prospective study in elderly female patients with constipation [113]. Constipation may also develop as a result of impaired rectal evacuation. Conditions responsible include prolapsing disorders and anismus. The latter is described as a functional disorder of the pelvic floor muscle in which straining or attempting to defecate leads to muscle contraction instead of relaxation, thereby causing a physiologic outlet obstruction and the inability to defecate [114].

The clinical assessment should include three elements: frequency of defecation, the consistency of the stool, and the presence of urgency.

In fact, patients with an evacuation disorder of the type seen in the solitary ulcer syndrome may make numerous visits to the bathroom where they may stay straining for hours. The physical exam should always include a neurological assessment. Unless the patients have symptoms suggestive of a mechanical cause, it is reasonable to treat the patient in general terms before carrying out sophisticated investigations such as sitz-marker transit time with ingestion of radio-opaque rings; scintigraphy with oral ^{111} Indium bound to diethylenetriaminepentacetic acid (DTPA) followed by a gamma scan at 72 and 96 hours; colonic manometry involving transanal insertion of open tipped or balloon probes; anal manometry particularly useful patients with anismus (high resting pressure at internal sphincter) and paradoxical contraction of the puborectalis muscle; balloon expulsion failure which is indicative of obstructed defecation; and MRI defecography. Treatment options can be medical or interventional. The former includes the management of any underlying disease, discontinuation of constipating drugs, dietary changes, and increased physical exercise. Biofeedback, neuromodulation, posterior tibial nerve stimulation, and retrograde (better than anterograde) irrigation are the available interventions. Currently, neuromodulation is the first choice of invasive treatment when the patient is still suffering despite medical and behavioral treatment. Elderly patients are generally not candidates for colectomy, which can be complicated by high rates of small bowel obstruction [115]. Furthermore, the reported 70% decrease in laxative consumption after colectomy may last for only a few years or less. In addition, about 10% of patients experience urgency and abdominal bloating after colectomy, unlike with neuromodulation [116].

Vascular Diseases of the Colon

Lower Gastrointestinal Bleeding

Elderly patients may develop a lower gastrointestinal (GI) bleeding due to common conditions, such as colorectal cancer, inflammatory bowel disease, hemorrhoids, infectious colitis, ischemia, radiation, and renal transplantation [117]. Uncommon conditions that can produce massive bleeding are coagulopathy, Dieulafoy's disease, aortointestinal fistula, rupture of a splenic artery aneurysm, microaneurysm of the superior hemorrhoidal artery, rupture of a pancreatic pseudocyst into the colon, angiosarcoma, and colorectal varices [118–121]. Regardless of the cause of the bleeding, a systematic approach to diagnosis and management is required to care for the elderly patient adequately. This includes stabilization of the patient and initial workup, followed by diagnostic studies (colonoscopy and angiography) to identify the source of bleeding, and finally intervention to stop the bleeding.

The importance of obtaining an accurate history cannot be overemphasized. For example, knowledge of prior abdominal aorta surgery may be critical. In addition to a detailed medical and surgical history, identifying risk factors for bleeding is essential in the history, such as the patient's use of anticoagulants or antiplatelet drugs. However, patients usually present with no history of lower GI bleed, and they frequently have no abdominal pain.

One should, obviously, perform a digital rectal examination and a limited rigid sigmoidoscopy with anoscopy as the initial examination. However, the yield from these diagnostic procedures is less than 10% [122]. The placement of a nasogastric tube can eliminate the stomach as a potential source, especially if clear bile is returned. The approach to the bleeding patient (after initial assessment is complete and the patient is believed to have a lower GI source for bleeding) is to proceed with colonoscopy or CT scan.

Classically, massive lower GI bleeding has been generally attributed to diverticular disease. The vasa recta, through their proximity with the diverticulum, can rupture either at the apex or at the neck as the vessel proceeds into the submucosa of the colon. The problem, however, is that most lower GI hemorrhage comes from the right side of the colon, where there are few or no diverticula. Evidence suggests that unexplained vigorous lower intestinal bleeding, even in the presence of known diverticulosis, is most likely due to an arteriovenous malformation (vascular ectasia, angiodysplasia) [123–125]. With the availability of CT-angiography and scinti-angiography, and the ability to identify preoperatively the site of bleeding, arteriovenous malformations have not uncommonly been observed in areas where diverticulosis is present. The current opinion is that lower GI bleeding mostly originates from vascular malformations. Since vascular malformations are more common in the right colon, we are assuming that the right colon is the most common site for lower GI hemorrhage. Bleeding associated with vascular malformations is usually less severe than that from diverticular hemorrhage.

The etiology of the vascular malformations has been unclear. However, new studies suggested that the vascular lesions are actually degenerative from an acquired and progressive dilatation of previously normal blood vessels as the result of the aging process [126].

Although Dieulafoy's lesion is most often a pinpoint nonulcerated arterial lesion, high in the gastric fundus, it has rarely been described as the source of bleeding in the colon, rectum, and anal canal.

Colorectal varices are a rare cause of lower GI bleed, almost always associated with cirrhosis, with resultant portal hypertension, or portal venous obstruction [127]. The condition has been reported in approximately 2.5% of those undergoing sclerotherapy for esophageal varices [128]. As few as 3.6% and as many as 56% of cirrhotic patients have been demonstrated to have concomitant rectal varices. Portal hypertension can also lead to stomal and parastomal varices, especially in an individual with sclerosing cholangitis and biliary cirrhosis as an extraintestinal manifestation of inflammatory bowel disease [129].

As implied from the foregoing, if the bleeding point is identified by means of angiography, tagged red cell scan, and endoscopy, appropriate therapy can be instituted: medical management, a local procedure, or resection, depending on the nature of the lesion and the patient's clinical course.

In the patient who continues to bleed and the source cannot be identified, blind left colectomy has historically been advocated, and later, right colectomy. Nonetheless, most surgeons today would rather perform subtotal colectomy with ileostomy. In spite of a 20% rebleeding rate after right hemicolectomy for angiographically demonstrated vascular malformations, some authors believe that the mortality risk of subtotal colectomy is greater than that of rebleeding [130]. Other authors reported an unacceptable mortality of 27% following subtotal colectomy [131]. On the other hand, rebleeding rate following subtotal colectomy with a mean follow-up of 1 year was 0% [132]. The rebleeding rate after segmental resection was 14% with a preoperative positive angiogram and 42% if the angiogram was negative. The latter patients also had an extremely high mortality rate (57%).

Nowadays, thanks to the availability of CT-angiogram and interventional radiology, surgeons should rarely have to resort to blind resections. Nonetheless, if the bleeding source is clearly demonstrated and cannot be controlled by some other means, a segmental colectomy is appropriate. Conversely, if the source of the bleeding is not identified with certainty, an emergency subtotal colectomy should be considered even in frail elderly patients. Finally, it is strongly recommended to perform a rigid sigmoidoscopy with the patient in stirrups prior to subtotal colectomy in order to definitively determine that the bleeding site is not the rectum or the anus.

Ischemic Colitis

Ischemic colitis is a term to describe a syndrome due to occlusive or nonocclusive vascular disease as it affects the large bowel. It is a very common disorder in elderly patients, being the most common form of intestinal ischemia and responsible for 1 in 2000 patient admissions [133]. It is a condition that usually is found in the aging population, with an increased incidence in women [134]. The following conditions produce their pathologic manifestations on an ischemic basis: arteriosclerosis, emboli, myocardial infarction, vasculitis, colorectal neoplasms, portal hypertension, strangulated hernia, volvulus, diabetes

unsafe [151]. Laparoscopic sigmoidopexy can be considered unless there are contraindications for pneumoperitoneum.

Cecal volvulus is much less common than sigmoid volvulus particularly in elderly subjects. There are three well-recognized types of cecal volvulus: the axial torsion type (type 1), loop type (type 2), and cecal bascule type (type 3). Eighty percent of all cecal volvulus cases are attributed to type 1 and 2 [152]. Low grade and colicky abdominal pain is usually the most common complaint. On X-ray or CT scan, the cecum and ascending colon can be displaced, the most common sites of displacement being the epigastrium followed by the left upper quadrant. Classic x-ray findings include a "coffee-bean" shape and visible mucosal folds at the site of obstruction with no gas distal to the point of obstruction. Colonoscopy has been successfully employed for detorsion of cecal volvulus. If attempts at colonoscopic detorsion fail, laparoscopy or laparotomy should be performed. In case if the viability of the bowel is compromised or if perforation is present, resection is mandated. If the bowel is viable, cecopexy may be performed. Cecopexy may be associated with high risk of recurrence.

References

1. Campbell FB, Samson RB. Carcinoma of the colon and rectum in geriatrics. J Am Geriatr Soc. 1953;1(8):557–60.
2. U.S. Census Bureau. 2010 census summary file 1. https://www.census.gov/prod/cen2010/doc/sf1.pdf. Accessed 10/18/2018.
3. Boddaert J, Raux M, Khiami F, Riou B. Perioperative management of elderly patients with hip fracture. Anesthesiology. 2014;121(6):1336–41.
4. Boakye D, Rillmann B, Walter V, Jansen L, Hoffmeister M, Brenner H. Impact of comorbidity and frailty on prognosis in colorectal cancer patients: a systematic review and meta-analysis. Cancer Treat Rev. 2018;64:30–9.
5. Ommundsen N, Nesbakken A, Wyller TB, et al. Post-discharge complications in frail older patients after surgery for colorectal cancer. Eur J Surg Oncol. 2018;44(10):1542–7.
6. Yeo H, Mao J, Abelson JS, et al. Development of a nonparametric predictive model for readmission risk in elderly adults after colon and rectal cancer surgery. J Am Geriatr Soc. 2016;64(11):e125–30.
7. Balducci L, Extermann M. Management of cancer in the older person: a practical approach. Oncologist. 2000;5(3):224–37.
8. Ommundsen N, Wyller TB, Nesbakken A, et al. Frailty is an independent predictor of survival in older patients with colorectal cancer. Oncologist. 2014;19(12):1268–75.
9. Fagard K, Leonard S, Deschodt M, et al. The impact of frailty on postoperative outcomes in individuals aged 65 and over undergoing elective surgery for colorectal cancer: a systematic review. J Geriatr Oncol. 2016;7(6):479–91.
10. Alibhai SMH. Comprehensive geriatric assessment in older patients with cancer: two steps forward? J Geriatr Oncol. 2013;4(3):205–7.
11. Montroni I, Ugolini G, Saur NM, et al. Personalized management of elderly patients with rectal cancer: expert recommendations of the European Society of Surgical Oncology, European Society of Coloproctology, International Society of Geriatric Oncology, and American College of Surgeons Commission on Cancer. Eur J Surg Oncol. 2018;44(11):1685–702.
12. Tarazona-Santabalbina FJ, Llabata-Broseta J, Belenguer-Varea Á, Álvarez-Martínez D, Cuesta-Peredo D, Avellana-Zaragoza JA. A daily multidisciplinary assessment of older adults undergoing elective colorectal cancer surgery is associated with reduced delirium and geriatric syndromes. J Geriatr Oncol. 2019;10(2):298–303.
13. Kristjansson SR, Nesbakken A, Jordhøy MS, et al. Comprehensive geriatric assessment can predict complications in elderly patients after elective surgery for colorectal cancer: a prospective observational cohort study. Crit Rev Oncol Hematol. 2010;76(3):208–17.
14. Hamaker ME, Vos AG, Smorenburg CH, et al. The value of geriatric assessments in predicting treatment tolerance and all-cause mortality in older patients with cancer. Oncologist. 2012;17(11):1439–49.
15. Ommundsen N, Wyller TB, Nesbakken A, et al. Preoperative geriatric assessment and tailored interventions in frail older patients with colorectal cancer: a randomized controlled trial. Color Dis. 2018;20(1):16–25.
16. Puts MTE, Hardt J, Monette J, et al. Use of geriatric assessment for older adults in the oncology setting: a systematic review. J Natl Cancer Inst. 2012;104(15):1133–63.
17. Bardram L, Funch-Jensen P, Jensen P, Crawford ME, Kehlet H. Recovery after laparoscopic colonic surgery with epidural analgesia, and early oral nutrition and mobilisation. Lancet. 1995;345(8952):763–4.
18. Bagnall NM, Malietzis G, Kennedy RH, Athanasiou T, Faiz O, Darzi A. A systematic review of enhanced recovery care after colorectal surgery in elderly patients. Color Dis. 2014;16(12):947–56.
19. Rumstadt B, Guenther N, Wendling P, et al. Multimodal perioperative rehabilitation for colonic surgery in the elderly. World J Surg. 2009;33(8):1757–63.

20. Wang Z, Tan KY, Tan P. Functional outcomes in elderly adults who have undergone major colorectal surgery. J Am Geriatr Soc. 2013;61(12):2249–50.
21. Gillis C, Li C, Lee L, et al. Prehabilitation versus rehabilitation: a randomized control trial in patients undergoing colorectal resection for cancer. Anaesthesiology. 2014;121(5):937–47.
22. Corman ML, Veidenheimer MC, Coller JA. Colorectal carcinoma: a decade of experience at the Lahey Clinic. Dis Colon Rectum. 1979;22(7):477–9.
23. Stewart RJ, Stewart AW, Turnbull PRG, Isbister WH. Sex differences in subsite incidence of large bowel cancer. Dis Colon Rectum. 1983;26(10):658.
24. Barone M, Prioletta M, Di Nuzzo D, Cipollone G, Camplese P, Mucilli F. Pulmonary metastasectomy in elderly colorectal cancer patients: a retrospective single center study. Updat Surg. 2016;68(4):357–67.
25. Hamaker ME, Schiphorst AH, Verweij NM, Pronk A. Improved survival for older patients undergoing surgery for colorectal cancer between 2008 and 2011. Int J Color Dis. 2014;29(10):1231–6.
26. Fugang W, Zhaopeng Y, Meng Z, Maomin S. Long-term outcomes of laparoscopy vs. open surgery for colorectal cancer in elderly patients: a meta-analysis. Mol Clin Oncol. 2017;7(5):771–6.
27. Volkmann R. Sammlung Klinischer Vortrage im Verbindung mit deutschen Klinikern. Chirurgie. 1874;29:1113–27.
28. Bueß G, Theiss R, Hutterer F, et al. Transanal endoscopic surgery of the rectum - testing a new method in animal experiments. Leber Magen Darm. 1983;13:73–7.
29. Atallah S, Albert M, Larach S. Transanal minimally invasive surgery: a giant leap forward. Surg Endosc. 2010;24(9):2200–5.
30. Popa D, Connolly TM, Barbon C, Bergamaschi R. Design defects can close a study. Color Dis. 2015;17(12):1121–2.
31. Hompes R, Rauh SM, Ris F, et al. Robotic transanal minimally invasive surgery for local excision of rectal neoplasms. Br J Surg. 2014;101(5):578–81.
32. Mason AY. Surgical access to the rectum – a transsphincteric exposure. Proc R Soc Med. 1970;63:91.
33. Bergamaschi R, Arnaud JP. Segmental resection of the rectum via a posterior approach in frail patients. Coloproctology. 1996;18(4):165–8.
34. Strauss AA, Strauss SF, Crawford RA, et al. Surgical diathermy of carcinoma of rectum: its clinical end results. JAMA. 1935;104:1480.
35. Fritsch A, Seidl W, Walzel C, et al. Palliative and adjunctive measures in rectal cancer. World J Surg. 1982;6(5):569–77.
36. Gordon-Watson C. Treatment of cancer of the rectum with radium by open operation. Proc R Soc Med. 1927;21(2):309–19.
37. Bergamaschi R, Manunta A, Arnaud JP. Endoscopic transanal resection for palliation of acutely obstructed rectal cancer in frail elderly patients. Tech Coloproctol. 1999;3:15–7.
38. Debras B, Bergamaschi R, Becouarn G, Arnaud JP. Absorption of glycine irrigating solution during endoscopic trans-anal resection of rectal tumors. Dis Colon Rectum. 1996;39(11):1245–8.
39. Kozak LJ, DeFrances CJ, Hall MJ. National hospital discharge survey: 2004 annual summary with detailed diagnosis and procedure data. Vital Health Stat 13. 2006;(165):1–209.
40. Sanders RS, Everhart JE, Donowitz M, et al. The burden of selected digestive diseases in the United States. Gastroenterology. 2002;122(5):1500–11.
41. Warner E, Crighton EJ, Moineddin R, et al. Fourteen-year study of hospital admissions for diverticular disease in Ontario. Can J Gastroenterol. 2007;21(2):97–9.
42. Etzioni DA, Mack TM, Beart RW, et al. Diverticulitis in the United States: 1998-2005: changing patterns of disease and treatment. Ann Surg. 2009;249(2):210–7.
43. Weizman AV, Nguyen GC. Diverticular disease: epidemiology and management. Can J Gastroenterol. 2011;25(7):385–9.
44. Bharucha AE, Parthasarathy G, Ditah I, et al. Temporal trends in the incidence and natural history of diverticulitis: a population-based study. Am J Gastroenterol. 2015;110(11):1589–96.
45. Sessa B, Galluzzo M, Ianniello S, Pinto A, Trinci M, Miele V. Acute perforated diverticulitis: assessment with multidetector computed tomography. Semin Ultrasound CT MR. 2016;37(1):37–48.
46. Hugues ES, Cuthbertson AM, Carden AB. The surgical management of acute diverticulitis. Med J Aust. 1963;50(1):780–2.
47. Hinchey EJ, Schaal PG, Richards GK. Treatment of perforated diverticular disease of the colon. Adv Surg. 1978;12:85–109.
48. Klarenbeek BR, de Korte N, van der Peet DL, Cuesta MA. Review of current classifications for diverticular disease and a translation into clinical practice. Int J Color Dis. 2012;27(2):207–14.
49. Vennix S, Morton DG, Hahnloser D, Lange JF, Bemelman WA. Research Committee of the European Society of Coloproctology. Systematic review of evidence and consensus on diverticulitis: an analysis of national and international guidelines. Color Dis. 2014;16(11):866–78.
50. Ambrosetti P. Acute left-sided colonic diverticulitis: clinical expressions, therapeutic insights, and role of computed tomography. Clin Exp Gastroenterol. 2016;9:249–57.
51. Mora López L, Flores Clotet R, Serra Aracil X, Montes Ortega N, Navarro SS. The use of the modified Neff classification in the management of acute diverticulitis. Rev Esp Enferm Dig. 2017;109(5):328–34.
52. Sartelli M, Moore FA, Ansaloni L, et al. A proposal for a CT driven classification of left colon acute diverticulitis. World J Emerg Surg. 2015;10:3.
53. Tursi A, Brandimarte G, Di Mario F, et al. Development and validation of an endoscopic

classification of diverticular disease of the colon: the DICA classification. Dig Dis. 2015;33(1):68–76.
54. Schoetz D. Diverticular disease of the colon. A century-old problem. Dis Colon Rectum. 1999;42(6):703–9.
55. Stollman N, Raskin JB. Diverticular disease of the colon. Lancet. 2004;363(9409):631–9.
56. Lamb MN, Kaiser AM. Elective resection versus observation after nonoperative management of complicated diverticulitis with abscess: a systematic review and meta-analysis. Dis Colon Rectum. 2014;57(12):1430–40.
57. Sallinen VJ, Mentula PJ, Leppäniemi AK. Nonoperative management of perforated diverticulitis with extraluminal air is safe and effective in selected patients. Dis Colon Rectum. 2014;57(7):875–81.
58. You K, Bendl R, Taut C, Sullivan R, Gachabayov M, Bergamaschi R. Study Group on Diverticulitis. Randomized clinical trial of elective resection versus observation in diverticulitis with extraluminal air or abscess initially managed conservatively. Br J Surg. 2018;105(8):971–9.
59. Feingold D, Steele SR, Lee S, et al. Practice parameters for the treatment of sigmoid diverticulitis. Dis Colon Rectum. 2014;57(3):284–94.
60. Constantinides VA, Tekkis PP, Senapati A. Association of Coloproctology of Great Britain Ireland. Prospective multicentre evaluation of adverse outcomes following treatment for complicated diverticular disease. Br J Surg. 2006;93(12):1503–13.
61. Janes SE, Meagher A, Frizelle FA. Management of diverticulitis. BMJ. 2006;332(7536):271–5.
62. Krukowski ZH, Matheson NA. Emergency surgery for diverticular disease complicated by generalized and faecal peritonitis: a review. Br J Surg. 1984;71(12):921–7.
63. Kronborg O. Treatment of perforated sigmoid diverticulitis: a prospective randomized trial. Br J Surg. 1993;80(4):505–7.
64. Zeitoun G, Laurent A, Rouffet F, et al. Multicentre randomized clinical trial of primary versus secondary sigmoid resection in generalized peritonitis complicating sigmoid diverticulitis. Br J Surg. 2000;87(10):1366–74.
65. Colcock BP. Surgical management of complicated diverticulitis. Surg Clin North Am. 1968;48(3):543–51.
66. Bahadrusingh AM, Longo WE. Colovaginal fistulas. Etiology and management. J Reprod Med. 2003;48(7):489–95.
67. Noecker CB. Perforation of sigmoid and small bowel into the uterus: secondary to diverticulitis of the sigmoid. Penn Med. 1929;32:496.
68. Small AJ, Young-Fadok TM, Baron TH. Expandable metal stent placement for benign colorectal obstruction. Surg Endosc. 2008;22(2):454–62.
69. Gachabayov M, Javadov M, Bergamaschi R. Blind colostomy. The case against. Surg Technol Int. 2019;34:137–8.
70. Wilks S. The morbid appearances in the intestines of Miss Bankes. Med Times Gaz. 1859;2:264.
71. Goligher JC, de Dombal FT, Watts JMCK, et al. Ulcerative colitis. London: Bailliére, Tindall and Cassell; 1968:2.
72. Nguyen GC, Bernstein CN, Benchimol EI. Risk of surgery and mortality in elderly-onset inflammatory bowel disease: a population-based cohort study. Inflamm Bowel Dis. 2017;23(2):218–23.
73. Welch CE, Hedberg SE. Colonic cancer in ulcerative colitis and idiopathic colonic cancer. JAMA. 1965;191:815.
74. Sachar DB, Auslander MO, Walfish JS. Aetiological theories of inflammatory bowel disease. Clin Gastroenterol. 1980;9:231.
75. Colombo F, Sahami S, de Buck Van Overstraeten A, et al. Restorative Proctocolectomy in elderly IBD patients: a multicentre comparative study on safety and efficacy. J Crohns Colitis. 2017;11(6):671–9.
76. Stryker SJ, Pemberton JH, Zinsmeister AR. Long-term results of ileostomy in older patients. Dis Colon Rectum. 1985;28(11):844–6.
77. Kim D, Taleban S. A comprehensive review of the diagnosis and pharmacological management of Crohn's disease in the elderly population. Drugs Aging. 2019; https://doi.org/10.1007/s40266-019-00672-x. [Epub ahead of print].
78. Olén O, Askling J, Sachs MC, et al. Mortality in adult-onset and elderly-onset IBD: a nationwide register-based cohort study 1964–2014. Gut. 2019. pii: gutjnl-2018-317572. https://doi.org/10.1136/gutjnl-2018-317572. [Epub ahead of print].
79. Keyashian K, Dehghan M, Sceats L, Kin C, Limketkai BN, Park KT. Comparative incidence of inflammatory bowel disease in different age groups in the United States. Inflamm Bowel Dis. 2019. pii: izz092. https://doi.org/10.1093/ibd/izz092. [Epub ahead of print].
80. Hawker PC, Gyde SN, Thompson H, et al. Adenocarcinoma of the small instestine complicating Crohn's disease. Gut. 1982;23(3):188–93.
81. Lavery IC, Jagelman DG. Cancer in the excluded rectum following surgery for inflammatory bowel disease. Dis Colon Rectum. 1982;25(6):522–4.
82. Pudipeddi A, Kariyawasam V, Haifer C, Baraty B, Paramsothy S, Leong RW. Safety of drugs used for the treatment of Crohn's disease. Expert Opin Drug Saf. 2019;26:1–11. https://doi.org/10.1080/14740338.2019.1612874. [Epub ahead of print]
83. Madoff RD, Mellgren A. One hundred years of rectal prolapse surgery. Dis Colon Rectum. 1999;42(4):441–50.
84. Gore RN, Levine MS, Laufer I, editors. Textbook of gastrointestinal radiology. Philadelphia: WB Saunders; 1994.
85. Jorge J, Wexner SD. Etiology and management of fecal incontinence. Dis Colon Rectum. 1993;36(1):77–97.
86. Bannister J, Abouzekry L, Read NW. Effect of aging on anorectal function. Gut. 1987;28(3):353–7.

87. Barrett JA, Brocklehurst JC, Kiff ES, Ferguson G, Faragher EB. Anal function in geriatric patients with faecal incontinence. Gut. 1989;30(9):1244–51.
88. Laurberg S, Swash M. Effects of aging on the anorectal sphincters and their innervation. Dis Colon Rectum. 1989;32(9):737–42.
89. Engel AF, Kamm MA, Sultan AH, Bartram CI, Nicholls RJ. Anterior anal sphincter repair in patients with obstetric trauma. Br J Surg. 1994;81(8): 1231–4.
90. Malouf AJ, Norton CS, Engel AF, Nicholls RJ, Kamm MA. Long-term results of overlapping anterior anal-sphincter repair for obstetric trauma. Lancet. 2000;355(9200):260–5.
91. Ogilvie H. Large-intestine colic due to sympathetic deprivation; a new clinical syndrome. Br Med J. 1948;2(4579):671–3.
92. Choung RS, Branda ME, Chitkara D, et al. Longitudinal direct medical costs associated with constipation in women. Aliment Pharmacol Ther. 2011;33(2):251–60.
93. Jones R, Lydeard S. Irritable bowel syndrome in the general population. BMJ. 1992;304(6819):87–90.
94. Sandler RS, Jordan MC, Shelton BJ. Demographic and dietary determinants of constipation in the US population. Am J Public Health. 1990;80(2):185–9.
95. Redmond JM, Smith GW, Barofsky I, Ratych RE, Goldsborough DC, Schuster MM. Physiological tests to predict long-term outcome of total abdominal colectomy for intractable constipation. Am J Gastroenterol. 1995;90(5):748–53.
96. Whitehead WE, Bharucha AE. Diagnosis and treatment of pelvic floor disorders: what's new and what to do. Gastroenterology. 2010;138(4):1231–5, 1235. e1–4.
97. Longstreth GF, Thompson WG, Chey WD, Houghton LA, Mearin F, Spiller RC. Functional bowel disorders. Gastroenterology. 2006;130(5):1480–91.
98. Agachan F, Chen T, Pfeifer J, Reissman P, Wexner SD. A constipation scoring system to simplify evaluation and management of constipated patients. Dis Colon Rectum. 1996;39(6):681–5.
99. Brown WJ, Mishra G, Lee C, Bauman A. Leisure time physical activity in Australian women: relationship with well being and symptoms. Res Q Exerc Sport. 2000;71(3):206–16.
100. Donald IP, Smith RG, Cruikshank JG, Elton RA, Stoddart ME. A study of constipation in the elderly living at home. Gerontology. 1985;31(2):112–8.
101. Kinnunen O. Study of constipation in a geriatric hospital, day hospital, old people's home and at home. Aging (Milano). 1991;3(2):161–70.
102. Bingham SA, Cummings JH. Effect of exercise and physical fitness on large intestinal function. Gastroenterology. 1989;97(6):1389–99.
103. Meshkinpour H, Selod S, Movahedi H, Nami N, James N, Wilson A. Effects of regular exercise in management of chronic idiopathic constipation. Dig Dis Sci. 1998;43(11):2379–83.
104. Karam SE, Nies DM. Student/staff collaboration: a pilot bowel management program. J Gerontol Nurs. 1994;20(3):32–40.
105. Smith B. Effect of irritant purgatives on the myenteric plexus in man and the mouse. Gut. 1968;9(2):139–43.
106. Riemann JF, Schmidt H. Ultrastructural changes in the gut autonomic nervous system following laxative abuse and in other conditions. Scand J Gastroenterol Suppl. 1982;71:111–24.
107. He CL, Burgart L, Wang L, et al. Decreased interstitial cell of cajal volume in patients with slow-transit constipation. Gastroenterology. 2000;118(1):14–21.
108. Wedel T, Roblick UJ, Ott V, et al. Oligoneuronal hypoganglionosis in patients with idiopathic slow-transit constipation. Dis Colon Rectum. 2002;45(1):54–62.
109. Riecken EO, Zeitz M, Emde C, et al. The effect of an anthraquinone laxative on colonic nerve tissue: a controlled trial in constipated women. Z Gastroenterol. 1990;28(12):660–4.
110. Leng-Peschlow E. Sennoside-induced secretion and its relevance for the laxative effect. Pharmacology. 1993;47(Suppl 1):14–21.
111. Ghadially FN, Walley VM. Melanoses of the gastrointestinal tract. Histopathology. 1994;25(3):197–207.
112. Campbell WL. Cathartic colon. Reversibility of roentgen changes. Dis Colon Rectum. 1983;26(7):445–8.
113. Baker JT, Harvey RF. Bowel habit in thyrotoxicosis and hypothyroidism. Br Med J. 1971;1(5744):322–3.
114. Kuijpers HC, Bleijenberg G. The spastic pelvic floor syndrome. A cause of constipation. Dis Colon Rectum. 1985;28(9):669–72.
115. Zutshi M, Hull TL, Trzcinski R, Arvelakis A, Xu M. Surgery for slow transit constipation: are we helping patients? Int J Color Dis. 2007;22(3):265–9.
116. Pikarsky AJ, Singh JJ, Weiss EG, Nogueras JJ, Wexner SD. Long-term follow-up of patients undergoing colectomy for colonic inertia. Dis Colon Rectum. 2001;44(2):179–83.
117. Stylianos S, Forde KA, Benvenisty AI, et al. Lower gastrointestinal hemorrhage in renal transplant recipients. Arch Surg. 1988;123(6):739–44.
118. Baig MK, Lewis M, Stebbing JF, et al. Multiple microaneurysms of the superior hemorrhoidal artery: unusual recurrent massive rectal bleeding: report of a case. Dis Colon Rectum. 2003;46(7):978–80.
119. Bradbury AW, Milne AA, Murie JA. Surgical aspects of Behçet's disease. Br J Surg. 1994;81(12):1712–21.
120. Cello JP, Wilcox CM. Evaluation and treatment of gastrointestinal tract hemorrhage in patients with AIDS. Gastroenterol Clin N Am. 1988;17(3):639–48.
121. Leitman IM, Paull DE, Shires GT 3rd. Evaluation and management of massive lower gastrointestinal hemorrhage. Ann Surg. 1989;209(2):175–80.
122. Richter JM, Christensen MR, Kaplan LM, et al. Effectiveness of current technology in the diagnosis and management of lower gastrointestinal hemorrhage. Gastrointest Endosc. 1995;41(2):93–8.

123. Alfidi RJ, Esselstyn CD, Tarar R, et al. Recognition and angio-surgical detection of arteriovenous malformations of the bowel. Ann Surg. 1971;174(4):573–82.
124. Baum S, Athanasoulis CA, Waltman AC. Angiographic diagnosis and control of large-bowel bleeding. Dis Colon Rectum. 1974;17(4):447–53.
125. Baum S, Athanasoulis CA, Waltman AC, et al. Angiodysplasia of the right colon: a cause of gastrointestinal bleeding. AJR Am J Roentgenol. 1977;129(5):789–94.
126. Boley SJ, Sammartano R, Adams A, et al. On the nature and etiology of vascular ectasias of the colon. Gastroenterology. 1977;72(4 Pt 1):650–60.
127. Iredale JP, Ridings P, McGinn FP, et al. Familial and idiopathic colonic varices: an unusual cause of lower gastrointestinal haemorrhage. Gut. 1992;33(9):1285–8.
128. Foutch PG, Sirak MV. Colonic variceal haemorrhage after endoscopic infection sclerosis of oesophageal varices: a report of three cases. Am J Gastroenterol. 1984;79(10):756–60.
129. Handschin AE, Weber M, Weishaupt D, et al. Contrast-enhanced three-dimensional magnetic resonance angiography for visualization of ectopic varices. Dis Colon Rectum. 2002;45(11):1541–4.
130. Boley SJ, Brandt LJ. Vascular ectasias of the colon—1986. Dig Dis Sci. 1986;31:26S.
131. Bender JS, Wiencek RG, Bouwman DL. Morbidity and mortality following total abdominal colectomy for massive lower gastrointestinal bleeding. Am Surg. 1991;57(8):536–40; discussion 540-1.
132. Parkes BM, Obeid FN, Sorensen VJ, et al. The management of massive lower gastrointestinal bleeding. Am Surg. 1993;59(10):676–8.
133. Elder K, Lashner BA, Al Solaiman F. Clinical approach to colonic ischemia. Cleve Clin J Med. 2009;76(7):410–09.
134. Parks TG. Ischaemic disease of the colon. Coloproctology. 1980;4:213.
135. Rubinstein E, Ibsen T, Rasmussen RB, et al. Formalin treatment of radiation-induced hemorrhagic proctitis. Am J Gastroenterol. 1986;81(1):44–5.
136. Boley SJ, Brandt LJ, Veith FJ. Ischemic disorders of the intestine. Curr Probl Surg. 1978;15(4):1–85.
137. Dirkx CA, Gerscovich EO. Sonographic findings in methamphetamine-induced ischemic colitis. J Clin Ultrasound. 1998;26(9):479–82.
138. Travis S, Davies DR, Creamer B. Acute colorectal ischaemia after anaphylactoid shock. Gut. 1991;32(4):444–6.
139. Jablonski M, Putzki H, Heymann H. Necrosis of the ascending colon in chronic hemodialysis patients: report of three cases. Dis Colon Rectum. 1987;30(8):623–5.
140. Rosenkrantz H, Bookstein JJ, Rosen RJ, et al. Postembolic colonic infarction. Radiology. 1982;142(1):47–51.
141. Paterno F, McGillicuddy EA, Schuster KM, Longo WE. Ischemic colitis: risk factors for eventual surgery. Am J Surg. 2010;200(5):646–50.
142. Schroeder T, Christoffersen JK, Andersen J, et al. Ischemic colitis complicating reconstruction of the abdominal aorta. Surg Gynecol Obstet. 1985;160(4):299–303.
143. Luna-Pérez P, Rodríguez-Ramírez SE. Formalin instillation for refractory radiation-induced hemorrhagic proctitis. J Surg Oncol. 2002;80(1):41–4.
144. Sahakitrungruang C, Thum-Umnuaysuk S, Patiwonqpaisarn A, et al. A novel treatment for haemorrhagic proctitis using colonic irrigation and oral antibiotic administration. Color Dis. 2011;13(5):e79–82.
145. Vernia P, Fracasso PL, Casale V, et al. Topical butyrate for acute radiation proctitis: randomised, crossover trial. Lancet. 2000;356(9237):1232–5.
146. Tjandra JJ, Sengupta S. Argon plasma coagulation is an effective treatment for refractory hemorrhagic radiation proctitis. Dis Colon Rectum. 2001;44(12):1759–65; discussion 1771.
147. Taïeb S, Rolachon A, Cenni J-C. Effective use of argon plasma coagulation in the treatment of severe radiation proctitis. Dis Colon Rectum. 2001;44(12):1766–71.
148. Buchi KN, Dixon JA. Argon laser treatment of hemorrhagic radiation proctitis. Gastrointest Endosc. 1987;33(1):27–30.
149. Bem J, Bem S, Singh A. Use of hyperbaric oxygen chamber in the management of radiation-related complications of the anorectal region: report of two cases and review of the literature. Dis Colon Rectum. 2000;43(10):1435–8.
150. Grossmann EM, Longo WE, Stratton MD, et al. Sigmoid volvulus in Department of Veterans Affairs medical centers. Dis Colon Rectum. 2000;43(3):414–8.
151. Daniels IR, Lamparelli MJ, Chave H, et al. Recurrent sigmoid volvulus treated by percutaneous endoscopic colostomy. Br J Surg. 2000;87(10):1419.
152. Lung BE, Yelika SB, Murthy AS, Gachabayov M, Denoya P. Cecal bascule: a systematic review of the literature. Tech Coloproctol. 2018;22(2):75–80.

Liver Transplant Surgery in the Elderly

22

Seigo Nishida

Liver transplant (LT) becomes a real practice for end-stage liver disease and liver failure patients. The results of liver transplantation have improved gradually over the years. Patient selection, anesthesia, liver transplant surgery, postoperative management, intensive care, and immunosuppression have significant improvement. The number of liver transplant performed in the United States steadily increased. UNOS data showed that the number of liver transplants in the United States was 1713 cases in 1988, 4519 cases in 1998, 6319 cases in 2008, and 8250 cases in 2018. General population becomes older. Liver transplant recipients in elderly have increased over the years. Percentage of the liver transplant over 65 years is also steadily increasing over the years (from 1.7% in 1988 to 21.8% in 2018) [1]. The older patients tend to have more comorbidities and higher surgical risk [2]. The chronological age is an important factor but not the best to assess the patient [2, 3]. Physiological age is more important [2, 3]. We reviewed the national data, publications, and our experience. We also discussed the donor, pretransplant evaluation and management, waiting list, liver transplant surgery, post liver transplant, immunosuppression, and future.

S. Nishida (✉)
Department of Surgery, Transplant Section, Westchester Medical Center and New York Medical College, Valhalla, NY, USA
e-mail: Seigo.Nishida@wmchealth.org

R. Latifi (ed.), *Surgical Decision Making in Geriatrics*,
https://doi.org/10.1007/978-3-030-47963-3_22

Increased Numbers of Older Recipients

United Network of Organ Sharing (UNOS) data showed that the number of liver transplant has increased steadily, and the age of the liver transplant recipients has increased [1]. UNOS data showed that the number of liver transplants in the United States was 1713 cases in 1988, 4519 cases in 1998, 6319 cases in 2008, and 8250 cases in 2018. Age group data showed that age group between 50 and 64 years old and age group 65 or more increased steadily [1]. Especially age group 65 or more has strongly increased (Figs. 22.1 and 22.2). UNOS data showed that the number of liver transplants of age group 65 or more in the United States was 29 cases in 1988, 322 cases in 1998, 621 cases in 2008, and 1795 cases in 2018 (Fig. 22.3). Percentage of the age group 65 or more also showed steady increase of this group (Fig. 22.4). UNOS data showed that the percentage of liver transplants of the age group 65 or more in the United States was 1.7% in 1988, 7.1% in 1998, 9.8% in 2008, and 21.8% in 2018 (Fig. 22.4). The waiting list of the age group 65 or more is also increasing [1–3]. Recent report showed about 23.8% of the liver transplant waiting list in 2017 is this group [1, 4].

The improvement of surgical technique and perioperative care gradually increased the indication of liver transplant for older liver transplant patients [5]. The aging population with hepatitis

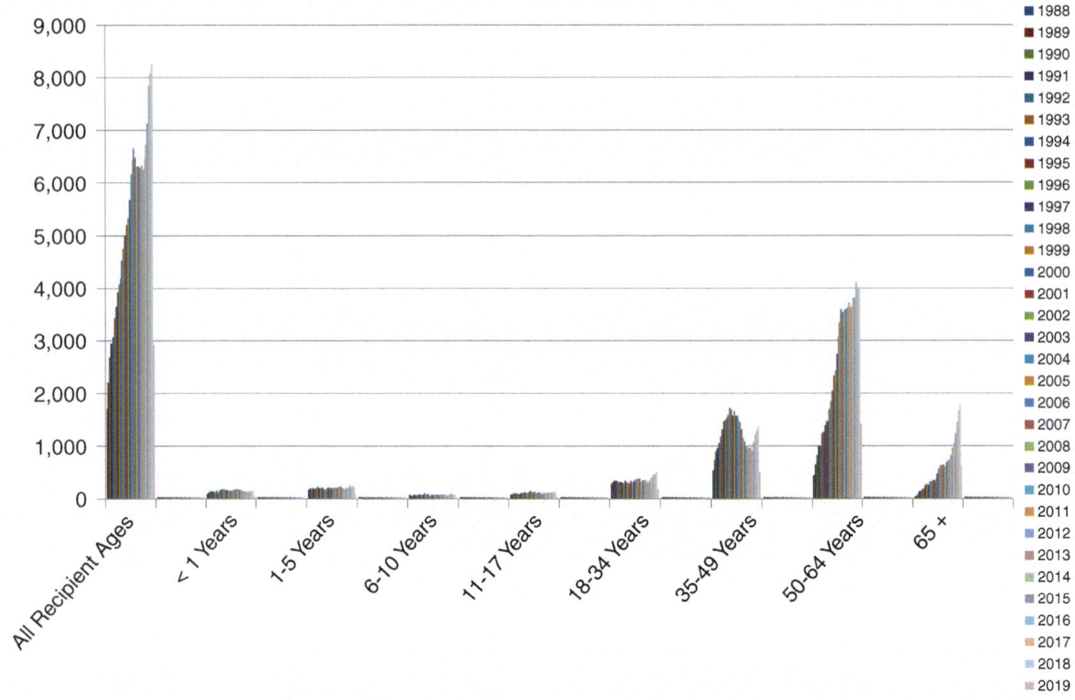

Fig. 22.1 Liver transplant between 1988 and 2019

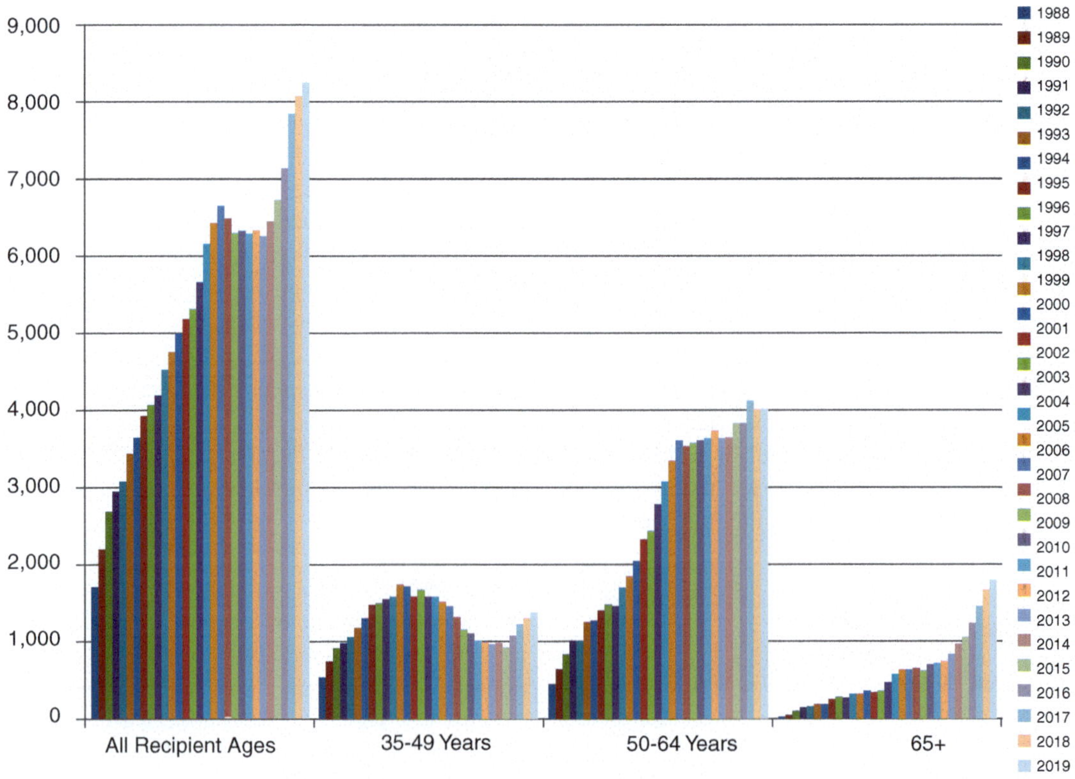

Fig. 22.2 Liver transplant age groups between 1988 and 2018

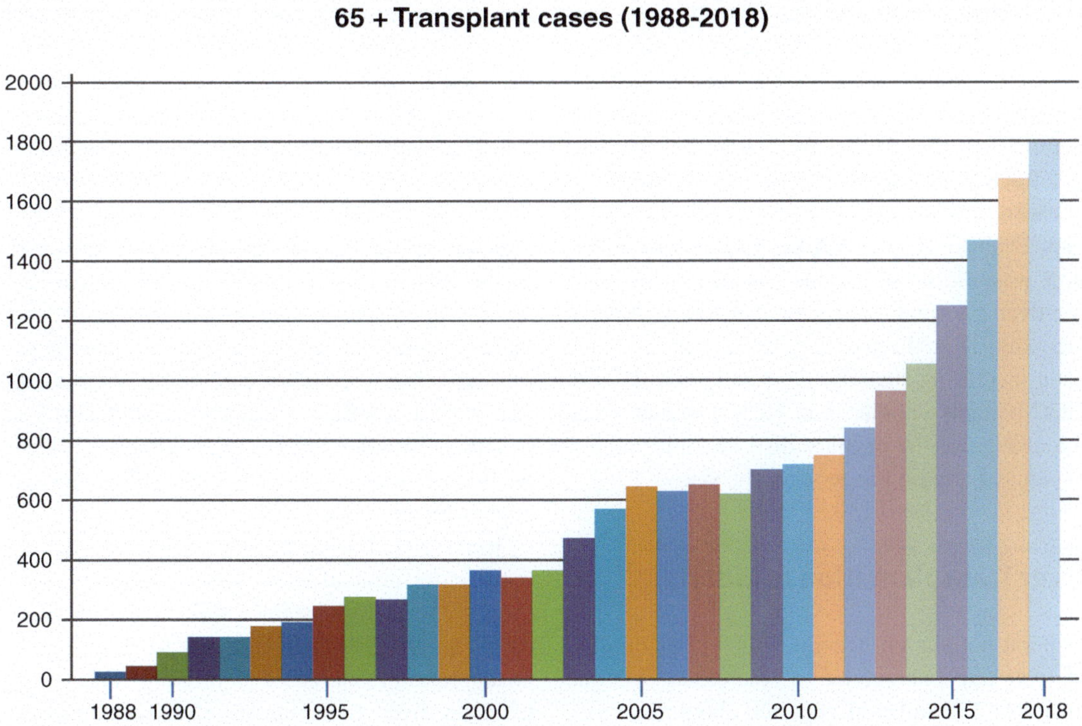

Fig. 22.3 Liver transplant 65+ between 1988 and 2018

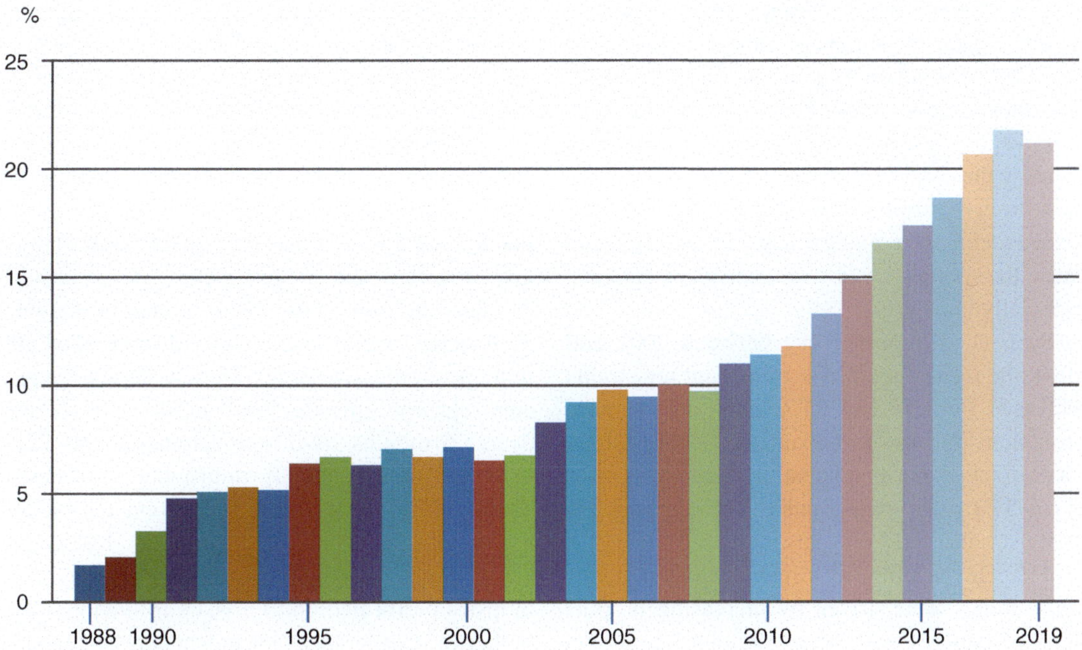

Fig. 22.4 65+ recipient percentages between 1988 and 2019

C, nonalcoholic steatohepatitis, and hepatocellular carcinoma may be the factor to increase the older patients [1–5]. The improvement of hepatitis C treatment has also change the patient populations [5]. Since the recent introduction of effective direct-acting antivirals (DAAs) for hepatitis C virus infection, recipient population may change. Before DAAs, HCV virus infection was the most common primary diagnosis for liver transplant recipient. Currently, NASH, alcoholic liver disease, and HCC patients have been increasing steadily. Especially NASH with HCC patients is older than HCV patients. Even the change of the primary disease, older recipients may increase further over the years.

Liver Transplantation in Elderly

General population becomes older. The number of liver transplantations has increased over the years [1]. Durand et al. reported about these topics in 2019 [2]. In general, older liver transplant patients have less chance of having rejection [2]. Especially patients with nonimmune conditions such as NASH and alcoholic cirrhosis do not require higher maintenance immunosuppression [2]. These patients may get benefits from lower immunosuppression. Dual or triple immunosuppressive therapy may not be needed and may increase the risk of infection and malignancy. We also observe similar outcome and agreed with this concept. More malignancy and infection can cause the mortalities in liver transplant patients, especially in elderly.

Su F et al. reported that between 2002 and 2014, the mean age of liver transplant registrants increased from 51.2 to 55.7 years, with a more prominent increase in hepatitis C virus-positive (50.9–57.9 years) than hepatitis C virus-negative (51.3–54.3 years) registrants [3]. The proportion of registrants aged ≥60 years increased from 19% to 41%. In hepatitis C virus-negative patients, aging trends were driven by increasing proportions of patients with hepatocellular carcinoma or nonalcoholic steatohepatitis. Among transplant registrants, increasing age was associated with increasing mortality before transplantation and decreasing likelihood of transplantation. Among transplant recipients, increasing age was associated with increasing posttransplantation mortality. There was little difference in 5-year transplant-related survival benefit between different age groups who had the same Model for End-Stage Liver Disease score. They concluded that dramatic aging of liver transplant registrants and recipients occurred from 2002 to 2014, driven by aging of the hepatitis C virus-positive cohort and increased prevalence of nonalcoholic steatohepatitis and hepatocellular carcinoma. Increasing age does not affect transplant-related survival benefit substantially because age diminishes both posttransplantation survival and waitlist survival approximately equally.

Benitez et al. from Spain reported prospective multicenter clinical trial of immunosuppressive drug withdrawal in stable adult liver transplant recipients [6]. They reported that lifelong immunosuppression increases morbidity and mortality in liver transplantation. Discontinuation of immunosuppressive drugs could lessen this burden, but the safety, applicability, and clinical outcomes of this strategy need to be carefully defined. They enrolled 102 stable liver recipients at least 3 years after transplantation in a single-arm multicenter immunosuppression withdrawal trial. Drugs were gradually discontinued over a 6–9-month period. The primary endpoint was the development of operational tolerance, defined as successful immunosuppressive drug cessation maintained for at least 12 months with stable graft function and no histopathologic evidence of rejection. Out of the 98 recipients evaluated, 57 rejected and 41 successfully discontinued all immunosuppressive drugs. In nontolerant recipients rejection episodes were mild and resolved over 5.6 months (two nontolerant patients still exhibited mild gradually improving cholestasis at the end of follow-up). In tolerant recipients no progressive clinically significant histological damage was apparent in follow-up protocol biopsies performed up to 3 years following drug withdrawal. Tolerance was independently associated with time since transplantation (odds ratio [OR] 1.353; P = 0.0001), recipient age (OR 1.073; P = 0.009), and male gender (OR 4.657;

$P = 0.016$). A predictive model incorporating the first two clinical variables identified subgroups of recipients with very high (79%), intermediate (30–38%), and very low (0%) likelihood of successful withdrawal. They concluded that when conducted at late time points after transplantation, immunosuppression withdrawal is successful in a high proportion of carefully selected liver recipients. A combination of clinical parameters could be useful to predict the success of this strategy. Additional prospective studies are now needed to confirm these results and to validate clinically applicable diagnostic biomarkers [6].

Author analyzed the graft survival and impact of specific risk factors on long-term graft survival outcome in primary liver transplant patients at single institute and reported at world transplant congress 2006 [7]. Purpose: Specific survival impact based on pretransplant disease etiology, donor factors, recipient factors, and intraoperative factors have not been addressed clearly on the long-term survival results. To evaluate the impact of these specific risk factors on long-term graft survival outcomes in liver transplant (LT), this retrospective study analyzed these factors on survival results with a long-term follow-up. Methods: Between June 1994 and December 2001, 1067 cadaveric adult whole LT were performed. Of 1067, 966 patients underwent primary LT. Median follow-up is 2199 days. Of 966, impact of donor factors (age, cytomegalovirus infection (CMV), presence of replaced hepatic artery), recipient factors (presence of HCV infection and tumor, CMV, age, sex, body weight, blood type, UNOS status, and length of the hospital stay), and surgical factors (cold and warm ischemia times, presence of portal vein thrombosis(PVT), intraoperative transfusion requirements of blood(PRBC), fresh frozen plasma(FFP) and platelet, duration of surgical procedure, type of arterial reconstruction, type of caval reconstruction, and type of biliary reconstruction) on graft survival were determined using univariate analysis. Cox proportional hazards modeling was further used to investigate the independent effects of multiple covariates on graft survival. Results: The overall 5 year patient and graft survival was 70% and 66%, respectively. Factors that significantly predict poor graft survival by univariate analysis were older donor age, recipient HCV infection, presence of tumor, presence of PVT, recipient CMV positive, amounts of FFP and PRBC, longer cold ischemia time, and longer hospital stay (Table 22.1). Factors that significantly predict poor graft survival by multivariate analysis were older donor age, recipient HCV infection, presence of tumor and PVT, and longer hospital stay (Table 22.2). Conclusion: Disease-related causes (HCV infection and presence of tumor), donor-related causes (older donor), and patient-related causes (PVT, longer hospital stay) represent the negative impact on the long-term graft survival. Specific surgical factors (caval, arterial, and biliary reconstruction) did not have the long-term survival impact. Donor age remained as a prognostic factor. If recipient were selected carefully, recipient age was not the independent prognostic factor for graft survival after liver transplantation.

Rana A et al. reported that no gains in long-term survival after liver transplantation over the past three decades [8]. The short-term survivals after orthotopic liver transplantation have improved due the improvement of surgical technique, introduction of calcineurin inhibitor immunosuppression, and improvement of patient management. Identified risk factors for patient survival are African American recipient, cold ischemia time, donor age, ETOH cirrhosis, hepatitis C, hospital admission, ICU admission, male gender, and recipient age. Those are similar findings with other reports [2–6]. Common cause of death in long-term survivors was malignancy, graft failure, and infection. Malignancy has been rising. Advanced recipient age had the most detrimental impact on long-term outcomes. Younger donors had the most beneficial impact. Long-term results of immunosuppression are the most common causes of death, especially malignancy and infection. Thus, most important thing is better long-term immunosuppression management. Those similar findings are also reported by others [2–5]. Adjustment of immunosuppression may be one of the important factors in long-term survival patients after liver transplantation.

Table 22.1 Univariate analysis for prognostic factor for graft survival after liver transplant

	Beta	Standard Error	*t*-value	Hazard Ratio	Wald Statist	*P*
Donor factors						
Age (yr)	0.008953	0.003014	2.970310	1.008993	8.822743	0.002977
CMV positive (yes/no)	−0.044282	0.114426	−0.386992	0.956684	0.149763	0.698765
Replaced hepatic artery (yes/no)	0.210779	0.119958	1.757107	1.234639	3.087424	0.078909
Recipient factors						
HCV infection (yes/no)	0.343779	0.104194	3.299423	1.410266	10.88619	0.000970
Tumor presence (yes/no)	0.537692	0.149788	3.589685	1.712051	12.88584	0.000332
CMV positive (yes/no)	0.302337	0.130044	2.324885	1.353017	5.405090	0.020084
Age(yr)	0.003242	0.004575	0.708623	1.003247	0.502146	0.478564
Gender (male/female)	−0.018986	0.107966	−0.175851	0.981193	0.030924	0.860412
Body weight (kg)	−0.002300	0.001606	−1.43210	0.997703	2.050908	0.152125
Blood type	−0.013324	0.018537	−0.718775	0.986765	0.516638	0.472285
UNOS status (1/2/3)	−0.000530	0.001091	−0.485657	0.999470	0.235862	0.627214
Length of hospital stay (day)	0.036814	0.005226	7.043732	1.037500	49.61415	0.000000
Operatlve factors						
Cold ischemic time (min)	0.000706	0.000348	2.029631	1.000706	4.119403	0.042402
Warm ischemic time (min)	0.004910	0.004694	1.046121	1.004922	1.094368	0.295513
Portal vein thrombosis (yes/no)	0.556528	0.145200	3.832831	1.744606	14.69059	0.000127
Amount of PRBC (unit)	0.016566	0.003550	4.666391	1.016704	21.77521	0.000003
Amount of **FFP** (unit)	0.008560	0.003227	2.652865	1.008597	7.037690	0.007985
Amount of pit (unit)	0.002731	0.003051	0.895279	1.002735	0.801525	0.370645
Operative time (min)	0.000261	0.000309	0.844134	1.000261	0.712562	0.398601
Arterial graft (yes/no)	−0.068743	0.066269	−1.03733	0.933567	1.076058	0.299589
Caval reconstruction (conventional piggyback)	0.338802	0.140215	2.416303	1.403265	5.838521	0.015685
Biliary reconstruction (R-oux-y/duct to duct)	0.093460	0.107387	0.870307	1.097966	0.757435	0.384139

Donor: Brain Death Donor, Donation After Cardiac Death Donor, and Live Donor

Donor age is a single independent prognostic factor after liver transplantation [9–15]. Recipient age is also a single independent prognostic factor after liver transplantation [3, 16]. When older donors are used, younger recipient do well. However, the transplant program tends to use older donor for older recipient. Gilbo et al. reported those issues. Older donors and recipients are increasingly considered for liver transplantation. Both donor and recipient age have a negative impact on outcomes. Large registry analyses show that older donors are frequently matched to older recipients. Whether age-related risks accumulate in a synergic negative effect on outcomes because of donor-recipient age matching is poorly understood. Gilbo et al. investigated the impact of donor-recipient age interaction on patient and death-censored graft survival in multivariate Cox regressions in 849 transplants (January 2000 to December 2015). Donors 70 years or older did not affect long-term patient or graft survival. Recipient age independently increased the risk of

Table 22.2 Multivariate analysis for prognostic factor for graft survival after liver transplant

	Beta	Standard Error	*t*-value	Hazard Ratio	Wald Statist.	*p*
Donor factors						
Donor age (yr)	0.007690	0.003081	2.496030	1.00772	6.23017	0.012564
Recipient factors						
PVT Presence (yes/no)	0.409647	0.151543	2.703170	1.506286	7.30713	0.006872
Presence of tumor	0.355031	0.169247	2.097713	1.426225	4.40040	0.035938
HCV infection (yes/no)	0.289715	0.106551	2.719013	1.336047	7.39303	0.006551
Recipient CMV positive	0.203998	0.131044	1.556708	1.226296	2.42334	0.119550
Hospital stay (day)	0.005049	0.001235	4.089408	1.005062	16.72326	0.000043
ICU stay (day)	0.022243	0.006309	3.525545	1.022492	12.42947	0.000423
Operative factors						
Cava (conventional)	0.111214	0.173019	0.642787	1.117635	0.41317	0.520367
Venous bypass(yes/no)	−0.090292	0.131215	−0.688125	0.913664	0.47352	0.491379
PRBC (unit)	0.008289	0.004407	1.881021	1.008324	3.53824	0.059978
FFP (unit)	−0.000653	0.003861	−0.169107	0.999347	0.02860	0.865714
Cold ischemic time(min)	0.000496	0.000361	1.372022	1.000496	1.88244	0.170066

death (hazard ratio [HR], 1.03; 95% confidence interval [CI], 1.02–1.05, $P < 0.0001$), but donor-recipient age interaction was non-influential. The negative impact of recipient age on patient survival was significant as early as 6 months after transplantation (HR, 1.06; 95% CI, 1.03–1.09; $P = 0.00008$). The adjusted risk of death was significant for patients aged 60 to 69 years (HR, 1.995; 95% CI, 1.40–2.85; $P < 0.0001$) and 70 years or older (HR, 2.001; 95% CI, 1.10–2.66; $P = 0.04$). In contrast, the risk of graft loss was not influenced by recipient age (HR, 1.02; 95% CI, 0.996–1.04; $P = 0.11$) or age interaction. They concluded that older livers can be safely used in older recipients without jeopardizing graft and patient survival if other risk factors are minimized [16]. Donation after cardiac death (DCD) donor has increased risk of primary nonfunctioning of the liver and higher incidence of biliary complications. Croome et al. reported that use of liver graft from donor 50 years or older had acceptable graft and patient survival [17]. Use of DCD graft has higher risk and older recipient can't tolerate those risks. Cautious selection of the older recipient for DCD organs is necessary. Similarly, use of live donor for low MELD patients may be the option with caution.

Pretransplant Evaluation of the Recipient and Selection of the Liver Transplant Candidate

In general, older patients have more comorbidities and cardiac risks [2–5]. Older patients were carefully selected to liver transplant candidacy because of the poor outcome after the liver transplant. But, there are recent reports of liver transplant in elderly patients with better outcomes [5].

Physical impairment, frailty, and old age in itself are associated with higher risk of poor outcome after liver transplantation. Careful Considerations in the selection of older candidates for liver transplantation may contribute the better outcome after liver transplantation. Durand et al. reported that with the average age of listing for liver transplantation increasing significantly in the last decade, 25% of patients receiving liver transplantation in the United States are over 65 years of age [2]. Durand et al. reported that above the standard evaluation, the selection of older candidates should pay particular attention to cardiovascular diseases, functional status, and the assessment of malignancy risk. Both ischemic (coronary artery disease) and nonischemic (cardiomyopathy, heart failure, arrhythmia, valvular heart disease) assessments become equally important during the evaluation in the older population. Age is a significant risk factor for all of these complications. Thus, at a minimum, initial transthoracic echocardiography, electrocardiography, noninvasive stress testing, and consultation with a cardiologist familiar with end-stage liver disease hemodynamics are important. If indicated, further cardiac evaluation using right side and left side cardiac catheterization are studied. Recent guideline emphasized a multidisciplinary approach to assess risks and prepare for peri- and postoperative cardiac complications. Cardiopulmonary exercise testing and other functional measures such as 6-minute walk distance also add prognostic data for candidates and correlate with the presence of frailty. Recent data have provided evidence that frailty predicts waitlist mortality independent of MELD score, and those older candidates are more likely to be frail with less physiological reserve. Although there are no universal standard criteria for frailty, and multiple tests are available, simple to perform performance-based tests of muscle function may have more clinical applications than imaging-based measures of sarcopenia. A recent frailty assessment called the Liver Frailty Index demonstrated that less than half of patients become "robust" after transplantation, supporting the need for pre- and posttransplant rehabilitation [2]. Durand et al. reported that extra hepatic malignancy directly increases with age, and most centers are reluctant to transplant patients with recent diagnoses of malignancy other than nonmelanoma skin or early, treated, less aggressive localized cancers, for fear of relapse following transplantation [2]. All candidates should undergo age and risk factor appropriate cancer screening prior to listing, such as colonoscopy, mammography, prostate screening, etc. Ongoing tobacco consumption should be strongly discouraged. Liver transplant candidates with a prior extra hepatic malignancy should have received definitive treatment with adequate tumor-free survival before liver transplant listing. Most programs would consider adequate tumor-free survival to be at least 1–5 years, depending again on the particular malignancy. As each patient often presents a unique clinical scenario, consultation with oncologists having specific experience in estimating the general risk of relapse, as well as additional risk with immunosuppressive therapy, is warranted particularly in the older population. Older recipients are at higher risk of developing de novo malignancy. However, the excess in cancer incidence compared to the general population is higher in the youngest recipients than in older recipients [2].

Waiting List and Accepting the Organ Offers

Increasing mortality while waiting in older patients and increasing risk after liver transplant surgery in older recipients are reality. Risk and benefit of the liver transplant in older recipient is difficult issues. Feng S et al. reported that Age and Waitlist Outcomes [3]. They analyzed the UNOS data.

Among 122,606 adults listed for transplantation between 2002 and 2014, 65,725 (53.6%) underwent transplantation, 21,009 (17.1%) died on the waiting list before undergoing transplantation, 11,124 (9.1%) were removed from the waitlist due to being too sick for transplant, 4823 (3.9%) were removed due to improved condition, 8080 (6.6%) were removed due to "other" factors, and 11,845 (9.7%) were still alive but had not been transplanted by June 30, 2014. Of the patients removed from the waiting list, 7185 (29.9%) died after removal by June 30, 2014. Waitlist outcomes by age group are analyzed. It showed that the proportion who dropped out due to being too sick to transplant increased from 6.3% in age group 18–49 years to 16.3% in age group ≥70 years. Conversely, the proportion who underwent transplantation progressively decreased with increasing age from 52.5% in age group 18–49 years to 44.1% in age group ≥70 years. Among registrants with HCC, the proportion of who died increased from 12.7% in age group 18–49 to 16.8% in age group ≥70 years, and the proportion who dropped out due to being too sick to transplant increased from 6.7% to 12.5%. The proportion who underwent transplantation progressively decreased with increasing age from 74% in age group 18–49 years to 61% in age group ≥70 years. Similar trends across age groups were observed in subgroups defined by MELD score ≤ 16 or > 16 at listing. In competing risk analyses, increasing age was associated with increased mortality before transplantation and decreased likelihood of transplantation. The associations between increasing age and increasing waitlist mortality were not substantially different in subgroups defined by MELD score and functional status. In other subgroup analyses, the association between increasing age and increasing waitlist mortality was substantially stronger in the HCV-negative (compared with HCV-positive), non-HCC (compared with HCC), and long-wait regions (compared to short-wait regions). Increasing age group was associated with decreasing MELD score at death, dropout, or transplantation, especially among patients without HCC, suggesting that older patients tolerate high MELD scores more poorly than younger patients. Causes of death on the waiting list were similar among all age groups.

Liver Transplant Surgery

Liver transplant surgery techniques have improved over the three decade [8]. It is still challenging for the high-risk patients. Selected older patients are safely transplanted; however, many attentions are necessary for the good outcome. Older patients have much comorbidity, especially cardiovascular issues. Pretransplant work-up for ischemic heart disease is mandatory [2–5]. High pulmonary artery pressure such as mean pulmonary artery pressure higher than 38 mmHg is risk factors for liver transplant. Anticoagulation, atrial fibrillation, and ischemic heart disease are also very common. Renal function is very important. In older patients, there are many issues including renal dysfunction, hepatorenal syndrome, indication of liver and kidney transplant, use of nephrotoxic medications, and immunosuppression. Neurological issues are also very common. They include hepatic encephalopathy, confusion, delirium, stroke, intracranial bleeding, central pontine myelinolysis, demyelination syndrome, posterior reversible encephalopathy syndrome, and neurotoxicity due to immunosuppression. Minimize the intraoperative bleeding, avoiding intraoperative complication, and good hemodynamic management are important to avoid morbidity and mortality. Shorter cold ischemia time, good quality of liver graft, and uneventful surgical course are ideal.

Posttransplant Management and Outcome

Watt et al. from Mayo clinic reported that although mortality rates following liver transplantation (LT) are well described, there is a lack of detailed, prospective studies determining patterns of, and risk factors for long-term mortality [18]. They analyzed the multicenter, prospectively obtained NIDDK Liver Transplantation Database of 798 transplant recipients from 1990 to 1994 (follow-up 2003). Overall, 327 recipients died. Causes of death >1 year: 28% hepatic, 22% malignancy, 11% cardiovascular, 9% infection, and 6% renal failure. Renal-related death increased dramatically over time. Risk factors for death >1 year (univariate): male gender, age/decade, pre-LT diabetes, post-LT diabetes, post-LT hypertension, post-LT renal insufficiency, retransplantation >1 year, pre-LT malignancy, alcoholic (ALD) disease, and metabolic liver disease, with similar risks noted for death >5 years. Hepatitis C, retransplantation, post-LT diabetes, hypertension, and renal insufficiency were significant risk factors for liver-related death. Cardiac deaths associated with age, male gender, ALD, cryptogenic disease, pre-LT hypertension, and post-LT renal insufficiency. Watt et al. summarized that the leading causes of late deaths after transplant were graft failure, malignancy, cardiovascular disease, and renal failure. Older age, diabetes, and renal insufficiency identified patients at highest risk of poor survival overall. Diligent management of modifiable post-LT factors including diabetes, hypertension, and renal insufficiency may impact long-term mortality.

Lower Immunosuppression for Nonimmune-Related Liver Disease Patients

Risk of developing chronic renal disease is higher in older liver transplant recipients [19]. Risk of developing de novo cancer is 10–15 times higher than in the general population [19]. Zhou et al. reported that using SRTR data (Oct 1987–Dec 2009), among 89,036 liver transplant recipients, 6834 recipients developed 9717 posttransplant malignancies. They focused on non-skin malignancies. A total of 3845 recipients suffered from 4854 de novo non-skin malignancies, including 1098 de novo hematological malignancies, 38 donor-related cases, and 3718 de novo solid-organ malignancies. Liver transplant recipients had more than 11 times elevated cancer risk compared with the general population.

Older Patients Have More Risks of Cardiac Complication, Infection, and Cancer Development

Recent report from Haugen et al. showed median length of stay decreased from 10 days in 2003–2006 to 9 days in 2013–2016. A stay of longer than 2 weeks for older liver transplant recipient decreased from 30.8% in 2003–2006 to 28.0% in 2013–2016. Graft survival in older liver transplant recipients also improved over time. One-year survival improved from 80% in 2003–2006 to 90% in 2013–2016, 3-year survival form 71% to 84%, and 5-year survival form 63% to 70% [4]. Based on the national study of 8627 older liver transplant recipients from 2003 to 2016, Haugen et al. showed the increase in number of liver transplant performed, improvements in length of stay, acute rejection, graft survival, and overall survival [4]. Ferman TJ et al. reported that liver transplant recipients older than 60 years old show executive and memory function improvement comparable to younger recipients [20].

Feng et al. reported about Age and Posttransplantation Outcomes [3]. They found that older patients had a significantly higher mortality after transplantation compared with younger patients both before and after adjusting for potential confounders. Compared with recipients aged 18–49 years, the adjusted hazard ratio increased to 1.16 for ages 50–59 years, 1.34 for ages 60–64 years, 1.61 for ages 65–69 years, and 1.87 for ages ≥70 years. Survival probability at 5 years posttransplant progressively declined from 0.78 in the 18–49-year age group to 0.62 in the ≥70 years age group. Other studies also showed 5-year survival is generally 10 to 20% lower in patients aged 60 or older to 70 years than in younger recipients [2, 8]. Watt et al. reported advanced age is significantly associated with the risk of death due to cardiovascular events and malignancy in liver transplant patients [18].

It is reported that older recipients have increased bone loss due to age and increased incidence of bone fractures after liver transplantation [2]. Durand et al. recommended that all patients should be advised to perform weight-bearing exercise, take vitamin D, and calcium supplements, and have bone density examination at least every 1–3 years, depending on other risk factors (steroid therapy, prior bone loss, family history). Treatment with antiresorptive therapy should be considered in all patients with osteoporosis, recent fracture, or declining bone density [2]. Older age in itself should not limit access to liver transplant in older patients.

Immunosuppression in the Elderly

In general, older patients do not require higher maintenance immunosuppression. Older patients may not tolerate well to modern immunosuppression due to side effects including neurological side effects, infection, and malignancy.

Recently, Rana et al. reported that there were no significant gains in unadjusted long-term outcomes among 1-year survivors over the past 30 years [8]. Cause of death analysis suggests malignancy after transplantation is a growing problem and preventing recurrent hepatitis C with direct-acting antivirals (DDAs) may only have a limited impact. Furthermore, rejection leading to graft failure and death had a rare occurrence (1.7% of long-term deaths) especially when compared with the sequelae of long-term immunosuppression: malignancy (16.4%), nonrejection graft failure (9.8%), and infection (10.5%) ($P < 0.001$). They concluded that there have been no appreciable improvements in long-term survival following liver transplantation among 1-year survivors. Long-term sequelae of immunosuppression, including malignancy and infection, are the most common causes of death. This study highlights the need for better long-term immunosuppression management.

Future

Careful selections of the donor and recipient have made improvement of outcome in early phase of liver transplant [2–5]. However, long-term outcome is not improved in the last three decade as reported by Rana et al. [8]. They reported that malignancy and infection are the most common causes of death [8]. Watt et al. reported that common causes of death were graft failure, malignancy, cardiovascular disease, and renal failure [20]. Better long-term immunosuppression management and better management of diabetes, hypertension, and renal insufficiency may improve the long-term outcome after liver transplantation.

References

1. Https://Optn.Transplant.Hrsa.Gov/Data/View-Data-Reports/National-Data/.
2. Durand F, Levitsky J, Cauchy F, Gilgenkrantz H, Soubrane O, Francoz C. Age and liver transplantation. J Hepatol. 2019;70(4):745–58. https://doi.org/10.1016/J.Jhep.2018.12.009. Epub 2018 Dec 18. Review.
3. Su F, Yu L, Berry K, Liou IW, Landis CS, Rayhill SC, Reyes JD, Ioannou GN. Aging of liver transplant registrants and recipients: trends and impact on wait-list outcomes, post-transplantation outcomes, and transplant-related survival benefit. Gastroenterology. 2016;150(2):441–53. https://doi.org/10.1053/J.Gastro.2015.10.043.E6; Quiz E16. Epub 2015 Oct 30.
4. CE H, CM H, Garonzik-Wang J, Pozo M, Warsame F, Mcadams-Demarco M, DL S. National trends in liver transplantation in older adults. J Am Geriatr Soc. 2018;66(12):2321–6. https://doi.org/10.1111/Jgs.15583. Epub 2018 Oct 16.
5. AI S, JN I, JL F, FJ D. Kidney and liver transplantation in the elderly. Br J Surg. 2016;103(2):E62–72. https://doi.org/10.1002/Bjs.10064. Epub 2015 Dec 14. Review.
6. Benítez C, Londoño MC, Miquel R, Manzia TM, Abraldes JG, Lozano JJ, Martínez-Llordella M, López M, Angelico R, Bohne F, Sese P, Daoud F, Larcier P, Roelen DL, Claas F, Whitehouse G, Lerut J, Pirenne J, Rimola A, Tisone G, Sánchez-Fueyo A. Prospective multicenter clinical trial of immunosuppressive drug withdrawal in stable adult liver transplant recipients. Hepatology. 2013;58(5):1824–35. https://doi.org/10.1002/Hep.26426. Epub 2013 Sep 19.
7. Nishida S, Levi DM, Moon JI, Selvaggi G, Kato T, Madariaga JR, et al. The impact of specific risk factors on long-term survival outcomes in adult liver transplant recipients. Wtc. Boston: Poster Presentation; 2006.
8. Rana A, Ackah RL, Webb GJ, Halazun KJ, Vierling JM, Liu H, Wu MF, Yoeli D, Kueht M, Mindikoglu AL, Sussman NL, Galván NT, Cotton RT, O'mahony CA, Goss IA. No gains in long-term survival after liver transplantation over the past three decades. Ann Surg. 2019;269(1):20–7. https://doi.org/10.1097/Sla.0000000000002650.
9. Halldorson JB, Bakthavatsalam R, Fix O, Reyes JD, Perkinsd-Meld JD. A simple predictor of post liver transplant mortality for optimization of donor/recipient matching. Am J Transplant. 2009;9:318–26.
10. Hoofnagle JH, Lombardero M, Zetterman RK, Lake J, Porayko M, Everhart J, et al. Donor age and outcome of liver transplantation. Hepatology. 1996;24:89–96.
11. Rana A, Hardy MA, Halazun KJ, Woodland DC, Ratner LE, Samstein B, et al. Survival outcomes following liver transplantation (soft) score: a novel method to predict patient survival following liver transplantation. Am J Transplant. 2008;8:2537–46.
12. Dutkowski P, Oberkofler CE, Slankamenac K, Puhan MA, Schadde E, Mullhaupt B, et al. Are there better guidelines for allocation in liver transplantation? A novel score targeting justice and utility in the model for end-stage liver disease era. Ann Surg. 2011;254:745–53, Discussion 753.
13. Blok JJ, Braat AE, Adam R, Burroughs AK, Putter H, Kooreman NG, et al. Validation of the donor risk index in orthotopic liver transplantation within the eurotransplant region. Liver Transpl. 2012;18:112–9.
14. Braat AE, Blok JJ, Putter H, Adam R, Burroughs AK, Rahmel AO, et al. The eurotransplant donor risk index in liver transplantation: Et-Dri. Am J Transplant. 2012;12:2789–96.
15. Feng S, Goodrich NP, Bragg-Gresham JL, Dykstra DM, Punch JD, Debroy MA, et al. Characteristics associated with liver graft failure: the concept of a donor risk index. Am J Transplant. 2006;6:783–90.
16. Gilbo N, Jochmans I, Sainz-Barriga M, Nevens F, Van Der Merwe S, Laleman W, Verslype C, Cassiman D, Verbeke L, Van Malenstein H, Roskams T, Pirenne J, Monbaliu D. Age Matching of elderly liver graft with elderly recipients does not have a synergistic effect on

long-term outcomes when both are carefully selected. Transplant Direct. 2019;5(4):E342. https://doi.org/10.1097/Txd.0000000000000883. Ecollection 2019 Apr.

17. Croome KP, Mathur AK, Lee DD, Moss AA, Rosen CB, Heimbach JK, Taner CB. Outcomes of donation after circulatory death liver grafts from donors 50 years or older: a multicenter analysis. Transplantation. 2018 Jul;102(7):1108–14.
18. KD W, RA P, WK K, JK H, MR C. Evolution of causes and risk factors for mortality post-liver transplant: results of the niddk long-term follow-up study. Am J Transplant. 2010;10(6):1420–7. https://doi.org/10.1111/J.1600-6143.2010.03126.X. Epub 2010 May 10.
19. Jie Zhou, Zhenhua HU, Qijun Zhang, Zhiwei LI, Jie Xiang, Sheng Yan, Jian Wu, Min Zhang, And Shusen Zheng. Spectrum of De Novo cancers and predictors in liver transplantation: analysis of the scientific registry of transplant recipients database. Plos One. 2016; 11(5): E0155179. Published Online 2016 May 12. https://doi.org/10.1371/Journal.Pone.0155179.
20. Ferman TJ, Keaveny AP, Schneekloth T, Heckman MG, Vargas E, Vasquez A, Rummans T, Taner CB, Niazi SK. Liver transplant recipients older than 60 years show executive and memory function improvement comparable to younger recipients. Psychosomatics. 2019;Pii: S0033-3182(19):30023–4. https://doi.org/10.1016/J.Psym.2019.01.008.

Renal Transplantation in the Elderly

23

Thomas Diflo

History

For thousands of years, there have been reports of transplantation of organs and tissues. The patron saints of physicians, the twin brother physicians Cosmas and Damian, transplanted the lower leg of a recently deceased Moor onto the gangrenous stump of a Roman church officer, with apparent good graft function [1], one of the miracles used to support their canonization.

In the 1930s, YY Voronoy, in the Soviet Union, used a temporary kidney transplant onto the brachial vessels of a patient who had gone into renal failure from mercury poisoning, in order to reverse the patient's anuria [2]. On December 23, 1953, Dr. Joseph Murray and his team at the Peter Bent Brigham Hospital began the modern era of transplantation when they performed the first successful living donor kidney transplant between two identical twins, the 23-year-old Herrick brothers [3].

Dr. Murray received the Nobel Prize in Medicine and Physiology in 1990 for this groundbreaking effort. For several years thereafter, given the small number of identical twin pairs, one of whom required transplantation, there were not many transplants performed. Once the immune response was better understood and appropriate medications to alter that response were discovered or developed, transplantation became a more routine therapy.

Throughout the 1960s and 1970s, the only immunosuppressive medications available were steroids and 6-mercaptopurine or its synthetic analog, azathioprine. Jean Borel, then a scientist at Sandoz, presented his studies of cyclosporine A (CyA) at the British Society of Immunology [4, 5], and Sir Roy Calne, at the University of Cambridge, began his experimental work with CyA in humans [6]. The introduction of CyA for transplantation was earthshaking, and 1-year patient and graft survival after kidney transplantation virtually doubled overnight.

Since then, there have been numerous other medications introduced which have improved graft function and survival, including tacrolimus, mycophenolic acid, sirolimus, and several monoclonal antibodies that work against specific portions of the immune response. Each of these introductions has allowed incremental improvements in patient and graft survival, but none have been as seminal as the introduction of cyclosporine.

In the 1960s and 1970s, transplantation was essentially unregulated. The lawful donation of tissues or organs was codified by the 1968 Uniform Anatomical Gift Act (UAGA) [7]. At the same time, a group convened at Harvard University to define brain death [8]. The 1980 Uniform Determination of Death Act defined

T. Diflo (✉)
Department of Surgery, New York Medical College, Westchester Medical Center, Valhalla, NY, USA
e-mail: Thomas.Diflo@wmchealth.org

R. Latifi (ed.), *Surgical Decision Making in Geriatrics*,
https://doi.org/10.1007/978-3-030-47963-3_23

death: "[A]n individual who has sustained either (1) irreversible cessation of circulatory or respiratory functions, or (2) irreversible cessation of all functions of the brain, including the brain stem, is dead" [7]. This allowed the procurement of viable organs from heart-beating cadavers.

In 1984 Senator Al Gore of Tennessee sponsored Public Law 98–507, the National Organ Transplant Act (NOTA) [9], which was passed into law. This Act established the Organ Procurement and Transplantation Network (OPTN), which is run by the Secretary of Health and Human Services. The law also established the Scientific Registry of Transplant Recipients (SRTR), as the central federal agency for data collection and analysis for all of the transplant programs in the country. In addition, prohibited was "the exchange of organs for transplantation for valuable consideration" [9] where "valuable consideration" is generally interpreted to mean money or property. The OPTN conducts the day-to-day management of transplantation, and, since its inception, the only OPTN has been the United Network for Organ Sharing (UNOS).

In subsequent years, there have been numerous additional laws, amendments, and changes to the rules that govern transplantation. From a practical point of view, all current regulatory policies are maintained by UNOS. The Unites States has been split up into 11 regions for the purpose of organ allocation and sharing. At the moment, except for some special circumstances, the kidneys are allocated first locally, then regionally, and then nationally.

The federal government, particularly its insurance endeavors, is represented by the Centers for Medicare and Medicaid Services (CMS). CMS's oversight of transplantation became more important after Public Law 92–603 amended the Social Security Act to provide Medicare coverage for most end-stage renal disease patients [10]. Once the Medicare program became responsible for dialysis and renal transplantation services, CMS developed a keen interest in the finances of transplantation and the performance of individual transplant centers.

The Scope of the Issue: Renal Failure and Transplantation in the Elderly

For the purposes of this chapter, we will define "elderly" as age 65 or greater. There is no doubt that transplantation is the preferred therapy for end-stage renal disease (ESRD) for those patients who can tolerate it, with significant improvements in overall health, quality of life, and huge financial advantages for the medical system. Worldwide, ESRD is a growing problem, particularly in the older population. In the United States, 50% of new ESRD patients annually are over the age of 65, and 33% are older than 70 [11, 12], and in Scandinavia, the incidence of new ESRD in elderly patients increased from 42% in 1990 to 65% in 2013 [13]. In the United States in 1992, only 4.2% of all kidney transplants were done on elderly patients, but this proportion increased to 17.2% in 2012 [13].

Patient and Graft Survival in the Elderly

One unfortunate, although perhaps not unexpected, fact about this is that both patient and graft survivals are inferior in the elderly when compared to younger patients. There is a significantly higher rate of death with a functioning graft, and cognitive impairment leads to more medication regimen nonadherence and rejection [14]. A study of overall mortality and death from infectious complications showed a linear increase in mortality over 10-year age increments but an *exponential* increase in deaths related to infection, with a fivefold increase in elderly recipients versus those younger than 30 [15]. Recipients over the age of 80 had a 30-day post-transplant mortality almost twice as high as those in their 60s (2.5% vs. 1.5%), as well as inferior 2-year survival (73% vs. 89%) [16]. A British study showed 4-year mortalities of 5.8% in patients younger than 50, 22% for those in their 60s, 32% in the 70s, and 45% in octogenarians [17]. All of this bad news needs to be tempered with the fact that 50% of transplant candidates over the age of 60 will die while on the waiting list [18].

While the results of transplantation for these patients are inferior to those of younger patients, it is perhaps more relevant to compare the results for patients who remain on the list to those who are transplanted. Transplant recipients over the age of 70 have a 41% lower overall risk of death compared to those who remain on dialysis on the list [19]. A French study showed similar results for patients over the age of 60 [20]. A European Consensus Statement from 2016 [21] concluded that selected elderly ESRD patients benefit significantly from transplantation, and age alone should not be a contraindication to transplantation. They did recommend careful screening for cardiovascular disease and cancer, assessing frailty, doing cognitive testing, and encouraging living donation.

Another interesting observation regarding elderly patients is the effects of an aging immune response [22]. As patients age, they become progressively less immunocompetent, which has salutary implications for the amount and degree of immunosuppression that they require.

While there are epidemiologic arguments both for and against transplanting elderly patients, perhaps it is more useful to assess what factors influence transplant success, positively and negatively. A recent study from the University of Minnesota showed strong predictors of graft loss in the elderly which included pre-transplant sensitization (panel reactive antibody [PRA] >10%), a history of congestive heart failure, delayed graft function, and acute cellular rejection [23]. A study from Johns Hopkins looked at a multivariate analysis of numerous pre-transplant patient factors and developed a risk prediction model which they then validated prospectively [24]. Perhaps one of the most important factors that need to be assessed before transplant is frailty.

Frailty

The importance of frailty in medical outcomes has been recognized since Fried's landmark "Frailty in Older Adults: Evidence for a Phenotype" [25]. There have been numerous studies of the effects of frailty on surgical and medical recovery, outcomes, and mortality. Clearly, not all elderly patients are frail, and not all frail patients are elderly. There is, however, an increasing incidence of frailty in cohorts of older patients.

There are interesting reciprocal effects of frailty and transplantation – frailty affects transplant outcomes, and transplantation influences frailty. A study from the Brigham and Women's Hospital noted that at least 15% of patients over the age of 65 are frail and that frailty in kidney recipients led to a twofold increase in the rate of delayed graft function, a significant increase in length of stay, and a similar increase in early readmissions [26, 27]. These are all causes for concern, but the increase in length of stay is perhaps the most alarming. Another study from Johns Hopkins showed that, in frail kidney recipients, each week of increased length of stay resulted in a 1.55-fold increased mortality risk [28]. On a more positive note, the same group has also looked at changes in frailty after renal transplantation and found that, while frailty worsens across the board shortly after transplantation, it generally improves to better than baseline by 3 months [28].

Evaluation for Transplantation

A recent survey of US transplant programs revealed that most programs do not have a specific age cutoff above which they will not evaluate a patient, and those that do have an age limit generally set it at 80 [29]. A French study on decision-making in whether to refer elderly patients for transplant evaluation created a scoring system based on such variables as age, diabetes mellitus, vascular disease, chronic obstructive pulmonary disease, cognitive deficits, and mobility. This generated a score from 0 to 56 points. Patients were then divided into quintiles, with Group I representing ≤6 points, up to Group V, representing ≥18 points. They found that patients in Group I had a post-transplant 3-year mortality of 30% and Group V had a 3-year mortality of 83% [30]. Such objective evaluation is useful in the decision whether to list a patient or not.

The basic requirements for transplant evaluation do not vary based on age. Demographic information, blood type, human leukocyte antigen (HLA) typing, HIV status, and hepatitis serologies are all required, as well as standard laboratory tests such as a complete blood count, comprehensive metabolic panel, prothrombin time, and aPTT. In order to be placed on the active transplant list, a patient must have a GFR of less than 20 mL/min/1.73m^3. In addition, the medical evaluation will include an EKG and chest X-ray and a PSA for men. Other routine evaluations are a colonoscopy for anyone over the age of 50 and Pap smear and mammography for all women.

The requirements for cardiac testing vary from center to center, but certainly all elderly potential transplant recipients will require at least an echocardiogram and a pharmacological stress test. Unless the potential recipient is in excellent physical condition, it is unlikely that he or she would be able to undergo treadmill testing. An alternative modality would be a dobutamine stress echo. After noninvasive testing, all elderly patients should be evaluated by a cardiologist, preferably one who has experience in working with a transplant team.

Whether to pursue cardiac catheterization for evaluation will vary from patient to patient. Certainly anyone with symptoms of angina or shortness of breath with moderate effort should be referred for catheterization. A positive result on noninvasive cardiac evaluation would also lead to a recommendation for catheterization. The issue of pursuing catheterization for asymptomatic patients who have longstanding diabetes mellitus is one that continues to be a topic of debate between cardiologists and transplant physicians. It is our practice to recommend catheterization for all such patients.

An additional layer of scrutiny that is worthwhile in the elderly candidates is evaluation by an anesthesiologist who is familiar with transplantation. He or she can help with risk stratification and recommend perioperative management strategies such as the use of Swan-Ganz pulmonary artery catheters at the time of surgery.

It is also useful to perform a non-contrast computed tomography (CT) scan of the abdomen and pelvis in all elderly candidates. Any patient with longstanding renal failure has an increased risk of renal malignancies, and the CT can help screen for these tumors. Probably the most useful aspect of these studies is in evaluating the degree and extent of calcification in the iliac arteries, which will help in preoperative planning. Any patient with overt peripheral vascular disease should also be evaluated by a vascular surgeon and have, at the minimum, noninvasive pulse volume recordings and arterial Dopplers. It is also worthwhile to screen these patients with a duplex ultrasound of the carotid arteries to assess the extent of cerebrovascular disease.

The CT scan, in combination with noninvasive maneuvers such as grip strength, can be helpful in evaluating sarcopenia in transplant candidates [31]. This leads to the wider consideration of measuring frailty in elderly candidates. As noted previously, frailty has significant deleterious effects in transplant patients. It appears appropriate to incorporate formal frailty evaluations in all elderly candidates, either by using Fried's schema [25] or some other validated frailty screening tool. Those patients who are deemed moderately or severely frail are likely not appropriate candidates for transplantation.

Cognitive dysfunction also has important negative effects on the outcomes after transplantation [14]. A recent study showed that all-cause graft loss was increased more than fivefold in recipients with moderate or severe cognitive impairment [32]. Therefore, it is appropriate that all elderly candidates be evaluated with (at least) a modified Mini-Mental Status Exam and/or consultation with a dedicated transplant psychiatrist.

Technical Considerations and the Perioperative Period

As mentioned before, patients with renal failure with longstanding hypertension and especially diabetes have a significant risk of atherosclerotic disease. Review of the abdominal-pelvic CT scans with transplant surgeons and radiologists is crucial for operative planning regarding placement of the renal graft, and which vessels are appropriate for arterial anastomosis. It has been our observation that even patients with heavily calcified common iliac arteries frequently have uninvolved sections of their external iliac arteries which will allow for safe anastomosis.

The preoperative evaluation by the cardiology team and the anesthesiologists will also help to dictate what type of hemodynamic monitoring is appropriate in the perioperative period. Those patients with significant cardiac disease may benefit from Swan-Ganz catheterization for the surgery and for several days afterward. Depending on the facilities available for postoperative care at one's institution, it is advisable that immediate postoperative care for elderly transplant recipients be provided in an intensive care setting, rather than a postoperative unit or floor bed.

Immunosuppression

Elderly patients tend to be less immunocompetent than younger recipients [22]. Most transplant centers use some sort of induction therapy at the time of transplantation in addition to steroids. In all cases, the induction consists of either an IL-2 receptor blocker (basiliximab [Simulect©, Novartis]), a rabbit anti-thymocyte globulin (Thymoglobulin©, Sanofi Genzyme), or another lymphocyte-depleting agent (alemtuzumab [Campath©, which is FDA approved for the treatment of B-cell CLL, but available for transplantation through a registry]). Perhaps not surprisingly, induction with alemtuzumab in elderly recipients is associated with increased graft loss and death [33], which speaks against use of this agent. In low immunological risk recipients, either no induction agent or the use of basiliximab is reasonable. In high-risk recipients, anti-thymocyte globulin, perhaps in a low-dose protocol, appears appropriate.

Standard postoperative immunosuppression in the United States consists of tacrolimus, a mycophenolic acid or sodium agent, and steroids. Given the increased chances of death from infection [15], careful monitoring of drug levels and rapid decrease in dosing of medications are crucial. Interestingly, it was the addition of mycophenolate to the tacrolimus regimen that had the greatest additive effect on graft survival, and it appears that steroids are of less importance. This has led to some recommending rapid weaning and elimination of steroids in the elderly recipient [11]. Of course, the specter of rejection and graft loss, as well as potential medication nonadherence, necessitates very close follow-up of the elderly transplant patient.

Outcomes

Several years ago, UNOS mandated that the "best" deceased donor kidneys be allocated to the youngest recipients, using the concept of maximizing life-years after transplant, which has had the reciprocal effect of allocating older, "worse" donor kidneys (so-called expanded criteria donors or ECDs) to older recipients. Intuitively, this is a reasonable argument, although it initially caused some controversy in the transplant community and in society at large. Nevertheless, it is interesting to look at the overall renal transplant outcomes in the elderly population.

Despite some of the dire results presented in this chapter, many elderly transplant recipients do quite well. Compared to continuing on dialysis, transplant recipients over the age of 70 have a 59% reduction of death rates [19]. Even elderly recipients of ECD kidneys had a 40% reduction in death rates compared to dialysis [34]. However, the transplanting of ECD kidneys to elderly recipients has had mixed results. A Eurotransplant study of preferentially allocating kidneys from donors over the age of 65 to elderly recipients resulted in a doubling of the number of such kidneys available for transplantation and a decrease in waiting time for the recipients. In addition, there was a significant decrease in cold ischemic time and delayed graft function, with no adverse effect on graft or patient survival [35]. A US study of age-matching in elderly donors and recipients showed lower rates of rejection (50% of the rate in a younger cohort) and improved death-censored survival [22]. Another US study showed that the results of transplanting ECD kidneys into recipients over the age of 70 were no different than through the use of standard-criteria kidneys [36]. On the other hand, an older US study showed that allocation of the highest-risk kidneys to elderly recipients had the worst results of all cohorts [37].

Recently at our center, we have begun to use kidneys from hepatitis C-positive donors in hepatitis C-negative recipients, with postoperative antiviral treatment as needed. We started with elderly recipients and had excellent results, which has led us to expand this program to younger recipients {unpublished data}. It is certainly worthwhile to consider the more widespread use of these kidneys in elderly recipients.

In all patients, the recipient of a living donor kidney is far and away better off than the recipient of almost any deceased donor kidney. Elderly recipients tend to receive fewer living donor transplants than younger patients. This may be partially related to the relative unavailability of living donors for these patients. A study out of UCLA assessed the outcomes of elderly recipients who had older living donors, over the age of 50. They found that living donation from this older donor group showed outcomes superior *to any deceased donor cohort.* In addition, the only recipient cohort that did better was elderly recipients of young living donors [37]. We have successfully used donors up to the age of 68 at our center. Given the excellent outcomes for elderly recipients of older living donor kidneys, it is certainly appropriate to consider such donor-recipient pairs.

Conclusions

Just as the population at large is aging, so too is the elderly population with renal failure. 50% of new ESRD patients annually are over the age of 65, and 33% are older than 70. Renal transplantation remains the optimal treatment for ESRD.

There are a number of additional considerations in evaluating elderly patients for transplantation. Long histories of hypertension and diabetes necessitate close attention to the possibilities of peripheral vascular and cardiovascular disease. All potential elderly candidates will require evaluation by a cardiologist, and many will require cardiac catheterization. Since frailty is a significant concern in the elderly population, all elderly candidates will require a frailty assessment. Similarly, objective cognitive assessment, preferably by a transplant psychiatrist, should be part of every elderly candidate's evaluation.

The relative decrease in immune function and the increased risk of post-transplant infectious death in elderly recipients require careful immunosuppressive management. It is inadvisable to use alemtuzumab for induction during transplantation, but basiliximab or low-dose antithymocyte globulin appears safe and effective. Postoperatively, immunosuppression with tacrolimus and mycophenolate, plus or minus steroids, has the best results, with close follow-up.

Although some of the data are contradictory, it appears that the use of ECD kidneys in elderly recipients is safe and effective, particularly in recipients over the age of 70. Any kidney, whether a standard-criteria or ECD, confers significant survival advantage over remaining on dialysis. Hepatitis C-positive kidneys can be safely and successfully transplanted into hepatitis C-negative recipients. Finally, while transplantation is the gold standard for the treatment of ESRD, living donation is the gold standard of transplantation. Elderly recipients have the best results with young living donor kidneys, but results with older donors are also excellent.

As the ESRD population ages, more and more elderly patients will present for transplantation. With careful evaluation, management, and follow-up, many of these patients can be transplanted safely and successfully.

References

1. https://www.wdl.org/en/item/3251/ A Verger's Dream: Saints Cosmas and Damian performing a miraculous cure by transplantation of a leg.
2. Starzl TE. My thirty-five year view of organ transplantation. History of Transplantation: Thirty-Five Recollections. UCLA Tissue Typing Laboratory 1991:145–79.
3. https://www.google.com/imgres?imgurl=https%3A%2F%2Fs3.amazonaws.com%2Fs3.timetoast.com%2Fpublic%2Fuploads%2Fphotos%2F4

190230%2FHerrick_brothers_122910.jpg%3F1474700015&imgrefurl=https%3A%2F%2Fwww.timetoast.com%2Ftimelines%2Fhistorical-events-from-1950-to-1955&docid=aIc9oXrIU8R9KM&tbnid=uj5fPDl1Dk_TOM%3A&vet=10ahUKEwiymaG-iY3bAhWqtVkKHQpaAL8QMwg6KAMwAw..i&w=609&h=480&client=firefox-b-1-ab&bih=781&biw=1413&q=herrick%20brothers&ved=0ahUKEwiymaG-iY3bAhWqtVkKHQpaAL8QMwg6KAMwAw&iact=mrc&uact=8.
4. Borel J-F. Jean-Francoişe Borel and the origin of the world's first billion dollar molecule. Transplantation. 2015 Oct;99(10):2012–4.
5. Borel JF, Feurer C, Magnee C, et al. Effects of the new antilymphocyte peptide Cyclosporin A in animals. Immunology. 1977;32:1017.
6. Calne R. Recollections from the Laboratory to the Clinic. History of Transplantation: Thirty-Five Recollections. UCLA Tissue Typing Laboratory 1991:227–241.
7. https://organdonor.gov/about-dot/laws/history.html.
8. American Medical Association. A definition of irreversible coma. JAMA. 1984;252(5):677–9.
9. 1984 Pub. L 98–507.
10. 1972 Pub. L 92–603.
11. Knoll GA. Kidney transplantation in the older adult. Am J Kidney Dis. Elsevier Inc. 2013;61(5):790–7.
12. https://www.usrds.org/adr.aspx.
13. Sørensen SS. Rates of renal transplantations in the elderly—data from Europe and the US. Transplantation Rev. Elsevier Inc. 2015 Oct 1;29(4):193–6.
14. Ponticelli C, Podestà MA, Graziani G. Renal transplantation in elderly patients. How to select the candidates to the waiting list? Transplantation Rev. Elsevier Inc. 2014;28(4):188–92.
15. Meier-Kriesche H-U, Ojo AO, Hanson JA, Kaplan B. Exponentially increased risk of infectious death in older renal transplant recipients. Kidney Int. Elsevier Masson SAS. 2001;59(4):1539–43.
16. Huang E, Poommipanit N, Sampaio MS, Kuo H-T, Reddy P, Gritsch HA, et al. Intermediate-term outcomes associated with kidney transplantation in recipients 80 years and older: an analysis of the OPTN/UNOS database. Transplantation. 2010 Nov;90(9):974–9.
17. Karim A, Farrugia D, Cheshire J, Mahboob S, Begaj I, Ray D, et al. Recipient age and risk for mortality after kidney transplantation in England. Transplantation. 2014 Apr;97(8):832–8.
18. Schold JD, Arrigain S, Flechner SM, Augustine JJ, Sedor JR, Wee A, et al. Dramatic secular changes in prognosis for kidney transplant candidates in the United States. Am J Transplant. 2018 Aug 14;19(2):414–24.
19. Rao PS, Merion RM, ASHBY VB, Port FK, Wolfe RA, Kayler LK. Renal transplantation in elderly patients older than 70 years of age: results from the scientific registry of transplant recipients. Transplantation. 2007 Apr;83(8):1069–74.
20. Savoye E, Tamarelle D, Chalem Y, Rebibou J-M, Tuppin P. Survival benefits of kidney transplantation with expanded criteria deceased donors in patients aged 60 years and over. Transplantation. 2007 Dec;84(12):1618–24.
21. Segall L, Nistor I, Pascual J, Mucsi I, Guirado L, Higgins R, et al. Criteria for and appropriateness of renal transplantation in elderly patients with end-stage renal disease. Transplantation. 2016 Oct;100(10):e55–65.
22. Tullius SG, Tran H, Guleria I, Malek SK, Tilney NL, Milford E. The combination of donor and recipient age is critical in determining host immunoresponsiveness and renal transplant outcome. Trans Meet Am Surg Assoc. 2010;128:275–89.
23. Faravardeh A, Eickhoff M, Jackson S, Spong R, Kukla A, Issa N, et al. Predictors of graft failure and death in elderly kidney transplant recipients. Transplantation. 2013 Dec;96(12):1089–96.
24. Grams ME, Kucirka LM, Hanrahan CF, Montgomery RA, Massie AB, Segev DL. Candidacy for kidney transplantation of older adults. J Am Geriatr Soc. 2012 Jan 12;60(1):1–7.
25. Fried LP. Frailty in older adults: evidence for a phenotype. J Gerontol. 2001;56A(3):M146–56.
26. Exterkate L, Slegtenhorst BR, Kelm M, Seyda M, Schuitenmaker JM, Quante M, et al. Frailty and transplantation. Transplantation. 2016 Apr;100(4):727–33.
27. Yamamoto A, Zhang JL, Rusinek H, Chandarana H, Vivier PH, Babb JS, et al. Quantitative evaluation of acute renal transplant dysfunction with low-dose three-dimensional MR renography. Radiology. 2011;260(3):781–9.
28. McAdams-DeMarco MA, Isaacs K, Darko L, Salter ML, Gupta N, King EA, et al. Changes in frailty after kidney transplantation. J Am Geriatr Soc. 2015;63(10):2152–7.
29. Mandelbrot DA, Fleishman A, Rodrigue JR, Norman SP, Samaniego M. Practices in the evaluation of potential kidney transplant recipients who are elderly: a survey of U.S. transplant centers. Clin Transpl. 2017;31(10):e13088–7.
30. Dusseux E, Albano L, Fafin C, Hourmant M, Guérin O, Couchoud C, et al. A simple clinical tool to inform the decision-making process to refer elderly incident dialysis patients for kidney transplant evaluation. Kidney Int Elsevier Masson SAS. 2015;88(1):121–9.
31. Ozkayar N, Altun B, Halil M, Kuyumcu ME, Arik G, Yesil Y, et al. Evaluation of sarcopenia in renal transplant recipients. Nephro Urol Mon. Kowsar. 2014;6(4):1–5.
32. Thomas AG, Ruck JM, Shaffer AA, Haugen CE, ScM HY, Warsame F, et al. Kidney transplant outcomes in recipients with cognitive impairment. Transplantation. 2019;103(7):1504–13.
33. Hurst FP, Altieri M, Nee R, Agodoa LY, Abbott KC, Jindal RM. Poor outcomes in elderly kidney transplant recipients receiving alemtuzumab induction. Am J Nephrol Karger Publishers. 2011;34(6):534–41.

34. Jay CL, Washburn K, Dean PG, Helmick RA, Pugh JA, Stegall MD. Survival benefit in older patients associated with earlier transplant with high KDPI kidneys. Transplantation. 2017 Apr;101(4):867–72.
35. Frei U, Noeldeke J, Machold-Fabrizii V, Arbogast H, Margreiter R, Fricke L, et al. Prospective age-matching in elderly kidney transplant recipients—a 5-year analysis of the Eurotransplant senior program. Am J Transplant. 2007; 0(0):071105081616014–???
36. Molnar MZ, Streja E, Kovesdy CP, Shah A, Huang E, Bunnapradist S, et al. Age and the associations of living donor and expanded criteria donor kidneys with kidney transplant outcomes. Am J Kidney Dis Elsevier Inc. 2012;59(6):841–8.
37. Gill J, Bunnapradist S, Danovitch GM, Gjertson D, Gill JS, Cecka M. Outcomes of kidney transplantation from older living donors to older recipients. Am J Kidney Dis. 2008;52(3):541–52.

Hepatopancreaticobiliary Surgery in the Elderly

24

Clara Angeles, Danny Lascano, and Gregory Veillette

Hepatopancreaticobiliary (HPB) surgery is a formidable challenge even in the young, healthy patient. These are typically long, difficult cases with the potential for catastrophic complications. Furthermore, the recovery from these surgeries places a huge physiologic and metabolic strain on the patient, which is oftentimes more demanding than the actual procedure. Therefore, the decision-making process involved in proceeding with HPB surgery in the elderly patient is extensive and extremely difficult. Failure to carefully select which patients to offer surgery will lead to significant harm and potentially death of the patient. The surgeon must have a good understanding of the patient's actual life expectancy, natural history of the disease process, likelihood of complications, and whether or not the patient can get through the surgical procedure and its potential complications.

Our commitment as surgeons and physicians is twofold: first to try and prolong life and second to improve/maintain the quality of life. When dealing with the elderly population, the bang for the buck may not be as large as compared to the younger population, in terms of prolongation of life. Therefore it is absolutely essential for the surgeon to understand both the natural history of the disease process with surgery and the natural history of the disease process without surgery and to carefully weigh the risk/benefit ratio for each particular patient.

One of the most important principles in the management of the elderly is the understanding that these patients lack the physiologic reserve that is typically seen in the younger patient. Therefore, the decision to operate or to not operate must be carefully calculated to consider the possible sequelae of each approach and determine which will offer the lowest strain to the patient. For example, the physiologic strain of a laparoscopic or open cholecystectomy is much less than a recurrent bout of pancreatitis or cholangitis from recurring choledocholithiasis. A common response from the surgeon when faced with operating on the elderly patient is "they are not going to tolerate the operation" or "they are too old for surgery." Unfortunately, oftentimes this statement is completely false and is not based on any objective data. This type of expectant management may suffice for certain types of surgical pathology; however, when dealing with benign and malignant disease processes of the liver, biliary tree, and pancreas, the outcomes of watchful waiting and/or nonoperative management can be devastating.

C. Angeles · D. Lascano
Department of Surgery, Westchester Medical Center, Valhalla, NY, USA

G. Veillette (✉)
Section of Hepatobiliary and Pancreatic Surgery, Department of Surgery, Westchester Medical Center, and New York Medical College, School of Medicine, Valhalla, NY, USA
e-mail: Gregory.Veillette@wmchealth.org

R. Latifi (ed.), *Surgical Decision Making in Geriatrics*,
https://doi.org/10.1007/978-3-030-47963-3_24

Physiologic Changes in the Elderly

Aging is associated with changes in the physiology and morphology of the organs in the immune and hepatopancreaticobiliary system. There are global variations that affect host defense mechanisms, and there are changes in each organ that increase the likelihood of certain pathologies.

Aging is associated with a loss of hepatocytes secondary to a decrease in hepatic blood flow [1]. This impairs the regeneration capability of the liver and results in an inability to recuperate from damage caused by resection, trauma, toxins, infections, and other insults. Aging itself is considered a risk factor associated with liver carcinogenesis and the development of hepatocellular carcinoma. Some genetic changes normal to the aging process, such as aberrant DNA methylation and shortening of telomere length, increase the risk of HCC development [2]. In addition, there are alterations in the structure of the bile ducts as well as a proliferation of the bile ducts that affects biliary and cholesterol metabolism [1, 2]. These changes result in an imbalance in the cholesterol saturation of bile which may be linked to a higher incidence of formation of cholesterol gallstones.

Gallbladder disease is common in the elderly. There are multiple physiologic changes in the motility and emptying of the gallbladder that make it prone to pathology. As explained above, aging is associated with a propensity to form gallstones and with greater probability of this causing disease [1, 2]. Multiple studies have also elucidated that in the elderly (animals and humans) there is a decreased sensitivity of the gallbladder to the enzyme cholecystokinin (CCK), its main stimulus for contraction [1, 3, 4]. Furthermore, there are increased serum concentrations of pancreatic polypeptide (PP), a hormone that inhibits the secretion of bile from the liver and gallbladder [1–4]. These all result in further formation of gallstones – due to stasis – and inability to utilize stored bile.

The pancreas undergoes a wide variety of pathological changes in the elderly. These are concerning, as they not only increase the likelihood of benign pathologies but can also lead to neoplastic degeneration. There is an increased fatty replacement and fibrosis, causing loss of volume of pancreatic cells, higher incidence of pancreatic steatosis, as well as a decreased in pancreatic exocrine function [5, 6]. Furthermore, with advancing age, there is lymphoplasmacytic infiltration, amyloid deposition, and development of intraepithelial neoplastic changes, making the risk of developing pancreatic cancer higher in the elderly [1, 5, 6].

Benign Biliary Disease in the Elderly

Biliary disease is very common in the general population. It is estimated that 18.8% of females and 9.5% of males have asymptomatic gallstones [4]. The prevalence of gallstones at 70 years of age increases to 24% for females and 15% for males and at 90 years of age to 35% and 24% for males and females, respectively [7]. Moreover, age is also a risk factor for developing symptoms from biliary disease. The risk of developing biliary symptoms for patients <65 years of age is approximately 15% at 10 years and 18% at both 15 and 20 years [4]. However, in older individuals, gallstone disease tends to be more virulent [1]. Older patients have increased risk of gangrene and perforation of the gallbladder. Other complications from gallstone disease such as choledocholithiasis, cholangitis, and pancreatitis are also more common and complex in the elderly [1, 7, 8].

After an initial episode of biliary colic, 20–40% of patients will experience recurrent episodes [9]. Approximately 14% will develop acute cholecystitis within 1 year, 5% will develop gallstone pancreatitis, and 5% will develop common bile duct stones [7, 9]. For this reason, the benefits of early cholecystectomy in symptomatic cholelithiasis have been clearly established in the general population [4, 7]. However, in older patients, the management of symptomatic cholelithiasis is handled differently. The decision to perform elective cholecystectomy is complicated by multiple competing risks, such as comorbidities, previous surgeries, medications, frailty, and fear. Bergman et al. analyzed the management of symptomatic gallstones in the elderly and found that there are marked differences in the management strategies between all age groups and specifically within the elderly group. In patients with symptomatic gallstones that were 65–74 years old, the incidence of surgery was 87.4%, compared to 63.5% for those 75–84 years of age, and only 22.1% for patients over 85 years old. He determined that with advancing age the tendency is toward a nonoperative management [9]. On the other hand, the number of patients requiring urgent admission also rose sharply with age, more than doubling when the youngest group was compared with the oldest group. Furthermore, open cholecystectomy was performed more often in the older groups, possibly in part because of greater disease severity or delayed operative management [10, 11]. This tendency for nonoperative management in the elderly could be leading to more urgent/emergent presentations with advancing age. If symptomatic cholelithiasis was treated similarly in all age groups, we may be able to decrease the incidence of emergency gallbladder surgery in the elderly.

Emergency Biliary Surgery in the Elderly

Acute cholecystitis, choledocholithiasis, cholangitis, and gallstone pancreatitis are typical indications for surgery; however in the elderly population, these conditions don't necessarily lead to a surgical intervention. Emergency gallbladder surgery has been associated with complication rates ranging from 44 to 66% and mortality rates as high as 10–19% [11]. Consequently, there is a tendency to treat the aging population nonoperatively. Acute cholecystitis in a patient over 65 years of age with multiple comorbidities is often treated (inappropriately) with a percutaneous cholecystostomy tube [8]. Cholangitis or choledocholithiasis is often managed by ERCP or percutaneous transhepatic biliary drainage [11]. Cholecystectomy is often delayed in gallstone pancreatitis. Some of these interventions may be denying older patients the appropriate timely surgical intervention.

Nowadays, the mortality associated with gallbladder surgery in this population is far lower than what has been historically reported. McKay et al. performed a population-based analysis of the morbidity and mortality of gallbladder surgery in the elderly and found that the 30-day mortality rate associated with gallbladder surgery in elderly patients was 1.3%. Among patients who underwent urgent/emergency surgery, the rate was only 1.6%, and it was 0.7% for patients who underwent elective surgery [11]. In addition, Tucker et al. examined administrative data from the American College of Surgeons National Surgical Quality Improvement Program for 23,582 patients who underwent cholecystectomy, comparing the use of open vs laparoscopic cholecystectomy. They found that elderly patients (>65 years) were more likely to undergo open cholecystectomy than younger patients. They also reported a higher 30-day postoperative mortality in older patients for both open and laparoscopic procedures (mortality rates of 2.7 and 0.7 versus 1.3 and 0.3%, respectively). They also elucidated an improved outcome in the laparoscopic group [12].

Symptomatic benign gallbladder disease should be treated early in elderly patients. The sequelae from nonoperative management of biliary disease (cholangitis, pancreatitis, cholecystitis) can be fatal in older individuals, and it is important to recognize the appropriate indications for surgery.

Gallbladder Cancer in the Elderly

Gallbladder cancer (GBC) is a disease of the elderly; the average age at diagnosis is 71 years of age [13]. The most significant risk factors for development of GBC are advanced age, female sex, cholelithiasis, porcelain gallbladder, gallbladder polyps, and obesity. Other less common risk factors are chronic infection with *Salmonella* species or *Helicobacter pylori* and anomalous pancreatobiliary duct junction [13]. GBCs are usually diagnosed on pathologic examination after cholecystectomy. Approximately 1 in 250, or 0.7% of laparoscopic cholecystectomies, will yield a pathologic diagnosis of incidental gallbladder cancer. The likelihood of malignancy increases in patients with polyps. Solitary polyps greater than 1 cm and sessile have the highest probability of cancer and should prompt surgery. For the remainder of GBC diagnosed preoperatively, the typical presenting symptoms are jaundice, abdominal fullness, and weight loss [13, 14].

Gallbladder cancer is rare and more common among women. The CDC reports an incidence of 1.4 cases per 100,000 women and 0.8 cases per 100,000 men. It carries a very poor prognosis with overall 5-year survival of 50% for Stage I cancers and 3% for Stage IV cancers [13]. Incidental gallbladder cancers are often diagnosed at an earlier stage and, hence, usually have improved survival.

Surgical resection offers the only potentially curative therapy for GBC. Current guidelines recommend cholecystectomy alone for a T1a lesion (cancer involves the lamina propria without invasion of the muscular layer). For T1b lesions or higher (cancer involving the muscular layer), an extended or radical cholecystectomy should be performed. This includes cholecystectomy with resection of liver segments IVb and V and a portal lymphadenectomy, to obtain negative margins. While extended resection has resulted in improved survival, recurrence rates are high and can occur at distant sites [13]. Since GBC has such a very poor prognosis, all efforts should be focused on early detection and aggressive resection. Although gallbladder cancer is a rare entity, elderly patients with benign, asymptomatic gallbladder diseases such as cholelithiasis and polyps should be thoroughly evaluated for their risk of developing GBC and offered an elective cholecystectomy, if warranted. Furthermore, those patients who present with resectable GBC should be offered radical resection, if feasible.

For advanced gallbladder cancer, the role of adjuvant therapy is receiving significant attention. This is important in the elderly population if they need to undergo a more limited resection [14]. Kasumova et al. set out to investigate the role of adjuvant therapy in T2 and T3 lesions based on the type of resection. In their study of 6825 individuals, patients who underwent extended cholecystectomy with adjuvant therapy had the longest median overall survival (22.4 and 23.3 months, respectively). Surprisingly, patients who underwent simple cholecystectomy with adjuvant therapy demonstrated a survival advantage relative to those who underwent extended cholecystectomy alone. These findings suggest that for select patients, including elderly patients, who are high-risk candidates for extensive surgery or for those in whom a complete curative resection is not possible, simple resection followed by adjuvant therapy should be considered [14].

Liver Cancer in the Elderly Patient

Primary liver cancer in the adult typically refers to either hepatocellular carcinoma (HCC, hepatoma) or cholangiocarcinoma (CCA), while secondary liver cancers include metastatic tumors such as colorectal cancer or neuroendocrine tumors. There are also benign solid (hemangioma, adenoma, and focal nodular hyperplasia) and cystic (simple, cystadenoma, infectious) lesions of the liver that may require surgical management.

Hepatocellular carcinoma (HCC) accounts for 85–90% of all primary liver cancers [15]; it ranks fifth among the most prevalent cancers worldwide and is the third most common cause of cancer-related death [15–17]. The main risk factors for developing HCC are chronic infection with hepatitis B virus (HBV), hepatitis C virus (HCV) infection, aflatoxin, hemochromatosis, and alcoholic and biliary cirrhosis [16]. The largest single risk factor for developing HCC worldwide is the presence of cirrhosis [16, 17]. HCC is more common in the elderly; in the United States, HCC incidence peaks above the age of 70 years [16]. In the coming years, there is expected to be an increase in the number of cases of HCC. This is due to a number of reasons: (1) the increased longevity of the population; (2) the rising incidence of liver cirrhosis unrelated to hepatitis viruses, such as nonalcoholic steatohepatitis (NASH) cirrhosis which can lead to HCC over a long period of time; and (3) some antiviral therapies, such as nucleoside analogs for chronic Hep B and interferon for chronic Hep C infections, have now been possibly linked to the development of HCC; their delayed effect will be seen in the coming years [16].

The management of HCC in the elderly is difficult, since the vast majority of HCCs occur within the cirrhotic liver. Typically, the armamentarium for the management of HCC includes liver transplantation, surgical resection, local-regional treatments (TACE, microwave, Y-90), chemotherapy, or palliative measures [17]. The general algorithm begins with consideration of curative options, either liver transplantation or surgical resection. Those patients with metastatic disease (bone, lung) should not be considered for these treatment options and should only be offered systemic chemotherapy, with local-regional treatment only for palliative purposes [17, 18]. Liver transplantation is considered in those patients who fall within the Milan criteria: single lesion less than 5 cm or up to three lesions less than 3 cm. Patients within Milan criteria are eligible to be listed for liver transplantation [19, 20]. However, once listed, the typical waiting times can be as long as 2–3 years, depending on the region in the country. Therefore, these patients need to undergo treatment with local-regional therapies to keep their tumor burden within Milan criteria until they can get a liver transplant. Patients who fall outside of Milan criteria at initial presentation, or while waiting for liver transplantation, are sometimes eligible to undergo down staging treatment to get their tumors to within Milan criteria, at which time they are eligible to be listed for liver transplantation [21, 22].

Liver transplantation is an enormous operation and therefore carries substantial surgical risk. Furthermore, there are many short- and long-term issues that come along with transplantation, mainly centered on the need for lifelong immunosuppression therapy. While liver transplant is feasible in the elderly population, it is usually not considered, given the significant risk of surgery, coupled with the huge scarcity of liver allografts. Therefore, the only potential curative option for HCC in the elderly patient is surgical resection.

Unfortunately, the vast majority of patients with HCC have underlying cirrhosis. This adds an extra layer of complexity to the surgical planning and decision-making. First, the cirrhotic liver does not have the regenerative or reparative ability of a non-cirrhotic liver [23–26]. Therefore, the amount of liver that can safely be resected is much less in the cirrhotic [24]. The use of liver volumes (from CT scan or MRI) to determine the future liver remnant and/or the use of ICG-clearance to determine total liver function are essential when considering surgical resection in the cirrhotic patient [27–29]. Second, many patients with cirrhosis have some degree of portal hypertension. Many times, the degree of portal hypertension alone may preclude liver surgery, but even if mild, evidence for portal hypertension (ascites, varices, hepatic encephalopathy) may signal the high risk for liver decompensation after resection [30, 31]. The use of scoring systems to assess the degree of liver dysfunction and portal hypertension are important adjuncts (MELD score and Childs-Pugh-Turcotte CPT score).

Cholangiocarcinoma is classically divided into two types: intrahepatic and extrahepatic. The extrahepatic lesions are divided into three subtypes: proximal (hilar or Klatskin tumors), middle, and distal. Surgery, with an R-0 resection, is the only curative option for all types of cholangiocarcinoma [32]. Intrahepatic cholangiocarcinoma is managed with liver resection, while distal cholangiocarcinoma requires a pancreaticoduodenectomy (Whipple procedure). The surgical management of middle and proximal cholangiocarcinoma depends on the extent of longitudinal spread along the biliary tree. While partial extrahepatic bile duct resection is necessary, a concomitant pancreaticoduodenectomy (for middle lesions) or liver resection (for hilar lesions) is often required [33]. Liver transplant has been studied for both intrahepatic and hilar cholangiocarcinoma. Currently, liver transplantation is not recommended for intrahepatic cholangiocarcinoma [34, 35] and is only recommended for a very select group of hilar cholangiocarcinoma [36].

Metastatic colorectal cancer (CRC) is the most common malignant lesion for which hepatectomy is considered in the United States [37]. The outcomes with aggressive surgery, as well as improvements in chemotherapy, have substantially increased survival in patients with advanced stage colorectal cancer [37–39]. The surgical management for metastatic CRC should be very closely orchestrated with medical oncology. Several factors have been shown to impact prognosis with hepatectomy for metastatic CRC. These include the number and size of the lesions, CEA level, time interval of development of liver metastases, and nodal status of the primary tumor [38]. However, the most important factor to consider when planning liver surgery for metastatic colorectal cancer is the ability to resect the entire tumor burden. Good quality, liver-dedicated, imaging is essential to assess tumor burden and to plan treatment. Unlike HCC, patients with metastatic colorectal cancer are typically not cirrhotic [39]. Nonetheless, it is critical to assess the liver for steatosis secondary to chemotherapy (CASH – chemotherapy-associated steatohepatitis). Hepatic steatosis is an injury pattern that can progress to fibrosis and even cirrhosis. Furthermore, similar to the cirrhotic liver, the injured fatty liver cannot regenerate and repair itself like a normal liver. Therefore, the detection of fatty damage to the liver is critical if considering any type of liver resection [40–44].

The technical aspects of liver surgery are similar for elderly patients, as they are for young patients. However, there are some essential considerations required when planning hepatectomy in the elderly. First, is the quality and status of the underlying liver. This will dictate how much liver can safely be resected. A normal, healthy liver can tolerate up to a 70–75% resection, while a cirrhotic liver may not tolerate even a 20–30% resection [45–46]. There is no magic number, and this must be individualized based on the degree of liver damage and dysfunction. There are several modalities to determine the underlying liver function and quality, including physical exam, serum liver function testing, cross-sectional imaging with volumetrics, liver biopsy, and ICG-clearance testing. The physical exam should be aimed at assessing for signs of liver disease (hepatomegaly, jaundice, cachexia, peripheral edema) and for signs of portal hypertension (ascites, splenomegaly, spider telangiectasia, superficial varices, encephalopathy, and gynecomastia). Serum liver function testing is the most basic investigation and should include markers for hepatocyte injury (AST, ALT), cholestasis (alkaline phosphatase, GGT), portal hypertension (platelet count, serum sodium), and true liver function (bilirubin, albumin, PT/INR). Also, if the history suggests risk factors, serum testing for viral hepatitis should be obtained. Liver-dedicated, triple-phase cross-sectional imaging with either CT scan or MRI should be performed on all patients. This will show evidence for any underlying liver abnormality (steatosis, cirrhosis, hepatomegaly) and for signs of portal hypertension (splenomegaly, varices, ascites, shunts). It is also critically important to assess the relationship of the tumor(s) to the inflow and outflow vessels, as well as surrounding structures (colon, diaphragm, stomach, duodenum) [46]. Liver volumes can be easily calculated using special radiologic software, which allows for determination of the future liver remnant based on the type and extent of resection [31]. If there are any concerns about the amount of liver to be resected, coupled with concerns about the quality of the underlying liver, a biopsy should be performed to rule out any pathology in the non-tumorous liver. Indocyanine green (ICG) clearance testing is very popular in Eastern countries and is an excellent test for preoperative assessment of liver function [29].

Once the adequate extent of resection is determined and deemed safe (in terms of future liver remnant), the hepatectomy should be done as meticulously as possible. The use of low CVP anesthesia during parenchymal transection is extremely beneficial in preventing excessive blood loss. The vast majority of liver resections (even in patients with cirrhosis) should be done with minimal to no blood transfusion requirements [47–50]. Knowledge of the patient's liver anatomy, careful dissection of the vasculature and biliary tree, and precise parenchymal transection will provide the best opportunity to avoid devascularization of remaining liver segments, excessive blood loss, and bile leaks.

One of the most critical components to successful liver surgery in the elderly is the

Table 24.1 Overview of the current literature

Author	Year	*n*	Morbidity (%)	Mortality (%)
Caratozzolo et al. [16]	2007	>70 $n = 51$ <70 $n = 93$	25% 10%, $p = 0.001$	0% 0%
Aldrighetti et al. [5]	2003	>70 $n = 32$ <70 $n = 95$	9% 22%, $p = 0.2$	0% 1%
Cho et al. [19]	2011	>70 $n = 75$ <70 $n = 75$	44% 33%, $p = 0.241$	0% 0%
Ferrero et al. [18]	2005	>70 $n = 64$ <70 $n = 177$	23% 42%, $p = 0.007$	3% 10%, $p = 0.113$
Cescon et al. [4]	2003	>70 $n = 23$ <70 $n = 99$	39% 32%, $p = 0.53$	0% 2%
Mann et al. [20]	2007	>70 $n = 49$ <70 $n = 142$	31% 19%, $p = 0.07$	0% 2%
Ijtsma et al. [13]	2008	>60 $n = 93$ <60 $n = 93$	47% 31%, $p = 0.024$	11% 2%, $p = 0.017$
Takahshi et al. [17]	2012	>80 $n = 21$ <80 $n = 410$	57% 46%, $p = 0.372$	5% 2%, $p = 0.332$
Adam et al. [12]	2010	>70 $n = 1624$ <70 $n = 6140$	32% 29%, $p < 0.001$	4% 2%, $p < 0.001$
Lee et al. [21]	2012	>70 $n = 61$ <70 $n = 90$	28% 14%, $p = 0.042$	
Menon et al. [22]	2006	>70 $n = 127$ <70 $n = 390$	31% 33%, $p = 0.22$	8% 5%, $p = 0.32$
Shirabe et al. [23]	2009	>80 $n = 43$ <80 $n = 307$	26% 22%, $p = 0.56$	0% 1%, $p = 0.99$

From Andert et al. [79], with permission

postoperative management [51]. Several studies have demonstrated that complex hepatobiliary surgery can be done safely in the elderly but with an overall higher morbidity and mortality (Table 24.1 and Fig. 24.1). Much of the morbidity, and subsequent mortality, is due to the inability of the elderly to tolerate the physiologic insults that are inherent to major liver surgery. Importantly, while the elderly patient may not necessarily be at any higher risk to develop certain complications, their ability to overcome a complication is what leads to further morbidity and increased mortality. For example, an 85-year-old male undergoes a right hepatectomy through a right subcostal or chevron incision. Postoperatively, he has significant pain that precludes him from taking deep breaths and clearing his secretions, leading to atelectasis, pneumonia, and a systemic inflammatory response syndrome. This scenario is extremely common after any major hepatobiliary surgery, and while the young patient could tolerate this without any difficulty, this could be a fatal event for the elderly patient.

The more common complications in the elderly include delirium, infection and cardiopulmonary decompensation [1, 47, 49, 51]. Therefore, the postoperative management must include cardiac optimization, careful fluid management, adequate pain control, avoidance of delirium-inducing medications, aggressive pulmonary toilet, prevention of aspiration, early and frequent ambulation, and good nutrition.

Pancreatic Cancer in the Elderly Patient

Pancreatic adenocarcinoma carries a dismal prognosis. Typically, these cancers present with painless jaundice (for pancreatic head/uncinate tumors), abdominal/back pain (for body and tail lesions), and unintentional weight loss. Unfortunately, by the time a patient with pancreatic cancer presents with symptoms, the majority of tumors are either locally advanced (involving the major visceral vessels) or metastatic [52, 53]. The median survival for untreated pancreatic

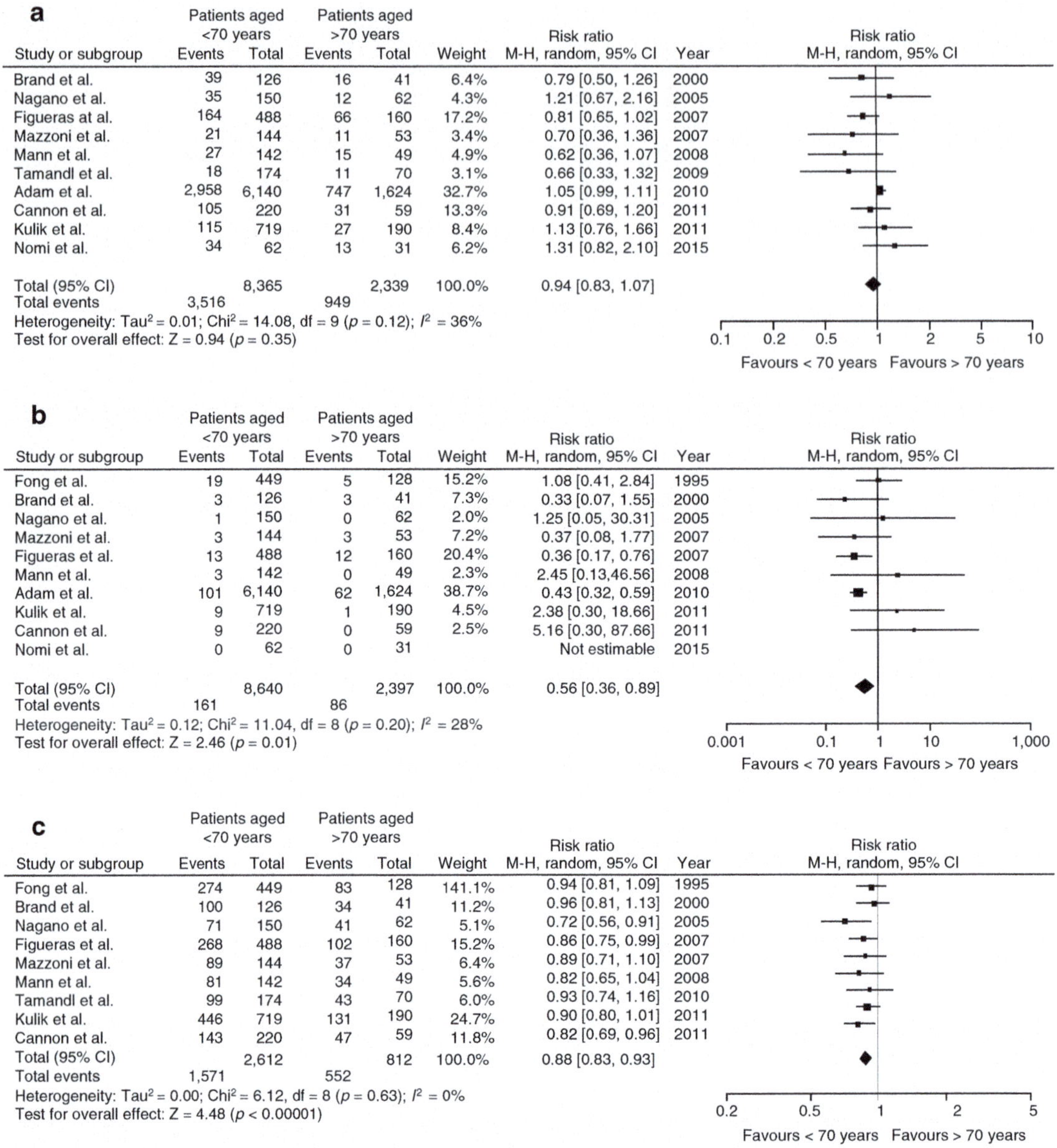

a

Study or subgroup	Patients aged <70 years Events	Total	Patients aged >70 years Events	Total	Weight	Risk ratio M-H, random, 95% CI	Year
Brand et al.	39	126	16	41	6.4%	0.79 [0.50, 1.26]	2000
Nagano et al.	35	150	12	62	4.3%	1.21 [0.67, 2.16]	2005
Figueras at al.	164	488	66	160	17.2%	0.81 [0.65, 1.02]	2007
Mazzoni et al.	21	144	11	53	3.4%	0.70 [0.36, 1.36]	2007
Mann et al.	27	142	15	49	4.9%	0.62 [0.36, 1.07]	2008
Tamandl et al.	18	174	11	70	3.1%	0.66 [0.33, 1.32]	2009
Adam et al.	2,958	6,140	747	1,624	32.7%	1.05 [0.99, 1.11]	2010
Cannon et al.	105	220	31	59	13.3%	0.91 [0.69, 1.20]	2011
Kulik et al.	115	719	27	190	8.4%	1.13 [0.76, 1.66]	2011
Nomi et al.	34	62	13	31	6.2%	1.31 [0.82, 2.10]	2015
Total (95% CI)		8,365		2,339	100.0%	0.94 [0.83, 1.07]	
Total events	3,516		949				

Heterogeneity: Tau^2 = 0.01; Chi^2 = 14.08, df = 9 (p = 0.12); I^2 = 36%
Test for overall effect: Z = 0.94 (p = 0.35)

b

Study or subgroup	Patients aged <70 years Events	Total	Patients aged >70 years Events	Total	Weight	Risk ratio M-H, random, 95% CI	Year
Fong et al.	19	449	5	128	15.2%	1.08 [0.41, 2.84]	1995
Brand et al.	3	126	3	41	7.3%	0.33 [0.07, 1.55]	2000
Nagano et al.	1	150	0	62	2.0%	1.25 [0.05, 30.31]	2005
Mazzoni et al.	3	144	3	53	7.2%	0.37 [0.08, 1.77]	2007
Figueras et al.	13	488	12	160	20.4%	0.36 [0.17, 0.76]	2007
Mann et al.	3	142	0	49	2.3%	2.45 [0.13,46.56]	2008
Adam et al.	101	6,140	62	1,624	38.7%	0.43 [0.32, 0.59]	2010
Kulik et al.	9	719	1	190	4.5%	2.38 [0.30, 18.66]	2011
Cannon et al.	9	220	0	59	2.5%	5.16 [0.30, 87.66]	2011
Nomi et al.	0	62	0	31		Not estimable	2015
Total (95% CI)		8,640		2,397	100.0%	0.56 [0.36, 0.89]	
Total events	161		86				

Heterogeneity: Tau^2 = 0.12; Chi^2 = 11.04, df = 8 (p = 0.20); I^2 = 28%
Test for overall effect: Z = 2.46 (p = 0.01)

c

Study or subgroup	Patients aged <70 years Events	Total	Patients aged >70 years Events	Total	Weight	Risk ratio M-H, random, 95% CI	Year
Fong et al.	274	449	83	128	141.1%	0.94 [0.81, 1.09]	1995
Brand et al.	100	126	34	41	11.2%	0.96 [0.81, 1.13]	2000
Nagano et al.	71	150	41	62	5.1%	0.72 [0.56, 0.91]	2005
Figueras et al.	268	488	102	160	15.2%	0.86 [0.75, 0.99]	2007
Mazzoni et al.	89	144	37	53	6.4%	0.89 [0.71, 1.10]	2007
Mann et al.	81	142	34	49	5.6%	0.82 [0.65, 1.04]	2008
Tamandl et al.	99	174	43	70	6.0%	0.93 [0.74, 1.16]	2010
Kulik et al.	446	719	131	190	24.7%	0.90 [0.80, 1.01]	2011
Cannon et al.	143	220	47	59	11.8%	0.82 [0.69, 0.96]	2011
Total (95% CI)		2,612		812	100.0%	0.88 [0.83, 0.93]	
Total events	1,571		552				

Heterogeneity: Tau^2 = 0.00; Chi^2 = 6.12, df = 8 (p = 0.63); I^2 = 0%
Test for overall effect: Z = 4.48 (p < 0.00001)

Fig. 24.1 (**a**) Postoperative morbidity for liver resection for CRLM in patients aged <70 years and patients aged >70 years. (**b**) Postoperative mortality for liver resection for CRLM in patients aged <70 years and patients aged >70 years. (**c**) Five-year survival for liver resection for CRLM in patients aged <70 years and patients aged >70 years. *CRLM* colorectal liver metastases, *M-H* Mantel-Haenszel. (From van Tuil et al. [80], with permission)

cancer is 8–12 months for locally advanced disease and 3–6 months for metastatic disease [53, 54]. Current chemotherapeutic regimens can offer several months of improvement in overall survival but must be weighed with the quality of life during the chemotherapy [55–57]. Only 10–20% of patients with pancreatic cancer have resectable disease at presentation. Despite huge improvements in mortality rates after pancreatic surgery, survival of pancreatic cancer remains poor, even with complete resection. The 5-year survival for a resected, node-negative pancreatic cancer is 25–40% but only 10–15% for a resected, node-positive tumor [52, 58, 59].

The surgical options for pancreatic cancer are based on the location of the tumor. In general, tumors of the pancreas either involve the head and uncinate process (which are to the right of the portal/mesenteric vein) or they involve the body and tail (which are to the left of the portal/mesenteric vein). Therefore, the vast majority of surgery for pancreatic lesions requires either resection of the head and uncinate process (pancreaticoduodenectomy) or resection of the body and tail (distal pancreatectomy). Sometimes, tumors in the body of the pancreas may extend over to the right side of the vessels. This may require extended resection or potentially a central/middle pancreatectomy. Enucleation of some pancreatic tumors is feasible but should never be offered for pancreatic adenocarcinoma [60, 61].

Pancreaticoduodenectomy (Whipple procedure) is the quintessential general surgical procedure. It involves resection of the duodenum, head and uncinate process of the pancreas, and the distal portion of the common bile duct. The shared blood supply and the intricate anatomic relationship of the pancreaticobiliary tree preclude isolated resection of a single structure. Furthermore, the intimate relationship of the duodenum and pancreas to the mesenteric vasculature (common hepatic artery, superior mesenteric artery, and vein) makes resection of these structures very treacherous. The reconstruction after resection requires establishing continuity of the pancreas, biliary tree, and stomach. This is accomplished by performing three separate anastomoses to the proximal jejunal limb, all of which have the potential for leakage [62]. In particular, the pancreatic anastomosis tends to be the most difficult and sinister. The soft nature of the pancreatic parenchyma, coupled with the small size of the pancreatic duct, makes for a high likelihood of leak (fistula). Leakage of pancreatic fluid can initiate autodigestion of local tissue, leading to intra-abdominal collection, sepsis, and massive bleeding [63, 64].

Historically, the Whipple procedure carried substantial morbidity, with mortality rates as high as 25%. Over the past 30 years, in experienced hands, the Whipple procedure has become a straightforward procedure, with acceptable mortality rates of only 2–3%. This significant improvement in mortality is the result mainly from mastery of the surgical technique, as well as improved postoperative management [62, 65]. However, despite this, the morbidity remains as high as 40%, with delayed gastric emptying and pancreatic leak as the most common complications [62–65]. Distal pancreatectomy does not require any type of enteric reconstruction, which makes it technically less demanding than pancreaticoduodenectomy and results in a lower mortality [66]. However, distal pancreatectomy still carries substantial morbidity, with pancreatic leak/fistula rates as high as 30–35% [66, 67].

The outcomes of pancreatic surgery in the elderly are acceptable, despite most studies showing a higher morbidity and mortality (Figs. 24.2 and

Study or Subgroup	≥80 Events	≥80 Total	<80 Events	<80 Total	Weight	Odds Ratio M-H, Fixed, 95% CI
Beltrame 2015	0	23	14	362	7.4%	0.51 [0.03, 8.84]
Belyaev 2013	9	76	43	1705	13.6%	5.19 [2.43, 11.09]
Gangl 2014	0	9	5	99	4.0%	0.90 [0.05, 17.64]
Hatzaras 2011	1	27	18	490	7.6%	1.01 [0.13, 7.85]
khan 2010	1	53	6	564	4.3%	1.79 [0.21, 15.14]
M. Lee 2010	4	74	27	703	20.5%	1.43 [0.49, 4.21]
Makary 2006	8	207	42	2491	26.0%	2.34 [1.09, 5.06]
Melis 2012	1	25	1	175	1.0%	7.25 [0.44, 119.76]
Oguro 2012	1	22	5	539	1.6%	5.09 [0.57, 45.48]
Sohn 1997	2	46	10	681	5.1%	3.05 [0.65, 14.35]
Tani 2009	0	25	8	310	5.4%	0.70 [0.04, 12.44]
Usuba et al 2014	0	10	2	50	3.6%	0.92 [0.04, 20.69]
Total (95% CI)		**597**		**8169**	**100.0%**	**2.22 [1.48, 3.31]**
Total events	27		181			

Heterogeneity: Chi² = 9.79, df = 11 (P= 0.55); I² = 0%
Test for overall effect: Z = 3.87 (P= 0.0001)

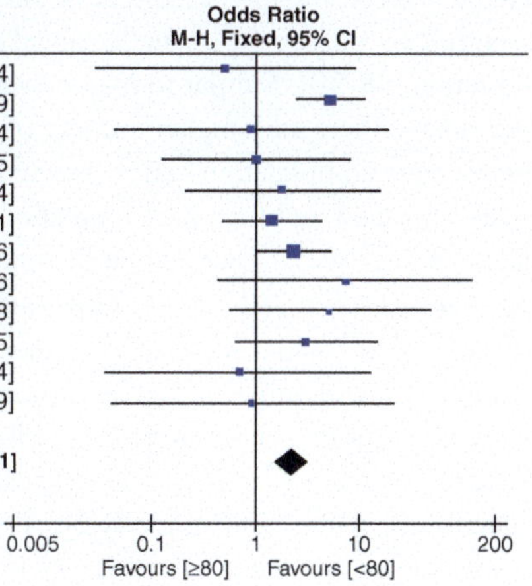

Fig. 24.2 Thirty-day mortality forest plot. Forest plot of comparison of 30-day mortality in patients aged 80 or over versus patients younger than 80. (From Kim et al. [81], with permission)

Study or Subgroup	≥80 Events	≥80 Total	<80 Events	<80 Total	Weight	Odds Ratio M-H, Fixed, 95% CI
Beltrame 2015	10	23	145	362	5.6%	1.15 [0.49, 2.70]
Hatzaras 2011	14	27	289	490	8.4%	0.75 [0.34, 1.63]
khan 2010	27	53	209	564	10.2%	1.76 [1.00, 3.10]
M. Lee 2010	35	74	359	703	20.8%	0.86 [0.53, 1.39]
Makary 2006	104	197	984	2491	39.3%	1.71 [1.28, 2.29]
Melis 2012	17	25	77	175	3.6%	2.70 [1.11, 6.60]
Oguro 2012	6	22	52	539	1.7%	3.51 [1.32, 9.37]
Sohn 1997	26	46	270	681	8.6%	1.98 [1.08, 3.62]
Usuba et al 2014	5	10	20	50	1.9%	1.50 [0.38, 5.86]
Total (95% CI)		**477**		**6055**	**100.0%**	**1.51 [1.25,1.83]**
Total events	244		2405			

Heterogeneity: Chi² = 15.08, df = 8 (P= 0.06); I² = 47%
Test for overall effect: Z = 4.30 (P < 0.0001)

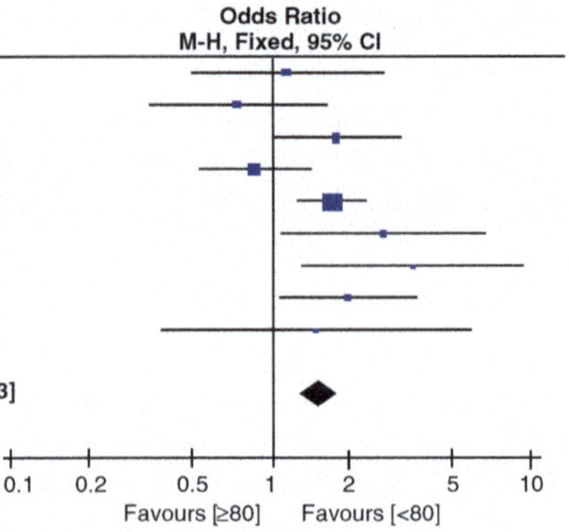

Fig. 24.3 Overall postoperative complication in octogenarians or older vs young patients forest plot. Forest plot of comparisons of overall complications after the operation in both age groups. (From Kim et al. [81], with permission)

Table 24.2 Summary of results of meta-analysis

Outcome	Number of studies	Participants aged 80 and over	Participants aged less than 80	Odds ratio fixed [95% CI]	Statistical difference
30-day mortality	12	597	8169	2.22 [1.48, 3.31]	$p < 0.001$
ASA ≥3	6	1191	11,442	2.33 [1.98, 2.74]	$p < 0.001$
Overall complications	9	477	6055	1.51 [1.25, 1.83]	$p < 0.001$
Postoperative pancreatic fistula	9	492	6940	0.99 [0.73, 1.34]	$p = 0.93$
Delayed gastric emptying	7	413	6266	1.77 [1.35, 2.31]	$p < 0.001$
Bile leak	5	351	4653	1.33 [0.77, 2.30]	$p = 0.30$
Postoperative hemorrhage	5	143	2526	1.44 [0.59, 3.53]	$p = 0.42$
Wound infection	4	339	4927	1.16 [0.79, 1.70]	$p = 0.45$
Reoperation	8	472	6581	1.28 [0.86, 1.89]	$p = 0.23$
Postoperative cardiac events	3	259	3291	3.24 [1.89, 5.56]	$p < 0.001$
Postoperative pneumonia	7	1416	14,684	1.72 [1.39, 2.13]	$p < 0.001$
Mean length of stay	5	765	7814	2.23 [1.36, 3.10]	$p < 0.001$

From Kim et al. [81], with permission

24.3). However, given the dismal prognosis of pancreatic cancer, the increased risk of surgery is outweighed by the survival benefit gained from tumor resection [68–70]. Similar to major liver surgery in the elderly, there are a higher number of cardiac and pulmonary complications after pancreatic surgery, with an equal number of pancreatic and biliary leaks (Table 24.2). Other patient factors that impact adverse outcomes after pancreaticoduodenectomy include American Society of Anesthesiologist class, low serum albumin levels, significant weight loss, and significant intraoperative blood transfusion requirement [71–76]. This suggests that poor functional status and underlying comorbidities contribute to the increased mortality in the elderly population, rather than age alone. Hence, a prehabilitation program may address both the frailty and nutritional aspects so as to optimize patients for surgery and improve their outcomes and may improve failure to rescue rates. Nakajima et al. show that of 76 patients who underwent prehabilitation had a decreased median length of stay of 23 days, in comparison to 30 days for those without prehabilitation [77]. This program was particularly unique in comparison to most American prehabilitation programs in that it incorporated leucine-rich essential amino acid supplements and supervised exercise therapy before their surgery. A small pilot study looked at the impact of prehabilitation on functional status and length of stay following pancreatic surgery and showed that mean length of stay was decreased to 12.6 days from 13.2 days [78].

Conclusion

Hepatopancreaticobiliary surgery can be done safely in the elderly patient. These are complex operations that require a multidisciplinary approach to the perioperative management of these patients. Careful preoperative planning, medical and nutritional optimization, meticulous surgical technique, and exceptional postoperative care are the key elements to successful outcomes. Age alone should never preclude the opportunity for elderly patients to undergo surgical procedures that can prolong life and/or improve quality of life.

References

1. Kahng KU, Roslyn JJ. Surgical issues for the elderly patient with hepatobiliary disease. Surg Clin N Am. 1994;74(2):345–73.
2. Wong CM, Ng IO. Molecular pathogenesis of hepatocellular carcinoma. Liver Int. 2008;28(2):160–74.
3. Flint A, Bradwejn J, Vaccarino F, Gutkowska J, Palmour R, Koszycki D. Aging and panicogenic response to cholecystokinin tetrapeptide: an examination of the cholecystokinin system. Neuropsychopharmacology. 2002;27(4):663.
4. Gracie WA, Ransohoff DF. The natural history of silent gallstones: the innocent gallstone is not a myth. N Engl J Med. 1982;307(13):798–800.
5. Matsuda Y. Age-related pathological changes in the pancreas. Front Biosci (Elite Ed). 2018;10:137–42.
6. Chantarojanasiri T, Hirooka Y, Ratanachu-ek T, Kawashima H, Ohno E, Goto H. Evolution of pancreas in aging: degenerative variation or early changes of disease? J Med Ultrason. 2015;42(2):177–83.
7. Parmar AD, Sheffield KM, Adhikari D, Davee RA, Vargas GM, Tamirisa NP, et al. PREOP-gallstones: a prognostic nomogram for the management of symptomatic cholelithiasis in older patients. Ann Surg. 2015;261(6):1184.
8. Pisano M, Ceresoli M, Cimbanassi S, Gurusamy K, Coccolini F, Borzellino G, et al. 2017 WSES and SICG guidelines on acute calculous cholecystitis in elderly population. World J Emerg Surg. 2019;14(1):10.
9. Bergman S, Al-Bader M, Sourial N, Vedel I, Hanna WC, Bilek AJ, et al. Recurrence of biliary disease following non-operative management in elderly patients. Surg Endosc. 2015;29(12):3485–90.
10. Gurusamy KS, Davidson BR. Surgical treatment of gallstones. Gastroenterol Clin. 2010;39(2):229–44.
11. McKay A, Katz A, Lipschitz J. A population-based analysis of the morbidity and mortality of gallbladder surgery in the elderly. Surg Endosc. 2013;27(7):2398–406.
12. Tucker JJ, Yanagawa F, Grim R, Bell T, Ahuja V. Laparoscopic cholecystectomy is safe but underused in the elderly. Am Surg. 2011;77(8):1014–20.
13. Hickman L, Contreras C. Gallbladder cancer: diagnosis, surgical management, and adjuvant therapies. Surg Clin. 2019;99(2):337–55.
14. Kasumova GG, Tabatabaie O, Najarian RM, Callery MP, Ng SC, Bullock AJ, et al. Surgical management of gallbladder cancer: simple versus extended cholecystectomy and the role of adjuvant therapy. Ann Surg. 2017;266(4):625–31.
15. Nordenstedt H, White DL, El-Serag HB. The changing pattern of epidemiology in hepatocellular carcinoma. Dig Liver Dis. 2010;42:S206–14.
16. Hamaoka M, Kobayashi T, Ishiyama K, Ohira M, Tahara H, Kuroda S, et al. Evaluation of the risk factors and prognostic factors of hepatectomy for hepatocellular carcinoma in patients aged 80 years or more. J Hepatobiliary Pancreat Sci. 2017;24(1):58–64.
17. Nishikawa H, Kimura T, Kita R, Osaki Y. Treatment for hepatocellular carcinoma in elderly patients: a literature review. J Cancer. 2013;4(8):635.
18. Kaibori M, Matsui K, Ishizaki M, et al. Hepatic resection for hepatocellular carcinoma in the elderly. J Surg Oncol. 2009;99(3):154–60.
19. Mazzaferro V, Regalia E, Doci R, Andreola S, Pulvirenti A, Bozzetti F, et al. Liver transplantation for the treatment of small hepatocellular carcinomas in patients with cirrhosis. N Engl J Med. 1996;334(11):693–700.
20. Mazzaferro V, Bhoori S, Sposito C, Bongini M, Langer M, Miceli R, Mariani L. Milan criteria in liver transplantation for hepatocellular carcinoma: an evidence-based analysis of 15 years of experience. Liver Transpl. 2011;17(S2):S44–57.
21. Yao FY, Ferrell L, Bass NM, Bacchetti P, Ascher NL, Roberts JP. Liver transplantation for hepatocellular carcinoma: comparison of the proposed UCSF criteria with the Milan criteria and the Pittsburgh modified TNM criteria. Liver Transpl. 2002;8(9):765–74.
22. Mazzaferro V, Llovet JM, Miceli R, Bhoori S, Schiavo M, Mariani L, et al. Predicting survival after liver transplantation in patients with hepatocellular carcinoma beyond the Milan criteria: a retrospective, exploratory analysis. Lancet Oncol. 2009;10(1):35–43.
23. Riehle KJ, Dan YY, Campbell JS, Fausto N. New concepts in liver regeneration. J Gastroenterol Hepatol. 2011;26:203–12.
24. Nagasue N, Yukaya H, Ogawa Y, Kohno H, Nakamura T. Human liver regeneration after major hepatic resection. A study of normal liver and livers with chronic hepatitis and cirrhosis. Ann Surg. 1987;206(1):30.
25. Schuppan D, Afdhal NH. Liver cirrhosis. Lancet. 2008;371(9615):838–51.
26. Koniaris LG, McKillop IH, Schwartz SI, Zimmers TA. Liver regeneration. J Am Coll Surg. 2003;197(4):634–59.

27. Chen MF, Hwang TL, Hung CF. Liver volumes: Human liver regeneration after major hepatectomy. A study of liver volume by computed tomography. Ann Surg. 1991;213(3):227.
28. Chuang YH, Ou HY, Lazo MZ, Chen CL, Chen MH, Weng CC, Cheng YF. Predicting post-hepatectomy liver failure by combined volumetric, functional MR image and laboratory analysis. Liver Int. 2018;38(5):868–74.
29. Olthof PB, Tomassini F, Huespe PE, Truant S, Pruvot FR, Troisi RI, et al. Hepatobiliary scintigraphy to evaluate liver function in associating liver partition and portal vein ligation for staged hepatectomy: liver volume overestimates liver function. Surgery. 2017;162(4):775–83.
30. Leung U, Simpson AL, Araujo RL, Gönen M, McAuliffe C, Miga MI, et al. Remnant growth rate after portal vein embolization is a good early predictor of post-hepatectomy liver failure. J Am Coll Surg. 2014;219(4):620–30.
31. Khan AS, Garcia-Aroz S, Ansari MA, Atiq SM, Senter-Zapata M, Fowler K, et al. Assessment and optimization of liver volume before major hepatic resection: current guidelines and a narrative review. Int J Surg. 2018;52:74–81.
32. Razumilava N, Gores GJ. Cholangiocarcinoma. Lancet. 2014;383(9935):2168–79.
33. Doherty B, Nambudiri VE, Palmer WC. Update on the diagnosis and treatment of cholangiocarcinoma. Curr Gastroenterol Rep. 2017;19(1):2.
34. Goldaracena N, Gorgen A, Sapisochin G. Current status of liver transplantation for cholangiocarcinoma. Liver Transpl. 2018;24(2):294–303.
35. De Vreede I, Steers JL, Burch PA, Rosen CB, Gunderson LL, Haddock MG, et al. Prolonged disease-free survival after orthotopic liver transplantation plus adjuvant chemoirradiation for cholangiocarcinoma. Liver Transpl. 2000;6:309–16.
36. Heimbach JK, Haddock MG, Alberts SR, Nyberg SL, Ishitani MB, Rosen CB, Gores GJ. Transplantation for hilar cholangiocarcinoma. Liver Transpl. 2004;10(S10):S65–8.
37. Fong Y, Cohen AM, Fortner JG, Enker WE, Turnbull AD, Coit DG, et al. Liver resection for colorectal metastases. J Clin Oncol. 1997;15(3):938–46.
38. Fong Y, Fortner J, Sun RL, Brennan MF, Blumgart LH. Clinical score for predicting recurrence after hepatic resection for metastatic colorectal cancer: analysis of 1001 consecutive cases. Ann Surg. 1999;230(3):309.
39. House MG, Ito H, Gönen M, Fong Y, Allen PJ, DeMatteo RP, et al. Survival after hepatic resection for metastatic colorectal cancer: trends in outcomes for 1,600 patients during two decades at a single institution. J Am Coll Surg. 2010;210(5):744–52.
40. Zorzi D, Laurent A, Pawlik TM, Lauwers GY, Vauthey JN, Abdalla EK. Chemotherapy-associated hepatotoxicity and surgery for colorectal liver metastases. Br J Surg. 2007;94(3):274–86.
41. Rao MS, Papreddy K, Abecassis M, Hashimoto T. Regeneration of liver with marked fatty change following partial hepatectomy in rats. Dig Dis Sci. 2001;46(9):1821–6.
42. Veteläinen R, van Vliet AK, van Gulik TM. Severe steatosis increases hepatocellular injury and impairs liver regeneration in a rat model of partial hepatectomy. Ann Surg. 2007;245(1):44.
43. Behrns KE, Tsiotos GG, DeSouza NE, Krishna MK, Ludwig J, Nagomey DM. Hepatic steatosis as a potential risk factor for major hepatic resection. J Gastrointest Surg. 1998;2(3):292–8.
44. Farrell GC, Larter CZ. Nonalcoholic fatty liver disease: from steatosis to cirrhosis. Hepatology. 2006;43(S1):S99–S112.
45. Lin TY, Chen CC. Metabolic function and regeneration of cirrhotic and non-cirrhotic livers after hepatic lobectomy in man. Ann Surg. 1965;162(6):959.
46. Fazakas J, Mandli T, Ther G, Arkossy M, Pap S, Füle B, et al. Evaluation of liver function for hepatic resection. Transplant Proc. 2006;38(3):798–800. Elsevier.
47. Melendez JA, Arslan V, Fischer ME, Wuest D, Jarnagin WR, Fong Y, Blumgart LH. Perioperative outcomes of major hepatic resections under low central venous pressure anesthesia: blood loss, blood transfusion, and the risk of postoperative renal dysfunction. J Am Coll Surg. 1998;187(6):620–5.
48. Smyrniotis V, Kostopanagiotou G, Theodoraki K, Tsantoulas D, Contis JC. The role of central venous pressure and type of vascular control in blood loss during major liver resections. Am J Surg. 2004;187(3):398–402.
49. Chen H, Merchant NB, Didolkar MS. Hepatic resection using intermittent vascular inflow occlusion and low central venous pressure anesthesia improves morbidity and mortality. J Gastrointest Surg. 2000;4(2):162–7.
50. Weber JC, Navarra G, Jiao LR, Nicholls JP, Jensen SL, Habib NA. New technique for liver resection using heat coagulative necrosis. Ann Surg. 2002;236(5):560.
51. Ekberg H, Tranberg KG, Andersson R, Jeppsson B, Bengmark S. Major liver resection: perioperative course and management. Surgery. 1986;100(1):1–8.
52. He C, Zhang Y, Cai Z, Lin X, Li S. Overall survival and cancer-specific survival in patients with surgically resected pancreatic head adenocarcinoma: a competing risk nomogram analysis. J Cancer. 2018;9(17):3156.
53. Cameron JL, Riall TS, Coleman J, Belcher KA. One thousand consecutive pancreaticoduodenectomies. Ann Surg. 2006;244(1):10.
54. Kamarajah SK, Burns WR, Frankel TL, Cho CS, Nathan H. Validation of the American Joint Commission on Cancer (AJCC) staging system for patients with pancreatic adenocarcinoma: a Surveillance, Epidemiology and End Results (SEER) analysis. Ann Surg Oncol. 2017;24(7):2023–30.
55. Neoptolemos JP, Stocken DD, Friess H, Bassi C, Dunn JA, Hickey H, et al. A randomized trial of chemora-

diotherapy and chemotherapy after resection of pancreatic cancer. N Engl J Med. 2004;350(12):1200–10.
56. Sultana A, Smith CT, Cunningham D, Starling N, Neoptolemos JP, Ghaneh P. Meta-analyses of chemotherapy for locally advanced and metastatic pancreatic cancer. J Clin Oncol. 2007;25(18):2607–15.
57. Oettle H, Neuhaus P, Hochhaus A, Hartmann JT, Gellert K, Ridwelski K, et al. Adjuvant chemotherapy with gemcitabine and long-term outcomes among patients with resected pancreatic cancer: the CONKO-001 randomized trial. JAMA. 2013;310(14):1473–81.
58. Balaban EP, Mangu PB, Khorana AA, Shah MA, Mukherjee S, Crane CH, et al. Locally advanced, unresectable pancreatic cancer: American Society of Clinical Oncology clinical practice guideline. J Clin Oncol. 2016;34(22):2654–68.
59. Bilimoria KY, Bentrem DJ, Ko CY, Ritchey J, Stewart AK, Winchester DP, Talamonti MS. Validation of the 6th edition AJCC pancreatic cancer staging system: report from the National Cancer Database. Cancer. 2007;110(4):738–44.
60. Rossi ML, Rehman AA, Gondi CS. Therapeutic options for the management of pancreatic cancer. World J Gastroenterol: WJG. 2014;20(32):11142.
61. Hidalgo M. Pancreatic cancer. N Engl J Med. 2010;362(17):1605–17.
62. Fernández-del Castillo C, Morales-Oyarvide V, McGrath D, Wargo JA, Ferrone CR, Thayer SP, et al. Evolution of the Whipple procedure at the Massachusetts General Hospital. Surgery. 2012;152(3):S56–63.
63. Adams DB. The pancreatic anastomosis: the danger of a leak, which anastomotic technique is better? J Gastrointest Surg. 2009;13(7):1182–3.
64. Reid-Lombardo KM, Farnell MB, Crippa S, Barnett M, Maupin G, Bassi C, Pancreatic Anastomotic Leak Study Group, et al. Pancreatic anastomotic leakage after pancreaticoduodenectomy in 1,507 patients: a report from the Pancreatic Anastomotic Leak Study Group. J Gastrointest Surg. 2007;11(11):1451–9.
65. Hüttner FJ, Fitzmaurice C, Schwarzer G, Seiler CM, Antes G, Büchler MW, Diener MK. Pylorus-preserving pancreaticoduodenectomy (pp Whipple) versus pancreaticoduodenectomy (classic Whipple) for surgical treatment of periampullary and pancreatic carcinoma. Cochrane Database Syst Rev.2016(2):CD006053.
66. Fahy BN, Frey CF, Ho HS, Beckett L, Bold RJ. Morbidity, mortality, and technical factors of distal pancreatectomy. Am J Surg. 2002;183(3):237–41.
67. Nathan H, Cameron JL, Goodwin CR, Seth AK, Edil BH, Wolfgang CL, et al. Risk factors for pancreatic leak after distal pancreatectomy. Ann Surg. 2009;250(2):277–81.
68. Beltrame V, Gruppo M, Pastorelli D, Pedrazzoli S, Merigliano S, Sperti C. Outcome of pancreaticoduodenectomy in octogenarians: single institution's experience and review of the literature. J Visc Surg. 2015;152(5):279–84.
69. Liang DH, Shirkey BA, Rosenberg WR, Martinez S. Clinical outcomes of pancreaticoduodenectomy in octogenarians: a surgeon's experience from 2007 to 2015. J Gastrointes Oncol. 2016;7(4):540.
70. Shamali A, De'Ath HD, Jaber B, Abuawad M, Barbaro S, Hamaday Z, Hilal MA. Elderly patients have similar short term outcomes and five-year survival compared to younger patients after pancreaticoduodenectomy. Int J Surg. 2017;45:138–43.
71. Tamirisa NP, Parmar AD, Vargas GM, Mehta HB, Kilbane EM, Hall BL, et al. Relative contributions of complications and failure to rescue on mortality in older patients undergoing pancreatectomy. Ann Surg. 2016;263(2):385.
72. Haigh PI, Bilimoria KY, DiFronzo LA. Early postoperative outcomes after pancreaticoduodenectomy in the elderly. Arch Surg. 2011;146(6):715–23.
73. Lee MK, DiNorcia J, Reavey PL, Holden MM, Genkinger JM, Lee JA, et al. Pancreaticoduodenectomy can be performed safely in patients aged 80 years and older. J Gastrointest Surg. 2010;14(11):1838–46.
74. Melis M, Marcon F, Masi A, Pinna A, Sarpel U, Miller G, et al. The safety of a pancreaticoduodenectomy in patients older than 80 years: risk vs. benefits. HPB. 2012;14(9):583–8.
75. Turrini O, Paye F, Bachellier P, Sauvanet A, Cunha AS, Le Treut YP, et al. Pancreatectomy for adenocarcinoma in elderly patients: postoperative outcomes and long term results: a study of the French Surgical Association. Eur J Surg Oncol. 2013;39(2):171–8.
76. Kim SY, Fink MA, Perini M, Houli N, Weinberg L, Muralidharan V, et al. Age 80 years and over is not associated with increased morbidity and mortality following pancreaticoduodenectomy. ANZ J Surg. 2018;88(5):E445–50.
77. Nakajima H, Yokoyama Y, Inoue T, Nagaya M, Mizuno Y, Kadono I, et al. Clinical benefit of preoperative exercise and nutritional therapy for patients undergoing hepato-pancreato-biliary surgeries for malignancy. Ann Surg Oncol. 2019;26(1):264–72.
78. Jones C, Kelliher L, Dickinson M, Brown R, Martin C, Kumar R. Impact of a prehabilitation program on functional status and length of stay following major pancreatic surgery with an ERAS pathway. Clin Nutr ESPEN. 2017;19:91.
79. Andert A, Lodewick T, Ulmer TF, et al. Liver resection in the elderly: a retrospective cohort study of 460 patients—feasible and safe. Int J Surg. 2016;28:126–30.
80. van Tuil T, Dhaif AA, te Riele WW, van Ramshorst B, van Santvoort HC. Systematic review and meta-analysis of liver resection for colorectal metastases in elderly patients. Dig surg. 2019;36:111–23.
81. Kim SY, Weinberg L, Christophi C, et al. The outcomes of pancreaticoduodenectomy in patients aged 80 or older: a systematic review and meta-analysis. HPB. 2017;19:475–82.

Plastic Surgery in the Elderly

25

Kaveh Alizadeh and Ashraf Elzanie

The longer I live, the more beautiful life becomes. – Frank Lloyd Wright

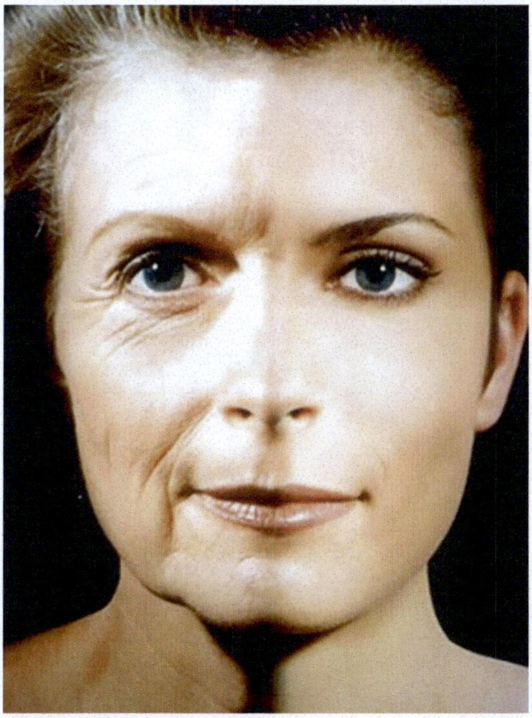

Introduction

As the population of the United States continues to age, there is an increasing demand for plastic surgery among the elderly population. The American Society of Plastic Surgeons reported a year to year 2% increase in procedures by patients over the age of 55 to 4.2 million cosmetic procedures in 2018 alone [1]. By 2050, the United States is expected to accelerate its elderly population growth of people 65 or older to 88.5 million people from its current 50 million today [2]. With age, there is an increasing demand for reconstructive surgeries such as wound care, tumor removal, and cancer reconstruction. ASPS statistics reveal a 29% increase in breast reconstruction alone with over 101,000 cases performed in 2018 [1]. The fastest rising trend in plastic surgery is the demand for post-bariatric body lifts with arm and lower truncal lift accounting for 5000% increase in numbers over the past two decades. With a third of Americans obese, there is no doubt these interventions will continue to rise. Furthermore, media and global access drive the celebrity culture and medical tourism which in turn have created a disproportionate demand for "minimally invasive" procedures that have made "60 is the new 40 years of age" culture and have contributed to the $40 billion plastic surgery global market. Therefore it is important to understand the plastic surgery principles surrounding the aging needs of the elderly

K. Alizadeh (✉) · A. Elzanie
Westchester Medical Center, Valhalla, NY, USA
e-mail: Kaveh.Alizadeh@wmchealth.org

R. Latifi (ed.), *Surgical Decision Making in Geriatrics*,
https://doi.org/10.1007/978-3-030-47963-3_25

population as it pertains to economic impact, improved outcomes, and the health of our society.

Biological Factors of Skin Aging

Over the past decade there has been tremendous amount of published literature focused on aging. With regard to plastic surgery, we are interested in factors that cause aging of the skin, muscle, bone, tendons, and peripheral nerves.

Extrinsic Factors

By far the most damaging external factor that can lead to "premature aging" is considered to be exposure to the sun. UVA and UVB rays of the sun can lead to free radicals which can directly damage the dermal layer, impair DNA repair of the skin, and suppress the immune mediated defense of the skin mediators. The buildup of damaged elastin at the dermal layer, combined with increased collagen breakdown and lack of repair, leads to uneven epidermal thickening and dermal thinning of the skin called solar elastosis which then further exacerbates the morphology of the skin with vessels that come close to the surface known as telangiectasia, and congregation of pigment-producing melanocyte cells called solar lentigo and dyschromia. Other factors that play a large role in the aging of the skin include dehydration, wind (primarily through dehydration), cold exposure, and smoking. Guyuron et al. studied the role of smoking in facial aging and found that a 5 year history of smoking can cause a noticeable difference in the appearance of aging in identical twins [3]. Similarly Ichibori and colleagues studied 67 pairs of Japanese monozygotic twins and found that cigarette smoke and sun exposure were the primary factors to worsen facial texturing [4] (Fig. 25.1).

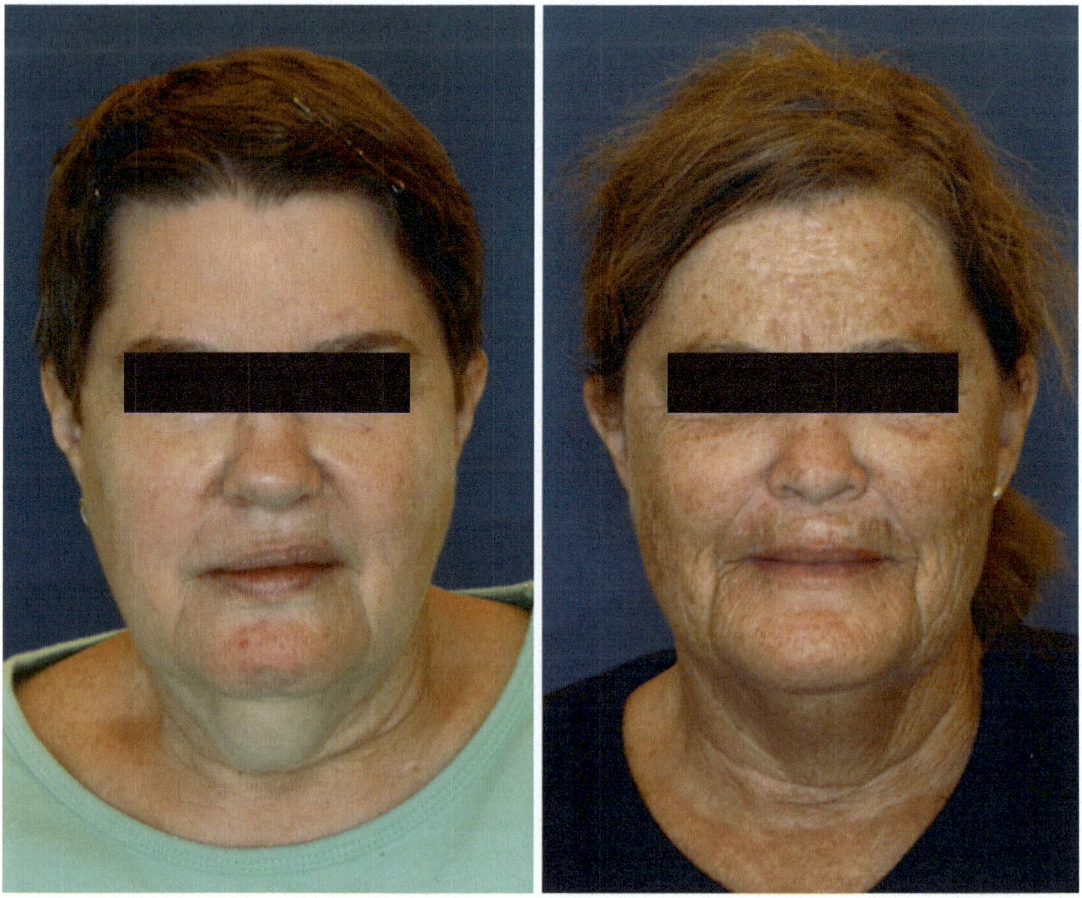

Fig. 25.1 Facial changes caused by smoking: a comparison between smoking and nonsmoking identical twins. (Reprinted with permission from Plast Reconstr Surg. 2009;123(4):1321–31)

Fig. 25.2 The aging of French actress Bridget Bardot over a 50 year period revealing the intrinsic and extrinsic forces at play in aging with loss of facial definition and diamond shape and descent of soft tissue in the neck

Another major noted factor for aging on our planet is the presence of gravity. Gravity places forces on our body which require a counterforce directed by our muscles, bones, tendons, and ligaments. Over time, with repetitive motion of these activities, a typical pattern of aging emerges in the face which has been well documented by investigators [5–7].

In a series of carefully reported studies of computed tomography scans of various age groups, Kahn and his team have shown that there is definitive changes in facial skeleton over time as the glabellar and maxillary angle in both the male and female subjects decrease with increasing age while there is a significant increase in pyriform aperture area from the young to the middle age group for both sexes. The pyriform and orbital aperture width and surface area increases significantly with age for both sexes. Mandibular length and height both decrease significantly for each sex while the mandibular angle significantly increases with age for both sexes. Coupled with Mendelson's work which showed increased descent of mid cheek soft tissue and loss of malar and mandibular ligament support, we can start to put together a predictable pattern of facial aging in all humans [8]. One cannot help but wonder about the circular nature of life since these facial patterns are very similar to the morphology of a baby's face (Fig. 25.2).

Intrinsic Factors

The ability of the skin to age well also depends on a multitude of factors including collagen which provides the structure foundation of the extracellular matrix (ECM) in the dermis. There are two major subtypes of collagen in the human skin: Type I collagen forms up to 90% of the skin while type III is responsible for about 10% of the dermis. El domiyati et al. have shown that human skin which primarily contains type I and III collagen can maintain its proportion well into the eighth decade of life unless interrupted by extrinsic factors [9], while Lovell and his team have suggested that the intrinsic effects of aging are primarily related to the inhibited synthesis of type I collagen [10]. Collagen, which is a major component of ECM, becomes fragmented and unevenly distributed due to increased activity of matrix metalloproteinases, and reactive oxygen species disrupt the normal signaling mechanism of transforming growth factor-β during aging.

The reduction in the amount of collagen hinders the mechanical interaction between fibroblasts and the ECM, and consequently leads to the deterioration of fibroblast function and further decrease in the amount of dermal collagen. Other ECM components, including elastic fibers, glycosaminoglycans (GAGs), and proteoglycans (PGs), are also negatively affected by aging. Dermal elastin which primarily stores energy for skin recoil and hyaluronic acid which gives its turgor decrease in intrinsically aged skin giving the typical appearance of dry skin that does not bounce. There is plenty of evidence that supports the loss of all these major components of the skin from ongoing protein glycation and inherent inflammatory damage [11–16].

There is also loss of muscle mass, and declining strength with age as nearly 30% of muscle mass is lost by the ninth decade of life [17]. This is likely due to decreased muscle build with lower body protein and increased expression of inflammatory factors leading to skeletal muscle catabolism with concomitant increased apoptosis and decreased mitochondrial function.

Age-related changes in bone involve reduced calcium and phosphate metabolism with resultant loss of mass and mineral content, increased marrow fat content, and altered response to growth factors and hormones [18]. Osteoblasts, osteoclasts, and osteocytes, like most other stem and progenitor cells, slow their activity leading to osteoporosis and skeletal fragility, increasing susceptibility to fractures [19]. The decline in structure and function of aged tendons results from degeneration of tenocytes and collagen fibers, accumulation of lipids and ground substance, and calcium deposits. Tenoblast metabolic activity also decreases which results in tensile strength loss, stiffness, and increased susceptibility to damage.

Within the peripheral nervous system, there is impaired regeneration of neurons which accumulate lipofuscin granules, with subsequent axonal loss, demyelination, and synapse number reduction and attenuated growth factor response [20]. These changes result in age-related declines in nerve conduction velocity, muscle strength, sensory discrimination, and autonomic responses and lead to poorer outcomes in reinnervation following peripheral nerve injury in the aging population [21] (Fig. 25.3).

Along with the molecular changes, there is a remarkable change that also happens in the facial soft tissue. Pesa and his colleagues have published a series of papers that have shown the intricate nature of the facial fat compartments and their relative contribution to aging over time. We now have a better understanding of why aging happens in such a predictable pattern in the periorbital area followed by fat involution in distinct compartments of the cheeks and lower face [23–25]. Even more fascinating is the knowledge about why men seem to age "slower" than women. MRI studies have shown that women tend to lose their diamond facial shape

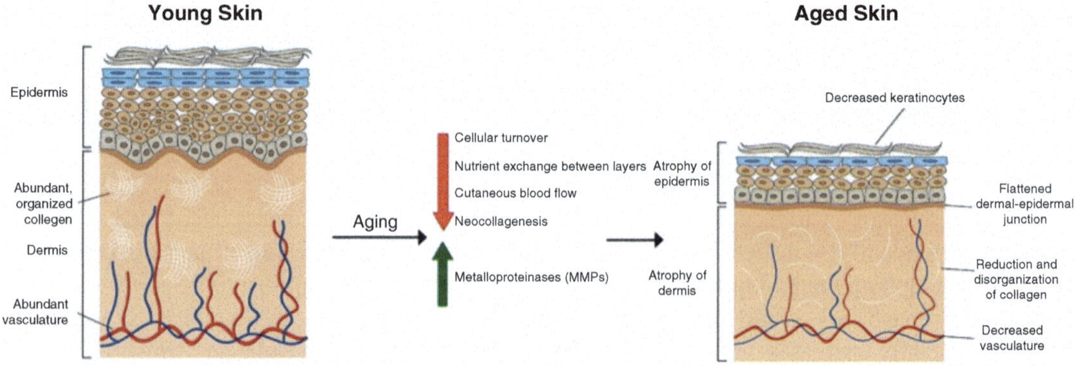

Fig. 25.3 Skin aging. Decreased cellular turnover and inefficient nutrient exchange between the different layers occurring with aging result in atrophy of both the epidermis and dermis. The decrease in collagen number and organization result from decreased production of collagen as well as increased breakdown by metalloproteinases. A reduction in the vasculature is also seen leading to inefficient cutaneous blood supply [22]

with time whereas men maintain their square shape as they age [26]. The result is that both men and women age with square-shaped faces as the bones resorb and soft tissue descend into jowls and lower neck. However, our brains perceive that women's faces lose their rejuvenated appeal as they lose their diamond shape (Table 25.1).

Physical Exam and Analysis

Armed with the knowledge of intrinsic and extrinsic factors of aging, it is important to evaluate the patients carefully based on these known parameters and document it carefully before one undertakes any surgery on the elderly. The Table 25.2 below shows what the surgeon needs to evaluate and document prior to operating on the geriatric patient.

Table 25.1 Changes in dermal extracellular components in the aging process

Photoaging	Components	Intrinsic Aging
Decreased and fragmented	Collagen	Decreased and fragmented
Abnormally accumulated (SE)	Elastic fiber	Decreased
Increased in SE region	Hyaluronic acid	Not changed
Increased	Total sulfated GAGs	Decreased
Increased in SE region	Versican	Not changed?
Not changed	Biglycan	Decreased
Decreased in SE region	Decorin	Not changed?

GAG glycosaminoglycan, *SE* solar elastosis
Source: Shin JW, Kwon SH, Choi JY, et al. Molecular Mechanisms of Dermal Aging and Antiaging Approaches. Int J Mol Sci. 2019;20(9):2126. Published 2019 Apr 29. https://doi:10.3390/ijms20092126

Planning for Plastic Surgery

It is important to note that although many of the changes of aging manifest themselves in the cosmetic evaluation, we also need to be aware of the physiological aspects of aging that can affect the surgical outcome in the elderly patient. The core values of plastic surgery dictate that the physician first address the functional impairments, next focus on form, and finally maximize the aesthetic benefit of the operation for the patient. There are a number of important considerations in planning reconstructive surgery for the geriatric patient.

Breast Reconstruction

With the aging of the world, the prevalence of breast cancer among elderly women is increasing. Nearly half of all women who are diagnosed with breast cancer are over the age of 65. As a result, more elderly women require breast reconstruction.

Table 25.2 A check list of all these factors must be considered when evaluating the progression of aging in facial surgery

Differential aging of the body part
Skin state and scarring history
Photo damage, dyschromia, and telangiectasia
Bony structure and previous fractures
Status of teeth and dental surgery
Fat vs muscle soft tissue cover
Quality vs quantity of facial rhytids
Thickness vs oiliness of the skin
Elasticity vs hyaluronic acid component of skin
Differential soft tissue descent of the face
Motor vs sensory deficits

Historically breast reconstruction after mastectomy in elderly women has been examined in the form of autologous vs implant-based techniques. The options for autologous approach in the elderly have included pedicled transverse rectus abdominis myocutaneous (TRAM) flaps or latissimus flaps vs free tissue transfer with abdominal tissue or other distant sites that limit the donor site at the cost of longer operative time and hospitalization. The non-autologous approach has involved use of expanders and permanent implant prosthesis and acellular dermal matrix (ADM) to provide a reliable aesthetic outcome. The latest trend in breast reconstruction involves a skin and nipple sparing mastectomy followed by use of a pre-pectoral implant and ADM for lower pole control [27]. These cases often require a secondary fat grafting to the upper pole to provide for better symmetry (Fig. 25.4).

There have been limited studies examining elderly women using a direct-to-implant reconstructive approach. The appeals of this approach in this particular population come in the form of a single surgical procedure, no tissue expansion needed and fewer follow-up visits. This study was a retrospective chart review over 4 years. A total of 24 breasts in 19 elderly patients were analyzed who underwent direct-to-implant reconstruction [28]. This was compared to a significantly younger control group population that underwent tissue expander with subsequent implant. Between the two groups there was no significant difference between wound complication rates or failed reconstruction. However, older individuals had significantly lower number of drain days, length of stay, readmissions, and postoperative visits.

Torabi and colleagues examined 339 patients that underwent deep inferior epigastric perforator (DIEP) microsurgical tissue transfer flaps (285 non elderly vs 54 elderly patients) with primary outcomes of complete flap loss, partial flap loss, or return to OR and were able to demonstrate a higher prevalence of medical comorbidities such as diabetes and hypertension with age [29]. When compared to the nonelderly population, elderly patients had a higher rate of complete flap loss and wound complications. More specifically, patients in the elderly cohort demonstrated an odds ratio of 10.92 for complete flap loss and had significantly higher rates of wound dehiscence (24.1% vs. 8.4%) vs the younger cohort. Success rates of free flap reconstruction in both cohorts were compared and found to be similar (99.6% in the elderly population compared to 96.3% in the nonelderly population). The study concluded that autologous free flap breast reconstruction remains a viable option for elderly patients although age is an independent risk factor for complete flap loss.

Santosa and colleagues published a high evidence-level multicenter prospective study which showed patients over the age of 60 did not have increased complications compared to their younger cohorts, although they tended to get more unilateral and delayed pedicled autologous options [30].

Hand and Microsurgery

It is important to preclude comorbidities such as vascular disease, smoking, and injury mechanism when analyzing the role of age in hand and microsurgery outcomes. Retrouvey took on this task when his team analyzed 284 patients that underwent finger reimplantation or revascularization with 32 of these patients over 60 years old [31]. Approximately half of the total population measured underwent revascularization alone while the other group underwent revascularization with reimplantation of the digit. When comparing the two populations, the older group had higher ASA scores and had more associated comorbidities. Overall a total of 88 patients had reimplanted digit failure or thrombosis of revascularization. Furthermore, this encompassed a total of 12 failures within the elderly group. When a multivariate logistic regression was performed, this showed that older patients did not experience major complications or failure rates when compared to younger age groups; however, they had higher rates of minor complications such as wound infections. Therefore, based on the data gathered from this study, ASA class and age should not preclude elderly patients from undergoing digital revascularization or reimplantation. The results of this

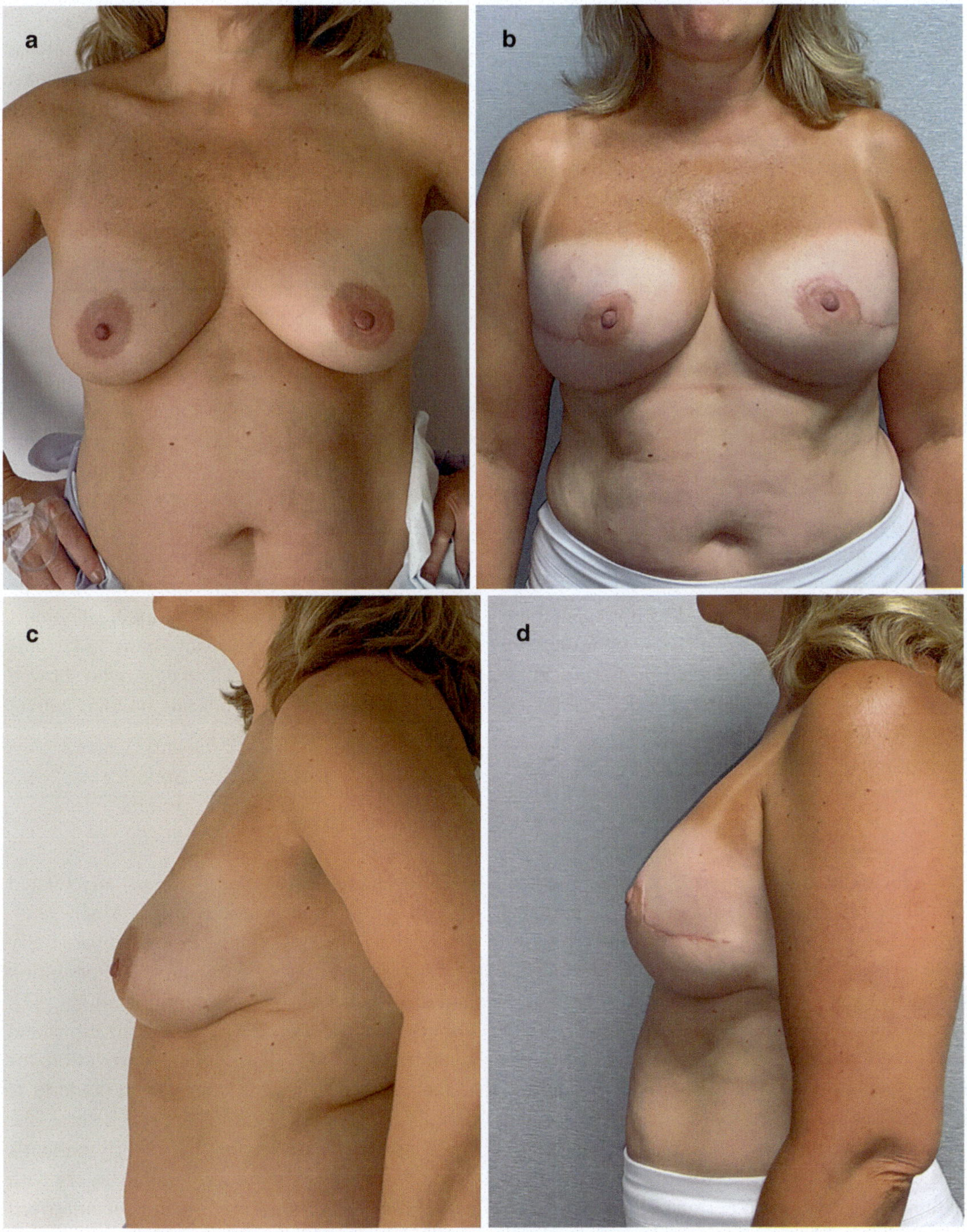

Fig. 25.4 (**a**) (front view) and (**b**) (side view) This patient was diagnosed with ductal cell carcinoma and required a skin and nipple sparing mastectomy followed by immediate implant reconstruction with acellular dermal matrix for lower pole support. (**c**) (front post-op) and (**d**). (Side view post-op) Patient is seen 4 months post reconstruction after having undergone an intermediary fat grafting session to the upper and medial poles to improve her projection and symmetry

study were corroborated by Barzin et al. in the *Journal of Hand Surgery* [32]. On the other hand (pun intended), Kwon et al. found that patients over age 70 had worse functional recovery than their younger cohorts with longer rehabilitation parameters [33]. Similarly, studies on tendon repair have found that older patients are more likely to require reoperation after flexor tendon repair [34, 35].

Microsurgical literature is more controversial with regard to the effect of aging on outcomes. A systematic review incorporated a total of 45 articles including 115 patients [36]. Majority of the included individuals were male and the mean age for this population was 73. Mean flap size was 598 cm^2 which ranged from 82 to 2500 cm^2. Only two of the 115 cases reported flap failure and the mortality rate for this population was 0.9 percent. Total complication rate for patients undergoing microvascular reconstruction was reported to be 22%. However, when stratified by flap type, there was no significant difference in complication rate. This study concluded that for microvascular reconstruction for complex scalp defects, age is not a risk factor for flap failure.

Üstün et al. [37] performed a systematic review and meta-analysis of free flaps in elderly patients. They found no difference in elderly versus young flap success rates or surgical complications; however, they did find significantly more medical complications and mortality in elderly patients. Jubbal's team analyzed 5951 cases of free tissue transfer and when controlling for comorbidities found that age itself was not significantly associated with complications. Age was significantly associated with increased mortality. As such, the authors recommended assessment of "physiological" age instead of chronological age in assessing patients for free tissue reconstruction [38].

Lower extremity reconstruction in the elderly is more complex due to peripheral artery disease and diabetes which makes it a better candidate for microsurgery where studies have found no difference in complication rates in patients >65; however, these patients required a higher rate of ICU admission [39–41].

Serletti et al. examined 100 patients >65 years retrospectively and found that chronological age did not predict flap complications [42]. However, higher ASA scores and length of operative time were significant predictors of postoperative surgical morbidity. Another literature review of head and neck free flap reconstruction found no difference in flap success, or mortality rate in elderly patients based on chronological age [43].

A large retrospective cohort study examined 211 patients 70 years or older at a single institution who underwent free tissue transfer. Although surgical complication rates were similar between the two groups (using rectus abdominis 32.7%, radial forearm 25.6%, jejunum 17.1%, fibula 18.1%, and latissimus dorsi 6.5%), there were significantly higher rates of medical complications in the octogenarian group when compared to septuagenarians. Furthermore, among this study's population it was identified that alcohol use, hypertension, and coronary artery disease were independent predictors of overall medical and surgical complications. Therefore, this study concluded that the elderly population must be carefully selected on the basis of medical comorbidities and overall functional status [44]. This was further illustrated by Chen et al. who showed a 37% surgical complication rate on their patient population over age 85 [45].

Facial Fractures

Facial fractures in the elderly are usually related to falls as opposed to trauma [46]. Atisha et al. found that elderly patients >64 years generally required significantly less operative intervention and also had fewer surgical complications [47]. Others have reported longer hospitalizations and rehabilitation [48]. The clinical correlation for management of these fractures in the elderly is usually a "watchful waiting" approach since one must also take into account the elderly patient's own wishes who may not want more surgery with plates and screws in their face for an improved aesthetic outcome.

Cosmetic Surgery

With the graying of the American population who want to keep looking young, there is an increased demand for cosmetic surgery among the elderly. A recent publication studies a total of

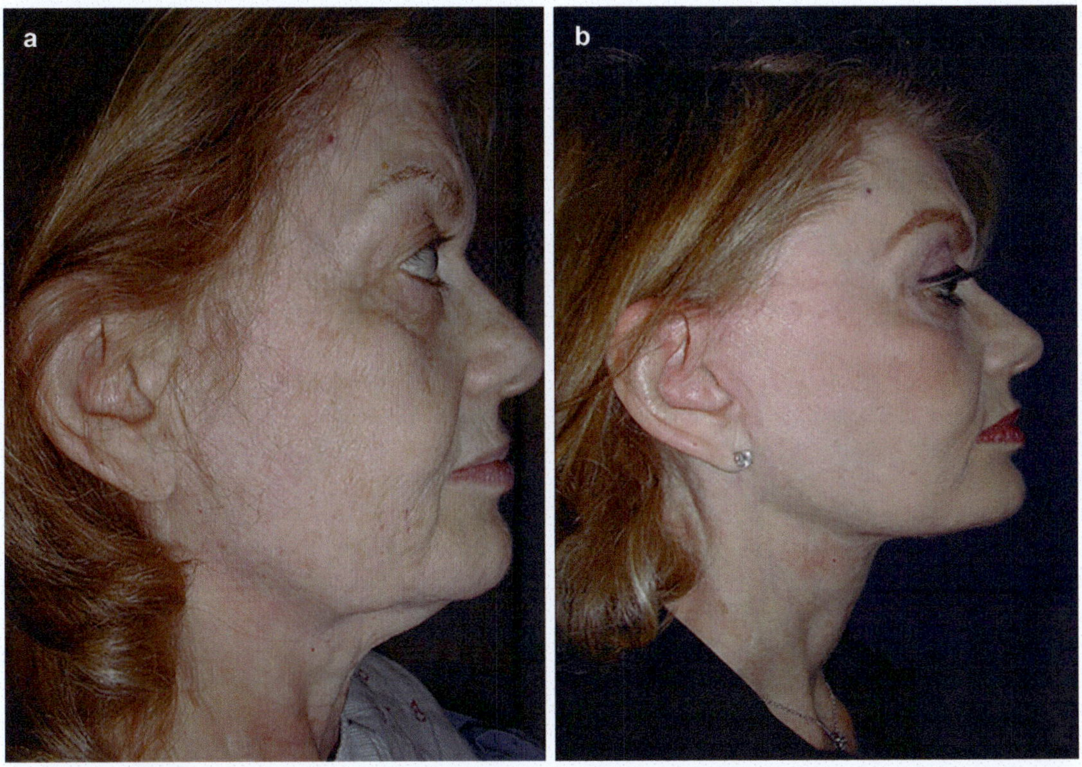

Fig. 25.5 (**a**) Sixty-year-old patient who presented for rhytidectomy and laser resurfacing to address her facial aging. (**b**) Same patient at age 70 returns 10 years post-op revealing increased facial skin aging with elastosis despite maintaining excellent skin contour of the neck

6786 elderly patients who underwent cosmetic procedures [49]. In the postoperative period, the overall complication rate was compared between elderly and younger populations and was found to have no statistical significant difference between the two groups. Furthermore, this was still true when stratified to octogenarians which showed no significant difference in complication rates among this population, although they had higher rates of postoperative infection and hematoma. This article concluded that there is no added risk to performing cosmetic procedures in the elderly, including octogenarians.

A single surgeon retrospective cohort study examined a total of 216 patients undergoing a face lift procedure. Patients were divided into an older cohort 65 and older with a total of 68 patients and a younger cohort under 65 years old with a total of 148 patients. Elderly patients' major and minor complication rates were not significantly different when compared to those in the younger age cohort with no mortality in either group. Based on these results, there was no significant difference in terms of outcomes measured from face lifts performed in the elderly and younger cohort [50].

Using a higher level of evidence, Karamanos et al. used NSQIP data to examine wound dehiscence from plastic surgery patients over a 5-year period and found that there was less than 1% dehiscence rate regardless of the age of the patient [51] (Fig. 25.5).

Abdominal Wall Reconstruction

Optimization of patients undergoing abdominal wall reconstruction for complex ventral hernias remains a complex issue. This particular cohort often presents with loss of abdominal domain, pain, difficulty performing routine physical activities in addition to being deconditioned, malnourished, and overweight. Furthermore, hernia repair and abdominal wall reconstruction of complex ventral hernias are major surgical procedures which induce further stress on the elderly population that may already have other comorbidities such as hypertension, cardiovascular disease, and diabetes. This cycle is initiated in the form of insulin resistance and proteolysis with further loss of lean body mass and the initiation of a catabolic state which leads to an impairment of immunologic function. Additional surgical adjuncts for abdominal wall closure such as implants and permanent mesh can act as a harbinger for chronic infection. However, based on previous literature there are preoperative modifiable risk factors that have been identified to reduce these effects. Obesity has been shown to be a significant risk factor for wound complications in the past. Previous studies have suggested that weight reduction with a BMI less than 40 is associated with better outcomes. A second modifiable risk factor is smoking cessation. With patients undergoing major reconstructive surgery, wound healing must be optimized. Negative effects of smoking in the form of reduction of blood and tissue oxygen tension levels as well as collagen formation lead to increased frequency of wound complications. Another modifiable risk factor includes glycemic control. It has been shown in previous studies that diabetics with a goal of HbA1c level of less than 6.5% preoperatively are associated with decreased rates of infection and other wound complications [52]. Sarcopenia is an additional modifiable risk factor that should be considered in this particular population. Sarcopenia when coupled with obesity has been shown to decrease overall survival as well as negative outcomes in the surgical as well as critical care settings. There are regimens available to counteract this detrimental disease process. Furthermore, these regimens are in the form of increased protein intake with combined resistance exercises which provide the best chance for preserving or regaining/maintaining functional status following complex hernia repair and abdominal wall reconstruction.

In terms of preoperative assessment of nutrition there is a lack of focus on preoperative nutritional optimization prior to surgery. It has been shown in previous studies that optimizing nutrition has been shown to decrease hospital stay, readmissions, and postoperative complications. There are a few tools available to clinicians to assess for nutritional status and who would benefit from nutritional intervention. One is in the form of body composition analysis which utilizes either bioelectrical impedance analysis or radiologic methods to determine body composition.

Giordano et al. used a propensity score and showed that abdominal wall reconstruction with acellular dermal matrix may result in more bulge/laxity rates in the elderly but does not significantly alter outcomes between populations over and under 65 years of age [53].

Pressure Ulcers

About 10% of elderly patients in hospitals will develop a pressure ulcer during hospitalization [54]. Much of this is related to chronic care protocols and procedures and lack of mobilization which would mean age should not be a factor. However, age and frailty are often intertwined, and Margolis et al. calculated the pressure ulcer probability of those >80 years were 4–20 times more likely to develop a pressure ulcer than the younger cohorts [55]. This makes it even more critical to make sure that the elderly patient is optimized from the nutritional and positioning standpoint while in an acute care setting.

Conclusion

Plastic surgery is always faithful to the principles of function, form, and cosmesis. The discipline is also unique in that the clinicians interact with all sexes, ages, and conditions from elective cosmetic surgery to acute posttraumatic reconstruction. With our novel understanding of the science of aging, it is imperative for the practitioner of plastic surgery to differentiate chronological age from frailty. The emerging published literature in the field shows that although one can safely operate on the elderly patient, the medical comorbidities must be thoroughly identified and discussed with the patients to make sure that the potential prolonged rehabilitation outcomes match the expectations and desires of the elderly patient. Armed with novel directions in the management of the elderly population, we can utilize "prehabilitation strategies" before major hospitalizations in order to improve outcomes, economic impact, and the health of our society.

References

1. ASPS 2018 statistics. Available at www.plasticsurgery.org.
2. US census bureau US department of commerce 2010.
3. Okada HC, Alleyne B, Varghai K, Kinder K, Guyuron B. Facial changes caused by smoking: a comparison between smoking and nonsmoking identical twins. Plast Reconstr Surg. 2013;132(5):1085–92. If we can get permission we can incorporate this photo as well form this study
4. Ichibori R, Fujiwara T, Tanigawa T, Kanazawa S, Shingaki K, Torii K, Tomita K, Yano K, Osaka Twin Research Group, Sakai Y, Hosokawa K. Objective assessment of facial skin aging and the associated environmental factors in Japanese monozygotic twins. J Cosmet Dermatol. 2014 Jun;13(2):158–63. https://doi.org/10.1111/jocd.12081.
5. Shaw RB Jr, Katzel EB, Koltz PF, Yaremchuk MJ, Girotto JA, Kahn DM, Langstein HN. Aging of the facial skeleton: aesthetic implications and rejuvenation strategies. Plast Reconstr Surg. 2011;127(1):374–83.
6. Mendelson B, Wong CH. Changes in the facial skeleton with aging: implications and clinical applications in facial rejuvenation. Aesthet Plast Surg. 2012;36(4):753–60. https://doi.org/10.1007/s00266-012-9904-3 . Epub 2012 May 12.
7. Shaw RB Jr, Kahn DM. Aging of the midface bony elements: a three-dimensional computed tomographic study. Plast Reconstr Surg. 2007;119(2):675–81; discussion 682-3.
8. Mendelson BC. Anatomic study of the retaining ligaments of the face and applications for facial rejuvenation. Aesthet Plast Surg. 2013;37(3):513–5. https://doi.org/10.1007/s00266-013-0066-8.
9. El-Domyati M, Attia S, Saleh F, Brown D, Birk E, Gasparro F, Ahmad H, Uitto J. Intrinsic aging vs. photoaging: a comparative histopathological, immunohistochemical, and ultrastructural study of skin:398–405First. Published: 04 October 2002 experimental dermatology
10. Type I and III collagen content and fibre distribution in normal human skin during ageing C.R. Lovellk.A . Smolenskiv.C. Duancen.D. Lights. Youngm. Dyson First published: October 1987.
11. Werth VP, Shi X, Kalathil E, Jaworsky C. Elastic fiber-associated proteins of skin in development and photoaging. Photochem Photobiol. 1996;63:308–13.
12. Watson REB, Griffiths CEM, Craven NM, Shuttleworth CA, Kielty CM. Fibrillin-rich microfibrils are reduced in photoaged skin. Distribution at the dermal-epidermal junction. J Invest Dermatol. 1999;112:782–7.
13. Yaar M, Gilchrest BA. Photoageing: mechanism, prevention and therapy. Br J Dermatol. 2007;157:874–87.
14. Farage MA, Miller KW, Elsner P, Maibach HI. Intrinsic and extrinsic factors in skin ageing: a review. Int J Cosmet Sci. 2008;30:87–95.
15. Varani J, Dame MK, Rittie L, Fligiel SE, Kang S, Fisher GJ, Voorhees JJ. Decreased collagen production in chronologically aged skin: roles of age-dependent alterations in fibroblast function and defective mechanical stimulation. Am J Pathol. 2006;168:1861–8.
16. Lapiere CM. The ageing dermis: the main cause for the appearance of 'old' skin. Br J Dermatol. 1990;122(Suppl. 35):5–1.
17. Ali S, Garcia JM. Sarcopenia, cachexia and aging: diagnosis, mechanisms and therapeutic options—a mini-review. Gerontology. 2014;60:294–305.
18. Boros K, Freemont T. Physiology of ageing of the musculoskeletal system. Best Pract Res Clin Rheumatol. 2017;31:203–17.
19. Ensrud KE. Epidemiology of fracture risk with advancing age. J Gerontol A Biol Sci Med Sci. 2013;68:1236–42.
20. Pannese E. Morphological changes in nerve cells during normal aging. Brain Struct Funct. 2011;216:85–9.
21. Kovacic U, Sketelj J, Bajrović FF. Chapter 26: age-related differences in the reinnervation after peripheral nerve injury. Int Rev Neurobiol. 2009;87:465–82.
22. Roh DS, Panayi AC, Bhasin S, Orgill DP, Sinha I. Implications of aging in plastic surgery. Plast Reconstr Surg Glob Open. 2019;7(1):e2085. Published online 2019 Jan 14. https://doi.org/10.1097/GOX.0000000000002085.
23. Rohrich RJ, Pessa JE. The fat compartments of the face: anatomy and clinical implications for cosmetic surgery. Plast Reconstr Surg. 2007;119(7):2219–27; discussion 2228-31.2008 May;121(5):1804-9.
24. Rohrich RJ, Pessa JE The retaining system of the face: histologic evaluation of the septal boundaries of the subcutaneous fat compartments.
25. JE P. SMAS fusion zones determine the subfascial and subcutaneous anatomy of the human face: fascial spaces, fat compartments, and models of facial aging. Aesthet Surg J. 2016;36(5):515–26. https://doi.org/10.1093/asj/sjv139. Epub 2016 Feb 23
26. Foissac R, Camuzard O, Piereschi S. High-resolution magnetic resonance imaging of aging upper face fat compartments. Plast Reconstr Surg. 2017;139(4):829–37.
27. Vidya R, Berna G, Sbitany H, et al. Prepectoral implant-based breast reconstruction: a joint consensus guide from UK, European and USA breast and plastic reconstructive surgeons. Ecancermedicalscience. 2019;13:927. Published 2019 May 7. https://doi.org/10.3332/ecancer.2019.927.

28. Knackstedt R, et al. Direct-to-implant breast reconstruction in women older than 65 years. Plast Reconstr Surg. 2018;141(4):1076. https://doi.org/10.1097/prs.0000000000004469.
29. Torabi R, Stalder MW, Tessler O. Assessing age as a risk factor for complications in autologous breast reconstruction. Plast Reconstr Surg. 2018;142(6):840e–6e.
30. Santosa KB, Qi J, Kim HM, et al. Effect of patient age on outcomes in breast reconstruction: results from a multicenter prospective study. J Am Coll Surg. 2016;223:745–54.
31. Retrouvey H, et al. The effect of increasing age on outcomes of digital revascularization or replantation. Plast Reconstr Surg. 2018:1. https://doi.org/10.1097/prs.0000000000005195.
32. Barzin A, Hernandez-Boussard T, Lee GK, et al. Adverse events following digital replantation in the elderly. J Hand Surg Am. 2011;36:870–4.
33. Kwon GD, Ahn BM, Lee JS, et al. The effect of patient age on the success rate of digital replantation. Plast Reconstr Surg. 2017;139:420–6.
34. Dy CJ, Daluiski A, Do HT, et al. The epidemiology of reoperation after flexor tendon repair. J Hand Surg Am. 2012;37:919–24.
35. Rigo IZ, Røkkum M. Predictors of outcome after primary flexor tendon repair in zone 1, 2 and 3. J Hand Surg Eur. 2016;41:793–801.
36. Sosin M, et al. Microsurgical scalp reconstruction in the elderly. Plast Reconstr Surg. 2015;135(3):856–66. https://doi.org/10.1097/prs.0000000000000959.
37. Üstün GG, Aksu AE, Uzun H, et al. The systematic review and meta-analysis of free flap safety in the elderly patients. Microsurgery. 2017;37:442–50.
38. Jubbal KT, Zavlin D, Suliman A. The effect of age on microsurgical free flap outcomes: an analysis of 5,591 cases. Microsurgery. 2017;8:858–64.
39. Xiong L, Gazyakan E, Wähmann M, et al. Microsurgical reconstruction for post-traumatic defects of lower leg in the elderly: a comparative study. Injury. 2016;47:2558–64.
40. Bhama PK, Patel SA, Khan U, et al. Head and neck free flap reconstruction in patients older than 80 years. J Reconstr Microsurg. 2014;30:523–30.
41. Goh CS, Kok YO, Yong CP, et al. Outcome predictors in elderly head and neck free flap reconstruction: a retrospective study and systematic review of the current evidence. J Plast Reconstr Aesthet Surg. 2018;71:719–28.
42. Serletti JM, Higgins JP, Moran S, et al. Factors affecting outcome in free-tissue transfer in the elderly. Plast Reconstr Surg. 2000;106:66–70.
43. Grammatica A, Piazza C, Paderno A, et al. Free flaps in head and neck reconstruction after oncologic surgery: expected outcomes in the elderly. Otolaryngol Head Neck Surg. 2015;152:796–802.
44. Garner WI. Free tissue transfer in the elderly: incidence of perioperative complications following microsurgical reconstruction of 197 septuagenarians and octogenarians. Yearbook of Plastic and Aesthetic Surgery. 2007;2007:274–5. https://doi.org/10.1016/s1535-1513(08)70243-1.
45. Chen XF, Chen YM, Gokavarapu S, et al. Free flap reconstruction for patients aged 85 years and over with head and neck cancer: clinical considerations for comprehensive care. Br J Oral Maxillofac Surg. 2017;55:793–7.
46. Falcone PA, et al. Maxillofacial fractures in the elderly. Plast Reconstr Surg. 1990;86(3):443–8. https://doi.org/10.1097/00006534-199009000-00009.
47. Atisha DM, Burr TV, Allori AC, et al. Facial fractures in the aging population. Plast Reconstr Surg. 2016;137:587–93.
48. Mundinger GS, Bellamy JL, Miller DT, et al. Defining population-specific craniofacial fracture patterns and resource use in geriatric patients: a comparative study of blunt craniofacial fractures in geriatric versus nongeriatric adult patients. Plast Reconstr Surg. 2016;137:386e–93e.
49. Yeslev M, et al. Safety of cosmetic procedures in elderly and octogenarian patients. Aesthet Surg J. 2015;35(7):864–73. https://doi.org/10.1093/asj/sjv053.
50. Gutowski Ka. The safety of Rhytidectomy in the elderly. Yearbook of Plastic and Aesthetic Surgery. 2013;2013:90–1. https://doi.org/10.1016/j.yprs.2011.11.005.
51. Karamanos E, Osgood G, Siddiqui A, et al. Wound healing in plastic surgery: does age matter? An American College of Surgeons National Surgical Quality Improvement Program study. Plast Reconstr Surg. 2015;135:876–81.
52. Joslyn NA, et al. Evidence-based strategies for the prehabilitation of the abdominal wall reconstruction patient. Plast Reconstr Surg. 2018;142:21s.
53. Giordano S, Schaverien M, Garvey PB, et al. Advanced age does not affect abdominal wall reconstruction outcomes using acellular dermal matrix: a comparative study using propensity score analysis. Am J Surg. 2017;213:1046–52.
54. Maklebust J, Magnan MA. Risk factors associated with having a pressure ulcer: a secondary data analysis. Adv Wound Care. 1994;7:25, 27–28, 31.
55. Margolis DJ, Bilker W, Knauss J, et al. The incidence and prevalence of pressure ulcers among elderly patients in general medical practice. Ann Epidemiol. 2002;12:321–5.

Drug Use in Elderly During Surgery: They Were Youngsters Once

26

James M. Feeney

As Americans live longer, growth in the number of older adults as a percentage of the population is unprecedented. In 2014, 14.5% (46.3 million) of the US population was aged 65 or older and is projected to reach 23.5% (98 million) by 2060 [1]. As people age, they tend toward increased contact with the medical community and acquire new diagnoses and treatments with increasing frequency. There are several reasons for this. People start a regimen of preventative care usually in middle age, mandating visits to a primary care physician more frequently than in their younger adulthood [2, 3]. Also, aging adults experience higher risk of chronic disease. In 2012, 60% of older adults managed two or more chronic conditions. According to the US Federal Interagency Forum on Aging-Related Statistics, the most common comorbidities include heart disease, cancer, chronic obstructive pulmonary disease (COPD), and diabetes mellitus.

These and other reasons combine to produce a higher rate of medication use in elderly patients, and that fact combined with the increasing rates of serious comorbidity and frailty complicates the performance of surgery in this population. Patients' experiences in the elective surgery realm demands planning surround perioperative optimization and medication use leading up to the day of surgery, multispecialty and multidisciplinary collaboration in the preoperative and postoperative phases, and close follow-up in the post-surgical, return to normal life phase. However, in the emergent setting, there is often not sufficient time to alter preoperative medication use or to mitigate chronic conditions optimally.

In this chapter, we will explore medication use in the elective perioperative period, including common pitfalls in preoperative and postoperative phases of care and including the resumption of preoperative care and home medication, with or without changes, as appropriate. However, first, we will briefly touch on the performance of emergent surgery in the geriatric population with multiple medications from chronic conditions.

Emergency Surgery in the Geriatric Population with Polypharmacy

Emergency surgery is increasingly common in the geriatric population, and therefore, surgery on the population with many medications is also increasingly common. In this sense, it is fairly straightforward: once the decision has been made that a patient needs an emergent operation, there are few indicators that surgery should be delayed or altered based on a patient's medication regimen. A few chronic conditions prohibit lifesav-

J. M. Feeney (✉)
Associate Professor of Surgery, Department of Surgery, Westchester Medical Center, and New York Medical College, School of Medicine, Valhalla, NY, USA
e-mail: James.Feeney@wmchealth.org

R. Latifi (ed.), *Surgical Decision Making in Geriatrics*,
https://doi.org/10.1007/978-3-030-47963-3_26

ing surgery, and so in the emergent setting, the optimization of medication regimens is a moot point. However few, there are medications and chronic conditions that may prohibit even emergent surgery. Of these comorbidities, the most common is anticoagulation.

Prior to undertaking emergency surgery, anticoagulation should be reversed prior to the initial incision. In the case of warfarin, prothrombin complex concentrate (PCC) rapidly corrects INR and allows for safe surgery. It can be relatively rapidly administered and is reasonably cost-efficient and safe from interactions and side effects. The half-life of warfarin is approximately 1 week, but it is quite variable in cases of increased physiologic stress. This means that the efficacy of the drug extends beyond the factors used to reverse the therapeutic effect, so bleeding complications should be monitored closely in the postoperative period. Although the postoperative bleeding complications need close monitoring, the conditions for which anticoagulation was necessary preoperatively are also still presumably present, and so the decision to restart anticoagulation should be considered carefully, especially in cases of spinal, intracranial, cardiac, and vascular procedures, where postoperative bleeding can have rapidly disastrous effects.

Direct oral anticoagulants include rivaroxaban (Xarelto), apixaban (Eliquis), and dabigatran (Pradaxa). They are increasing in popularity in the past several years as prophylaxis against stroke. There are reversal agents for dabigatran (idarucizumab or Praxbind) and rivaroxaban and apixaban (andexanet alfa or Andexxa); however they are expensive ($3500–$49,500 per dose). Data for the use of PCC to reverse DOACs is very limited, although it has become commonplace in trauma settings. In many instances, because the half-lives of these agents are short, and in the cases when there is no reversal agent, delaying the operation, if prudent, may suffice.

Contrary to the case with anticoagulation, timely recognition of adrenal insufficiency may present an opportunity to improve the time to the operating room in patients presenting with surgical causes of sepsis or septic shock. Adrenal insufficiency is prevalent in 20% percent of patients admitted to ICU through the emergency department, and rapid treatment of adrenal insufficiency may color the decision to operate. Insufficient patients appear to be more severely ill, in shock, or with end-organ dysfunction, and these assessments may lead to a decision for lifesaving surgery even though they represent potential complications of an undiagnosed preexisting condition. Often, these patients have a history of glucocorticoid use for other conditions. Even if intermittent, past glucocorticoid use may predispose patients to adrenal insufficiency in times of increased physiologic stress. Treatment is fairly straightforward, with few added risks, and involves physiologic doses of glucocorticoids. In patients with circulatory collapse refractory to volume or pressors or in patients requiring pressors for no other apparent cause, adrenal insufficiency should be suspected and treated empirically where appropriate, especially after the use of etomidate for induction of anesthesia or intubation. These actions completely ameliorate the effect of the adrenal insufficiency and can allow surgery to proceed unabated. There is more discussion on treatment and recognition of adrenal insufficiency in the sections that follow.

Pharmacologic Preoperative Risk Mitigation in the Elective Geriatric Population

Perioperative Cardiac Risk Stratification and the Use of Perioperative Beta Blockade

Patients with known or suspected coronary artery disease may benefit from perioperative cardiac risk stratification. In patients with known CAD, the risk stratification involves a determination of whether the operation planned is high, intermediate, or low risk. Following that, the determination as to the patient's status is made, depending on the severity of the patient's disease. These factors help to determine the course of the anesthesia and surgery, but it should be plainly stated that there is no such thing as "cardiac clearance" for surgery; use of this term should be abandoned.

There are several methods of perioperative cardiac risk stratification, but perhaps the one backed by the largest data set is the NSQIP database risk model. It uses five factors to predict cardiac risk, with good predictive ability (C statistic 0.874) [4]. The five factors, based on multivariate regression analysis and validated in over 250,000 patients, are age, ASA class, creatinine, functional status, and the type of operation.

After patients are risk stratified, the decision for further testing or intervention should be pursued [5]. The American College of Cardiology and the American Heart Association Joint Task Force issued a guideline in 2014 that details the algorithm to pursue further testing or intervention prior to surgery, and that guideline is provided as Fig. 26.1. From a perioperative risk mitigation standpoint, it is essential to continue most home medications up to and including the morning of surgery and then to resume home medications as early as possible postoperatively. Occasionally, the perioperative evaluation process leads to the discovery of significant disease burdens in previously undiagnosed patients. In these patients, it is wise to stabilize newly discovered chronic disease, if at all possible, prior to surgical intervention for other problems. This includes the use of beta adrenergic receptor blockade, which in times past was added for the purpose of preventing perioperative cardiac events prior to surgery.

There is some data that suggests that, in fact, patients with a revised cardiac risk index >3 may benefit from addition of a beta adrenergic receptor antagonist prior to surgery, insomuch as there is a reduction in perioperative AMI. However, current evidence suggests that addition of beta adrenergic receptor antagonist to patients not chronically using this class of drug is usually unwarranted in the elective setting and more often results in increased mortality and stroke rates persisting up to 1 year after the combination of non-cardiac surgery and perioperative treatment with beta adrenergic receptor blockade [6]. Prior opinion and published literature supported the widespread use of beta adrenergic receptor antagonists, even in patients without known heart disease; however, currently, the data does not support the use of these medications prophylactically in the perioperative period, except in patients already prescribed these medications as part of their home regimen. In cases of newly discovered or uncontrolled congestive heart failure, the addition of beta adrenergic receptor blockade may be part of a regimen change intended toward future treatment, but the protective effects of beta blockers preoperatively are a myth that has been all but debunked.

Liver Disease

Patients with known liver disease, like those with known cardiac disease, should have their physiologic status optimized prior to elective surgery. This may include adjustment of chronic home medications to improve synthetic function or in the case of coagulopathy from liver disease supplementing with exogenous clotting factors at the time of surgery. Patients are typically stratified using the model for end-stage liver disease (MELD) scores or Childs-Turcotte-Pugh scores, and reducing these scores preoperatively may prevent postoperative complication rated [7]. Childs-Turcotte-Pugh class is easy to calculate from a few data points and thus has advantages; however, the MELD score is linearly correlated with outcomes postoperatively and therefore is a bit more precise [7]. The optimization in the preoperative period may involve addition or increases in the doses of potassium sparing or other diuretics, paracentesis, albumin infusion, and optimization of sodium intake. Patients with liver disease metabolize and/or excrete medications differently, and the postoperative care, covered later, should be done in a multidisciplinary approach to avoid overdose or toxicity.

Anticogulation and Elective Surgery

Anticoagulation is perhaps the most common medication that can influence patients' intraoperative and postoperative courses. Mitigating bleeding risk involves correcting or reversing anticoagulation as close to the onset of surgery as

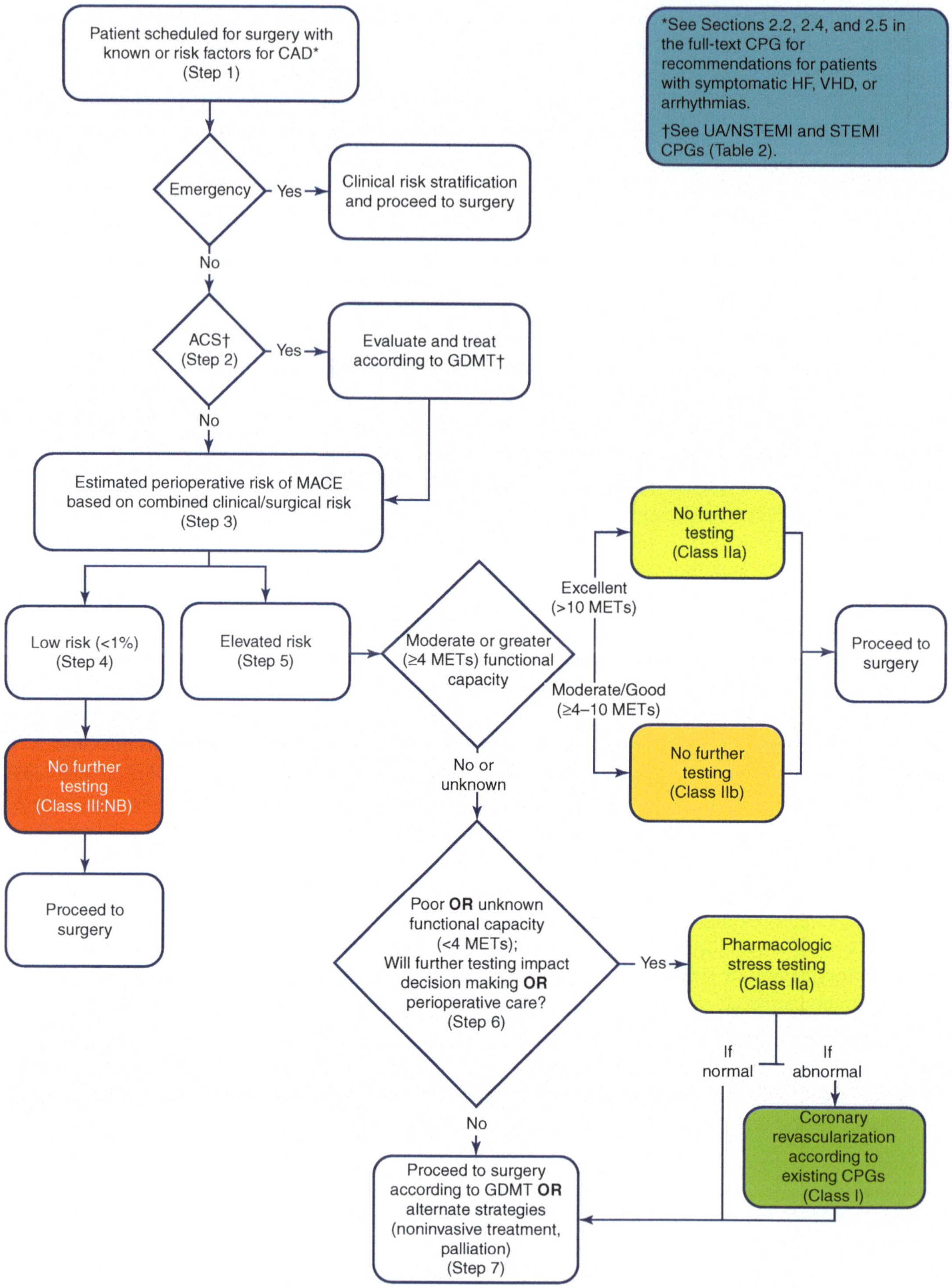

Fig. 26.1 Stepwise approach to perioperative cardiac assessment for CAD. (From Fleisher et al. [5] with permission)

practical and then restarting the medication once safe. There are four main classes of medications with anticoagulant properties: antiplatelet medications, heparins, warfarin, and direct oral anticoagulants (DOACs).

Heparins are, perhaps, the simplest to deal with in the perioperative period. Heparin injections should be discontinued 24 hours preoperatively and partial thromboplastin time (PTT) checked immediately before surgery. For this reason, many advocate switching warfarin daily dosing to injections with heparins preoperatively to protect against thrombosis or stroke in the immediate preoperative period, although the benefit of this strategy is not proven.

Antiplatelet agents include COX inhibitors, like aspirin; glycoprotein IIb/IIIa inhibitors, like abciximab, tirofiban, and eptifibatide; phosphodiesterase inhibitors like dipyridamole; and protease-activated receptor-1 antagonists (vorapaxar). Most of these medications, with the notable exceptions of aspirin and clopidogrel, are prescribed and used in the treatment of acute coronary syndrome, although a full familiarity with the mechanisms of actions is still required preoperatively for any patients taking these medications. Antiplatelet agents should be discontinued for at least 7 days prior to elective operative intervention, and if the risk of discontinuation outweighs the benefit of the elective surgery (e.g., in the case of recent coronary stenting or AMI), then the elective case should be postponed until the antiplatelet agents can be safely stopped.

Direct oral anticoagulants were initially approved by the US Food and Drug Administration in November 2011. The promises of a safer therapeutic window and an improved mitigation of stroke risk with less monitoring and fewer drug interactions made the drugs instantly appealing. However, the risks of intractable bleeding were hard to ignore, and physicians tempered their enthusiasm for the positive aspects of the drugs with caution for the bleeding risks [8–10]. Comparatively, the risks seemed similar to warfarin; however, most of the data in early phase III clinical trials dealt with incidence and prevalence of hemorrhage, not with outcomes [11]. Warfarin, the erstwhile gold standard for outpatient anticoagulation, has a narrow therapeutic index, and its efficacy is influenced by several other medications or even by changes in dietary habits [12, 13]. Once taken, warfarin is quickly absorbed through the gastrointestinal tract, and within 90 min it reaches maximal blood concentrations [14, 15]. The terminal half-life of warfarin is approximately 1 week [14]. However, its effective half-life is found to be between 36 and 42 h [15]. As a result of the narrow therapeutic window and long half-life, toxicity is a common finding and leads to increases in mortality and morbidity, as well as further exposure to transfusion and other blood products (e.g., cryoprecipitate, fibrinogen, or factors) as attempts to reverse coagulopathy become protocolized across institutions. As intensive as the management of warfarin can be, it was heretofore seen as a risk worth taking to mitigate the stroke risks that are so well characterized by the various stroke risk scoring systems, for example, CHADS2 [12, 13].

Concerns for the safety of patients prescribed warfarin led to the creation of DOACs as a novel therapeutic class. DOACs include the direct thrombin inhibitor, dabigatran (Pradaxa), and the direct factor Xa inhibitors rivaroxaban (Xarelto), apixaban (Eliquis), and edoxaban (Savaysa). Rather than inhibiting the synthetic pathway of coagulation factors as warfarin does, DOACs instead influence procoagulant enzymatic activity [16]. DOACs exhibit a linear relationship between plasma levels and anticoagulant effect and have similar pharmacodynamic properties to heparins, taking between just 1 and 4 h to reach peak plasma levels and having relatively short half-lives of 5–17 h [16].

Overall, DOACs have been advocated as preferred outpatient anticoagulants for their rapid onset, short half-life, and limited number of food and drug interactions [12, 16]. Because of the short half-life of DOACs, the manufacturers recommend no preoperative switching of anticoagulants (e.g., from DOACs to heparins) [17]; the industry suggests that in cases of elective surgery, it is enough for the patient to simply stop taking DOACs 24–48 h before operative intervention [16]. Additionally, though there is a lack of established antidotes for DOACs, in the case of emergency, according to the manufacturers, allowing for natural elimination is often sufficient [16]. There are antidotes for dabigatran, rivaroxaban, and apixaban, but these are extremely expensive, currently costing upward of $49,500 per dose.

Warfarin, however, has well-established protocols for reversal, most commonly including prothrombin concentrate complex (PCC), and vitamin K, vitamin K alone, or fresh frozen plasma and vitamin K infusion. For elective surgery, discontinuation of warfarin should be at least 1 week prior to the operative intervention, to allow for the synthesis of clotting factors to resume normally. For routinely anticoagulated patients, coagulation factors should be checked immediately preoperatively, and coagulopathy should be treated or the case delayed for any abnormalities [18].

Postoperative Considerations

Certainly, postoperative pharmacologic considerations carry a vast opportunity for untoward outcomes when treating geriatric patients. Geriatric patients may have organ dysfunction that is masked in the preoperative phase, and not well represented biochemically, for example, underlying liver, kidney, or heart dysfunction that is not manifested in daily life but becomes apparent postoperatively after significant physiologic stress. The old adage "start low and go slow" applies to treatments started in geriatric patients postoperatively, although there are several special circumstances that may require intervention and are unique to or more common in geriatric populations.

Delirium

Delirium is defined as an acutely altered and fluctuating mental status, with altered level of consciousness and disorientation [19]. It is quite common and often undiagnosed in the geriatric population, with an incidence ranging from 9% to 87% [19]. Risk factors for the development of delirium include older age, dementia, psychopathological symptoms, medical comorbidities, frailty, and functional impairment. Additionally, different operations lead to delirium at different rates, with low-risk procedures including cataract surgery (4%) and high-risk procedures including vascular surgery (36%) [20]. The risk factors are additive, and therefore, patients can be assessed and identified as having a higher risk for delirium postoperatively, so that supportive and environmental steps can be taken to mitigate the delirium, as much as possible [21].

The diagnosis of delirium is usually established with the use of assessment scales or tools, one of the most common being the confusion assessment method-intensive care unit (CAM-ICU) tool. This scoring system has well-established reliability and validity for assessing delirium [22]. Other popular tools include the Mini-Mental Status Exam, Informant Questionnaire for Cognitive Decline in the Elderly, and the Memorial Delirium Assessment Scale [21].

Treatment of delirium consists of addressing routine metabolic causes, including electrolyte abnormalities, glucose, oxygenation, and ventilation. Additionally, routine sources of sepsis should be sought and ruled out, including urosepsis, pneumonia, line sepsis, and surgical site infection. If found, these should be addressed quickly.

Finally, many common medications may cause or exacerbate delirium in elderly patients, including cimetidine, corticosteroids, diphenhydramine, belladonna, promethazine, warfarin, opiates, benzodiazepines, and antiparkinsonian medications. Additionally, ethanol withdrawal or withdrawal from other substances may often be confused with or exacerbate delirium. Careful history should identify patients at risk for ethanol withdrawal, and treatment for withdrawal should be accompanied by treatment with thiamine, to reduce the effects of Korsakoff's psychosis [21].

Treatment includes optimizing environmental and supportive measures, in addition to pharmacologic management. Environmental supports, such as hearing and vision aids nearby, attention to early resumption of enteral nutrition, sleep hygiene, patients out of bed, tubes and catheters removed, avoidance of dehydration and hypovolemia, family involvement and interaction, and attention to electrolyte abnormalities, have been demonstrated to reduce delirium from 15% to 10% [23].

Pharmacologic management included haloperidol, which is the treatment of choice for delirium [24], which is considered superior to benzodiazepines both for the avoidance of side effects attributed to benzodiazepines as well as superior outcomes in symptom management. Atypical antipsychotics are not superior to haloperidol but may be preferred for ease of administration, preferred route, or length of half-life [25]. Usually, loading doses of 2–5 mg repeated every 15 min while agitation persists. After the delirium is controlled, scheduled antipsychotic medication is prescribed over the next few days to prevent relapses [21].

Potential side effects of treatment of delirium that require monitoring include extrapyramidal side effects and prolonged QT syndrome. Daily EKC should be obtained to follow corrected QT (cQT) intervals, and if found to be greater than 440 ms in males and 460 ms in females, the haloperidol should be discontinued. Corrected QT interval is used because it is heart rate-independent [26].

Extrapyramidal side effects include acute dystonia, akathisia, drug-induced Parkinsonism, and Tardive dyskinesia. Acute dystonia typically occurs within minutes of atypical antipsychotic medication administration. It is characterized by painful convulsive movements of the neck, tongue, and body [26]. Usual treatment is anticholinergic drugs, including benzhexol or benztropine intramuscularly. Akathisia is a very distressing side effect that occurs usually days to weeks after taking antipsychotic drugs and is characterized by difficulty in keeping one's legs in place. Treatment includes reducing antipsychotic dose or beta adrenergic receptor blockers such as propranolol. Drug-induced Parkinsonism presents identically to Parkinson's disease. It includes muscle stiffness, pill rolling tremor, and bradykinesia. It typically begins several months after antipsychotic drug treatment. It is treated with benzhexol or other anticholinergic medications. Tardive dyskinesia is the most difficult side effect of atypical antipsychotics to treat. It typically begins years after chronic treatment with antipsychotic drugs. It is characterized by irregular movements of the tongue and face. The prognosis is usually poor [26].

In summary, delirium is a common and often underdiagnosed complication of the medical care of the geriatric patient, particularly in the postoperative patient. Diagnosing delirium starts with a search for the underlying cause and appropriate treatment. Environmental and supportive measures have modest success in reducing delirium, and treatment with atypical antipsychotics is the mainstay. Side effects of antipsychotics are often profound, so the duration of treatment should be limited as much as possible.

Adrenal Insufficiency

Adrenal insufficiency is a difficult problem to diagnose and is often not immediately obvious to treating physicians. The benefits of treating relative adrenal insufficiency in septic shock, sepsis, subarachnoid hemorrhage, and critical illness are well documented [27–29]. Especially well studied is the effect of treating relative adrenal insufficiency in cardiac surgery patients. Patients have less dependence of vasopressors and improved clinical outcome. In sepsis, the Surviving Sepsis campaign recommends treating vasopressor dependence after fluid resuscitation with empiric corticosteroids. This represents a paradigm shift over decades ago, when steroids were thought to be universally detrimental.

Patients chronically taking corticosteroids as outpatients should be suspected of relative adrenal insufficiency, even without biochemical proof of glandular dysfunction. These patients should be treated for relative adrenal insufficiency as a matter of routine [30, 31]. Early treatment with steroids clearly reduces mortality and decreases vasopressor use [31, 32], and in patients with hemodynamic instability, shock, and vasopressor requirement, adrenal insufficiency should be considered and treated empirically, preferably within the first hour after vasopressor use is required [33].

Etomidate is a popular drug for the induction of anesthesia, especially in the cardiac surgery population, because it does not depress myocardial activity [34]. Patients requiring vasopressors who were induced using etomidate should be strongly suspected of having adrenal insufficiency and should be treated empirically [34]. Many authors historically have raised concerns with the use of corticosteroids in the postoperative setting, due to the negative effects of corticosteroids on wound healing. However, the doses used in the postoperative setting are usually physiologic (no more than normal secretory levels), compared with large doses used when the negative wound healing effects were first described. Additionally, the effects of corticosteroids are minimal when compared to the effects of persistent hypotension, shock, poor oxygen delivery, and acidosis that proceed from untreated adrenal insufficiency. Considering that adrenal insufficiency is present in nearly 20% of ICU patients, it is important to attune ourselves to the diagnosis and treatment of this comorbidity, because the consequences of untreated adrenal insufficiency are disastrous.

Conclusion

Pharmacologic management of the geriatric patient is fraught with pitfalls and complications. The polypharmacy patient is difficult to balance and difficult to manage in the postoperative setting. Preoperative planning for the most comorbid patients is helpful in the postoperative setting, and attention to the most common postoperative complications can prevent life-threatening complications from affecting our patients and their families. Careful attention to the geriatric patient and to the management of preexisting organ dysfunction can help secure a successful outcome for the geriatric patient, even the most comorbid, most frail, and most difficult to manage of these patients.

References

1. Colby SL, Ortman JM. Projections of the size and composition of the U.S. population: 2014 to 2060, current population reports, P25–1143, U.S. Washington, DC: Census Bureau; 2014.
2. Ward BW, Schiller JS, Goodman RA. Multiple chronic conditions among US adults: a 2012 update. Prev Chronic Dis. 2014;11:130389.
3. Federal Interagency Forum on Aging-Related Statistics. Older Americans 2012: key indicators of well-being. Washington, DC: U.S. Government Printing Office; 2012.
4. Gupta PK, Gupta H, Sundaram A, Kaushik M, Fang X, Miller WJ, Esterbrooks DJ, Hunter CB, Pipinos II, Johanning JM, Lynch TG, Forse RA, Mohiuddin SM, Mooss AN. Development and validation of a risk calculator for prediction of cardiac risk after surgery. Circulation. 2011;124(4):381.
5. Fleisher LA, Fleischmann KE, Auerbach AD. 2014 ACC/AHA guideline on perioperative cardiovascular evaluation and management of patients undergoing noncardiac surgery: a report of the American College of Cardiology/American Heart Association Task Force on practice guidelines. J Am Coll Cardiol. 2014;64(22):e77–137.
6. Wijeysundera DN, Duncan D, Nkonde-Price C, Virani SS, Washam JB, Fleischmann KE, Fleisher LA. ACC/AHA Task Force Members. Perioperative beta blockade in noncardiac surgery: a systematic review for the 2014 ACC/AHA guideline on perioperative cardiovascular evaluation and management of patients undergoing noncardiac surgery: a report of the American College of Cardiology/American Heart Association Task Force on practice guidelines. Circulation. 2014;130(24):2246–64.
7. Friedman LS. Surgery in the patient with liver disease. Trans Am Clin Climatol Assoc. 2010;121:192–204. discussion 205
8. Cotton BA, McCarthy JJ, Holcomb JB. Acutely injured patients on dabigatran. N Engl J Med. 2011;365(21):2039–40.
9. Cano E, Miyares MA. Clinical challenges in a patient with dabigatran-induced fatal hemorrhage. Am J Geriatr Pharmacother. 2012;10:160–3.
10. Chen BC, Viny AD, Garlich FM, Basciano P, Howland MA, Smith SW, et al. Hemorrhagic complications associated with dabigatran use. Clin Toxicol (Phila). 2012;50:854–7.
11. Hankey GJ, Stevens SR, Piccini JP, Lokhnygina Y, Mahaffey KW, Halperin JL, et al. ROCKET AF Steering Committee and Investigators Intracranial hemorrhage among patients with atrial fibrillation anticoagulated with warfarin or rivaroxaban: the rivaroxaban once daily, oral, direct factor Xa inhibition compared with vitamin K antagonism for prevention of stroke and embolism trial in atrial fibrillation. Stroke. 2014;45(5):1304–12.

12. Dossett LA, Riesel JN, Griffin MR, Cotton BA. Prevalence and implications of preinjury warfarin use. Arch Surg. 2011;146:565–70.
13. Gordon JL, Fabian TC, Lee MD, Dugdale M. Anticoagulant and antiplatelet medications encountered in emergency surgery patients: a review of reversal strategies. J Trauma Acute Care Surg. 2013;75(3):475–86.
14. Di Bartolomeo S, Marino M, Valent F, De Palma R. Effects of anticoagulant and antiplatelet drugs on the risk for hospital admission for traumatic injuries: a case-control and population-based study. J Trauma Acute Care Surg. 2014;76:437–42.
15. Obeng-Gyasi S, Loor MM, Samotowka MA, Moorman ML. Management of dabigatran-induced anticoagulation in trauma and acute care surgery patients. J Trauma Acute Care Surg. 2012;73(5):1064–9.
16. Dempfle CE. Direct oral anticoagulants-pharmacology, drug interactions, and side effects. Semin Hematol. 2014;51(2):89–97.
17. Hirsh J, Dalen JE, Anderson DR, Poller L, Bussey H, Ansell J, et al. Oral anticoagulants: mechanism of action, clinical effectiveness, and optimal therapeutic range. Chest J. 2001;119:8S–21S.
18. Feeney JM, Santone E, DiFiori M, Kis L, Jayaraman V, Montgomery SC. Compared to warfarin, direct oral anticoagulants are associated with lower mortality in patients with blunt traumatic intracranial hemorrhage: a TQIP study. J Trauma Acute Care Surg. 2016 Nov;81(5):843–8.
19. Demeure MJ, Fain MJ. The elderly surgical patient and postoperative delirium. J Am Coll Surg. 2006 Nov;203(5):752–7.
20. Robinson TN, Eiseman B. Postoperative delirium in the elderly: diagnosis and management. Clin Interv Aging. 2008;3(2):351–5.
21. Inouye SK. Delirium in older persons. N Engl J Med. 2006;354(11):1157–65.
22. Ely EW, Inouye SK, Bernard GR, Gordon S, Francis J, May L, Truman B, Speroff T, Gautam S, Margolin R, Hart RP, Dittus R. Delirium in mechanically ventilated patients: validity and reliability of the confusion assessment method for the intensive care unit (CAM-ICU). JAMA. 2001;286(21):2703–10.
23. Inouye SK, Bogardus ST Jr, Charpentier PA, Leo-Summers L, Acampora D, Holford TR, Cooney LM Jr. A multicomponent intervention to prevent delirium in hospitalized older patients. N Engl J Med. 1999;340(9):669–76.
24. Kalisvaart KJ, de Jonghe JF, Bogaards MJ, Vreeswijk R, Egberts TC, Burger BJ, Eikelenboom P, van Gool WA. Haloperidol prophylaxis for elderly hip-surgery patients at risk for delirium: a randomized placebo-controlled study. J Am Geriatr Soc. 2005;53(10):1658–66.
25. Han CS, Kim YK. A double-blind trial of risperidone and haloperidol for the treatment of delirium. Psychosomatics. 2004 Jul-Aug;45(4):297–301.
26. Jacobi J, Fraser GL, Coursin DB, Riker RR, Fontaine D, Wittbrodt ET, Chalfin DB, Masica MF, Bjerke HS, Coplin WM, Crippen DW, Fuchs BD, Kelleher RM, Marik PE, Nasraway SA Jr, Murray MJ, Peruzzi WT, Lumb PD, Task Force of the American College of Critical Care Medicine (ACCM) of the Society of Critical Care Medicine (SCCM), American Society of Health-System Pharmacists (ASHP), American College of Chest Physicians. Clinical practice guidelines for the sustained use of sedatives and analgesics in the critically ill adult. Crit Care Med. 2002;30(1):119–41.
27. Weant KA, Sasaki-Adams D, Dziedzic K, Ewend M. Acute relative adrenal insufficiency after aneurismal subarachnoid hemorrhage. Neurosurgery. 2008;63:645–50.
28. de Jong MF, Beishuizen A, Spijkstra JJ, Groeneveld ABJ. Relative adrenal insufficiency as a predictor of disease severity, mortality, and beneficial effects of corticosteroid treatment in septic shock. Crit Care Med. 2007;35:1896–903. https://doi.org/10.1097/01.CCM.0000275387.51629.ED.
29. Marik PE, Pastores SM, Annane D, Meduri GU, Sprung CL, Arlt W. Recommendations for the diagnosis and management of corticosteroid insufficiency in critically ill adult patients: consensus statements from an international task force by the American College of Critical Care Medicine. Crit Care Med. 2008;36:1937–49.
30. Gotoh S, Nishimura N, Takahashi O, Shiratsuka H, Horinouchi H, Ono H, Uchiyama N, Chohnabayashi N. Adrenal function in patients with community-acquired pneumonia. Eur Respir J. 2008;31:1268–73.
31. Nguyen HB, Jaehne AK, Jayaprakash N, et al. Early goal-directed therapy in severe sepsis and septic shock: insights and comparisons to ProCESS, ProMISe, and ARISE. Crit Care. 2016;20(1):160. Published 2016 Jul 1.
32. Dellinger RP, Levy MM, Carlet JM, Bion J, Parker MM, Jaeschke R, et al. Surviving Sepsis Campaign: international guidelines for management of severe sepsis and septic shock: 2008. Crit Care Med. 2008;36(1):296–327.
33. Rivers EP, Blake HC, Dereczyk B, Ressler JA, Talos EL, Patel R, et al. Adrenal dysfunction in hemodynamically unstable patients in the emergency department. Acad Emerg Med. 1999;6(6):626–30.
34. Mohammad Z, Afessa B, Finkielman JD. The incidence of relative adrenal insufficiency in patients with septic shock after the administration of etomidate. Crit Care. 2006;10(4):R105.

Urologic Surgery in the Elderly

27

John L. Phillips, Nikhil Gopal, Jason Elyaguov, and Muhammad S. Choudhury

Urology consistently ranks among the top three specialties in the United States in terms of the total volume of older adults seen in clinical practice. Only ophthalmology and cardiology outrank urology in terms of the total volume of geriatric care provided in the specialty [1]. In the future, increased population growth among older adults will lead to even higher incidence of urological disorders requiring surgical intervention [2]. Moreover, the incidence and prevalence rates substantially increase with advancing age for some of the most common urologic conditions such as urologic cancers (prostate, bladder, kidney), benign prostatic hyperplasia (BPH), urinary incontinence, and pelvic organ prolapse [3].

Age-related physiological, functional, cognitive, and associated comorbidities present a unique challenge in elderly patients who need urological surgery. A thorough preoperative evaluation is crucial for stratifying risks for a particular surgery, identifying patients' specific intraoperative and postoperative needs based on their comorbidities and for maximally optimizing patients before the procedure [4].

Multiple assessment tools have been created to aid the health care provider in evaluating the elderly patient. Comprehensive geriatric assessment (CGA) is a commonly used tool which is multidimensional and interdisciplinary and assists in determining medical, psychological, and functional capabilities of an elderly person. This evaluation includes comorbid conditions and disease severity, medication review, nutritional status, basic activities of daily living (ADL), instrumental activities of daily living (IADL), psychological assessment with testing (mini-mental status), mood testing using the depression scale, social assessment, and environmental assessment [5].

With this evaluation elderly patients can be categorized into three groups: fit, vulnerable, and frail [6]. Fit elderly patients can be treated like their younger counterpart, vulnerable elderly may be able to undergo standard therapy after appropriate medical care, and frail elderly will be at increased risk. Details about the evaluation of elderly patients requiring surgical intervention can be found in Chaps. 1 and 2.

Urologic Cancer in the Elderly

More than 50% of all cancers and over 70% of all cancer-related deaths in the United States occur in patients who are >65 years [7]. This is especially true in the case of the three most common urologic cancers, prostate, bladder, and kidney, which are much more prevalent in older adults. The median age at diagnosis for

J. L. Phillips · N. Gopal · J. Elyaguov
M. S. Choudhury (✉)
Department of Urology, New York Medical College, Valhalla, NY, USA
e-mail: Muhammad_choudhury@nymc.edu

R. Latifi (ed.), *Surgical Decision Making in Geriatrics*,
https://doi.org/10.1007/978-3-030-47963-3_27

prostate cancer is 66 years, and 70% of deaths due to prostate cancer occur in men aged 75 years or older [8]. The median age at diagnosis for bladder cancer is 73 years, and 30% of all bladder cancer are diagnosed in individuals between the age of 75–84 years [9, 10]. Similarly, the median age at diagnosis for kidney cancer, the third most common urologic cancer, is 64 years, and 24.2% is diagnosed in patient over the age of ≥74 years [11].

The challenge of managing older patients with urologic cancer is to accurately assess whether the expected benefits of treatment surpasses the risks in a population with decreased life expectancy, competing comorbidities, and decreased tolerance to stress. In this section, we primarily review the surgical management of common urologic cancers in elderly patient population. We have not reviewed the management of advanced cancer as it is beyond the scope of this chapter.

Localized Prostate Cancer

Prostate cancer, the most frequently diagnosed male cancer in the USA and one of three most common causes of cancer-related deaths, is predominantly a disease of older men, with a median age at diagnosis of 66 years with 70% of death due to prostate cancer occurring in men aged 75 years or older [8].

Treatment decision for localized prostate cancer (c T1–3, No,Mo disease) in older men is challenging because of the wide risk spectrum of prostate cancer and competing causes of death from comorbidities in this patient population. The decision to treat older men with prostate cancer should take into account the risk of dying from the cancer (which depends on its grade and stage/risk category), the risk of dying from another cause (which depends on the severity of comorbidities than on age), potential treatment risks, and risk of developing prostate cancer–associated complications that might interfere with existing comorbidity, and lastly patients' preference [8].

The importance of risk category in prostate cancer was highlighted in a population-based, nationwide Swedish study of a large cohort of 117,328 patients with prostate cancer which showed that mortality risk at 15 years was independent of patients' age at diagnosis but directly linked to the National Comprehensive Cancer Network (NCCN) risk groups, with 10% (Low risk), 20% (Intermediate risk), and 35–40% (High risk) [12].

In this study, death from causes other than prostate cancer was mainly linked to comorbidities, but the aggressiveness of prostate cancer outweighed the comorbid conditions as a risk of death for the intermediate- and high-risk groups. The presence of comorbidities was the strongest predictor of death (from causes other than prostate cancer) in men with localized prostate cancer in this study from the United States [13]. Age was a less significant factor in the study. Therefore, in elderly patients with comorbidities, risk-stratified approach to the management of localized prostate cancer is critical.

Active Surveillance

Active surveillance has been introduced as a treatment option in select men with localized prostate cancer to reduce over-treatment by identifying those who will likely benefit from definitive therapy, while men with true favorable risk prostate cancer are spared curative interventions and its adverse effects [14, 15]. Active surveillance for localized prostate cancer involves actively monitoring the course of the disease with the goal to intervene with curative intent if the cancer progresses.

Large, prospective active surveillance cohorts with predefined follow-up programs and criteria for recommending curative therapy have achieved excellent 98%–99.6% (95% CI 98.6–99) 10-year cause-specific survival [16–19]. Follow-up in these cohorts include regular Prostate Specific Antigen (PSA) testing, digital rectal examinations, and sequential surveillance biopsies.

The Protect study, the first randomized, controlled trial comparing curative interventions (i.e., radical prostatectomy and radiotherapy) to an active monitoring strategy, revealed no survival difference between the three treatment strategies and the 10-year cause-specific survival in men on active monitoring was 98.8% [20].

Active surveillance is preferred for men with very low-risk prostate cancer and life expectancy >20 years and for men with low-risk prostate cancer and life expectancy ≥10 years. Patients with favorable intermediate-risk prostate cancer may also be considered for active surveillance [14, 15].

The advantage of active surveillance is that approximately 2/3 of eligible men will avoid treatment and potential side effects and maintain quality of life/normal activates. The disadvantage of active surveillance is that about 1/3 of men will require treatment, although treatment delays do not seem to impact cure rate. Other disadvantages include chance of missed opportunity for cure despite very low and the need for periodic prostate biopsy [14, 15].

Active surveillance is especially suited for older men with favorable-risk prostate cancer, limited life expectancy, and competing comorbidities.

Observation

Observation strategy in prostate cancer involves monitoring the course of the disease with the expectation to deliver palliative therapy (usually Androgen Deprivation Therapy – ADT) for the development of symptoms or change in exam or PSA levels that suggest symptoms are imminent.

In contrast to active surveillance, in observation strategy no curative treatment is planned. Observation strategy allows elderly frail men to avoid complications of unnecessary treatment.

Radical Prostatectomy

Radical prostatectomy is appropriate for any patient whose tumor is clinically confined to the prostate, who has a life expectancy of 10 years or more, and who has no serious comorbid conditions that would contraindicate an elective operation, and who prefers a surgical option after considering alternatives [14, 15]. Radical prostatectomy is a highly effective treatment for localized prostate cancer.

A low 15-year prostate cancer–specific mortality of 12% was reported in patients who underwent radical prostatectomy (5% for patients with low-risk disease), although it is unclear whether the favorable prognosis is due to the effectiveness of the procedure or the low lethality of cancers detected in the PSA era [21]. Radical prostatectomy improves life expectancy in older patients with few comorbidities in intermediate or high-grade disease [22]. Some patients at high- or very-high-risk prostate cancer may benefit from radical prostatectomy. In an analysis of 842 men with Gleason scores 8–10 at biopsy who underwent radical prostatectomy, predictors of unfavorable outcome included PSA level over 10 ng/mL, clinical stage T2b or higher, Gleason score 9–10, a higher number of biopsy cores with high grade cancer, and over 50% core involvement [23].

Even in a high-risk disease, age has a minor effect on cancer-specific mortality after radical prostatectomy, 9.6% for patients aged 70 years or older vs 9.2% for patients younger than 70 years at 10-year follow-up [24].

The two most important complications after radical prostatectomy, urinary incontinence and erectile dysfunction, are adversely affected by increasing age. A large study of 8295 patients with normal continence and international index of erectile function (IIEF) >18 who underwent radical prostatectomy between January 2009 and July 2013 showed 1-year continence rate of 93.2% in men <65 years of age compared to 86.5% in men >75 years of age. Additionally, 1-year potency rates were 59.3% in men <65 years of age versus 31.3% in men >75 years. In multivariate analysis, older age showed a significant negative effect in both functional outcomes [25]. Other large series have similarly showed the negative effect of age on sexual and urinary function [26, 27].

The 30-day mortality after radical prostatectomy increases with age, although only 1% of men aged 70–79 years die during this time period [28]. The same study showed risks of death and postoperative complications are more dependent on comorbidities than age.

Robotic-assisted radical prostatectomy is now widely used in the United States and many parts of the world. Two randomized trials found a lower rate of blood transfusion with robotic approach [29, 30]. The rate of positive surgical margin is similar between two approaches. Data from a prospective RCT found no difference in margin status between open and robotic approaches. Ten percent of patients in the open and 15% of patients in the robotic group had a positive surgical margin ($P = 0.21$) [29]. Multiple studies have found no statistical difference in the rates of continence or preservation of erectile function after open, robotic, or perineal radical prostatectomy [29–32].

Radiation therapy for localized prostate cancer is an acceptable treatment option [14, 15] and is often used in elderly patients because it is considered less invasive and has acceptable clinical outcome and health-related Quality of Life (QOL) [33].

Cryotherapy has been advocated as a possible treatment for elderly men with localized prostate cancer who may not be candidate for more invasive therapy [34].

Bladder Cancer

Bladder cancer, the second most common urologic cancer, is also more prevalent in the elderly. The incidence increases with age, the median age at diagnosis is 73 years, and individuals aged 75–84 years account for the largest percentage (30%) of new cases [35]. In 2017, there were 79,030 new cases and 16,870 deaths related to bladder cancer in the United States [35]. Approximately 75% of these newly diagnosed patients have non-muscle-invasive bladder cancer and their primary risk is frequent recurrence in the bladder. The remaining 25% have muscle-invasive bladder cancer at presentation and their primary risk is disease progression, metastases, and death.

Non-muscle-Invasive Bladder Cancer (NMIBC)

NMIBC involves the bladder mucosa and may extend to the submucosal layer of the bladder wall. These patients usually present with gross, painless hematuria, and the initial treatment across all age group is trans-urethral resection of bladder tumor (TURBT). It is a safe and effective procedure for endoscopic removal of all visible tumor and can be performed under regional or general anesthesia. Risks related to the procedure (e.g., bladder perforation, hematuria) are similar across all age groups, and any additional risks, in general, is the reflection of associated comorbidities in elderly patients.

Following initial TURBT, 40–70% of these patients will suffer recurrence in the bladder within 5 years, requiring repeated TURBT. Intravesical therapy is the main stay of treatment for NMIBC to reduce the risk of recurrence and disease progression. The use of intravesical therapy following TURBT is determined by tumor grade, stage, multiplicity, size, time, and rate of recurrence [36]. The commonly used intravesical agents are BCG, Mitomycin, Adriamycin, Epirubicin, etc.

Age and Intravesical Therapy

There is concern that patients older than 70 years of age has lower response rate to intravesical BCG therapy. A retrospective Phase II database review from a multicenter trial found that overall response to either BCG or BCG plus interferon was lower in patients aged 80 years or over compared with younger patients [37]. At 24-month follow-up, absolute response was reduced by 22% in patients aged 80 years or over compared with patients aged 61–70 years (39% vs 61%). Age remained a significant variable for decreased response rate after controlling for multiple other relevant variables.

In 2007, Herr from Memorial Sloan Kettering Cancer Center reported his 20-year experience with intravesical BCG therapy in NMIBC [38]. Outcome measures included initial response to BCG and tumor-free recurrence. When the series was stratified by age, >70 vs <70, a small but significant difference was seen in the tumor-free recurrence, favoring patients younger than 70. Increased age seemed to confer a less-durable response to BCG, earlier recurrence, and shorter cancer-free survival time.

A Canadian study of 238 patients, showed 2-year Progression free Survival (PFS) of 87% among patients <75 years vs 65% in patients >75 year ($P < 0.001$). On multivariable analysis, age was an independent risk factor for progression but recurrence free survival was similar among age strata [39].

In general, NMIBC is treated in elderly patients similarly to their younger counterpart with initial transurethral resection of bladder tumor followed by intravesical therapy when appropriate. It appears that response rate to intravesical therapy is lower in elderly patients.

Muscle-Invasive Bladder Cancer

Approximately 25% of bladder cancer patients present de novo with muscle-invasive disease. These patients are at risk of disease spread to regional lymph nodes and systemic metastases. Radical cystectomy and urinary diversion is the most curative treatment option in this group of patients [36].

Radical cystectomy is a complex surgical procedure associated with complications, morbidity, mortality, and extended hospital stays, even in experienced hands [40, 41]. Radical cystectomy poses even more of a challenge in elderly patients with multiple comorbidities.

Concerns have been raised that older bladder cancer patients are not offered curative treatment such as radical cystectomy as often, because clinicians lack quantitative and reliable estimates of competing mortality risks when considering treatments for bladder cancer [42]. Two separate studies which used SEER (Surveillance, Epidemiology and End Results) Medicare database showed that older patients aged >75 are less likely to be treated with radical cystectomy [43, 44], the most curative treatment option for muscle-invasive bladder cancer.

This study evaluated Surveillance, Epidemiology and End Results (SEER) data and found that individuals aged 75 years or over with muscle-invasive bladder cancer had a higher prevalence of cardiac disease, prior cancer diagnosis, chronic anemia, and poor American Society of Anesthesiologist (ASA) physical status classification [44].

Similarly, other studies have shown elderly patients are more likely to have more comorbidities, higher complication rate, and longer hospital stay, which likely influences the clinician's decision to use radical cystectomy in the elderly [45, 46].

Patients over 75 years often have longer hospital stays (5 days vs 4 days, $p = 0.03$) and higher minor complications rate (72% vs 51%, $p = 0.04$) than younger patients [45].

A large study involving 5207 patients with muscle-invasive bladder cancer who were treated with radical cystectomy found 30- and 90-day mortality rate of 5.2 and10.6%, respectively. When broken down by age 90 mortality increased with increasing age at 6.4% for 65–69 years, 10.1% for 70–79 years, and 14.8% for patients >80 years ($p < 0.001$) respectively. The 90-day mortality rates also increased with increasing Charlson comorbidity index (CCI, 0, 1, 2, and 3) 6.3, 10.3, 12.6 and 15.9% $p < 0.001$) [46].

Other studies have shown that performance status rather than age or comorbidities appear to correlate better with outcome after the treatment of muscle-invasive bladder cancer in elderly patients. A study evaluated 152 patients aged 70 years or older, including 33 (22%) patients over 80 years, treated for muscle-invasive disease. In total, 106 patients underwent cystectomy either as primary or secondary treatment. Median survival for the entire cohort was 22 months. Karnofsky performance status (KPS) was assessed before definitive treatment and was then correlated with outcomes. The authors found an overall 4-year survival rate of 14% for those with KPS below 80% compared with 33% for those with KPS above 80%. In a multivariable analysis that included age, KPS, marital status, treatment type, and disease stage, the only independent predictor of overall survival was the KPS score.

Patients with KPS score below 80% had a 1.8 fold increase in the risk of death from any cause compared with those with a better performance status. This finding held true regardless of the definitive treatment chosen, radical cystectomy, radiation therapy, or chemotherapy [47].

That being said, many older patients appear to tolerate cystectomy well. In a series from Memorial Sloan Kettering Cancer Center, studying cystectomy in octogenarian, radical cystectomy in older patients with bladder cancer provided similar disease control and survival outcome with risks of high grade perioperative morbidity comparable to those in younger patients [48].

However, the selection of elderly "fit" patients for radical cystectomy remains a challenge for the clinicians. A publicly available universal surgical risk calculator (http://riskcalculator.facs.org/) was developed using standardized clinical data from the American College of Surgeons National Surgical Quality Improvement Program. Twenty-one preoperative factors are used to estimate postoperative mortality, morbidity, and specific complications with excellent performance [49]. This tool may assist physicians in counselling patients for radical cystectomy.

Comprehensive geriatric assessment (CGA) is another tool which may assist clinicians in counselling patients who need radical cystectomy. The CGA assesses many areas of health of the elderly including function, comorbidities, socioeconomics, cognition, emotion, medication, nutrition, dementia, fall risk, etc. [50]. This tool has been utilized and validated showing its effectiveness in identifying the true health status of these patients [51].

Laparoscopic/Robot-Assisted Radical Cystectomy Versus Open Radical Cystectomy for Elderly Population

As elderly patients potentially experience a higher incidence of complications following radical cystectomy, laparoscopic/robot-assisted radical cystectomy has been proposed as a way of reducing complications, length of stay, etc., in patients with muscle-invasive bladder cancer.

In 2018, an open-level, randomized, phase 3, non-inferiority trial compared robot-assisted radical cystectomy versus open radical cystectomy in patients with bladder cancer (RAZOR) included 302 patients from 15 medical centers in the United States, 150 of whom underwent robotic cystectomy and 152 underwent open cystectomy. The median age for the robotic cystectomy group was 70 years and for the open cystectomy group was 67 years. Two-year progression-free survival in the robotic cystectomy group was 72.3% (95% CI 64.3–78.8) and 71.6% in the open cystectomy group (95% CI 63.6–78.2). Adverse events occurred in 101 (67%) of 150 patients in robotic cystectomy group, and 105 (69%) of 152 patients in the open cystectomy group. The most common adverse event was urinary tract infection, 35% in robotic cystectomy group, and 26% in the open cystectomy group. Postoperative ileus developed in 22% of the robotic cystectomy group and 20% in the open cystectomy group [52].

A study involving a group of elderly patients aged 80–94 years of age who underwent robotic-assisted radical cystectomy reported 2-year recurrence-free survival rate of 73%, disease specific survival of 74%, and overall survival of 61% [53].

However, a study from Memorial Sloan Kettering Cancer Center failed to identify a large advantage for robotic-assisted cystectomy over standard open cystectomy in terms of 90-day complication rate, hospital stay, pathologic outcomes, and 3- and 6-month quality-of-life outcome [54].

Surgical treatment of muscle-invasive bladder cancer in elderly patients is a challenge, but appropriately selected patient tolerate radical cystectomy well with acceptable morbidity and mortality risk. Robot-assisted radical cystectomy may be an option in elderly patients.

Bladder-sparing therapies with endoscopic resection, adjuvant chemotherapy, and/or radiation therapy have been used in older patients with muscle-invasive bladder cancer to avoid the risks of radical cystectomy and in patients who refuse extirpative surgery. Some studies have shown similar survival rate for more conservative therapy compared with radical cystectomy [55].

Renal Cancer

Renal cancer, the third most common urologic cancer, is a disease more common in older adults. In 2017, approximately 63,990 new cases were diagnosed and 14,400 deaths were related to renal cancer in the United States [35]. The median age at diagnosis was 64 years, and about 75% of these cases were diagnosed in patients 55 years or older, of which 26.8% were in patients between the age of 65 and74 years and another 21.9% in patients 75 years or older. Of the newly diagnosed renal cancer patients, 65% were confined to the kidney, 16% presented with regional lymph node involvement, and 16% with distant metastases [56]. An increased use of cross-sectional imaging since the mid-1970s has led to rising incidence of incidentally found, asymptomatic, lower stage, small renal masses and over 50% of the newly diagnosed renal cancer are asymptotic and found incidentally in imaging studies done for unrelated complaints [57–59]. Moreover, in the United States, over 80% of cancer patients aged ≥65 years have at least one comorbidity [60].

A small renal mass (SRM) is defined as an incidentally detected, contrast-enhancing solid or cystic lesion that is ≤4 cm, consistent with clinical stage T1a renal cell cancer, and tumor size appears to be directly related to the risk of malignancy and the presence of high-grade pathology [61, 62]. Small renal masses (SRM) are benign in 15–20%, potentially aggressive in 10% of cases, and the remaining 70%, although malignant, exhibiting indolent biologic behavior [63–65]. Given the substantial proportion of benign lesions and malignant tumors with indolent biologic behavior in this patient population, there is increasing enthusiasm for active surveillance (AS) as a management strategy especially in older patients with comorbidities.

Active Surveillance (AS)

In selecting renal cancer patient for AS, careful consideration should be given to patient factors, tumor factors, and patient preference. Patient factors which should be considered when contemplating AS include age, life expectancy <5 years, high comorbidities, excessive perioperative risk, poor functional status, marginal renal function, and patient preference to avoid treatment risks. In addition, tumor factors which require consideration include tumor size <4 cm, tumor growth <5 mm per year, non-infiltrative in imaging studies, low complexity, and favorable histology, if renal mass biopsy is performed [61].

In the case of AS for SRM, patients are followed with periodic imaging studies such as renal sonogram, CAT Scan, or MRI; and surgical intervention is considered if the growth rate of the tumor is high (>0.5 cm/year), the tumor increases in size to >4 cm, hematuria is present, or the patient prefers it [61, 62].

In a review of 20 separate active surveillance cohort studies, cancer-specific survival (CSS) of 92.3–100% and overall survival (OS) of 42.8–97.6% were reported [62]. Although, the median follow-up in these studies were short ranging between 22 and 47.6 months.

Thermal Ablation (TA)

Minimally invasive surgical techniques such as cryoablation and radiofrequency ablation have been used in older patients with comorbidities as an alternative to active surveillance in renal cancer patients, who are either not candidates or unwilling to have major surgical procedure such as partial or radical nephrectomy.

Guidelines generally recommend considering percutaneous thermal ablation as an option for clinical T1a renal masses <3 cm. Both cryoablation and radiofrequency ablations are options and demonstrate no significant difference in complications, local recurrences, metastatic progression, or cancer-specific survival [61, 66, 67].

A systematic review and meta-analysis comparing thermal ablation (n = 3974) and partial nephrectomy ($n = 2519$) showed all-cause mortality and cancer-specific mortality were higher among patients undergoing thermal ablation but there was no significant difference in local recurrence rate or risk of metastases [68].

In this study, the review of complications revealed decreased blood loss, transfusion rate, shorter hospital stay, and improved renal function with thermal ablation when compared to partial nephrectomy.

Radical and Partial Nephrectomy

Historically, radical nephrectomy has been the standard of care for localized renal cancer. There has been concern about the effect of radical nephrectomy on renal function especially in older patients with compromised renal function. Elderly patients with decreased glomerular filtration rate are at increased risk of mortality, cardiovascular event, and hospitalization [69].

The reported 3-year probability of GFR <60 ml/min in 65% of patients after radical nephrectomy versus 20% after partial nephrectomy ($p < 0.001$) highlights the importance of preserving renal functional mass when treating patients with renal tumor [70]. Moreover, both radical and partial nephrectomy provide comparable cancer-specific and recurrence-free survival [71]. In terms of complication rate and mortality, both techniques are almost identical, except for perioperative bleeding ($P < 0.001$) and urinary fistula ($p < 0.001$), which are more frequent after partial nephrectomy [72].

The adoption of newer surgical techniques, such as laparoscopic and robotic assisted surgery, has decreased complications, blood loss, transfusion rate, pain, and hospital stays after radical and partial nephrectomy [73–75].

Therefore, when elderly patients who require surgical intervention for renal tumor, especially those with compromised renal function, partial nephrectomy should be considered in appropriately selected patients, preferably using minimally invasive approach. However, radical nephrectomy will continue to be a treatment option in patients with large tumor, complex tumor, intra-renal tumor, and tumor extension into the renal venous system.

Benign Prostatic Hyperplasia

Benign prostatic hyperplasia (BPH) and its potential for bothersome lower urinary tract symptoms (LUTS) is one of the most commonly treated conditions in a urologist's practice. Long established as a disease of age, BPH can be seen in men as early as their 40s and is as prevalent as nearly 90% by the eighth decade of life [76].

A detailed medical history assessing voiding habits can determine the extent of bladder outlet obstruction (BOO); however, distinguishing certain lifestyle habits, such as caffeine intake or sleep behaviors, is important in identifying all possible contributors to LUTS. The International Prostate Symptom Score (IPSS) was developed by the American Urological Association (AUA) as a patient questionnaire that measures seven parameters such as frequency, urgency, nocturia, intermittency, slow stream, feeling of incomplete voiding, and having to push to void in a scale of 0–5 reflecting difficulty with urination [77]. Out of 35, an IPSS score of 20 or higher is considered severely symptomatic. A thorough physical assessment including a digital rectal exam should be performed, and a transrectal ultrasound to more accurately measure the prostate size can be considered. A prostate volume of 30 ml or greater is when BPH and concomitant LUTS can begin to manifest [78].

Uroflowmetry is an objective tool often performed in the office to measure the urinary flow and voided volume. A maximum flow rate (Qmax) of 10–15 ml/sec or less likely indicates some form of urinary obstruction. A post-void residual (PVR) via bladder ultrasound is subsequently measured, in which over 50–100 ml or over of urine retained is considered high [79]. Both of these parameters, however, may reflect inherent bladder pathology and detrusor activity [80]. In such cases, a urodynamic study (UDS) may be done to show potential detrusor over- or under-activity that contributes to LUTS. UDS thus proves especially useful in determining optimal medical and surgical therapy.

First-line therapy for BPH involves medical management initially with alpha-blockers such as tamsulosin, doxazosin, alfuzosin, and terazosin to open the prostatic urethra for easier voiding. Those with selective α1-adrenoceptor blockade are better tolerated, reducing the most common side effects of hypotension, dizziness, and asthenia. Medication targeting the androgen-dependent growth of the prostate has also been developed, notably via 5a-reductase inhibitors such as finasteride and dutasteride. These medications are most effective starting at prostate sizes of 40–50 ml or greater [81]. When combining the two classes of drugs, symptom scores and urinary flow rates improve significantly [82]. Thus, since the 1980s, the advent and success of medical therapy in treating BPH has allowed more men to delay surgical intervention, a major consideration in the older patient.

Though dual medication treatment has been shown to prevent the progression of disease, multiple signs of worsening obstructive uropathy would prompt a urologist to encourage surgery. Signs of bladder decompensation include increasing post-void residual urine, bladder trabeculations, and diverticulae. In addition, stasis of urine and incomplete voiding can lead to the development of bladder calculi as well as recurrent urinary infections. Any changes in the bladder can be seen during an office cystoscopy, which the surgeon may consider prior to the procedure to determine which approach would be most appropriate and what results may be expected postoperatively. Finally, acute urinary retention, with or without kidney injury, is often a hard indication for prostate surgery.

Once surgery is decided, preoperative lab work should include a basic metabolic panel and urine analysis with culture. If microhematuria is detected (>3 RBC/hpf), the urologist should evaluate for other potential sources of blood in the urine beyond the prostate. CT scan imaging, preferably with a delayed contrast study, as well as cystoscopy, would offer a comprehensive workup in ruling out malignancy, the gravest concern.

Furthermore, serum prostate specific antigen (PSA) should be checked in those whose life expectancy is greater than 10 years, as prostate cancer can rarely manifest as LUTS. Special attention should be brought to those on 5a-reductase inhibitors, which can reduce PSA by 50% on such treatment [83]. Though PSA can elevate with BPH, and consideration of PSA density and PSA velocity would help judge likelihood of benign disease, a TRUS-guided biopsy of the prostate is definitive for formal tissue diagnosis.

Most of the procedures to be described involve an endoscopic approach and are thus considered minimally invasive surgeries (MIS). As a result, they carry lower surgical risk compared to open procedures and have minimal recovery time. These are therefore attractive options in the elderly population who may have multiple comorbidities and other clinical limitations. Since there are various interventions available, surgeon experience and patient characteristics ultimately help determine the optimal approach.

Monopolar Transurethral Resection of the Prostate

In monopolar transurethral resection of the prostate (TURP), a wired loop distributes a current through the prostate gland to an electrode in the grounding pad on the patient. The current best performs in a non-ionic solution, commonly glycine, to maximize cutting potential. Employed through a cysto-resectoscope, the loop resects and cauterizes tissue down to the prostatic capsule. All of the prostatic chips are then collected via an Ellik evacuator and sent as specimen to pathology. Special attention is made to achieve hemostasis throughout the case. A catheter is then placed at the conclusion of the procedure and can be removed in 24 hours provided hemostasis is maintained and continuous bladder irrigation is no longer required. As an endoscopic procedure, monopolar TURP does necessitate dorsal lithotomy position, which may prove difficult in older patients with limited mobility and fragile or contracted extremities. Nevertheless, it

adequate management and treatment of this condition in the elderly.

Urinary incontinence can broadly be divided into stress, urgency, overflow, and mixed UI. Stress UI is leakage of urine with increased abdominal process; if mild, leakage will occur only with forceful coughing, sneezing, or running; but if severe, leakage can occur with simply standing. Urgency UI is leakage of urine preceded by a strong sensation of bladder fullness with concomitant need to void (so-called urge). Overflow UI occurs when pressure within a distended bladder overcomes the closure pressure of the urethral sphincter, resulting in leakage, simulating an overflowing receptacle. Finally, mixed UI refers to a combination of any of the three previously described types of incontinence [3].

The etiology of UI in the geriatric population is multifactorial and may be explained through factors both intrinsic and extrinsic to the genitourinary system. With aging, parts of the bladder smooth muscle become replaced with collagen, resulting in decreased bladder capacity and decreased contractility. This can manifest clinically in urinary urgency or frequency that may lead to urge UI and/or diminished emptying of the bladder that may result in overflow UI. The apoptosis of muscle with age can also be seen in the urethral sphincter and pelvic muscles that support the urethra, resulting in decreased sphincter tone and urethral hypermobility that can manifest in stress UI. Elderly patients are more likely to have medical comorbidities with concomitant medications that may contribute to UI. For instance, a patient with osteoarthritis and congestive heart failure may experience increased urine production at night due to translocation of fluid from lower extremities when supine. Due to limitation of mobility with arthritis, the urinary urgency experienced by the patient may evolve into urge UI [102]. Geriatricians often refer to UI as a geriatric syndrome, exemplifying its often systemic etiology [103].

The multifaceted nature of the etiology of UI in the elderly is also reflected in its management, ranging from behavioral to pharmaceutical and finally surgical. While this section is focused primarily on surgical treatments, the discussion of operative indications will involve consideration of nonsurgical options. Unlike other disease states such as benign prostatic enlargement (BPE), treatment for UI does not always need to proceed in a stepwise fashion from conservative to surgical intervention. Rather, the specific nature and severity of UI as well as the overall health of the patient need to be taken into account when deciding on treatment, with the focus being on maximizing quality of life. It is important to remember that chronological age does not always equate to functional age; that is, a 65-year-old independent male only on a baby aspirin who runs half a mile each day will be treated very differently than a 65year-old wheelchair-bound male with DM and CHF who recently suffered a stroke. The distinction between these two individuals is captured within the concept of frailty [104]. It is defined as individuals with impaired cognition, motor function, physical abilities, nutrition, and endurance. In short, they have decreased reserve to recover from an emotional or physical insult. Frailty is correlated with increased postoperative complications and thus should be adequately assessed preoperatively.

The use of catheters in the management of incontinence: Prior to discussing the management of specific types of UI, a brief discussion of the role of urinary catheters in the management of UI is warranted. Particularly in frail individuals not amenable to operative intervention, practitioners, patients, and caregivers may consider them a simple, yet effective way to minimize the morbidity of incontinence-associated dermatitis [100] and the financial and emotional cost of using multiple adult diapers daily. Urinary catheters can be divided into either external or indwelling, with the difference being that external urinary catheters are placed around as opposed to inside the urethral meatus. The benefit of external or condom catheters (Fig. 27.1) are that they divert urine from the patient's body without the risk of catheter-associated urinary tract infections (CAUTI) caused by indwelling urethral catheters [105]. However, condom catheters require patients to adequately empty their urine; therefore, while it may reduce the perceived leakage from overflow UI, it does not

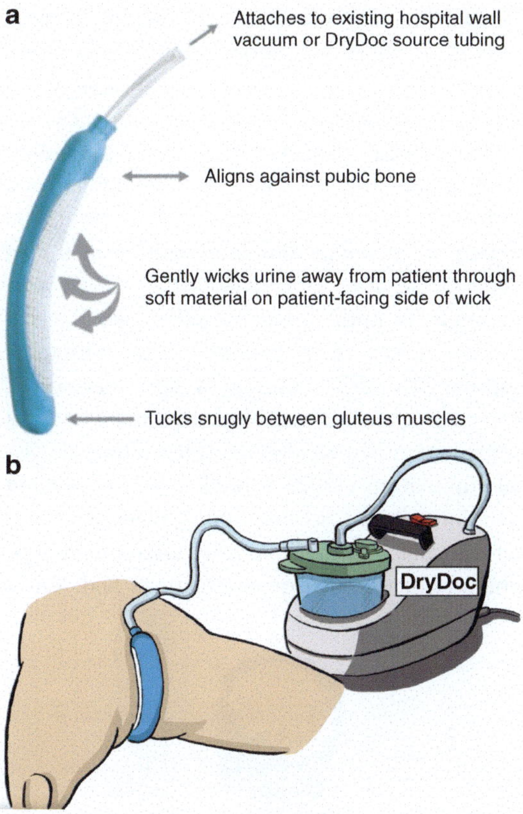

Fig. 27.1 Female external urinary catheter (PureWick™). Top image showing components of catheter taken from incontinencesource.com. Bottom image showing the placement of catheter taken from video entitled "Purewick Demonstration" from PureWick Inc. on youtube.com. Catheter is tucked between gluteus muscle posteriorly and aligns to pubic bones anteriorly. In this way, the catheter is positioned over the urethral meatus. Urine collects at the base of the catheter which is attached to a suction cannister to provide negative pressure for the translocation of urine away from catheter and into tubing. (Courtesy of Bard. © 2019 BD. Used with permission)

alleviate the urine trapped within the distended bladder, which may lead to urinary tract infections and upper tract urinary damage. Thus, patients with overflow UI merit an indwelling urethral catheter. This example illustrates the usefulness of obtaining a bladder scan in elderly patients that present with moderate–to-severe urinary incontinence in order to rule out overflow UI [3].

In summary, external urinary catheters may be an appropriate option for the management of severe urinary incontinence, particularly in a frail elderly patient, with the benefit of reducing reliance on adult diapers and skin breakdown associated with urine. Indwelling catheters should generally be discouraged due to risk of CAUTI which can be associated with significant morbidity and mortality. The one situation where indwelling urinary catheters should be employed is in the case of overflow UI, and for this reason, elderly patients with moderate–to-severe UI should have a bladder scan to rule out elevated PVR.

Male and Female Urge Urinary Incontinence (UUI)

Urge urinary incontinence (UUI) or the involuntary loss of urine preceded by a strong desire to void can be viewed as one endpoint of the spectrum of the overactive bladder (OAB) syndrome complex. Prevalence of OAB increases with age due to both physiological changes within the bladder (i.e., decreased bladder capacity with smooth muscle remodeling into collagen) and the development of medical conditions that can cause urinary frequency and urgency (i.e., diuretics for CHF or autonomic neuropathy associated with diabetes mellitus) [3, 106]. Particularly with the development of UUI, behavioral management such as avoidance of bladder irritants (coffee, alcohol, chocolates, etc.) can often be ineffective as a sole therapy. Pharmacologic therapy for OAB consists of anticholinergics and ß3 agonists. While anticholinergics such as oxybutynin have been shown to be effective in reducing urinary leakage and improving health care–related quality of life, they are not without side effects. In particular, increased risk of dementia has been associated with higher burden of anticholinergic medications, which can be particularly problematic in a frail elderly patient [107]. While ß3 agonists do not carry the cognitive risks of anticholinergics, their pro-hypertensive effects limit their use in patients with hypertension, a common condition in the elderly. Thus, due to inadequate and even harmful effects of nonsurgical options for UUI in the elderly, surgical interventions are particularly appealing as a first-line treatment.

The intravesical detrusor administration of onabotulinum toxin A (Botox) represents one such surgical modality for UUI. Treatment doses range from 100 to 200 units, with the drug inhibiting the release of acetylcholine to motor neurons of the bladder, resulting in decreased smooth muscle spasticity. Acetylcholine inhibition appears to also occur in afferent neurons, producing decreased sensation of fullness. Both afferent and efferent effects of Botox combine to produce reduced urge and urge incontinence [108]. However, Botox can produce adverse effects such as urinary retention requiring temporary catheterization, urinary tract infection, and/or rarely diaphragmatic inhibition affecting respiration.

Studies have shown an association of old age, male sex, multiple comorbidities, and preoperative post-void residual >100 cc with increased adverse outcomes after intradetrusor Botox injections [109]. In regard to the first named variable, when stratifying a geriatric female population into the frail elderly and the non-frail elderly, Liao et al. demonstrated that the non-frail elderly group behaved similar to young patients in terms of postoperative complications. In particular, the frail elderly group had higher rates of elevated PVR (>150 cc) and urinary retention as compared to the two other groups. Furthermore, while frail elderly patients had significantly improved UUI and OAB symptoms post-injection, they had reduced long-term success [110]. Decreased success in this population may be due to multifaceted nature of urinary incontinence as a byproduct of poor functional status and comorbidities. Thus, functional as opposed to chronological age appears to correlate more strongly with worsened outcomes after Botox injection.

Botox injection represents an important therapeutic armamentarium for incontinence due to OAB in the elderly, but particularly in the frail elderly, patients and their caregivers should be counseled about an increased risk of adverse effects such as urinary retention that may require a temporary indwelling catheter or clean intermittent catheterization. Furthermore, realistic expectations need to be made about treatment effect, particularly in an elderly patient with reduced functional status and/or medical comorbidities. Especially as compared to a younger and/or healthier patient, Botox injection may just reduce as opposed to eliminate incontinence.

Sacral neuromodulation represents another surgical modality for the treatment of patients with urge urinary incontinence. This technique consists of percutaneous implantation of a lead electrode at the S3 foramen (Fig. 27.2). The electrode is initially connected to an external stimulator, and if the patient's OAB symptoms improve by 50% during a test period of 1–4 weeks, an implantable stimulator is placed in the patient's gluteal pocket. Resulting electrical stimulation via the lead electrode is thought to inhibit both preganglionic bladder motor nerves and interneurons of afferent nerves, which reduces detrusor contractions and sensa-

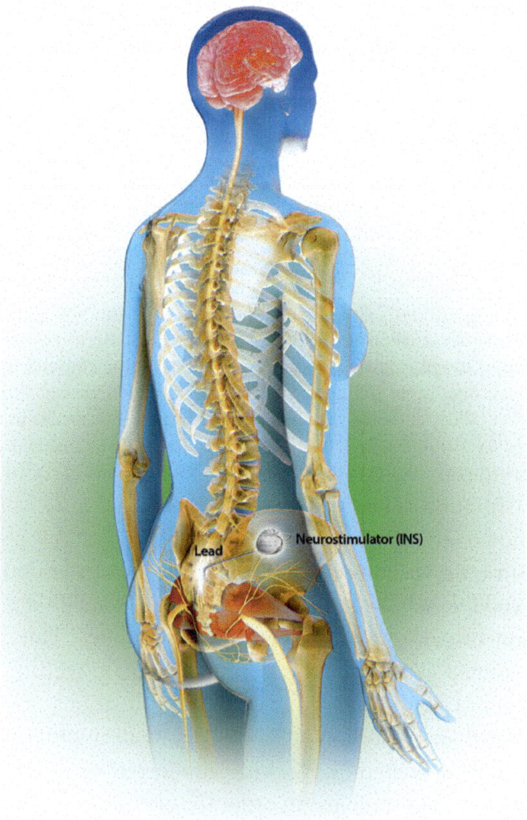

Fig. 27.2 Sacral neuromodulation. Note the placement of percutaneously placed lead at S3 foramen. Degree of stimulation controlled externally by the patient. (Courtesy of InterStim by Medtronic, Fridley, MN, USA)

tion of fullness, both of which contribute to urge urinary incontinence [111].

In regard to the geriatric population, sacral neuromodulation has only been investigated in women older than 55 years. Similar to the findings of Botox, although an improved benefit in urinary symptoms was seen post-intervention, older women had a decreased complete dry rate as compared to their younger counterparts. In addition to age, medical comorbidities were negatively associated with response post-sacral neuromodulation [112]. As with Botox injection, elderly females with multiple comorbidities and likely multifactorial etiology of urinary incontinence should be counseled that sacral neuromodulation may only reduce as opposed to eliminate incontinence. Other important preoperative considerations are that patients should have sufficient manual dexterity and cognitive function to titrate settings of the stimulator to optimize urinary symptoms. Sacral neuromodulation was also traditionally contraindicated in patients with significant neurological or orthopedic disease who may require high-resolution MRIs as these imaging modalities are contraindicated in the setting of implanted pulse generator [111]. However, the newer stimulators are compatible with MRI.

UUI is a common cause of urinary incontinence in elderly males and females, and due to first-line behavioral and pharmacological options generally being ineffective and/or harmful, minimally invasive surgery should be a consideration in these individuals. Both intravesical Botox instillation and sacral neuromodulation have shown efficacy in the geriatric population, but particularly with the frail elderly, preoperative expectations of reduction as opposed to elimination of urinary incontinence should be stressed. Patients should be informed that they are at higher risk for urinary retention and urinary tract infection after Botox injection, the former of which may require temporary catheterization. Patients with dementia and poor dexterity should not undergo sacral neuromodulation; furthermore, this surgical modality should be limited to only females, as results of this therapy on older males have not been published. Adequate selection and preoperative counseling of patients will ensure success with these surgical interventions.

Male Stress Urinary Incontinence

According to the Incontinence Society, <10% of male urinary incontinence is attributed to stress incontinence [113]. The majority of stress incontinence seen in older males occurs as a result of radical prostatectomy, with the etiology purported to be damage to the intrinsic urethral sphincter; disruption of nerve endings to the bladder; loss of urethral support; and/or reduction in length of membranous urethra [114]. Although stress incontinence is commonly seen after radical prostatectomy, between 84% and 97% of patients fully regain continence by 12 months postoperatively. However, about 30% of patients undergoing a prostatectomy will require either adjuvant or salvage radiation for disease recurrence, with incontinence rates of 33–42% after radiation [115]. Given that approximately 70,000 patients underwent a radical prostatectomy in 2010 [116], a not-insignificant proportion of the geriatric population will experience prolonged postoperative urinary leaking that can be a severe impediment to their quality of life, especially since many of these patients had no symptoms prior to their surgery and/or radiation.

As mentioned above, one theory for post-prostatectomy stress incontinence involves loss of urethral support; if this theory is correct, then increasing periurethral muscle support should reduce urinary incontinence. This is the rationale for many urologists recommending pelvic floor muscle therapy (PFMT) (Kegel exercises) for patients with urinary leakage after radical prostatectomy. However, evidence has been mixed on the therapy's efficacy. One study demonstrated a significant improvement in continence after 3 months in patients undergoing PFMT as compared to a control group, but the relative benefit decreased at 12-month follow-up [117]. Thus, PFMT may simply reduce time to urinary continence only in those patients that may have otherwise regained urinary continence by

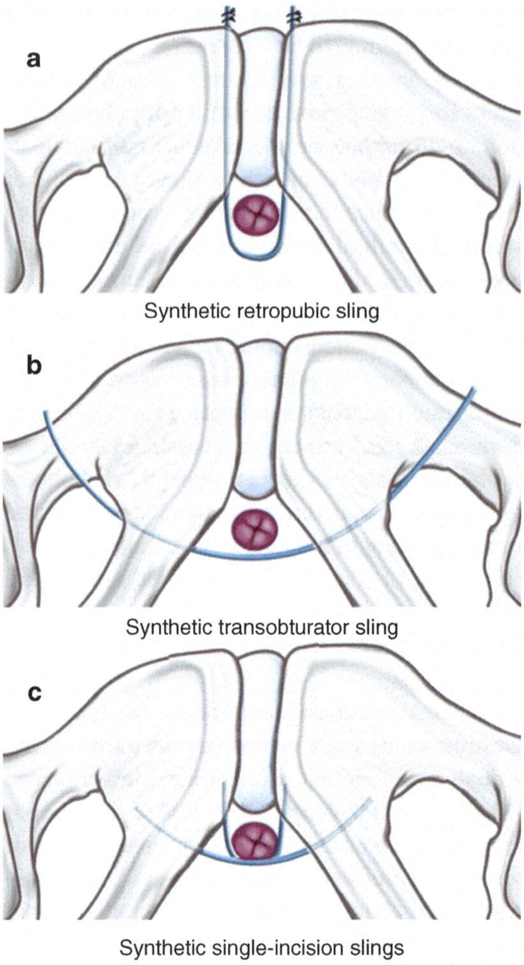

Fig. 27.4 Schematic showing the placement of the three different urethral slings in females. All the slings are placed in the mid-urethra but differ in the mechanism of anchoring. The retropubic sling is anchored by the endopelvic fascia. The transobturator sling is anchored by the obturator internus and externus as well as surrounding fascia. The single-incision (or mini-sling) is anchored only by the obturator internus [134]. Note that, unlike the transobturator sling, the mini-sling does not pierce through the obturator membrane, which should result in reduced likelihood of injury to the obturator nerve which translates into theoretical reduced thigh/groin pain postoperatively. (From Baggish and Karram [135], with permission)

thral coaptation [128]. Unlike the more traditional pubovaginal sling or Burch colposuspension procedures, mid-urethral sling placement does not require an abdominal incision and is thus less invasive with reduced postoperative recovery; it is able to be performed as an outpatient procedure. The TOMUS randomized trial comparing retropubic vs. transobturator slings demonstrated that both had equivalent efficacy in relief of urinary leakage through both subjective and objective measures. However, both had different rare serious side effects, with the retropubic route having a higher rate of bladder perforation, while the transobturator route resulting in more chronic leg pain [130]. Both prospective and retrospective cohorts comparing older women (>70 years) to younger counterparts with pure stress urinary incontinence demonstrate that, while older females enjoy similar improvement in stress incontinence, they have a higher incidence of postoperative urinary tract infections and de novo urge urinary incontinence [131]. The recent introduction of the "mini-sling" may diminish the thigh pain associated with traditional transobturator devices; indeed, recent trials have shown a favorable efficacy to side effect profile [132, 133]. However, further trials are needed, particularly in the elderly, to precisely define the role of the mini-sling in the management of stress incontinence (Fig. 27.4) [134]. As discussed earlier, urinary incontinence in the geriatric population should be thought of as a syndrome with multifactorial etiologies. With decreased bladder compliance as collagen replaces smooth muscle with aging, as the bladder fills with urine postoperatively after mid-urethral sling repair, its decreased capacity with concomitant urges become apparent. As always, adequate preoperative counseling to both the patient and family is important to accurately illustrate expected postoperative outcomes. Surgery for stress incontinence is a choice, rather than a necessity, but should be an informed choice for the patient.

References

1. Drach GW, Griebling TL. Geriatric urology. J AM Geriatric Soc. 2003;51:5355–8.
2. Takao T, Tsujimura A, Kiuchi H, et al. Urological surgery in patients aged 80 years and older: a

30 year retrospective clinical study. Int J Urol. 2008;15:789–93.
3. Griebling T. Chapter 88: Aging and geriatric urology. Campbell-Walsh Urology. 11th ed; 2016. p. 2083–2102.e11.
4. Kabarriti AE, Pietzak EJ, Drach GW, Guzzo TJ. Surgical considerations for the geriatric urology patient. Urol Pract. 2016;3:153–63.
5. Wieland D, Hirth V. Comprehensive geriatric assessment. Cancer Control. 2003;10:454–62.
6. Balducci L, Extermann M. Management of the frail person with advanced cancer. Crit Rev Oncol Hematol. 2000;33:143–8.
7. Noone Am HN, Krapcho M, Miller D, Brest A, Yum, Ruhl J, et al. SEER cancers statistics review, 1975–2015: National Cancer Institute. Bethesda, MD, based on November 2017 SEER data submission posted to the SEER website April 2018. Available at: https://seer.cancer.gov/csr/1975 2015/.
8. Droz J, Aapro M, Balducci L, Boyle H, Broeck TV et al. Management of prostate cancer in older patients: updated recommendation of a working group of the International Society of Geriatric Oncology. www.thelancet.com/oncology Vol. 15 August 2014.
9. Surveillance, Epidemiology and End Research Program. SEER stat fact sheets: bladder cancer [Internet]. Bethesda (MD): National Cancer Institute; 2016 [cited 2016 May 1]. Available from https://seer.cancer.gov/statfacts/html/urinb.html.
10. Canadian Cancer Society's Advisory Committee on Cancer Statistics. Canadian Cancer Statistics 2015. Toronto, ON: Canadian Cancer Society; 2015.
11. SEER cancer gov cancer of the kidney and renal pelvis – SEER statfact sheets. Available from: http://seer.cancer.gov/statfact/html/kidrp.html. Accessed 6 Mar 2013.
12. Rider JR, Sandin F, Andren O, Wiklund P, Hugosson J, et al. Long term outcomes among noncuratively treated men according to prostate cancer risk category in a nationwide, population-based study. Eur Urol. 2013;63:88–96.
13. Albertsen PC, Moore DF, Shih W, Lin Y, Li H, et al. Impact of comorbidity on survival among men with localized prostate cancer. J Clin Oncol. 2011;29:1335–41.
14. NCCN Clinical Practice Guideline in Oncology: Prostate Cancer: Version 2.2019 – April 17, 2019. NCCN.org.
15. Clinical localized Prostate Cancer: AUA/ASTRO/SUO Guideline: April 2017. American Urological Association.
16. Thomsen FB, Jakobsen H, Langkidde NC, Borre M, Jakobsen EB, et al. Active surveillance for localized prostate cancer: nationwide observational study. J U. 2019;201:520–7.
17. Klotz L, Vesprini D, Sethkavalan P, et al. Long-tern follow-up of a large active surveillance cohort of patients with prostate cancer. J Clin Oncol. 2015;33:272.
18. Tosoian JJ, Manawala M, Epstein JI, et al. Intermediate and long-term outcomes from a prospective active-surveillance program for favorable-risk prostate cancer. J Clin Oncol. 2015;33:3379.
19. Bakhorst LP, Valdagni R, Rannikko A, et al. A decade of active surveillance in the PRIAS study: an update and evaluation of the criteria used to recommend a switch to active treatment. Eur Urol. 2016;70:954.
20. Hamdy FC, Donovan JL, Lane JA, et al. 10 year outcomes after monitoring, surgery or radiotherapy for localized prostate cancer. N Engl J Med. 2016;375:141–5.
21. Stephenson AJ, Kattan MW, Eastham JA, et al. Prostate cancer specific mortality after radical prostatectomy for patients treated in the prostate-specific antigen era. J Clin Oncol. 2009;27:4300–5.
22. Alibhai SM, Naglie G, Nam R, Trachtenberg J, Krahn MD. Do older men benefit from curative therapy of localized prostate cancer? J Clin Oncol. 2003;21:3318–27.
23. Pieroazio PM, Ross AE, Liu BM, et al. Preoperative characteristics of high-Gleason disease predictive of favorable pathological and clinical outcomes at radical prostatectomy. BJU Int. 2012;100:112–1128.
24. A Briganti, M Spahn, S Joniau, et al.; for the European Multicenter Prostate Cancer Clinical and Translational Research Group (EMPaCT). Impact of age and comorbidities on long-term survival of patients with high-risk prostate cancer treated with radical prostatectomy: a multi-institutional competing-risks analysis. Eur Urol. 2013;63:693–701.
25. Mandel P, Graefen M, Michl U, et al. The effect of age on functional outcomes after radical prostatectomy. Urol Oncol. 2015;33:203.e11–8.
26. Catalona WJ, Carvalhal GF, Mager DE, et al. Potency, continence and complication rates in 1,870 consecutive radical retropubic prostatectomies. J Urol. 1999;162:433–8.
27. Shikanov S, Desai V, Razmaria A, et al. Robotic radical prostatectomy for elderly patients: probability of achieving continence and potency 1 year after surgery. J Urol. 2010;183:1803–7.
28. Alibhai SM, Leach M, Tomlinson G, et al. 30-day mortality and major complications after radical prostatectomy: influence of age and comorbidity. J Natl Cancer Inst. 2005;97:1525–32.
29. Yaxley JW, Coughlin GD, Chambers SK, et al. Robot-assisted laparoscopic prostatectomy versus open radical retropubic prostatectomy: early outcomes from a randomised controlled phase 3 study. Lancet. 2016;388:1057.
30. Wallerstedt A, Tyritzis SI, Thorsteinsdottir T, et al. Short-term results after robot-assisted laparoscopic radical prostatectomy compared to open radical prostatectomy. Eur Urol. 2015;67:660.
31. Krambeck AE, DiMarco DS, Rangel LJ, et al. Radical prostatectomy for prostatic adenocarcinoma: a matched comparison of open retropubic and robot-assisted techniques. BJU Int. 2009;103:448.

32. Alemozaffar M, Sanda M, Yecies D, et al. Benchmarks for operative outcomes of robotic and open radical prostatectomy: results from the health professionals follow-up study. Eur Urol. 2015;67(3):432.
33. Nauiki S, Ishidoya S, Kawamura S, et al. Quality of life among elderly men treated for prostate cancer with either radical prostatectomy or external beam radiation therapy. J Cancer Res Clin Oncol. 2010;136:379–86.
34. Dhar N, Ward JF, Cher ML, et al. Primary full-gland prostate cryoablation in older men (> age 75 years): results form 860 patients tracked with the COLD registry. BJU Int. 2011;108:508–12.
35. Siegel RL, Miller KD, Jemal A. Cancer statistics. CA Cancer J Clin. 2017;2017:67.
36. AUA NMIBC guideline Diagnosis and treatment of non-muscle invasive bladder cancer: AUA/SUO Joint guideline 2016.
37. Joudi FN, Smith BJ, O'Donnell MA, et al. The impact of age on the response of patients with superficial bladder cancer to intravesical immunotherapy. J Urol. 2006;175:1634–9.
38. Herr HW. Age and outcome of superficial bladder cancer treated with Bacille Calmette-Guérin therapy. Urology. 2007;70:65–8.
39. Margel D, Alkhateeb SS, Finelli A, Fleshner N. Diminished efficacy of Bacille Calmette-Guerin among elderly patients with non-muscle invasive bladder cancer. Urology. 2011;78:848–54.
40. Shabsigh A, Korets R, Vora KC, Brooks CM, Cronin AM, Savage C, et al. Defining early morbidity of radical cystectomy for patients with bladder cancer using a standardized reporting methodology. Eur Urol. 2009;55:164–74.
41. Guancial EA, Roussel B, Bergsma DP, Bylund KC, Sahasrabudhe D, Messing E, et al. Bladder cancer in the elderly patient challenges and solutions. Clin Interv Aging. 2015;10:939–49.
42. Erlich A, Zlotta A. Treatment of bladder cancer in the elderly. Investig Clin Urol. 2016;57 supp 1:526–35.
43. Chamie K, Hu B, Devere White RW, Ellison LM. Cystectomy in the elderly: does the survival benefit in younger patients translate to the octogenarians? BJU Int. 2008;102:284–90.
44. Prout GR Jr, Welsey MN, Yancik R, Ries LA, Havlik RJ, Edwards BK. Age and comorbidity impact surgical therapy in older bladder carcinoma patients: a population based study. Cancer. 2005;104:1638–47.
45. Daneshmand S, Ahmadi H, Schuckman AK, Mitra AP, Cai J, Miranda G, et al. Enhanced recovery protocol after radical cystectomy for bladder cancer. J Urol. 2014;192:50–5.
46. Schiffman J, Gandaglia G, Larcher A, Sun M, Tian Z, Shariat SF, et al. Contemporary 90-day mortality rate after radical cystectomy in the elderly. Eur J Surg Oncol. 2014;40:1738–45.
47. Weizer AZ, Joshi D, Daignault S, Kinnaman M, Hussian M, Montie JE, Zhang Y, Lee CT. Performance status is a predictor of overall survival of elderly patients with muscle invasive bladder cancer. J Urol 2007;177:1287–1293. [Pub Med 17382715].
48. Donat SM, Siegrist T, Cronin A, Savage C, Milowski MI, Herr HW. Radical cystectomy in octgenariaus: does morbidity outweigh the potential survival benefits? J Urol. 2010;183:2171–7.
49. Bilimoria KY, Liu Y, Paruch JL, Zhou L, Kmiecik TE, Ko CY, et al. Development and evaluation of the universal ACS NSQIP surgical risk calculator: a decision aid and informed consent tool for patients and surgeons. J Am Coll Surg. 2013;217:833–42.
50. Hurria A, Cirrincione CT, Muss HB, Kornblith AB, Barry W, Artz AS, et al. Implementing a geriatric assessment in cooperative group clinical cancer trials: CALGB 360401. J Clin Oncol. 2011;29:1290–6.
51. Isbarn H, Jeldres C, Zini L, Perrotte P, Baillargeon-Gagne S, Capitanio U, et al. A population based assessment of perioperative mortality after cystectomy for bladder cancer. J Urol. 2009;182:70–7.
52. Parek DJ, Reis IM, Cartle EP, Gonzalgo MI, Woods ME, et al. Robot-assisted radical cystectomy versus open radical cystectomy in patients with bladder cancer (RAZOR): an open-label, randomized, phase 3, non-inferiority trial. Lancet. 2018;391: 2525–36.
53. Nguyen DP, Al Hussein Al Awamlh B, Charles Osterberg E, Chrystal J, Flynn T, Lee DJ, et al. Postoperative complications and short-term oncological outcomes of patients aged ≥80 years undergoing robot-assisted radical cystectomy. World J Urol. 2015;33:1315–21.
54. Bochner BH, Sjoberg DD, Laudone VP. A randomized trial of robot-assisted laparoscopic radical cystectomy. N Engl J Med. 2014;371: 389–90.
55. Martini T, Mayr R, Wehrberger, et al. Comparison of radical cystectomy with conservative treatment in geriatric (>1+80) patient with muscle-invasive bladder cancer. Int. Brza J Urol. 2013;39:622–30.
56. SEER 18 2010–2014 https:seer.cancer.gov statfacts html kidrp.html.
57. Jayson M, Sanders H. Increased incidence of serendipitously discovered renal cell carcinoma. Urology. 1998;51:203–5.
58. Kane CJ, Mallinm K, Ritchey J, Cooperberg MR, Carroll PR. Renal cell cancer stage migration: analysis of the national cancer database. Cancer. 2008;113:78–83.
59. Hollingsworth JM, Miller DC, Daignault S, Hollenback JM, et al. Rising incidence of small renal masses: a need to reassess treatment effect. J Natl Cancer Inst. 2006;98:1331–4.
60. Yanick R, Ganz PA, Varricchio CG, Conley B. Perspectives on comorbidity and cancer in older patients: approaches to expand the knowledge base. J Clin Oncol. 2001;19(4):1147–51.
61. Campbell S, Uzzo RG, Allaf ME, et al. Renal mass and localized renal cancer: AUA guideline. J Urol. 2017;198:520–9.

62. Ritasu BT, Kutikov A, Uzzo RG, et al. Active surveillance for small renal masses: when less is more. Eur Urol Focus. 2016;2:660–8.
63. Kutikov A, Fossett LK, Ramchandani P, et al. Incidence of benign pathologic findings at partial nephrectomy for solitary renal mass presumed to be renal cell carcinoma on preoperative imaging. Urology. 2006;68:737–40.
64. Thompson RH, Kurta JM, Kaag M, et al. Tumor size is associated with malignant potential in renal cell carcinoma cases. J Urol. 2009;181:2033–6.
65. Frank I, Blute ML, Cheville JC, Lohse CM, Weaver AL, Zincke H. Solid renal tumors: an analysis of pathological features related to tumor size. J Urol. 2003;170:2217–20.
66. Ljungberg B, Bensalah K, Canfield S, et al. EAU guidelines on renal cell carcinoma: 2014 update. Eur Urol. 2015;67:913–24.
67. Finelli A, Ismaila N, Bro B, et al. Management of small renal masses: American Society of Clinical Oncology Clinical Practice Guideline. J Clin Oncol. 2017;35:668–80.
68. Rivero JR, De La Cerda JIII, Wang H, et al. Partial nephrectomy versus thermal ablation for clinical stageT1 renal masses: systematic review and meta-analysis of more than 3900 patients. J Vasc Int Rad. 2018;29:18–29.
69. Go AS, Chertow GM, Fan D, McCuloch CE, Hsu C. Chronic kidney disease and the risks of death, cardiovascular events and hospitalization. N Engl J Med. 2004;351(13):1296–305.
70. Huang WC, Levey AS, Serio AM, Snyder M, et al. Chronic kidney disease after nephrectomy in patients with renal cortical tumour: a retrospective cohost study. Lancet Oncl. 2006;7:735.
71. Liangyou G, Zin M, Hongzhao L, Luyao C, Youngpeng X, et al. Comparison of oncologic outcomes between partial and radical nephrectomy for localized renal cell carcinoma: a systematic review and meta analysis. Surg Oncol. 2016 Dec;25(4):385.
72. Van Poppel H, Da Pozzo L, Albrecht W, et al. A prospective randomized EORTC intergroup phase 3 study comparing the complications of elective nephron-sparing surgery and radical nephrectomy for low stage renal cell carcinoma. Eur Urol. 2007;51(6):1606–15.
73. Al-Qudah HS, Rodriguez AR, Sexton WJ. Laparoscopic management of kidney cancer: updated review. Cancer Control. 2007;14(3):218–30.
74. Portis AJ, Yan Y, Landman J, et al. Long-term follow-up after laparoscopic radical nephrectomy. J Urol. 2002;167(3):1257–62.
75. Xia L, Wang X, Xu T, Guzzo TJ. Systematic review and meta-analysis of comparative studies reporting perioperative outcomes of robot-assisted partial nephrectomy versus open partial nephrectomy. J Endourol. 2017;31(9):893–909.
76. Berry SJ, Coffey DS, Walsh PC, et al. The development of human benign prostatic hyperplasia with age. J Urol. 1984;132(3):474–9.
77. Abrams P, Chapple C, Khoury S, et al. Evaluation and treatment of lower urinary tract symptoms in older men. J Urol. 2009;181(4):1779–87.
78. Jacobsen SJ, Jacobson DJ, Girman CJ, et al. Natural history of prostatism: risk factors for acute urinary retention. J Urol. 1997;158(2):481–7.
79. McConnell JD, Barry MJ, Bruskewitz RC. Benign prostatic hyperplasia: diag- nosis and treatment. Agency for Health Care Policy and Research. Clin Pract Guidel Quick Ref Guide Clin. 1994;8:1–17.
80. Osman NI, Chapple CR, Abrams P, et al. Detrusor underactivity and the underactive bladder: a new clinical entity? A review of current terminology, definitions, epidemiology, aetiology, and diagnosis. Eur Urol. 2014;65(2):389–98.
81. McConnell JD, Bruskewitz R, Walsh P, et al. The effect of finasteride on the risk of acute urinary retention and the need for surgical treatment among men with benign prostatic hyperplasia. Finasteride Long-Term Efficacy and Safety Study Group. N Engl J Med. 1998;338(9):557–63.
82. McConnell JD, Roehrborn CG, Bautista OM, et al. The long-term effect of doxazosin, finasteride, and combination therapy on the clinical progression of benign prostatic hyperplasia. N Engl J Med. 2003;349(25):2387–98.
83. Guess HA, Heyse JF, Gormley GJ. The effect of finasteride on prostate-specific antigen in men with benign prostatic hyperplasia. Prostate. 1993;22(1):31–7.
84. Masumori N, Furuya R, Tanaka Y, et al. The 12-year symptomatic outcome of transurethral resection of the prostate for patients with lower urinary tract symptoms suggestive of benign prostatic obstruction compared to the urodynamic findings before surgery. BJU Int. 2010;105:1429–33.
85. Ahyai SA, Gilling P, Kaplan SA, et al. Meta-analysis of functional outcomes and complications following transurethral procedures for lower urinary tract symptoms resulting from benign prostatic enlargement. Eur Urol. 2010;58:384–97.
86. Mayer EK, Kroeze SG, Chopra S, et al. Examining the 'gold standard': a comparative critical analysis of three consecutive decades of monopolar transurethral resection of the prostate (TURP) outcomes. BJU Int. 2012;110:1595–601.
87. Omar MI, Lam TB, Alexander CE, et al. Systematic review and meta-analysis of the clinical effectiveness of bipolar compared with monopolar transurethral resection of the prostate (TURP). BJU Int. 2014;113:24–35.
88. Mamoulakis C, Ubbink DT, de la Rosette JJ. Bipolar versus monopolar transurethral resection of the prostate: a systematic review and meta- analysis of randomized controlled trials. Eur Urol. 2009;56:798–809.
89. Tan A, Liao C, Mo Z, et al. Meta-analysis of holmium laser enucleation versus transurethral resection of the prostate for symptomatic prostatic obstruction. Br J Surg. 2007;94:1201–8.

28 Gynecologic Surgery in the Elderly

Tana Pradhan, Cara Grimes, Gizelka David-West, Alessandra Marino, and Stephanie Twomey

Aging is an extraordinary process where you become the person you always should have been —David Bowie

According to the National Institutes of Health, approximately 8.5% of the world's population is aged 65 and older, and this percentage is expected to double by 2050 [1]. It is important to note that the ratio of males to females decreases in the aging population due to the longer life expectancy of women. This is expected to result in a phenomenon referred to as "the excess of women," making the care of problems in the geriatric female patient paramount.

Gynecologic surgeries in the geriatric patient are performed for either benign or malignant disease process which includes pelvic organ dysfunction with the preoperative evaluation being tailored to the reason for surgery. The risk assessment of a patient undergoing a debulking surgery for ovarian cancer will be different than that of a patient requiring a simple vulvectomy. Our goal in this chapter is to cover the most common gynecologic conditions requiring surgery in elderly women and review points to consider when making the decision to take a geriatric woman for gynecologic surgery. We will also present the top considerations for providing perioperative care and uncover barriers that may be encountered in this unique population.

Gynecologic Conditions in Older Adult Women

Overall the most common gynecologic complaints specific to the aging female are atrophic vaginitis, post-menopausal bleeding, pelvic floor disorders, and vulvar dermatitis. A pelvic mass in the post-menopausal women is less common, however, when encountered often warrants exploratory surgery and therefore will be covered in this chapter. Certainly, each clinical presentation has a host of differential diagnoses, and only a subset of these conditions warrant surgical evaluation.

Older women with atrophic vaginitis may complain of dryness, dyspareunia, soreness, pain,

T. Pradhan (✉) · G. David-West · A. Marino
S. Twomey
Department of Obstetrics and Gynecology, Division of Gynecologic Oncology, New York Medical College/Westchester Medical Center, Valhalla, NY, USA
e-mail: tanashah.pradhan@wmchealth.org

C. Grimes
Department of Obstetrics and Gynecology, Advanced Urogynecology and Female Pelvic Medicine and Reconstructive Surgery, New York Medical College/Westchester Medical Center, Valhalla, NY, USA

R. Latifi (ed.), *Surgical Decision Making in Geriatrics*,
https://doi.org/10.1007/978-3-030-47963-3_28

Table 28.1 Gynecologic conditions and treatments in older women

Condition	Presenting symptoms	Treatment approach
Vulvar/vaginal atrophy	Dryness, itching, postcoital bleeding	Moisturizers, lubricants, vaginal estrogen therapy
Vulvar/vaginal dysplasia	Itching, visible lesion, bleeding or discharge	*Conservative management:* Topical immunomodulator *Surgical management:* Wide local excision of vulva Laser ablation of vagina Upper vaginectomy
Vulvar cancer	Visible lesion with or without pain, bleeding, discharge Inguinal lymphadenopathy	*Depends on stage of cancer:* Vulvectomy with or without groin node dissection Radiation therapy Chemotherapy
Endometrial hyperplasia	Vaginal bleeding	*Depends on type of hyperplasia:* Ranges from hormonal therapy to surgical intervention with simple hysterectomy
Endometrial cancer	Most common: painless post-menopausal bleeding	*Conservative management:* For patient who are not surgical candidates, hormonal therapy is an option *Surgical management:* Surgery with hysterectomy, BSO, and pelvic lymph node dissection is mainstay of treatment
Ovarian cancer	Bloating, loss of appetite, weight loss, pelvic pain, nausea, and vomiting	*Depends on stage of cancer:* Surgical debulking and staging Chemotherapy
Pelvic floor disorders	Vaginal bulge or pressure, urinary or fecal incontinence, defecatory dysfunction (splinting, straining, incomplete evacuation), incomplete bladder emptying	*Conservative management:* Behavioral therapy, pelvic floor physical therapy, medication therapy, pessaries *Surgical management:* Midurethral sling, reconstructive procedure (retains vaginal function), obliterative procedure (lose vaginal function)

or urinary symptoms which prompt a gynecologic evaluation (Table 28.1). On exam there may be evidence of petechiae, cervix flush to the vagina or loss of rugae especially if there is severe atrophy. There may even be post-menopausal bleeding as a result. With advancing age (or premature estrogen deprivation, e.g., oophorectomy or cancer-related treatment), the loss of estrogen causes the dryness of the genitourinary tissues which predominantly have estrogen receptors. Typically, a trial of vaginal lubricants and/or topical estrogen, if not contraindicated, is prescribed as long as no bleeding or genital lesions are noted.

Menopause is the cessation of menses for 12 months as a result of loss of depletion of ovarian follicles. Any bleeding after this time period warrants a gynecologic evaluation. Most older women have painless bleeding, and only approximately 10–20% of these cases result in the diagnosis of an endometrial cancer. However, the incidence of cancer is on the rise due to increasing rates of obesity in the United States which results in increased peripheral estrogens with resultant stimulation of the endometrial lining and development of preinvasive or invasive disease. Other causes of post-menopausal bleeding include endometrial or cervical polyps, atrophic vaginitis, exogenous hormone use, trauma, atrophic endometritis, or other genitourinary cancers.

Although pelvic masses are a less common gynecologic condition in the elderly woman, the incidence of ovarian cancer is highest in women age 65 and over and may require radical surgery. Due to the low overall incidence of ovarian cancer in the general population and the lack of a cost-effective screening tool, screening of the general population is not recommended. Women may present with vague symptoms such as weight loss, abdominal bloating, early satiety, or nausea. Additionally, women may have mild abdominal pain, urinary frequency, or urgency. There has been some effort to create a symptom index to predict ovarian cancer (e.g., symptoms present for less than 1 year and greater than 12 days per month) [2]. However, healthcare providers must pay attention to the constellation of nonspecific symptoms as mentioned above with their frequency and initiated prompt referral to a gynecologist or a gynecologic oncologist for a comprehensive assessment. Unfortunately 75% of women diagnosed with epithelial ovarian cancer have advanced stage disease due to lack of effective screening, vague symptomatology causing diagnostic delays, and the anatomic location of the ovaries making significant symptoms apparent only with evidence of large tumors and metastatic disease.

Vulvar dermatitis or vulvar lesions can also present with post-menopausal bleeding and vulvar pain. Up to 1 in 30 elderly women may develop vulvar lichen sclerosis, a benign chronic inflammatory skin condition which can be associated with squamous cell cancer at a frequency of approximately 4.5%. This condition is diagnosed based on clinical history and exam, along with vulvar biopsy for pathologic diagnosis. Lichen sclerosis is primarily treated with topical steroid creams and ointments as well as interlesion steroid injections. Patients often present with vulvar irritation and pruritis, which can progress to erosions and fissures due to chronic scratching. Surgical intervention for this condition is considered if cancer is found associated with lichen sclerosis, or often a last resort once the patient has failed all medical management options.

Pelvic floor disorders include urinary incontinence, fecal incontinence, defecatory dysfunction, and pelvic organ prolapse. Pelvic floor disorders are more common with advancing age and result in a decreased quality of life leading women to seek medical attention. Specifically, urinary incontinence is considered a geriatric syndrome which is a highly prevalent multifactorial health condition that can have substantial morbidity and is associated with adverse outcomes of aging in older adults including disability, nursing home admission, and mortality [3]. Approximately 6% of nursing home admissions of older women can be attributed directly to urinary incontinence resulting in an estimated cost of $3 billion per year [4]. Further, meta-analyses found that a median of 58% of nursing home residents suffer from urinary incontinence [5].

Initial treatment for many pelvic floor disorders is conservative and includes behavioral therapy and pelvic floor muscle training. Second line treatments include anticholinergic medications for urgency incontinence and pessaries for pelvic organ prolapse or stress incontinence. There is concern about significant associations identified between anticholinergic medication use and increased risk of cognitive impairment and dementia [6]. Finally, surgical procedures exist to treat pelvic organ prolapse, and the options vary depending on a patient's comorbidities and goals of treatment.

Perioperative Considerations in the Older Woman

After a directed assessment, the gynecologic surgeon may consider operative intervention for further diagnosis or definitive treatment. The decision to proceed with surgery in the elderly woman involves careful consideration of disease outcomes, surgical approach and frailty assessment. Even though advanced age is a risk factor for significant surgical morbidity and mortality, age itself should not preclude a woman from having indicated gynecologic surgery. The preoperative assessment should be comprehensive and useful to detect potential sequelae in the postoperative period.

Common indications for surgery in the elderly women are (1) post-menopausal bleeding, (2) endometrial cancer, (3) pelvic masses suspicious for ovarian cancer, (4) vulvar precancer or cancer, and (5) pelvic organ prolapse or incontinence.

For post-menopausal bleeding, the procedure is commonly hysteroscopy with dilation and curettage to evaluate and sample the endometrium for polyps, hyperplasia or to diagnose malignancy. This procedure is typically an ambulatory procedure and can be performed under sedation if needed. If hyperplasia or malignancy is discovered, hysterectomy may be warranted.

For endometrial cancer, standard of care is a hysterectomy with removal of tubes and ovaries and possible retroperitoneal lymph node evaluation including pelvic and para-aortic lymph nodes. Laparoscopic or robotic approaches for hysterectomy require positioning in steep Trendelenburg position and insufflation with the need of maintaining pneumoperitoneum which may be difficult for extreme obesity or significant cardiopulmonary standpoint. Recently sentinel lymph node algorithms with minimally invasive surgery commonly robotic surgery have been adopted as standard of care for endometrioid endometrial lesions. Sentinel lymph node algorithms reduce the rates of lymphedema and reduce time under anesthesia required for complete lymphadenectomy in those tumors believed to have little risk for nodal disease. Finally, for women who are suboptimal surgical candidates for laparoscopic/robotic or laparotomy approaches, a vaginal hysterectomy with bilateral salpingo-oophorectomy can suffice to remove the primary site of disease. Since endometrial cancer is typically found at clinical stage 1 (85% of cases), hysterectomy with removal of tubes and ovaries can provide excellent disease control.

Conversely, for ovarian cancer, initial surgical cytoreduction via a midline incision laparotomy is the standard of care for suspected advanced disease (Stages III and IV). Ovarian cancer surgeries can be minimally invasive if the disease is confined to the ovary isolated to 1 or 2 quadrants of disease or recurrent disease. Palliative surgery for bowel obstruction is another common indication for operative management either in the initial or recurrent setting. Surgical procedures include radical oophorectomy and hysterectomy, pelvic and para-aortic lymph node dissection, total omentectomy, bowel resection commonly rectosigmoid resection, diaphragm and peritoneal stripping, partial hepatectomy, distal pancreatectomy, and splenectomy. The goal of cytoreduction is debulking to no gross residual disease to afford the longest overall survival.

Most of the gynecologic data on frailty assessment has recently been in the setting of a suspected advanced epithelial ovarian cancer diagnosis. Surgery is the cornerstone of treatment of ovarian cancer, however, there is data showing equivalent survival between neoadjuvant chemotherapy and upfront debulking surgery. There is a strong bias toward upfront surgical management in the United States due to tumor resistance and higher rates of survival in the United States than reported in international studies. Frailty in the older woman would influence the decision to pursue neoadjuvant therapy if there are potentially modifiable factors that reduce the risk of morbidity and mortality.

A frailty index has been explored as a means of predicting outcomes in advanced ovarian cancer. To date more than 50% of studies on frailty are in the male population, and little data exists on the preoperative assessment of older women with a suspected cancer diagnosis. In retrospective analysis of older advanced ovarian cancer patients, frail patients were more likely to be older, higher BMI, higher ASA score, have less surgical complexity, and higher rates of having residual disease [7]. They were also three times more likely to have an Intensive Care Unit Admission within 30 days of surgery and a non-home discharge (two times more likely than non-frail patients). Even though age was a risk factor, it was frailty that was predictive of adverse outcomes for patient undergoing laparotomy [8].

In endometrial cancer, minimally invasive surgery is the cornerstone of treatment, however, few studies look at age and frailty in this population. In the landmark Phase III trial LAP2 which established minimally invasive surgery as standard of care, a recent secondary analysis was performed looking at advanced age and complications each of the trial groups, open versus laparoscopic (including robotic) surgery. In the study population 31% were 70 and older but only 6.6% were 80 and older [9]. Overall complications such as prolonged hospitalization, readmission, and death were still lower in the minimally invasive group regardless of age [10].

Surgery for vulvar disorders includes wide local excision of the vulva, simple vulvectomy, and radical vulvectomy with inguinal lymphadenectomy. The postoperative morbidity of these surgeries increases as they become more radical, with the risk of complex wound failure being as high as 85% and the risk of lower extremity lymphedema up to 70% [11]. With such high risk for morbidity, the preoperative evaluation of the geriatric patient must be multifaceted to ensure the right candidate proceeds with the right surgical intervention. Over the years, the more radical vulvectomy surgeries with full inguinal lymphadenectomy have been modified to reduce morbidity by decreasing the size of the vulvar incision and by employing sentinel lymph node dissection for early stage vulvar cancers. Such modifications in technique have lowered wound complication rates to 30% and lymphedema rates to approximately 20% [12]. Additionally, for patients who are poor surgical candidates or with unresectable vulvar tumors, external beam pelvic radiation therapy with concurrent radiosensitizing chemotherapy is utilized as either definitive therapy or as neoadjuvant therapy to then decrease the morbidity of surgical resection.

The two main pelvic floor disorders that are often corrected surgically are stress incontinence and pelvic organ prolapse.

Advances in anti-incontinence surgery over the last 25 years have resulted in a minimally invasive same day midurethral sling surgery that is appropriate for the majority of women with simple stress incontinence. This allows a safe and effective procedure in an outpatient setting that reduces risk exposure. Age-related risks of incontinence surgery are not well delineated in the medical literature, but overall, sling outcome studies that stratify women by age sometimes find increased risks of postoperative complications (urgency, recurrent UTIs) with equal or slightly less effective success outcomes [13].

Pelvic organ prolapse can be surgically treated via the abdominal route (usually through a minimally invasive approach) or a vaginal approach. A common consideration when approaching surgery for pelvic organ prolapse is to balance risk and benefits associated with reconstructive versus obliterative procedures to correct pelvic organ prolapse. Reconstructive procedures in which a woman retains functional use of her vagina can be native tissue repairs which involve suturing tissues to ligaments to provide support and plicating the fibromuscular tissue of the vagina to strengthen and support her tissues or involve the use of grafts/mesh to augment native tissue support. For elderly patients with multiple comorbidities who do not desire future penetrative vaginal intercourse, obliterative procedures can be considered. In these procedures a woman is left with a foreshortened and narrowed vagina. Obliterative procedures offer the lowest morbidity and highest success rates but at a cost of no more penetrative sexual activity.

Sung et al. analyzed data from the Nationwide Inpatient Sample (NIS). The overall risk of death following urogynecologic procedures was low at 0.04%, but elderly patients undergoing urogynecologic procedures were at higher risk for inhospital death and perioperative complications, even after adjusting for comorbidities. Elderly women 80 years and over who underwent obliterative procedures vs reconstructive procedures had a lower risk of complications compared with those who underwent reconstructive procedures for prolapse (17.0% vs 24.7%, $P < 0.01$) [14].

Future Considerations

In the surgical literature, enhanced recovery pathways (ERAS) have become increasingly adopted for both open and minimally invasive gynecologic surgery. These pathways offer a series of perioperative considerations which emphasize returning the patient back to its physiologic homeostasis. For example, certain strategies used in ERAS protocols are: addressing pain by limiting narcotics and treating pain preoperatively; early to immediate refeeding; and avoiding excessive intravenous fluids. While older women may be excellent candidates to include in these pathways, they have been excluded in initial studies. There are some feasibility studies looking at older women in ERAS protocols, but increased efforts to include this population are necessary. [15]

The Utility of Comprehensive Geriatric Assessment: First Step Toward Better Postoperative HRQoL

Comprehensive geriatric assessment (CGA) is defined as a multidisciplinary diagnostic and therapeutic process that identifies medical, psychosocial, and functional limitations of a frail elderly, aimed to improve the outcome [25]. The components of the CGA are shown in Table 29.2. A recent Cochrane review on the impact of CGA on elderly admitted to surgical services found that older people who received CGA had a lower risk of dying and was more likely to return to the same location they lived in before hospital admission [26]. The application of CGA even in medical inpatients and community-dwelling older people has been shown to improve mortality at 36-month follow-up and increases the chance of living independently at home [27]. CGA has been shown to decrease the length of stay, rates of institutionalization, and readmission, and improves functional status in oncology [28], vascular [29], orthopedic [30], and elective general surgery [31].

Table 29.2 Components of comprehensive geriatric assessment (CGA)

Domain	Variables
Medical	Comorbid conditions and disease severity Medication review Nutritional status
Mental health	Cognition Mood and anxiety Fear
Functional capacity	Basic activities of daily living Gait and balance Activity/exercise status Instrumental activities of daily living
Social circumstances I	Support from family or friends Social network such as visitors or daytime activities Eligibility for being offered care resources
Environment	Home comfort, facilities, and safety Potential use of telehealth Transport facilities Accessibility to local resources

With the demographic shift, higher proportion of elderly will undergo emergency general surgery (EGS) procedures. EGS comprises illnesses that require surgical intervention urgently. A third of EGS patients undergo surgical intervention, and 28.8% of those were elderly undergoing major emergency general surgery in the United States [32]. Importantly, while elective surgery rates decrease drastically in patients over the age of 75 years, rates of emergency interventions increase. There is a rising demand for establishing the CGA by geriatrician lead team in emergency departments. There are some heterogeneous studies in the literature, but well-designed RCTs are lacking [33].

Outcomes Measures of Postoperative Health-Related Quality of Life (HRQoL) in Elderly

While there is an overall agreement of HRQoL, less clear are a number of questions such as what is it that we are measuring, or even more obvious, how do we define a good quality of life after surgery vs another patient who is not doing that well? Surprisingly, the universal outcome measure (or a generally agreed-upon measure) does not exist [34]. Hence, this is an area of ongoing intensive research. So far the general functional outcomes such as activities of daily living (ADLs) and independent activities of daily living (IADLs) are used to assess the QOL in elderly patients [35]. Both ADLs and IADLs provide a measure of independence, which matters to the elderly when it comes to postoperative QOL. Functional capacity takes into account both these basic activities of daily living (ADLs), eating, bathing, dressing, toileting, and walking; and IADLs, shopping, banking, and housekeeping [36]. Evaluation of ADLs and IADLs are considered outcomes of long-term recovery, and the data about the impact of surgery on them is available but scant. Surgical stress, frailty, sarcopenia, hormonal dysregulation, and prolonged bed rest all act in concert and dramatically reduce muscle mass and functional capacity, which leads to impaired ADLs and IADLs.

The short-term outcome measure of HRQoL is the "recovery from surgery," and this aspect of geriatric care has an immediate impact on the clinicians and is an important indicator of the quality of healthcare. The new quality payment program (QPP), 2017, has been implemented by Centers for Medicare & Medicaid Services (CMS) with the goal of disposing payment with high-quality care in the hospitals [37]. The short-term (early) recovery is divided into three phases: early, intermediate, or late postoperative recovery.

Early postoperative recovery phase has been defined as first seven postoperative days which are influenced by pain, nausea, perioperative medications, and delirium [38, 39]. The intermediate phase has been defined as the first 28 or 60 days which are influenced by pain, anxiety and depression, physical impairment, and cognitive dysfunction [40, 41]. The late postoperative recovery phase has been defined as the first 6 weeks to 3 months [42].

This postoperative functional decline has been labeled as "hospitalization associated disability" [43]. Prolonged bed rest, polypharmacy, overuse of urinary catheterization, physical restraints, and restricted diets contribute to the feeling of loss of independence in elderly [44]. Table 29.3 shows the validated tools to measure long- and short-term recovery [45–47]. The limitation here is that there are plenty of scales available, but none of them is specific for the elderly.

Table 29.3 Validated tools to assess short-term and long-term recovery

Tools to assess long-term recovery	Description
1. Katz Index of Activities of Daily Living (measure ADL) [45]	Eating, bathing, dressing, toileting, walking
2. Lawton's instrumental activities of daily living scale (IADL) [46]	Shopping, banking, housekeeping
Tools to assess short-term recovery [47]	
1. Post-anesthesia recovery score	Composite score in PACU on inpatients
2. Post-discharge surgical recovery scale (PSR)	Composite score on day 4 after ambulatory surgery
3. 24-Hour Functional Ability Questionnaire (24hFAQ)	Composite score after 24 hours of ambulatory surgery
4. Post-anesthesia short-term quality of life tool (PASQOL)	Composite score on day 7 after ambulatory surgery
5. Surgical recovery index (SRI)	Composite score in inpatients on day 7, 14, 21 and 28
6. Functional recovery index (FRI)	Composite score after day 1, 3, 5 and 7 of ambulatory surgery
7. Postoperative quality of recovery score (PQRS)	Dichotomous score noted baseline at 15 and 45 min then day 1 and 3 and 3 months on inpatients
8. Surgical recovery scale (SRS	Composite score in inpatients on day 3, 7, 30, and 60

Interventions to Improve the Postoperative Health-Related Quality of Life (HRQoL)

Lawrence et al. prospectively examined the functional independence after major abdominal surgery in the elderly. They observed that cognitive status improves by 3 weeks (after a mild decline), ADLs improve by 6 weeks to 3months, and IADLs take 3–6 months to improve to baseline [48]. Many times in our day to day surgical practice, we have seen that eventful short-term recovery affects long-term recovery, and the negative outcomes may even logarithmically potentiate [49]. So logically, improving short-term recovery will enhance long-term recovery. However, the independent interventions targeting long-term outcome measure are also important.

The interventions can be divided into three phases: (1) prehospital intervention; (2) inhospital intervention; and (3) posthospital intervention. Importantly, prehospital, inhospital, and posthospital interventions represent the entire continuum of care.

Prehospital Intervention

In 2012, American College of Surgeons, National Surgical Quality Improvement

Program (ACS NSQIP), and the American Geriatrics Society (AGS) created the best practices guidelines to optimize perioperative care of the geriatric surgical patient [50]. They emphasize that in addition to obtaining complete history and performing detailed physical examination, elderly should be screened for depression, postoperative risk of delirium, alcohol or substance abuse, nutritional status, risk factors for postoperative pulmonary complications, baseline frailty score, patient's social support system, and capacity to understand the anticipated surgery. The intervention should be done in the form of preoperative cardiac evaluation according to the American College of Cardiology/American Heart Association for patients undergoing noncardiac surgery along with appropriate preoperative diagnostic tests. The complete documentation of medication history, polypharmacy, functional status, and history of falls is of paramount importance. Addressing the concern in any of the abovementioned domain will translate into an intervention/s to improve geriatric surgical care.

Prehabilitation is a very effective intervention performed during the preoperative period and has been shown to improve functional status before and after surgery. This program which ranges from 6 to 8 weeks employs (1) exercise techniques targeting specific muscles or joints by walking, cycling, or swimming to increase VO2max and anaerobic threshold (AT); (2) patient education about surgery and the recovery process; (3) dietary counseling to improve preoperative nutritional status: and (4) stress reduction under direct supervision of psychologist by relaxation with breathing exercises at home [51, 52].

The evidence of postoperative improvement after prehabilitation is encouraging from the initial studies, but there is heterogeneity of the data, wide inclusion and exclusion criteria. Some patients benefit more from the other techniques, which are based on their physiological reserve and type of disease process. So more dedicated studies with well-defined outcome measures are needed [53, 54].

Inhospital Interventions

The literature and guidelines are replete with the inhospital intervention, and all of them emphasize evidence-based geriatric care. The guidelines on optimal perioperative management by ACS NSQIP and AGS provide a sound platform [55]. They emphasize on employing regional anesthetic techniques with multimodal or opioid-sparing analgesia to prevent cognitive and functional decline. The judicious use of intravenous fluids along with appropriate hemodynamic management is one of the important interventions in immediate perioperative care. Correct medical reconciliation of all indicated and current medications is very important. Preventing delirium [56], postoperative nausea vomiting [57], pulmonary complications [58], falls [59], UTI [60], pressure ulcers [61], and functional decline [62] in elderly should be addressed on admission and should be part of all quality improvement projects in the hospitals.

Post-discharge Interventions

Discharge planning involves formulating a comprehensive aftercare plan for patients in whom they are briefed about their disposition, dosage, side effects, and major drug interactions of new medication started in the hospital. It also involves arranging for special care needs such as diet, intravenous medication, wound care, or physiotherapy. The patients are given follow-up appointments with contact numbers for clinic visits [63]. The aim is to reduce hospital length of stay and unplanned readmission to hospital [64]. A Cochrane database review on the effects of individualized discharge plans showed that it leads to reductions in hospital length of stay and readmission rates for elderly admitted to hospitals. However, the authors did emphasize that its impact on mortality, health outcomes, and cost still remains uncertain [65].

Enhanced recovery after surgery (ERAS) programs are a series of evidence-based perioperative interventions that support recovery by reducing the physical stress reactions after sur-

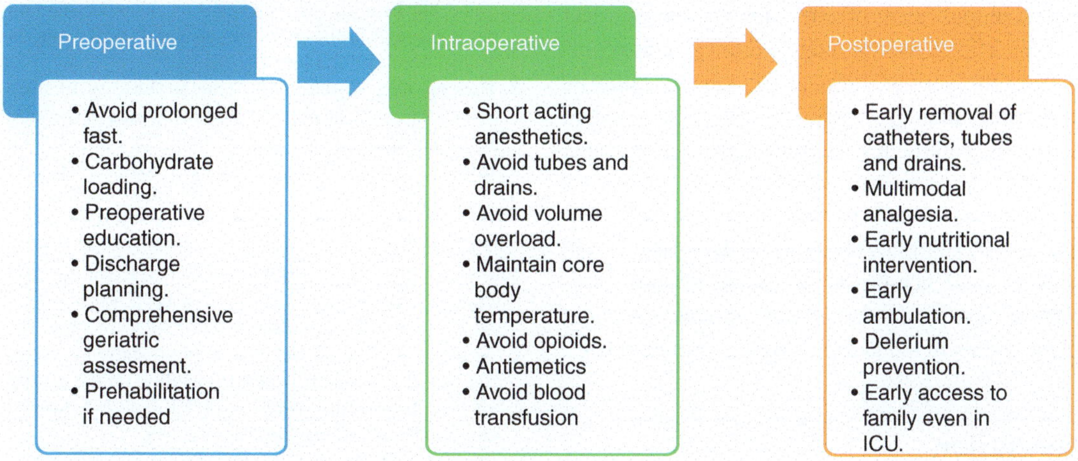

Fig. 29.2 An illustration of a typical ERAS protocol. ICU intensive care unit

gery [66]. This series of the perioperative protocol to enhance recovery was first applied in 1995 by a Danish surgeon, Henrik Kehlet for colectomy [67]. This has now been applied to other major operations and even medium-size surgeries. The series of perioperative, intraoperative, and postoperative interventions described above constitute ERAS protocol, depicted in Fig. 29.2 [68]. The studies have shown comparable efficacy of implementing ERAS protocol on postoperative outcomes in younger and elderly patients. Kisialeuski et al., compared outcomes after implementing ERAS protocol in elderly with non-elderly population after laparoscopic colorectal surgery. The older group had higher ASA scores, but there was no difference in time to first bowel movement, hospital length of stay, or the number of perioperative complications [69]. In a Danish study on fast track hip and knee replacement of patients above 85 years, most patients were able to leave the hospital in 3 days [70]. The same group also emphasized the need for anemia management both pre- and postoperatively in this patient group. The underlying message being that individualizing the ERAS protocol will further enhance the outcomes.

The improved longevity has led to the demographic shift in trauma and acute care surgical services. Many preoperative and intraoperative ERAS interventions outlined in Fig. 29.2 are not entirely feasible for trauma patients. Some examples being the reduced duration of fasting, discharge planning, CGA, avoiding blood transfusion, etc.

In a systematic review, Paduraru et al. concluded that implementing ERAS protocol was beneficial in elderly who underwent emergency surgery with a reduction in postoperative complications, hospitalization, and readmission rates [71].

This systematic review of heterogeneous studies, four cohort studies, and one randomized controlled trial evaluated the impact of ERAS after emergency surgery calls for more in order to evaluate this concept and to make more tangible recommendations.

ACS NSQIP Geriatric Surgery Pilot Project: Spreading the Message

The role of and the importance of teamwork and communication in surgery have been established. Evidence demonstrates the positive effects on both technical skills and patient outcomes [72]. In order to improve geriatric patient outcomes, the ACS has partnered with John A. Hartford Foundation to develop a geriatric surgery quality initiative program and created Coalition for Quality in Geriatric Surgery (CQGS) Project [73]. This is a 4-year initiative to define the processes, resources, and infrastructures necessary to provide optimal care of the older adult surgical patient. The CQGS has engaged the groups representing the various surgical services, anesthesia, geriatrics, nursing, social work, pharmacy, patient advocacy, emergency medicine, physical therapy, community resources, advocacy and regulatory organizations, and, perhaps most importantly, patients and families themselves. This project will provide evidence-based standards of care, uniform system for data collection, and measurement of outcomes in the elderly surgical population.

Conclusion

For the elderly, living independently and with dignity, postoperatively is one of the most important aspects in deciding to undergo surgery. Postoperative quality of life should be the most important patient-centered outcome measure in this fastest-growing segment of the world population. Aging, frailty, and high burden of comorbidity are the risk factors for poor functional recovery. It is important to be able to identify elderly who are at greater risk for poor postoperative functional outcomes. The comprehensive geriatric assessment offers a great advantage and has shown to improve outcomes. Measuring and documenting ADLs and IADLs is very important to understand asses and improve QOL. A validated HRQoL tool specific to the elderly population is needed. The postoperative QOL is affected by preoperative, intraoperative, and postoperative course in the hospital. American College of Surgeon developed the optimal perioperative management of the geriatric patient. These applications of these guidelines along with ERAS protocol has shown to improve the outcomes and hence the quality of life. The importance of good clinical judgment and experience is of paramount importance.

References

1. Colloca G, Santoro M, Gambassi G. Age-related physiologic changes and perioperative management of elderly patients. Surg Oncol. 2010;19:124–30. https://doi.org/10.1016/j.suronc.2009.11.011.
2. Fried TR, Bradley EH, Towle VR, et al. Understanding the treatment preferences of seriously ill patients. N Engl J Med. 2002;346(14):1061–6.
3. Budenz CL, Cosetti MK, Coelho DH, et al. The effects of cochlear implantation on speech perception in older adults. J Am Geriatr Soc. 2011;59:446–53. https://doi.org/10.1111/j.1532-5415.2010.03310.x.
4. St. Clair SF, et al. Hip and knee arthroplasty in the geriatric population. Clin Geriatr Med. 2006;22(3):515–33.
5. Carr LK, Walsh PJ, Abraham VE, et al. Favorable outcome of pubovaginal slings for geriatric women with stress incontinence. J Urol. 1997;157(1):125–8.X–1B.
6. Milicevic S. The impact of benign prostatic hyperplasia surgical treatment with turp method on the quality of life. Acta Inform Med. 2011;19(3):142–5. https://doi.org/10.5455/aim.2011.19.142-145.
7. Manku K, Bacchetti P, Leung JM. Prognostic significance of postoperative in-hospital complications in elderly patients, I: long-term survival. Anesth Analg. 2003;96(2):583–9.
8. Slade MS, Simmons RL, Yunis E, et al. Immunodepression after major surgery in normal patients. Surgery. 1975;78:363–72.
9. Banz VM, Jakob SM, Inderbitzin D. Review article: improving outcome after major surgery: pathophysiological considerations. Anesth Analg. 2011;112:1147–55.
10. Kelley K, Bluthe R, Dantzer R, et al. Cytokine-induced sickness behavior. Brain Behav Immun. 2003;17(Suppl 1):S112–8.
11. The WHOQOL Group. The World Health Organization Quality of Life Assessment (WHOQOL). Development and psychometric properties. Soc Sci Med. 1998;46:1569–85.
12. Sharma D. Health-related quality of life and its assessment in GI Surgery. Indian J Surg. 2004;66:323–35.
13. Selim AJ, Rogers W, Fleishman JA, et al. Updated U.S. population standard for the Veterans RAND 12-item Health Survey (VR-12). Qual Life Res. 2009;18(1):43–52.
14. Moriarty DG, Zack MM, Kobau R. The Centers for Disease Control and Prevention's Healthy Days Measures—population tracking of perceived physical and mental health over time. Health Qual Life Outcomes. 2003;1:37.
15. Campolina AG, Ciconelli RM. SF-36 and the development of new assessment tools for quality of life [in Portuguese]. Acta Reumatol Port. 2008;33(2):127–33.
16. Whynes DK, TOMBOLA Group. Correspondence between EQ-5D health state classifications and EQ VAS scores. Health Qual Life Outcomes. 2008;6:94. https://doi.org/10.1186/1477-7525-6-94. ISSN 1477-7525.
17. Richardson J, Iezzi A, Khan MA, Maxwell A. Validity and reliability of the Assessment of Quality of Life (AQoL)-8D multi-attribute utility instrument. Patient. 2014;356:85–96. https://doi.org/10.1007/s40271-013-0036-x.

18. Hajebrahimi S, Corcos J, Lemieux CM. International consultation on incontinence questionnaire short form: comparison of physician versus patient completion and immediate and delayed self-administration. Urology. 2004;63:1076–8.
19. The Criteria Committee of the New York Heart Association Nomenclature and criteria for diagnosis of diseases of the heart and blood vessels. Boston: Little Brown; 1964.
20. Aaronson NK, Ahmedzai S, Bergman B, et al. The European Organization for Research and Treatment of Cancer QLQ-C30: a quality of life instrument for use in international clinical trials in oncology. J Natl Cancer Inst. 1993;85:365–76.
21. Oken MM, Creech RH, Tormey DC, et al. Toxicity and response criteria of the Eastern Cooperative Oncology Group. Am J Clin Oncol. 1982;5:649.
22. Bjelland I, Dahl AA, Haug TT, Neckelmann D. The validity of the Hospital Anxiety and Depression Scale; an updated review. J Psychosom Res. 2002;52:69–77. https://doi.org/10.1016/S0022-3999(01)00296-3.
23. Bottomley A, Jones D, Claassens L. Patient-reported outcomes: assessment and current perspectives of the guidelines of the Food and Drug Administration and the reflection paper of the European Medicines Agency. Eur J Cancer. 2009;45(3):347–53. https://doi.org/10.1016/j.ejca.2008.09.032.
24. Marquis P, Caron M, Emery MP, et al. The Role of Health-Related Quality of Life Data in the Drug Approval Processes in the US and Europe: A Review of Guidance Documents and Authorizations of Medicinal Products from 2006 to 2010. Pharm Med. 2011;25(3):147–60.
25. Stuck AE, Siu AL, Wieland GD, et al. Comprehensive geriatric assessment: a meta-analysis of controlled trials. Lancet. 1993;342:1032.
26. Eamer G, Taheri A, Chen SS, et al. Comprehensive geriatric assessment for older people admitted to a surgical service. Cochrane Database Syst Rev. 2018;1(1):CD012485. Published 2018 Jan 31. https://doi.org/10.1002/14651858.CD012485.pub2.
27. Stuck AE, Siu AL, Wieland GD, Adams J, Rubenstein LZ. Comprehensive geriatric assessment: a meta-analysis of controlled trials. Lancet. 1993;342:1032–6.
28. Feng MA, McMillan DT, Crowell K, et al. Geriatric assessment in surgical oncology: a systematic review. J Surg Res. 2015;193:265–72.
29. Partridge JS, Harari D, Martin FC, et al. The impact of pre-operative comprehensive geriatric assessment on postoperative outcomes in older patients undergoing scheduled surgery: a systematic review. Anaesthesia. 2014;69:8–16.
30. Hawley S, Javaid MK, Prieto-Alhambra D, et al. Judge at the REFReSH study group. Clinical effectiveness of orthogeriatric and fracture liaison service models of care for hip fracture patients: population-based longitudinal study. Age Ageing. 2016;45:236–42.
31. Harari D, Hopper A, Dhesi J, et al. Proactive care of older people undergoing surgery ('POPS'): designing, embedding, evaluating and funding a comprehensive geriatric assessment service for older elective surgical patients. Age Ageing. 2007;36:190–6.
32. Scott JW, Olufajo OA, Brat GA, et al. Use of national burden to define operative emergency general surgery. JAMA Surg. 2016;151:e160480.
33. Vilches-Moraga A, Fox J. Geriatricians and the older emergency general surgical patient: proactive assessment and patient centred interventions. Salford-POP-GS. Aging Clin Exp Res. 2018;30(3):277–82. https://doi.org/10.1007/s40520-017-0886-5.
34. Deiner S, Silverstein J. Long term outcomes in elderly surgical patients. Mt Sinai J Med. 2012;79(1):95–106.
35. Gazala S, Tul Y, Wagg A, et al. Quality of life and long-term outcomes of octo- and nonagenarians following acute care surgery: a cross sectional study. World J Emerg Sur. 2013;8:23. https://doi.org/10.1186/1749-7922-8-23.
36. Applegate WB, Blass JP, Williams TF. Instruments for the functional assessment of older patients. Current concepts in geriatrics. N Engl J Med. 1990;322(17):1207–15.
37. Centers for Medicare & Medicaid Services Quality Payment Program: Delivery system reform, Medicare payment reform and MACRA. Available at: http://www.cms.gov/Medicare/Quality-Initiatives-Patient-Assessment-Instruments/ValueBased-Programs/MACRA-MIPS-and-APMs/Quality-Payment-Program-QPP.html. Accessed 16 June 2019.
38. Oakes CL, Ellington KJ, Oakes KJ, Olson RL, Neill KM, Vacchiano CA. Assessment of postanesthesia short-term quality of life: a pilot study. AANA J. 2002;70:267–73.
39. Krenk L, Rasmussen LS. Postoperative delirium and postoperative cognitive dysfunction in the elderly – what are the differences? Minerva Anestesiol. 2011;77:742–9.
40. Talamini MA, Stanfield CL, Chang DC, et al. The surgical recovery index. Surg Endosc. 2004;18:596–600.
41. Paddison JS, Sammour T, Kahokehr A, et al. Development and validation of the Surgical Recovery Scale (SRS). J Surg Res. 2011;167:85–91.
42. Myles PS, Reeves MD, Anderson H, et al. Measurement of quality of recovery in 5672 patients after anaesthesia and surgery. Anaesth Intensive Care. 2000;28:276–80.
43. Covinsky KE, Pierluissi E, Johnston CB. Hospitalization associated disability: "she was probably able to ambulate, but I'm not sure". JAMA. 2011;306:1782–93.
44. Sourdet S, Lafont C, Rolland Y, et al. Preventable iatrogenic disability in elderly patients during hospitalization. J Am Med Dir Assoc. 2015;16:674–81.
45. Katz S, Ford AB, Moskowitz RW, Jackson BA, Jaffe MW. Studies of illness in the aged. The index of ADL: a standardized measure of biological and psychosocial function. JAMA. 1963;185:914–9. [PubMed: 14044222]
46. Lawton MP, Brody EM. Assessment of older people: self-maintaining and instrumental activities of daily living. The Gerontologist. 1969;9:179–86.

Surgical Decision-Making in the Elderly with Serious Surgical Illness: The Role of Palliative Care

30

Vincent Finbarr Blood, Matthew K. McIntyre, and Christian A. Bowers

Our reluctance to honestly examine the experience of aging and dying has increased the harm we inflict on people and denies the basic comforts they most need.
Lacking a coherent view of how people might live successfully all the way to the very end, we have allowed ourselves to be controlled by the imperatives of medicine, technology and strangers.
Atul Gawande (Surgeon and Author), Being Mortal: Medicine and What Matters in the End

Surgeons constantly face new challenges as knowledge and research along with technology are continuously evolving to surgically achieve optimal patient outcomes. Historically, surgeons have led by example through translational science, implementing new technological innovations, or advancing surgical techniques. Surgeons have followed the same time-honored leadership blueprint in the application of palliative care to the geriatric surgical patient (GSP). Palliative care is unique with its humanistic battlefield that not only seeks to holistically understand our patients and their families but also our colleagues and perhaps, most importantly, ourselves. This challenge is to the very essence of our souls, values, and culture, and not to surgical technique or technology.

We must behave and think differently to change this dogma lest we lose our purpose as surgeons: to treat our older surgical patients with compassion and dignity within a comprehensive patient-based approach. We must have dynamic vision and foresight of what is important to the GSP as defined not just by their pathology but by their personal, psychosocial, clinical situation, and overall quality of life (QoL). This perspective permits Surgical Palliative Care (SPC) to demonstrate integrity to the GSP's entire situation to determine which course honors and follows the patient's values best. This unique and vulnerable patient population needs us to be a different type of hero than the traditional paintings of the famous general surgeons that line our country's oldest and most prestigious surgery departments. In this chapter, we attempt to understand the intersection of Geriatric Medicine (GM), surgery, and SPC. Barriers to SPC will be

V. F. Blood (✉)
Department of Surgery, Westchester Medical Center, Valhalla, NY, USA
e-mail: Vincent.Blood@wmchealth.org

M. K. McIntyre
School of Medicine, New York Medical College, Valhalla, NY, USA

C. A. Bowers
Department of Neurosurgery, Westchester Medical Center, New York Medical College, Valhalla, NY, USA

R. Latifi (ed.), *Surgical Decision Making in Geriatrics*,
https://doi.org/10.1007/978-3-030-47963-3_30

acknowledged and discussed. Solutions and tools to assist the surgeon in this difficult transformation and decision-making will be postulated.

Understanding Each Other: Geriatric Medicine, Palliative Care, and Surgery – A Prerequisite for SPC Decision-Making

A crucial first step in integrating SPC into an effective interdisciplinary team caring for the older age patient with serious surgical illness is an accurate understanding of the other disciplines involved, primarily GM and palliative care (PC). Why does each entity act and think in the way that they do? Mutual understanding and exchange of ideas through effective communication is a key component to working as a team to reach the moral and ethical common goal of optimizing patient-focused care.

Cultural differences across disciplines and an unwillingness to engage in meaningful conversations contribute to isolation and limitations in interdisciplinary care. Ideally, the objective is a patient-focused, comprehensive interdisciplinary care team utilized for establishing palliative care decisions based on open discussions. A multidisciplinary team differs from an interdisciplinary team because in an interdisciplinary team different disciplines assume differing levels of control depending on the clinical and personal situation, whereas in a multidisciplinary team one discipline is in charge of all decision-making and overall care. Surgeons may interpret this as loss of control and ownership of their patient.

The number of older adults undergoing procedures for serious surgical illness is growing, as half of all US surgical procedures are in patients over 65, but their mortality rate continues to decrease [1, 2]. It is thought that over half of all procedures performed in the USA are in those over 65, and each individual in this group is likely to require at least one surgery [3]. Therefore, surgeons frequently manage older patients, but PC is under-recognized despite having demonstrated its efficacy, and the demand for SPC is constantly increasing [4].

SPC implementation has been problematic due to the various heterogenous clinical scenarios inherent to the GSP. The advanced patient care options available permits various surgical subspecialists to push the envelope for what GSP can tolerate. However, these treatment advances may come at an inconspicuous cost as GSPs are less likely to receive palliative or hospice care in their final year of life when compared to medical patients [5]. This is unfortunate since SPC patients have better pain management, higher satisfaction with care, increased QoL, and reduced healthcare costs [6, 7].

Given the remarkable advanced surgical care now available, the pertinent question is shifting to "not can we operate but, rather should we?" What will the patient's functional status be and how does that coincide with their wishes or advanced directive? Surgeons must shift to a comprehensive patient-focused, interdisciplinary care model that considers patient and family goals, values, and potential outcomes instead of what the technical surgical possibilities are. The nuances of the technical mastery of a Whipple, for example, are lost on the patient and his family if the patient becomes impaired and struggles with a protracted hospital course full of suffering that ultimately ends in their death anyways.

What Makes the Geriatric Patient Unique?

The GSP is unique from a personal, psychosocial standpoint and a clinical perspective. Many of these patients have chronic illness, comorbidities, and decreased physiologic reserve that portend worse postoperative outcomes. Polypharmacy complicates medical care in the elderly and is frequent as 39% of patients over 65 years take more than five medications daily [8]. Furthermore, older patients frequently have baseline impaired activities of daily living (ADLs), decreased activity and independence, varying degrees of frailty, and impaired cognition [9]. The already complicated informed decision-making process is further complicated by the frequent psychological and emotional impairments prevalent in older patients. Appropriately, family and caregivers are asked to assist, care for, and make important healthcare decisions, but then indirect decision-making and surrogacy issues can further cloud the surgical decision-making process. The surgeon must accordingly recognize that they are treating a "care unit" consisting of patient, family members, and caregivers [10]. This "care unit" must be communicated with, educated, and involved in decision-making at all phases of the patient's course. The GSP's wishes, goals of care (GOC), and social situation are key elements to the decision-making process, and families should be involved early, and frequently, to establish a relationship with the physician in the event the patient becomes incapacitated or their wishes are unknown.

Additionally, given that over 1.7 million American geriatric patients are living in nursing homes or hospice, the social aspect of the patient's life must be considered in any surgical decision [11]. For example, if the patient's baseline functional status is poor, as evidenced by hospice dwelling, then shared decision-making must be utilized *before* an intervention is planned. This can be especially challenging in new or acute surgical scenarios such as traumatic injury where the care is very protocolized and time is precious. This highlights the need for established GOC and realistic conversations months or years before an acute scenario arises, something that is not possible for a trauma or acute care surgeon to do. We must therefore rely on our primary care colleagues to routinely have and document these conversations in an easily accessible database. Patients should also be encouraged to share their goals with family members who will be entrusted to make surrogate decisions on their behalf should the situation arise.

Geriatric Patients and Barriers to Palliative Care

Firstly, PC is frequently viewed as a terminal event with end of life care, but it offers so much more to patient, family, and the surgeon. This narrow perspective deprives patients of potential benefits and impedes optimal care. Secondly, a potentially dangerous and harmful false dichotomy exists: (1) palliative and symptom-based care aimed at increasing QoL from a physical, psychological, and spiritual perspective and (2) curative- and disease-focused care. These two areas are often seen as mutually exclusive. However, these two goals can and should be carried out in parallel with a high level of coordination because a single clinician operating in a vacuum will struggle and typically fail to achieve this goal. The lead clinician should be determined by whether palliative or curative treatment is the goal.

Alarmingly, focus of care decisions are often heavily influenced by providers while excluding patient wishes, GOC, and without discussing pertinent end-of-life issues. A frequent misconception is that SPC is a primary care process rather than an adjunctive, parallel system of care. In GSPs, the SPC process can be longer, more dynamic, and subject to change depending on numerous dynamic variables such as chronic illness, worsening comorbidities, and psychosocial personal change. Over time, the weight of each factor can change also as priorities shift and different decisions become appropriate. SPC and PC focusing on shared common ground (QoL, control over one's life, care for patient and family, collaborative care etc.) will help to integrate these disciplines and deliver the best care.

What Is Palliative Care?: Definitions, Principles, and Proper Use

PC is an essential part of total care offering a wide range of treatment and support for these patients and their families and is not just a place to send patients who we believe can no longer benefit from surgery for cure or where surgery is no longer a realistic, viable treatment option (Table 30.1). Nor is it equivalent to giving up or withholding care. It is not an "on" or "off" switch, but rather a spectrum that assumes a larger and larger role of the care as a patient's demise becomes more and more certain.

Table 30.1 Principles of palliative care

Basis is a relief of pain and suffering (physical, emotional, and spiritual)	End of life care is one important part of palliative care not it's "only" role
Improving quality of life	Treatment of the "whole" patient (patient-focused model)
Multidisciplinary and integration into an interdisciplinary team	Continuous reassessment and change of objectives if necessary
Curative treatment and symptom-based palliation are not mutually exclusive	Symptom management
Early education, effective communication, and mutual understanding	Caring for the caretaker is essential
Early establishment of patient's wishes, feelings, and goals of care	Palliative care extends beyond death and includes the bereavement process

Jerant et al. acknowledged that PC must be individualized to unique patient populations, especially in assisted living situations [12]. They examined barriers to PC in the older age patient and proposed a model to overcome them. This "TLC" model consisting of timing, teamwork, longitudinal care, collaborative, and comprehensive care was shown to improve palliative care interventions. PC is caring for the critically ill patient with advanced/terminal disease and their care unit with an "interdisciplinary" team whose primary objective is to increase QoL and decrease pain and suffering.

The graphs in Fig. 30.1 represent PC implementation models to help understand the PC team's interrelationship with other disciplines. The first model is a dated and ineffective way to use PC, albeit the most common interpretation by the public and non-PC providers. The key to effective PC is early involvement with patient and family as well as early integration and formation of the interdisciplinary care team.

The above points give credence to the effective, proper use of PC models as are summarized in Fig. 30.1b, c. As discussed, early coordinated involvement is critical to effective PC and decision-making. Curative treatment and palliative treatment (both medical and surgical) should be carried out simultaneously and in parallel. The situation is dynamic. Constant reassessment, education, communication, and changing care goals to meet the situation are required within the patient's clinical course and personal choices. The balance of treatment aimed at disease (curative) versus symptoms (palliative) as well as supportive care will vary according to time, place, and individualization of the patient's situation. Services offered can be surgical, medical, and nonmedical. They are aimed at curative treatment when appropriate and always strive to decrease pain and suffering and increase QoL. With the integration of palliative care, transition to different phases of care will be smoother with early and effective implementation as shown when these models are used.

Shared decision-making is a vitally important PC principle as it centers on the patient, promotes interaction, and offers a degree of control over life to the patient and family. The PC team assists in decision-making and should be involved in end of life or de-escalation of care scenarios, fitting with the second PC implementation model. These situations are dynamic and patients and families frequently do change their minds, and early PC involvement will help everyone get a better end result of increasing the patient's QoL and decreasing pain and suffering and to clarify goals of care and expectations for treatment and prognosis. SPC necessitates the formation of an interdisciplinary care team in order to provide the highest level of care. Formation, acceptance, and integration of an interdisciplinary team will require major changes in work flow from the current practice for many.

Surgical Palliative Care (SPC): The Concept of SPC as a Paradox Is a Potentially Harmful Misconception

> Before undertaking surgery, the surgeon should consider the whole man, his life, history, habits, constitutional idiosyncrasies, previous ailments, interactions of his mind, embed and body.
> John Hunter (Father of Modern Surgery)

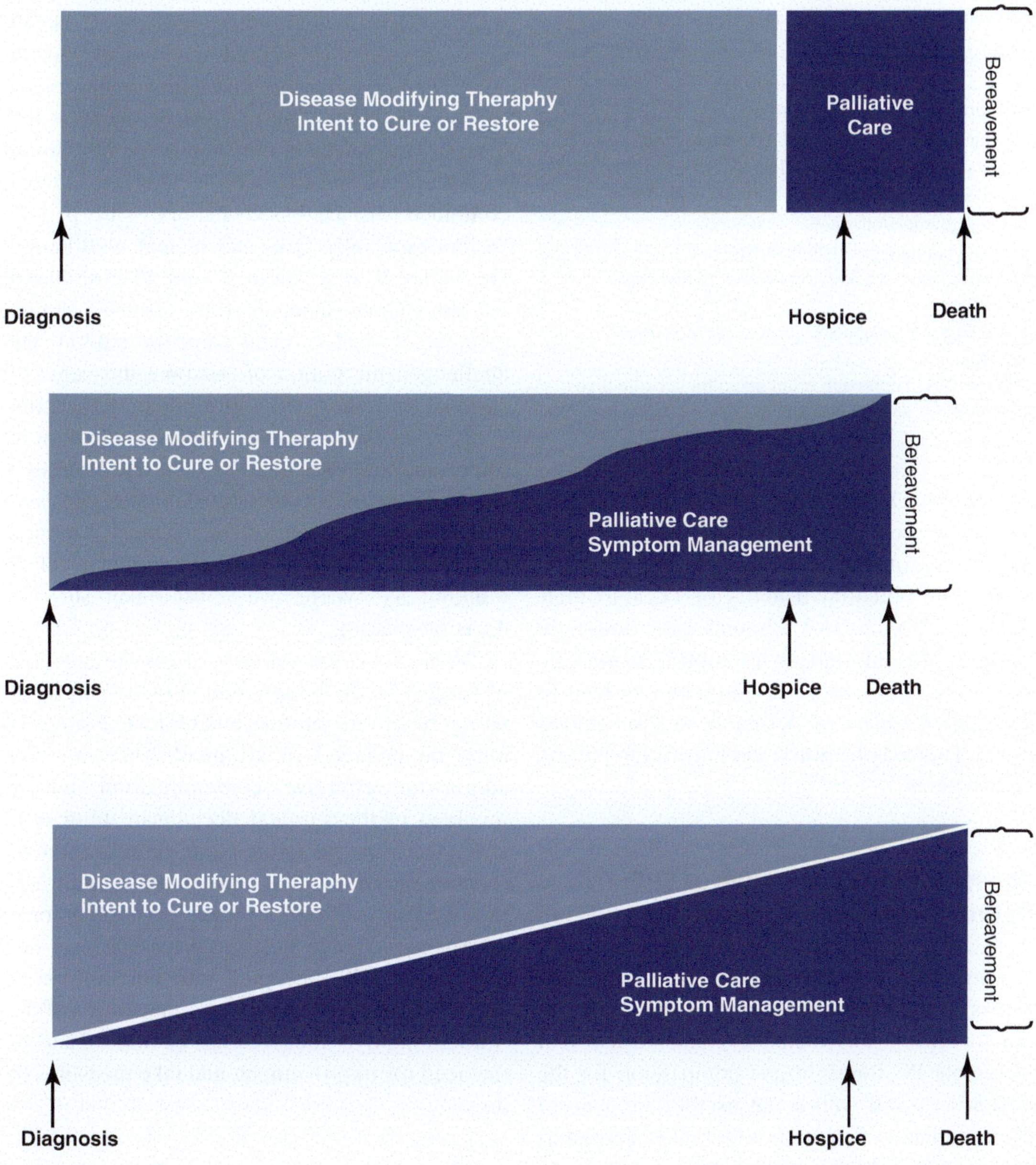

Fig. 30.1 (**a–c**) Diagrams of models of palliative care – demonstrating correct and improper use

Surgical Palliative Care

SPC is PC applied to the advanced, seriously ill, or terminal surgical patient, while palliative surgery is surgery performed to ameliorate symptoms without hope for curing pathology and is part of SPC. SPC has not been adopted widely or easily due to many current barriers impeding a surgeon's acceptance and implementation of SPC into their arsenal of overall patient care.

Barriers to SPC

SPC is not paradoxical, but is a valuable treatment option and potential supportive tool for

Table 30.2 Barriers to surgical palliative care

Surgical culture
Surgical personality and training
Definitions of success and failure
Working with an interdisciplinary care team
Understanding of surgical ethics
Psychology of the surgeon – patient relationship
Miscommunication
Misunderstanding of palliative care
Paucity of research and corresponding high-level evidence
Uncertainty of prognosis, clinical and personal situation
Disease-based approach often given bigger role than patient-based factors and model

the surgeon (Table 30.2). The second PC implementation model demands that surgeons play a major role on the interdisciplinary team and permit the potential benefits of PC into their practice. These GSP clinical situations are complicated and fluid, so attempting to manage everything alone as a solo surgeon can lead to the consequences of acting as a "misguided hero". This may be grave and harmful to patient and surgeon.

The Historical Perspective of SPC: Return to a Surgical Tradition

Historically, palliative surgery was ubiquitous across surgical disciplines since surgery's modern foundation and their early teachings reveal incredible PC foresight and compassion for the patient's overall well-being as they recognized the importance of a patient's QoL [13]. PS examples include Billroth performing a gastrectomy to relieve a patient of a horrible QoL with intractable vomiting, decreased nutrition, and an inability to eat [14]. Cardiothoracic surgery's simple incision through stenotic heart valves to relieve severe congestive heart failure symptoms is another [15]. Neurosurgical PS interventions include cordotomy for providing pain relief in a terminal patient [16].

It is important to reflect that Balfour Mount, a urologist, is credited with establishing the term palliative care in the early 1970s. His early writings described the principles of PC as we know them today [17]. Mount was extremely concerned and disheartened with surgery's narrow-minded focus on surgical procedures and disease. He rejected this purely patho-centric model of care and stressed the need for one that combined disease-based factors with patient factors especially QoL and patient well-being. He believed that surgical and disease-based variables alone disregarded the most important element of total surgical care, the patient. He challenged the notion of outcome measures of survival and short-term morbidity as the only goals of surgical care. The reassessment of more important outcomes such as QoL and leading a good life devoid of, or with minimal, pain and suffering are prescient and on the mark. Mount's challenge to his colleagues and higher level of thinking is where we must take surgical decision-making.

We must be compelled to stress the wonders of surgery in the larger context of patient well-being, function, comfort, and QoL. The surgeon must be engaged in all phases of care and administer effective communication to all involved in the surgical decision-making process. The surgeon's role is to guide treatment recommendations and support the patient and family. This is "true" courage and a surgeon's purest, most noble destiny. Why SPC has not been more fully accepted into our culture as surgeons and in practice is concerning. The time is long overdue. We must be honest, accept the need for improvement, and take measures to do so.

Surgical Palliative Care "Areas for Improvement"

Surgical Culture: The Surgical Personality

What is a surgical personality and does it affect one's ability to incorporate PC into surgery and a surgical decision-making framework? The "surgical personality" is a hardworking, self-critical, strong minded, independent, decisive, action oriented, hands on individual with an absolute obsession on achieving positive results. This personality is stereotypical because these traits that attract people to become surgeons are the required characteristics in order to successfully endure the rigorous training and become competent surgeons [18]. However, these same necessary "survival" traits are not conducive to fitting in and working in a PC setting. Uncertainty runs rampant in surgical decision-making in relation to SPC in older adults with serious surgical illness. Uncertainty of prognosis, intricate personal and family dynamics, highly charged emotional situations, decreased function, frailty, dependence on others, and complex psychological issues are just some of the many challenges in SPC. For surgeons, perceived loss of control of "our patient," unwillingness or reluctance to accept other colleagues' opinions, and intense feelings also make it difficult to reconcile, but surgeons must undergo introspection and change if they want the best outcome for their patients.

Surgical Training and Defining Success and Failure

Surgical training reinforces and even selects for the surgical personality. There is very little praise in training, but plenty of criticism and failure is defined by death, and complications are questioned and taken personally. What went wrong? What did you do? These frequent questions after failure may illicit many strong emotions including guilt, self-blame, self-loathing, shame, and sadness. Surgeons are deeply affected by what surgical culture defines as success and failure, so it is critical to make a reassessment of outcomes and what our perception of success and failure are. In SPC, success can be a death, as long as the patient's goals and wishes were achieved and followed, respectively. Palliation can be a success by supporting your patient, so they may live the best life they can as defined by them until the end. A patient centric model of living life with quality until the end is as much of a success as curative treatment. The reality of incurable disease, unrealistic expectations, and death often bring to light our limitations and are seen as failure.

The Psychology of the Surgeon: Patient Relationship

The geriatric patient confronting serious surgical illness often views the surgeon as a hero and savior, and a rescue is expected, even when impossible. We can give our all to care for our patients even when there is no cure or surgical treatment but must not permit the hero/savior physician expectation to cause the surgeon to feel a sense of failure or frustration at not being able to cure or fix the patient because providing PC is frequently the best possible outcome [19]. When the surgeon feels like they have nothing to offer, this may be the time they actually have the most to offer their patient at his or her most difficult phase of life. Be there, show interest, listen, talk, support, be compassionate, and show presence. This may be the most valuable gift we have. We have ourselves. To offer ourselves is the ultimate gift to our patient even if there are no associated RVUs or CPT codes. We must learn to laugh and cry with our patients. Each patient and family will react differently. There is no right answer other than patient individualization.

The psychological phenomenon of transference may be experienced by the patient after clinical deterioration or bad news. A recent systematic review by Srinivasa et al. showed that patient complications effect surgeon personal and professional well-being and identified four main themes to these occupational hazards. (1) surgeons have feelings of anxiety, guilt, shame, and others that interfere with their personal and professional lives, (2) surgeons lack coping strategies and can turn to substance abuse, (3) talking with trusted colleges is seen as weakness, and (4) these complications affect future practice [20].

A Distinct Surgical Ethics

Ethics are foundational and play a daily major role in every SPC interaction. Patient autonomy and capacity are particularly emphasized given the prevalence and significance of control and dependence issues. Issues of surrogacy including advance directives add additional complexity, but indirect surrogate decision-making should reflect what the patient would want given the totality of the circumstances. Strained patient family dynamics may also be a confounder due to complex relationships and differing opinions. Finally, human bias, including the surgeon's, may factor into the situation, but surgeons must educate and communicate effectively with surrogates so as to do what the patient would likely want. The interdisciplinary team must support and help navigate this complex dynamic.

Dr. Miles Little proposed a distinctly surgical ethics that overlaps with the goals of SPC and combines within the broader medical ethics [21]. He describes five distinct ethical principles of surgery within the surgeon patient relationship: First, rescue, the surgeon attempts to cure and save the patient and, if successful, is comfortable. Second, proximity requires self-awareness and is critical to understand how intimate and sacred the patient-surgeon bond is, and SPC must account for this. The ethical principles of ordeal and aftermath are the third and fourth principles where the surgeon must be present and guide the patient after the realization of the enormous physical and mental toll surgery has taken on them. Finally, presence is the fifth and final ethical principle. It is the only principle experienced solely by the surgeon and consists of being there for the patient in mind, body, and soul regardless of the clinical or personal course. Surgeons must never abandon their patient, and distancing behavior must be avoided [22]. Surgeons must always be present to support and demonstrate compassion, and the effect on family members, caregivers, and caretakers alike cannot be overstated. The serious consequences this experience may have on caregivers need further inquiry [23].

Surgical Palliative Care Summary

Traditional surgical culture focuses on disease-specific or surgical outcomes, and their communication to patients is easy. Although challenging, we must explore our own personalities, training, psychology, ethics, and overall surgical culture. Our self-awareness will guide us through the SPC decision process which is complex, dynamic, and high-stakes emotionally and physically. We must help the patient care unit make the best decision possible within the context of the individual patient situation and using all the information we have available.

Surgical Palliative Care Decision-Making: A Paradigm and Framework to Improve Care

Introduction: SPC Decision-Making

Figure 30.2 depicts the framework of SPC surgical decision-making that can be used as an adjunct to assist the entire interdisciplinary team, not to solve the complicated clinical scenarios by following an algorithm, but should be included in the assessment of all patients to foster a cognizance of the possibility of SPC for all geriatric surgical patients. Methods to foster and improve integration of PC into surgical and trauma ICUs must be thought of at each decision point with the individual well-being as the shared objective of all and the center [24]. Within this context for decision-making, the timing and physical location of the patient are critical factors. The patient's physical location (nursing home, skilled nursing facility, etc.) and the acuity of the surgical decision-making are critical as well with regard to an emergent situation or an outpatient elective procedure being considered.

The Surgical Palliative Care Decision Points

The first and most important decision point is (Decision 1 – Fig. 30.2) whether or not to initially involve a SPC team with older ill patients. Uncertainty of prognosis, complex family dynamics, lack of understanding, complex decision-making, decreased capacity, and many other factors often make SPC involvement a necessity to best care, even when many surgeons may feel initially that formal SPC interdisciplinary care teams may be unwarranted because they

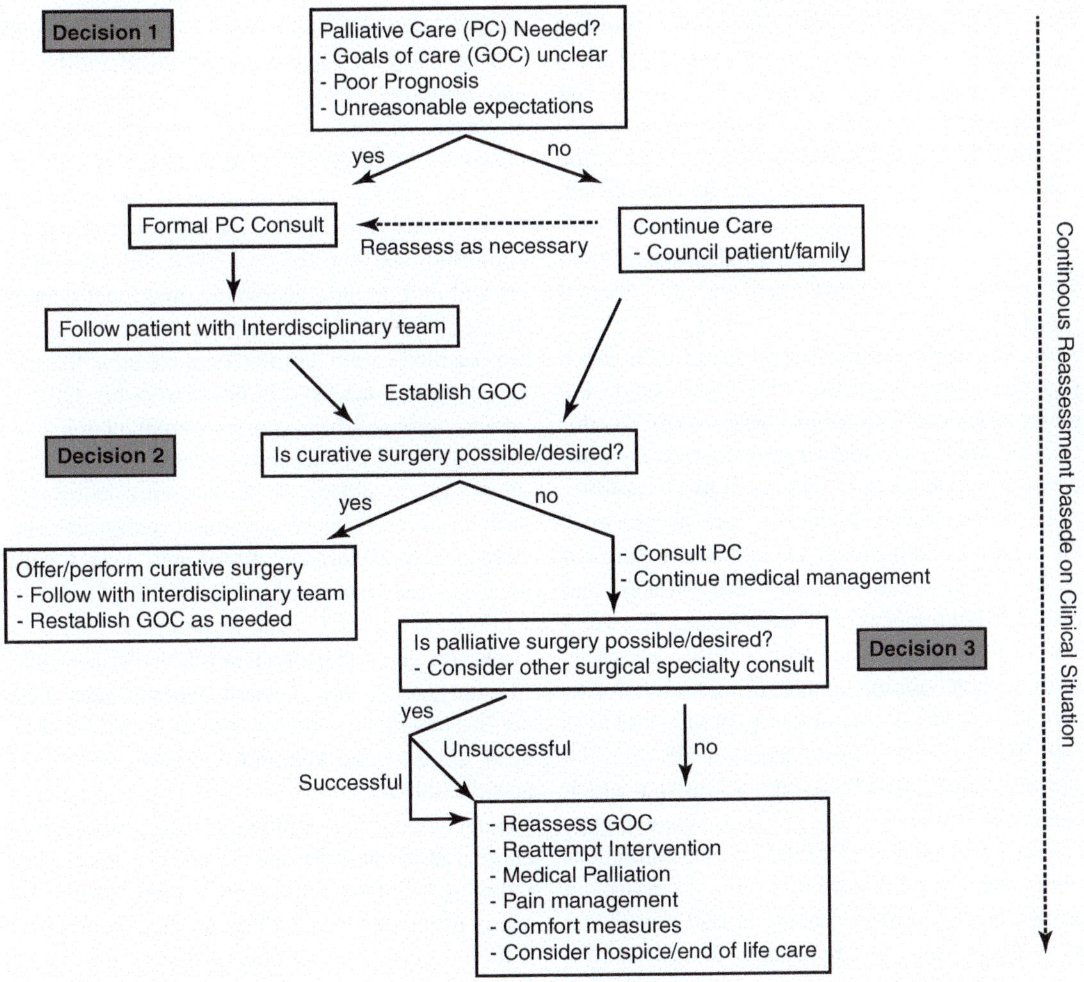

Fig. 30.2 SPC decision-making

can handle it themselves. SPC initiation does not have any requisite criteria for initiation although SPC guidelines for triggers to SPC involvement, as a means of increasing utilization and improving care, have been proposed [25]. Others have proposed protocols to accomplish essential SPC goals at reasonable time intervals [26, 27]. However, earlier is better so as to avoid the older implementation model of only engaging SPC after a decision has been made to not pursue further aggressive measures.

Decision 2 of Fig. 30.2 is whether curative surgery is the correct option for the individual patient. Traditionally, SPC is initiated when the decision "no" has been selected, but this is too late for optimal SPC utilization. Even if curative surgery is pursued, palliative treatments and support can also continue in parallel. Palliative surgery (decision 3) must account for not only the risk benefit discussion but also must factor preoperative optimization for what are typically very sick patients. The interdisciplinary care team is critical at all of the points in Fig. 30.2, and establishing the GOC is crucial, and GOC are readjusted to focus on QoL and relief of pain and suffering. Decisions concerning disposition and social factors are critical and must be discussed thoroughly as a change in settings can be disruptive and patients must feel in control and not abandoned.

Following Decision 3 of the SPC model is the critical decision to transition to end of life (EOL) care. The physician must effectively communicate the reasons for the transition and the process of EOL care while respecting the patient and family wishes. It should be a shared decision giving the care unit a role and sense of control of the situation. EOL goals of care must be established, support provided, and symptom management made a priority [28]. After death, the bereavement phase must be acknowledged and all affected cared for.

In summary, palliation has historically been part of a surgeon's job description and has been a major American College of Surgeons (ACS) educational goal for residency programs. Many educational initiatives have addressed this area from, "The Workforce on Palliative Care in Surgery," the creation of a Resident's Handbook in Surgical Palliative Care to ongoing educational and research conferences and workshops. The ACS has established PC guidelines [25]. Surgeons must lead by building on this strong foundation to restore the surgical tradition of palliation and SPC [19]. The educational component for SPC must be a priority, and it must focus on teaching that which can prove elusive to many brilliant technical physicians but is vital in SPC: the art of communication. There are models and aids to assist in cultivating this skill. "Breaking of Bad News" and EOL discussions are true tests of the ability of a surgeon to communicate an understanding of SPC and decision-making. Aids like the SPIKES Model give us guidance and structure to such difficult conversations [29], but it must be formally studied for trainees to learn it [30].

Models and Quantitative Tools to Aid in Decision-Making

As outlined elsewhere in this book, there is an emerging area of research across surgery into the effect of frailty, i.e., a reduced physiologic reserve, on surgical outcomes. Intuitively, we all know that the more comorbidities a patient has, the worse they are likely to do the more invasive and longer a procedure is. However, measures of frailty actually quantify this effect, and, as one might expect, frailty has the potential to better predict outcomes and augment more traditional scoring systems for the prediction of morbidity and mortality.

Based on large datasets, over 150 different measures of frailty have been developed; however the most commonly used is the modified frailty index (mFI) [31]. The mFI is a set of 11 comorbidities that are each assigned 1 point if present and include history of functional dependence, impaired sensorium, diabetes, hypertension on medication, chronic or acute lung disease, myocardial infarction, congestive heart failure, angina/prior cardiac surgery/percutaneous cardiac intervention, transient ischemic attacks, cardiovascular accidents or stroke with neurologic deficits, and peripheral vascular disease/rest pain. Across several surgical disciplines, a high mFI has been associated with increased risk for death, complications, and other morbidities [32, 33]. However, given that this is a relatively new field of study, it is still unclear which index best applies to a given situation. For example, Ondeck et al. showed that age and American society of anesthesiologists (ASA) scale better predicted adverse outcomes following hip arthroplasty compared to the mFI and Charlson comorbidity index (CCI) [34], while conversely this same group found that the CCI best predicted outcomes following spinal tumor surgery [35]. Conversely, Bateni et al. showed that the ASA, mFI, and CCI have limited predictive value for stage 4 cancer patients with bowel obstruction [36]. Therefore, while measures of frailty have promise to be used in predictive models of outcomes, several obstacles remain: (1) each disease state (with its given intrinsic mortality) may be best predicted by a different frailty index; (2) there are a dearth of prospective studies in this field; (3) for a given disease, outcomes may be best predicted by some other non-frailty variable; however, more disease-specific study is needed; and (4) frailty indices have been validated for standardized endpoints such as mortality and postoperative complications; however, it is unclear whether they predict other endpoints such as long-term outcomes, discharge home, or other QoL measures.

As alluded to above, the study of surgical outcomes frequently focuses on in-hospital complications and mortality; however it fails to identify other areas that matter to patient well-being. Traditional scoring systems such as the SOFA, APACHE, and intracranial hemorrhage score have all been developed to predict in-hospital outcomes especially mortality. The ethical issue with improving in-hospital survival at whatever cost is that it leaves the possibility that we will be discharging patients to a fate that is worse than death. One example of this is the decompressive craniectomy (DC) trials for malignant MCA infarction in elderly patients which showed that DC saved lives, but at the cost of creating significantly more extremely disabled elderly patients [37]. As a result, patients deserve the information to provide informed consent to our interventions including the quantitative risk of never being able to function in society again or the inability to participate in the activities they care about most, particularly relevant in the geriatric population, who have fewer years of life left and tend to be less interested in permanent significant disability than younger patients. Additionally, clinicians have a poor ability to predict short-term mortality and are frequently overly optimistic about prognosis [38]. The scoring systems discussed above are an attempt to account for a patient's physiologic reserve to better estimate outcomes. Future studies must include long-term outcomes below in order for our patients to truly provide informed consent.

These somewhat overlooked areas of surgical research include patient pain, depression, discharge location (home, nursing home, rehab), and long-term measures of QoL. For example, a recent systematic review found no randomized trials investigating QoL following cardiac surgery among the elderly but did show, in lower-quality studies, that QoL decreases among 8–19% of patients following cardiac surgery [39]. While this is only one example, it highlights the need for more study on endpoints that matter to patients such as the ability to return to work, pain, reintegration with society, and feeling like a burden on their families. One of the reasons these endpoints are underresearched is that it is logistically difficult to objectively measure them. Pain, for example, is fluid and changes by the day, thus making it difficult to reliably quantify. Another reason these endpoints are understudied is that we are not evaluated on them. Hospitals and physicians are heavily scrutinized for their mortality and complication rate; however long-term QoL measures are often overlooked as a quality measure.

While a somewhat abstract concept, measuring QoL is an important endpoint that should be used more frequently in surgical research. Common measures of health-related QoL is the Short Form 12 (SF-12) and Short Form 36 (SF-36) which ask questions regarding pain and whether health (including emotional health) interfered with working, socializing, accomplishments, etc. The major drawback to this type of scoring system is that the patient must be cognitively present and able to complete the survey in order for it to be useful, and thus it may be less accurate among nursing home residents [40]. Another common approximation of patient QoL is the mini-mental status exam (MMSE), which can be useful to predict postoperative complications. For example, the MMSE have been shown to predict postoperative delirium and long-term cognition following cardiac surgery [41]. Other scoring systems designed for the PC practitioner have also been developed to predict short-term mortality among terminal patients. One of the most common, the Palliative Prognostic Index (PPI) includes variables such as oral intake, edema, dyspnea at rest, delirium, and the palliative performance scale. A PPI >4 is associated with a predicted lifespan shorter than 6 weeks [42, 43]. While these tools help to predict short-term survival, they are limited in their long-term predictive value, thus making their use as predictors of long-term patient-centric outcomes limited.

In summary, while several systems for approximating frailty and predicting outcomes exist, their widespread implementation into preoperative decision algorithms is dependent on validation for each disease state and determination of their predictive value for non-traditional endpoints such as QoL, return to work, and pain. These endpoints are what patients care most about and should be included in future studies examining long-term outcomes of our interventions. However, as noted, prospective collection of these endpoints is difficult to obtain and difficult to use in research. Consequently, they have been largely ignored in favor of measuring mortality. As a result, mortality is the most frequently studied endpoint in SPC [7, 44]. The goal of future research should be to determine an objective measure of long-term outcomes that could be used to establish goals of care including the decision to operate. Therefore, as a means of providing information for informed consent, we have an ethical obligation to study these endpoints and develop objective tools to predict them.

Surgical Palliative Care Research and Ways to Improve the Evidence: Where Things Stand Today

The American College of Surgeons in 2003 identified seven key areas for the research and implementation of palliative care in surgery (Table 30.3). An excellent systematic review by Lilley et al. examined the state of the research in achieving these goals in surgical palliative care. They found only 25 suitable articles in the field of surgical palliative care have been published and only 9 of which are randomized clinical trials [7]. For purposes of brevity, we will only focus on surgical decision-making; however we encourage the reader to examine the Lilley article for more information.

Of the three studies that have investigated SPC and surgical decision-making, two by Miner et al. examined the role for palliative surgery among advanced cancer patients. First, in 2004, they showed that palliative procedures (including for obstruction, neurologic symptoms, pain, dyspnea, and jaundice) provided symptom improvement or resolution in 80% in 30 days. The median survival was 194 days from surgery; however the procedures were associated with 29% morbidity and 11% mortality within 30 days [45]. In 2011, they went on to describe how the palliative triangle of patient symptoms, values, and GOC can be used to carefully select patients for palliative surgery. Using this triangle they showed a 91% symptom improvement or resolution, lower postoperative morbidity (20%) and mortality (4%), and prolonged survival to 212 days, on average [46]. Similarly, Tan et al. showed that among patients undergoing colorectal surgery, those managed by a dedicated geriatric surgical team focusing on preoperative evaluation/rehabilitation and functional recovery had a high rate of return to functional status postoperatively [47].

Table 30.3 7 Domains of surgical palliative care

Surgical decision-making (disease/procedure focused)
Patient decision-making (patient focused)
End-of-life decision-making (recognition and the process)
Symptom management
Communication
Process of care
Surgical education

Research on surgical decision-making among elderly patients is lacking. This is true for both palliative and curative surgery. At a minimum, surgeons should incorporate GOC discussions into all conversations with patients and their families pre- and postoperatively. The research outlined above is a start; however there is a tremendous paucity of work in this field. Specifically, there is a need for surgical decision algorithms that incorporate palliative care concepts and predict mortality, QoL, and patient factors.

Gaps in knowledge must focus on patient function and QoL measures as well as effects on caregivers. Recent studies have emphasized that caregivers are subject to long-term effects of a loved one being hospitalized including post-traumatic stress disorder and lower QoL that can extend for over a year after hospitalization [48, 49]. Physicians have a role in preventing some of these adverse effects through counseling and providing family-centered counseling, which may even reduce ICU length of stay [50]. In summary, we must ask the right questions that will help us and our patients arrive at the best decision for the patient in the broad context of the entirety of their situation. Of major relevance, the effects on the caregiver's well-being are often overlooked, and future research is needed to establish ways of better caring for the caregiver and incorporating their opinions into the decision-making process.

Our approach to methodology along with studying the pertinent outcome measures and questions is key. The longitudinal nature of older patient's care can cross many care settings and physicians and can include the opinions and biases of multiple caretakers. Collecting data is difficult due to the patient's inability to participate in feedback secondary to decreased cognitive or overall function, frailty, loss to follow-up, distorted, missing data, and even death. Alternate means of gaining data may be needed including behavioral observation or data obtained by proxies given that randomized control trials are difficult in this setting [51, 52].

SPC decisions are of great complexity. Decision science examines all techniques and issues of such a decision-making process. Collaboration with PC as evidence grows may help ensure the decisions are made in the context of patient and family desires [53]. Best evidence and a structured decision process which requires study is needed to ensure that all relevant, critical SPC issues are addressed and help enable patients and families to be informed so as to make the best decision in accordance with the patient's wishes and values.

Conclusion

> Cure some, Treat often, Comfort all.
> Hippocrates

Through the lens of the surgical geriatric patient with serious illness, we examined barriers, proposed aids, and solutions, demonstrated a basis for these tools and where future inquiry can optimize SPC. SPC deserves its place in the field of surgery. In the end, it is about us and our patients. It is both that simple and complex. Peace for our seriously ill geriatric surgical patients can be achieved with dignity and humanity. A focus on the patient, their QoL, wishes, and "entirety" as people is necessary and of great importance to this "success." Achieving acceptable QoL, patient well-being, and dying devoid of pain and suffering is the means to this peace for patient, family, and surgeon. We must acknowledge we are human with all that entails: weakness, strength, fear, courage, and hope. It is our job to do the best we can with the objective data available and, most importantly, incorporate it into a patient-centric, interdisciplinary, and comprehensive model of care. We must give our geriatric surgical patients with serious and advanced disease the best life possible and an acceptable QoL as defined by the patient to the very end. A prolonged death filled with angst, pain, and suffering of the physical, emotional, and spiritual type must be accepted as wrong, harmful, and, ultimately, eradicated. While daunting, this will truly make us the heroes our patients often see us as. This chapter is meant to inspire and support surgeons in the pursuit of this most humanistic and pure form of care. The time for this paradigm shift in surgical culture is now. Good luck to all.

References

1. Chibbaro S, Di Rocco F, Makiese O, et al. Neurosurgery and elderly: analysis through the years. Neurosurg Rev. 2011;34(2):229–33.
2. Maldaner N, Sarnthein J, Bozinov O, Regli L, Neidert MC. Neurosurgery in octogenarians: a prospective study of perioperative morbidity, mortality, and complications in elderly patients. World Neurosurg. 2018;110:e287–95.
3. Yang R, Wolfson M, Lewis MC. Unique aspects of the elderly surgical population: an anesthesiologist's perspective. Geriatr Orthop Surg Rehabil. 2011;2(2):56–64.
4. Kavalieratos D, Corbelli J, Zhang D, et al. Association between palliative care and patient and caregiver outcomes: a systematic review and meta-analysis. JAMA. 2016;316(20):2104–14.
5. Olmsted CL, Johnson AM, Kaboli P, Cullen J, Vaughan-Sarrazin MS. Use of palliative care and hospice among surgical and medical specialties in the veterans health administration. JAMA Surg. 2014;149(11):1169–75.
6. Lilley EJ, Williams KJ, Schneider EB, et al. Intensity of treatment, end-of-life care, and mortality for older patients with severe traumatic brain injury. J Trauma Acute Care Surg. 2016;80(6):998–1004.

7. Lilley EJ, Khan KT, Johnston FM, et al. Palliative care interventions for surgical patients a systematic review. JAMA Surg. 2016;151(2):172–83.
8. Charlesworth CJ, Smit E, Lee DSH, Alramadhan F, Odden MC. Polypharmacy among adults aged 65 years and older in the United States: 1988–2010. J Gerontol Ser A Biol Sci Med Sci. 2015;70(8):989–95.
9. Sue K, Mazzotta P, Grier E. Palliative care for patients with communication and cognitive difficulties. Can Fam Physician. 2019;65(Suppl 1):S19–24.
10. Swartz K, Collins LG. Caregiver care. Am Fam Physician. 2011;83(11):1309–17.
11. US Department of Health and Human Services. Long-term care providers and services users in the United States, 2015–2016. 2019.
12. Jerant AF, Azari RS, Nesbitt TS, Meyers FJ. The TLC model of palliative care in the elderly: preliminary application in the assisted living setting. Ann Fam Med. 2004;2(1):54–60.
13. Dunn GP, Editor-in-chief F. Role of general surgery in the palliative care of patients with cancer – Oxford Medicine.
14. Ellis H. A history of surgery. London: Greenwich Medical Media; 2001.
15. Steiner JM, Cooper S, Kirkpatrick JN. Palliative care in end-stage valvular heart disease. Heart. 2017;103(16):1233–7.
16. Romanelli P, Esposito V, Adler J. Ablative procedures for chronic pain. Neurosurg Clin N Am. 2004;15(3):335–42.
17. Mount B. Whole person care: beyond psychosocial and physical needs. Am J Hosp Palliat Med. 1993;10(1):28–37.
18. Cassell J. Expected miracles: surgeons at work. Philadelphia: Temple University Press; 1991.
19. Dunn P, Dunn GP, Dunn P. Surgery, palliative care, and the American College of Surgeons. Ann Palliat Med. 2015;4(1):5.
20. Srinivasa S, Gurney J, Koea J. Potential consequences of patient complications for surgeon well-being: a systematic review. JAMA Surg. 2019;154(5):451–7.
21. Little M. The fivefold root of an ethics of surgery. Bioethics. 2002;16(3):183–201.
22. Tarpley MJ, Tarpley JL. Spiritual dimensions of surgical palliative care. Surg Clin North Am. 2011;91(2):305–15.
23. Thomsen KT, Guldin MB, Nielsen MK, Ollars CL, Jensen AB. A process evaluation of systematic risk and needs assessment for caregivers in specialised palliative care. BMC Palliat Care. 2017;16(1):1–8.
24. Mosenthal AC, Weissman DE, Curtis JR, et al. Integrating palliative care in the surgical and trauma intensive care unit. Crit Care Med. 2012;40(4):1199–206.
25. TQIP, ACS. Palliative Care Best Practices Guidelines.
26. Hui D, Hannon BL, Zimmermann C, Bruera E. Improving patient and caregiver outcomes in oncology: team-based, timely, and targeted palliative care. CA Cancer J Clin. 2018;68(5):356–76.
27. Luckett T, Phillips J, Agar M, Virdun C, Green A, Davidson PM. Elements of effective palliative care models: a rapid review. BMC Health Serv Res. 2014;14(1):1–22.
28. Peschman J, Brasel KJ. End-of-life care of the geriatric surgical patient. Surg Clin North Am. 2015;95(1):191–202.
29. Baile WF, Buckman R, Lenzi R, Glober G, Beale EA, Kudelka AP. SPIKES-A six-step protocol for delivering bad news: application to the patient with cancer. Oncologist. 2000;5(4):302–11.
30. Lord L, Clark-Carter D, Grove A. The effectiveness of communication-skills training interventions in end-of-life noncancer care in acute hospital-based services: a systematic review. Palliat Support Care. 2016;14(4):433–44.
31. Knoop V, Costenoble A, Vella Azzopardi R, Vermeiren S, Debain A, Jansen B, et al. The operationalization of fatigue in frailty scales: a systematic review. Ageing Res Rev. 2019;53:100911.
32. Seib CD, Rochefort H, Chomsky-Higgins K, et al. Association of patient frailty with increased morbidity after common ambulatory general surgery operations. JAMA Surg. 2018;153(2):160–8.
33. Velanovich V, Antoine H, Swartz A, Peters D, Rubinfeld I. Accumulating deficits model of frailty and postoperative mortality and morbidity: its application to a national database. J Surg Res. 2013;183(1):104–10.
34. Ondeck NT, Bohl DD, Bovonratwet P, et al. Predicting adverse outcomes after total hip arthroplasty: a comparison of demographics, the American Society of Anesthesiologists class, the modified Charlson comorbidity index, and the modified frailty index. J Am Acad Orthop Surg. 2018;26(20):735–43.
35. Lakomkin N, Zuckerman SL, Stannard B, Montejo J, Sussman ES, Virojanapa J, et al. Preoperative risk stratification in spine tumor surgery: a comparison of the modified charlson index, frailty index, and ASA score. Spine. 2019;44(13):E782–E787.
36. Bateni SB, Bold RJ, Meyers FJ, Canter DJ, Canter RJ. Comparison of common risk stratification indices to predict outcomes among stage IV cancer patients with bowel obstruction undergoing surgery. J Surg Oncol. 2018;117(3):479–87.
37. Das S, Mitchell P, Ross N, Whitfield PC. Decompressive Hemicraniectomy in the treatment of malignant middle cerebral artery infarction: a meta-analysis. World Neurosurg. 2019;123:8–16.
38. White N, Reid F, Harris A, Harries P, Stone P. A systematic review of predictions of survival in palliative care: how accurate are clinicians and who are the experts? PLoS One. 2016;11(8):1–20.
39. Abah U, Dunne M, Cook A, et al. Does quality of life improve in octogenarians following cardiac surgery? A systematic review. BMJ Open. 2015;5(4):e006904.
40. Andresen EM, Gravitt GW, Aydelotte ME, Podgorski CA. Limitations of the SF-36 in a sample of nursing home residents. Age Ageing. 1999;28(6):562–6.

41. Price CC, Garvan C, Hizel LP, Lopez MG, Billings FT. Delayed recall and working memory MMSE domains predict delirium following cardiac surgery. J Alzheimers Dis. 2017;59(3):1027–35.
42. Stone CA, Tiernan E, Dooley BA. Prospective validation of the palliative prognostic index in patients with Cancer. J Pain Symptom Manag. 2008;35(6):617–22.
43. Liu Y, Su L, Wang Y, Liu S, Dong B. The application of the palliative prognostic index in predicting the life expectancy of patients in palliative care: a systematic review and meta-analysis. Aging Clin Exp Res. 2018;30(12):1417–28.
44. Lilley EJ, Cooper Z, Schwarze ML, Mosenthal AC. Palliative care in surgery: defining the research priorities. Ann Surg. 2018;267(1):66–72.
45. Miner TJ, Brennan MF, Jaques DP. A prospective, symptom related, outcomes analysis of 1022 palliative procedures for advanced cancer. Ann Surg. 2004;240(4):719–27.
46. Miner TJ, Cohen J, Charpentier K, McPhillips J, Marvell L, Cioffi WG. The palliative triangle. Arch Surg. 2011;146(5):517.
47. Tan KY, Tan P, Tan L. A collaborative transdisciplinary "geriatric surgery service" ensures consistent successful outcomes in elderly colorectal surgery patients. World J Surg. 2011;35(7):1608–14.
48. Alfheim HB, Hofsø K, Småstuen MC, Tøien K, Rosseland LA, Rustøen T. Post-traumatic stress symptoms in family caregivers of intensive care unit patients: a longitudinal study. Intensive Crit Care Nurs. 2019;50:5–10.
49. Alfheim HB, Småstuen MC, Hofsø K, Tøien K, Rosseland LA, Rustøen T. Quality of life in family caregivers of patients in the intensive care unit: a longitudinal study. Aust Crit Care. 2018:1–7.
50. Goldfarb MJ, Bibas L, Bartlett V, Jones H, Khan N. Outcomes of patient- and family-centered care interventions in the ICU. Crit Care Med. 2017;45(10):1751–61.
51. Houska A, Loučka M. Patients' autonomy at the end of life: a critical review. J Pain Symptom Manag. 2019;57(4):835–45.
52. Hall A, Rowland C, Grande G. How should end-of-life advance care planning discussions be conducted according to patients and informal carers? A qualitative review of reviews. J Pain Symptom Manag. 2019;58(2):311–35.
53. Bakitas M, Lyons KD, Hegel MT, et al. Effects of a palliative care intervention on clinical outcomes. JAMA. 2009;302(7):741–9.

31 Family Involvement: What Does a Loved One Want at the End of Life?

Patrice L. Anderson

EVERYONE dies. Death is not an inherent failure........... The vital goal almost all people want from medicine is not having a good death but having as good a life as possible all the way to the very end.

Atul Gawande [1].

Everyone dies, and one of the hardest things anyone may ever have to do is to make end-of-life decisions for someone they care about. If a patient is incapacitated and no longer able to decide for themselves, others must decide for them. Sometimes patients specify future healthcare desires in legal documents. However, without documentation physicians must often rely on surrogates. As our population gets older, end-of-life care has become one of the most prevalent and important issues in healthcare. This chapter will explore how end-of-life decisions are made, who should be making them, and how the healthcare team can support patients and family members through the end-of-life process?

P. L. Anderson (✉)
Department of Surgery, New York Medical College, School of Medicine and Westchester Medical Center, Valhalla, NY, USA
e-mail: PatriceLynn.Anderson@wmchealth.org

R. Latifi (ed.), *Surgical Decision Making in Geriatrics*,
https://doi.org/10.1007/978-3-030-47963-3_31

How End-of-Life Care Decisions Are Made and Who Should Make Them

When a patient is no longer able to speak for themselves, others must decide for them. This daunting task is much easier if there is written documentation as to the patient's wishes. The Patient Self-Determination Act encourages all patients to make and document choices about the types and extent of medical care they would like to accept or refuse should they become unable to make these decisions in the future due to illness [1]. This federal law went into effect December 1991, and it requires hospitals, nursing homes, and hospices to advise patients about their right to accept or refuse medical/surgical care and to execute an advanced directive [1].

Written Guidance

Advanced directives enable both physicians and patient's loved ones to make educated, informed decisions about a patient's healthcare when a patient does not have decision-making capacity, and they make provisions for healthcare decisions in the future. These are fluid documents that can be changed and adjusted by the patient, as his/her medical circumstances and goals of care change [2]. As an extension of patient autonomy and self-determination, these legal documents allow a patient to detail their medical wishes, values, and preferences about end-of-life care. The two main types of advanced directives are the living will and the durable power of attorney for healthcare.

A *living will* is a document wherein a patient provides goals of care and treatment preferences to guide future healthcare decision-making [3]. Here a patient may clearly specify which procedures they would want if they were unable to voice an opinion. *A durable power of attorney for healthcare* is a document that assigns a healthcare proxy who is someone to make medical decisions for a patient when they are incapacitated. A durable power of attorney can also be referred to as a representative, an agent, or a proxy. This is usually someone the patient trusts will act in accordance with their values and wishes.

Many states have their own advanced directive forms.

Though not an advanced directive, the National Physician's Orders For Life-Sustaining Treatment (POLST) Paradigm is another way for patients to establish their end-of-life wishes. A POLST form has a set of specific medical orders that a seriously ill patient can fill in and have their physician sign. This form addresses wishes in an emergency such as cardiopulmonary resuscitation and intubation. This form is most appropriately used by patients who are likely to have a predictable medical crisis based on a terminal diagnosis. Per the POLST.org website: "The POLST decision–making process and resulting medical orders are intended for patients who are considered to be at risk of a life–threatening clinical event because they have a serious, life–limiting medical condition, which may include advanced frailty" [4]. Emergency personnel, like EMTs and paramedics, are unable to use advanced directives to guide their care. They are required to provide every possible treatment to keep a patient alive and transport them to the emergency room. However, emergency personnel can follow a POLST form because once this form is signed by a physician, it is a medical order.

Table 31.1 The POLST Paradigm varies by state

POLST (Physician Orders for Life-Sustaining Treatment)
POST (Physician Orders for Scope of Treatment)
MOLST (Medical Orders for Life-Sustaining Treatment)
MOST (Medical Orders for Scope of Treatment)
TPOPP (Transportable Physician Orders for Patient Preference)
COLST (Clinician Order for Life Sustaining Treatment)
DMOST (Delaware Medical Orders for Scope of Treatment)
IPOST (Iowa Physician Orders for Scope of Treatment)
TOPP (Transportable Orders for Patient Preferences)
AzPOLST (Arizona Provider Orders for Life-Sustaining Treatment
LaPOST (Louisiana Physician Orders for Scope of Treatment)
OkPOLST (Oklahoma Physician Orders for Life-Sustaining Treatment)
PAPOLST (Pennsylvania Orders for Life-Sustaining Treatment)
WyoPOLST (Wyoming Providers Orders for Life-Sustaining Treatment)
SAPO (State Authorized Portable Orders)
SMOST (Summary of Physician Orders for Scope of Treatment)

The POLST program, or similar programs which go by different names, can be found at the POLST.org website. This paradigm exists at some level in 50 states. The POLST Paradigm is known by different names in different states (Table 31.1).

When There Is No Written or Direct Verbal Guidance: Surrogacy

Default surrogate consent statues, available in the majority of states, allow physicians to consult either a designated individual or a group of individuals who will be tasked with making medical decisions for patients who are unable to do so for themselves [5, 6]. Surrogate laws were developed to help protect the wishes of incapacitated patients that have no advanced directives. Most states have developed a hierarchy model, or priority list, indicating which immediate family members may serve as a surrogate, for example, (1) a spouse (unless legally separated), (2) an adult child, (3) a parent, and (4) an adult sibling.

Many states have broadened the definition of a surrogate to include nonfamily members incorporating friends and domestic partners [5–7]. Additionally, some states have developed a mechanism for "unbefriended" patients. These are patients whose identity is unknown or patients who have no one to speak for them. In this situation, physicians, often in conjunction with ethics committees or other physicians, make end-of-life decisions, incorporating their medical knowledge and the best interests standard [6, 7]. If making life-ending decisions is too difficult and overwhelming for a loved one, they have the option of completely deferring decision-making to the treating physician or team. Most families, however, decide to use a consensus model. This style of decision-making is a collaborative process among family surrogates and multiple members of the patient's treatment team, each of whom has a different insight on how to best honor a patient's goals of life [6].

Several different approaches to surrogacy decision-making at the end of life have been proposed. One is called *substituted judgment* [6, 8, 9]. In this method, a surrogate is asked to put him- or herself in the place of the patient and to decide as if they are that person, to extrapolate from the patient's known personal values, and to think like the patient would think. The surrogate is often encouraged to pull from previous discussions with the patient and to act as an informant as to the patient's wishes without relying on their own personal preferences. This can be difficult as there is often discordance between what a patient would want and what their surrogate would do. A systematic review by Shalowitz and colleagues evaluated 40 years of research on surrogate decision-making. They identified 16 studies that examined 151 hypothetical scenarios. Their goal was to assess how accurately surrogates were able to predict a patient's treatment choice. They concluded that both the patient designated and default surrogates predicted the patient's treatment preferences with only 68% accuracy [9]. This discrepancy may stem from the fact that surrogates are unable to remove themselves completely from the situation and the decisions surrogates make are not substantially different than those they would have made for themselves. Another reason surrogates may not adequately represent a patient's treatment preference is because these preferences are fluid and change. As a patient's health status changes, so too do their priorities and wishes with regard to life-sustaining treatment over time [10, 11].

Questions we can ask that may help surrogates make the best decision possible are depicted in Table 31.2 [8]. Substituted judgment supports a patient's autonomy, but it may also help the surrogate psychologically. Families feel a significant burden when making life and death decisions for a loved one. Emphasizing substituted judgment may alleviate some of this by framing the decision based on what the patient would want rather than what the surrogate would want to do.

A second approach to surrogacy decision-making is the *best interests standard*. In this case healthcare management decisions are made in accordance with societal norms and beliefs about what is best for a dying patient [6, 8, 12]. This approach is designed to incorporate a patient's quality of life based on what most people would want.

Table 31.2 Questions that may help surrogates make end-of-life decisions

Has the patient ever spoken about what they want at the end of their life?
Has the patient expressed an opinion about how someone else has been treated that could be used to extrapolate in the situation?
What have been the patient's values in life?
What has given their life meaning?
Would the patient want the most days of life, or would the patient want fewer days with a better quality of life?

Framing the End-of-Life Discussion

Often, family members or surrogates are overwhelmed with choices when a loved one nears the end of their life and the patient can no longer speak for themselves. The concepts and terminology with which they must grapple are often new, confusing, and frightening. This can be a time of incomparable sadness and unparalleled stress that has the potential to affect family members long after the patient dies. The decision-making burden can be profound. The hope to improve the experience of a dying patient and their family has become a widespread, international resolve. The number of publications over the last 50 years on "end-of-life care" has gone up exponentially (Fig. 31.1) as have the available resources for patients, families, and medical personnel (Table 31.3).

The framing of end-of-life discussions, the terminology used, the compassion shown, and the explanations provided can have an enormous effect on improving the experience of death for the surrogates, the family, the friends, and ultimately the patient. Family members often struggle with the subtle differences between aggressive treatment, stopping curative therapy, do not resuscitate (DNR), withholding life sustaining treatment, no escalation of care, withdraw of care, and comfort measures. This is an area where the medical community can substantially decrease the burden felt by families after someone dies if they allow families to voice their concerns and work through the stumbling blocks with them.

What Does "Do Everything" Really Mean?

When faced with a loved one actively dying, family members often request that physicians to "do

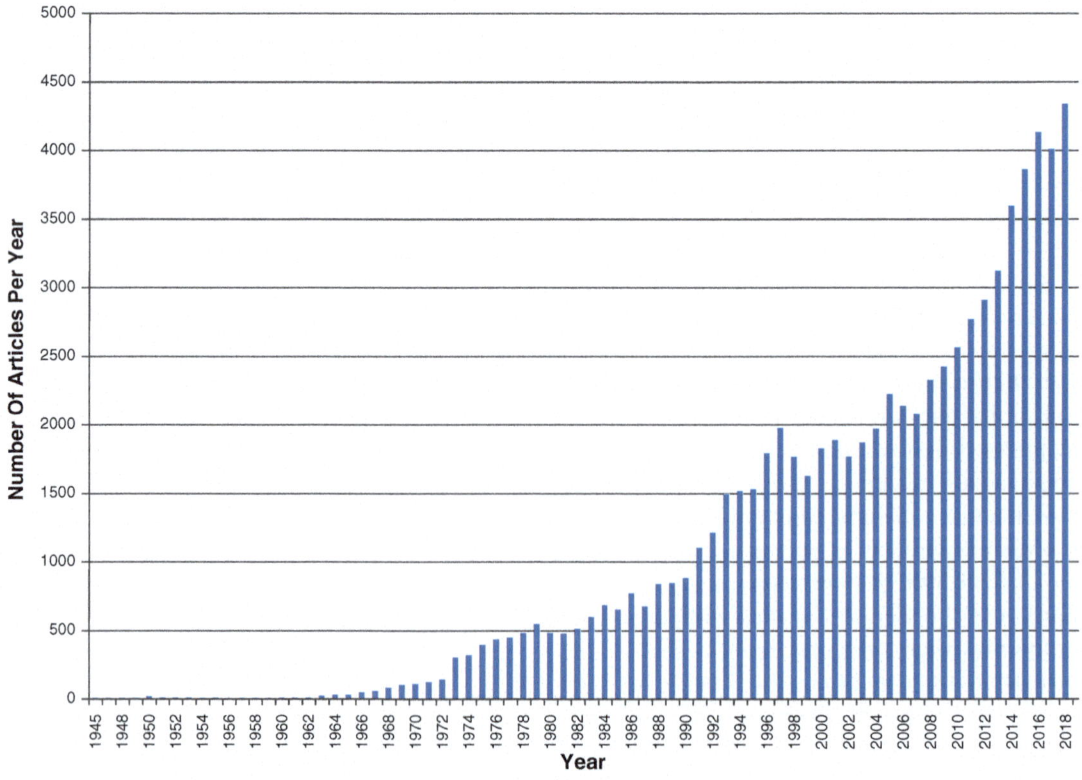

Fig. 31.1 Number of articles on "end-of-life care" published per year on PubMed since 1945

Table 31.3 Some of the end-of-life care resources available for patients, families, and physicians

American Society of Clinical Oncology
National Palliative Care Research Center
Promoting Excellence and End-of-Life Care
End-of-Life Nursing Education Consortium
AMDA Foundation
Association for Death Education and Counseling
Center to Advance Palliative Care
The Education on Palliative and End-of-Life Care Project
Aging, Health And Healing
National POLST Paradigm
The Coalition to Transform Advanced Care
Supportive Care Matters
What Matters Now
CancerSource
Aging With Dignity
Cancer Resources
Journey of Hearts: A Healing Placed in Cyberspace
US Living Will Registry
Veterans Guide to Long-Term Care
National Association of Social Workers
Hospice And Palliative Nurse Association
Hospice Foundation of America
American Academy of Hospice And Palliative Medicine
National Coalition for Cancer Survivorship
National Family Caregivers Association
MyDirectives
Gift From Within – PTSD Resources for Survivors And Caregivers
American Medical Association – Education for Physicians on End-of-Life Care Project
Open Society Institute – Project on Death in America
Robert Wood Johnson Foundation – Last Rites Campaign

everything." This stems from a deep-seated need to feel like they have done everything possible to help their loved ones. As healthcare professionals, it is our responsibility to reframe this request and focus end-of-life discussions on advocating for what is ultimately in accordance with what the patient would want instead of just doing everything. The medical equivalent of doing everything is a full code. These extraordinary measures are meant for critically ill patients as a temporizing measure for the body to have time to regain normal function. Full codes are not intended to be used for someone who is dying and will never regain a good quality of life. Coding these patients may only delay the inevitable and the patient will still die. In fact, this has the potential of increasing the patient's and the family's suffering without changing the outcome. It is imperative that we help the families understand the balance between the burden of treatment and the benefits to the patient's quality of life, not their quantity of life [13].

Stopping Aggressive Treatment Is Not the Same as "Doing Nothing"

If there is a "do everything," that implies that there must be a "do nothing." An integral part of end-of-life discussions should include dispelling the misconceived notion that stopping aggressive treatment means to "do nothing". We do not stop caring for the patient. It is the exact opposite. We listen to what the patients want. In 1927, when speaking at the Harvard Medical School, Dr. Peabody summarized this well when he said that "one of the essential qualities of the clinician is interest in humanity, for the secret of the care of the patient is caring for the patient" [14]. Often families need to be reassured that appropriate symptom management will continue in accordance with the patient's wishes even after curative treatment has stopped. Too often, the end of a patient's life can be negatively impacted by interventions meant to cure them.

Withdrawal of Care Is Not the Same as "Killing Someone"

The decision to withhold or withdraw life-saving treatment can be emotionally challenging, and many families equate this to abandonment, "giving up," or "killing someone." If aggressive care is incapable of extending a patient's life in a meaningful way in accordance with the patient's desired quality of life, stopping this type of treatment is appropriate, ethical, and humane. In fact, withdrawing some treatments and shifting the goals of care from cure to comfort allows physicians to concentrate solely on relief of symptoms, and patients often feel better. It is important for physicians to recognize that families go through a process in which they must transition from hoping for a cure to the acceptance that death is imminent. This process can be devastating and hard. Various family members often make this transition at different times and usually after the patient themselves have accepted their death. It is important that as physicians we approach this time with empathy and compassion.

"Allow a Natural Death" (AND) May Seem More Acceptable Than Do Not Resuscitate (DNR)

For a medical professional, a do not resuscitate order is equivalent to allowing a natural death. It means that if a patient's heart stops or they stop breathing, we will not employ extraordinary measures to prolong their life. For some families "do not resuscitate" has a negative connotation because it implies that we are electively not doing something we are capable of doing. Many have postulated that allowing a natural death may seem more compassionate to families, and this could ultimately make this decision easier [15, 16]. In the end, the result for the patient is the same – there will be no extraordinary measures taken to prolong life. For the family, however, allowing death to occur naturally may be easier to accept.

What Is Quality End-of-Life Care, and Is There Such a Thing as a "Good Death"?

When a patient's illness defies treatment and their goals of life can no longer be met, focus must move from cure to comfort. This is when the medical community needs to make every effort to facilitate a "good death" or a dignified death. What does this mean? In 1997 the Institute of Medicine defined a good death as "one that is free from avoidable distress and suffering, for patients, family and caregivers; in general accord with the patient's and family's wishes; and reasonably consistent with clinical, cultural and ethical standards" [17]. "Dying with dignity" has been defined as physical comfort, autonomy, meaningfulness, usefulness, preparedness, and interpersonal connection [18]. For each person their definition of a "good" death is shaped by individual hopes, fears, and beliefs. There is no standard "dignified death" just as there is no standard "good death." It is our role to help families figure out this very personalized definition and honor it.

A recent systematic review identified and critically appraised tools measuring the quality of death and dying [19]. The review highlighted that though there are an abundance of tools available to assess the quality of death and dying, because this is such a personal matter, we may never develop a single measurement tool [19]. Instead of trying to find a tool, concentrating on topics revealed in several studies could help ensure a quality end of life or good death. Themes that are important to patients at the end of life are summarized in Table 31.4 [19–23].

Table 31.4 Themes important to patients at the end of life

The dying process: dying during sleep, quickly, where, when, how
Adequate pain and symptom management: not suffering, dying with no pain, no thirst, no anxiety, calm sleep, no dyspnea
Emotional well-being: emotional support, peace, psychological comfort, whole person well-being
Family: family support, surrounded by loved ones, family acceptance of death, desire not to be a burden on the family, connectedness with loved ones
Dignity: respect as an individual, maintaining independence, autonomy
Life completion: saying goodbye, acceptance of death, preparedness
Religion and spirituality: comfort, faith, clergy support
Treatment preferences: not prolonging life unnecessarily, reassurance that all available life-saving treatments have been exhausted, a sense of control over treatment choices
Quality of life: maintaining pleasure
Relationship with the help care team: trust, support, and comfort from the entire treatment team

Expanding the Overall Treatment Plan to Include Support for Surrogate Decision-Makers

Often surrogates struggle with trying to make an informed decision which adequately reflects their loved one's values, preferences, and end-of-life goals. Making life-sustaining treatment decisions for a loved one can be very traumatic. As our population ages, many of these decisions are now being made in our ICUs. Whether it is a patient-designated healthcare proxy or a default surrogate, a designee may experience psychological and physical stress including anxiety, fear, depression, fatigue, posttraumatic stress symptoms, and "post-intensive care unit syndrome" [24–29]. Patient age, lack of information, and patient illness severity are some of the predictors for psychological and physical distress in family members [24, 30].

Like patient-centered care at the end of life, there is also evidence that family-centered care is of great importance as well [24, 31]. This type of care incorporates the emotional needs of family members into the overall treatment plan of the patient. End-of-life family conferences should be specifically designed to help families work through difficult decisions that need to be made with effective communication, respect, collaboration, and support [32, 33]. These discussions need to start early, and they should include patient's priorities, goals of care, appropriateness of treatment, disease trajectory, and possibility for quality survival. These conversations often include difficult truths and prognostic estimates that may facilitate family decision-making that has the potential to maximize quality of life over quantity. Patient's families need to have the opportunity to discuss what they think the patient may want, to express their grief and sadness, to alleviate their feelings of any guilt, and to understand the ultimate goals of care. To avoid rushed, last-minute, middle of the night decisions, aspects of care that should be addressed during these goal-oriented family meetings should include many aspects of care (Table 31.5) [2].

Table 31.5 Aspects of care that should be addressed during goal-oriented family meetings

Respiratory distress and requirement of a ventilator
Cardiopulmonary resuscitation
What to do with pacemakers/defibrillators
Hemodialysis
Enteral feeding access or parenteral nutrition
Use of antibiotics
Utilization of IV hydration
Life-sustaining medications
Sedation and pain management
DNR/AND orders
Organ tissue donation
POLST and MOLST forms

What Happens If the Medical Staff and the Family Members Disagree on a Treatment Plan

On occasion family members request interventions that a physician may find "futile" or "medically inappropriate." This can be a challenging ethical situation. In 2015 the American Thoracic Society (ATS), the American Association for Critical Care Nurses (AACN), the American College of Chest Physicians (ACCP), the European Society for Intensive Care Medicine (ESICM), and the Society of Critical Care Medicine (SCCM) jointly published a policy on how to respond to requests for potentially inappropriate treatments in intensive care units [34]. This multi-society statement was designed to provide clinicians with tools for managing surrogates of critically ill patients. They emphasized early proactive communication and collaborative decision-making, utilizing "potentially inappropriate" care instead of "futile" care when speaking with family members, and they provided steps for resolution of conflict regarding potentially inappropriate treatments [34].

More recently, in 2019, the American Medical Association Principles of Medical Ethics [3] suggested that the physician:

1. Review the patient's goals of care and desired quality of life with the family.
2. Reassure the family that medically appropriate interventions, specifically symptom management, will be provided.

3. Negotiate a mutually agreed upon plan of care that includes both the physicians clinical judgment and the family's understanding of the patient's wishes.
4. Seek guidance from an ethics committee if the family continues to request treatment that the physician feels is not medically warranted.

It Takes a Village

Providing exceptional care for dying patients takes an extensive treatment team including: physicians, palliative care providers, nursing staff, social work, clergy, and the patient's family. This end-of-life shared decision model strives to find a cohesive consensus between patient autonomy, family needs, and medical opinion.

Conclusion

The reality is that every one of us will die one day. As healthcare providers sometimes all we can do is ensure patient autonomy and fulfill their wishes and those of their loved ones. There are countless technical and scholarly articles about end-of-life care, but perhaps in the end – it is about honoring a dying patient and their family by treating them with openness, respect, compassion, dignity, advocacy, and commitment to the highest quality of care. Even with all we know, and all we can do as medical professionals, this may be where we make the greatest impact on our patients and their loved ones. In the end everyone should get to die how they lived, on their own terms.

References

1. Gawande A. Quantity and quality of life. JAMA. 2016;315(3):267.
2. H.R.4449 – 101st Congress (1989–1990): Patient Self Determination Act of 1990 [Internet]. Congress.gov. 2019 [cited 1 June 2019]. Available from: https://www.congress.gov/bill/101st-congress/house-bill/4449.
3. Advance Care Planning: Healthcare Directives [Internet]. National Institute on Aging. 2019 [cited 20 May 2019]. Available from: https://www.nia.nih.gov/health/advance-care-planning-healthcare-directives.
4. Code of Medical Ethics: Caring for patients at the end of life [Internet]. American Medical Association. 2019 [cited 21 July 2019]. Available from: https://www.ama-assn.org/delivering-care/ethics/code-medical-ethics-caring-patients-end-life.
5. POLST A, Planning A, Directives A, Form P, FAQs P, Programs S et al. Home – POLST [Internet]. POLST. 2019 [cited 26 May 2019]. Available from: https://polst.org/.
6. If There is No Advance Directive or Guardian, Who Makes Medical Treatment Choices? [Internet]. Americanbar.org. 2019 [cited 1 May 2019]. Available from: https://www.americanbar.org/groups/law_aging/publications/bifocal/vol_37/issue_1_october2015/hospitalist_focus_group/.
7. Kohn NA, Blumenthal JA. Designating health care decision-makers for patients without advance directives: a psychological critique. Georgia Law Review 2008;42:979. Available at SSRN: https://ssrn.com/abstract=1008494.
8. Bifocal, A Journal of the ABA Commission on Law and Aging [Internet]. Americanbar.org. 2019 [cited 1 June 2019]. Available from: http://americanbar.org/publications/bifocal/vol_36/issue_1_october2014/default_surrogate_consent_statutes.html.
9. Understanding Healthcare Decisions at the End of Life [Internet]. National Institute on Aging. 2019 [cited 31 May 2019]. Available from: https://www.nia.nih.gov/health/understanding-healthcare-decisions-end-life.
10. Shalowitz D, Garrett-Mayer E, Wendler D. The accuracy of surrogate decision makers. Arch Intern Med. 2006;166(5):493.
11. Danis M. Stability of choices about life-sustaining treatments. Ann Intern Med. 1994;120(7):567.
12. Emanuel L. Advance directives. Stability of patients' treatment choices. Arch Intern Med. 1994;154(2):209–17.
13. Torke A, Alexander G, Lantos J. Substituted judgment: the limitations of autonomy in surrogate decision making. J Gen Intern Med. 2008;23(9):1514–7.
14. Quill T. Discussing treatment preferences with patients who want "everything". Ann Intern Med. 2009;151(5):345.
15. Peabody F. The care of the patient. JAMA. 1927;88(12):877.
16. Barnato A, Arnold R. The effect of emotion and physician communication behaviors on surrogates' life-sustaining treatment decisions. Crit Care Med. 2013;41(7):1686–91.
17. Venneman S, Narnor-Harris P, Perish M, Hamilton M. "Allow natural death" versus "do not resuscitate": three words that can change a life. J Med Ethics. 2008;34(1):2–6.
18. Field M, Cassel C. Approaching death. Washington, D.C.: National Academy Press; 1997.
19. Proulx K, Jacelon C. Dying with dignity: The good patient versus the good death. Am J Hosp Palliat Med. 2004;21(2):116–20.
20. Kupeli N, Candy B, Tamura-Rose G, Schofield G, Webber N, Hicks S, et al. Tools measuring quality of death, dying, and care, completed after death: sys-

tematic review of psychometric properties. Patient. 2018;12(2):183–97.
21. Emanuel E, Emanuel L. The promise of a good death. Lancet. 1998;351:SII21–9.
22. Meier E, Gallegos J, Thomas L, Depp C, Irwin S, Jeste D. Defining a good death (successful dying): literature review and a call for research and public dialogue. Am J Geriatr Psychiatry. 2016;24(4):261–71.
23. Singer P, Martin D, Kelner M. Quality end-of-life care. JAMA. 1999;281(2):163.
24. Wong A. Having a good death in our ICU | EUROASIA [Internet]. Esicm.org. 2019 [cited 30 May 2019]. Available from: https://www.esicm.org/blog/?p=1908.
25. [Internet]. Atsjournals.org. 2019 [cited 30 June 2019]. Available from: https://www.atsjournals.org/doi/pdf/10.1164/rccm.200409-1295OC.
26. Davidson J, Jones C, Bienvenu O. Family response to critical illness. Crit Care Med. 2012;40(2):618–24.
27. Mercadante S, Gregoretti C, Cortegiani A. Palliative care in intensive care units: why, where, what, who, when, how. BMC Anesthesiol. 2018;18(1):106.
28. Pochard F, Azoulay E, Chevret S, Lemaire F, Hubert P, Canoui P, et al. Symptoms of anxiety and depression in family members of intensive care unit patients: ethical hypothesis regarding decision-making capacity. Crit Care Med. 2001;29(10):1893–7.
29. Pochard F, Darmon M, Fassier T, Bollaert P, Cheval C, Coloigner M, et al. Symptoms of anxiety and depression in family members of intensive care unit patients before discharge or death. A prospective multicenter study. J Crit Care. 2005;20(1):90–6.
30. Schmidt M, Azoulay E. Having a loved one in the ICU. Curr Opin Crit Care. 2012;18(5):540–7.
31. A communication strategy and brochure for relatives of patients dying in the ICU. N Engl J Med. 2007;357(2):203.
32. Fumis R, Ranzani O, Martins P, Schettino G. Emotional disorders in pairs of patients and their family members during and after ICU stay. PLoS One. 2015;10(1):e0115332.
33. Wiegand D, Grant M, Cheon J, Gergis M. Family-centered end-of-life care in the ICU. J Gerontol Nurs. 2013;39(8):60–8.
34. Bosslet G, Pope T, Rubenfeld G, Lo B, Truog R, Rushton C, et al. An official ATS/AACN/ACCP/ESICM/SCCM policy statement: responding to requests for potentially inappropriate treatments in intensive care units. Am J Respir Crit Care Med. 2015;191(11):1318–30.

Spiritual and Religious Considerations in the Care of the Elderly

32

Husham Abdelrahman, Mohammad Asim, and Ayman El-Menyar

Religiosity and spirituality are important elements of well-being among the elder population worldwide. Although both are often approached as similar concepts, in reality, they are not identical but have common characteristics. According to the 2000/2001 Gallup polls, nearly two-thirds of elderly US citizens (above 50 years) noted religion as "very important" and attended regularly church or synagogue. This could explain the increasing interest to understand this important and complex aspect of care. Notably elderly people are being reported more religious and spiritual than the younger population. For instance, in the USA >90% of elderly are considered as religious or spiritual, while only 6–10% are atheists. It has been suggested that the well-being, inner peace, personal meaning, belonging and connectedness, and belief in life hereafter would help to guide the actions and are considered important coping factors among the elderly. Moreover, terminal illness would prompt and provoke the religiosity and/or spirituality needs especially upon approaching the end of life [1]. An earlier study interviewed 205 advanced cancer patients about religiousness and spiritual support. Although spiritual support is associated with improved quality of life, such needs were not fulfilled by the religious communities or health-care system [2]. Overall, the majority of elderly patients relied on religion, at least to a moderate degree, in enduring health issues and adverse societal circumstances.

H. Abdelrahman · M. Asim
Department of Surgery, Trauma Surgery, Hamad General Hospital, Doha, Qatar

A. El-Menyar (✉)
Department of Surgery, Trauma Surgery, Hamad General Hospital, Doha, Qatar

Clinical Medicine, Weill Cornell Medical College, Doha, Qatar

What Does Religiosity and/or Spirituality Mean?

Although both terms are often used interchangeably, the difference between religiosity and spirituality is beyond this text. However, for the sake of simplicity and understanding, "Spirituality involves humans' search for the purpose of life, while religion is associated with an organized entity with specific fundamental beliefs, rituals, and practices about a higher power or *God* [3]." For instance, in different faiths like Christianity, Islam, Jewish, Buddhism, Hindu, and the sub-affiliations (Protestant, Catholics, Eastern Orthodox, Mormon, and others), it defines relationships and responsibilities with certain ritualistic practices and attendance in a worship place.

There are three main dimensions of religiousness which include the organizational (social worship practices), non-organizational (praying/meditation/reciting and watching holy books), and subjective (or intrinsic) religiosity

R. Latifi (ed.), *Surgical Decision Making in Geriatrics*,
https://doi.org/10.1007/978-3-030-47963-3_32

(motivating/drives behavior and influence decision-making).

Both religiosity and spirituality are important consideration for the care of the elderly; however with respect to subjective health status, spirituality might be a crucial explanatory factor for good health [4]. It is important to note that spirituality could be associated with religiosity for those who have faith in God but for atheists it will be unrelated [3]. Notably, spiritual commitment, belief, attitude, and practice of people may influence their perception of health status, interaction with their ailments, decision-making, wishes, and preferences [5].

How to Measure Religiosity and Spirituality?

There are various measures to assess religious involvement such as organizational religiousness, non-organizational religiousness, intrinsic religiosity, self-rated religiousness, and observer-rated religiousness. On the other hand, the measures of spirituality includes self-rated spirituality, observer-rated spirituality, and daily spiritual experiences [6–12].

Spirituality can be described in several ways, and its perception could vary for different individuals and thus difficult to measure or capture in activities (practices) or personal inquiry (attitudes and experiences) [7]. Often the religiosity is measured within the spirituality scope as if the latter being considered broader and that is also confusing.

A valid measure for assessment of spirituality is the Spirituality Index of Well-Being (SIWB) [9]. This instrument contains 20 items with 2 subscales involving religious and existential well-being which demonstrated associations between health and well-being in the outpatient settings for geriatric primary care. The Spiritual Needs Questionnaire (SpNQ) is a common measure for the assessment of spiritual needs of patients with chronic pain conditions and cancer [10]. Moreover, the Spiritual Needs Assessment for Patients (SNAP) is another instrument consisting of 23 items intended to measure the psychosocial, spiritual, and religious needs [11].

The Problem of Being Hospitalized Among the Elderly

To further define the problem of the sick elderly patients, it is very important to consider the stressors of being hospitalized in the elder population. In simple words, it is the the fear for loss. These stressors include fear of disease seriousness and nearness to dying, abandoning social roles, dependence, confronting the unknown, and threatened sense of control and adequacy along with unpleasant physical symptoms, effects of hospitalization which restrict mobility, and limited stimulation which assaults patient's sense of competence [13]. Based on the current literature, religious and spiritual beliefs could help in explaining the meaning of existential questions that arise with respect to a sickness and the possibility of death [14]. However, there exists a great challenge for health professionals to help provide spiritual nursing care to patients experiencing sickness [15]. This is attributed to the fact that health-care professionals are often inexperienced and unaware to deal with the religious and spiritual beliefs of the patients and relatives.

The Importance of Religious and Spiritual Beliefs in Clinical Practice

Religious and spiritual beliefs are considered as essential factors both socially and psychologically in health determination, perception, and response to treatment. In accordance to the World Health Organization (WHO), the definition of health refers to the "state of complete physical, mental and social well-being and not merely the absence of disease or infirmity" [16]. That is why spiritual well-being is integral to the evaluation of health-related quality of life [17]. Religion and spirituality are assumed to provide hope and courage in difficult situations as well. Even among socially isolated elderly residing in institutions, the personal meaning, individual value of religion, and access to spiritual resources are important components of well-being [18].

The effects of religious and spiritual beliefs can be summarized in three psychological benefits. First is a *positive attitude* about life, stress, and disease that may improve resilience and outcomes both in short- and long-term basis. Second, they provide *a sense of meaning or purpose* that delineate behaviors, support, and relationships. Third, they develop an *optimized tolerance and coping* to illness, hospitalization, and disability.

Religious commitment is beneficial for well-being and prevention of both mental and physical illnesses, enhance recovery, and empower coping abilities [19]. Health-promoting practices such as avoidance of smoking (tobacco), heavy alcohol use, or other substance abuse resulted in healthier and longer life, e.g., Mormons and Seventh-Day Adventist have higher life expectancy (2–4 years) than in the Federal Republic of Germany [20]. Elderly people who have such community networks are less likely to neglect themselves. There appears to be a strong link between religiosity and successful adaptation to the challenges of aging [21]. An earlier study suggested that there is a need to enhance the skills and capacity of self-care among elder population which strongly correlates with decline in self-neglect and increases their well-being [22]. Studies also showed that spiritual support by medical team in critically sick patients influences medical decision and improves quality of life especially at the end of life with greater hospice utilization and less aggressive medical interventions [23].

Is Religiosity and/or Spirituality an Alternative or Complementary to Medical/Surgical Treatment?

Historically religiosity and spirituality were somewhat considered the same (synonymous), before the era of modern and scientific-based practice, where for a while it was separated nearly completely. However, recently the interest was renewed, and currently they are considered as important adjuncts to complement and supplement the different medical management approaches.

That's particularly true considering the reality of conventional treatment, where fears, side effects, previous unpleasant encounters, and the need for supportive personal approaches are by large the most common explanation for their choice. The overall belief is that religiosity is associated with better physical, mental, psychological, social, and functional health as well as less chronic pain experience [24].

Why People Seek Alternative or Adjunct Therapies?

There are many causes that limit the conventional treatment, such as side effects, previous negative experiences, looking for more personal support, and a holistic sense of health. One should not think of religious or spiritual healing as a substitute for standard medical care; but active religious commitment is beneficial for prevention, coping, and even recovering from both mental and physical illnesses as previously stated [25]. Furthermore, they are very helpful in rehabilitation and disability acceptance [25]. An earlier study showed that spirituality and religiousness are associated with the utilization of complementary and alternative medicine (CAM) among adult population [26]. Another concept is the integral medicine which emphasizes that human beings have various emotional, mental, and spiritual dimensions to treat which are necessary for the diagnosis and management of disease and improvement of well-being [27]. Therefore, the medical team should be aware of the implications and consider the sociocultural, spiritual, and psychological parameters in the patients. Therefore, integral medicine is the next step of integrative medicine with multidimensional approach involving curing from physical to psychological and cultural to spiritual in addition to conventional and alternative medicine.

Patient's Perspective on Clinical Management and Decision

The potentially harmful impact of the health exterminators could be expressed in attitude issues with respect to inflexibility, anxiety, reluctance, and looking for religious rituals like praying and chanting as a substitution to lifesaving interventions and therapies. The religious devotion might enhance excessive guilt, inflexibility, and anxiety. Patients with mental illness are more vulnerable of developing religious preoccupations and delusions. Certain religious groups discourage mental and physical health care, including potentially lifesaving therapies (e.g., blood transfusions, treatment of life-threatening infections, insulin therapy) [28], and may substitute religious rituals (e.g., praying, chanting, lighting candles). Some religious communities are more rigid and may isolate elder population from nonparticipating family members and the broader social community [29]. The elderly people often have distinct spiritual needs that may overlap with but are not the same as psychologic needs. Therefore, identification of patients' spiritual needs might help to understand and provide the desired support in terms of spiritual counseling, involvement in religious activities, and socialization with the specific religious community members [30].

Caregiver's Perspective on Religiosity and Spirituality

In a study among caregivers of patients with terminal cancer, the investigators found that caregivers with a strong personal religious faith and many social contacts were better able to cope with the stresses of caregiving during a 2-year period [31]. However, there has been less focus on the role that religion/spirituality might play in coping with stress of informal caregiving of older persons who need assistance with activities of daily living. Caregiving for elder population may be influenced by religion/spirituality beliefs as those caregivers with firm religious beliefs are more involved and motivated to provide care among eldely and mentally ill relatives [32]. Moreover the relationship quality might influence caregiving as poor quality of relationship between caregiver and care recipient may be considered as a risk factor for subjective burden or higher levels of overload [31, 33].

Interestingly, the religious beliefs of individuals lead to different health-care practices which may vary depending upon a specific religion. There are four major pathways which could explain the influence of spirituality/religion on health which included *health behaviors* (specific diet or prohibition of alcohol and smoking), *social support*, *psychological states* (religious people have more optimism and faith which help in coping with stress), and *"psi" influences* (belief in supernatural laws), and they might have indirect effect on health [34].

Notably, in light of perennial discussions about the relationship between science and faith, most physicians apply medical science while maintaining a belief that God intervenes in patients' health [35]. This also indicates a way that religious characteristics may influence the care of patients in clinical contexts like end-of-life care in which some patients and families articulate hopes for miracles [36]. This is what the secular physician may not notice or may ignore, but the religious physician may emphasize or exaggerate. In particular, those patients who have stronger religious beliefs seek out religious physicians, and those patients who are less inclined toward religion are more comfortable with secular physicians. Therefore, clinicians and patients who shared strong religious concordance showed preference for religion and spirituality [35]. Some studies suggested that there is less empirical basis for prediction about the association of religious beliefs or activity with favorable health outcomes [37].

Religiosity and/or Spirituality Consideration for Treatment of Elderly Patients

Taking a history of spiritual/religious belief is a sensitive issue and should be performed with great attention and care. Although there are several validated screening tools for assessing the two concepts, it might be more appropriate to use a generic approach without forcing or intruding way before introducing the concepts into the inquiry process unless as previously mentioned to be self-reported or clear and obvious. One of the approaches to do that is to ask a personal question about the "most important coping mechanism" and from there to carry further clarifications about why is it valid or not to consider religion and spirituality. Also to quantify these effects, the proposed questions would be of great help: "whether their spiritual beliefs are an important part of their life?" and "How these beliefs influence the way they take care of themselves and choices?" and "whether they participate in a religious community or activity and would like their needs to be fulfilled."

Furthermore, the religiosity and/or spirituality of the treating physicians and caregiver can potentially impact the process of care. The important considerations should bring in the discussion the concerns and wishes, understanding the requirements of the elderly as well as coping with the stresses in dealing with terminal, complex, and end-of-life issues. The inquiry about these issues (or spiritual beliefs) is relevant when the patient self-report to be religious. Among patients with palliative care for frailty, severe distress, or critical or near end of life, their religious needs are obvious, and justify the call for support of spiritual services in order to provide compassionate holistic care [38]. It is more likely that patients may encounter different opinions with regard to their beliefs and medical treatment in relation to the religious beliefs of their physicians.

The Evidence of Religiosity and/or Spirituality in Health Care

The current scientific methodology primarily based on physical measures to determine the evidence behind spirituality and religion in disease or health influence; thus it is instrumentally incorrect to rely on it completely. Religiosity and spirituality are more likely to be associated with higher social support especially in terms of religious organizational activities, alleviation of depressive symptoms, enhanced cognitive function, and greater cooperativeness [6, 39]. Intrinsic religiosity is associated with better physical functioning and lower depression scores among elderly survivors of cancer [40, 41]. Although there exists a link between religiosity and physical and/or mental functioning, still a strong scientific evidence is lacking.

As long as the absence of evidence is not evidence of absence, it needs to be interpreted carefully. Therefore, the health-care professional should not ignore the conceptualization of spirituality and religiosity even if it has not been well defined and understood. As part of the commitment to patient-centered care, the physician should respect the patient's beliefs, request for pastoral care, and religious service for terminally ill patients. As a matter of fact, in such patients the physician rather should embrace the spiritual approach and utilize efficiently, to the maximum, all related health benefit through social and psychosocial support that enhances recovery and helps patients to cope with their disease, health-related decisions, and acceptance of therapies [42]. The long-term effects expected are satisfaction, trust on the care system, monetary savings related to reduced length of stay, and return to the premorbid status [2]. Nevertheless the spiritual support is underutilized both by the religious communities and by the medical team which can potentially improve the patient's quality of life [2]. Therefore, the spiritual support should be considered at the level of training for the clergy services, particularly at the community level and to the medical team.

Religious Reasons Informing End-of-Life Controversies

Do Not Resuscitate (DNR) and Withdrawal Treatment

The End of Life (EoL) remains a challenging issue to be addressed among patients and their relatives in different geographic locations around the world with diverse beliefs and religious doctrine and commitment [43]. While it might be clear from the technical point of the view by the physician that the futility of care and lack of evidence about provision of care would result in a clinically acceptable outcome, sometimes this might not be clear to the other parties involved such as the patient and relatives or guidance clergy, religious or advisory bodies. The patient or more often the relatives would have different point of views, and in the absence of a guidance like a living will or predictive or evidence of patient's previously expressed preferences, DNR becomes an ethical dilemma. In some occasions the call for legal considerations and decision might be further modified by different state laws and definition of these important concepts as well as patient beliefs and religion [44]. Withdrawing treatment is referred to as an active, life-shortening treatment which is considered unacceptable. However, withholding treatment is a passive intervention and so deemed permissible under certain circumstances [45].

Although there exists an important link between religiosity and/or spirituality with health, however, the strong evidence pertaining to the influence of the different dimensions is lacking due to methodological weakness, incompatibility, and several confounding issues. Considering this, the renewing interest on the linkage did not address how physicians should develop adequate cultural and religious conceptions, overcome cultural stereotyping, and practice spirit of culturally sensitive medical care [44]. Beyond doubt there is a need to respect patient wish for religious services and rather embrace that as complementary to support health in the holistic view. At the end, religion commonly informs ethical judgment and underlies the public discourse and that's particularly true in communities that are considered religious [46]. Overall, the religious and cultural beliefs for predicting personal willingness toward withdrawal treatment or DNR or mercy killing are far more complicated and need better understanding of the aspects possibly influencing cultural or religious convictions.

Advanced Directives and Living Will

Advance directives refer to the communication or conversation and/or documentation between the physician and the patient focusing on clarifying the patient's wishes and values and projecting the type of medical treatment the patient would want to opt under certain circumstances in which the patient becomes incapacitated to make a decision [47]. It also relates to a surrogate decision-making either specified by the advance directives or the law which typically designates in order the spouse, adult child, parent, adult sibling, grandparent, or friend to assume this role [48]. The expression can be either one of the three as follows: the living will (LW), the durable power of attorney for health care (DPA), or the conversation with a physician or others. In many institutions, assessments of physicians are required to determine the patient incapacity of decision-making [49] and open the door for considering alternative consent like emergency, surrogate, and therapeutic privilege among others according to local laws and regulations.

Role of Hospital Clergy/Chaplain

There is certain public opinion which considers the community clergy as a symbolic moral voice, and their attitude and opinions are a key influence in the formulation of the beliefs and acceptance of ethical dilemmas such as DNR and withdrawal treatment. Interestingly, when comparing the community clergy with the hospital chaplain, the former would typically be expected to be traditional in their views of religion and its influence on health, while the latter mostly has extensive training in the mental, social, and spiritual needs of hospitalized patients particularly among the elderly and supporting the end-of-life care [50]. A recent study showed that the majority (80%) of clergy believe under certain circumstances the terminally ill patients should be allowed to die; on the other hand, they outrightly disregarded the idea of physician-aid in dying or physician-assisted suicide [51]

Moreover, it is highly recommended to include hospice clergy in health-care teams, if the patient was receptive to spiritual support-linked care by religious community and clergy at patient community with all the positive support of the community which can be called upon to help for patient convalescence like meals, transportation, escort, and housekeeping. It also helps to reduce alienation and isolation and increase sense of belonging and coping with illness and its related limitations for a fast recovery. Therefore, the interplay between religious beliefs, medical knowledge, and trust needs to be established within the general population and among patients facing a life-threatening illness.

Balancing Faith and Health Care

Religion is far more beyond a set of beliefs and practices which cannot be ascertained instrumentally; rather it encompasses a spiritual way of being existed in the world [46]. Over the past decades, clinicians were trying to understand and balance the health care by restoring spiritual basis of medicine, which is supposed to have a strong link with health-care practice [52]. It has been suggested that the spiritual support provided by a clinical team results in improved hospice utilization and quality of life in patients approaching end of life [53]. However, surrounding issues of trust, the role of miracles, and caution of prognostication are some of the factors hindering the partnership between religion and medicine [51]. Hospital clergy services; include the importance of listening, understanding grief, being a supportive presence for the family, and displaying appropriate bedside manners which provide spiritual support at the end of life. Clergy training focusing on the health-care education and end-of-life care could help in overcoming issues of medical disbelief in terminally ill patients [54]. Therefore, there is a strong need to strengthen the relationships between clergy and clinicians for establishing a holistic approach of patient care.

In conclusion, religiosity and spirituality are important elements of well-being among elder population worldwide. The level of religious participation among elder population is greater than that in any other age group. There are various measures to assess religious involvement such as organizational religiousness, non-organizational religiousness, intrinsic religiosity, self-rated religiousness, and observer-rated religiousness. In patients with palliative care for frailty, cancer, or critical or near end of life, the religious needs are obvious and justify the call for the support of spiritual services in order to provide compassionate and holistic care. On the other hand, the religious devotion might enhance excessive guilt, inflexibility, and anxiety. Therefore, it is important to take into consideration the potential benefits and the possible drawbacks of involving religiosity and spirituality in elderly patients seeking care for health issues and surgery.

References

1. Alcorn SR, Balboni MJ, Prigerson HG, Reynolds A, Phelps AC, Wright AA, Block SD, Peteet JR, Kachnic LA, Balboni TA. "If God wanted me yesterday, I wouldn't be here today": religious and spiritual themes in patients' experiences of advanced cancer. J Palliat Med. 2010;13(5):581–8. https://doi.org/10.1089/jpm.2009.0343.
2. Balboni TA, Vanderwerker LC, Block SD, Paulk ME, Lathan CS, Peteet JR, Prigerson HG. Religiousness and spiritual support among advanced cancer patients and associations with end-of-life treatment preferences and quality of life. J Clin Oncol. 2007;25(5):555–60.
3. Tanyi RA. Towards clarification of the meaning of spirituality. J Adv Nurs. 2002;39(5):500–9.
4. Daaleman TP, Perera S, Studenski SA. Religion, spirituality, and health status in geriatric outpatients. Ann Fam Med. 2004;2(1):49–53.
5. Jaul E, Zabari Y, Brodsky J. Spiritual background and its association with the medical decision of, DNR at terminal life stages. Arch Gerontol

Geriatr. 2014;58(1):25–9. https://doi.org/10.1016/j.archger.2013.08.004.
6. Koenig HG, George LK, Titus P. Religion, spirituality, and health in medically ill hospitalized older patients. J Am Geriatr Soc. 2004;52(4):554–62.
7. Arrey AE, Bilsen J, Lacor P, Deschepper R. Spirituality/religiosity: a cultural and psychological resource among Sub-Saharan African migrant women with HIV/AIDS in Belgium. PLoS One. 2016;11(7):e0159488. https://doi.org/10.1371/journal.pone.0159488.
8. Ellison CW. Spiritual well-being: conceptualization and measurement. J Psychol Theol. 1983;11(4):330–40.
9. Monod S, Brennan M, Rochat E, Martin E, Rochat S, Büla CJ. Instruments measuring spirituality in clinical research: a systematic review. J Gen Intern Med. 2011;26(11):1345–57. https://doi.org/10.1007/s11606-011-1769-7.
10. Büssing A, Balzat HJ, Heusser P. Spiritual needs of patients with chronic pain diseases and cancer – validation of the spiritual needs questionnaire. Eur J Med Res. 2010;15(6):266–73.
11. Sharma RK, Astrow AB, Texeira K, Sulmasy DP. The Spiritual Needs Assessment for Patients (SNAP): development and validation of a comprehensive instrument to assess unmet spiritual needs. J Pain Symptom Manag. 2012;44(1):44–51. https://doi.org/10.1016/j.jpainsymman.2011.07.008.
12. Austin P, Macdonald J, MacLeod R. Measuring spirituality and religiosity in clinical settings: a scoping review of available instruments. Religions. 2018;9(3):70. https://doi.org/10.3390/rel9030070.
13. Kornfeld DS. The hospital environment: its impact on the patient. Adv Psychosom Med. 1972;8:252–70.
14. Bousso RS, Poles K, Serafim Tde S, de Miranda MG. [Religious beliefs, illness and death: family's perspectives in illness experience]. Rev Esc Enferm USP. 2011;45(2):397–403.
15. Emblen J, Pesut B. Strengthening transcendent meaning: a model for the spiritual nursing care of patients experiencing suffering. J Holist Nurs. 2001;19(1):42–56.
16. World Health Organization. About World Health Organization. Constitution. Available at: http://www.who.int/governance/eb/constitution/en/. Accessed 7 May 2019.
17. Bulkley J, McMullen CK, Hornbrook MC, Grant M, Altschuler A, Wendel CS, Krouse RS. Spiritual well-being in long-term colorectal cancer survivors with ostomies. Psychooncology. 2013;22(11):2513–21. https://doi.org/10.1002/pon.3318.
18. Fry PS. Religious involvement, spirituality and personal meaning for life: existential predictors of psychological wellbeing in community-residing and institutional care elders. Aging Mental Health. 2000;4:375–87. doi.org/10.1080/713649965
19. Slade M. Mental illness and well-being: the central importance of positive psychology and recovery approaches. BMC Health Serv Res. 2010;10:26. https://doi.org/10.1186/1472-6963-10-26.
20. Grundmann E. Cancer morbidity and mortality in USA Mormons and seventh-day Adventists. Arch Anat Cytol Pathol. 1992;40(2–3):73–8.
21. Helm HM, Hays JC, Flint EP, Koenig HG, Blazer DG. Does private religious activity prolong survival? A six-year follow-up study of 3,851 older adults. J Gerontol A Biol Sci Med Sci. 2000;55(7):M400–5.
22. Mardan H, Hamid T, Redzuan M, Ibrahim R. Correlate of self-care and self-neglect among community-dwelling older adults. Iran J Nurs Midwifery Res. 2014;19(7 Suppl 1):S71–6.
23. Balboni TA, Balboni M, Enzinger AC, Gallivan K, Paulk ME, Wright A, Steinhauser K, VanderWeele TJ, Prigerson HG. Provision of spiritual support to patients with advanced cancer by religious communities and associations with medical care at the end of life. JAMA Intern Med. 2013;173(12):1109–17. https://doi.org/10.1001/jamainternmed.2013.903.
24. Cassileth BR. Unorthodox cancer medicine. Cancer Investig. 1986;4(6):591–8.
25. Underwood-Gordon L, Peters DJ, Bijur P, Fuhrer M. Roles of religiousness and spirituality in medical rehabilitation and the lives of persons with disabilities. A commentary. Am J Phys Med Rehabil. 1997;76(3):255–7.
26. Ellison CG, Bradshaw M, Roberts CA. Spiritual and religious identities predict the use of complementary and alternative medicine among US adults. Prev Med. 2012;54(1):9–12. https://doi.org/10.1016/j.ypmed.2011.08.029.
27. Ross CL. Integral healthcare: the benefits and challenges of integrating complementary and alternative medicine with a conventional healthcare practice. Integr Med Insights. 2009;4:13–20.
28. Greil A, McQuillan J, Benjamins M, Johnson DR, Johnson KM, Heinz CR. Specifying the effects of religion on medical helpseeking: the case of infertility. Soc Sci Med. 2010;71(4):734–42. https://doi.org/10.1016/j.socscimed.2010.04.033.
29. Atashzadeh-Shoorideh F, Zakaryaee NS, Fani M. The barriers and facilitators in providing spiritual care for parents who have children suffering from cancer. J Family Med Prim Care. 2018;7(6):1319–26. https://doi.org/10.4103/jfmpc.jfmpc_76_18.
30. Blando J. Spirituality, religion, and counseling. Couns Hum Dev. 2006;39(2):1.
31. Kim Y, Given BA. Quality of life of family caregivers of cancer survivors: across the trajectory of the illness. Cancer. 2008;112(11 Suppl):2556–68. https://doi.org/10.1002/cncr.23449.
32. Chang BH, Noonan AE, Tennstedt SL. The role of religion/spirituality in coping with caregiving for disabled elders. Gerontologist. 1998;38(4):463–70.
33. Draper BM. Risk factors for stress in elderly caregivers. Int J Geriat Psych. 1996;11:227–31.
34. Oman D, Thoresen CE. 'Does religion cause health?': differing interpretations and diverse meanings. J

Health Psychol. 2002;7(4):365–80. https://doi.org/10.1177/1359105302007004326.
35. Curlin FA, Sellergren SA, Lantos JD, Chin MH. Physicians' observations and interpretations of the influence of religion and spirituality on health. Arch Intern Med. 2007;167(7):649–54.
36. Sulmasy DP. Spiritual issues in the care of dying patients: "... it's okay between me and god". JAMA. 2006;296(11):1385–92.
37. Sloan RP, Bagiella E. Claims about religious involvement and health outcomes. Ann Behav Med. 2002;24(1):14–21.
38. Chattopadhyay S. Religion, spirituality, health and medicine: why should Indian physicians care? J Postgrad Med. 2007;53:262–6.
39. Yeung WJ, Chan Y. The positive effects of religiousness on mental health in physically vulnerable populations: a review on recent empirical studies and related theories. Int J Psych Rehab. 2007;11(2):37–52.
40. Caplan LS, Sawyer P, Holt C, Allman RM. Religiosity and function among community-dwelling older adult survivors of cancer. J Relig Spiritual Aging. 2013;25(4):311–25.
41. Wei D, Liu XY, Chen YY, Zhou X, Hu HP. Effectiveness of physical, psychological, social, and spiritual intervention in breast cancer survivors: an integrative review. Asia Pac J Oncol Nurs. 2016;3(3):226–32.
42. Nichols SW. Examining the impact of spiritual care in long-term care. Omega (Westport). 2013;67(1–2):175–84.
43. Rubulotta F, Rubulotta G, Santonocito C, Ferla L, Celestre C, Occhipinti G, Ramsay G. End-of-life care is still a challenge for Italy. Minerva Anestesiol. 2010;76(3):203–8.
44. Schweda M, Schicktanz S, Raz A, Silvers A. Beyond cultural stereotyping: views on end-of-life decision making among religious and secular persons in the USA, Germany, and Israel. BMC Med Ethics. 2017;18(1):13. https://doi.org/10.1186/s12910-017-0170-4.
45. Barilan YM. Human dignity, human rights, and responsibility. The new language of global bioethics and biolaw. Cambridge: MIT Press; 2012.
46. Sloan RP, Bagiella E, VandeCreek L, Hover M, Casalone C, Jinpu Hirsch T, Hasan Y, Kreger R, Poulos P. Should physicians prescribe religious activities? N Engl J Med. 2000;342(25):1913–6.
47. Lum HD, Sudore RL. Advance care planning and goals of care communication in older adults with cardiovascular disease and multi-morbidity. Clin Geriatr Med. 2016;32(2):247–60. https://doi.org/10.1016/j.cger.2016.01.011.
48. Vig EK, Starks H, Taylor JS, Hopley EK, Fryer-Edwards K. Surviving surrogate decision-making: what helps and hampers the experience of making medical decisions for others. J Gen Intern Med. 2007;22(9):1274–9.
49. DeMartino ES, Dudzinski DM, Doyle CK, Sperry BP, Gregory SE, Siegler M, Sulmasy DP, Mueller PS, Kramer DB. Who decides when a patient can't? Statutes on alternate decision makers. N Engl J Med. 2017;376(15):1478–82. https://doi.org/10.1056/NEJMms1611497.
50. Flannelly KJ, Weaver AJ, Smith WJ, Oppenheimer JE. A systematic review on chaplains and community-based clergy in three palliative care journals: 1990–1999. Am J Hosp Palliat Care. 2003;20(4):263–8.
51. Balboni MJ, Sullivan A, Smith PT, Zaidi D, Mitchell C, Tulsky JA, Sulmasy DP, VanderWeele TJ, Balboni TA. The views of clergy regarding ethical controversies in care at the end of life. J Pain Symptom Manag. 2018;55(1):65–74.e9. https://doi.org/10.1016/j.jpainsymman.2017.05.009.
52. Puchalski CM. The role of spirituality in health care. Proc (Bayl Univ Med Cent). 2001;14(4):352–7.
53. Balboni TA, Paulk ME, Balboni MJ, Phelps AC, Loggers ET, Wright AA, Block SD, Lewis EF, Peteet JR, Prigerson HG. Provision of spiritual care to patients with advanced cancer: associations with medical care and quality of life near death. J Clin Oncol. 2010;28(3):445–52. https://doi.org/10.1200/JCO.2009.24.8005.
54. Balboni MJ, Sullivan A, Enzinger AC, Smith PT, Mitchell C, Peteet JR, Tulsky JA, VanderWeele T, Balboni TA. U.S. clergy religious values and relationships to end-of-life discussions and care. J Pain Symptom Manag. 2017;53(6):999–1009. https://doi.org/10.1016/j.jpainsymman.2016.12.346.

Surgery in the Elderly with Mental Health Issues

33

Faisal Jehan and Rifat Latifi

The demographics of the US population are undergoing a dramatic shift as the population is not only growing but aging as well. Since 1980, the geriatric population has increased by 21% [1]. Over the past three decades, the US population has been growing at a rate of 39%; however, those over the age bracket of 65 and 85 have almost doubled and tripled during this time [2]. This rightward shift of the population histogram has been attributed to the increased life expectancy due to the effective screening measures/diagnostic tools and advancements in the management of diseases. The second factor contributing to such a shift is the aging "baby boomers" [3]. With continued advancements in the healthcare and standard of living, life expectancy is further estimated to increase. It is predicted that by the next 20 years, those over the age of 65 will comprise 1/5th of the total population [4]. This rapidly growing new subset of population has peculiar needs and poses a great impact on our health. For instance, 9.6 million emergency department (ED) visits were attributed to age 65 and above from 2009 to 2010, accounting for a rate of 511 per1000 persons, and it increased with age [5, 6]. Similarly, in the year 2010, adults aged 85 and over accounted for only 2% of the US population but disproportionately accounted for 9% of hospital discharges [6]. In order to provide quality care, the healthcare system must adjust promptly to be able to take optimal care of this new predominant subset of the population.

Surgery in the Elderly

As the US population continues to age, along with the increase of E.D. visits and hospitalization, elderly patients will also require increased surgical procedures and resources [7]. Elderly patients have proportionately higher rates per population, compared to their younger counterparts, of inpatient as well as outpatient surgical and nonsurgical procedures [3, 7, 8]. In 2006, about 35% of the total inpatient procedures while 32% of outpatient procedures were performed in elderly patients [8, 9]. Data from the National Health Statistics Reports has demonstrated that out of the 45,963 procedures, 16,238 were performed in patients aged 65 years and over [9]. With such growing proportion, to achieve optimal surgical outcomes, focused research on this subgroup is warranted to allow better understanding of the specific needs of these patients. Surgical management of elderly patients is far more challenging compared to young patients for numerous important reasons. First and foremost, increasing age-related comorbidities, hyper-

F. Jehan
Department of Surgery, Westchester Medical Center and New York, Medical College, Valhalla, NY, USA

R. Latifi (✉)
Department of Surgery, New York Medical College, School of Medicine and Westchester Medical Center Health Network, Valhalla, NY, USA
e-mail: rifat.latifi@wmchealth.org

R. Latifi (ed.), *Surgical Decision Making in Geriatrics*,
https://doi.org/10.1007/978-3-030-47963-3_33

tension, dementia, heart diseases, degenerative joint diseases, may require preoperative optimization and complicate the postoperative care. Second, the acute surgical stress in elderly patients with baseline limited physiologic reserves to withstand stressors severely impedes healing and the recovery from surgery. Third, the unreliable clinical signs and symptoms and the attenuated inflammatory and immune response in the elderly further complicates the course, leading to a delay in recognition and appropriate management of postoperative complications. The unpredictable response of sedatives, anesthetics, and narcotics and the drug–drug interaction with patient home medications are also critical challenges in elderly patients.

Mental Health Issues and Epidemiology

Mental health has been defined by the World Health Organization (WHO) as "a state of well-being in which every individual realizes his or her own potential, can cope with the normal stresses of life, can work productively and fruitfully, and is able to make a contribution to her or his community" [10, 11]. Unfortunately, however, every one in three individuals faces a disruption in their mental health at some point of their life [11]. Mental illness refers to a group of conditions involving changes in emotion, thinking, behavior, or difficulty/distress functioning in social, work, or family activities. Mental health issues and neuropsychiatric disorders comprise about 14% of the global burden of disease [12]. In 2013, five different mental illnesses made into the top 20 causes of global burden of disease, leading to the most years lived with disability [13]. Major depression was second, anxiety disorders was seventh, schizophrenia was eleventh, dysthymia was sixteenth, and bipolar disorder was seventeenth on the list. According to the 2016 data from the National Survey on Drug Use and Health (NSDUH) by the Substance Abuse and Mental Health Services Administration (SAMHSA), approximately 44.7 million adults had some kind of mental illness while 4.2% (10.4 million) of all US adults had severe mental illness [14].

The demographic shift as the population is living longer and improved survival comes at a cost of greater proportion living with morbidities and disabilities increasing the prevalence of mental illness [15]. Mental illness has evolved as a major contributor to the overall morbidity and disability globally [12].

Classification of Mental Health Diseases

Mental health illness encompasses a wide array of disorders. For the sake of simplicity, we have categorized broadly some of the major mental illnesses in Table 33.1.

Table 33.1 Major mental illnesses

S. No	Category	Disorders
1.	Mood disorders	Depression Bipolar disorder Mania Dysthymia Cyclothymia
2.	Psychotic disorders	Acute psychotic disorder Schizophrenia Schizoaffective
3.	Anxiety disorders	Generalized anxiety disorder Obsessive-compulsive disorder (OCD) Panic disorder Post-traumatic stress disorder (PTSD) Social phobia (social anxiety disorder) Specific phobia
4.	Personality disorders	Paranoid personality disorder Schizoid personality disorder Schizotypal personality disorder Antisocial personality disorder Borderline personality disorder Histrionic personality disorder Narcissistic personality disorder Avoidant personality disorder Dependent personality disorder Obsessive-compulsive personality disorder

Table 33.1 (continued)

S. No	Category	Disorders
5.	Eating disorders	Anorexia nervosa Bulimia nervosa Binge eating disorder
6.	Developmental disorders	Intellectual disability disorder Autism spectrum disorder Attention deficit hyperkinetic disorder (ADHD)
7.	Cognitive disorders	Dementia Delirium
8.	Drug abuse	Alcohol and drug abuse and withdrawals

Mental Illness in the Elderly

As the proportion of elderly population continues to increase as a result of improved survival from previously lethal conditions and the aging baby boomers, a larger subset of the elderly have to live with morbidities, disabilities, and some form of mental illness [16]. As stated previously, between 2015 and 2050, it is estimated that the number of people over the age 60 will increase from 900 million to 2 billion [17]. Elderly patients by virtue of their diminished physiologic reserves and decreased stress coping capacity are not only at a higher risk for mental illness but also have worse outcomes [18]. Elderly individuals have unique physical and mental health challenges that need to be timely identified and addressed. Major depression, dementia (Alzheimer's and vascular), anxiety disorders, late-life schizophrenia, and drug abuse disorders especially alcohol are common mental health problems that affect the elderly. These can be severely debilitating in terms of quality of life if they remain unrecognized or untreated in the elderly.

Elderly patients who live alone, have no family support (friends/relatives), are socioeconomically and geographically disadvantaged have experienced a traumatic brain injury or those who are suffering from a progressive or chronic illness or pain syndrome are at high risk for mental illness and depression. Screening of at-risk elderly individuals and early diagnosis and appropriate pharmacological and psychosocial treatment is pivotal in the management of these patients.

Every one in five individuals over the age of 60 suffer from some form a mental or neurological illness [17, 19, 20]. Mental and neurological disorders contribute 6.6% of all disability-adjusted life years (DALYs) among people aged 60 and over. Furthermore, these mental and neurological disorders account for 17.4% of years lived with disability (YLDs) in the elderly group. The two most common mental and neurological disorders in individuals aged 60 and over are depression and dementia. Depression affects about 7% while dementia affects 5% of the world's elderly population. Similarly, anxiety disorders affect 3.8% while substance abuse affects about 1% of the world elderly population. Substance abuse in the elderly predisposes them to self-harm. Substance abuse problems are particularly difficult to diagnose in the elderly people and often remain undiagnosed and untreated. Mental health issues in the elderly often culminate in the cases of self-harm or suicide. About 25% of deaths from self-inflicting injuries worldwide are in the elderly sub-group of the population. The elderly individuals have the highest suicide rate of any age group [21]. Compared to an overall rate of 11 per 100,000 for all ages, men aged 85 years or older have a suicide rate of 45.23 per 100,000 [21].

Risk Factors for Mental Health Issues Among the Elderly

Elderly patients by virtue of their age-dependent decline in physiological reserves, additional comorbidities, and decreased stress coping mechanisms are at an increased risk for developing mental health problems. Older adults come across different life stressors at different points in their lives. However, life stressors are not only more common in the later part of left, but elderly individuals have a significantly less ability to deal with those stressors due to the decline in intellectual, cognitive, and functional abilities. For example, elderly individuals are more likely to experience grief or bereavement events due to loss of a significant other, retirement from job, or a change in their social and economic status. In addition, progressive chronic medical conditions, pain syndromes, reduced agility, and increasing frailty and delayed return to baseline health from illness predispose them to require placement in a long-term facility care (subacute rehabilitation or skilled nursing facility centers). This results in the separation of the elderly from the family; the increased loneliness and isolation predispose them to develop mental health issues like depression and anxiety. Mental health and physical health are interconnected and dependent on each other. For example, depression is thought to be associated with the increased risk of coronary heart disease, osteoporosis, diabetes complications, cancer incidence and progression, and cancer mortality [22–25]. Similarly, a meta-analysis has shown the association between positive emotional well-being and favorable prognosis of physical illness [26]. On the other hand, patients with chorionic debilitating illnesses like cancer have an increased risk of developing mental illness [27].

Abuse is also common in the elderly. Abuse can be in any form including physical, verbal, psycho-emotional, financial, and sexual abuse. Abuse can also be in the form of abandonment, neglect, and nutritional deprivation. Data from the World Health Organization suggest that 1 in 6 older individual experience some form of elder abuse [17]. Elder abuse is a major stressor in the elderly and not only leads to physical injuries but is also associated with serious life-long mental health consequences, including depression, anxiety, post-traumatic stress disorder, and suicide.

Dementia and Depression Among the Elderly as Public Health Issues

Dementia

Although normal aging is associated with some slowing of mental ability, usually it is not severe enough to interfere with the function of everyday life. On the other hand, dementia is a decline in mental ability severe enough to interfere with the performance of daily life activities. Dementia, although more common in the elderly, is not a normal part of aging. Dementia is defined as a chronic and progressive decline in memory, cognition, thinking, behavior, and emotional response that significantly impairs the ability of an individual to perform the activities of daily living. An estimated 50 million people worldwide have dementia, and the total number is projected to increase to 82 million in 2030 and 152 million in 2050 [17]. Dementia is not only a physical and social burden on the patient, family, and the society, but also has an enormous economic effect on the healthcare system. The direct costs of medical treatment and requirement of long-term care and the indirect costs due to loss of productivity and dependence associated with dementia are staggering. In addition, the emotional, physical, and socioeconomic stress to the family and caregivers is unimaginable.

Depression

Depression is one of the most prevalent mental health problems in the elderly. It is estimated that about 7% of the elderly population suffer from depression [17]. Depression is associated with a decline in physical, emotional, mental, and social functioning and accounts for 5.7% of years lost to disability (YLDs) in the elderly over the age of 60. Depressive symptoms and their effects on functioning also complicate the treatment of other chronic diseases and further worsen the outcomes [28]. Due to overlapping symptoms among normal aging, age-related chronic diseases, and depression, it often remains underdiagnosed and the treatment is often overlooked. Depression in the elderly is associated with increased functional decline compared to those with chronic medical conditions such as lung disease, hypertension, or diabetes. Depression in the elderly is also associated with increased health resource utilization by increased emergency department visits and hospitalization and longer hospital length of stays.

Challenges in the Assessment and Diagnosis in the Elderly

Elderly patients most commonly present with atypical signs and symptoms of altered mental state, anorexia, hypothermia instead of fever, lack of localized pain, and decline in functional status after any organic illness [29]. These symptoms are similar to the presentation of mental illness in the elderly. In addition, the classical features of the mental diseases are seldom seen in elderly patients. For instance, elderly patients usually do not meet the full criteria for depression or anxiety and often present with vague somatic symptoms. This makes prompt diagnosis and detection of treatable mental illness a challenge in the elderly. Increasing frailty in the elderly, decreased mobility, and other age-related medical comorbidities further complicate the detection of mental disorders in these patients. Frailty, immobility, impaired cognition, and incontinence which are often referred to as geriatric syndromes may mimic or mask the psychopathology, making diagnosis in this subset further challenging.

Patient-related factors is another important barrier in the diagnosis of mental health problems in the elderly. Elderly patients have multiple comorbidities and are more likely to report somatic symptoms of pain, fatigue, and sensorium and psychological symptoms are usually underreported and are masked by these somatic symptoms. This leads to missed diagnosis and a wrong perception of lower prevalence of mental diseases in the elderly. Sometimes mental illness is considered as a part of normal aging in the society, and this stereotype further lowers the actual prevalence of mental illness in the elderly. Patients taking multiple medications can have drug-related side effects which can make the differentiation of psychiatric illness in elderly patients. Because of the lack of geriatric specialists, primary care physicians carry much of the burden for diagnosis of mental disorders in the elderly population. However, their familiarity with geriatric-specific syndromes and mental health illness in the elderly is only minimal. Reported data from the literature has consistently shown that the ability of primary care physicians to recognize mental illness and referrals for proper treatment is below the standard of care. For instance, a study that interviewed primary care providers demonstrated that only 55% of internists felt confident in diagnosing depression, while only 35% of the total felt confident in prescribing antidepressants to elderly patients. Due to the lack of geriatric mental health specialists and the complex referral system if one is available, it is estimated that up to 63% of the elderly population above 65 years with a mental disorder do not have an adequate access for the diagnosis and treatment of mental health illness. Barriers to timely and correct diagnosis and treatment of mental illness can be broadly classified in three groups as shown in Table 33.2.

Table 33.2 Barriers to identification and treatment of mental illness in the elderly

S. No	Category	Barriers
1.	Patient barriers	Preference for primary care Tendency to emphasize somatic problems Reluctance to disclose psychological symptoms
2.	Provider barriers	Lack of awareness of manifestation of mental disorders Complexity of treatment Reluctance to inform patients of a diagnosis
3.	System barriers	Time pressures Reimbursement policies Complex referral process Lack of geriatric specialists

Preparing the Elderly Patient with Mental Health Issues for Surgery

In order to achieve optimal outcomes in elderly patients with mental illness undergoing surgery, a robust preoperative optimization program, vigilant intraoperative care, and close postoperative follow-up are required. A multidisciplinary patient-centered team approach is indispensable for the preparation of elderly patients with mental health issues. It starts with the patient's primary care provider and will require the involvement of psychiatrist, geriatrician, anesthesiologist, cardiologist, and surgeon to plan the optimal approach and time and delineate the postoperative care of patient.

Preoperative Screening and Optimization

The preoperative phase is by far one of the most critical phases to determine the outcomes of elderly patients with mental illness after surgery. A structured preoperative screening and optimization is required to adequately prepare this complex subset of patients for the stress of surgery.

Screening and Specialist Referral

Preoperative assessment by a multidisciplinary team and geriatric experts can improve outcomes after surgery for the elderly, who are more likely than the younger patients to develop preventable postoperative adverse events [30]. Data across different surgical specialties has consistently shown that patients who received preoperative interdisciplinary intervention before surgery have fewer complications and shorter length of stay. The surgeon has the ultimate responsibility for a surgical patient; therefore, the surgeon should ensure that all the appropriate preoperative evaluations and interventions required are performed so that the patient can make informed decisions and receive the highest quality care. The primary care physician of the patient, the geriatric psychiatrist, and the surgeon should form a triangle of patient advocacy and further determine if the patient would require further consults A cardiologist referral for preoperative cardiac evaluation according to the American College of Cardiology/American Heart Association algorithm for patients undergoing noncardiac surgery. Identify patients with chronic obstructive/restrictive pulmonary diseases who are at risk of postoperative pulmonary complications and implement appropriate strategies for prevention. Patient baseline frailty score, functional status, and history of falls should be documented.

Screening for Cognitive Impairment/Dementia

A detailed history and a cognitive assessment should be performed for all patients with a known history of cognitive impairment or dementia, mental illness, strong family history of dementia, or history of transient ischemic attacks or strokes. A cognitive assessment can be easily performed using tools such as the Mini-Cog (Tables 33.3 and 33.4) [31]. Based on the Mini-Cog evaluation, if

Table 33.3 Cognitive assessment with the Mini-Cog: 3 Item Recall and Clock Draw

1. GET THE PATIENT'S ATTENTION, THEN SAY:
"I am going to say three words that I want you to remember now and later. The words are: banana, sunrise, chair. Please say them for me now." Give the patient 3 tries to repeat the words. If unable after 3 tries, go to next item.
2. SAY ALL THE FOLLOWING PHRASES IN THE ORDER INDICATED:
"Please draw a clock in the space below. Start by drawing a large circle. Put all the numbers in the circle and set the hands to show 11:10 (10 past 11)." If subject has not finished clock drawing in 3 minutes, discontinue and ask for recall items.
3. SAY: "What were the three words I asked you to remember?"

From Borson et al. [31], with permission

Table 33.4 Interpretation of the Mini-Cog

SCORING:
3 item recall (0 to 3 points): 1 point for each correct word Clock draw (0 or 2 points): 0 points for abnormal clock 2 points for normal clock
A NORMAL CLOCK HAS ALL OF THE FOLLOWING ELEMENTS:
All numbers 1 to 12, each only once, are present in the correct order and direction (clockwise) inside the circle. Two hands are present, one pointing to 11 and one pointing to 2.
ANY CLOCK MISSING ANY OF THESE ELEMENTS IS SCORED ABNORMAL. REFUSAL TO DRAW A CLOCK IS SCORED ABNORMAL.
Total score of 0, 1, or 2 suggests possible impairment. Total score of 3, 4, or 5 suggests no impairment.

From Borson et al. [31], with permission

the patient has evidence of cognitive impairment, coordinate with the patient primary care physician and consider patient evaluation by a geriatric mental health specialist for further evaluation and optimization. Obtain a detailed history from the patient's primary care and family members to better determine the functional and cognitive decline over time. Patients with significant decline over a short period of time or those with recent decline will specially benefit from the evaluation by a geriatric mental health specialist. Postoperative decline in cognition and delirium are common postoperative complications. In order to correctly identify these complications and quantify the change from baseline, preoperative accurate documentation of the patient's preoperative cognitive and functional status is very important. Patients with preexisting cognitive impairment have an increased risk of postoperative delirium [32]. An episode of delirium in the elderly after surgery is associated with increased hospital length of stay, postoperative functional decline, and increased mortality [32, 33].

Screening for Depression

Screening of depression in elderly patients is recommended specially in those with risk factors like poor health status, cognitive impairment, living alone, divorced, diagnosis of a chronic illness, and socioeconomically challenged [34]. Other risk factors for depression among geriatric patients include female sex, disability, bereavement, sleep disturbance, and prior depression diagnosis [35]. Simple tools like the Patient Health Questionnaire-2 (PHQ-2, Table 33.5) is used [36]. Patients who screen positive on the PHQ-2 will require a referral to the geriatric psychiatrist for detailed evaluation. Preoperative depression has been associated with worse outcomes after surgery. For instance, preoperative depression has been shown to be associated with increased mortality and longer length of stay after cardiac surgery [37]. Similarly, the difficulty to differentiate between psychotic pain and distress in depression from somatic pain has been shown to be associated with higher pain perception and increased requirements of postoperative analgesia [38, 39].

Table 33.5 Screening for depression with the Patient Health Questionnaire-2 (PHQ-2)

ASK THE PATIENT THE FOLLOWING QUESTIONS:
1. "In the past 12 months, have you ever had a time when you felt sad, blue, depressed, or down for most of the time for at least 2 weeks?"
2. "In the past 12 months, have you ever had a time, lasting at least 2 weeks, when you didn't care about the things that you usually cared about or when you didn't enjoy the things that you usually enjoyed?"
If the patient answers YES to either question, then further evaluation by a primary care physician, geriatrician, or mental health specialist is recommended.

From Li et al. [36], with permission

Screening for Alcohol and Substance Abuse

All elderly patients undergoing surgery should be screened for alcohol dependence and substance abuse using the CAGE questionnaire (Cut down, Annoyed, Guilty, Eye-opener) [40]. Patients who screen positive should be motivated to see a substance abuse specialist for preoperative abstinence or medical detoxification. The patient should be educated on different platforms for help. Daily multivitamins with B6 and B12 and oral or parental thiamine are recommended in the perioperative period for patients with alcohol abuse disorder. In addition, strict monitoring in the perioperative period and the possible use of prophylactic medications to prevent withdrawals is also warranted in these patients.

A national survey during 2005–2006 showed that about 60% of older adults (50 years and older) were using alcohol [41]. The prevalence of binge-drinking among elderly men and women was 14.5% and 3.3%, respectively [42]. Different studies have shown the association of preoperative alcohol abuse and dependence and increased rates of postoperative complications, including but not limited to pneumonia, sepsis, wound infection, and disruption, as well as mortality [43].

Determining Capacity and Informed Consent

Assessing mental capacity in vulnerable patients is a necessary function of respecting patient autonomy. Assessing the patient's decision-making capacity, not often easy in the elderly, is critical in determining his or her ability to provide informed surgical consent. A detailed discussion about the need for surgery, indications, benefits, risks, and alternatives to surgery should be held with the patient. The patient can also include their loved one in this discussion too. After the discussion, the patient should be able to describe in their own words the salient features of the consent, risks, and benefits. In order to be able to sign the consent, the following criteria must be meant in order to determine decision-making capacity.

1. The patient acknowledges his or her medical condition, treatment options, and the likely outcomes.
2. The patient understands the relevant information communicated by the physician.
3. The patient can clearly indicate his or her treatment choice.
4. The patient can engage in a rational discussion about the treatment options.

If any of the above requirements is not met or if unable to determine capacity, a psychiatric consult should be placed for detailed evaluation.

Goals of Care Discussion

Goals of care discussion before surgical procedures is of great value and builds a trust between the physician and the patient. It is especially relevant in patients with multiple comorbidities requiring high-risk surgeries. The discussion about a curative surgery vs. a palliative surgery should also be discussed with the patient upfront. In all these discussions, treatment goals, and plans, the surgeon should always empathetically understand patients' wishes and expectations and respect them in all circumstances. These discussions about patients' preferences and goals of care should be clearly documented in the medical records. The surgeon should give a realistic expectation of the postoperative course and possible complications, and should preemptively arrange for any postoperative needs that need that may come up, e.g. the need for ostomy nurse, home healthcare, long-term IV antibiotics or need for dialysis. If relevant, include the discussion of possible functional decline and need for rehabilitation or nursing home care during the informed consent process.

Involve the Family

The involvement of family in the treatment of elderly patients with mental health issues is critical in the management of these patients. The determination of the patient's family and social support systems to minimize stressors and make a smooth postoperative transition of the patient back to his family. If the patient lacks social support or family preoperative involvement of physical therapy, social worker and case manager to determine the needs of the patients should be encouraged.

Advance Directives and Assigning Health Care Proxy

The surgeon should also carry a discussion with the patient about advance directives and designate a health care proxy or surrogate decision maker, in case the patient at some stage in their treatment lacks capacity. This discussion should include dependence on feeding tubes, tracheostomies, and ventilator. These documents should be placed in the medical chart. A study of deceased elderly individuals showed that about one-third of elderly individuals required medical decision making; however, they lacked decision-making capacity toward the end of life [44].

Multidisciplinary Team Approach

The optimal surgical preparation of patient with mental illness require multidisciplinary approach involving the primary care physician (PCP), the surgeon, the anesthesiologist, and the geriatric psychiatrist. The early involvement of all these pillars will help identify the areas that can be optimized before surgery and decrease postoperative adverse events.

Polypharmacy

Before surgery, a complete review and documentation of all the medications should be performed. This should include a complete list of all prescription and nonprescription agents (over-the-counter pain medications, vitamins, eye drops) and herbal products. The correct dosage and duration should be confirmed with the pharmacy, and if needed, a pharmacist should be involved as well. The risk for adverse drug reactions and drug–drug interactions can be minimized by identifying medications that can be discontinued before surgery or can be avoided, and by dose-reducing or substituting to another medication with safer profile. Polypharmacy is considered as a major risk factor for developing postoperative delirium. Avoid sedatives like benzodiazepines and consider reducing benzodiazepines when possible if the patient is already taking benzodiazepines. For sleep, the non-sedative hypnotics like melatonin and readjusting the sleep-awake cycle should be recommended. Avoid using narcotics; however, adequately control the pain with non-narcotic medications to reduce the risk for developing postoperative delirium. Other drugs with strong anticholinergic or sedative effects like the histamine (H1) antagonists and tricyclic antidepressants should be avoided if possible. Critical medications like antiplatelets/anticoagulants should be discontinued for the minimum possible duration after discussion with the primary care physician and cardiologist and resumed immediately postoperatively when possible, to reduce perioperative risks of adverse events (cardiac, stroke, etc.). The ACC/AHA guidelines should be followed for perioperative beta blockers and statins [45]. Preoperative statins should be started as soon as possible before surgery for patients who have known vascular disease, elevated low-density lipoprotein cholesterol, or ischemia on thallium testing. For patients undergoing noncardiac surgery, who are currently taking statins, statin therapy should be continued. Statin use may also be considered for patients undergoing vascular and intermediate risk surgery. For medications that are cleared renally, doses of medications should be adjusted based on glomerular filtration rate. If a patient develops a postoperative acute kidney injury, all the nephrotoxic medications should be discontinued. Continuous monitoring for polypharmacy and potential adverse interactions and discontinuation of non-essential medications preoperatively are recommended, if possible. The addition of new medications should be kept to a minimum postoperatively.

Intraoperative Care of Elderly Patients with Mental Health Issues

Aspiration Risk

Elderly patients with mental illness may be under the effect of psychotropic medications and are at an increased risk of aspiration. Care should be taken during the preoperative period and during intubation. The optimal NPO time should be determined by the anesthesia.

Correct Dosing

For elderly patients with mental illness who are taking other sedatives/tranquilizers, the anesthetic dose should be adjusted accordingly by the anesthesiologist.

Postoperative Care

Delirium

As discussed earlier elderly patients at risk of developing delirium should be identified before surgery, and these risk factors should be clearly documented in the patient chart. For patients at risk for postoperative delirium, the administration of benzodiazepines, antihistamines, and other sedative medications should be avoided if possible. Early identification of patients who develop postoperative delirium and minimizing the causative agents will help improve patient outcomes.

Pain Control

The use of non-narcotics (NSAIDs, acetaminophen, gabapentin) for pain control should be encouraged. Acetaminophen is considered to have the safest risk profile in elderly patients. NSAIDs can be used for a shorter period postoperatively if the creatinine is normal. Adequate hydration is recommended while using NSAIDs. Local anesthetic patches, regional nerve blocks, and pain catheters can provide adequate pain control without the use of narcotics and sedatives. Narcotics increases the risk of delirium in postoperative patients and should be avoided if possible.

Disposition: Intensive Care Unit (ICU) Versus Floor

Patients may require an initial ICU stay overnight for hemodynamic monitoring. Early discussion should be held with the ICU team and a low threshold kept for ICU. Certain large centers are advocating the need for geriatric intensive care units (GICUs) for the management of sick geriatric patients requiring critical care. Some studies have shown improved outcomes for patients managed at geriatric ICUs.

Discharge Planning and Care

Psychosocial Support

Patients should receive continuous psychosocial support and should be followed by a psychiatrist/mental health specialist in the hospital to titrate psychotropic medications.

Physical and Occupational Therapy (PT/OT)

Patients should be mobilized by physical therapy as soon as possible after surgery. Early physical and psychological rehabilitation is the key to prevent postoperative complications.

Family Involvement and Education

Early family involvement in the postoperative care of patient is pivotal for the early recovery after surgery in geriatric patients. Family education about the postoperative care and the needs post discharge should be initiated immediately postoperatively.

Discharge Goals

Discharge goals should be delineated, and a comprehensive discharge plane given to the patient. Patient and family should be educated about home medications and precautions and the signs and symptoms to watch after surgery. If a patient is going to a rehabilitation center or a skilled nursing facility, patient's plan should be clearly communicated to the accepting center.

Follow-Up

Early follow-up with the surgeon, primary care physician, and psychiatrist to track recovery and identify any potential complications is recommended.

Summary

As the US population is aging, the longevity comes at a cost of living with increased lifelong morbidities and disabilities, making the elderly susceptible to mental health problems. Clinical presentation of elderly patients with mental disorders may be different from that of other adults, making timely diagnosis and detection of treatable illness more difficult. Elderly patients are more likely to undergo inpatient and outpatient surgical procedures. Therefore, early identification of elderly patients with mental illness undergoing surgery and a multidisciplinary team approach to their preoperative optimization and timely interventions and resource allocation will help improve clinical outcomes of these patients.

References

1. Joseph B, Hassan A. Geriatric trauma patients: what is the difference? Curr Surg Rep. 2016;4(1):1.
2. Hobbs F, Damon BL. Sixty-five plus in the United States: US Department of Commerce, Bureau of the Census; 1996.
3. Chow WB, Rosenthal RA, Merkow RP, Ko CY, Esnaola NF. Optimal preoperative assessment of the geriatric surgical patient: a best practices guideline from the American College of Surgeons National Surgical Quality Improvement Program and the American Geriatrics Society. J Am Coll Surg. 2012;215(4):453–66.
4. Vincent GK, Velkoff VA. The next four decades: The older population in the United States: 2010 to 2050: US Department of Commerce, Economics and Statistics Administration, US; 2010.
5. Albert M, McCaig LF, Ashman JJ. Emergency department visits by persons aged 65 and over: United States, 2009–2010: US Department of Health and

Human Services, Centers for Disease Control and ...; 2013.
6. Levant S, Chari K, DeFrances CJ. Hospitalizations for patients aged 85 and over in the United States, 2000–2010: US Department of Health and Human Services, Centers for Disease Control; 2015.
7. Etzioni DA, Liu JH, O'Connell JB, Maggard MA, Ko CY. Elderly patients in surgical workloads: a population-based analysis. Am Surg. 2003;69(11):961.
8. Cullen KA, Hall MJ, Golosinskiy A. Ambulatory surgery in the United States, 2006; 2009.
9. DeFrances CJ, Lucas CA, Buie VC, Golosinskiy A. 2006 National hospital discharge survey. Natl Health Stat Rep. 2008;5(July):1–20.
10. Vigo D, Thornicroft G, Atun R. Estimating the true global burden of mental illness. Lancet Psychiatry. 2016;3(2):171–8.
11. Ginn S, Horder J. "One in four" with a mental health problem: the anatomy of a statistic. BMJ. 2012;344:e1302.
12. Prince M, Patel V, Saxena S, Maj M, Maselko J, Phillips MR, Rahman A. No health without mental health. Lancet. 2007;370(9590):859–77.
13. Vos T, Barber RM, Bell B, Bertozzi-Villa A, Biryukov S, Bolliger I, Charlson F, Davis A, Degenhardt L, Dicker D. Global, regional, and national incidence, prevalence, and years lived with disability for 301 acute and chronic diseases and injuries in 188 countries, 1990–2013: a systematic analysis for the Global Burden of Disease Study 2013. Lancet. 2015;386(9995):743–800.
14. Ahrnsbrak R, Bose J, Hedden S, Lipari R, Park-Lee E. Key substance use and mental health indicators in the United States: Results from the 2016 National Survey on Drug Use and Health. Center for Behavioral Health Statistics and Quality, Substance Abuse and Mental Health Services Administration: Rockville, MD, USA; 2017.
15. Lozano R, Naghavi M, Foreman K, Lim S, Shibuya K, Aboyans V, Abraham J, Adair T, Aggarwal R, Ahn SY. Global and regional mortality from 235 causes of death for 20 age groups in 1990 and 2010: a systematic analysis for the Global Burden of Disease Study 2010. Lancet. 2012;380(9859):2095–128.
16. Ortman JM, Velkoff VA, Hogan H. An aging nation: the older population in the United States: United States Census Bureau, Economics and Statistics Administration, US; 2014.
17. Mental health of older adults. World Health Organization. 2017. Available from: https://www.who.int/news-room/fact-sheets/detail/mental-health-of-older-adults. Access Date: 25 June 2019
18. Smyer MA, Qualls SH. Aging and mental health. Oxford, UK: Blackwell Publishing; 1999.
19. Heine C, Browning CJ. Mental health and dual sensory loss in older adults: a systematic review. Front Aging Neurosci. 2014;6:83.
20. American Association of Geriatric Psychiatry. Geriatrics and mental health—the facts. 2008. Available at: http://www.aagponline.org/prof/facts_mh.asp. Accessed 25 June 2018.
21. Centers for Disease Control and Prevention National Center for Injury Prevention and Control. WISQARS™ (Web-based Injury Statistics Query and Reporting System). https://www.cdc.gov/injury/wisqars/index.html. 2017.
22. Michelson D, Stratakis C, Hill L, Reynolds J, Galliven E, Chrousos G, Gold P. Bone mineral density in women with depression. N Engl J Med. 1996;335(16):1176–81.
23. Glassman AH, Shapiro PA. Depression and the course of coronary artery disease. Am J Psychiatr. 1998;155(1):4–11.
24. Spiegel D, Giese-Davis J. Depression and cancer: mechanisms and disease progression. Biol Psychiatry. 2003;54(3):269–82.
25. Satin JR, Linden W, Phillips MJ. Depression as a predictor of disease progression and mortality in cancer patients: a meta-analysis. Cancer. 2009;115(22):5349–61.
26. Lamers SM, Bolier L, Westerhof GJ, Smit F, Bohlmeijer ET. The impact of emotional well-being on long-term recovery and survival in physical illness: a meta-analysis. J Behav Med. 2012;35(5):538–47.
27. Somerset W, Stout SC, Miller AH, Musselman D. Breast cancer and depression. Oncology (Williston Park). 2004;18(8):1021–34; discussion 35–6, 47–8.
28. Chapman DP, Perry GS, Strine TW. PEER REVIEWED: The vital link between chronic disease and depressive disorders. Prev Chronic Dis. 2005;2(1).
29. Limpawattana P, Phungoen P, Mitsungnern T, Laosuangkoon W, Tansangworn N. Atypical presentations of older adults at the emergency department and associated factors. Arch Gerontol Geriatr. 2016;62:97–102.
30. McDonald SR, Heflin MT, Whitson HE, TO D, Lidsky ME, Liu P, Poer CM, Sloane R, Thacker JK, White HK. Association of integrated care coordination with postsurgical outcomes in high-risk older adults: the Perioperative Optimization of Senior Health (POSH) initiative. JAMA Surg. 2018;153(5):454–62.
31. Borson S, Scanlan J, Brush M, Vitaliano P, Dokmak A. The mini-cog: a cognitive 'vital signs' measure for dementia screening in multi-lingual elderly. Int J Geriatr Psychiatry. 2000;15(11):1021–7.
32. Ansaloni L, Catena F, Chattat R, Fortuna D, Franceschi C, Mascitti P, Melotti R. Risk factors and incidence of postoperative delirium in elderly patients after elective and emergency surgery. Br J Surg. 2010;97(2):273–80.
33. Rudolph JL, Inouye SK, Jones RN, Yang FM, Fong TG, Levkoff SE, Marcantonio ER. Delirium: an independent predictor of functional decline after cardiac surgery. J Am Geriatr Soc. 2010;58(4):643–9.
34. Cole MG, Dendukuri N. Risk factors for depression among elderly community subjects: a systematic review and meta-analysis. Am J Psychiatr. 2003;160(6):1147–56.

35. Steffens DC, Fisher GG, Langa KM, Potter GG, Plassman BL. Prevalence of depression among older Americans: the Aging, Demographics and Memory Study. Int Psychogeriatr. 2009;21(5):879–88.
36. Li C, Friedman B, Conwell Y, Fiscella K. Validity of the Patient Health Questionnaire 2 (PHQ-2) in identifying major depression in older people. J Am Geriatr Soc. 2007;55(4):596–602.
37. Contrada RJ, Boulifard DA, Hekler EB, Idler EL, Spruill TM, Labouvie EW, Krause TJ. Psychosocial factors in heart surgery: Presurgical vulnerability and postsurgical recovery. Health Psychol. 2008;27(3):309.
38. De Cosmo G, Congedo E, Lai C, Primieri P, Dottarelli A, Aceto P. Preoperative psychologic and demographic predictors of pain perception and tramadol consumption using intravenous patient-controlled analgesia. Clin J Pain. 2008;24(5):399–405.
39. Taenzer P, Melzack R, Jeans ME. Influence of psychological factors on postoperative pain, mood and analgesic requirements. Pain. 1986;24(3):331–42.
40. Hinkin CH, Castellon SA, Dickson-Fuhrman E, Daum G, Jaffe J, Jarvik L. Screening for drug and alcohol abuse among older adults using a modified version of the CAGE. Am J Addict. 2001;10(4):319–26.
41. Blazer DG, Wu L-T. The epidemiology of substance use and disorders among middle aged and elderly community adults: national survey on drug use and health. Am J Geriatr Psychiatry. 2009;17(3):237–45.
42. Blazer DG, Wu L-T. The epidemiology of at-risk and binge drinking among middle-aged and elderly community adults: National Survey on Drug Use and Health. Am J Psychiatr. 2009;166(10):1162–9.
43. Nath B, Li Y, Carroll JE, Szabo G, Tseng JF, Shah SA. Alcohol exposure as a risk factor for adverse outcomes in elective surgery. J Gastrointest Surg. 2010;14(11):1732–41.
44. Silveira MJ, Kim SY, Langa KM. Advance directives and outcomes of surrogate decision making before death. N Engl J Med. 2010;362(13):1211–8.
45. Woolger JM. Preoperative testing and medication management. Clin Geriatr Med. 2008;24(4):573–83.

Part IV

Teaching Geriatric Surgery

34 How Do We Teach Medical Students to Care for the Elderly

Anitha Srinivasan

Geriatrics as a Component of Surgical Education

In the United States, the geriatric population is rising due to improved standards of living and better medical care. The population aged 65 and above is expected to double to 84 million from 2010 to 2050 [1]. With this comes the demand for medical and surgical care of this population. In surgical care there is a projected demand for vascular surgery to increase by 31% and general surgery by 18%. These patients need well-informed surgical providers with educational exposure to geriatric care to achieve positive results. This will become an unmet need in the future generation of healthcare unless it is addressed in today's medical education [2].

History and Current State of Geriatrics as Part of Surgical Education

The need for geriatric care and its awareness among specialty trainees was recognized in the mid-1990s [2]. However there still exists a lack of formalized training as part of surgical education and recognition of its importance toward caring for the future generations [3]. There are currently ongoing programs to educate surgical residents in training to orient them toward care of geriatric patients in surgery, e.g., Chief Resident Immersion Training program for care of the elderly (CRIT at Boston University) [4]. However this may not be the ideal teaching moment in a formative surgical career [5]. Given that geriatric patients are going to be a significant part of surgery in the future, it is important that the students understand what to expect if they choose surgery as their field of practice. There is enough rationale to inculcate geriatric surgical education in the medical school curriculum as part of the surgery teaching [6]. This, we hope, will lead to a well-informed medical graduate who chooses a career in surgery knowing the needs of the future population they will be serving.

The scope of teaching the medical students must include a holistic approach to create an awareness that is embedded within their decision-making in surgery [7]. Every geriatric patient presentation must not pose a new challenge to the future surgeon. Instead a well-trained surgeon should be capable of informed decision-making for the best outcome in that particular case, due to their training in geriatric surgery from an early stage of their careers. It is such knowledge one hopes to gain from surgical student education regarding geriatrics that will lead to the

A. Srinivasan (✉)
Metropolitan Hospital Breast Health Center, Department of Surgery, New York Medical College, School of Medicine, Valhalla, NY, USA

Metropolitan Hospital, New York, NY, USA
e-mail: anitha.srinivasan@nychhc.org

R. Latifi (ed.), *Surgical Decision Making in Geriatrics*,
https://doi.org/10.1007/978-3-030-47963-3_34

unconscious selection of the right choices for the patient with knowledge of evidence-based practices in geriatrics.

The Components of Care of the Geriatric Surgical Patient and What a Student Should Learn and Expect

Here is a suggested format for introducing the surgical student to geriatric care in surgical disease (Table 34.1).

- Assessment and preoperative preparation
- Operative care and postoperative inhospital care
- Long-term outcomes, functional, and ethical considerations

Introductory Lectures

The student starts the rotation with the general introductory lectures in surgery and trauma. In addition, there should be a lecture given by the geriatric medical doctor, a surgeon who practices surgery in the elderly, and by a palliative care specialist. This will familiarize students with theoretical aspects of geriatric care related to surgery.

Understanding the Role of Multidisciplinary Care in Geriatric Surgery: Simulate Creation of an Ideal Team of Care for the Elderly Patient

The components of a geriatric surgical team must include the participation of a geriatric care team which includes specialists that address medical and palliative care components of the elderly patient's surgical condition. Often a social worker is essential to ensure smooth transitions of care and safe discharges and creation of goals of care. Family involvement and caregiver contribution is absolutely essential to decision-making and evaluation of an elderly patient in surgery. Role play of such scenarios with relevant case discussions will be informative for the students. Here the students are following the model of a flipped classroom acting as specialists in a geriatric care scenario. They create a checklist for the assessment of a geriatric patient and the necessary team members and present it to the class justifying the role of each team member.

Table 34.1 Format for introducing the surgical student to geriatric care in surgical disease

Assessment and preoperative preparation	Operative care and postoperative inhospital care	Long-term outcomes, functional, and ethical considerations
Introductory lectures	The art of multidisciplinary discussion in surgical decision-making	Surgical clinic shadowing and experience to asses post-op patients
Understanding the role of multidisciplinary care in geriatric surgery – simulate creation of an ideal team of care for the elderly patient	Observation of operative cases involving geriatric patients	ICU rotation in surgery and witness end-of-life discussions and understanding ethics committee
Role play and group discussion to simulate a clinic preoperative assessment	Shadowing the geriatrician and understanding postoperative care of the elderly patient	Shadowing rehabilitation post-surgery
Observation of the preplanning/assessment of the geriatric patient in the clinic setting	Understanding the role of a palliative care physician in the surgical setting	
Simulation case scenarios – discuss rationale for decision in each case	Postoperative care and follow-up of the patient	

Role Play and Group Discussion to Simulate a Clinic Preoperative Assessment

In routine surgical preoperative assessment, a surgeon focuses on the technique of choice, method, timing of surgery, and the expected outcome and discusses the same with the patient regarding risks and benefits. In the geriatric patient, before embarking on such a discussion, one has to first asses the following: the frailty of the patient [8, 9], role of delirium expected post-operation, and overall effect of dementia in the patient. Students should be tested with various scenarios of patients presenting for different surgery needs and asked to assess them for their surgical risk, benefit summary, and the goal of the surgery [10, 11]. Simulation case scenarios such as the following must be set to educate the student and have them familiarized with the issues in planning surgery in the elderly population.

Simulation Case Scenarios: Discuss Rationale for Decision in Each Case

- Case #1: Asymptomatic inguinoscrotal hernia in the elderly – discuss in immobile, mobile, functional patient, and highly functional athletic elderly
- Case#2: Elderly patient seeking cosmetic surgery – minor procedure under local anesthesia – injections, fillers, liposuction, or abdominoplasty in an elderly woman
- Case#3: Independently living mobile patient with previous attacks of diverticulitis – currently asymptomatic with cardiac comorbidity
- Case#4: Patient with gallstones, New Hampshire resident with significant medical comorbidities, independently living patient with one previous attack
- Case#5: NH resident immobile patient with incidental breast mass that was biopsied and found to be a hormone-positive ductal carcinoma. Next steps

Observation of the Preplanning/ Assessment of the Geriatric Patient in the Clinic Setting

The medical student in the surgery clinic plays a key role in the history-taking and observation of the clinical assessment by the surgeon. If this is done in consort with the geriatrician in the clinic setting, the student should witness and understand the role each and grasp the decision-making skill involved.

The Art of Multidisciplinary Discussion in Surgical Decision-Making

The decision to operate or not: Often surgeons play an authoritative role in decision-making regarding surgery and type of interventions. However, this role has to be collaborative in the care of the elderly and has to be adapted to a multidisciplinary approach with input from the entire care team and family. The goals maybe different from the one a surgeon would routinely make in the case of a non-elderly patient who does not exhibit signs of frailty or poor healing.

Ask the students to do a group discussion and role play of a geriatric patient visit who is seeking a surgical opinion (Fig. 34.1). Change the cases with varying diagnoses and the functional and frailty level of the patient. Ask the student to explain the rationale for the decision to operate or avoid surgery. Also ask the student to discuss how to elicit the patient goals of the surgical outcome. This is when the student should familiarize themselves with the need for multidisciplinary care in the decision-making process [5, 7]. The importance of the involvement of a geriatrician, the patient's primary care physician, the patient's wishes and healthcare designee, and long-term goals of care all have to play a role in this preoperative, decision-making process (Fig. 34.2). The geriatric portion of the medical student curriculum should use small-group instructional methods consistent with adult learning principles that

Fig. 34.1 Teaching methodologies for geriatric surgery medical student education

Fig. 34.2 Optimal geriatric surgical care

include practice-based learning, case-based learning, patient simulation using Objective Structured Video Examination (OSVE), and didactic sessions. Faculty instruction is a shared responsibility between geriatricians and general surgeons [2]. Pre- and postexposure and teaching assessments can be used to asses knowledge gain in the subject prior to direct patient involvement in surgery [10].

Observation of Operative Cases Involving Geriatric Patients

The students will gain a lot of knowledge by observation and assisting in the surgery of the elderly. Here an added emphasis is placed on anesthesia preparation, techniques, and the aim to minimize time under anesthesia. Cardiac assessments including risks posed by pacemakers and AEDs are taken into consideration as is the proclivity to venous thromboembolism. Positioning of the patient and the collaborative approach to a speedy surgery and smooth recovery is to be noted in the operating room environment.

Shadowing the Geriatrician and Understanding Postoperative Care of the Elderly Patient

A geriatrician is able to view the patient's course from an alternate window than that of a surgeon. He or she takes into consideration various factors in pre- and postoperative care of the elderly surgical patient [9]. The role of patient wishes, goals, medications, functionality, frailty, and unexpected events due to surgery (e.g., DVT, Delirium) are often in the purview of the geriatrician. Ideal surroundings and the environment of care post-operation are always of utmost importance in a speedy recovery and return to functionality. Preventing delays and prolonged hospitalizations in elderly, and its importance is often exemplified in the geriatrician's practice and follow-up.

Understanding the Role of a Palliative Care Physician in the Surgical Setting

A Palliative care physician provides insight into the limitations and resetting of goals of care according to the surgical findings, procedures, and any altered diagnosis encountered. In these cases, the palliative care physician often establishes the tenet that discussing issues with the family and placing patient wishes first is the primary aim. Has the intervention and/or future plan for interventions answer the critical question of, "have you improved patient's life and established the basis of 'do no harm' in every particular case"? The palliative care team often involves discussions with caregivers to understand decisional capacity of the patient and advanced directives and the role of a Do Not Resuscitate (DNR) order as the case may be. The role of palliation is indispensable in the care of an elderly surgical patient. Often there are decisions that are made that need review and reconsideration based on the surgical disease and findings, including the possibility of a malignancy in the diagnosis.

Postoperative Care and Follow-Up of the Patient

Depending on the patient's outcome and procedure performed, the student should be exposed to and participate in rounding in the intensive care unit and gain knowledge into the geriatric issues at hand. These may include medication management, reconciliation, drug interactions, renal dosing, signs of delirium, and worsening dementia. Often the nursing care of the ICU plays a key role in an elderly surgical patient's recovery – the prevention of pressure ulcers, the importance of turning and positioning, and the pulmonary toilet are keys in preventing hospital-acquired infections. The key roles of physical therapy in combating deconditioning, which is accelerated in the elderly population postoperatively and the occupational therapist in reaching functional levels of independence are essential. A multidisciplinary care plan, rounding, and joint decision-making should be witnessed by the student for full understating of the process. The student must familiarize himself or herself with hospital-acquired infections in the setting of postoperative recovery such as a ventilator-associated event (VAE), CLABSI, CAUTI, etc. and apply it to the relevance of the case observed and followed.

Surgical Clinic Shadowing and Experience to Assess Postoperative Patients

With the intended outcome for a successful hospital discharge achieved in the elderly, every effort must be made for the student to follow the patient and observe a typical postoperative clinic visit. This is essential in understanding the rate of wound healing and the return of functionality in the elderly patient. There is data suggesting the need for depression screening in the elderly after a major surgery, especially in post-cardia surgery. Mobility, quality of life, and independence in daily activities can be assessed at a postoperative outpatient visit.

ICU Rotation in Surgery and Witness End-of-Life Discussions and Understanding Ethics Committee

Some outcomes maybe less than ideal and unanticipated which lead to prolonged ICU stays, the need for reestablishing goals of care and, at times, the need to consider imitation or withdrawal of care. Such cases are done after an extensive evaluation of goals of care, futility of care, and the patient's directives in consort with the family discussions. Some cases may require the intervention of neurology to establish brain death protocol, which should be introduced in concept to the medical student. In some instances the patient-stated directives of care and the family and caregiver discussions may not align requiring the intervention of an ethics committee. The composition, role, and interaction of such an ethics committee are important for the student to comprehend the end-of-life discussion issues. The interested medical student should at every opportunity be allowed to shadow or observe, with the family's permission, on end-of-life discussions and also ethics committee meetings. The role of organ transplantation and donor establishment rules should be discussed with every instance of end-of-life withdrawal of care and explained to the student.

Shadowing Rehabilitation Post-surgery in Elderly Patients

An ideal education in elderly surgical care should include the role of an active rehabilitation unit and understanding its function the patient recovery. The student should follow-up the patient's path in an acute rehabilitation unit if available within the hospital. For example, witnessing a joint replacement in an elderly patient is incomplete if a student does not appreciate the crucial role of physical and occupational therapy in recovery of the patient. The student also completely grasps the surgical outcome when they visit and witness the work done in an offsite rehabilitation center which may vary for each elderly patient. Such opportunities if allowable by schedule will greatly enrich and ensure completeness of education about the surgical care of the elderly. The education of geriatric surgical patient care should thus be embarked upon during the early years of clinical training the medical student and ensure a comprehensive exposure to the various multidisciplinary aspects of care.

References

1. Ortman JM, Velkoff VA, Hogan H. An aging nation: the older population in the United States. Population estimates and projections. May 2014. https://www.census.gov/prod/2014pubs/p25-1140.pdf.
2. Solomon DH, Burton JR, Lundebjerg NE, Eisner J. The new frontier: increasing geriatrics expertise in surgical and medical specialties. J Am Geriatr Soc. 2000;48(6):702–4.
3. Barbas AS, Haney JC, Henry BV, Heflin MT, Lagoo SA. Development and implementation of a formalized geriatric surgery curriculum for general surgery

residents. Gerontol Geriatr Educ. 2014;35(4):380–94. https://doi.org/10.1080/02701960.2013.879444.
4. Association of Directors of Geriatrics Academic Programs (ADGAP). Programs. http://adgap.americangeriatrics.org/adgap-programs/crit/.
5. Hagiwara Y, Pagan-Ferrer J, Sanchez-Reilly S. Impact of an early-exposure geriatrics curriculum in an accelerated baccalaureate-MD program. Gerontol Geriatr Educ. 2018:1–6. https://doi.org/10.1080/02701960.2018.1464919.
6. Foguem C, Manckoundia P. Experience of medical students mentoring in an acute geriatric unit: use of evidence-based medicine. Eur J Intern Med. 2019;60:e23–5. https://doi.org/10.1016/j.ejim.2018.11.008.
7. Camp ME, Jeon-Slaughter H, Johnson AE, Sadler JZ. Medical student reflections on geriatrics: moral distress, empathy, ethics and end of life. Gerontol Geriatr Educ. 2018;39(2):235–48. https://doi.org/10.1080/02701960.2017.1391804.
8. Griebling TL. Re: Report of a simplified frailty score predictive of short-term postoperative morbidity and mortality. J Urol. 2016;196(3):789–90. https://doi.org/10.1016/j.juro.2016.06.040.
9. Revenig LM, Canter DJ, Kim S, Liu Y, Sweeney JF, Sarmiento JM, et al. Report of a simplified frailty score predictive of short-term postoperative morbidity and mortality. J Am Coll Surg. 2015;220(5):904–11.e1. https://doi.org/10.1016/j.jamcollsurg.2015.01.053.
10. Webb TP, Duthie E Jr. Geriatrics for surgeons: infusing life into an aging subject. J Surg Educ. 2008;65(2):91–4. https://doi.org/10.1016/j.jsurg.2007.12.004.
11. Williams BC, Fitzgerald JT. Brief report: brief instrument to assess geriatrics knowledge of surgical and medical subspecialty house officers. J Gen Intern Med. 2006;21(5):490–3. https://doi.org/10.1111/j.1525-1497.2006.00433.x.

Index

R. Latifi (ed.), *Surgical Decision Making in Geriatrics*,
https://doi.org/10.1007/978-3-030-47963-3

O

P

The manufacturer's authorised representative in the EU is Springer Nature Customer Service Centre GmbH, Europaplatz 3, 69115 Heidelberg, Germany. If you have any concerns regarding our products, please contact ProductSafety@springernature.com

Printed and bound by CPI Group (UK) Ltd, Croydon, CR0 4YY
30/09/2026
02225705-0008